W9-BNJ-743

PAUL A. FOERSTER

Precalculus
with Trigonometry
Concepts and Applications

Key Curriculum Press
Innovators in Mathematics Education

Project Editor: Anna Werner
Project Administrator: Christopher David
Consulting Editors: Christian Aviles-Scott, Cindy Clements, Stacey Miceli, Abby Tanenbaum
Editorial Assistants: Halo Golden, Michael Hyett, Beth Masse
Reviewers: Bill Medigovich, San Francisco, CA; Calvin Rouse, Laney College, Oakland, CA; Wayne Nirode, Troy High School, Troy, OH; Mary Anne Molnar, Northern Valley Regional High School, Demarest, NJ
Multicultural Reviewer: Swapna Mukhopadhyay, San Diego State University, San Diego, CA
Accuracy Checkers: Dudley Brooks, Cavan Fang
Production Editor: Jennifer Strada
Copyeditor: Elliot Simon
Manager, Editorial Production: Deborah Cogan
Production Director: Diana Jean Parks
Production Coordinator and Cover Designer: Jenny Somerville
Text Designer: Adriane Bosworth
Art Editor and Photo Researcher: Laura Murray
Art and Design Coordinator: Caroline Ayres
Illustrators: Jon Cannell, Michael Meister, Bill Pasini, Matthew Perry, Charlene Potts, Rose Zgodzinski
Technical Art: Lineworks, Inc.
Photography: Lonnie Duka, Index Stock, and Hillary Turner Photography
Student Models: Nizhoni Ellenwood, Sean Parks, and Cynthia Tram
Cover Photo Credit: Duomo/Corbis
Composition and Prepress: GTS Graphics
Printer: Quebecor World Book Services

Executive Editor: Casey FitzSimons
Publisher: Steven Rasmussen

Key Curriculum Press
1150 65th Street
Emeryville, CA 94608
editorial@keypress.com
http://www.keypress.com

Printed in the United States of America
10 9 8 7 6 5 4 3 2 1 06 05 04 03 02
ISBN 1-55953-391-9

Photograph credits appear on the last two pages of the book.

To my students, for whom the text was written.

To my wife, Peggy, who gives me love and encouragement.

Field Testers

Gary Anderson, Seattle Academy of Arts and Sciences, Seattle, WA
David Badal, Bentley Upper School, Lafayette, CA
Brenda Batten, Thomas Heyward Academy, Ridgeland, SC
Jim Bigelow, Redlands East Valley High School, Redlands, CA
Amarjit Chadda, Los Altos High School, Los Altos, CA
Michelle Clawson, Hamlin High School, Hamlin, TX
Melissa Cleavelin, St. Mary's Academy, Denver, CO
John Crotty, Career High School, New Haven, CT
Deborah Davies, University School of Nashville, Nashville, TN
Carol DeCuzzi, Audubon Junior/Senior High School, Audubon, NY
Louise Denney, Troup High School, LaGrange, GA
Autie Doerr, Anderson High School, Austin, TX
Paul A. Foerster, Alamo Heights High School, San Antonio, TX
Kevin Fox, Millburn High School, Millburn, NJ
Bill Goldberg, Viewmont High School, Bountiful, UT
Donna Gunter, Anderson High School, Austin, TX
Teresa Ann Guthie-Sarver, Melbourne High School, Melbourne, FL
Karen Hall, Skyview High School, Vancouver, WA
Rick Kempski, Nashoba Regional High School, Bolton, MA
Ken Koppelman, Montgomery High School, Santa Rosa, CA
Stella Leland, Caney Creek High School, Conroe, TX
Dan Lotesto, Riverside University High School, Milwaukee, WI
Bill Miron, Millburn High School, Millburn, NJ
Deb Prantl, Georgetown Day School, Washington, DC
John Quintrell, Montgomery High School, Santa Rosa, CA
Louise Schumann, Redlands High School, Redlands, CA
Luis Shein, Bentley Upper School, Lafayette, CA
Gregory J. Skufca, Melbourne High School, Melbourne, FL
Chris Sollars, Alamo Heights High School, San Antonio, TX
Judy Stringham, Martin High School, Arlington, TX
Nancy Suazo, Española Valley High School, Española, NM
Rickard Swinth, Montgomery High School, Santa Rosa, CA
Susan Thomas, Alamo Heights High School, San Antonio, TX
Anne Thompson, Hawken School, Gates Mills, OH
Julie Van Horn, Manitou Springs High School, Manitou Springs, CO
Barbara Whalen, St. John's Jesuit High School, Toledo, OH
Margaret B. Wirth, J. H. Rose High School, Greenville, NC
Isabel Zsohar, Alamo Heights High School, San Antonio, TX
Richard J. Zylstra, Timothy Christian School, Elmhurst, IL

Author's Acknowledgments

Special thanks go to Debbie Davies and her students at The University School in Nashville, Tennessee, who gave consistent comments and encouragement by field testing the text in its formative stage, and who drafted parts of the *Instructor's Guide.* Thanks also to Karen Hall and her students at Skyview High School in Vancouver, Washington, who also participated in the initial field test.

Thanks to other teachers and reviewers whose comments helped spot ways to make the text even more useful for the education of our students. In particular, thanks to Brenda Batten of Thomas Heyward Academy in South Carolina; Judy Stringham of Martin High School in Texas; Bill Goldberg of Viewmont High School in Utah; Anne Thompson of the Hawken School in Ohio; Richard Zylstra of Timothy Christian School in Illinois; Barbara Whalen of St. John's Jesuit High School in Ohio; John Quintrell, Rick Swinth, and Ken Koppelman of Montgomery High School in California; Louise Schumann of Redlands High School in California; Deb Prantl of Georgetown Day School in the District of Columbia; and Gary Anderson of the Seattle Academy of Arts and Sciences in Washington.

The readability of the text is enhanced by the fact that much of the wording of the original edition came directly from the mouths of students as they learned to grasp the ideas. Thanks go to my own students Nancy Carnes, Susan Curry, Brad Foster, Stacey Lawrence, Kelly Sellers, Ashley Travis, and Debbie Wisneski, whose class notes formed the basis for many sections.

Above all, thanks to my colleagues Susan Thomas, Isabel Zsohar, and Chris Sollars, and their students at Alamo Heights High School in San Antonio, who supplied daily input so that the text could be complete, correct, and teachable.

Finally, thanks to the late Richard V. Andree and his wife Josephine for allowing their children, Phoebe Small and Calvin Butterball, to make occasional appearances in my texts.

Paul A. Foerster

About the Author

Paul Foerster enjoys teaching mathematics at Alamo Heights High School in San Antonio, Texas, which he has done since 1961. After earning a bachelor's degree in chemical engineering, he served four years in the U.S. Navy. Following his first five years at Alamo Heights, he earned a master's degree in mathematics. He has published five textbooks, based on problems he wrote for his own students to let them see more realistically how mathematics is applied in the real world. In 1983 he received the Presidential Award for Excellence in Mathematics Teaching, the first year of the award. He raised three grown children with the late Jo Ann Foerster, and he also has two grown stepchildren through his wife Peggy Foerster, as well as three grandchildren. Paul plans to continue teaching for the foreseeable future, relishing the excitement of the ever-changing content and methods of the evolving mathematics curriculum.

Foreword

by John Kenelly, Clemson University

Déjà vu! Foerster has done it again. *Precalculus with Trigonometry: Concepts and Applications* succeeds as *Calculus: Concepts and Applications* succeeds, but in a far more difficult arena. In calculus, technology enhances and removes the necessity for many shopworn drills, but modernizing precalculus is a far larger challenge. The course must prepare students for a wide variety of instructional approaches, so the selection and structuring of its topics is a formidable task. Educational leadership is required and Foerster gives us that leadership. Again the author's blend of the best of the past and the promise of the future is rich.

What can develop prerequisite skills and enroll the precalculus course in the reform movement? Why, Foerster's approach indeed! The formerly stale trigonometric identities are included but now in the exciting context of harmonic analysis. What fun; mathematics with music! We can finally go beyond Klein's ancient article "The Sine of G Major" to a broad and varied collection of motivating topics. Understanding elementary functions is still fundamental, and Foerster brings the topic exciting coverage through his careful integration of data analysis, modeling, and statistics. Mathematical modeling enthusiasts, including myself, will review these sections with special pleasure.

Great texts have to motivate students as well as instructors. Foerster succeeds here as only a gifted author can. The problem sets are especially valuable. This book will be on every teacher's shelf even if it is not in his or her classroom. Where else but Foerster do you find 200 million people jumping off tables to create a sinusoidal wave around the world! Where else will you find proportionality relations between length, area, and volume linked with Stanley Langley's unsuccessful transition between flying models and full-size airplanes in 1896? Every section is loaded with problems in application contexts that will enrich all our classes.

The information age is an unrelenting driving force in educational reform, but some faculty have not taken on the responsibility of developing students' mathematical skills for these new conditions. Foerster's brilliant solution to the difficulty of finding an effective approach to the problematic precalculus course will please all of us, students and instructors alike.

John Kenelly has been involved with the Advanced Placement Calculus program for over 30 years. He was Chief Reader and later Chair of the AP Calculus Committee when Paul Foerster was grading the AP exams in the 1970s, and he was instrumental in getting the reading sessions moved to Clemson University when they outgrew the prior facilities. He is a leader in the development of the graphing calculator and in pioneering its use in college and school classrooms. His organization of Technology Intensive Calculus for Advanced Placement sessions following recent AP readings has allowed calculus instructors to share ideas for implementing the changes in calculus that have been made inevitable by the advent of technology. He currently serves as Treasurer of the Mathematical Association of America.

Contents

A Note to the Student from the Author

I wrote this book to make available to students and other teachers the things I have learned in over 40 years of teaching. I have organized it to take full advantage of technology, particularly graphing calculators, while still giving students experience with pencil-and-paper and mental computations. It embodies the best parts of my earlier text, *Precalculus with Trigonometry: Functions and Applications*, rewritten to incorporate advances and experience accumulated since its publication.

In previous courses you have learned about various kinds of functions. These functions tell how the value of one variable quantity is related to the value of another, such as the distance you have driven as a function of time you have been driving. In this course you will expand your knowledge to include transformations that can be performed on functions, and functions that vary periodically. You will learn how to decide just which kind of function will fit best for a set of real-world data that is inherently scattered. You will get reinforcement of the fact that variables really *vary*, not just stand for unknown constants. Finally, you will be introduced to the rate at which a variable quantity varies, thus laying the foundation for calculus.

You will have the opportunity to learn mathematics four ways—algebraically, numerically, graphically, and verbally. Thus, in whichever of these areas your talents lie, you will have an opportunity to excel. For example, if you are a verbal person, you can profit by reading the text, explaining clearly the methods you use, and writing in the journal you will be asked to keep. If your talents are visual, you will have ample opportunity to learn from the shapes of graphs you will plot on the graphing calculator. The calculator will also allow you to do numerical computations, such as regression analysis, that would be too time-consuming to do with just pencil and paper.

One thing to bear in mind is that mathematics is not a spectator sport. You must learn by doing, not just by watching others. You will have a chance to participate in cooperative groups, learning from your classmates as you work on the Explorations that introduce you to new concepts and techniques. The Do These Quickly problems at the beginning of each Problem Set ask you to recall quickly things you may have forgotten. There are some problems, marked by a shaded star, that will prepare you for a topic in a later section. In addition to the sample Chapter Test at the end of each chapter, there are Review Problems keyed to each section of that chapter. You may rehearse for a test on just those topics that have been presented, and you can check your answers in the back of the book. The Concept Problems give you a chance to apply your knowledge to new and challenging situations. So keep up with your homework to help ensure your success.

Many times you will see applications to real-world problems as a motivation for studying a particular topic. At other times you may see no immediate use for a topic. Learn the topic anyway and learn it well. Mathematics has a structure that you must discover, and the big picture may become clear only after you have unveiled its various parts. The more you understand about mathematics, the more deeply you will be able to understand other subjects, sometimes where there is no obvious connection, such as in theology, history, or law. Remember, what you know, you may never use. But what you don't know, you'll *definitely* never use.

In conclusion, let me wish you the best as you embark on this course. Keep a positive attitude and you will find that mastering mathematical concepts and techniques can give you a sense of accomplishment that will make the course seem worthwhile, and maybe even fun.

Paul A. Foerster
Alamo Heights High School
San Antonio, Texas

Precalculus

with Trigonometry

Concepts and Applications

CHAPTER

1

Functions and Mathematical Models

If you shoot an arrow into the air, its height above the ground depends on the number of seconds since you released it. In this chapter you will learn ways to express quantitatively the relationship between two variables such as height and time. You will deepen what you have learned in previous courses about functions and the particular relationships that they describe—for example, how height depends on time.

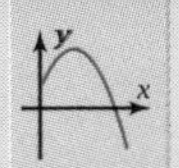

Mathematical Overview

In previous courses you have studied linear functions, quadratic functions, exponential functions, power functions, and others. In precalculus mathematics you will learn general properties that apply to all kinds of functions. In particular you will learn how to transform a function so that its graph fits real-world data. You will gain this knowledge in four ways.

Graphically The graph on the right shows the graph of a quadratic function. The y-variable could represent the height of an arrow at various times, x, after its release into the air. For larger time values, the quadratic function shows that y is negative. These values may or may not be reasonable in the real world.

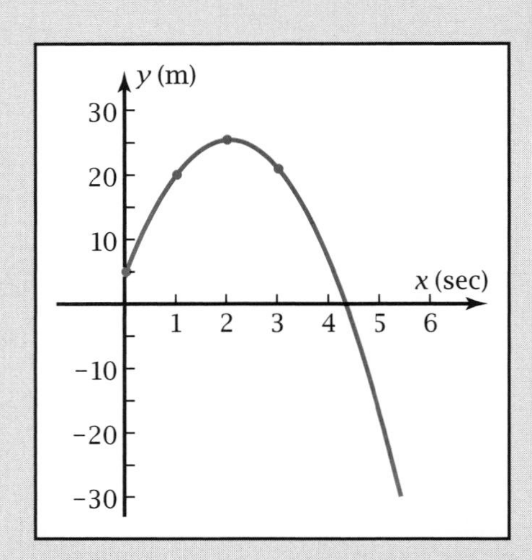

Algebraically The equation of the function is:

$$y = -4.9x^2 + 20x + 5$$

Numerically This table shows corresponding x- and y-values, which satisfy the equation of the function.

x (sec)	y (m)
0	5.0
1	20.1
2	25.4
3	20.9

Verbally *I have learned that when the variables in a function stand for things in the real world, the function is being used as a mathematical model. The coefficients in the equation of the function $y = -4.9x^2 + 20x + 5$ also show a connection to the real world. For example, the −4.9 coefficient is a constant that is a result of the gravitational acceleration and 5 reflects the initial height of the arrow.*

1-1 Functions: Algebraically, Numerically, Graphically, and Verbally

If you pour a cup of coffee, it cools more rapidly at first, then less rapidly, finally approaching room temperature. Since there is *one and only one* temperature at any one given time, the temperature is called a **function** of time. In this course you'll refresh your memory about some kinds of functions you have studied in previous courses. You'll also learn some new kinds of functions, and you'll learn properties of functions so that you will be comfortable with them in later calculus courses. In this section you'll see that you can study functions in four ways.

OBJECTIVE Work with functions that are defined algebraically, graphically, numerically, or verbally.

You can show the relationship between coffee temperature and time **graphically.** Figure 1-1a shows the temperature, y, as a function of time, x. At $x = 0$ the coffee has just been poured. The graph shows that as time goes on, the temperature levels off, until it is so close to room temperature, 20°C, that you can't tell the difference.

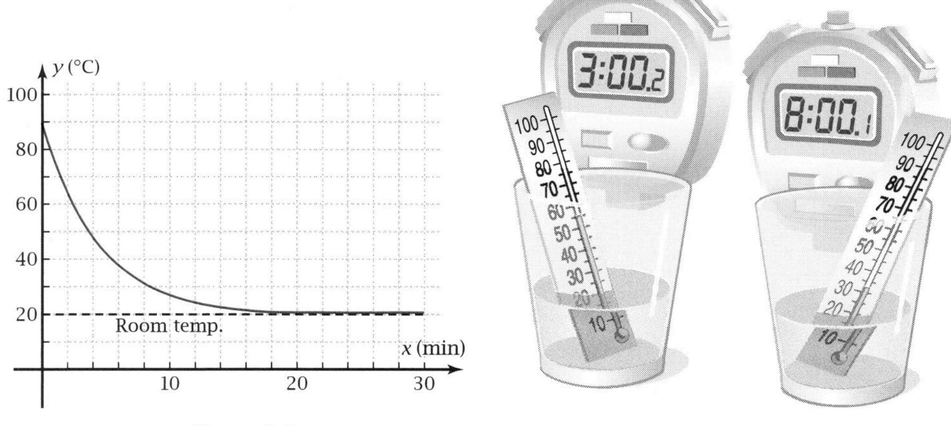

Figure 1-1a

This graph might have come from numerical data, found by experiment. It actually came from an **algebraic** equation, $y = 20 + 70(0.8)^x$.

From the equation, you can find **numerical** information. If you enter the equation into your grapher, then use the table feature, you can find these temperatures, rounded to 0.1 degree.

x (min)	y (°C)
0	90
5	42.9
10	27.5
15	22.5
20	20.8

Functions that are used to make predictions and interpretations about something in the real world are called **mathematical models.** Temperature is the **dependent variable** because the temperature of the coffee depends on the time it has been cooling. Time is the **independent variable.** You cannot change

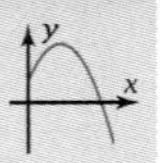

time simply by changing coffee temperature! Always plot the independent variable on the horizontal axis and the dependent variable on the vertical axis.

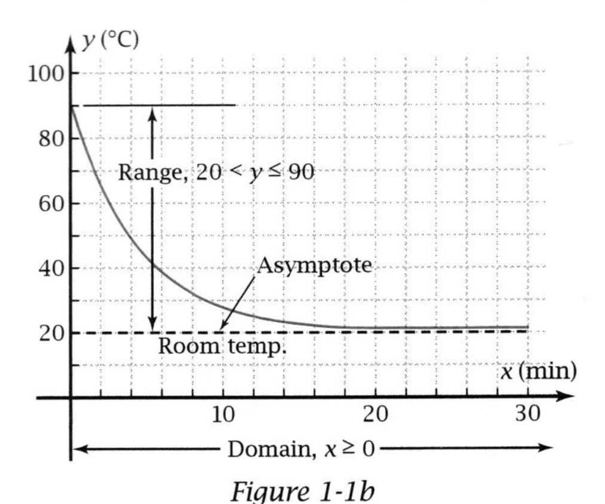

Figure 1-1b

The set of values the independent variable of a function can have is called the **domain.** In the coffee cup example, the domain is the set of nonnegative numbers, or $x \geq 0$. The set of values of the dependent variable corresponding to the domain is called the **range** of the function. If you don't drink the coffee (which would end the domain), the range is the set of temperatures between 20°C and 90°C, including the 90°C but not the 20°C, or $20 < y \leq 90$. The horizontal line at 20°C is called an **asymptote.** The word comes from the Greek *asymptotos,* meaning "not due to coincide." The graph gets arbitrarily close to the asymptote but never touches it. Figure 1-1b shows the domain, range, and asymptote.

This example shows you how to describe a function **verbally.**

▶ ***EXAMPLE 1*** The time it takes you to get home from a football game is related to how fast you drive. Sketch a reasonable graph showing how this time and speed are related. Tell the domain and range of the function.

Solution It seems reasonable to assume that the time it takes *depends on* the speed you drive. So you must plot time on the *vertical* axis and speed on the *horizontal* axis.

To see what the graph should look like, consider what happens to the time as the speed varies. Pick a speed value and plot a point for the corresponding time (Figure 1-1c). Then pick a faster speed. Because the time will be shorter, plot a point closer to the horizontal axis (Figure 1-1d).

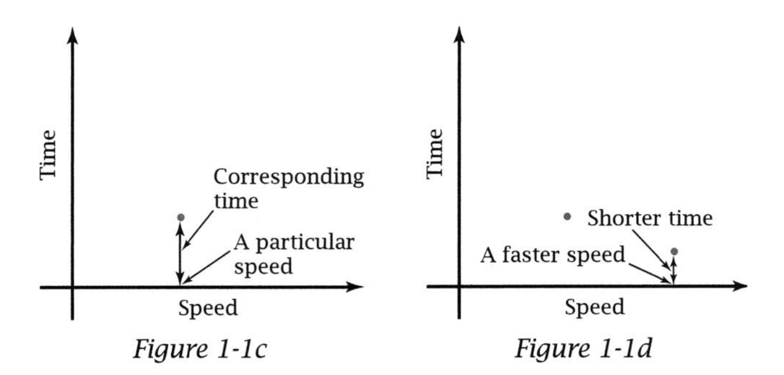

Figure 1-1c *Figure 1-1d*

For a slower speed, the time will be longer. Plot a point farther from the horizontal axis (Figure 1-1e). Finally, connect the points with a smooth curve, since it is possible to drive at any speed within the speed limit. The graph never touches either axis, as shown in Figure 1-1f. If speed were zero, you would never get home. The length of time would be infinite. Also, no matter how fast you drive, it will always take you some time to get home. You cannot arrive instantaneously.

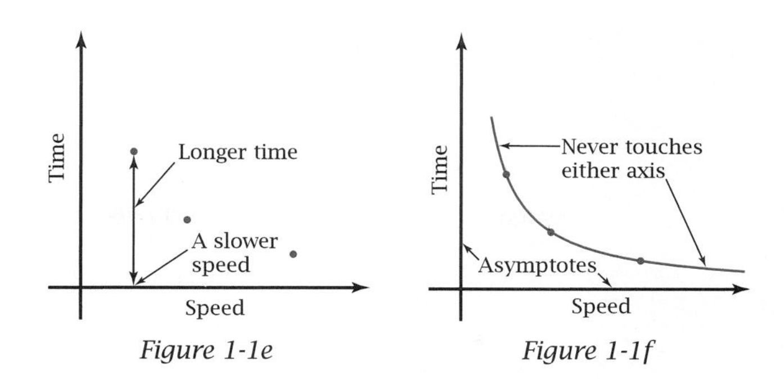

Figure 1-1e

Figure 1-1f

Domain: 0 ≤ speed ≤ speed limit

Range: time ≥ minimum time at speed limit ◀

This problem set will help you see the relationship between variables in the real world and functions in the mathematical world.

Problem Set 1-1

Calc

1. *Archery Problem 1:* An archer climbs a tree near the edge of a cliff, then shoots an arrow high into the air. The arrow goes up, then comes back down, going over the cliff and landing in the valley, 30 m below the top of the cliff. The arrow's height, y, in meters above the top of the cliff depends on the time, x, in seconds, since the archer released it. Figure 1-1g shows the height as a function of time.

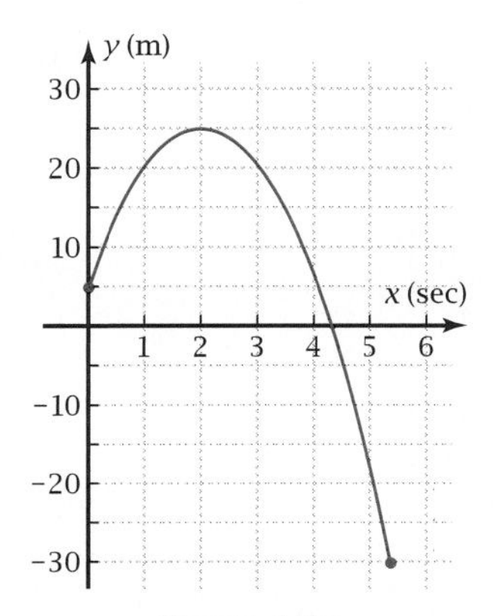

Figure 1-1g

a. What was the approximate height of the arrow at 1 second? At 5 seconds? How do you explain the fact that the height is negative at 5 seconds?

b. At what *two* times was the arrow at 10 m above the ground? At what time does the arrow land in the valley below the cliff?

c. How high was the archer above the ground at the top of the cliff when she released the arrow?

d. Why can you say that altitude is a *function* of time? Why is time *not* a function of altitude?

e. What is the domain of the function? What is the corresponding range?

2. *Gas Temperature and Volume Problem:* When you heat a fixed amount of gas, it expands, increasing its volume. In the late 1700s, French chemist Jacques Charles used numerical measurements of the temperature and volume of a gas to find a quantitative relationship between these two variables. Suppose that these temperatures and volumes had been recorded for a fixed amount of oxygen.

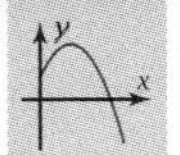

Pietro Longhi's painting, The Alchemists, *depicts a laboratory setting from the middle of the 18th century.*

T (°C)	V (l)
0	9.5
50	11.2
100	12.9
150	14.7
200	16.4
250	18.1
300	19.9

a. On graph paper, plot V as a function of T. Choose scales that go at least from $T = -300$ to $T = 400$.

b. You should find, as Charles did, that the points lie in a straight line! Extend the line backward until it crosses the T-axis. The temperature you get is called *absolute zero*, the temperature at which, supposedly, all molecular motion stops. Based on your graph, what temperature in degrees Celsius is absolute zero? Is this the number you recall from science courses?

c. Extending a graph beyond all given data, as you did in 2b, is called **extrapolation.** "*Extra-*" means "beyond," and "*-pol-*" comes from "pole," or end. Extrapolate the graph to $T = 400$ and predict what the volume would be at 400°C.

d. Predict the volume at $T = 30$°C. Why do you suppose this prediction is an example of **interpolation**?

e. At what temperature would the volume be 5 liters? Which do you use, interpolation or extrapolation, to find this temperature?

f. Why can you say that the volume is a *function* of temperature? Is it also true that the temperature is a function of volume? Explain.

g. Considering volume to be a function of temperature, write the domain and the range for this function.

h. See if you can write an algebraic equation for V as a function of T.

i. In this problem, the temperature is the independent variable and the volume is the dependent variable. This implies that you can change the volume by changing the temperature. Is it possible to change the *temperature* by changing the *volume,* such as you would do by pressing down on the handle of a tire pump?

3. *Mortgage Payment Problem:* People who buy houses usually get a loan to pay for most of the house and pay on the resulting *mortgage* each month. Suppose you get a \$50,000 loan and pay it back at \$550.34 per month with an interest rate of 12% per year (1% per month). Your balance, B dollars, after n monthly payments is given by the algebraic equation

$$B = 50{,}000(1.01^n) + \frac{550.34}{0.01}(1 - 1.01^n)$$

a. Make a table of your balances at the end of each 12 months for the first 10 years of the mortgage. To save time, use the table feature of your grapher to do this.

b. How many months will it take you to pay off the entire mortgage? Show how you get your answer.

c. Plot on your grapher the graph of B as a function of n from $n = 0$ until the mortgage is paid off. Sketch the graph on your paper.

d. True or false: "After half the payments have been made, half the original balance remains to be paid." Show that your conclusion agrees with your graph from part c.

e. Give the domain and range of this function. Explain why the domain contains only *integers*.

4. *Stopping Distance Problem:* The distance your car takes to stop depends on how fast you are going when you apply the brakes. You may recall from driver's education that it takes more than twice the distance to stop your car if you double your speed.

a. Sketch a reasonable graph showing your stopping distance as a function of speed.

b. What is a reasonable domain for this function?

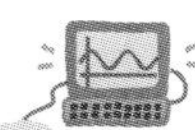

c. Consult a driver's manual, the Internet, or another reference source to see what the stopping distance is for the maximum speed you stated for the domain in part b.

d. When police investigate an automobile accident, they estimate the speed the car was going by measuring the length of the skid marks. Which are they considering to be the independent variable, the speed or the length of the skid marks? Indicate how this would be done by drawing arrows on your graph from part a.

5. *Stove Heating Element Problem:* When you turn on the heating element of an electric stove, the temperature increases rapidly at first, then levels off. Sketch a reasonable graph showing temperature as a function of time. Show the horizontal asymptote. Indicate on the graph the domain and range.

6. In mathematics you learn things in four ways—algebraically, graphically, numerically, and verbally.

a. In which of the five problems above was the function given algebraically? Graphically? Numerically? Verbally?

b. In which problem or problems of the five above did you go from verbal to graphical? From algebraic to numerical? From numerical to graphical? From graphical to algebraic? From graphical to numerical? From algebraic to graphical?

1-2 Kinds of Functions

In the last section you learned that you can describe functions algebraically, numerically, graphically, or verbally. A function defined by an algebraic equation often has a descriptive name. For instance, the function $y = -x^2 + 5x + 3$ is called *quadratic* because the highest power of x is x squared and *quadrangle* is one term for a square. In this section you will refresh your memory about verbal names for algebraically defined functions and see what their graphs look like.

OBJECTIVE Make connections among the algebraic equation for a function, its name, and its graph.

Definition of Function

If you plot the function $y = -x^2 + 5x + 3$, you get a graph that rises and then falls, as shown in Figure 1-2a. For any x-value you pick, there is only *one* y-value. This is not the case for all graphs. For example, in Figure 1-2b there are places where the graph has more than one y-value for the same x-value. Although the two variables are related, the relation is not called a function.

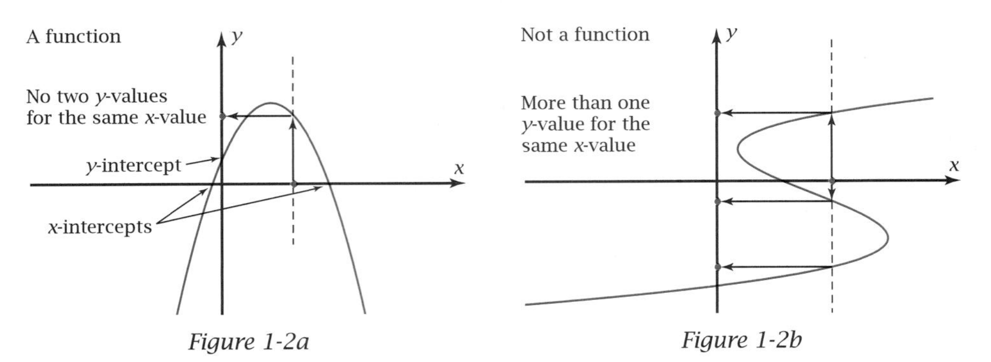

Figure 1-2a *Figure 1-2b*

Each point on a graph corresponds to an **ordered pair** of numbers, (x, y). A **relation** is any set of ordered pairs. A **function** is a set of ordered pairs for which each value of the independent variable (often x) in the domain has only *one* corresponding value of the dependent variable (often y) in the range. So Figures 1-2a and 1-2b are both graphs of relations, but only Figure 1-2a is the graph of a function.

The ***y*-intercept** of a function is the value of y when $x = 0$. It gives the place where the graph crosses the y-axis (Figure 1-2a). An ***x*-intercept** is a value of x for which $y = 0$. Functions can have more than one x-intercept.

$f(x)$ Terminology

You should recall $f(x)$ notation from previous courses. It is used for y, the dependent variable of a function. With it, you show what value you substitute for x, the independent variable. For instance, to substitute 4 for x in the quadratic function, $f(x) = x^2 + 5x + 3$, you would write:

$$f(4) = 4^2 + 5(4) + 3 = 39$$

The symbol $f(4)$ is pronounced "f of 4" or sometimes "f at 4." You must recognize that the parentheses mean *substitution* and *not* multiplication.

This notation is also useful if you are working with more than one function of the same independent variable. For instance, the height and velocity of a falling object both depend on time, t. So you could write the equations for the two functions this way:

$$h(t) = -4.9t^2 + 10t + 70 \quad \text{(for the height)}$$

$$v(t) = -9.8t + 10 \quad \text{(for the velocity)}$$

In $f(x)$, the variable x or any value substituted for x is called the **argument** of the function. It is important for you to distinguish between f and $f(x)$. The symbol f is the *name* of the function. The symbol $f(x)$ is the *y-value* of the function. For instance, if f is the square root function, then $f(x) = \sqrt{x}$ and $f(9) = \sqrt{9} = 3$.

Names of Functions

Functions are named for the operation performed on the independent variable. Here are some kinds of functions you may recall from previous courses, along with their typical graphs. In these examples, the letters a, b, c, m, and n are real numbers; the letter x stands for the independent variable; and y or $f(x)$ stands for the dependent variable.

- **Polynomial function,** *Figure 1-2c*

  General equation: $f(x) = a_n x^n + a_{n-1} x^{n-1} + \cdots + a_1 x + a_0$, where n is a nonnegative integer

  Verbally: $f(x)$ is a polynomial function of x. (If $n = 3$, f is a cubic function. If $n = 4$, f is a quartic function.)

  Features: The graph crosses the x-axis up to n times and has up to $n - 1$ vertices (points where the function changes direction). The domain is all real numbers.

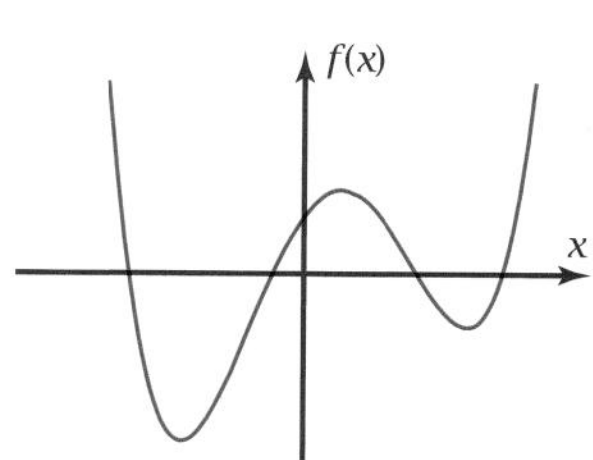
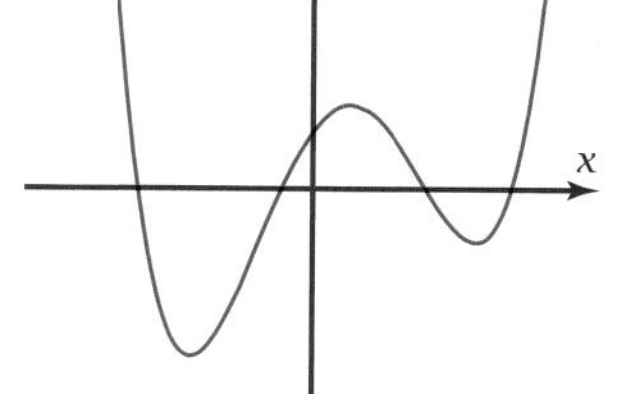

Figure 1-2c

- **Quadratic function,** *Figure 1-2d* (another special case of a polynomial function)

  General equation: $f(x) = ax^2 + bx + c$

  Verbally: $f(x)$ varies quadratically with x, or $f(x)$ is a quadratic function of x.

  Features: The graph changes direction at its one vertex. The domain is all real numbers.

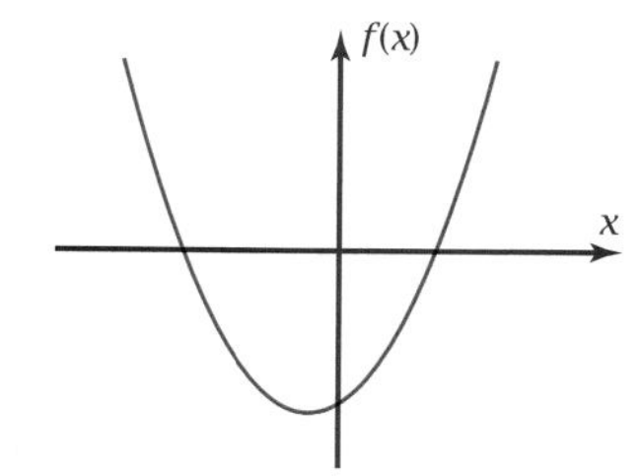

Figure 1-2d

- **Linear function,** *Figure 1-2e* (special case of a polynomial function)

  General equation: $f(x) = ax + b$ (or $f(x) = mx + b$)

  Verbally: $f(x)$ varies linearly with x, or $f(x)$ is a linear function of x.

  Features: The straight-line graph, $f(x)$, changes at a constant rate as x changes. The domain is all real numbers.

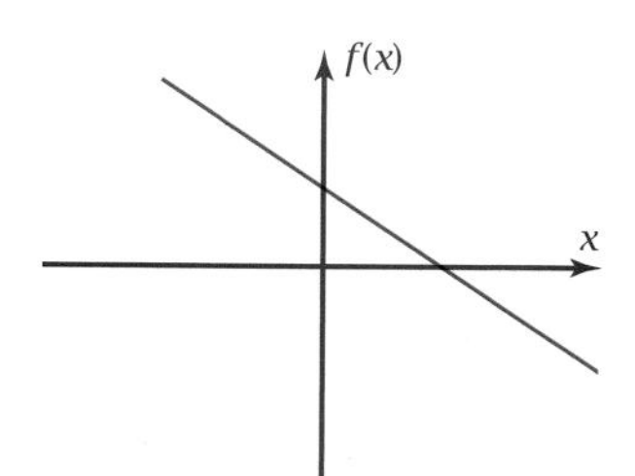

Figure 1-2e

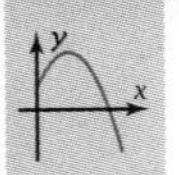

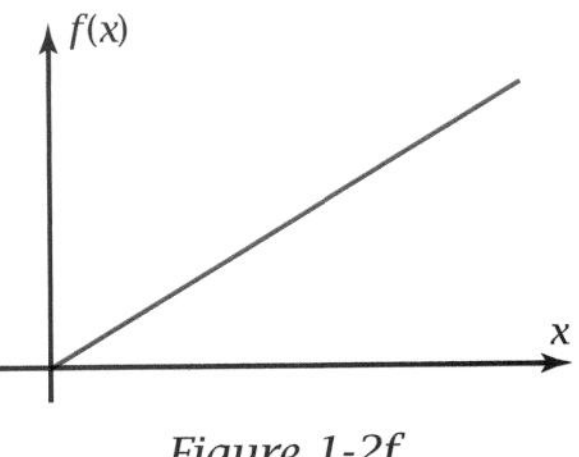

Figure 1-2f

- **Direct variation function,** *Figure 1-2f* (a special case of a linear, power, or polynomial function)

  General equation: $f(x) = ax$ (or $f(x) = mx + 0$, or $f(x) = ax^1$)

  Verbally: $f(x)$ varies directly with x, or $f(x)$ is directly proportional to x.

  Features: The straight-line graph goes through the origin. The domain is $x \geq 0$ (as shown) for most real-world applications.

- **Power function,** *Figure 1-2g* (a polynomial function if b is a nonnegative integer)

  General equation: $f(x) = ax^b$ (a *variable* with a *constant* exponent)

  Verbally: $f(x)$ varies directly with the bth power of x, or $f(x)$ is directly proportional to the bth power of x.

  Features: The graph contains the origin if b is positive. The domain is nonnegative real numbers if b is positive and positive real numbers if b is negative.

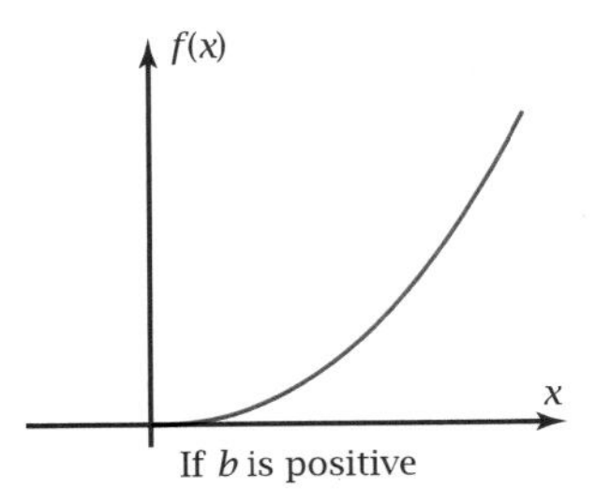

If b is positive

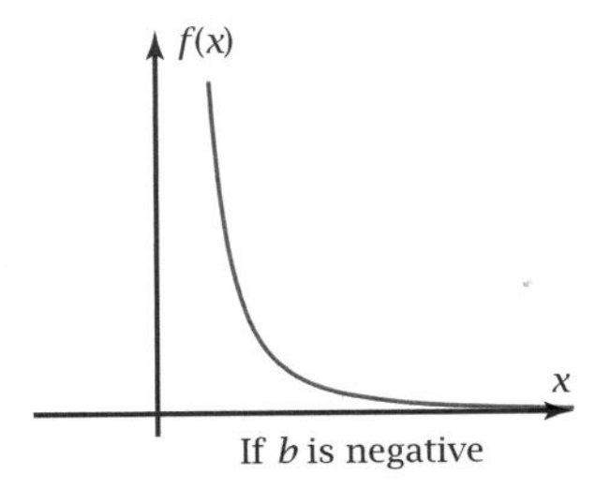

If b is negative

Figure 1-2g

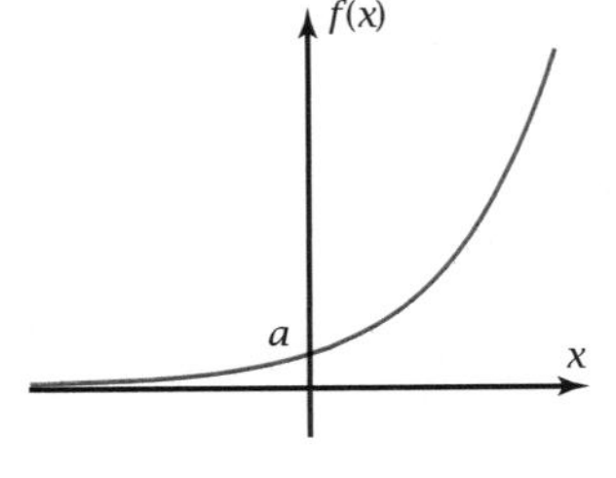

Figure 1-2h

- **Exponential function,** *Figure 1-2h*

  General equation: $f(x) = a \cdot b^x$ (a *constant* with a *variable* exponent)

  Verbally: $f(x)$ varies exponentially with x, or $f(x)$ is an exponential function of x.

  Features: The graph crosses the y-axis at $f(0) = a$ and has the x-axis as an asymptote.

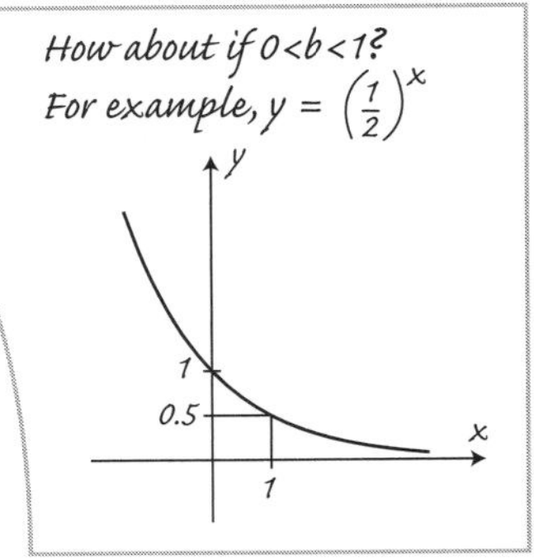

f(x)

x

If n is odd

Figure 1-2i

- **Inverse variation function,** *Figure 1-2i* (special case of power function)

  General equation: $f(x) = \dfrac{a}{x}$,

  or

  $f(x) = \dfrac{a}{x^n}$ ($f(x) = ax^{-1}$, or $f(x) = ax^{-n}$)

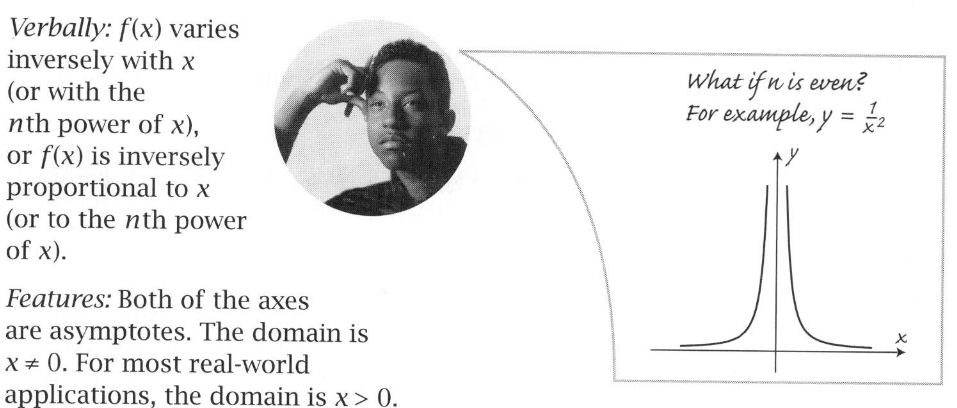

Verbally: $f(x)$ varies inversely with x (or with the nth power of x), or $f(x)$ is inversely proportional to x (or to the nth power of x).

Features: Both of the axes are asymptotes. The domain is $x \neq 0$. For most real-world applications, the domain is $x > 0$.

- **Rational algebraic function,** *Figure 1-2j*

General equation: $f(x) = \dfrac{p(x)}{q(x)}$, where p and q are polynomial functions.

Verbally: $f(x)$ is a rational function of x.

Features: A rational function has a discontinuity (asymptote or missing point) where the denominator is 0; may have horizontal or other asymptotes.

Figure 1-2j

Restricted Domains and Boolean Variables

Suppose that you want to plot a graph using only part of your grapher's window. For instance, let the height of a growing child between ages 3 and 10 be given by $y = 3x + 26$, where x is in years and y is in inches. The domain here is $3 \leq x \leq 10$.

Your grapher has **Boolean variable** capability. A Boolean variable, named for George Boole, an Irish logician and mathematician (1815–1864), equals 1 if a given condition is true and 0 if that condition is false. For instance, the compound statement

$$(x \geq 3 \text{ and } x \leq 10)$$

equals 1 if $x = 7$ (which *is* between 3 and 10) and equals 0 if $x = 2$ or $x = 15$ (neither of which is between 3 and 10). To plot a graph in a **restricted domain,** divide the equation by the appropriate Boolean variable. For the equation above, enter

$$y_1 = 3x + 26\ /\ (x \geq 3 \text{ and } x \leq 10)$$

If x is between 3 and 10, inclusive, the $3x + 26$ is divided by 1, which leaves it unchanged. If x is not between 3 and 10, inclusive, the $3x + 26$ is divided by 0 and the grapher plots nothing. Here is an example.

▶ ***EXAMPLE 1*** Plot the graph of $f(x) = -x^2 + 5x + 3$ in the domain $0 \leq x \leq 4$. What kind of function is this? Give the range. Find a pair of real-world variables that could have a relationship described by a graph of this shape.

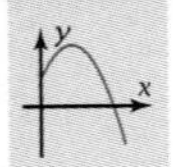

Solution Enter $y_1 = -x^2 + 5x + 3 \,/\, (x \geq 0 \text{ and } x \leq 4)$ Divide by a Boolean variable.

The graph in Figure 1-2k shows the restricted domain.

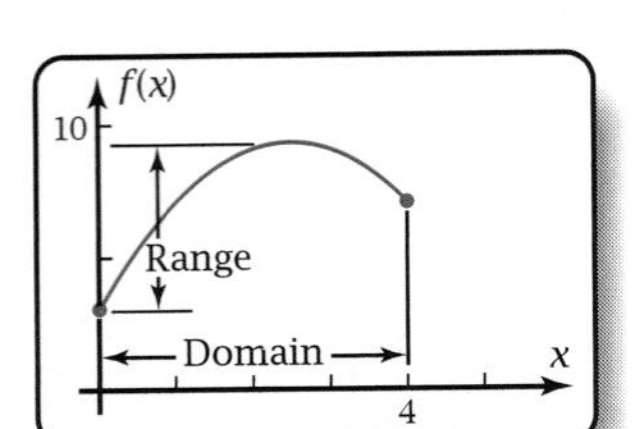

Figure 1-2k

The function is quadratic because $f(x)$ equals a second-degree expression in x.

The range is $3 \leq f(x) \leq 9.25$. You can find this interval by tracing to the left endpoint of the graph where $f(0) = 3$ and to the high point where $f(2.5) = 9.25$. (At the right endpoint, $f(4) = 7$, which is between 3 and 9.25.)

The function could represent the relationship between something that rises for a while then falls, such as a punted football's height as a function of time or (if $f(x)$ is multiplied by 10) the grade you could make on a test as a function of the number of hours you study for it. (The grade could be lower for longer times if you stay up too late and thus are sleepy during the test.) ◀

TECHNIQUE: Restricted Domain and Boolean Variables

A **Boolean variable** is a variable that has a given condition attached to it. If the condition is true, the variable equals 1. If the condition is false, the variable is 0.

To plot a function in a **restricted domain,** divide any one term of the function's equation by the appropriate Boolean variable.

▶ EXAMPLE 2 As children grow older, their height and weight are related. Sketch a reasonable graph to show this relation and describe it. Tell what kind of function has a graph like the one you drew.

Solution Weight depends on height, so weight is on the vertical axis, as shown on the graph in Figure 1-2l. The graph curves upward because doubling the height more than doubles the weight. Extending the graph sends it through the origin, but the domain starts beyond the origin at a value greater than zero, since a person never has zero height or weight. The graph stops at the person's adult height and weight. A power function has a graph like this.

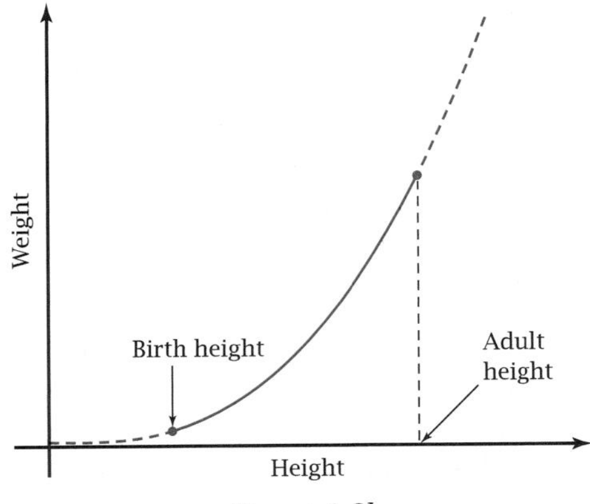

Figure 1-2l

◀

Problem Set 1-2

For Problems 1–4

a. Plot the graph on your grapher using the domain given. Sketch the result on your paper.

b. Give the range of the function.

c. Name the kind of function.

d. Describe a pair of real-world variables that could be related by a graph of this shape.

1. $f(x) = 2x + 3$ domain: $0 \le x \le 10$
2. $f(x) = 0.2x^3$ domain: $0 \le x \le 4$
3. $g(x) = \frac{12}{x}$ domain: $0 < x \le 10$
4. $h(x) = 5 \times 0.6^x$ domain: $-5 \le x \le 5$

For Problems 5–18

a. Plot the graph using a window set to show the entire graph, when possible. Sketch the result.

b. Give the y-intercept and any x-intercepts and locations of any vertical asymptotes.

c. Give the range.

5. Quadratic (polynomial) function $f(x) = -x^2 + 4x + 12$ with the domain $0 \le x \le 5$
6. Quadratic (polynomial) function $f(x) = x^2 - 6x + 40$ with the domain $0 \le x \le 8$
7. Cubic (polynomial) function $f(x) = x^3 - 7x^2 + 4x + 12$ with the domain $-1 \le x \le 7$
8. Quartic (polynomial) function $f(x) = x^4 + 3x^3 - 8x^2 - 12x + 16$ with the domain $-3 \le x \le 3$
9. Power function $f(x) = 3x^{2/3}$ with the domain $0 \le x \le 8$
10. Power function $f(x) = 0.3x^{1.5}$ with the domain $0 \le x \le 9$
11. Linear function $f(x) = -0.7x + 4$ with the domain $-3 \le x \le 10$
12. Linear function $f(x) = 3x + 6$ with the domain $-5 \le x \le 5$
13. Exponential function $f(x) = 3 \times 1.3^x$ with the domain $-5 \le x \le 5$
14. Exponential function $f(x) = 20 \times 0.7^x$ with the domain $-5 \le x \le 5$
15. Inverse square variation function $f(x) = \frac{25}{x^2}$ with the domain $x > 0$
16. Direct variation function $f(x) = 5x$ with the domain $x \ge 0$
17. Rational function $y = \frac{x - 2}{(x - 4)(x + 1)}$ with the domain $-3 \le x \le 6$, $x \ne 4$, $x \ne -1$
18. Rational function $y = \frac{x^2 - 2x - 2}{x - 3}$ with the domain $-2 \le x \le 6$, $x \ne 3$

For Problems 19–28, name a kind of function that has the graph shown.

19.

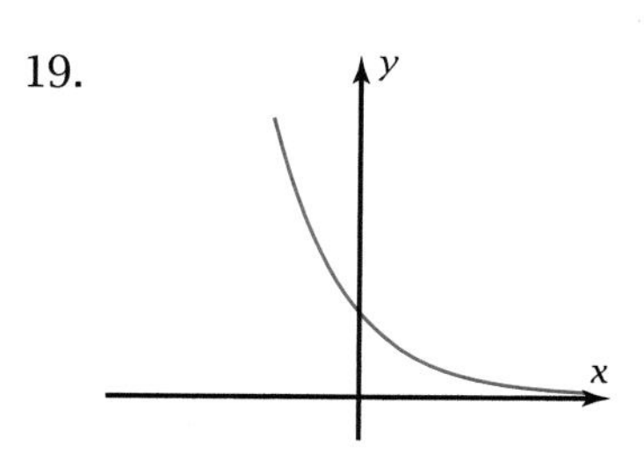

20.

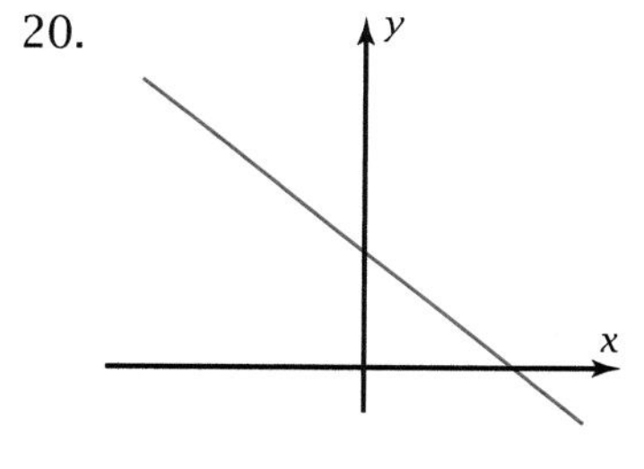

21.

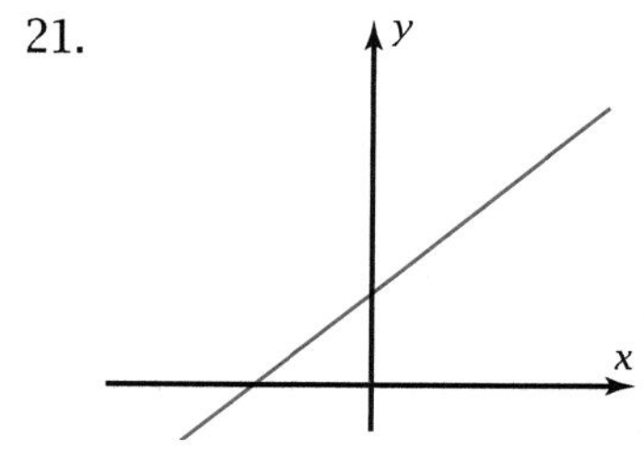

22.

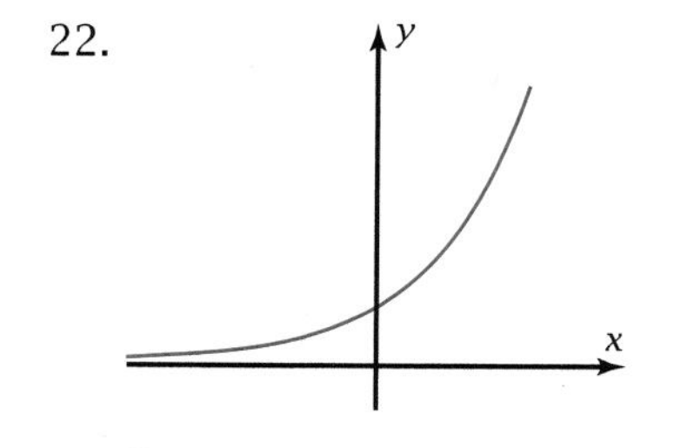

23.

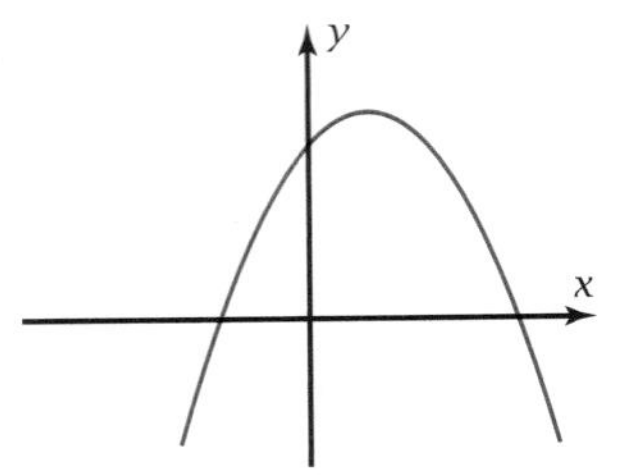

24.

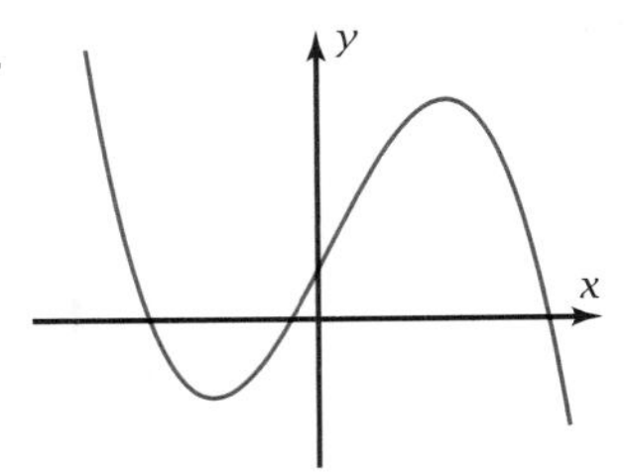

25.

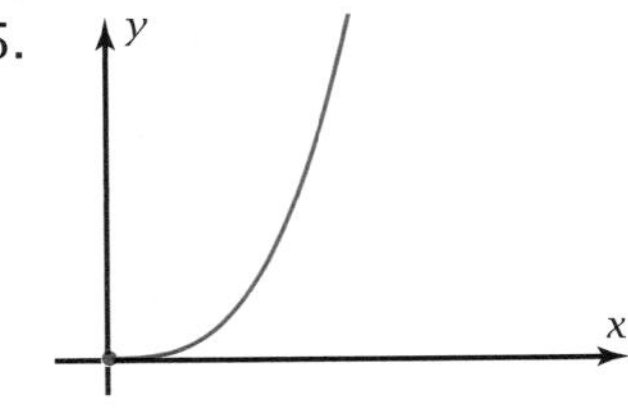

26.

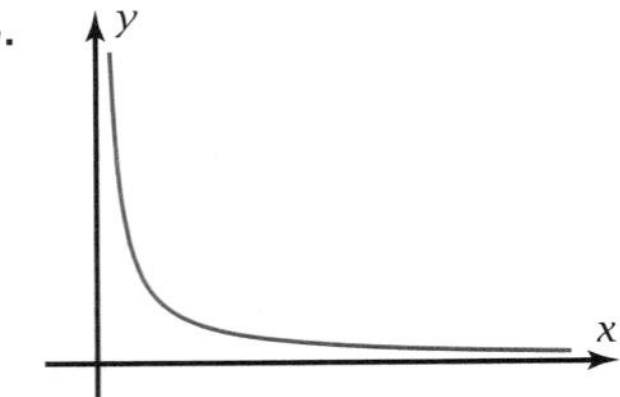

27.

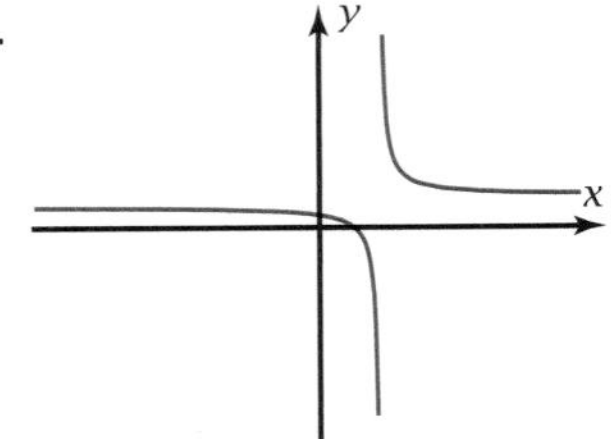

28.

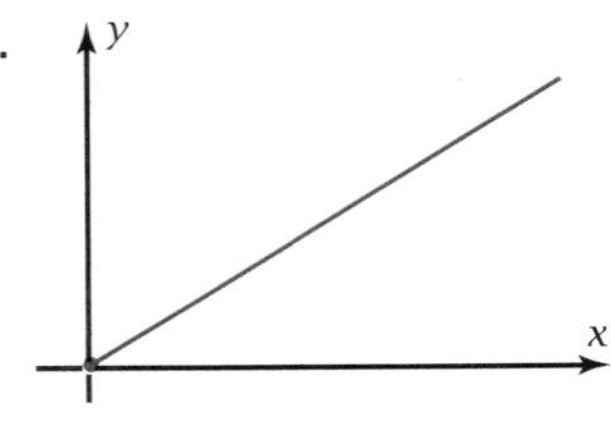

For Problems 29–32

a. Sketch a reasonable graph showing how the variables are related.

b. Identify the type of function it could be (quadratic, power, exponential, and so on).

29. The weight and length of a dog.

30. The temperature of a cup of coffee and the time since the coffee was poured.

31. The purchase price of a home in a particular neighborhood based on the price of the lot plus a certain fixed amount per square foot for the house. The cost is a function of the area, measured in square feet.

32. The height of a punted football as a function of the number of seconds since it was kicked.

For Problems 33–38, tell whether or not the relation graphed is a function. Explain how you made your decision.

33.

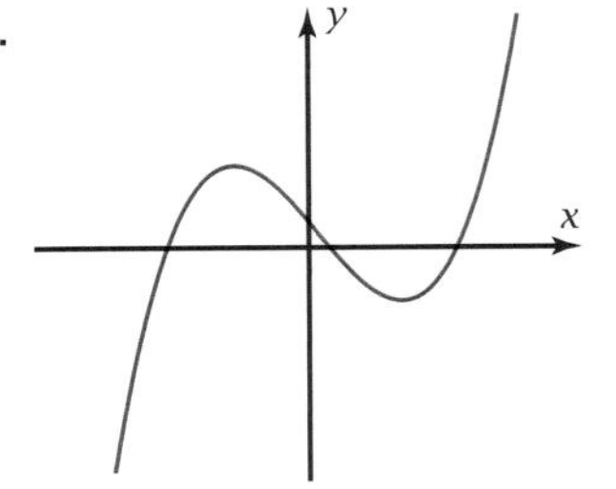

34.

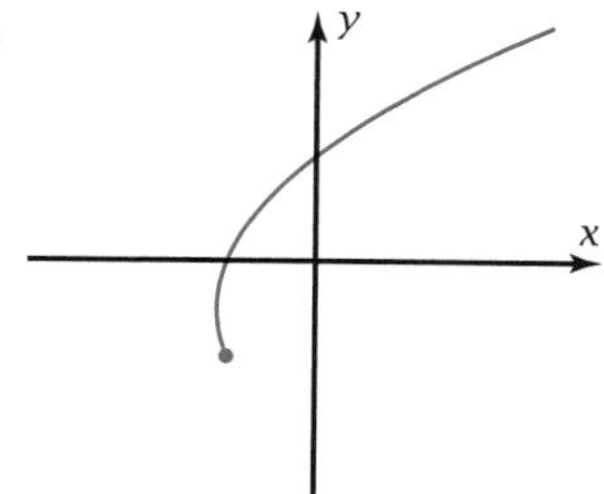

35.

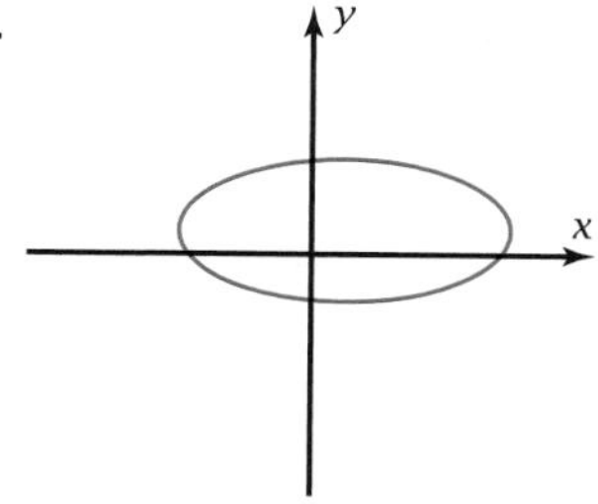

36.

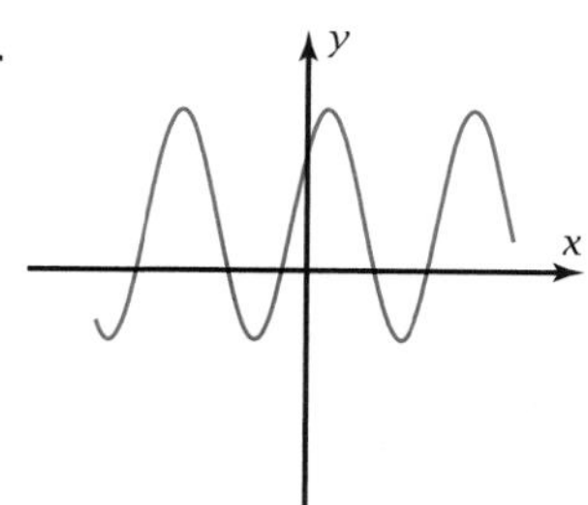

37.

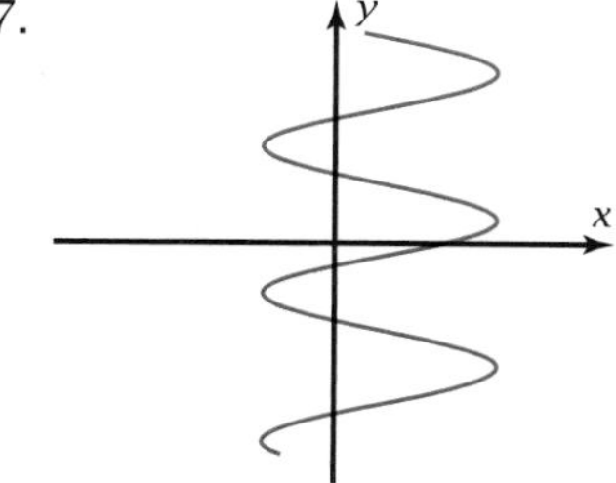

38.

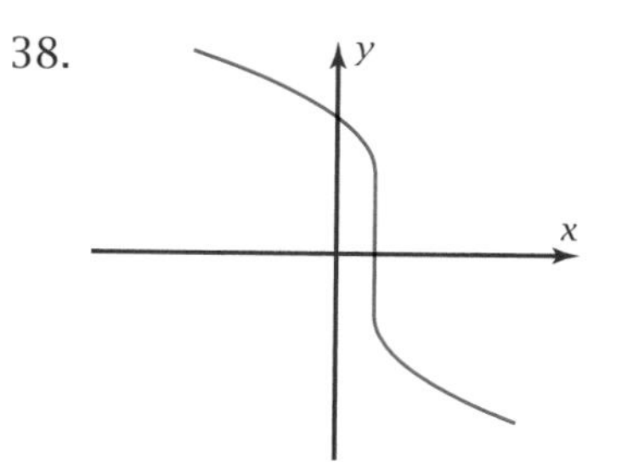

39. *Vertical Line Test Problem:* There is a graphical way to tell whether or not a relation is a function. It is called the **vertical line test.**

Property: The Vertical Line Test

If any vertical line cuts the graph of a relation in more than one place, then the relation is not a function.

Figure 1-2m illustrates the test.

a. Based on the definition of function, explain how the vertical line test distinguishes between relations that are functions and relations that are not.

b. Sketch the graphs in Problems 33 and 35. On your sketch, show how the vertical line test tells you that the relation in 33 is a function but the relation in 35 is not.

40. Explain why a function can have more than one x-intercept but only one y-intercept.

41. What is the *argument* of the function $y = f(x - 2)$?

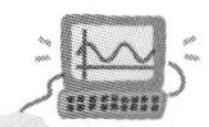

42. *Research Problem 1:* Look up George Boole on the Internet or another reference source. Describe several of Boole's accomplishments that you discover. Include your source.

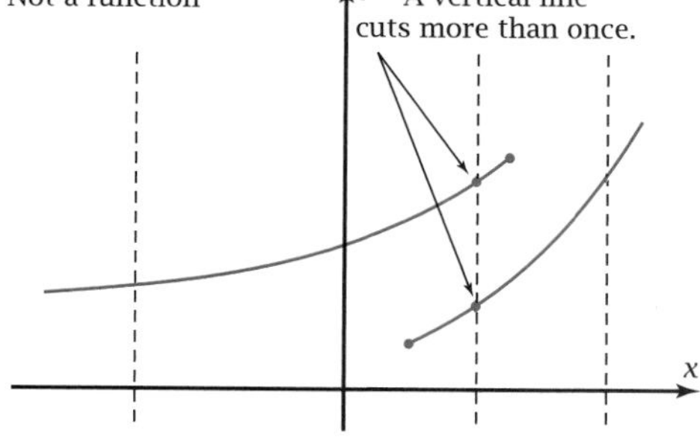

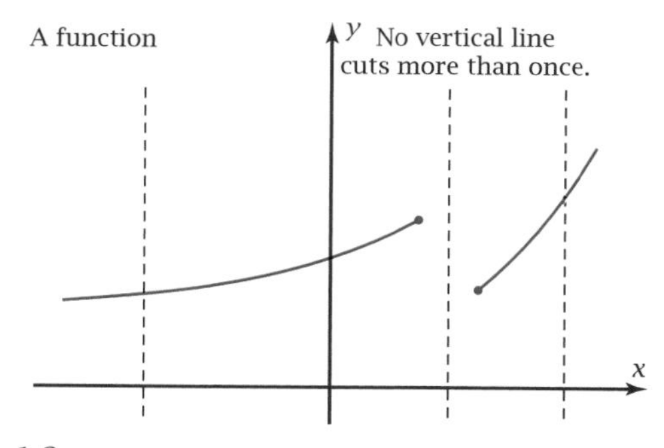

Figure 1-2m

1-3 Dilation and Translation of Function Graphs

Each of the two graphs below shows the unit semicircle and a transformation of it. The left graph shows the semicircle **dilated** (magnified) by a factor of 5 in the x-direction and by a factor of 3 in the y-direction. The right graph shows the unit semicircle **translated** by 4 units in the x-direction and by 2 units in the y-direction.

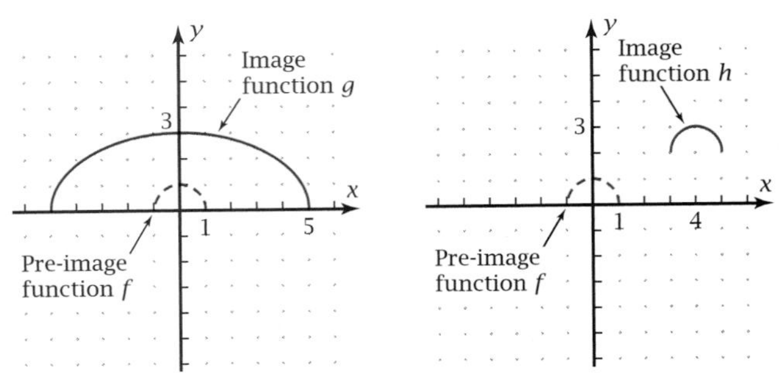

Figure 1-3a

The transformed functions, g and h, in Figure 1-3a are called **images** of the function f. The original function, f, is called the **pre-image.** In this section you will learn how to transform the equation of a function so that its graph will be dilated and translated by given amounts in the x- and y-directions.

OBJECTIVE Transform a given pre-image function so that the result is a graph of the image function that has been dilated by given factors and translated by given amounts.

Dilations

To get the vertical dilation in the left graph of Figure 1-3a, multiply each y-coordinate by 3. Figure 1-3b shows the image, $y = 3f(x)$.

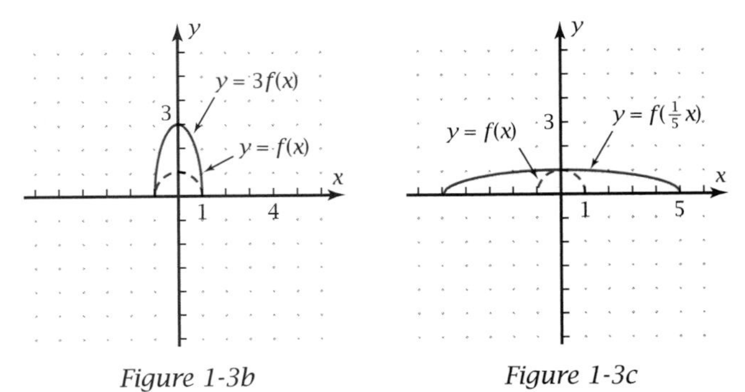

Figure 1-3b *Figure 1-3c*

The horizontal dilation is trickier. Each value of the argument must be 5 times what it was in the pre-image. Substituting u for the argument of f, you get

$$y = f(u)$$

$$x = 5u$$

$$\frac{1}{5}x = u$$

Figure 1-3c shows the graph of the image, $y = f\left(\frac{1}{5}x\right)$.

Putting the two transformations together gives the equation for $g(x)$.

$$g(x) = 3f\left(\frac{1}{5}x\right)$$

EXAMPLE 1 The equation of the pre-image function in Figure 1-3a is $f(x) = \sqrt{1 - x^2}$. Confirm on your grapher that $g(x) = 3f\left(\frac{1}{5}x\right)$ is the transformed image function

a. By direct substitution into the equation

b. By using the grapher's built-in variables feature

Solution a. $g(x) = 3\sqrt{1 - (x/5)^2}$ Substitute $x/5$ as the argument of f. Multiply the entire expression by 3.

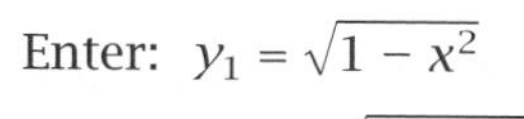

Enter: $y_1 = \sqrt{1 - x^2}$

$y_2 = 3\sqrt{1 - (x/5)^2}$

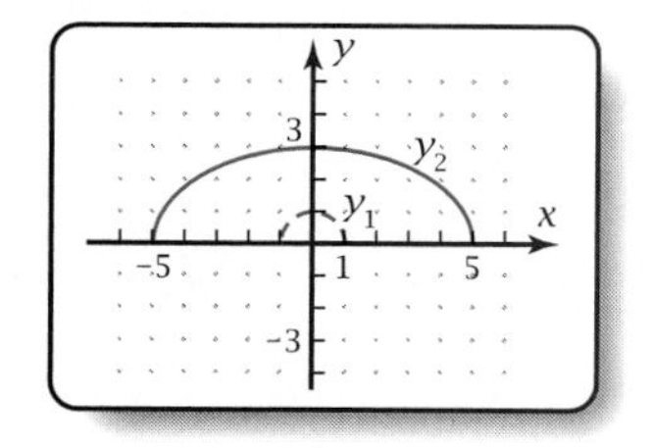

Figure 1-3d

The graph in Figure 1-3d shows a dilation by 5 in the x-direction and 3 in the y-direction. Use the grid-on feature to make the grid points appear. Use equal scales on the two axes so the graphs have the correct proportions. If the window is **friendly** in the x-direction (that is, integer values of x are grid points), then the graph will go all the way to the x-axis.

b. Enter: $y_3 = 3y_1(x/5)$ y_1 is the function *name* in this format, not the function value.

This graph is the same as the graph of y_2 in Figure 1-3d. ◀

You may ask, "Why do you *multiply* by the y-dilation and *divide* by the x-dilation?" You can see the answer by using y for $g(x)$ and dividing both sides of the equation by 3:

$$y = 3f\left(\frac{1}{5}x\right)$$

$$\frac{1}{3}y = f\left(\frac{1}{5}x\right)$$ Divide both sides by 3 $\left(\text{or multiply by } \frac{1}{3}\right)$.

You actually divide by *both* dilation factors, y by the y-dilation and x by the x-dilation.

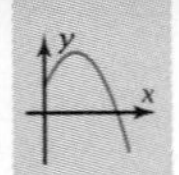

Translations

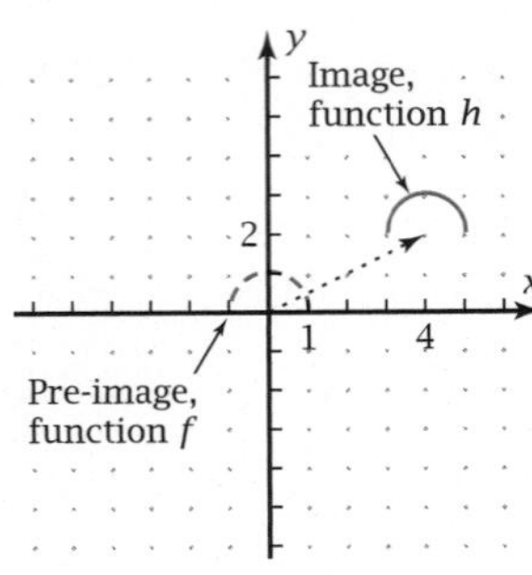

Figure 1-3e

The translations in Figure 1-3a that transform $f(x)$ to $h(x)$ are shown again in Figure 1-3e. To figure out what translation has been done, ask yourself, "To where did the point at the origin move?" As you can see, the center of the semicircle, initially at the origin, has moved to (4, 2). So there is a horizontal translation of 4 units and a vertical translation of 2 units.

To get a vertical translation of 2 units, add 2 to each y-value:

$$y = 2 + f(x)$$

To get a horizontal translation of 4 units, note that what was happening at $x = 0$ in function f has to be happening at $x = 4$ in function h. Again, substituting u for the argument of f gives

$$h(x) = 2 + f(u)$$

$$x = u + 4$$

$$x - 4 = u$$

$$h(x) = 2 + f(x - 4)$$ Substitute $x - 4$ as the argument of f.

▶ **EXAMPLE 2** The equation of the pre-image function in Figure 1-3e is $f(x) = \sqrt{1 - x^2}$. Confirm on your grapher that $g(x) = 2 + f(x - 4)$ is the transformed image function by

a. Direct substitution into the equation

b. Using the grapher's built-in variables feature

Solution

a. $g(x) = 2 + \sqrt{1 - (x - 4)^2}$ Substitute $x - 4$ for the argument. Add 2 to the expression.

Enter: $y_1 = \sqrt{1 - x^2}$

$y_2 = 2 + \sqrt{1 - (x - 4)^2}$

The graph in Figure 1-3f shows an x-translation of 4 and y-translation of 2.

b. Enter: $y_3 = 2 + y_1(x - 4)$

The graph is the same as that for y_2 in Figure 1-3f. ◀

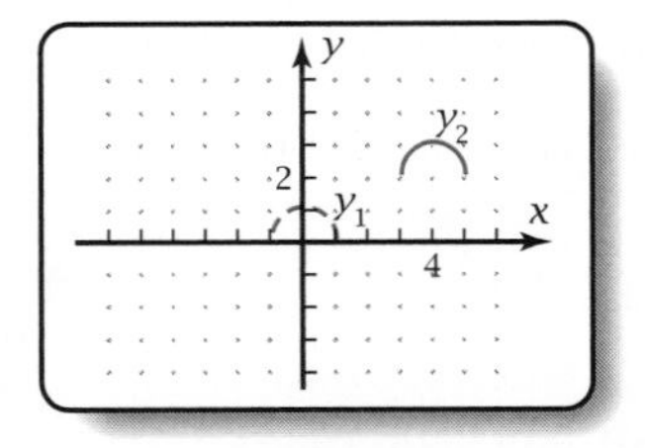

Figure 1-3f

Again, you may ask, "Why do you subtract an x-translation and add a y-translation?" The answer again lies in associating the y-translation with the y variable. You actually subtract *both* translations:

$$y = 2 + f(x - 4)$$

$$y - 2 = f(x - 4)$$ Subtract 2 from both sides.

The reason for writing the transformed equation with y by itself is to make it easier to calculate the dependent variable, either by pencil and paper or on your grapher.

This box summarizes the dilations and translations of a function and its graph.

PROPERTY: Dilations and Translations

The function g given by

$$\frac{1}{a}g(x) = f\left(\frac{1}{b}x\right) \quad \text{or, equivalently,} \quad g(x) = a\,f\left(\frac{1}{b}x\right)$$

represents a **dilation** by a factor of a in the y-direction and by a factor of b in the x-direction.

The function given by

$$h(x) - c = f(x - d) \quad \text{or, equivalently,} \quad h(x) = c + f(x - d)$$

represents a **translation** of c units in the y-direction and a translation of d units in the x-direction.

Note: If the function is only dilated, the x-dilation is the number you can substitute for x to make the argument equal 1. If the function is only translated, the x-translation is the number to substitute for x to make the argument equal 0.

▶ ***EXAMPLE 3*** The three graphs in Figure 1-3g show three different transformations of the pre-image graph to image graphs $y = g(x)$. Explain verbally what transformations were done. Write an equation for $g(x)$ in terms of the function f.

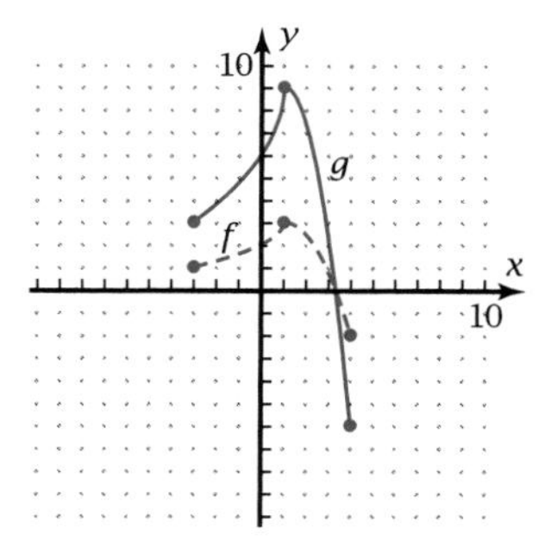

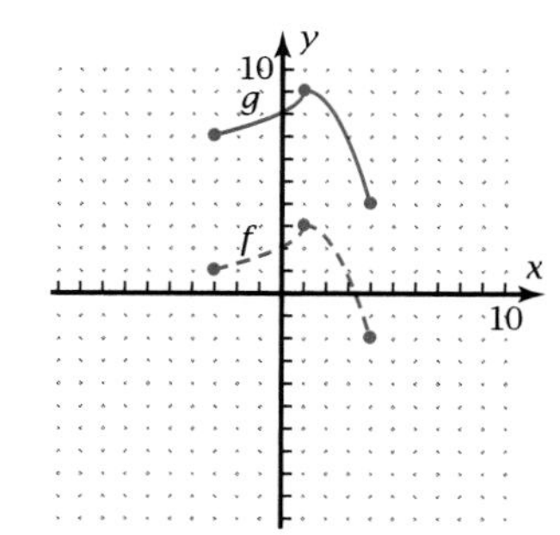

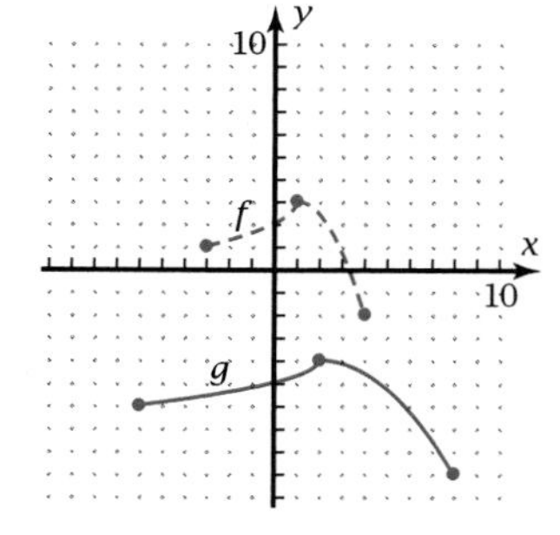

Figure 1-3g

Solution Left graph: Vertical dilation by a factor of 3

Equation: $g(x) = 3f(x)$

Note: Each point on the graph of g is 3 times as far from the x-axis as the corresponding point on the graph of f. Note that the vertical dilation moved points above the x-axis farther up and moved points below the x-axis farther down.

Middle graph: Vertical translation by 6 units

Equation: $g(x) = 6 + f(x)$

Note: The vertical dilation moved all points on the graph of f up by the same amount, 6 units. Also note that the fact that $g(1)$ is three times $f(1)$ is purely coincidental and is not true at other values of x.

Right graph: Horizontal dilation by a factor of 2 and vertical translation by −7

Equation: $g(x) = -7 + f\left(\frac{1}{2}x\right)$

Note: Each point on the graph of g is twice as far from the y-axis as the corresponding point on the graph of f. The horizontal dilation moved points to the right of the y-axis farther to the right and moved points to the left of the y-axis farther to the left. ◀

Problem Set 1-3

Do These Quickly

From now on there will be ten short problems at the beginning of most problem sets. Some of the problems are intended for review of skills from previous sections or chapters. Others are to test your general knowledge. Speed is the key here, not detailed work. Try to do all ten problems in less than five minutes.

Q1. $y = 3x^2 + 5x - 7$ is a particular example of a —?— function.

Q2. Write the general equation of a power function.

Q3. Write the general equation of an exponential function.

Q4. Calculate the product: $(x - 7)(x + 8)$

Q5. Expand: $(3x - 5)^2$

Q6. Sketch the graph of a relation that is not a function.

Q7. Sketch the graph of $y = \frac{2}{3}x + 4$.

Q8. Sketch an isosceles triangle.

Q9. Find 30% of 3000.

Q10. Which one of these is not the equation of a function?

A. $y = 3x + 5$ B. $f(x) = 3 - x^2$

C. $g(x) = |x|$ D. $y = \pm\sqrt{x}$

E. $y = 5x^{2/3}$

For Problems 1–6, let $f(x) = \sqrt{9 - x^2}$.

a. Write the equation for $g(x)$ in terms of x.

b. Plot the graphs of f and g on the same screen. Use a friendly window with integers from about −10 to 10 as grid points. Use the same scale on both axes. Sketch the result.

c. Describe how $f(x)$ was transformed to get $g(x)$.

1. $g(x) = 2f(x)$
2. $g(x) = -3 + f(x)$
3. $g(x) = f(x - 4)$
4. $g(x) = f\left(\frac{1}{3}x\right)$
5. $g(x) = 1 + f\left(\frac{1}{2}x\right)$
6. $g(x) = \frac{1}{2}f(x + 3)$

For Problems 7–12

a. Describe how the pre-image function f (dashed) was transformed to get the graph of image g (solid).

b. Write an equation for $g(x)$ in terms of the function f.

7.

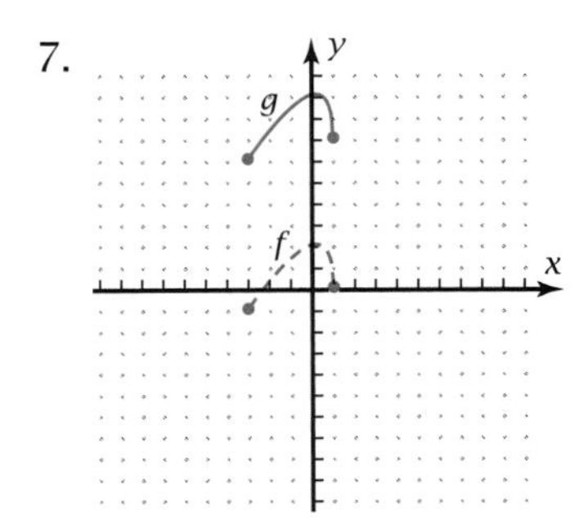

8.

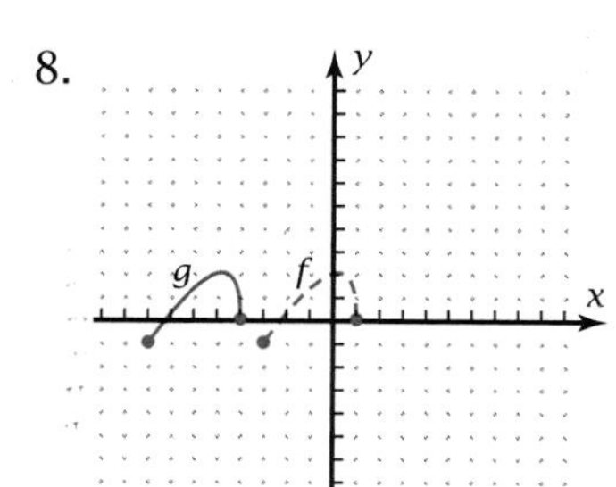

9.

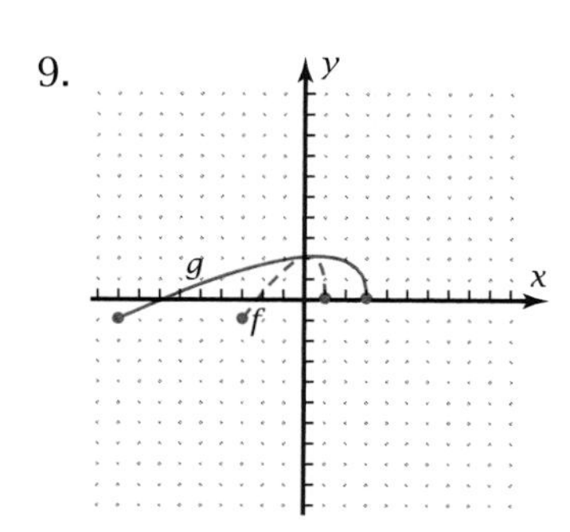

10.

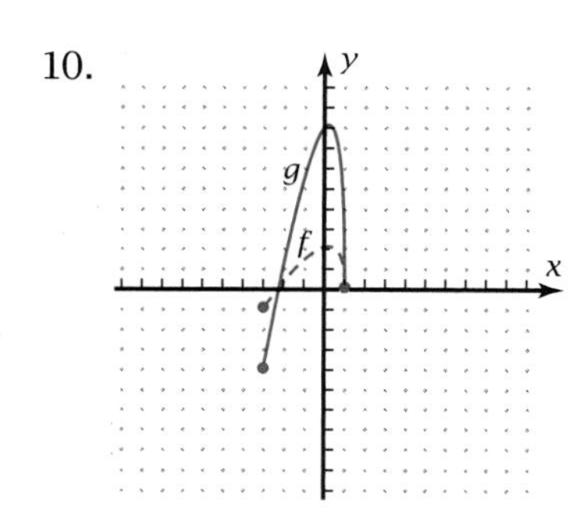

11.

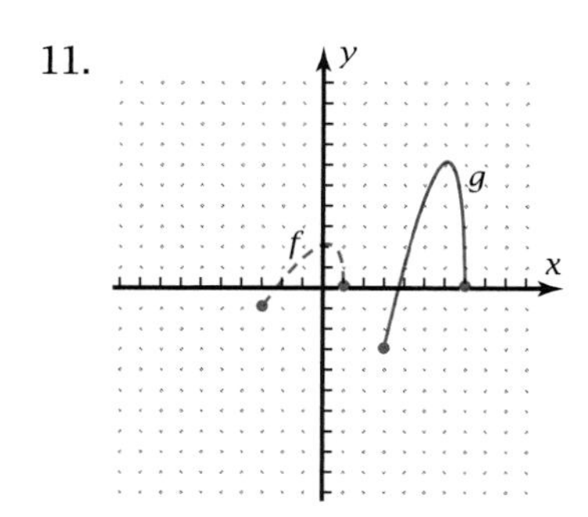

12.

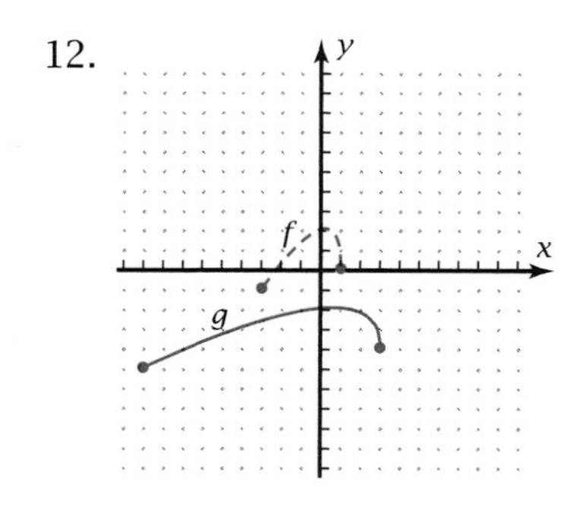

13. The equation of f in Problem 11 is $f(x) = 4\sqrt{1-x} + 2.25(x-1)$. Enter this equation and the equation for $g(x)$ into your grapher and plot the graphs. Does the result agree with the figure from Problem 11?

14. The equation of f in Problem 12 is $f(x) = 4\sqrt{1-x} + 2.25(x-1)$. Enter this equation and the equation for $g(x)$ into your grapher and plot the graphs. Does the result agree with the figure from Problem 12?

Figure 1-3h shows the graph of the pre-image function f. For Problems 15–20

a. Sketch the graph of the image function g on a copy of Figure 1-3h.

b. Identify the transformation(s) that are done.

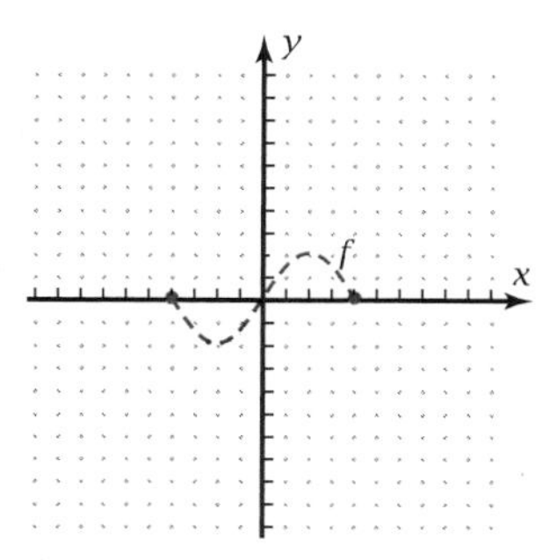

Figure 1-3h

15. $g(x) = f(x+9)$

16. $g(x) = f\left(\frac{1}{2}x\right)$

17. $g(x) = 5f(x)$

18. $g(x) = 4 + f(x)$

19. $g(x) = 5f(x+9)$

20. $g(x) = 4 + f\left(\frac{1}{2}x\right)$

The wind's force on the sail translates the windsurfer over the water.

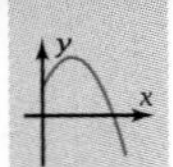

1-4 Composition of Functions

If you drop a pebble into a pond, a circular ripple extends out from the drop point (Figure 1-4a). The radius of the circular ripple is a function of time. The area enclosed by the circular ripple is a function of the radius. So area is a function of time through this chain of functions:

> Area depends on radius. Radius depends on time.

In this case, the area is a **composite function** of time. In this section you will learn some of the mathematics of composite functions.

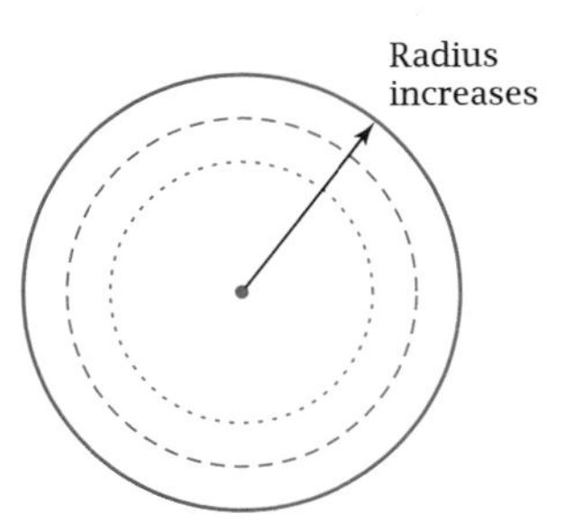

Figure 1-4a

OBJECTIVE Given two functions, graph and evaluate the composition of one function with the other.

Suppose that the radius of the ripple is increasing at a constant rate of 8 inches per second. Let $R(t)$ be the number of inches of the radius as a function of t seconds after the drop. Then you can write

$$R(t) = 8t$$

Recall that the area of a circle is πr^2, where r is the radius. So you can write

$$A(R(t)) = \pi(8t)^2 = 64\pi t^2$$

The symbol $A(R(t))$ is pronounced "A of R of t." Function R is called the **inside function** because you apply the rule that defines function R first to t. Function A is called the **outside function** and is applied to the value of the inside function at t. The symbol $A(R(t))$ can also be written $A \circ R(t)$ or $(A \circ R)(t)$. You can use the first set of parentheses to remind yourself that the entire symbol $A \circ R$ is the name of the function.

▶ **EXAMPLE 1** Let $f(x) = 2^x$ and let $g(x) = \sqrt{x}$. Find

a. $f(g(3))$.

b. $g(f(3))$. Show that it does *not* equal $f(g(3))$.

c. $f(f(3))$.

d. An algebraic equation that expresses $f(g(x))$ explicitly in terms of x.

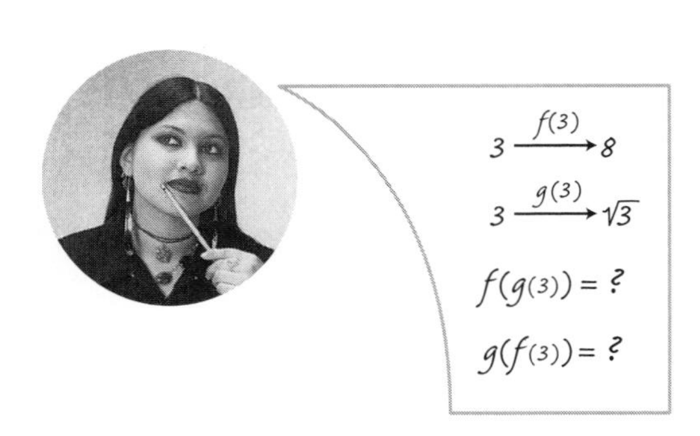

Solution

a. $f(g(3)) = f(\sqrt{3}) = 2^{\sqrt{3}} = 3.3219...$ Evaluate the inside function first. Then apply the outside function.

b. $g(f(3)) = g(2^3) = g(8) = \sqrt{8} = 2.8284...$, which does not equal 3.3219...

c. $f(f(3)) = f(2^3) = f(8) = 2^8 = 256$ Composition of f with itself.

d. $f(g(x)) = f(\sqrt{x}) = 2^{\sqrt{x}}$ ◀

Domain and Range of a Composite Function

If the domains of two functions are restricted, then the domain and range of the composition of those two functions are also restricted. Example 2 shows you why this happens and how to find the domain and range.

▶ **EXAMPLE 2** Let $g(x) = x - 3$, for $2 \le x \le 7$. See Section 1-2 for plotting in a restricted domain.

Let $f(x) = -2x + 8$, for $1 \le x \le 5$.

The graphs of g and f are shown in Figure 1-4b (left and right, respectively).

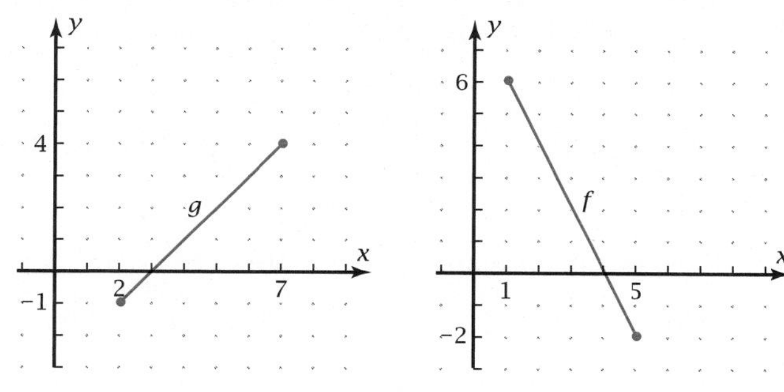

Figure 1-4b

a. Find $f(g(5))$. Show how you found this by sketching the graphs.

b. Try to find $f(g(8))$. Explain why this value is undefined.

c. Try to find $f(g(2))$. Explain why this value is undefined even though 2 is in the domains of both f and g.

d. Plot f, g, $f \circ g$, and $g \circ f$ on your grapher.

e. Find an equation that expresses $f(g(x))$ explicitly in terms of x. Find the domain and range of $f \circ g$ algebraically and show that they agree with the graph in part d.

f. Find the domain of $g \circ f$ and show that it agrees with the graph in part d.

Solution

a. By graph (Figure 1-4c, left) or by algebra, find that $g(5) = 2$. Substitute 5 for x in function g.

By graph (Figure 1-4c, right) or by algebra, find that $f(2) = 4$. Substitute $g(5)$ for x in function f.

Therefore, $f(g(5)) = f(2) = 4$. Evaluate the inside function first.

Figure 1-4c shows substituting 5 for x in function g and getting 2, then substituting 2 for x in function f and getting 4 for the final answer.

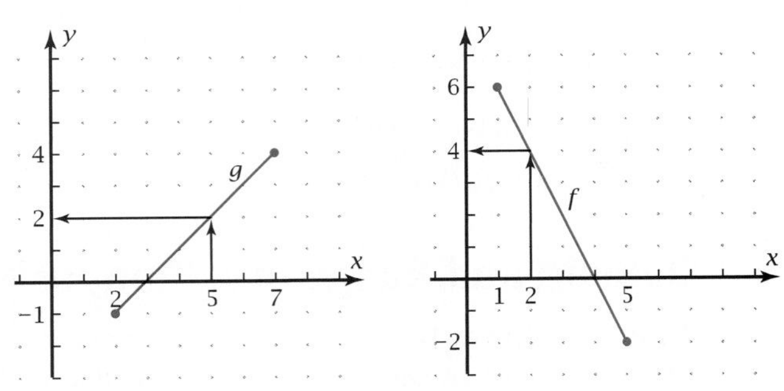

Figure 1-4c

b. $f(g(8))$ is undefined because $g(8)$ is undefined. In other words, 8 is outside the domain of g.

c. $g(2) = -1$.

$f(-1)$ is undefined because -1 is outside the domain of f.

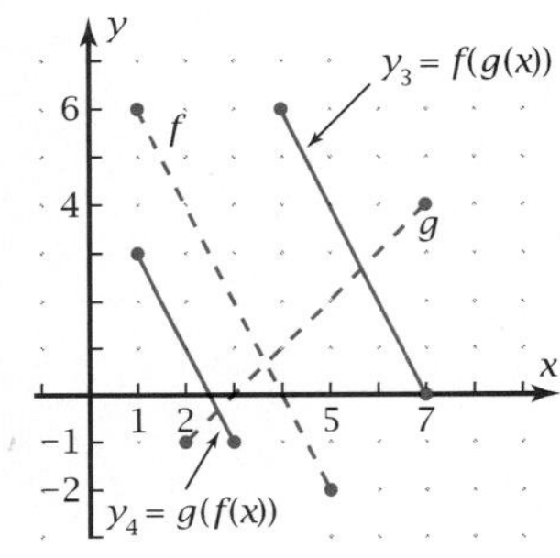

Figure 1-4d

d. $y_1 = -2x + 8$ / $(x \geq 1$ and $x \leq 5)$

$y_2 = x - 3$ / $(x \geq 2$ and $x \leq 7)$

$y_3 = y_1(y_2(x))$ — Use the $f(x)$ capability of your grapher.

$y_4 = y_2(y_1(x))$

The graphs are shown in Figure 1-4d.

e. $f(g(x)) = f(x - 3)$ — Substitute $x - 3$ for $g(x)$ in $f(g(x))$.

$f(g(x)) = -2(x - 3) + 8$ — Substitute $x - 3$ for x in $f(x) = -2x + 8$.

$f(g(x)) = -2x + 14$ — Simplify.

To calculate the domain algebraically, first note that $g(x)$ must be within the domain of f.

$1 \leq g(x) \leq 5$ — Write $g(x)$ in the domain of f.

$1 \leq x - 3 \leq 5$ — Substitute $x - 3$ for $g(x)$.

$4 \leq x \leq 8$ — Add 3 to all three members of the inequality.

Next, note that x must also be in the domain of g, specifically, $2 \leq x \leq 7$. The domain of $f \circ g$ is the *intersection* of these two intervals.

Number-line graphs will help you visualize the intersection.

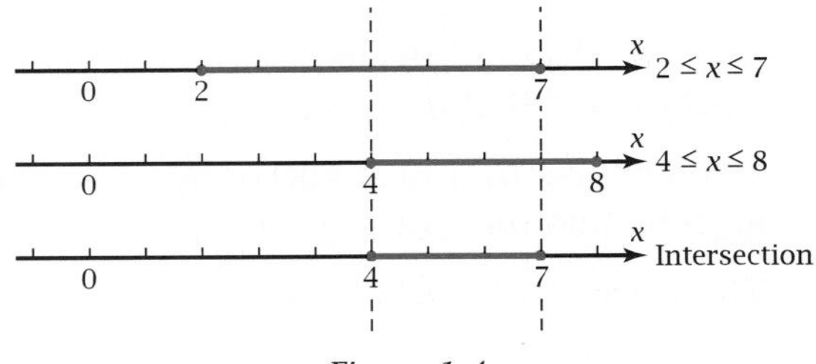

Figure 1-4e

Therefore, the domain of $f \circ g$ is $4 \le x \le 7$. The range of $f \circ g$ may be found from the graph or, in this case, by substituting 4 and 7 into the equation of the composed function.

$$f(g(4)) = -2(4) + 14 = 6$$

$$f(g(7)) = -2(7) + 14 = 0$$

Therefore, the range of $f \circ g$ is $0 \le y \le 6$.

f. $2 \le f(x) \le 7$ — $f(x)$ must be within the domain of g.

$2 \le -2x + 8 \le 7$ — Substitute $-2x + 8$ for $f(x)$.

$-6 \le -2x \le -1$ — Subtract 8 from all three members of the inequality.

$3 \ge x \ge 0.5$ — Divide by -2. The order reverses when you divide by a negative number.

$0.5 \le x \le 3$ and $1 \le x \le 5$ — x must also be in the domain of f.

Therefore, the domain of $g \circ f$ is $1 \le x \le 3$, which agrees with Figure 1-4d. ◀

DEFINITION AND PROPERTIES: Composite Function

The **composite function** $f \circ g$ (pronounced "f composition g") is the function

$$(f \circ g)(x) = f(g(x))$$

The domain of $f \circ g$ is the set of all values of x in the domain of g for which $g(x)$ is in the domain of f.

Domain of g — Range of g — Range of f — Domain of f — x_1 — x_2 — $g(x_1)$ — $g(x_2)$ — $f(g(x_1))$ — $f(g(x_2))$ is undefined because $g(x_2)$ is not in the domain of f

Function g is called the **inside function** and is performed on x first.

Function f is called the **outside function** and is performed on $g(x)$, the value assigned to x by g.

Note: Horizontal dilations and translations are examples of composite functions because they are performed on x.

Problem Set 1-4

Do These Quickly

Q1. What transformation of f is represented by $g(x) = 3f(x)$?

Q2. What transformation of f is represented by $h(x) = 5 + f(x)$?

Q3. If g is a horizontal translation of f by -4 spaces, then $g(x) =$ —?—.

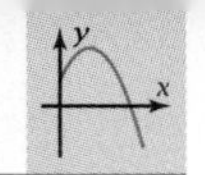

Q4. If p is a vertical dilation of f by a factor of 0.2, then $p(x) =$ —?—.

Q5. Why is $y = 3x^5$ *not* an exponential function, even though it has an exponent?

Q6. Write the general equation of a quadratic function.

Q7. For what value of x will the graph of $y = \frac{x-3}{x-5}$ have a discontinuity?

Q8. Simplify: x^3x^5

Q9. Find 40% of 300.

Q10. Which of these is a horizontal dilation by a factor of 2?

A. $g(x) = 2f(x)$ C. $g(x) = f(0.5x)$

B. $g(x) = 0.5f(x)$ D. $g(x) = f(2x)$

1. *Flashlight Problem:* You shine a flashlight, making a circular spot of light with radius 5 cm on a wall. Suppose that, as you back away from the wall, the radius of the spot of light increases at a rate of 7 cm/sec.
   a. Write an equation for $R(t)$, the radius of the spot of light t sec after you started backing away.
   b. Find the radius of the spot of light when $t = 4$ and when $t = 10$ sec. Use the answers to find the area of the wall illuminated at these two times.
   c. The area, $A(t)$, illuminated by the light is a function of the radius, and the radius is a function of t. Write an equation for the composite function $A(R(t))$. Simplify.
   d. At what time t will the area illuminated be 50,000 cm^2?

2. *Bacterial Culture Problem:* When you grow a culture of bacteria in a petri dish, the area of the culture is a function of the number of bacteria present. Suppose that the area of the culture, $A(t)$, measured in square millimeters, is given by this function of time, t, measured in hours:

   $$A(t) = 9(1.1^t)$$

   a. Find the area at times $t = 0$, $t = 5$, and $t = 10$. Is the area changing at an increasing rate or

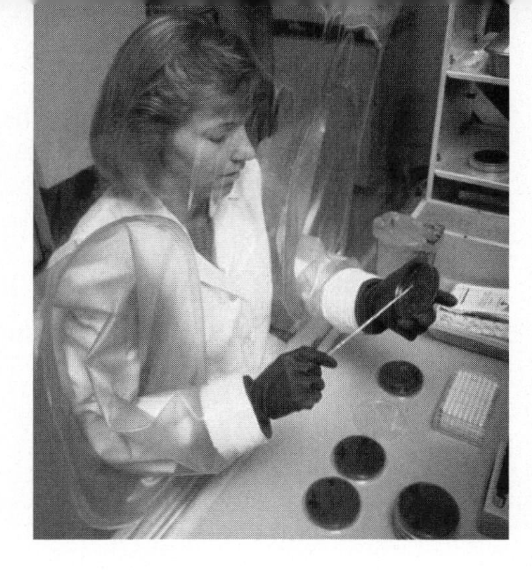

at a decreasing rate? How do the values of $A(0)$, $A(5)$, and $A(10)$ allow you to answer this question?
   b. Assume that the bacteria culture is circular. Use the results of part a to find the radius of the culture at these three times. Is the radius changing at an increasing rate or at a decreasing rate?
   c. The radius of the culture is a function of the area. Write an equation for the composite function $R(A(t))$.
   d. The radius of the petri dish is 30 mm. The culture is centered in the dish and grows uniformly in all directions. What restriction does this fact place on the domain of t for the function $R \circ A$?

3. *Two Linear Functions Problem 1:* Let f and g be defined by

   $f(x) = 9 - x \qquad 4 \le x \le 8$

   $g(x) = x + 2 \qquad 1 \le x \le 5$

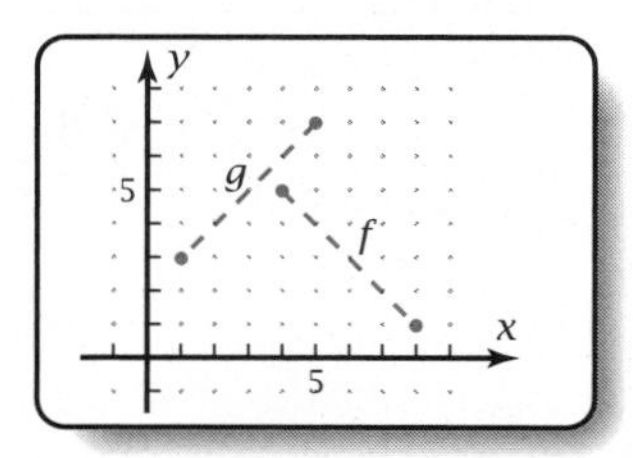

Figure 1-4f

   a. The graphs of f and g are shown in Figure 1-4f. Plot these graphs on your grapher, making sure they agree with the figure.

On the same screen, plot the graphs of $y = f(g(x))$ and $y = g(f(x))$. Sketch the graphs, showing the domains of each. You may use a copy of Figure 1-4f.

b. Find $f(g(3))$. Show on your sketch from part a the *two* steps by which you can find this value directly from the graphs of f and g.

c. Explain why $f(g(6))$ is undefined. Explain why $f(g(1))$ is undefined, even though $g(1)$ *is* defined.

d. Find an equation for $f(g(x))$ explicitly in terms of x. Simplify as much as possible.

e. Find the domains of both $f \circ g$ and $g \circ f$ algebraically. Do they agree with your graphs in part a?

f. Find the range of $f \circ g$ algebraically. Does it agree with your graph in part a?

g. Find $f(f(5))$. Explain why $g(g(5))$ is undefined.

4. *Quadratic and Linear Function Problem:* Let f and g be defined by

$f(x) = -x^2 + 8x - 4 \qquad 1 \le x \le 6$

$g(x) = 5 - x \qquad 0 \le x \le 7$

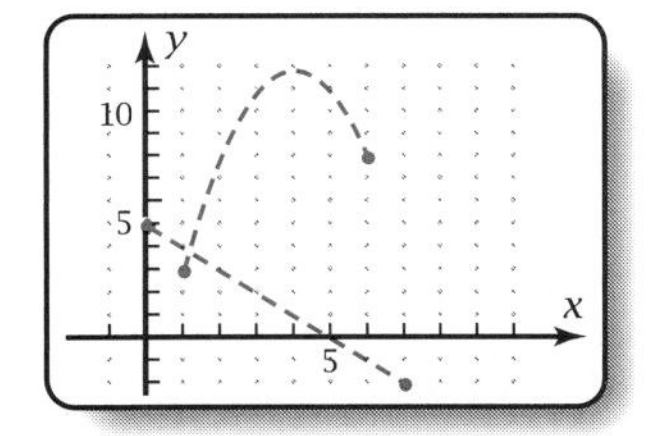

Figure 1-4g

a. The graphs of f and g are shown in Figure 1-4g. Plot these graphs on your grapher. On the same screen, plot the graph of $y = f(g(x))$. Sketch the resulting graph, showing the domain and range. You may use a copy of Figure 1-4g.

b. Show that $f(g(3))$ is defined, but $g(f(3))$ is undefined.

c. Find the domain of $f \circ g$ algebraically. Show that it agrees with your graph in part a.

d. Find an equation for $f(g(x))$ explicitly in terms of x. Simplify as much as possible. Plot the graph of this equation on the same screen as your graph in part a, observing the domain in part c. Does the graph coincide with the one you plotted in part a?

5. *Square Root and Linear Function Problem:* Let f and g be defined by

$f(x) = 3\sqrt{x - 4}$ where the values of x make $f(x)$ a real number

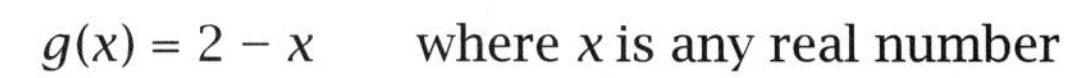

$g(x) = 2 - x$ where x is any real number

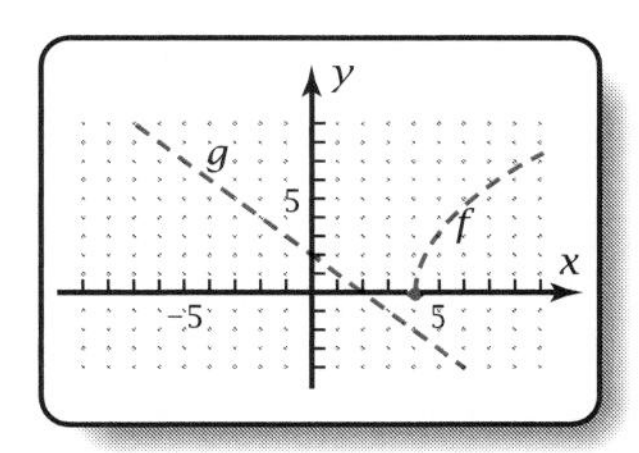

Figure 1-4h

a. Plot the graphs of f and g. Use a friendly x-window that includes $x = 4$ as a grid point. Do the graphs agree with Figure 1-4h?

b. Explain why the domain of f is $x \ge 4$.

c. Plot $f \circ g$ on the same screen as in part a. Sketch the function.

d. Find an equation for $f(g(x))$ explicitly in terms of x. Simplify as much as possible. Explain why the domain of $f \circ g$ is $x \le -2$.

e. Find an equation for $g(f(x))$. What is the domain of $g \circ f$?

6. *Square Root and Quadratic Function Problem:* Let f and g be defined by

$f(x) = \sqrt{x - 5}$ where the values of x make $f(x)$ a real number

$g(x) = x^2 - 4$ where x is any real number

a. Plot the graphs of f, g, and $f \circ g$ on the same screen. Use a friendly x-window of about $-10 \le x \le 10$ that includes the integers as grid points. Sketch the result.

b. Find the domain of $f \circ g$. Explain why the domain of $f \circ g$ has positive and negative numbers, whereas the domain of f has only positive numbers.

c. Show algebraically that $f(g(4))$ is defined, but $f(g(1))$ is undefined.

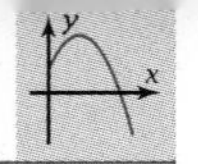

d. Show algebraically that $g(f(6))$ is defined, but $g(f(3))$ is undefined.

e. Plot the graph of $g \circ f$. Make sure that the window has negative and positive y-values.

f. The graph of $g \circ f$ appears to be a straight line. By finding an equation for $g(f(x))$ explicitly in terms of x, determine whether it is true.

7. *Square and Square Root Functions:* Let f and g be defined by

$f(x) = x^2$ where x is any real number

$g(x) = \sqrt{x}$ where the values of x make $g(x)$ a real number

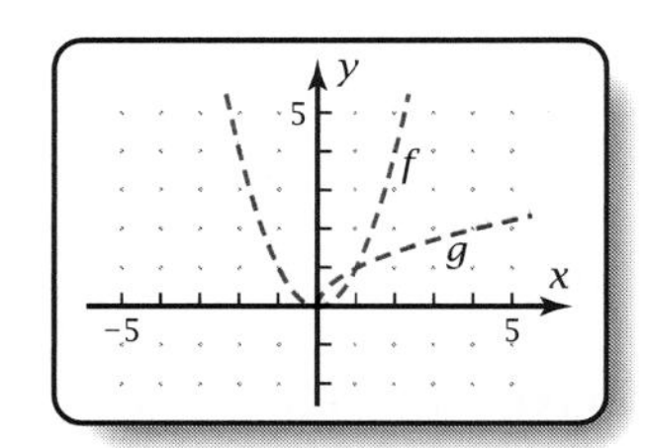

Figure 1-4i

a. Find $f(g(3))$, $f(g(7))$, $g(f(5))$, and $g(f(8))$. What do you notice in each case? Make a conjecture: "For all values of x, $f(g(x)) = $ —?—, and $g(f(x)) = $ —?—."

b. Test your conjecture by finding $f(g(-9))$ and $g(f(-9))$. Does your conjecture hold for negative values of x?

c. Plot $f(x)$, $g(x)$, and $f(g(x))$ on the same screen. Use approximately equal scales on both axes, as shown in Figure 1-4i. Explain why $f(g(x)) = x$, but only for nonnegative values of x.

d. Deactivate $f(g(x))$, and plot $f(x)$, $g(x)$, and $g(f(x))$ on the same screen. Sketch the result.

e. Explain why $g(f(x)) = x$ for nonnegative values of x, but $-x$ (the opposite of x) for negative values of x. What other familiar function has this property?

8. *Linear Function and Its Inverse Problem:* Let f and g be defined by

$$f(x) = \frac{2}{3}x - 2$$

$$g(x) = 1.5x + 3$$

a. Find $f(g(6))$, $f(g(-15))$, $g(f(10))$, and $g(f(-8))$. What do you notice in each case?

b. Plot the graphs of f, g, $f \circ g$, and $g \circ f$ on the same screen. How are the graphs of $f \circ g$ and $g \circ f$ related? How are the graphs of $f \circ g$ and $g \circ f$ related to their "parent" graphs f and g?

c. Show algebraically that $f(g(x))$ and $g(f(x))$ both equal x.

d. Functions f and g in this problem are called *inverses* of each other. Whatever f "does to x," g "undoes." Let $h(x) = 5x - 7$. Find an equation for the function that is the inverse of h.

For Problems 9 and 10, find what transformation will transform f (dashed graph) into g (solid graph).

9.

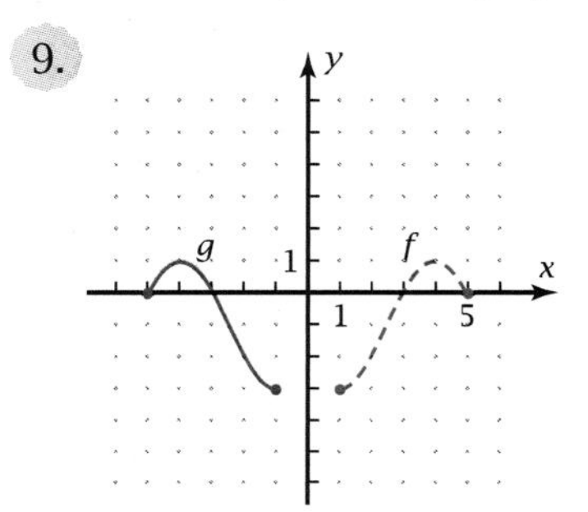

10.

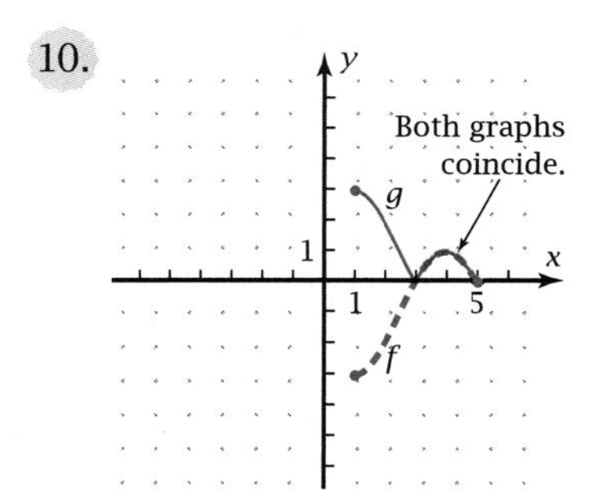

11. *Horizontal Translation and Dilation Problem:* Let f, g, and h be defined by

$f(x) = x^2$ for $-2 \le x \le 2$

$g(x) = x - 3$ for all real values of x

$h(x) = \frac{1}{2}x$ for all real values of x

a. $f(g(x)) = f(x - 3)$. What transformation of function f is equivalent to the composite function $f \circ g$?

b. $f(h(x)) = f\left(\frac{1}{2}x\right)$. What transformation of function f is equivalent to the composite function $f \circ h$?

c. Plot the graphs of f, $f \circ g$, and $f \circ h$. Sketch the results. Do the graphs confirm your conclusions in part a and part b?

1-5 Inverse of Functions

Figure 1-5a shows the graph of a quadratic function. The solid part could represent the price you pay for a pizza as a function of its diameter. There could be a fixed amount charged for the cooking and service, plus an amount for the pizza itself, which varies with the square of its diameter. Figure 1-5b shows the variables interchanged so that price is on the horizontal axis. The new relation is called the **inverse** of the original function.

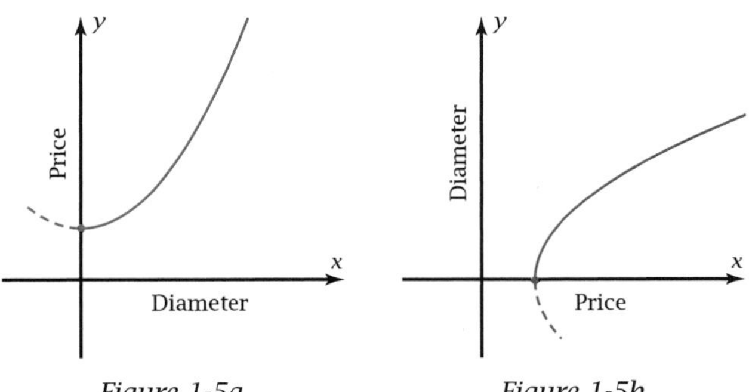

Figure 1-5a *Figure 1-5b*

If the domain included the dashed part of the graph in Figure 1-5a, then the inverse relation would not be a function. As shown in Figure 1-5b, there would be places where there are two y-values for the same x-value. In this section you will study relations formed by inverting functions, as well as other transformations.

OBJECTIVE Given a function, find its inverse relation, and tell whether or not the inverse relation is a function.

Example 1 shows you how to find and plot the graph of the inverse relation if you are given the equation of the original function.

▶ **EXAMPLE 1** Given $y = 0.5x^2 + 2$,

a. Write the equation for the inverse relation by transforming the equation so that y is in terms of x.

b. Plot the function and its inverse on the same screen, using equal scales for the two axes. Explain why the inverse relation is not a function.

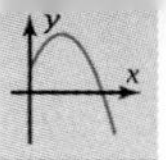

c. Plot the line $y = x$ on the same screen. How are the graph of the function and its inverse relation related to this line?

Solution

a. $x = 0.5y^2 + 2$ — Interchange x and y to get the inverse relation.

$y^2 = 2x - 4$

$y = \pm\sqrt{2x - 4}$ — Take the square root of each side.

b. Enter the equations this way:

$y_1 = 0.5x^2 + 2$

$y_2 = \sqrt{(2x - 4)}$

$y_3 = -\sqrt{(2x - 4)}$

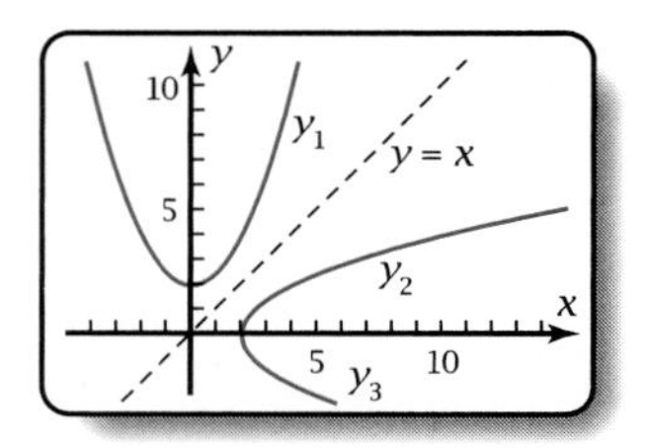

Figure 1-5c

Figure 1-5c shows the graph. The inverse relation is not a function because there are places where there are two different y-values for the same x-value.

c. Figure 1-5c also shows the graph of $y = x$. The two graphs are reflections of each other across this line. ◀

There is a simple way to plot the graph of the inverse of a function with the help of **parametric equations.** These are equations in which both x and y are expressed as a function of a third variable, t. You'll learn more about these equations in later chapters. Example 2 shows how you can use the parametric mode on your grapher to graph inverse relations.

▶ EXAMPLE 2 Plot the graph of $y = 0.5x^2 + 2$ for x in the domain $-2 \le x \le 4$ and its inverse using parametric equations. What do you observe about the domain and range of the function and its inverse?

Solution Put your grapher in parametric mode. Then on the y= menu, enter

$x_{1t} = t$

$y_{1t} = 0.5t^2 + 2$ — Because $x = t$, this is equivalent to $y = 0.5x^2 + 2$.

$x_{2t} = 0.5t^2 + 2$

$y_{2t} = t$ — For the inverse, interchange the equations for x and y.

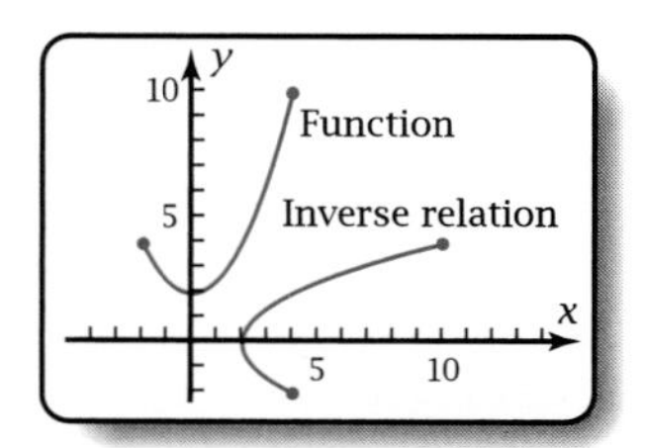

Figure 1-5d

Set the window with a t-range of $-2 \le t \le 4$. Use a convenient t-step, such as 0.1. The result is shown in Figure 1-5d.

The range of the inverse relation is the same as the domain of the function, and vice versa. So range and domain are interchanged. ◀

If the inverse of a function is also a function, then the original function is **invertible.** The symbol $f^{-1}(x)$, pronounced "f inverse of x," is used for the y-value in the inverse function.

▶ EXAMPLE 3 Let $f(x) = 3x + 12$.

a. Find the equation of the inverse of f. Plot function f and its inverse on the same screen.

b. Explain how you know that f is an invertible function.

c. Show algebraically that the composition of f^{-1} with f is $f^{-1}(f(x)) = x$.

Solution

a. Let $y = 3x + 12$.

$x = 3y + 12$ Interchange the variables to invert the function.

$y = \frac{1}{3}x - 4$ Transform the equation so that y is in terms of x.

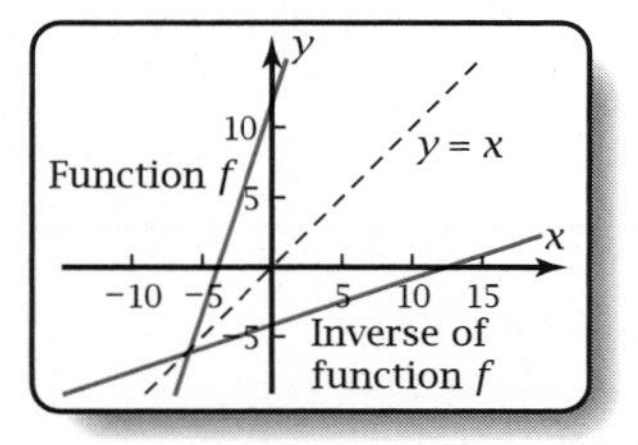

Figure 1-5e

Figure 1-5e shows f and its inverse on the same screen. The two graphs are mirror images across the line $y = x$.

b. The inverse of the function $f(x) = 3x + 12$ is also a function because no values of x have more than one corresponding value of y.

c. $f^{-1}(f(x)) = \frac{1}{3}(3x + 12) - 4 = x + 4 - 4 = x$ Substitute $3x + 12$ for $f(x)$.

$\therefore f^{-1}(f(x)) = x$, Q.E.D. ◀

Note: The three-dot mark $\therefore$ stands for "therefore." The letters Q.E.D. stand for the Latin words *quod erat demonstrandum,* meaning "which was to be demonstrated." Note also that for any invertible function,

$$f^{-1}(f(x)) = x \quad \text{and} \quad f(f^{-1}(x)) = x$$

There is a quick way to tell whether or not a function is invertible. Figure 1-5f shows the air pressure in a punctured tire as a function of time. Because there are no two times at which the pressure is the same, there will be no two pressures for which the time is the same. This function is called a **one-to-one function.** The inverse of this function is shown in Figure 1-5g. You can tell from the graph that it is a function. Functions that are strictly decreasing, as in Figure 1-5f, or strictly increasing are one-to-one functions, which is a guarantee for invertibility.

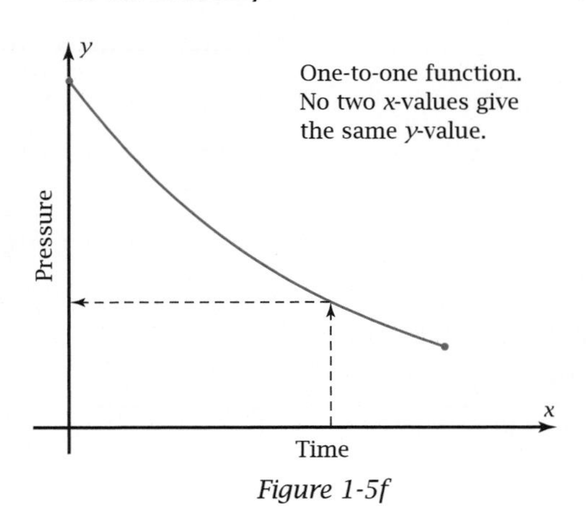

Figure 1-5f

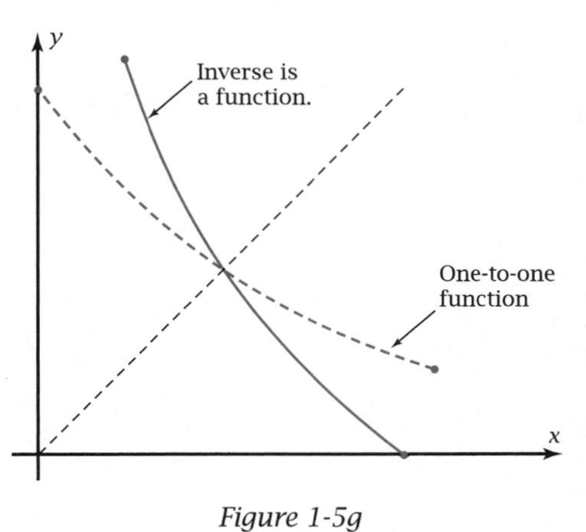

Figure 1-5g

Figure 1-5h shows the height of a ball as a function of time. It is not a one-to-one function because there are two different times at which the height is the same. You can tell from Figure 1-5i that the function is not invertible because the inverse relation has two y-values for the same x-value.

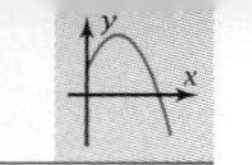

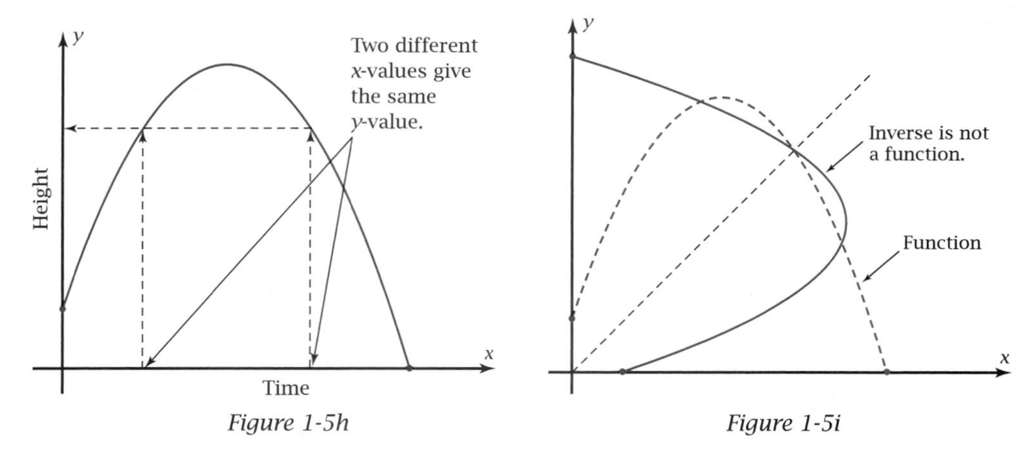

Figure 1-5h *Figure 1-5i*

This box summarizes the information of this section regarding inverses of functions.

DEFINITIONS AND PROPERTIES: *Function Inverses*

- The **inverse** of a relation in two variables is formed by interchanging the two variables.
- If the inverse of function f is also a function, then f is **invertible.**
- If f is invertible and $y = f(x)$, then you can write the inverse of f as $y = f^{-1}(x)$.
- To plot the graph of the inverse of a function, either

  Interchange the variables, solve for y, and plot the resulting equation(s), or

  Use parametric mode, as in Example 2.
- If f is invertible, then the compositions of f and f^{-1} are

  $f^{-1}(f(x)) = x$, provided x is in the domain of f and $f(x)$ is in the domain of f^{-1}

  $f(f^{-1}(x)) = x$, provided x is in the domain of f^{-1} and $f^{-1}(x)$ is in the domain of f
- A one-to-one function is invertible. Strictly increasing or strictly decreasing functions are one-to-one functions.

Problem Set 1-5

Do These Quickly

Q1. In the composite function $m(d(x))$, function d is called the —?— function.

Q2. In the composite function $m(d(x))$, function m is called the —?— function.

Q3. Give another symbol for $m(d(x))$.

Q4. If $f(x) = 2x$ and $g(x) = x + 3$, find $f(g(1))$.

Q5. Find $g(f(1))$ for the functions in Q4.

Q6. Find $f(f(1))$ for the functions in Q4.

Q7. $|3 - 5| =$ —?—

Q8. Is $f(x) = 2^x$ an exponential function or a power function?

Q9. If $f(x) = 2^x$, find $f(0)$.

Q10. If $f(x) = 2^x$, find an equation for $g(x)$, a horizontal translation of $f(x)$ by -3 units.

For Problems 1–4, sketch the line $y = x$ and sketch the inverse relation on a copy of the given figure. Be sure that the inverse relation is a reflection of the function graph across $y = x$. Tell whether or not the inverse relation is a function.

1.

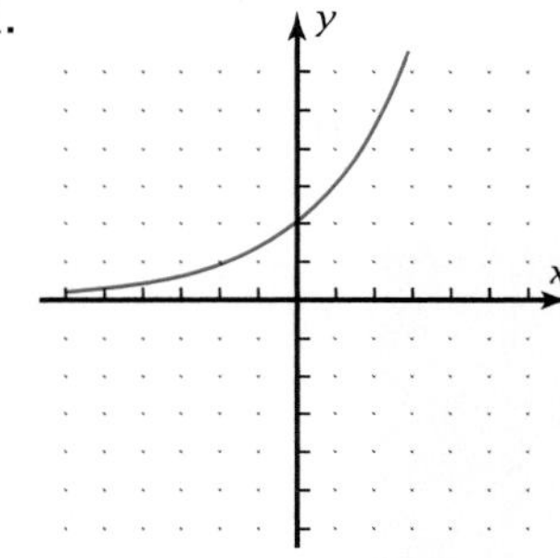

2.

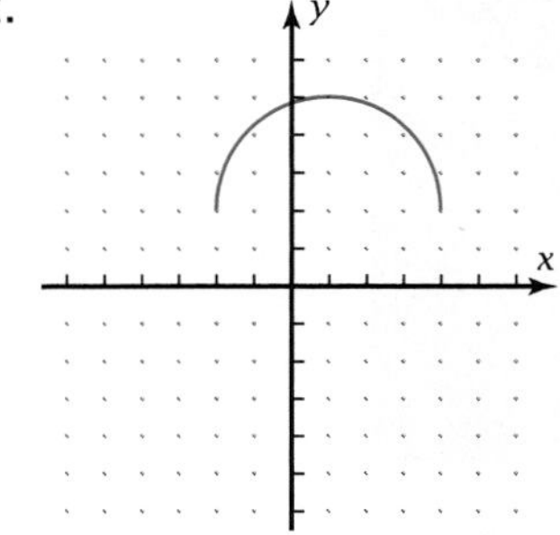

3.

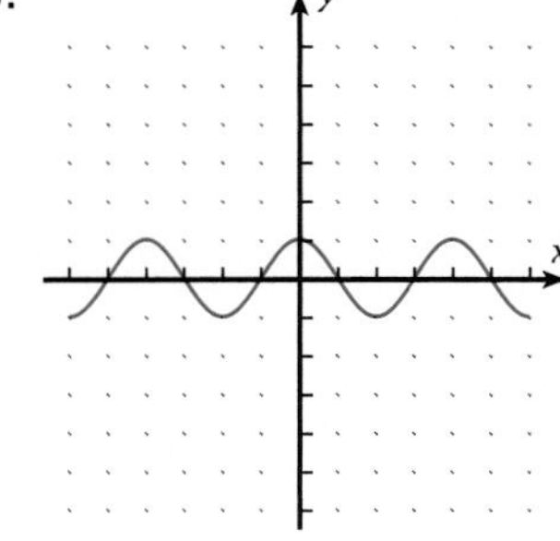

4.

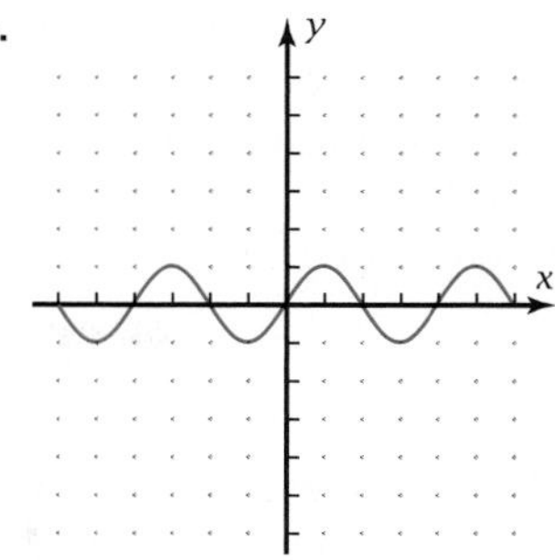

For Problems 5–16, plot the function in the given domain using parametric mode. On the same screen, plot its inverse relation. Tell whether or not the inverse relation is a function. Sketch the graphs.

5. $f(x) = 2x - 6, \quad -1 \le x \le 5$
6. $f(x) = -0.4x + 4, \quad -7 \le x \le 10$
7. $f(x) = -x^2 + 4x + 1, \quad 0 \le x \le 5$
8. $f(x) = x^2 - 2x - 4, \quad -2 \le x \le 4$
9. $f(x) = 2^x \quad x$ is any real number
10. $f(x) = 0.5^x \quad x$ is any real number
11. $f(x) = -\sqrt{3 - x}, \quad -6 \le x \le 3$
12. $f(x) = \sqrt[3]{x}, \quad -1 \le x \le 8$
13. $f(x) = \dfrac{1}{x - 3}, \quad -2 \le x \le 8$
14. $f(x) = \dfrac{x}{x + 1}, \quad -6 \le x \le 4$
15. $f(x) = x^3, \quad -2 \le x \le 1$
16. $f(x) = 0.016x^4, \quad -4 \le x \le 5$

For Problems 17–20, write an equation for the inverse relation by interchanging the variables and solving for y in terms of x. Then plot the function and its inverse on the same screen, using function mode. Sketch the result, showing that the function and its inverse are reflections across the line $y = x$. Tell whether or not the inverse relation is a function.

17. $y = 2x - 6$
18. $y = -0.4x + 4$
19. $y = -0.5x^2 - 2$
20. $y = 0.4x^2 + 3$
21. Show that $f(x) = \frac{1}{x}$ is its own inverse function.
22. Show that $f(x) = -x$ is its own inverse function.

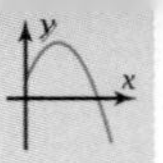

23. *Cost of Owning a Car Problem:* Suppose you have fixed costs (car payments, insurance, and so on) of \$300 per month and operating costs of \$0.25 per mile you drive. The monthly cost of owning the car is given by the linear function

$$c(x) = 0.25x + 300$$

where x is the number of miles you drive the car in a given month and $c(x)$ is the number of dollars per month you spend.

a. Find $c(1000)$. Explain the real-world meaning of the answer.
b. Find an equation for $c^{-1}(x)$, where x now stands for the number of dollars you spend instead of the number of miles you drive. Explain why you can use the symbol c^{-1} for the inverse relation. Use the equation of $c^{-1}(x)$ to find $c^{-1}(437)$ and explain its real-world meaning.
c. Plot $y_1 = c(x)$ and $y_2 = c^{-1}(x)$ on the same screen, using function mode. Use a window of $0 \le x \le 1000$ and use equal scales on the two axes. Sketch the two graphs, showing how they are related to the line $y = x$.

24. *Deer Problem:* The surface area of a deer's body is approximately proportional to the $\frac{2}{3}$ power of the deer's weight. (This is true because the area is proportional to the square of the length and the weight is proportional to the cube of the length.) Suppose that the particular equation for area as a function of weight is given by the power function

$$A(x) = 0.4x^{2/3}$$

where x is the weight in pounds and $A(x)$ is the surface area measured in square feet.

a. Find $A(50)$, $A(100)$, and $A(150)$. Explain the real-world meaning of the answers.
b. True or false: "A deer twice the weight of another deer has a surface area twice that of the other deer." Give numerical evidence to support your answer.
c. Find an equation for $A^{-1}(x)$, where x now stands for area instead of weight.
d. Plot A and A^{-1} on the same screen using function mode. Use an x-window of $0 \le x \le 250$. How are the two graphs related to the line $y = x$?

25. *Braking Distance Problem:* The length of skid marks, $d(x)$ feet, left by a car braking to a stop is a direct square power function of x, the speed in miles per hour when the brakes were applied. Based on information in the *Texas Drivers Handbook* (1997), $d(x)$ is given approximately by

$$d(x) = 0.057x^2 \qquad \text{for } x \ge 0$$

The graph of this function is shown in Figure 1-5h.

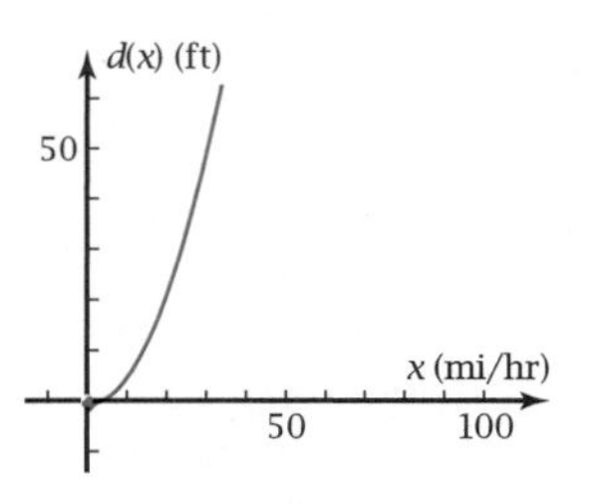

Figure 1-5h

a. When police officers investigate automobile accidents, they use the length of the skid marks to calculate the speed of the car at the time it started to brake. Write an equation for the inverse function, $d^{-1}(x)$, where x is now the length of the skid marks. Explain why you need to take only the positive square root.

b. Find $d^{-1}(200)$. What does this number represent in the context of this problem?

c. Suppose that the domain of function d started at -20 instead of zero. With your grapher in parametric mode, plot the graphs of function d and its inverse relation. Use the window shown in Figure 1-5h with a t-range of $-20 \le t \le 70$. Sketch the result.

d. Explain why the inverse of d in part c is not a function. What relationship do you notice between the domain and range of d and its inverse?

26. *Discrete Function Problem:* Figure 1-5i shows a function that consists of a **discrete** set of points. Show that the function is one-to-one and thus is invertible, although the function is increasing in some parts of the domain and decreasing in other parts.

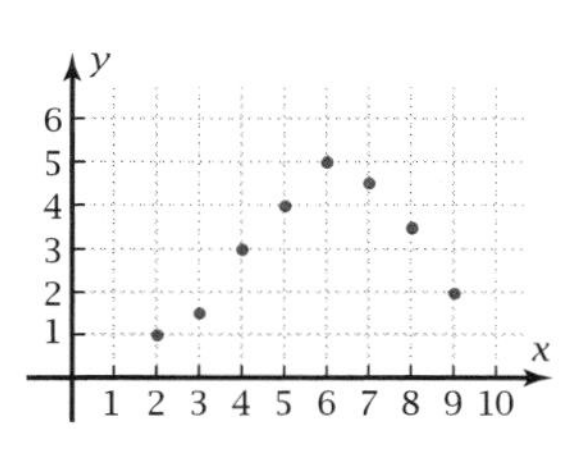

Figure 1-5i

27. *Tabular Function Problem:* A function is defined by the following table of values. Tell whether or not the function is invertible.

x	y
50	1000
60	2000
70	2500
80	2000
90	1500

28. *Horizontal Line Test Problem:* The vertical line test of Problem 39 in Section 1-2 helps you see graphically that a relation is a function if no vertical line crosses the graph more than once. The **horizontal line test** allows you to tell whether or not a function is invertible. Sketch two graphs, one for an invertible function and one for a noninvertible function, that illustrate the following test.

> **Property: The Horizontal Line Test**
>
> If a horizontal line cuts the graph of a function in more than one place, then the function is not invertible because it is not one-to-one.

1-6 Reflections, Absolute Values, and Other Transformations

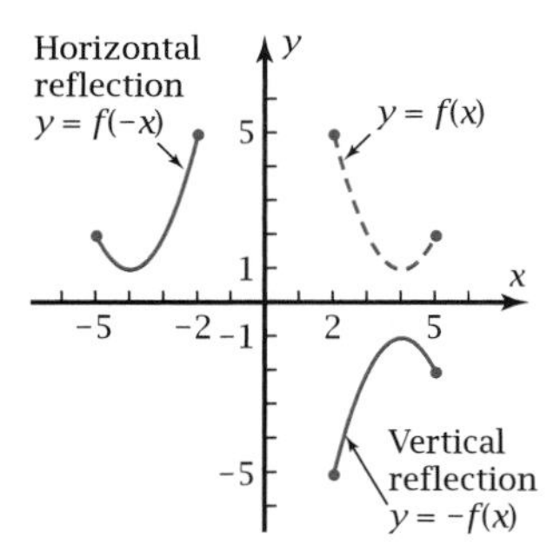

Figure 1-6a

In Section 1-3 you learned that if $y = f(x)$, then multiplying x by a constant causes a horizontal dilation. Suppose that the constant is -1. Each x-value will be $1/(-1)$ or -1 times what it was in the pre-image. Figure 1-6a shows that the resulting image is a horizontal reflection of the graph across the y-axis. The new graph is the same size and shape, just a mirror image of the original. Similarly, a vertical dilation by a factor of -1 reflects the graph vertically across the x-axis.

In this section you will learn special transformations of functions that reflect the graph in various ways. You will also learn what happens when you take the absolute value of a function or of the independent variable x. Finally, you will learn about odd and even functions.

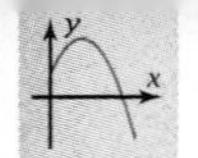

OBJECTIVE Given a function, transform it by reflection and by applying *absolute value* to the function or its argument.

Reflections Across the *x*-axis and *y*-axis

Example 1 shows you how to plot the graphs in Figure 1-6a.

► **EXAMPLE 1** The pre-image function $y = f(x)$ in Figure 1-6a is $f(x) = x^2 - 8x + 17$, where $2 \le x \le 5$.

a. Write an equation for the reflection of this pre-image function across the *y*-axis.

b. Write an equation for the reflection of this pre-image function across the *x*-axis.

c. Plot the pre-image and the two reflections on the same screen.

Solution

a. A reflection across the *y*-axis is a horizontal dilation by a factor of −1. So,

$y = f(-x) = (-x)^2 - 8(-x) + 17$ — Substitute $-x$ for x.

$y = x^2 + 8x + 17$

Domain: $2 \le -x \le 5$

$-2 \ge x \ge -5$ or $-5 \le x \le -2$ — Multiply all three sides of the inequality by −1. The inequalities reverse.

b. $y = -f(x)$ — For a reflection across the *x*-axis, find the opposite of $f(x)$.

$y = -x^2 + 8x - 17$ — The domain remains $2 \le x \le 5$.

c. $y_1 = x^2 - 8x + 17 \,/\, (x \ge 2 \text{ and } x \le 5)$ — Divide by a Boolean variable to restrict the domain.

$y_2 = x^2 + 8x + 17 \,/\, (x \ge -5 \text{ and } x \le -2)$

$y_3 = -x^2 + 8x - 17 \,/\, (x \ge 2 \text{ and } x \le 5)$

The graphs are shown in Figure 1-6a.

You can check the algebraic solutions by plotting $y_4 = y_1(-x)$ and $y_5 = -y_1(x)$. Graph the equations using thick style. You should find that the graphs overlay y_2 and y_3. ◄

PROPERTY: Reflections Across the Coordinate Axes

$g(x) = -f(x)$ is a vertical reflection of function f across the *x*-axis.

$g(x) = f(-x)$ is a horizontal reflection of function f across the *y*-axis.

Absolute Value Transformations

Suppose you shoot a basketball. While in the air, it is above the basket level sometimes and below it at other times. Figure 1-6b shows $y = f(x)$, the **displacement** from the level of the basket as a function of time. If the ball is above the basket, its displacement is positive; if the ball is below the basket, it's negative.

Figure 1-6b

Distance, however, is the magnitude (or size) of the displacement, which is never negative. Distance equals the **absolute value** of the displacement. The light solid graph in Figure 1-6c shows $y = g(x) = |f(x)|$. Taking the absolute value of $f(x)$ retains the positive values of y and reflects the negative values vertically across the x-axis.

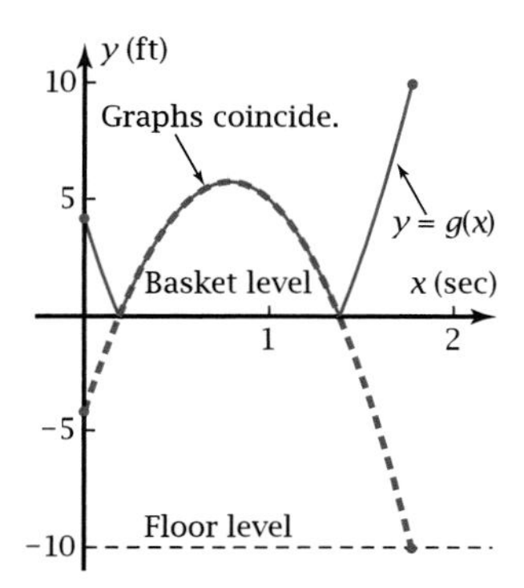

Figure 1-6c

Figure 1-6d shows what happens for $g(x) = f(|x|)$, where you take the absolute value of the argument. For positive values of x, $|x| = x$; so $g(x) = f(x)$ and the graphs coincide. For negative values of x, $|x| = -x$, and so $g(x) = f(-x)$, a horizontal reflection of the positive part of function f across the y-axis. Notice that the graph of f for the negative values of x is not a part of the graph of $f(|x|)$.

The equation for $g(x)$ can be written this way:

$$g(x) = \begin{cases} f(x) & \text{if } x \geq 0 \\ f(-x) & \text{if } x < 0 \end{cases}$$

Because there are two different rules for $g(x)$ in different "pieces" of the domain, g is called a **piecewise function** of x.

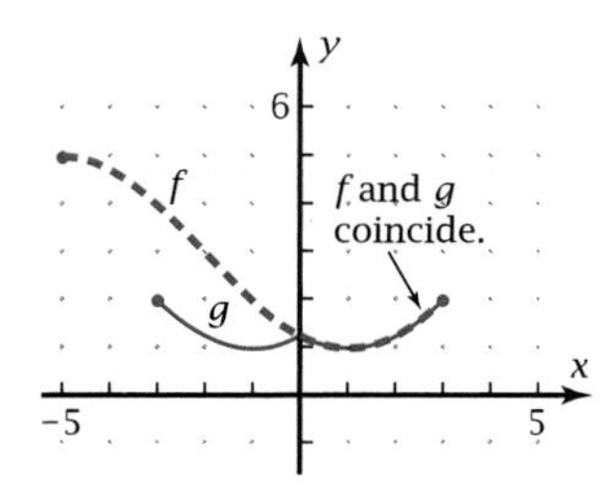

Figure 1-6d

PROPERTY: Absolute Value Transformations

The transformation $g(x) = |f(x)|$

- Reflects f across the x-axis if $f(x)$ is negative.
- Leaves f unchanged if $f(x)$ is nonnegative.

The transformation $g(x) = f(|x|)$

- Leaves f unchanged for nonnegative values of x.
- Reflects the part of the graph for positive values of x to the corresponding negative values of x.
- Eliminates the part of f for negative values of x.

Even Functions and Odd Functions

Figure 1-6e shows the graph of $f(x) = -x^4 + 5x^2 - 1$, a polynomial function with only even exponents. (The number 1 equals $1x^0$, which has an even exponent.) Figure 1-6f shows the graph of $f(x) = -x^3 + 6x$, a polynomial function with only odd exponents.

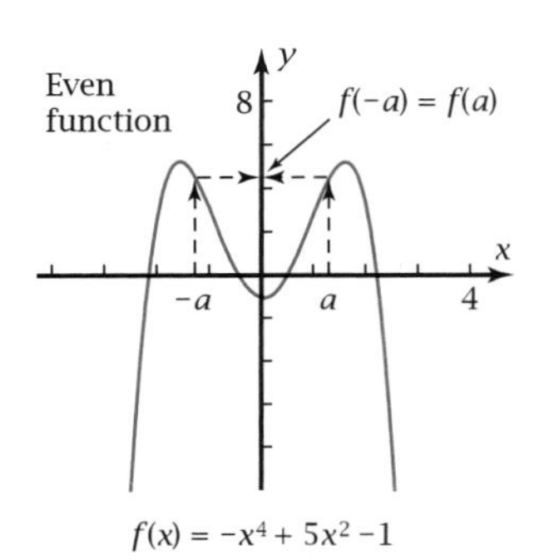

Figure 1-6e

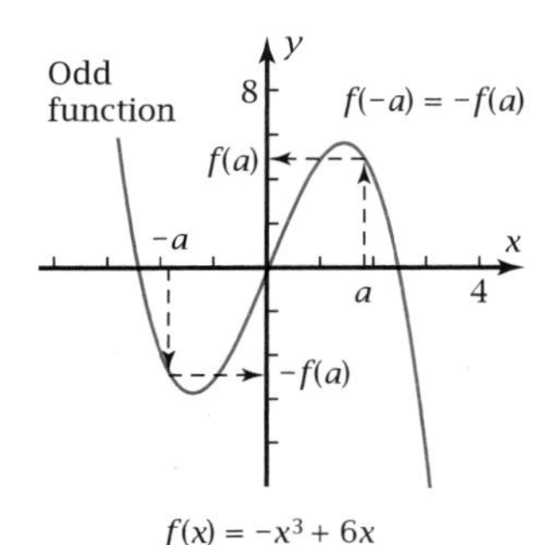

Figure 1-6f

Reflecting the even function $f(x) = -x^4 + 5x^2 - 1$ horizontally across the y-axis leaves the graph unchanged. You can see this algebraically given the property of powers with even exponents.

$$f(-x) = -(-x)^4 + 5(-x)^2 - 1 \quad \text{Substitute } -x \text{ for } x.$$

$$f(-x) = -x^4 + 5x^2 - 1 \quad \text{Negative number raised to an even power.}$$

$$f(-x) = f(x)$$

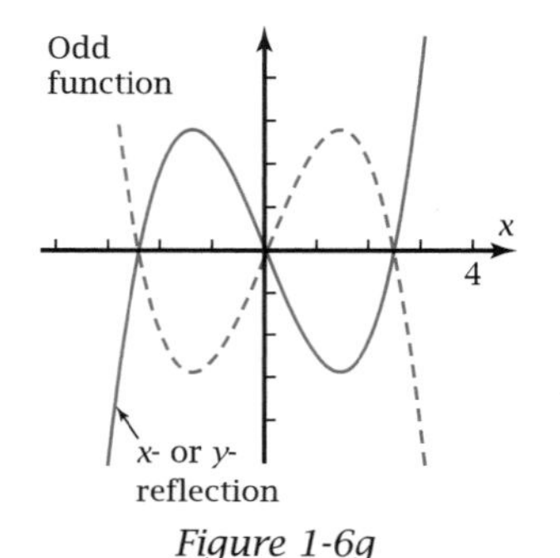

Figure 1-6g

Figure 1-6g shows that reflecting the odd function $f(x) = -x^3 + 6x$ horizontally across the y-axis has the same effect as reflecting it vertically across the x-axis. Algebraically,

$$f(-x) = -(-x)^3 + 6(-x) \quad \text{Substitute } -x \text{ for } x.$$

$$f(-x) = x^3 - 6x \quad \text{Negative number raised to an odd power.}$$

$$f(-x) = -f(x)$$

Any function having the property $f(-x) = f(x)$ is called an **even function.** Any function having the property $f(-x) = -f(x)$ is called an **odd function.** These names apply even if the equation for the function does not have exponents.

DEFINITION: Even Function and Odd Function

The function f is an **even function** if and only if $f(-x) = f(x)$ for all x in the domain.

The function f is an **odd function** if and only if $f(-x) = -f(x)$ for all x in the domain.

Note: For odd functions, reflection across the y-axis gives the same image as reflection across the x-axis. For even functions, reflection across the y-axis is the same as the pre-image. Most functions do not possess the property of oddness or evenness. As you will see later in the course, there are some functions that have these properties even though there are no exponents in the equation!

Problem Set 1-6

Do These Quickly 

Q1. If $f(x) = 2x$, then $f^{-1}(x) = $ —?—.

Q2. If $f(x) = x - 3$, then $f^{-1}(x) = $ —?—.

Q3. If $f(x) = 2x - 3$, then $f^{-1}(x) = $ —?—.

Q4. If $f(x) = x^2$, write the equation for the inverse relation.

Q5. Explain why the inverse relation in Problem Q4 is not a function.

Q6. If $f(x) = 2^x$, then $f^{-1}(8) = $ —?—.

Q7. If the inverse relation for function f is also a function, then f is called —?—.

Q8. Write the definition of a one-to-one function.

Q9. Give a number x for which $|x| = x$.

Q10. Give a number x for which $|x| = -x$.

For Problems 1–4, sketch on copies of the figure the graphs of

a. $g(x) = -f(x)$
b. $h(x) = f(-x)$
c. $a(x) = |f(x)|$
d. $v(x) = f(|x|)$

1.

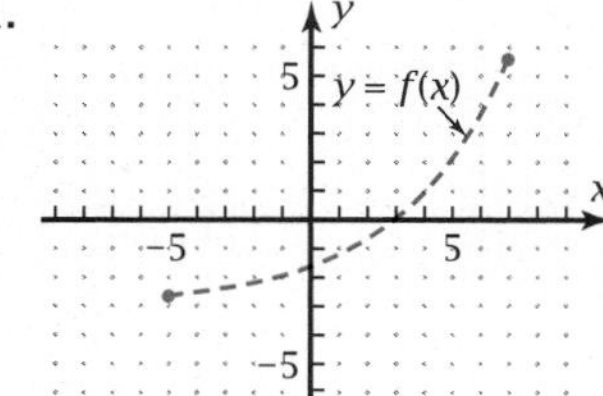

2.

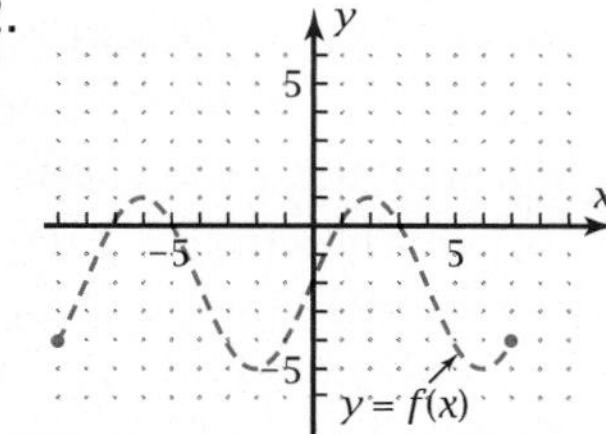

3.

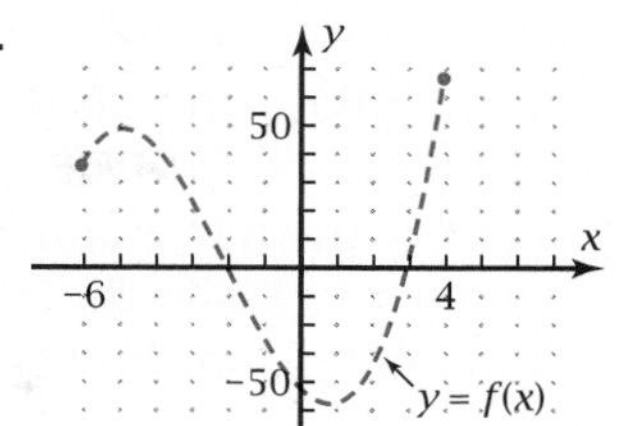

4.

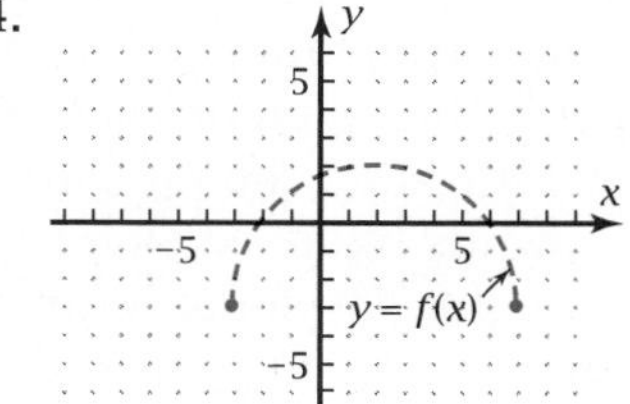

5. The equation for the function in Problem 3 is $f(x) = x^3 + 6x^2 - 13x - 42$ for $-6 \le x \le 4$. Plot the function as y_1 on your grapher. Plot $y_2 = y_1(|x|)$ using thick style. Does the result confirm your answer to Problem 3d?

6. The equation for the function in Problem 4 is $f(x) = -3 + \sqrt{25 - (x-2)^2}$ for $-3 \le x \le 7$. Plot the function as y_1 on your grapher. Plot $y_2 = y_1(|x|)$ using thick style. Does the result confirm your answer to Problem 4d?

7. *Absolute Value Transformations Problem:* Figure 1-6h shows the graph of $f(x) = 0.5(x-2)^2 - 4.5$ in the domain $-2 \le x \le 6$.

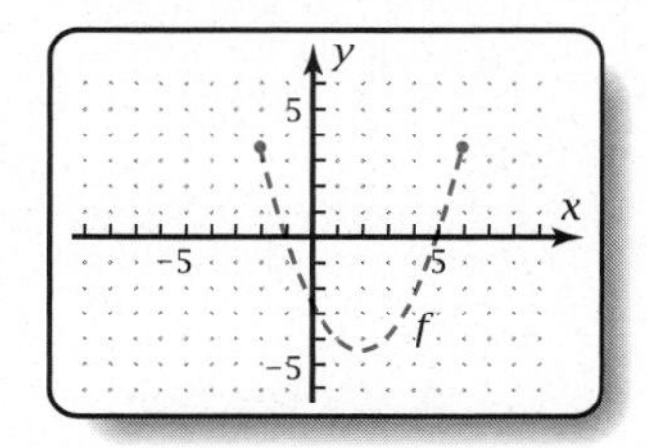

Figure 1-6h

a. Plot the graph of $y_1 = f(x)$. On the same screen, plot $y_2 = |f(x)|$ using thick style. Sketch the result and describe how this transformation changes the graph of f.

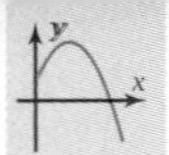

b. Deactivate y_2. On the same screen as y_1, plot the graph of $y_3 = f(|x|)$ using thick style. Sketch the result and describe how this transformation changes the graph of f.

c. Use the equation of function f to find the value of $|f(3)|$ and the value of $f(|-3|)$. Show that both results agree with your graphs in parts a and b. Explain why -3 is in the domain of $f(|x|)$ although it is not in the domain of f itself.

d. Figure 1-6i shows the graph of a function g, but you don't know the equation. On a copy of this figure, sketch the graph of $y = |g(x)|$, using the conclusion you reached in part a. On another copy of this figure, sketch the graph of $y = g(|x|)$, using the conclusion you reached in part b.

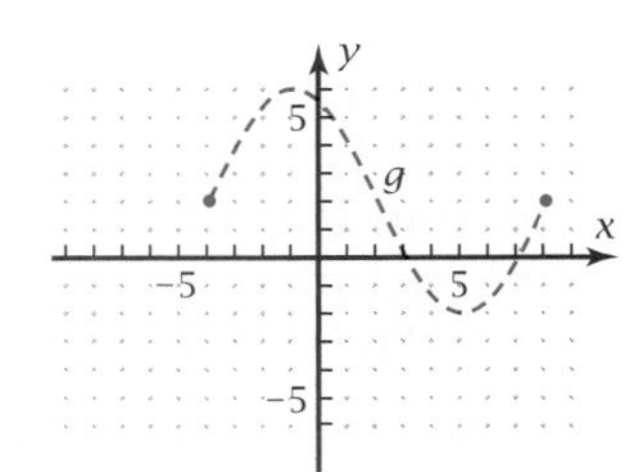

Figure 1-6i

8. *Displacement vs. Distance Absolute Value Problem:* Calvin's car runs out of gas as he is going uphill. He continues to coast uphill for a while, stops, then starts rolling backward without applying the brakes. His displacement, y, in meters, from a gas station on the hill as a function of time, x, in seconds, is given by

$$y = -0.1x^2 + 12x - 250$$

a. Plot the graph of this function. Sketch the result.

b. Find Calvin's displacement at 10 seconds and at 40 seconds. What is the real-world meaning of his negative displacement at 10 seconds?

c. What is Calvin's distance from the gas station at times $x = 10$ and $x = 40$? Explain why both values are positive.

d. Define Calvin's distance from the gas station. Sketch the graph of distance versus time.

e. If Calvin keeps moving as indicated in this problem, when will he pass the gas station as he rolls back down the hill?

9. *Even Function and Odd Function Problem:* Figure 1-6j shows the graph of the even function $f(x) = -x^4 + 5x^2 + 1$. Figure 1-6k shows the graph of the odd function $g(x) = -x^3 + 6x$.

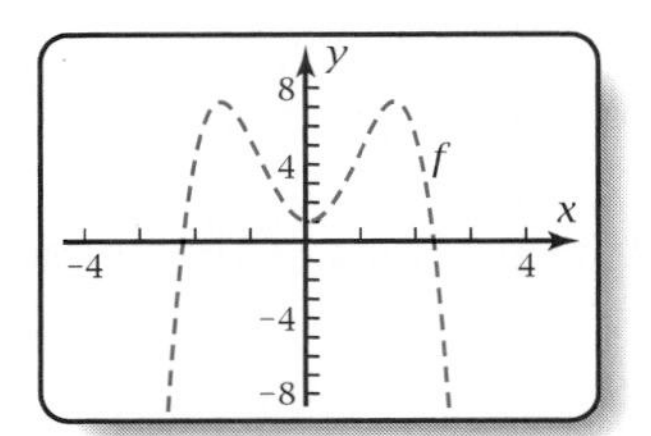

Figure 1-6j

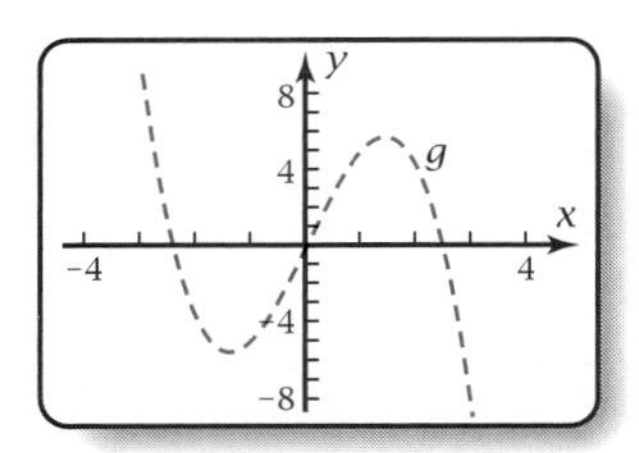

Figure 1-6k

a. On the same screen, plot $y_1 = f(x)$ and $y_2 = f(-x)$. Use thick style for y_2. Based on the properties of negative numbers raised to even powers, explain why the two graphs are identical.

b. Deactivate y_1 and y_2. On the same screen, plot $y_3 = g(x)$, $y_4 = g(-x)$, and $y_5 = -g(x)$. Use thick style for y_5. Based on the properties of negative numbers raised to odd powers, explain why the graphs of y_4 and y_5 are identical.

c. Even functions have the property $f(-x) = f(x)$. Odd functions have the property $f(-x) = -f(x)$. Figure 1-6l shows two functions, h and j, but you don't know the equation of either function. Tell which function is an even function and which is an odd function.

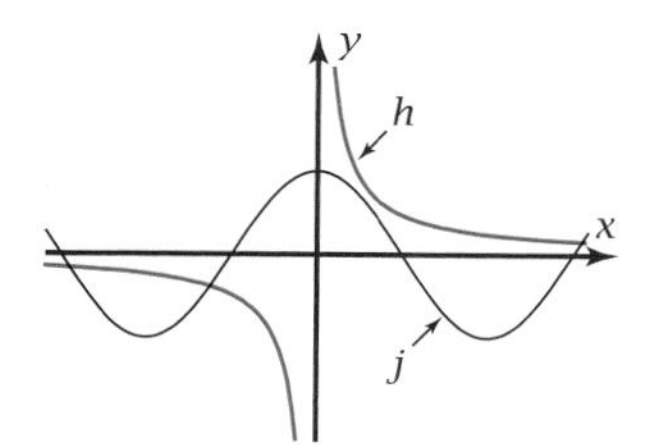

Figure 1-6l

d. Let $e(x) = 2^x$. Sketch the graph. Based on the graph, is function e an odd function, an even function, or neither? Confirm your answer algebraically by finding $e(-x)$.

10. *Absolute Value Function—Odd or Even?* Plot the graph of $f(x) = |x|$. Sketch the result. Based on the graph, is function f an odd function, an even function, or neither? Confirm your answer algebraically by finding $f(-x)$.

11. *Step Discontinuity Problem 1:* Figure 1-6m shows the graph of

$$f(x) = \frac{|x|}{x}$$

The graph has a **step discontinuity** at $x = 0$, where $f(x)$ jumps instantaneously from -1 to 1.

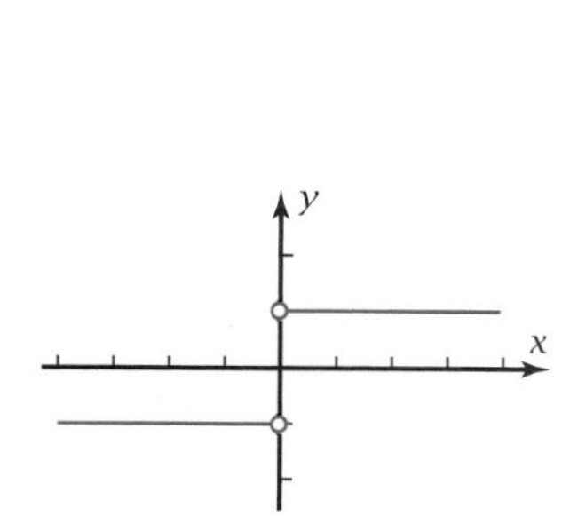

Figure 1-6m

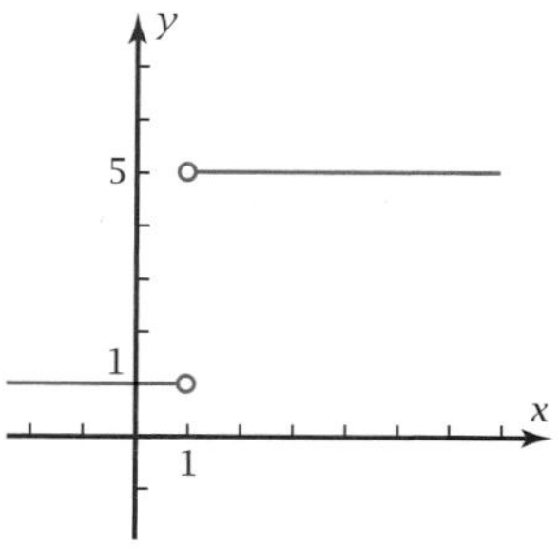

Figure 1-6n

a. Plot the graph of $y_1 = f(x)$. Use a friendly window that includes $x = 0$ as a grid point. Does your graph agree with the figure?

b. Figure 1-6n is a vertical dilation of function f with vertical and horizontal translations. Enter an equation for this function as y_2, using operations on the variable y_1. Use a friendly window that includes $x = 1$ as a grid point. When you have duplicated the graph in Figure 1-6n, write an equation for the transformed function in terms of function f.

c. Figure 1-6o shows the graph of the quadratic function $y = (x - 3)^2$ to which something has been added or subtracted to give it a step discontinuity of 4 spaces at $x = 5$. Find an equation of the function. Verify that your equation is correct by plotting on your grapher.

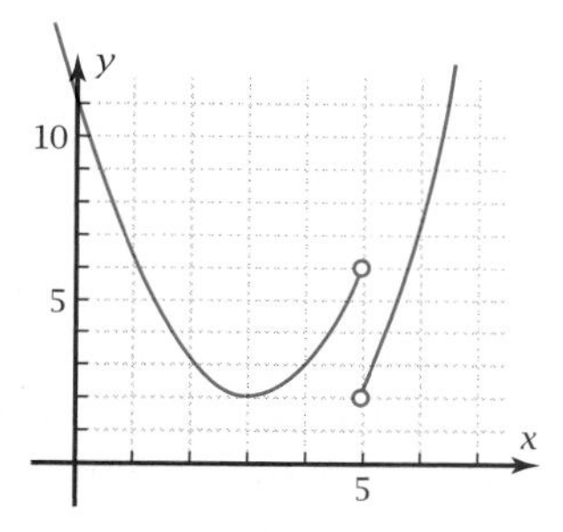

Figure 1-6o

12. *Step Functions—The Postage Stamp Problem:* Figure 1-6p shows the graph of the **greatest integer function,** $f(x) = [x]$. In this function, $[x]$ is the greatest integer less than or equal to x. For instance, $[3.9] = 3$, $[5] = 5$, and $[-2.1] = -3$.

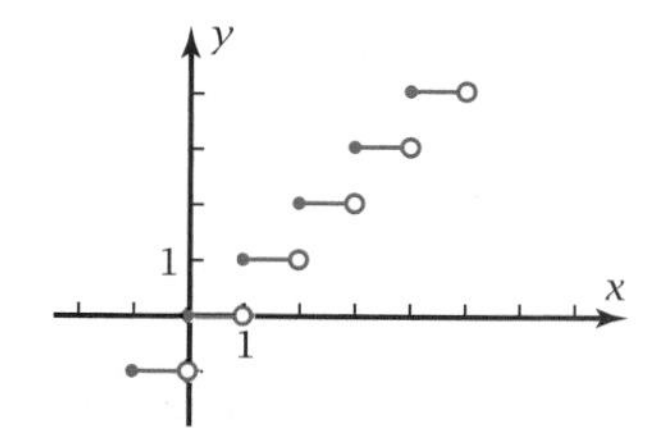

Figure 1-6p

a. Plot the greatest integer function using dot style so that points will not be connected. Most graphers use the symbol int(x) for $[x]$. Use a friendly x-window that has the integers as grid points. Trace to $x = 2.9$, $x = 3$, and $x = 3.1$. What do you find for the three y-values?

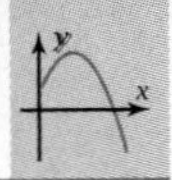

b. In the year 2001, the postage for a first-class letter was 34 cents for weights up to one ounce and 23 cents more for each additional ounce or fraction of an ounce. Sketch the graph of this function.

c. Using a transformation of the greatest integer function, write an equation for the postage as a function of the weight. Plot it on your grapher. Does the graph agree with the one you sketched in part b?

d. First-class postage rates apply only until the letter reaches the weight at which the postage would exceed \$3.25. What is the domain of the function in part c?

13. *Piecewise Functions—Weight Above and Below Earth's Surface Problem:* When you are above the surface of Earth, your weight is inversely proportional to the square of your distance from the center of Earth. This happens because the farther away you are, the weaker the gravitational force is between Earth and you. When you are below the surface of Earth, your weight is directly proportional to your distance from the center. At the center you would be "weightless" because Earth's gravity would pull you equally in all directions. Figure 1-6q shows the graph of the weight function for a 150-pound person. The radius of Earth is about 4000 miles. The weight is called a **piecewise function** of the distance because it is given by different equations in different "pieces" of the domain. Each piece is called a **branch** of the function. The equation of the function can be written

$$y = \begin{cases} ax & \text{if } 0 \le x \le 4000 \\ \dfrac{b}{x^2} & \text{if } x \ge 4000 \end{cases}$$

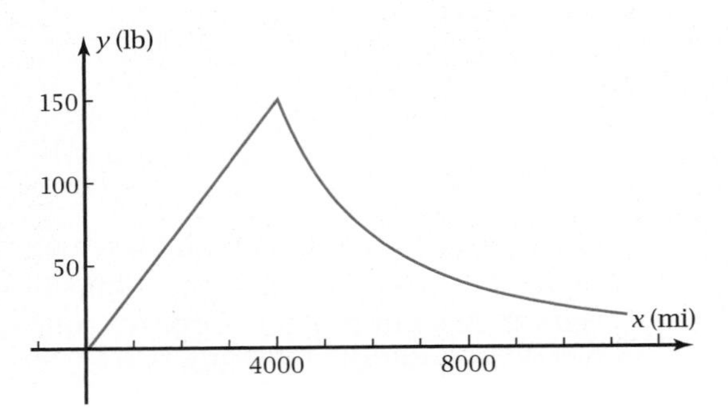

Figure 1-6q

a. Find the values of a and b that make $y = 150$ when $x = 4000$ for each branch.

b. Plot the graph of f. Use Boolean variables to restrict the domain of the graph.

c. Find y if $x = 3000$ and if $x = 5000$.

d. Find the two distances from the center at which the weight would be 50 pounds.

1-7 Precalculus Journal

In this chapter you have been learning mathematics graphically, numerically, and algebraically. An important ability to develop for any subject you study is to *verbalize* what you have learned, both orally and in writing. To gain verbal practice, you should start a **journal.** In it you will record topics you have studied and topics about which you are still unsure. The word *journal* comes from the same word as the French *jour*, meaning "day." *Journey* has the same root and means "a day's travel." Your journal will give you a written record of your travel through mathematics.

OBJECTIVE Start writing a journal in which you can record things you have learned about precalculus mathematics and questions you have concerning concepts about which you are not quite clear.

You should use a bound notebook or a spiral notebook with large index cards for pages. This way your journal will hold up well under daily use. Researchers use such notebooks to record their findings in the laboratory or in the field. Each entry should start with the date and a title for the topic. A typical entry might look like this.

Topic: Inverse of a Function 9/15

I've learned that you invert a function by interchanging the variables. Sometimes an inverse is a relation that is not a function. If it is a function, the inverse of $y = f(x)$ *is* $y = f^{-1}(x)$. *At first, I thought this meant* $\frac{1}{f(x)}$, *but after losing 5 points on a quiz, I realized that wasn't correct. The graphs of* f *and* f^{-1} *are reflections of each other across the line* $y = x$, *like this:*

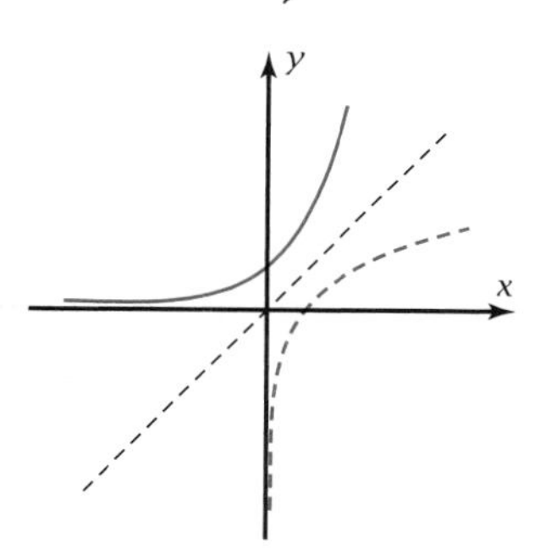

Problem Set 1-7

1. Start a journal for recording your thoughts about precalculus mathematics. The first entry should include such things as these:
   - Sketches of graphs from real-world information
   - Familiar kinds of functions from previous courses
   - How to dilate, translate, reflect, compose, and invert the graph of a function
   - How you feel about what you are learning, such as its potential usefulness to you
   - Any difficulties or misconceptions you had but overcame
   - Any topics about which you are still unsure

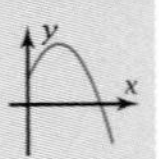

1-8 Chapter Review and Test

In this chapter you saw how you can use functions as mathematical models algebraically, graphically, numerically, and verbally. Functions describe a relationship between two variable quantities, such as distance and time for a moving object. Functions defined algebraically are named according to the way the independent variable appears in the equation. If x is an exponent, the function is an exponential function, and so forth. You can transform the graphs of functions by dilating and translating in the x- and y-directions. Some of these transformations reflect the graph across the x- or y-axis or the line $y = x$. A good understanding of functions will prepare you for later courses in calculus, in which you will learn how to find the rate of change of y as x varies.

You may continue your study of precalculus mathematics either with periodic functions in Chapters 2 through 6 or with fitting of functions to real-world data starting with Chapter 7.

The Review Problems below are numbered according to the sections of this chapter. Answers are provided at the back of the book. The Concept Problems allow you to apply your knowledge to new situations. Answers are not provided, and, in some chapters, you may be required to do research to find answers to open-ended problems. The Chapter Test is more like a typical classroom test your instructor might give you. It has a calculator part and a noncalculator part, and the answers are *not* provided.

Review Problems

R1. *Punctured Tire Problem:* For parts a–d, suppose that your car runs over a nail. The tire's air pressure, y, measured in psi (pounds per square inch), decreases with time, x, measured in minutes, as the air leaks out. A graph of pressure versus time is shown in Figure 1-8a.

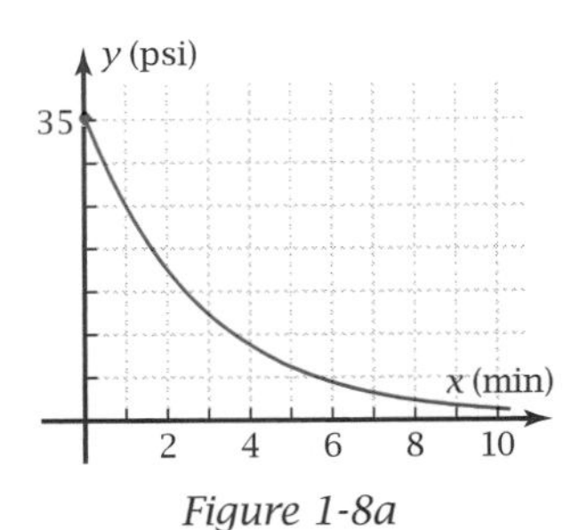

Figure 1-8a

a. Find graphically the pressure after 2 min. Approximately how many minutes can you drive before the pressure reaches 5 psi?

b. The algebraic equation for the function in Figure 1-8a is

$$y = 35 \times 0.7^x$$

Make a table of numerical values of pressure for times of 0, 1, 2, 3, and 4 min.

c. Suppose the equation in part b gives reasonable answers until the pressure drops to 5 psi. At that pressure, the tire comes loose from the rim and the pressure drops to zero. What is the domain of the function described by this equation? What is the corresponding range?

d. The graph in Figure 1-8a gets closer and closer to the x-axis but never quite touches. What special name is given to the x-axis in this case?

e. *Earthquake Problem 1:* Earthquakes happen when rock plates slide past each other. The stress between plates that builds up over a number of years is relieved by the quake in a few seconds. Then the stress starts building up again. Sketch a reasonable graph showing stress as a function of time.

In 1995, a magnitude 7.2 earthquake struck the region of Kobe and Osaka in south-central Japan.

R2. For parts a–e, name the kind of function for each equation given.

a. $f(x) = 3x + 7$

b. $f(x) = x^3 + 7x^2 - 12x + 5$

c. $f(x) = 1.3^x$

d. $f(x) = x^{1.3}$

e. $f(x) = \dfrac{x - 5}{x^2 - 2x + 3}$

f. Name a pair of real-world variables that could be related by the function in part a.

g. If the domain of the function in part a is $2 \le x \le 10$, what is the range?

h. In a flu epidemic, the number of people infected depends on time. Sketch a reasonable graph of the number of people infected as a function of time. What kind of function has a graph that most closely resembles the one you drew?

i. For Figures 1-8b through 1-8d, what kind of function has the graph shown?

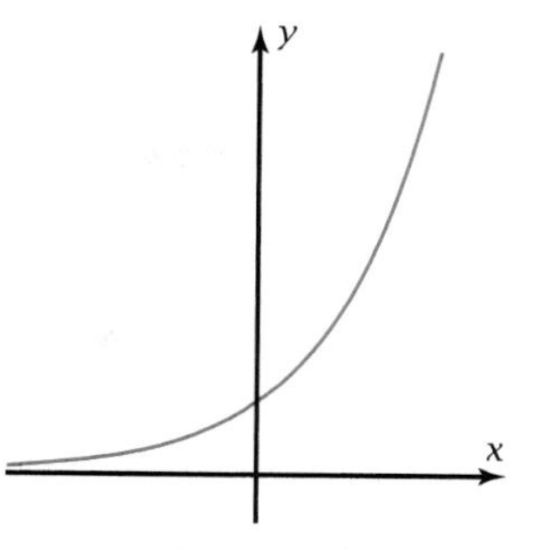

Figure 1-8b

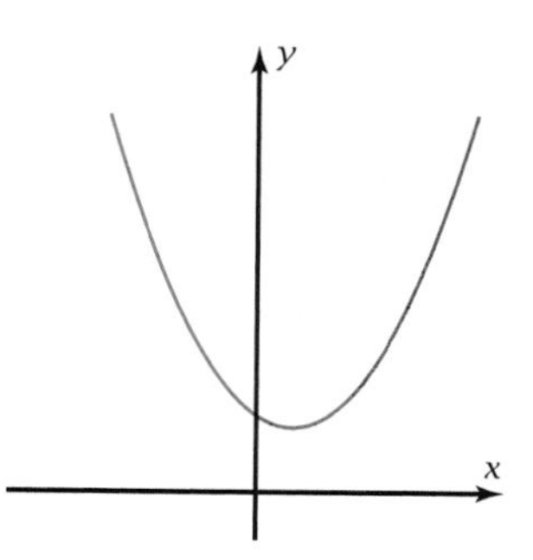

Figure 1-8c

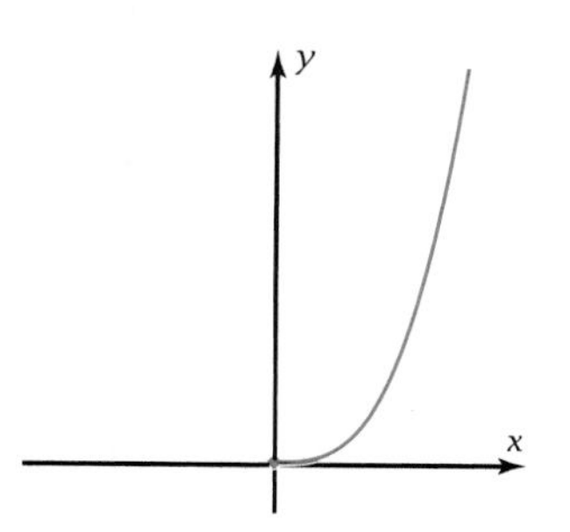

Figure 1-8d

j. Explain how you know that the relation in Figure 1-8e is a function, but the relation in Figure 1-8f is not a function.

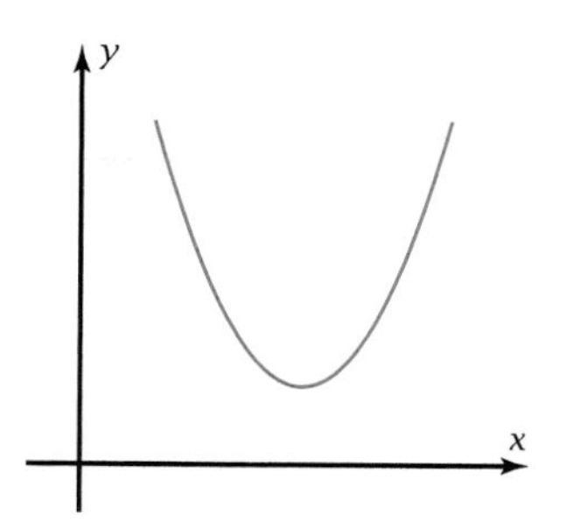

Figure 1-8e

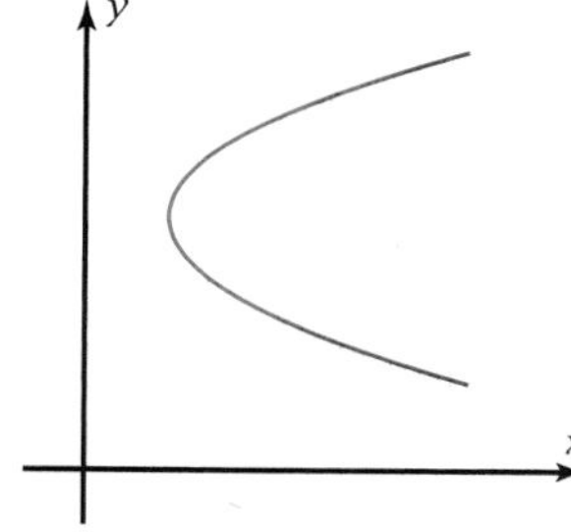

Figure 1-8f

R3. a. For functions f and g in Figure 1-8g, identify how the pre-image function f (dashed) was transformed to get the image function g (solid). Write an equation for $g(x)$ in terms of x given that the equation of f is

$$f(x) = \sqrt{4 - x^2}$$

Confirm the result by plotting the image and the pre-image on the same screen.

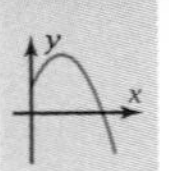

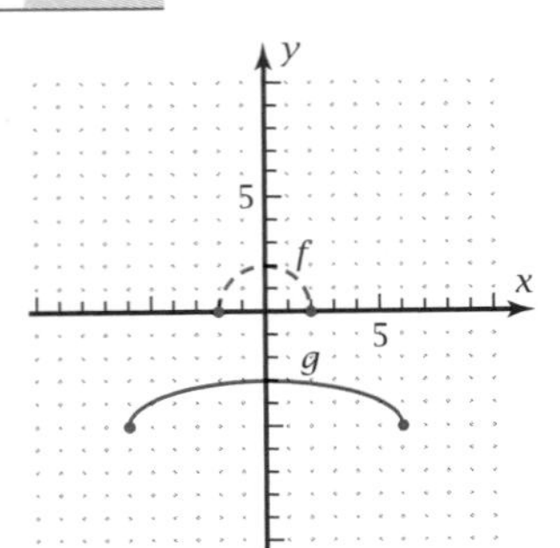

Figure 1-8g

Figure 1-8h

b. If $g(x) = 3f(x - 4)$, explain how function f was transformed to get function g. Using the pre-image in Figure 1-8h, sketch the graph of g on a copy of this figure.

R4. *Height and Weight Problem:* For parts a–e, the weight of a growing child depends on his or her height, and the height depends on age. Assume that the child is 20 inches when born and grows 3 inches per year.

a. Write an equation for $h(t)$ (in inches) as a function of t (in years).

b. Assume that the weight function W is given by the power function $W(h(t)) = 0.004\ h(t)^{2.5}$. Find $h(5)$ and use the result to calculate the predicted weight of the child at age 5.

c. Plot the graph of $y = W(h(t))$. Sketch the result.

d. Assume the height of the child increases at a constant rate. Does the weight also seem to increase at a constant rate? Explain how you arrived at your answer.

e. What is a reasonable domain for t for the function $W \circ h$?

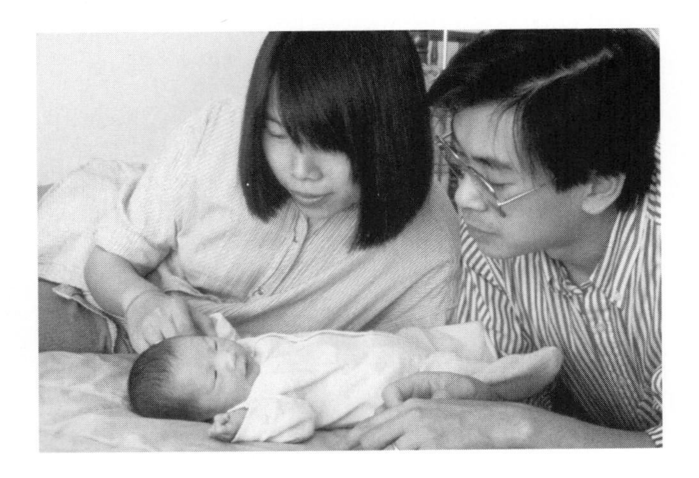

Two Linear Functions Problem 2: For parts f–i, let functions f and g be defined by

$f(x) = x - 2 \quad$ for $4 \le x \le 8$

$g(x) = 2x - 3 \quad$ for $2 \le x \le 6$

f. Plot the graphs of f, g, and $f(g(x))$ on the same screen. Sketch the results.

g. Find $f(g(4))$.

h. Show that $f(g(3))$ is undefined, even though $g(3)$ is defined.

i. Calculate the domain of the composite function $f \circ g$ and show that it agrees with the graph you plotted in part f.

R5. Figure 1-8i shows the graph of $f(x) = x^2 + 1$ in the domain $-1 \le x \le 2$.

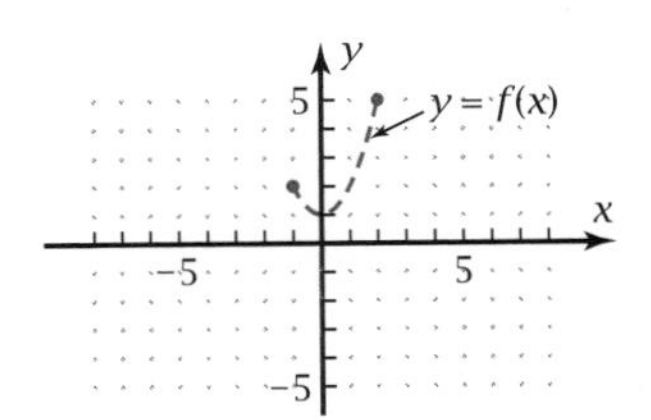

Figure 1-8i

a. On a copy of the figure, sketch the graph of the inverse relation. Explain why the inverse is not a function.

b. Plot the graphs of f and its inverse relation on the same screen using parametric equations. Also plot the line $y = x$. How are the graphs of f and its inverse relation related to the line $y = x$? How are the domain and range of the inverse relation related to the domain and range of function f?

c. Write an equation for the inverse of $y = x^2 + 1$ by interchanging the variables. Solve the new equation for y in terms of x. How does this solution reveal that there are two different y-values for some x-values?

d. On a copy of Figure 1-8j, sketch the graph of the inverse relation. What property does the function graph have that allows you to conclude that the function is invertible?

What are the vertical lines at $x = -3$ and at $x = 3$ called?

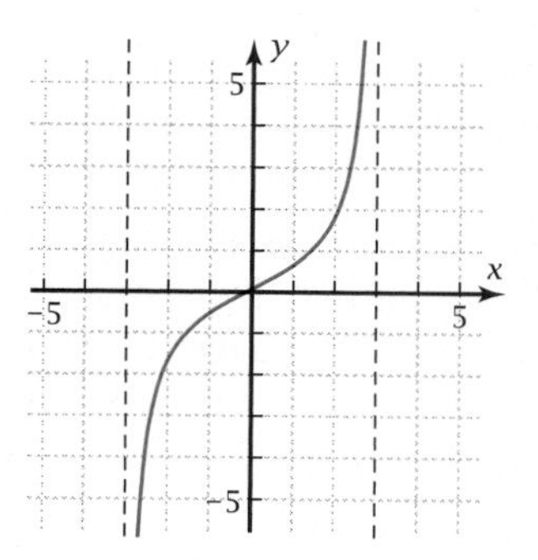

Figure 1-8j

e. *Spherical Balloon Problem:* Recall that the volume, $V(x)$, measured in in.^{3}, of a sphere is given by

$$V(x) = \frac{4}{3}\pi x^3$$

where x is the radius of the sphere in inches. Find $V(5)$. Find x if $V(x) = 100$. Write an equation for $V^{-1}(x)$. Explain why both the V function and the V^{-1} function are examples of power functions. Describe a kind of problem for which the V^{-1} equation would be more useful than the V equation.

f. Sketch the graph of a one-to-one function. Explain why it is invertible.

R6. a. On four copies of $y = f(x)$ in Figure 1-8k, sketch the graphs of these four functions: $y = -f(x)$, $y = f(-x)$, $y = |f(x)|$, and $y = f(|x|)$.

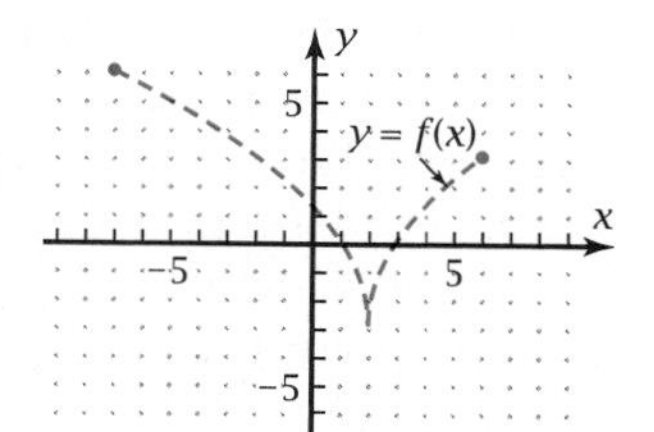

Figure 1-8k

b. Function f in part a is defined piecewise:

$$f(x) = \begin{cases} 3\sqrt{x-2} - 3 & \text{if } 2 \le x \le 6 \\ 3\sqrt{3-x} - 3 & \text{if } -7 \le x \le 2 \end{cases}$$

Plot the two branches of this function as y_1 and y_2 on your grapher. Does the graph agree with Figure 1-8k? Plot $y = f(|x|)$ by plotting $y_3 = y_1(|x|)$ and $y_4 = y_2(|x|)$. Does the graph agree with your answer to the corresponding portion of part a?

c. Explain why functions with the property $f(-x) = -f(x)$ are called *odd* functions and functions with the property $f(-x) = f(x)$ are called *even* functions.

d. Plot the graph of

$$f(x) = 0.2x^2 - \frac{|x-3|}{x-3}$$

Use a friendly window that includes $x = 3$ as a grid point. Sketch the result. Name the feature that appears at $x = 3$.

R7. In Section 1-7 you started a precalculus journal. In what ways do you think that keeping this journal will help you? How could you use the completed journal at the end of the course? What is your responsibility throughout the year to ensure that writing the journal has been a worthwhile project?

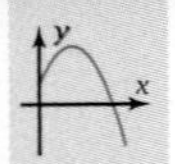

Concept Problems

C1. *Four Transformation Problem:* Figure 1-8l shows a pre-image function f (dashed) and a transformed image function g (solid). Dilations and translations were performed in both directions to get the g graph. Figure out what the transformations were. Write an equation for $g(x)$ in terms of f. Let $f(x) = x^2$ with domain $-2 \le x \le 2$. Plot the graph of g on your grapher. Does your grapher agree with the figure?

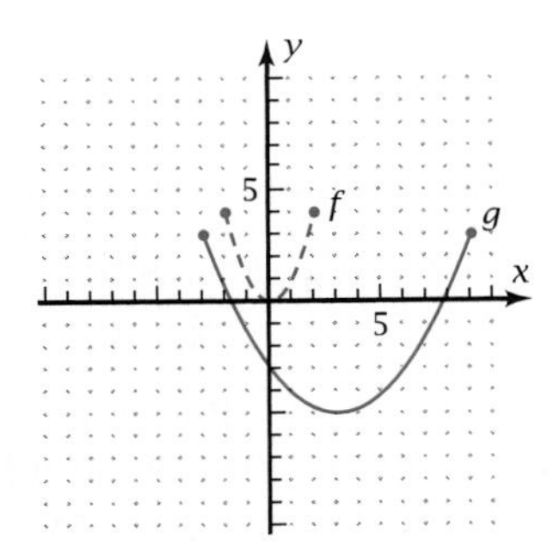

Figure 1-8l

C2. *Sine Function Problem:* If you enter $y_1 = \sin(x)$ on your grapher and plot the graph, the result resembles Figure 1-8m. (Your grapher should be in "radian" mode.) The function is called the **sine function** (pronounced like "sign"), which you will study starting in Chapter 2.

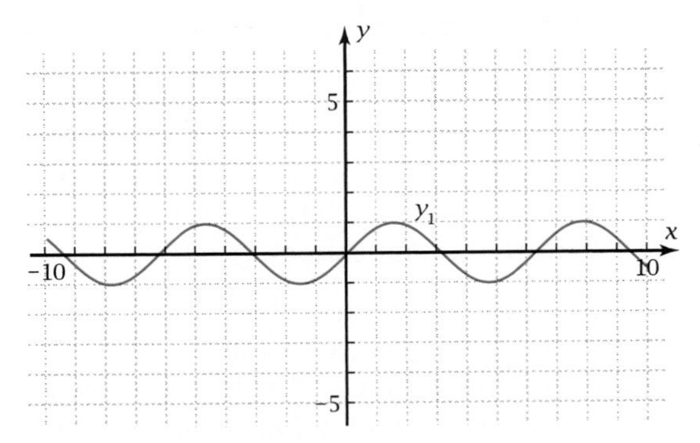

Figure 1-8m

a. The sine function is an example of a **periodic function.** Why do you think this name is given to the sine function?

b. The **period** of a periodic function is the difference in x positions from one point on the graph to the point where the graph first starts repeating itself. Approximately what does the period of the sine function seem to be?

c. Is the sine function an odd function, an even function, or neither? How can you tell?

d. On a copy of Figure 1-8m, sketch a vertical dilation of the sine function graph by a factor of 5. What is the equation for this dilated function? Check your answer by plotting the sine graph and the transformed image graph on the same screen.

e. Figure 1-8n shows a two-step transformation of the sine graph in Figure 1-8m. Name the two transformations. Write an equation for the transformed function and check your answer by plotting both functions on your grapher.

f. Let $f(x) = \sin x$. What transformation would $g(x) = \sin\left(\frac{1}{2}x\right)$ be? Confirm your answer by plotting both functions on your grapher.

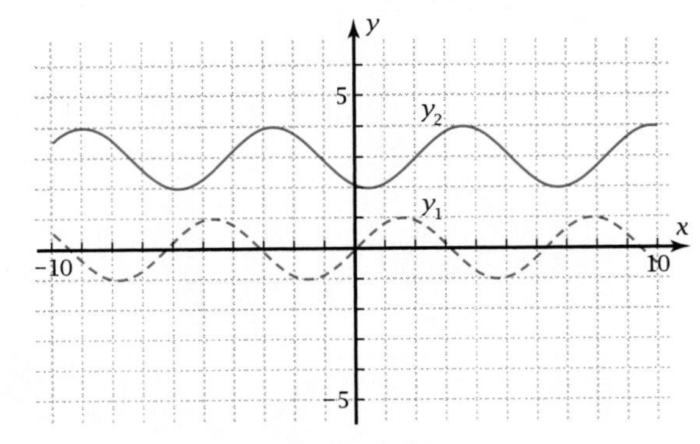

Figure 1-8n

Chapter Test

PART 1: No calculators allowed (T1–T11)

For Problems T1–T4, name the type of function that each of the graphs shows.

T1.

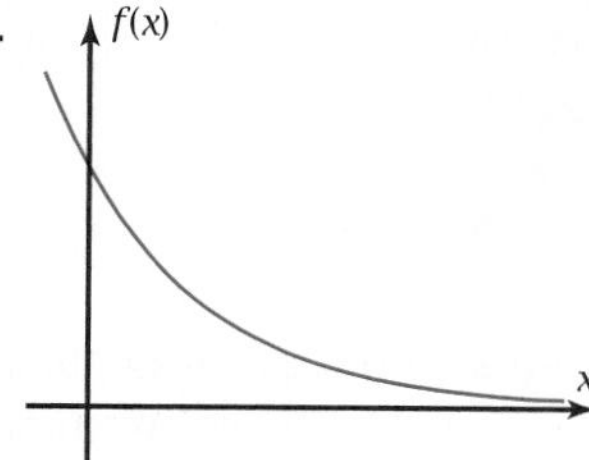

T2.

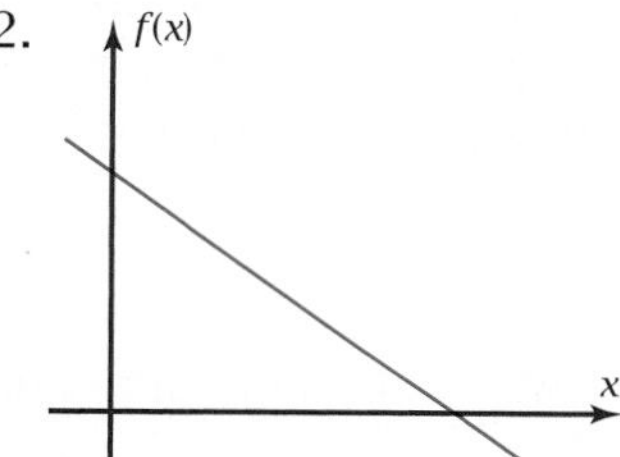

T3.

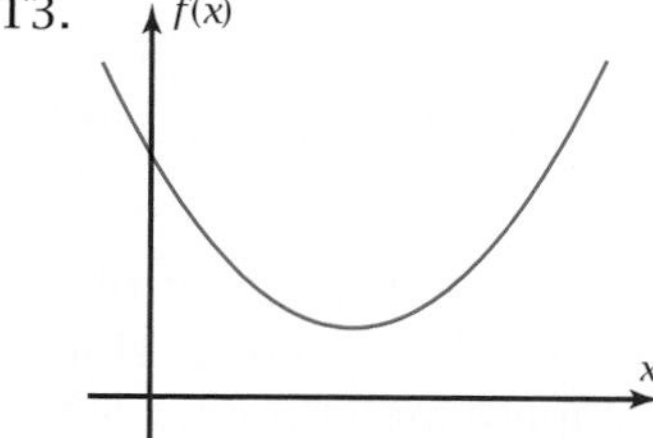

T4.

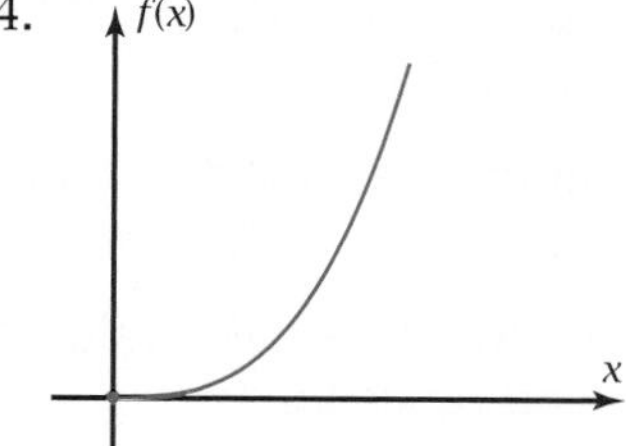

T5. Which of the functions in T1–T4 are one-to-one functions? What conclusion can you make about a function that is not one-to-one?

T6. When you turn on the hot water faucet, the time the water has been running and the temperature of the water are related. Sketch a reasonable graph.

For Problems T7 and T8, tell whether the function is odd, even, or neither.

T7.

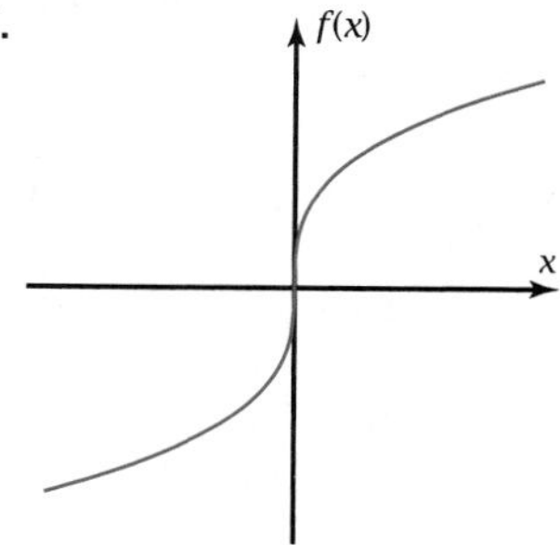

T8.

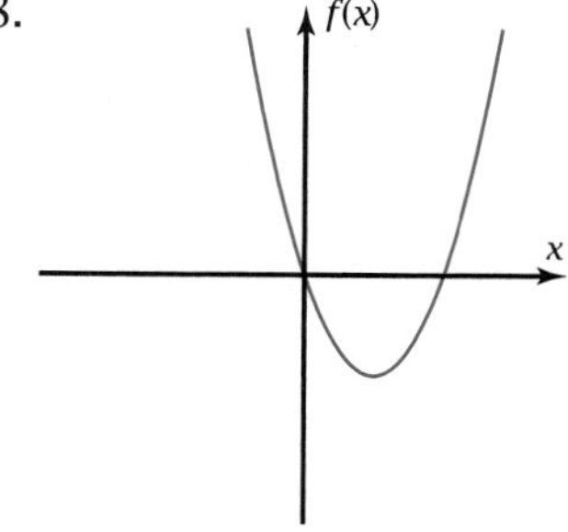

For Problems T9–T11, describe how the graph of f (dashed) was transformed to get the graph of g (solid). Write an equation for $g(x)$ in terms of f.

T9.

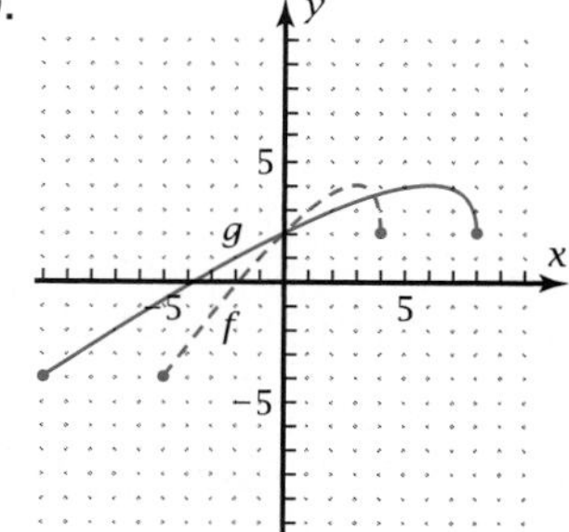

T10.

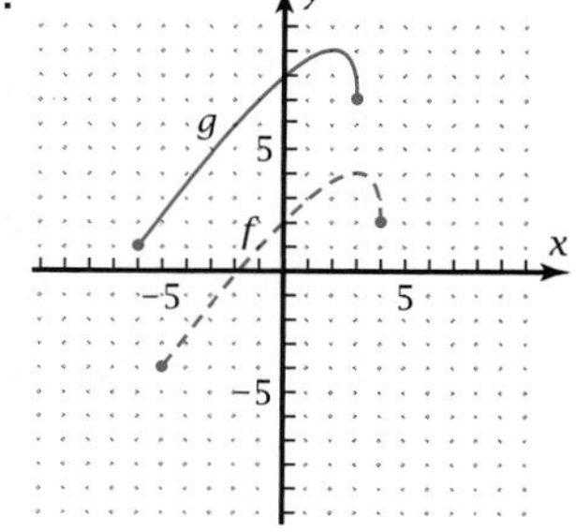

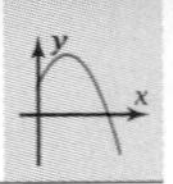

T11.

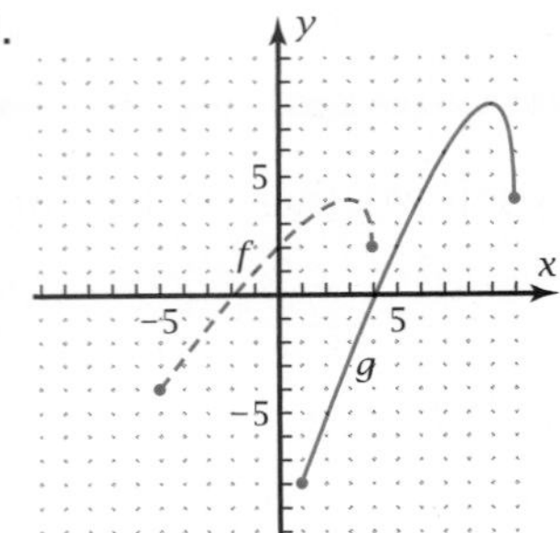

PART 2: Graphing calculators allowed (T12–T28)

T12. Figure 1-8o shows the graph of a function, $y = f(x)$. Give the domain and the range of f.

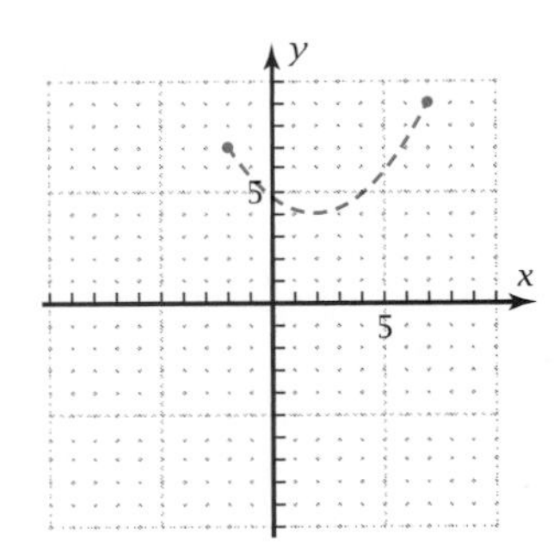

Figure 1-8o

For Problems T13–T16, sketch the indicated transformations on copies of Figure 1-8o. Describe the transformations.

T13. $y = \frac{1}{2}\, f(x)$

T14. $y = f\left(\frac{2}{3}x\right)$

T15. $y = f(x + 3) - 4$

T16. The inverse relation of $f(x)$

T17. Explain why the inverse relation in T16 is not a function.

T18. Let $f(x) = \sqrt{x}$. Let $g(x) = x^2 - 4$. Find $f(g(3))$. Find $g(f(3))$. Explain why $f(g(1))$ is not a real number, even though $g(1)$ is a real number.

T19. Use the absolute value function to write a single equation for the discontinuous function in Figure 1-8p. Check your answer by plotting it on your grapher.

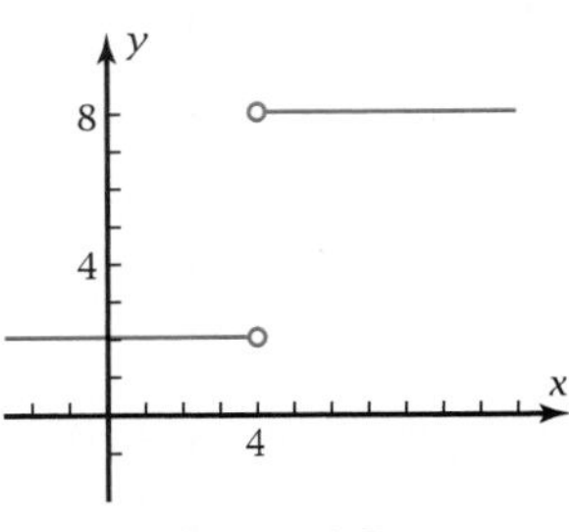

Figure 1-8p

Wild Oats Problem: Problems T20–T27 refer to the competition of wild oats, a kind of weed, with the wheat crop. Based on data in A. C. Madgett's book *Applications of Mathematics: A Nationwide Survey,* submitted to the Ministry of Education, Ontario (1976), the percent loss in wheat crop, $L(x)$, is approximately

$$L(x) = 3.2x^{0.52}$$

where x is the number of wild oat plants per square meter of land.

T20. Describe how $L(x)$ varies with x. What kind of function is L?

T21. Find $L(150)$. Explain verbally what this number means.

T22. Suppose the wheat crop reduces to 60% of what it would be without the wild oats. How many wild oats per square meter are there?

T23. Let $y = L(x)$. Find an equation for $y = L^{-1}(x)$. For what kind of calculations would the $y = L^{-1}(x)$ equation be more useful than $y = L(x)$?

T24. Find $L^{-1}(100)$. Explain the real-world meaning of the answer.

T25. Based on your answer to T24, what would be a reasonable domain and range for L?

T26. Plot $y_1 = L(x)$ and $y_2 = L^{-1}(x)$ on the same screen. Use equal scales for the two axes. Use the domain and range from T24. Sketch the results with the line $y = x$.

T27. How can you tell that the inverse relation is a function?

T28. What did you learn as a result of this test that you didn't know before?

CHAPTER

2

Periodic Functions and Right Triangle Problems

Ice-skater Michelle Kwan rotates through many degrees during a spin. Her extended hands come back to the same position at the end of each rotation. Thus the position of her hands is a periodic function of the angle through which she rotates. This angle is not restricted to the 0°–180° angles you studied in geometry. In this chapter you will learn about some special periodic functions.

Mathematical Overview

In Chapter 1 you studied various types of functions from previous courses and how these functions can be mathematical models of the real world. In this chapter you will study functions for which the y-values repeat at regular intervals. You will study these periodic functions in four ways.

Algebraically $\cos\theta = \dfrac{u}{r}$

(θ is the Greek letter theta.)

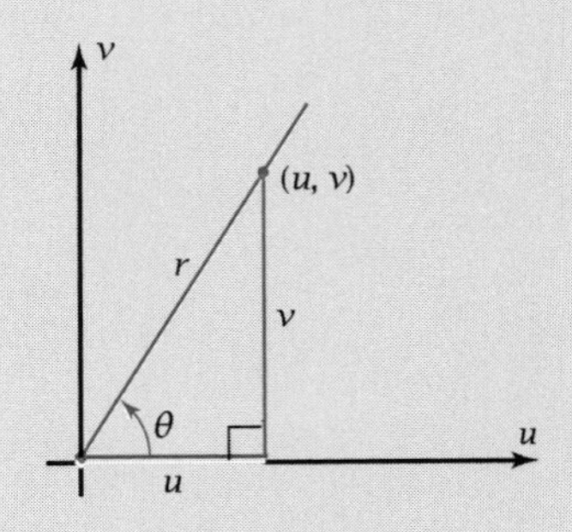

Numerically

θ	$y = \cos\theta$
0°	1
30°	0.8660...
60°	0.5
90°	0

Graphically This is the graph of a cosine function. Here, y depends on the angle, θ, which can take on negative values and values greater than 180°.

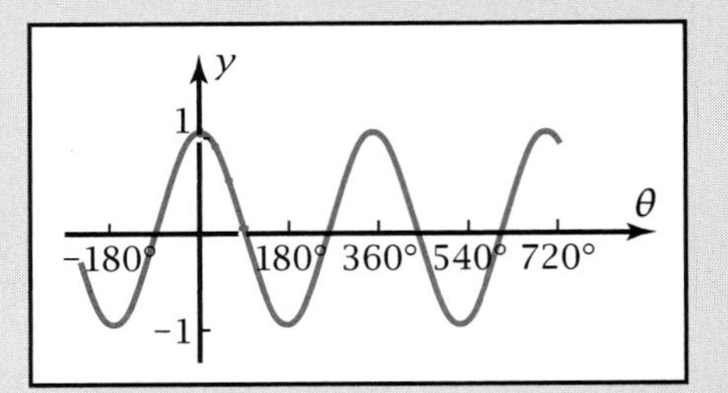

Verbally *I have learned that cosine, sine, tangent, cotangent, secant, and cosecant are ratios of sides of a right triangle. If the measure of an angle is allowed to be greater than 90° or to be negative, then these functions become periodic functions of the angle. I still have trouble remembering which ratio goes with which function.*

2-1 Introduction to Periodic Functions

As you ride a Ferris wheel, your distance from the ground depends on the number of degrees the wheel has rotated (Figure 2-1a). Suppose you start measuring the number of degrees when the seat is on a horizontal line through the axle of the wheel. The Greek letter θ (theta) often stands for the measure of an angle through which an object rotates. A wheel rotates through 360° each revolution, so θ is not restricted. If you plot θ, in degrees, on the horizontal axis and the height above the ground, y, in meters, on the vertical axis, the graph looks like Figure 2-1b. Notice that the graph has repeating y-values corresponding to each revolution of the Ferris wheel.

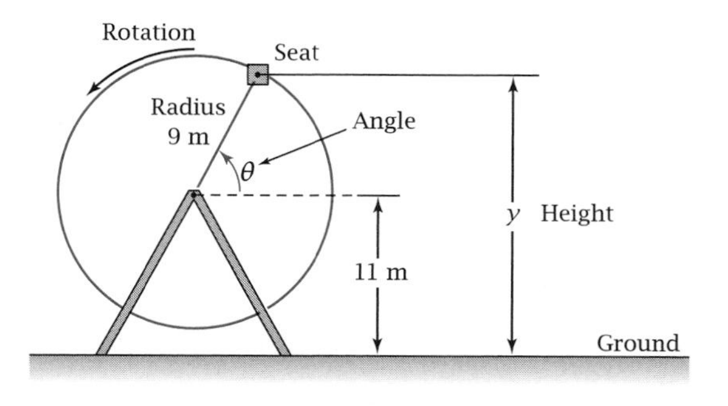

Figure 2-1a

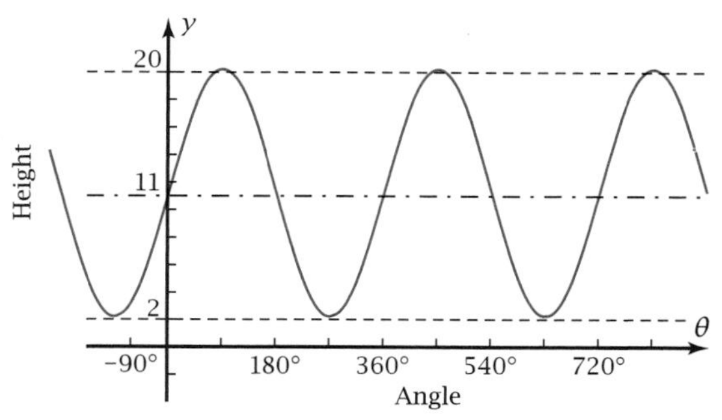

Figure 2-1b

OBJECTIVE Find the function that corresponds to the graph in Figure 2-1b and graph it on your grapher.

Exploratory Problem Set 2-1

1. The graph in Figure 2-1c is the **sine function** (pronounced like "sign"). Its abbreviation is sin, and it is written $\sin(\theta)$ or $\sin\theta$. Plot $y_1 = \sin(x)$ on your grapher (using x instead of θ). Use the window shown, and make sure your grapher is in **degree mode.**

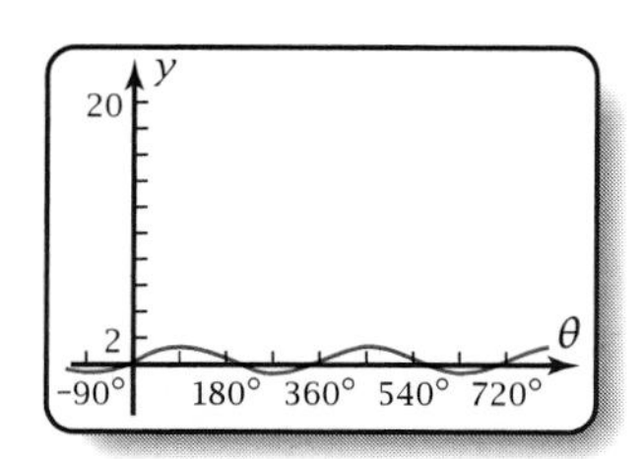

Figure 2-1c

2. The graphs in Figures 2-1b and 2-1c are called **sinusoids** (pronounced like "sinus," a skull cavity). What two transformations must you perform on the sinusoid in Figure 2-1c to get the sinusoid in Figure 2-1b?

3. Enter in your grapher an appropriate equation for the sinusoid in Figure 2-1b as y_2. Verify that your equation gives the correct graph.

4. Explain how an angle can have measure of more than 180°. Explain the real-world significance of the negative values of θ in Figures 2-1b and 2-1c.

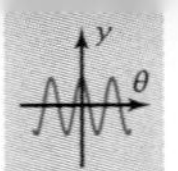

2-2 Measurement of Rotation

In the Ferris wheel problem of Section 2-1, you saw that you can use an angle to measure an amount of rotation. In this section you will extend the concept of an angle to angles that are greater than 180° and to angles whose measures are negative. You will learn why functions such as your height above the ground are periodic functions of the angle through which the Ferris wheel turns.

OBJECTIVE Given an angle of any measure, draw a picture of that angle.

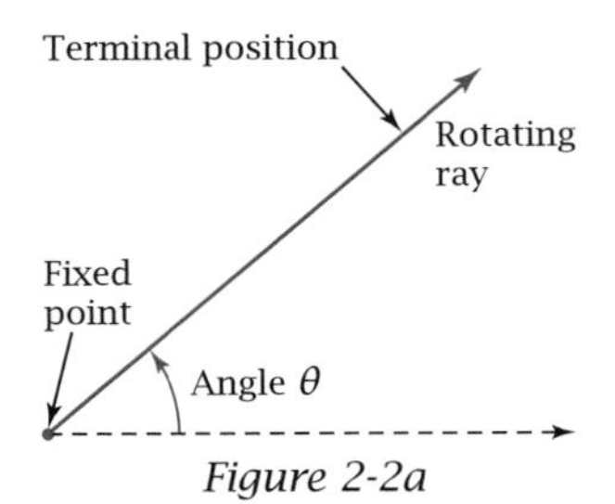

Figure 2-2a

An angle as a measure of rotation can be as large as you like. For instance, a figure skater might spin through an angle of thousands of degrees. To put this idea into mathematical terms, consider a ray with a fixed starting point. Let the ray rotate through a certain number of degrees, θ, and come to rest in a terminal (or final) position, as shown in Figure 2-2a.

So that the terminal position is uniquely determined by the angle measure, a **standard position** is defined. The initial position of the rotating ray is along the positive horizontal axis in a coordinate system, with its starting point at the origin. Counterclockwise rotation to the terminal position is measured with positive degrees, and clockwise rotation is measured with negative degrees.

DEFINITION: Standard Position of an Angle

An angle is in **standard position** in a Cartesian coordinate system if

- Its vertex is at the origin.
- Its initial side is along the positive horizontal axis.
- It is measured *counterclockwise* from the horizontal axis if the angle measure is positive and *clockwise* from the horizontal axis if the angle measure is negative.

Several angles in standard position are shown in Figure 2-2b. Because x and y are used elsewhere, the axes are labeled u and v (v for vertical).

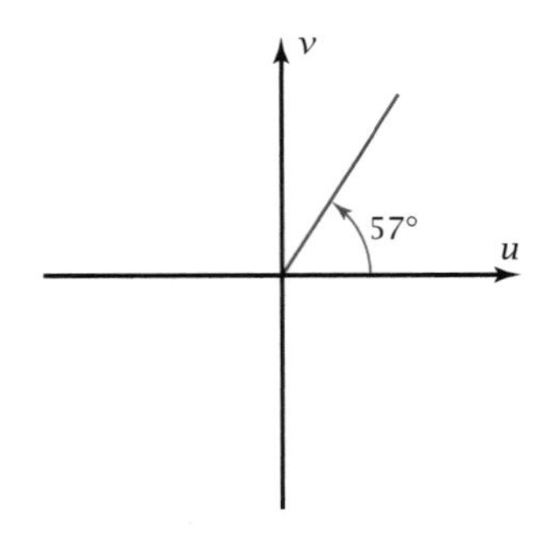

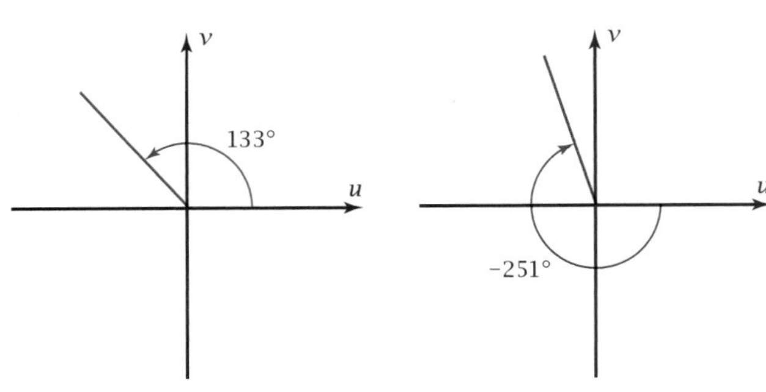

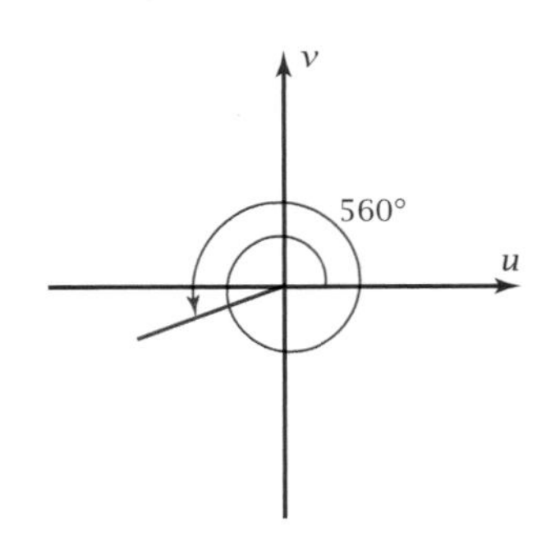

Figure 2-2b

The same position can have several corresponding angle measurements. For instance, the 493° angle terminates in the same position as 133° after one full revolution (360°) more. The −227° angle terminates there as well, by rotating clockwise instead of counterclockwise. Figure 2-2c shows these three **coterminal angles.**

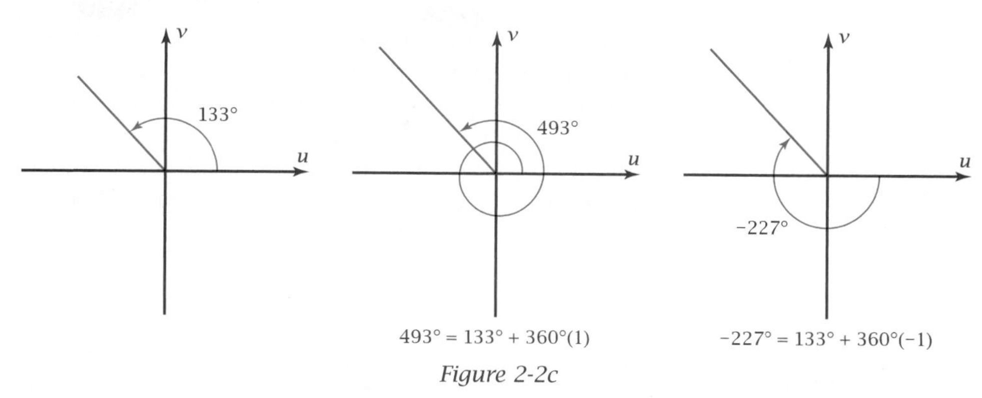

Figure 2-2c

Letters such as θ may be used for the measure of an angle or for the angle itself. Other Greek letters are often used as well: α (alpha), β (beta), γ (gamma), ϕ (phi) (pronounced "fye" or "fee"), ω (omega).

You might recognize some of the letters on this Greek restaurant sign.

DEFINITION: Coterminal Angles

Two angles in standard position are **coterminal** if and only if their degree measures differ by a multiple of 360°. That is, ϕ and θ are coterminal if and only if

$$\phi = \theta + n360°$$

where n stands for an integer.

Note: Coterminal angles have *terminal* sides that *coincide,* hence the name.

To draw an angle in standard position, you can find the measure of the positive acute angle between the horizontal axis and the terminal side. This angle is called the **reference angle.**

DEFINITION: Reference Angle

The **reference angle** of an angle in standard position is the *positive, acute* angle between the horizontal axis and the terminal side.

Note: Reference angles are always measured *counterclockwise.* Angles whose terminal sides fall on one of the axes do not have reference angles.

Example 1 shows how to find reference angles for angles terminating in each of the four quadrants.

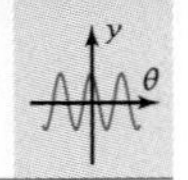

▶ EXAMPLE 1 Sketch angles of 71°, 133°, 254°, and 317° in standard position and calculate the measure of each reference angle.

Solution To calculate the measure of the reference angle, sketch an angle in the appropriate quadrant, then look at the geometry to figure out what to do.

Figure 2-2d shows the four angles along with their reference angles. For an angle between 0° and 90° (in Quadrant I), the angle and the reference angle are equal. For angles in other quadrants, you have to calculate the positive acute angle between the *x*-axis and the terminal side of the angle. If the angle is not between 0° and 360°, you can first find a coterminal angle that is between these values. From there on, it is an "old" problem like Example 1.

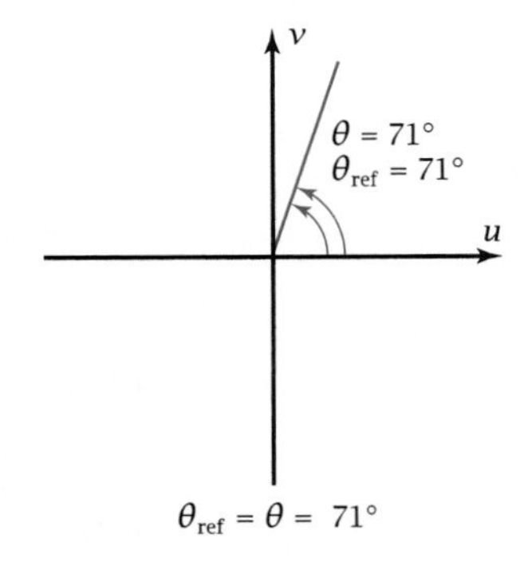

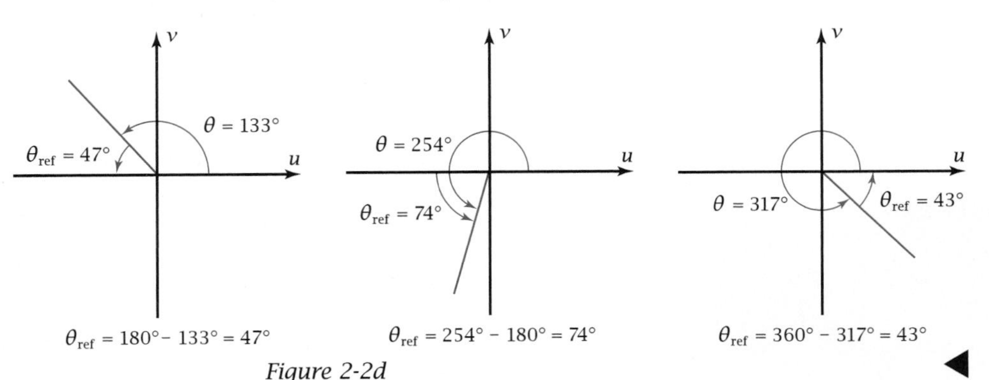

Figure 2-2d

◀

▶ EXAMPLE 2 Sketch an angle of 4897° in standard position, and calculate the measure of the reference angle.

Solution

$$\frac{4897}{360} = 13.6027\ldots$$ Divide 4897 by 360 to find the number of whole revolutions.

This number tells you that the terminal side makes 13 whole revolutions plus another 0.6027... revolution. To find out which quadrant the angle falls in, multiply the decimal part of the revolutions by 360 to find the number of degrees. The answer is θ_c, a coterminal angle to θ between 0° and 360°.

$$\theta_c = (0.6027\ldots)(360) = 217°$$ Do the computations without rounding.

Sketch the 217° angle in Quadrant III, as shown in Figure 2-2e.

Figure 2-2e

From the figure, you should be able to see that

$$\theta_{ref} = 217° - 180° = 37°$$

Two things are important for you to remember as you draw the reference angle. First, it is always between the terminal side and the *horizontal* axis (never the vertical axis). Second, the reference angle sometimes goes from the axis to the terminal side and sometimes from the terminal side to the axis. To figure out which way it goes, recall that the reference angle is *positive.* Thus it always goes in the *counterclockwise* direction.

Problem Set 2-2

Do These Quickly

5 min

Q1. A function that repeats its values at regular intervals is called a —?— function.

Describe the transformations in Q2–Q5.

Q2. $g(x) = 5\,f(x)$

Q3. $g(x) = f(3x)$

Q4. $g(x) = 4 + f(x)$

Q5. $g(x) = f(x - 2)$

Q6. If $f(x) = 2x + 6$, then $f^{-1}(x) =$ —?—.

Q7. How many degrees are there in two revolutions?

Q8. Sketch the graph of $y = 2^x$.

Q9. 40 is 20% of what number?

Q10. $x^{20}/x^5 =$

A. x^{15}
B. x^4
C. x^{25}
D. x^{100}
E. None of these

For Problems 1–20, sketch the angle in standard position, mark the reference angle, and find its measure.

1. 130°
2. 198°
3. 259°
4. 147°
5. 342°
6. 21°
7. 54°
8. 283°
9. −160°
10. −220°
11. −295°
12. −86°
13. 98.6°
14. 57.3°
15. −154.1°
16. −273.2°
17. 5481°
18. 7321°
19. −2746°
20. −3614°

For Problems 21–26, the angles are measured in degrees, minutes, and seconds. There are 60 minutes (60′) in a degree and 60 seconds (60″) in a minute. To find 180° − 137°24′, you calculate 179°60′ − 137°24′. Sketch each angle in standard position, mark the reference angle, and find its measure.

21. 145°37′
22. 268°29′
23. 213°16′
24. 121°43′
25. 308°14′51″
26. 352°16′44″

For Problems 27 and 28, sketch a reasonable graph of the function, showing how the dependent variable is related to the independent variable.

27. A girl jumps up and down on a trampoline. Her distance from the ground depends on time.

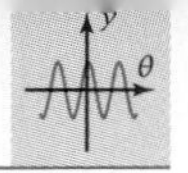

28. The pendulum in a grandfather clock swings back and forth. The distance from the end of the pendulum to the left side of the clock depends on time.

For Problems 29 and 30, write an equation for the image function, g (solid graph), in terms of the pre-image function f (dashed graph).

29.

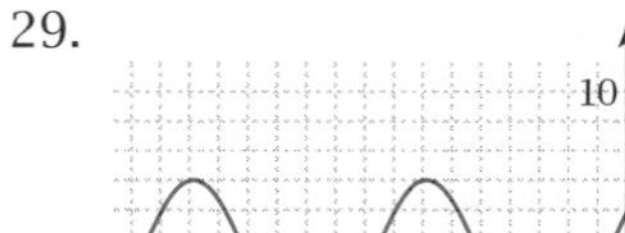

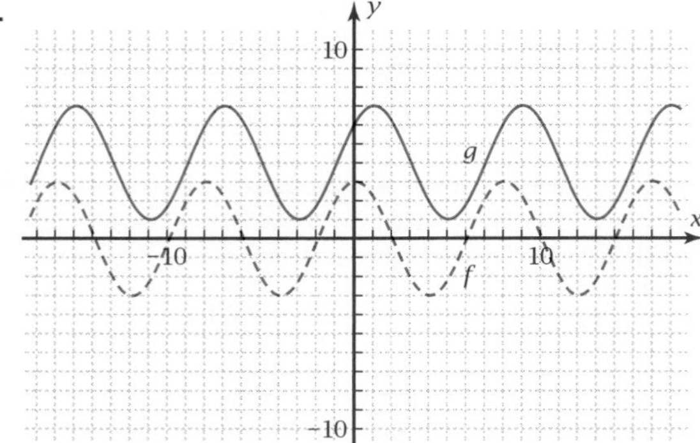

30.

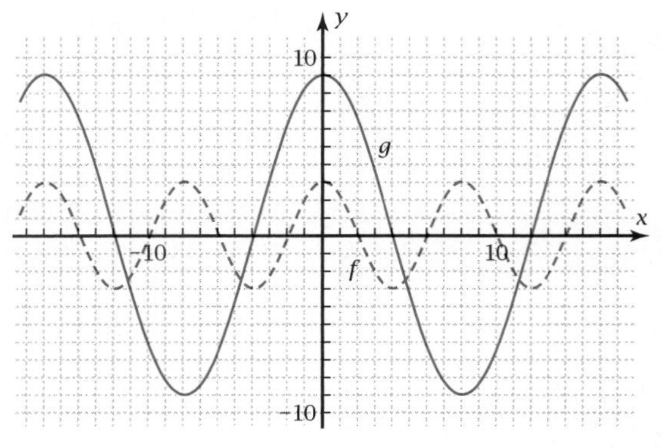

2-3 Sine and Cosine Functions

From previous mathematics courses, you may recall working with sine, cosine, and tangent for angles in a right triangle. Your grapher has these functions in it. If you plot the graph of $y = \sin\theta$ and $y = \cos\theta$, you get the periodic functions shown in Figure 2-3a. Each graph is called a sinusoid, as you learned in Section 2-1. To get these graphs, you may enter the equations in the form $y = \sin x$ and $y = \cos x$, and use degree mode.

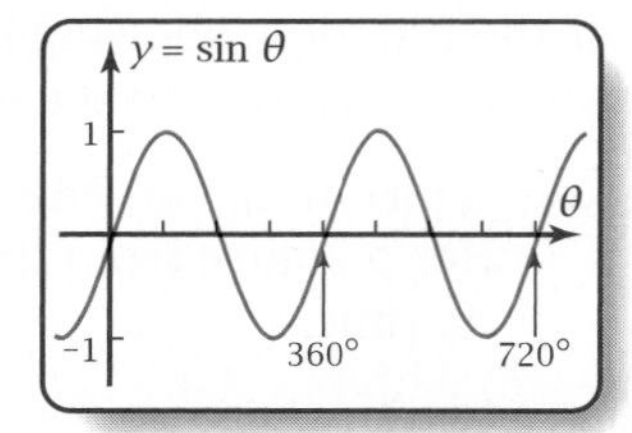

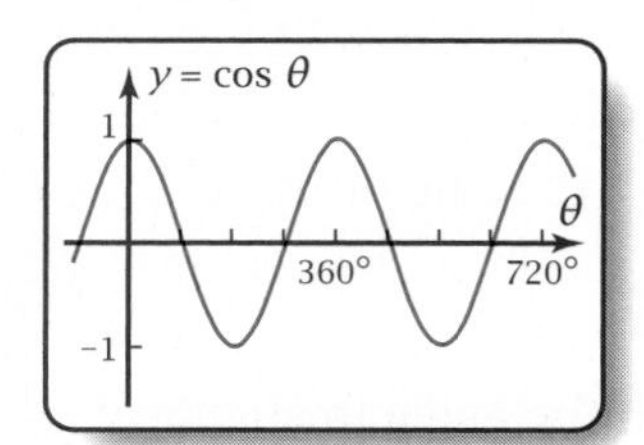

Figure 2-3a

In this section you will see how the reference angles of Section 2-2 let you extend the right triangle definitions of sine and cosine to include angles of any measure. You will also see how these definitions lead to sinusoids.

OBJECTIVE Extend the definitions of sine and cosine for any angle.

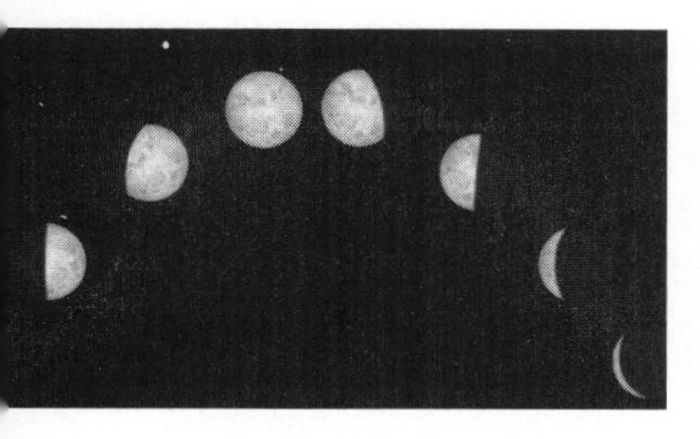

Periodicity is quite common. The phases of the moon are just one example of a periodic phenomenon.

A **periodic function** is a function whose values repeat at regular intervals. The graphs in Figure 2-3a are examples. The part of the graph from a point to the point where the graph starts repeating itself is called a **cycle.** For a periodic function, the **period** is the difference between the horizontal coordinates corresponding to one cycle. Figure 2-3b shows an example. The sine and cosine functions complete a cycle every 360°, as you can see in Figure 2-3a. So the period of these functions is 360°. A horizontal translation of one period makes the pre-image and image graphs identical.

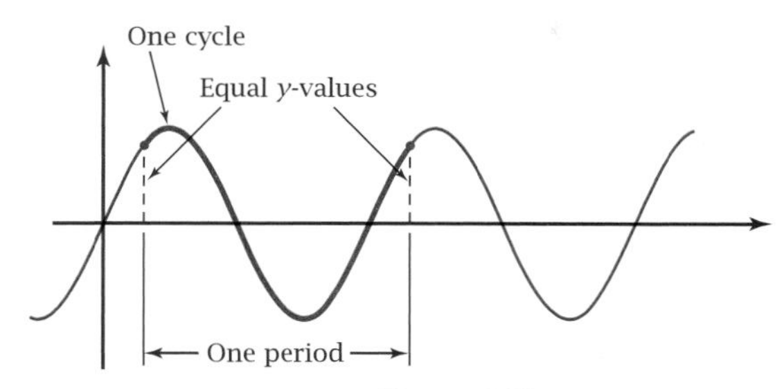

Figure 2-3b

> ***DEFINITION: Periodic Function***
>
> The function f is a **periodic function** of x if and only if there is a number p for which $f(x - p) = f(x)$ for all values of x in the domain.
>
> If p is the smallest such number, then p is called the **period** of the function.

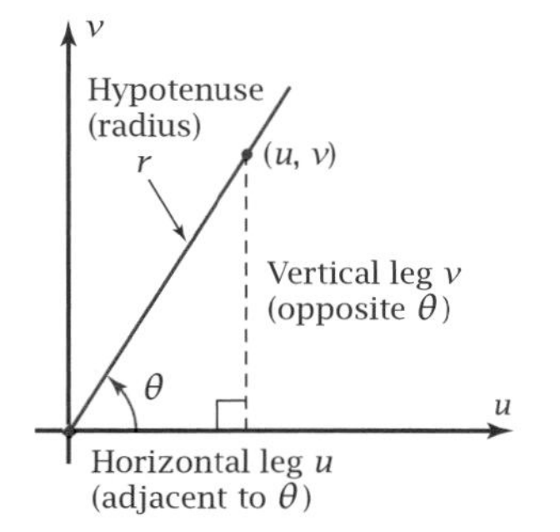

Figure 2-3c

To see why the sine and cosine graphs are periodic, consider an angle θ in standard position, terminating in Quadrant I in a uv-coordinate system (Figure 2-3c). If you pick a point (u, v) on the terminal side and draw a perpendicular from it to the horizontal axis, you form a right triangle.

The right triangle definition of these **trigonometric functions** for acute angle θ is

$$\sin \theta = \frac{\text{opposite leg}}{\text{hypotenuse}} \qquad \cos \theta = \frac{\text{adjacent leg}}{\text{hypotenuse}}$$

The word **trigonometry** comes from the Greek roots for "triangle" (*trigon-*) and "measurement" (*-metry*). You can define sine and cosine as ratios of the lengths of the sides of a right triangle. By the properties of similar triangles, these ratios depend only on the measure of θ, not on where the point (u, v) lies on the terminal side. Figure 2-3d illustrates this fact.

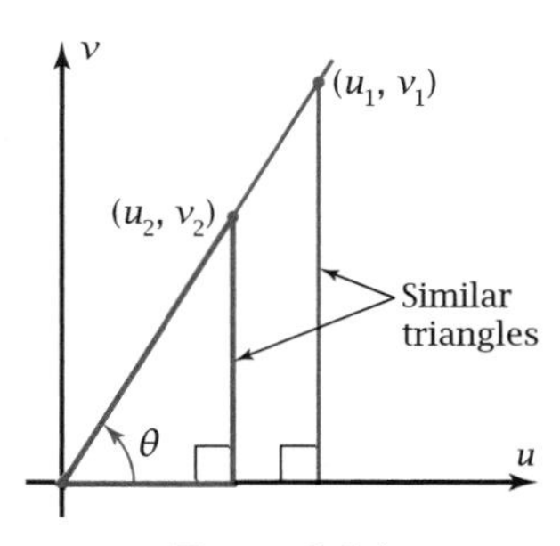

Figure 2-3d

As you can see from Figure 2-3c, with θ in standard position, the length of the opposite leg of the right triangle is equal to the vertical coordinate of

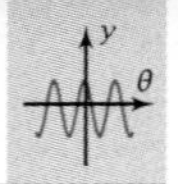

point (u, v) and the length of the adjacent leg is equal to the horizontal coordinate. The hypotenuse is equal to r (for "radius"), the distance from the origin to point (u, v). As shown in Figure 2-3e, r is always positive because it is the radius of a circle. Because for any angle in standard position you can pick an arbitrary point on its terminal side, the following definitions apply to any size angle.

DEFINITION: Sine and Cosine Functions of Any Size Angle

Let (u, v) be a point r units from the origin on the terminal side of a rotating ray. If θ is the angle in standard position to the ray, then

$$\sin\theta = \frac{v}{r} \qquad \cos\theta = \frac{u}{r}$$

Note: You can write the symbols $\sin\theta$ and $\cos\theta$ as $\sin(\theta)$ and $\cos(\theta)$. In this form you will recognize them as examples of the $f(x)$ notation, where sin and cos are the names of the functions and θ is the independent variable or argument.

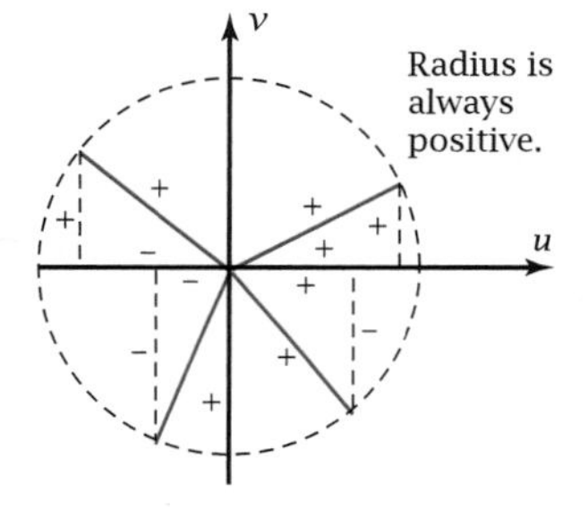

Figure 2-3e

From Figure 2-3e you can determine that $\sin\theta$ will be positive in Quadrant I and Quadrant II, and it will be negative in Quadrant III and Quadrant IV. Similarly, you can determine the sign of $\cos\theta$ in each of the different quadrants: positive in Quadrants I and IV and negative in the other two quadrants.

Now, imagine θ increasing as the ray rotates counterclockwise around the origin. As shown in Figure 2-3f, the vertical v-coordinate of point (u, v) increases when θ is in Quadrant I, decreases when θ is in Quadrant II, becomes negative when θ is in Quadrant III, and is negative but increasing toward zero when θ is in Quadrant IV.

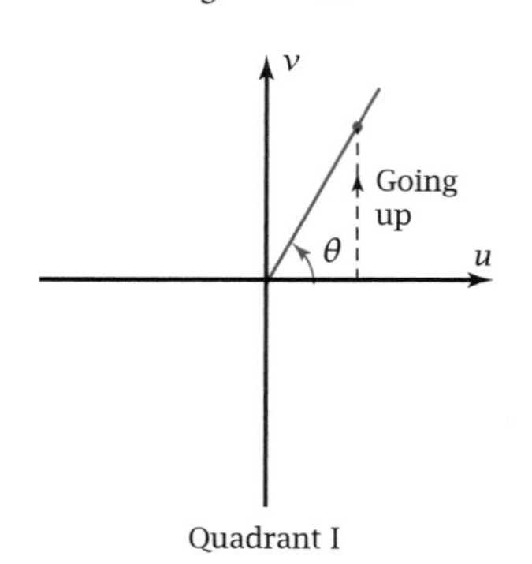

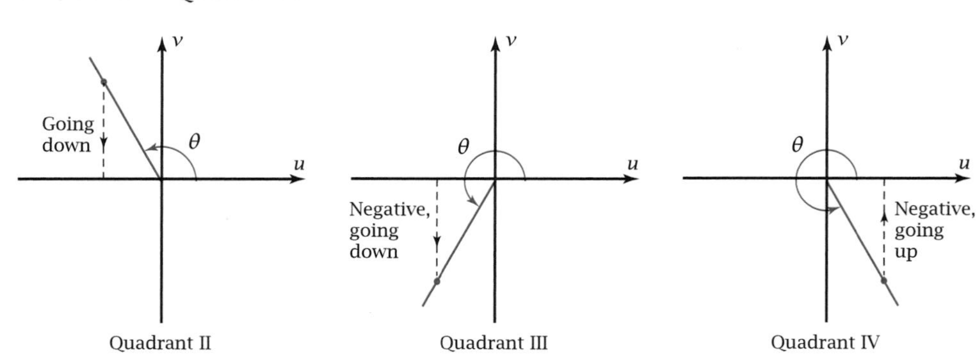

Figure 2-3f

When θ is $0°$, $v = 0$, and so $\sin 0° = 0$. When θ is $90°$, v and r are equal, so $\sin 90° = 1$. Similarly, $\sin 180° = 0$ and $\sin 270° = -1$. From the time θ gets back around to $360°$, the sine values start repeating. So by allowing angles greater than $90°$, you can see that the sine is a periodic function of the angle. The increasing and decreasing pattern agrees with the sine graph in Figure 2-3a.

You can repeat this analysis using the cosine function. By examining the values of the horizontal coordinate, u, you will see that $\cos 0° = 1$ and that $\cos \theta$ decreases as θ increases through Quadrant I, becomes negative in Quadrant II, increases to 0 in Quadrant III, and increases back to 1 by the end of Quadrant IV, where $\theta = 360°$. This is the behavior shown in the cosine graph of Figure 2-3a.

If you have computer software for animation available, such as The Geometer's Sketchpad®, you can show dynamically how the rotating ray in the uv-coordinate system generates sinusoids in the θy-coordinate system.

► EXAMPLE 1 Draw a 147° angle in standard position. Mark the reference angle and find its measure. Then find cos 147° and $\cos \theta_{ref}$. Explain the relationship between the two cosine values.

Solution Draw the angle and its reference angle, as shown in Figure 2-3g.

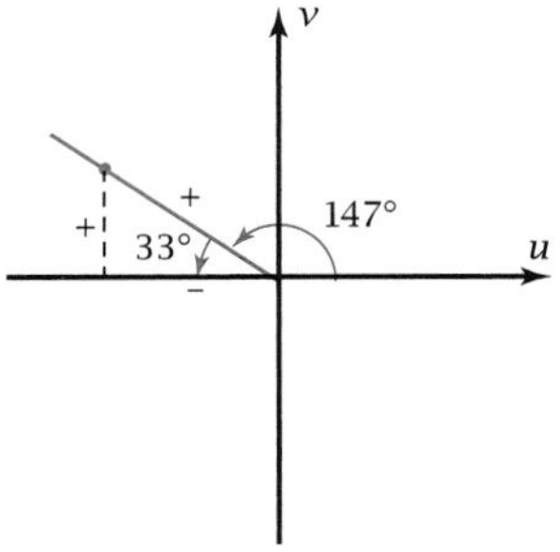

$\theta_{ref} = 180° - 147° = 33°$ Because θ_{ref} and 147° must add up to 180°.

$\cos 147° = -0.8386\ldots$ and

$\cos 33° = 0.8386\ldots$ By calculator.

Figure 2-3g

Both cosine values have the same magnitude. Cos 147° is negative because the horizontal coordinate of a point in Quadrant II is negative. The radius, r, is always considered to be positive because it is the radius of a circle traced as the ray rotates. ◄

Note: When you write the cosine of an angle in degrees, such as cos 147°, you must write the degree sign. Writing cos 147 without the degree sign has a different meaning, as you will see when you learn about angles in radians in the next chapter.

► EXAMPLE 2 The terminal side of angle θ in standard position contains the point $(u, v) = (8, -5)$. Sketch the angle in standard position. Use the definitions of sine and cosine to find $\sin \theta$ and $\cos \theta$.

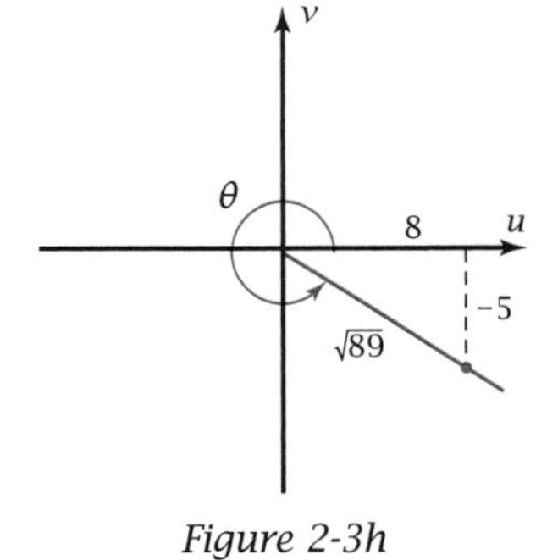

Figure 2-3h

Solution Draw the angle as in Figure 2-3h. Mark the 8 as the u-coordinate and the -5 as the v-coordinate.

$r = \sqrt{8^2 + (-5)^2} = \sqrt{89}$ Use the Pythagorean theorem to find r. Show $\sqrt{89}$ on the figure.

$\sin \theta = \dfrac{-5}{\sqrt{89}} = -0.5299\ldots$ and

$\cos \theta = \dfrac{8}{\sqrt{89}} = 0.8479\ldots$ By the definitions. ◄

Figure 2-3i shows the **parent sine function,** $y = \sin\theta$. You can plot sinusoids with other proportions and locations by transforming this parent graph. Example 3 shows you how to do this.

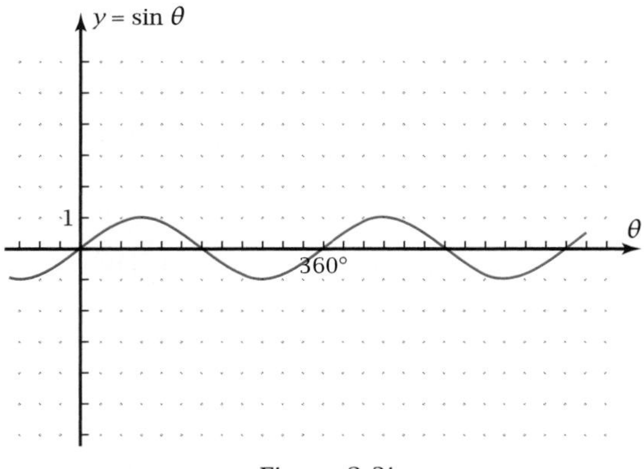

Figure 2-3i

► ***EXAMPLE 3*** Let $y = 4\sin\theta$. What transformation of the parent sine function is this? On a copy of Figure 2-3i, sketch the graph of this image sinusoid. Check your sketch by plotting the parent sinusoid and the transformed sinusoid on the same screen.

Solution The transformation is a vertical dilation by a factor of 4.

Find places where the pre-image function has high points, low points, or zeros. Multiply the y-values by 4 and plot the resulting **critical points** (Figure 2-3j, left side). Sketch a smooth curve through the critical points (Figure 2-3j, right side). The grapher will confirm that your sketch is correct.

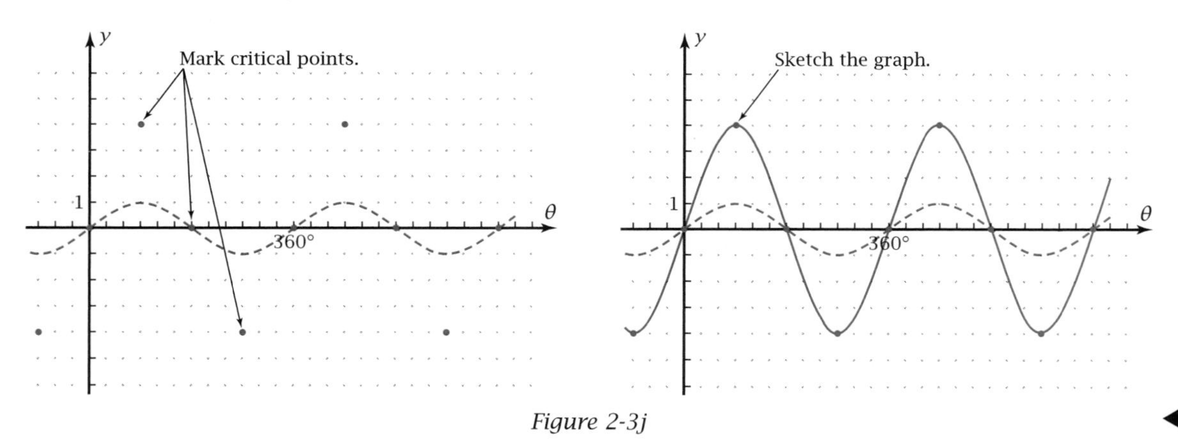

Figure 2-3j ◄

Problem Set 2-3

Do These Quickly

Q1. Write the general equation of an exponential function.

Q2. The equation $y = 3x^{1.2}$ represents a particular —?— function.

Q3. Find the reference angle for 241°.

Q4. Name these Greek letters: α, β, γ, ϕ

Q5. What transformation is the image function $y = (x - 3)^5$ of the pre-image $y = x^5$?

Q6. Find x if $5 \log 2 = \log x$.

Q7. Sketch a reasonable graph showing the distance of your foot from the pavement as a function of the distance your bicycle has traveled.

Q8. $3.7^0 =$ —?— (3.7 with a zero exponent, not 3.7 degrees.)

Q9. What is the value of 5! (five factorial)?

Q10. What percent of 300 is 60?

For Problems 1–6, sketch the angle in standard position in a *uv*-coordinate system, and then mark and find the measure of the reference angle. Find the sine or cosine of the angle and its reference angle. Explain the relationship between them.

1. sin 250°
2. sin 320°
3. cos 140°
4. cos 200°
5. cos 300°
6. sin 120°

For Problems 7–14, use the definition of sine and cosine to write $\sin\theta$ and $\cos\theta$ for angles whose terminal side contains the given point.

7. (7, 11)
8. (4, 1)
9. (−2, 5)
10. (−6, 9)
11. (4, −8)
12. (8, −3)
13. (−24, −7) (Surprising?!)
14. (−3, −4) (Surprising?!)

Figure 2-3k shows the parent function graphs $y = \sin\theta$ and $y = \cos\theta$. For Problems 15–20, give the transformation of the parent function represented by the equation. Sketch the transformed graph on a copy of Figure 2-3k. Confirm your sketch by plotting both graphs on the same screen on your grapher.

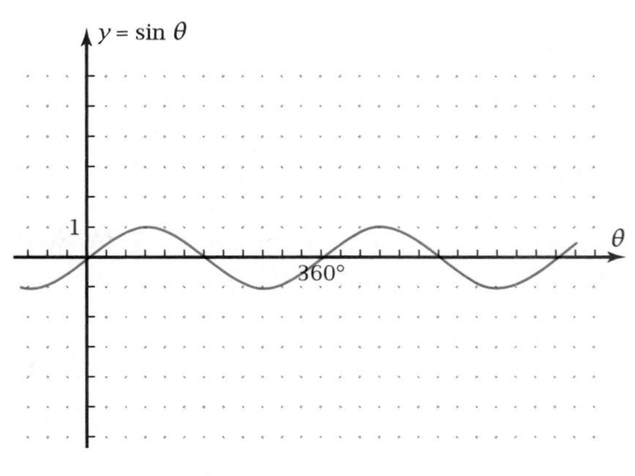

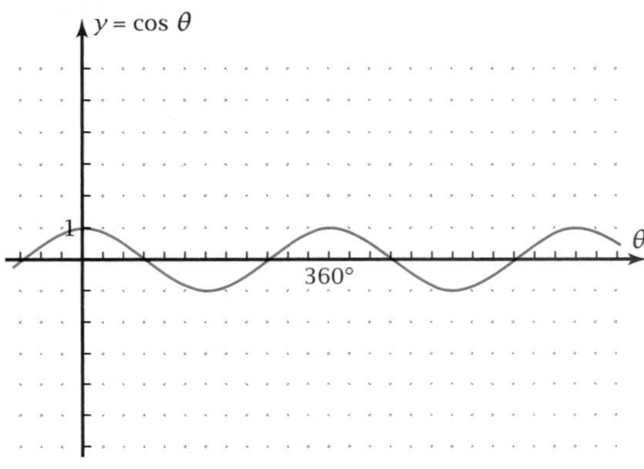

Figure 2-3k

15. $y = \sin(\theta - 60°)$
16. $y = 4 + \sin\theta$
17. $y = 3\cos\theta$
18. $y = \cos\frac{1}{2}\theta$
19. $y = 3 + \cos 2\theta$
20. $y = 4\cos(\theta + 60°)$
21. Draw the *uv*-coordinate system. In each quadrant, put a + or a − sign to show whether $\cos\theta$ is positive or negative when θ terminates in that quadrant.
22. Draw the *uv*-coordinate system. In each quadrant, put a + or a − sign to show whether $\sin\theta$ is positive or negative when θ terminates in that quadrant.
23. *Functions of Reference Angles Problem:* This property relates the sine and cosine of an angle to the sine and cosine of the reference angle. Give numerical examples to show that the property is true for both sine and cosine.

Property: Sine and Cosine of a Reference Angle

$\sin\theta_{\text{ref}} = |\sin\theta|$ and $\cos\theta_{\text{ref}} = |\cos\theta|$

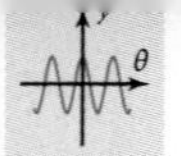

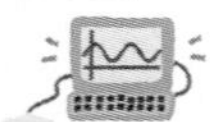

24. *Construction Problem 1:* For this problem, use pencil and paper or a computer graphing program such as The Geometer's Sketchpad. Construct a right triangle with one horizontal leg 8 cm long and an acute angle of 35° with its vertex at one end of the 8-cm leg. Measure the hypotenuse and the other leg. Use these measurements to calculate the values of sin 35° and cos 35° from the definitions of sine and cosine. How well do the answers agree with the values you get directly by calculator? While keeping the angle equal to 35°, increase the sides of the right triangle. Calculate the values of sin 35° and cos 35° in the new triangle. What do you find?

2-4 Values of the Six Trigonometric Functions

In Section 2-3 you recalled the definitions of sine and cosine of an angle and saw how to extend these definitions to include angles beyond the range of 0° to 90°. With the extended definitions, $y = \sin\theta$ and $y = \cos\theta$ are periodic functions whose graphs are called sinusoids. In this section you will define four other trigonometric functions. You will learn how to evaluate all six trigonometric functions approximately by calculator and exactly, in special cases, using the definitions. In the next section you will use what you have learned to calculate unknown side and angle measures in right triangles.

OBJECTIVE Be able to find values of the six trigonometric functions approximately, by calculator, for any angle and exactly for certain special angles.

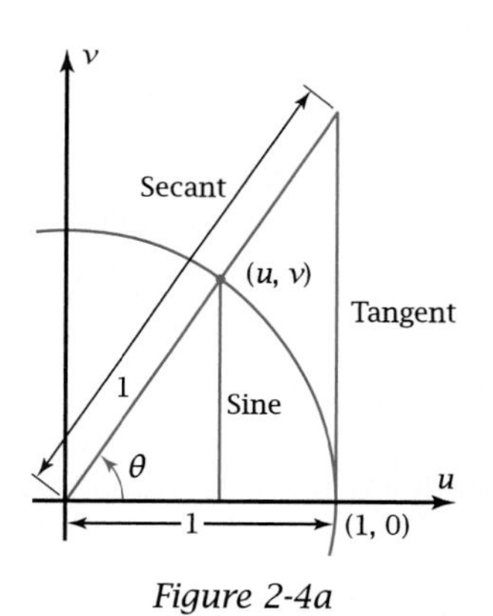

Figure 2-4a

Figure 2-4a shows a unit circle (radius is 1 unit) centered at the origin in a *uv*-coordinate system. A ray at an angle θ intersects the circle at the point (u, v). Two right triangles are shown with θ as an acute angle. The smaller triangle has its vertical side from (u, v) to the *u*-axis. The larger triangle has its vertical side tangent to the circle at the point (1, 0). The hypotenuse of the larger triangle is a secant line because it cuts through the circle.

Recall that $\sin\theta$ in a right triangle equals the ratio of the opposite leg to the hypotenuse. Because the hypotenuse of the smaller triangle is 1 (the radius of the circle), $\sin\theta = v$, the vertical coordinate of the point (u, v). The ratio of the opposite leg to the adjacent leg is also a function of θ. Because the adjacent leg of θ in the larger right triangle is 1, this ratio equals the length of the tangent line segment. The ratio of the opposite leg to the adjacent leg is called the *tangent* of θ, abbreviated as $\tan\theta$. Similarly, the length of the secant line segment (the hypotenuse of the larger triangle) equals the ratio of the hypotenuse to the adjacent leg. This ratio is called the *secant* of θ and abbreviated as $\sec\theta$.

There are two more basic trigonometric functions: the *cotangent* ($\cot\theta$) and the *cosecant* ($\csc\theta$) functions. These are defined as the ratios of the adjacent leg to the opposite leg and of the hypotenuse to the opposite leg, respectively. This table presents the definitions of the six trigonometric functions, both in right triangle form and in the more general coordinate form. You should memorize these definitions.

DEFINITIONS: The Six Trigonometric Functions

Let (u, v) be a point r units from the origin on the terminal side of a rotating ray. If θ is the angle in standard position to the ray, then the following hold.

Right Triangle Form	**Coordinate Form**
$\sin\theta = \dfrac{\text{opposite}}{\text{hypotenuse}}$	$\sin\theta = \dfrac{\text{vertical coordinate}}{\text{radius}} = \dfrac{v}{r}$
$\cos\theta = \dfrac{\text{adjacent}}{\text{hypotenuse}}$	$\cos\theta = \dfrac{\text{horizontal coordinate}}{\text{radius}} = \dfrac{u}{r}$
$\tan\theta = \dfrac{\text{opposite}}{\text{adjacent}}$	$\tan\theta = \dfrac{\text{vertical coordinate}}{\text{horizontal coordinate}} = \dfrac{v}{u}$
$\cot\theta = \dfrac{\text{adjacent}}{\text{opposite}}$	$\cot\theta = \dfrac{\text{horizontal coordinate}}{\text{vertical coordinate}} = \dfrac{u}{v}$
$\sec\theta = \dfrac{\text{hypotenuse}}{\text{adjacent}}$	$\sec\theta = \dfrac{\text{radius}}{\text{horizontal coordinate}} = \dfrac{r}{u}$
$\csc\theta = \dfrac{\text{hypotenuse}}{\text{opposite}}$	$\csc\theta = \dfrac{\text{radius}}{\text{vertical coordinate}} = \dfrac{r}{v}$

Notes: The cotangent, secant, and cosecant functions are reciprocals of the tangent, cosine, and sine functions, respectively. The relationship between each pair of functions, such as cotangent and tangent, is called the **reciprocal property** of trigonometric functions, which you will fully explore in Chapter 4.

When you write the functions in a column in the order $\sin\theta$, $\cos\theta$, $\tan\theta$, $\cot\theta$, $\sec\theta$, $\csc\theta$, the functions and their reciprocals will have this pattern:

Approximate Values by Calculator

Example 1 shows how you can find approximate values of all six trigonometric functions using your calculator.

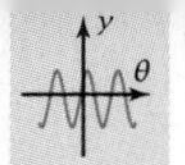

EXAMPLE 1 Evaluate by calculator the six trigonometric functions of 58.6°. Round to four decimal places.

Solution You can find sine, cosine, and tangent directly on the calculator.

$$\sin 58.6° = 0.8535507\ldots \approx 0.8536$$

Note preferred usage of the ellipsis and the ≈ sign.

$$\cos 58.6° = 0.5210096\ldots \approx 0.5210$$

$$\tan 58.6° = 1.6382629\ldots \approx 1.6383$$

The other three functions are the reciprocals of the sine, cosine, and tangent functions. Notice that the reciprocals follow the pattern described earlier.

$$\cot 58.6° = \frac{1}{\tan 58.6°} = 0.61040260\ldots \approx 0.6104$$

$$\sec 58.6° = \frac{1}{\cos 58.6°} = 1.91935031\ldots \approx 1.9194$$

$$\csc 58.6° = \frac{1}{\sin 58.6°} = 1.17157643\ldots \approx 1.1716$$

Exact Values by Geometry

If you know a point on the terminal side of an angle, you can calculate the values of the trigonometric functions exactly. Example 2 shows you the steps.

EXAMPLE 2 The terminal side of angle θ contains the point (−5, 2). Find *exact* values of the six trigonometric functions of θ. Use radicals if necessary, but no decimals.

Solution

- Sketch the angle in standard position (Figure 2-4b).
- Pick a point on the terminal side, (−5, 2) in this instance, and draw a perpendicular.
- Mark lengths on the sides of the reference triangle, using the Pythagorean theorem.

$$r = \sqrt{(-5)^2 + 2^2} = \sqrt{29}$$

$$\sin\theta = \frac{\text{vertical}}{\text{radius}} = \frac{2}{\sqrt{29}}$$

$$\cos\theta = \frac{\text{horizontal}}{\text{radius}} = \frac{-5}{\sqrt{29}} = -\frac{5}{\sqrt{29}}$$

$$\tan\theta = \frac{\text{vertical}}{\text{horizontal}} = \frac{2}{-5} = -\frac{2}{5}$$

$$\cot\theta = \frac{1}{\tan\theta} = -\frac{5}{2}$$

$$\sec\theta = \frac{1}{\cos\theta} = -\frac{\sqrt{29}}{5}$$

$$\csc\theta = \frac{1}{\sin\theta} = \frac{\sqrt{29}}{2}$$

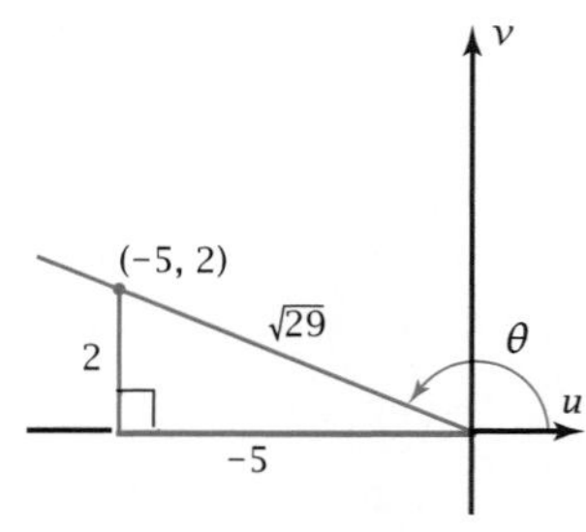

Figure 2-4b

Note: You can use the proportions of the sides in the 30°-60°-90° triangle and the 45°-45°-90° triangle to find exact function values for angles whose reference angle is a multiple of 30° or 45°. Figure 2-4c shows these proportions.

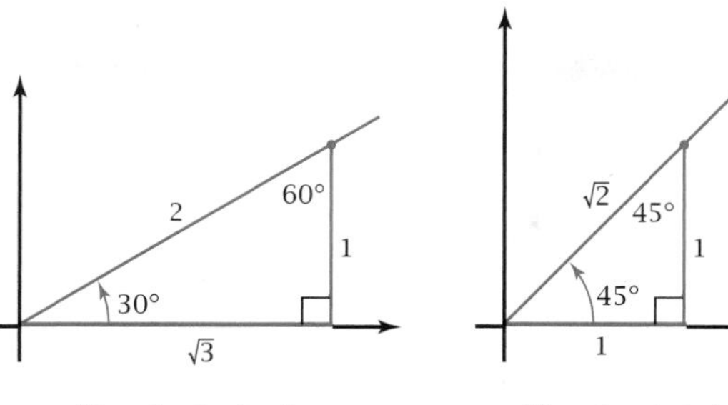

Figure 2-4c

▶ **EXAMPLE 3** Find *exact* values (no decimals) of the six trigonometric functions of 300°.

Solution Sketch an angle terminating in Quadrant IV and a reference triangle (Figure 2-4d).

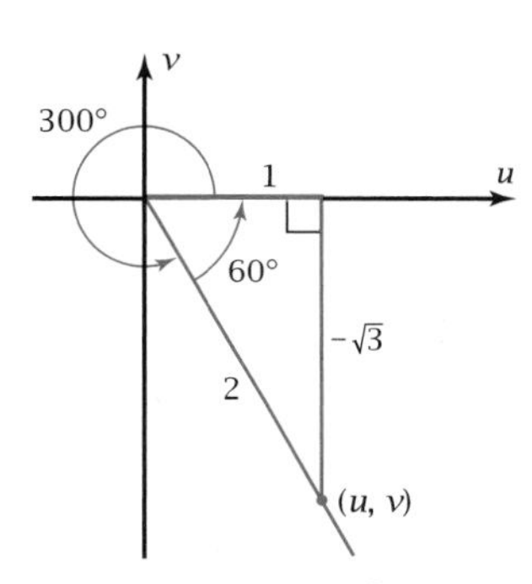

Figure 2-4d

$$\sin\theta = \frac{-\sqrt{3}}{2} = -\frac{\sqrt{3}}{2}$$ Use the *negative* square root because v is negative.

$$\cos\theta = \frac{1}{2}$$

$$\tan\theta = \frac{-\sqrt{3}}{1} = -\sqrt{3}$$ Do any obvious simplification.

$$\cot\theta = \frac{1}{\tan\theta} = -\frac{1}{\sqrt{3}}$$ Use the reciprocal relationship.

$$\sec\theta = \frac{1}{\cos\theta} = \frac{2}{1} = 2$$

$$\csc\theta = \frac{1}{\sin\theta} = \frac{2}{-\sqrt{3}} = -\frac{2}{\sqrt{3}}$$ ◀

Example 4 shows how to find the function values for an angle that terminates on a quadrant boundary.

▶ **EXAMPLE 4** Without a calculator, evaluate the six trigonometric functions for an angle of 180°.

Solution Figure 2-4e shows a 180° angle in standard position. The terminal side falls on the negative side of the horizontal axis. Pick any point on the terminal side, such as (−3, 0). Note that although the u-coordinate of the point is negative, the distance r from the origin to the point is positive because it is the radius of a circle. The vertical coordinate, v, is 0.

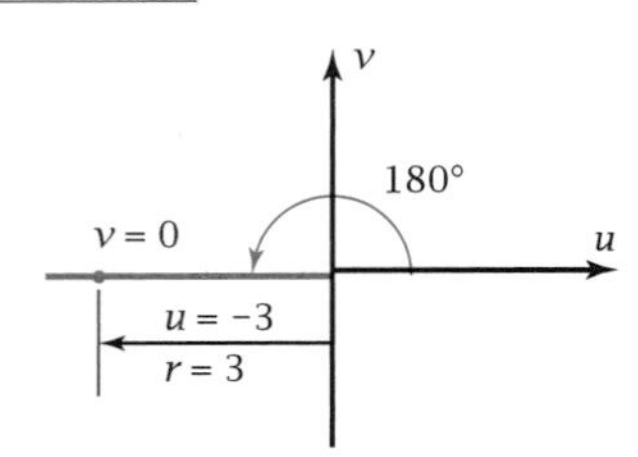

Figure 2-4e

$\sin 180° = \dfrac{\text{vertical}}{\text{radius}} = \dfrac{0}{3} = 0$ Use "vertical, radius" rather than "opposite, hypotenuse."

$\cos 180° = \dfrac{\text{horizontal}}{\text{radius}} = \dfrac{-3}{3} = -1$

$\tan 180° = \dfrac{\text{vertical}}{\text{horizontal}} = \dfrac{0}{-3} = 0$ Do the obvious simplification.

$\cot 180° = \dfrac{1}{\tan 180°} = \dfrac{1}{0}$ No value. Undefined because of division by zero.

$\sec 180° = \dfrac{1}{\cos 180°} = \dfrac{1}{-1} = -1$

$\csc 180° = \dfrac{1}{\sin 180°} = \dfrac{1}{0}$ No value. ◀

Problem Set 2-4

Do These Quickly

Problems Q1–Q5 concern the right triangle in Figure 2-4f.

Q1. Which side is the leg opposite θ?

Q2. Which side is the leg adjacent to θ?

Q3. Which side is the hypotenuse?

Q4. $\cos \theta =$ —?—

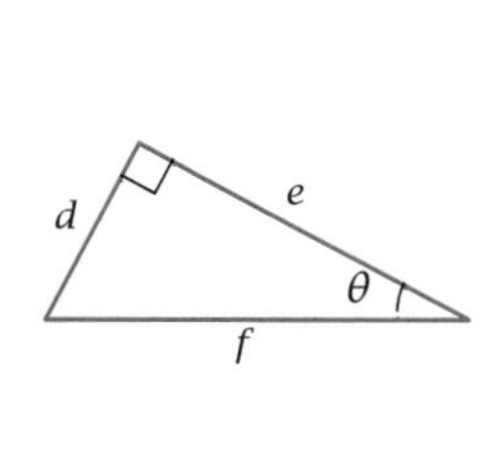

Figure 2-4f

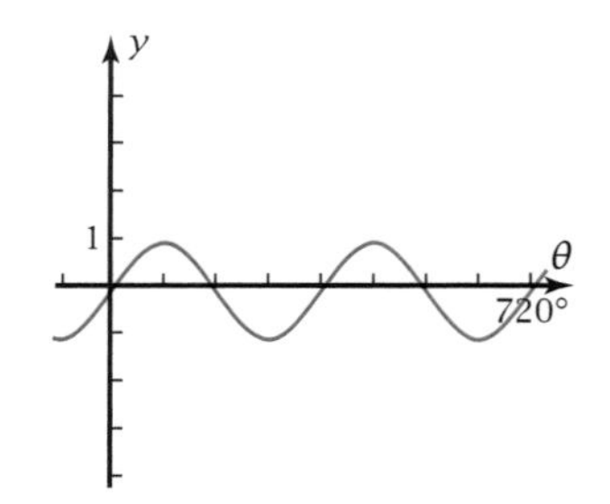

Figure 2-4g

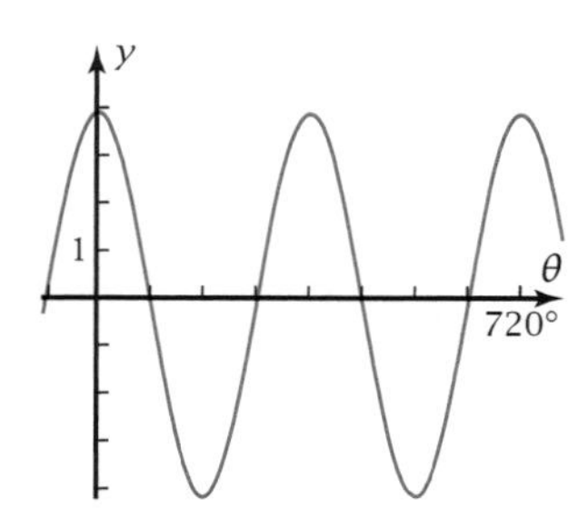

Figure 2-4h

Q5. $\sin \theta =$ —?—

Q6. Write an equation for the sinusoid in Figure 2-4g.

Q7. Write an equation for the sinusoid in Figure 2-4h.

Q8. How was the parent function transformed to get the sinusoid in Figure 2-4h?

Q9. Sketch the graph of $y = -x^2$.

Q10. A one-to-one function is

A. Always increasing

B. Always decreasing

C. Always positive

D. Always negative

E. Always invertible

For Problems 1–6, find a decimal approximation for the given function value. Round the answer to four decimal places.

1. $\cot 38°$
2. $\cot 140°$
3. $\sec 238°$
4. $\sec (-53°)$
5. $\csc (-179°)$
6. $\csc 180°$ (Surprise?)

For Problems 7–10, find the exact values (no decimals) of the six trigonometric functions of an angle θ in standard position whose terminal side contains the given point.

7. $(4, -3)$ 8. $(-12, 5)$

9. $(-5, -7)$ 10. $(2, 3)$

For Problems 11–14, if θ terminates in the given quadrant and has the given function value, find the exact values (no decimals) of the six trigonometric functions of θ.

11. Quadrant II, $\sin\theta = \frac{4}{5}$

12. Quadrant III, $\cos\theta = -\frac{1}{3}$

13. Quadrant IV, $\sec\theta = 4$

14. Quadrant I, $\csc\theta = \frac{13}{12}$

For Problems 15–20, find the exact values of the six trigonometric functions of the given angle.

15. $60°$ 16. $135°$

17. $-315°$ 18. $330°$

19. $180°$ 20. $-270°$

For Problems 21–32, find the exact value (no decimals) of the given function. Try to do this *quickly,* from memory or by visualizing the figure in your head.

21. $\sin 180°$ 22. $\sin 225°$

23. $\cos 240°$ 24. $\cos 120°$

25. $\tan 315°$ 26. $\tan 270°$

27. $\cot 0°$ 28. $\cot 300°$

29. $\sec 150°$ 30. $\sec 0°$

31. $\csc 45°$ 32. $\csc 330°$

33. Find all values of θ from 0° through 360° for which
    a. $\sin\theta = 0$ b. $\cos\theta = 0$
    c. $\tan\theta = 0$ d. $\cot\theta = 0$
    e. $\sec\theta = 0$ f. $\csc\theta = 0$

34. Find all values of θ from 0° through 360° for which
    a. $\sin\theta = 1$ b. $\cos\theta = 1$
    c. $\tan\theta = 1$ d. $\cot\theta = 1$
    e. $\sec\theta = 1$ f. $\csc\theta = 1$

For Problems 35–42, find the exact value (no decimals) of the given expression. Note that the expression $\sin^2\theta$ means $(\sin\theta)^2$ and similarly for other functions. You may check your answers using your calculator.

35. $\sin 30° + \cos 60°$ 36. $\tan 120° + \cot(-30°)$

37. $\sec^2 45°$ 38. $\cot^2 30°$

39. $\sin 240° \csc 240°$ 40. $\cos 120° \sec 120°$

41. $\tan^2 60° - \sec^2 60°$ 42. $\cos^2 210° + \sin^2 210°$

43. From geometry recall that **complementary angles** have a sum of 90°.
    a. If $\theta = 23°$, what is the complement of θ?
    b. Find cos 23° and find sin (complement of 23°). What relationship do you notice?
    c. Based on what you've discovered, what do you think the prefix "co-" stands for in the names cosine, cotangent, and cosecant?

44. *Pattern in Sine Values Problem:* Find the exact values of sin 0°, sin 30°, sin 45°, sin 60°, and sin 90°. Make all the denominators equal to 2 and all the numerators radicals. Describe the pattern you see.

45. *Sketchpad Project—Sinusoids:* Figure 2-4i shows a unit circle in a *uv*-plane and a sinusoid whose *y*-values equal the *v*-values of the point *P*, where a rotating ray cuts the unit circle. Draw these figures using a program

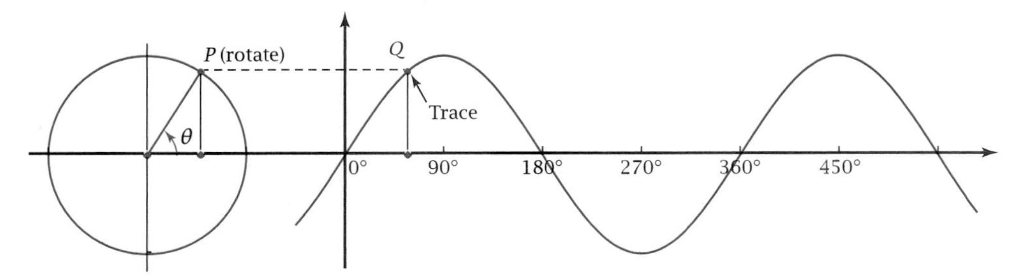

Figure 2-4i

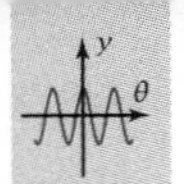

such as The Geometer's Sketchpad so that when you move the intersection point P around the circle, point Q traces the sinusoid. Write a paragraph explaining what you learned about the geometrical relationship between the angle θ rotating around a unit circle and the number θ plotted along the horizontal axis.

46. *Journal Problem:* Update your journal, writing about concepts you have learned since the last entry and concepts about which you are still unsure.

2-5 Inverse Trigonometric Functions and Triangle Problems

You have learned how to evaluate trigonometric functions for specific angle measures. Next you'll learn to use function values to find the angles. You'll also learn how to find unknown measures in a right triangle.

OBJECTIVE Given two sides of a right triangle or a side and an acute angle, find measures of the other sides and angles.

Inverses of Trigonometric Functions

In order to find the measure of an angle when its function value is given, you could press the appropriate inverse function keys on your calculator. For instance, if you know that $\cos\theta = 0.8$, you press

$$\cos^{-1} 0.8 = 36.8698...^\circ$$

The symbol $\cos^{-1}$ is the familiar inverse function terminology of Chapter 1. You say "inverse cosine of 0.8." Note that it does *not* mean the −1 power of cos, which is the reciprocal of cosine:

$$(\cos x)^{-1} = \frac{1}{\cos x}$$

Some calculators avoid this difficulty by using the symbol acos x, where the "a" can be thought of as standing for "angle." You can call $\cos^{-1} x$ or acos x "an angle whose cosine is x."

Trigonometric functions are periodic, so they are not one-to-one functions. There are many angles whose cosine is 0.8 (Figure 2-5a). However, for each trigonometric function there is a **principal branch** of the function that *is* one-to-one and includes angles between 0° and 90°. The calculator is programmed to give the one angle on the principal branch. The symbol $\cos^{-1} 0.8$ means the one angle on the principal branch whose cosine is 0.8. The inverse of the cosine function on the principal branch is a function denoted $\cos^{-1} x$.

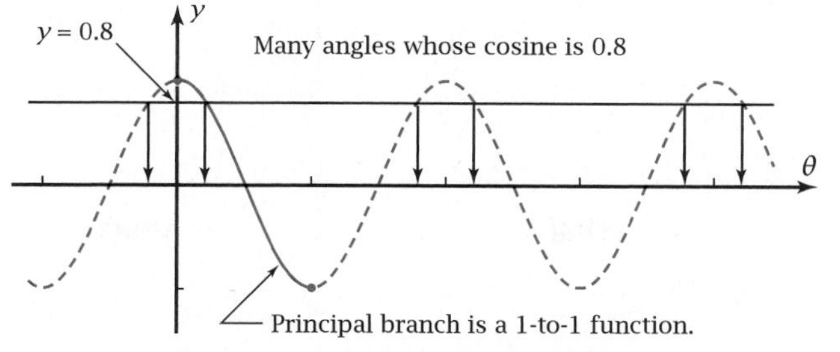

Figure 2-5a

Later, in Section 4-6, you will learn more about the principal branches of all six trigonometric functions. For the triangle problems of this section, all of the angles will be acute. So the value the calculator gives you is the value of the angle you want.

The definitions of the inverse trigonometric functions are presented in the box.

DEFINITIONS: Inverse Trigonometric Functions

If x is the value of a trigonometric function at angle θ, then the **inverse trigonometric functions** can be defined on limited domains.

$\theta = \sin^{-1} x$ means $\sin\theta = x$ and $-90° \le \theta \le 90°$

$\theta = \cos^{-1} x$ means $\cos\theta = x$ and $0° \le \theta \le 180°$

$\theta = \tan^{-1} x$ means $\tan\theta = x$ and $-90° < \theta < 90°$

Notes:

- Words: "The angle θ is the angle on the principal branch whose sine (and so on) is x."
- Pronunciation: "Inverse sine of x," and so on, never "sine to the negative 1."
- The symbols $\sin^{-1} x$, $\cos^{-1} x$, and $\tan^{-1} x$ are used only for the value the calculator gives you, not for other angles that have the same function values. The symbols $\cot^{-1} x$, $\sec^{-1} x$, and $\csc^{-1} x$ are similarly defined.
- **Caution:** The symbol $\sin^{-1} x$ does *not* mean the reciprocal of $\sin x$.

Right Triangle Problems

Trigonometric functions and inverse trigonometric functions often come up in applications, such as the right triangle problems.

▶ ***EXAMPLE 1*** Suppose you have the job of measuring the height of the local water tower. Climbing makes you dizzy, so you decide to do the whole job at ground level. You find that from a point 47.3 m from the base of the tower you must look up at an angle of 53° (angle of elevation) to see the top of the tower (Figure 2-5b). How high is the tower?

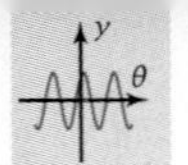

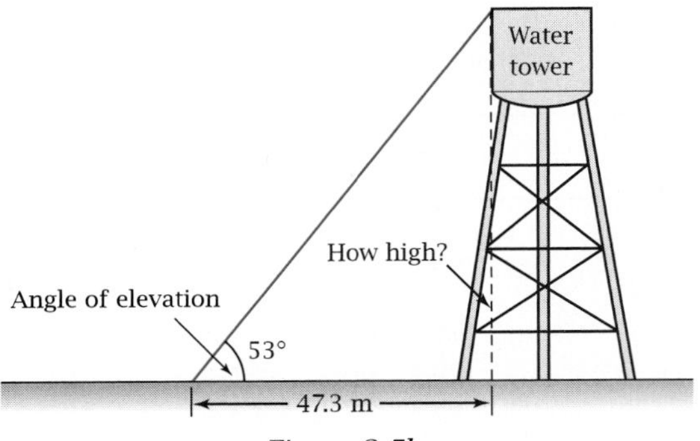

Figure 2-5b

Solution Sketch an appropriate right triangle, as shown in Figure 2-5c. Label the known and unknown information.

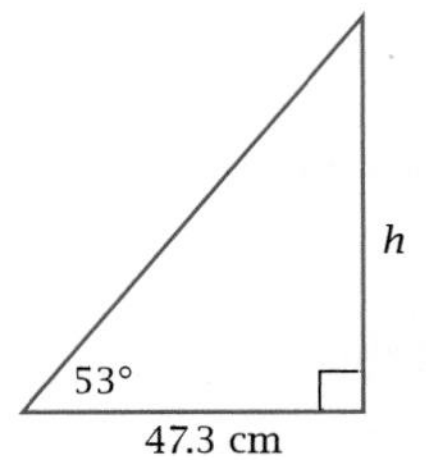

Figure 2-5c

$$\frac{h}{47.3} = \tan 53°$$ Write a ratio for tangent.

$$h = 47.3 \tan 53° = 62.7692\ldots$$ Solve for h.

The tower is about 62.8 m high. Write the real-world answer. ◀

Note that the angle must always have the degree sign, even during the computations. The symbol tan 53 has a different meaning, as you will learn when you study *radians* in the next chapter.

▶ EXAMPLE 2 A ship is passing through the Strait of Gibraltar. At its closest point of approach, Gibraltar radar determines that the ship is 2400 m away. Later, the radar determines that the ship is 2650 m away (Figure 2-5d).

a. By what angle θ did the ship's bearing from Gibraltar change?

b. How far did the ship travel between the two observations?

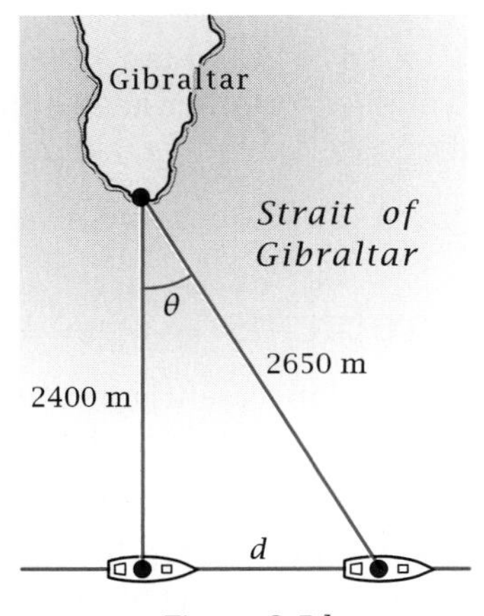

Figure 2-5d

Solution

a. Draw the right triangle and label the unknown angle θ. By the definition of cosine,

$$\cos\theta = \frac{\text{adjacent}}{\text{hypotenuse}} = \frac{2400}{2650}$$

$$\theta = \cos^{-1}\frac{2400}{2650} = 25.0876...^\circ$$ Take the inverse cosine to find θ.

The angle is about 25.09°.

b. Label the unknown side d, for distance. By the definition of sine,

$$\frac{d}{2650} = \sin 25.0876...^\circ$$ Use the unrounded angle measure that is in your calculator.

$$d = 2650\sin 25.0876...^\circ = 1123.6102...$$

The ship traveled about 1124 m. ◀

Problem Set 2-5

Do These Quickly

Problems Q1–Q6 refer to the right triangle in Figure 2-5e.

Q1. $\sin\theta =$ —?—

Q2. $\cos\theta =$ —?—

Q3. $\tan\theta =$ —?—

Q4. $\cot\theta =$ —?—

Q5. $\sec\theta =$ —?—

Q6. $\csc\theta =$ —?—

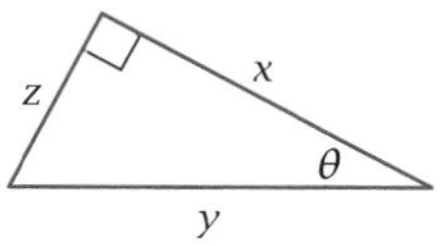

Figure 2-5e

Q7. $\sin 60^\circ =$ —?— (No decimals!)

Q8. $\cos 135^\circ =$ —?— (No decimals!)

Q9. $\tan 90^\circ =$ —?— (No decimals!)

Q10. The graph of the periodic function $y = \cos\theta$ is called a —?—.

For Problems 1–6, evaluate the inverse trigonometric function for the given value.

1. Find $\sin^{-1} 0.3$. Explain what the answer means.
2. Find $\cos^{-1} 0.2$. Explain what the answer means.
3. Find $\tan^{-1} 7$. Explain what the answer means.
4. Explain why $\sin^{-1} 2$ is undefined.
5. Find $\cos(\sin^{-1} 0.8)$. Explain, based on the Pythagorean theorem, why the answer is a rational number.
6. Find $\sin(\cos^{-1} 0.28)$. Explain, based on the Pythagorean theorem, why the answer is a rational number.
7. *Principal Branches of Sine and Cosine Problem:* Figure 2-5f shows the principal branch of $y = \sin\theta$ as a solid line on the sine graph. Figure 2-5g shows the principal branch of $y = \cos\theta$ as a solid line on the cosine graph.

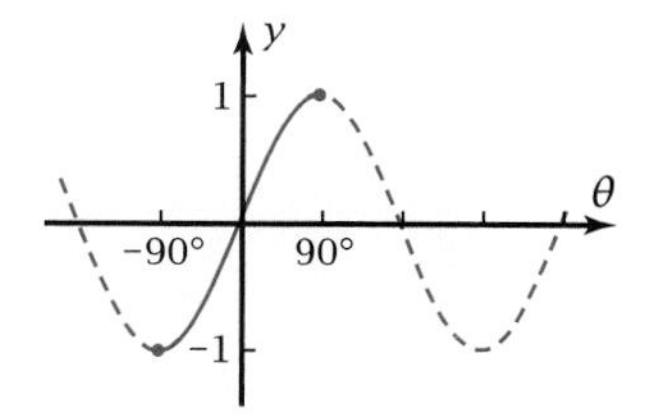

Figure 2-5f

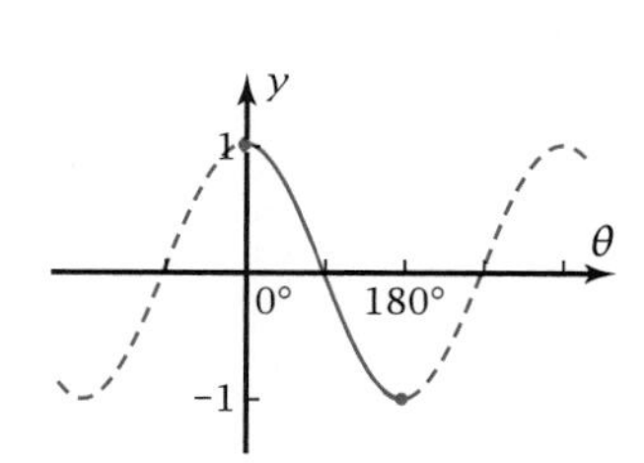

Figure 2-5g

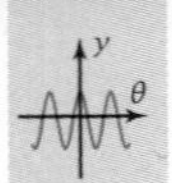

a. Why is neither the entire sine function nor the entire cosine function invertible?

b. What are the domains of the principal branches of cosine and sine? What property do these principal branches have that makes them invertible functions?

c. Find $\theta = \sin^{-1}(-0.9)$. Explain why the answer is a negative number.

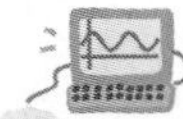

8. *Construction Problem 2:* Draw a right triangle to scale, with one leg 8 cm long and the adjacent acute angle 34°. Draw on paper with ruler and protractor or on the computer with a program such as The Geometer's Sketchpad:

a. Measure the opposite leg and the hypotenuse correct to the nearest 0.1 cm.

b. Calculate the lengths of the opposite leg and the hypotenuse using appropriate trigonometric functions. Show that your measured values and the calculated values agree within 0.1 cm.

9. *Ladder Problem 1:* Suppose you have a ladder 6.7 m long.

a. If the ladder makes an angle of 63° with the level ground when you lean it against a vertical wall, how high up the wall is the top of the ladder?

b. Your cat is trapped on a tree branch 6.5 m above the ground. If you place the ladder's top on the branch, what angle does the bottom of the ladder make with the level ground?

10. *Flagpole Problem:* You must order a new rope for the flagpole. To find out what length of rope is needed, you observe that the pole casts a shadow 11.6 m long on the ground. The angle of elevation of the Sun is 36° at this time of day (Figure 2-5h). How tall is the pole?

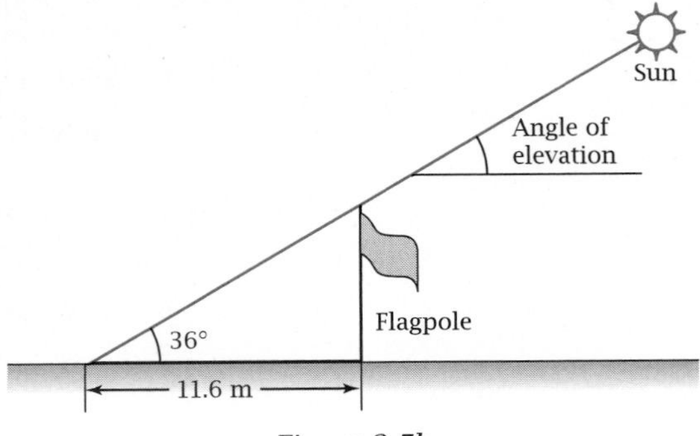

Figure 2-5h

11. *Tallest Skyscraper Problem:* The Petronas Twin Towers in Kuala Lumpur, Malaysia, are two of the world's tallest skyscrapers. The towers reach 451.9 m above the ground. Suppose that at a particular time the towers cast shadows on the ground 950 m long. What is the angle of elevation of the Sun at this time?

12. *The Grapevine Problem:* Interstate 5 in California enters the San Joaquin Valley through a mountain pass called the Grapevine. The road descends from an altitude of 3000 ft above sea level to 500 ft above sea level in a slant distance of 6 mi.

a. Approximately what angle does the roadway make with the horizontal?

b. What assumption must you make about how the road slopes?

13. *Grand Canyon Problem:* From a point on the North Rim of the Grand Canyon, a surveyor measures an angle of depression of 1.3° to a point on the South Rim (Figure 2-5i). From an aerial photograph she determines that the horizontal distance between the two points is 10 mi. How many feet is the South Rim below the North Rim?

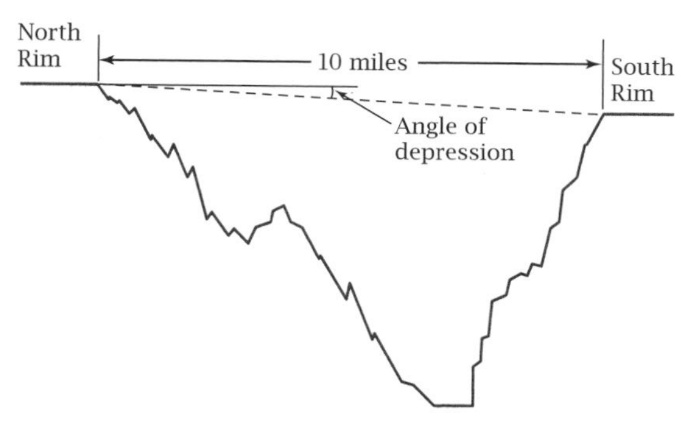

Figure 2-5i

14. *Airplane Landing Problem:* Commercial airliners fly at an altitude of about 10 km. Pilots start descending the airplanes toward the airport when they are far away so that they will not have to dive at a steep angle.

    a. If the pilot wants the plane's path to make an angle of 3° with the ground, at what horizontal distance from the airport must she start descending?

    b. If she starts descending when the plane is at a horizontal distance of 300 km from the airport, what angle will the plane's path make with the horizontal?

    c. Sketch the actual path of the plane just before and just after it touches the ground.

15. *Radiotherapy Problem:* A doctor may use a beam of gamma rays to treat a tumor that is 5.7 cm beneath the patient's skin. To avoid damaging a vital organ, the radiologist moves the source over 8.3 cm (Figure 2-5j).

    a. At what angle θ to the patient's skin must the radiologist aim the source to hit the tumor?

    b. How far will the beam travel through the patient's body before reaching the tumor?

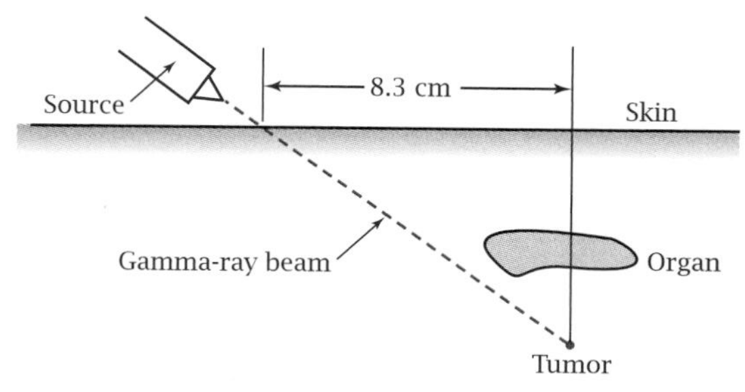

Figure 2-5j

16. *Triangular Block Problem:* A block bordering Market Street is a right triangle (Figure 2-5k). You take 125 paces on Market Street and 102 paces on Pine Street as you walk around the block.

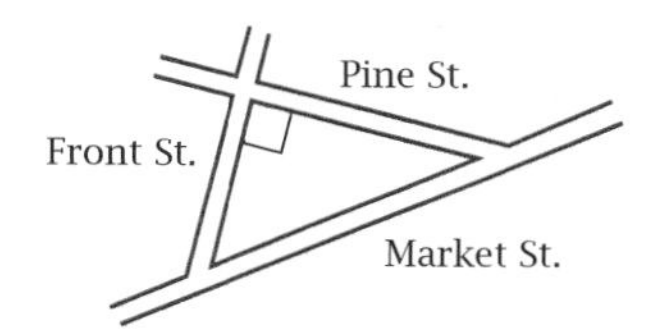

Figure 2-5k

    a. At what angle do Pine and Market Streets intersect?

    b. How many paces must you take on Front Street to complete the trip?

17. *Surveying Problem 1:* When surveyors measure land that slopes significantly, the slant distance they measure is longer than the horizontal distance they must draw on the map. Suppose that the distance from the top edge of the Cibolo Creek bed to the edge of the water is 37.8 m (Figure 2-5l). The land slopes downward at an angle of 27.6° to the horizontal.

    a. What is the horizontal distance from the top of the creek bed to the edge of the creek?

    b. How far below the level of the surrounding land is the surface of the water in the creek?

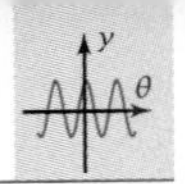

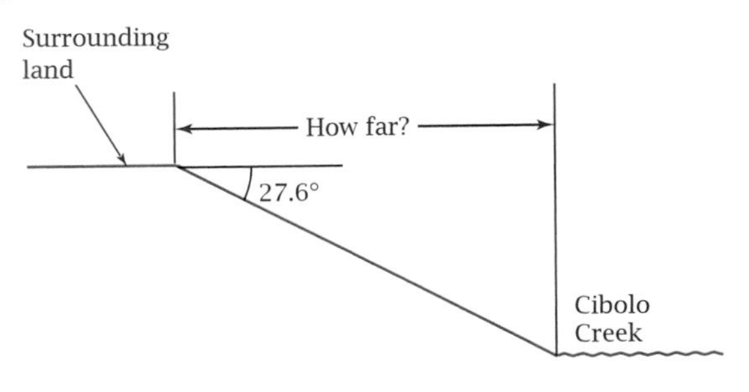

Figure 2-5l

18. *Highland Drive Problem:* One of the steeper streets in the United States is the 500 block of Highland Drive on Queen Anne Hill in Seattle. To measure the slope of the street, Tyline held a builder's level so that one end touched the pavement. The pavement was 14.4 cm below the level at the other end. The level itself was 71 cm long (Figure 2-5m).

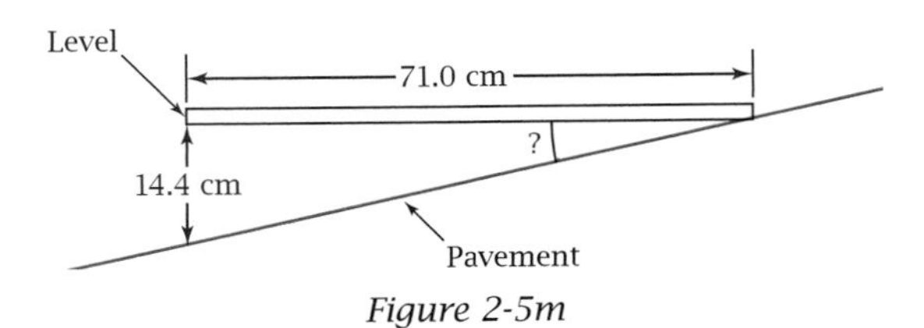

Figure 2-5m

a. What angle does the pavement make with the level?
b. A map of Seattle shows that the horizontal length of this block of Highland Drive is 365 ft. How much longer than 365 ft is the *slant* distance up this hill?
c. How high does the street rise up in this block?

19. *Submarine Problem 1:* As a submarine at the surface of the ocean makes a dive, its path makes a 21° angle with the surface.

a. If the submarine goes for 300 m along its downward path, how deep will it be? What horizontal distance is it from its starting point?
b. How many meters must it go along its downward path to reach a depth of 1000 m?

20. *Planet Diameter Problem:* You can find the approximate diameter of a planet by measuring the angle between the lines of sight to the two sides of the planet (Figure 2-5n).

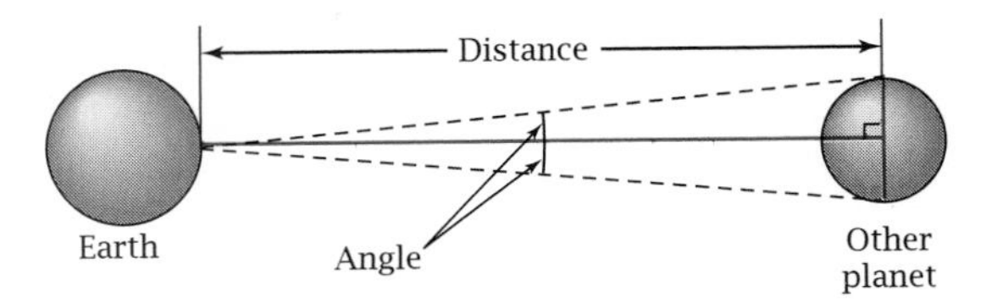

Figure 2-5n

a. When Venus is closest to Earth (25,000,000 mi), the angle is 0°1′2.5″ (zero degrees, 1 minute, 2.5 seconds). Find the approximate diameter of Venus.
b. When Jupiter is closest to Earth (390,000,000 mi), the angle is 0°0′46.9″. Find the approximate diameter of Jupiter.

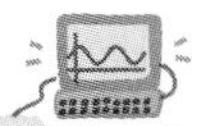

c. Check an encyclopedia, an almanac, or the Internet to see how close your answers are to the accepted diameters.

21. *Window Problem:* Suppose you want the windows of your house built so that the eaves completely shade them from the sunlight in the summer and the sunlight completely fills the windows in the winter. The eaves have an overhang of 3 ft (Figure 2-5o).

a. How far below the eaves should the top of a window be placed for the window to receive full sunlight in midwinter, when the Sun's noontime angle of elevation is 25°?

b. How far below the eaves should the bottom of a window be placed for the window to receive no sunlight in midsummer, when the Sun's angle of elevation is 70°?

c. How tall will the windows be if they meet both requirements?

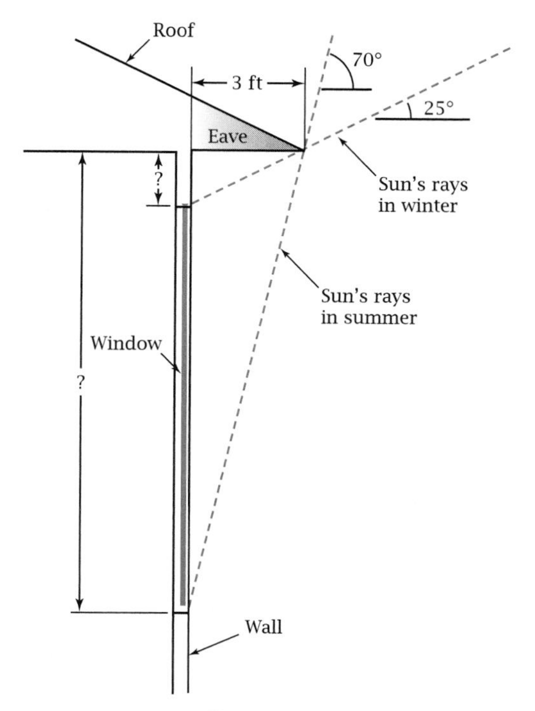

Figure 2-5o

22. *Grand Piano Problem 1:* A 28-in. prop holds the lid open on a grand piano. The base of the prop is 55 in. from the lid's hinge.

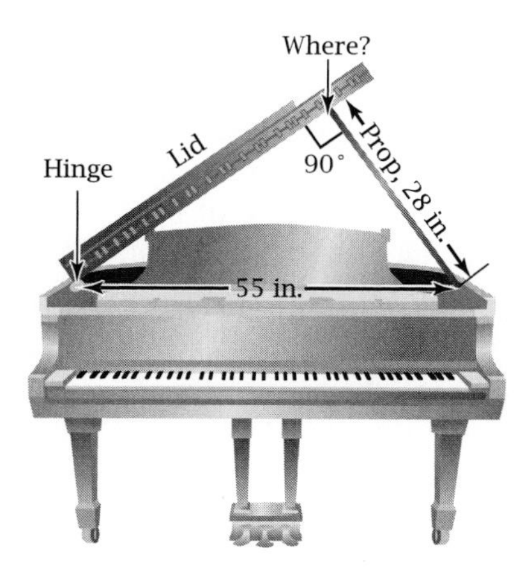

a. What angle does the lid make with the piano top when the prop is placed perpendicular to the lid?

b. Where should the prop be placed on the lid to make the right angle in part a?

c. The piano also has a shorter (13-in.) prop. Where on the lid should this prop be placed to make a right angle with the lid?

23. *Handicap Ramp Project:* In this project you will measure the angle a typical handicap access ramp makes with the horizontal.

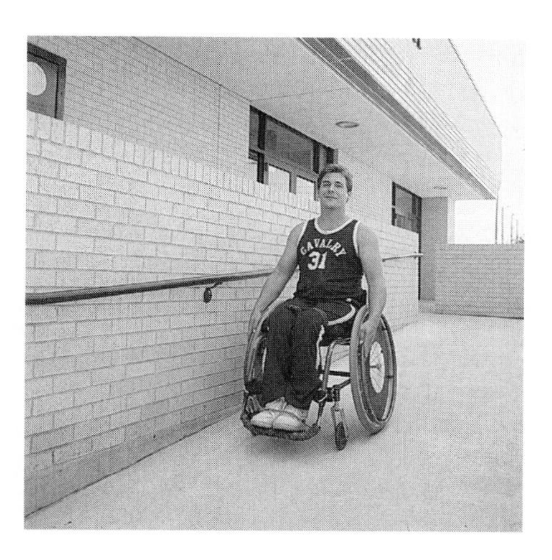

a. Based on handicap access ramps you have seen, make a conjecture about the angle these ramps make with the horizontal. Each member of your group should make his or her own conjecture.

b. Find a convenient handicap access ramp. With a level and ruler, measure the run and the rise of the ramp. Use these numbers to calculate the angle the ramp makes with the horizontal.

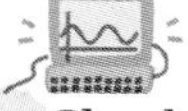

c. Check on the Internet or use another reference source to find the maximum angle a handicap access ramp is allowed to make with the horizontal.

24. *Pyramid Problem:* The Great Pyramid of Cheops in Egypt has a square base 230 m on each side. The faces of the pyramid make an angle of 51°50′ with the horizontal (Figure 2-5p).

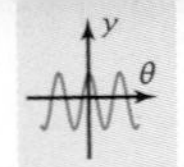

a. How tall is the pyramid?

b. What is the shortest distance you would have to climb to get to the top?

c. Suppose that you decide to make a model of the pyramid by cutting four isosceles triangles out of cardboard and gluing them together. How large should you make the base angles of these isosceles triangles?

d. Show that the ratio of the distance you calculated in part b to one-half the length of the base of the pyramid is very close to the golden ratio, $\frac{\sqrt{5}+1}{2}$.

e. See Martin Gardner's article in the June 1974 issue of *Scientific American* for other startling relationships among the dimensions of this pyramid. Also, see Public Broadcasting System's Web site *(www.pbs.org)* and click on Nova: Secrets of Lost Empires for photographs and information about the pyramids.

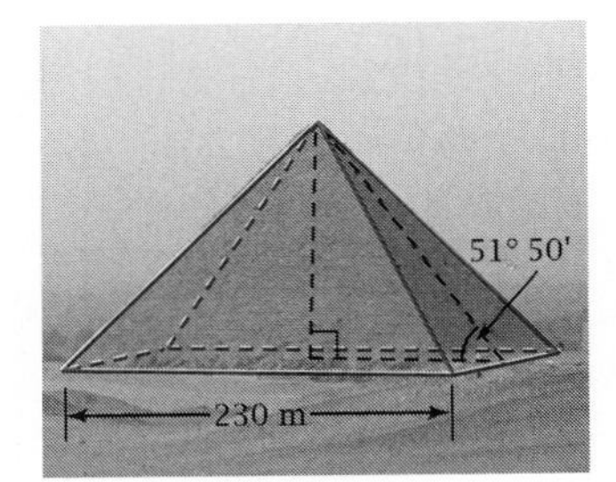

Figure 2-5p

2-6 Chapter Review and Test

This chapter introduced you to periodic functions. You saw how these functions occur in the real world. For instance, your distance from the ground as you ride a Ferris wheel changes periodically. The sine and cosine functions are periodic. By dilating and translating these functions, you can get sinusoids of different proportions, with their critical points located at different places. You also learned about the other four trigonometric functions and their relationship to sine and cosine. Finally, you learned how to use these six functions and their inverses to find unknown sides and angles in right triangles.

Review Problems

R0. Update your journal with what you have learned since the last entry. Include such things as

- How angles can have measures that are negative or greater than 180°, and reference angles
- The definitions of sine, cosine, tangent, cotangent, secant, and cosecant
- Why sine and cosine graphs are periodic
- Inverse trigonometric functions used to find angles
- Applications to right triangle problems

R1. *Hose Reel Problem:* You unwind a hose by turning the crank on a hose reel mounted to the wall (Figure 2-6a, facing page). As you crank, the distance your hand is above the ground is a periodic function of the angle through which the reel has rotated (Figure 2-6b, solid graph). The distance, y, is measured in feet, and the angle, θ, is measured in degrees.

a. The dashed graph in Figure 2-6b is the pre-image function $y = \sin \theta$. Plot this sine function graph on your grapher. Does the result agree with Figure 2-6b?

b. The solid graph in Figure 2-6b is a dilation and translation of $y = \sin \theta$. Figure out what the two transformations are, and write an equation for the function. When you plot the transformed graph on your grapher, does the result agree with Figure 2-6b?

c. What is the name for the periodic graphs in Figure 2-6b?

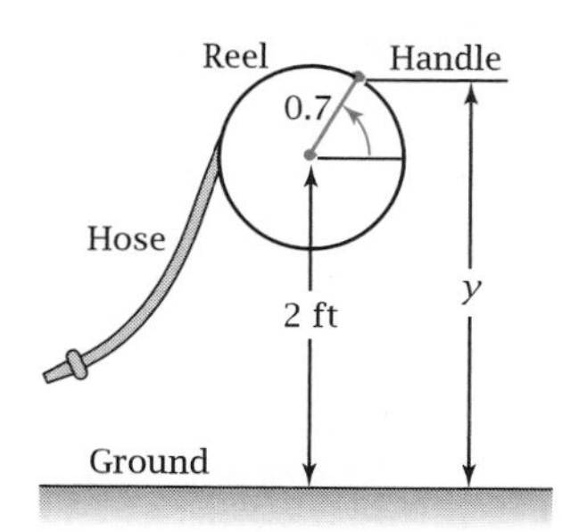

Figure 2-6a

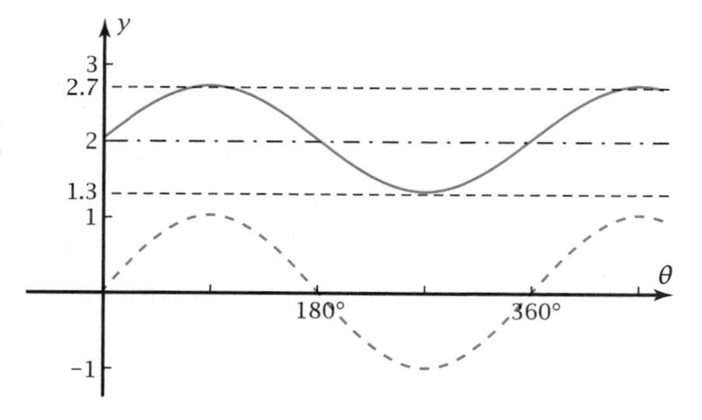

Figure 2-6b

R2. For each angle measure, sketch an angle in standard position. Mark the reference angle and find its measure.

a. 110° b. −79°

c. 2776°

R3. a. Find $\sin \theta$ and $\cos \theta$ given that the terminal side of θ contains the point $(u, v) = (-5, 7)$.

b. Find decimal approximations for sin 160° and cos 160°. Draw a 160° angle in standard position in a *uv*-coordinate system and mark the reference angle. Explain why sin 160° is positive but cos 160° is negative.

c. Sketch the graphs of the parent sinusoids $y = \cos \theta$ and $y = \sin \theta$.

d. In which two quadrants on a *uv*-coordinate system is $\sin \theta$ negative?

e. For $y = 4 + \cos 2\theta$, what are the transformations of the parent function graph $y = \cos \theta$? Sketch the graph of the transformed function.

R4. a. Find a decimal approximation for csc 256°.

b. Find exact values (no decimals) of the six trigonometric functions of 150°.

c. Find the exact value of $\sec \theta$ if $\theta_{ref} = 45°$ and θ terminates in Quadrant III.

d. Find the exact value of $\cos \theta$ if the terminal side of θ contains the point $(-3, 5)$.

e. Find the exact value of $\sec(-120°)$.

f. Find the exact value of $\tan^2 30° - \csc^2 30°$.

g. Explain why tan 90° is undefined.

R5. a. Find a decimal approximation for $\theta = \cos^{-1} 0.6$. What does the answer mean?

b. *Galleon Problem:* Imagine that you are on a salvage ship in the Gulf of Mexico. Your sonar system has located a sunken Spanish galleon at a slant distance of 683 m from your ship, with an angle of depression of 28°.

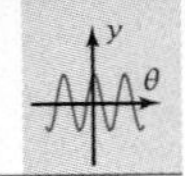

i. How deep is the water at the location of the galleon?
ii. How far must your ship go to be directly above the galleon?
iii. Your ship moves horizontally toward the galleon. After 520 m, what is the angle of depression?
iv. How could the crew of a fishing vessel use the techniques of this problem while searching for schools of fish?

Concept Problems

C1. *Tide Problem 1:* The average depth of the water at the beach varies with time due to the motion of the tides. Figure 2-6c shows the graph of depth, in feet, versus time, in hours, for a particular beach. Find the four transformations of the parent cosine graph that would give the sinusoid shown. Write an equation for this particular sinusoid, assuming 1 degree represents 1 hour. Confirm your answer by grapher.

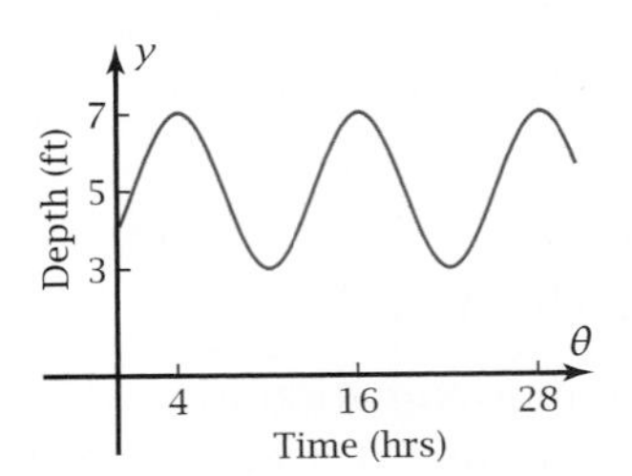

Figure 2-6c

C2. Figure 2-6d shows three cycles of the sinusoid $y = 10 \sin \theta$. The horizontal line $y = 3$ cuts each cycle at two points.

a. Estimate graphically the six values of θ where the line intersects the sinusoid.

b. Calculate the six points in part a numerically, using the intersection feature of your grapher.

c. Calculate the six points in part a algebraically, using the inverse sine function.

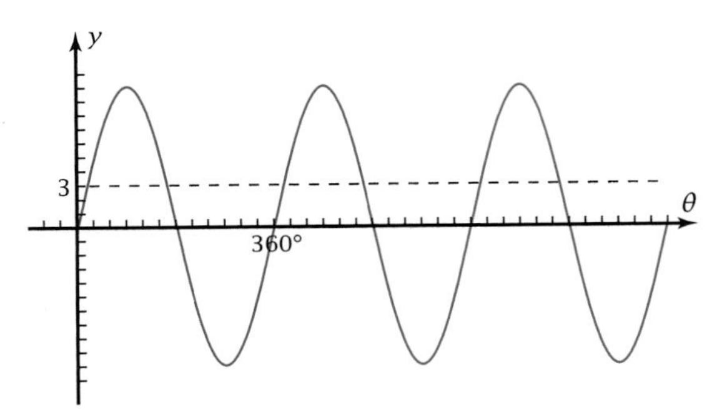

Figure 2-6d

C3. On your grapher, make a table of values of $\cos^2 \theta + \sin^2 \theta$ for each 10°, starting at $\theta = 0°$. What is the pattern in the answers? Explain why this pattern applies to any value of θ.

C4. A ray from the origin of a *uv*-coordinate system starts along the positive *u*-axis and rotates around and around the origin. The **slope** of the ray depends on the angle measure through which the ray has rotated.

a. Sketch a reasonable graph of the slope as a function of the angle of rotation.

b. What function on your grapher is the same as the one you sketched in part a?

Chapter Test

PART 1: No calculators allowed (T1–T8)

T1. Sketch an angle θ in standard position whose terminal side contains the point (3, −4). Show the reference angle. Find the exact values of the six trigonometric functions of θ.

T2. Sketch an angle of 120° in standard position. Show the reference angle and its measure. Find the exact values of the six trigonometric functions of 120°.

T3. Sketch an angle of 225° in standard position. Show the reference angle and its measure. Find the exact values of the six trigonometric functions of 225°.

T4. Sketch an angle of 180° in standard position. Pick a point on the terminal side, and write the horizontal coordinate, vertical coordinate, and radius. Use these numbers to find the exact values of the six trigonometric functions of 180°.

T5. The number of hairs on a person's head and his or her age are related. Sketch a reasonable graph.

T6. The distance between the tip of the "second" hand on a clock and the floor depends on time. Sketch a reasonable graph.

T7. Only one of the functions in Problems T5 and T6 is periodic. Which one is that?

T8. Figure 2-6e shows the graph of $y = \sin\theta$ (dashed) and its principal branch (solid). Explain why the $y = \sin\theta$ function is not invertible but the function defined by its principal branch is invertible.

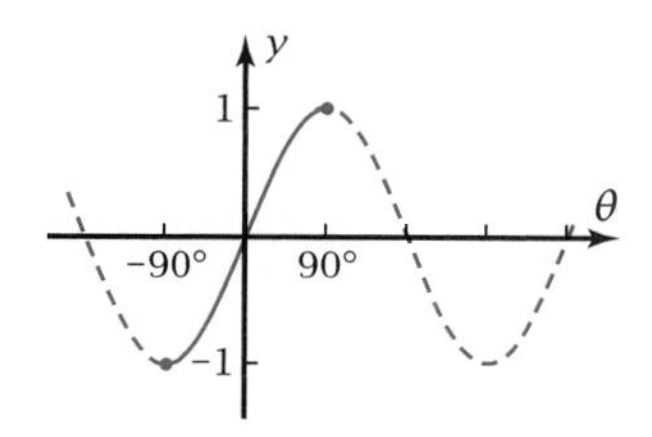

Figure 2-6e

PART 2: Graphing calculators allowed (T9–T22)

For Problems T9–T12, use your calculator to find each value.

T9. sec 39°

T10. cot 173°

T11. csc 191°

T12. $\tan^{-1} 0.9$. Explain the meaning of the answer.

T13. Calculate the measure of side x.

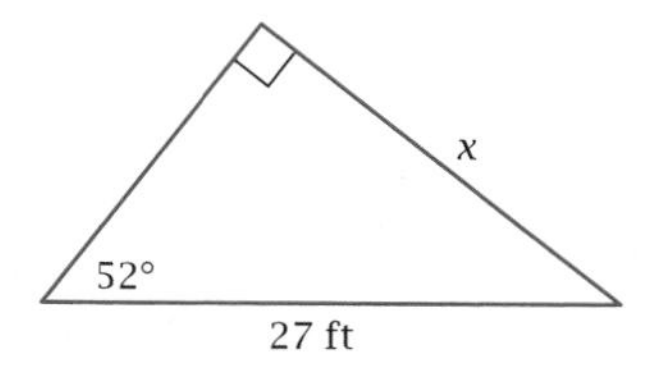

T14. Calculate the measure of side y.

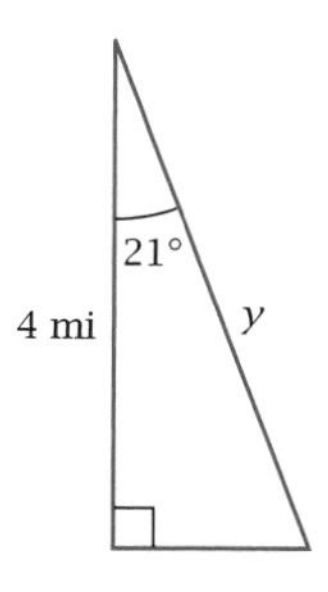

T15. Calculate the measure of angle B.

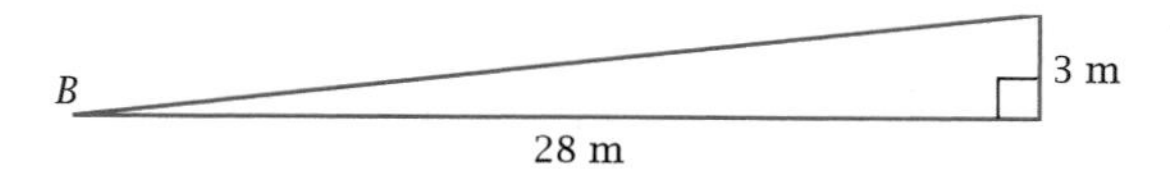

T16. Calculate the measure of side z.

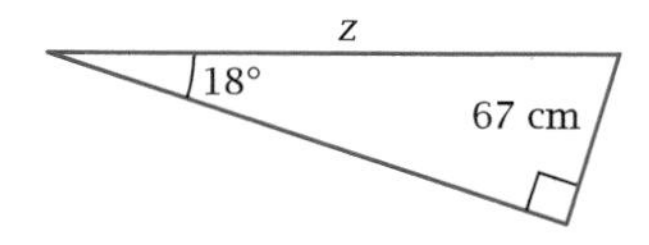

T17. Calculate the measure of angle A.

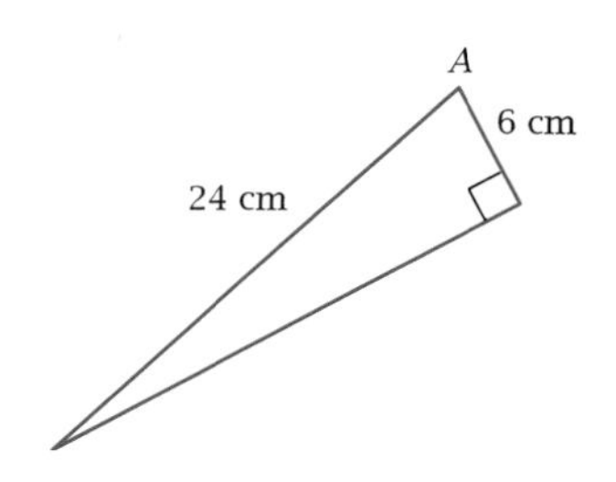

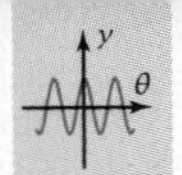

Buried Treasure Problem 1: For Problems T18–T20, use Figure 2-6f of a buried treasure. From the point on the ground at the left of the figure, sonar detects the treasure at a slant distance of 19.3 m, at an angle of 33° with the horizontal.

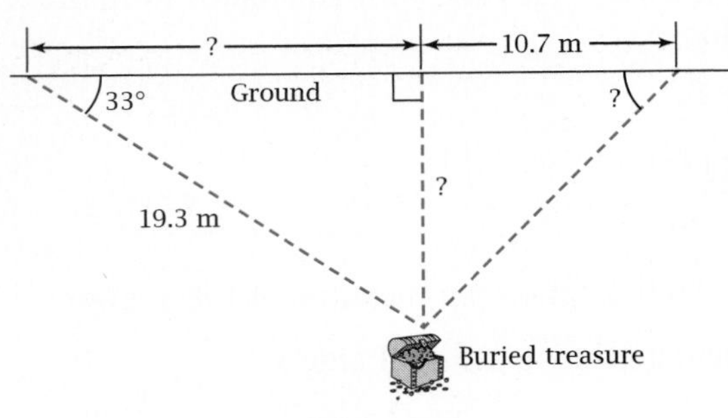

Figure 2-6f

T18. How far must you go from the point on the left to be directly over the treasure?

T19. How deep is the treasure below the ground?

T20. If you keep going to the right 10.7 m from the point directly above the treasure, at what angle would you have to dig to reach the buried treasure?

T21. In Figure 2-6g, the solid graph shows the result of *three* transformations to the parent function $y = \cos\theta$ (dashed). Write the equation for the transformed function. Check your results by grapher.

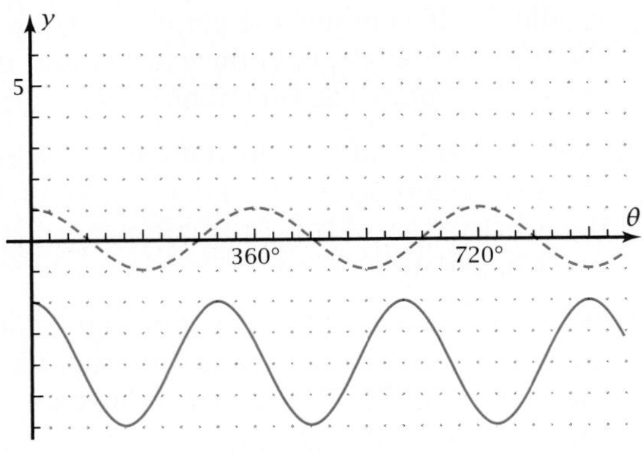

Figure 2-6g

T22. What did you learn from this test that you didn't know before?

Applications of Trigonometric and Circular Functions

CHAPTER 3

Stresses in the earth compress rock formations and cause them to buckle into sinusoidal shapes. It is important for geologists to be able to predict the depth of a rock formation at a given point. Such information can be very useful for structural engineers as well. In this chapter you'll learn about the circular functions, which are closely related to trigonometric functions. Geologists and engineers use these functions as mathematical models to make calculations for such wavy rock formations.

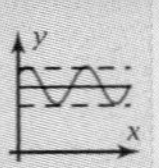

Mathematical Overview

So far you've learned about transformations and sinusoids. In this chapter you'll combine what you've learned so that you can write a particular equation for a sinusoid that fits any given conditions. You will approach this in four ways.

Graphically The graph shows a sinusoid that is a cosine function transformed through vertical and horizontal translations and dilations. The independent variable is x rather than θ so that you can fit sinusoids to situations that do not involve angles.

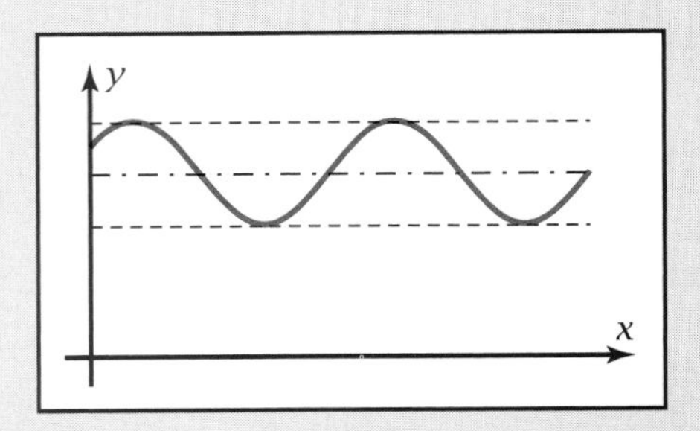

Algebraically Particular equation: $y = 7 + 2\cos\frac{\pi}{3}(x - 1)$

Numerically

x	y
1	9
2	8
3	6
4	5

Verbally *The circular functions are just like the trigonometric functions except that the independent variable is an arc of a unit circle instead of an angle. Angles in radians form the link between angles in degrees and numbers of units of arc length.*

3-1 Sinusoids: Amplitude, Period, and Cycles

Figure 3-1a shows a dilated and translated sinusoid and some of its geometric features. In this section you will learn how these features relate to transformations you've already learned.

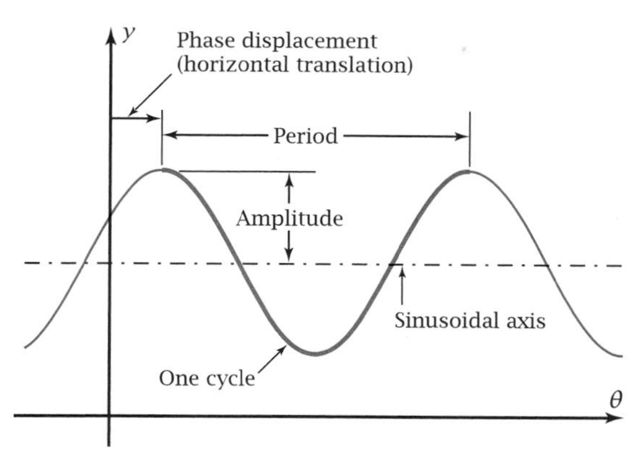

Figure 3-1a

OBJECTIVE Learn the meanings of *amplitude, period, phase displacement,* and *cycle of a sinusoidal graph.*

Exploratory Problem Set 3-1

1. Sketch one **cycle** of the graph of the parent sinusoid $y = \cos\theta$, starting at $\theta = 0°$. What is the **amplitude** of this graph?

2. Plot the graph of the transformed cosine function $y = 5\cos\theta$. What is the amplitude of this function? What is the relationship between the amplitude and the vertical dilation of a sinusoid?

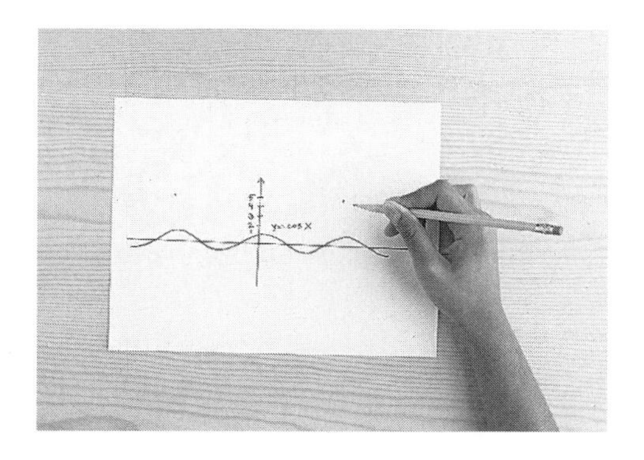

3. What is the period of the transformed function? What is the period of the parent function $y = \cos\theta$?

4. Plot the graph of $y = \cos 3\theta$. What is the period of this transformed function? How is the 3 related to the transformation? How could you calculate the period using the 3?

5. Plot the graph of $y = \cos(\theta - 60°)$. What transformation is caused by the 60°?

6. The $(\theta - 60°)$ in Problem 5 is called the **argument** of the cosine. The **phase displacement** is the value of θ that makes the argument equal 0. What is the phase displacement for this function? How is the phase displacement related to the horizontal translation?

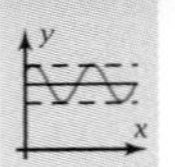

7. Plot the graph of $y = 6 + \cos \theta$. What transformation is caused by the 6?

8. The **sinusoidal axis** runs along the middle of the graph of a sinusoid. It is the dashed centerline shown in Figure 3-1a. What transformation of the $y = \cos x$ function does the location of the sinusoidal axis indicate?

9. Suppose that $y = 6 + 5 \cos 3(\theta - 60°)$. What are the amplitude, period, phase displacement, and sinusoidal axis of the graph? Check by plotting on your grapher.

10. Update your journal with things you have learned in this problem set.

3-2 General Sinusoidal Graphs

In Section 3-1, you encountered the terms *period, amplitude, cycle, phase displacement,* and *sinusoidal axis.* They are often used to describe horizontal and vertical translation and dilation of sinusoids. In this section you'll make the connection between the new names and these transformations. By so doing you will be able to fit an equation to any given sinusoid. This will help you use sinusoidal functions as mathematical models for real-world applications such as the variation of average daily temperature with time of the year.

OBJECTIVE Given any one of these sets of information about a sinusoid, find the other two.

- The equation
- The graph
- The amplitude, period or frequency, phase displacement, and sinusoidal axis

Recall from Chapter 2 that the period of a sinusoid is the number of degrees per cycle. The reciprocal of the period, or the number of cycles per degree, is called the **frequency.** It is convenient to use the frequency when the period is very short. For instance, the alternating electrical current in the United States has a frequency of 60 cycles per second.

You can see how the general sinusoidal equations allow for all four transformations.

DEFINITION: General Sinusoidal Equation

$y = C + A \cos B(\theta - D)$ or $y = C + A \sin B(\theta - D)$, where

- $|A|$ is the amplitude (A is the vertical dilation, which can be positive or negative)
- B is the reciprocal of the horizontal dilation
- C is the location of the sinusoidal axis (vertical translation)
- D is the phase displacement (horizontal translation)

The period can be calculated from the value of B. Since $\frac{1}{B}$ is the horizontal dilation, and since the parent cosine and sine functions have periods of 360°, the period equals $\frac{1}{|B|}(360°)$. Use the absolute value sign, since dilations can be positive or negative.

PROPERTY: Period and Frequency of a Sinusoid

For general equations $y = C + A\cos B(\theta - D)$ or $y = C + A\sin B(\theta - D)$,

$$\text{period} = \frac{1}{|B|}(360°) \quad \text{and} \quad \text{frequency} = \frac{1}{\text{period}} = \frac{|B|}{360°}$$

Next you'll use these properties and the general equation to graph sinusoids and find their equations.

Background: Concavity, Points of Inflection, and Upper and Lower Bounds

A smoothly curved graph can have a **concave** (hollowed-out) side and a **convex** (bulging) side, as Figure 3-2a shows for a typical sinusoid. In calculus, for reasons you will learn, mathematicians usually refer to the concave side. Figure 3-2a also shows regions where the concave side of the graph is up or down. A **point of inflection** occurs where a graph stops being concave one way and starts being concave the other way. The word originates from the British spelling, *inflexion,* which means "not flexed."

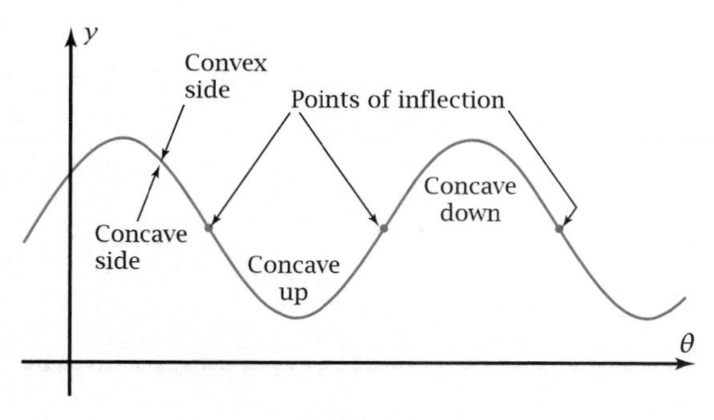

Figure 3-2a

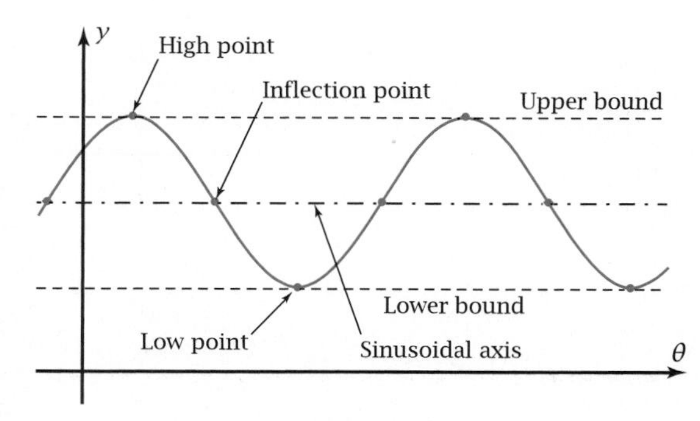

Figure 3-2b

As you can see from Figure 3-2b, the sinusoidal axis goes through the points of inflection. The lines through the high points and the low points are called the **upper bound** and the **lower bound,** respectively. The high points and low

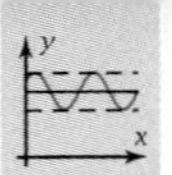

points are called **critical points** because they have a "critical" influence on the size and location of the sinusoid. Note that it is a quarter of a cycle between a critical point and the next point of inflection.

▶ ***EXAMPLE 1*** Suppose that a sinusoid has a period of 12° per cycle, an amplitude of 7 units, a phase displacement of −4° with respect to the parent cosine function, and a sinusoidal axis 5 units below the θ-axis. Without using your grapher, sketch this sinusoid and then find an equation for it. Verify with your grapher that your equation and the sinusoid you sketched agree with each other.

Solution First, draw the sinusoidal axis at $y = -5$, as in Figure 3-2c. (The long-and-short dashed line is used by draftspersons for centerlines.) Use the amplitude of 7 to draw the upper and lower bounds 7 units above and 7 units below the sinusoidal axis.

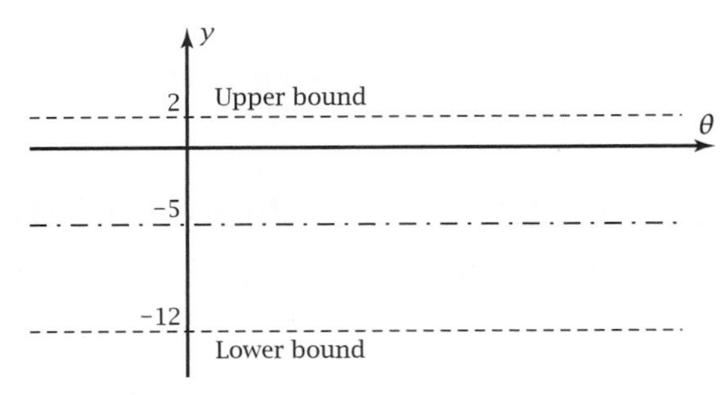

Figure 3-2c

Next, find some critical points on the graph (Figure 3-2d). Start at $\theta = -4°$, because that is the phase displacement, and mark a high point on the upper bound. (Cosine starts a cycle at a high point since $\cos 0° = 1$.) Then use the period of 12° to plot the ends of the next two cycles.

$$-4° + 12° = 8°$$

$$-4° + 2(12°) = 20°$$

Mark some low critical points halfway between consecutive high points.

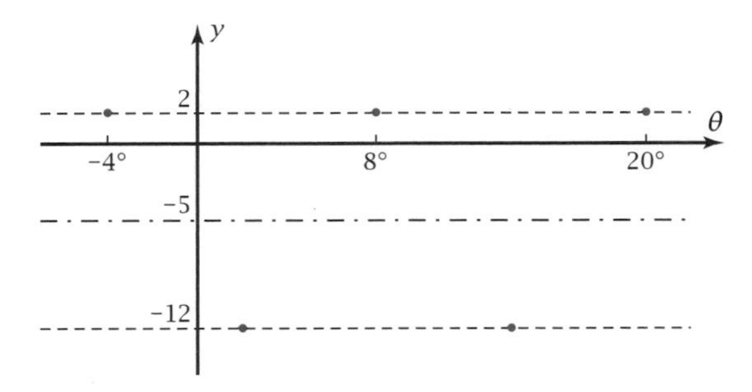

Figure 3-2d

Now mark the points of inflection (Figure 3-2e). They come on the sinusoidal axis, halfway between consecutive high and low points.

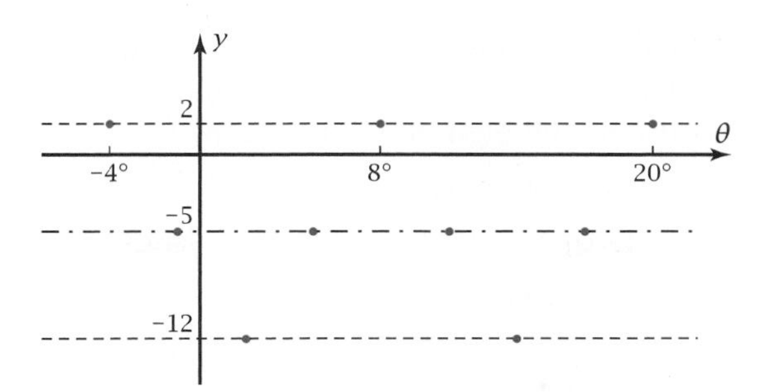

Figure 3-2e

Finally, sketch the graph in Figure 3-2f by connecting the critical points and points of inflection with a smooth curve. Be sure the graph is rounded at the critical points and that it changes concavity at the points of inflection.

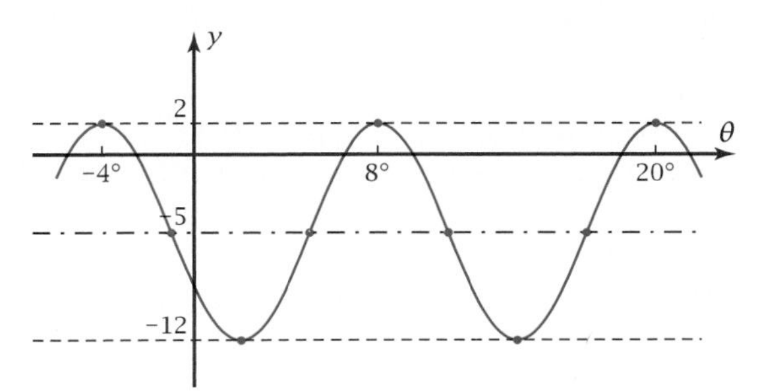

Figure 3-2f

Since the period of this sinusoid is 12° and the period of the parent cosine function is 360°, the horizontal dilation is

$$\text{dilation} = \frac{12°}{360°} = \frac{1}{30}$$

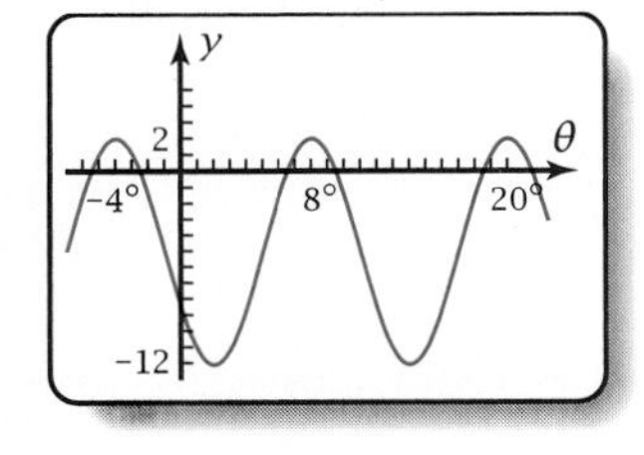

Figure 3-2g

The coefficient B in the sinusoidal equation is the reciprocal of $\frac{1}{30}$, namely, 30. Thus a particular equation is

$$y = -5 + 7\cos 30(\theta + 4°)$$

Plotting the graph on your grapher confirms that this equation produces the correct graph (Figure 3-2g). ◀

▶ EXAMPLE 2 For the sinusoid in Figure 3-2h, give the period, frequency, amplitude, phase displacement, and sinusoidal axis location. Write a particular equation of the sinusoid. Check your equation by plotting it on your grapher.

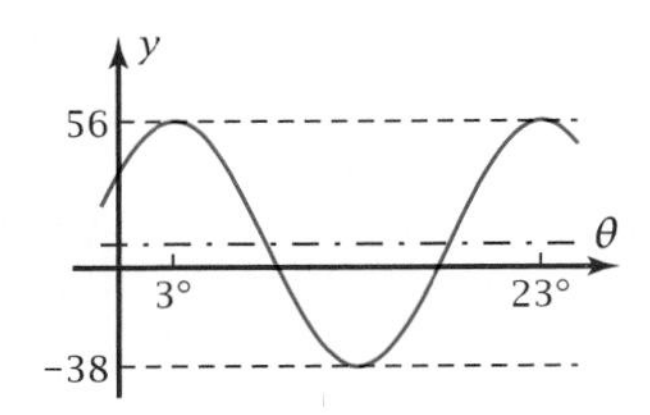

Figure 3-2h

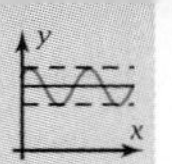

Solution As you will see later, you can use either the sine or the cosine as the pre-image function. Here, use the cosine function, because its "first" cycle starts at a high point and two of the high points are known.

- To find the period, look at the cycle shown on Figure 3-2h. It starts at 3° and ends at 23°, so the period is 23° − 3° = 20°.
- The frequency is the reciprocal of the period, $\frac{1}{20}$ cycle per degree.
- The sinusoidal axis is halfway between the upper and lower bounds. So $y = \frac{1}{2}(-38 + 56) = 9$.
- The amplitude is the distance between the upper or lower bound and the sinusoidal axis.

  $$A = 56 - 9 = 47$$

- Using the cosine function as the parent function, the phase displacement is 3°. (You could also use 23° or −17°.)
- The horizontal dilation is $\frac{20°}{360°}$, so $B = \frac{360°}{20°} = 18$, since it is the reciprocal of the horizontal dilation. So a particular equation is

  $$y = 9 + 47 \cos 18(\theta - 3°)$$

Plotting the corresponding graph on your grapher confirms that the equation is correct. ◀

You can find an equation of a sinusoid when only part of a cycle is given. The next example shows you how to do this.

▶ **EXAMPLE 3** Figure 3-2i shows a quarter-cycle of a sinusoid. Write a particular equation and check it by using your grapher.

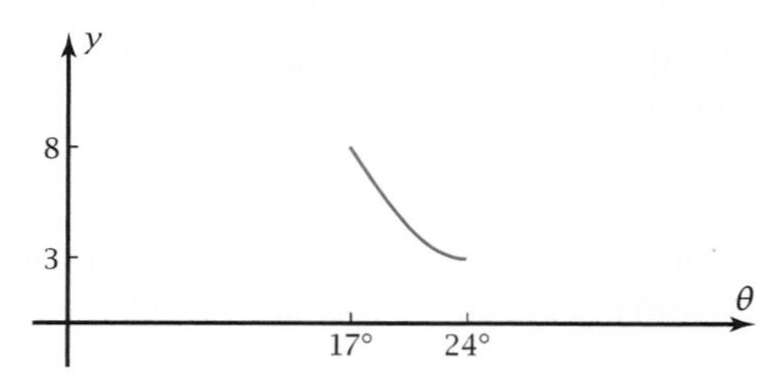

Figure 3-2i

Solution Imagine the entire cycle from the part of the graph that is shown. You can tell that a low point is at $\theta = 24°$ because the graph appears to level out there. So the lower bound is at $y = 3$. The point at $\theta = 17°$ must be an inflection point on the sinusoidal axis at $y = 8$ since the graph is a quarter-cycle. So the amplitude is 8 − 3, or 5. Sketch the lower bound, the sinusoidal axis, and the upper

bound. Next, locate a high point. Each quarter-cycle covers $(24° - 17°) = 7°$, so the critical points and points of inflection are spaced 7° apart. Thus a high point is at $\theta = 17° - 7° = 10°$. Then sketch at least one complete cycle of the graph (Figure 3-2j).

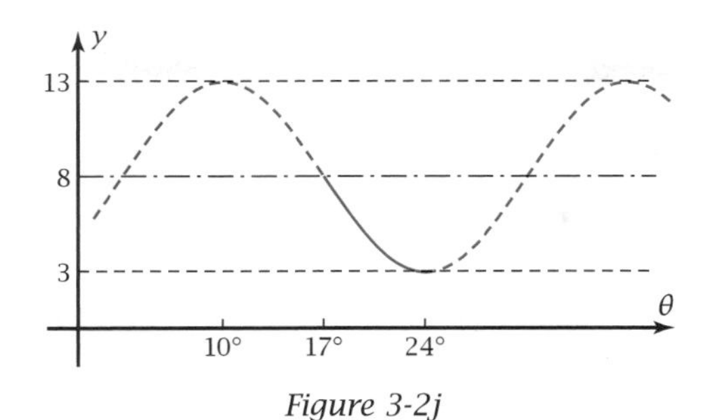

Figure 3-2j

The period is $4(7°) = 28°$ because a quarter of the period is 7°. The horizontal dilation is $\frac{28°}{360°} = \frac{7}{90}$.

The coefficient B in the sinusoidal equation is the reciprocal of this horizontal dilation. Using the techniques of Example 2, a particular equation is

$$y = 8 + 5 \cos \tfrac{90}{7}(\theta - 10°)$$

Plotting the graph on your grapher shows that the equation is correct. ◀

Note that in all of the examples so far *a particular equation* is used. There are many equivalent forms for the equation, depending on which cycle you pick for the "first" cycle and whether you use the parent sine or cosine functions. The next example shows some possibilities.

▶ **EXAMPLE 4** For the sinusoid in Figure 3-2k, write a particular equation using

a. Cosine, with a phase displacement other than 10°

b. Sine

c. Cosine, with a negative vertical dilation factor

d. Sine, with a negative vertical dilation factor

Confirm on your grapher that all four equations give the same graph.

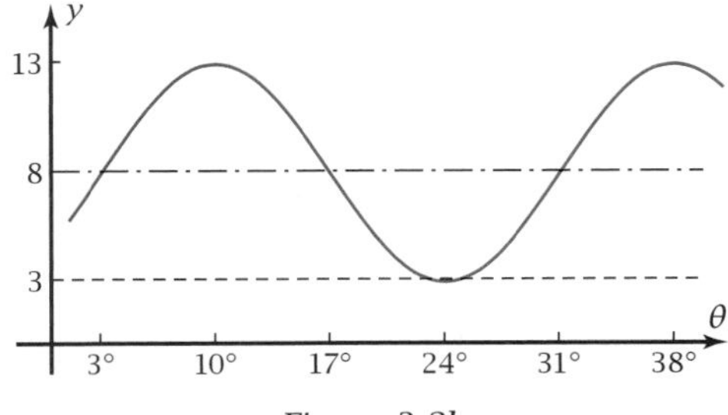

Figure 3-2k

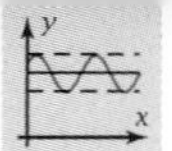

Solution a. Notice that the sinusoid is the same one as in Example 3. To find a different phase displacement, look for another high point. A convenient one is at $\theta = 38°$. All of the other constants remain the same. So another particular equation is

$$y = 8 + 5\cos\tfrac{90}{7}(\theta - 38°)$$

b. The graph of the parent sine function starts at a point of inflection on the sinusoidal axis, while *going up*. Two possible starting points appear in Figure 3-2k, one at $\theta = 3°$, another at $\theta = 31°$.

$$y = 8 + 5\sin\tfrac{90}{7}(\theta - 3°) \quad \text{or} \quad y = 8 + 5\sin\tfrac{90}{7}(\theta - 31°)$$

c. Changing the vertical dilation factor from 5 to −5 causes the sinusoid to be reflected in the sinusoidal axis. So if you use −5, the "first" cycle starts as a *low* point instead of a high point. The most convenient low point in this case is at $\theta = 24°$.

$$y = 8 - 5\cos\tfrac{90}{7}(\theta - 24°)$$

d. With a negative dilation factor, the sine function starts a cycle at a point of inflection, going *down*. One such point is shown in Figure 3-2k at $\theta = 17°$.

$$y = 8 - 5\sin\tfrac{90}{7}(\theta - 17°)$$

Plotting these four graphs on your grapher reveals only one image. The graphs are superimposed on one another. ◀

Problem Set 3-2

Do These Quickly

Problems Q1–Q5 refer to Figure 3-2l.

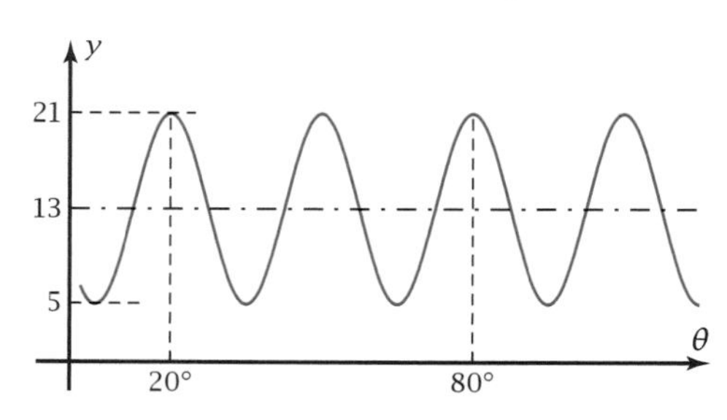

Figure 3-2l

Q1. How many cycles are there between $\theta = 20°$ and $\theta = 80°$?

Q2. What is the amplitude?

Q3. What is the period?

Q4. What is the vertical translation?

Q5. What is the horizontal translation (for cosine)?

Q6. Find the exact value (no decimals) of sin 60°.

Q7. Find the approximate value of sec 71°.

Q8. Find the approximate value of $\cot^{-1} 4.3$.

Q9. Find the measure of the larger acute angle of a right triangle with legs 11 ft and 9 ft.

Q10. Expand: $(3x - 5)^2$

For Problems 1–4, find the amplitude, period, phase displacement, and sinusoidal axis location. Without using your grapher, sketch the graph by locating critical points. Then check your graph using your grapher.

1. $y = 7 + 4\cos 3(\theta + 10°)$
2. $y = 3 + 5\cos\frac{1}{4}(\theta - 240°)$
3. $y = -10 + 20\sin\frac{1}{2}(\theta - 120°)$
4. $y = -8 + 10\sin 5(\theta + 6°)$

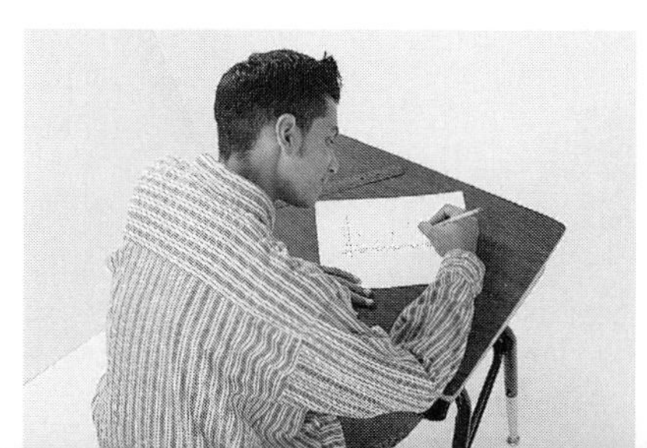

For Problems 5–8,

a. Find a particular equation for the sinusoid using cosine or sine, whichever seems easier.

b. Give the amplitude, period, frequency, phase displacement, and sinusoidal axis location.

c. Use the equation to calculate y for the given values of θ. Show that the result agrees with the given graph for the first value.

5. $\theta = 60°$ and $\theta = 1234°$

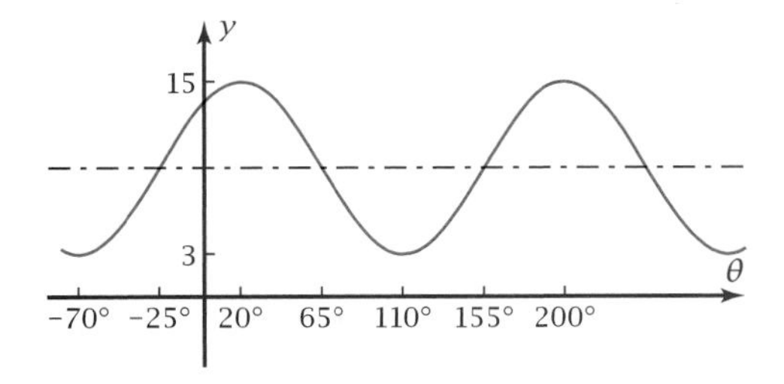

6. $\theta = 10°$ and $\theta = 453°$

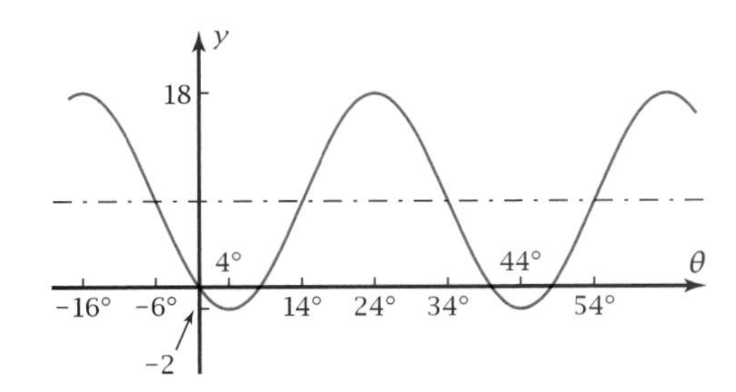

7. $\theta = 70°$ and $\theta = 491°$

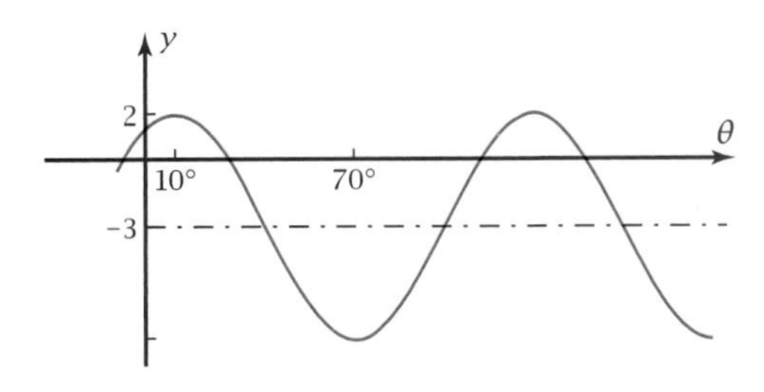

8. $\theta = 8°$ and $\theta = 1776°$

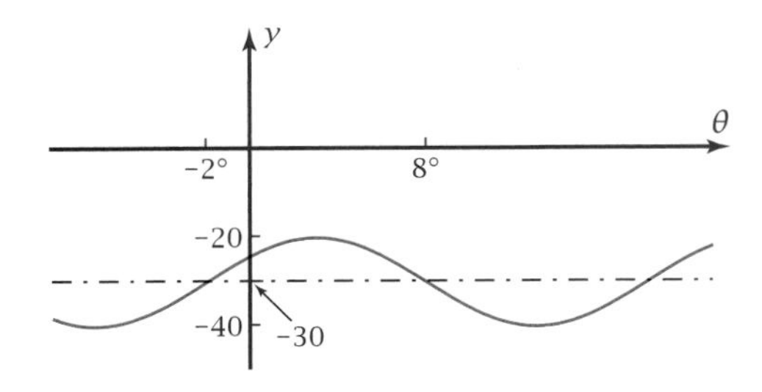

For Problems 9–14, find a particular equation of the sinusoid that is graphed.

9.

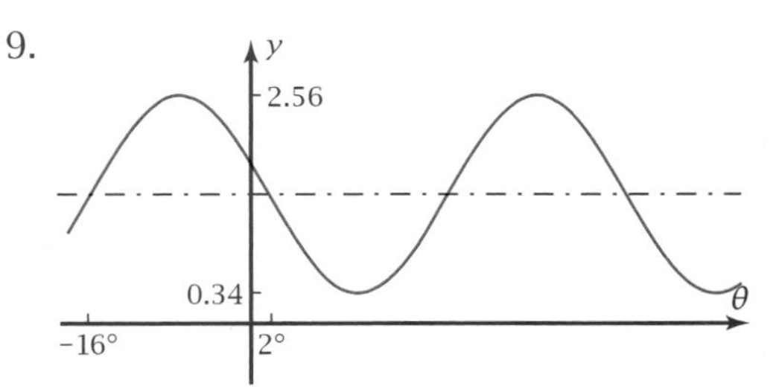

10.

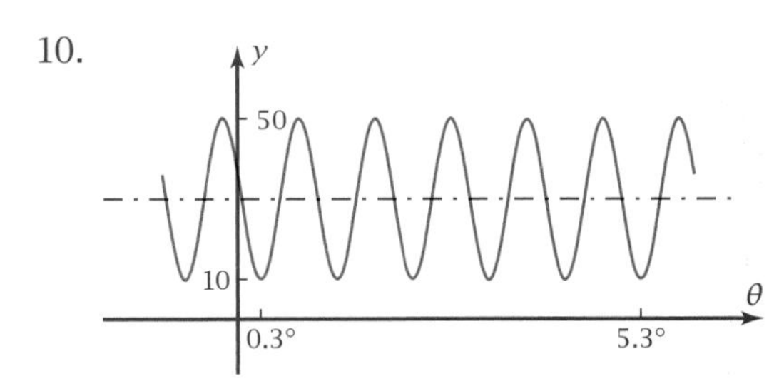

11.

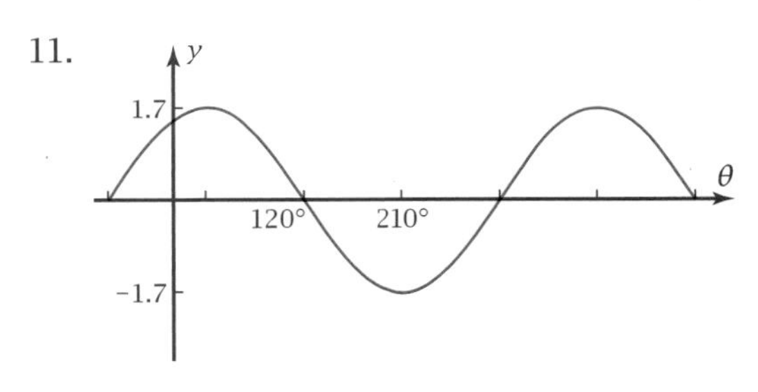

12.

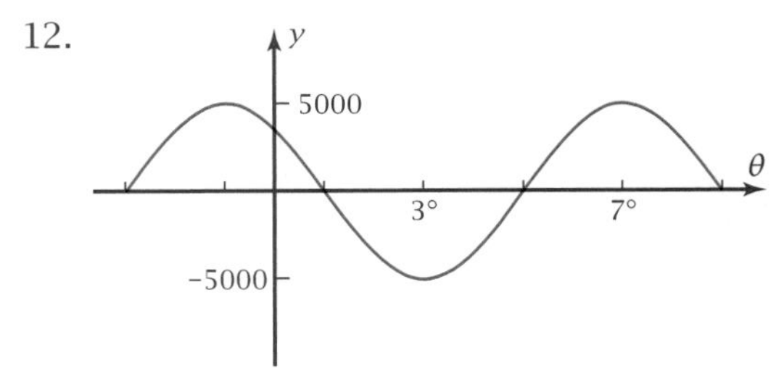

13.

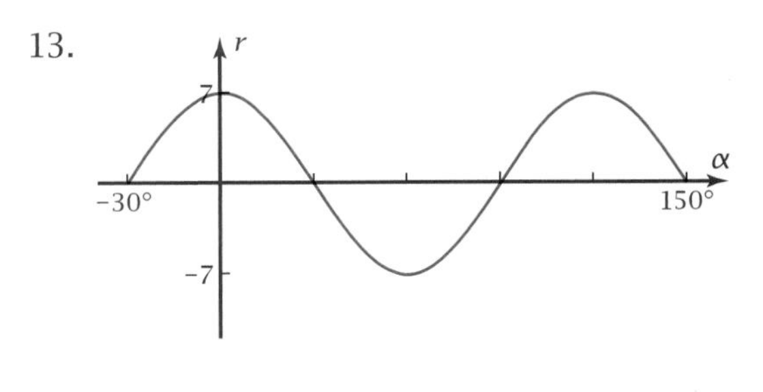

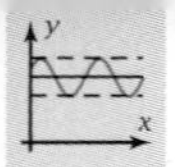

14.

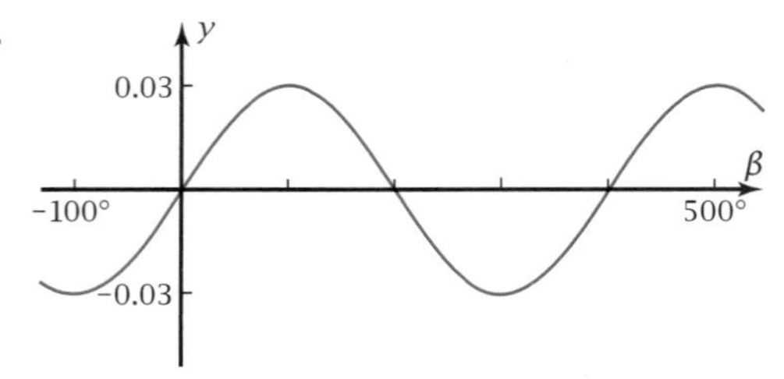

In Problems 15 and 16, a half-cycle of a sinusoid is shown. Find a particular equation.

15.

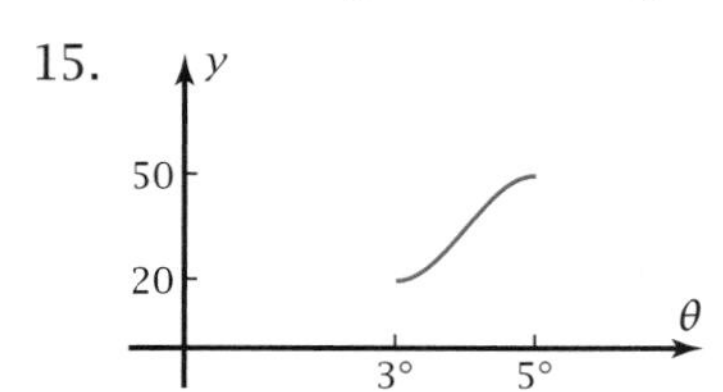

16.

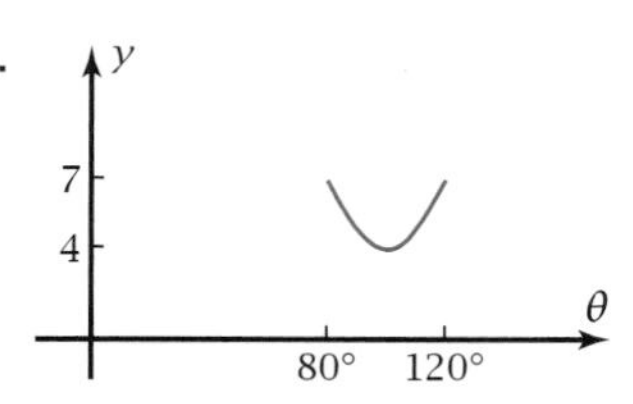

In Problems 17 and 18, a quarter-cycle of a sinusoid is shown. Find a particular equation.

17.

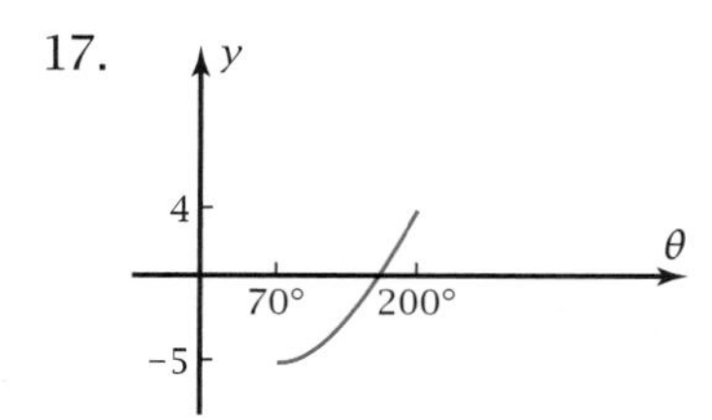

18.

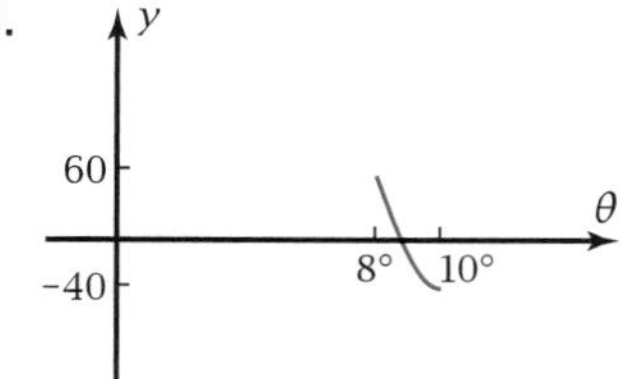

19. If the sinusoid in Problem 17 is extended to $\theta = 300°$, what is the value of y? If the sinusoid is extended to $\theta = 5678°$, is the point on the graph above or below the sinusoidal axis? How far?

20. If the sinusoid in Problem 18 is extended back to $\theta = 2.5°$, what is the value of y? If the sinusoid is extended to $\theta = 328°$, is the point on the graph above or below the sinusoidal axis? How far?

For Problems 21 and 22, sketch the sinusoid described and write a particular equation for it. Check the equation on your grapher to make sure it produces the graph you sketched.

21. The period equals 72°, amplitude is 3 units, phase displacement (for $y = \cos x$) equals 6°, and the sinusoidal axis is at $y = 4$ units.

22. The frequency is $\frac{1}{10}$ cycle per degree, amplitude equals 2 units, phase displacement (for $y = \cos x$) equals −3°, and the sinusoidal axis is at $y = -5$ units.

For Problems 23 and 24, write four different particular equations for the given sinusoid, using

a. Cosine as the parent function with positive vertical dilation

b. Cosine as the parent function with negative vertical dilation

c. Sine as the parent function with positive vertical dilation

d. Sine as the parent function with negative vertical dilation

Plot all four equations on the same screen, thus confirming that the graphs are the same.

23.

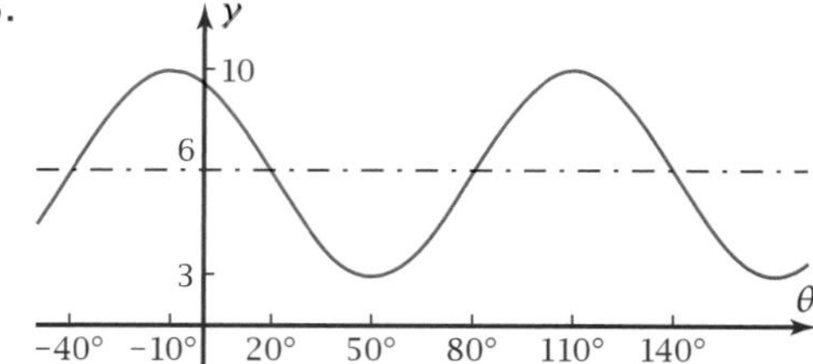

24.

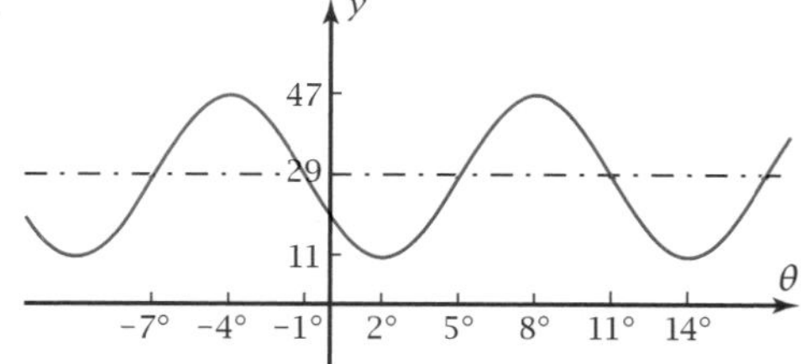

25. *Frequency Problem:* The unit for the period of a sinusoid is degrees per cycle. The unit for the frequency is cycles per degree.

a. Suppose that a sinusoid has period $\frac{1}{60}$ degree/cycle. What would the frequency be? Why might people prefer to speak of the frequency of such a sinusoid rather than the period of this sinusoid?

b. For $y = \cos 300\theta$, what is the period? What is the frequency? How can you calculate the frequency quickly, using the 300?

26. *Inflection Point Problem:* Sketch the graph of a function that has high and low critical points. On the sketch, show

a. A point of inflection

b. A region where the graph is concave up

c. A region where the graph is concave down

27. *Horizontal vs. Vertical Transformations Problem:* In the function

$$y = 3 + 4\cos 2(\theta - 5°)$$

the 3 and 4 are the vertical transformations, but the 2 and −5 are the reciprocal and opposite of the horizontal transformations.

a. Show that you can transform the given equation to

$$\frac{y-3}{4} = \cos\left(\frac{\theta - 5°}{1/2}\right)$$

b. Examine the equation in part a for the transformations that are applied to the x- and y-variables. What is the form of these transformations?

c. Why is the original form of the equation more useful than the form in part a?

28. *Journal Problem:* Update your journal with things you have learned about sinusoids. In particular, explain how the amplitude, period, phase displacement, frequency, and sinusoidal axis location are related to the four constants in the general sinusoidal equation. What is meant by *critical points, concavity,* and *points of inflection*?

3-3 Graphs of Tangent, Cotangent, Secant, and Cosecant Functions

If you enter tan 90°, your calculator will give you an error message. This happens because tangent is defined as a quotient. On a unit circle, a point on the terminal side of a 90° angle has horizontal coordinate zero and vertical coordinate 1. Division of a nonzero number by zero is undefined, which you'll see leads to **vertical asymptotes** at angle values for which division by zero would occur. In this section you'll also see that the graphs of the tangent, cotangent, secant, and cosecant functions are **discontinuous** where the function value would involve division by zero.

OBJECTIVE Plot the graphs of the tangent, cotangent, secant, and cosecant functions, showing their behavior when the function value is undefined.

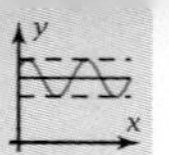

You can plot cotangent, secant, and cosecant by using the fact that they are reciprocals of tangent, cosine, and sine, respectively.

$$\cot\theta = \frac{1}{\tan\theta} \qquad \sec\theta = \frac{1}{\cos\theta} \qquad \csc\theta = \frac{1}{\sin\theta}$$

Figure 3-3a shows the graphs of $y = \tan\theta$ and $y = \cot\theta$, and Figure 3-3b shows the graphs $y = \sec\theta$ and $y = \csc\theta$, all as they might appear on your grapher. If you use a friendly window that includes multiples of 90° as grid points, you'll see that the graphs are discontinuous. Notice that the graphs go off to infinity (positive or negative) at odd or even multiples of 90°, exactly at the places where the functions are undefined.

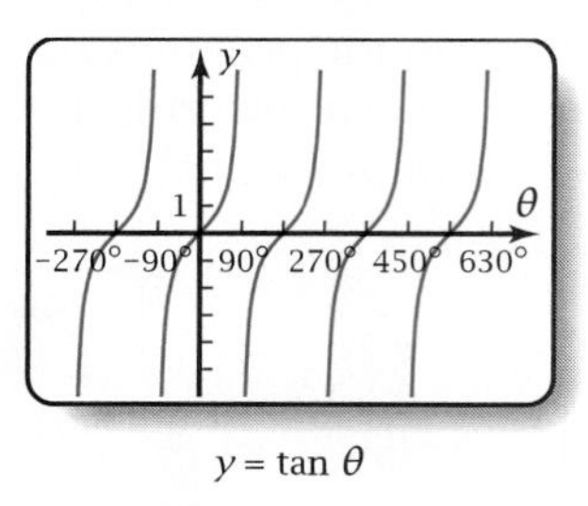

$y = \tan\theta$

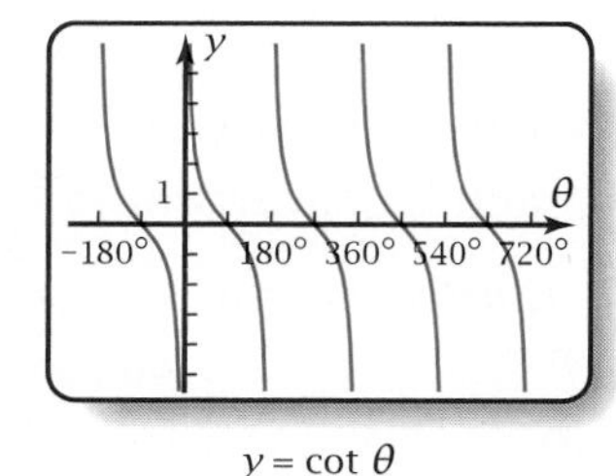

$y = \cot\theta$

Figure 3-3a

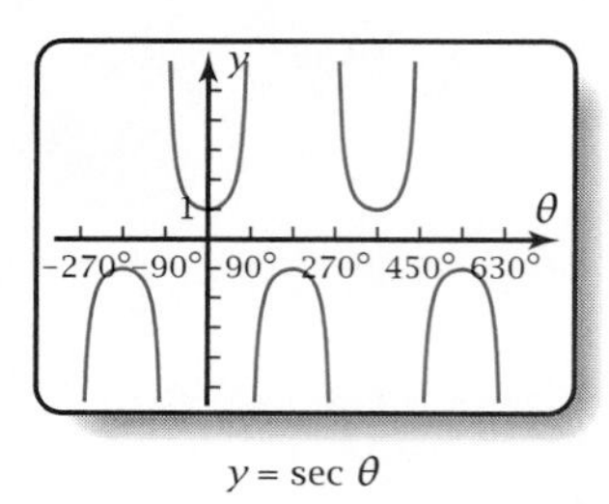

$y = \sec\theta$

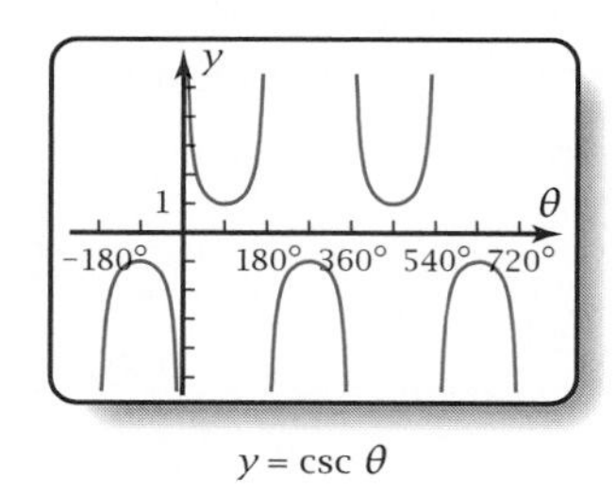

$y = \csc\theta$

Figure 3-3b

To see why the graphs have these shapes, it helps to look at transformations performed on the parent cosine and sine graphs.

▶ ***EXAMPLE 1*** Sketch the graph of the sine function $y = \sin\theta$. Use the fact that $\csc\theta = \frac{1}{\sin\theta}$ to sketch the graph of the cosecant function. Show how the asymptotes of the cosecant function are related to the graph of the sine function.

Solution Sketch the sine graph as in Figure 3-3c. Where the value of the sine function is zero, the cosecant function will be undefined because of division by zero. Draw vertical asymptotes at these values of θ.

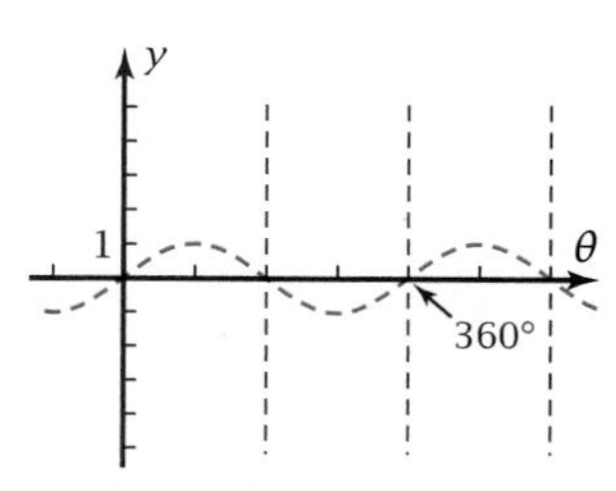

Figure 3-3c

Where the sine function equals 1 or −1, so does the cosecant function. This is because the reciprocal of 1 is 1 and the reciprocal of −1 is −1. Mark these points as in Figure 3-3d. As the sine gets smaller, the cosecant gets bigger. For instance, the reciprocal of 0.2 is 5. The reciprocal of −0.5 is −2. Sketch the graph consistent with these facts, as in Figure 3-3d. ◀

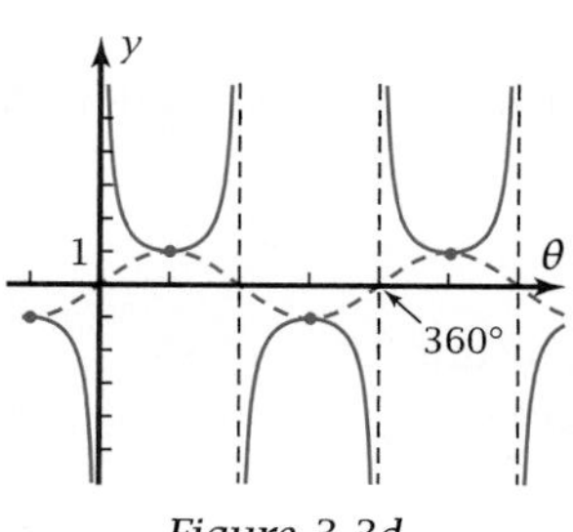

Figure 3-3d

To understand why the graphs of the tangent and cotangent functions have the shapes in Figure 3-3a, it helps to examine how these functions are related to the sine and the cosine functions. By definition,

$$\tan\theta = \frac{v}{u}$$

Dividing the numerator and the denominator by r gives

$$\tan\theta = \frac{v/r}{u/r}$$

By the definitions of sine and cosine, the numerator equals $\sin\theta$ and the denominator equals $\cos\theta$. As a result, these **quotient properties** are true.

PROPERTIES: Quotient Properties for Tangent and Cotangent

$$\tan\theta = \frac{\sin\theta}{\cos\theta} \quad \text{and} \quad \cot\theta = \frac{\cos\theta}{\sin\theta}$$

The quotient properties let you construct the tangent and cotangent graphs from the sine and cosine.

▶ EXAMPLE 2 On paper, sketch the graphs of $y = \sin x$ and $y = \cos x$. Use the quotient property to sketch the graph of $y = \cot x$. Show the asymptotes and points where the graph crosses the θ-axis.

Solution Draw the graphs of the sine and the cosine functions (dashed and solid, respectively) as in Figure 3-3e. Because $\cot\theta = \frac{\cos\theta}{\sin\theta}$, show the asymptotes where $\sin\theta = 0$, and show the θ-intercepts where $\cos\theta = 0$.

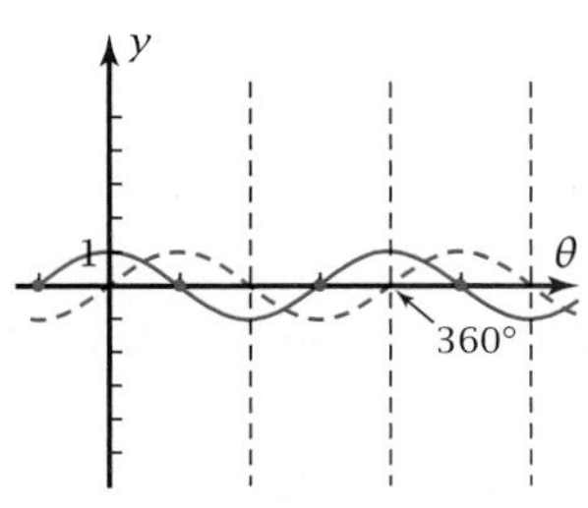

Figure 3-3e

At $\theta = 45°$, and wherever else the graphs of the sine and the cosine functions intersect each other, $\frac{\cos\theta}{\sin\theta}$ will equal 1. Wherever sine and cosine are opposites of each other, $\frac{\cos\theta}{\sin\theta}$ will equal −1. Mark these points as in Figure 3-3f. Then draw the cotangent graph through the marked points, consistent with the asymptotes. The final graph is in Figure 3-3g.

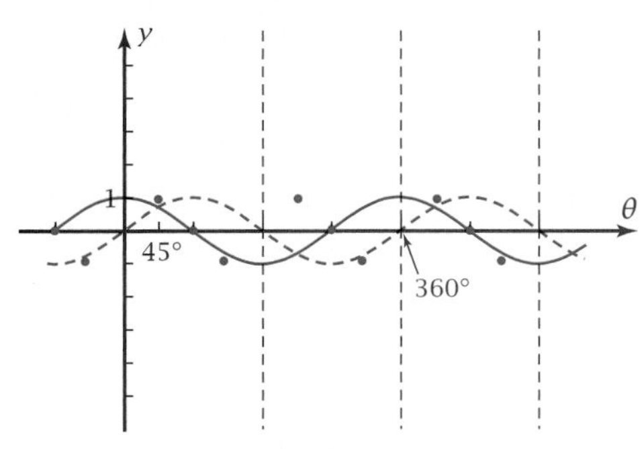

Figure 3-3f

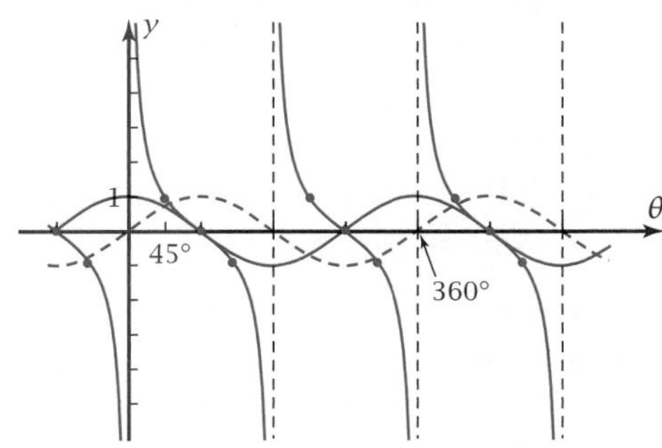

Figure 3-3g

◀

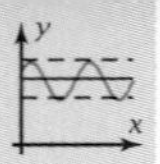

Problem Set 3-3

Do These Quickly

5 min

Problems Q1–Q7 refer to the equation $y = 3 + 4\cos 5(\theta - 6°)$.

Q1. The graph of the equation is called a —?—.

Q2. The amplitude is —?—.

Q3. The period is —?—.

Q4. The phase displacement with respect to $y = \cos\theta$ is —?—.

Q5. The frequency is —?—.

Q6. The sinusoidal axis is at $y =$ —?—.

Q7. The lower bound is at $y =$ —?—.

Q8. What kind of function is $y = x^5$?

Q9. What kind of function is $y = 5^x$?

Q10. The "If..." part of the statement of a theorem is called the

A. Conclusion B. Hypothesis
C. Converse D. Inverse
E. Contrapositive

---

1. *Secant Function Problem*
   a. Sketch two cycles of the parent cosine function $y = \cos\theta$. Use the fact that $\sec\theta = \frac{1}{\cos\theta}$ to sketch the graph of $y = \sec\theta$.
   b. How can you locate the asymptotes in the secant graph by looking at the cosine graph? How does your graph compare with the secant graph in Figure 3-3b?
   c. Does the secant function have critical points? If so, find some of them. If not, explain why not.
   d. Does the secant function have points of inflection? If so, find some of them. If not, explain why not.

2. *Tangent Function Problem*
   a. Sketch two cycles of the parent function $y = \cos\theta$ and two cycles of the parent function $y = \sin\theta$ on the same axes.
   b. Explain how you can use the graphs in part a to locate the θ-intercepts and the vertical asymptotes of $y = \tan\theta$.
   c. Mark the asymptotes, intercepts, and other significant points on your sketch in part a. Then sketch the graph of $y = \tan\theta$. How does the result compare with Figure 3-3a?
   d. Does the tangent function have critical points? If so, find some of them. If not, explain why not.
   e. Does the tangent function have points of inflection? If so, find some of them. If not, explain why not.

3. *Quotient Property for Tangent Problem:* Plot these three graphs on the same screen on your grapher. Explain how the result confirms the quotient property for tangent.

   $y_1 = \sin\theta$
   $y_2 = \cos\theta$
   $y_3 = y_1/y_2$

4. *Quotient Property for Cotangent Problem:* On the same screen on your grapher, plot these three graphs. Explain how the result confirms the quotient property for cotangent.

   $y_1 = \sin\theta$
   $y_2 = \cos\theta$
   $y_3 = y_2/y_1$

5. Without referring to Figure 3-3a, sketch quickly the graphs of $y = \tan\theta$ and $y = \cot\theta$.

6. Without referring to Figure 3-3b, sketch quickly the graphs of $y = \sec\theta$ and $y = \csc\theta$.

7. Explain why the period of the $y = \tan\theta$ and $y = \cot\theta$ functions is only 180° instead of 360°, like the other four trigonometric functions.

8. Explain why it is meaningless to talk about the amplitude of the tangent, cotangent, secant, and cosecant functions.

9. What is the domain of $y = \sec\theta$? What is its range?

10. What is the domain of $y = \tan\theta$? What is its range?

For Problems 11–14, what are the dilation and translation caused by the constants in the equation? Plot the graph on your grapher and show that these transformations are correct.

11. $y = 2 + 5\tan 3(\theta - 5°)$
12. $y = -1 + 3\cot 2(\theta - 30°)$
13. $y = 4 + 6\sec \frac{1}{2}(\theta + 50°)$
14. $y = 3 + 2\csc 4(\theta + 10°)$
15. *Rotating Lighthouse Beacon Problem:* Figure 3-3h shows a lighthouse 500 m from the shore.

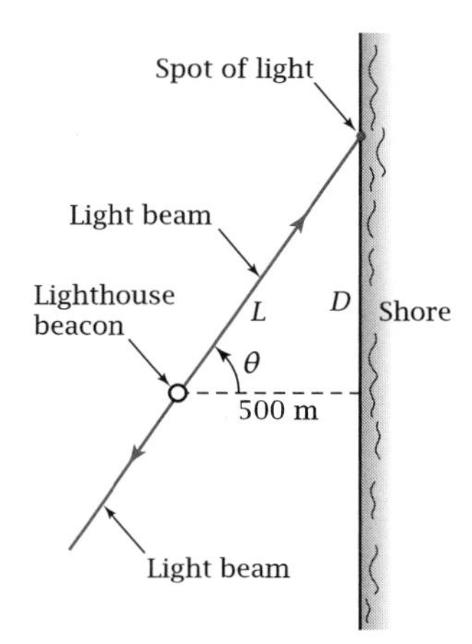

Figure 3-3h

A rotating light on the top of the lighthouse sends out light beams in opposite directions. As the beacon rotates through an angle θ, a spot of light moves along the shore. Let L be the length of the light beam from the lighthouse to the spot on the shore. Let D be the *directed* distance along the shore to the spot of light; that is, D is positive to the right if you're on the shore facing the lighthouse and negative to the left.

a. Plot the graphs of D and L as functions of θ. Use a θ-range of 0° to 360° and a y-range of −2000 to 2000. Sketch your results.
b. When $\theta = 55°$, where does the spot of light hit the shore? How long is the light beam?
c. What is the first positive value of θ for which the light beam will be 2000 m long?
d. Explain why D is negative for $90° < \theta < 180°$ and for $270° < \theta < 360°$ and why L is negative for $90° < \theta < 270°$.
e. Explain the physical significance of the asymptote in each graph at $\theta = 90°$.

3-4 Radian Measure of Angles

With your calculator in degree mode, press sin 60°. You get

$$\sin 60° = 0.866025403\ldots$$

Now change to **radian** mode and press $\sin\left(\frac{\pi}{3}\right)$. You get the same answer!

$$\sin\left(\frac{\pi}{3}\right) = 0.866025403\ldots$$

In this section you will learn what radians are and how to convert angle measures between radians and degrees. The radian measure of angles allows you to expand on the concept of trigonometric functions, as you'll see in the next section. Through this expansion of trigonometric functions, you can model real-world phenomena where independent variables represent distance, time, or any other quantity, not just an angle measure in degrees.

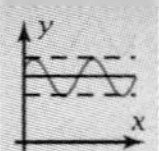

OBJECTIVE

- Given an angle measure in degrees, convert it to radians, and vice versa.
- Given an angle measure in radians, find trigonometric function values.

Excerpt from an old Babylonian cuneiform text

The degree as a unit of angular measure came from ancient mathematicians, probably Babylonians. It is assumed that they divided a revolution into 360 parts we call degrees because there were approximately 360 days in a year and they used the base-60 (sexagesimal) number system. There is another way to measure angles, called radian measure. This mathematically more natural unit of angular measure is derived by wrapping a number line around a unit circle (a circle of radius 1 unit) in a *uv*-coordinate system, as shown in Figure 3-4a. This way, each point on the number line corresponds to a point on the perimeter of the circle.

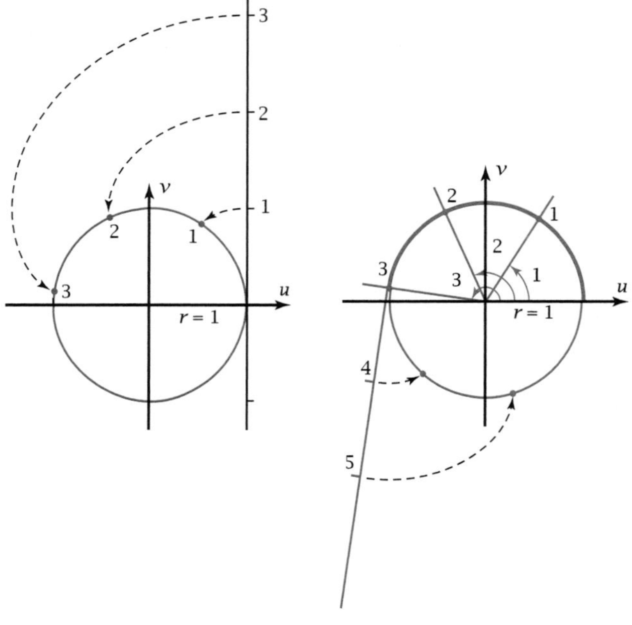

Figure 3-4a

If you draw rays from the origin to the points 1, 2, and 3 on the circle (right side of Figure 3-4a), the corresponding central angles have radian measures of 1, 2, and 3, respectively.

But, you may ask, what happens if the same angle is in a larger circle? Would the same radian angle measurement correspond to it? How would you calculate the radian angle measurement in this case? Figures 3-4b and 3-4c answer these questions. Figure 3-4b shows an angle of 1, in radians, and the arcs it **subtends** (cuts off) on circles of radius 1 unit and *x* units. The arc subtended on the unit circle has a length of 1 unit. By the properties of similar geometric figures, the arc subtended on the circle

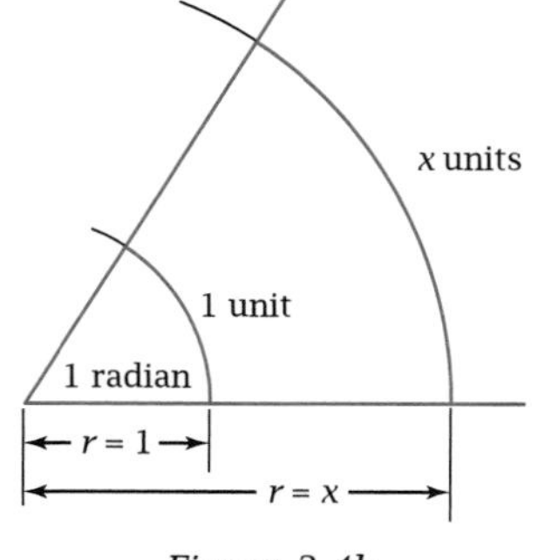

Figure 3-4b

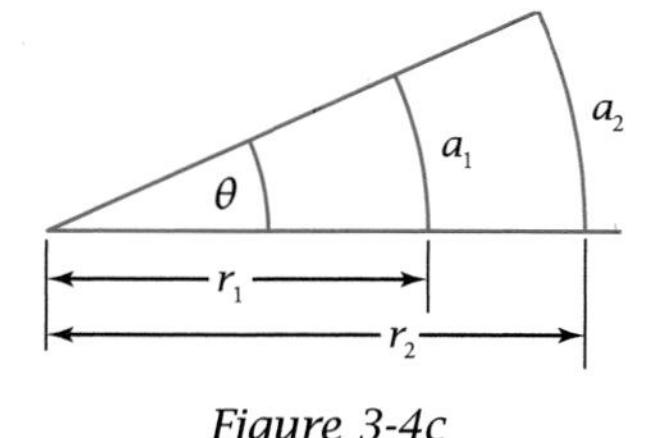

Figure 3-4c

of radius x has length of x units. So 1 radian subtends an arc of length equal to the radius of the circle.

For any angle measure, the arc length and radius are proportional $\left(\frac{a_1}{r_1} = \frac{a_2}{r_2}\right.$, as shown in Figure 3-4c$\left.\right)$, and their quotient will be a unitless number that uniquely corresponds to and describes the angle. So, in general, the radian measure of an angle equals the length of the subtended arc divided by the radius.

DEFINITION: Radian Measure of an Angle

$$\text{radian measure} = \frac{\text{arc length}}{\text{radius}}$$

Note: One radian subtends an arc whose length equals the radius of the circle (Figure 3-4b).

For the work that follows, it is important to distinguish between the name of the angle and the measure of that angle. Measures of θ will be written this way:

θ is the *name* of the angle.

$m°(\theta)$ = degree measure of angle θ.

$m^R(\theta)$ = radian measure of angle θ.

Since the circumference of a circle is $2\pi r$, and r for a unit circle is 1, the wrapped number line in Figure 3-4a divides the circle into 2π units (a little more than six parts). So there are 2π radians in a complete revolution. Since there are also 360° in a complete revolution, you can convert degrees to radians, or the other way around, by setting up these proportions.

$$\frac{m^R(\theta)}{m°(\theta)} = \frac{2\pi}{360°} = \frac{\pi}{180°} \quad \text{or} \quad \frac{m°(\theta)}{m^R(\theta)} = \frac{360°}{2\pi} = \frac{180°}{\pi}$$

Solving for $m^R(\theta)$ and $m°(\theta)$, respectively, will give you

$$m^R(\theta) = \frac{\pi}{180°} m°(\theta) \quad \text{and} \quad m°(\theta) = \frac{180°}{\pi} m^R(\theta)$$

These equations lead to this procedure for accomplishing the objective of this section.

PROCEDURE: Radian–Degree Conversion

- To find the radian measure of θ, multiply the degree measure by $\frac{\pi}{180°}$.
- To find the degree measure of θ, multiply the radian measure by $\frac{180°}{\pi}$.

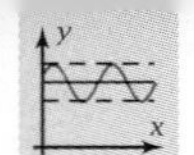

▶ EXAMPLE 1 Convert 135° to radians.

Solution In order to keep the units straight, write each quantity as a fraction with the proper units. If you have done the work correctly, certain units will cancel, leaving the proper units for the answer.

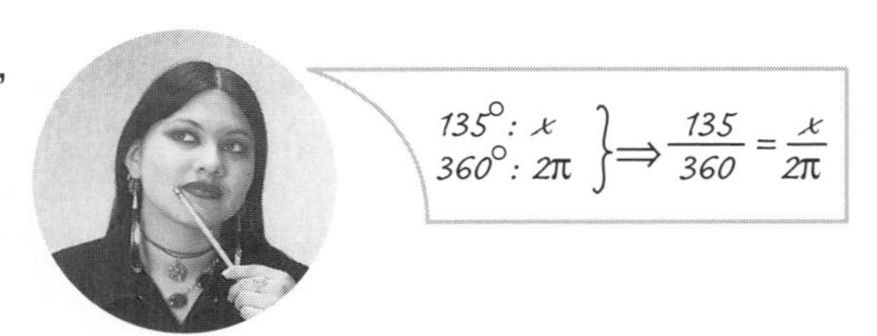

$$m^R(\theta) = \frac{135 \text{ degrees}}{1} \cdot \frac{\pi \text{ radians}}{180 \text{ degrees}} = \frac{3}{4}\pi = 2.3561\ldots \text{ radians}$$ ◀

Note:

- If the *exact* value is called for, leave the answer as $\frac{3}{4}\pi$. If not, you have the choice of writing the answer as a multiple of π or converting to a decimal.
- The procedure for canceling units, as shown in Example 1, is called **dimensional analysis.** You will use this procedure throughout your study of mathematics.

▶ EXAMPLE 2 Convert 5.73 radians to degrees.

Solution

$$\frac{5.73 \text{ radians}}{1} \cdot \frac{180 \text{ degrees}}{\pi \text{ radians}} = 328.3048\ldots^\circ$$ ◀

▶ EXAMPLE 3 Find tan 3.7.

Solution Unless the argument of a trigonometric function has the degree sign, it is assumed to be a measure in radians. (That is why it has been important for you to include the degree sign up till now.) Set your calculator to radian mode and enter tan 3.7.

$$\tan 3.7 = 0.6247\ldots$$ ◀

▶ EXAMPLE 4 Find the radian measure and the degree measure of an angle whose sine is 0.3.

Solution

$\sin^{-1} 0.3 = 0.3046\ldots$ radians Set your calculator to radian mode.

$\sin^{-1} 0.3 = 17.4576\ldots^\circ$ Set your calculator to degree mode. ◀

To check if these answers are in fact the same, you could convert one to the other.

$$0.3046\ldots \cdot \frac{180}{\pi} = 17.4576\ldots$$ Use the 0.3046... already in your calculator, without rounding off.

Radian Measures of Some Special Angles

It will help you later in calculus to concentrate on concepts and methods to be able to recall quickly the radian measures of certain special angles, such as those whose degree measures are multiples of 30° and 45°. By the technique of Example 1,

$$30° \longrightarrow \frac{\pi}{6} \text{ radians, or } \frac{1}{12} \text{ revolution}$$

$$45° \longrightarrow \frac{\pi}{4} \text{ radians, or } \frac{1}{8} \text{ revolution}$$

If you remember these two, you can find others quickly by multiplication. For instance,

$$60° \longrightarrow 2(\pi/6) = \frac{\pi}{3} \text{ radians, or } \frac{1}{6} \text{ revolution}$$

$$90° \longrightarrow 3(\pi/6) \text{ or } 2(\pi/4) = \frac{\pi}{2} \text{ radians, or } \frac{1}{4} \text{ revolution}$$

$$180° \longrightarrow 6(\pi/6) \text{ or } 4(\pi/4) = \pi \text{ radians, or } \frac{1}{2} \text{ revolution}$$

For 180°, you can simply remember that a full revolution is 2π radians, so half a revolution is π radians.

Figure 3-4d shows the radian measures for some special first-quadrant angles. Figure 3-4e shows radian measures for larger angles that are $\frac{1}{4}$, $\frac{1}{2}$, $\frac{3}{4}$, and 1 revolution. The box summarizes this information.

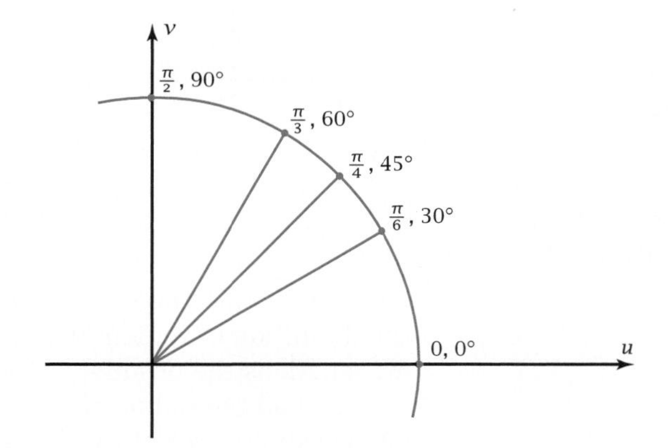

Figure 3-4d

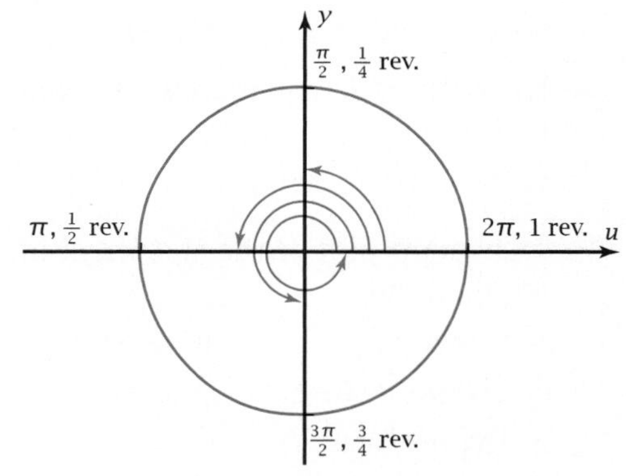

Figure 3-4e

PROPERTY: Radian Measures of Some Special Angles

Degrees	Radians	Revolutions
30°	$\pi/6$	$\frac{1}{12}$
45°	$\pi/4$	$\frac{1}{8}$
60°	$\pi/3$	$\frac{1}{6}$
90°	$\pi/2$	$\frac{1}{4}$
180°	π	$\frac{1}{2}$
360°	2π	1

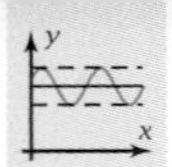

EXAMPLE 5 Find the exact value of $\sec \dfrac{\pi}{6}$.

Solution $\sec \dfrac{\pi}{6} = \sec 30° = \dfrac{1}{\cos 30°} = \dfrac{1}{\sqrt{3}/2} = \dfrac{2}{\sqrt{3}}$ Recall how to find cos 30° exactly.

Problem Set 3-4

Do These Quickly

Q1. Sketch the graph of $y = \tan \theta$.

Q2. Sketch the graph of $y = \sec \theta$.

Q3. What is the first positive value of θ at which $y = \cot \theta$ has a vertical asymptote?

Q4. What is the first positive value of θ for which $\csc \theta = 0$?

Q5. What is the exact value of tan 60°?

Q6. What transformation of function f is represented by $g(x) = 3\,f(x)$?

Q7. What transformation of function f is represented by $h(x) = f(10x)$?

Q8. Write the general equation of a quadratic function.

Q9. $3^{2005} \div 3^{2001} =$ —?—.

Q10. The "then" part of the statement of a theorem is called the

A. Converse B. Inverse

C. Contrapositive D. Conclusion

E. Hypothesis

---

1. *Wrapping Function Problem:* Figure 3-4f shows a unit circle in a uv-coordinate system. Suppose you want to use the angle measure in radians as the independent variable. Imagine the x-axis from an xy-coordinate system placed tangent to the circle. Its origin, $x = 0$, is at the point $(u, v) = (1, 0)$. Then the x-axis is wrapped around the circle.

   a. Show where the points $x = 1$, 2, and 3 on the number line map onto the circle.

   b. On the sketch from part a, show angles in standard position of 1, 2, and 3 radians.

   c. Explain how the length of the arc of a unit circle subtended by a central angle of the circle is related to the radian measure of that angle.

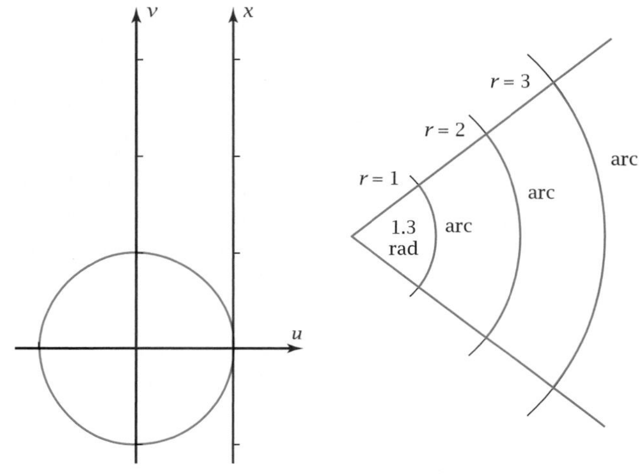

Figure 3-4f Figure 3-4g

2. *Arc Length and Angle Problem:* As a result of the definition of radians, you can calculate the arc length as the product of the angle in radians and the radius of the circle. Figure 3-4g (above) shows arcs of three circles subtended by a central angle of 1.3 radians. The circles have radii of 1, 2, and 3 centimeters.

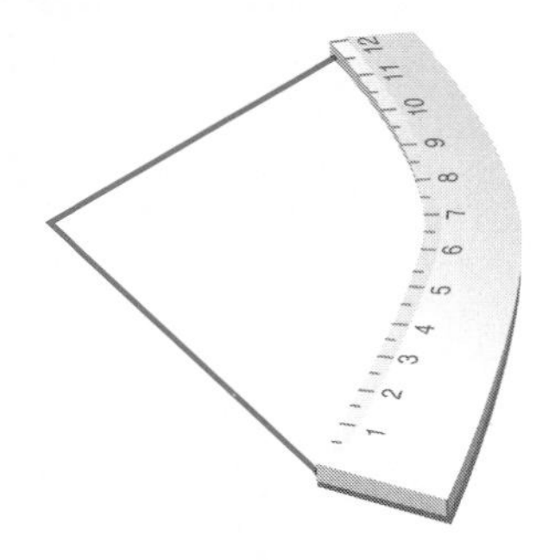

a. How long would the arc of the 1-cm circle be if you measured it with a flexible ruler?

b. Find how long the arcs are on the 2-cm circle and on the 3-cm circle using the properties of similar geometrical figures.

c. On a circle of radius r meters, how long would an arc be that is subtended by an angle of 1.3 radians?

d. How could you find *quickly* the length a of an arc of a circle of radius r meters that is subtended by a central angle of measure θ radians? Write a formula for the arc length.

For Problems 3–10, find the *exact* radian measure of the angle (no decimals).

3. 60°
4. 45°
5. 30°
6. 180°
7. 120°
8. 450°
9. −225°
10. 1080°

For Problems 11–14, find the radian measure of the angle in decimal form.

11. 37°
12. 54°
13. 123°
14. 258°

For Problems 15–24, find the *exact* degree measure of the angle given in radians (no decimals). Use the most time-efficient method.

15. $\frac{\pi}{10}$ radians
16. $\frac{\pi}{2}$ radians
17. $\frac{\pi}{6}$ radians
18. $\frac{\pi}{4}$ radians
19. $\frac{\pi}{12}$ radians
20. $\frac{2\pi}{3}$ radians
21. $\frac{3\pi}{4}$ radians
22. π radians
23. $\frac{3\pi}{2}$ radians
24. $\frac{5\pi}{6}$ radians

For Problems 25–30, find the degree measure in decimal form of the angle given in radians.

25. 0.34 radians
26. 0.62 radians
27. 1.26 radians
28. 1.57 radians
29. 1 radian
30. 3 radians

For Problems 31–34, find the function value (decimal form) for the angle in radians.

31. sin 5
32. cos 2
33. tan (−2.3)
34. sin 1066

For Problems 35–38, find the radian measure (decimal form) of the angle.

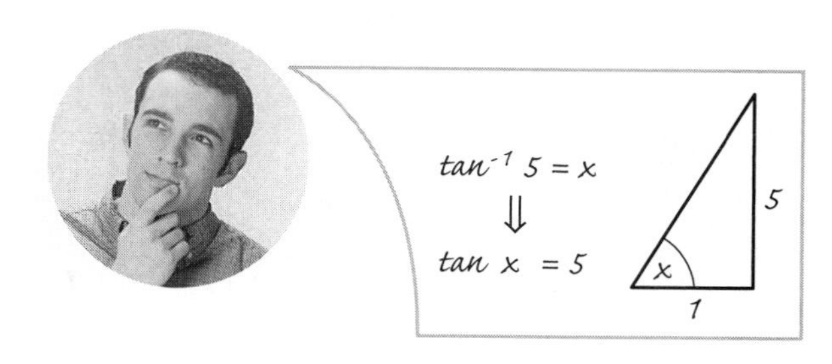

35. $\sin^{-1} 0.3$
36. $\tan^{-1} 5$
37. $\cot^{-1} 3$
38. $\csc^{-1} 1.001$

For Problems 39–44, find the *exact* value of the indicated function (no decimals). Note that since the degree sign is not used, the angle is assumed to be in radians.

39. $\sin \frac{\pi}{3}$
40. $\cos \pi$
41. $\tan \frac{\pi}{6}$
42. $\cot \frac{\pi}{2}$
43. $\sec 2\pi$
44. $\csc \frac{\pi}{4}$

For Problems 45–48, find the *exact* value of the expression (no decimals).

45. $\sin \frac{\pi}{2} + 6 \cos \frac{\pi}{3}$
46. $\csc \frac{\pi}{6} \sin \frac{\pi}{6}$
47. $\cos^2 \pi + \sin^2 \pi$
48. $\tan^2 \frac{\pi}{3} - \sec^2 \frac{\pi}{3}$

For Problems 49 and 50, write a particular equation for the sinusoid graphed.

49.

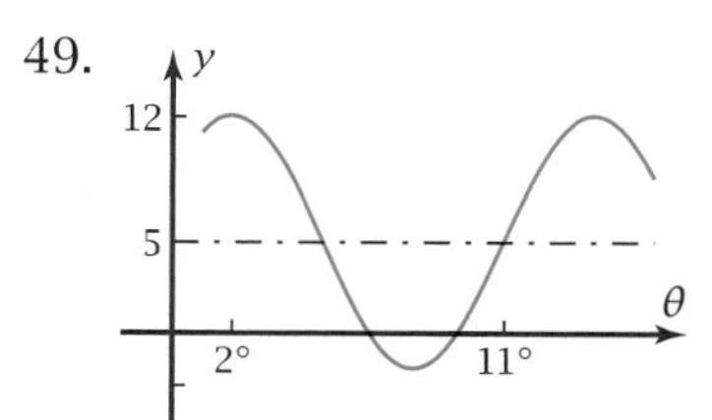

50.

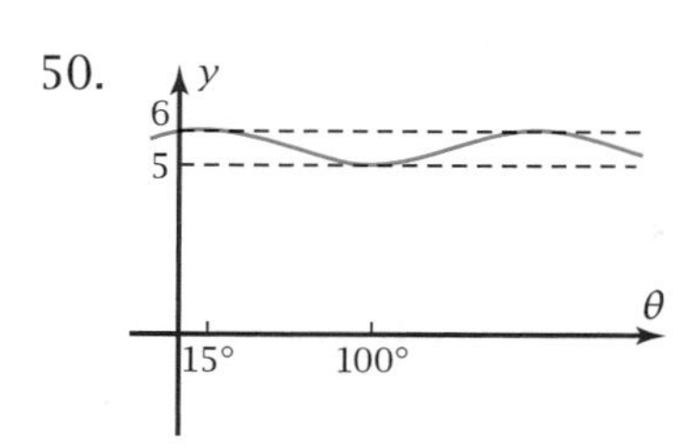

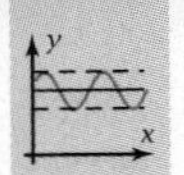

For Problems 51 and 52, find the length of the side marked x in the right triangle.

51.

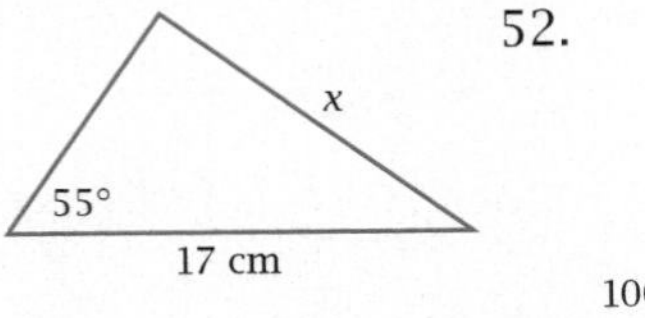

52.

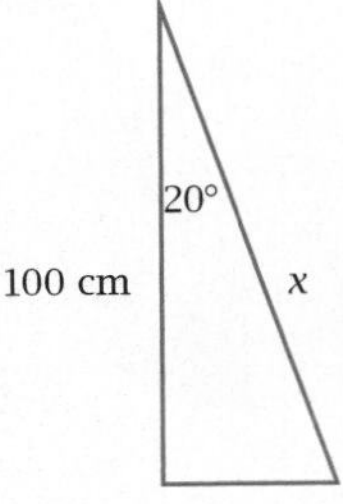

For Problems 53 and 54, find the degree measure of angle θ in the right triangle.

53.

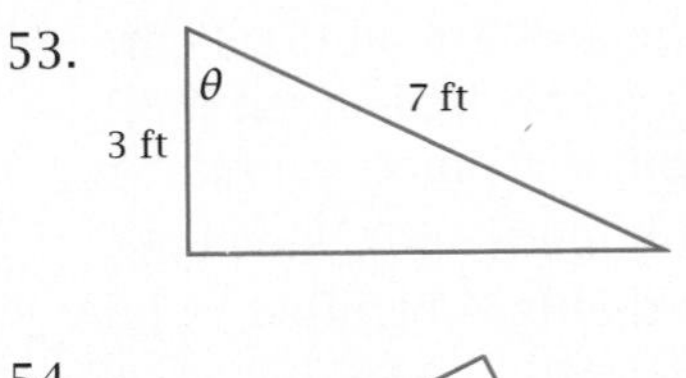

54.

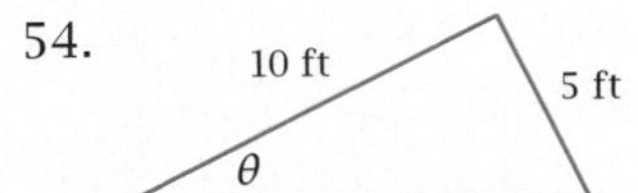

3-5 Circular Functions

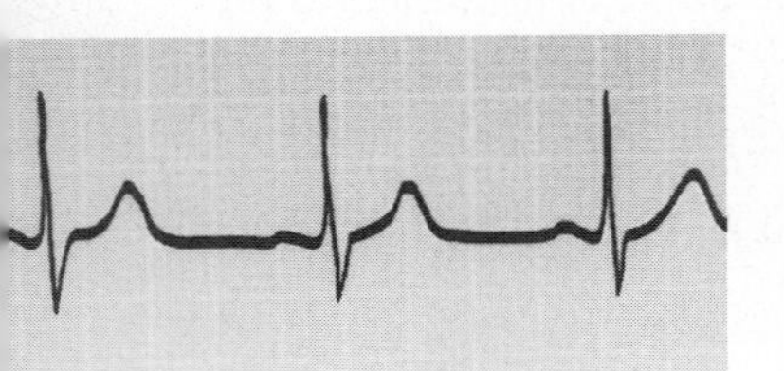

The normal human EKG (electrocardiogram) is periodic.

In many real-world situations, the independent variable of a periodic function is time or distance, with no angle evident. For instance, the normal daily high temperature varies periodically with the day of the year. In this section you will define **circular functions,** which are periodic functions whose independent variable is a real number without any units. These functions, as you will see, are identical to trigonometric functions in every way except for their argument. Circular functions are more appropriate for real-world applications. They also have some advantages in later courses in calculus, for which this course is preparing you.

OBJECTIVE Learn about the circular functions and their relationship to trigonometric functions.

Two cycles of the graph of the parent cosine function are completed in 720° (Figure 3-5a, left) or in 4π units (Figure 3-5a, right), because 4π radians corresponds to two revolutions.

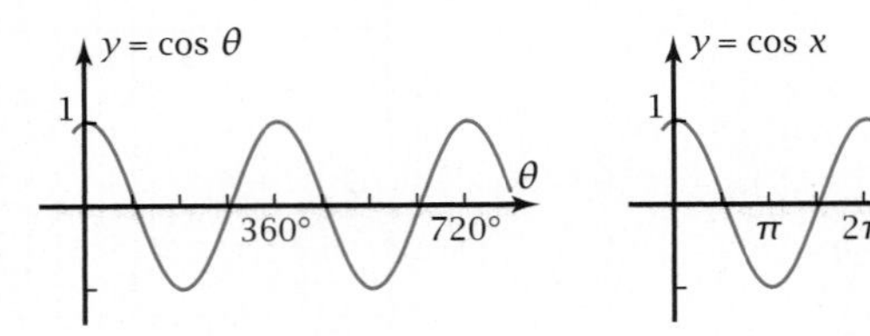

Figure 3-5a

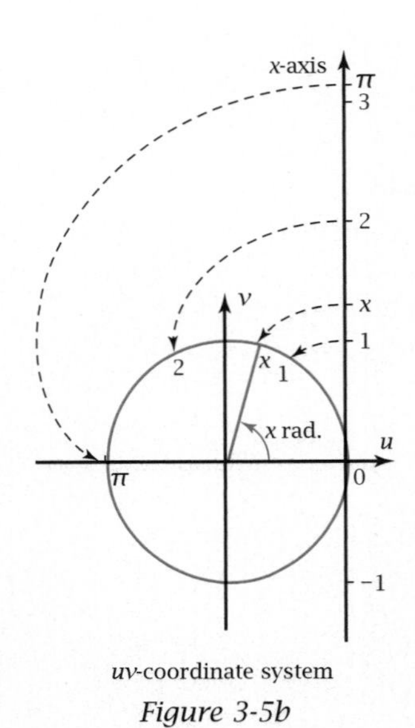

uv-coordinate system

Figure 3-5b

To see how the independent variable can represent a real number, imagine the x-axis from an xy-coordinate system lifted out and placed tangent to a unit circle in a uv-coordinate system at the point (1, 0) (Figure 3-5b). The origin of the x-axis is at the point $(u, v) = (1, 0)$. Then wrap the x-axis around the unit circle. As the axis wraps, $x = 1$ maps onto an angle of 1 radian, $x = 2$ maps onto 2 radians, $x = \pi$ maps onto π radians, and so on.

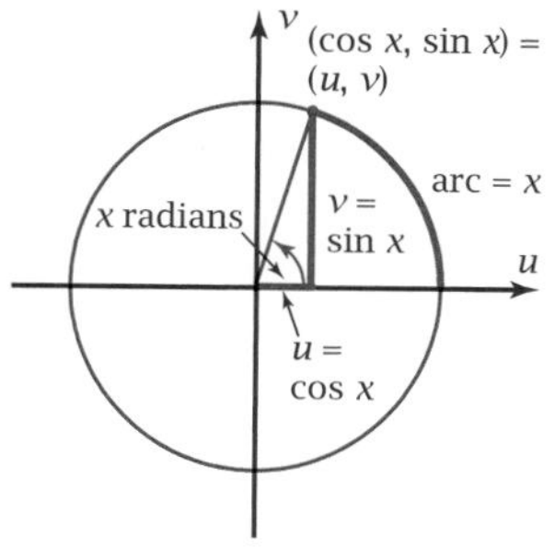

Figure 3-5c

The distance x on the x-axis is equal to the arc length on the unit circle. This arc length is equal to the radian measure for the corresponding angle. Thus the functions $\sin x$ and $\cos x$ for a number x on the x-axis are the same as the sine and cosine of an angle of x radians.

Figure 3-5c shows an arc of length x on a unit circle, *with* the corresponding angle being shown. The arc is in **standard position** on the unit circle, with its initial point at (1, 0) and its terminal point at (u, v). The sine and cosine of x are defined in the same way as for the trigonometric functions.

$$\cos x = \frac{\text{horizontal coordinate}}{\text{radius}} = \frac{u}{1} = u$$

$$\sin x = \frac{\text{vertical coordinate}}{\text{radius}} = \frac{v}{1} = v$$

The name *circular function* comes from the fact that x equals the length of an arc on a unit circle. The other four circular functions are defined as ratios of sine and cosine.

DEFINITION: *Circular Functions*

If (u, v) is the terminal point of an arc of length x in standard position on a unit circle, then the **circular functions** of x are defined as follows:

$$\sin x = v \qquad \cos x = u$$

$$\tan x = \frac{\sin x}{\cos x} \qquad \cot x = \frac{\cos x}{\sin x}$$

$$\sec x = \frac{1}{\cos x} \qquad \csc x = \frac{1}{\sin x}$$

Circular functions are equivalent to trigonometric functions in radians. This equivalency provides an opportunity to expand the concept of trigonometric functions. You have seen trigonometric functions first defined using the angles of a right triangle and later expanded to include all angles. From now on, the concept of trigonometric functions includes circular functions and can have both degrees and radians as arguments. The way the two kinds of trigonometric functions are distinguished is by their arguments. If the argument is measured in degrees, Greek letters represent them (for example, $\sin \theta$). If the argument is measured in radians, they are represented by letters from the Roman alphabet (for example, $\sin x$).

► ***EXAMPLE 1*** Plot the graph of $y = 4 \cos 5x$ on your grapher, in radian mode. Find the period graphically and algebraically. Compare your results.

Solution Figure 3-5d shows the graph.

By tracing the graph, the first high point beyond $x = 0$ is between $x = 1.25$ and $x = 1.3$. So graphically the period is between 1.25 and 1.3.

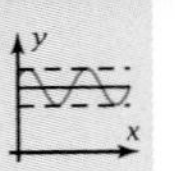

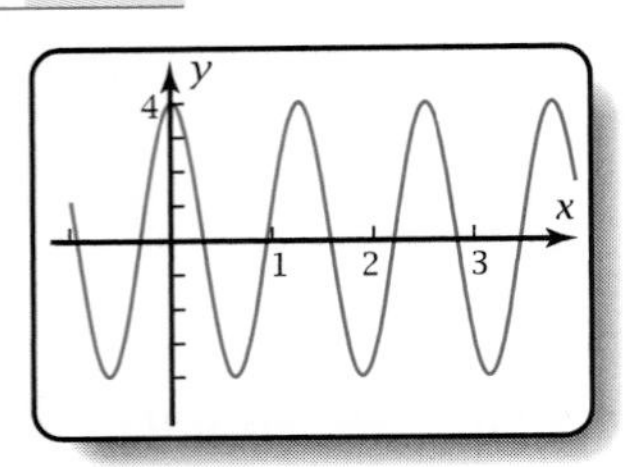

Figure 3-5d

To find the period algebraically, recall that the 5 in the argument of the cosine function is the reciprocal of the horizontal dilation. The period of the parent cosine function is 2π, since there are 2π radians in a complete revolution. Thus the period of the given function is

$$\frac{1}{5}(2\pi) = 0.4\pi = 1.2566\ldots$$

The answer found graphically is close to this exact answer. ◀

▶ EXAMPLE 2 Find a particular equation for the sinusoid function in Figure 3-5e. Notice x on the horizontal axis, not θ, indicating the angle is measured in radians. Confirm your answer by plotting the equation on your grapher.

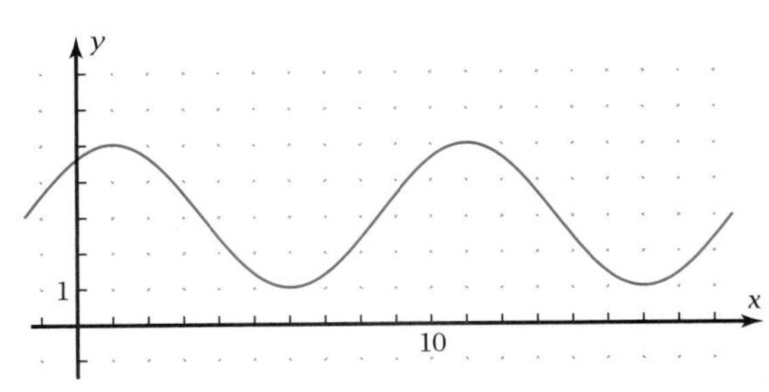

Figure 3-5e

Solution

$y = C + A \cos B(x - D)$ — Write the general sinusoidal equation, using x instead of θ.

- Sinusoidal axis is at $y = 3$. So $C = 3$. — Get A, B, C, and D using information from the graph.
- Amplitude $= 2$. So $A = 2$.
- Period $= 10$. — From one high point to the next is $11 - 1 = 10$.
- Dilation $= \frac{10}{2\pi} = \frac{5}{\pi}$, so $B = \frac{\pi}{5}$. — B is the reciprocal of the horizontal dilation.
- Phase displacement $= 1$ (for $y = \cos x$). So $D = 1$. — Cosine starts a cycle at a high point.

$y = 3 + 2 \cos \frac{\pi}{5}(x - 1)$ — Write the particular equation.

Plotting this equation in radian mode confirms that it is correct. ◀

▶ EXAMPLE 3 Sketch the graph of $y = \tan \frac{\pi}{6}x$.

Solution In order to graph the function, you need to identify its period, the locations of its inflection points, and its asymptotes.

$\text{Period} = \frac{6}{\pi} \cdot \pi = 6$ — Horizontal dilation is the reciprocal of $\frac{\pi}{6}$. The period of the tangent is π.

For this function, the points of inflection are also the x-intercepts, or where the value of the function equals zero. So

$$\frac{\pi}{6}x = 0, \pm\pi, \pm 2\pi, \ldots$$

$$x = 0, \pm 6, \pm 12, \ldots$$

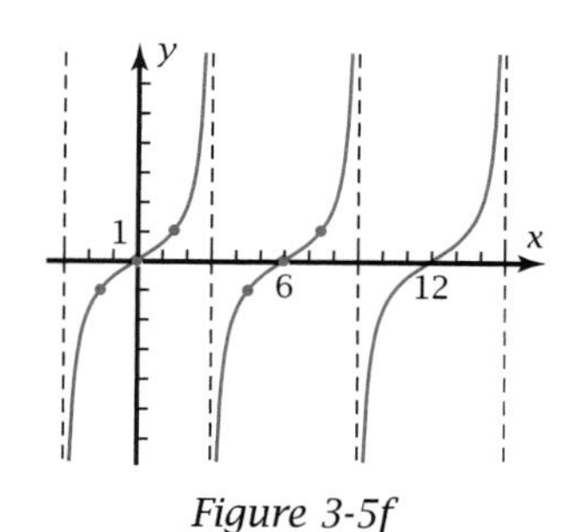

Figure 3-5f

Asymptotes are where the function is undefined. So

$$\frac{\pi}{6}x = -\frac{\pi}{2}, \frac{\pi}{2}, \frac{3\pi}{2}, \frac{5\pi}{2}, \ldots$$

$$x = -3, 3, 9, 15, \ldots$$

Recall that halfway between a point of inflection and an asymptote the tangent equals 1 or −1. The graph in Figure 3-5f shows these features. ◀

Note that in the graphs of circular functions the number π appears either in the equation as a coefficient of x or in the graph as a scale mark on the x-axis.

Problem Set 3-5

Do These Quickly (5 min)

Q1. How many radians in 180°?

Q2. How many degrees in 2π radians?

Q3. How many degrees in 1 radian?

Q4. How many radians in 34°?

Q5. Find sin 47°.

Q6. Find sin 47.

Q7. Find the period of $y = 3 + 4\cos 5(\theta - 6°)$.

Q8. Find the upper bound for y for the sinusoid in Q7.

Q9. How long does it take you to go 300 miles at an average speed of 60 mi/h?

Q10. Write 5% as a decimal.

For Problems 1–4, find the exact arc length of a unit circle subtended by the given angle (no decimals).

1. 30°
2. 60°
3. 90°
4. 45°

For Problems 5–8, find the exact degree measure of the angle that subtends the given arc length of a unit circle.

5. $\frac{\pi}{3}$ units
6. $\frac{\pi}{6}$ units
7. $\frac{\pi}{4}$ units
8. $\frac{\pi}{2}$ units

For Problems 9–12, find the exact arc length of a unit circle subtended by the given angle in radians.

9. $\frac{\pi}{2}$ radians
10. π radians
11. 2 radians
12. 1.467 radians

For Problems 13–16, evaluate the trigonometric function in decimal form.

13. tan 1
14. sin 2
15. sec 3
16. cot 4

For Problems 17–20, find the inverse circular function in decimal form.

17. $\cos^{-1} 0.3$
18. $\tan^{-1} 1.4$
19. $\csc^{-1} 5$
20. $\sec^{-1} 9$

For Problems 21–24, find the exact value of the circular function (no decimals).

21. $\sin \frac{\pi}{3}$
22. $\cos \frac{\pi}{4}$
23. $\tan \frac{\pi}{6}$
24. $\csc \pi$

For Problems 25–28, find the period, amplitude, phase displacement, and sinusoidal axis location. Use these features to sketch the graph. Confirm your graph by plotting the sinusoids on your grapher.

25. $y = 3 + 2\cos\frac{\pi}{5}(x - 4)$

26. $y = -4 + 5\sin\frac{2\pi}{3}(x + 1)$

27. $y = 2 + 6\sin\frac{\pi}{4}(x + 1)$

28. $y = 5 + 4\cos\frac{\pi}{3}(x - 2)$

For Problems 29–32, find the period and critical points or points of inflection, and sketch the graph.

29. $y = \cot\frac{\pi}{4}x$

30. $y = \tan 2\pi x$

31. $y = 2 + \sec x$

32. $y = 3\csc x$

For Problems 33–42, find a particular equation for the circular function graphed.

33.

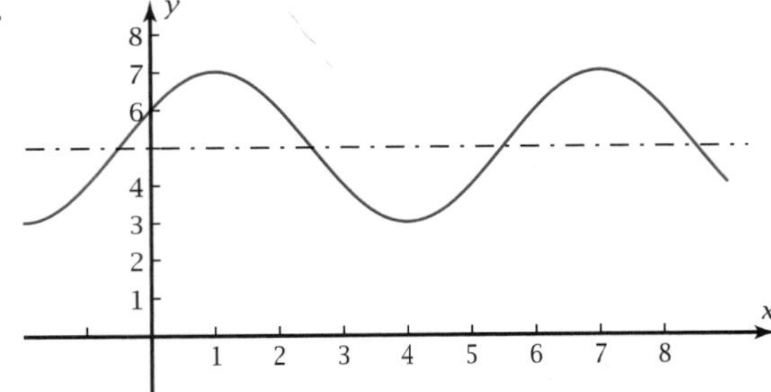

34.

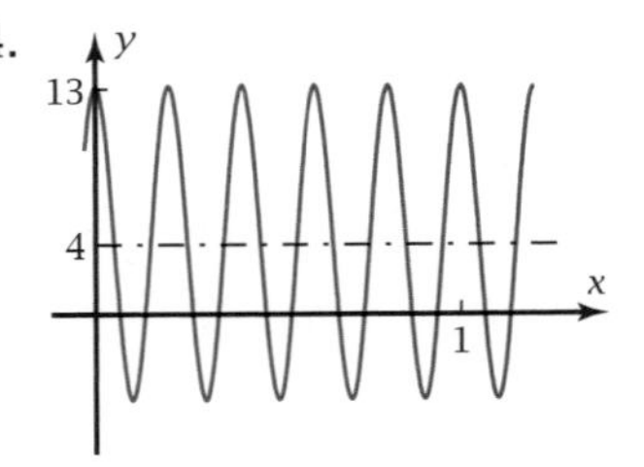

35.

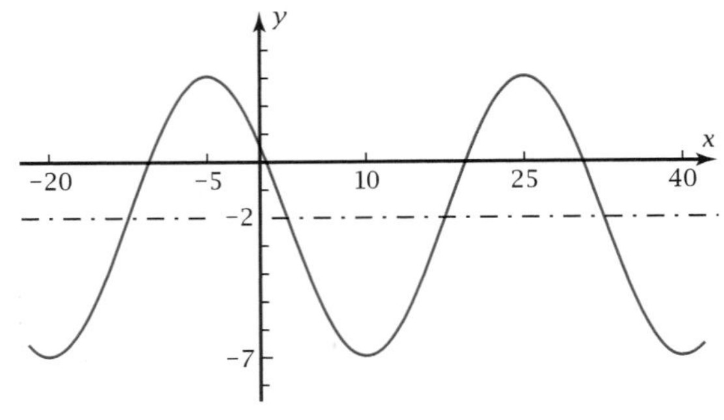

36.

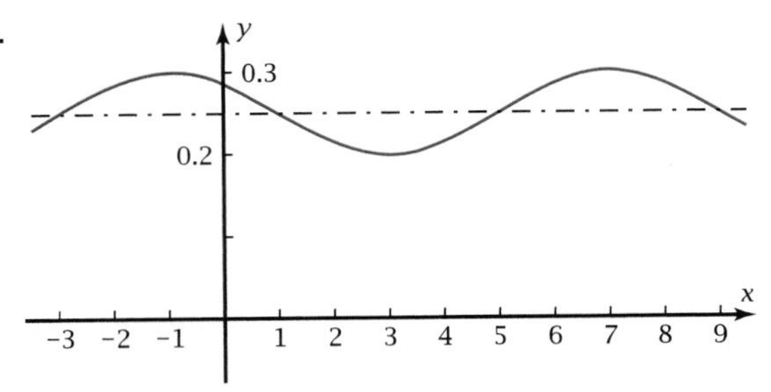

37.

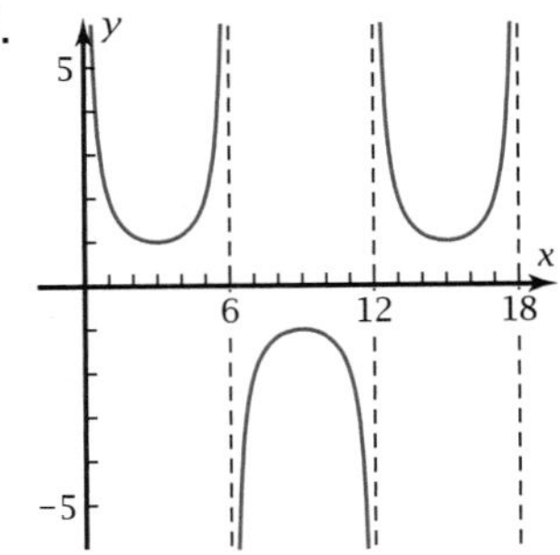

38.

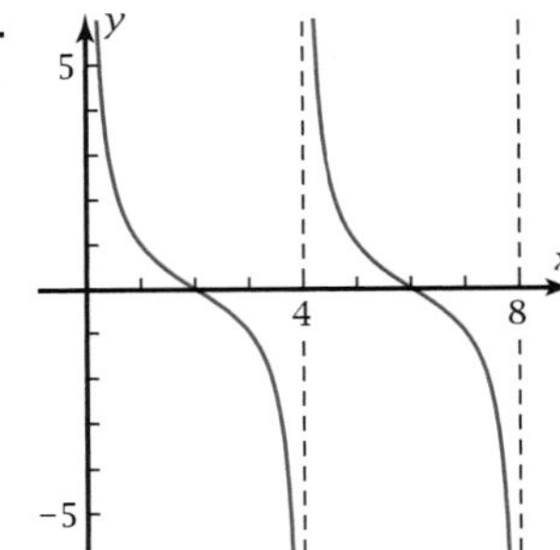

39.

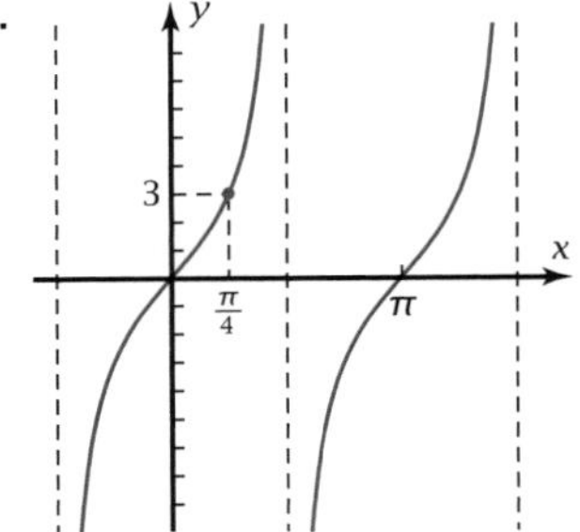

40.

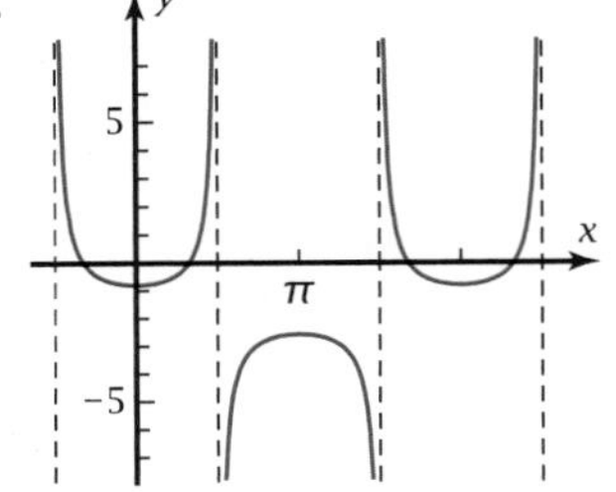

41.

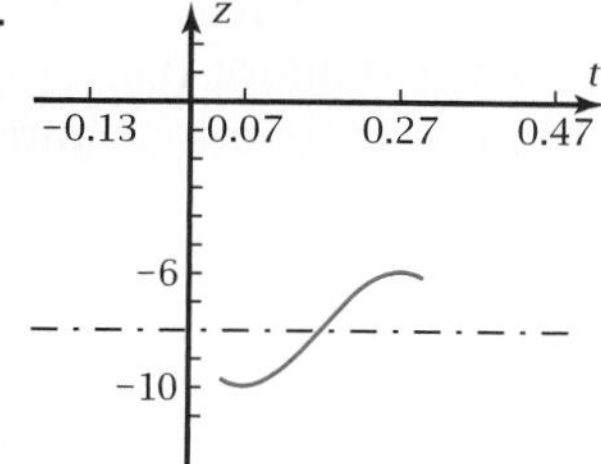

42.

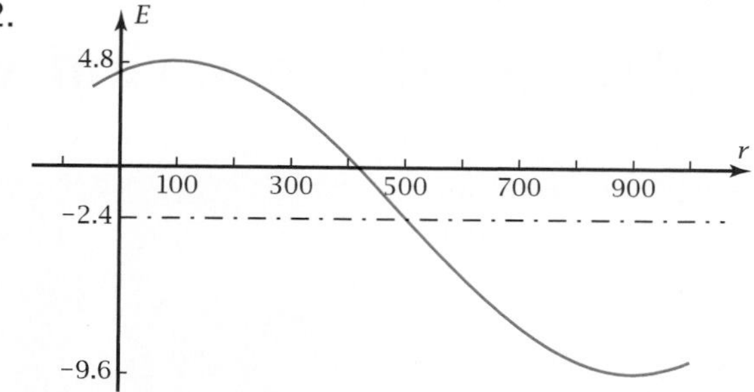

43. For the sinusoid in Problem 41, find the value of z at $t = 0.4$ on the graph. If the graph is extended to $t = 50$, is the point on the graph above or below the sinusoidal axis? How far above or below?

44. For the sinusoid in Problem 42, find the value of E at $r = 1234$ on the graph. If the graph is extended to $r = 10{,}000$, is the point on the graph above or below the sinusoidal axis? How far above or below?

45. *Sinusoid Translation Problem:* Figure 3-5g shows the graphs of $y = \cos x$ (dashed) and $y = \sin x$ (solid). Note that the graphs are congruent to each other (if superimposed, they coincide), differing only in horizontal translation.

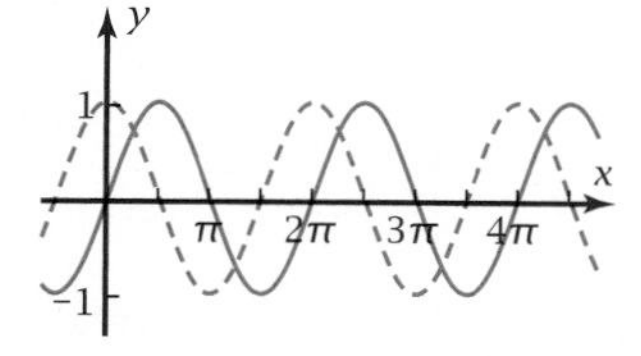

Figure 3-5g

a. What translation would make the cosine graph coincide with the sine graph? Complete the equation: $\sin x = \cos(\text{—?—})$.

b. Let $y = \cos(x - 2\pi)$. What effect would this translation have on the cosine graph?

c. Name a positive and a negative translation that would make the sine graph coincide with itself.

d. Explain why $\sin(x - 2\pi n) = \sin x$ for any integer n. How is the 2π related to the sine function?

46. *The Inequality $\sin x < x < \tan x$ Problem:* In this problem you will examine the $\sin x < x < \tan x$ inequality. Figure 3-5h shows angle AOB in standard position, with subtended arc AB of length x on a unit circle.

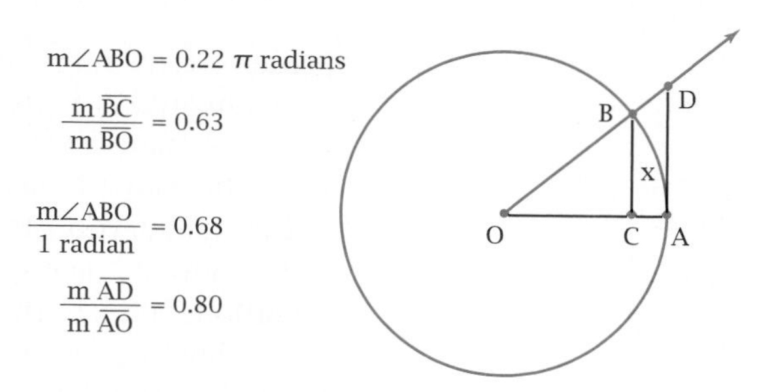

Figure 3-5h

a. Based on the definition of radians, explain why x is also the radian measure of angle AOB.

b. Based on the definitions of sine and tangent, explain why BC and AD equal $\sin x$ and $\tan x$, respectively.

c. From Figure 3-5h it appears that $\sin x < x < \tan x$. Make a table of values to show numerically that this inequality is true even for values of x very close to zero.

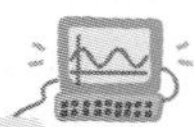

d. Use The Geometer's Sketchpad to construct Figure 3-5h. By moving point B around the circle, observe the relationship between the values of $\sin x$, x, and $\tan x$ whenever arc AB is in the first quadrant.

47. *Circular Function Comprehension Problem:* For circular functions such as $\cos x$, the independent variable, x, represents the length of an arc of a unit circle. For other functions you have studied, such as the quadratic function $y = ax^2 + bx + c$, the independent variable, x, stands for a distance along a horizontal number line, the x-axis.

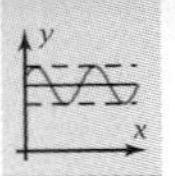

a. Explain how the concept of wrapping the x-axis around the unit circle links the two kinds of functions.

b. Explain how angle measures in radians link the circular functions to the trigonometric functions.

48. *Journal Problem:* Update your journal with things you have learned about the relationship of trigonometric functions and circular functions.

3-6 Inverse Circular Relations: Given y, Find x

A major reason for finding the particular equation of a sinusoid is to use it to evaluate y for a given x or to calculate x when you are given y. Functions are used this way to make predictions in the real world. For instance, you can express the time of sunrise as a function of the day of the year. With this equation, you can predict the time of sunrise on a given day by pressing the right keys on a calculator. Predicting the day(s) on which the Sun rises at a given time is more complicated. In this section you will learn graphical, numerical, and algebraic ways to find x for a given y-value.

Radar speed guns use inverse relations to calculate the speed of a car from time measurements.

OBJECTIVE Given the equation of a circular function or trigonometric function and a particular value of y, find specified values of x or θ:

- Graphically
- Numerically
- Algebraically

The Inverse Cosine Relation

The symbol $\cos^{-1} 0.3$ means the inverse cosine function evaluated at 0.3, a particular arc or angle whose cosine is 0.3. By calculator, in radian mode,

$$\cos^{-1} 0.3 = 1.2661\ldots$$

The **inverse cosine relation** includes all arcs or angles whose cosine is a given number. The term that you'll use in this text is **arccosine,** abbreviated **arccos.** So arccos 0.3 means any arc or angle whose cosine is 0.3, not just the function value. Figure 3-6a

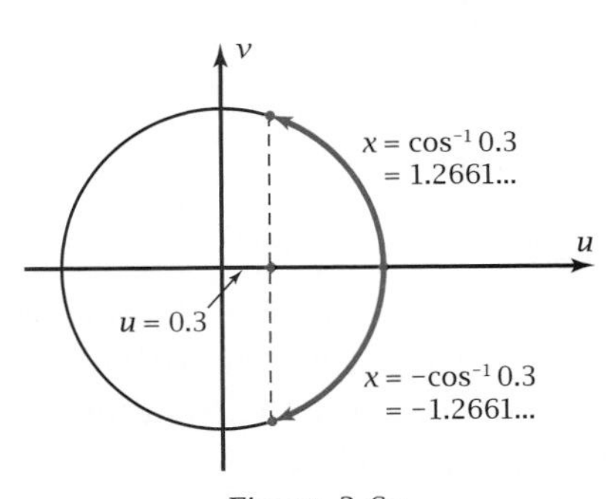

Figure 3-6a

shows that both 1.2661... and −1.2661... have cosines equal to 0.3. So −1.2661... is also a value of arccos 0.3.

The **general solution** for the arccosine of a number is written this way:

$$\arccos 0.3 = \pm\cos^{-1} 0.3 + 2\pi n \qquad \text{General solution for arccos 0.3}$$

where n stands for an integer. The ± sign tells you that both the value from the calculator and its opposite are values of arccos 0.3. The $2\pi n$ tells you that any arc that is an integer number of revolutions added to these values is also a value of arccos 0.3. If n is a negative integer, a number of revolutions is being subtracted from these values. Notice there are infinitely many such values.

The arcsine and arctangent relations will be defined in Section 4-4 in connection with solving more general equations.

DEFINITION: *Arccosine, the Inverse Cosine Relation*

$$\arccos x = \pm\cos^{-1} x + 2\pi n$$

Verbally: Inverse cosines come in opposite pairs with all their coterminals.

Note: The function value $\cos^{-1} x$ is called the **principal value** of the inverse cosine relation. This is the value the calculator is programmed to give you. In Section 4-6, you will learn why certain quadrants are picked for these inverse function values.

Finding *x* When You Know *y*

Figure 3-6b shows a sinusoid with a horizontal line drawn at $y = 5$. The horizontal line cuts the part of the sinusoid shown at six different points. Each point corresponds to a value of x for which $y = 5$. The next examples show how to find the values of x by three methods.

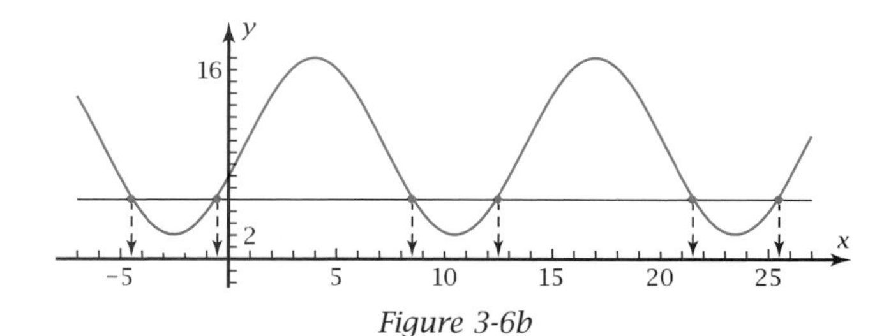

Figure 3-6b

▶ ***EXAMPLE 1*** Find **graphically** the six values of x for which $y = 5$ for the sinusoid in Figure 3-6b.

Solution On the graph, draw lines from the intersection points down to the x-axis (Figure 3-6b). The values are:

$$x \approx -4.5, -0.5, 8.5, 12.5, 21.5, 25.5$$

◀

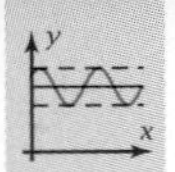

▶ EXAMPLE 2 Find **numerically** the six values of x in Example 1. Show that the answers agree with those found graphically in Example 1.

Solution

$y_1 = 9 + 7\cos\frac{2\pi}{13}(x - 4)$ — Write the particular equation using the techniques of Section 3-5.

$y_2 = 5$ — Plot a horizontal line at $y = 5$.

$x \approx 8.5084\ldots$ and $x \approx 12.4915\ldots$ — Use the intersect or solver feature on your grapher to find two adjacent x-values.

$x \approx 8.5084\ldots + 13(-1) = -4.4915\ldots$ — Add multiples of the period to find other x-values.

$x \approx 12.4915\ldots + 13(-1) = -0.5085\ldots$

$x \approx 8.5084\ldots + 13(1) = 21.5084\ldots$

$x \approx 12.4915\ldots + 13(1) = 25.4915\ldots$

These answers agree with the answers found graphically in Example 3. ◀

Note that the ≈ sign is used for answers found numerically because the solver or intersect feature on most calculators gives only approximate answers.

▶ EXAMPLE 3 Find **algebraically** (by calculation) the six values of x in Example 1. Show that the answers agree with those in Examples 1 and 2.

Solution

$9 + 7\cos\frac{2\pi}{13}(x - 4) = 5$ — Set the two functions equal to one another.

$\cos\frac{2\pi}{13}(x - 4) = -\frac{4}{7}$ — Simplify the equation by peeling away the constants.

$\frac{2\pi}{13}(x - 4) = \arccos\left(-\frac{4}{7}\right)$ — Take the arccosine of both sides.

$x = 4 + \frac{13}{2\pi}\arccos\left(-\frac{4}{7}\right)$ — Rearrange the equation to isolate x.

$x = 4 + \frac{13}{2\pi}\left(\pm\cos^{-1}\left(-\frac{4}{7}\right) + 2\pi n\right)$ — Substitute for arccosine.

$x = 4 \pm \frac{13}{2\pi}\cos^{-1}\left(-\frac{4}{7}\right) + 13n$ — Distribute the $\frac{13}{2\pi}$ to both terms.

$x = 4 \pm 4.5084\ldots + 13n$

$x = 8.5084\ldots + 13n$ or $-0.5084\ldots + 13n$

$x = -4.4915\ldots, -0.5084\ldots, 8.5084\ldots, 12.4915\ldots,$
$21.5084\ldots,$ or $25.4915\ldots$ — Let n be $0, \pm1, \pm2$.

These agree with the graphical and numerical solutions in Examples 1 and 2. ◀

Note that in the $13n$ term, the 13 is the period. The $13n$ in the general solution for x means that you need to add multiples of the period to the values of x you get for the inverse function. Note, too, that you can put the $8.5084... + 13n$ and $-0.5084... + 13n$ into the y= menu of your grapher and make a table of values. For most graphers you will have to use x in place of n.

Note also that the algebraic solution gets all the values at once rather than one at a time numerically.

Problem Set 3-6

Do These Quickly 5 min

Q1. What is the period of the circular function $y = \cos 4x$?

Q2. What is the period of the trigonometric function $y = \cos 4\theta$?

Q3. How many degrees in $\frac{\pi}{6}$ radians?

Q4. How many radians in 45°?

Q5. Sketch the graph of $y = \sin \theta$.

Q6. Sketch the graph of $y = \csc \theta$.

Q7. Find the smaller acute angle of a right triangle with legs of 3 miles and 7 miles.

Q8. $x^2 + y^2 = 9$ is the equation of a(n) —?—.

Q9. What is the general equation of an exponential function?

Q10. Functions that repeat themselves at regular intervals are called —?— functions.

For Problems 1–4, find the first five positive values of the inverse circular relation.

1. arccos 0.9
2. arcccos 0.4
3. arccos (−0.2)
4. arccos (−0.5)

For the circular sinusoids graphed in Problems 5–10,

a. Estimate graphically the x-values shown for the indicated y-value.

b. Find a particular equation of the sinusoid.

c. Find the x-values in part a numerically, using the equation.

d. Find the x-values in part a algebraically.

5. $y = 6$

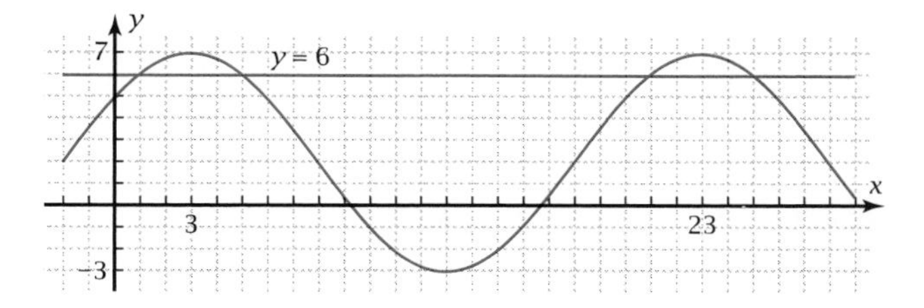

6. $y = 5$

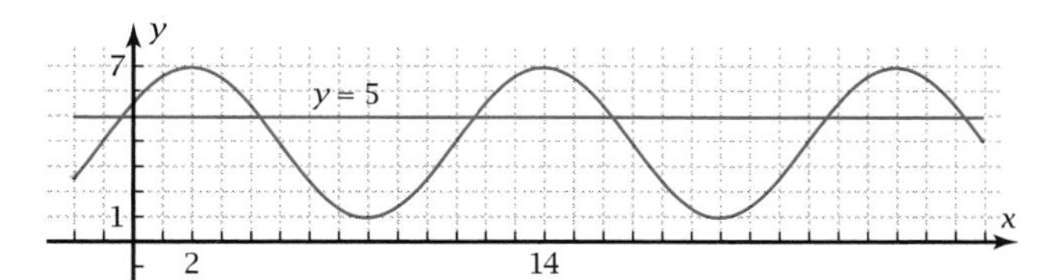

7. $y = -1$

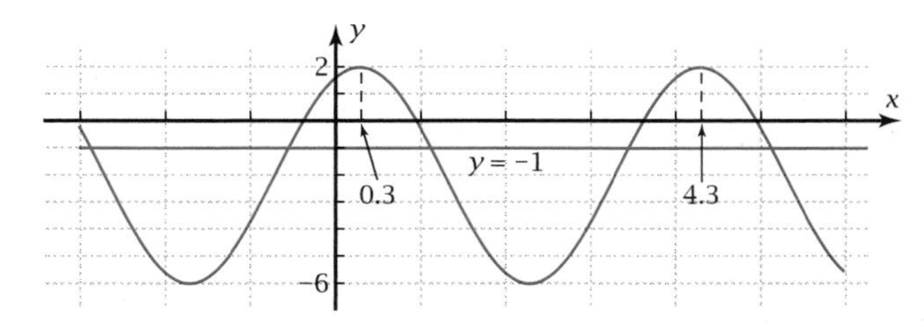

8. $y = -2$

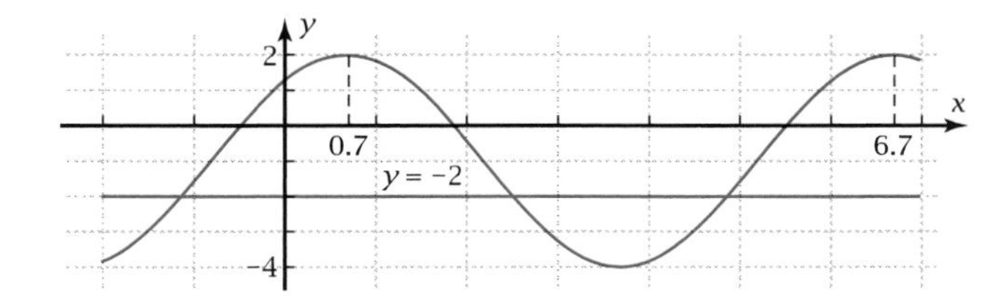

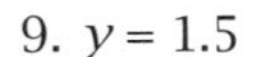

9. $y = 1.5$

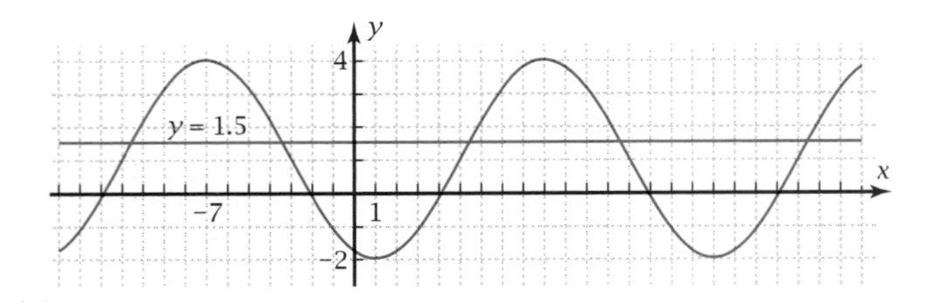

10. $y = -4$

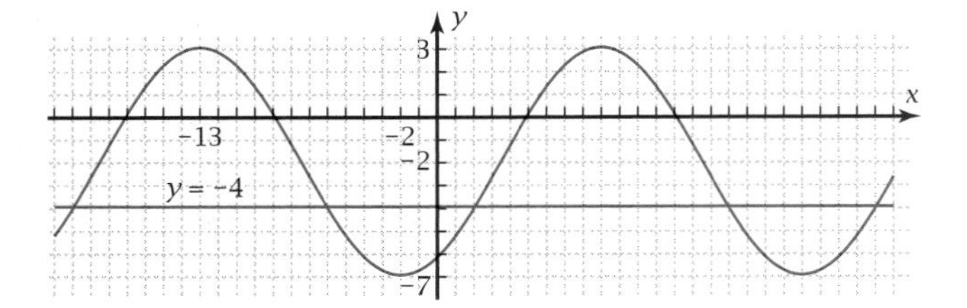

11. $y = 3$

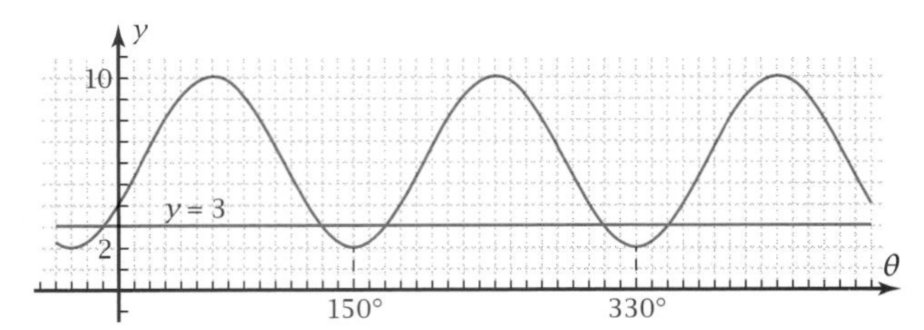

12. $y = 5$

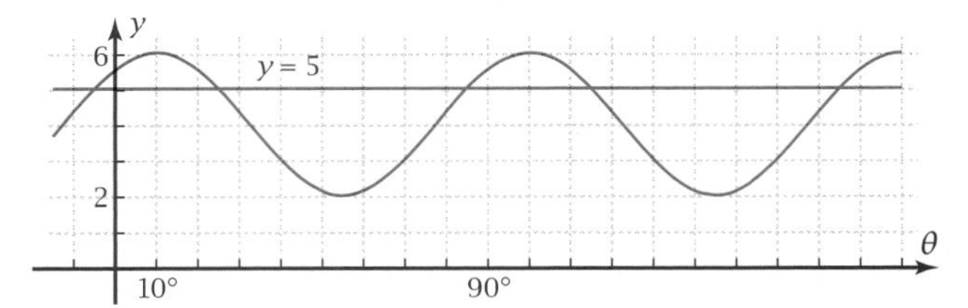

For the trigonometric sinusoids graphed in Problems 11 and 12,

a. Estimate graphically the first three positive values of θ for the indicated y-value.
b. Find a particular equation for the sinusoid.
c. Find the θ-values in part a numerically, using the equation.
d. Find the θ-values in part a algebraically.

13. Figure 3-6c shows the parent cosine function $y = \cos x$.
    a. Find algebraically the six values of x shown, for which $\cos x = -0.9$.
    b. Find algebraically the first value of x greater than 200 for which $\cos x = -0.9$.

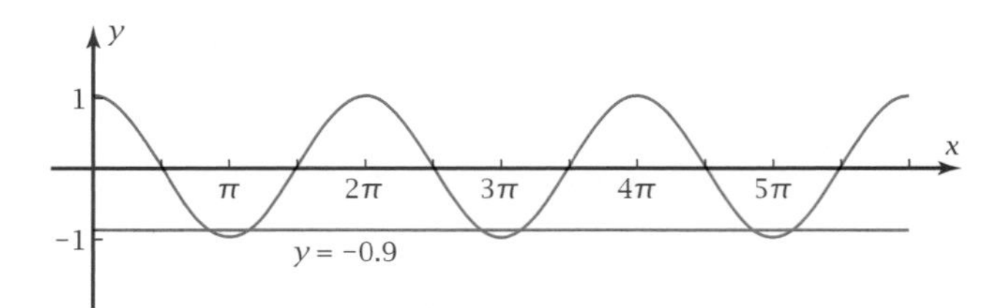

Figure 3-6c

3-7 Sinusoidal Functions as Mathematical Models

Chemotherapy medication destroys red blood cells as well as cancer cells. The red cell count goes down for a while and then comes back up again. If a treatment is taken every three weeks, then the red cell count resembles a periodic function of time (Figure 3-7a). If such a

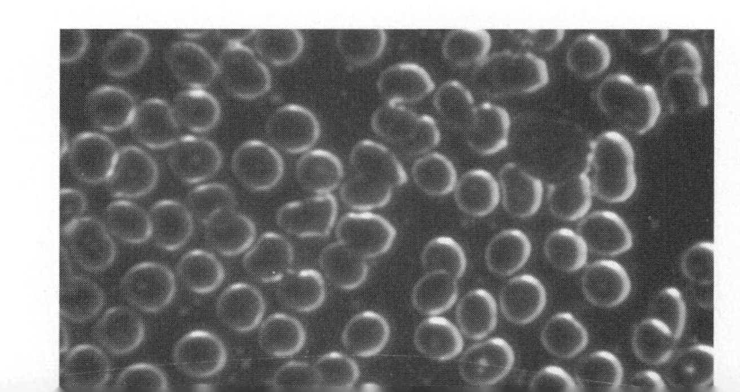

function is regular enough, you can use a sinusoidal function as a mathematical model.

In this section you'll start with a verbal description of a periodic phenomenon, interpret it graphically, get an algebraic equation from the graph, and use the equation to find numerical answers.

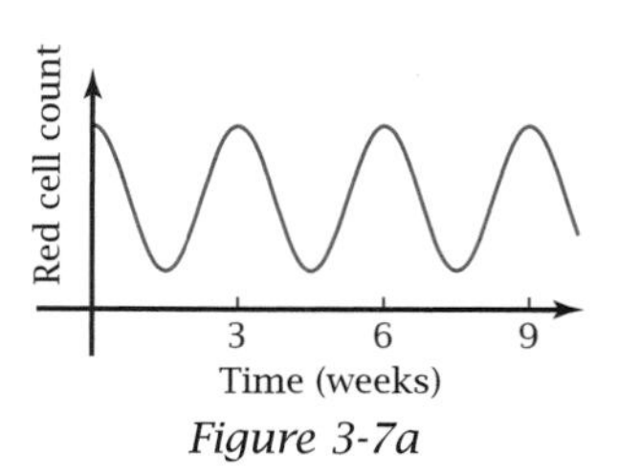

Figure 3-7a

OBJECTIVE Given a verbal description of a periodic phenomenon, write an equation using sine or cosine functions and use the equation as a mathematical model to make predictions and interpretations about the real world.

▶ EXAMPLE 1

Waterwheel Problem: Suppose that the waterwheel in Figure 3-7b rotates at 6 revolutions per minute (rpm). Two seconds after you start a stopwatch, point P on the rim of the wheel is at its greatest height, $d = 13$ ft, above the surface of the water. The center of the waterwheel is 6 ft above the surface.

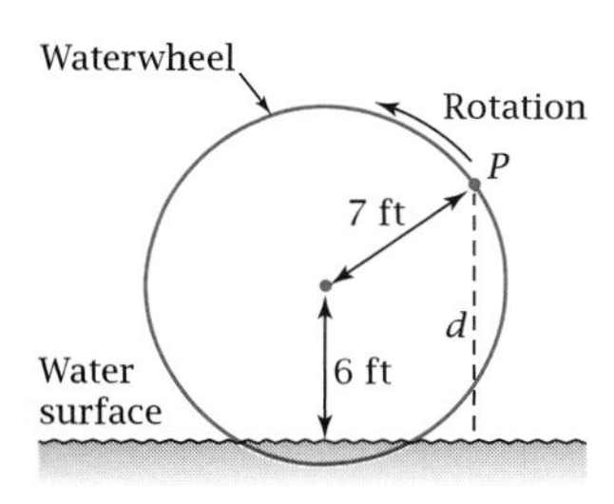

Figure 3-7b

a. Sketch the graph of d as a function of t, in seconds, since you started the stopwatch.

b. Assuming that d is a sinusoidal function of t, write a particular equation. Confirm by graphing that your equation gives the graph you sketched in part a.

c. How high above or below the water's surface will P be at time $t = 17.5$ sec? At that time, will it be going up or down?

d. At what time t was point P first emerging from the water?

Solution

a. From what's given, you can tell the location of the sinusoidal axis, the "high" and "low" points, and the period.

Sketch the sinusoidal axis at $d = 6$ as shown in Figure 3-7c.

Sketch the upper bound at $d = 6 + 7 = 13$ and the lower bound at $d = 6 - 7 = -1$.

Sketch a high point at $t = 2$.

Because the waterwheel rotates at 6 rpm, the period is $\frac{60}{6} = 10$ sec. Mark the next high point at $t = 2 + 10 = 12$.

Mark a low point halfway between the two high points, and mark the points of inflection on the sinusoidal axis halfway between each consecutive high and low.

Sketch the graph through the critical points and the points of inflection. The finished sketch is shown in Figure 3-7c.

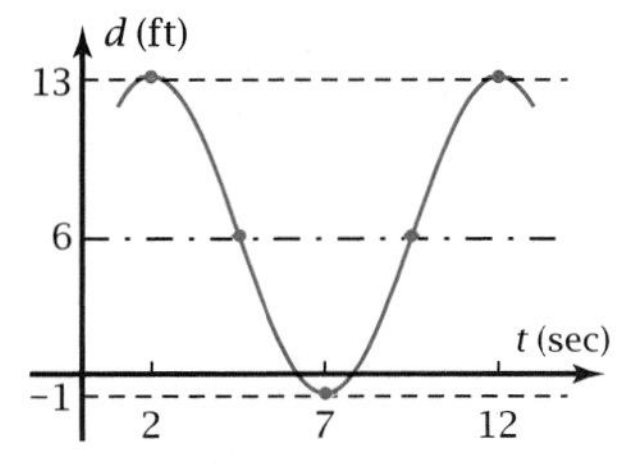

Figure 3-7c

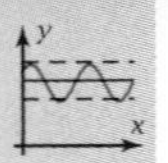

b. $d = C + A \cos B(t - D)$ — Write the general equation. Use d and t for the variables.

From the graph, $C = 6$ and $A = 7$.

$D = 2$ — Cosine starts a cycle at a high point.

Horizontal dilation $\dfrac{10}{2\pi} = \dfrac{5}{\pi}$ — Period of this sinusoid is 10. Period of the circular cosine function is 2π.

$B = \dfrac{\pi}{5}$ — B is the reciprocal of the horizontal dilation.

$\therefore d = 6 + 7 \cos \frac{\pi}{5}(t - 2)$ — Write the particular equation.

Plotting on your grapher confirms that the equation is correct (Figure 3-7d).

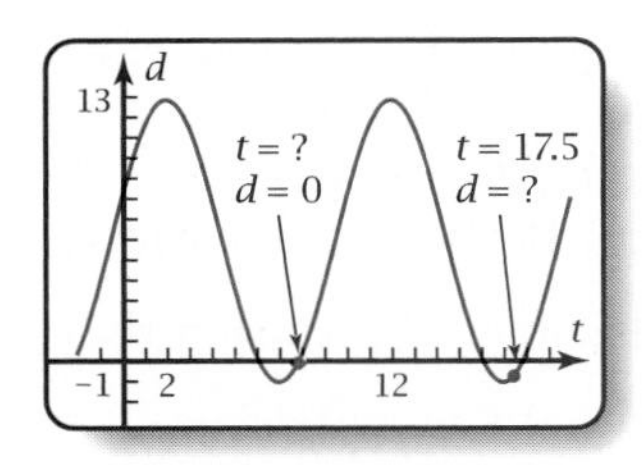

Figure 3-7d

c. Set the window on your grapher to include 17.5. Then trace or scroll to this point (Figure 3-7d). From the graph, $d = -0.6573... \approx -0.7$ ft and is going up.

d. Point P is either submerging into or emerging from the water when $d = 0$. At the first zero shown in Figure 3-7d, the point is going into the water. At the next zero, the point is emerging. Using the intersect, zeros, or solver feature of your grapher, you'll find the point is at

$$t = 7.8611... \approx 7.9 \text{ sec}$$

◀

Note that it is usually easier to use the cosine function for these problems, because its graph starts a cycle at a high point.

Problem Set 3-7

Do These Quickly

Problems Q1–Q8 concern the circular function $y = 4 + 5 \cos \frac{\pi}{6}(x - 7)$.

Q1. The amplitude is —?—.

Q2. The period is —?—.

Q3. The frequency is —?—.

Q4. The sinusoidal axis is at $y =$ —?—.

Q5. The phase displacement with respect to the parent cosine function is —?—.

Q6. The upper bound is at $y =$ —?—.

Q7. If $x = 9$, then $y =$ —?—.

Q8. The first three positive x-values at which low points occur are —?—, —?—, and —?—.

Q9. Two values of $x = \arccos 0.5$ are —?— and —?—.

Q10. If $y = 5 \cdot 3^x$, then adding 2 to the value of x multiplies the value of y by —?—.

---

1. *Steamboat Problem:* Mark Twain sat on the deck of a river steamboat. As the paddle wheel turned, a point on the paddle blade moved so that its distance, d, from the water's surface was a sinusoidal function of time. When Twain's stopwatch read 4 sec, the point was at its highest, 16 ft above the water's surface. The wheel's diameter was 18 ft, and it completed a revolution every 10 sec.
   a. Sketch the graph of the sinusoid.
   b. What is the lowest the point goes? Why is it reasonable for this value to be negative?

c. Find a particular equation for distance as a function of time.

d. How far above the surface was the point when Mark's stopwatch read 17 sec?

e. What is the first positive value of t at which the point was at the water's surface? At that time, was the point going into or coming out of the water? How can you tell?

f. "Mark Twain" is a pen name used by Samuel Clemens. What is the origin of that pen name? Give the source of your information.

2. *Fox Population Problem:* Naturalists find that populations of some kinds of predatory animals vary periodically with time. Assume that the population of foxes in a certain forest varies sinusoidally with time. Records started being kept at time $t = 0$ years. A minimum number of 200 foxes appeared when $t = 2.9$ years. The next maximum, 800 foxes, occurred at $t = 5.1$ years.

   a. Sketch the graph of this sinusoid.

   b. Find a particular equation expressing the number of foxes as a function of time.

   c. Predict the fox population when $t = 7, 8, 9,$ and 10 years.

   d. Foxes are declared an endangered species when their population drops below 300. Between what two nonnegative values of t were the foxes first endangered?

   e. Show on your graph in part a that your answer is correct.

3. *Bouncing Spring Problem:* A weight attached to the end of a long spring is bouncing up and down (Figure 3-7e). As it bounces, its distance from the floor varies sinusoidally with time. Start a stopwatch. When the stopwatch reads 0.3 sec, the weight first reaches a high point 60 cm above the floor. The next low point, 40 cm above the floor, occurs at 1.8 sec.

   a. Sketch the graph of this sinusoidal function.

   b. Find a particular equation for distance from the floor as a function of time.

   c. What is the distance from the floor when the stopwatch reads 17.2 sec?

   d. What was the distance from the floor when you started the stopwatch?

   e. What is the first positive value of time when the weight is 59 cm above the floor?

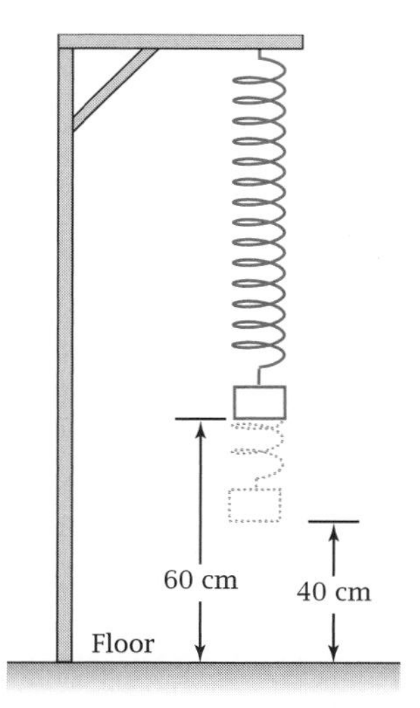

Figure 3-7e

4. *Rope Swing Problem:* Zoey is at summer camp. One day she is swinging on a rope tied to a tree branch, going back and forth alternately over land and water. Nathan starts a stopwatch. When $x = 2$ seconds, Zoey is at one end of her swing, $y = -23$ feet from the river bank (see Figure 3-7f). When $x = 5$ seconds, she is then at the other end of her swing, $y = 17$ feet from the

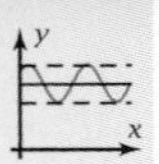

river bank. Assume that while she is swinging, y varies sinusoidally with x.

a. Sketch the graph of y versus x and write the particular equation.

b. Find y when $x = 13.2$ sec. Was Zoey over land or over water at this time?

c. Find the first positive time when Zoey was directly over the river bank ($y = 0$).

d. Zoey lets go of the rope and splashes into the water. What is the value of y for the end of the rope when it comes to rest? What part of the mathematical model tells you this?

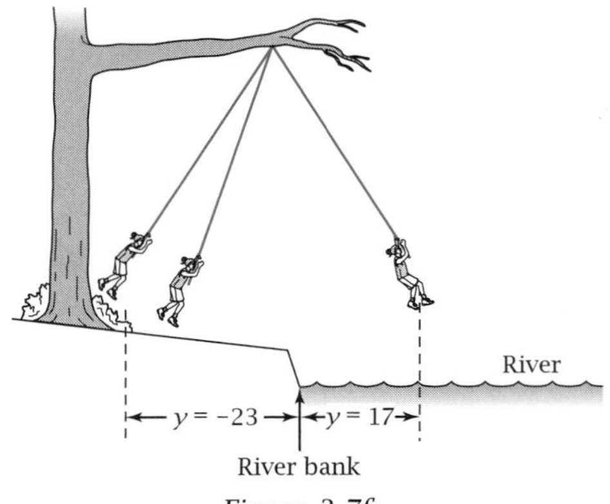

Figure 3-7f

5. *Roller Coaster Problem 1:* A theme park is building a portion of a roller coaster track in the shape of a sinusoid (Figure 3-7g). You have been hired to calculate the lengths of the horizontal and vertical support beams to use.

   a. The high and low points of the track are separated by 50 m horizontally and 30 m vertically. The low point is 3 m below the ground. Let y be the distance (in meters) a point on the track is above the ground. Let x be the horizontal distance (in meters) a point on the track is from the high point. Find a particular equation for y as a function of x.

   b. The vertical support beams are spaced 2 m apart, starting at the high point and ending just before the track goes below ground. Make a table of values of the lengths of the beams.

   c. The horizontal beams are spaced 2 m apart, starting at ground level and ending just below the high point. Make a table of values of horizontal beam lengths.

   d. The builder must know how many vertical and horizontal beams to order. In the most time-efficient way, find the total length of the vertical beams and the total length of the horizontal beams.

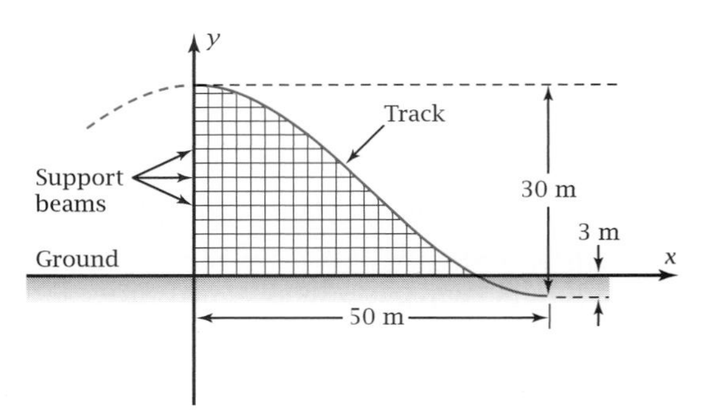

Figure 3-7g

6. *Buried Treasure Problem 2:* Suppose you seek a treasure that is buried in the side of a mountain. The mountain range has a sinusoidal vertical cross section (Figure 3-7h). The valley to the left is filled with water to a depth of 50 m, and the top of the mountain is 150 m above the water level. You set up an x-axis at water level and a y-axis 200 m to the right of the deepest part of the water. The top of the mountain is at $x = 400$ m.

   a. Find a particular equation expressing y for points on the *surface* of the mountain as a function of x.

   b. Show algebraically that the sinusoid in part a contains the origin (0, 0).

c. The treasure is located beneath the surface at the point (130, 40), as shown in Figure 3-7h. Which would be a shorter way to dig to the treasure, a horizontal tunnel or a vertical tunnel? Show your work.

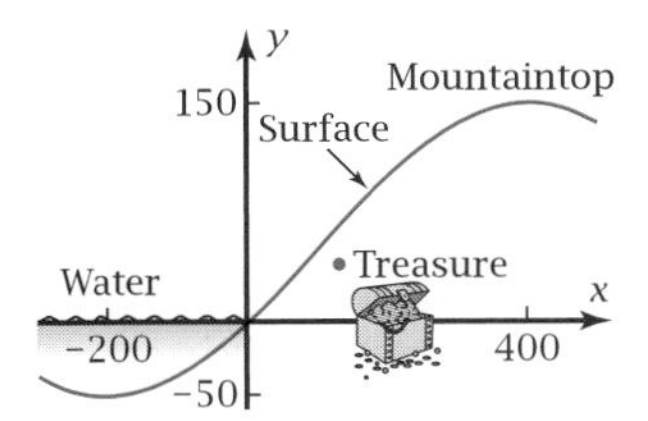

Figure 3-7h

7. *Sunspot Problem:* For several hundred years, astronomers have kept track of the number of solar flares, or "sunspots," that occur on the surface of the Sun. The number of sunspots in a given year varies periodically, from a minimum of about 10 per year to a maximum of about 110 per year. Between 1750 and 1948 there were exactly 18 complete cycles.

a. What is the period of a sunspot cycle?

b. Assume that the number of sunspots per year is a sinusoidal function of time and that a maximum occurred in 1948. Find a particular equation for the number of sunspots per year as a function of the year.

c. How many sunspots will there be in the year 2020? This year?

d. What is the first year after 2020 in which there will be about 35 sunspots? What is the first year after 2020 in which there will be a maximum number of sunspots?

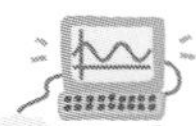

e. Find out how closely the sunspot cycle resembles a sinusoid by looking on the Internet or in another reference.

8. *Tide Problem 2:* Suppose that you are on the beach at Port Aransas, Texas, on August 2. At 2:00 p.m., at high tide, you find that the depth of the water at the end of a pier is 1.5 m. At 7:30 p.m., at low tide, the depth of the water is 1.1 m deep. Assume that the depth varies sinusoidally with time.

a. Find a particular equation for depth as a function of time that has elapsed since 12:00 midnight at the beginning of August 2.

b. Use your mathematical model to predict the depth of the water at 5:00 p.m. on August 3.

c. At what time does the first low tide occur on August 3?

d. What is the earliest time on August 3 that the water will be 1.27 m deep?

e. A high tide occurs because the Moon is pulling the water away from Earth slightly, making the water a bit deeper at a given point. How do you explain the fact that there are *two* high tides each day at most places on Earth? Provide the source for your information.

9. *Shock-Felt-Round-the-World Problem:* Suppose that one day all 200+ million people in the United States climb up on tables. At time $t = 0$ they all jump off. The resulting shock wave starts the earth vibrating at its fundamental period of 54 minutes. The surface first moves

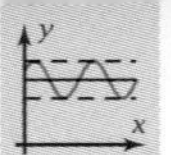

down from its normal position and then moves up an equal distance above its normal position (Figure 3-7i). Assume that the amplitude is 50 m.

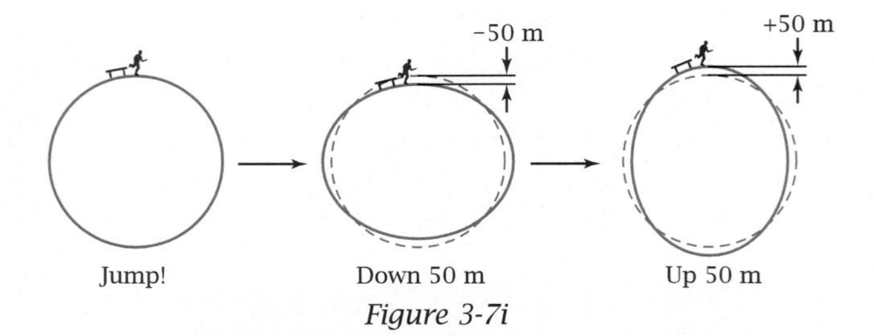

Figure 3-7i

a. Sketch the graph of displacement of the surface from its normal position as a function of time elapsed since the people jumped.

b. At what time will the surface be its farthest above normal position?

c. Find a particular equation expressing displacement above normal position as a function of time since the jump.

d. What is the displacement when $t = 21$?

e. What are the first three positive times at which the displacement is −37 m?

10. *Island Problem:* Ona Nyland owns an island several hundred feet from the shore of a lake. Figure 3-7j shows a vertical cross section through the shore, lake, and island. The island was formed millions of years ago by stresses that caused the earth's surface to warp into the sinusoidal pattern shown. The highest point on the shore is at $x = -150$ feet. From measurements on and near the shore (solid part of the graph), topographers find that an equation of the sinusoid is

$$y = -70 + 100 \cos \frac{\pi}{600}(x + 150)$$

where x and y are in feet. Ona consults you to make predictions about the rest of the graph (dotted).

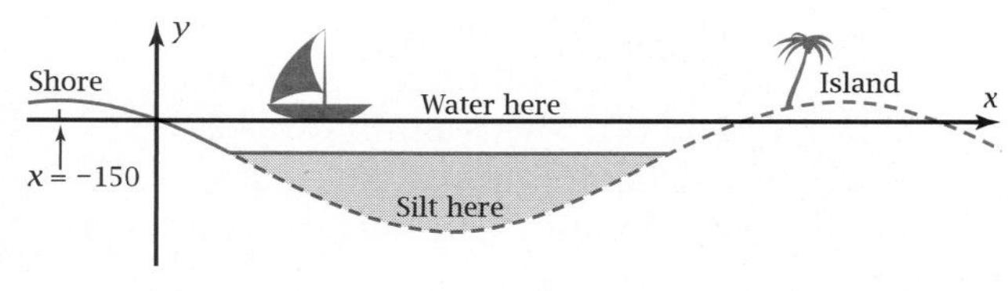

Figure 3-7j

a. What is the highest the island goes above the water level in the lake? How far from the y-axis is this high point? Show how you got your answers.

b. What is the deepest the sinusoid goes below the water level in the lake? How far from the y-axis is this low point? Show how you got your answers.

c. Over the centuries silt has filled the bottom of the lake so that the water is only 40 feet deep. That is, the silt line is at $y = -40$ feet. Plot the graph. Use a friendly window for x and a window with a suitable range for y. Then find graphically the range of x-values between which Ona would expect to find silt if she goes scuba diving in the lake.

d. If Ona drills an offshore well at $x = 700$ feet, through how much silt would she drill before she reaches the sinusoid? Describe how you got your answer.

e. The sinusoid appears to go through the origin. Does it actually do this, or does it just miss? Justify your answer.

f. Find algebraically the range of x-values between which the island is at or above the water level. How wide is the island, from the water on one side to the water on the other?

11. *Pebble-in-the-Tire Problem:* As you stop your car at a traffic light, a pebble becomes wedged between the tire treads. When you start moving again, the distance between the pebble and the pavement varies sinusoidally with the distance you have gone. The period is the circumference of the wheel. Assume that the diameter of the wheel is 24 in.

a. Sketch the graph of this sinusoidal function.

b. Find a particular equation for the function. (It is possible to get an equation with *zero* phase displacement.)

c. What is the pebble's distance from the pavement when you have gone 15 in.?

d. What are the first two distances you have gone when the pebble is 11 in. from the pavement?

12. *Oil Well Problem:* Figure 3-7k shows a vertical cross section through a piece of land. The y-axis is drawn coming out of the ground at

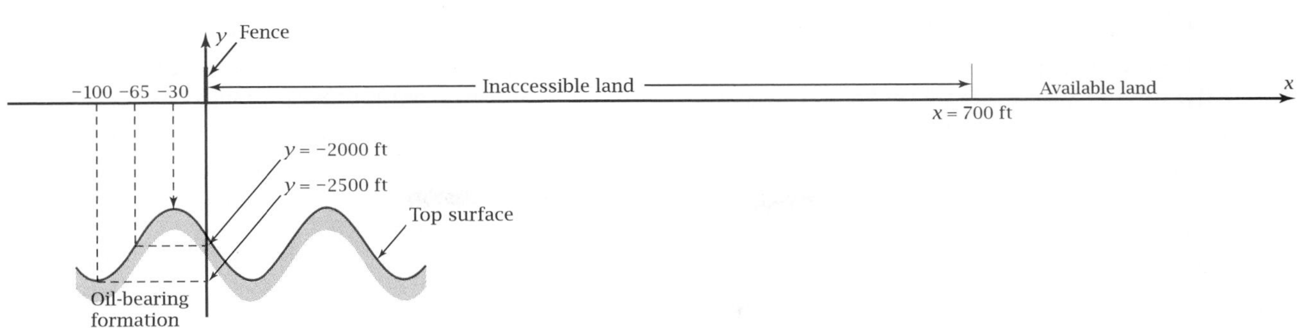

Figure 3-7k

the fence bordering land owned by your boss, Earl Wells. Earl owns the land to the left of the fence and is interested in getting land on the other side to drill a new oil well. Geologists have found an oil-bearing formation below Earl's land that they believe to be sinusoidal in shape. At $x = -100$ feet, the top surface of the formation is at its deepest, $y = -2500$ feet. A quarter-cycle closer to the fence, at $x = -65$ feet, the top surface is only 2000 feet deep. The first 700 feet of land beyond the fence is inaccessible. Earl wants to drill at the first convenient site beyond $x = 700$ feet.

a. Find a particular equation for y as a function of x.

b. Plot the graph on your grapher. Use a friendly window for x of about [−100, 900]. Describe how the graph confirms that your equation is correct.

c. Find graphically the first interval of x-values in the available land for which the top surface of the formation is no more than 1600 feet deep.

d. Find algebraically the values of x at the ends of the interval in part c. Show your work.

e. Suppose that the original measurements had been slightly inaccurate and that the value of y shown at −65 feet was at $x = -64$ instead. Would this fact make much difference in the answer to part c? Use the most time-efficient method to reach your answer. Explain what you did.

13. *Sound Wave Problem:* The hum you hear on some radios when they are not tuned to a station is a sound wave of 60 cycles per second.

Bats use ultrasonic sounds (20–100 kHz) for their communication and navigation, sounds that are undetectable by the human ear.

a. Is the 60 cycles per second the period, or is it the frequency? If it is the period, find the frequency. If it is the frequency, find the period.

b. The *wavelength* of a sound wave is defined as the distance the wave travels in a time equal to one period. If sound travels at 1100 ft/sec, find the wavelength of the 60-cycle-per-second hum.

c. The lowest musical note the human ear can hear is about 16 cycles per second. In order to play such a note, the pipe on an organ must be exactly half as long as the wavelength. What length of organ pipe would be needed to generate a 16-cycle-per-second note?

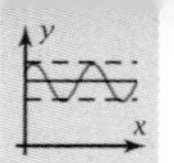

14. *Sunrise Project:* Assume that the time of sunrise varies sinusoidally with the day of the year. Let t be the time of sunrise. Let d be the day of the year, starting with $d = 1$ on January 1.

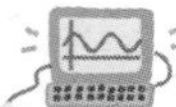

a. On the Internet or from an almanac, find for your location the time of sunrise on the longest day of the year, June 21, and on the shortest day of the year, December 21. If you choose, you may use the data for San Antonio, 5:34 a.m. and 7:24 a.m., CST, respectively. The phase displacement for cosine will be the value of d at which the Sun rises the latest. Use the information to find a particular equation for time of sunrise as a function of the day number.

b. Calculate the time of sunrise today at the location used for the equation in part a. Compare the answer to your data source.

c. What is the time of sunrise on your birthday, taking daylight saving time into account if necessary?

d. What is the first day of the year on which the Sun rises at 6:07 a.m. in the locality in part a?

e. In the northern hemisphere, Earth moves faster in wintertime, when it is closer to the Sun, and slower in summertime, when it is farther from the Sun. As a result, the actual high point of the sinusoid occurs later than predicted, and the actual low point occurs earlier than predicted (Figure 3-7l). A representation of the actual graph can be plotted by putting in a phase displacement that *varies.* See if you can duplicate the graph in Figure 3-7l on your grapher. Is the modified graph a better fit for the actual sunrise data for the locality in part a?

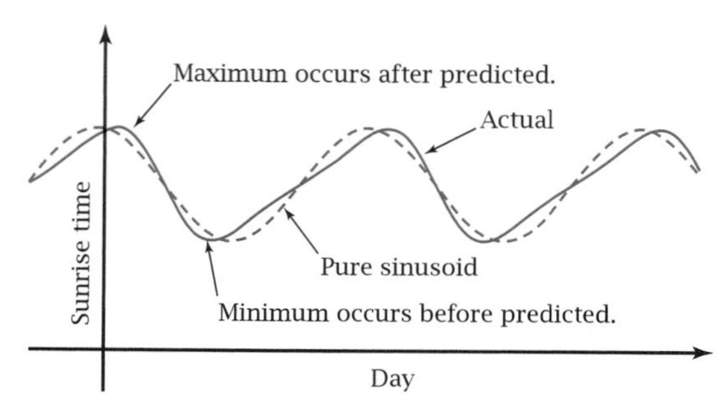

Figure 3-7l

15. *Variable Amplitude Pendulum Project:* If there were no friction, the displacement of a pendulum from its rest position would be a sinusoidal function of time,

$$y = A \cos Bt$$

To account for friction, assume that the amplitude A decreases exponentially with time,

$$A = a \cdot b^t$$

Make a pendulum by tying a weight to a string hung from the ceiling or some other convenient place (see Figure 3-7m).

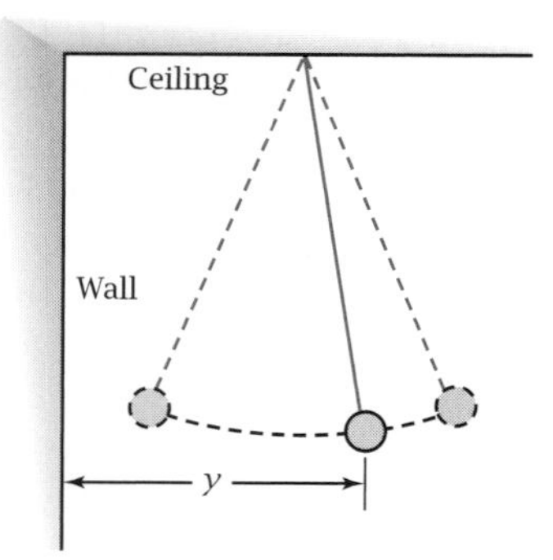

Figure 3-7m

Find its period by measuring the time for 10 swings and dividing by 10. Record the amplitude when you first start the pendulum and measure it again after 30 seconds. From these measurements, find the constants a, b, and B and write a particular equation for the position of the pendulum as a function of time. Test your equation by using it to predict the displacement of the pendulum at time $t = 10$ seconds and seeing if the pendulum really is where you predicted it to be at that time. Write an entry in your journal describing this experiment and its results.

t	d
0	10
1	42.2
2	14.9
3	41.2
4	23.3
5	32.2
6	32.1
7	24.3

3-8 Chapter Review and Test

In this chapter you learned how to graph trigonometric functions. The sine and cosine functions are continuous sinusoids, while other trigonometric functions are discontinuous, having vertical asymptotes at regular intervals. You also learned about circular functions, which you can use to model real-world phenomena mathematically, and you learned the way radians provide a link between these circular functions and trigonometric funtions.

Review Problems

R0. Update your journal with what you have learned since the last entry. Include such things as

- The one most important thing you have learned as a result of studying this chapter
- The graphs of the six trigonometric functions
- How the transformations of sinusoidal graphs relate to function transformations in Chapter 1
- How the circular and trigonometric functions are related
- Why circular functions are usually more appropriate as mathematical models than are trigonometric functions.

R1. a. Sketch the graph of a sinusoid. On the graph, show the difference in meaning between a cycle and a period. Show the amplitude, the phase displacement, and the sinusoidal axis.

b. In $y = 3 + 4 \cos 5(\theta - 10°)$, what name is given to the quantity $5(\theta - 10°)$?

R2. a. Without using your grapher, show that you understand the effects of the constants in a sinusoidal equation by sketching the graph of $y = 3 + 4 \cos 5(\theta - 10°)$. Give the amplitude, period, sinusoidal axis location, and phase displacement.

b. Using the cosine function, find a particular equation for the sinusoid in Figure 3-8a. Find another particular equation using the sine function. Show that the equations are equivalent to each other by plotting them on the same screen. What do you observe about the two graphs?

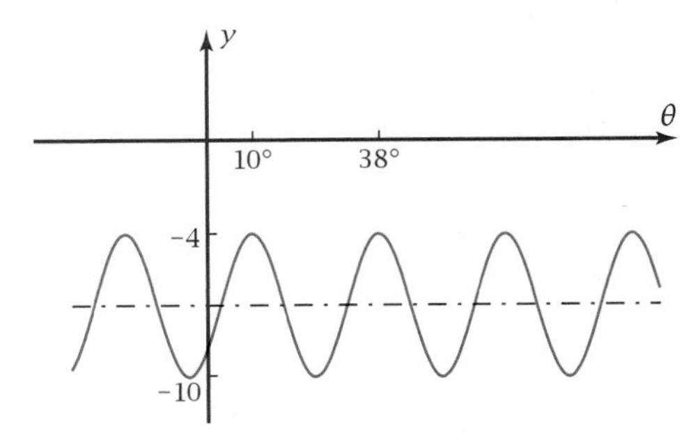

Figure 3-8a

c. A quarter-cycle of a sinusoid is shown in Figure 3-8b. Find a particular equation for it.

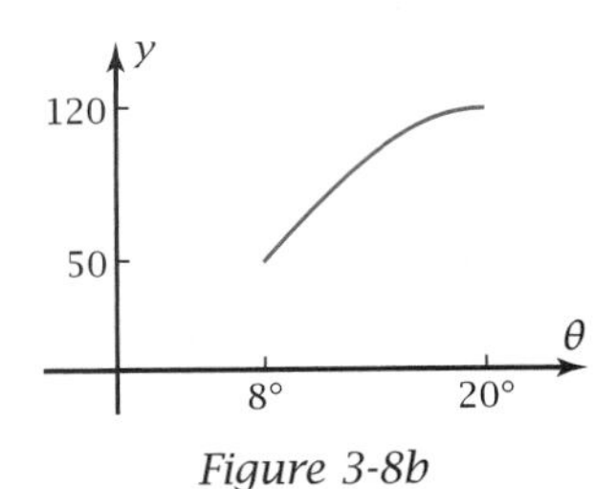

Figure 3-8b

d. At what value of θ shown in Figure 3-8b does the graph have a point of inflection? At what point does the graph have a critical point?

e. Find the frequency of the sinusoid shown in Figure 3-8b.

R3. a. Sketch the graph of $y = \tan \theta$.

b. Explain why the period of tangent is 180° rather than 360° like sine and cosine.

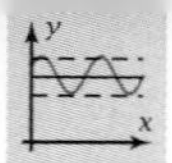

c. Plot the graph of $y = \sec \theta$ on your grapher. Explain how you did this.

d. Use the relationship between sine and cosecant to explain why the cosecant function has vertical asymptotes at $\theta = 0°$, $180°$, $360°$, ...

e. Explain why the graph of cosecant has high and low points but no points of inflection. Explain why the graph of cotangent has points of inflection but no high or low points.

f. For $y = 2 + 0.4 \cot \frac{1}{3}(\theta - 40°)$, give the vertical and horizontal dilations and the vertical and horizontal translations. Then plot the graph to confirm that your answers are right. What is the period of this function? Why is it not meaningful to talk about its amplitude?

R4. a. How many radians in 30°? in 45°? in 60°? Write the answers exactly, in terms of π.

b. How many degrees in an angle of 2 radians? Write the answer as a decimal.

c. Find cos 3 and cos 3°.

d. Find the radian measure of $\cos^{-1} 0.8$ and $\csc^{-1} 2$.

e. How long is the arc of a circle subtended by a central angle of 1 radian if the radius of the circle is 17 units?

R5. a. Draw a unit circle in a uv-coordinate system. In this coordinate system, draw an x-axis vertically with its origin at the point $(u, v) = (1, 0)$. Show where the points $x = 1$, 2, and 3 units map onto the unit circle as the x-axis is wrapped around it.

b. How long is the arc of a unit circle subtended by a central angle of 60°? Of 2.3 radians?

c. Find sin 2° and sin 2.

d. Find the value of the inverse trigonometric function $\cos^{-1} 0.6$.

e. Find the exact values (no decimals) for the circular functions $\cos \frac{\pi}{6}$, $\sec \frac{\pi}{4}$, and $\tan \frac{\pi}{2}$.

f. Sketch the graphs of the parent circular functions $y = \cos x$ and $y = \sin x$.

g. Explain how to find the period of the circular function $y = 3 + 4 \sin \frac{\pi}{10}(x - 2)$ from the constants in the equation. Sketch the graph. Confirm by your grapher that your sketch is correct.

h. Find a particular equation for the circular function sinusoid for which half a cycle is shown in Figure 3-8c.

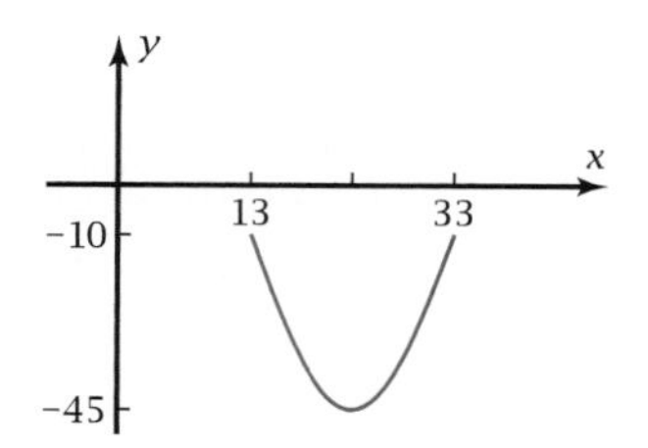

Figure 3-8c

R6. a. Find the general solution for the inverse circular relation arccos 0.8.

b. Find the first three positive values of the inverse circular relation arccos 0.8.

c. Find the least value of arccos 0.1 that is greater than 100.

d. For the sinusoid in Figure 3-8d, find the four values of x shown for which $y = 2$
- Graphically, to one decimal place
- Numerically, by finding the particular equation and plotting the graph
- Algebraically, using the particular equation

e. What is the next positive value of x for which $y = 2$, beyond the last positive value shown in Figure 3-8d?

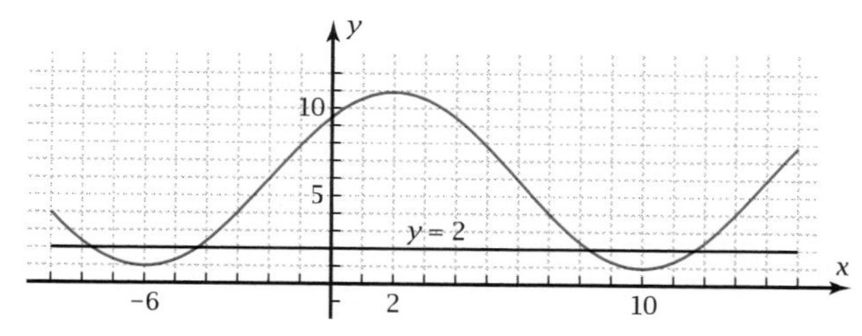

Figure 3-8d

R7. *Porpoising Problem:* Assume that you are aboard a research submarine doing submerged training exercises in the Pacific Ocean. At time $t = 0$ you start porpoising (alternately deeper and then shallower). At time $t = 4$ min you are at your deepest, $y = -1000$ m. At time $t = 9$ min you next reach your shallowest, $y = -200$ m. Assume that y varies sinusoidally with time.

a. Sketch the graph of y versus t.

b. Find an equation expressing y as a function of t.

c. Your submarine can't communicate with ships on the surface when it is deeper than $y = -300$ m. At time $t = 0$, could your submarine communicate? How did you arrive at your answer?

d. Between what two nonnegative times is your submarine first unable to communicate?

Concept Problems

C1. *Pump Jack Problem:* An oil well pump jack is shown in Figure 3-8e. As the motor turns, the walking beam rocks back and forth, pulling the rod out of the well and letting it go back into the well. The connection between the rod and the walking beam is a steel cable that wraps around the cathead. As time goes on, the distance d from the ground to point P, where the cable connects to the rod, varies periodically.

a. As the walking beam rocks, its angle θ with the ground varies sinusoidally with time. The angle goes from a minimum of -0.2 radians to a maximum of 0.2 radians. How many degrees corresponds to this range of angles?

b. The radius of the circular arc on the cathead is 8 ft. What arc length on the cathead corresponds to the range of angles in part a?

c. The distance, d, between the cable-to-rod connector and the ground varies sinusoidally with time. What is the amplitude of the sinusoid?

d. Suppose that the pump is started at time $t = 0$ sec. One second later, P is at its highest point above the ground. P is at its next low point 2.5 sec after that. When the walking beam is horizontal, point P is 7 ft above the ground. Sketch the graph of this sinusoid.

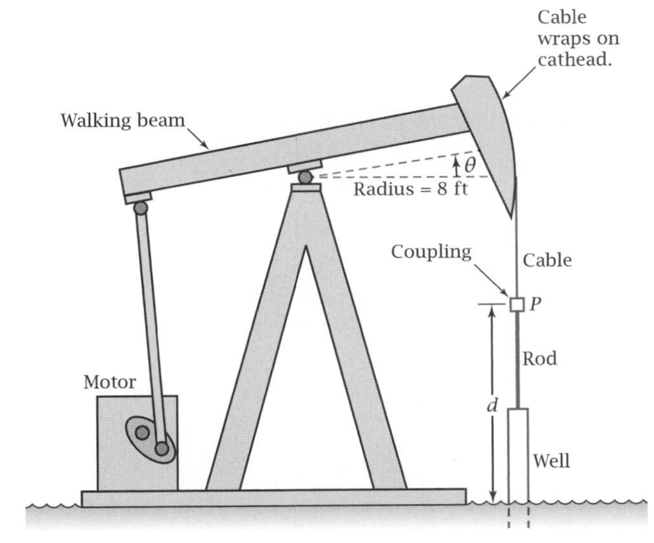

Figure 3-8e

e. Find a particular equation expressing d as a function of t.

f. How far is P above the ground when $t = 9$?

g. For how long does P stay more than 7.5 ft above the ground on each cycle?

h. True or false? "The angle is always the independent variable in a periodic function."

C2. *Inverse Circular Relation Graphs:* In this problem you'll investigate the graphs of the

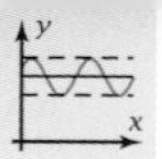

inverse sine and inverse cosine functions and the general inverse sine and cosine relations from which they come.

a. On your grapher, plot the inverse circular function $y = \sin^{-1} x$. Use a friendly window with a range for x of about $[-10, 10]$ that includes $x = 1$ and $x = -1$ as grid points. Use the same scales on both the x- and y-axes. Sketch the result.

b. The graph in part a is only for the inverse sine *function*. The entire inverse sine *relation*, $y = \arcsin x$, can be plotted by putting your grapher in **parametric** mode. In this mode, both x and y are functions of a third variable, usually t. Enter the parametric equations this way:

$$x = \sin t$$
$$y = t$$

Plot the graph using a t-range the same as the x-range in part a. Sketch the graph.

c. Describe how the graphs in part a and part b are related to each other.

d. Explain algebraically how the parametric functions in part b and the function $y = \sin^{-1} x$ are related.

e. Find a way to plot the ordinary sine function, $y = \sin x$, on the same screen, as in part b. Use a different style for this graph so that you can distinguish it from the other. The result should look like the graphs in Figure 3-8f.

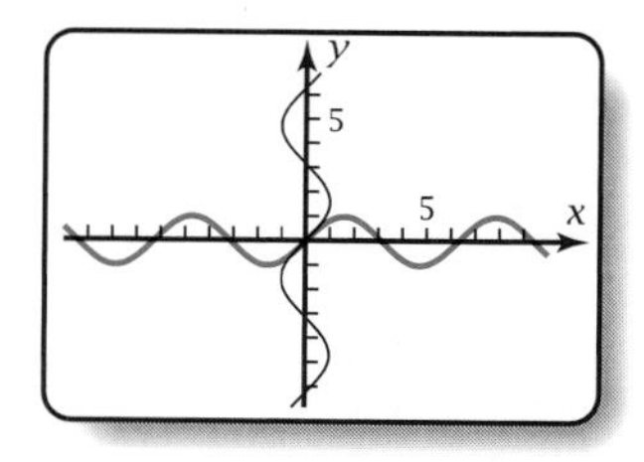

Figure 3-8f

f. How are the two graphs in Figure 3-8f related to each other? Find a geometric transformation of the sine graph to get the arcsine graph.

g. Explain why the arcsine graph in Figure 3-8f is not a *function* graph but the principal value of the inverse sine you plotted in part a *is* a function graph.

h. Using the same scales as in part b, plot the graphs of $y = \cos x$ and the inverse cosine relation. Sketch the result. Do the two graphs have the same relationship as those in Figure 3-8f?

i. Repeat part h for the inverse tangent function.

j. Write an entry in your journal telling what you have learned from this problem.

C3. *Angular Velocity Problem:* A figure skater is spinning with her arms outstretched. She turns through an angle of 12 radians in 2 seconds, which means that her **angular velocity** is $\omega = 6$ radians per second.

a. Her fingertip travels in a circle of radius 80 cm. How far does her fingertip travel in the 2 seconds? What is the **linear velocity** of her fingertip in centimeters per second? How can the linear velocity be calculated easily using just the angular velocity and the radius of rotation?

b. The end of her shoulder is 18 cm from her axis of rotation. Explain why it is correct to say that her shoulder has both the same velocity as and a smaller velocity than her fingertip, depending on which kind of velocity you use.

c. As the skater draws in her arms, she is increasing her angular velocity to 17 radians per second. The tip of her elbow is now 23 cm from her axis of rotation. What is her elbow's angular velocity?

d. Write in your journal about angular velocity. Explain the numerical relationship that exists between angular and linear velocity if you measure angles in radians and the fact that different points on the same rotating object can have the same angular velocity but different linear velocities.

Chapter Test

PART 1: No calculators allowed (T1–T7)

T1. Figure 3-8g shows an x-axis drawn tangent to a unit circle in a uv-coordinate system. On a copy of this figure, show approximately where the point $x = 2.3$ maps onto the unit circle when the x-axis is wrapped around the circle.

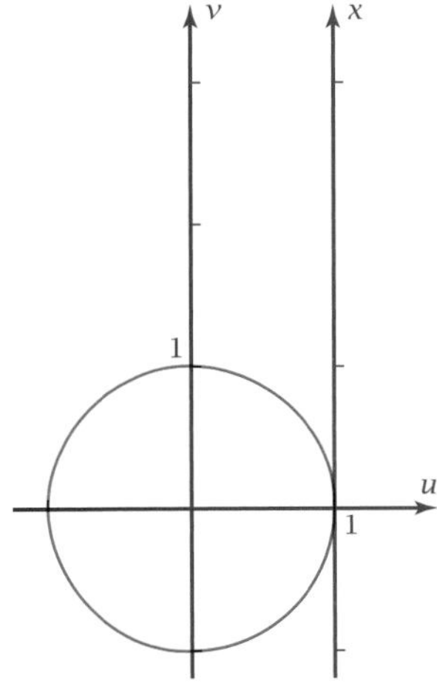

Figure 3-8g

T2. Sketch an angle of 2.3 radians on the copy of Figure 3-8g.

T3. What are the steps needed to find a decimal approximation for the degree measure of an angle of 2.3 radians? In what quadrant would this angle terminate?

T4. Give the exact number of radians in 120° (no decimals).

T5. Give the exact number of degrees in $\frac{\pi}{5}$ radians (no decimals).

T6. Give the period, amplitude, vertical translation, and phase displacement of this circular function:

$$f(x) = 3 + 4\cos\frac{\pi}{5}(x - 1)$$

T7. Sketch at least two cycles of the sinusoid in T6.

PART 2: Graphing calculators allowed (T8–T17)

T8. A long pendulum hangs from the ceiling. As it swings back and forth, its distance from the wall varies sinusoidally with time. At time $x = 1$ second it is at its closest point, $y = 50$ cm. Three seconds later it is at its farthest point, $y = 160$ cm. Sketch the graph.

T9. Figure 3-8h shows a half-cycle of a circular function sinusoid. Find a particular equation for this sinusoid.

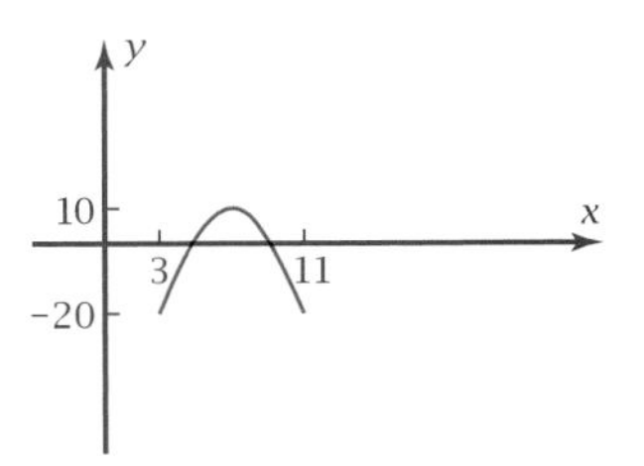

Figure 3-8h

For Problems T10–T16, Figure 3-8i shows the depth of the water at a point near the shore as it varies due to the tides. A particular equation relating d, in feet, to t, in hours after midnight on a given day, is

$$d = 3 + 2\cos\frac{\pi}{5.6}(t - 4)$$

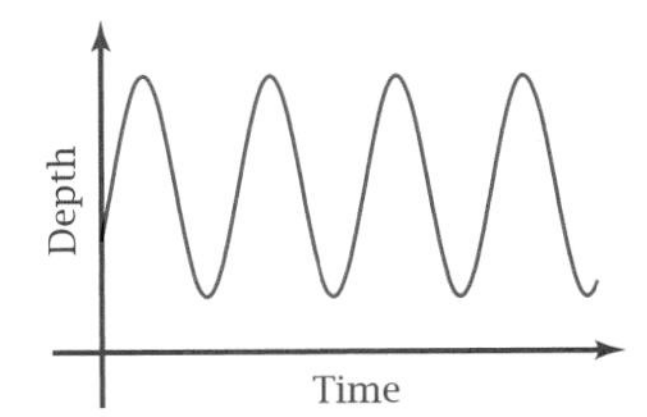

Figure 3-8i

T10. Find a time at which the water is its deepest. How deep is this?

T11. At what time after the time you found in T10 is the water its next shallowest? How deep is it at that time?

T12. What does t equal at 3:00 p.m.? How deep is the water at that time?

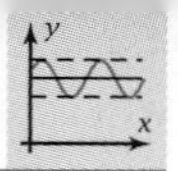

T13. Plot the graph of the sinusoid in Figure 3-8i on your grapher. Use a friendly window with an x-range (actually, t) of about [0, 50] and an appropriate window for y (actually, d).

T14. By tracing your graph in T13, find, approximately, the first interval of nonnegative times for which the water is less than 4.5 feet deep.

T15. Set your grapher's table mode to begin at the later time from T14, and set the table increment at 0.01. Find to the nearest 0.01 hours the latest time at which the water is still less than 4.5 feet deep.

T16. Solve algebraically for the first positive time at which the water is exactly 4.5 feet deep.

T17. What did you learn from this sample test that you did not know before?

Trigonometric Function Properties, Identities, and Parametric Functions

CHAPTER 4

The Foucault pendulum shown in the photograph provided physical proof in the middle of the 19th century that Earth is rotating around its axis. Pendulums can have many different paths in different planes. A pendulum's path is a two-dimensional curve that you can describe by x- and y-displacements from its rest position as functions of time. You can predict a pendulum's position at any given time using parametric functions. Pythagorean properties of trigonometric functions allow you to conclude whether the path of a pendulum is an ellipse or a circle.

Mathematical Overview

There are three kinds of algebraic properties relating trigonometric functions of one argument. For example, the graphs of $y = \cos^2 x$ and $y = \sin^2 x$ are sinusoids. If you add these two functions together, the result is always 1. In this chapter you'll learn properties you can apply to proving identities and solving trigonometric equations. You'll gain this knowledge in four ways.

Algebraically Pythagorean property: $\cos^2 x + \sin^2 x = 1$

Graphically

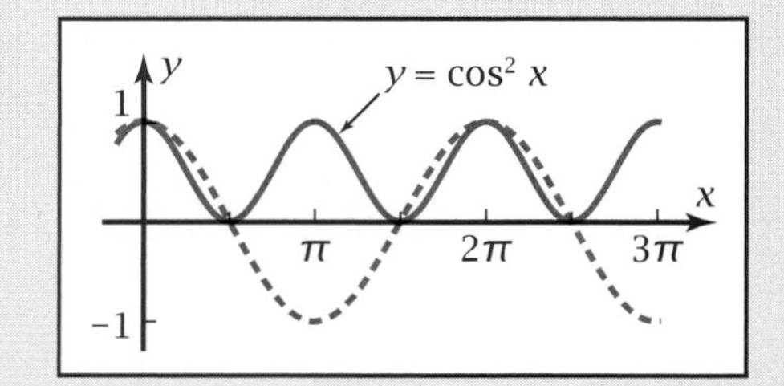

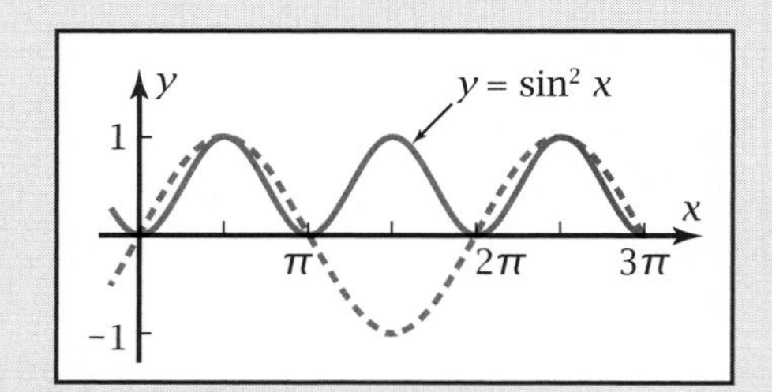

Numerically $\cos^2 37° + \sin^2 37° = 0.6378... + 0.3621... = 1$

Verbally *I can use the Pythagorean, reciprocal, and quotient properties to transform one trig expression to another form. I can check my answer graphically by plotting the original expression and the transformed one. I can also check the answer numerically by making tables of both expressions.*

4-1 Introduction to the Pythagorean Property

Figure 4-1a shows the graphs of $y = \cos^2 x$ (on the left) and $y = \sin^2 x$ (on the right). Both graphs are sinusoids, as you will prove in the next section. In this section you'll learn that the sum of the two functions always equals 1.

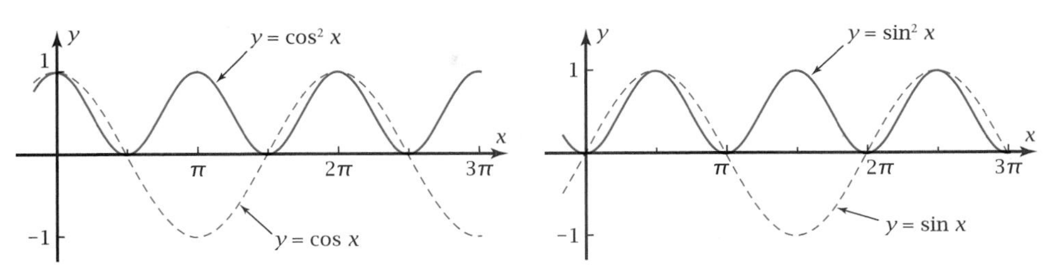

Figure 4-1a

OBJECTIVE Investigate the sum of the squares of the cosine and sine of the same argument.

Exploratory Problem Set 4-1

1. If you enter $\cos^2 0.7$ and $\sin^2 0.7$ into your calculator, you get these numbers:

$$\cos^2 0.7 = 0.5849835715$$

$$\sin^2 0.7 = 0.4150164285$$

Without using your calculator, do the addition. What do you notice?

2. Enter $y_1 = (\cos(x))^2$ and $y_2 = (\sin(x))^2$ into your grapher. (This is how your grapher recognizes $\cos^2 x$ and $\sin^2 x$.) Enter $y_3 = y_1 + y_2$ and then make a table of values of the three functions for each 0.1 radian, starting at 0. What do you notice about y_3?

3. Plot the three functions on the same screen. Do the y_1 and y_2 graphs agree with Figure 4-1a? How does the relationship between y_1 and y_2 give you graphical evidence that $\cos^2 x + \sin^2 x$ is equal to 1, no matter what x is?

4. Remake the table of Problem 2 with your grapher in degree mode. Does your conclusion in Problem 3 apply to trigonometric functions independent of whether x is measured in degrees or radians?

5. Figure 4-1b shows a unit circle in a uv-coordinate system and an angle of 50° in standard position. Use the definitions of cosine and sine to explain why $\cos 50° = u$ and $\sin 50° = v$.

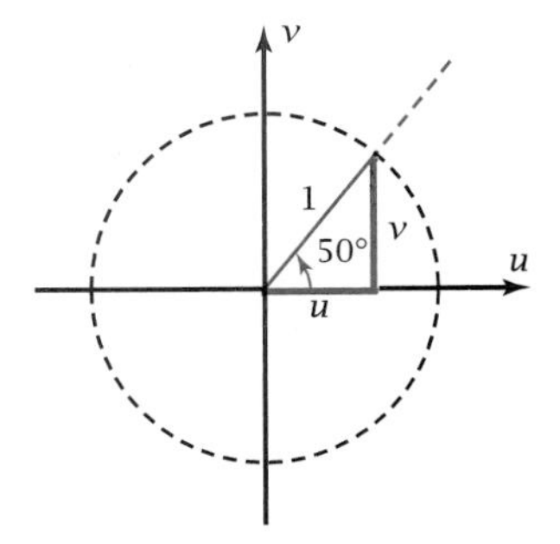

Figure 4-1b

6. Show numerically that $\cos^2 50° + \sin^2 50° = 1$. Explain graphically why this **Pythagorean property** is true.

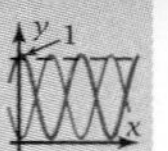

4-2 Pythagorean, Reciprocal, and Quotient Properties

In Section 4-1, you discovered the Pythagorean property

$$\cos^2 x + \sin^2 x = 1$$

for all values of x. You also know that secant, cosecant, and cotangent are reciprocals of cosine, sine, and tangent, respectively. In this section you will prove these properties algebraically, along with the quotient properties, such as

$$\tan x = \frac{\sin x}{\cos x}$$

OBJECTIVE Derive algebraically three kinds of properties expressing relationships among trigonometric functions.

Since the properties you'll learn in this section apply to all trigonometric functions, the argument x will be used both for degrees and for radians.

Reciprocal Properties

In Chapter 2, in order to find values of the secant, cosecant, and tangent functions, you took advantage of the fact that each is the reciprocal of one of the functions on your grapher. For instance,

$$\sec x = \frac{1}{\cos x}$$

because, for right triangles,

$$\sec x = \frac{\text{hypotenuse}}{\text{adjacent leg}} \quad \text{and} \quad \cos x = \frac{\text{adjacent leg}}{\text{hypotenuse}}$$

This relationship between secant and cosine is called a **reciprocal property.** As you can see from the graphs in Figure 4-2a, each y-value for the secant graph is the reciprocal of the corresponding y-value for the cosine graph. For instance, because $\cos\left(\frac{\pi}{3}\right) = \frac{1}{2}$, it follows that $\sec\left(\frac{\pi}{3}\right) = 2$. As you saw in Section 3-3, the asymptotes for the graph of the secant function occur at

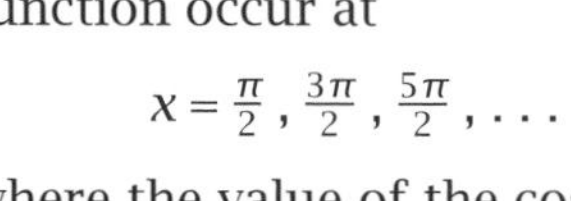

$$x = \frac{\pi}{2}, \frac{3\pi}{2}, \frac{5\pi}{2}, \ldots$$

where the value of the cosine function is zero.

Secant
Cosine

Figure 4-2a

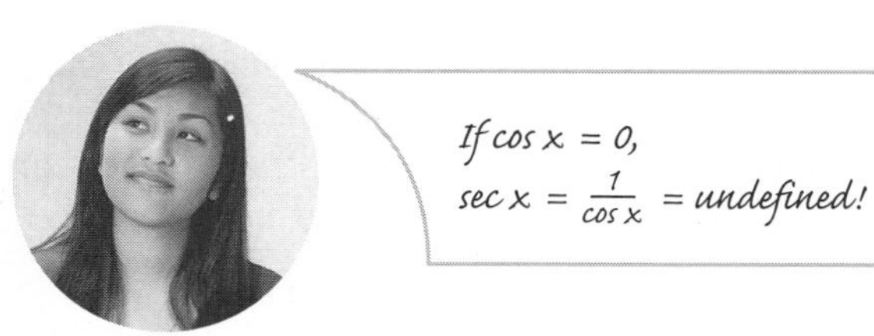

This box summarizes the three reciprocal properties.

PROPERTIES: The Reciprocal Properties

$$\sec x = \frac{1}{\cos x} \qquad \csc x = \frac{1}{\sin x} \qquad \cot x = \frac{1}{\tan x}$$

The domain excludes those values of x that produce a denominator equal to zero.

Quotient Properties

If you divide $\sin x$ by $\cos x$, an interesting result appears.

$$\frac{\sin x}{\cos x} = \frac{\frac{\text{opposite leg}}{\text{hypotenuse}}}{\frac{\text{adjacent leg}}{\text{hypotenuse}}}$$ Definition of sine and cosine.

$$= \frac{\text{opposite leg}}{\text{hypotenuse}} \cdot \frac{\text{hypotenuse}}{\text{adjacent leg}}$$ Multiply the numerator by the reciprocal of the denominator.

$$= \frac{\text{opposite leg}}{\text{adjacent leg}}$$ Simplify.

$$= \tan x$$ Definition of tangent.

$$\therefore \tan x = \frac{\sin x}{\cos x}$$ Transitivity and reflexivity.

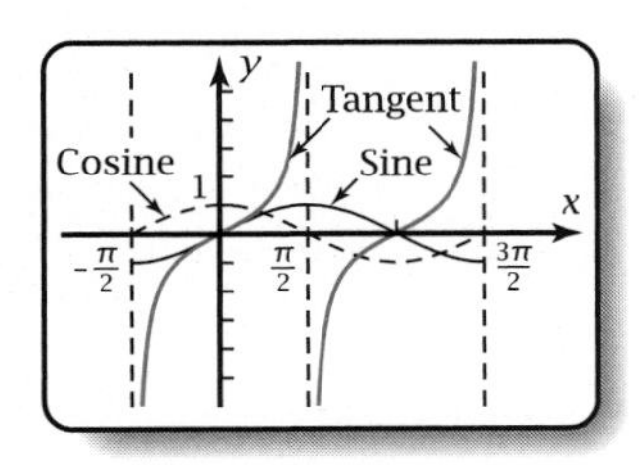

Figure 4-2b

This relationship is called a **quotient property.** If you plot

$y_1 = \sin(x)/\cos(x)$

$y_2 = \tan x$

the graphs will be superimposed (Figure 4-2b).

Because cotangent is the reciprocal of tangent, another quotient property is

$$\cot x = \frac{\cos x}{\sin x}$$

Each of these quotient properties can be expressed in terms of secant and cosecant. For instance,

$$\tan x = \frac{\sin x}{\cos x}$$

$$= \frac{\frac{1}{\csc x}}{\frac{1}{\sec x}}$$ Use the reciprocal properties for sine and cosine.

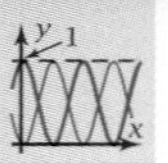

$$= \frac{\sec x}{\csc x} \quad \text{Simplify.}$$

$$\therefore \tan x = \frac{\sec x}{\csc x}$$

Again using the reciprocal property for cotangent,

$$\cot x = \frac{\csc x}{\sec x}$$

This box records the two quotient properties in both of their forms. The properties apply unless a denominator equals zero.

PROPERTIES: The Quotient Properties

$$\tan x = \frac{\sin x}{\cos x} = \frac{\sec x}{\csc x} \qquad \text{Domain: } x \neq \frac{\pi}{2} + \pi n \text{, where } n \text{ is an integer.}$$

$$\cot x = \frac{\cos x}{\sin x} = \frac{\csc x}{\sec x} \qquad \text{Domain: } x \neq \pi n \text{, where } n \text{ is an integer.}$$

Pythagorean Properties

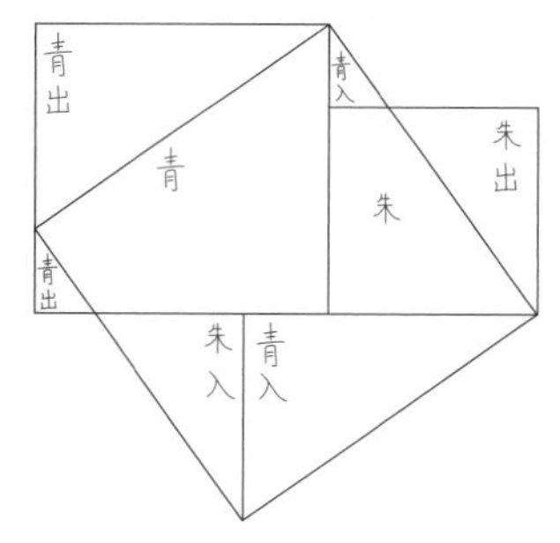

This diagram shows one possible reconstruction of the geometrical figure in Liu Hui's 3rd-century A.D. proof of the Pythagorean theorem.

Figure 4-2c shows an arc of length x in standard position on a unit circle in a uv-coordinate system. By the Pythagorean theorem, point (u, v) at the endpoint of arc x has the property

$$u^2 + v^2 = 1$$

This property is true even if x terminates in a quadrant where u or v is negative, since squares of negative numbers are the same as the squares of their absolute values.

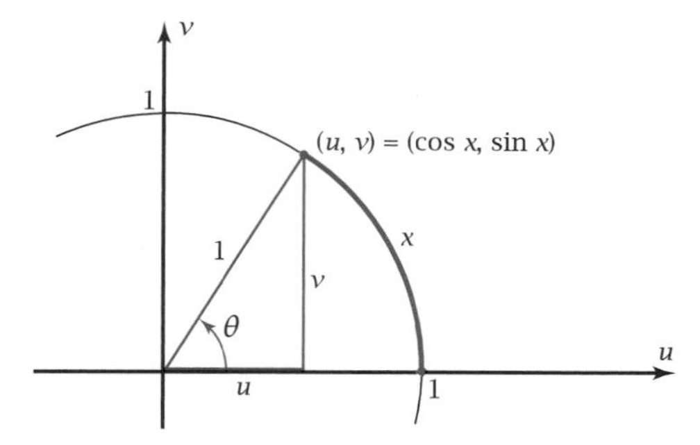

Figure 4-2c

By the definitions of cosine and sine,

$$u = \cos x \qquad \text{and} \qquad v = \sin x \qquad u = \frac{u}{1} \text{ and } v = \frac{v}{1}$$

Substitution into $u^2 + v^2 = 1$ gives the Pythagorean property for sine and cosine.

$$\cos^2 x + \sin^2 x = 1$$

Two other Pythagorean properties can be derived from this one.

$\cos^2 x + \sin^2 x = 1$	Start with the Pythagorean property for cosine and sine.
$\dfrac{\cos^2 x}{\cos^2 x} + \dfrac{\sin^2 x}{\cos^2 x} = \dfrac{1}{\cos^2 x}$	Divide both sides of the equation by $\cos^2 x$.
$1 + \tan^2 x = \sec^2 x$	$\dfrac{\sin^2 x}{\cos^2 x} = \left(\dfrac{\sin x}{\cos x}\right)^2 = \tan^2 x$ and $\dfrac{1}{\cos^2 x} = \sec^2 x$

Dividing by $\sin^2 x$ instead of by $\cos^2 x$ results in the property

$$\cot^2 x + 1 = \csc^2 x$$

This box records the three Pythagorean properties.

PROPERTIES: The Three Pythagorean Properties

$\cos^2 x + \sin^2 x = 1$	Domain: All real values of x.
$1 + \tan^2 x = \sec^2 x$	Domain: $x \neq \frac{\pi}{2} + \pi n$, where n is an integer.
$\cot^2 x + 1 = \csc^2 x$	Domain: $x \neq \pi n$, where n is an integer.

Problem Set 4-2

Do These Quickly

Q1. What is the exact value of cos 30°?

Q2. What is the exact value of $\sin\left(\frac{\pi}{4}\right)$?

Q3. What is the exact value of tan 60°?

Q4. What is the exact value of $\cot\left(\frac{\pi}{2}\right)$?

Q5. Write cos 57° in decimal form.

Q6. Write sin 33° in decimal form.

Q7. Write sec 81° in decimal form.

Q8. Write csc 9° in decimal form.

Q9. From the answers to Q5–Q8, what relationship exists between the values of sine and cosine? How about between secant and cosecant?

Q10. The period of the circular function $y = 3 + 4\cos 5(x - 6)$ is

A. 3 B. 4
C. 5 D. 6
E. None of these

1. What is the reciprocal property for sec x?
2. Explain why $\cot x \cdot \tan x = 1$.
3. Write tan x in terms of sin x and cos x.
4. Show how you can transform the reciprocal property $\cot x = \frac{\cos x}{\sin x}$ algebraically to express cot x in terms of sec x and csc x.
5. Explain geometrically why the property $\cos^2 x + \sin^2 x = 1$ is called a Pythagorean property.
6. By appropriate operations on the Pythagorean property $\cos^2 x + \sin^2 x = 1$, derive the Pythagorean property $\cot^2 x + 1 = \csc^2 x$.
7. Sketch the graph of the trigonometric function $y = \sin\theta$. On the same axes, sketch the graph of $y = \csc\theta$ using the fact that the y-value for $\csc\theta$ is the reciprocal of the corresponding y-value for $\sin\theta$. Where do the asymptotes occur in the graph of the cosecant function?

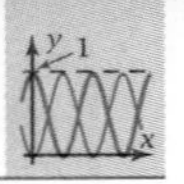

8. On your grapher, make a table with columns showing the values of the trigonometric expressions $\tan^2 \theta$ and $\sec^2 \theta$ for 0°, 15°, 30°, ... What relationship do you notice between the two columns? How do you explain this relationship? How do you explain what happens at 90°?
9. Show algebraically that $\sin^2 x = 1 - \cos^2 x$.
10. Show algebraically that $\cot^2 x = \csc^2 x - 1$.
11. Many trigonometric properties involve the number 1. Use these properties to write *six* trigonometric expressions that equal 1.
12. Use the Pythagorean properties to write expressions equivalent to each of the following expressions.

    a. $\sin^2 x$ b. $\cos^2 x$ c. $\tan^2 x$

    d. $\cot^2 x$ e. $\sec^2 x$ f. $\csc^2 x$

13. *Duality Property of Trigonometric Functions:* The cosine of an angle is the *sine* of the *complement* of that angle. The two functions that satisfy this property are called **cofunctions** of each other. For instance, $\cos 70° = \sin 20°$, as you can check by your calculator. Each of the properties of this section has a **dual,** a property in which each function in the original property has been replaced by its cofunction. For example,

$$\tan x = \frac{\sin x}{\cos x} \quad \longrightarrow \quad \cot x = \frac{\cos x}{\sin x}$$

Show that each of the properties in this section has a dual that is also a valid property. Explain how this duality property can help you memorize the properties.

4-3 Identities and Algebraic Transformation of Expressions

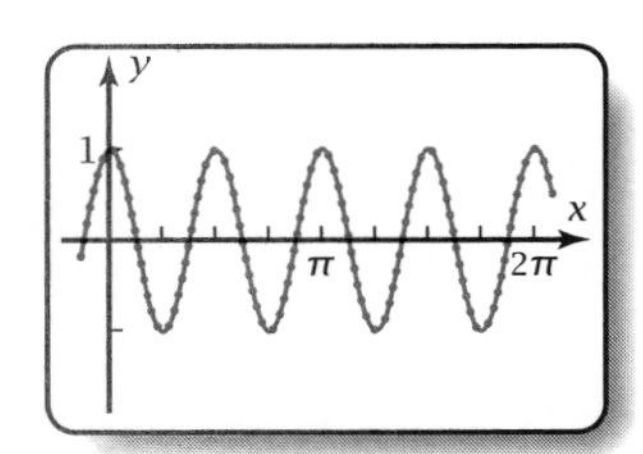

Figure 4-3a

Figure 4-3a shows the graphs of these functions:

$$y_1 = \cos^2 x - \sin^2 x$$

$$y_2 = 1 - 2\sin^2 x$$

The first graph is a thin, solid line, and the second is a thick, dotted line. As you can see, the two graphs are identical. The equation

$$\cos^2 x - \sin^2 x = 1 - 2\sin^2 x$$

is called an **identity** because the two sides of the equation represent identical numbers for all values of x for which the expressions are defined. In this section you will gain fluency with the properties from the last section by using them to transform one trigonometric expression to another one, such as the left side of the identity to the right side.

OBJECTIVE Given a trigonometric expression, transform it to an equivalent form that is perhaps simpler or more useful.

Transformations

Here are examples for transforming one expression to another.

▶ ***EXAMPLE 1*** Transform $\sin x \cot x$ into $\cos x$.

Solution Your thought process should be: "The product $\sin x \cot x$ has two factors, and the result has only one factor. Can I convert one of the factors into a fraction and cancel?"

$\sin x \cot x$	Start by writing the given expression.
$= \sin x \cdot \dfrac{\cos x}{\sin x}$	Substitute using the quotient properties to get $\cos x$ into the expression.
$= \cos x$	Simplify.
$\therefore \sin x \cot x = \cos x$, Q.E.D.	Use the transitive property for completeness. ◀

▶ ***EXAMPLE 2*** Transform $\cos^2 x - \sin^2 x$ into $1 - 2\sin^2 x$.

Solution Your thought process should be:

- The result has only sines in it, so I need to get rid of cosines.
- Since the expressions involve *squares* of functions, I'll think of the *Pythagorean* properties.
- I can write the Pythagorean property $\cos^2 x + \sin^2 x = 1$ as $\cos^2 x = 1 - \sin^2 x$.

$\cos^2 x - \sin^2 x$	Start by writing the given expression.
$= (1 - \sin^2 x) - \sin^2 x$	Substitute $1 - \sin^2 x$ for $\cos^2 x$ using the Pythagorean property.
$= 1 - 2\sin^2 x$	Combine like terms.
$\therefore \cos^2 x - \sin^2 x = 1 - 2\sin^2 x$, Q.E.D.	Use the transitive property. ◀

Identities

To prove that a given trigonometric equation is an identity, start with the expression on one side of the equation and transform it to the other. The only difference between this example and the previous ones is that you may pick either side of the equation to work on.

▶ ***EXAMPLE 3*** Prove algebraically that $(1 + \cos x)(1 - \cos x) = \sin^2 x$ is an identity.

Proof

$(1 + \cos x)(1 - \cos x)$	Start with one member of the equation, usually the more complicated one.
$= 1 - \cos^2 x$	Complete the multiplication.

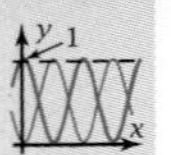

$= \sin^2 x$ — Look for familiar expressions. Since the functions are squared, think Pythagorean!

$\therefore (1 + \cos x)(1 - \cos x) = \sin^2 x$, Q.E.D. — Use the transitive property. ◀

Note:

- Start by writing "Proof." This word tells the reader of your work that you have stopped *stating* the problem and started *solving* it. Writing "Proof" also gets your pencil moving! Sometimes you don't see how to prove something until you actually start doing it.
- It is tempting to *start* with the given equation, then work on *both* sides until you have a statement that is obviously true, such as $\cos x = \cos x$. What this actually does is to prove the *converse* of what you were asked to prove. That is, "*If* the identity is true, *then* the reflexive axiom, such as $y = y$, is true." This is circular reasoning. It is dangerous because you might actually "prove" something that is false by taking an irreversible step, such as squaring both sides of the equation.
- You can never prove graphically or numerically that an identity is true for all values of x. However, you can confirm the validity of an identity for a set of values graphically by plotting both sides and showing that the graphs coincide, or numerically by generating a table of values. Figure 4-3a at the beginning of this section shows you an example of a graphical verification.

▶ EXAMPLE 4 Prove algebraically that $\cot A + \tan A = \csc A \sec A$ is an identity.

Proof

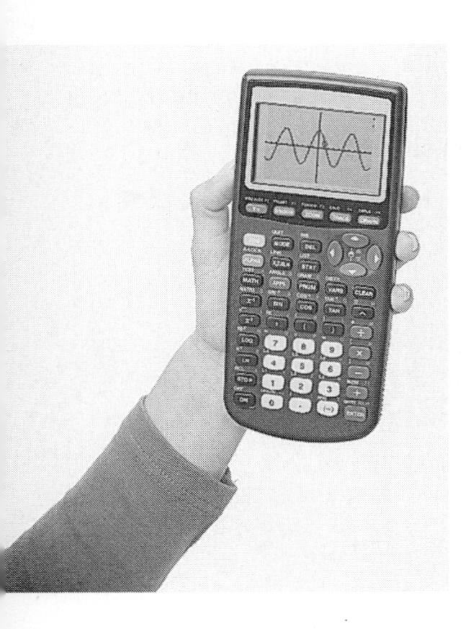

$\cot A + \tan A$	Pick one member of the equation to start with.
$= \dfrac{\cos A}{\sin A} + \dfrac{\sin A}{\cos A}$	The result has only one term. Try making fractions to add together.
$= \dfrac{\cos^2 A + \sin^2 A}{\sin A \cos A}$	Find a common denominator and add the fractions.
$= \dfrac{1}{\sin A \cos A}$	Simplify the numerator by recognizing the Pythagorean property.
$= \dfrac{1}{\sin A} \cdot \dfrac{1}{\cos A}$	The result has *two* factors, so *make* two factors.
$= \csc A \sec A$	Use the reciprocal properties to get the $\csc A$ and $\sec A$ that appear in the answer.
$\therefore \cot A + \tan A = \csc A \sec A$, Q.E.D.	Use the transitive property. ◀

Note: Avoid the temptation to use a shortcut where you write only cos or sec. These are the *names* of the functions, not the values of the functions. Equality applies to numbers, not to names.

▶ EXAMPLE 5 Prove algebraically that $\dfrac{1 - \cos B}{\sin B} = \dfrac{\sin B}{1 + \cos B}$ is an identity. Confirm graphically on a reasonable interval.

Proof

$$\frac{\sin B}{1 + \cos B}$$

Start with the more complicated side of the equation (binomial denominator).

$$= \frac{\sin B}{1 + \cos B} \cdot \frac{1 - \cos B}{1 - \cos B}$$

Multiply by a clever form of 1 (see the note following this example).

$$= \frac{\sin B\,(1 - \cos B)}{1 - \cos^2 B}$$

Multiply the denominator but not the numerator. You want $1 - \cos B$ in your result.

$$= \frac{\sin B\,(1 - \cos B)}{\sin^2 B}$$

Make the denominator have *one* term by recognizing the Pythagorean property.

$$= \frac{1 - \cos B}{\sin B}$$

Cancel $\sin B$ in the numerator with one $\sin B$ in the denominator.

$$\therefore \frac{1 - \cos B}{\sin B} = \frac{\sin B}{1 + \cos B}, \quad \text{Q.E.D.}$$

Enter $y_1 = (1 - \cos(x))/\sin(x)$ and $y_2 = \sin(x)/(1 + \cos(x))$. Plot the graphs using different styles, such as dashed for one and solid or path style for the other. Figure 4-3b shows the result, using a friendly window for x containing multiples of π as grid points.

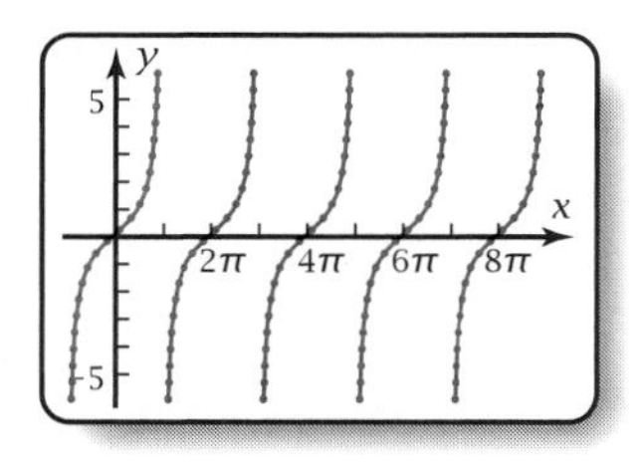

Figure 4-3b

◀

Note: There are two ways to think up the form of 1 to multiply by in the second line of Example 5. First, the expressions $(1 + \cos B)$ and $(1 - \cos B)$ are **conjugate binomials,** or conjugates. When you multiply conjugates together, you get a difference of two squares (no middle term). This allows you to use the Pythagorean property in the next step. Second, you want the quantity $(1 - \cos B)$ in the result. So you *put* it there by multiplying by a form of 1 containing it.

This box summarizes useful techniques from these examples.

PROCEDURE: Transforming Trigonometric Expressions and Proving Identities

1. Start by writing the given expression or, for an identity, by picking the side of the equation you wish to start with and writing it down. Usually it is easier to start with the more complicated side.
2. Look for *algebraic* things to do.
   a. If there are two terms and you want only one term, then
      i. Add fractions, or
      ii. Factor something out.
   b. Multiply by a clever form of 1 in order to
      i. Multiply a numerator or denominator by its conjugate binomial, or
      ii. Get a desired expression into the numerator or denominator.
   c. Do any obvious calculations (distribute, square, multiply polynomials, and so on).
   d. Factor out an expression you want to appear in the result.
3. Look for *trigonometric* things to do.
   a. Look for familiar trigonometric expressions you can transform.
   b. If there are *squares* of functions, think of Pythagorean properties.
   c. Reduce the number of different functions, transforming them to ones you want in the result.
   d. Leave expressions you want unchanged.
4. Keep looking at the result and thinking of ways you can get closer to it.

Problem Set 4-3

Do These Quickly

Q1. Write the Pythagorean property for cosine and sine.

Q2. Write the quotient property for tangent in terms of sine and cosine.

Q3. Write the quotient property for tangent in terms of secant and cosecant.

Q4. Write the reciprocal property for secant.

Q5. Why does $(\tan x)(\cot x)$ equal 1?

Q6. Sketch the graph of the parent cosine function $y = \cos x$.

Q7. Sketch the graph of the parent sine function $y = \sin\theta$.

Q8. What is the vertical dilation for $y = 2 + 3\cos 4(x - 5)$?

Q9. The reference angle for 260° is —?—.

Q10. $y = 3(1.06^x)$ is an example of a(n) —?— function.

For Problems 1–26, show the steps in transforming the expression on the left to the one on the right.

1. $\cos x \tan x$ to $\sin x$
2. $\csc x \tan x$ to $\sec x$
3. $\sec A \cot A \sin A$ to 1
4. $\csc B \tan B \cos B$ to 1
5. $\sin^2\theta \sec\theta \csc\theta$ to $\tan\theta$
6. $\cos^2\alpha \csc\alpha \sec\alpha$ to $\cot\alpha$
7. $\cot R + \tan R$ to $\csc R \sec R$
8. $\cot D \cos D + \sin D$ to $\csc D$
9. $\csc x - \sin x$ to $\cot x \cos x$
10. $\sec x - \cos x$ to $\sin x \tan x$
11. $\tan x(\cot x \cos x + \sin x)$ to $\sec x$
12. $\cos x(\sec x + \cos x \csc^2 x)$ to $\csc^2 x$
13. $(1 + \sin B)(1 - \sin B)$ to $\cos^2 B$
14. $(\sec E - 1)(\sec E + 1)$ to $\tan^2 E$
15. $(\cos\phi - \sin\phi)^2$ to $1 - 2\cos\phi\sin\phi$
16. $(1 - \tan\phi)^2$ to $\sec^2\phi - 2\tan\phi$
17. $(\tan n + \cot n)^2$ to $\sec^2 n + \csc^2 n$
18. $(\cos k - \sec k)^2$ to $\tan^2 k - \sin^2 k$
19. $\dfrac{\csc^2 x - 1}{\cos x}$ to $\cot x \csc x$
20. $\dfrac{1 - \cos^2 x}{\tan x}$ to $\sin x \cos x$
21. $\dfrac{\sec^2\theta - 1}{\sin\theta}$ to $\tan\theta \sec\theta$
22. $\dfrac{1 + \cot^2\theta}{\sec^2\theta}$ to $\cot^2\theta$

23. $\dfrac{\sec A}{\sin A} - \dfrac{\sin A}{\cos A}$ to $\cot A$
24. $\dfrac{\csc B}{\cos B} - \dfrac{\cos B}{\sin B}$ to $\tan B$
25. $\dfrac{1}{1 - \cos x} + \dfrac{1}{1 + \cos x}$ to $2\csc^2 x$
26. $\dfrac{1}{\sec D - \tan D} + \dfrac{1}{\sec D + \tan D}$ to $2\sec D$

For Problems 27–36, prove algebraically that the given equation is an identity.

27. $\sec x(\sec x - \cos x) = \tan^2 x$
28. $\tan x(\cot x + \tan x) = \sec^2 x$
29. $\sin x(\csc x - \sin x) = \cos^2 x$
30. $\cos x(\sec x - \cos x) = \sin^2 x$
31. $\csc^2\theta - \cos^2\theta \csc^2\theta = 1$
32. $\cos^2\theta + \tan^2\theta \cos^2\theta = 1$
33. $(\sec\theta + 1)(\sec\theta - 1) = \tan^2\theta$
34. $(1 + \sin\theta)(1 - \sin\theta) = \cos^2\theta$
35. $(2\cos x + 3\sin x)^2 + (3\cos x - 2\sin x)^2 = 13$
36. $(5\cos x - 4\sin x)^2 + (4\cos x + 5\sin x)^2 = 41$
37. Confirm that the equation in Problem 33 is an identity by plotting the two graphs.
38. Confirm that the equation in Problem 34 is an identity by plotting the two graphs.

39. Confirm that the equation in Problem 35 is an identity by making a table of values.

40. Confirm that the equation in Problem 36 is an identity by making a table of values.

41. Prove that the equation $\cos x = 1 - \sin x$ is *not* an identity.

42. Prove that the equation $\tan^2 x - \sec^2 x = 1$ is *not* an identity.

Problems 43–54 involve more complicated algebraic techniques. Prove that each equation is an identity.

43. $\sec^2 A + \tan^2 A \sec^2 A = \sec^4 A$

44. $\cos^4 t - \sin^4 t = 1 - 2\sin^2 t$

45. $\dfrac{1}{\sin x \cos x} - \dfrac{\cos x}{\sin x} = \tan x$

46. $\dfrac{\sin x}{\csc x} + \dfrac{\cos x}{\sec x} = 1$

47. $\dfrac{1}{1 + \cos p} = \csc^2 p - \csc p \cot p$

48. $\dfrac{\cos x}{\sec x - 1} - \dfrac{\cos x}{\tan^2 x} = \cot^2 x$

49. $\dfrac{1 + \sin x}{1 - \sin x} = 2\sec^2 x + 2\sec x \tan x - 1$

50. $\sin^3 z \cos^2 z = \sin^3 z - \sin^5 z$

51. $\sec^2 \theta + \csc^2 \theta = \sec^2 \theta \csc^2 \theta$

52. $\sec \theta + \tan \theta = \dfrac{1}{\sec \theta - \tan \theta}$

53. $\dfrac{1 - 3\cos x - 4\cos^2 x}{\sin^2 x} = \dfrac{1 - 4\cos x}{1 - \cos x}$

54. $\dfrac{\sec^2 x - 6\tan x + 7}{\sec^2 x - 5} = \dfrac{\tan x - 4}{\tan x + 2}$

55. *Journal Problem:* Update your journal with what you have learned recently about transforming trigonometric expressions algebraically. In particular, show how you can use the three kinds of properties from Section 4-2 to transform an expression to a different form.

4-4 Arcsine, Arctangent, Arccosine, and Trigonometric Equations

In Section 3-6, you learned how to solve equations that reduce to the form

$$\cos(x) = a, \qquad \text{where } a \text{ is a constant}$$

You learned that the general solution is

$$x = \arccos(a) = \pm\cos^{-1}(a) + 2\pi n$$

where n is an integer representing a number of cycles or revolutions. In this section you will solve trigonometric equations involving sine or tangent rather than just cosine, and where the argument may be in degrees or in radians.

OBJECTIVE Find algebraically or numerically the solutions for equations involving circular or trigonometric sines, cosines, and tangents of one argument.

Arcsine, Arctangent, and Arccosine

You recall from Section 3-6 that $\arccos x$ means any of the angles whose cosine is x. $\text{Arcsin}\, x$ and $\arctan x$ have the analogous meaning for sine and tangent.

Within any one revolution there are two values of the inverse trigonometric relation for any given argument. Figure 4-4a shows how to find the values of arcsin $\frac{3}{5}$, arccos $\frac{3}{5}$, and arctan $\frac{3}{5}$. Sketch a reference triangle with appropriate sides of 3 and 5, then look for a reference triangle in another quadrant for which the sides have the ratio $\frac{3}{5}$. You find the general solution by adding integer numbers of revolutions, $360n°$ or $2\pi n$ radians.

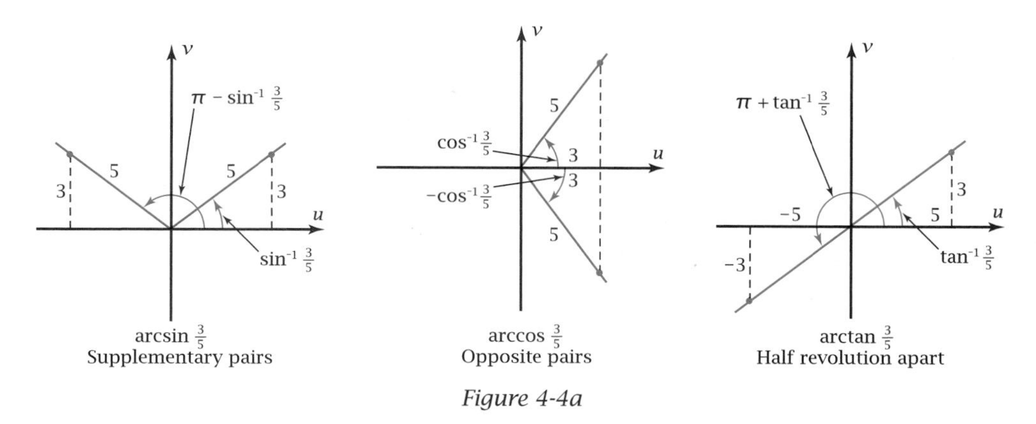

Figure 4-4a

Note that the second angle involves a reflection of the reference triangle across an axis. To remember *which* axis requires only recall of the definition of the trigonometric functions.

arcsine: $\sin x = \dfrac{\text{vertical coordinate}}{\text{radius}}$ — Reflect across the vertical axis (Figure 4-4a, left).

arccosine: $\cos x = \dfrac{\text{horizontal coordinate}}{\text{radius}}$ — Reflect across the horizontal axis (Figure 4-4a, middle).

arctangent: $\tan x = \dfrac{\text{vertical coordinate}}{\text{horizontal coordinate}}$ — Reflect across *both* axes (Figure 4-4a, right).

EXAMPLE 1 Solve the equation $10 \sin (x - 0.2) = -3$ algebraically for x in the domain $[0, 4\pi]$. Verify the solutions graphically.

Solution

$10 \sin (x - 0.2) = -3$ — Write the given equation.

$\sin (x - 0.2) = -0.3$ — Reduce the equation to the function (argument) equal to a constant form.

$x - 0.2 = \arcsin (-0.3)$ — Take the arcsine of both sides.

$x = 0.2 + \arcsin (-0.3)$ — Isolate x.

$x = 0.2 + \sin^{-1} (-0.3) + 2\pi n$ — Supplementary pairs.

or

$x = -0.1046\ldots + 2\pi n$ or $3.6462\ldots + 2\pi n$

$S = \{3.6462\ldots, 6.1784\ldots, 9.9294\ldots, 12.4616\ldots\}$

Choose the values of n that give solutions in the domain.

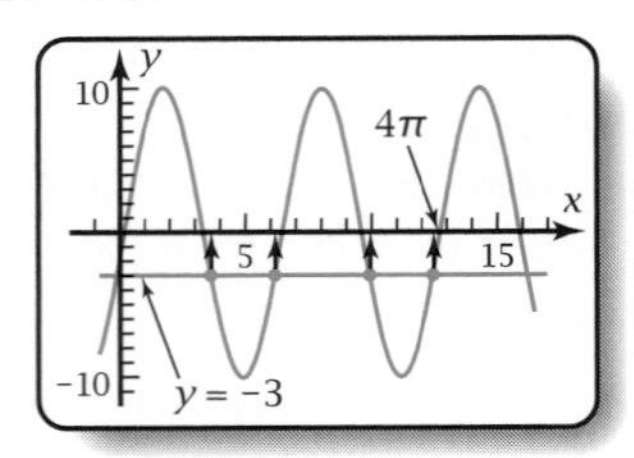

Figure 4-4b

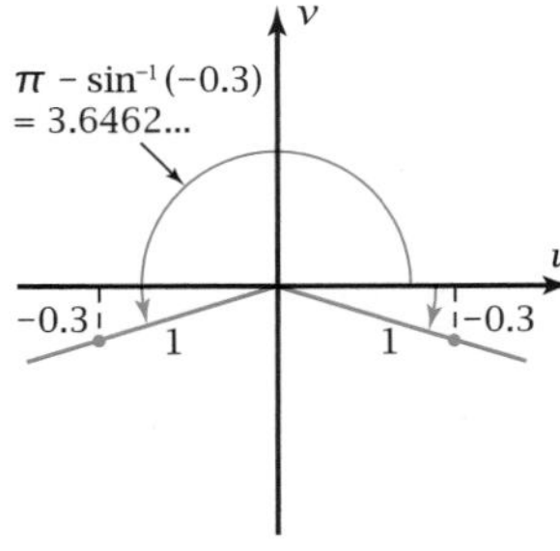

Figure 4-4c

The graph in Figure 4-4b shows $y_1 = 10 \sin (x - 0.2)$ with the line $y_2 = -3$. Use the intersect feature of your grapher to show that the lines do intersect at the points in the solution set. (Some intersections are out of the domain.) ◀

Note:

- You can enter $y_1 = 0.2 + \sin^{-1}(-0.3) + 2\pi n$ and $y_2 = 0.2 + (\pi - \sin^{-1}(-0.3)) + 2\pi n$ in your grapher, using x in place of n, and use the table feature to find the particular values.
- The function value $\sin^{-1}(-0.3) = -0.1046\ldots$ terminates in Quadrant IV. The other value is the supplement of this number. Figure 4-4c shows that subtracting $-0.1046\ldots$ from π gives an angle in Quadrant III, where the other value must be if its sine is negative.

$$\pi - (-0.1046\ldots) = \pi + 0.1046\ldots = 3.6462\ldots$$

Interval Notation

A compact way to write a domain such as $0 \le x \le 4\pi$ is $[0, 4\pi]$. This set of values of x is called the **closed interval** from $x = 0$ to $x = 4\pi$. The **open interval** from $x = 0$ to $x = 4\pi$ is written $(0, 4\pi)$, and means $0 < x < 4\pi$. The symbol $\in$ from set terminology is used to show that x is an "element of" or is in a given interval. So you can write the domain for the closed interval

$$x \in [0, 4\pi]$$

which is pronounced "x is an element of the closed interval from 0 to 4π." The various interval notations are summarized in this box.

DEFINITIONS: Interval Notation

WRITTEN	*MEANING*	*NAME*
$x \in [0, 4\pi]$	$0 \le x \le 4\pi$	Closed interval
$x \in (0, 4\pi)$	$0 < x < 4\pi$	Open interval
$x \in [0, 4\pi)$	$0 \le x < 4\pi$	Half-open interval
$x \in (0, 4\pi]$	$0 < x \le 4\pi$	Half-open interval

▶ **EXAMPLE 2** Solve the equation $4 \tan 2\theta = -5$ algebraically for the first three positive values of θ. Verify the solutions graphically.

Solution

$$4 \tan 2\theta = -5$$

$$\tan 2\theta = -1.25$$

$$2\theta = \arctan(-1.25) = \tan^{-1}(-1.25) + 180n^\circ$$

Write the general solution (half a revolution apart).

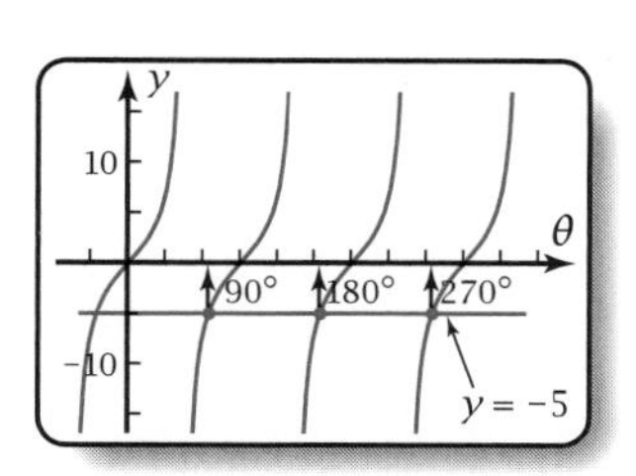

Figure 4-4d

$$\theta = \frac{1}{2}\tan^{-1}(-1.25) + 90n°$$

Solve for θ; divide *both* terms on the right side of the equation by 2.

$$\theta = -25.6700...° + 90n°$$

$$S = \{64.32...°, 154.32...°, 244.32...°\}$$

Choose the values of n that give the first three positive answers.

The graph in Figure 4-4d shows $y = 4\tan 2\theta$ and $y = -5$, with intersections at the three positive values that are in the solution set. ◀

To find the quadrant for the "other" angle, try sketching a *uv*-diagram. Figure 4-4e shows the reference triangle reflected across both axes, making the two angles a half-revolution apart.

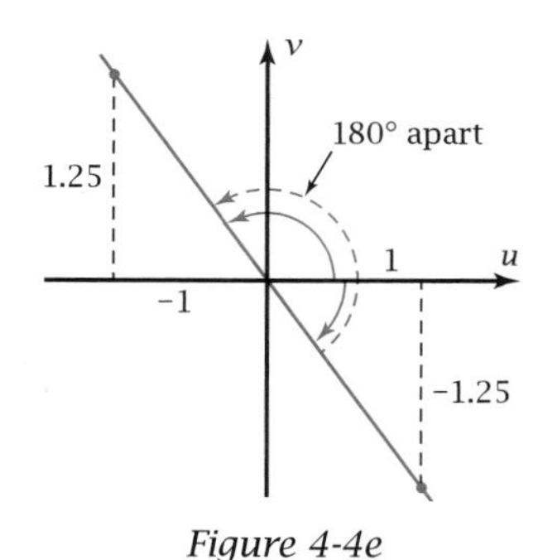

Figure 4-4e

PROPERTIES: General Solutions for Arcsine, Arccosine, and Arctangent

Let A stand for the argument of the inverse sine, cosine, or tangent.

$$\theta = \arcsin A = \sin^{-1} A + 360n° \quad \text{or} \quad (180° - \sin^{-1} A) + 360n°$$

$$x = \arcsin A = \sin^{-1} A + 2\pi n \quad \text{or} \quad (\pi - \sin^{-1} A) + 2\pi n$$

Verbally: Inverse sines come in supplementary pairs (plus coterminals).

Visually: Reflect the reference triangle across the *vertical* axis.

$$\theta = \arccos A = \pm\cos^{-1} A + 360n°$$

$$x = \arccos A = \pm\cos^{-1} A + 2\pi n$$

Verbally: Inverse cosines come in opposite pairs (plus coterminals).

Visually: Reflect the reference triangle across the *horizontal* axis.

$$\theta = \arctan A = \tan^{-1} A + 180n°$$

$$x = \arctan A = \tan^{-1} A + \pi n$$

Verbally: Inverse tangents come in pairs a half revolution apart.

Visually: Reflect the reference triangle across *both* axes.

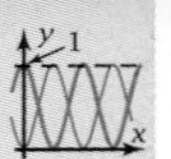

Quadratic Forms

You may need to use the quadratic formula or factoring if you have to solve algebraically an equation that has squares of trigonometric functions.

EXAMPLE 3 Solve algebraically: $\cos^2 \theta + \sin \theta + 1 = 0$, $\theta \in [-90°, 270°)$.

Solution

$\cos^2 \theta + \sin \theta + 1 = 0$

$(1 - \sin^2 \theta) + \sin \theta + 1 = 0$ — Use the Pythagorean property to change all terms to *one* function.

$\sin^2 \theta - \sin \theta - 2 = 0$ — Put the equation in $ax^2 + bx + c = 0$ form.

No solution

~~$\sin \theta = 2$~~ or $\sin \theta = -1$ — Use the quadratic formula. Discard impossible solutions.

$\theta = \arcsin(-1) = -90° + 360n°$ — $180° - (-90°) = 270°$ is coterminal with $-90°$.

$S = \{-90°\}$ — The 270° when $n = 1$ is out of the domain (half *open* interval).

Note that in this case you could have factored to solve the quadratic equation.

$(\sin \theta - 2)(\sin \theta + 1) = 0$

$\sin \theta - 2 = 0$

or $\sin \theta + 1 = 0$

$\sin \theta = 2$

or $\sin \theta = -1$

If a product is zero, then one of its factors has to be zero!

Numerical Solutions

Some trigonometric equations cannot be solved algebraically. This is true if the variable appears both transcendentally (in the argument of the function) and algebraically (not in the argument), such as

$0.2x + \sin x = 2$

There is no algebraic solution because you cannot transform the equation to f(argument) = constant. In other cases, the algebraic solution may be difficult to find. In such cases, a numerical solution with the help of graphs is appropriate.

EXAMPLE 4 Solve $0.2x + \sin x = 2$ for all real values of x.

Solution The graph in Figure 4-4f shows $y_1 = 0.2x + \sin x$ and $y_2 = 2$ intersecting at the three points $x \approx 7.0$, 9.3, and 12.1.

$S = \{6.9414..., 9.2803..., 12.1269...\}$

Use the intersect or solver feature on your grapher.

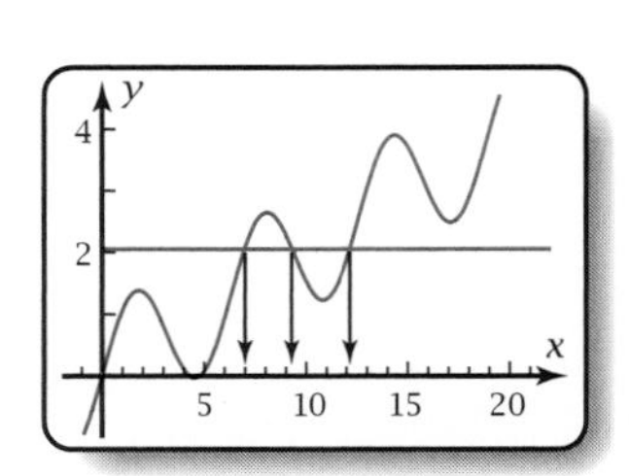

Figure 4-4f

Problem Set 4-4

Do These Quickly

Q1. Write the particular equation of the sinusoid with amplitude 2, period 120°, sinusoidal axis at $y = 5$, and phase displacement 17° (for cosine).

Q2. Sketch a reasonable graph for the time of sunset as a function of the day of the year.

Q3. Sketch the graph of $y = \sec x$.

Q4. What is the exact value (no decimals) of $\cos 30°$?

Q5. What is the exact value (no decimals) of $\sin \frac{\pi}{4}$?

Q6. Sketch the reference angle for 260°.

Q7. Right triangle XYZ has right angle Y. Side x is opposite angle X, and so on. Find $\csc X$.

Q8. Find the degree measure of the acute angle $\cot^{-1} 3$.

Q9. What is the value of n if $\log 32 = n \log 2$?

Q10. The graph of $y = 3x^2 + 2x - 7$ is called a(n) —?—.

For Problems 1–10,

a. Find the general solution for θ or x.

b. Find the particular solutions that are in the given domain.

1. $\theta = \arcsin 0.7$ $\qquad \theta \in [0°, 720°]$
2. $\theta = \arcsin (-0.6)$ $\qquad \theta \in [0°, 720°]$
3. $x = \arcsin (-0.2)$ $\qquad x \in [0, 4\pi]$
4. $x = \arcsin 0.9$ $\qquad x \in [0, 4\pi]$
5. $\theta = \arctan (-4)$ $\qquad \theta \in [0°, 720°]$
6. $\theta = \arctan 0.5$ $\qquad \theta \in [0°, 720°]$
7. $x = \arctan 10$ $\qquad x \in [0, 4\pi]$
8. $x = \arctan (-0.9)$ $\qquad x \in [0, 4\pi]$
9. $\theta = \arccos 0.2$ $\qquad \theta \in [0°, 720°]$
10. $x = \arccos (-0.8)$ $\qquad x \in [0, 4\pi]$
11. Confirm graphically that the answers to Problem 1b are correct.
12. Confirm graphically that the answers to Problem 8b are correct.
13. Explain why there are *no* solutions for $x = \arccos 2$ but there *are* solutions for $x = \arctan 2$.
14. Explain why there are *no* solutions for $\theta = \arcsin 3$ but there *are* solutions for $\theta = \arctan 3$.

For Problems 15–20, solve the equation in the given domain.

15. $\tan \theta + \sqrt{3} = 0$ $\qquad \theta \in [0°, 720°]$
16. $2 \cos \theta + \sqrt{3} = 0$ $\qquad \theta \in [0°, 720°]$
17. $2 \sin (\theta + 47°) = 1$ $\qquad \theta \in [-360°, 360°]$
18. $\tan (\theta - 81°) = 1$ $\qquad \theta \in [-180°, 540°]$
19. $3 \cos \pi x = 1$ $\qquad x \in [0, 6]$
20. $5 \sin \pi x = 2$ $\qquad x \in [-2, 4]$
21. Confirm graphically that the answers to Problem 17 are correct.
22. Confirm graphically that the answers to Problem 18 are correct.
23. Figure 4-4g shows the graph of $y = 2 \cos^2 \theta - \cos \theta - 1$. Calculate algebraically the θ-intercepts in the domain $0° \le \theta \le 720°$, and show that they agree with the graph.

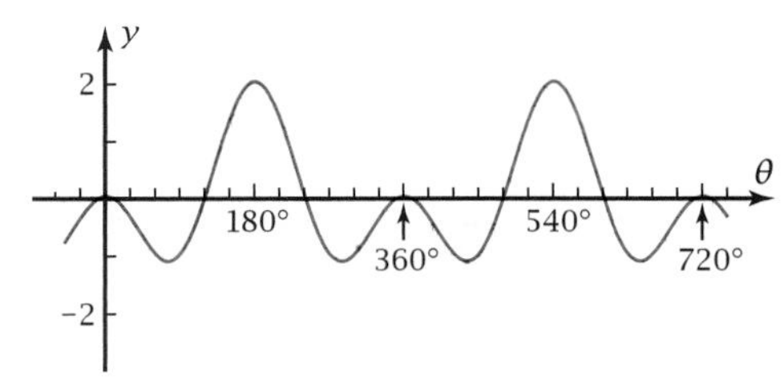

Figure 4-4g

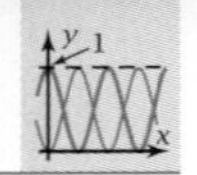

24. Figure 4-4h shows the graph of $y = 2\sin^2\theta - 3\sin\theta + 1$. Calculate algebraically the θ-intercepts in the domain $0° \le \theta \le 720°$, and show that they agree with the graph.

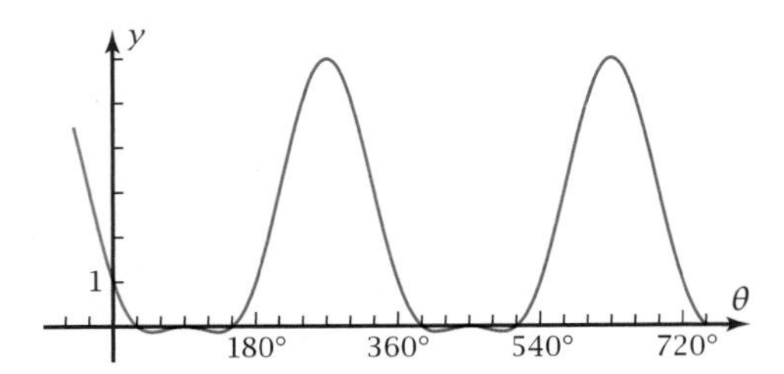

Figure 4-4h

25. Figure 4-4i shows the graph of $y = 2\sin^2\theta - 3\sin\theta - 2$. Calculate algebraically the θ-intercepts in the domain $0° \le \theta \le 720°$, and show that they agree with the graph.

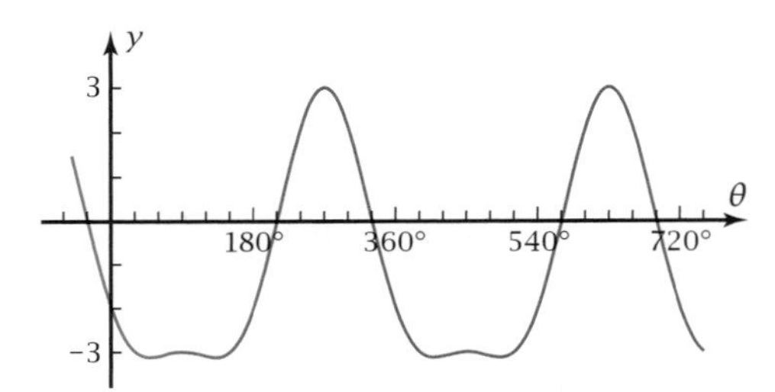

Figure 4-4i

26. Figure 4-4j shows the graph of $y = \cos^2\theta + 5\cos\theta + 6$. Calculate the θ-intercepts algebraically. Tell why the results you got agree with the graph.

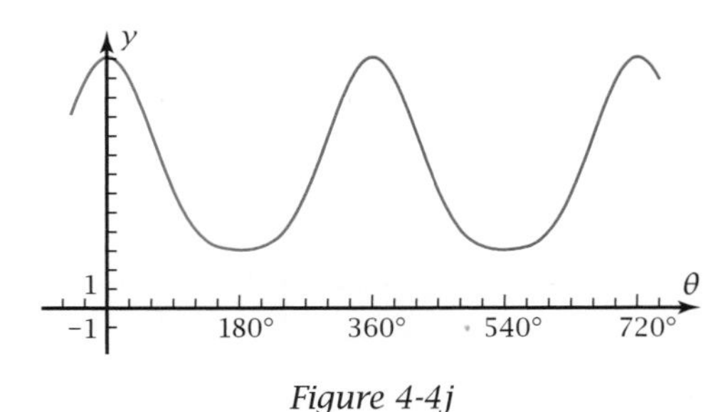

Figure 4-4j

For Problems 27–30,

a. Solve the equation graphically using the intersect feature of your grapher.

b. Solve the equation algebraically, confirming the graphical solution.

27. $3\cos^2\theta = 2\cos\theta \qquad \theta \in [0°, 360°)$

28. $\tan^2\theta = 2\tan\theta \qquad \theta \in [0°, 360°)$

29. $4\cos^2 x + 2\sin x = 3 \qquad x \in [0, 2\pi)$

30. $5\sin^2 x - 3\cos x = 4 \qquad x \in [0, 2\pi)$

31. *Rotating Beacon Problem:* Figure 4-4k shows a rotating beacon on a lighthouse 500 yards offshore. The beam of light shines out of both sides of the beacon, making a spot of light that moves along the beach with a displacement y, measured in yards, from the point on the beach that is closest to the lighthouse.

a. Write an equation for y in terms of θ.

b. The beacon rotates with an angular velocity of 5 degrees per second. Let t be the time, in seconds, since the beam was perpendicular to the beach (that is, $y = 0$). By appropriate substitution, write y as a function of t.

c. A house on the beach is at a displacement $y = 600$. Find the first four positive values of t when the spot of light illuminates the house.

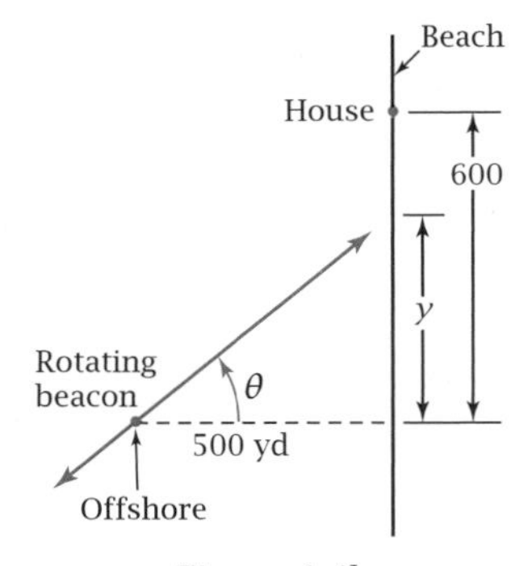

Figure 4-4k

32. *Numerical Solution of Equation Problem 1:* Figure 4-4l shows the graphs of $y = x$ and $y = \cos x$ as they might appear on your grapher.

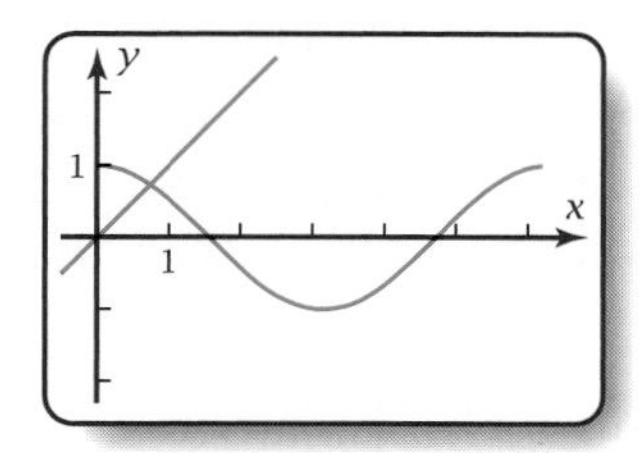

Figure 4-4l

a. Read from the graph a value of x for which $\cos x \approx x$.

b. Solve numerically to find a more precise value of x in part a.

c. Are there other values of x for which $\cos x = x$? How did you reach your conclusion?

d. Explain why the equation $\cos x = x$ cannot be solved algebraically.

33. *Numerical Solution of Equation Problem 2:* Figure 4-4m shows $y = x$ and $y = \tan \pi x$.

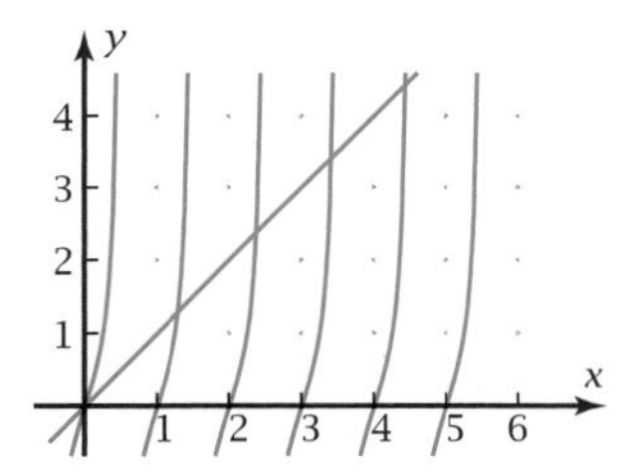

Figure 4-4m

a. Read from the graph the first three values of x for which $\tan \pi x = x$.

b. Solve numerically to find the three precise values in part a.

c. Explain why the equation $\tan \pi x = x$ cannot be solved algebraically.

34. *Numerical Solution of Equation Problem 3:* Figure 4-4n shows $y = x$ and $y = 5 \sin \frac{\pi}{2}x$.

a. Find numerically the greatest value of x for which $5 \sin \frac{\pi}{2}x = x$.

b. Find numerically the next-to-greatest value of x for which $5 \sin \frac{\pi}{2}x = x$. The zoom feature of your grapher may help.

c. Explain why the equation $5 \sin \frac{\pi}{2}x = x$ cannot be solved algebraically.

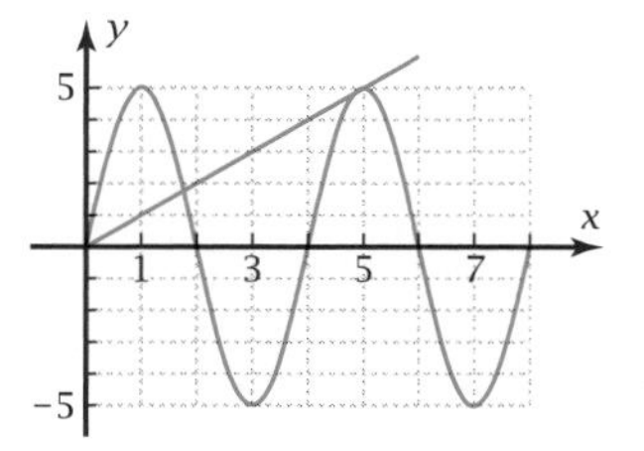

Figure 4-4n

35. *Trigonometric Inequality Problem 1:* Figure 4-4o shows the region of points that satisfy the **trigonometric inequality**

$$y \le 2 + 3 \cos \frac{\pi}{6}x$$

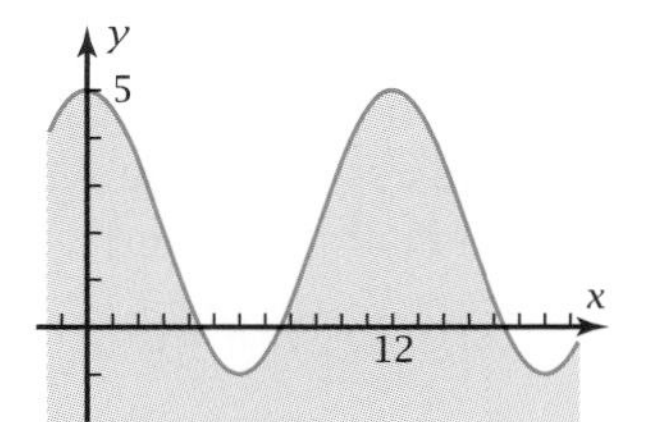

Figure 4-4o

a. Duplicate the figure on your grapher. Use the appropriate style to shade the region below the boundary curve.

b. On the same screen, plot the graph of $y \ge 0$. Use the appropriate style so that the grapher will shade the region. Sketch the **intersection** of the two regions.

c. Find the interval of x-values centered at $x = 12$ in which both inequalities are satisfied.

36. *Trigonometric Inequality Problem 2:* Figure 4-4p shows the region of points that satisfy the trigonometric inequality

$$y \le 5 \sin \frac{\pi}{4}x$$

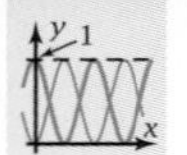

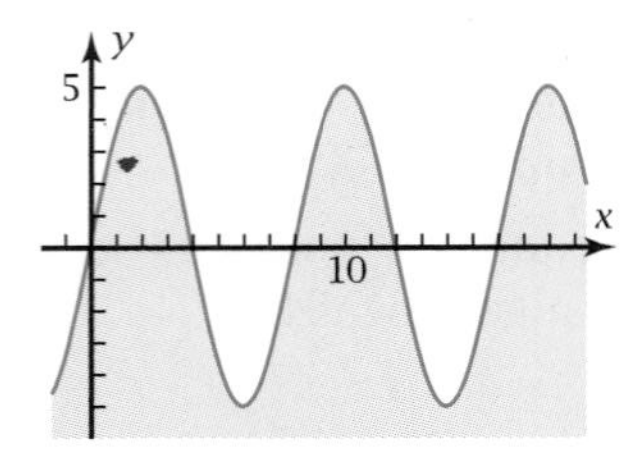

Figure 4-4p

a. Duplicate the figure on your grapher. Use the appropriate style to shade the region.

b. On the same screen, plot the region of points that satisfy the inequality $y \geq 0.3x$. Sketch the intersection of the two regions for $x \geq 0$.

c. Find all intervals of x-values for $x \geq 0$ for which both inequalities are satisfied.

37. *Surprise Problem:* Try solving this equation algebraically. Show how to interpret the results graphically. In particular, what do the graphs of the two sides look like?

$$\frac{1 - \sin x}{\cos x} = \frac{\cos x}{1 + \sin x}$$

4-5 Parametric Functions

If two related variables x and y both depend on a third, independent variable t, the pair of equations in x and t and y and t is called a **parametric function.** For instance, if a pendulum is swinging in a rotating path as in Figure 4-5a, then both the x- and y-coordinates of the pendulum bob depend on the time, t. In this section you'll apply the Pythagorean properties to prove that the graphs of certain parametric functions are ellipses or hyperbolas.

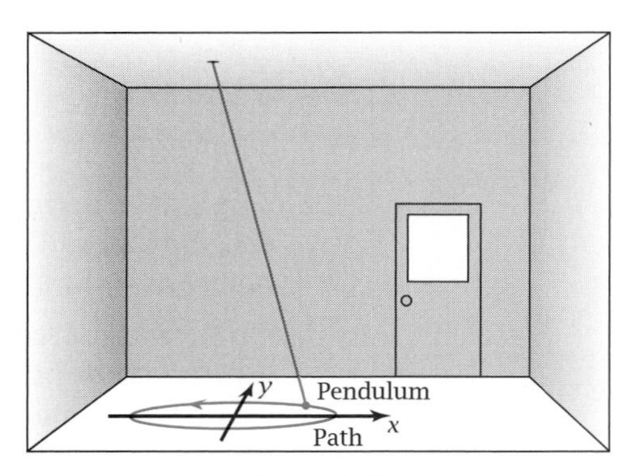

Figure 4-5a

OBJECTIVE Given equations of a parametric function, plot the graph and make conclusions about the geometrical figure that results.

Suppose that the pendulum illustrated in Figure 4-5a has two displacements, measured in centimeters from its rest position. Its displacement is x in the direction parallel to the wall with the door and y in the direction perpendicular to that wall. Neglecting the effects of air friction on the pendulum, x and y will be sinusoidal functions of time t, measured in seconds. For instance, the equations might be

$$x = 30 \cos \frac{\pi}{1.5}t$$

$$y = 20 \sin \frac{\pi}{1.5}t$$

These are the **parametric equations** for the position of the moving pendulum. The independent variable t is called the **parameter.** (The prefix *para-* is a Greek word meaning "beside" or "near" (as in *parallel*), and the suffix *-meter* means "measure.") Because x and y are often functions of time, the variable t is usually used for the parameter.

This example shows you how to plot a pair of parametric equations when both are sinusoids, as they would be in the pendulum example. So that you may more easily see some properties of parametric functions, some equations will be graphed in degree mode.

▶ EXAMPLE 1 Plot the graph of the following parametric function in degree mode.

$$x = 5 \cos t$$
$$y = 7 \sin t$$

Solution Set your grapher to parametric mode and enter the two equations. Choose a window that uses equal scales on both axes. Because the amplitudes of x and y are 5 and 7, respectively, the window will have to be at least −5 to 5 in the x-direction and −7 to 7 in the y-direction. Set the t-range for at least 0° to 360°. Use a t-step of 5°. Figure 4-5b shows the graph. ◀

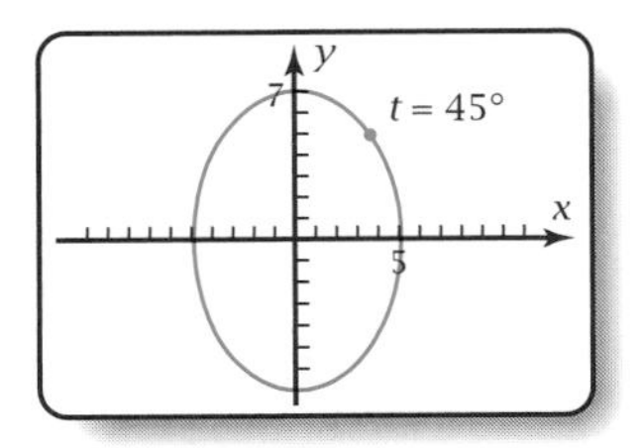

Figure 4-5b

Note that the parameter t is *not* an angle in standard position. Figure 4-5b shows that when $t = 45°$, the angle is considerably larger than 45°. In Chapter 13, you'll learn geometrical properties of parametric functions that reveal how the angle is related to points on the path.

Pythagorean Properties to Eliminate the Parameter

You can sometimes discover properties of a graph by eliminating the parameter, thus getting a single Cartesian equation with only x and y. The next example shows you how.

▶ EXAMPLE 2 For the parametric function $x = 5 \cos t$, $y = 7 \sin t$ in Example 1, eliminate the parameter to get a Cartesian equation relating x and y. Describe the graph.

Solution Since $\cos^2 t + \sin^2 t = 1$, you can eliminate the parameter by solving the given equations for $\cos t$ and $\sin t$, squaring both sides of each equation and then adding.

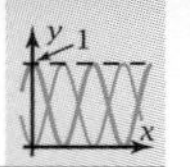

$$x = 5\cos t \Rightarrow \frac{x}{5} = \cos t \Rightarrow \left(\frac{x}{5}\right)^2 = \cos^2 t$$

$$y = 7\sin t \Rightarrow \frac{y}{7} = \sin t \Rightarrow \left(\frac{y}{7}\right)^2 = \sin^2 t$$

$$\left(\frac{x}{5}\right)^2 + \left(\frac{y}{7}\right)^2 = \cos^2 t + \sin^2 t$$

Add the two equations, left side to left side, right side to right side.

$$\left(\frac{x}{5}\right)^2 + \left(\frac{y}{7}\right)^2 = 1$$

Use the Pythagorean property for cosine and sine.

The path is an **ellipse.**

The unit circle $x^2 + y^2 = 1$ is dilated horizontally by 5 and vertically by 7.

▶ **EXAMPLE 3** Plot the graph of the parametric equations

$$x = 6 + 5\cos t$$
$$y = -3 + 7\sin t$$

Describe the effect of the constants 6 and −3 on the graph.

Solution The graph in Figure 4-5c shows that the 6 and the −3 are horizontal and vertical translations, respectively, and give the coordinates of the center of the ellipse.

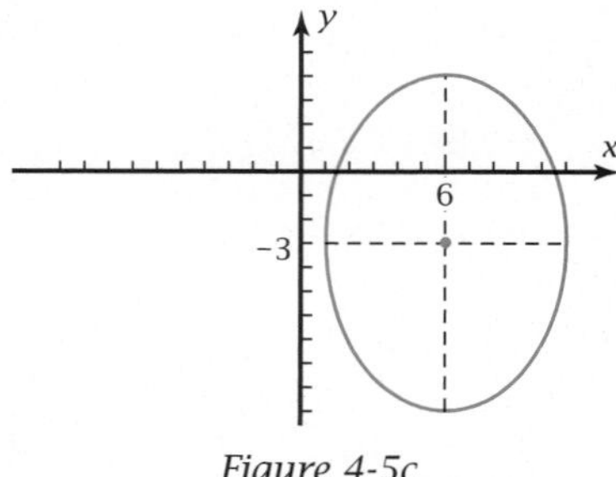

Figure 4-5c

Parametric Equations from Graphs

From the previous examples you can generalize the parametric equations of an ellipse.

> **PROPERTY: Parametric Equations for an Ellipse**
>
> The general parametric equations for an ellipse are
>
> $$x = h + a\cos t$$
> $$y = k + b\sin t$$
>
> where a and b are called the x- and y-radii, respectively, and h and k are the coordinates of the center. If $a = b$, the figure is a circle.

Note: The coefficients a and b are also the horizontal and vertical dilations of the **unit circle**

$$x = \cos t$$
$$y = \sin t$$

Also, the constants h and k are the horizontal and vertical translations, respectively, of the center of the unit circle.

If you know this property, you can use parametric functions to plot pictures of solid objects such as cones and cylinders on your grapher. The next example shows you how to do this.

▶ EXAMPLE 4 Figure 4-5d shows the outlines of a cylinder. Duplicate this figure on your grapher by finding parametric equations of ellipses to represent the bases and then drawing lines to represent the walls.

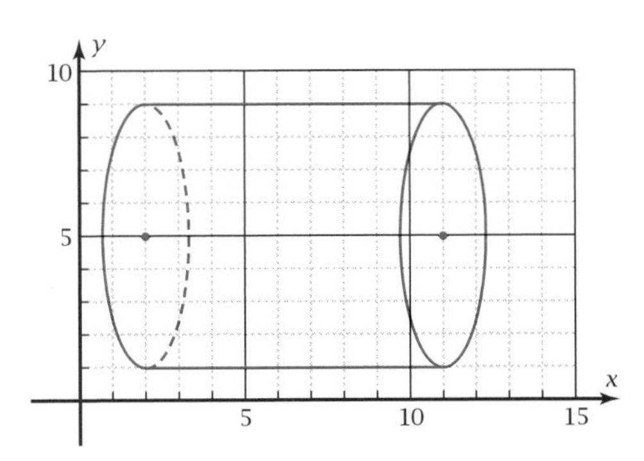

Figure 4-5d

Solution The bases are ellipses because the cylinder is shown in perspective. The right ellipse is centered at the point (11, 5), with an x-radius of about 1.3 and a y-radius of 4. The left ellipse is congruent with the right one, with the center at the point (2, 5). Half of the left ellipse is hidden by the cylinder.

$$x_1 = 11 + 1.3\cos t$$
$$y_1 = 5 + 4\sin t$$

Parametric equations for the right ellipse.

$$x_2 = 2 + 1.3\cos t$$
$$y_2 = 5 + 4\sin t$$

Parametric equations for the left ellipse.

Plot these functions on your grapher in parametric mode, using degrees, with equal scales on the two axes. Use a t-range of 0° to 360° to get a complete revolution for each ellipse. Use the draw command to draw lines from the point (2, 9) to the point (11, 9) and from the point (2, 1) to the point (11, 1), representing the walls of the cylinder. ◀

Problem Set 4-5

Do These Quickly

Q1. What is the Pythagorean property for cosine and sine?

Q2. What is the Pythagorean property for secant and tangent?

Q3. If $\cos^{-1} x = 1.2$, what is the general solution for $\arccos x$?

Q4. If $\sin^{-1} x = 56°$, what is the general solution for $\arcsin x$?

Q5. For right triangle ABC, if B is the right angle, then $\sin A$ = —?—.

Q6. For right triangle ABC in Q5, side a^2 = —?— in terms of sides b and c.

Q7. If $y = \cos B\theta$ has a period of 180°, what does B equal?

Q8. What is the period of the parent sine function $y = \sin x$?

Q9. If an angle has measure $\frac{\pi}{6}$ radians, what is its degree measure?

Q10. The exact value of $\cos \frac{\pi}{4}$ is

A. 0 B. $\frac{1}{\sqrt{2}}$ C. $\frac{1}{2}$

D. $\frac{\sqrt{3}}{2}$ E. 1

Problems 1 and 2 show you the relationship among x, y, and t in parametric functions. For each problem,

a. Make a table of x- and y-values for a range of t-values. Include negative values of t.

b. Plot the points (x, y) on graph paper and connect them with a line or smooth curve.

c. Confirm that your graph is correct by plotting it on your grapher using parametric mode.

1. $x = 3t + 1$
   $y = 2t - 1$

2. $x = 1 + t^2$
   $y = t + 2$

For Problems 3–6,

a. Plot the graph on your grapher. Sketch the results.

b. Use the Pythagorean property for cosine and sine to eliminate the parameter t.

c. Explain how you know that the graph is an ellipse or a circle.

3. $x = 3 \cos t$
   $y = 5 \sin t$

4. $x = 6 \cos t$
   $y = 6 \sin t$

5. $x = 5 + 7 \cos t$
   $y = 2 + 3 \sin t$

6. $x = 4 + 3 \cos t$
   $y = -1 + 6 \sin t$

The truncated cylindrical tower of the Museum of Modern Art in San Francisco, California, has an elliptical cross section.

Problems 7–14 show solid three-dimensional figures. The ellipses represent circular bases of the solids. The dashed lines represent hidden edges.

a. Write parametric equations for the ellipses.

b. Make the figure on your grapher using draw commands for the straight lines.

7. Cone

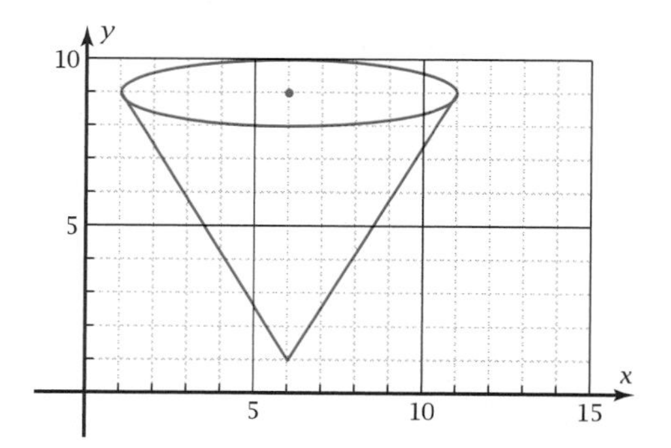

8. Cone

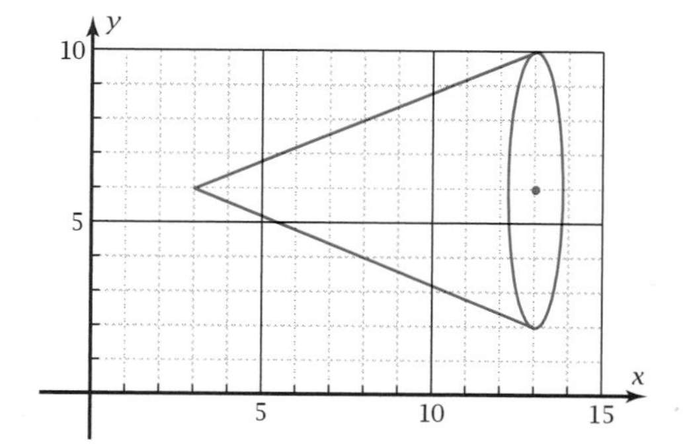

9. Cylinder

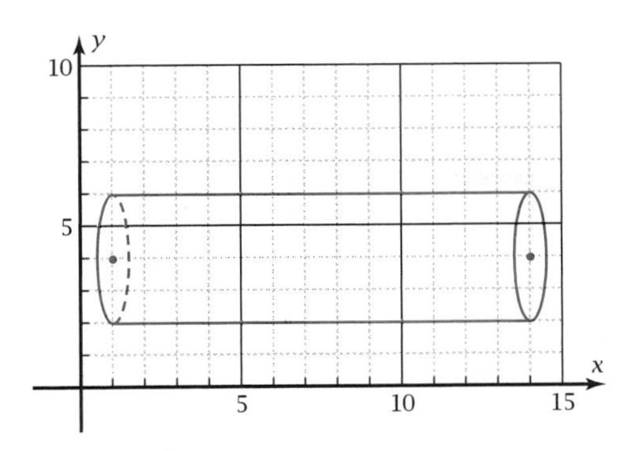

10. Cylinder

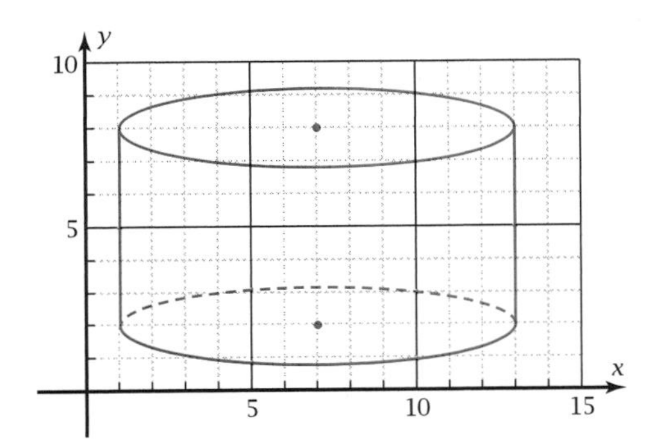

11. Frustum of a cone

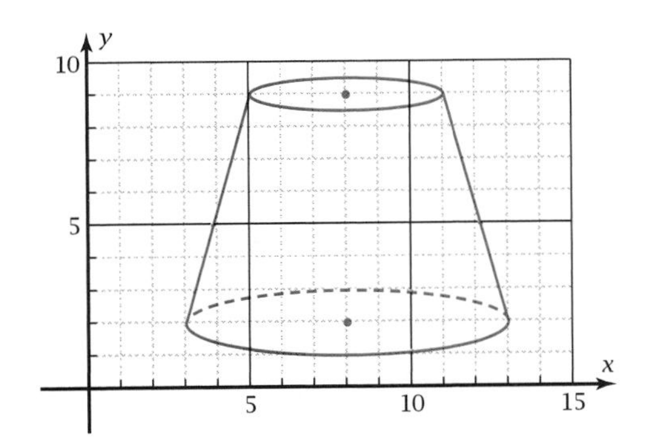

12. Two-napped cone

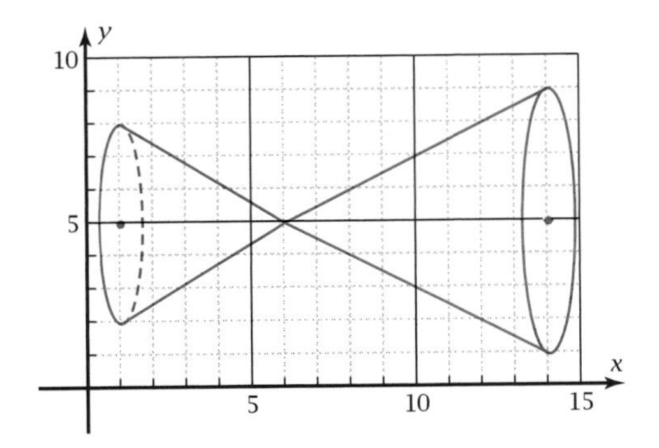

13. Hemisphere
Include the equation of the semicircle.

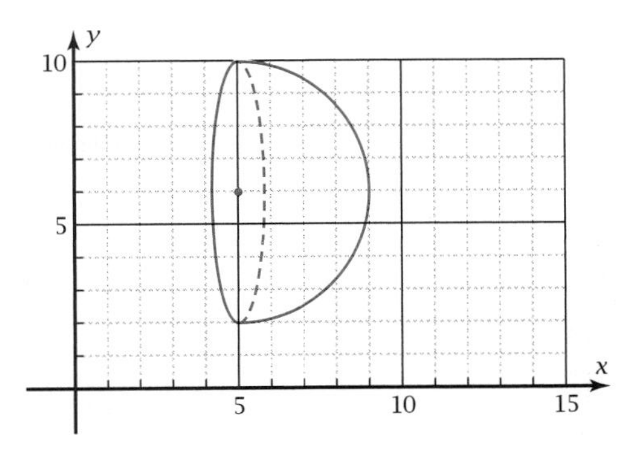

14. Hemisphere
Include the equation of the semicircle.

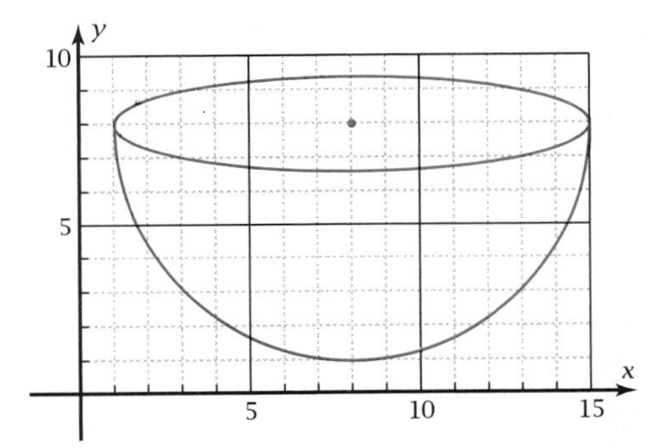

15. *Projectile Problem:* If a ball is thrown through the air, its motion in the horizontal and vertical directions is modeled by two different physical laws. Horizontally, the ball moves at a constant rate if you ignore air resistance. Vertically, the ball accelerates downward due to gravity. Let x be the ball's horizontal displacement from its starting point and y be the vertical displacement above its starting point. Suppose a ball is thrown with a horizontal velocity of 20 m/sec and an initial upward velocity of 40 m/sec. The parametric equations for its position (x, y) at time t seconds are

$$x = 20t$$
$$y = 40t - 4.9t^2$$

The graph of x and y as functions of t is shown in Figure 4-5e.

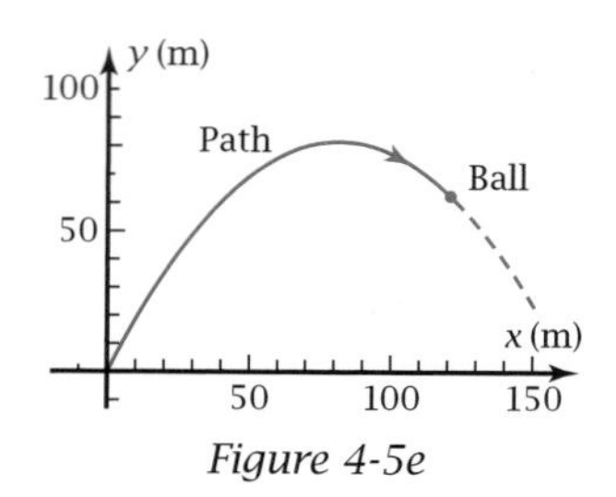

Figure 4-5e

a. What is the position of the ball at time $t = 3$ sec?

b. When is the ball at a horizontal distance of $x = 100$ m? How high is it at that time?

c. At what *two* times is the ball 30 m above the ground? Find x at these times.

d. A fence 2 m high is at $x = 160$ m. According to this parametric function, will the ball go over the fence, hit the fence, or hit the ground before reaching the fence? How can you tell?

e. Eliminate the parameter t, showing that y is a quadratic function of x.

16. *Parametric Function Domain Problem:* Sometimes when you eliminate the parameter, the Cartesian function has a domain different from the parametric function. In this problem you will investigate the parametric function

$$x_1 = 3\cos^2 t$$
$$y_1 = 2\sin^2 t$$

a. Set your grapher to radian mode. Plot the graph using a t-range of $[-2\pi, 2\pi]$ and a window that includes positive and negative values of x and y. Sketch the result.

b. Based on your graph, make a conjecture about what geometric figure the graph is.

c. Eliminate the parameter with the help of the Pythagorean properties. Solve the equation for y in terms of x; that is, find $y = f(x)$. Does the Cartesian equation confirm or refute your answer to part b?

d. You can plot the Cartesian equation in part c in parametric mode this way:

$$x_2 = t$$
$$y_2 = f(t)$$

The $y = f(t)$ is the Cartesian equation you found in part c, with t in place of x. Plot the graph. Compare the x-domain and y-range of the Cartesian equation to those of the parametric equations. Describe your observations.

Problems 17–20 involve parametric functions that have interesting graphs. Plot the graph on your grapher and sketch the result. Use radian mode.

17. *Asteroid Problem:* $x = 8\cos^3 t$
$y = 8\sin^3 t$

This curve is also called a hypocycloid of 4 cusps.

18. *Cycloid Problem:* $x = t + \sin t$
$y = 1 - \cos t$

19. *Conchoid of Nicomedes Problem:*

$$x = \tan t + 5\sin t$$
$$y = 1 + 5\cos t$$

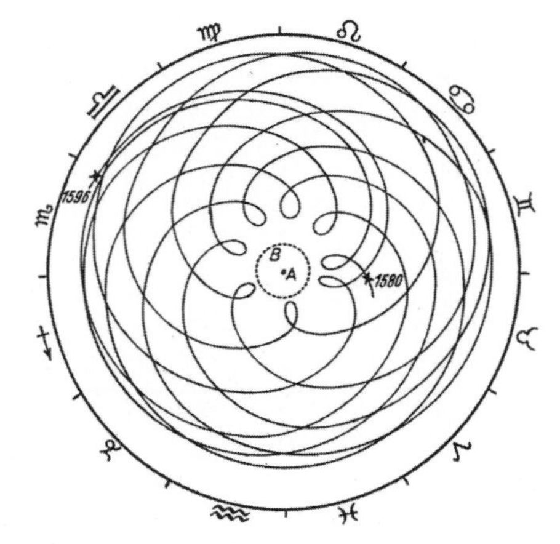

The motion of Mars. Adapted from Johannes Kepler's Astronomia Nova *(1609).*

20. *Involute of a Circle Problem 1:*

$$x = \cos t + t \sin t$$
$$y = \sin t - t \cos t$$

21. *Graphs of Inverse Trigonometric Relations by Parametrics:* Figure 4-5f shows the graph of the relation $y = \arcsin x$. Note that this is *not* a function, because there is more than one value of y corresponding to the same value of x. In this problem you will learn how to duplicate the graph on your grapher.

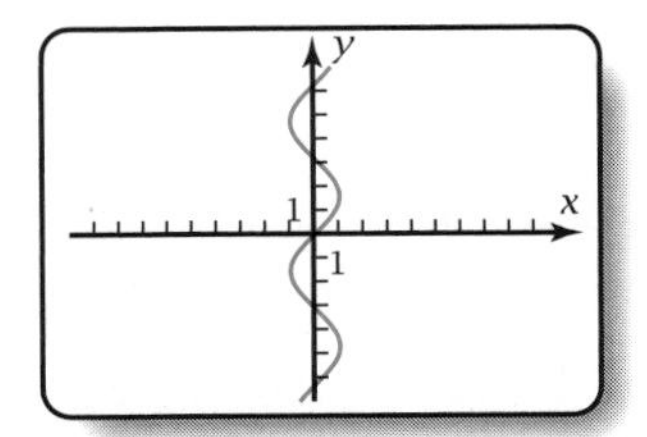

Figure 4-5f

a. With your grapher in function mode, plot $y = \sin^{-1} x$ (radian mode). Use a friendly window with $x = -1$ and $x = 1$ as grid points, a y-range of at least $-7 \le y \le 7$, and approximately equal scales on both axes. Why does the grapher show only part of the graph in Figure 4-5f?

b. Set your grapher in parametric mode. Enter the parametric equations

$$x_1 = \sin t$$
$$y_1 = t$$

Use a t-range as large as the y-window. Describe the results. Based on the definition of arcsine, explain why these parametric equations generate the entire inverse sine relation graph.

c. With the first parametric equations still active, enter the equations

$$x_2 = t$$
$$y_2 = \sin t$$

Use a t-range as large as the x-window. Sketch the resulting graphs. How are the two graphs related to each other?

d. Repeat parts b and c for $y = \arccos x$ and $y = \cos x$.

e. Repeat parts b and c for $y = \arctan x$ and $y = \tan x$.

4-6 Inverse Trigonometric Relation Graphs

You have learned that an inverse trigonometric relation, such as arcsin 0.4, has many values; but when you enter $\sin^{-1} 0.4$ into your calculator, it gives you only one of those values. In this section you'll learn which value the calculator has been programmed to give. You'll also learn how to calculate exact values of inverse trigonometric functions.

OBJECTIVE

- Plot graphs of inverse trigonometric functions and relations.
- Find exact values of functions of inverse trigonometric functions.

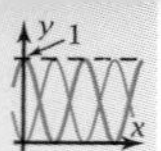

Graphs and Principal Branches

Figure 4-6a shows $y = \tan^{-1} x$, the inverse trigonometric *function.* It is a reflection of one branch of the graph of $y = \tan x$ through the line $y = x$ (Figure 4-6b).

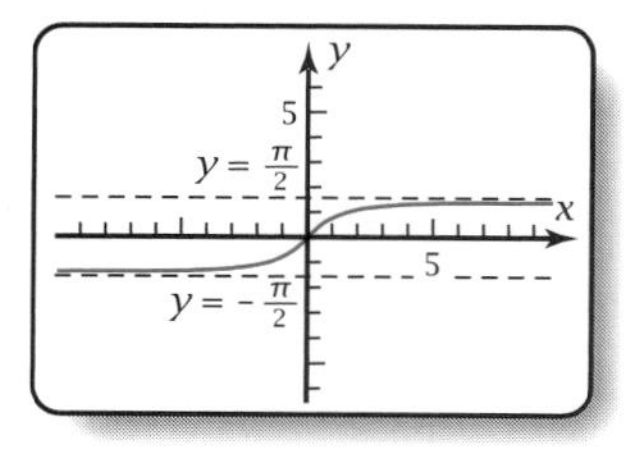

Figure 4-6a

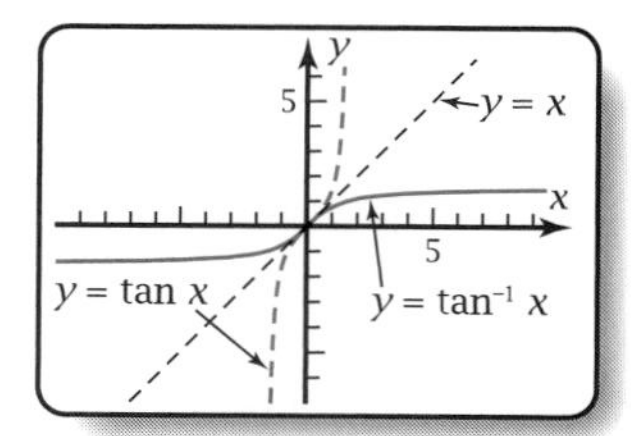

Figure 4-6b

The inverse tangent function is called the **principal branch** of the inverse tangent relation, $y = \arctan x$. Using parametric mode, you can plot $y = \arctan x$. Enter

$$x = \tan t$$
$$y = t$$

The definition of arctangent tells you that $y = \arctan x$ if and only if $x = \tan y$. So if you set $y = t$, you can think of parametric mode as an x= menu. The graph will have all branches of $y = \arctan x$ that fit in the window you have chosen (Figure 4-6c).

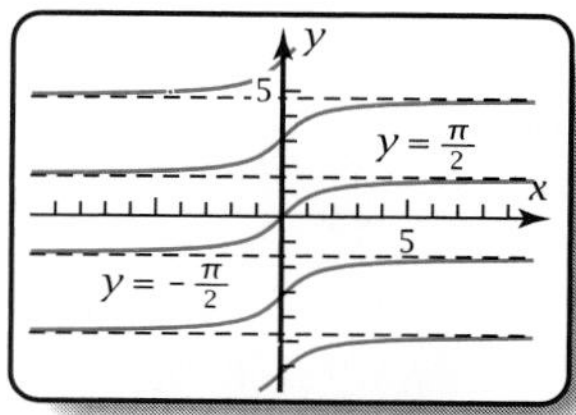

Figure 4-6c

The inverse circular function, $y = \tan^{-1} x$, is defined by designating one of the branches of arctangent to be the principal branch. Do this by restricting the range of the arctangent to meet the criteria in this box.

Criteria for Selecting Principal Branches for Inverse Trigonometric Functions

1. It must be a *function.*
2. It must use the *entire domain* of the inverse trigonometric relation.
3. It should be one *continuous* graph, if possible.
4. It should be *centrally located,* near the origin.
5. If there is a choice between two possible branches, use the *positive* one.

The ranges of the other five inverse trigonometric functions are also defined to meet these criteria. Figure 4-6d shows the results of applying these criteria to all six inverse trigonometric relations. The highlighted portion of each graph shows the inverse trigonometric function. The rest of each graph shows more of the inverse trigonometric relation. Notice the ranges of y that give the principal branches.

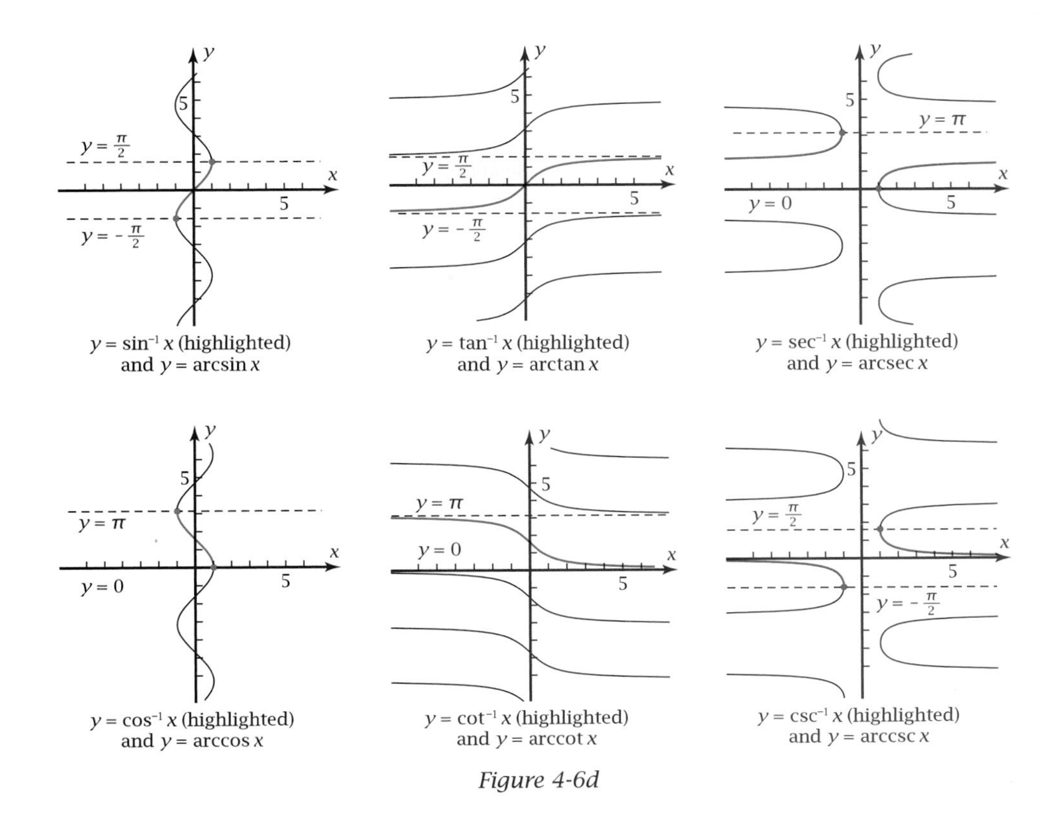

$y = \sin^{-1} x$ (highlighted) and $y = \arcsin x$

$y = \tan^{-1} x$ (highlighted) and $y = \arctan x$

$y = \sec^{-1} x$ (highlighted) and $y = \text{arcsec}\, x$

$y = \cos^{-1} x$ (highlighted) and $y = \arccos x$

$y = \cot^{-1} x$ (highlighted) and $y = \text{arccot}\, x$

$y = \csc^{-1} x$ (highlighted) and $y = \text{arccsc}\, x$

Figure 4-6d

DEFINITIONS: Ranges and Domains of Inverse Trigonometric Functions

FUNCTION	*RANGE (NUMERICALLY)*	*RANGE (VERBALLY)*	*DOMAIN*
$y = \sin^{-1} x$	$y \in [-\frac{\pi}{2}, \frac{\pi}{2}]$	Quadrants I and IV	$x \in [-1, 1]$
$y = \cos^{-1} x$	$y \in [0, \pi]$	Quadrants I and II	$x \in [-1, 1]$
$y = \tan^{-1} x$	$y \in \left(-\frac{\pi}{2}, \frac{\pi}{2}\right)$	Quadrants I and IV	$x \in (-\infty, \infty)$
$y = \cot^{-1} x$	$y \in (0, \pi)$	Quadrants I and II	$x \in (-\infty, \infty)$
$y = \sec^{-1} x$	$y \in [0, \pi]$ and $y \neq \frac{\pi}{2}$	Quadrants I and II	$\lvert x \rvert \geq 1$
$y = \csc^{-1} x$	$y \in [-\frac{\pi}{2}, \frac{\pi}{2}]$ and $y \neq 0$	Quadrants I and IV	$\lvert x \rvert \geq 1$

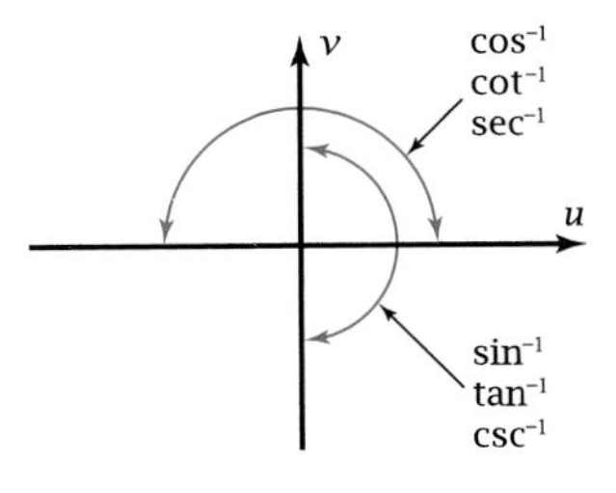

Figure 4-6e

To remember these ranges, it might help you to think of y as an angle in radians in a uv-coordinate system (Figure 4-6e). If the argument is positive, the arc or angle terminates in Quadrant I. If the argument is negative, the arc or angle terminates in Quadrant II or Quadrant IV, depending on which inverse function it is. None of the inverse functions terminates in Quadrant III.

If you could perpendicularly project an image of the railing in this spiral staircase onto the wall behind it, you would get a curve that resembles the graph of arccosine.

Exact Values of Inverse Circular Functions

Recall that it is possible to find exact trigonometric and circular function values for certain special angles or arcs. For instance, $\cos\left(\frac{\pi}{6}\right) = \frac{\sqrt{3}}{2}$. It is also possible to find exact values of expressions involving inverse trigonometric functions.

▶ ***EXAMPLE 1*** Evaluate $\tan\left(\sin^{-1}\left(-\frac{2}{3}\right)\right)$ geometrically to find the exact value. Check your answer numerically.

Solution Draw an angle in standard position whose sine is $-\frac{2}{3}$. The angle terminates in Quadrant IV because the range of the inverse sine function is Quadrants I and IV. Draw the reference triangle, as shown in Figure 4-6f, and find the third side. Then use the definition of tangent.

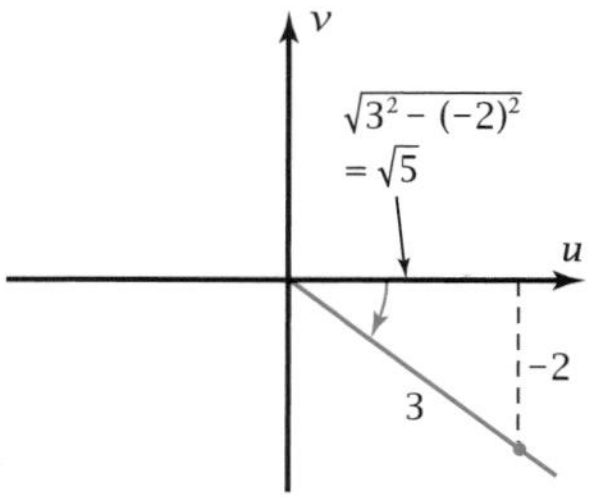

$$\tan\left(\sin^{-1}\left(-\tfrac{2}{3}\right)\right) = \frac{-2}{\sqrt{5}}$$

Figure 4-6f

Check: When you evaluate $\frac{-2}{\sqrt{5}}$, you get $-0.8844...$, which agrees with $\tan\left(\sin^{-1}\left(-\frac{2}{3}\right)\right) = \tan(-0.7297...) = -0.8844....$ ◀

▶ ***EXAMPLE 2*** Evaluate $y = \sin(\cos^{-1} x)$ geometrically to find the answer in radical form. Set your answer to y, and plot it together with the original equation on the same screen to confirm that your answer is correct.

Solution Figure 4-6g shows the two possible quadrants for $\cos^{-1} x$.

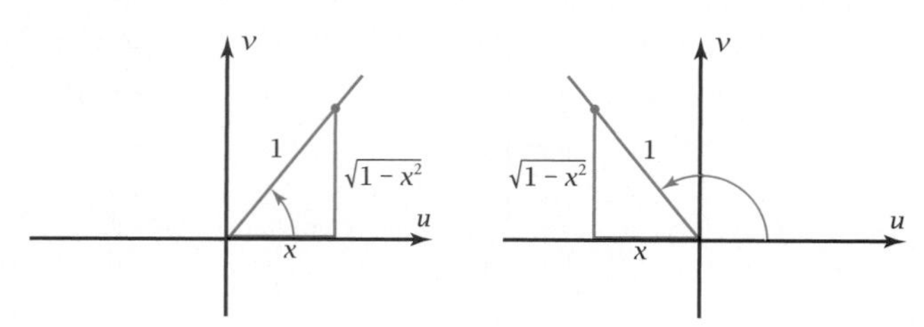

Figure 4-6g

By the definition of cosine, you can label the horizontal leg of the reference triangle as x and the radius as 1. The third side is given by the Pythagorean theorem. Note that in both Quadrants I and II, the third side is *positive.* So you use the positive square root in both cases. By the definition of sine,

$$y = \sin(\cos^{-1} x) = \frac{\sqrt{1 - x^2}}{1} = \sqrt{1 - x^2}$$

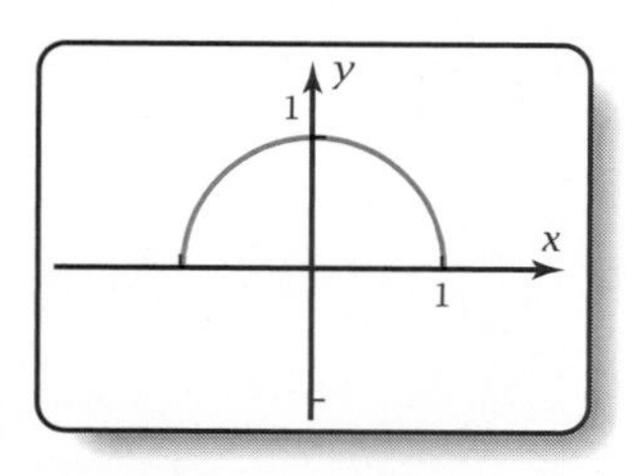

Figure 4-6h

Figure 4-6h shows that the two graphs are both semicircles of radius 1. ◀

The Composite of a Function and Its Inverse Function

If you apply the techniques of Examples 1 and 2 to a function and its inverse function, an interesting property reveals itself. Example 3 shows how this is done.

▶ EXAMPLE 3 Evaluate $y = \cos(\cos^{-1} x)$. Explain why the answer is reasonable. Set your answer equal to y, and plot it together with the original function on the same screen to verify that the answer is correct.

Solution $\cos^{-1} x$ means "the angle whose cosine is x." So by definition,

$$y = \cos(\cos^{-1} x) = x$$

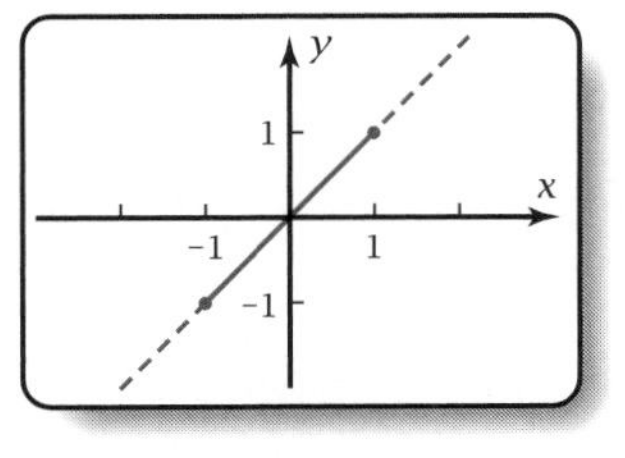

Figure 4-6i

Draw a *uv*-diagram as in the earlier examples if you need further convincing.

Figure 4-6i shows both the $y = x$ and the $y = \cos(\cos^{-1} x)$ graphs. Note that the graph of $\cos(\cos^{-1} x)$ has domain $[-1, 1]$ because the inverse cosine function is defined only for those values of x. ◀

Example 3 illustrates a general property of the function of an inverse function, which you may recall from Section 1-5.

PROPERTY: The Composite of a Function and Its Inverse Function

$$f(f^{-1}(x)) = x \qquad \text{and} \qquad f^{-1}(f(x)) = x$$

provided x is in the *range* of the outside function and in the *domain* of the inside function.

In Problem 23 of Problem Set 4-6, you will prove this property. To illustrate the restrictions in the box,

$$\cos^{-1}(\cos 10) = \cos^{-1}(-0.8390...)$$
$$= 2.5663..., \text{ not } 10. \qquad \text{10 is not in the range of } \cos^{-1}.$$

$$\cos(\cos^{-1} 3) \text{ is undefined, not } 3. \qquad \text{3 is not in the domain of } \cos^{-1}.$$

In the first case, 10 is not in the range of the inverse cosine function (principal branch). In the second case, 3 is not in the domain of the inverse cosine function.

Problem Set 4-6

Do These Quickly

Q1. The function $y = 5 + 6 \cos 7(x - 8)$ is a horizontal translation of $y = \cos x$ by —?— units.

Q2. The sinusoid in Q1 is a vertical translation of $y = \cos x$ by —?— units.

Q3. The sinusoid in Q1 is a vertical dilation of $y = \cos x$ by —?— units.

Q4. The sinusoid in Q1 is a horizontal dilation of $y = \cos x$ by —?— units.

Q5. The period of the sinusoid in Q1 is —?—.

Q6. If $f(x) = x^3$, then the inverse function $f^{-1}(x) =$ —?—.

Q7. What geometric figure is the graph of the parametric functions $x = 3 \cos t$ and $y = 5 \sin t$?

Q8. Write the Pythagorean property that involves tangent.

Q9. Without your grapher, evaluate $\cos \pi$.

Q10. Given $A = \arcsin x$, write the general solution for A in terms of $\sin^{-1} x$.

---

1. With your grapher in function mode, plot the graphs of $y = \sin^{-1} x$, $y = \cos^{-1} x$, and $y = \tan^{-1} x$. Use a friendly window that includes $x = 1$ and $x = -1$ as grid points. How do the graphs compare with the graphs shown in Figure 4-6d? Specifically, does each graph have the same y-range as shown for the principal branch?
2. With your grapher in parametric mode, plot the graphs of $y = \arcsin x$, $y = \arccos x$, and $y = \arctan x$. Use a friendly window that includes $x = 1$ and $x = -1$ as grid points. Use equal scales on both axes. To make the graph fill the screen, the t-range should be the same as the y-range. How do the graphs compare with the graphs shown in Figure 4-6d?

For Problems 3 and 4, with your grapher in parametric mode, plot on the same screen the two graphs. Use a window for y with a range of at least $[-7, 7]$ and a window for x that makes the scales on the two axes come out the same. Use different styles for the two graphs. Describe what you can do to show that these two graphs are reflections across the line $y = x$.

3. $y = \arcsin x$ and $y = \sin x$
4. $y = \arctan x$ and $y = \tan x$

For Problems 5–14, calculate the exact value of the inverse function geometrically. Assume the principal branch in all cases. Check your answers by direct calculation.

5. $\tan\left(\cos^{-1} \frac{4}{5}\right)$
6. $\cos\left(\tan^{-1} \frac{4}{3}\right)$
7. $\sin\left(\tan^{-1} \frac{5}{12}\right)$
8. $\sec\left(\sin^{-1} \frac{15}{17}\right)$
9. $\cos\left(\sin^{-1}\left(-\frac{8}{17}\right)\right)$
10. $\cot\left(\csc^{-1}\left(-\frac{13}{12}\right)\right)$
11. $\sec\left(\cos^{-1} \frac{2}{3}\right)$ (Surprise?)
12. $\tan(\cot^{-1} 4)$ (Surprise?)
13. $\cos(\cos^{-1} 3)$
14. $\sec(\text{arcsec}\ 0)$
15. Explain why $\cos(\cos^{-1} 3)$ in Problem 13 does *not* equal 3.
16. Explain why $\sec(\text{arcsec}\ 0)$ in Problem 14 does *not* equal 0.

For Problems 17–22, evaluate the function geometrically to find the answer in radical form. Set your answer equal to y, and plot it together with the original equation on the same screen to show that your answer is correct.

17. $y = \cos(\sin^{-1} x)$
18. $y = \tan(\sin^{-1} x)$
19. $y = \sin(\tan^{-1} x)$
20. $y = \cos(\tan^{-1} x)$
21. $y = \sin(\sin^{-1} x)$
22. $y = \tan(\tan^{-1} x)$
23. *Composite of a Function and Its Inverse Problem:* In Problems 21 and 22, you found that $\sin(\sin^{-1} x) = x$ and that $\tan(\tan^{-1} x) = x$. These are examples of a general property of functions and their inverse functions, to which you were introduced in Chapter 1. In this problem you will prove the property.
    a. Prove that $f^{-1}(f(x)) = x$ by letting $y = f(x)$, applying the definition of f^{-1} and using a clever substitution.
    b. Prove that $f(f^{-1}(x)) = x$ by letting $y = f^{-1}(x)$, applying the definition of f^{-1} and using a clever substitution.
24. *Interpretation Problem—Composite of a Function and Its Inverse:* In Problem 23, you proved that the composite function of a function and its inverse function is equal to x. In this problem you will see some surprises!
    a. Explain why the graph of $y = \tan(\tan^{-1} x)$ in Problem 22 is equivalent to $y = x$ for all values of x but the graph of $y = \sin(\sin^{-1} x)$ in Problem 21 is equivalent to $y = x$ for only certain values of x.
    b. Figure 4-6j shows the result of plotting on your grapher

    $$y = \sin^{-1}(\sin x)$$

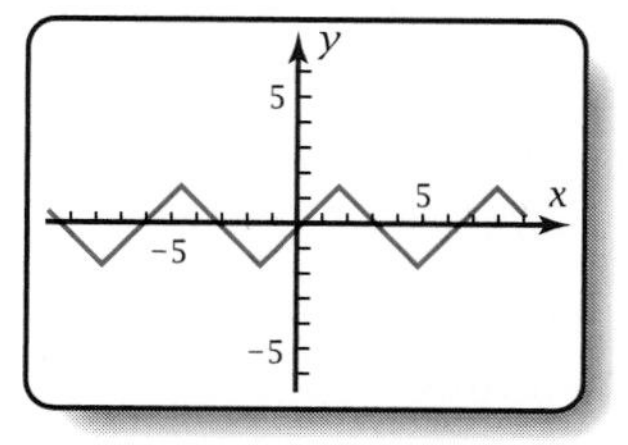

Figure 4-6j

Why is the graph "saw-toothed" instead of linear?

c. Plot $y = \cos^{-1}(\cos x)$. Sketch the result. Explain why the graph is *not* equivalent to $y = x$.

d. Plot $y = \tan^{-1}(\tan x)$. Use dot style rather than connected style. Sketch the result. Explain why the graph is *not* equivalent to $y = x$.

25. *Tunnel Problem:* Scorpion Gulch and Western Railway is preparing to build a new line through Rolling Mountains. They have hired you to do some calculations for tunnels and bridges needed on the line (Figure 4-6k).

You set up a Cartesian coordinate system with its origin at the entrance to a tunnel through Bald Mountain. Your surveying crew finds that the mountain rises 250 m above the level of the track and that the next valley goes down 50 m below the level of the track. The cross section of the mountain and valley is roughly sinusoidal, with a horizontal distance of 700 m from the top of the mountain to the bottom of the valley (Figure 4-6k).

a. Write a particular equation expressing the vertical distance y, in meters, from the track to the surface of the mountain or valley as a function of x, in meters, from the tunnel entrance. You can find the constants A, B, and C from the given information. Finding the phase displacement D requires that you substitute the other three constants and the coordinates (0, 0) for (x, y), then solve for D.

b. How long will the tunnel be? How long will the bridge be?

c. The railway company thinks it might be cheaper to build the line if it is raised by 20 m. The tunnel will be shorter, and the bridge will be longer. Find the new values of x at the beginning and end of the tunnel and at the beginning and end of the bridge. How long will each be under these conditions?

26. *Journal Problem:* Update your journal with things you have learned since the last entry. Include such things as

- How to plot graphs of *inverse trigonometric relations*
- How the ranges of the *inverse trigonometric functions* are chosen
- How to calculate values of inverse trigonometric functions geometrically

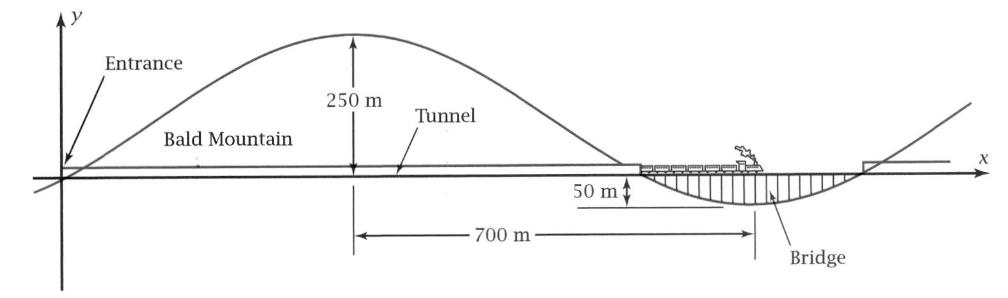

Figure 4-6k

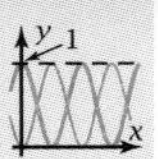

4-7 Chapter Review and Test

In this chapter you've learned how to transform trigonometric expressions and solve equations using the Pythagorean, quotient, and reciprocal properties. The Pythagorean properties help to show that certain parametric function graphs are circles or ellipses. The parametric functions let you plot graphs of inverse trigonometric relations. Analyzing these graphs and identifying the principal branches give more meaning to the values the calculator gives for the inverse trigonometric functions.

Review Problems

R0. Update your journal with what you have learned in this chapter. Include such topics as

- Statements of the three kinds of properties
- How to prove that a trigonometric equation is an identity
- How to solve conditional trigonometric equations algebraically, numerically, and graphically
- What a parametric function is, how to graph it, and how to eliminate the parameter to get a Cartesian equation
- How to graph inverse trigonometric relations and find ranges for inverse trigonometric functions

R1. Figure 4-7a shows a unit circle and an angle θ in standard position.

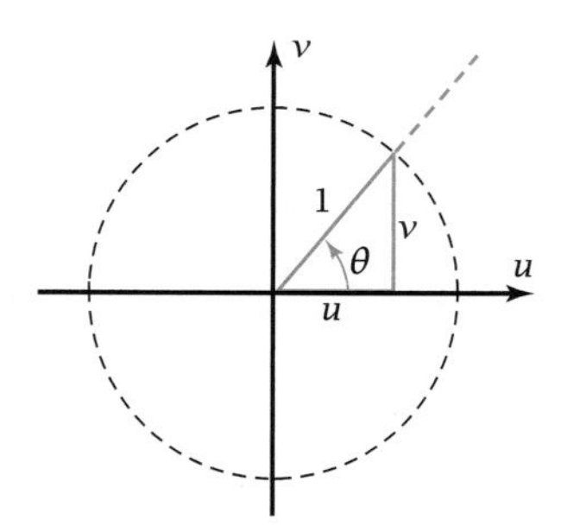

Figure 4-7a

a. Explain why $u^2 + v^2 = 1$.

b. Explain why $u = \cos\theta$ and $v = \sin\theta$.

c. Explain why $\cos^2\theta + \sin^2\theta = 1$.

d. Give a numerical example that confirms the property in part c.

e. Plot on the same screen $y_1 = \cos^2\theta$ and $y_2 = \sin^2\theta$. Sketch the graphs. How do the graphs support the Pythagorean property $\cos^2\theta + \sin^2\theta = 1$?

R2. a. Write equations expressing $\tan x$ and $\cot x$ in terms of $\sin x$ and $\cos x$.

b. Write equations expressing $\tan x$ and $\cot x$ in terms of $\sec x$ and $\csc x$.

c. Write three equations in which the product of two trigonometric functions equals 1.

d. Make a table of values showing numerically that $\cos^2 x + \sin^2 x = 1$.

e. Write equations expressing

i. $\sin^2 x$ in terms of $\cos x$

ii. $\tan^2 x$ in terms of $\sec x$

iii. $\csc^2 x$ in terms of $\cot x$

f. Sketch the graph of the parent function $y = \cos x$. On the same set of axes, sketch the graph of $y = \sec x$ using the fact that secant is the reciprocal of cosine.

R3. a. Transform $\tan A \sin A + \cos A$ to $\sec A$. What values of A are excluded from the domain?

b. Transform $(\cos B + \sin B)^2$ to $1 + 2\cos B \sin B$. What values of B are excluded from the domain?

c. Transform $\frac{1}{1 + \sin C} + \frac{1}{1 - \sin C}$ to $2\sec^2 C$. What values of C are excluded from the domain?

d. Prove that the equation $\csc D(\csc D - \sin D) = \cot^2 D$ is an identity. What values of A are excluded from the domain?

e. Prove that the equation $(3 \cos E + 5 \sin E)^2 + (5 \cos E - 3 \sin E)^2 = 34$ is an identity.

f. Show that the two expressions in part b are equivalent by plotting each on your grapher.

g. Make a table of values to show that the equation in part e is an identity.

R4. a. Find the general solution for $\theta = \arcsin 0.3$.

b. Solve $1 + \tan 2\pi(x + 0.6) = 0$ algebraically for the first four positive values of x. Confirm graphically that your solutions are correct.

c. Solve $(2 \cos \theta - 1)(2\sin \theta + \sqrt{3}) = 0$ algebraically in the domain $\theta \in [0°, 540°]$. Confirm graphically that your solutions are correct.

R5. a. Plot the graph of this parametric function on your grapher. Sketch the result.

$$x = -2 + 5 \cos t$$
$$y = 1 + 3 \sin t$$

b. Use the Pythagorean property for cosine and sine to eliminate the parameter in part a.

c. How can you conclude from the answer to part b that the graph is an ellipse? Where is the center of the ellipse? What are the x- and y-radii?

d. Figure 4-7b shows a solid cone in perspective. Write parametric equations for the ellipse that represents the circular base of the cone. Draw the cone on your grapher.

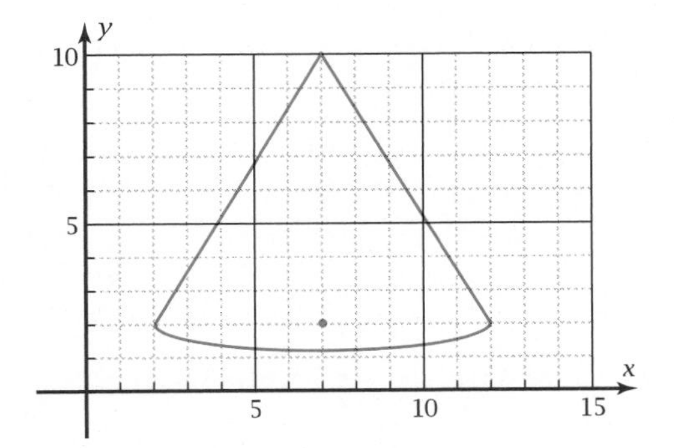

Figure 4-7b

R6. a. Using parametric mode on your grapher, duplicate the graph of the circular relation $y = \arccos x$ shown in Figure 4-7c.

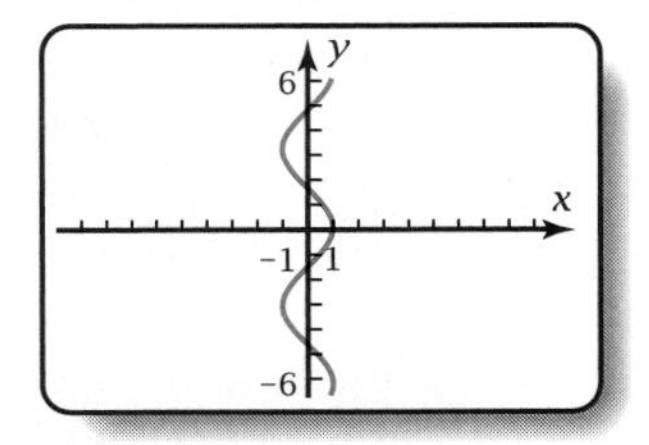

Figure 4-7c

b. Sketch the graph of $y = \cos^{-1} x$, the principal branch of $y = \arccos x$. Explain the specifications used for selecting this principal branch. What is the range of this inverse cosine function?

c. How is the graph of $y = \arccos x$ related to the graph of $y = \cos x$?

d. Find geometrically the exact value (no decimals) of $\sin(\tan^{-1} 2)$. Check the answer by direct calculation.

e. Write an equation for $y = \tan(\cos^{-1} x)$ that does not involve trigonometric or inverse trigonometric functions. Confirm your answer by plotting it together with the given function on the same screen. Sketch the result.

f. Prove that $\cos(\cos^{-1} x) = x$.

g. Show on a *uv*-diagram the range of values of the functions $\sin^{-1}$ and $\cos^{-1}$.

h. Explain why the prefix *arc-* is appropriate in the names *arccos*, *arcsin*, and so on.

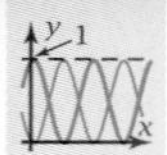

Concept Problems

C1. *Pendulum Problem:* Figure 4-7d shows a pendulum hanging from the ceiling. The pendulum bob traces out a counterclockwise circular path of radius 20 cm (which appears elliptical because it is drawn in perspective). At any time t, in seconds, since the pendulum was started in motion, it is over the point (x, y) on the floor, where x and y are in centimeters. The pendulum makes a complete cycle in 3 sec.

Léon Foucault demonstrates his pendulum and the rotation of Earth at the Pantheon in Paris (1851).

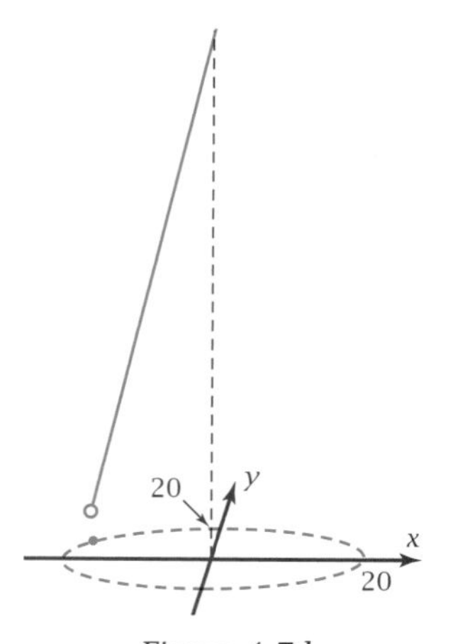

Figure 4-7d

a. Assuming that the pendulum bob was at the point (20, 0) at time $t = 0$, write parametric equations for the circular path it traces.

b. Where is the pendulum at time $t = 5$ sec?

c. Find the first three times when the pendulum bob has a y-coordinate of 10 cm. What are the x-coordinates at each of these times?

d. Explain how this problem ties together all of the topics in this chapter.

C2. Prove that each of these equations is an identity.

a. $$\frac{1 + \sin x + \cos x}{1 + \sin x - \cos x} = \frac{1 + \cos x}{\sin x}$$

b. $$\frac{1 + \sin x + \cos x}{1 - \sin x + \cos x} = \frac{1 + \sin x}{\cos x}$$

C3. *Square of a Sinusoid Problem:* Figure 4-7e shows the graphs of $y_1 = \cos x$ (dashed) and $y_2 = \cos^2 x$ (solid). The squared graph seems to be sinusoidal.

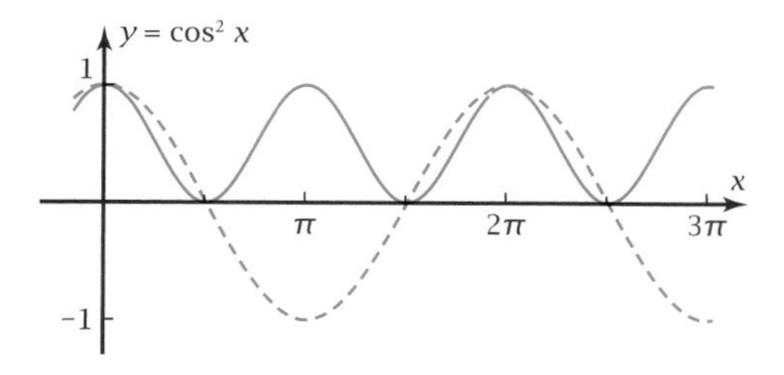

Figure 4-7e

a. Assuming that $y_2 = \cos^2 x$ is a sinusoid, find its period, amplitude, sinusoidal axis location, and phase displacement from the $y = \cos x$ function.

b. Write the particular equation of the sinusoid you described in part a.

c. Give numerical and graphical evidence that the sinusoid in part b is identical to $y_2 = \cos^2 x$.

Chapter Test

PART 1: No calculators allowed (T1–T8)

T1. Write the Pythagorean property for cosine and sine.

T2. Write a quotient property involving cosine and sine.

T3. Write the reciprocal property for cotangent.

T4. Write the reciprocal property for secant.

T5. The value of $\sin^{-1} 0.5$ is 30°. Write the general solution for $\theta = \arcsin 0.5$.

T6. The value of $\tan^{-1} \sqrt{3}$ is $\frac{\pi}{3}$. Write the general solution for $x = \arctan \sqrt{3}$.

T7. Explain why the range of $y = \cos^{-1} x$ is $[0, \pi]$ but the range of $y = \sin^{-1} x$ is $[-\frac{\pi}{2}, \frac{\pi}{2}]$.

T8. Find geometrically the exact value of $\cos(\tan^{-1} 2)$.

PART 2: Graphing calculators allowed (T9–T19)

T9. Transform $(1 + \sin A)(1 - \sin A)$ to $\cos^2 A$. What values of A are excluded from the domain?

T10. Prove that $\tan B + \cot B = \csc B \sec B$ is an identity. What values of B must be excluded from the domain?

T11. Multiply the numerator and denominator of

$$\frac{\sin C}{1 + \cos C}$$

by the conjugate of the denominator. Show that the result is equivalent to

$$\frac{1 - \cos C}{\sin C}$$

T12. Plot the graphs of both expressions in T11 to confirm that the two expressions are equivalent. Sketch the graphs. What values of C are excluded from the domain?

T13. With your calculator in degree mode, find the value of $\cos^{-1} 0.6$. Show the angle in a *uv*-coordinate system.

T14. Find another angle between 0° and 360° whose cosine is 0.6. Show it on the *uv*-coordinate system in T13.

T15. Write the general solution for the inverse trigonometric relation $\theta = \arccos 0.6$. Show how you can use the ± sign to simplify writing this solution.

T16. Find the fifth positive value of θ for which $\cos\theta = 0.6$. How many revolutions, n, do you have to make to get to that value of θ in the *uv*-coordinate system?

T17. Find algebraically the general solution for

$$4\tan(\theta - 25°) = 7$$

T18. Write parametric equations for this ellipse (Figure 4-7f).

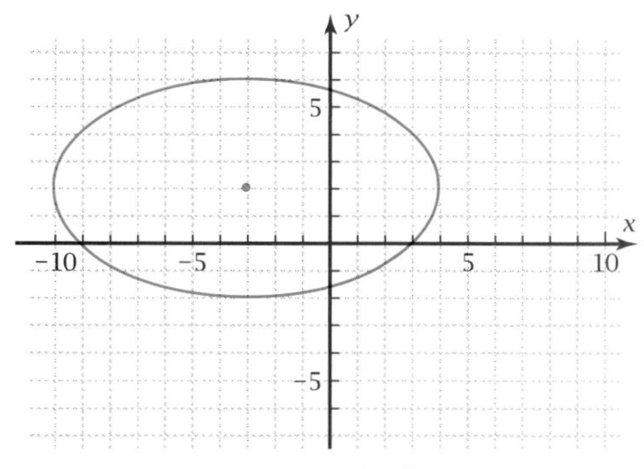

Figure 4-7f

T19. Write the parametric equations you use to plot $y = \arctan x$ (Figure 4-7g).

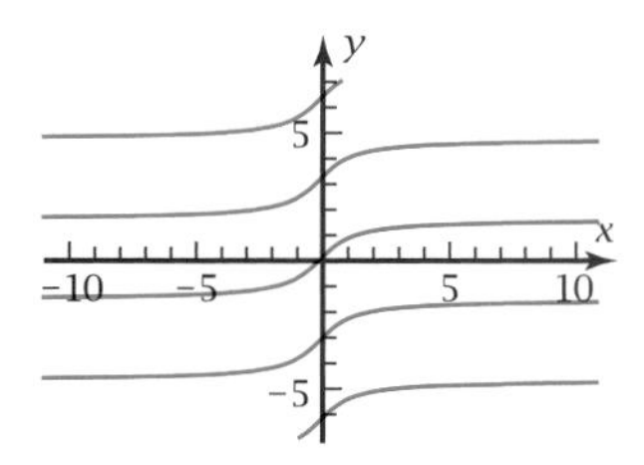

Figure 4-7g

T20. What did you learn as a result of working this test that you did not know before?

CHAPTER

5

Properties of Combined Sinusoids

When two vehicles are going nearly the same speed on the highway, the combined sound of their engines sometimes seems to pulsate. The same thing happens when two airplane engines are going at slightly different speeds. The phenomenon is called *beats.* Using the concept of beats, a vibrato sound can be generated on a piano by tuning two strings for the same note at slightly different frequencies. In this chapter you'll learn about combinations of sinusoids so you can analyze these harmonic phenomena.

Mathematical Overview

In Chapter 4 you learned the Pythagorean, quotient, and reciprocal properties of the trigonometric functions. Each of these properties involves functions of *one* argument. In this chapter you'll learn properties in which functions of *different* arguments appear. These properties allow you to analyze more complicated periodic functions that are sums or products of sinusoids. You'll learn this in four ways.

Graphically A variable-amplitude periodic function

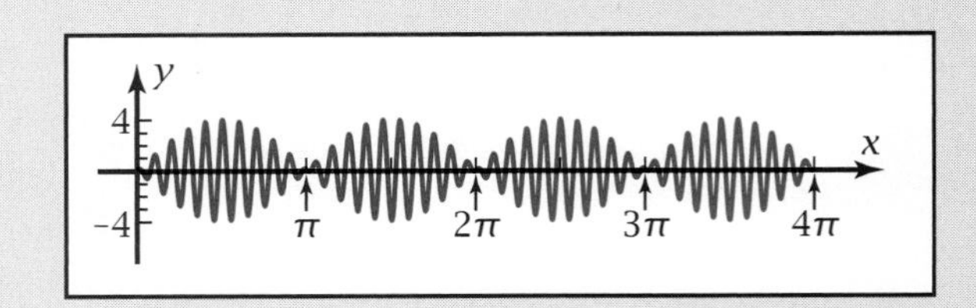

Algebraically You can represent this graph by either a product of two sinusoids or a sum of sinusoids.

$$y = 4 \sin x \cdot \cos 20x \qquad \text{or} \qquad y = 2 \sin 21x + 2 \sin 19x$$

Numerically If $x = 1$, the y-value of either function equals 1.3735....

Verbally *I learned at first that if the amplitude varies, the combined graph is a product of two sinusoids. I was surprised to find out later that you can also write it as a sum of two sinusoids. For the product, the sinusoids have much different periods. For the sum, they have nearly equal periods. I use the sum and product properties to transform one expression into the other. I also learned that this property explains how AM and FM radio work, but I'll need to do more research to fully understand it.*

5-1 Introduction to Combinations of Sinusoids

You probably know that music, like any other sound, is transmitted by waves. A "pure" musical note can be represented by a sine or cosine graph. The frequency of the note is represented by the period of the graph, and the loudness of the note is represented by the amplitude of the graph. Figure 5-1a represents two musical notes of the same frequency that are played at the same time, $y = 3 \cos \theta$ (dotted) and $y = 4 \sin \theta$ (dashed). The solid graph represents the sound wave formed by adding these two sounds, $y = 3 \cos \theta + 4 \sin \theta$. In this section you will explore this combined wave graph and show that it, too, is a sinusoid.

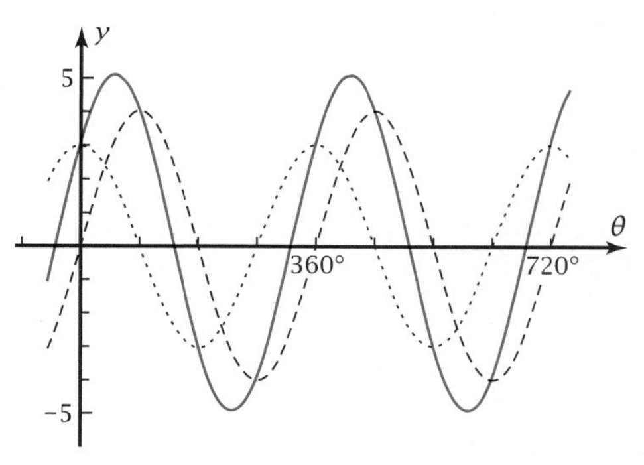

Figure 5-1a

OBJECTIVE Investigate graphs formed by sums of sines and cosines.

Exploratory Problem Set 5-1

1. Plot the graphs of $y_1 = 3 \cos \theta$, $y_2 = 4 \sin \theta$, and $y_3 = y_1 + y_2$ on the same screen. Do your graphs agree with the ones in Figure 5-1a?
2. The y_3 graph in Problem 1 seems to be a sinusoid. Estimate graphically its period, amplitude, and phase displacement (for $y = \cos x$).
3. Find numerically the amplitude and phase displacement of y_3 by using the MAX feature of your grapher to find the first high point. Do the results confirm your estimates in Problem 2?
4. Plot the sinusoid $y_4 = A \cos (\theta - D)$, where A and D are the amplitude and phase displacement you found numerically in Problem 3. Use a different style for this graph so you can distinguish it from the graph of y_3. Does the y_3 graph really seem to be a sinusoid?
5. Make a table of values of y_3 and y_4 for various values of θ. Do the values confirm or refute the conjecture that y_3 is a sinusoid?
6. See if you can find the A and D constants in Problem 4 algebraically, using the factors 3 and 4 from the equations for y_1 and y_2.
7. Substitute two different angles for θ and D and show that $\cos (\theta - D)$ does *not* equal $\cos \theta - \cos D$.
8. Based on your observation in Problem 7, what property of multiplication and subtraction does *not* apply to the operations of cosine and subtraction?
9. Update your journal with things you have learned in this problem set.

5-2 Composite Argument and Linear Combination Properties

In Problem Set 5-1 you saw a graph that represented a sound wave generated by two different musical instruments playing the same note. The equation was

$$y = 3\cos\theta + 4\sin\theta$$

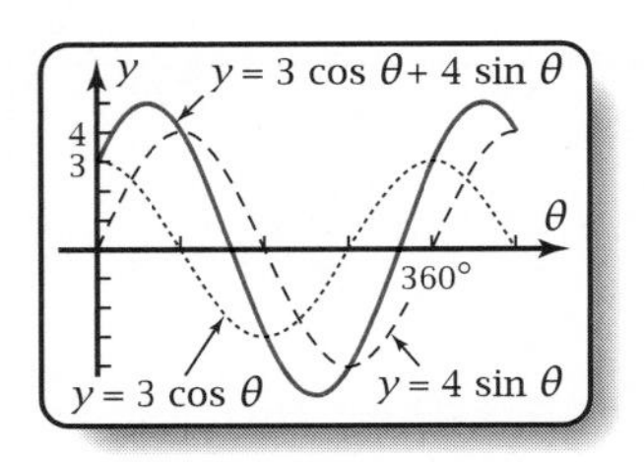

Figure 5-2a

Figure 5-2a shows that although the individual sound waves have amplitudes of 3 and 4, the combined sound wave does not have amplitude 3 + 4, or 7.

In this section you'll learn algebraic ways to find the amplitude and phase displacement of such a **linear combination** of cosine and sine, that is, an equation in the form of $y = a\cos\theta + b\sin\theta$. You'll do this with the help of the **composite argument property,** by which you can express $\cos(A - B)$ in terms of cosines and sines of A and B.

OBJECTIVE Derive a composite argument property expressing $\cos(A - B)$ in terms of cosines and sines of A and B, and use it to express a linear combination of cosine and sine as a single cosine with a phase displacement.

Linear Combination Property

The graph of $y = 3\cos\theta + 4\sin\theta$ shown in Figure 5-2a is a sinusoid. (You'll prove it algebraically in Problem 32 of Problem Set 5-2.) You can write its equation in the form

$$y = A\cos(\theta - D)$$

where A is the amplitude and D is the phase displacement for $y = \cos x$. If you plot the graph and then use the maximum feature on your grapher, you will find that

$$A = 5 \quad \text{and}$$

$$D = 53.1301\ldots°$$

When you "mix" sound, the principles of linear combination of sound waves apply.

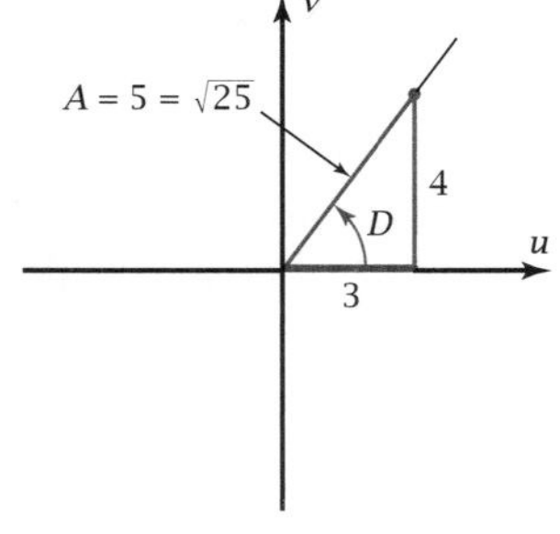

Figure 5-2b

Actually, D is the angle in standard position with $u = 3$ (the coefficient of cosine) and $v = 4$ (the coefficient of sine), as shown in Figure 5-2b. Once you derive the composite argument property mentioned in the objective, you can prove that the amplitude A is the length of the hypotenuse of the reference triangle for angle D. You find A using the Pythagorean theorem and D by applying the concept of arctangent.

$$A = \sqrt{3^2 + 4^2} = 5$$

$$D = \arctan \frac{4}{3} = 53.1301...° + 180n° = 53.1301...°$$

Choose $n = 0$ so D terminates in Quadrant I.

So $y = 5 \cos (\theta - 53.1301...°)$ is equivalent to $y = 3 \cos \theta + 4 \sin \theta$. The graphical solution confirms this.

► ***EXAMPLE 1*** Express $y = -8 \cos \theta + 3 \sin \theta$ as a single cosine with a phase displacement.

Solution Sketch angle D in standard position with $u = -8$ and $v = 3$ (the coefficients of cosine and sine, respectively), as shown in Figure 5-2c.

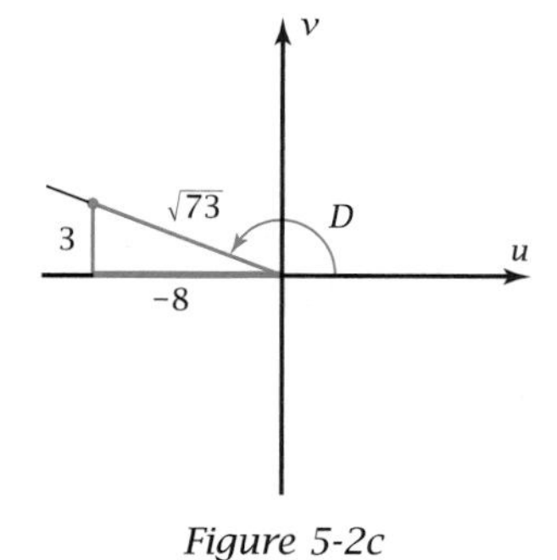

Figure 5-2c

$$A = \sqrt{(-8)^2 + 3^2} = \sqrt{73}$$ Find A by the Pythagorean theorem.

$$D = \arctan \frac{3}{-8} = -20.5560...° + 180n°$$ Find D using the definition of arctangent.

$$= 159.4439...°$$ Choose $n = 1$ to place D in the correct quadrant.

$$\therefore y = \sqrt{73} \cos (\theta - 159.4439...°)$$ ◄

PROPERTY: Linear Combination of Cosine and Sine with Equal Periods

$$b \cos x + c \sin x = A \cos (x - D)$$

where

$$A = \sqrt{b^2 + c^2} \quad \text{and} \quad D = \arctan \tfrac{c}{b}$$

The quadrant for $D = \arctan \frac{c}{b}$ depends on the signs of b and c and may be determined by sketching D in standard position. The hypotenuse of the reference triangle is A.

Composite Argument Property for Cosine $(A - B)$

In Section 5-1 you found that the cosine function does *not* distribute over addition or subtraction. Consider this proof by counterexample,

$$\cos (58° - 20°) = \cos 38° = 0.7880...$$

$$\cos 58° - \cos 20° = 0.5299... - 0.9396... = -0.4097...$$

$$\therefore \cos (58° - 20°) \neq \cos 58° - \cos 20°$$

However, you *can* express $\cos (58° - 20°)$ exactly in terms of sines and cosines of 58° and 20°. The result is

$$\cos (58° - 20°) = \cos 58° \cos 20° + \sin 58° \sin 20°$$

Both sides equal 0.7880....

Next you'll see how to generalize the results for any angles A and B. Figure 5-2d shows angles A and B in standard position and shows their difference, angle $(A - B)$. The coordinates of the points where the initial and terminal sides of angle $(A - B)$ cut the unit circle are

$$(\cos A, \sin A) \qquad \text{and} \qquad (\cos B, \sin B)$$

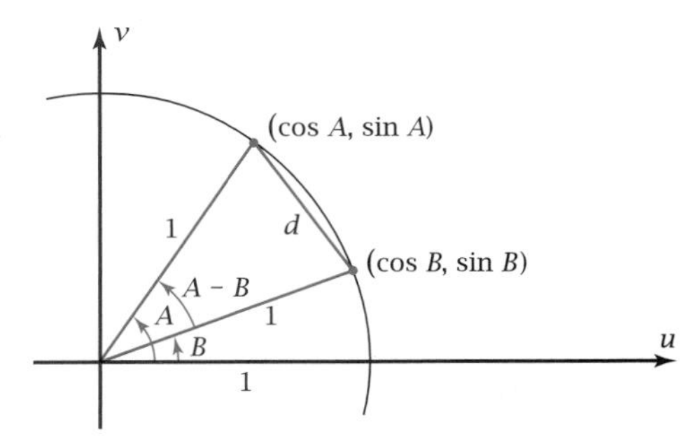

Figure 5-2d

The chord between the initial and terminal points has length d. The length d can be written using the **distance formula** (which you derived in earlier algebra study from the Pythagorean theorem).

$$d^2 = (\cos A - \cos B)^2 + (\sin A - \sin B)^2$$

Distance formula for points (u_1, v_1) and (u_2, v_2): $d^2 = (u_2 - u_1)^2 + (v_2 - v_1)^2$

$$= \cos^2 A - 2 \cos A \cos B + \cos^2 B + \sin^2 A - 2 \sin A \sin B + \sin^2 B$$

Expand the squares.

$$= (\cos^2 A + \sin^2 A) + (\cos^2 B + \sin^2 B) - 2 \cos A \cos B - 2 \sin A \sin B$$

Commute and associate the squared terms.

$$= 1 + 1 - 2 \cos A \cos B - 2 \sin A \sin B$$

Use the Pythagorean property for cosine and sine.

$$\therefore d^2 = 2 - 2 \cos A \cos B - 2 \sin A \sin B$$

Now consider Figure 5-2e, which shows angle $(A - B)$ rotated into standard position. The coordinates of the terminal and initial points of the angle in this position are

$$(\cos (A - B), \sin (A - B)) \qquad \text{and} \qquad (1, 0)$$

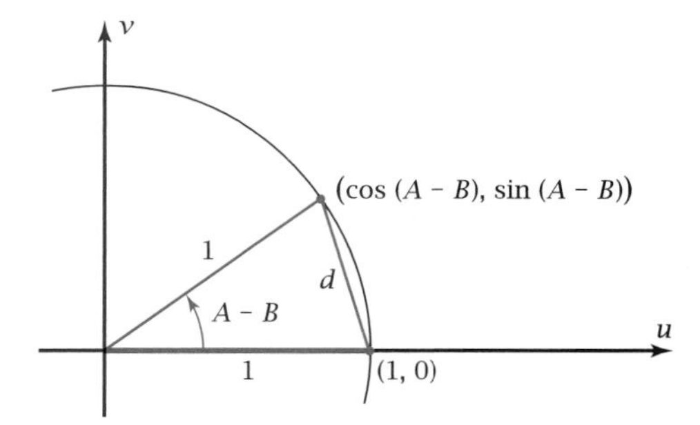

Figure 5-2e

The chord still has length d. By the distance formula and subsequent algebra,

$$d^2 = (\cos(A-B) - 1)^2 + (\sin(A-B) - 0)^2$$

$$= \cos^2(A-B) - 2\cos(A-B) + 1 + \sin^2(A-B)$$

Expand the squares.

$$= (\cos^2(A-B) + \sin^2(A-B)) + 1 - 2\cos(A-B)$$

Notice the Pythagorean property.

$$\therefore d^2 = 2 - 2\cos(A-B)$$

Equate the two expressions for d^2 to get

$$2 - 2\cos(A-B) = 2 - 2\cos A\cos B - 2\sin A\sin B$$

$$-2\cos(A-B) = -2\cos A\cos B - 2\sin A\sin B$$

$$\cos(A-B) = \cos A\cos B + \sin A\sin B$$

This is the property that was illustrated by numerical example with $A = 58°$ and $B = 20°$. You might remember this property most easily by expressing it verbally: "Cosine of first angle minus second angle equals cosine (first angle) times cosine (second angle) plus sine (first angle) times sine (second angle)."

PROPERTY: Composite Argument Property for cos (A − B)

$$\cos(A-B) = \cos A\cos B + \sin A\sin B$$

Cosine of the difference of two angles is cosine of first times cosine of second plus sine of first times sine of second.

▶ EXAMPLE 2 Express $7\cos(\theta - 23°)$ as a linear combination of $\cos\theta$ and $\sin\theta$.

Solution

$$7\cos(\theta - 23°) = 7(\cos\theta\cos 23° + \sin\theta\sin 23°)$$

Apply the composite argument property.

$$= 7\cos\theta\cos 23° + 7\sin\theta\sin 23°$$

$$= (7\cos 23°)\cos\theta + (7\sin 23°)\sin\theta$$

Associate the constant factors.

$$7\cos(\theta - 23°) = 6.4435\ldots\cos\theta + 2.7351\ldots\sin\theta$$

A linear combination of $\cos\theta$ and $\sin\theta$. ◀

Algebraic Solution of Equations

You can use the linear combination property to solve certain trigonometric equations algebraically.

▶ *EXAMPLE 3* Solve $-2\cos x + 3\sin x = 2$ for x in the domain $x \in [-2\pi, 2\pi]$. Verify the solution graphically.

Solution

$$-2\cos x + 3\sin x = 2$$ Write the given equation.

Transform the left side, $-2\cos x + 3\sin x$, into the form $A\cos(x - D)$. Draw angle D in standard position (Figure 5-2f).

$$A = \sqrt{(-2)^2 + (3)^2} = \sqrt{13}$$ Use the Pythagorean theorem to calculate A.

$$D = \arctan\frac{-3}{2} = -0.9827\ldots + \pi n = 2.1587\ldots$$

Use $n = 1$ for the proper arctangent value.

$$\therefore \sqrt{13}\cos(x - 2.1587\ldots) = 2$$ Rewrite the equation using $A\cos(x - D)$.

$$\cos(x + 2.1587\ldots) = \frac{2}{\sqrt{13}}$$

$$x - 2.1587\ldots = \arccos\left(\frac{2}{\sqrt{13}}\right)$$

$$x = 2.1587\ldots \pm 0.9827\ldots + 2\pi n$$

Rewrite the equation and evaluate the arccosine.

$$x = 3.141\ldots + 2\pi n \quad \text{or} \quad 1.1760\ldots + 2\pi n$$

Evaluate $2.1587\ldots + 0.9827\ldots$ and $2.1587\ldots - 0.9827.$

$$S = \{-5.1071\ldots, -3.1415\ldots, 1.1760\ldots, 3.1415\ldots\}$$

Pick values of n to get x in the domain.

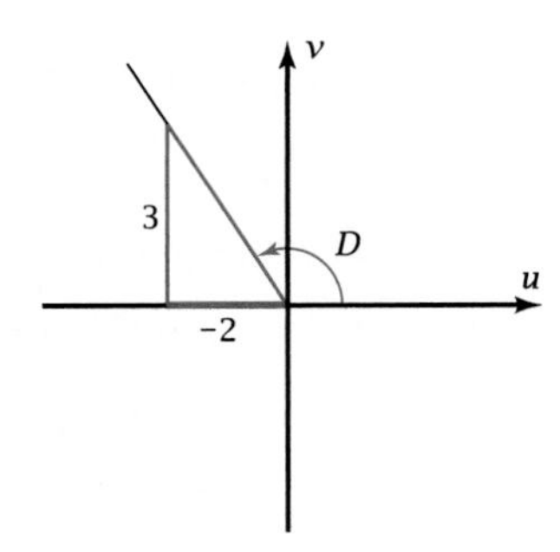

Figure 5-2f

Figure 5-2g on the next page shows the graphs of $y = -2\cos x + 3\sin x$ and the line $y = 2$. Note that the graph is a sinusoid, as you discovered algebraically while solving the equation. By using the intersect feature, you can see that the four solutions are correct and that they are the only solutions in the $[-2\pi, 2\pi]$ domain. The graph also shows the phase displacement of 2.1587... and the amplitude of $\sqrt{13}$, which equals approximately 3.6. ◀

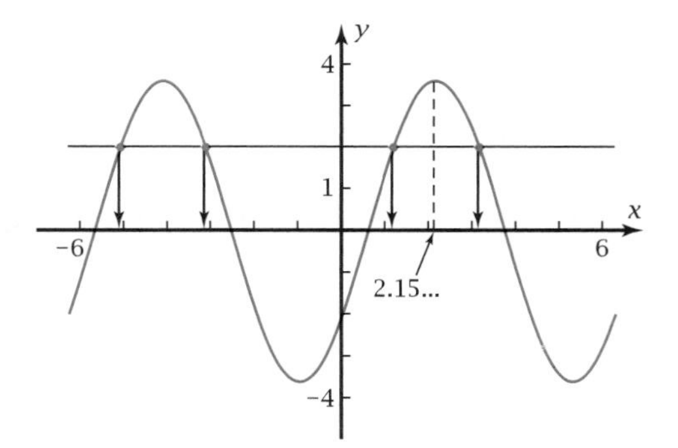

Figure 5-2g

Problem Set 5-2

Do These Quickly 5 min

Q1. State the Pythagorean property for secant and tangent.

Q2. State the reciprocal property for cosecant.

Q3. State the quotient property for cotangent in terms of sine and cosine.

Q4. Is $\cos^2 x = 1 - \sin^2 x$ an identity?

Q5. Is $\cot x \tan x = 1$ an identity?

Q6. Is $\cos x \sin x = 1$ an identity?

Q7. Find the exact value (no decimals) of $\cos \frac{\pi}{4}$.

Q8. Find the exact value (no decimals) of $\tan 30°$.

Q9. Find the first three positive angles $\theta = \arccos 0.5$.

Q10. Factor: $x^2 - 5x - 6$

For Problems 1–12, write the linear combination of cosine and sine as a single cosine with a phase displacement.

1. $y = 12 \cos \theta + 5 \sin \theta$
2. $y = 4 \cos \theta + 3 \sin \theta$
3. $y = -7 \cos \theta + 24 \sin \theta$
4. $y = -15 \cos \theta + 8 \sin \theta$
5. $y = -8 \cos \theta - 11 \sin \theta$
6. $y = -7 \cos \theta - 10 \sin \theta$
7. $y = 6 \cos \theta - 6 \sin \theta$
8. $y = \cos \theta - \sin \theta$
9. $y = \sqrt{3} \cos \theta + \sin \theta$
10. $y = (\sqrt{6} + \sqrt{2}) \cos \theta + (\sqrt{6} - \sqrt{2}) \sin \theta$ (Surprising result?)
11. $y = -3 \cos x + 4 \sin x$ (radian mode)
12. $y = -5 \cos x - 12 \sin x$ (radian mode)
13. Confirm by graphing that your answer to Problem 1 is correct.

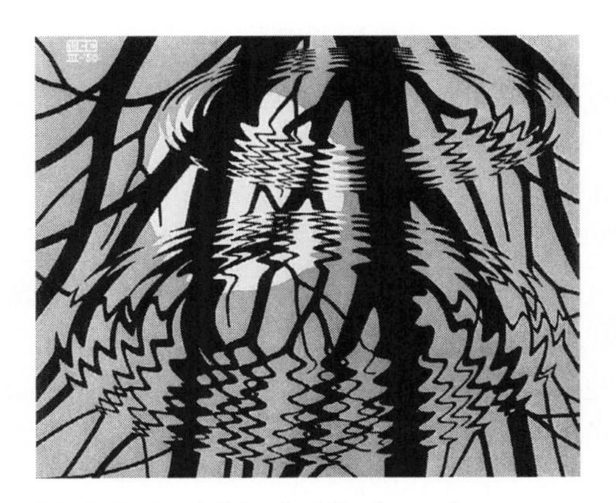

M. C. Escher's Rippled Surface *shows the reflection of branches in the water waves, which have a sinusoidal pattern.* (*M. C. Escher's* Rippled Surface

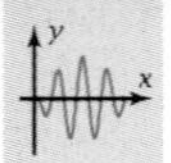

14. Confirm by graphing that your answer to Problem 2 is correct.

15. Express the circular function $y = \cos 3x + \sin 3x$ as a single cosine with a phase displacement. What effect does the "3" have on your work?

16. Figure 5-2h shows a cosine graph and a sine graph. Find equations for these two sinusoids. Then find an equation for the sum of the two sinusoids as a single cosine with a phase displacement. Verify your answers by plotting them on your grapher.

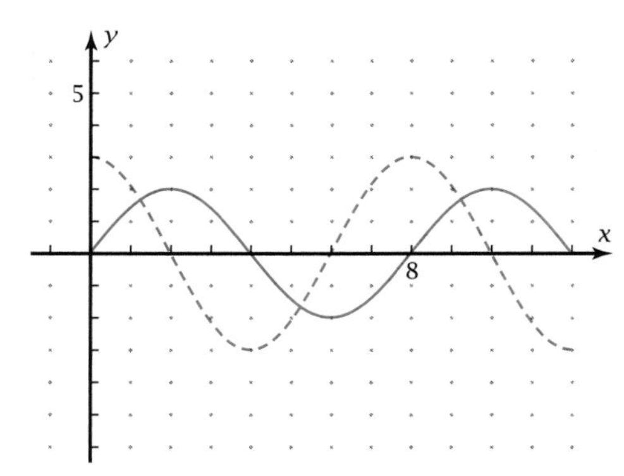

Figure 5-2h

17. Prove by counterexample that cosine does not distribute over subtraction. That is, give a numerical example to show that $\cos(A - B) \neq \cos A - \cos B$.

18. Make a table of values to show numerically that $\cos(A - B) = \cos A \cos B + \sin A \sin B$.

For Problems 19–22, express each equation as a linear combination of cosine and sine.

19. $y = 10 \cos(\theta - 30°)$. Confirm graphically that your answer is correct.

20. $y = 20 \cos(\theta - 60°)$. Confirm graphically that your answer is correct.

21. $y = 5 \cos(3\theta - 150°)$

22. $y = 8 \cos(2\theta - 120°)$

For Problems 23–26, solve the equation algebraically. Use the domain $x \in [0, 2\pi]$ or $\theta \in [0°, 360°]$.

23. $5 \cos \theta + 7 \sin \theta = 3$

24. $2 \cos x + 5 \sin x = 4$

25. $-8 \cos x - 3 \sin x = 5$

26. $7 \cos \theta - 4 \sin \theta = 6$

27. Use the composite argument property to show that this equation is an identity:

$$\cos 2\theta = \cos 5\theta \cos 3\theta + \sin 5\theta \sin 3\theta$$

Use the result to solve this equation for $\theta \in [0°, 360°]$.

$$\cos 5\theta \cos 3\theta + \sin 5\theta \sin 3\theta = 0.3$$

28. *Musical Note Problem:* The Nett sisters, Cora and Clara, are in a band. Each one is playing the note A. Their friend Tom is standing at a place where the notes arrive exactly a quarter cycle out of phase. If x is time in seconds, the function equations of Cora's and Clara's notes are

Cora: $y = 100 \cos 440\pi x$

Clara: $y = 150 \sin 440\pi x$

a. The sound Tom hears is the sum of Cora's and Clara's sound waves. Write an equation for this sound as a single cosine with a phase displacement.

b. The amplitudes 100 and 150 measure the loudness of the two notes Cora and Clara are playing. Is this statement true or false? "Tom hears a note 250 units loud, the sum of 100 and 150." Explain how you reached your answer.

c. The frequency of the A being played by Cora and Clara is 220 cycles per second. Explain how you can figure this out from the two equations. Is the following true or false? "The note Tom hears also has a frequency of 220 cycles per second."

29. *Cofunction Property for Cosines and Sines Problem:*

a. Show that $\cos 70° = \sin 20°$.

b. Use the composite argument property and the definition of complementary angles to show in general that $\cos(90° - \theta) = \sin\theta$.

c. What does the prefix *co-* mean in the name *cosine*?

30. *Even Property of Cosine Problem:*

a. Show that $\cos(-54°) = \cos 54°$.

b. You can write $\cos(-54°)$ as $\cos(0° - 54°)$. Use the composite argument property to show algebraically that $\cos(-\theta) = \cos\theta$.

c. Recall that functions with the property $f(-x) = f(x)$ are called *even* functions. Show why this name is picked by letting $f(x) = x^6$ and showing that $f(-x) = f(x)$.

31. *Composite Argument Property Derivation Problem:* Derive the property

$$\cos(A - B) = \cos A \cos B + \sin A \sin B$$

Try to do this on your own, looking at the text only long enough to get you started again if you get stuck.

32. *Linear Combination of Cosine and Sine Derivation Problem:* In this problem you'll see how to prove the linear combination property.

a. Use the composite argument property to show that

$$A\cos(\theta - D) =$$
$$(A\cos D)\cos\theta + (A\sin D)\sin\theta$$

b. Let $A\cos D = b$, and let $A\sin D = c$. Square both sides of each equation to get

$$A^2\cos^2 D = b^2$$

$$A^2\sin^2 D = c^2$$

Explain why $A^2 = b^2 + c^2$.

c. Explain why $D = \arccos\frac{b}{A}$ and $D = \arcsin\frac{c}{A}$, and thus why $D = \arctan\frac{c}{b}$.

33. *Journal Problem:* Update your journal with what you have learned since the last entry. In particular, explain what the composite argument property is and how you can use it to prove that a sum of cosine and sine with equal periods is a single cosine with the same period and a phase displacement.

5-3 Other Composite Argument Properties

In Section 5-2 you used the composite argument property for cosine to show that a linear combination of cosine and sine with equal periods is a single cosine with the same period but with a different amplitude and a phase displacement. In this section you'll learn composite argument properties for sine and tangent involving $(A + B)$ as well as $(A - B)$. You'll learn the cofunction properties and odd–even function properties that allow you to derive these new composite argument properties quickly from $\cos(A - B)$.

OBJECTIVE For trigonometric functions f, derive and learn properties for:

- $f(-x)$ in terms of $f(x)$
- $f(90° - \theta)$ in terms of functions of θ or $f\left(\frac{\pi}{2} - x\right)$ in terms of functions of x
- $f(A + B)$ and $f(A - B)$ in terms of functions of A and functions of B

The Odd–Even Properties

If you take the functions of opposite angles or arcs, interesting patterns emerge.

$\sin(-20°) = -0.3420...$ and $\sin 20° = 0.3420...$

$\cos(-20°) = 0.9396...$ and $\cos 20° = 0.9396...$

$\tan(-20°) = -0.3639...$ and $\tan 20° = 0.3639...$

These numerical examples illustrate the fact that sine and tangent are odd functions and cosine is an even function. Figure 5-3a shows graphically why these properties apply for any value of θ.

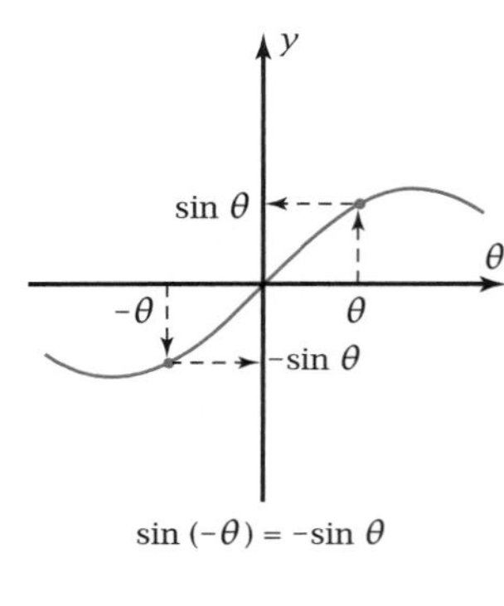

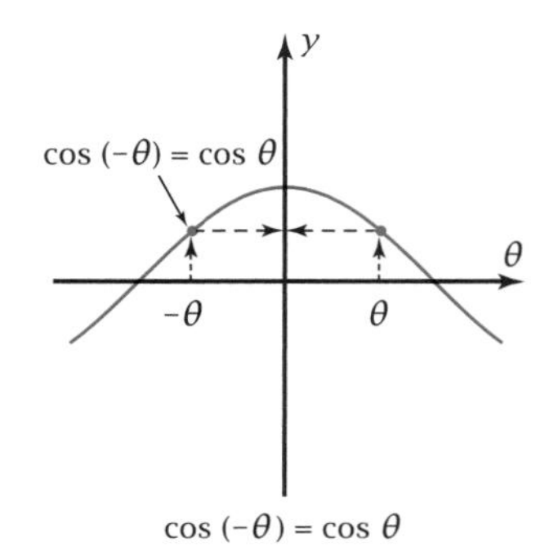

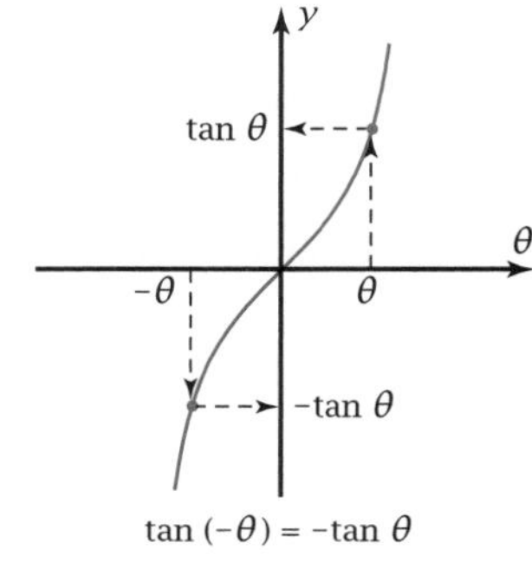

Figure 5-3a

The reciprocals of the functions have the same **parity** (oddness or evenness) as the original functions.

PROPERTIES: Odd and Even Functions

Cosine and its reciprocal are *even* functions. That is,

$\cos(-x) = \cos x$ and $\sec(-x) = \sec x$

Sine and tangent, and their reciprocals, are *odd* functions. That is,

$\sin(-x) = -\sin x$ and $\csc(-x) = -\csc x$

$\tan(-x) = -\tan x$ and $\cot(-x) = -\cot x$

The Cofunction Properties: Functions of $(90° - \theta)$ or $\left(\frac{\pi}{2} - x\right)$

The angles 20° and 70° are **complementary** angles because they add up to 90°. (The word comes from "complete," since the two angles *complete* a right angle.) The angle 20° is the complement of 70°, and the angle 70° is the complement of 20°. An interesting pattern shows up if you take the function and the cofunction of complementary angles.

$$\cos 70° = 0.3420\ldots \quad \text{and} \quad \sin 20° = 0.3420\ldots$$

$$\cot 70° = 0.3639\ldots \quad \text{and} \quad \tan 20° = 0.3639\ldots$$

$$\csc 70° = 1.0641\ldots \quad \text{and} \quad \sec 20° = 1.0641\ldots$$

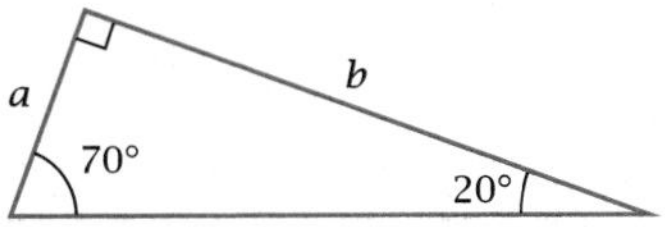

a
b
70°
20°
c

Figure 5-3b

You can verify these patterns by using the right triangle definitions of the trigonometric functions. Figure 5-3b shows a right triangle with acute angles 70° and 20°. The opposite leg for 70° is the adjacent leg for 20°. Thus,

$$\cos 70° = \frac{\text{adjacent leg}}{\text{hypotenuse}} = \frac{a}{c} \quad \text{and} \quad \sin 20° = \frac{\text{opposite leg}}{\text{hypotenuse}} = \frac{a}{c}$$

$$\therefore \cos 70° = \sin 20°$$

The prefix *co-* in the names cosine, cotangent, and cosecant comes from the word *complement.* In general, the cosine of an angle is the *sine* of the *complement* of that angle. The same property is true for cotangent and cosecant, as you can verify with the help of Figure 5-3b.

The cofunction properties are true regardless of the size of the angle or arc. For instance, if θ is 234°, then the complement of θ is $90° - 234°$, or $-144°$.

$$\cos 234° = -0.5877\ldots \quad \text{and}$$

$$\sin(90° - 234°) = \sin(-144°) = -0.5877\ldots$$

$$\therefore \cos 234° = \sin(90° - 234°)$$

Note that it doesn't matter which of the two angles you consider to be "the angle" and which one you consider to be "the complement." It is just as true, for example, that

$$\sin 20° = \cos(90° - 20°)$$

The cofunction properties for the trigonometric functions are summarized verbally as:

The cosine of an angle equals the *sine* of the *complement* of that angle.

The cotangent of an angle equals the *tangent* of the *complement* of that angle.

The cosecant of an angle equals the *secant* of the *complement* of that angle.

PROPERTIES: Cofunction Properties for Trigonometric Functions

When working with degrees:

$\cos\theta = \sin(90° - \theta)$ and $\sin\theta = \cos(90° - \theta)$

$\cot\theta = \tan(90° - \theta)$ and $\tan\theta = \cot(90° - \theta)$

$\csc\theta = \sec(90° - \theta)$ and $\sec\theta = \csc(90° - \theta)$

When working with radians:

$\cos x = \sin(\frac{\pi}{2} - x)$ and $\sin x = \cos(\frac{\pi}{2} - x)$

$\cot x = \tan(\frac{\pi}{2} - x)$ and $\tan x = \cot(\frac{\pi}{2} - x)$

$\csc x = \sec(\frac{\pi}{2} - x)$ and $\sec x = \csc(\frac{\pi}{2} - x)$

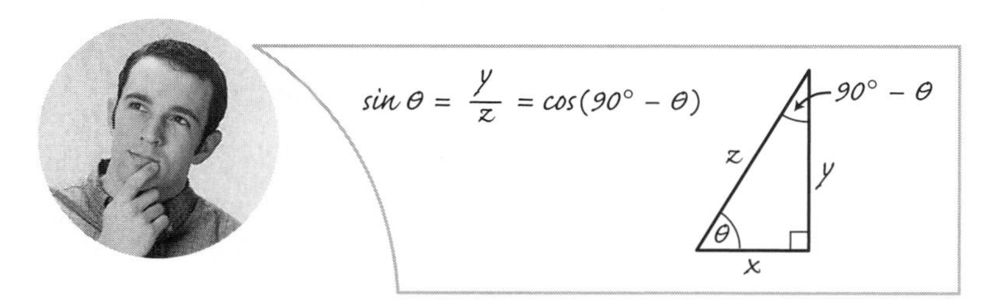

The Composite Argument Property for cos (A + B)

You can write the cosine of a *sum* of two angles in terms of functions of those two angles. You can transform the cosine of a sum to a cosine of a difference with some insightful algebra and the odd–even properties.

$\cos(A + B)$

$= \cos[A - (-B)]$ — Change the sum into a difference.

$= \cos A\cos(-B) + \sin A\sin(-B)$ — Use the composite argument property for $\cos(A - B)$.

$= \cos A\cos B + \sin A(-\sin B)$ — Cosine is an even function. Sine is an odd function.

$= \cos A\cos B - \sin A\sin B$

$\therefore \cos(A + B) = \cos A\cos B - \sin A\sin B$

The only difference between this property and the one for $\cos(A - B)$ is the minus sign between the terms on the right side of the equation.

The Composite Argument Properties for sin (A − B) and sin (A + B)

You can derive composite argument properties for $\sin(A - B)$ with the help of the cofunction property.

$$\sin(A - B) = \cos[90° - (A - B)]$$ Transform to a cosine using the cofunction property.

$$= \cos[(90° - A) + B]$$ Distribute the minus sign, then associate $(90° - A)$.

$$= \cos(90° - A)\cos B - \sin(90° - A)\sin B$$ Use the composite argument property for $\cos(A + B)$.

$$= \sin A \cos B - \cos A \sin B$$ Use the cofunction property the other way around.

$$\therefore \sin(A - B) = \sin A \cos B - \cos A \sin B$$

The composite argument property for $\sin(A + B)$ is

$$\sin(A + B) = \sin A \cos B + \cos A \sin B$$

You can derive it by writing $\sin(A + B)$ as $\sin[A - (-B)]$ and using the same reasoning as for $\cos(A + B)$.

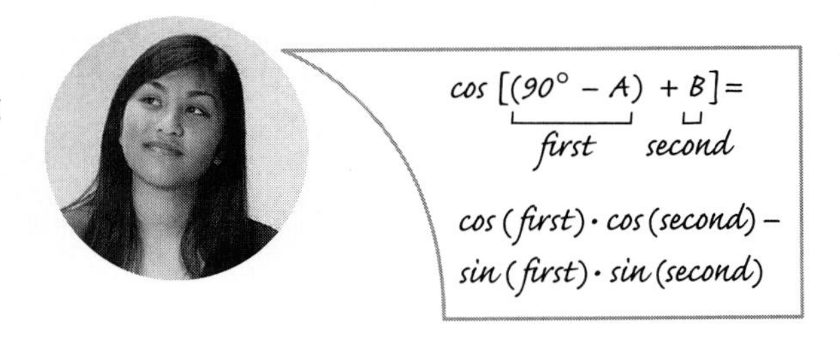

The Composite Argument Properties for tan (A − B) and tan (A + B)

You can write the tangent of a composite argument in terms of tangents of the two angles. This requires factoring out a "common" factor that isn't actually there!

$$\tan(A - B) = \frac{\sin(A - B)}{\cos(A - B)}$$ Use the quotient property for tangent to "bring in" sines and cosines.

$$= \frac{\sin A \cos B - \cos A \sin B}{\cos A \cos B + \sin A \sin B}$$ Use the composite argument properties for $\sin(A - B)$ and $\cos(A - B)$.

$$= \frac{\cos A \cos B\left(\dfrac{\sin A \cos B}{\cos A \cos B} - \dfrac{\cos A \sin B}{\cos A \cos B}\right)}{\cos A \cos B\left(\dfrac{\cos A \cos B}{\cos A \cos B} + \dfrac{\sin A \sin B}{\cos A \cos B}\right)}$$ Factor out $\cos A \cos B$ in the numerator and denominator to put cosines in the minor denominators.

$$= \frac{\dfrac{\sin A}{\cos A} - \dfrac{\sin B}{\cos B}}{1 + \dfrac{\sin A \sin B}{\cos A \cos B}}$$ Cancel all common factors.

$$= \frac{\tan A - \tan B}{1 + \tan A \tan B}$$ Use the quotient property to get only tangents.

$$\therefore \tan(A - B) = \frac{\tan A - \tan B}{1 + \tan A \tan B}$$

You can derive the composite argument property for $\tan(A + B)$ by writing $\tan(A + B)$ as $\tan[A - (-B)]$ and then using the fact that tangent is an odd function. The result is

$$\tan(A - B) = \frac{\tan A + \tan B}{1 - \tan A \tan B}$$

This table summarizes the composite argument properties for cosine, sine, and tangent. Like the composite argument properties for $\cos(A - B)$ and $\cos(A + B)$, notice that the signs between the terms change when you compare $\sin(A - B)$ and $\sin(A + B)$ or $\tan(A - B)$ and $\tan(A + B)$.

PROPERTIES: *Composite Argument Properties for Cosine, Sine, and Tangent*

$$\cos(A - B) = \cos A \cos B + \sin A \sin B$$

$$\cos(A + B) = \cos A \cos B - \sin A \sin B$$

$$\sin(A - B) = \sin A \cos B - \cos A \sin B$$

$$\sin(A + B) = \sin A \cos B + \cos A \sin B$$

$$\tan(A - B) = \frac{\tan A - \tan B}{1 + \tan A \tan B}$$

$$\tan(A + B) = \frac{\tan A + \tan B}{1 - \tan A \tan B}$$

Algebraic Solution of Equations

You can use the composite argument properties to solve certain trigonometric equations algebraically.

▶ ***EXAMPLE 1*** Solve the equation for x in the domain $x \in [0, 2\pi]$. Verify the solutions graphically.

$$\sin 5x \cos 3x - \cos 5x \sin 3x = \frac{1}{2}$$

Solution

$$\sin 5x \cos 3x - \cos 5x \sin 3x = \frac{1}{2}$$ Write the given equation.

$$\sin(5x - 3x) = \frac{1}{2}$$ Use the composite argument property for $\sin(A - B)$.

$$\sin 2x = \frac{1}{2}$$

$$2x = \arcsin \frac{1}{2}$$

$$= 0.5235\ldots + 2\pi n \quad \text{or} \quad (\pi - 0.5235\ldots) + 2\pi n$$

Use the definition of arcsine to write the general solution.

$$x = 0.2617... + \pi n \quad \text{or} \quad 1.3089... + \pi n$$

$$S = \{0.2618..., 1.3089..., 3.4033..., 4.4505...\}$$

Use $n = 0$ and $n = 1$ to get the solutions in the domain. ◀

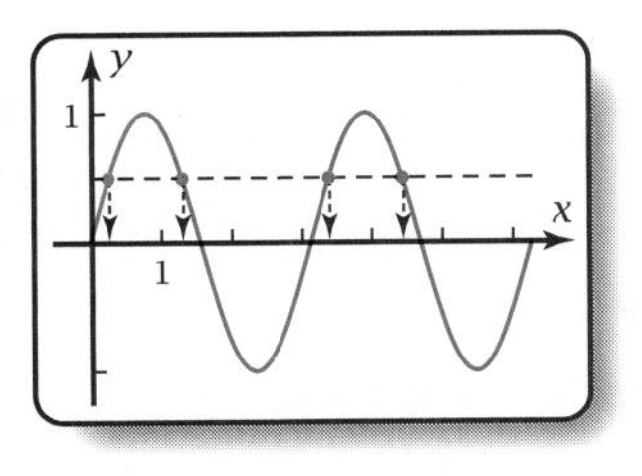

Figure 5-3c

In this case, the answers turn out to be simple multiples of π. See if you can figure out why $x = \frac{\pi}{12}, \frac{5\pi}{12}, \frac{13\pi}{12}$, and $\frac{17\pi}{12}$.

Figure 5-3c shows the graphs of $y = \sin 5x \cos 3x - \cos 5x \sin 3x$ and the line $y = 0.5$. Note that the graph of $y = \sin 5x \cos 3x - \cos 5x \sin 3x$ is equivalent to the sinusoid $y = \sin 2x$. By using the intersect feature, you can see that the four solutions are correct and that they are the only solutions in the domain $x \in [0, 2\pi]$.

Problem Set 5-3

Do These Quickly

Q1. If one value of $\arcsin x$ is 30°, find another positive value of $\arcsin x$, less than 360°.

Q2. The value 30 is what percentage of 1000?

Q3. $2^{10} =$ —?—

Q4. $\cos 7 \cos 3 + \sin 7 \sin 3 = \cos$ —?—

Q5. What is the amplitude of the sinusoid $y = 8 \cos \theta + 15 \sin \theta$?

Q6. If $A \cos(\theta - D) = 8 \cos \theta + 15 \sin \theta$, then D could equal —?—.

Q7. $\tan^2 47° - \sec^2 47° =$ —?—

Q8. $\log 3 + \log 4 = \log$ —?—

Q9. $\tan x = \dfrac{\sec x}{\csc x}$ is called a —?— property.

Q10. Sketch the graph of an exponential function with base between 0 and 1.

---

1. Prove by counterexample that $\sin(A + B) \neq \sin A + \sin B$.
2. Prove by counterexample that the operation tangent does not distribute over addition.
3. Show by numerical example that $\tan(A - B) = \dfrac{\tan A - \tan B}{1 + \tan A \tan B}$.
4. Show by numerical example that $\sin(A - B) = \sin A \cos B - \cos A \sin B$.
5. Make a table of values to confirm that $\cos(-x) = \cos x$.
6. Make a table of values to confirm that $\tan(-x) = -\tan x$.
7. Confirm graphically that $\cot \theta = \tan(90° - \theta)$.
8. Confirm graphically that $\cos \theta = \sin(90° - \theta)$.
9. *Odd–Even Property Geometrical Proof Problem:* Figure 5-3d shows angles of θ and $-\theta$ in standard position in a uv-coordinate system. The u-coordinates of the points where the angles cut a unit circle are equal. The v-coordinates of these points are opposites of each other.

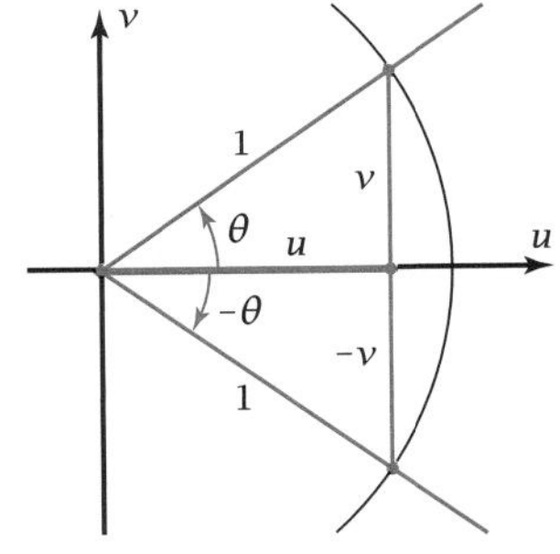

Figure 5-3d

a. Based on the definition of sine, explain why $\sin(-\theta) = -\sin \theta$.

b. Based on the definition of cosine, explain why $\cos(-\theta) = \cos\theta$.

c. Based on the definition of tangent, explain why $\tan(-\theta) = -\tan\theta$.

d. Based on the reciprocal properties, explain why secant, cosecant, and cotangent have the same odd–even properties as cosine, sine, and tangent, respectively.

10. *Odd–Even Property Proof:* Recall that $y = -f(x)$ is a vertical reflection of $y = f(x)$ across the x-axis and that $y = f(-x)$ is a horizontal reflection across the y-axis.

   a. Figure 5-3e shows $y = \sin x$. Sketch the graph resulting from a reflection of $y = \sin x$ across the y-axis, and sketch another graph resulting from a reflection of $y = \sin x$ across the x-axis. Based on the results, explain why the sine is an *odd* function.

   b. Figure 5-3f shows $y = \cos x$. Sketch the graph resulting from a reflection of $y = \cos x$ across the y-axis. From the result, explain why cosine is an *even* function.

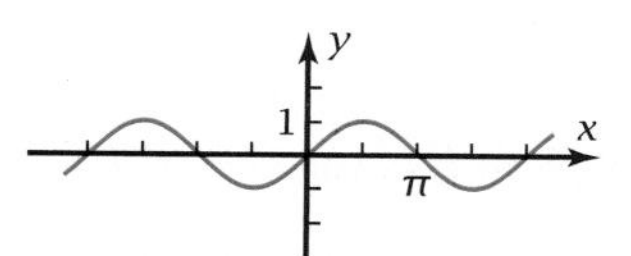

Figure 5-3e

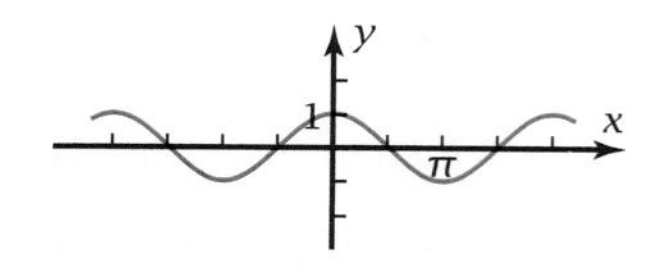

Figure 5-3f

For Problems 11–20, use the composite argument properties to show that the given equation is an identity.

11. $\cos(\theta - 90°) = \sin\theta$

12. $\cos(x - \frac{\pi}{2}) = \sin x$

13. $\sin(x - \frac{\pi}{2}) = -\cos x$

14. $\sec(\theta - 90°) = \csc\theta$ (Be clever!)

15. $\sin(\theta + 60°) - \cos(\theta + 30°) = \sin\theta$

16. $\sin(\theta + 30°) + \cos(\theta + 60°) = \cos\theta$

17. $\sqrt{2}\cos(x - \frac{\pi}{4}) = \cos x + \sin x$

18. $(\cos A\cos B - \sin A\sin B)^2 + (\sin A\cos B + \cos A\sin B)^2 = 1$

19. $\sin 3x\cos 4x + \cos 3x\sin 4x = \sin 7x$

20. $\cos 10x\cos 6x + \sin 10x\sin 6x = \cos 4x$

For Problems 21–26, use the composite argument properties to transform the left side of the equation to a *single* function of a composite argument. Then solve the equation algebraically to get

a. The general solution for x or θ

b. The particular solutions for x in the domain $x \in [0, 2\pi)$ or for θ in the domain $\theta \in [0°, 360°)$

21. $\cos x\cos 0.6 - \sin x\sin 0.6 = 0.9$

22. $\sin\theta\cos 35° + \cos\theta\sin 35° = 0.5$

23. $\sin 3\theta\cos\theta - \cos 3\theta\sin\theta = 0.5\sqrt{2}$

24. $\cos 3x\cos x + \sin 3x\sin x = -1$

25. $\dfrac{\tan 2x - \tan x}{1 + \tan 2x\tan x} = \sqrt{3}$

26. $\dfrac{\tan\theta + \tan 27°}{1 - \tan\theta\tan 27°} = 1$

Exact Function Value Problems: Figure 5-3g shows angles A and B in standard position in a uv-coordinate system. For Problems 27–32, use the information in the figure to find *exact* values (no decimals!) of the following. Check your answers by calculating A and B, adding or subtracting them, and finding the function values directly.

27. $\cos(A - B)$

28. $\cos(A + B)$

29. $\sin(A - B)$

30. $\sin(A + B)$

31. $\tan(A - B)$

32. $\tan(A + B)$

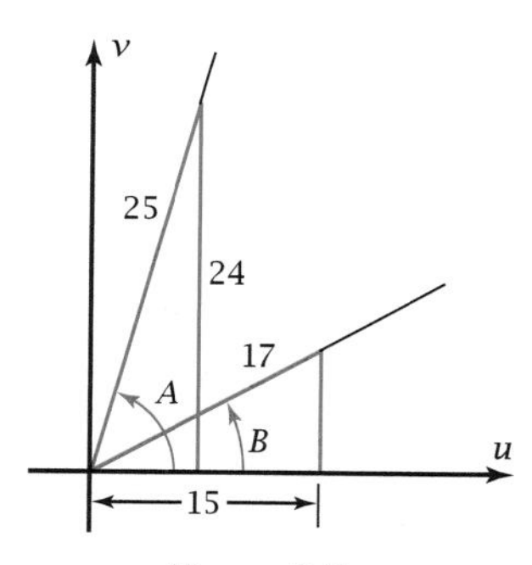

Figure 5-3g

For Problems 33 and 34, use the composite argument properties with exact values of functions of special angles (such as 30°, 45°, 60°) to show that these numerical expressions are exact values of $\sin 15°$ and $\cos 15°$. Confirm numerically that the values are correct.

33. $\sin 15° = \dfrac{\sqrt{6} - \sqrt{2}}{4}$

34. $\cos 15° = \dfrac{\sqrt{6} + \sqrt{2}}{4}$

For Problems 35 and 36, use the exact values of $\sin 15°$ and $\cos 15°$ from Problems 33 and 34 and the cofunction properties to find exact values (no decimals) of these expressions.

35. $\sin 75°$

36. $\cos 75°$

For Problems 37 and 38, use the values of $\sin 15°$ and $\cos 15°$ from Problems 33 and 34, with appropriate simplification, to show that these numerical expressions are exact values of $\tan 15°$ and $\cot 15°$.

37. $\tan 15° = 2 - \sqrt{3}$

38. $\cot 15° = 2 + \sqrt{3}$

39. *Cofunction Property for the Inverse Sine Function Problem:* In this problem you will prove that $\cos^{-1} x$ is the complement of $\sin^{-1} x$.
   a. Let $\theta = 90° - \sin^{-1} x$. Use the composite argument property to prove that $\cos \theta = x$.
   b. From part a, it follows that $\theta = \arccos x$, the inverse trigonometric *relation.* Use the fact that $-90° \le \sin^{-1} x \le 90°$ to show that θ is in the interval $[0°, 180°]$.
   c. How does part b allow you to conclude that θ is $\cos^{-1} x$, the inverse trigonometric *function*?

40. *Cofunction Properties for the Inverse Circular Functions Problem:* Use the cofunction properties for the inverse circular functions to calculate these values. Show that each answer is in the range of the inverse cofunction.
   a. $\cos^{-1}(-0.4)$
   b. $\cot^{-1}(-1.5)$
   c. $\csc^{-1}(-2)$

Properties: Cofunction Properties for the Inverse Circular Functions

$\cos^{-1} x = \frac{\pi}{2} - \sin^{-1} x$

$\cot^{-1} x = \frac{\pi}{2} - \tan^{-1} x$

$\csc^{-1} x = \frac{\pi}{2} - \sec^{-1} x$

Triple Argument Properties Problems: The composite argument properties have sums of *two* angles or arcs. It is possible to derive **triple argument properties** for three angles or arcs. Derive properties expressing the given function in terms of $\sin A$, $\sin B$, $\sin C$, $\cos A$, $\cos B$, and $\cos C$ (start by associating two of the three angles).

41. $\cos(A + B + C)$

42. $\sin(A + B + C)$

5-4 Composition of Ordinates and Harmonic Analysis

In Section 5-3 you learned that a sum of two sinusoids with equal periods is another sinusoid with the same period, as shown in Figure 5-4a. A product of sinusoids with equal periods is also a sinusoid, as shown in Figure 5-4b.

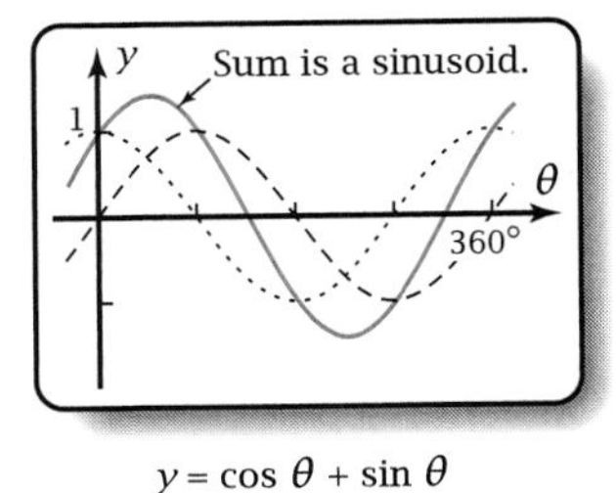

$y = \cos \theta + \sin \theta$

Figure 5-4a

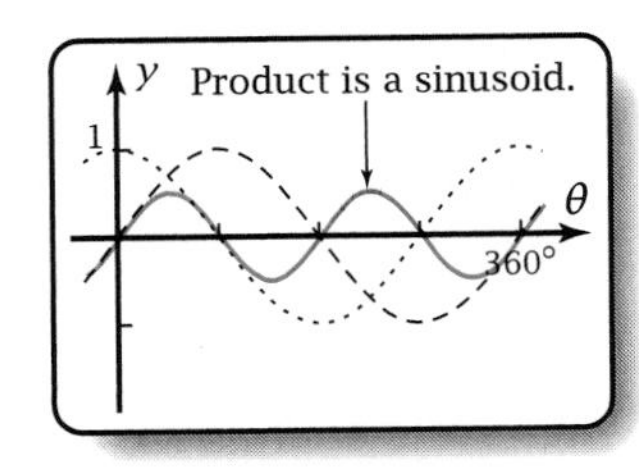

$y = \sin \theta \cdot \cos \theta$

Figure 5-4b

Figure 5-4c shows the result of adding two sinusoids with unequal periods, which might happen, for example, if two musical notes of different frequency are played at the same time. In this section you will learn about **composition of ordinates,** by which sinusoids are added or multiplied, and **harmonic analysis,** by which you reverse the process to find the parent sinusoids.

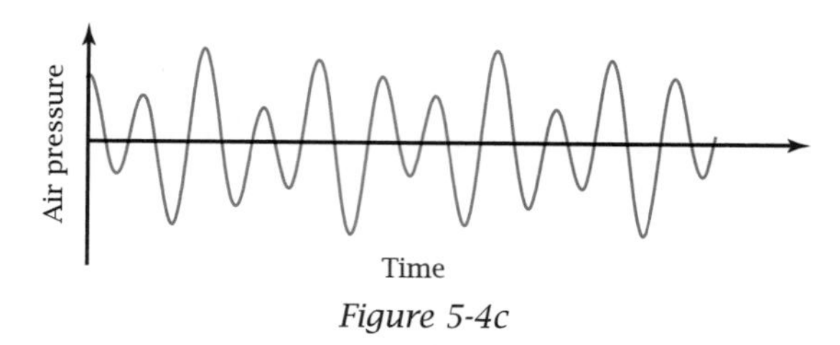

Figure 5-4c

OBJECTIVE

- Given two sinusoids, form a new graph by adding or multiplying ordinates (y-coordinates).
- Given a graph formed by adding or multiplying two sinusoids, find the equations of the two sinusoids.

Sum of Two Sinusoids with Unequal Periods

Example 1 shows you how to sketch a graph composed of two sinusoids whose periods are not equal.

▶ EXAMPLE 1

Figure 5-4d shows the graphs of $y_1 = 3\cos\theta$ and $y_2 = \sin 4\theta$.

On a copy of this figure, sketch the graph of

$$y = 3\cos\theta + \sin 4\theta$$

Then plot the function on your grapher. How well does your sketch match the accurate graph?

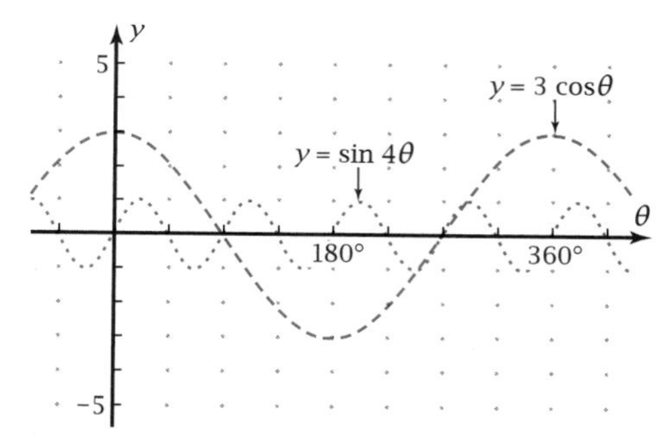

Figure 5-4d

Solution

Look for high points, low points, and zeros on the two graphs. At each θ-value you chose, estimate the ordinate of each graph and add them together. Put a dot on the graph paper at that value of θ with the appropriate ordinate. Figure 5-4e shows how you might estimate the ordinate at one particular point.

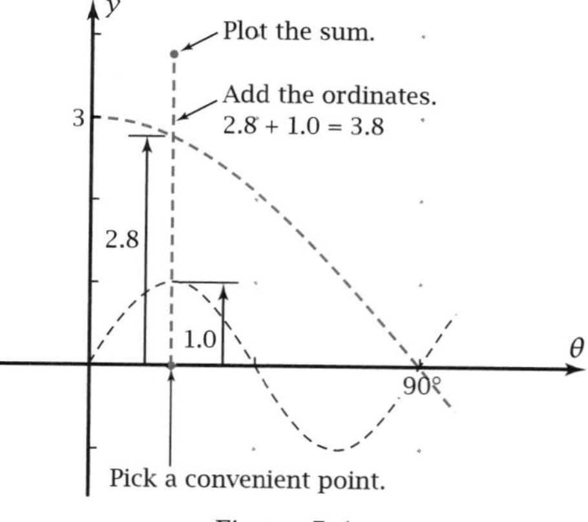

Figure 5-4e

Figure 5-4f shows the dots plotted at places where the auxiliary graphs have critical points or zeros. Once you see the pattern, you can connect the dots with a smooth curve, as in Figure 5-4g.

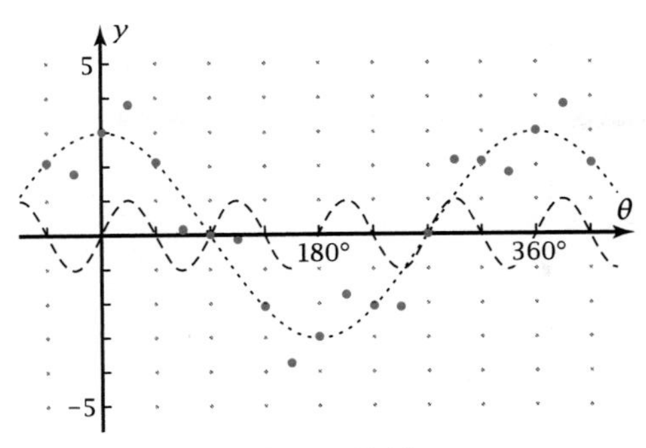

Figure 5-4f

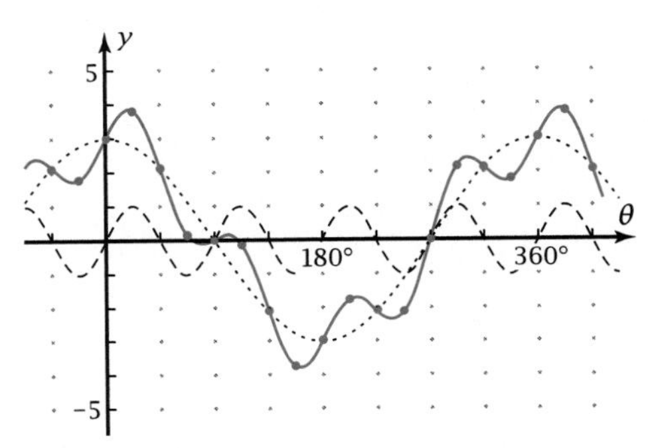

Figure 5-4g

Plotting the equation on your grapher confirms that your sketch is correct. ◀

In Example 1, adding $3 \cos \theta$ to $\sin 4\theta$, which has a smaller period, produces different vertical translations at different points. The graph of $y = 3 \cos \theta$ has a **variable sinusoidal axis**, which passes through the points of inflection of the composed graph.

Product of Two Sinusoids with Unequal Periods

Example 2 shows you how to combine the two sinusoids of Example 1 by multiplying instead of adding.

▶ ***EXAMPLE 2*** Figure 5-4d shows the graphs of $y_1 = 3 \cos \theta$ and $y_2 = \sin 4\theta$.

On a copy of Figure 5-4d, sketch the graph of

$$y = 3 \cos \theta \cdot \sin 4\theta$$

Then plot the figure on your grapher. How well does your sketch match the accurate graph?

Solution The thought process is the same as for Example 1, but this time you multiply the ordinates instead of adding them. Figure 5-4h shows the results.

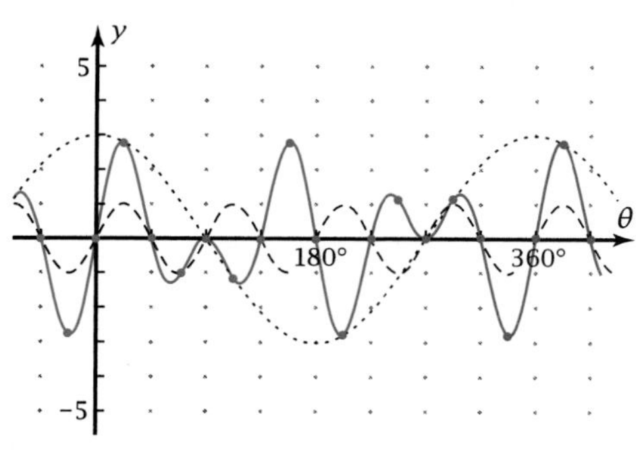

Figure 5-4h

Plotting the equation on your grapher confirms that your sketch is correct. ◀

In Example 2, multiplying $\sin 4\theta$ by $3 \cos \theta$ gives $\sin 4\theta$ a different dilation at different points. The function $y = 3 \cos \theta \sin 4\theta$ seems to have a **variable amplitude.** Because $y = \sin 4\theta$ has an amplitude of 1, the composed graph touches the graph of $y = 3 \cos \theta$ wherever the graph of $y = \sin 4\theta$ has a high point. The graph of $y = 3 \cos \theta$ forms an **envelope** for the composed graph. Note that at each place where either graph crosses the θ-axis, the composed graph also crosses the axis.

PROPERTIES: Sums and Products of Sinusoids with Unequal Periods

If two sinusoids have greatly different periods, then

- Adding two sinusoids produces a function with variable sinusoidal axis.
- Multiplying two sinusoids produces a function with a variable amplitude.

Harmonic Analysis: The Reverse of Composition of Ordinates

You can use the properties of sums and products of sinusoids with different periods to help you "decompose" a complicated graph into the two sinusoids that formed it. The procedure is used, for example, by technicians when their ship's sonar detects a complicated wave pattern and they want to find out if the sounds are being generated by another ship or by a whale.

▶ EXAMPLE 3 The function in Figure 5-4i is a sum or a product of two sinusoids. Find a particular equation, and confirm your answer by plotting the equation on your grapher.

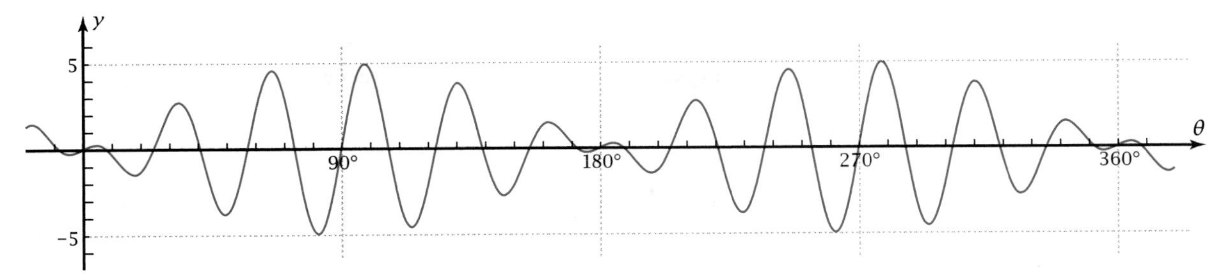

Figure 5-4i

Solution Your thought process should be:

1. The *amplitude* varies, so it is a *product* of sinusoids.
2. Sketch the larger-period sinusoid, forming an **envelope curve** that touches high and low points, as shown in Figure 5-4j. Where the envelope curve becomes negative, it touches low points on the given curve.

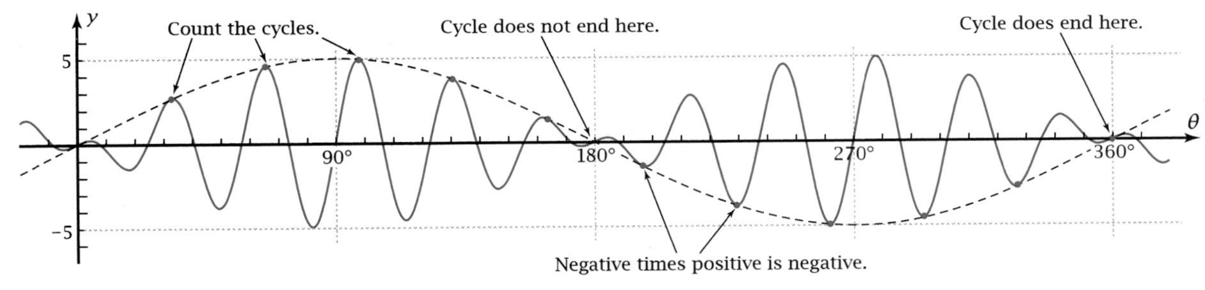

Figure 5-4j

3. The larger-period sinusoid has equation $y = 5 \sin \theta$.
4. Let the amplitude of the smaller-period sinusoid be 1. Count the number of cycles the given graph makes in one cycle of the longer-period envelope, 11 in this case. To help you count cycles, look for points where the given curve is **tangent** to the envelope curve.
5. The smaller-period sinusoid appears to be a cosine. (The value of the composed graph is 0, not 1, at $\theta = 0°$ because the larger graph is zero there.) So its equation is $y = \cos 11\theta$.
6. Write the equation for the composed function: $y = 5 \sin \theta \cdot \cos 11\theta$
7. Plot the graph (Figure 5-4k) to check your answer. Use a window with a range for x, such as $[-20, 200]$, that is small enough to separate the cycles and includes the ends where characteristic behavior occurs. If the plotted graph doesn't agree with the given graph, go back and check your work. ◀

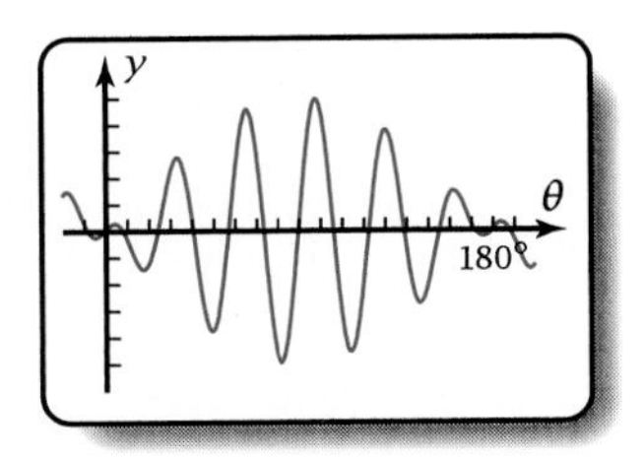

Figure 5-4k

The next example shows how to do the harmonic analysis if the argument is in radians.

▶ EXAMPLE 4 The function in Figure 5-4l is a sum or a product of two sinusoids. Find the particular equation, and confirm your answer by plotting the equation on your grapher.

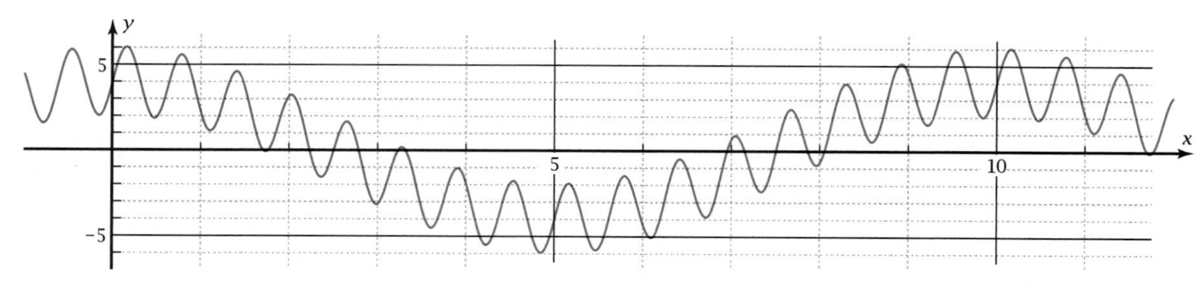

Figure 5-4l

Solution Your thought process should be:

1. The *sinusoidal axis* varies, so it is a *sum* of sinusoids.
2. Sketch the larger-period sinusoid as a sinusoidal axis through the points of inflection halfway between high points and low points (Figure 5-4m).

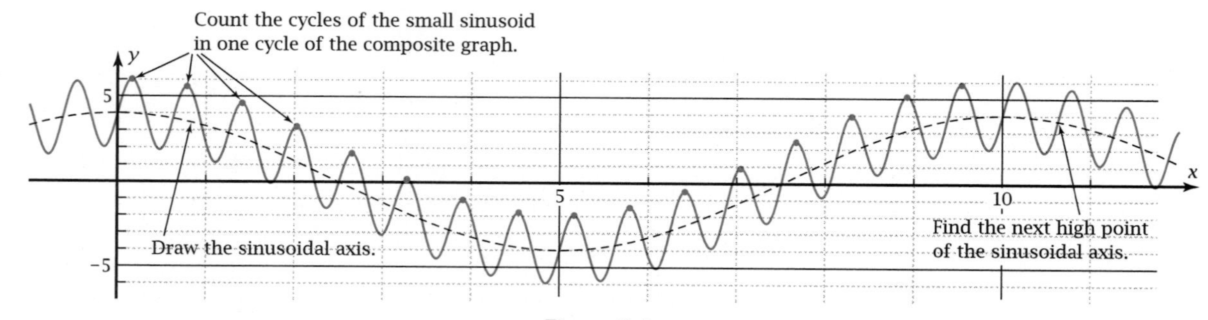

Figure 5-4m

3. The larger-period sinusoid has amplitude 4 and period 10 and starts at a high point, so its equation is $y = 4\cos\frac{\pi}{5}x$.

   The coefficient of x is 2π divided by the period.

4. The smaller-period sinusoid has amplitude 2. Because it is on the varying sinusoidal axis at $x = 0$, its function is sine. It makes 16 cycles in one period of the larger-period sinusoid, $0 \le x < 10$. So its equation is $y = 2\sin\frac{16\pi}{5}x$.
5. The equation of the composed function is $y = 4\cos\frac{\pi}{5}x + 2\sin\frac{16\pi}{5}x$.
6. Plotting on your grapher in radian mode confirms that the equation is correct. ◀

Problem Set 5-4

Do These Quickly

5 min

Q1. If $\cos A = 0.6$, $\sin A = 0.8$, $\cos B = \dfrac{1}{\sqrt{2}}$, and $\sin B = \dfrac{-1}{\sqrt{2}}$, then $\cos(A - B) =$ —?—.

Q2. In general, $\cos(x + y) =$ —?— in terms of cosines and sines of x and y.

Q3. $\sin 5 \cos 3 + \cos 5 \sin 3 = \sin$(—?—).

Q4. $\sin(90° - \theta) = \cos$(—?—).

Q5. How do you tell that sine is an *odd* function?

Q6. $\cos(x + x) =$ —?— in terms of $\cos x$ and $\sin x$.

Q7. If the two legs of a right triangle are 57 and 65, find the tangent of the smallest angle.

Q8. The period of the sinusoid $y = 5 + 7\cos\frac{\pi}{4}(x - 6)$ is —?—.

Q9. $x = 3 + 2\cos t$ and $y = 5 + 4\sin t$ are parametric equations for a(n) —?—.

Q10. If $g(x) = f(x - 7)$, what transformation is done on function f to get function g?

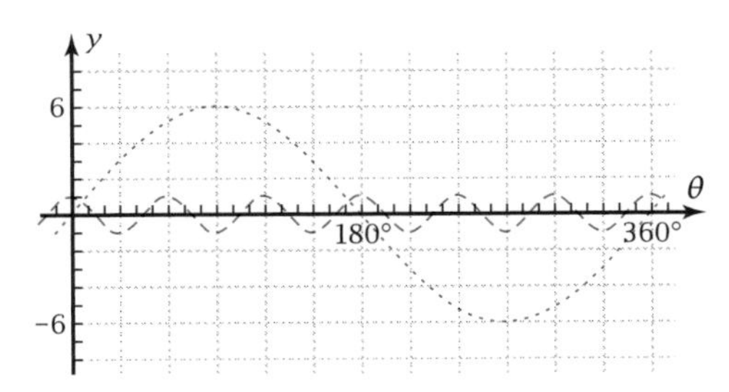

Figure 5-4n

1. *Product of Two Sinusoids Problem:* Figure 5-4n shows the graphs of

$$y_1 = 6\sin\theta \qquad \text{and} \qquad y_2 = \cos 6\theta$$

a. Which graph corresponds to which function?

b. Without using your grapher, sketch on a copy of Figure 5-4n the graph of

$$y = 6\sin\theta \cdot \cos 6\theta$$

c. On your grapher, plot the graph of $y_3 = y_1 \cdot y_2$. Use a window with a range for x of at least [0°, 360°] and an appropriate range for y. How closely does your composed graph in part b resemble the actual graph on your grapher?

d. The graph of $y = A\cos 6\theta$ is a sinusoid with a fixed amplitude A. How would you describe the graph of $y = 6\sin\theta \cdot \cos 6\theta$?

The Tacoma Narrows Bridge in Tacoma, Washington, collapsed on November 7, 1940, due to its resonance with the wind's frequency.

2. *Sum of Two Sinusoids Problem:* Figure 5-4n shows the graphs of

$$y_1 = 6\sin\theta \qquad \text{and} \qquad y_2 = \cos 6\theta$$

a. Without using your grapher, sketch on a copy of Figure 5-4n the graph of

$$y = 6\sin\theta + \cos 6\theta$$

b. On your grapher, plot the graph of $y_3 = y_1 + y_2$. Use a window with a range for x of at least [0°, 360°] and an appropriate range for y. How closely does your composed graph in part a resemble the actual graph on your grapher?

c. The graph of $y = C + \cos 6\theta$ is a sinusoid with a fixed vertical translation C. How would you describe the graph of $y = 6\sin\theta + \cos 6\theta$?

Harmonic Analysis Problems: For Problems 3–12, find the particular equation of the graph shown. Each is the sum or the product of sinusoids with unequal periods. Make note of whether or not the argument is in degrees or radians.

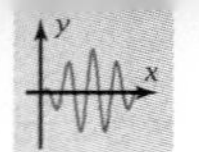

3.

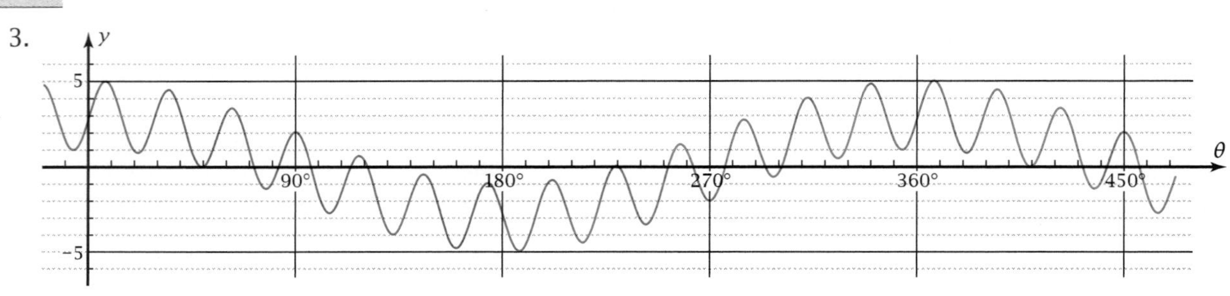
y
5
−5
θ
90°
180°
270°
360°
450°

4.

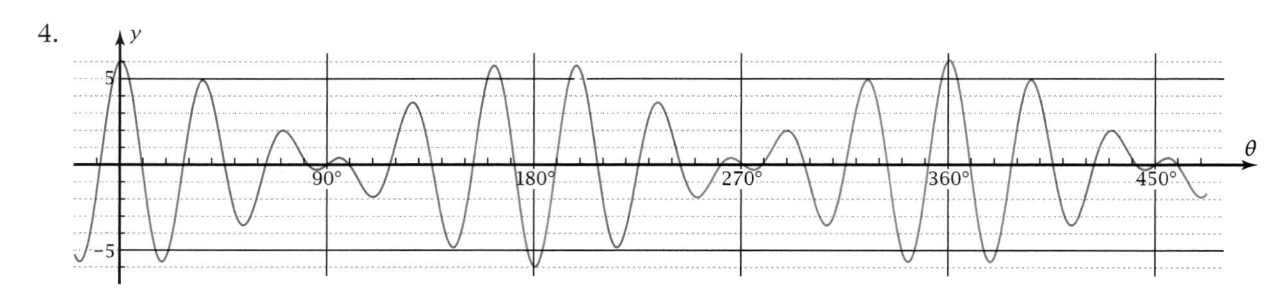
y
5
−5
θ
90°
180°
270°
360°
450°

5.

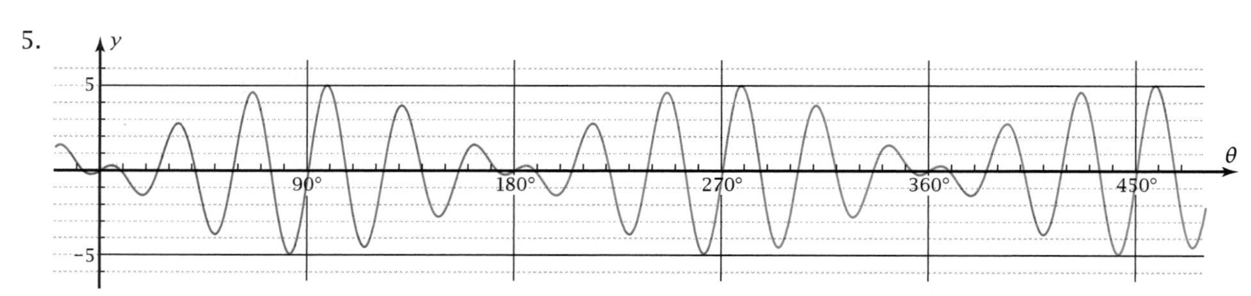
y
5
−5
θ
90°
180°
270°
360°
450°

6.

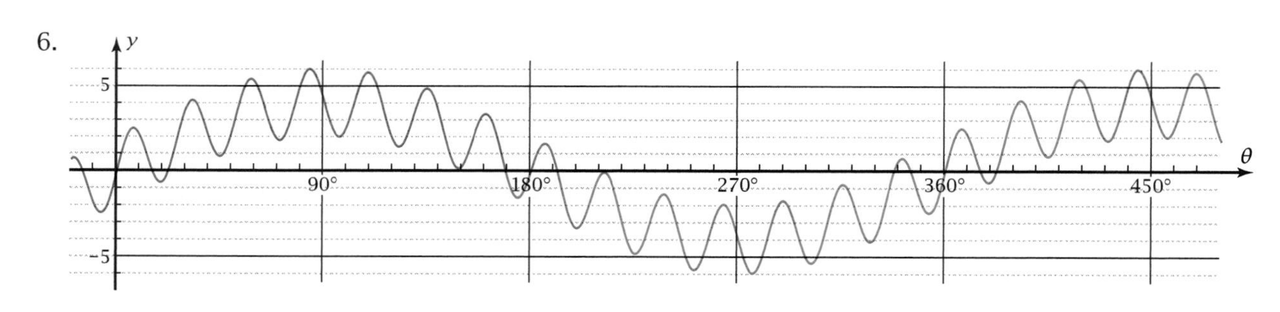
y
5
−5
θ
90°
180°
270°
360°
450°

7.

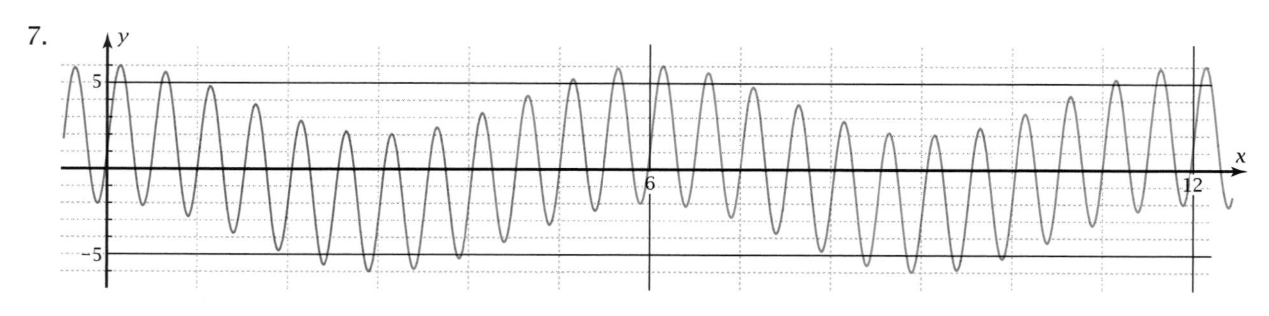
y
5
−5
x
6
12

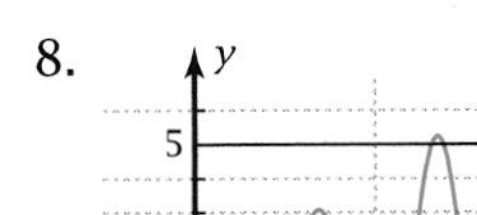

8.

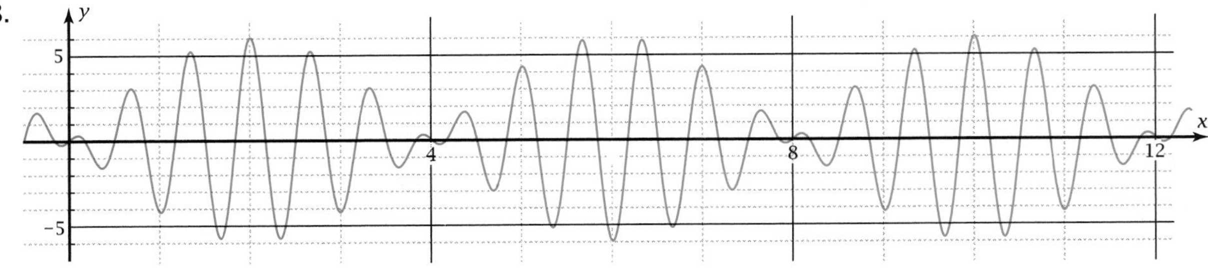

9.

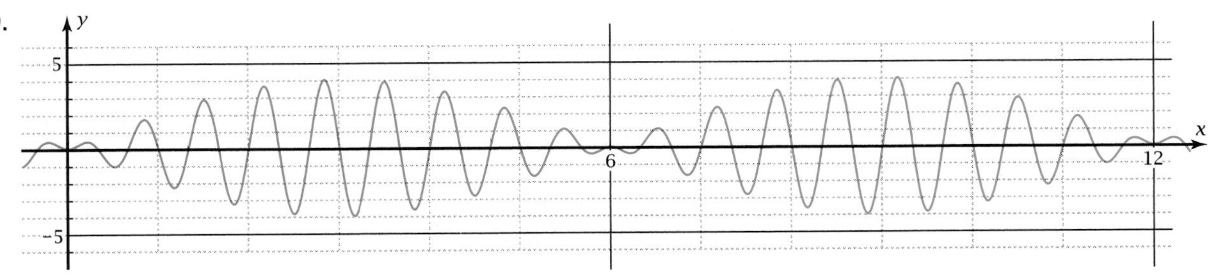

10.

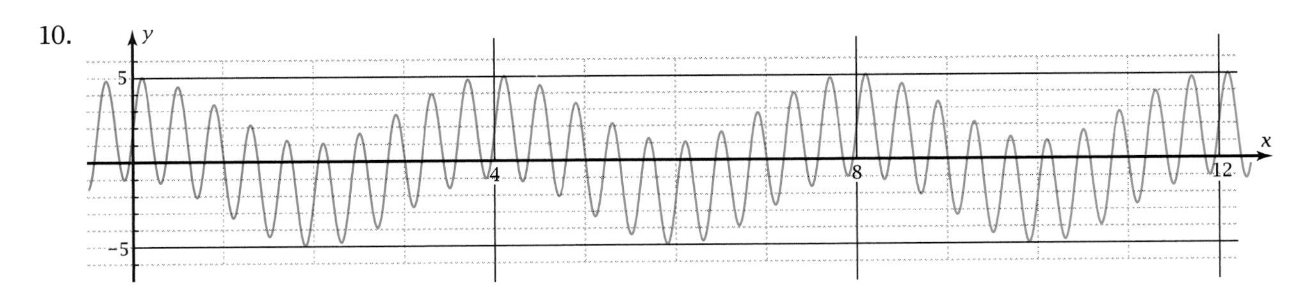

11.

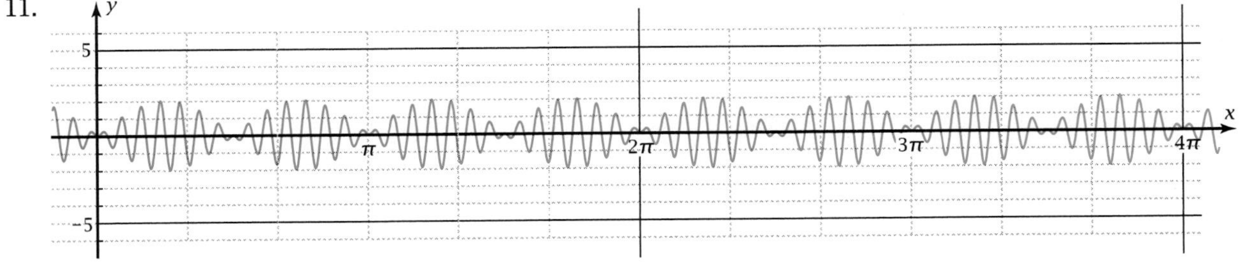

12. A combination of *three* sinusoids!

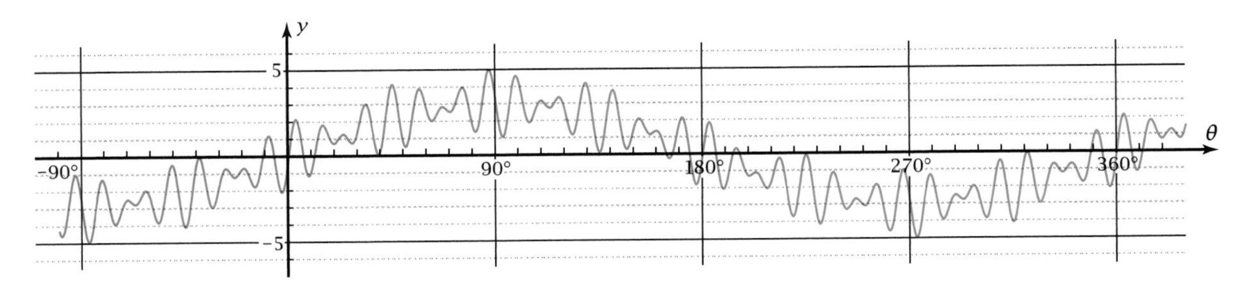

13. *Submarine Sonar Problem:* The sonar on a surface ship picks up sound being generated by equipment on a submarine. By performing harmonic analysis on the sound wave, it is possible to identify the national origin of the submarine.

    a. Figure 5-4o shows the pattern of a sound wave, where x is time, in seconds. Find an equation for this graph. Observe that the longer-wave sinusoidal axis completes *three* cycles before the wave pattern starts repeating itself.

    b. What is the period of the longer sinusoid? What is the period of the shorter sinusoid?

    c. Recall that the frequency of a sinusoid is the reciprocal of the period. What are the frequencies of the two sinusoids in part b?

    d. United States submarines have electrical generators that rotate at 60 cycles per second and other electrical generators that rotate at 400 cycles per second. Based on your results in part c, could the sound have been coming from the generators on a U.S. submarine?

14. *Sunrise Project:* The angle of elevation of the Sun at any time of day on any day of the year is a sum of two sinusoids. The first sinusoid is caused by the rotation of Earth about its own axis. In San Antonio this sinusoid has a period of 1 day and an amplitude of 61° (the complement of the 29° latitude on which San Antonio lies). The sinusoid reaches a minimum at midnight on any day. The other sinusoid is caused by the rotation of Earth around the Sun. That sinusoid has a period of 365.25 days and an amplitude of 23.5° (equal to the tilt of Earth's axis with respect to the ecliptic plane). This second sinusoid reaches a minimum at day −10 (late December). Part of the graph of this composed function is shown in Figure 5-4p. On this graph, y is the angle of elevation on day x. Note that $x = 1$ at the *end* of day 1, $x = 2$ at the *end* of day 2, etc.

    a. Write a particular equation of this composed graph. Use radians as the argument of the function. Use the equation to show that the maximum angle of elevation on day 1 ($x = 0.5$) is lower than the maximum angle on day 30. Confirm that this is true by direct measurement on Figure 5-4p.

    b. Find the maximum angle of elevation on the longest day of the year, June 21.

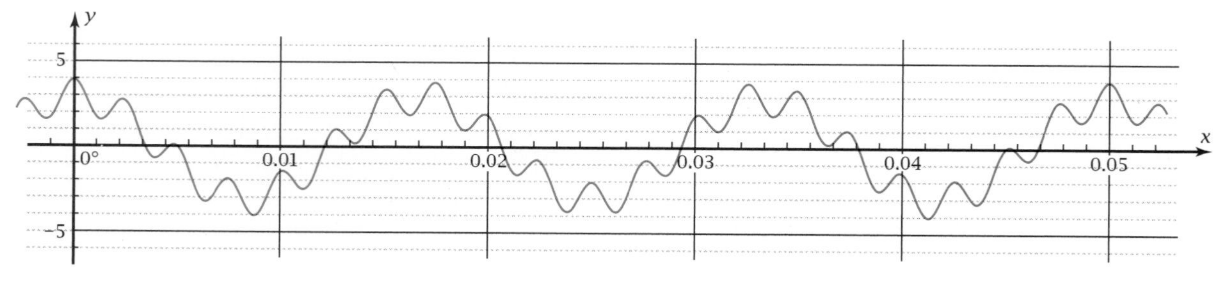

Figure 5-4o

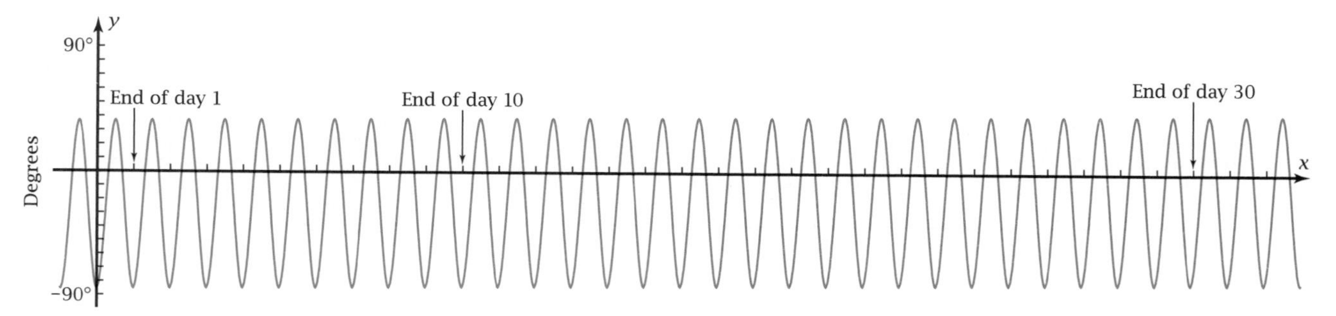

Figure 5-4p

Mariners' astrolabe from around 1585. An astrolabe measured the elevation of the Sun or another star to predict the time of sunrise and sunset.

c. Sunrise occurs at the time when the angle of elevation equals zero, as it is increasing. Find the time of sunrise on day 1 by solving numerically for the value of x close to 0.25 when $y = 0$. Then convert your answer to hours and minutes. Show that the Sun rises earlier on day 30 than it does on day 1. How much earlier?

d. How do you interpret the parts of the graph that are *below* the x-axis?

15. *Journal Problem:* Update your journal with things you have learned since the last entry. In particular, mention how you decide whether a composed graph is a sum or a product of sinusoids. You might also mention how comfortable you are getting with the amplitudes and periods of sinusoidal graphs and whether and how your confidence with these concepts has grown since you first encountered them in Chapter 2.

5-5 The Sum and Product Properties

Figure 5-5a shows the graphs of two sinusoids with very different periods. Figure 5-5b shows two sinusoids with nearly equal periods. The product of the two sinusoids in Figure 5-5a is the variable-amplitude function in Figure 5-5c. Surprisingly, the *sum* of the two nearly equal sinusoids in Figure 5-5b also produces the wave pattern in the third figure! The y-values add up where the waves are in phase (i.e., two high points coincide) and cancel out where the waves are out of phase. This phenomenon is heard as what's called *beats*. For example, two piano strings for the same note can be tuned at slightly different frequencies to produce a vibrato effect. This is what piano tuners listen for.

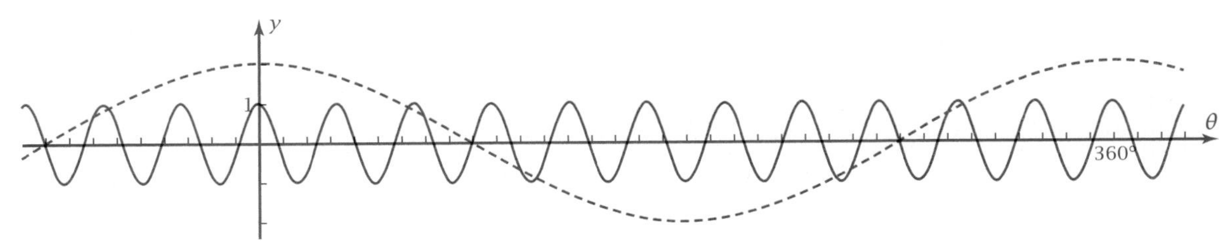

$y_1 = \cos 11\theta$ and $y_2 = 2\cos\theta$

Figure 5-5a

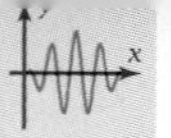

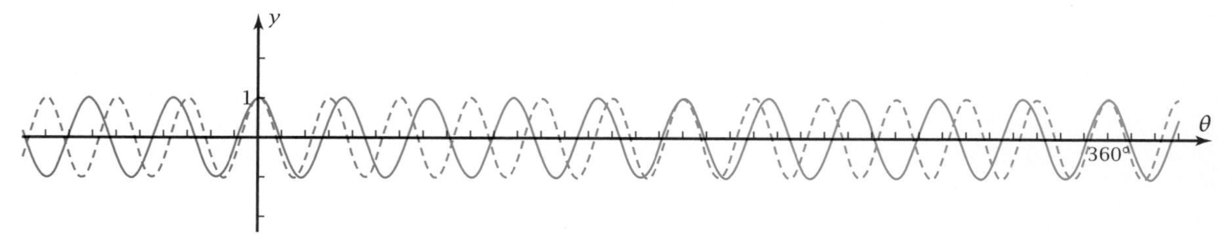

$y_1 = \cos 12\theta$ and $y_2 = \cos 10\theta$

Figure 5-5b

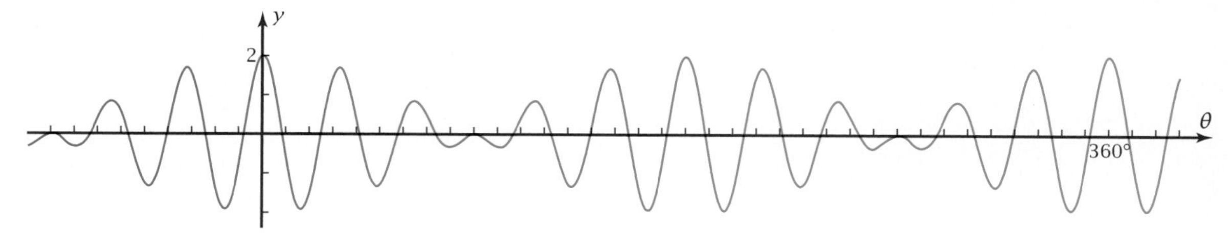

$y = 2\cos 11\theta \cdot \cos\theta$ or $y = \cos 12\theta + \cos 10\theta$

Figure 5-5c

In this section you will learn algebraic properties that allow you to prove that the sum graph and product graph are equivalent.

OBJECTIVE Transform a sum of two sinusoids to a product of two sinusoids, and vice versa.

Product to Sum

From Figure 5-5c, it appears that

$$\cos 12\theta + \cos 10\theta = 2\cos 11\theta \cdot \cos\theta$$

To see why, write 12θ as $(11\theta + \theta)$ and 10θ as $(11\theta - \theta)$ and use composite argument properties.

$$\cos 12\theta = \cos(11\theta + \theta) = \cos 11\theta\cos\theta - \sin 11\theta\sin\theta$$

$$\cos 10\theta = \cos(11\theta - \theta) = \cos 11\theta\cos\theta + \sin 11\theta\sin\theta$$

Add the two equations to get

$$\cos 12\theta + \cos 10\theta = 2\cos 11\theta\cos\theta$$

If you subtract the equations, you get

$$\cos 12\theta - \cos 10\theta = -2\sin 11\theta\sin\theta$$

Using similar steps, you can derive these properties in general. Use A and B in place of the 11θ and θ and you'll get **sum and product properties** expressing the sum or difference of two cosines as a product of two cosines or sines or the other way around.

$$\cos(A + B) + \cos(A - B) = 2\cos A\cos B$$

$$\cos(A + B) - \cos(A - B) = -2\sin A\sin B$$

Two other sum and product properties come from adding or subtracting the composite argument properties for sine.

$$\sin(A+B) = \sin A \cos B + \cos A \sin B$$
$$\underline{\sin(A-B) = \sin A \cos B - \cos A \sin B}$$

$$\sin(A+B) + \sin(A-B) = 2\sin A \cos B$$ By adding the two equations.

$$\sin(A+B) - \sin(A-B) = 2\cos A \sin B$$ By subtracting the two equations.

This table summarizes the four properties.

PROPERTIES: Sum and Product Properties—Product to Sum

$$2\cos A \cos B = \cos(A+B) + \cos(A-B)$$
$$-2\sin A \sin B = \cos(A+B) - \cos(A-B)$$
$$2\sin A \cos B = \sin(A+B) + \sin(A-B)$$
$$2\cos A \sin B = \sin(A+B) - \sin(A-B)$$

It is probably easier to derive these properties as you need them than it is to memorize them.

▶ EXAMPLE 1 Transform $2\sin 13^\circ \cos 48^\circ$ to a sum (or difference) of functions with *positive* arguments. Demonstrate numerically that the answer is correct.

Solution Sine multiplied by cosine appears in the *sine* composite argument properties. So the answer will be the sum of two sines. If you have not memorized the sum and product properties, you would write

$$\sin(13^\circ + 48^\circ) = \sin 13^\circ \cos 48^\circ + \cos 13^\circ \sin 48^\circ$$
$$\underline{\sin(13^\circ - 48^\circ) = \sin 13^\circ \cos 48^\circ - \cos 13^\circ \sin 48^\circ}$$

$$\sin 61^\circ + \sin(-35^\circ) = 2\sin 13^\circ \cos 48^\circ$$ *Add* the equations.

$$2\sin 13^\circ \cos 48^\circ = \sin 61^\circ - \sin 35^\circ$$ Use the symmetric property of equality. Also, sine is an *odd* function.

Check: 0.3010... = 0.3010... ◀

Sum to Product

Example 2 shows you how to reverse the process and transform a sum of two sinusoids into a product.

▶ EXAMPLE 2 Transform $\cos 7\theta - \cos 3\theta$ to a product of functions with positive arguments.

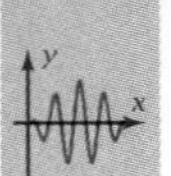

Solution First, think of writing $\cos 7\theta$ and $\cos 3\theta$ as cosines with composite arguments, then use appropriate calculations to find out what those two arguments are.

Let $\cos 7\theta = \cos(A + B)$ and let $\cos 3\theta = \cos(A - B)$.

$$A + B = 7\theta$$
$$A - B = 3\theta$$
$$2A = 10\theta$$
$$A = 5\theta$$

Substituting 5θ for A in either equation, you get $B = 2\theta$. Now, put these values of A and B into the composite argument properties for cosine.

$$\cos 7\theta = \cos(5\theta + 2\theta) = \cos 5\theta \cos 2\theta - \sin 5\theta \sin 2\theta$$
$$\cos 3\theta = \cos(5\theta - 2\theta) = \cos 5\theta \cos 2\theta + \sin 5\theta \sin 2\theta$$
$$\cos 7\theta - \cos 3\theta = -2 \sin 5\theta \sin 2\theta$$

The arguments have no negative signs, so you need no further transformations. ◀

From the algebraic steps in Example 2, you can see that A equals half the sum of the arguments. You can also tell that B equals half the difference of the arguments. So a general property expressing a difference of two cosines as a product is

$$\cos x - \cos y = -2 \sin \tfrac{1}{2}(x + y) \sin \tfrac{1}{2}(x - y)$$

You can also write the other three sum and product properties in this form. The results are in this box. Again, do not try to memorize the properties. Instead, derive them from the composite argument properties, as in Example 2, or look them up when you need to use them.

PROPERTIES: Sum and Product Properties—Sum to Product

$$\sin x + \sin y = 2 \sin \tfrac{1}{2}(x + y) \cos \tfrac{1}{2}(x - y)$$
$$\sin x - \sin y = 2 \cos \tfrac{1}{2}(x + y) \sin \tfrac{1}{2}(x - y)$$
$$\cos x + \cos y = 2 \cos \tfrac{1}{2}(x + y) \cos \tfrac{1}{2}(x - y)$$
$$\cos x - \cos y = -2 \sin \tfrac{1}{2}(x + y) \sin \tfrac{1}{2}(x - y)$$

Note: Both functions on the "sum" side are always the same.

▶ **EXAMPLE 3** Figure 5-5d shows a periodic trigonometric function with a variable sinusoidal axis. Using harmonic analysis, find a particular equation for this function as a sum of two sinusoids. Then transform the sum into a product. Confirm graphically that both equations produce the function in Figure 5-5d.

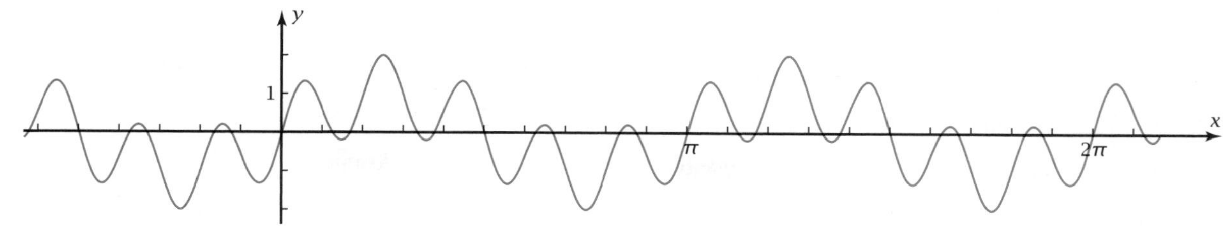

Figure 5-5d

Solution On a copy of Figure 5-5d, sketch the sinusoidal axis. Figure 5-5e shows the result.

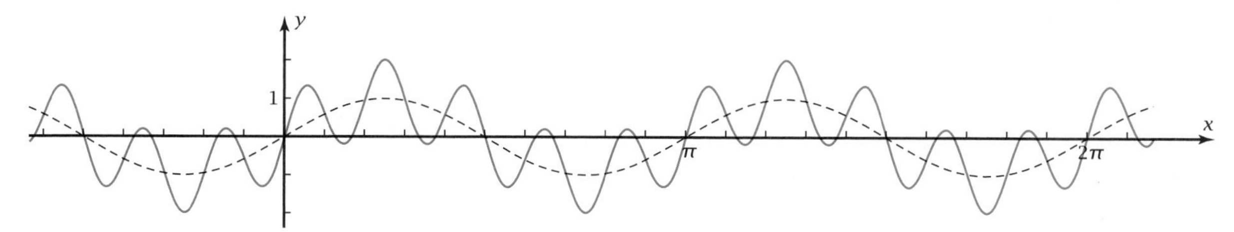

Figure 5-5e

By the techniques of Section 5-4,

$$y = \sin 2x + \sin 10x$$

To transform this equation into a product, write

$$y = 2\sin \tfrac{1}{2}(2x + 10x)\cos \tfrac{1}{2}(2x - 10x)$$ Use the property from the box, or derive it.

$$y = 2\sin 6x \cos(-4x)$$

$$y = 2\sin 6x \cos 4x$$ Cosine is an *even* function.

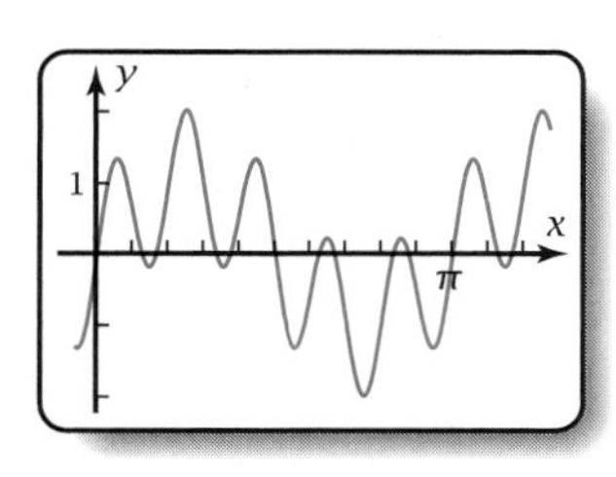

Figure 5-5f

Both equations give the graph shown in Figure 5-5f, which agrees with Figure 5-5e. ◀

Problem Set 5-5

Do These Quickly

Q1. Find values of x and y if $x + y = 20$ and $x - y = 12$.

Q2. Find values of x and y if $x + y = A$ and $x - y = B$.

Q3. Sketch the graph of the parent trigonometric sinusoid $y = \cos\theta$.

Q4. Sketch the graph of the parent circular sinusoid $y = \sin x$.

Q5. How many degrees are there in $\frac{\pi}{4}$ radians?

Q6. How many radians are there in 180°?

Q7. What is the exact value (no decimals) of $\cos\left(\frac{\pi}{6}\right)$?

Q8. What is the exact value (no decimals) of tan 30°?

Q9. $\cos(3x + 5x) =$ —?— in terms of functions of $3x$ and $5x$.

Q10. Find in decimal degrees $\theta = \cot^{-1}\left(\frac{3}{7}\right)$.

Transformation Problems: For Problems 1–8, transform the product into a sum or difference of sines or cosines with *positive* arguments.

1. $2 \sin 41° \cos 24°$
2. $2 \cos 73° \sin 62°$
3. $2 \cos 53° \cos 49°$
4. $2 \sin 29° \sin 16°$
5. $2 \cos 3.8 \sin 4.1$
6. $2 \cos 2 \cos 3$
7. $2 \sin 3x \sin 7.2$
8. $2 \sin 8x \cos 2x$

For Problems 9–16, transform the sum or difference to a product of sines and/or cosines with *positive* arguments.

9. $\cos 46° + \cos 12°$
10. $\cos 56° - \cos 24°$
11. $\sin 2 + \sin 6$
12. $\sin 3 - \sin 8$
13. $\cos 2.4 - \cos 4.4$
14. $\sin 1.8 + \sin 6.4$
15. $\sin 3x - \sin 8x$
16. $\cos 9x + \cos 11x$

Graphing Problems: For Problems 17–20, use harmonic analysis to find an equation of the given graph as a product or sum of sinusoids. Then transform the product into a sum or the sum into a product. Confirm graphically that both of your equations produce the given graph.

17.

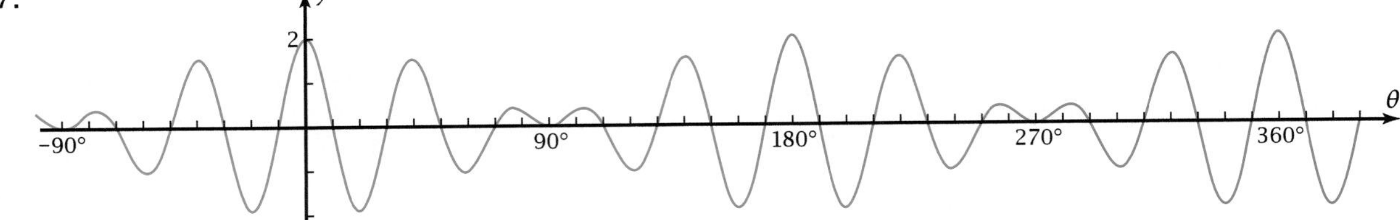

18.

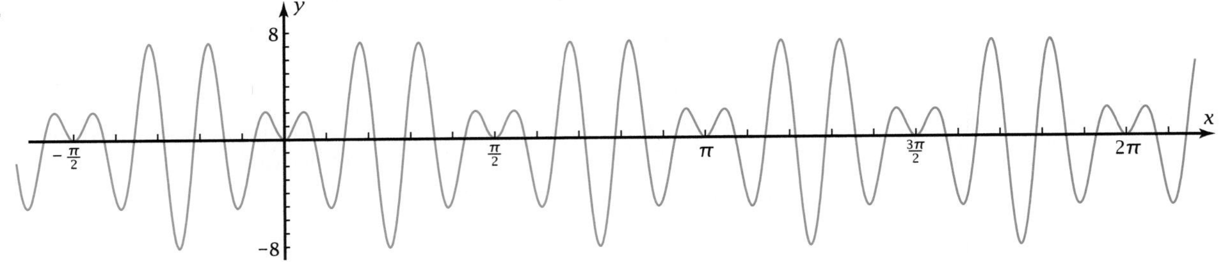

19.

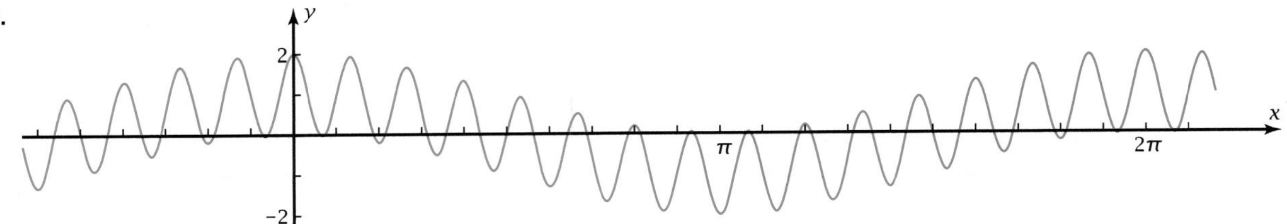

20.

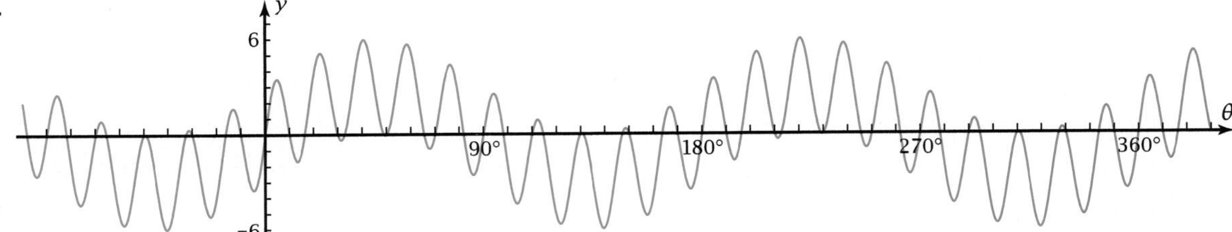

Algebraic Solution of Equations Problems 1: You can use the sum and product properties to find algebraically the exact solutions of certain equations. For Problems 21–24, solve the equation by first transforming it to a product equal to zero and then setting each factor equal to zero. Use the domain $\theta \in [0°, 360°]$ or $x \in [0, 2\pi]$.

21. $\sin 3x - \sin x = 0$
22. $\sin 3\theta + \sin \theta = 0$
23. $\cos 5\theta + \cos 3\theta = 0$
24. $\cos 5x - \cos x = 0$

Identities Problems: For Problems 25–30, prove that the given equation is an identity.

25. $\cos x - \cos 5x = 4 \sin 3x \sin x \cos x$
26. $\dfrac{\sin 5x + \sin 7x}{\cos 5x + \cos 7x} = \tan 6x$
27. $\cos x + \cos 2x + \cos 3x = \cos 2x(1 + 2 \cos x)$
28. $\sin (x + y) \sin (x - y) = \sin^2 x - \sin^2 y$
29. $\cos (x + y) \cos (x - y) = \cos^2 x - \sin^2 y$
30. $\sin (x + y) \cos (x - y) = \frac{1}{2} \sin 2x + \frac{1}{2} \sin 2y$
31. *Piano Tuning Problem:* Note A on the piano has a frequency of 220 cycles/sec. Inside the piano there are two strings for this note. When the note is played, the hammer attached to the A key hits both strings and starts them vibrating. Suppose that one of the two A strings is tuned to 221 cycles/sec and the other is tuned to 219 cycles/sec.

 a. The combined sound of these two notes is the sum of the two sound waves. Write an equation for the combined sound wave, where the independent variable is t, in seconds. Use the (undisplaced) cosine function for each sound wave, and assume that each has an amplitude of 1.
 b. Use the sum in part a and transform it to a product of two sinusoids.
 c. Explain why the combined sound is equivalent to a sound with a frequency of 220 cycles/sec and an amplitude that varies. What is the frequency of the variable amplitude? Describe how the note sounds.

32. *Car and Truck Problem:* Suppose that you are driving an 18-wheeler tractor-trailer truck along the highway. A car pulls up alongside your truck and then moves ahead very slowly. As it passes, the combined sounds of the car and truck engines pulsate louder and softer. The combined sound can be either a sum of the two sinusoids or a product of them.
 a. The tachometer says that your truck's engine is turning at 3000 revolutions per minute (rev/min), which is equivalent to 50 revolutions per second (rev/sec). The pulsations come and go once a second, which means that the amplitude sinusoid has a period of *two* seconds. Write an equation for the sound intensity as a product of two sinusoids. Use cosine for each, with independent variable t seconds. Use 2 for the amplitude of the larger sinusoid and 1 for the smaller.
 b. Use the product in part a and transform it to a sum of two sinusoids.
 c. At what rate is the car's engine rotating?
 d. Explain why the period of the amplitude sinusoid is *two* seconds, not one second.

33. *AM/FM Radio Project:* AM ("amplitude modulation") radio works by having a sound wave of a relatively low frequency (long period) cause variations in the amplitude of a "carrier wave" that has a very high frequency (VHF). So the sound wave is multiplied by the carrier wave. An example of the resulting wave pattern is shown in Figure 5-5g. In this project you will find equations of the two waves that were multiplied to form this graph. Then you will see how you can form the same wave pattern by *adding* two waves of nearly equal frequency (FM, or frequency modulation radio).
 a. By harmonic analysis, find the equations of the two circular function sinusoids that were multiplied to form the graph

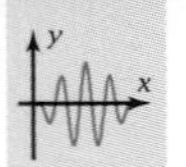

in Figure 5-5g. Which equation is the sound wave, and which is the carrier wave?

b. You can add two sinusoids of nearly equal period to form the same wave pattern as in Figure 5-5g. Use the sum and product properties to find equations for these two sinusoids.

c. Confirm by graphing that the product of the sinusoids in part a and the sum of the sinusoids in part b give the same wave pattern as in Figure 5-5g.

d. The scale on the t-axis in Figure 5-5g is in milliseconds. Find the frequency of the carrier wave in kilocycles per second. (This number, divided by 10, is what appears on an AM radio dial.)

e. Research to find the name used in radio waves for "cycles per second."

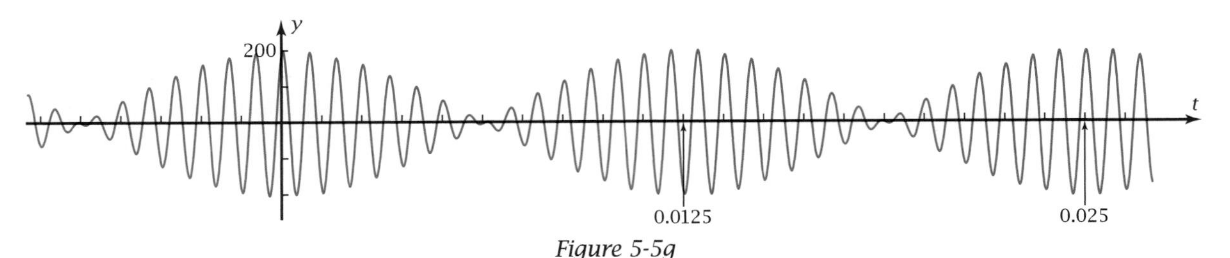

Figure 5-5g

5-6 Double and Half Argument Properties

If you write cos $2x$ as cos $(x + x)$ and use the composite argument property, you can get a **double argument property** expressing cos $2x$ in terms of sines and cosines of x. By performing algebraic operations on the double argument properties, you can derive similar **half argument properties.** In this section you'll learn these properties and show that the product of two sinusoids with equal periods is a sinusoid with half the period and half the amplitude.

OBJECTIVE

- Prove that a product of sinusoids with equal periods is also a sinusoid.
- Derive properties for $\cos 2A$, $\sin 2A$, and $\tan 2A$ in terms of functions of A.
- Derive properties for $\cos \frac{1}{2}A$, $\sin \frac{1}{2}A$, and $\tan \frac{1}{2}A$ in terms of functions of A.

Product of Sinusoids with Equal Periods

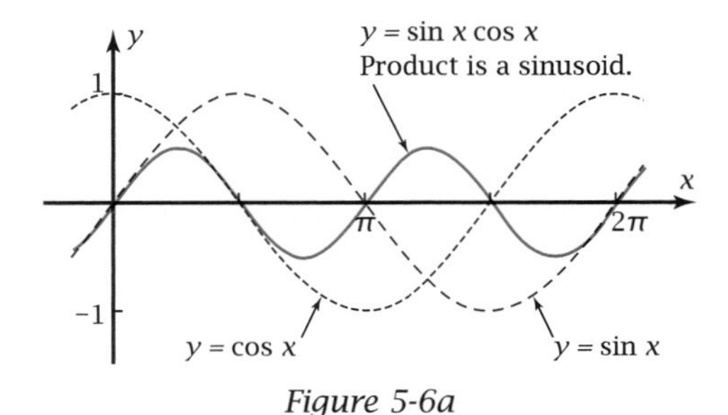

Figure 5-6a

Figure 5-6a shows that $y = \sin x \cos x$ appears to be a sinusoid with half the period and half the amplitude of the parent sine and cosine functions. Its equation would be $y = 0.5 \sin 2x$. This calculation shows that this is true.

$\frac{1}{2} \sin 2x = \frac{1}{2} \sin (x + x)$ — Write $2x$ as $x + x$.

$= \frac{1}{2}(\sin x \cos x + \cos x \sin x)$ — Use the composite argument property.

$= \frac{1}{2}(2 \sin x \cos x)$

$= \sin x \cos x$

$\therefore \sin x \cos x = \frac{1}{2} \sin 2x$ — The product of sine and cosine with equal periods is a sinusoid.

The square of cosine and the square of sine have a similar property. Figure 5-6b gives graphical evidence.

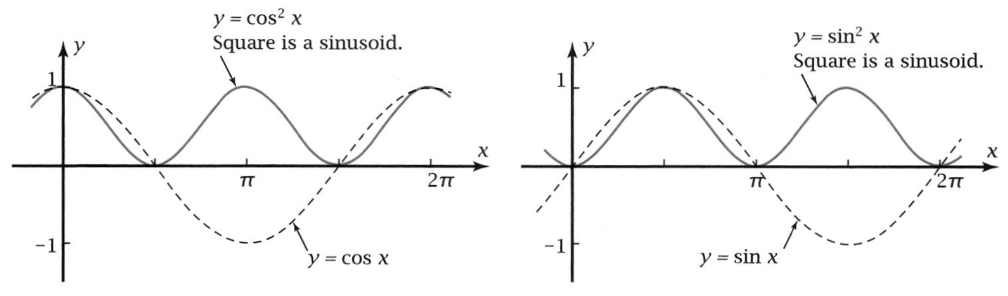

Figure 5-6b

Both $y = \cos^2 x$ and $y = \sin^2 x$ are sinusoids with amplitude $\frac{1}{2}$, period π, and sinusoidal axis $y = \frac{1}{2}$. Notice that the two graphs are a half-cycle out of phase. (This is why $\cos^2 x + \sin^2 x = 1$.) This table summarizes these conclusions, fulfilling the first part of this section's objective.

PROPERTIES: Products and Squares of Cosine and Sine

$\sin x \cos x = \frac{1}{2} \sin 2x$	Product of sine and cosine property.
$\cos^2 x = \frac{1}{2} + \frac{1}{2} \cos 2x$	Square of cosine property.
$\sin^2 x = \frac{1}{2} - \frac{1}{2} \cos 2x$	Square of sine property.

Note: You can observe that the product of two sinusoids with equal periods, equal amplitudes, and zero vertical translation and dilation is a sinusoid with one half the period and one half the amplitude.

Double Argument Properties

Multiplying both sides of the equation $\sin x \cos x = \frac{1}{2} \sin 2x$ by 2 gives an equation expressing $\sin 2x$ in terms of the sine and cosine of x.

$\sin 2x = 2 \sin x \cos x$ — Double argument property for sine.

Here's how the double argument property for cosine can be derived.

$$\cos 2x = \cos(x + x)$$
$$= \cos x \cos x - \sin x \sin x$$
$$= \cos^2 x - \sin^2 x$$

This table summarizes the double argument properties. You can prove the double argument property for tangent in Problem 31 of Problem Set 5-6.

PROPERTIES: Double Argument Properties

Double Argument Property for Sine

$$\sin 2A = 2 \sin A \cos A$$

Double Argument Properties for Cosine

$$\cos 2A = \cos^2 A - \sin^2 A$$
$$\cos 2A = 2\cos^2 A - 1$$
$$\cos 2A = 1 - 2\sin^2 A$$

Double Argument Property for Tangent

$$\tan 2A = \frac{2 \tan A}{1 - \tan^2 A}$$

► ***EXAMPLE 1*** If $\cos x = 0.3$, find the exact value of $\cos 2x$. Check your answer numerically by finding the value of x, doubling it, and finding the cosine of the resulting argument.

Solution

$\cos 2x = 2\cos^2 x - 1$ — Double argument property for cosine in terms of cosine alone.

$= 2(0.3)^2 - 1$

$= -0.82$

Check:

$x = \cos^{-1} 0.3 = 1.2661...$

$\cos 2(1.2661...) = \cos 2.5322... = -0.82$ — The answer is correct. ◄

Note that for the properties to apply, all that matters is that one argument is twice the other argument. Example 2 shows this.

► ***EXAMPLE 2*** Write an equation expressing $\cos 10x$ in terms of $\sin 5x$.

Solution

$\cos 10x = \cos 2(5x)$ — Transform to double argument.

$= 1 - 2\sin^2 5x$ — Use the double argument property for cosine involving only sine.

$\therefore \cos 10x = 1 - 2\sin^2 5x$ ◄

EXAMPLE 3 Do Example 2 directly from the composite argument property for cosine.

Solution

$$\cos 10x = \cos(5x + 5x)$$ Transform to composite argument.

$$= \cos 5x \cos 5x - \sin 5x \sin 5x$$ Use the composite argument property for cosine.

$$= \cos^2 5x - \sin^2 5x$$

$$= (1 - \sin^2 5x) - \sin^2 5x$$ Use the Pythagorean property.

$$= 1 - 2\sin^2 5x$$

Half Argument Properties

The square of cosine property and the square of sine property state

$$\cos^2 x = \tfrac{1}{2} + \tfrac{1}{2}\cos 2x \quad \text{and} \quad \sin^2 x = \tfrac{1}{2} - \tfrac{1}{2}\cos 2x$$

The argument x on the left is half the argument $2x$ on the right. Substituting A for $2x$ leads to

$$\cos^2 \tfrac{1}{2}A = \tfrac{1}{2}(1 + \cos A) \quad \text{and} \quad \sin^2 \tfrac{1}{2}A = \tfrac{1}{2}(1 - \cos A)$$

Taking the square roots gives **half argument** properties for $\cos \frac{1}{2}A$ and $\sin \frac{1}{2}A$.

$$\cos \tfrac{1}{2}A = \pm\sqrt{\tfrac{1}{2}(1 + \cos A)} \quad \text{and} \quad \sin \tfrac{1}{2}A = \pm\sqrt{\tfrac{1}{2}(1 - \cos A)}$$

The way you can determine whether to choose the positive or negative sign of the ambiguous plus/minus sign is by looking at the quadrant in which $\frac{1}{2}A$ terminates (*not* the quadrant in which A terminates). For instance, if $A = 120°$, then half of A is 60°, which terminates in the first quadrant, as shown in the left graph of Figure 5-6c. In this case, both $\sin \frac{1}{2}A$ and $\cos \frac{1}{2}A$ are positive. If $A = 480°$ (which is coterminal with 120°), then half A is 240°, which terminates in Quadrant III, as shown in the right graph of Figure 5-6c. In this case, you'll choose the negative signs for $\cos \frac{1}{2}A$ and $\sin \frac{1}{2}A$. The signs of sine and cosine don't have to be the same. If half of A falls in Quadrant II or Quadrant IV, the signs of $\cos \frac{1}{2}A$ and $\sin \frac{1}{2}A$ are opposites.

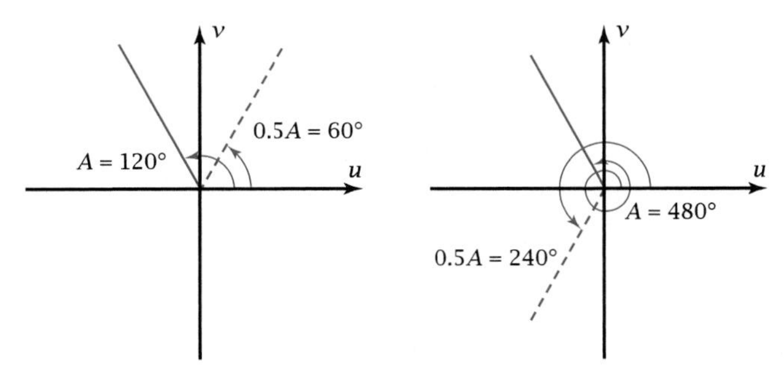

Figure 5-6c

You can derive a property for $\tan \frac{1}{2}A$ by dividing the respective properties for sine and cosine. In Problem 32 of Problem Set 5-6, you will see how this is done. The result is remarkable because you can drop the radical sign and the ambiguous plus/minus sign.

The half argument properties are summarized in this table.

PROPERTIES: Half Argument Properties

Half Argument Property for Sine

$$\sin \tfrac{1}{2}A = \pm\sqrt{\tfrac{1}{2}(1 - \cos A)}$$

Half Argument Property for Cosine

$$\cos \tfrac{1}{2}A = \pm\sqrt{\tfrac{1}{2}(1 + \cos A)}$$

Half Argument Properties for Tangent

$$\tan \tfrac{1}{2}A = \pm\sqrt{\frac{1 - \cos A}{1 + \cos A}} = \frac{1 - \cos A}{\sin A} = \frac{\sin A}{1 + \cos A}$$

Note: The ambiguous plus/minus sign is determined by the quadrant where $\frac{1}{2}A$ terminates.

Example 4 shows you how to calculate the functions of half an angle and twice an angle using the properties and how to verify the result by direct computation.

► ***EXAMPLE 4*** If $\cos A = \frac{15}{17}$ and A is in the open interval (270°, 360°),

a. Find the exact value of $\cos \frac{1}{2}A$.

b. Find the exact value of $\cos 2A$.

c. Verify your answers numerically by calculating the values of $2A$ and $\frac{1}{2}A$ and taking the cosines.

Solution Sketch angle A in standard position, as shown in Figure 5-6d. Pick a point on the terminal side with horizontal coordinate 15 and radius 17. Draw a line perpendicular to the horizontal axis. By the Pythagorean theorem and by noting that angle A terminates in Quadrant IV, you can determine that the vertical coordinate of the point is −8.

a.

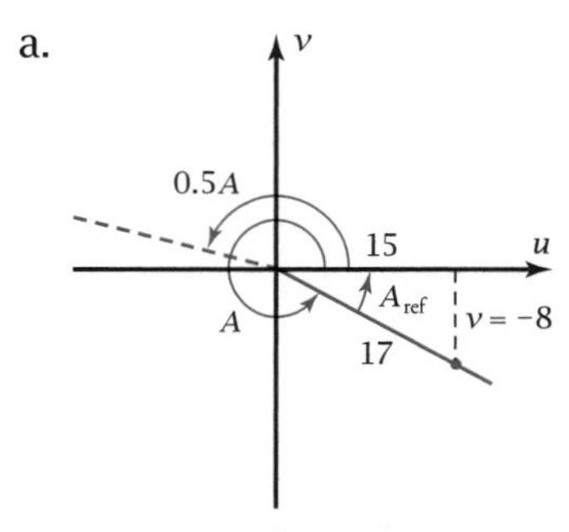

Figure 5-6d

$270° < A < 360°$ Write the given range as an inequality.

$135° < \frac{1}{2}A < 180°$ Divide by 2 to find the quadrant in which half A terminates.

Therefore, $\frac{1}{2}A$ terminates in Quadrant II, where cosine is negative.

$\cos \frac{1}{2}A = -\sqrt{\frac{1}{2}(1 + \frac{15}{17})} = -\frac{4}{\sqrt{17}}$ Use the half argument property with the minus sign.

b. $\cos 2A = 2\cos^2 A - 1 = \frac{450}{289} - 1 = \frac{161}{289}$ Use the form of cos 2A involving the *given* function value.

c. $A = \arccos \frac{15}{17} = \pm 28.0724...° + 360n°$

Use the definition of arccosine to write the general solution for A.

$A = -28.0724...° + 360° = 331.9275...°$

Write the particular solution. See Figure 5-6d.

$\frac{1}{2}A = 165.9637...°$ and $2A = 663.8550...°$

$\cos 165.9637...° = -0.9701... = -\frac{4}{\sqrt{17}}$

$\cos 663.8550...° = 0.5570... = \frac{161}{289}$

The answers are correct. ◀

Problem Set 5-6

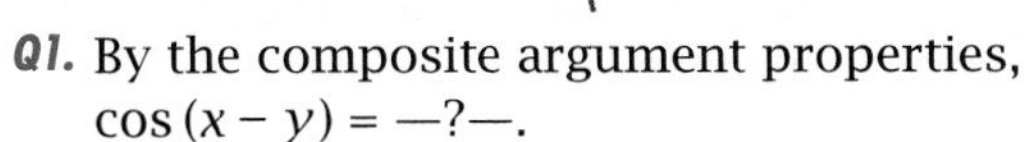

Q1. By the composite argument properties, $\cos(x - y) = $ —?—.

Q2. By the composite argument properties, $\sin x \cos y - \cos x \sin y = $ —?—.

Q3. True or false: $\tan(x + y) = \tan x + \tan y$

Q4. $\log x + \log y = \log$ (—?—)

Q5. The equation $3(x + y) = 3x + 3y$ is an example of the —?— property for multiplication over addition.

Q6. Find the amplitude of the sinusoid $y = 2\cos\theta + 5\sin\theta$.

Q7. Find the phase displacement for $y = \cos x$ of $y = 2\cos\theta + 5\sin\theta$.

Q8. Find the measure of the smaller acute angle of a right triangle with legs 13 and 28 cm.

Q9. What is the period of the circular function $y = \sin 5x$?

Q10. The graph of the parametric function $x = 5\cos t$ and $y = 4\sin t$ is a(n) —?—.

---

1. Explain why $\cos 2x$ does *not* equal $2\cos x$ by considering the differences in the graphs of $y_1 = \cos 2x$ and $y_2 = 2\cos x$.
2. Prove by a numerical counterexample that $\tan \frac{1}{2}x \neq \frac{1}{2}\tan x$.

For Problems 3–6, illustrate by numerical example that the double argument property is true by making a table of values.

3. $\sin 2x = 2\sin x \cos x$
4. $\cos 2x = \cos^2 x - \sin^2 x$
5. $\cos 2x = 1 + 2\cos^2 x$
6. $\tan 2x = \frac{2\tan x}{1 - \tan^2 x}$

For Problems 7–10, illustrate by numerical example that the half argument property is true by making a table of values.

7. $\sin \frac{1}{2}A$ for $A \in [0, 180°]$

8. $\cos \frac{1}{2}A$ for $A \in [0°, 180°]$
9. $\cos \frac{1}{2}A$ for $A \in [360°, 540°]$
10. $\sin \frac{1}{2}A$ for $A \in [180°, 360°]$

Sinusoid Problems: For Problems 11–16, the graph of each function is a sinusoid.

a. Plot the graph of the given function.
b. From the graph, find the equation of the sinusoid.
c. Verify algebraically that the equation is sinusoidal.

11. $y = 6 \sin x \cos x$
12. $y = 8 \cos^2 x$
13. $y = 10 \sin^2 x$
14. $y = \cos x + \sin x$
15. $y = \cos^2 3x$
16. $y = 12 \cos 5x \sin 5x$

17. *Sinusoid Conjecture Problem 1:* In this section you have proved that certain graphs that look like sinusoids really *are* sinusoids. Figure 5-6e shows the graphs of

$y_1 = 2 + \cos x$

$y_2 = 4 + \sin x$

$y_3 = (2 + \cos x)(4 + \sin x)$

The graphs of y_1 and y_2 are sinusoids with vertical displacements. Is y_3 a sinusoid? If so, find a particular equation for the graph. If not, explain why not.

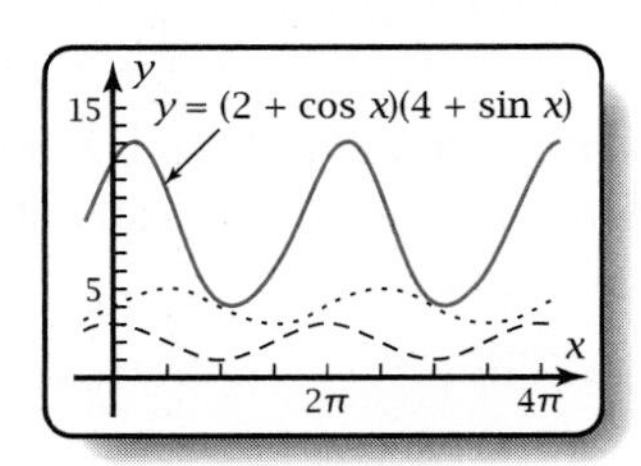

Figure 5-6e

18. *Sinusoid Conjecture Problem 2:* Figure 5-6f shows the graphs of

$y_1 = \cos \theta$

$y_2 = \sin (\theta - 30°)$

$y_3 = (\cos \theta)(\sin (\theta - 30°))$

The graph y_2 is a sinusoid with a horizontal displacement. Is y_3 a sinusoid? If so, find a particular equation for the graph. If not, explain why not.

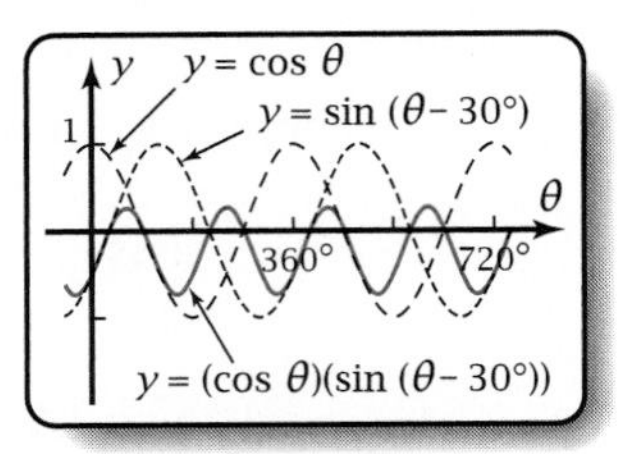

Figure 5-6f

19. *Half Argument Interpretation Problem:* Figure 5-6g shows the two functions

$y_1 = \cos \frac{1}{2}\theta$ and

$y_2 = \sqrt{\frac{1}{2}(1 + \cos \theta)}$

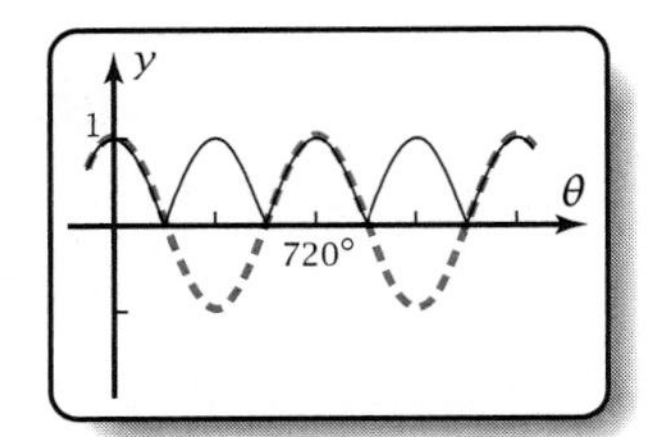

Figure 5-6g

a. Which graph is y_1 and which is y_2?
b. The half argument property for cosine contains the ambiguous ± sign. For what intervals of θ should you use the positive sign? The negative sign?
c. What one transformation can you do to the y_1 equation to make its graph identical to the graph of y_2?
d. Explain why $\sqrt{n^2} = |n|$, not just n. Use the result, and the way you derived the half argument property for cosine, to explain the origin of the ± sign in the half argument property for cosine.

20. *Terminal Position of $\frac{1}{2}A$ Problem:* You have seen that the half argument properties can involve a different sign for different coterminal angles. Is the same thing true for the double argument properties? For instance, 30° and 390° are coterminal, as in Figure 5-6h, but $\frac{1}{2}(30°)$ and $\frac{1}{2}(390°)$ are not coterminal. Show that the same thing does *not* happen for 2(30°) and 2(390°).

Show that if any two degree measures A and B are coterminal, then $2A$ and $2B$ are also coterminal.

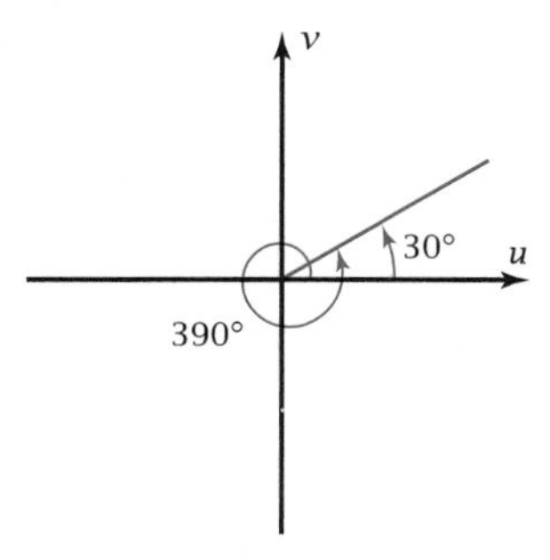

Figure 5-6h

Exact Values Problems: For Problems 21–26:

a. Use the double and half argument properties to find the *exact* values of the functions, using radicals and fractions if necessary.

b. Show that your answers are correct by finding the measure of A and then evaluating the functions directly.

21. If $\cos A = \frac{3}{5}$ and $A \in (0°, 90°)$, find $\sin 2A$ and $\cos \frac{1}{2}A$.

22. If $\cos A = \frac{3}{5}$ and $A \in (270°, 360°)$, find $\cos 2A$ and $\sin \frac{1}{2}A$.

23. If $\cos A = -\frac{3}{5}$ and $A \in (180°, 270°)$, find $\sin 2A$ and $\cos \frac{1}{2}A$.

24. If $\cos A = -\frac{3}{5}$ and $A \in (90°, 180°)$, find $\cos 2A$ and $\sin \frac{1}{2}A$.

25. If $\cos A = \frac{3}{5}$ and $A \in (630°, 720°)$, find $\sin 2A$ and $\cos \frac{1}{2}A$.

26. If $\cos A = -\frac{3}{5}$ and $A \in (450°, 540°)$, find $\cos 2A$ and $\sin \frac{1}{2}A$.

27. *Sine Double Argument Property Derivation Problem:* Starting with $\sin 2x = \sin (x + x)$, derive the property $\sin 2x = 2 \sin x \cos x$.

28. *Cosine Double Argument Properties Derivation Problem:*

    a. Starting with $\cos 2x = \cos (x + x)$, derive the property $\cos 2x = \cos^2 x - \sin^2 x$.

    b. Using the Pythagorean properties, prove that $\cos 2x = 2 \cos^2 x - 1$.

    c. Using the Pythagorean properties, prove that $\cos 2x = 1 - 2 \sin^2 x$.

29. *Sine Times Cosine Is a Sinusoid Problem:* Using the double argument properties, prove algebraically that $y = \sin x \cos x$ is a sinusoid.

30. *Squares of Cosine and Sine Are Sinusoids Problem:* Using the double argument properties, prove algebraically that $y = \cos^2 x$ and $y = \sin^2 x$ are sinusoids.

31. *Double Argument Property for Tangent:* The table of properties in this section lists this property expressing $\tan 2A$ in terms of $\tan A$.

$$\tan 2A = \frac{2 \tan A}{1 - \tan^2 A}$$

    a. Show graphically that the property is true on an interval of your choice. Write a few sentences explaining what you did and your results. List any domain restrictions on the argument A.

    b. Derive this double argument property algebraically by starting with the appropriate composite argument property for tangent.

    c. Derive this property again, this time by starting with the quotient property for $\tan 2A$ and substituting the double argument properties for sine and cosine. You will need to use some insightful algebraic operations to transform the resulting sines and cosines back into tangents.

32. *Half Argument Property for Tangent:*

    a. Based on the quotient properties, tell why

$$\tan \tfrac{1}{2}A = \pm\sqrt{\frac{1 - \cos A}{1 + \cos A}}$$

    b. Starting with the property in part a, derive another form of the half argument property for tangents given in the table on page 210,

$$\tan \tfrac{1}{2}A = \pm\frac{\sin A}{1 + \cos A}$$

    First multiply under the radical sign by 1 in the form

$$\frac{1 + \cos A}{1 + \cos A}$$

    c. Confirm graphically that the answer to part b is true. Write a few sentences explaining what you did and your results. Based on the graphs, explain why only the positive sign applies and never the negative sign.

Algebraic Solution of Equations Problems 2: For Problems 33–38, solve the equation algebraically, using the double argument or half argument properties appropriately to transform the equation to a suitable form.

33. $4 \sin x \cos x = \sqrt{3},\ x \in [0, 2\pi]$

34. $\cos^2 \theta - \sin^2 \theta = -1,\ \theta \in [0°, 360°]$

35. $\cos^2 \theta = 0.5,\ \theta \in [0°, 360°]$

36. $\dfrac{2 \tan x}{1 - \tan^2 x} = \sqrt{3},\ x \in [0, 2\pi]$

37. $\sqrt{\tfrac{1}{2}(1 + \cos x)} = \tfrac{1}{2}\sqrt{3},\ x \in [0, 4\pi]$

38. $\sqrt{\tfrac{1}{2}(1 - \cos \theta)} = 1,\ \theta \in [0°, 720°]$

Identity Problems: For Problems 39–44, prove that the given equation is an identity.

39. $\sin 2x = \dfrac{2 \tan x}{1 + \tan^2 x}$

40. $\cos 2y = \dfrac{1 - \tan^2 y}{1 + \tan^2 y}$

41. $\sin 2\phi = 2 \cot \phi \sin^2 \phi$

42. $\tan \beta = \dfrac{1 - \cos 2\beta}{\sin 2\beta}$

43. $\sin^2 5\theta = \tfrac{1}{2}(1 - \cos 10\theta)$

44. $\cos^2 3x = \tfrac{1}{2}(1 + \cos 6x)$

5-7 Chapter Review and Test

In this chapter you have extended the study of trigonometric function properties that you started in Chapter 4. Specifically, you learned properties such as $\cos(x - y) = \cos x \cos y - \sin x \sin y$ that apply to functions of more than one argument. You saw that these properties allow you to analyze graphs that are composed of sums and products of sinusoids. Sometimes the composed graph was another sinusoid, and sometimes it was a periodic function with a varying amplitude or a varying sinusoidal axis. Finally, you applied these properties to derive double and half argument properties that express, for instance, $\sin 2A$ and $\sin \tfrac{1}{2}A$ in terms of functions of A.

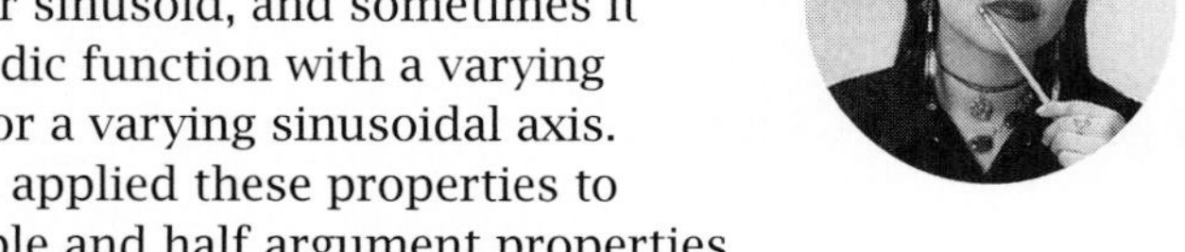

For your future reference, the box on page 216 lists the ten kinds of properties you have learned in Chapters 4 and 5.

Review Problems

R0. Update your journal with what you have learned in this chapter. For example, include

- Products and sums of sinusoids with equal periods
- Products and sums of sinusoids with much different periods
- Harmonic analysis of graphs composed of two sinusoids

- Transformations between sums and products of sinusoids
- Double argument and half argument properties to prove certain products are sinusoids

R1. Figure 5-7a shows the graph of $y = 5\cos\theta + 12\sin\theta$.

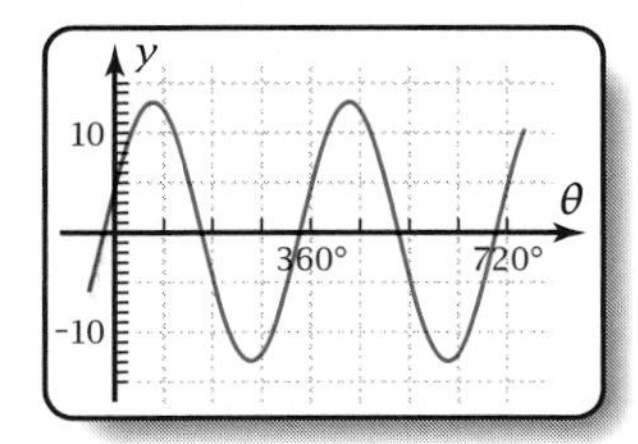

Figure 5-7a

a. The graph in Figure 5-7a is a sinusoid. Estimate its amplitude and its phase displacement with respect to the parent cosine curve, and then write an equation for the displaced sinusoid.

b. Plot your equation in part a and the equation $y = 5\cos\theta + 12\sin\theta$ on the same screen. How well do the two graphs agree with each other and with Figure 5-7a?

R2. Figure 5-7b shows the graphs of $y_1 = \cos(\theta - 60°)$ and $y_2 = \cos\theta - \cos 60°$ as they might appear on your grapher.

a. Without actually plotting the graphs, identify which is y_1 and which is y_2. Explain how you chose them. From the graphs, how can you conclude that the cosine function does *not* distribute over subtraction?

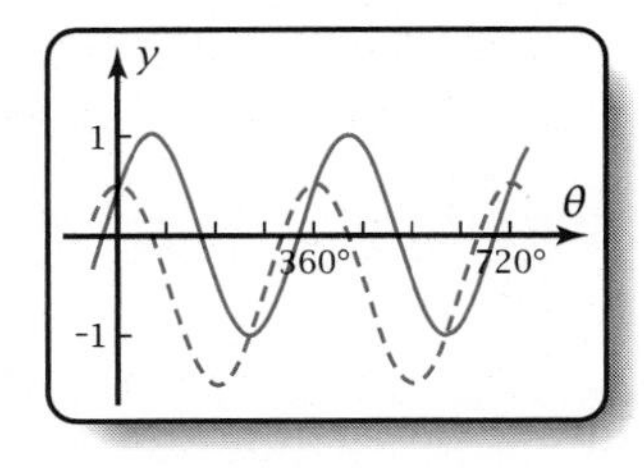

Figure 5-7b

b. Write $\cos(\theta - 60°)$ as a linear combination of sine and cosine. Verify graphically that your answer is correct.

c. Write the trigonometric expression $8\cos\theta + 15\sin\theta$ as a single sinusoid with a phase displacement for $y = \cos x$. Plot the graphs of the given expression and your answer. Explain how the two graphs confirm that your answer is correct.

d. Write the circular function expression $-9\cos x + 7\sin x$ as a single sinusoid with a phase displacement for $y = \cos x$. Make a table of values of the given expression and your answer. Explain how the numbers in the table indicate that your answer is correct.

e. Figure 5-7c shows the graphs of $y_1 = 4\cos x + 3\sin x$ and $y_2 = 2$ as they might appear on your grapher. Solve the equation $4\cos x + 3\sin x = 2$ algebraically for $x \in [0, 2\pi]$. Show that the solutions agree with the graph.

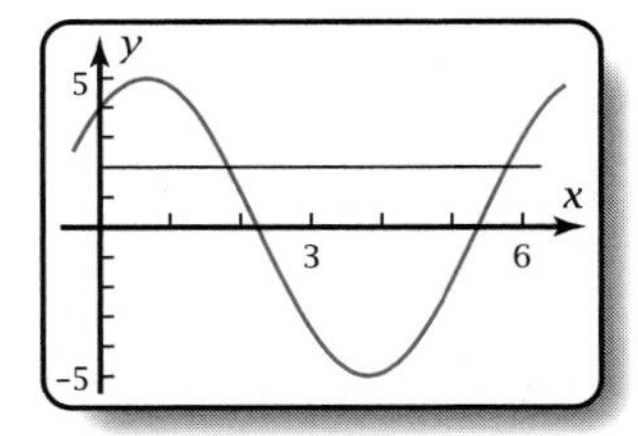

Figure 5-7c

R3. a. Express $\sin(-x)$, $\cos(-x)$, and $\tan(-x)$ in terms of the same function of (positive) x.

b. Express $\sin(x + y)$ in terms of sines and cosines of x and y.

c. Express $\cos(x + y)$ in terms of sines and cosines of x and y.

d. Express $\tan(x - y)$ in terms of $\tan x$ and $\tan y$.

e. Express $\cos(90° - \theta)$ in terms of $\sin\theta$.

f. Express $\cot(\frac{\pi}{2} - x)$ in terms of $\tan x$. What restrictions are there on the domain?

g. Express $\csc(\frac{\pi}{2} - x)$ in terms of $\sec x$. What restrictions are there on the domain?

h. Sketch $y = \tan x$, and thus show graphically that $\tan(-x) = -\tan x$. What restrictions are there on the domain?

PROPERTIES: Summary of Trigonometric Function Properties

Reciprocal Properties

$\cot x = \dfrac{1}{\tan x}$ or $\tan x \cot x = 1$

$\sec x = \dfrac{1}{\cos x}$ or $\cos x \sec x = 1$

$\csc x = \dfrac{1}{\sin x}$ or $\sin x \csc x = 1$

Quotient Properties

$\tan x = \dfrac{\sin x}{\cos x} = \dfrac{\sec x}{\csc x}$

$\cot x = \dfrac{\cos x}{\sin x} = \dfrac{\csc x}{\sec x}$

Pythagorean Properties

$\cos^2 x + \sin^2 x = 1$

$1 + \tan^2 x = \sec^2 x$

$\cot^2 x + 1 = \csc^2 x$

Odd–Even Function Properties

$\sin(-x) = -\sin x$ (odd function)

$\cos(-x) = \cos x$ (even function)

$\tan(-x) = -\tan x$ (odd function)

$\cot(-x) = -\cot x$ (odd function)

$\sec(-x) = \sec x$ (even function)

$\csc(-x) = -\csc x$ (odd function)

Cofunction Properties

$\cos(90° - \theta) = \sin\theta$, $\cos(\frac{\pi}{2} - x) = \sin x$

$\cot(90° - \theta) = \tan\theta$, $\cot(\frac{\pi}{2} - x) = \tan x$

$\csc(90° - \theta) = \sec\theta$, $\csc(\frac{\pi}{2} - x) = \sec x$

Linear Combination of Cosine and Sine

$b \cos x + c \sin x = A \cos(x - D)$, where

$A = \sqrt{b^2 + c^2}$ and $D = \arctan\frac{c}{b}$

Composite Argument Properties

$\cos(A - B) = \cos A \cos B + \sin A \sin B$

$\cos(A + B) = \cos A \cos B - \sin A \sin B$

$\sin(A - B) = \sin A \cos B - \cos A \sin B$

$\sin(A + B) = \sin A \cos B + \cos A \sin B$

$\tan(A - B) = \dfrac{\tan A - \tan B}{1 + \tan A \tan B}$

$\tan(A + B) = \dfrac{\tan A + \tan B}{1 - \tan A \tan B}$

Sum and Product Properties

$2 \cos A \cos B = \cos(A + B) + \cos(A - B)$

$2 \sin A \sin B = -\cos(A + B) + \cos(A - B)$

$2 \sin A \cos B = \sin(A + B) + \sin(A - B)$

$2 \cos A \sin B = \sin(A + B) - \sin(A - B)$

$\cos x + \cos y = 2 \cos \frac{1}{2}(x + y) \cos \frac{1}{2}(x - y)$

$\cos x - \cos y = -2 \sin \frac{1}{2}(x + y) \sin \frac{1}{2}(x - y)$

$\sin x + \sin y = 2 \sin \frac{1}{2}(x + y) \cos \frac{1}{2}(x - y)$

$\sin x - \sin y = 2 \cos \frac{1}{2}(x + y) \sin \frac{1}{2}(x - y)$

Double Argument Properties

$\sin 2x = 2 \sin x \cos x$

$\cos 2x = \cos^2 x - \sin^2 x = 1 - 2\sin^2 x$

$= 2\cos^2 - 1$

$\tan 2x = \dfrac{2 \tan x}{1 - \tan^2 x}$

$\sin^2 x = \frac{1}{2}(1 - \cos 2x)$

$\cos^2 x = \frac{1}{2}(1 + \cos 2x)$

Half Argument Properties

$\sin \frac{1}{2}x = \pm\sqrt{\frac{1}{2}(1 - \cos x)}$

$\cos \frac{1}{2}x = \pm\sqrt{\frac{1}{2}(1 + \cos x)}$

$\tan \frac{1}{2}x = \pm\sqrt{\dfrac{1 - \cos x}{1 + \cos x}}$

$= \dfrac{\sin x}{1 + \cos x} = \dfrac{1 - \cos x}{\sin x}$

i. Figure 5-7d shows the graph of $y = \cos 3x \cos x + \sin 3x \sin x$ and the line $y = 0.4$. Solve the equation $\cos 3x \cos x + \sin 3x \sin x = 0.4$ algebraically for $x \in [0, 2\pi]$. Explain how the graph confirms your solutions.

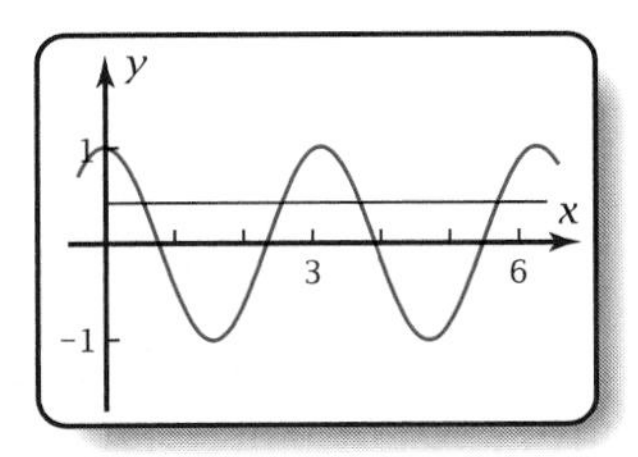

Figure 5-7d

R4. Figure 5-7e shows a sinusoid and a linear function.

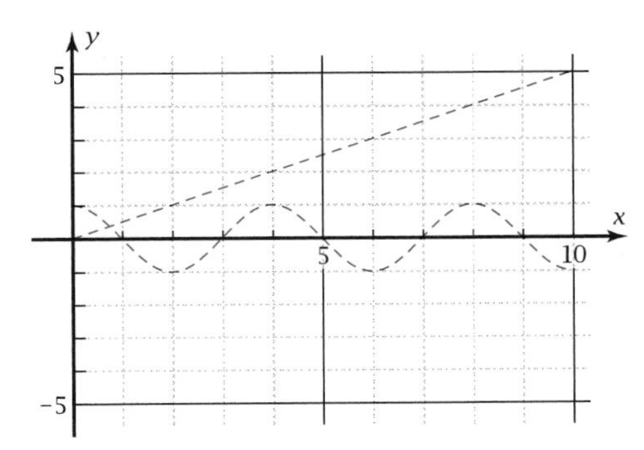

Figure 5-7e

a. On a copy of the figure, sketch the graph of the function composed by *adding* ordinates.

b. On another photocopy of Figure 5-7e, sketch the graph composed by *multiplying* the ordinates.

c. Confirm your answers to parts a and b by finding equations of the two parent graphs, plotting them and the composed graphs on your grapher and comparing them with your sketches.

d. Describe verbally the two composed graphs in part c.

e. By harmonic analysis, find a particular equation for the composed function in Figure 5-7f. Check your answer by plotting on your grapher.

f. By harmonic analysis, find a particular equation for the composed function in Figure 5-7g. Check your answer by plotting on your grapher.

R5. a. Transform $\cos 13° \cos 28°$ to a sum (or difference) of sines or cosines with positive arguments.

b. Transform $\sin 5 - \sin 8$ to a product of sines and cosines with positive arguments.

c. Figure 5-7h shows the graph of $y = 4 \sin x \sin 11x$. Transform the expression on the right-hand side of this equation to a

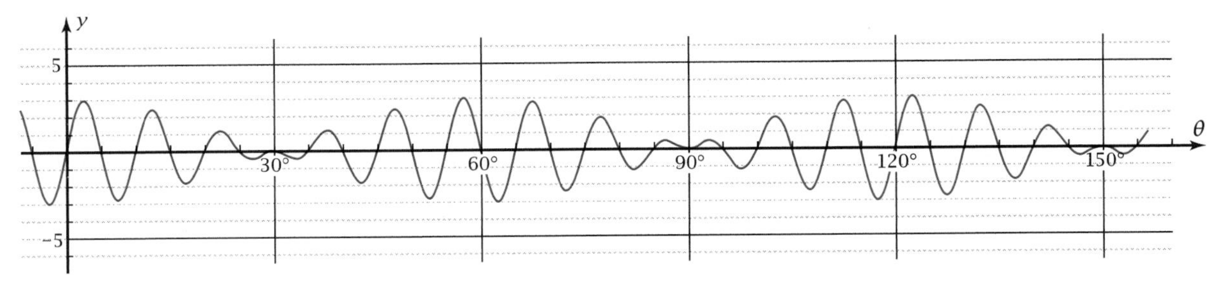

Figure 5-7f

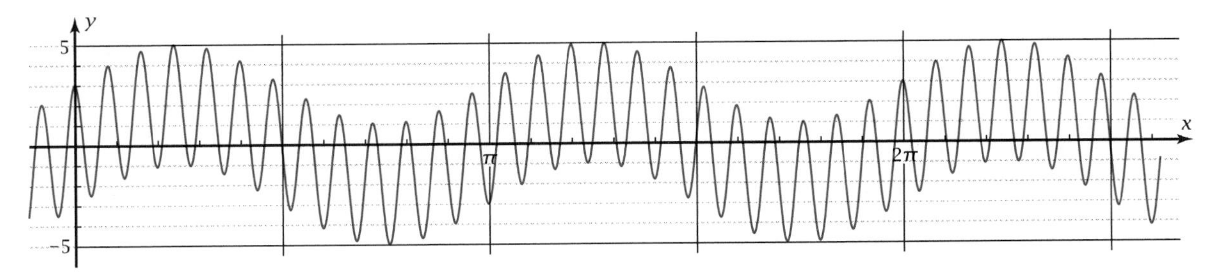

Figure 5-7g

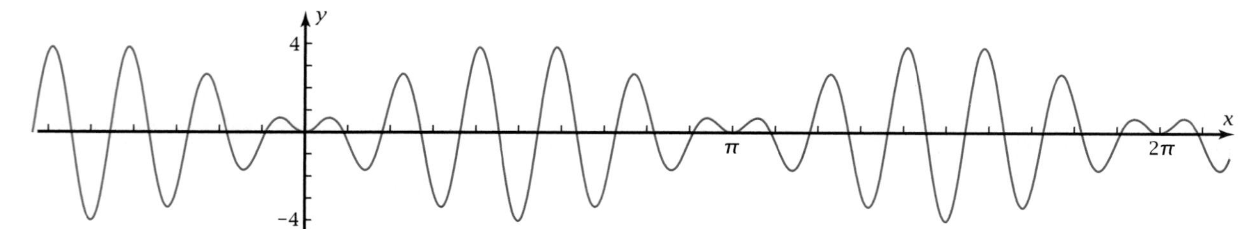

Figure 5-7h

sum (or difference) of sinusoids whose arguments have positive coefficients. Check graphically that your answer and the given equation both agree with Figure 5-7h.

d. Solve $2 \sin 3\theta + 2 \sin \theta = 1$ algebraically for $\theta \in [0°, 360°]$.

e. Use the sum and product properties to prove that $\cos(x + \frac{\pi}{3}) \cos(x - \frac{\pi}{3}) = \cos^2 x - \frac{3}{4}$ is an identity.

R6. Figure 5-7i shows the graph of $y = \sin^2 3x$.

a. The function is a sinusoid. Find its equation from the graph. Confirm algebraically that your equation is correct by using the double argument properties.

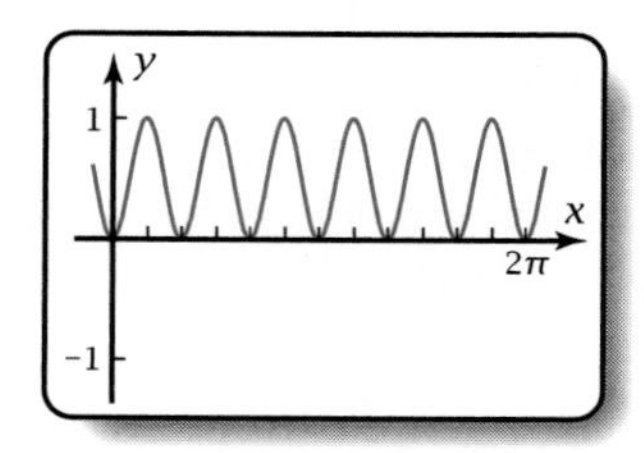

Figure 5-7i

b. Give a numerical counterexample to prove that $\cos 2x \neq 2 \cos x$.

c. Write an equation expressing $\cos 2x$ in terms of $\cos x$ alone. Make a table of values to demonstrate that the equation is an identity.

d. Write an equation expressing $\tan 2x$ in terms of $\tan x$ alone. What restrictions are there on the domain of x?

e. Suppose that $A \in (360°, 450°)$ and $\sin A = \frac{24}{25}$. Find the exact values (no decimals) of $\cos 2A$ and $\cos \frac{1}{2}A$. Then calculate the measure of A (no round-off), and find decimal values of $\cos 2A$ and $\cos \frac{1}{2}A$ directly. How do these answers compare with the exact values?

f. Prove that the equation $\sin 2A = 2 \tan A \cos^2 A$ is an identity for all the values where both sides are defined. What is the definition of this identity?

g. Solve the equation

$$\sqrt{\tfrac{1}{2}(1 - \cos \theta)} = 0.5 \qquad \theta \in [0°, 720°]$$

Apply the half argument property to the left side of the equation. What kind of solution did you just give? Plot the graph of the expression on the left side and the graph of $y = 0.5$. How many solutions are there? Have you found all of them using both methods?

Concept Problems

C1. *Exact Value of sin 18° Project:* You have learned how to find exact values of functions of multiples of 30° and 45°. By the composite argument property, you found an exact value for $\sin 15°$ by writing it as $\sin(45° - 30°)$. In this problem you will combine trigonometric properties with algebraic techniques and some ingenuity to find an exact value of $\sin 18°$.

a. Use the double argument property for sine to write an equation expressing $\sin 72°$ in terms of $\sin 36°$ and $\cos 36°$.

b. Transform the equation in part a so that $\sin 72°$ is expressed in terms of $\sin 18°$ and $\cos 18°$. You should find that the *sine* form of the double argument property for $\cos 36°$ works best.

c. Recall by the cofunction property that $\sin 72° = \cos 18°$. Replace $\sin 72°$ in your equation from part b with $\cos 18°$. If you have done everything correctly, the $\cos 18°$ should disappear from the equation, leaving a *cubic* (third-degree) equation in $\sin 18°$.

d. Solve the equation in part c for $\sin 18°$. It may help to let $x = \sin 18°$ and solve for x. If you rearrange the equation so that the right side is 0, you should find that $(2x - 1)$ is a factor of the left side. You can find the other factor by long division or synthetic substitution. To find the exact solutions, recall the multiplication property of zero and the quadratic formula.

e. You should have *three* solutions for the equation in part d. Only one of these solutions is possible. *Which* solution?

f. A pattern shows up for some exact values of $\sin \theta$:

$$\sin 15° = \frac{\sqrt{6} - \sqrt{2}}{4}$$

$$\sin 18° = \frac{\sqrt{5} - 1}{4} = \frac{\sqrt{5} - \sqrt{1}}{4}$$

$$\sin 30° = \frac{1}{2} = \frac{2}{4} = \frac{\sqrt{4} - \sqrt{0}}{4}$$

See if you can extend this pattern to sines of other angles!

C2. *A Square Wave Function and Fourier Series Project:* Figure 5-7j shows the graph of

$$y = \cos x - \tfrac{1}{3}\cos 3x$$

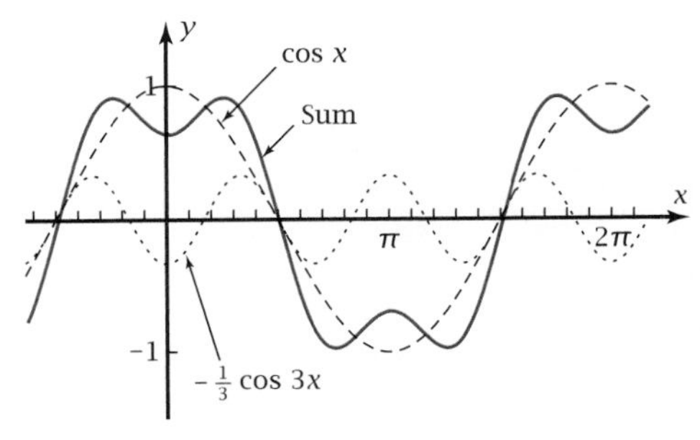

Figure 5-7j

The smaller sinusoid pulls the larger one closer to the x-axis at some places and pushes it farther away at others. The terms in the given equation form this **partial sum** of a **Fourier series.**

$$y = \cos x - \tfrac{1}{3}\cos 3x + \tfrac{1}{5}\cos 5x - \tfrac{1}{7}\cos 7x + \tfrac{1}{9}\cos 9x - \tfrac{1}{11}\cos 11x$$

The more terms that are added to the partial sum, the more the result looks like a **square wave.** For instance, the graph of the partial sum with 11 terms is shown in Figure 5-7k.

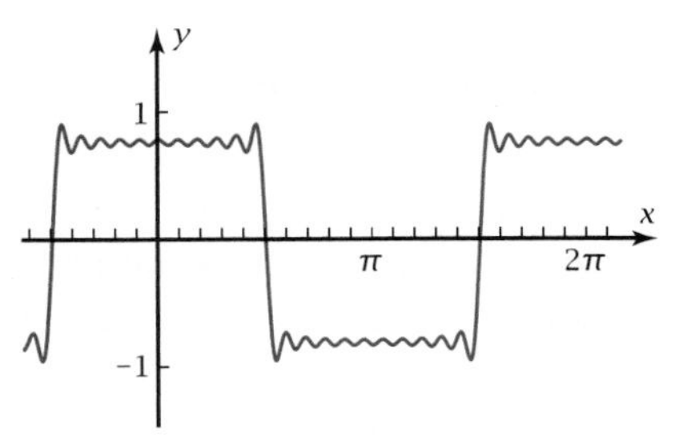

Figure 5-7k

Jean-Baptiste-Joseph Fourier, French mathematician (1768–1830)

a. Plot the graph of the partial sum with 11 terms and show that it is similar to the graph in Figure 5-7k. Find the sum and sequence commands that will allow you to write the equation without typing in all of the terms.

b. If you plot the partial sum with 12 terms, some of the high points and low points in Figure 5-7k are reversed. So if you average the 11-term and the 12-term sums, you get a better square-wave pattern. Do this on your grapher. (If you perform some clever calculations first, you can find a relatively easy way to add one term to the 11-term sum that gives the average without having to compute the 12-term sum.)

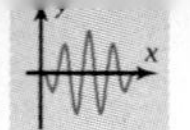

Electrical generators can produce alternating currents with the sawtooth wave pattern shown in Figure 5-7k.

c. Plot the average of the 50-term sum and the 51-term sum. How close does the graph come to a "square wave"?

d. When the square wave is on the high portion of its graph, there is an "axis" about which it oscillates. Find the y-value of this axis. Explain how you found your answer. How is the answer related to π?

e. Plot the 10th partial sum of the Fourier sine series

$$y = \sin x + \tfrac{1}{2}\sin 2x + \tfrac{1}{3}\sin 3x + \cdots + \tfrac{1}{10}\sin 10x$$

From your graph, figure out why the result is called a **sawtooth wave pattern.**

Chapter Test

PART 1: No calculators (T1–T9)

T1. What are the amplitude and period of the sinusoid $y = 6\cos x + 7\sin x$?

T2. The graph of $y = \cos(x - 2)$ is a sinusoid with a phase displacement of 2. Given that $\cos 2 \approx -0.42$ and $\sin 2 \approx 0.91$, write y as a linear combination of $\cos x$ and $\sin x$.

T3. What are the amplitude and period of the sinusoid $y = 2\sin\theta\cos\theta$?

T4. The graph of $y = 5\cos x + \sin 8x$ is periodic but not a sinusoid. Based on the form of the equation, describe in words what the graph would look like.

T5. The process of finding the two sinusoids that have been added or multiplied to form a more complicated graph is called —?—.

T6. Is tangent an odd or even function? Write the property in algebraic form.

T7. According to the cofunction property, $\cos 13° =$ —?—.

T8. The expression $\cos 9 + \cos 5$ can be transformed to $2\cos x\cos y$. What do x and y equal?

T9. You can write the double argument property for cosine in the form $\cos x = 1 - 2\sin^2\frac{1}{2}x$. Show algebraically how you can transform this equation into the half argument property for sine.

PART 2: Graphing calculators are allowed (T10–T19)

T10. Figure 5-7m shows the graph of the linear combination of two sinusoids with equal periods,

$$y = -4\cos\theta + 3\sin\theta$$

Write the equation as a single cosine with a phase displacement. Show that the calculated phase displacement agrees with the graph in the figure.

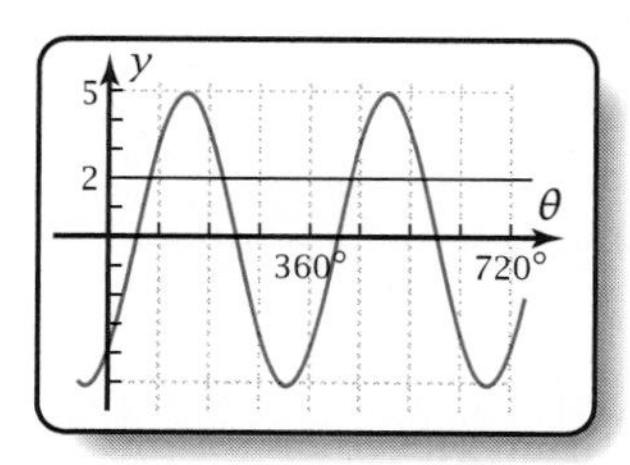

Figure 5-7m

T11. Solve the equation $-4\cos\theta + 3\sin\theta = 2$ algebraically for $\theta \in [0°, 720°]$. Show that the solutions agree with the four shown in Figure 5-7m.

T12. In Problem T2 you transformed

$$y = \cos(x - 2) \quad \text{to}$$

$$y = \cos x \cos 2 + \sin x \sin 2$$

Plot both graphs on the same screen. Explain in writing how your graphs confirm that the two equations are equivalent.

T13. Figure 5-7n shows the graph of $y = \cos^2 \theta$ as it might appear on your grapher. From the graph, find the equation of this sinusoid. Then use the double argument properties to prove algebraically that your answer is correct.

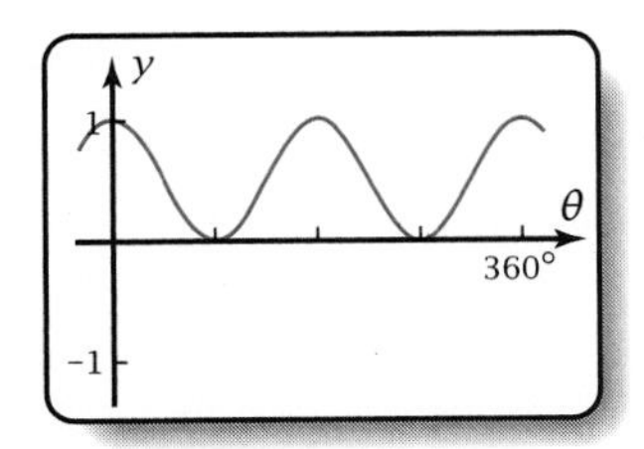

Figure 5-7n

T14. Figure 5-7o shows graphs of two sinusoids with different periods. On a copy of the figure, sketch the graph of the *sum* of these two sinusoids. Then figure out equations for each sinusoid and plot the result on your grapher. How well does the sketch agree with the plot?

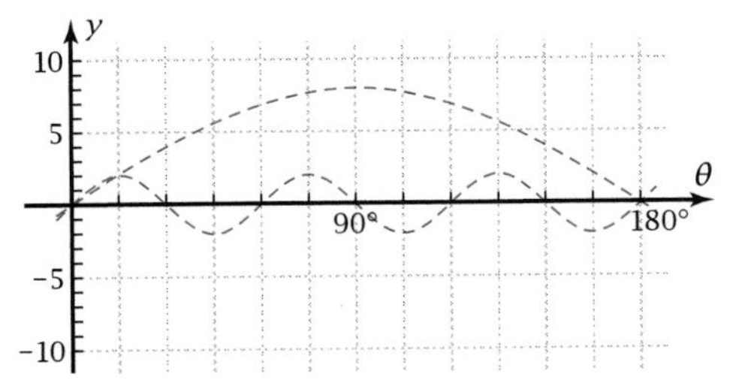

Figure 5-7o

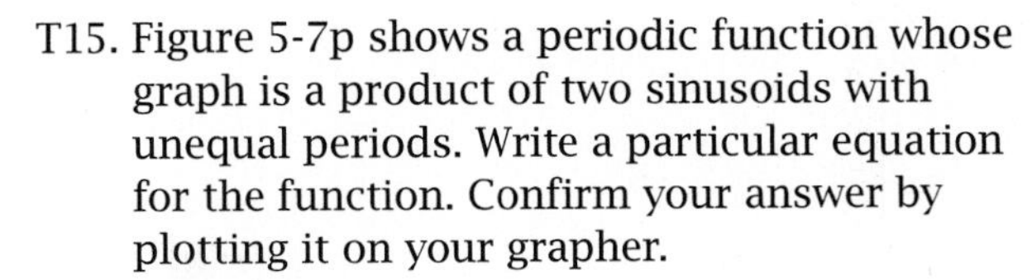

T15. Figure 5-7p shows a periodic function whose graph is a product of two sinusoids with unequal periods. Write a particular equation for the function. Confirm your answer by plotting it on your grapher.

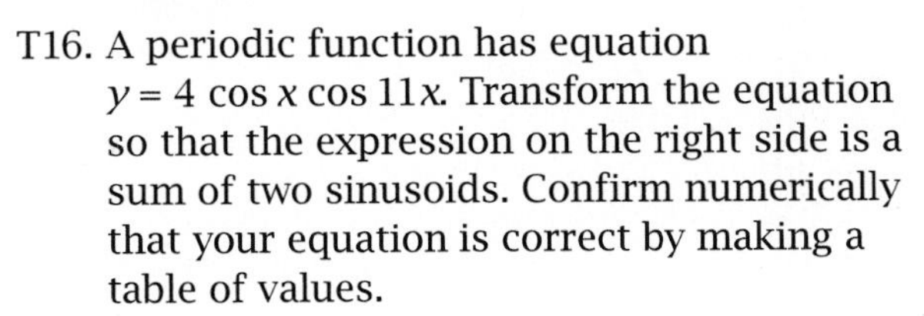

T16. A periodic function has equation $y = 4 \cos x \cos 11x$. Transform the equation so that the expression on the right side is a sum of two sinusoids. Confirm numerically that your equation is correct by making a table of values.

T17. Suppose that A is an angle between 0° and 90° and that $\cos A = \frac{15}{17}$. What does $\sin(90° - A)$ equal? What property can you use to find this answer quickly?

T18. For angle A in Problem T17, find the exact value (no decimals) of $\sin 2A$ and $\cos \frac{1}{2}A$. Then find the measure of A, and calculate $\sin 2A$ and $\cos \frac{1}{2}A$ directly. How do the answers compare?

T19. What did you learn from this test that you did not know before?

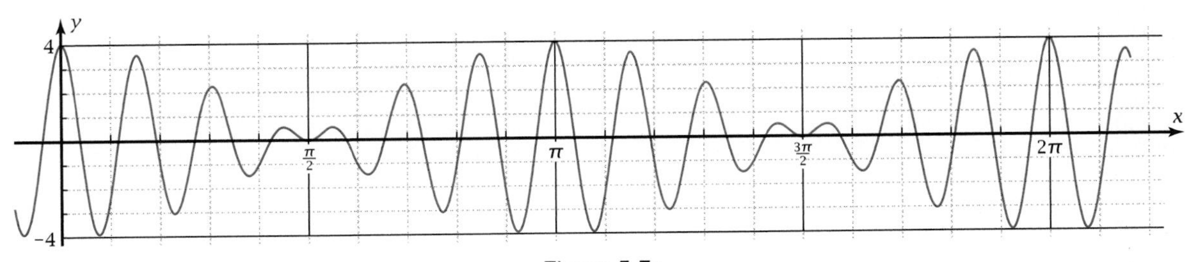

Figure 5-7p

CHAPTER

6

Triangle Trigonometry

Surveyors measure irregularly shaped tracts of land by dividing them into triangles. To calculate area, side lengths, and angles, they must extend their knowledge of right triangle trigonometry to include triangles that have no right angle. In this chapter you'll learn how to do these calculations.

Mathematical Overview

The Pythagorean theorem describes how to find the hypotenuse of a right triangle if you know the lengths of the two legs. If the angle formed by the two given sides is not a right angle, you can use the law of cosines (an extension of the Pythagorean theorem) to find side lengths or angle measures. If two angles and a side opposite one of the angles or two sides and an angle opposite one of the sides are given, you can use the law of sines. The area can also be calculated from side and angle measures. These techniques give you a way to analyze vectors, which are quantities, such as velocity, that have both direction and magnitude. You will learn about these techniques in four ways.

Graphically Make a scale drawing of the triangle using the given information, and measure the length of the third side.

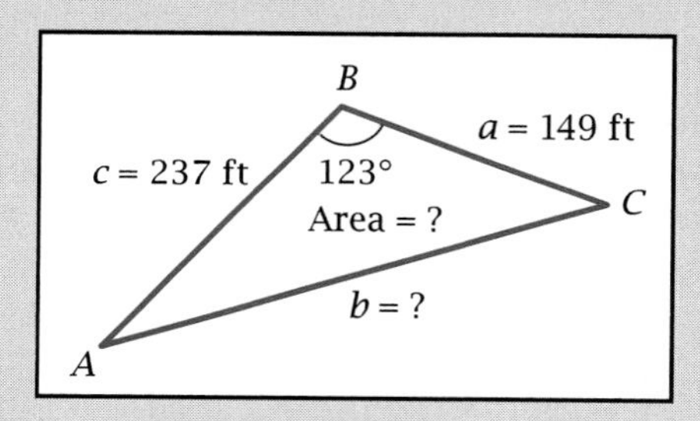

Algebraically Law of cosines: $b^2 = a^2 + c^2 - 2ac \cos B$

Area of a triangle: $A = \frac{1}{2}ac \sin B$

Numerically The length of side b: $b = \sqrt{149^2 + 237^2 - 2(149)(237) \cos 123°}$

$= 341.81\ldots$

The area of the triangle: $A = \frac{1}{2}(237)(149) \sin 123° = 14807.98\ldots$

$\approx 14{,}808 \text{ ft}^2$

Verbally *If I know two sides and the included angle, I can use the law of cosines to find the third side. I can also find the area from two sides and the included angle. If I know three sides, I can use the law of cosines "backward" to find the angle between any two of the sides.*

6-1 Introduction to Oblique Triangles

You already know how to find unknown sides and angles in right triangles using trigonometric functions. In this section you'll be introduced to a way of calculating the same kind of information if none of the angles of the triangle is a right angle. Such triangles are called **oblique triangles.**

OBJECTIVE Given two sides and the included angle of a triangle, find by direct measurement the third side of the triangle.

Exploratory Problem Set 6-1

1. Figure 6-1a shows five triangles. Each one has sides of 3 cm and 4 cm. They differ in the size of the angle included between the two sides. Measure the sides and angles. Do you agree with the given measurements in each case?
2. Measure a, the third side of each triangle. Find the third side if A were 180° and if A were 0°. Record your results in table form.
3. Store the data from Problem 2 in lists on your grapher. Make a connected plot of the data on your grapher.
4. The plot looks like a half-cycle of a sinusoid. Find the equation of the sinusoid that has the same low and high points and plot it on the same screen. Does the data really seem to follow a sinusoidal pattern?
5. By the Pythagorean theorem, $a^2 = 3^2 + 4^2$ if A is 90°. If A is *less* than 90°, side a is less than 5, so it seems you must *subtract* something from $3^2 + 4^2$ to get the value of a^2. See if you can find *what* is subtracted!
6. What did you learn from this problem set that you did not know before?

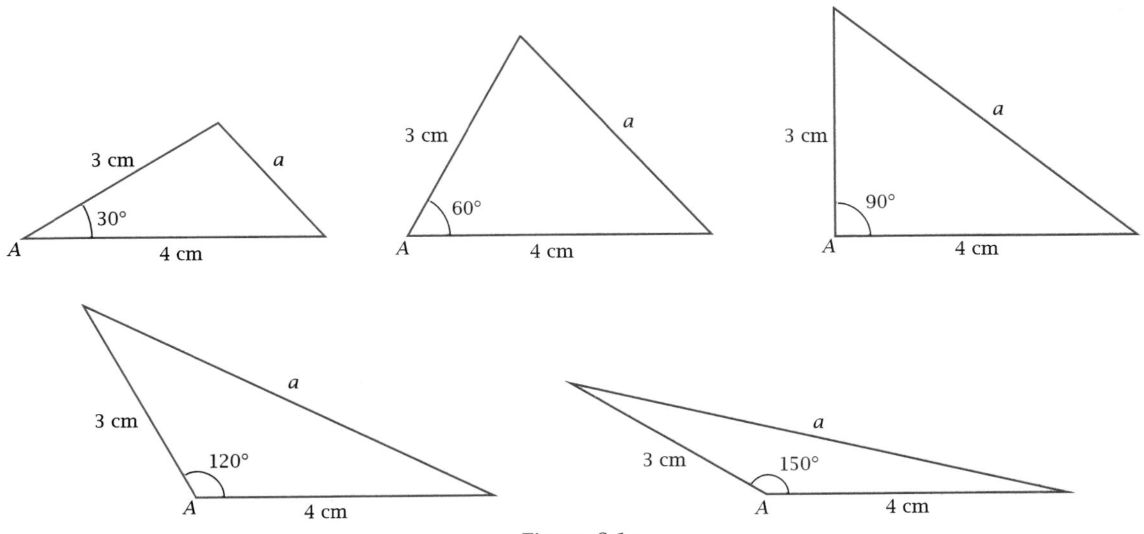

Figure 6-1a

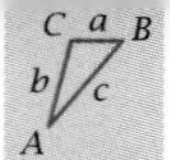

6-2 Oblique Triangles: Law of Cosines

In Problem Set 6-1 you measured the third sides of triangles for which two sides and the included angle were known. Think of three of the triangles with included angles 60°, 90°, and 120°.

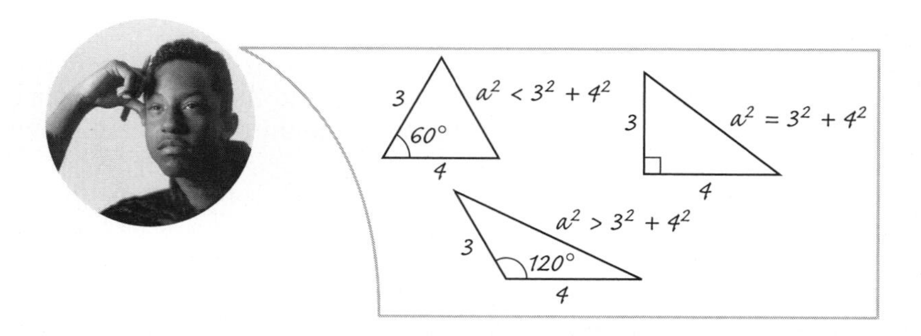

For the right triangle in the middle, you can get the third side, a, using the Pythagorean theorem.

$$a^2 = 3^2 + 4^2$$

For the 60° triangle on the left, the value of a^2 is less than $b^2 + c^2$. For the 120° triangle on the right, a^2 is greater than $b^2 + c^2$.

The equation you'll use to find the exact length of the third side from the measures of two sides and the included angle is called the **law of cosines** (since it involves the cosine of the angle). In this section you'll see why the law of cosines is true and how to use it.

OBJECTIVES

- Given two sides and the included angle of a triangle, derive and use the law of cosines for finding the third side.
- Given three sides of a triangle, find an angle.

Derivation of the Law of Cosines

Suppose that the lengths of two sides, b and c, of $\triangle ABC$ are known, as is the measure of the included angle, A (Figure 6-2a, left side). You must find the third side, a.

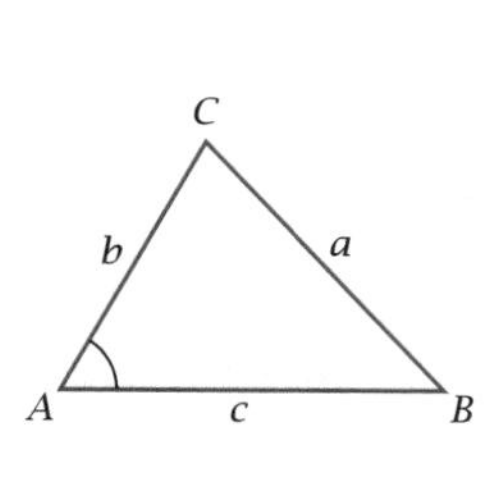

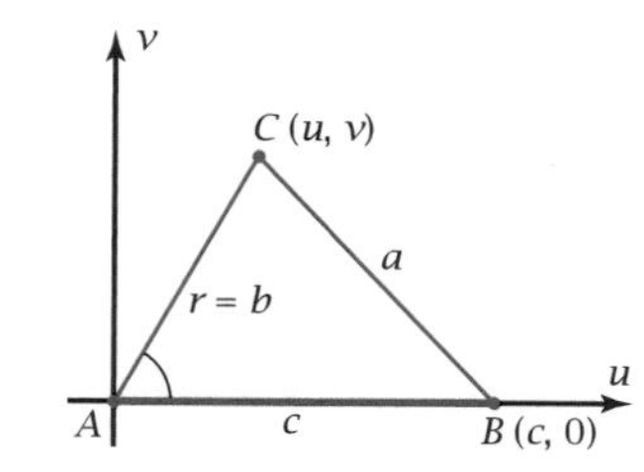

Figure 6-2a

If you construct a uv-coordinate system with angle A in standard position, as on the right in Figure 6-2a, then vertices B and C have coordinates $B(c, 0)$ and $C(u, v)$. By the distance formula,

$$a^2 = (u - c)^2 + (v - 0)^2$$

By the definitions of cosine and sine,

$$\frac{u}{b} = \cos A \Rightarrow u = b \cos A$$

$$\frac{v}{b} = \sin A \Rightarrow v = b \sin A$$

Substituting these values for u and v and completing the appropriate algebraic operations gives

$$a^2 = (u - c)^2 + (v - 0)^2$$

$$a^2 = (b \cos A - c)^2 + (b \sin A - 0)^2$$

$$a^2 = b^2 \cos^2 A - 2bc \cos A + c^2 + b^2 \sin^2 A$$ Calculate the squares.

$$a^2 = b^2(\cos^2 A + \sin^2 A) - 2bc \cos A + c^2$$ Factor b^2 from the first and last terms.

$$a^2 = b^2 + c^2 - 2bc \cos A$$ Use the Pythagorean property.

PROPERTY: The Law of Cosines

In triangle ABC with sides a, b, and c,

$$a^2 = b^2 + c^2 - 2bc \cos A$$

side² + side² − 2 (side) (side) cosine of included angle = (third side)²

Notes:

- If the angle is 90°, the law of cosines reduces to the Pythagorean theorem, because cos 90° is zero.
- If A is obtuse, $\cos A$ is negative. So you are subtracting a negative number from $b^2 + c^2$, giving the larger value for a^2, as you found out through Problem Set 6-1.
- You should not jump to the conclusion that the law of cosines gives an easy way to *prove* the Pythagorean theorem. Doing so would involve circular reasoning, because the Pythagorean theorem (in the form of the distance formula) was used to *derive* the law of cosines.

- A capital letter is used for the vertex, the angle at that vertex, or the measure of that angle, whichever is appropriate. If confusion results, you can use the symbols from geometry, such as m∠A for the measure of angle A.

Applications of the Law of Cosines

You can use the law of cosines to calculate either a side or an angle. In each case, there are different parts of a triangle given. Watch for what these "givens" are.

▶ EXAMPLE 1 In △PMF, angle $M = 127°$, side $p = 15.78$ ft, and side $f = 8.54$ ft. Find the third side, m.

Solution First, sketch the triangle and label the sides and angles, as shown in Figure 6-2b. (It does not need to be accurate, but it must have the right relationship among sides and angles.)

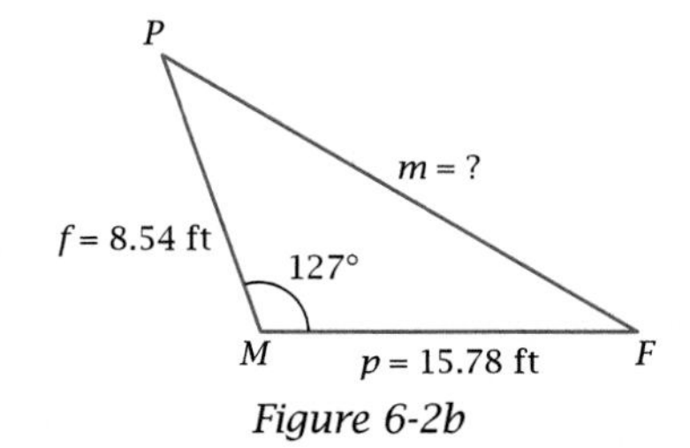

Figure 6-2b

$m^2 = 8.54^2 + 15.78^2 - 2(8.54)(15.78) \cos 127°$ Use the law of cosines for side m.

$m^2 = 484.1426...$

$m = 22.0032... \approx 22.0$ ft ◀

▶ EXAMPLE 2 In △XYZ, side $x = 3$ m, side $y = 7$ m, and side $z = 9$ m. Find the measure of the largest angle.

Solution Sketch the triangle.

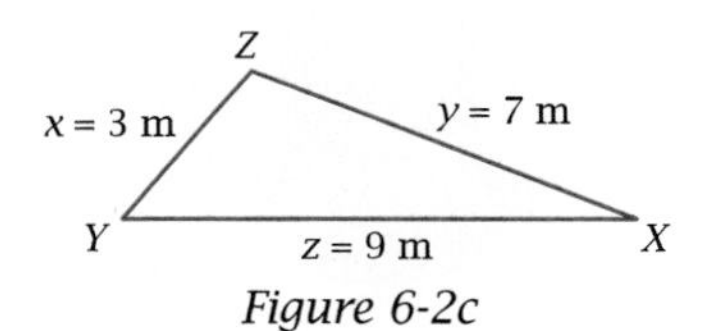

Figure 6-2c

Recall from geometry that the largest side is opposite the largest angle, in this case, Z. Use the law of cosines with this angle and the two sides that include it.

$9^2 = 7^2 + 3^2 - 2 \cdot 7 \cdot 3 \cos Z$

$81 = 49 + 9 - 42 \cos Z$

$\dfrac{81 - 49 - 9}{-42} = \cos Z$

$-0.5476... = \cos Z$

$Z = \arccos(-0.5476...) = \cos^{-1}(-0.5476...) = 123.2038...$
$\approx 123.2°$ ◀

Note that arccos (−0.5476...) = $\cos^{-1}$ (−0.5476...), because there is only one value of an arccosine between 0° and 180°, the range of angles possible in a triangle.

▶ EXAMPLE 3 Suppose that the measures of the sides in Example 2 had been $x = 3$ m, $y = 7$ m, and $z = 11$ m. What is the measure of angle Z in this case?

Solution Write the law of cosines for side z, the side that is across from angle Z.

$$11^2 = 7^2 + 3^2 - 2 \cdot 7 \cdot 3 \cos Z$$

$$121 = 49 + 9 - 42 \cos Z$$

$$\frac{121 - 49 - 9}{-42} = \cos Z$$

$$-1.5 = \cos Z$$

$$z^2 = x^2 + y^2 - 2xy \cos Z$$

No such triangle. cos Z must be in the range [−1, 1]. ◀

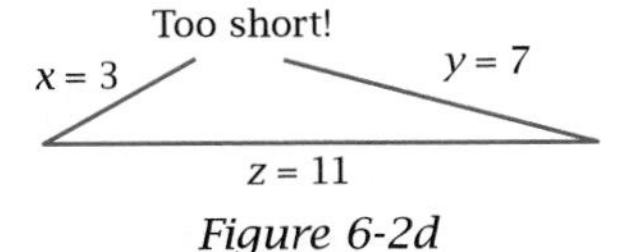

Figure 6-2d

The geometrical reason why there is no answer in Example 3 is that no two sides of a triangle can add up to less than the third side. Figure 6-2d illustrates this fact. The law of cosines signals this inconsistency algebraically by giving you a cosine outside the interval [−1, 1].

Problem Set 6-2

Do These Quickly (5 min)

Problems Q1–Q6 refer to right triangle QUI (Figure 6-2e).

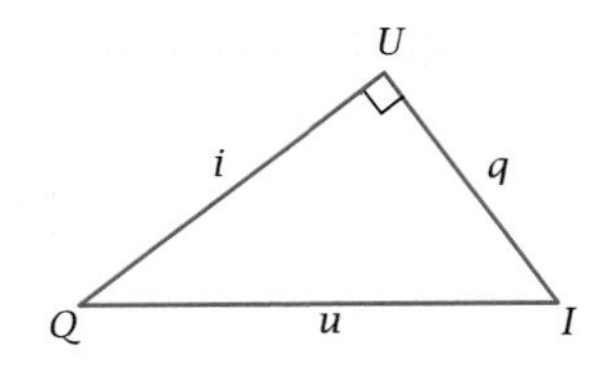

Figure 6-2e

Q1. $\cos Q =$ —?—

Q2. $\tan I =$ —?—

Q3. $\sin U =$ —?—

Q4. In terms of side u and angle I, what does i equal?

Q5. In terms of sides u and q, what does i equal?

Q6. In terms of the inverse tangent function, angle $Q =$ —?—.

Q7. The graph of $y = 5 \cos \theta + \sin 12\theta$ is periodic with a varying —?—.

Q8. In terms of cosines and sines of 53° and 42°, $\cos(53° - 42°) =$ —?—.

Q9. What transformation of $y = \cos x$ is expressed by $y = \cos 5x$?

Q10. Express $\sin 2x$ in terms of $\sin x$ and $\cos x$.

For Problems 1–4, find the length of the specified side.

1. Side r in $\triangle RPM$, if $p = 4$ cm, $m = 5$ cm, and $R = 51°$.
2. Side d in $\triangle CDE$, if $c = 7$ in., $e = 9$ in., and $D = 34°$.
3. Side r in $\triangle PQR$, if $p = 3$ ft, $q = 2$ ft, and $R = 138°$.
4. Side k in $\triangle HJK$, if $h = 8$ m, $j = 6$ m, and $K = 172°$.

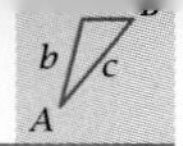

For Problems 5–12, find the measure of the specified angle.

5. Angle U in $\triangle UMP$, if $u = 2$ in., $m = 3$ in., and $p = 4$ in.
6. Angle G in $\triangle MEG$, if $m = 5$ cm, $e = 6$ cm, and $g = 8$ cm.
7. Angle T in $\triangle BAT$, if $b = 6$ km, $a = 7$ km, and $t = 12$ km.
8. Angle E in $\triangle PEG$, if $p = 12$ ft, $e = 22$ ft, and $g = 16$ ft.
9. Angle Y in $\triangle GYP$, if $g = 7$ yd, $y = 5$ yd, and $p = 13$ yd.
10. Angle N in $\triangle GON$, if $g = 6$ mm, $o = 3$ mm, and $n = 12$ mm.
11. Angle O in $\triangle NOD$, if $n = 1475$ yd, $o = 2053$ yd, and $d = 1428$ yd.
12. Angle Q in $\triangle SQR$, if $s = 1504$ cm, $q = 2465$ cm, and $r = 1953$ cm.

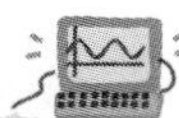

13. *Accurate Drawing Problem 1:* Using computer software such as The Geometer's Sketchpad, or using ruler and protractor, construct $\triangle RPM$ from Problem 1. Then measure side r. Does the measured value agree with the calculated value in Problem 1 within 0.1 cm, or ±0.1 cm?

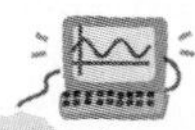

14. *Accurate Drawing Problem 2:* Using computer software such as The Geometer's Sketchpad, or using ruler, compass, and protractor, construct $\triangle MEG$ from Problem 6. Construct the longest side, 8 cm, first. Then draw an arc or circle of radius 5 cm from one endpoint and an arc of radius 6 cm from the other endpoint. The third vertex is the point where the arcs intersect. Measure angle G. Does the measured value agree with the calculated value in Problem 6 within one degree, or ±1°?

15. *Fence Problem:* Mattie works for a fence company. She has the job of pricing a fence to go across a triangular lot at the corner of Alamo and Heights streets, as shown in Figure 6-2f. The streets intersect at an angle of 65°. The lot extends 200 ft from the intersection along Alamo and 150 ft from the intersection along Heights.
    a. How long will the fence be?
    b. How much will it cost her company to build it if fencing costs \$3.75 per foot?
    c. What price should she quote to the customer if the company is to make a 35% profit?

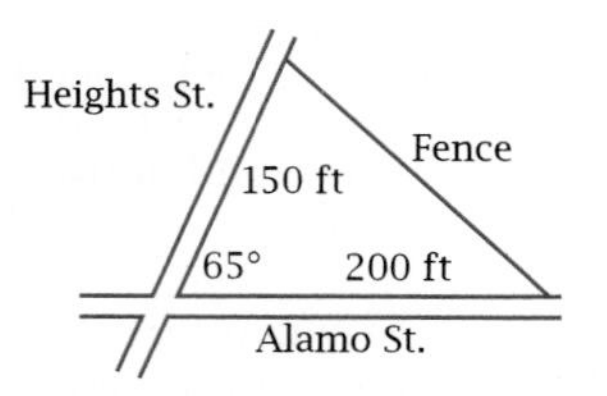

Figure 6-2f

16. *Flight Path Problem:* Sam flies a helicopter to drop supplies to stranded flood victims. He will fly from the supply depot, S, to the drop point, P. Then he will return to the helicopter's base at B, as shown in Figure 6-2g. The drop point is 15 mi from the supply depot. The base is 21 mi from the drop point. It is 33 miles between the supply depot and the base. Because the return flight to the base will be made after dark, Sam wants to know in what direction to fly. What is the angle between the two paths at the drop point?

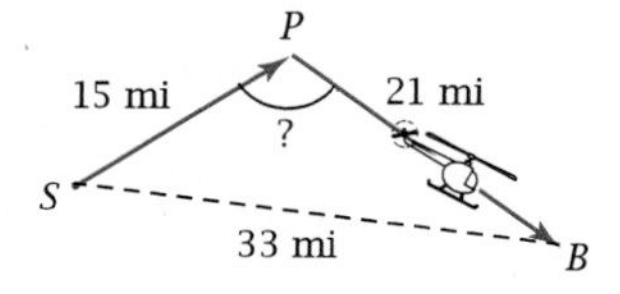

Figure 6-2g

17. *Derivation of the Law of Cosines Problem:* Figure 6-2h shows $\triangle XYZ$ with angle Z in standard position. The sides that include angle Z are 4 units and 5 units long. Find the coordinates of points X and Y in terms of the 4, the 5, and angle Z. Then use the distance formula and appropriate algebra and trigonometry to show:

$$z^2 = 5^2 + 4^2 - 2 \cdot 5 \cdot 4 \cdot \cos Z$$

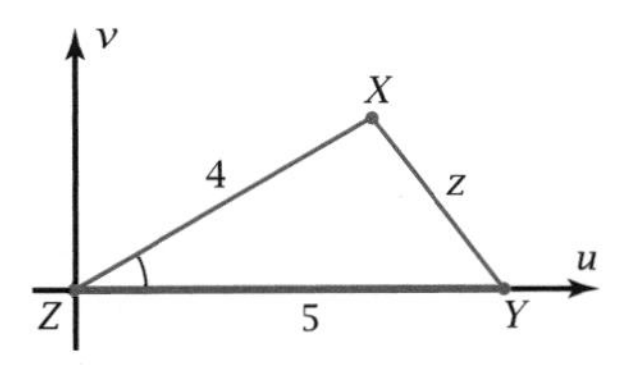

Figure 6-2h

18. *Acute, Right, or Obtuse Problem:* The law of cosines states that in $\triangle XYZ$,

$$x^2 = y^2 + z^2 - 2yz \cos X$$

a. Explain how the law of cosines allows you to make a quick test to see if angle X is acute, right, or obtuse, as shown:

> **Property: Test for the Size of an Angle in a Triangle**
>
> In $\triangle XYZ$:
>
> If $x^2 < y^2 + z^2$, then X is an acute angle.
>
> If $x^2 = y^2 + z^2$, then X is a right angle.
>
> If $x^2 > y^2 + z^2$, then X is an obtuse angle.

b. Without using your calculator, find whether angle X is acute, right, or obtuse if $x = 7$ cm, $y = 5$ cm, and $z = 4$ cm.

6-3 Area of a Triangle

Recall from geometry that the area of a triangle equals half the product of the base and the altitude. In this section you'll learn how to find this area from two side lengths and the included angle measure. This is the same information you use in the law of cosines to calculate the length of the third side.

OBJECTIVE Given the measures of two sides and the included angle, find the area of the triangle.

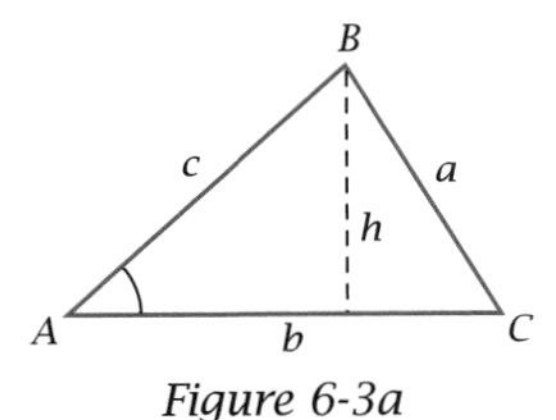

Figure 6-3a

Figure 6-3a shows $\triangle ABC$ with base b and altitude h.

$\text{Area} = \frac{1}{2}bh$ From geometry, area equals half of base times height.

$\text{Area} = \frac{1}{2}b(c \sin A)$ Because $\sin A = h/c$.

$\text{Area} = \frac{1}{2}bc \sin A$

PROPERTY: Area of a Triangle

In $\triangle ABC$,

$$\text{Area} = \frac{1}{2}bc \sin A$$

Verbally: The area of a triangle equals half the product of two of its sides and the sine of the included angle.

▶ **EXAMPLE 1** In $\triangle ABC$, side $a = 13$ in., side $b = 15$ in., and angle $C = 71°$. Find the area of the triangle.

Solution Sketch the triangle to be sure you're given two sides and the *included* angle (Figure 6-3b).

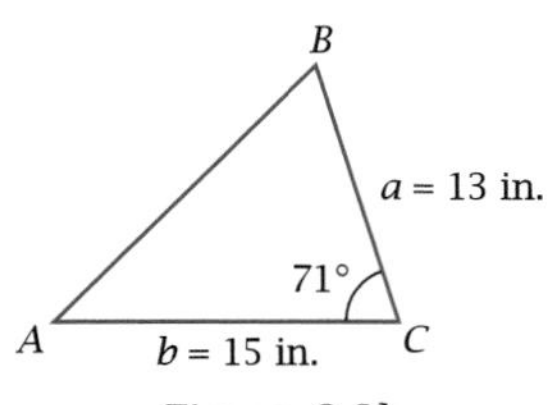

Figure 6-3b

$$\text{Area} = \frac{1}{2}(13)(15)\sin 71°$$
$$= 92.1880\ldots \approx 92.19 \text{ in.}^2$$

◀

▶ **EXAMPLE 2** Find the area of $\triangle JDH$ if $j = 5$ cm, $d = 7$ cm, and $h = 11$ cm.

Solution

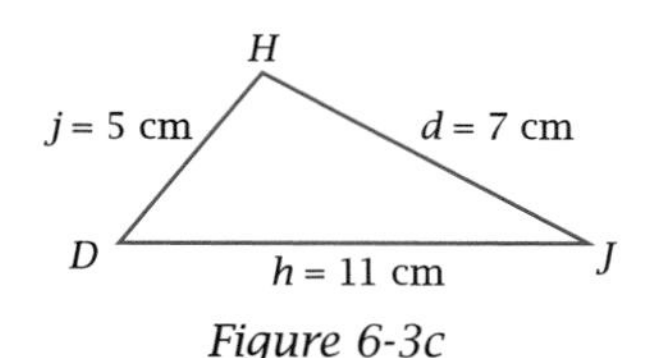

Figure 6-3c

Sketch the triangle to give yourself a picture of what has to be done.

$$h^2 = j^2 + d^2 - 2jd\cos H$$ Use the law of cosines to calculate an angle.

$$\cos H = \frac{j^2 + d^2 - h^2}{2jd}$$ Solve for $\cos H$.

$$\cos H = \frac{5^2 + 7^2 - 11^2}{2(5)(7)} = -0.6714\ldots$$

$$H = \arccos(-0.6714\ldots) = \cos^{-1}(-0.6714\ldots) = 132.1774\ldots°$$

Store the answer, without rounding, for use in the next part.

$$\text{Area} = \frac{1}{2}(5)(7)\sin 132.1774\ldots° = 12.9687\ldots \approx 12.97 \text{ cm}^2$$

◀

Hero's Formula

It is possible to find the area of a triangle directly from the lengths of three sides, as given in Example 2. The method uses Hero's formula, after Hero of Alexandria, who lived about 100 B.C.

Hero of Alexandria

PROPERTY: Hero's Formula

In $\triangle ABC$, the area is given by

$$\text{Area} = \sqrt{s(s-a)(s-b)(s-c)}$$

where s is the semiperimeter (half the perimeter), $\frac{1}{2}(a + b + c)$.

EXAMPLE 3 Find the area of $\triangle JDH$ in Example 2 using Hero's formula. Confirm that you get the same answer.

Solution

$$s = \tfrac{1}{2}(5 + 7 + 11) = 11.5$$

$$\text{Area} = \sqrt{11.5(11.5 - 5)(11.5 - 7)(11.5 - 11)} = \sqrt{168.1875}$$

$$= 12.9687\ldots \approx 12.97 \text{ cm}^2$$

Which agrees with Example 2. ◀

Problem Set 6-3

Do These Quickly

Problems Q1–Q5 refer to Figure 6-3d.

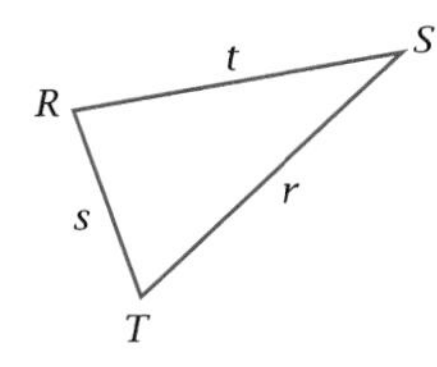

Figure 6-3d

Q1. State the law of cosines using angle R.

Q2. State the law of cosines using angle S.

Q3. State the law of cosines using angle T.

Q4. Express $\cos T$ in terms of sides r, s, and t.

Q5. Why do you need only the function, $\cos^{-1}$, not the relation, arccos, when using the law of cosines?

Q6. When you multiply two sinusoids with much different periods, you get a function with a varying —?—.

Q7. What is the first step in proving that a trigonometric equation is an identity?

Q8. Which trigonometric functions are *even* functions?

Q9. If θ is in standard position, then $\dfrac{\text{horizontal coordinate}}{\text{radius}}$ is the definition of —?—.

Q10. In the composite argument properties, $\cos(x + y) =$ —?—.

For Problems 1–4, find the area of the indicated triangle.

1. $\triangle ABC$, if side $a = 5$ ft, side $b = 9$ ft, and angle $C = 14°$.

2. $\triangle ABC$, if side $b = 8$ m, side $c = 4$ m, and angle $A = 67°$.

3. $\triangle RST$, if side $r = 4.8$ cm, side $t = 3.7$ cm, and angle $S = 43°$.

4. $\triangle XYZ$, if side $x = 34.19$ yd, side $z = 28.65$ yd, and angle $Y = 138°$.

For Problems 5–7, use Hero's formula to calculate the area of the triangle.

5. $\triangle ABC$, if side $a = 6$ cm, side $b = 9$ cm, and side $c = 11$ cm.

6. $\triangle XYZ$, if side $x = 50$ yd, side $y = 90$ yd, and side $z = 100$ yd.

7. $\triangle DEF$, if side $d = 3.7$ in., side $e = 2.4$ in., and side $f = 4.1$ in.

8. *Comparison of Methods Problem:* Reconsider Problems 1 and 7.
   a. For $\triangle ABC$ in Problem 1, calculate the third side using the law of cosines. Store the answer without round-off. Then find the area using Hero's formula. Do you get the same answer as in Problem 1?
   b. For $\triangle DEF$ in Problem 7, calculate the measure of angle D using the law of cosines. Store the answer without round-off. Then find the area using the area formula as in Example 2. Do you get the same answer as in Problem 7?

9. *Hero's Formula and Impossible Triangles Problem:* Suppose someone tells you that $\triangle ABC$ has sides $a = 5$ cm, $b = 6$ cm, and $c = 13$ cm.
   a. Explain why there is no such triangle.
   b. Apply Hero's formula to the given information. How does Hero's formula allow you to detect that there is no such triangle?

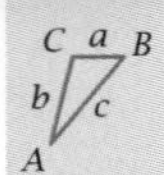

10. *Lot Area Problem:* Sean works for a real estate company. The company has a contract to sell the triangular lot at the corner of Alamo and Heights streets (Figure 6-3e). The streets intersect at an angle of 65°. The lot extends 200 ft from the intersection along Alamo and 150 ft from the intersection along Heights.

    a. Find the area of the lot.

    b. Land in this area is valued at $35,000 per acre. An acre is 43,560 square feet. How much is the lot worth?

    c. The real estate company will earn a commission of 6% of the sales price. If the lot sells for what it is worth, how much will the commission be?

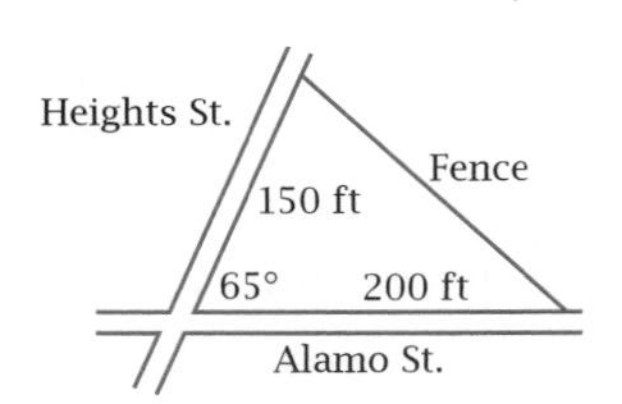

Figure 6-3e

11. *Variable Triangle Problem:* Figure 6-3f shows angle θ in standard position in a uv-coordinate system. The fixed side of the angle is 3 units long and the rotating side is 4 units long. As the angle increases, the area of the triangle shown in the figure is a function of θ.

    a. Write the area as a function of θ.

    b. Make a table of values of area for each 15° from 0° through 180°.

    c. Is this statement true or false? "The area is an increasing function of θ for all angles from 0° through 180°." Give evidence to support your answer.

    d. Find the domain of θ for which this statement is true: "The area is a sinusoidal function of θ." Explain why the statement is false outside this domain.

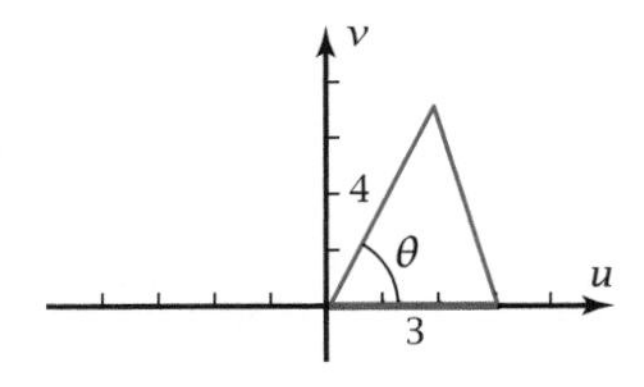

Figure 6-3f

12. *Unknown Angle Problem:* Suppose you need to construct a triangle with one side 14 cm, another side 11 cm, and a given area.

    a. What *two* possible values of the included angle will produce an area of 50 cm^2?

    b. Show that there is only *one* possible value of the included angle if the area were 77 cm^2.

    c. Show algebraically that there is *no* possible value of the angle if the area were 100 cm^2.

13. *Derivation of the Area Formula Problem:* Figure 6-3g shows $\triangle XYZ$ with angle Z in standard position. The two sides that include angle Z are 4 and 5 units long. Find the altitude h in terms of the 4 and angle Z. Then show that the area of the triangle is given by

    Area $= \frac{1}{2}(5)(4)\sin Z$

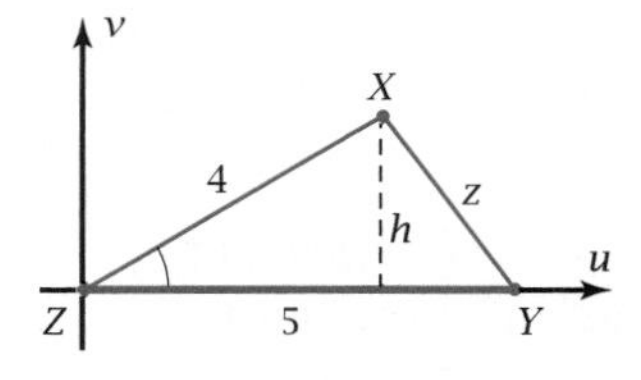

Figure 6-3g

6-4 Oblique Triangles: Law of Sines

Because the law of cosines involves all three sides of a triangle, you must know at least two of the sides to use it. In this section you'll learn the law of sines that lets you calculate a side of a triangle if only one side and two angles are given.

OBJECTIVE Given the measure of an angle, the length of the side opposite this angle, and one other piece of information about the triangle, find the other side and angle measures.

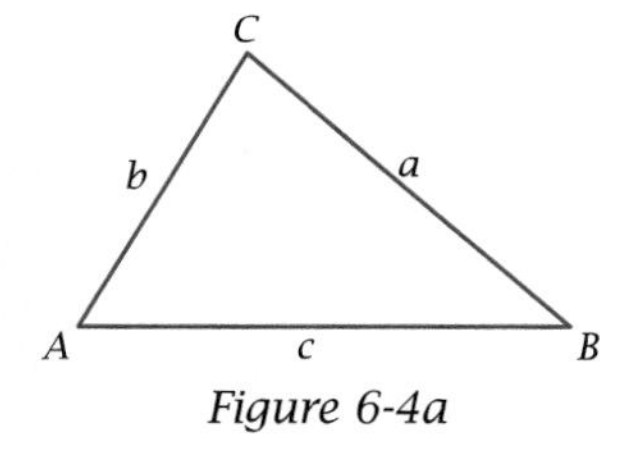

Figure 6-4a

Figure 6-4a shows $\triangle ABC$. In the previous section you found that the area is equal to $\frac{1}{2}bc \sin A$. The area is constant no matter which pair of sides and included angle you use.

$$\tfrac{1}{2}bc \sin A = \tfrac{1}{2}ac \sin B = \tfrac{1}{2}ab \sin C \qquad \text{Set the areas equal.}$$

$$bc \sin A = ac \sin B = ab \sin C \qquad \text{Multiply by 2.}$$

$$\frac{bc \sin A}{abc} = \frac{ac \sin B}{abc} = \frac{ab \sin C}{abc} \qquad \text{Divide by } abc.$$

$$\frac{\sin A}{a} = \frac{\sin B}{b} = \frac{\sin C}{c}$$

This final relationship is called the **law of sines.** If three nonzero numbers are equal, then their reciprocals are equal. So you can write the law of sines in another algebraic form.

$$\frac{a}{\sin A} = \frac{b}{\sin B} = \frac{c}{\sin C}$$

PROPERTY: The Law of Sines

In $\triangle ABC$,

$$\frac{\sin A}{a} = \frac{\sin B}{b} = \frac{\sin C}{c} \quad \text{and} \quad \frac{a}{\sin A} = \frac{b}{\sin B} = \frac{c}{\sin C}$$

Verbally: Within any given triangle, the ratio of the sine of an angle to the length of the side opposite that angle is constant.

Because of the different combinations of sides and angles for any given triangle, it is convenient to revive some terminology from geometry. The acronym SAS stands for "side, angle, side." This means that as you go around the perimeter of the triangle, you are given the length of a side, the measure of an angle, and the length of a side, in that order. So SAS is equivalent to knowing two sides and the included angle, the same information that is used in the law of cosines and in the area formula. Similar meanings are attached to ASA, AAS, SSA, and SSS.

Given AAS, Find the Other Sides

Example 1 shows you how to calculate two side lengths from the third side and two of the angles.

► **EXAMPLE 1** In $\triangle ABC$, angle $B = 64°$, angle $C = 38°$, and side $b = 9$ ft. Find the lengths of sides a and c.

Solution First, draw a picture, as in Figure 6-4b.

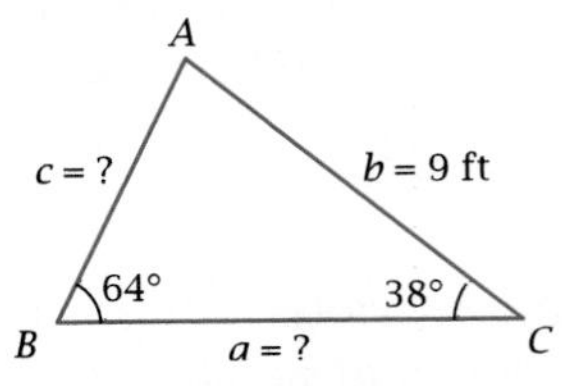

Case: AAS

Figure 6-4b

Because you know the angle opposite side c but not the angle opposite side a, it's easier to start with finding side c.

$$\frac{c}{\sin 38°} = \frac{9}{\sin 64°}$$ Use the appropriate parts of the law of sines.

$$c = \frac{9 \sin 38°}{\sin 64°}$$ Multiply both sides by $\sin 38°$ to isolate c on the left.

$$c = 6.1648\ldots$$

To find a by the law of sines, you need the measure of A, the opposite angle.

$$A = 180° - (38° + 64°) = 78°$$ The sum of the internal angles in a triangle equals 180°.

$$\frac{a}{\sin 78°} = \frac{9}{\sin 64°}$$ Use the appropriate parts of the law of sines with a in the numerator.

$$a = \frac{9 \sin 78°}{\sin 64°}$$

$$a = 9.7946\ldots$$

$$\therefore a \approx 9.79 \text{ ft} \quad \text{and} \quad c \approx 6.16 \text{ ft}$$ ◀

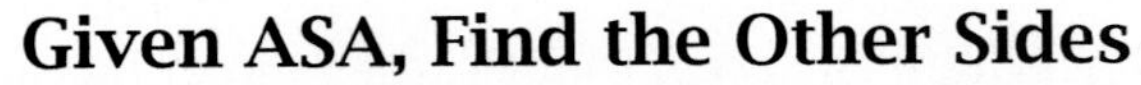

Given ASA, Find the Other Sides

Example 2 shows you how to calculate side lengths if the given side is included between the two given angles.

▶ **EXAMPLE 2** In $\triangle ABC$, side $a = 8$ m, angle $B = 64°$, and angle $C = 38°$. Find sides b and c.

Solution First, draw a picture (Figure 6-4c). The picture reveals that in this case you do not know the angle opposite the given side. So you calculate this angle first. From there on, it is a familiar problem, similar to Example 1.

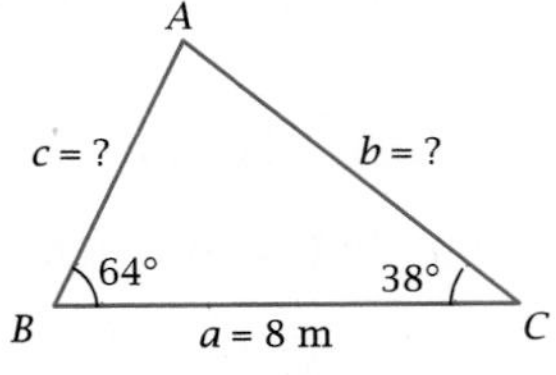

Case: ASA

Figure 6-4c

$$A = 180° - (38° + 64°) = 78°$$

$$\frac{b}{\sin 64°} = \frac{8}{\sin 78°}$$ Use the appropriate parts of the law of sines.

$$b = \frac{8 \sin 64°}{\sin 78°} = 7.3509...$$

$$\frac{c}{\sin 38°} = \frac{8}{\sin 78°}$$ Use the appropriate parts of the law of sines.

$$c = \frac{8 \sin 38°}{\sin 78°} = 5.0353...$$

$\therefore b \approx 7.35$ m and $c \approx 5.04$ m ◀

Law of Sines for Angles

You can use the law of sines to find an unknown angle of a triangle. However, you must be careful because there are *two* values of the inverse sine relation between 0° and 180°, either of which could be the answer. For instance, $\arcsin 0.8 = 53.1301...°$ and $126.9698...°$; both could be angles of a triangle. Problem 11 in Problem Set 6-4 shows you how to handle this situation.

Problem Set 6-4

Do These Quickly (5 min)

Q1. State the law of cosines for $\triangle PAF$ involving angle P.

Q2. State the formula for the area of $\triangle PAF$ involving angle P.

Q3. Write two values of $\theta = \sin^{-1} 0.5$ that lie between 0° and 180°.

Q4. If $\sin \theta = 0.372...$, then $\sin(-\theta) =$ —?—.

Q5. $\cos\left(\frac{\pi}{6}\right) =$

A. $\frac{1}{\sqrt{3}}$ B. $\frac{1}{2}$ C. $\frac{2}{\sqrt{3}}$

D. $\frac{\sqrt{3}}{2}$ E. $\sqrt{3}$

Q6. A(n) —?— triangle has no equal sides and no equal angles.

Q7. A(n) —?— triangle has no right angle.

Q8. State the Pythagorean property for cosine and sine.

Q9. $\cos 2x = \cos(x + x) =$ —?— in terms of cosines and sines of x.

Q10. The amplitude of $y = 3 + 4\cos 5(x - 6)$ is —?—.

---

1. In $\triangle ABC$, angle $A = 52°$, angle $B = 31°$, and side $a = 8$ cm. Find side b and side c.
2. In $\triangle PQR$, angle $P = 13°$, angle $Q = 133°$, and side $q = 9$ in. Find side p and side r.
3. In $\triangle AHS$, angle $A = 27°$, angle $H = 109°$, and side $a = 120$ yd. Find side h and side s.
4. In $\triangle BIG$, angle $B = 2°$, angle $I = 79°$, and side $b = 20$ km. Find side i and side g.
5. In $\triangle PAF$, angle $P = 28°$, side $f = 6$ m, and angle $A = 117°$. Find side a and side p.
6. In $\triangle JAW$, angle $J = 48°$, side $a = 5$ ft, and angle $W = 73°$. Find side j and side w.
7. In $\triangle ALP$, angle $A = 85°$, side $p = 30$ ft, and angle $L = 87°$. Find side a and side l.
8. In $\triangle LOW$, angle $L = 2°$, side $o = 500$ m, and angle $W = 3°$. Find side l and side w.

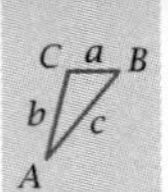

9. *Island Bridge Problem:* Suppose that you work for a construction company that is planning to build a bridge from the land to a point on an island in a lake (Figure 6-4d). The only two places on the land to start the bridge are point X and point Y, 1000 m apart. Point X has better access to the lake but is farther from the island than point Y. To help decide between x and y, you need the precise lengths of the two possible bridges. From X you measure a 42° angle to the point on the island, and from Y you measure 58°.

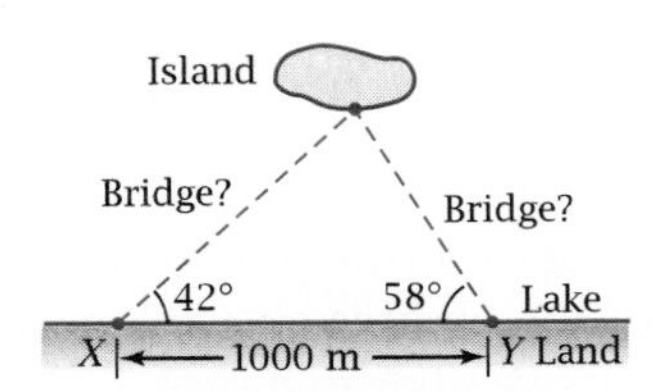

Figure 6-4d

a. How long would each bridge be?

b. If constructing the bridge costs \$370 per meter, how much could be saved by constructing the shorter bridge?

c. How much could be saved by constructing the shortest possible bridge (if that were okay)?

10. *Walking Problem 1:* Amos walks 800 ft along the sidewalk next to a field. Then he turns at an angle of 43° to the sidewalk and heads across the field (Figure 6-4e). When he stops, he looks back at the starting point, finding a 29° angle between his path across the field and the direct route back to the starting point.

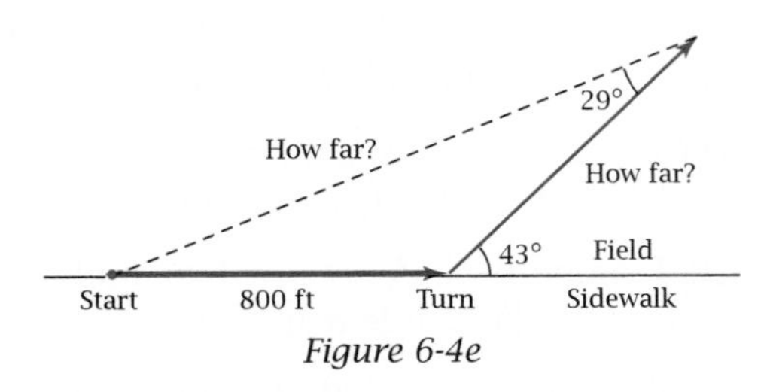

Figure 6-4e

a. How far across the field did Amos walk?

b. How far does he have to walk to go directly back to the starting point?

c. Amos walks 5 ft/sec on the sidewalk but only 3 ft/sec across the field. Which way is quicker for him to return to the starting point—by going directly across the field, or by retracing the original route?

11. *Law of Sines for Angles Problem:* You can use the law of sines to find an unknown angle measure, but the technique is risky. Suppose that $\triangle ABC$ has sides 4 cm, 7 cm, and 10 cm, as shown in Figure 6-4f.

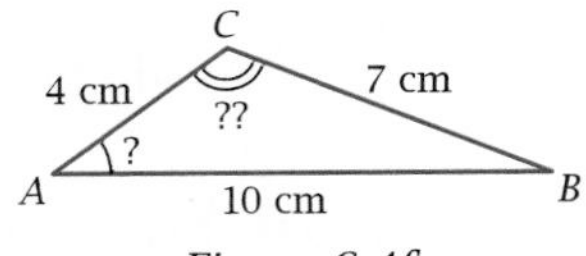

Figure 6-4f

a. Use the law of cosines to find the measure of angle A.

b. Use the answer to part a (don't round) and the law of *sines* to find the measure of angle C.

c. Find the measure of angle C again, using the law of *cosines* and the given side measures.

d. Your answers to parts b and c probably do not agree. Show that you can get the correct answer from your work with the law of sines in part b by considering the *general* solution for arcsine.

e. Why is it dangerous to use the law of sines to find an angle measure but not dangerous to use the law of cosines?

12. *Accurate Drawing Problem 3:* Using computer software such as The Geometer's Sketchpad, or using ruler and protractor with pencil and paper, construct a triangle with base 10.0 cm and base angles of 40° and 30°. Measure the length of the side opposite the 30° angle. Then calculate its length using the law of sines. Your measured value should be within 0.1 cm, or ±0.1 cm, of the calculated value.

13. *Derivation of the Law of Sines Problem:* Derive the law of sines. If you cannot do it from memory, consult the text long enough to get started. Then try doing it on your own.

6-5 The Ambiguous Case

From one end of a long segment, you draw an 80-cm segment at an angle of 26°. From the other end of the 80-cm segment, you draw a 50-cm segment, completing a triangle. Figure 6-5a shows the two possible triangles you might create.

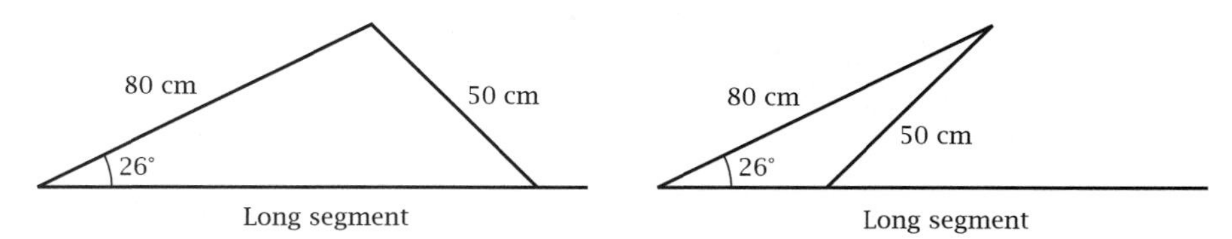

Figure 6-5a

Figure 6-5b shows why there are two possible triangles. A 50-cm arc from the upper vertex cuts the long segment in two places. Each point could be the third vertex of the triangle.

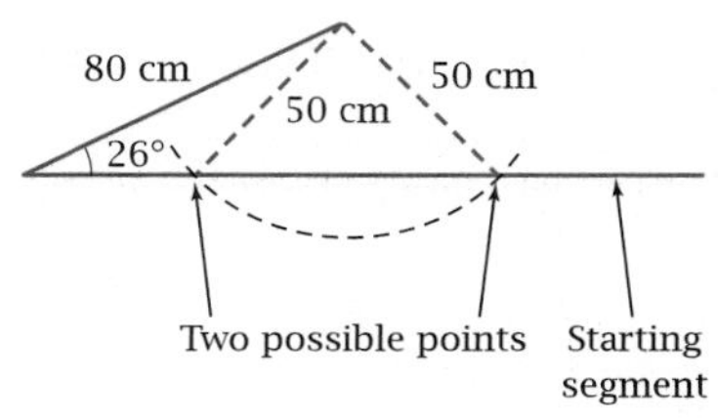

Figure 6-5b

As you go around the perimeter of the triangle in Figure 6-5b, the given information is a side, another side, and an angle (SSA). Because there are two possible triangles that have these specifications, SSA is called the **ambiguous case.**

OBJECTIVE Given two sides and an angle not contained by them in a triangle, calculate the possible values of the third side.

▶ EXAMPLE 1 In $\triangle XYZ$, side $x = 50$ cm, side $z = 80$ cm, and angle $X = 26°$, as in Figure 6-5a. Find the possible values of side y.

Solution Sketch a triangle and label the given sides and angle (Figure 6-5c).

Figure 6-5c

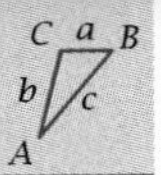

Using the law of sines to find y would require several steps. Here is a shorter way, using the law of cosines.

$$50^2 = y^2 + 80^2 - 2 \cdot y \cdot 80 \cdot \cos 26°$$ Write the law of cosines for the known angle, x.

This is a quadratic equation in the variable y. You can solve it using the quadratic formula.

$$y^2 - (160 \cos 26°)y + 6400 - 2500 = 0$$ Make one side equal zero.

$$y^2 + (-160 \cos 26°)y + 3900 = 0$$ Get the form: $ay^2 + by + c = 0$

$$y = \frac{160 \cos 26° \pm \sqrt{(-160 \cos 26°)^2 - 4 \cdot 1 \cdot 3900}}{2 \cdot 1}$$

Use the quadratic formula: $y = \dfrac{-b \pm \sqrt{b^2 - 4ac}}{2a}$

$$y = 107.5422\ldots \text{ or } 36.2648\ldots$$

$$y \approx 107.5 \text{ cm or } 36.3 \text{ cm}$$ ◀

You may be surprised if you use different lengths for side x in Example 1. Figures 6-5d and 6-5e show this side as 90 cm and 30 cm, respectively, instead of 50 cm. In the first case, there is only *one* possible triangle. In the second case, there is none.

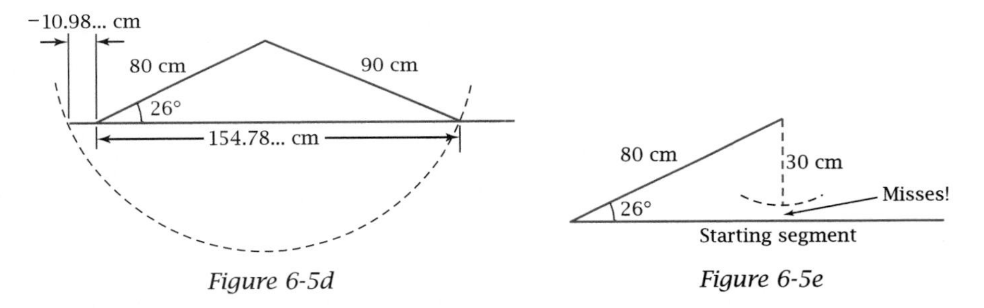

Figure 6-5d *Figure 6-5e*

The quadratic formula technique of Example 1 detects both of these results. For 30 cm, the discriminant, $b^2 - 4ac$, equals $-1319.5331\ldots$, meaning there are no real solutions for the equation and thus no triangle. For 90 cm,

$$y = 154.7896\ldots \text{ or } -10.9826\ldots$$

Although $-10.9826\ldots$ cannot be the side measure for a triangle, it does equal the *directed* distance to the point where the arc would cut the starting segment if this segment were extended back the other direction.

Problem Set 6-5

Do These Quickly

Problems Q1–Q6 refer to the triangle in Figure 6-5f.

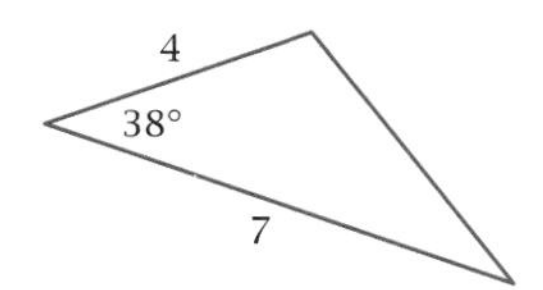

Figure 6-5f

Q1. The acronym SAS stands for —?—.

Q2. Find the length of the third side.

Q3. What method did you use in Problem Q2?

Q4. Find the area of this triangle.

Q5. The largest angle in this triangle is opposite the —?— side.

Q6. The sum of the angle measures in this triangle is —?—.

Q7. Find the amplitude of the sinusoid $y = 4\cos x + 3\sin x$.

Q8. The period of the circular function $y = 3 + 7\cos\frac{\pi}{8}(x - 1)$ is

A. 16 B. 8 C. $\frac{\pi}{8}$ D. 7 E. 3

Q9. The value of the inverse circular function $x = \sin^{-1} 0.5$ is —?—.

Q10. A value of the inverse circular relation $x = \arcsin 0.5$ between $\frac{\pi}{2}$ and 2π is —?—.

For Problems 1–8, find the possible lengths of the indicated side.

1. In $\triangle ABC$, angle $B = 34°$, side $a = 4$ cm, and side $b = 3$ cm. Find side c.
2. In $\triangle XYZ$, angle $X = 13°$, side $x = 12$ ft, and side $y = 5$ ft. Find side z.
3. In $\triangle ABC$, angle $B = 34°$, side $a = 4$ cm, and side $b = 5$ cm. Find side c.
4. In $\triangle XYZ$, angle $X = 13°$, side $x = 12$ ft, and side $y = 15$ ft. Find side z.
5. In $\triangle ABC$, angle $B = 34°$, side $a = 4$ cm, and side $b = 2$ cm. Find side c.
6. In $\triangle XYZ$, angle $X = 13°$, side $x = 12$ ft, and side $y = 60$ ft. Find side z.
7. In $\triangle RST$, angle $R = 130°$, side $r = 20$ in., and side $t = 16$ in. Find side s.
8. In $\triangle OBT$, angle $O = 170°$, side $o = 19$ m, and side $t = 11$ m. Find side b.

For Problems 9–12, use the law of sines to find the indicated angle measure. You must determine beforehand whether there are two possible angles or just one.

9. In $\triangle ABC$, angle $A = 19°$, side $a = 25$ mi, and side $c = 30$ mi. Find angle C.
10. In $\triangle HSC$, angle $H = 28°$, side $h = 50$ mm, and side $c = 20$ mm. Find angle S.
11. In $\triangle XYZ$, angle $X = 58°$, side $x = 9.3$ cm, and side $z = 7.5$ cm. Find angle Z.
12. In $\triangle BIG$, angle $B = 110°$, side $b = 1000$ yd, and side $g = 900$ yd. Find angle G.
13. *Radio Station Problem:* Radio station KROK plans to broadcast rock music to people on the beach near Ocean City (O.C. in Figure 6-5g). Measurements show that Ocean City is 20 miles from KROK, at an angle of 50° north of west. KROK's broadcast range is 30 miles.

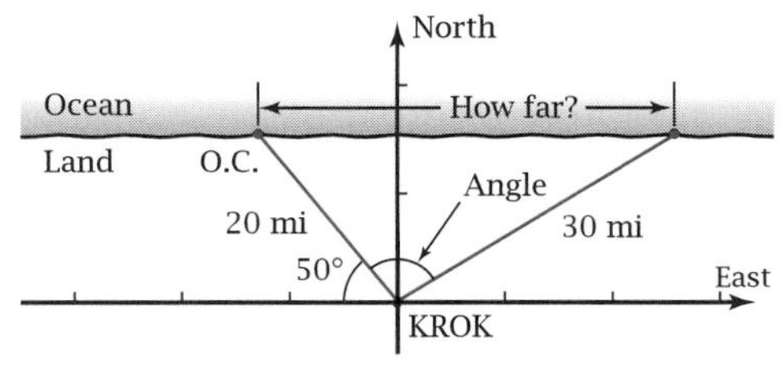

Figure 6-5g

a. Use the law of cosines to calculate how far along the beach to the east of Ocean City people can hear KROK.

b. There are two answers to part a. Show that *both* answers have meaning in the real world.

c. KROK plans to broadcast only in an angle between a line from the station through

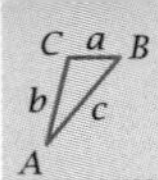

Ocean City and a line from the station through the point on the beach farthest to the east of Ocean City that people can hear the station. What is the measure of this angle?

14. *Six SSA Possibilities Problem:* Construct $\triangle XYZ$. Figure 6-5h shows six possibilities if angle X, and sides x and y are given. For each case, explain the relationship among x, y, and the quantity $y \sin X$.

a.
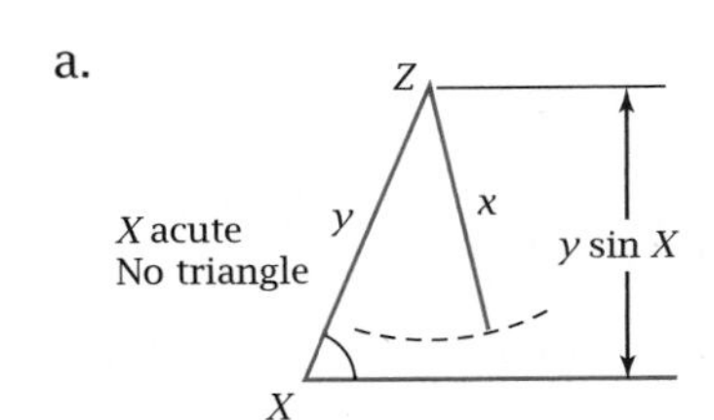

b.
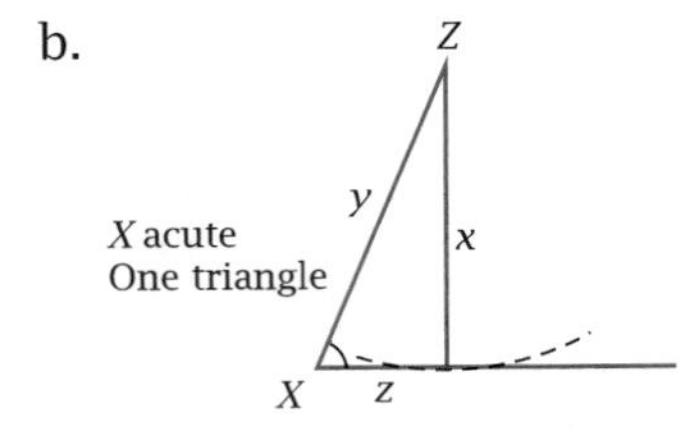

c.
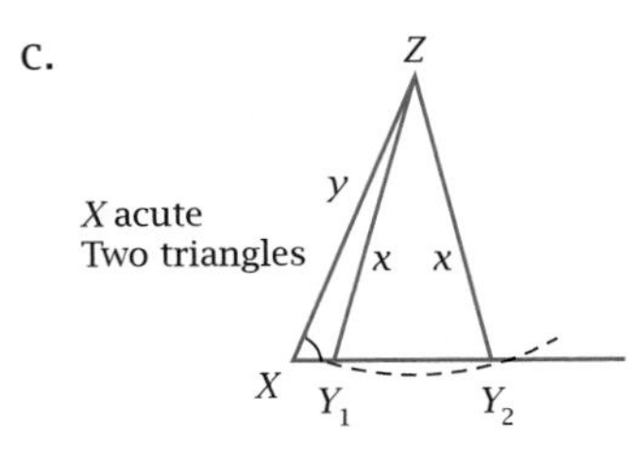

e.
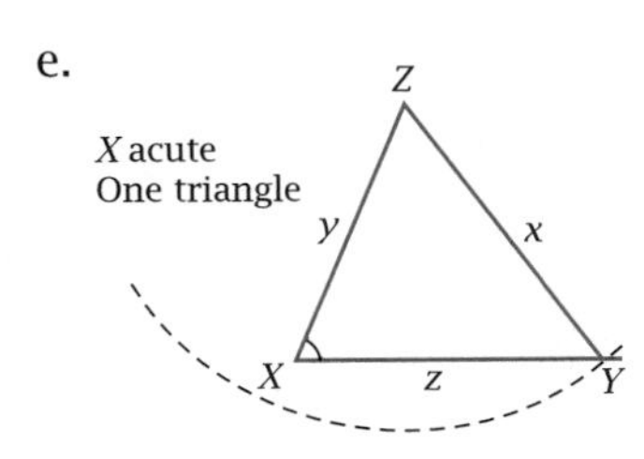

e.
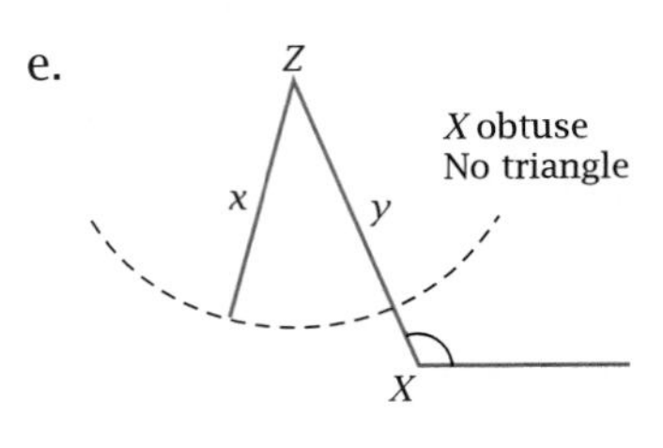

f.
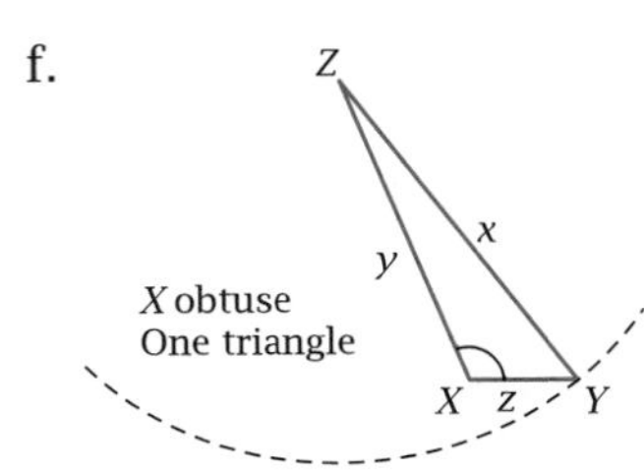

Figure 6-5h

6-6 Vector Addition

Suppose you start at the corner of a room and walk 10 ft at an angle of 70° to one of the walls (Figure 6-6a). Then you turn 80° clockwise and walk another 7 ft. If you had walked straight from the corner to your stopping point, how far and in what direction would you have walked?

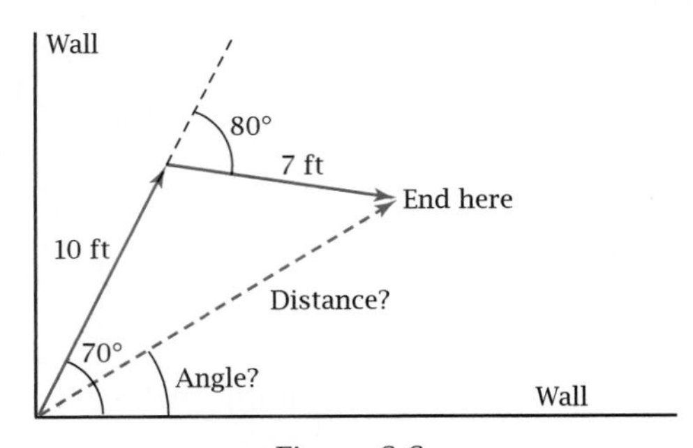

Figure 6-6a

The two motions described are called **displacements.** They are **vector quantities** that have both **magnitude** (size) and **direction** (angle). Vector quantities are represented by directed line segments called **vectors.** A quantity such as distance, time, or volume that has no direction is called a **scalar quantity.**

OBJECTIVE Given two vectors, add them to find the resultant vector.

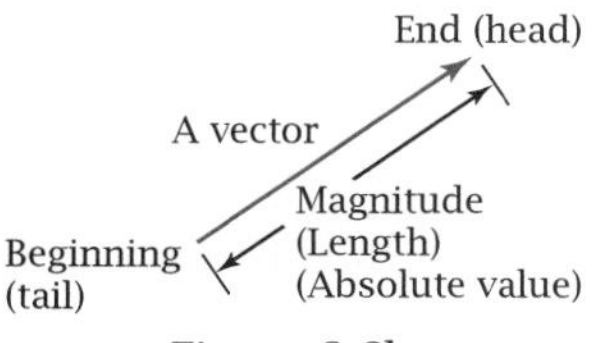

Figure 6-6b

The *length* of the directed line segment represents the *magnitude* of the vector quantity, and the *direction* of the segment represents the vector's direction. An arrowhead on a vector distinguishes the end (its **head**) from the beginning (its **tail**), as shown in Figure 6-6b.

A typhoon's wind speed can reach up to 150 miles per hour.

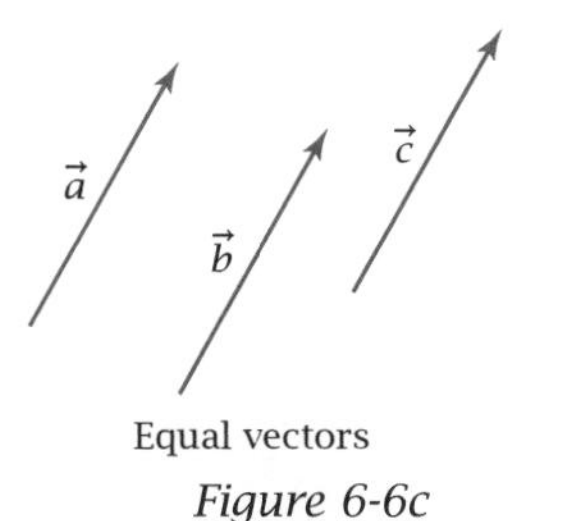

Figure 6-6c

A variable used for a vector has a small arrow over the top of it, like this, $\vec{x}$, to distinguish it from a scalar. The magnitude of the vector is also called its **absolute value** and is written $|\vec{x}|$. Vectors are **equal** if they have the same magnitude and the same direction. Vectors $\vec{a}$, $\vec{b}$, and $\vec{c}$ in Figure 6-6c are equal vectors, even though they start and end at different places. So you can **translate** a vector without changing its value.

DEFINITION: Vector

A **vector** $\vec{v}$ is a directed line segment.

The absolute value, or magnitude, of a vector, $|\vec{v}|$, is a scalar quantity equal to its length.

Two vectors are equal if and only if they have the same magnitude and the same direction.

▶ **EXAMPLE 1** You start at the corner of a room and walk as in Figure 6-6a. Find the displacement that results from the two motions.

Solution Draw a diagram showing the two given vectors and the displacement that results, $\vec{x}$ (Figure 6-6d). They form a triangle with sides 10 and 7 and included angle 100° (that is, 180° − 80°).

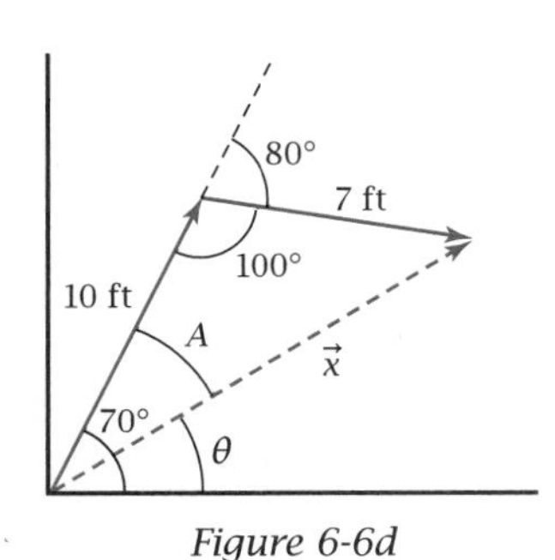

Figure 6-6d

$|\vec{x}|^2 = 10^2 + 7^2 - 2(10)(7)\cos 100° = 173.3107\ldots$ Use the law of cosines.

$|\vec{x}| = 13.1647\ldots$ Store without rounding for use later.

$7^2 = 10^2 + 13.16\ldots^2 - 2(10)(13.16\ldots)\cos A$ Use the law of cosines to find angle A.

$$\cos A = \frac{10^2 + 13.16\ldots^2 - 7^2}{2(10)(13.16\ldots)} = 0.8519\ldots$$

$A = 31.5770\ldots°$

$\theta = 70° - 31.5770\ldots° = 38.423\ldots°$

The vector representing the resultant displacement is approximately 13.2 ft at an angle of 38.4° to the wall. ◀

This example leads to the geometrical definition of vector addition. The **sum** of two vectors goes from the beginning of the first vector to the end of the second, representing the resultant displacement. Because of this, the sum of two vectors is also called the **resultant vector.**

DEFINITION: Vector Addition

The *sum* $\vec{a} + \vec{b}$ is the vector from the beginning of $\vec{a}$ to the end of $\vec{b}$ if the tail of $\vec{b}$ is placed at the head of $\vec{a}$.

Example 2 shows how to add two vectors that are not yet head-to-tail, using *velocity* vectors, for which the magnitude is the scalar *speed.*

▶ EXAMPLE 2 A ship near the coast is going 9 knots at an angle of 130° to a current of 4 knots. (A knot, kt, is a nautical mile per hour, slightly more than a regular mile per hour). What is the ship's resultant velocity with respect to the current?

Solution Draw a diagram showing two vectors 9 and 4 units long, tail-to-tail, making an angle of 130° with each other, as shown in Figure 6-6e. Translate one of the vectors so that the two vectors are head-to-tail. Draw the resultant vector, $\vec{v}$, from the beginning (tail) of the first to the end (head) of the second.

Figure 6-6e

In its new position, the 4-kt vector is parallel to its original position. The 9-kt vector is a transversal cutting two parallel lines. So the angle between the vectors forming the triangle shown in Figure 6-6e is the supplement of the given 130° angle, namely, 50°. From here on the problem is like Example 1. See if you can get these answers.

$$|\vec{v}| = 7.1217\ldots$$

$$A = 25.4838\ldots°$$

$$\theta = 130° - 25.4838\ldots° = 104.4161\ldots°$$

$$\therefore \vec{v} \approx 7.1 \text{ kt at } 104.4° \text{ to the current}$$ ◀

Vector Addition by Components

Suppose that an airplane is climbing with a horizontal velocity of 300 mph and a vertical velocity of 170 mph (Figure 6-6f). Let $\vec{i}$ and $\vec{j}$ be **unit vectors** in the horizontal and vertical directions, respectively. This means that each of the vectors has a magnitude of 1 mph.

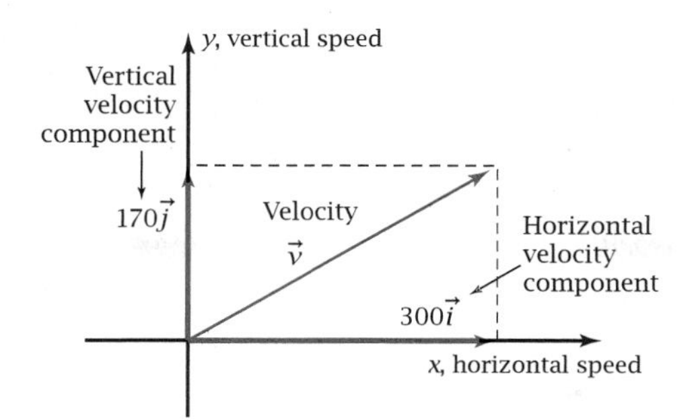

Figure 6-6f

You can write the resultant velocity vector $\vec{v}$ as the sum

$$\vec{v} = 300\,\vec{i} + 170\,\vec{j}$$

The $300\,\vec{i}$ and $170\,\vec{j}$ are called the horizontal and vertical **components** of $\vec{v}$. A product of a scalar, 300, and the unit vector $\vec{i}$ is a vector in the same direction as the unit vector but 300 times as long.

▶ **EXAMPLE 3** Vector $\vec{a}$ has magnitude 3 and direction 143° from the horizontal (Figure 6-6g). **Resolve** $\vec{a}$ into horizontal and vertical components.

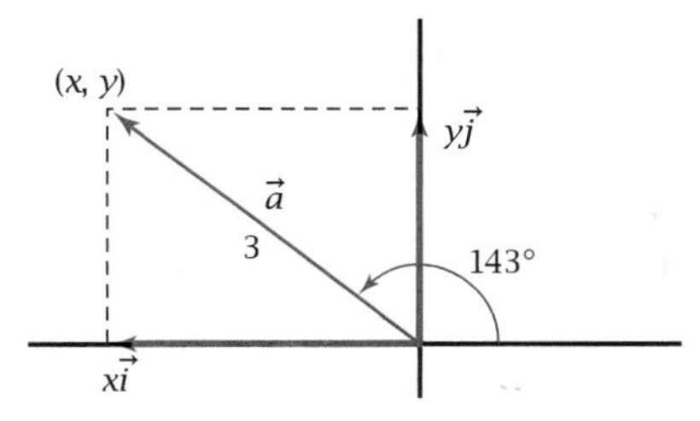

Figure 6-6g

Solution Let (x, y) be the point at the head of $\vec{a}$. By the definitions of cosine and sine,

$$\frac{x}{3} = \cos 143° \qquad \text{and} \qquad \frac{y}{3} = \sin 143°$$

$$\therefore x = 3\cos 143° = -2.3959\ldots \qquad \text{and} \qquad y = 3\sin 143° = 1.8054\ldots$$

$$\therefore \vec{a} \approx -2.396\,\vec{i} + 1.805\vec{j}$$

Note that multiplying a vector by a *negative* number, such as $-2.396\vec{i}$ in Example 3, gives a vector that points in the *opposite* direction. From Example 3 you can reach the following conclusion. ◀

PROPERTY: Components of a Vector

If $\vec{v}$ is a vector in the direction θ in standard position, then

$$\vec{v} = x\vec{i} + y\vec{j}$$

where $x = |\vec{v}|\cos\theta$ and $y = |\vec{v}|\sin\theta$.

Components make it easy to add two vectors. As shown in Figure 6-6h, if $\vec{r}$ is the resultant of $\vec{a}$ and $\vec{b}$, then the components of $\vec{r}$ are the sums of the components of $\vec{a}$ and $\vec{b}$. Because the two horizontal components have the same direction, you can add them simply by adding their coefficients. The same is true for the vertical components.

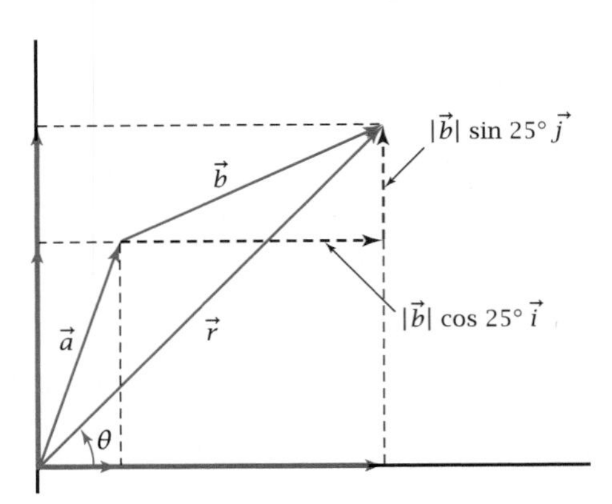

Figure 6-6h

▶ EXAMPLE 4 Vector $\vec{a}$ is 5 at 70°, and vector $\vec{b}$ is 6 at 25° (Figure 6-6h). Find the resultant, $\vec{r}$, as:

a. The sum of two components

b. A magnitude and a direction angle

Solution

a. $\vec{r} = \vec{a} + \vec{b}$

$= (5\cos 70°)\vec{i} + (5\sin 70°)\vec{j} + (6\cos 25°)\vec{i} + (6\sin 25°)\vec{j}$ Write the components.

$= (5\cos 70° + 6\cos 25°)\vec{i} + (5\sin 70° + 6\sin 25°)\vec{j}$ Combine like terms.

$= 7.1479\ldots\vec{i} + 7.2341\ldots\vec{j}$

$\approx 7.15\vec{i} + 7.23\vec{j}$ Round the *final* answer.

b. $|\vec{r}| = \sqrt{(7.1479\ldots)^2 + (7.2341\ldots)^2} = 10.1698\ldots$ By the Pythagorean theorem.

$\theta = \arctan\dfrac{7.2341\ldots}{7.1479\ldots} = 45.3435\ldots° + 180n° = 45.3435\ldots°$ Pick $n = 0$.

$\therefore \vec{r} \approx 10.17$ at $45.34°$ Round the *final* answer. ◀

Navigation Problems

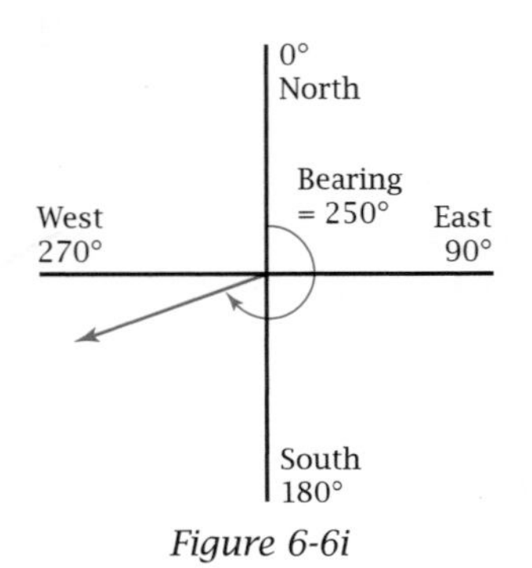

Figure 6-6i

A **bearing** is an angle measured clockwise from north, used universally by navigators for a velocity or a displacement vector. Figure 6-6i shows a bearing of 250°.

These sailors continue the tradition of one of the first seafaring people. Pacific Islanders read the waves and clouds to determine currents and predict weather.

▶ ***EXAMPLE 5*** Victoria walks 90 m due south (bearing 180°), then turns and walks 40 m more along a bearing of 250° (Figure 6-6j).

a. Find her resultant displacement vector from the starting point.

b. What is the starting point's bearing from the place where Victoria stops?

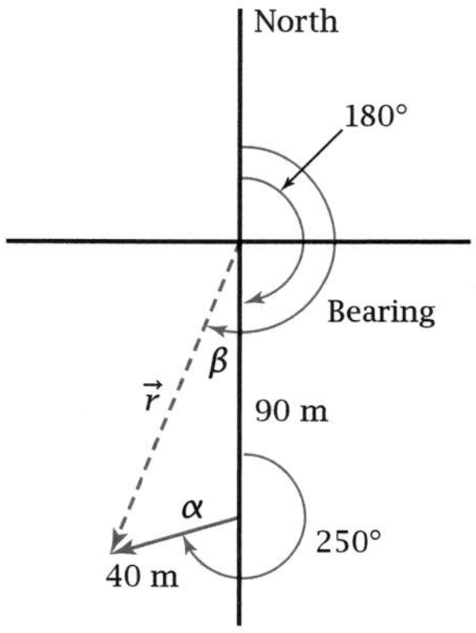

North
180°
Bearing
β
$\vec{r}$
90 m
α
250°
40 m

Figure 6-6j

Solution a. The resultant vector, $\vec{r}$, goes from the beginning of the first vector to the end of the second. Angle α is an angle in the resulting triangle.

$\alpha = 360° - 250° = 110°$

$|\vec{r}|^2 = 90^2 + 40^2 - 2(90)(40)\cos 110° = 12162.5450\ldots$ Use the law of cosines.

$|\vec{r}| = 110.2839\ldots$ Store without rounding.

To find the bearing, first calculate angle β in the resulting triangle.

$$\cos\beta = \frac{90^2 + (110.2839\ldots)^2 - 40^2}{2(90)(110.2839\ldots)} = 0.9401\ldots$$ Use the law of cosines.

$\beta = 19.9272\ldots°$

Bearing $= 180° + 19.9272\ldots° = 199.9272\ldots°$ See Figure 6-6j.

$\therefore \vec{r} \approx 110.3$ m at a bearing of 199.9°

b. The bearing from the ending point to the starting point is the opposite of the bearing from the starting point to the ending point. To find the opposite, add 180° to the original bearing.

Bearing $= 199.9272\ldots° + 180° = 379.9272\ldots°$

Because this bearing is greater than 360°, find a coterminal angle by subtracting 360°.

Bearing $= 379.9272\ldots° \approx 19.9°$ ◀

Problem Set 6-6

Do These Quickly

Q1. $\cos 90° =$

A. 1 B. 0 C. −1 D. $\frac{1}{2}$ E. $\frac{\sqrt{3}}{2}$

Q2. $\tan\frac{\pi}{4} =$

A. 1 B. 0 C. −1 D. $\frac{1}{2}$ E. $\frac{\sqrt{3}}{2}$

Q3. In $\triangle FED$, the law of cosines states that $f^2 =$ —?—.

Q4. A triangle has sides 5 ft and 8 ft and included angle 30°. What is the area of the triangle?

Q5. For $\triangle MNO$, $\sin M = 0.12$, $\sin N = 0.3$, and side $m = 24$ cm. How long is side n?

Q6. Finding equations of two sinusoids that are combined to form a graph is called —?—.

Q7. If $\sin\theta = \frac{5}{13}$ and angle θ is in Quadrant II, what is $\cos\theta$?

Q8. If $\theta = \csc^{-1}\left(\frac{11}{7}\right)$, then $\theta = \sin^{-1}$(—?—).

Q9. The equation $y = 3 \cdot 5^x$ represents a particular —?— function.

Q10. What transformation is applied to $f(x)$ to get $g(x) = f(3x)$?

For Problems 1–3, translate one vector so that the two vectors are head-to-tail, and then use appropriate triangle trigonometry to find $|\vec{a} + \vec{b}|$ and the angle the resultant vector makes with $\vec{a}$ (Figure 6-6k).

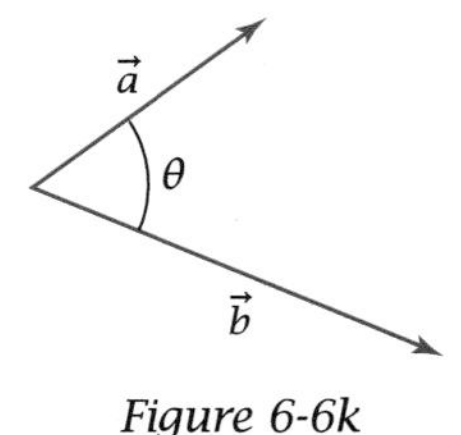

Figure 6-6k

1. $|\vec{a}| = 7$ cm, $|\vec{b}| = 11$ cm, and $\theta = 73°$
2. $|\vec{a}| = 8$ ft, $|\vec{b}| = 2$ ft, and $\theta = 41°$
3. $|\vec{a}| = 9$ in., $|\vec{b}| = 20$ in., and $\theta = 163°$
4. *Velocity Problem:* A plane flying with an air speed of 400 mi/hr crosses the jet stream, which is blowing at 150 mi/hr. The angle between the two velocity vectors is 42° (Figure 6-6l). The plane's actual velocity with respect to the ground is the vector sum of these two velocities.

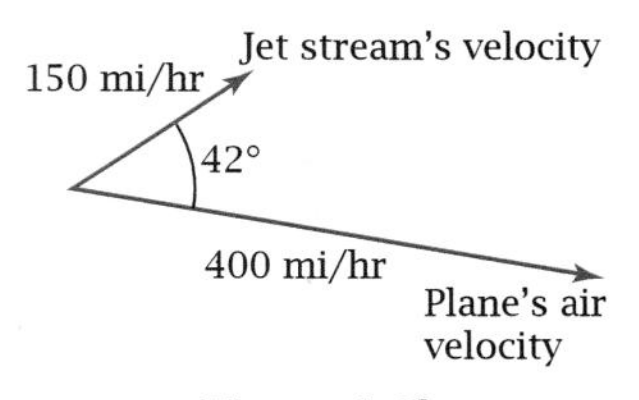

Figure 6-6l

   a. What is the actual ground speed? Why is it *less* than 400 mi/hr + 150 mi/hr?
   b. What angle does the plane's ground velocity vector make with its 400-mi/hr air velocity vector?

5. *Displacement Vector Problem:* Lucy walks on a bearing of 90° (due east) for 100 m and then on a bearing of 180° (due south) for 180 m.
   a. What is her bearing from the starting point?
   b. What is the starting point's bearing from her?
   c. How far along the bearing in part b must Lucy walk to go directly back to the starting point?

6. *Swimming Problem:* Suppose you swim across a stream that has a 5-km/hr current.

   a. Find your actual velocity vector if you swim perpendicular to the current at 3 km/hr.
   b. Find your speed through the water if you swim perpendicular to the current but your resultant velocity makes an angle of 34° with the direction you are heading.
   c. If you swim 3 km/hr, can you make it straight across the stream? Explain.

For Problems 7–10, resolve the vector into horizontal and vertical components.

7.

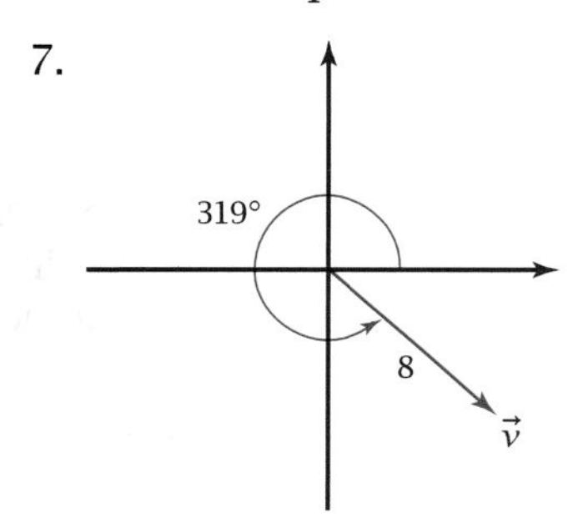

8.

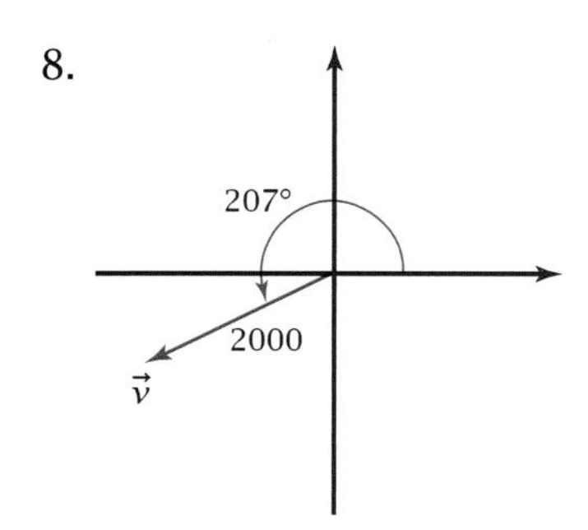

9.

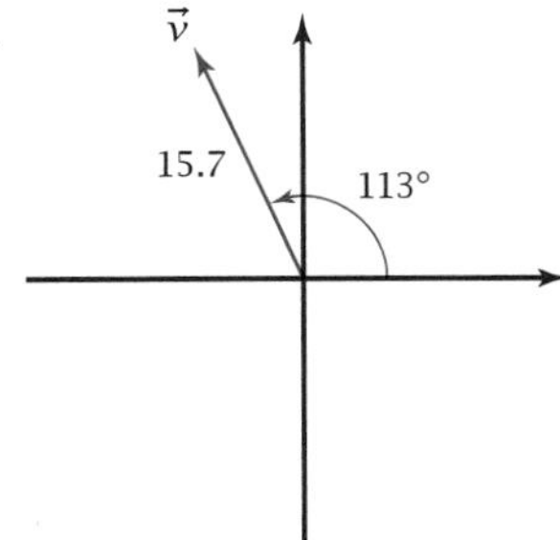

10.

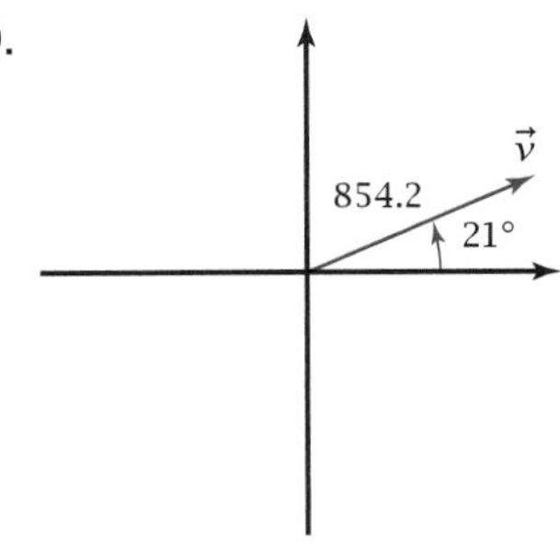

11. If $\vec{r}$ = 21 units at $\theta = 70°$ and $\vec{s}$ = 40 units at $\theta = 120°$, find $\vec{r} + \vec{s}$:

    a. As a sum of two components

    b. As a magnitude and direction

12. If $\vec{u}$ = 12 units at $\theta = 60°$ and $\vec{v}$ = 8 units at $\theta = 310°$, find $\vec{u} + \vec{v}$:

    a. As a sum of two components

    b. As a magnitude and direction

13. A ship sails 50 mi on a bearing of 20° and then turns and sails 30 mi on a bearing of 80°. Find the resultant displacement vector as a distance and a bearing.

14. A plane flies 30 mi on a bearing of 200° and then turns and flies 40 mi on a bearing of 10°. Find the resultant displacement vector as a distance and a bearing.

15. A plane flies 200 mi/hr on a bearing of 320°. The air is moving with a wind speed of 60 mi/hr on a bearing of 190°. Find the plane's resultant velocity vector (speed and bearing) by adding these two velocity vectors.

16. A scuba diver swims 100 ft/min on a bearing of 170°. The water is moving with a current of 30 ft/min on a bearing of 115°. Find the diver's resultant velocity (speed and bearing) by adding these two velocity vectors.

17. *Spaceship Problem 1:* A spaceship is moving in the plane of the Sun, the Moon, and Earth. It is being acted upon by three forces (Figure 6-6m). The Sun pulls with a force of 90 newtons at 40°. The Moon pulls with a force of 50 newtons at 110°. Earth pulls with a force of 70 newtons at 230°. What is the resultant force as a sum of two components? What is the magnitude of this force? In what direction will the spaceship move as a result of these forces?

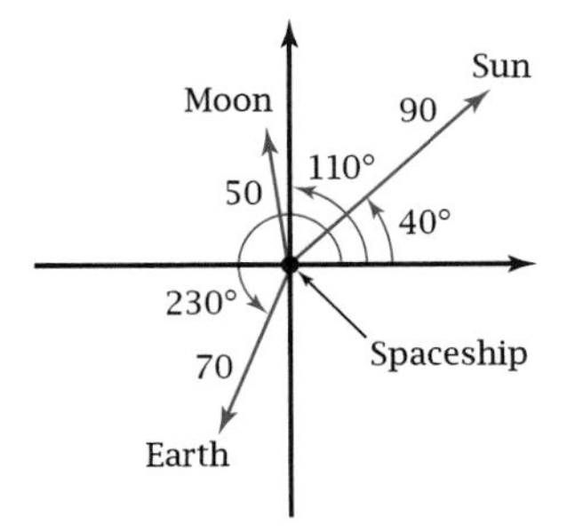

Figure 6-6m

Problems 18–22 refer to vectors $\vec{a}$, $\vec{b}$, and $\vec{c}$ in Figure 6-6n.

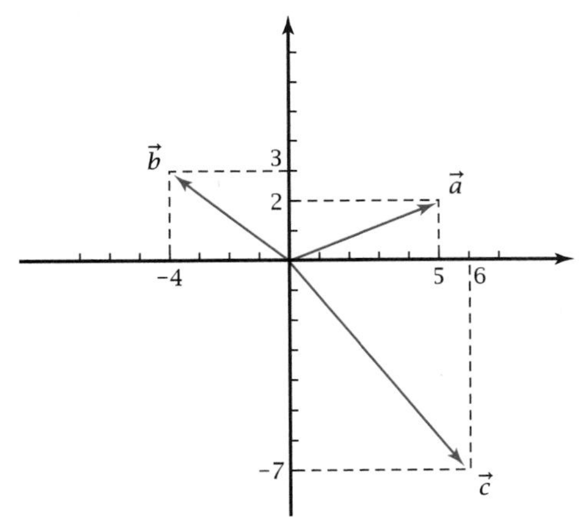

Figure 6-6n

18. *Commutativity Problem:*

    a. On graph paper, plot $\vec{a} + \vec{b}$ by translating $\vec{b}$ so that its tail is at the head of $\vec{a}$.

    b. On the same axes, plot $\vec{b} + \vec{a}$ by translating $\vec{a}$ so that its tail is at the head of $\vec{b}$.

    c. How does your figure show that vector addition is **commutative**?

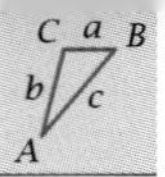

19. *Associativity Problem:* Show that vector addition is **associative** by plotting on graph paper $(\vec{a} + \vec{b}) + \vec{c}$ and $\vec{a} + (\vec{b} + \vec{c})$.
20. *Zero Vector Problem:* Plot on graph paper the sum $\vec{a} + (-\vec{a})$. What is the magnitude of the resultant vector? Can you assign a direction to the resultant vector? Why is the resultant called the **zero vector**?
21. *Closure Under Addition Problem:* How can you conclude that the set of vectors is **closed** under addition? Why is the existence of the zero vector necessary to ensure closure?
22. *Closure Under Multiplication by a Scalar Problem:* How can you conclude that the set of vectors is closed under multiplication by a scalar? Is the existence of the zero vector necessary to ensure closure in this case?
23. Look up the origin of the word *scalar.* Tell the source of your information.

6-7 Real-World Triangle Problems

Previously in this chapter you have encountered some real-world triangle problems in connection with learning the law of cosines, the law of sines, the area formula, and Hero's formula. You were able to tell which technique to use by the section of the chapter in which the problem appeared. In this section you will encounter such problems without having those external clues.

OBJECTIVE Given a real-world problem, identify a triangle, and use the appropriate technique to calculate unknown side lengths and angle measures.

Surveying instrument

To accomplish this objective, it helps to formulate some conclusions about which method is appropriate for a given set of information. Some of these conclusions are contained in this table.

PROCEDURES: *Triangle Techniques*

Law of Cosines

- Usually you can use it to find the third side from two sides and the included angle (SAS).
- You can also use it in reverse to find an angle if you know three sides (SSS).
- You can use it to find *both* values of the third side in the ambiguous SSA case.
- You *can't* use it if you know only *one* side because it involves all three sides.

(continued)

Procedures: Triangle Techniques, continued

Law of Sines

- Usually you can use it to find a side when you know an angle, the opposite side, and another angle (ASA or AAS).
- You can also use it to find an angle, but there are *two* values of the arcsine between 0° and 180° that could be the answer.
- You *can't* use it for the SSS case because you must know at least one angle.
- You *can't* use it for the SAS case because the side *opposite* the angle is unknown.

Area Formula

- You can use it to find the area from two sides and the included angle (SAS).

Hero's Formula

- You can use it to find the area from three sides (SSS).

Problem Set 6-7

Do These Quickly

Q1. For $\triangle ABC$, write the law of cosines involving angle B.

Q2. For $\triangle ABC$, write the law of sines involving angles A and C.

Q3. For $\triangle ABC$, write the area formula involving angle A.

Q4. Sketch $\triangle XYZ$, given x, y, and angle X, showing how you can draw *two* possible triangles.

Q5. Draw a sketch showing a vector sum.

Q6. Draw a sketch showing the components of $\vec{v}$.

Q7. Write $\vec{a} + \vec{b}$ if $\vec{a} = 4\vec{i} + 7\vec{j}$ and $\vec{b} = -6\vec{i} + 8\vec{j}$.

Q8. $\cos \pi =$

A. 1 B. 0 C. −1 D. $\frac{1}{2}$ E. $\frac{\sqrt{3}}{2}$

Q9. By the composite argument properties, $\sin(A - B) =$ —?—.

Q10. What is the phase displacement for $y = 7 + 6\cos 5(\theta + 37°)$ with respect to the parent cosine function?

1. *Studio Problem:* A contractor plans to build an artist's studio with a roof that slopes differently on the two sides (Figure 6-7a). On one side, the roof makes an angle of 33° with the horizontal. On the other side, which has a window, the roof makes an angle of 65° with the horizontal. The walls of the studio are planned to be 22 ft apart.

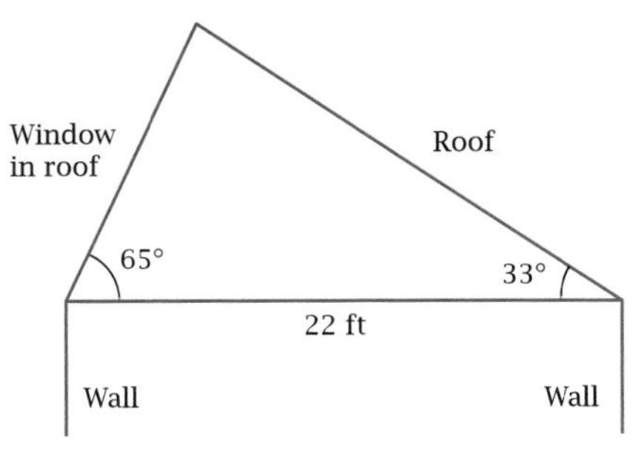

Figure 6-7a

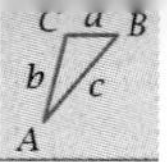

a. Calculate the lengths of the two parts of the roof.

b. How many square feet of paint will be needed for each triangular end of the roof?

2. *Detour Problem:* Suppose that you are the pilot of an airliner. You find it necessary to detour around a group of thundershowers, as shown in Figure 6-7b. You turn your plane at an angle of 21° to your original path, fly for a while, turn, and then rejoin your original path at an angle of 35°, 70 km from where you left it.

a. How much farther did you have to go because of the detour?

b. What is the area of the region enclosed by the triangle?

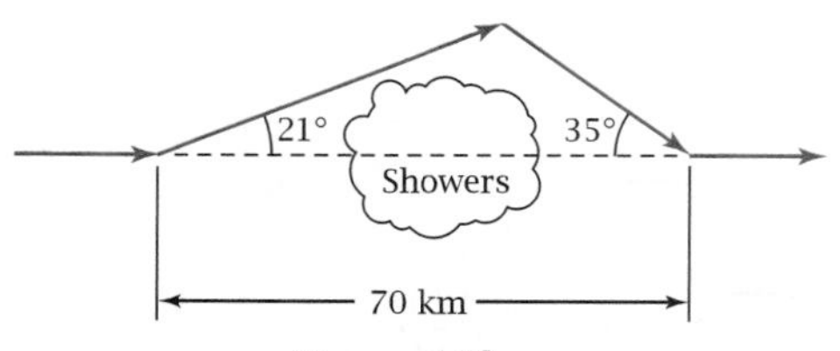

Figure 6-7b

3. *Pumpkin Sale Problem:* Scorpion Gulch Shelter is having a pumpkin sale for Halloween. They will display the pumpkins on a triangular region in the parking lot, with sides of 40 feet, 70 feet, and 100 feet. Each pumpkin takes about 3 square feet of space.

a. About how many pumpkins can the shelter display?

b. Find the measure of the middle-sized angle.

4. *Underwater Research Lab Problem:* A ship is sailing on a path that will take it directly over an occupied research lab on the ocean floor. Initially, the lab is 1000 yards from the ship on a line that makes an angle of 6° with the surface (Figure 6-7c). When the ship's slant distance has decreased to 400 yards, it can contact people in the lab by underwater telephone. Find the *two* distances from the ship's starting point at which it is at a slant distance of 400 yards from the lab.

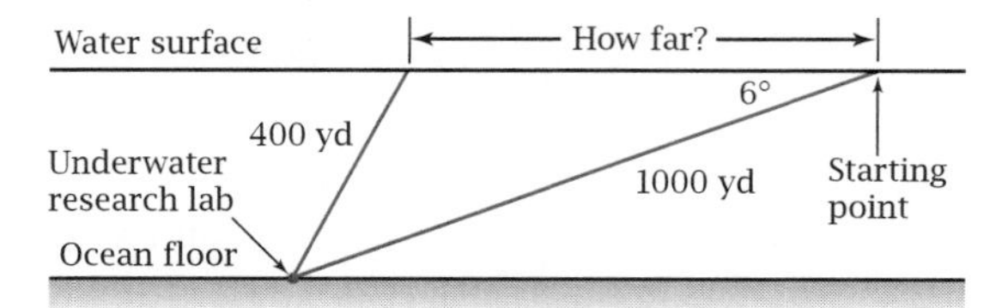

Figure 6-7c

5. *Truss Problem:* A builder has specifications for a triangular truss to hold up a roof. The horizontal side of the triangle will be 30 ft long. An angle at one end of this side will be 50°. The side to be constructed at the other end will be 20 ft long. Use the law of sines to find the angle opposite the 30-ft side. How do you interpret the results?

6. *Mountain Height Problem:* A surveying crew has the job of measuring the height of a mountain (Figure 6-7d). From a point on level ground they measure an angle of elevation of 21.6° to the top of the mountain. They move 507 m closer and find that the angle of elevation is now 35.8°. How high is the mountain? (You might have to calculate some other information along the way!)

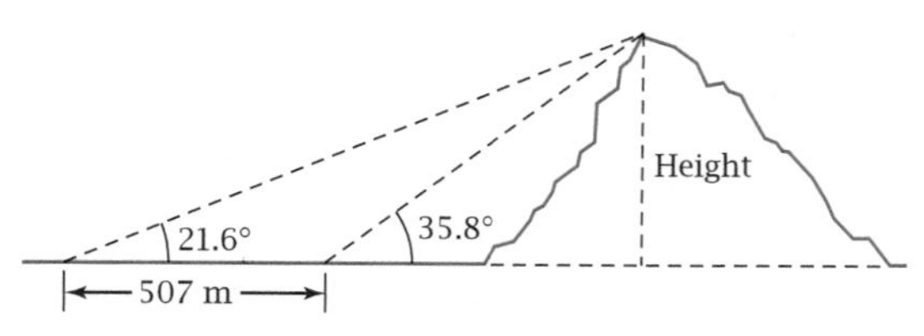

Figure 6-7d

7. *Rocket Problem:* An observer 2 km from the launching pad observes a rocket ascending vertically. At one instant, the angle of elevation is 21°. Five seconds later, the angle has increased to 35°.

Space shuttle on launch pad at Cape Kennedy, Florida

   a. How far did the rocket travel during the 5-sec interval?
   b. What was its average speed during this interval?
   c. If the rocket keeps going vertically at the same average speed, what will be the angle of elevation 15 sec after the *first* sighting?

8. *Grand Piano Problem 2:* The lid on a grand piano is held open by a 28-in. prop. The base of the prop is 55 in. from the lid's hinges, as shown in Figure 6-7e. At what possible distances along the lid could you place the end of the prop so that the lid makes a 26° angle with the piano?

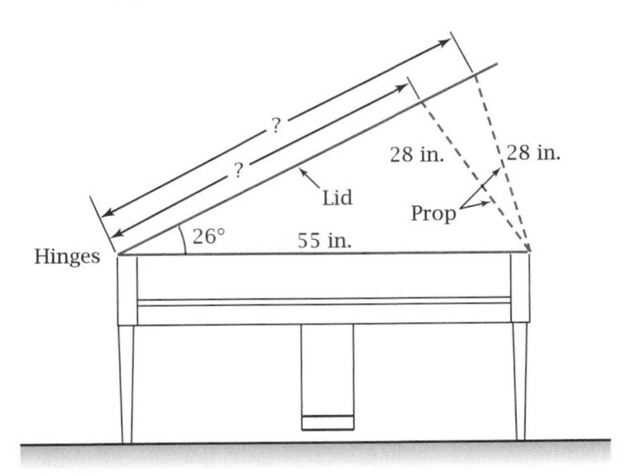

Figure 6-7e

9. *Airplane Velocity Problem:* A plane is flying through the air at a speed of 500 km/hr. At the same time, the air is moving 40 km/hr with respect to the ground at an angle of 23° with the plane's path. The plane's ground speed is the vector sum of the plane's air velocity and the wind velocity. Find the plane's ground speed if it is flying
   a. Against the wind
   b. With the wind

10. *Canal Barge Problem:* In the past, it was common to pull a barge with tow ropes on opposite sides of a canal (Figure 6-7f). Assume that one person exerts a force of 50 lb at an angle of 20° with the direction of the canal. The other person pulls at an angle of 15° to the canal with just enough force so that the resultant vector is directly along the canal. Find the force, in pounds, with which the second person must pull and the magnitude of the resultant force vector.

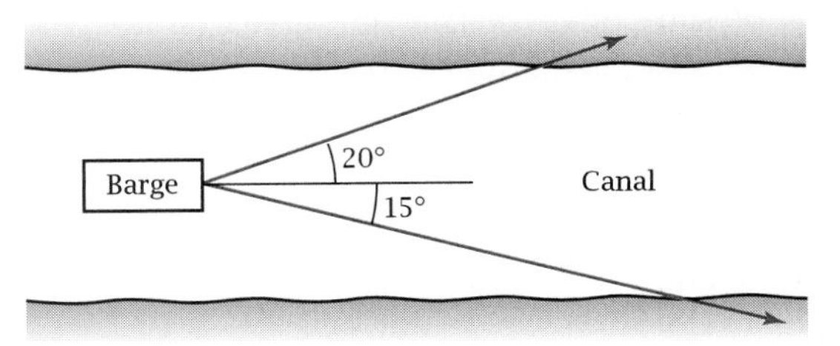

Figure 6-7f

11. *Airplane Lift Problem:* When an airplane is in flight, the air pressure creates a force vector, called the *lift,* that is perpendicular to the wings. When the plane banks for a turn, this lift vector may be resolved into horizontal and vertical components. The vertical component has magnitude equal to the plane's weight (this is what holds the plane up). The horizontal component is a *centripetal* force that makes the plane go on its curved path. Suppose that a jet plane weighing 500,000 lb banks at an angle θ (Figure 6-7g).
    a. Make a table of values of lift and horizontal component for each 5° from 0° through 30°.
    b. Based on your answers to part a, why can a plane turn in a *smaller* circle when it banks at a *greater* angle?

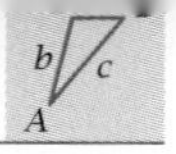

c. Why does a plane fly *straight* when it is *not* banking?

d. If the maximum lift the wings can sustain is 600,000 lb, what is the maximum angle at which the plane can bank?

e. What *two* things might happen if the plane tried to bank at an angle steeper than in part d?

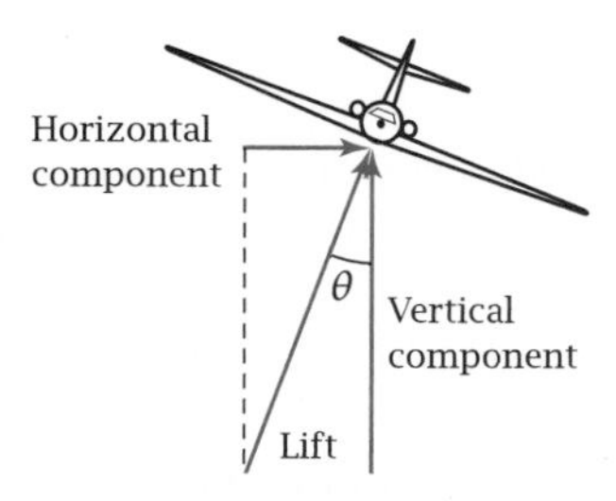

Figure 6-7g

12. *Ship's Velocity Problem:* A ship is sailing through the water in the English Channel with a velocity of 22 knots, as shown in Figure 6-7h. The current has a velocity of 5 knots on a bearing of 213°. The actual velocity of the ship is the vector sum of the ship's velocity and the current's velocity. Find the ship's actual velocity.

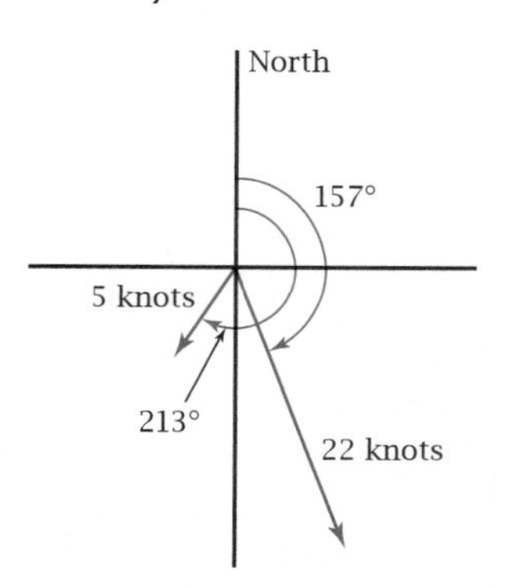

Figure 6-7h

13. *Wind Velocity Problem:* A navigator on an airplane knows that the plane's velocity through the air is 250 km/hr on a bearing of 237°. By observing the motion of the plane's shadow across the ground, she finds to her surprise that the plane's ground speed is only 52 km/hr and that its direction is along a bearing of 15°. She realizes that the ground velocity is the vector sum of the plane's velocity and the wind velocity. What wind velocity would account for the observed ground velocity?

14. *Space Station Problem 1:* Ivan is in a space station orbiting Earth. He has the job of observing the motion of two communications satellites.

 a. As Ivan approaches the two satellites, he finds that one of them is 8 km away, the other is 11 km away, and the angle between the two (with Ivan at the vertex) is 120°. How far apart are the satellites?

 b. A few minutes later, Satellite No. 1 is 5 km from Ivan and Satellite No. 2 is 7 km from Ivan. At this time, the two satellites are 10 km apart. At which of the three space vehicles does the largest angle of the resulting triangle occur? What is the measure of this angle? What is the area of the triangle?

 c. Several orbits later, only Satellite No. 1 is visible, while Satellite No. 2 is near the opposite side of Earth (Figure 6-7i). Ivan determines that angle A is 37.7°, angle B is 113°, and the distance between him and Satellite No. 1 is 4362 km. Correct to the nearest kilometer, how far apart are Ivan and Satellite No. 2?

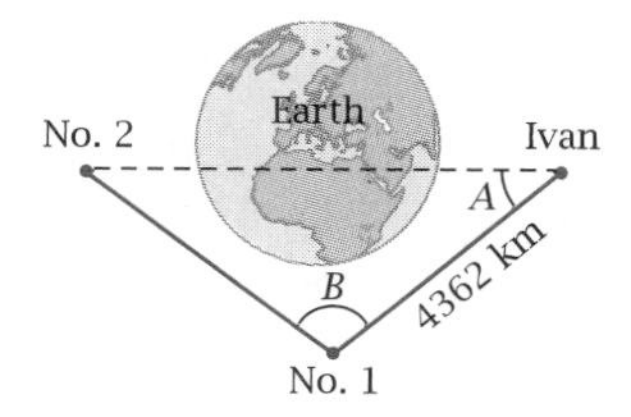

Figure 6-7i

The International Space Station is a joint project of the United States, the Russian Federation, Japan, the European Union, Canada, and Brazil. Construction began in 1998 and probably will be completed in 2006.

15. *Visibility Problem:* Suppose that you are aboard a plane destined for Hawaii. The pilot announces that your altitude is 10 km. You decide to calculate how far away the horizon is. You draw a sketch as in Figure 6-7j and realize that you must calculate an *arc length.* You recall from geography that the radius of Earth is about 6400 km. How far away is the horizon along Earth's curved surface? Surprising?

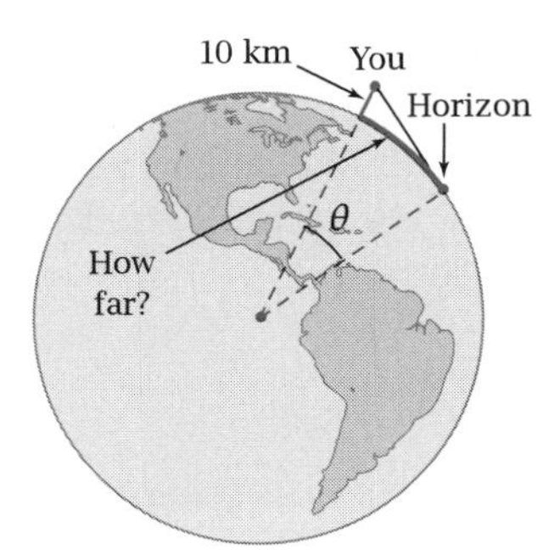

Figure 6-7j

16. *Hinged Rulers Problem:* Figure 6-7k shows a meterstick (100-cm ruler) with a 60-cm ruler attached to one end by a hinge. The other ends of both rulers rest on a horizontal surface. The hinge is pulled upward so that the meterstick makes an angle of θ with the surface.

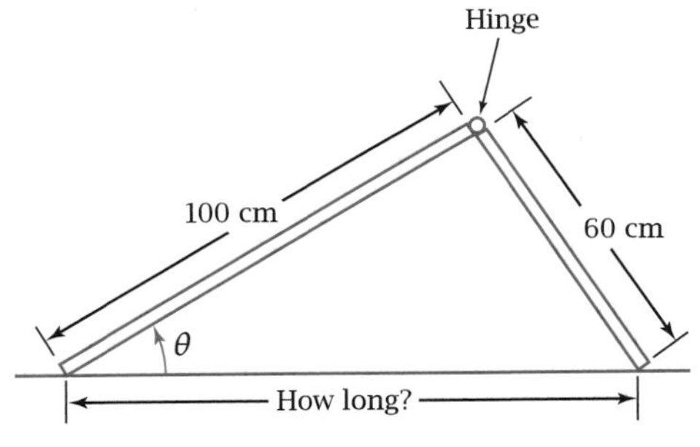

Figure 6-7k

   a. Find the two possible distances between the ruler ends if $\theta = 20°$.
   b. Show that there is *no* possible triangle if $\theta = 50°$.
   c. Find the value of θ that gives just *one* possible distance between the ends.

17. *Surveying Problem 2:* A surveyor measures the three sides of a triangular field and gets 114 m, 165 m, and 257 m.
   a. What is the measure of the largest angle of the triangle?
   b. What is the area of the field?

18. *Surveying Problem 3:* A field has the shape of a quadrilateral that is *not* a rectangle. Three sides measure 50 m, 60 m, and 70 m, and two angles measure 127° and 132° (Figure 6-7l).

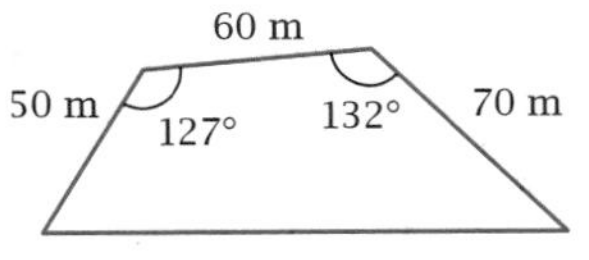

Figure 6-7l

   a. By dividing the quadrilateral into two triangles, find its area. You may have to find some sides and angles first.
   b. Find the length of the fourth side.
   c. Find the measures of the other two angles.

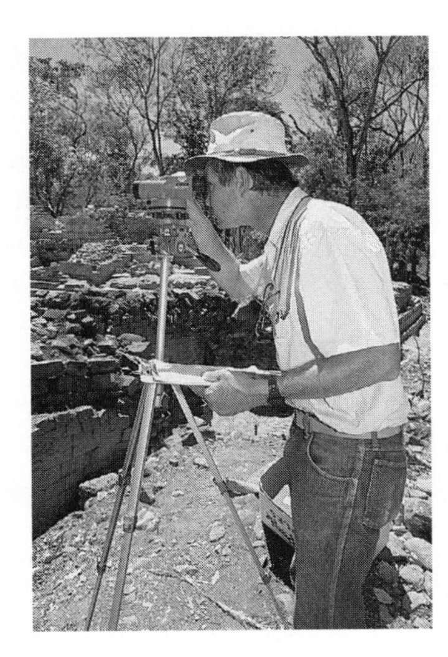

19. *Surveying Problem 4:* Surveyors find the area of an irregularly shaped tract of land by taking "field notes." These notes consist of the length of each side and information for finding each angle measure. For this problem, starting at one vertex, the tract is divided into triangles. For the first triangle, two sides and the included angle are known (Figure 6-7m), so you can calculate its area. To calculate the area of the next triangle, you must recognize that one of its sides is also the *third* side of the *first* triangle and that one of its angles is an angle of the polygon (147° in Figure 6-7m) *minus* an angle of the first triangle. By calculating this side and angle and using the

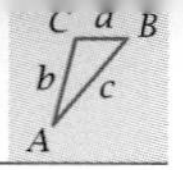

next side of the polygon (15 m in Figure 6-7m), you can calculate the area of the second triangle. The areas of the remaining triangles are calculated in the same manner. The area of the tract is the sum of the areas of the triangles.

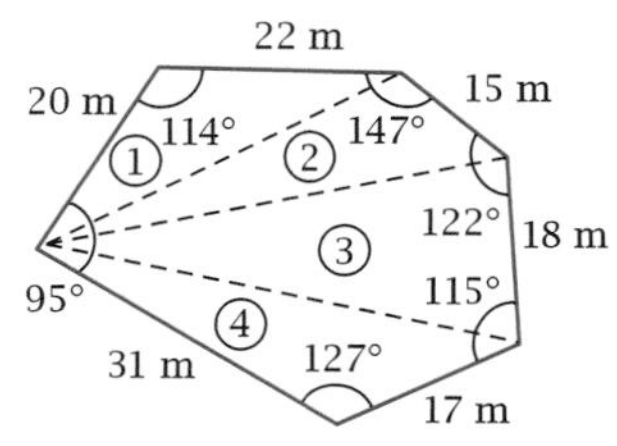

Figure 6-7m

a. Write a program for calculating the area of a tract using the technique described. The input should be the sides and angles of the polygon. The output should be the area of the tract.

b. Use your program to calculate the area of the triangle in Figure 6-7m. If you get approximately 1029.69 m^2, you can assume that your program is working correctly.

c. Show that the last side of the polygon is 30.6817 m long, which is close to the measured value of 31 m.

d. The polygon in Figure 6-7m is a **convex polygon** because none of the angles measures more than 180°. Explain why your program might give wrong answers if the polygon were *not* convex.

6-8 Chapter Review and Test

In this chapter you returned to the analysis of triangles started in Chapter 2. You expanded your knowledge of trigonometry to include oblique triangles as well as right triangles. You learned techniques to find side and angle measures for various sets of given information. These techniques are useful for real-world problems, including analyzing vectors.

Review Problems

R0. Update your journal with things you learned in this chapter. Include topics such as the law of cosines and sines, the area formulas, how these are derived, and when it is appropriate to use them. Also include how triangle trigonometry is applied to vectors.

R1. Figure 6-8a shows triangles of sides 4 cm and 5 cm, with a varying included angle θ. The length of the third side (dashed) is a function of angle θ. The five values of θ shown are 30°, 60°, 90°, 120°, and 150°.

a. Measure the length of the third side (dashed) for each triangle.

b. How long would the third side be if the angle were 180°? If it were 0°?

c. If $\theta = 90°$, you can calculate the length of the dashed line by means of the Pythagorean theorem. Does your measured length in part a agree with this calculated length?

d. If y is the length of the dashed line, the law of cosines states that

$$y = \sqrt{5^2 + 4^2 - 2 \cdot 5 \cdot 4 \cdot \cos \theta}$$

Plot the data from parts a and b and this equation for y on the same screen. Does the data seem to fit the law of cosines? Does the graph seem to be part of a sinusoid? Explain.

R2. a. Sketch a triangle with sides representing 50 ft and 30 ft and an included angle of 153°. Find the length of the third side.

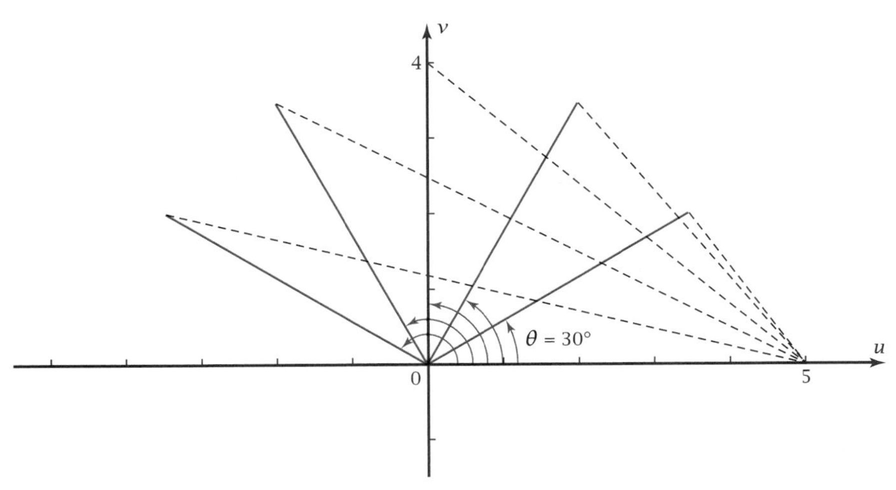

Figure 6-8a

b. Sketch a triangle with sides representing 8 m, 5 m, and 11 m. Calculate the measure of the largest angle.

c. Suppose you want to construct a triangle with sides 3 cm, 5 cm, and 10 cm. Explain why this is geometrically impossible. Show how computation of an angle using the law of cosines leads to the same conclusion.

d. Sketch $\triangle DEF$ with angle D in standard position in a uv-coordinate system. Find the coordinates of points E and F in terms of sides e and f, and angle D. Use the distance formula to prove that you can calculate side d using

$$d^2 = e^2 + f^2 - 2ef \cos D$$

R3. a. Sketch a triangle with sides representing 50 ft and 30 ft and an included angle of 153°. Find the area of the triangle.

b. Sketch a triangle with sides representing 8 mi, 11 mi, and 15 mi. Find the measure of one angle and then use it to find the area of the triangle. Calculate the area again using Hero's formula. Show that the answers are the same.

c. Suppose that two sides of a triangle are 10 yd and 12 yd and that the area is 40 yd^2. Find the two possible values of the included angle between these two sides.

d. Sketch $\triangle DEF$ with side d horizontal. Draw the altitude from vertex D to side d. What does this altitude equal in terms of side e and angle F? By appropriate geometry, show that the area of the triangle is

$$\text{Area} = \tfrac{1}{2}de \sin F$$

R4. a. Sketch a triangle with one side 6 in., the angle opposite that side equal to 39°, and another angle, 48°. Calculate the length of the side opposite the 48° angle.

b. Sketch a triangle with one side representing 5 m and its two adjacent angles measuring 112° and 38°. Find the length of the longest side of the triangle.

c. Sketch a triangle with one side 7 cm, a second side 5 cm, and the angle opposite the 5-cm side equal to 31°. Find the *two* possible values of the angle opposite the 7-cm side.

d. Sketch $\triangle DEF$ and show sides d, e, and f. Write the area three ways: in terms of angle D, in terms of angle E, and in terms of angle F. Equate the areas and then perform calculations to derive the three-part equation expressing the law of sines.

R5. Figure 6-8b shows a triangle with sides 5 cm and 8 cm and angles θ and ϕ, not included by these sides.

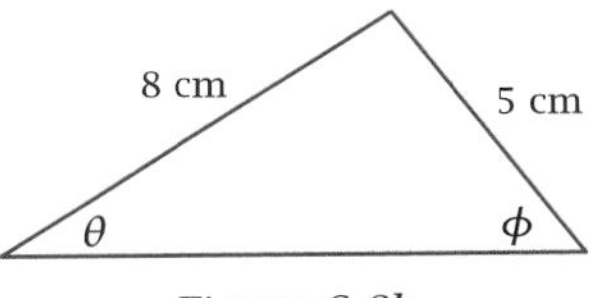

Figure 6-8b

a. If $\theta = 22°$, calculate the *two* possible values of the length of the third side.

b. If $\theta = 85°$, show algebraically that there is *no* possible triangle.

c. Calculate the value of θ for which there is exactly *one* possible triangle.

d. If $\phi = 47°$, calculate the *one* possible value of the third side of the triangle.

R6. a. Vectors $\vec{a}$ and $\vec{b}$ make an angle of 174° when placed tail-to-tail (Figure 6-8c). The magnitudes of the vectors are $|\vec{a}| = 6$ and $|\vec{b}| = 10$. Find the magnitude of the resultant vector $\vec{a} + \vec{b}$ and the angle this resultant makes with $\vec{a}$ when they are placed tail-to-tail.

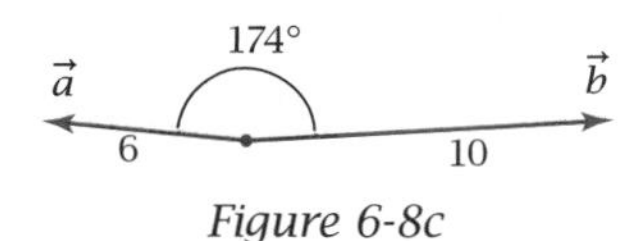

Figure 6-8c

b. Suppose that $\vec{a} = 5\vec{i} + 3\vec{j}$ and $\vec{b} = 7\vec{i} - 6\vec{j}$. Find the sum vector $\vec{a} + \vec{b}$ as sums of components. Then find the vector again as a magnitude and an angle in standard position.

c. A ship moves west (bearing of 270°) for 120 miles and then turns and moves on a bearing of 130° for another 200 miles. How far is the ship from its starting point? What is the ship's bearing from its starting point?

d. A plane flies through the air at 300 km/hr on a bearing of 220°. Meanwhile, the air is moving at 60 km/hr on a bearing of 115°. Find the plane's resultant ground velocity as a sum of two components, where unit vector $\vec{i}$ points north and $\vec{j}$ points east. Then find the plane's resultant ground speed and the bearing on which it is actually moving.

R7. *Airport Problem:* Figure 6-8d shows Nagoya Airport and Tokyo Airport 260 km apart. The ground controllers at Tokyo Airport monitor planes within a 100-km radius of the airport.

a. Plane 1 is 220 km from Nagoya Airport at an angle of 32° to the straight line between the airports. How far is Plane 1 from Tokyo Airport? Is it really out of range of Tokyo Ground Control, as suggested by Figure 6-8d?

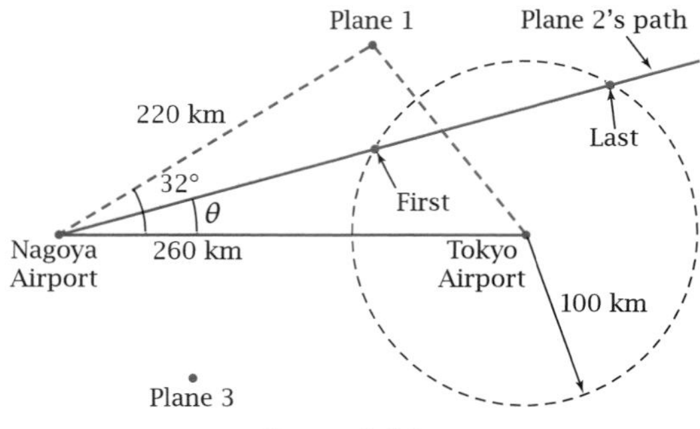

Figure 6-8d

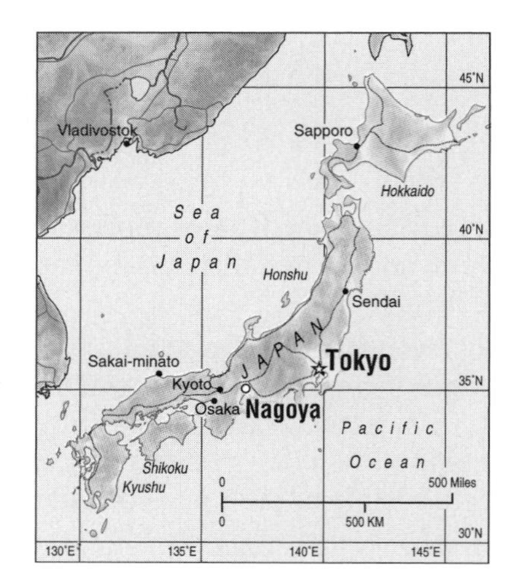

b. Plane 2 is going to take off from Nagoya Airport and fly past Tokyo Airport. Its path will make an angle of θ with the line between the airports. If $\theta = 15°$, how far will Plane 2 be from Nagoya Airport when it first comes within range of Tokyo Ground Control? How far from Nagoya Airport is it when it is last within range? Store both of these distances in your calculator, without rounding.

c. Show that if $\theta = 40°$, Plane 2 is never within range of Tokyo Ground Control.

d. Calculate the value of θ for which Plane 2 is within range of Tokyo Ground Control at just *one* point. How far from Nagoya Airport is this point? Store the distance in your calculator, without rounding.

e. Show numerically that the square of the distance in part d is exactly equal to the product of the two distances in part b. What theorem from geometry expresses this fact?

f. Plane 3 (Figure 6-8d) reports that it is being forced to land in a field! Nagoya Airport and Tokyo Airport report that the angles between Plane 3's position and the line between the airports are 35° and 27°, respectively. Which airport is Plane 3 closer to? How much closer?

Helicopter Problem: The rotor on a helicopter creates an upward force vector (Figure 6-8e). The vertical component of this force (the lift) balances the weight of the helicopter and keeps it up in the air. The horizontal component (the thrust) makes the helicopter move forward. Suppose that the helicopter weighs 3000 lb.

g. At what angle will the helicopter have to tilt forward to create a thrust of 400 lb?

h. What will be the magnitude of the total force vector?

i. Explain why the helicopter can hover over the same spot by judicious choice of the tilt angle.

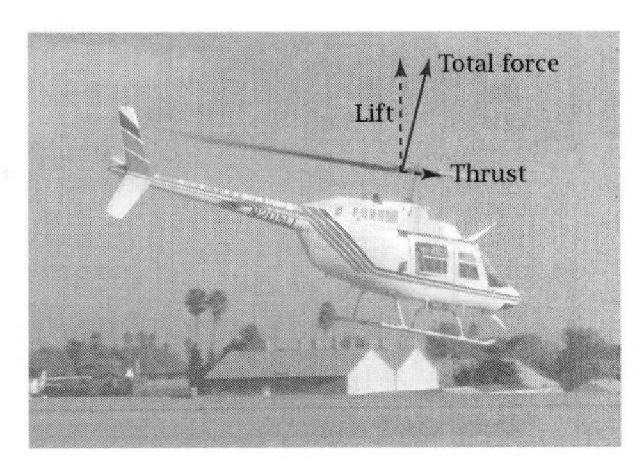

Figure 6-8e

Concept Problems

C1. *Essay Problem:* Research the contributions of different cultures to trigonometry. Use the following resources or others you might find on the Web or in your local library: Eli Maor, *Trigonometric Delights* (Princeton: Princeton University Press, 1998); David Blatner, *The Joy of π* (New York: Walker Publishing Co., 1997). Write an essay about what you have learned.

C2. *Reflex Angle Problem:* Figure 6-8f shows quadrilateral *ABCD*, in which angle *A* is a **reflex angle** measuring 250°. The resulting figure is called a **nonconvex polygon.** Note that the diagonal from vertex *B* to *D* lies *outside* the figure.

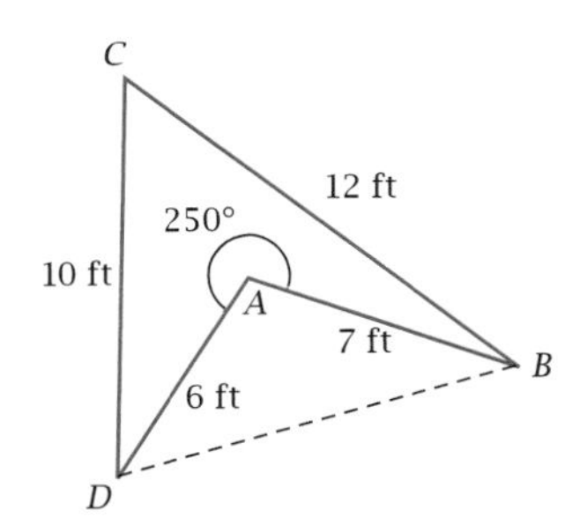

Figure 6-8f

a. Find the measure of angle *A* in △*ABD.* Next, calculate the length *DB* using the sides 6 ft and 7 ft shown in Figure 6-8f. Then calculate *DB directly,* using the 250° measure of angle *A.* Do you get the same answer? Explain why or why not.

b. Calculate the area of △*ABD* using the nonreflex angle you calculated in part a. Then calculate the area of this triangle directly using the 250° measure of angle *A.* Do you get the same answer for the area? Explain why or why not.

c. Use the results in part a to find the area of △*BCD.* Then find the area of quadrilateral *ABCD.* Explain how you can find this area *directly* using the 250° measure of angle *A.*

C3. *Angle of Elevation Experiment:* Construct an **inclinometer** that you can use to measure angles of elevation. One way to do this is to hang a piece of wire, such as a straightened paper clip, from the hole in a protractor, as shown in Figure 6-8g. Then tape a straw to the protractor so that you can sight a distant object more accurately. As you view the top of a building

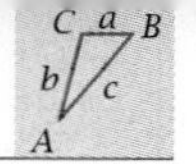

or tree along the straight edge of the protractor, gravity holds the paper clip vertical, allowing you to determine the angle of elevation. Use your apparatus to measure the height of a tree or building using the techniques of this chapter.

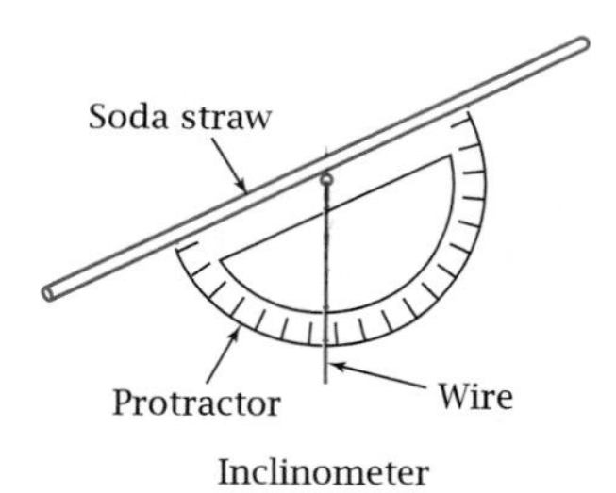

Figure 6-8g

C4. *Euclid's Problem:* This problem comes from Euclid's *Elements.* Figure 6-8h shows a circle with a secant line and a tangent line.

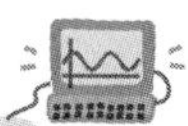

a. Sketch a similar figure using a dynamic geometry program, such as The Geometer's Sketchpad, and measure the lengths of the secant segments, $\overline{PQ}$ and $\overline{PR}$, and the tangent segment $\overline{PS}$; by varying the radius of the circle and the angle QPO, see if it is true that

$$PS^2 = PQ \cdot PR$$

b. Using the trigonometric laws and identities you've learned, prove that the equation in part a is a true statement.

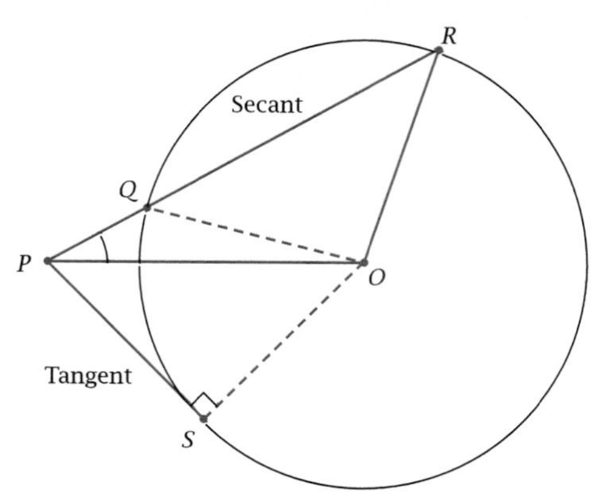

Figure 6-8h

Chapter Test

PART 1: No calculators (T1–T9)

To answer T1–T3, refer to Figure 6-8i.

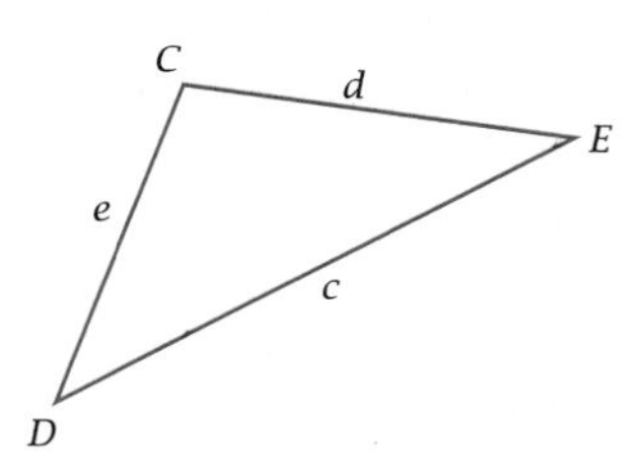

Figure 6-8i

T1. Write the law of cosines involving angle D.

T2. Write the law of sines (either form).

T3. Write the area formula involving sides d and e.

T4. Explain why you cannot use the law of cosines for the triangle in Figure 6-8j.

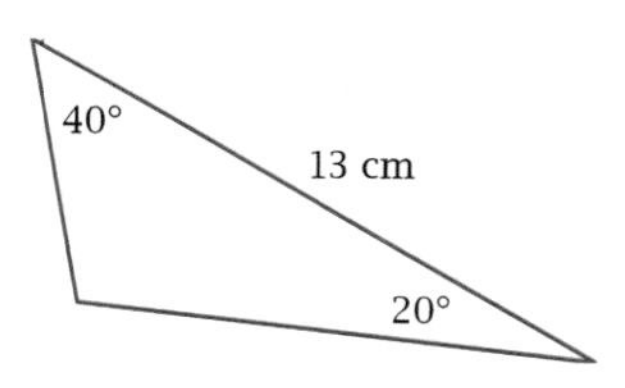

Figure 6-8j

T5. Explain why you cannot use the law of sines for the triangle in Figure 6-8k.

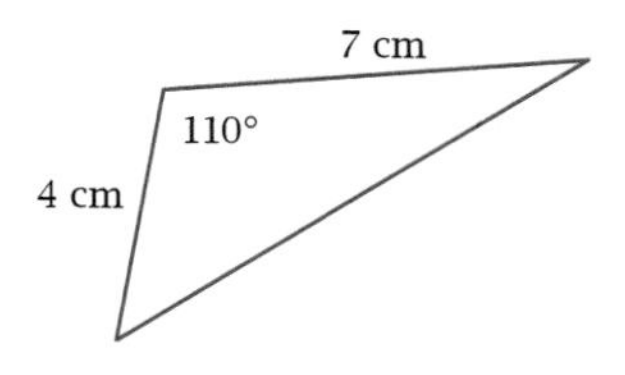

Figure 6-8k

T6. Explain why there is *no* triangle with the side measurements given in Figure 6-8l.

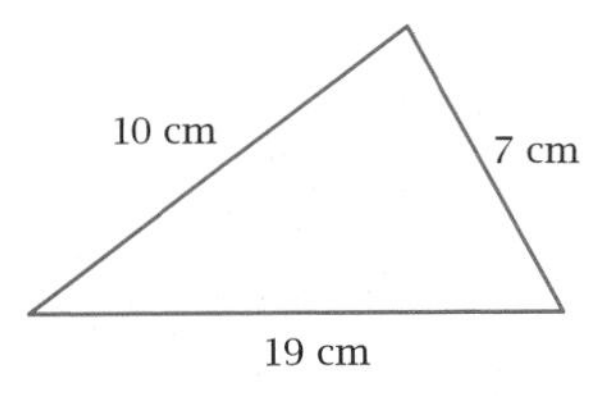

Figure 6-8l

T7. Explain why you can use the inverse cosine *function*, $\cos^{-1}$, when you are finding an angle of a triangle by the law of cosines but must use the inverse sine *relation*, arcsin, when you are finding an angle of a triangle by the law of sines.

T8. Sketch the vector sum $\vec{a} + \vec{b}$ (Figure 6-8m).

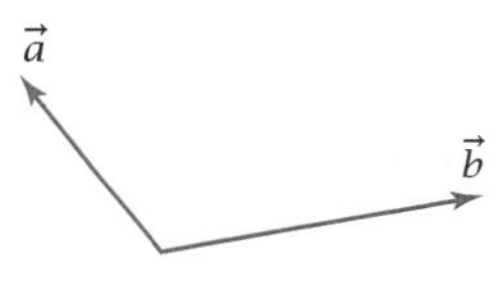

Figure 6-8m

T9. Sketch vector $\vec{v} = 3\vec{i} - 5\vec{j}$ and its components in the x- and y-directions.

PART 2: Graphing calculators are allowed (T10–T21)

T10. Construct a triangle with sides 7 cm and 5 cm and an included angle 24°. Measure the third side.

T11. Calculate the length of the third side in Problem T10. Does the measurement in Problem T10 agree with this calculated value?

T12. Sketch a triangle with a base representing 50 ft and two base angles of 38° and 47°. Calculate the measure of the third angle.

T13. Calculate the length of the shortest side of the triangle in Problem T12.

T14. Sketch a triangle. Make up lengths for the three sides that give a *possible* triangle. Calculate the size of the largest angle. Store the answer without rounding.

T15. Find the area of the triangle in Problem T14. Use the angle you calculated in Problem T13. Store the answer without rounding.

T16. Use Hero's formula to calculate the area of your triangle in Problem T14. Does it agree with your answer to Problem T15?

T17. Figure 6-8n shows a circle of radius 3 cm. Point P is 5 cm from the center. From P, a secant line is drawn at an angle of 26° to the line connecting the center to P. Use the law of cosines to calculate the two unknown lengths marked a and b in the figure.

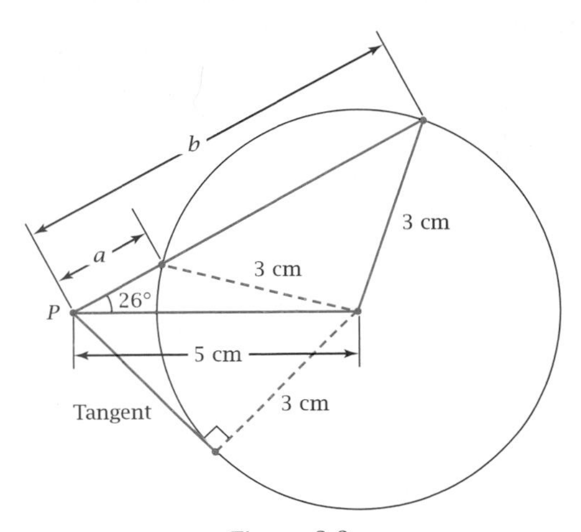

Figure 6-8n

T18. Recall that the radius of a circle drawn to the point of tangency is perpendicular to the tangent. Use this fact to calculate the length of the tangent segment from point P in Figure 6-8n.

T19. Show numerically that the product of your two answers in Problem T17 equals the square of the tangent length in Problem T18. This geometrical property appears in Euclid's *Elements.*

T20. For vector $\vec{v} = 3\vec{i} - 5\vec{j}$, calculate the magnitude. Calculate the direction as an angle in standard position.

T21. What did you learn from this test that you did not know before?

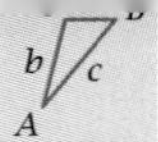

6-9 Cumulative Review, Chapters 1–6

These problems constitute a 2- to 3-hour "rehearsal" for your examination on Chapters 1–6.

1. Write the general equation of a quadratic function.
2. Name the transformations applied to function f to get g, when $g(x) = 5f(3x)$.
3. In Figure 6-9a, name the transformations applied to function f to get function h. Write an equation for $h(x)$ in terms of function f.

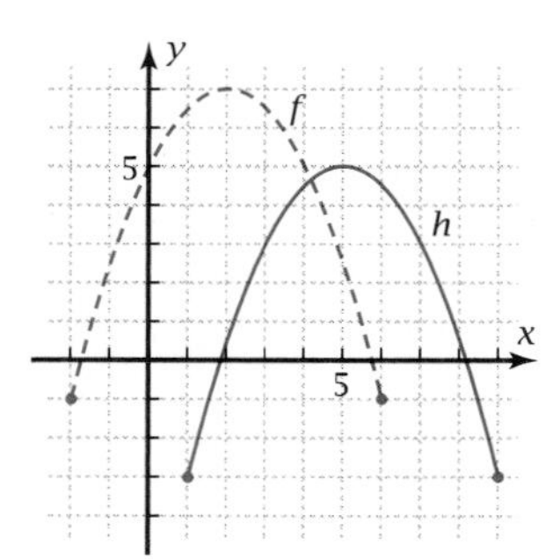

Figure 6-9a

4. On a copy of Figure 6-9b, sketch the graph of $y = f^{-1}(x)$.

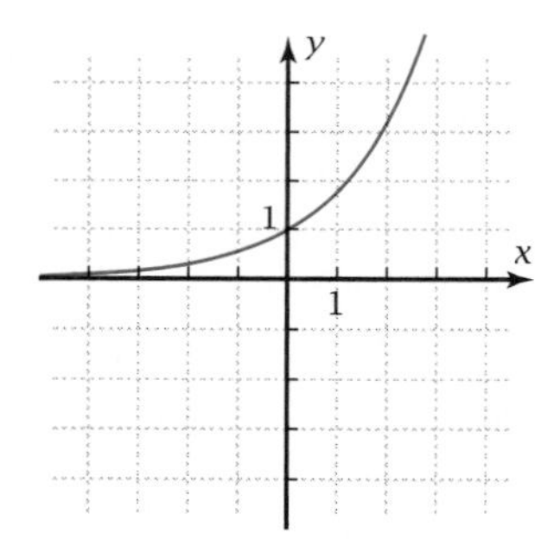

Figure 6-9b

5. If $f(-x) = -f(x)$ for all x in the domain, then f is a(n) —?— function.
6. Write an equation for $g(x)$ in terms of $f(x)$ that has *all* of these features:
   - A horizontal dilation by a factor of 2
   - A vertical dilation by a factor of 3
   - A horizontal translation by 4 units
   - A vertical translation by 5 units
7. *Satellite Problem 1, Part 1:* A satellite is in orbit around Earth. From where you are on Earth's surface, the straight-line distance to the satellite (*through* Earth, at times) is a periodic function of time. Sketch a reasonable graph.

8. Sketch an angle of $-213°$ in standard position. Mark its reference angle and find the measure of the reference angle.
9. The terminal side of angle θ contains the point (12, −5) in the uv-coordinate system. Write the *exact* values (no decimals) of the six trigonometric functions of θ.
10. Write the exact value (no decimals) of $\sin 240°$.
11. Draw 180° in standard position. Explain why $\cos 180° = -1$.
12. Sketch the graph of the parent sinusoidal function $y = \sin \theta$.
13. What special name is given to the kind of periodic function you graphed in Problem 12?
14. How many radians are there in 360°? 180°? 90°? 45°?

15. How many degrees are there in 2 radians?

16. Sketch a graph showing a unit circle centered at the origin of a uv-coordinate system. Sketch an x-axis tangent to the circle, going vertically through the point $(u, v) = (1, 0)$. If the x-axis is wrapped around the unit circle, show that the point $(2, 0)$ on the x-axis corresponds to an angle of 2 radians.

17. Sketch the graph of the parent sinusoidal function $y = \cos x$.

18. For $y = 3 + 4 \cos 5(x + 6)$, find:

    a. The horizontal dilation

    b. The vertical dilation

    c. The horizontal translation

    d. The vertical translation

19. For sinusoids, list the special names given to:

    a. The horizontal dilation

    b. The vertical dilation

    c. The horizontal translation

    d. The vertical translation

20. Write a particular equation for the sinusoid in Figure 6-9c.

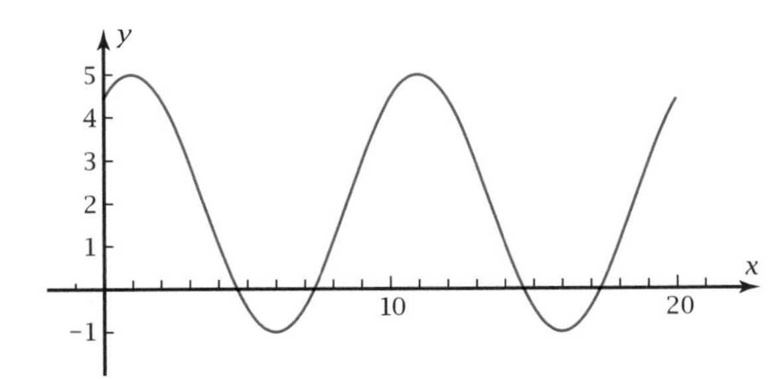

Figure 6-9c

21. If the graph in Problem 20 were plotted on a wide-enough domain, predict y for $x = 342.7$.

22. For the sinusoid in Problem 20, find algebraically the first three positive values of x if $y = 4$.

23. Show graphically that your three answers in Problem 22 are correct.

24. *Satellite Problem 1, Part 2:* Assume that in Problem 7, the satellite's distance varies sinusoidally with time. Suppose that the satellite is closest, 1000 miles from you, when $t = 0$ minutes. Half a period later, when $t = 50$ minutes, it is at its maximum distance from you, 9000 miles. Write a particular equation for distance, in thousands of miles, as a function of time.

25. There are three kinds of properties that involve just *one* argument. Write the name of each kind of property, and give an example of each.

26. Use the properties in Problem 25 to prove that this equation is an identity. What restrictions are there on the domain of x?

$$\sec^2 x \sin^2 x + \tan^4 x = \frac{\sin^2 x}{\cos^4 x}$$

27. Other properties involve functions of a composite argument. Write the composite argument property for $\cos(x - y)$. Then express this property verbally.

28. Show numerically that $\cos 34° = \sin 56°$.

29. Use the property in Problem 27 to prove that the equation $\cos(90° - \theta) = \sin \theta$ is an identity. How does this explain the result in Problem 28?

30. Show that the function

$$y = 3 \cos \theta + 4 \sin \theta$$

is a sinusoid by finding algebraically the amplitude and phase displacement and writing y as a single sinusoid.

31. The function

$$y = 12 \sin \theta \cos \theta$$

is equivalent to the sinusoid $y = 6 \sin 2\theta$. Prove algebraically that this is true by applying the composite argument property to $\sin 2\theta$.

32. Write the double argument property expressing $\cos 2x$ in terms of $\sin x$ alone. Use this property to show algebraically that the graph of $y = \sin^2 x$ is a sinusoid.

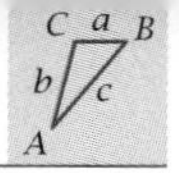

33. Find a particular equation for the function in Figure 6-9d.

34. Find a particular equation for the function in Figure 6-9e.

35. Transform the function

$$y = 2\cos 20\theta \cos\theta$$

to a sum of two cosine functions.

36. Find the periods of the two sinusoids in the equation given in Problem 35 and the periods of the two sinusoids in the answer. What can you tell about the periods of the two sinusoids in the given equation and about the periods of the sinusoids in the answer?

37. Find the (one) value of the inverse trigonometric function $\theta = \tan^{-1} 5$.

38. Find the general solution for the inverse trigonometric relation $x = \arcsin 0.4$.

39. Use parametric functions to create the graph of $y = \arccos x$, as shown in Figure 6-9f.

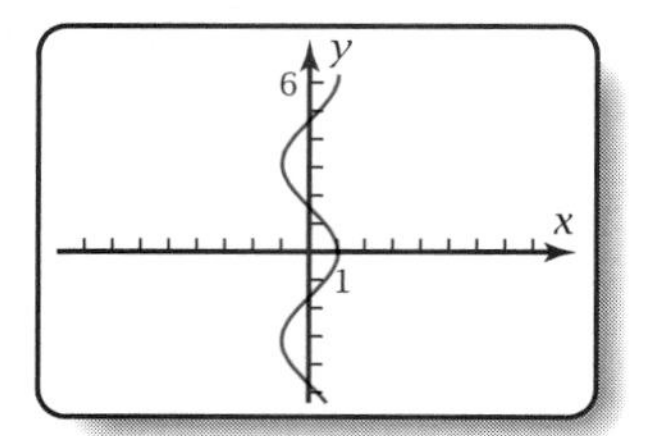

Figure 6-9f

40. The inverse trigonometric function $y = \cos^{-1} x$ is the principal branch of $y = \arccos x$. Define the domain and range of $y = \cos^{-1} x$.

41. Find the first four positive values of θ, if $\theta = \arctan 2$.

42. State the law of cosines.

43. State the law of sines.

44. State the area formula for a triangle given two sides and the included angle.

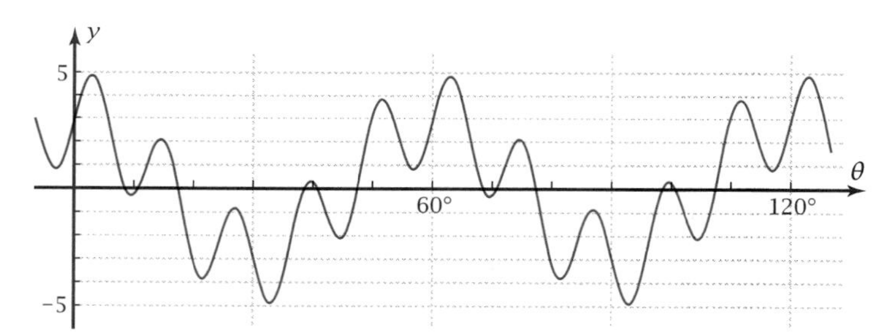

Figure 6-9d

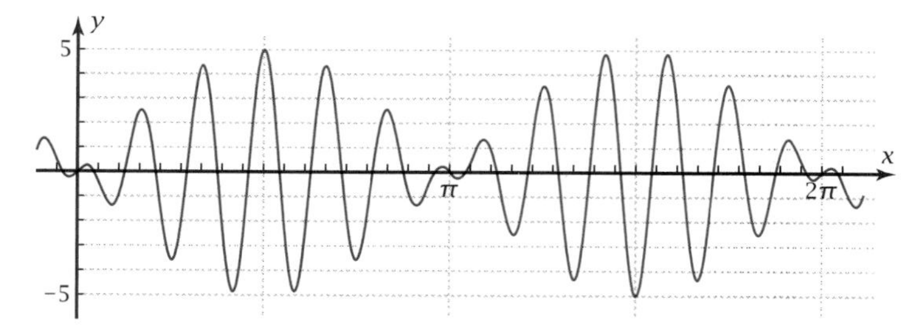

Figure 6-9e

45. If a triangle has sides 6 ft, 7 ft, and 12 ft, find the measure of the largest angle.

46. Find the area of the triangle in Problem 45 using Hero's formula.

47. Vector $\vec{a} = -3\vec{i} + 4\vec{j}$. Vector $\vec{b} = 5\vec{i} + 12\vec{j}$.

    a. Find the resultant vector $\vec{a} + \vec{b}$ in terms of its components.

    b. Find the magnitude and angle in standard position of the resultant vector.

    c. Sketch a figure to show $\vec{a}$ and $\vec{b}$ added geometrically, head-to-tail.

    d. Is this true or false? "$|\vec{a} + \vec{b}| = |\vec{a}| + |\vec{b}|$." Explain why your answer is reasonable.

48. *Satellite Problem 1, Part 3:* In Problem 24 you assumed that the distance between you and the satellite was a sinusoidal function of time. In this problem you will get a more accurate mathematical model.

    a. Use the law of cosines and the distances in Figure 6-9g to find y as a function of angle x radians.

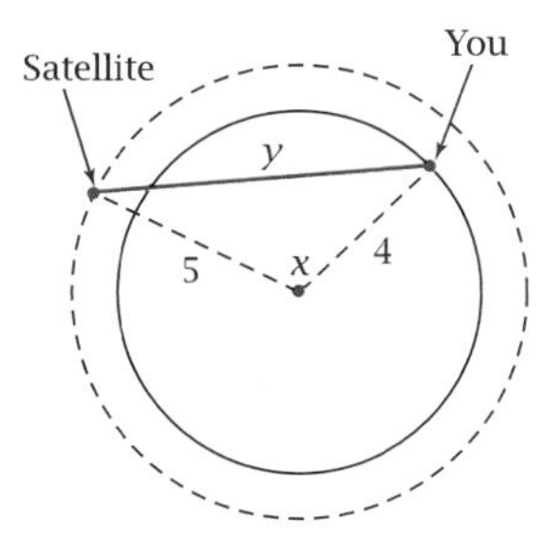

Figure 6-9g

    b. Use the fact that it takes 100 minutes for the satellite to make one orbit to get the equation for y as a function of time t. Assume that angle $x = 0$ when time $t = 0$.

    c. Plot the equation from part b and the equation from Problem 24 on the same screen, thus showing that the functions have the same high points, low points, and period but that the equation from part b is *not* a sinusoid.

49. What do you consider to be the *one* most important thing you have learned so far as a result of taking precalculus?

CHAPTER

7

Properties of Elementary Functions

If a mother rat is twice as long as her offspring, then the mother's weight is about eight times the baby's weight. But the mother rat's skin area is only about four times the baby's skin area. So the baby rat must eat more than the mother rat in proportion to its body weight to make up for the heat loss through its skin. In this chapter you'll learn how to use functions to model and explain situations like this.

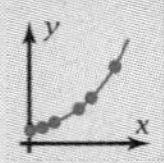

Mathematical Overview

In this chapter you will extend what you have already learned about some of the more familiar functions in algebra, as well as some you may not have yet encountered. These functions are

- Linear
- Quadratic
- Power
- Exponential
- Logarithmic
- Logistic

You will study these functions in four ways.

Algebraically You can define each of these functions algebraically, for example, the logarithmic function

$y = \log_b x$ if and only if $b^y = x$

Numerically You can find interesting numerical relationships between the values of x and y variables. Exponential functions exhibit the add–multiply property: as a result of adding a constant to x, the corresponding y-value is multiplied by a constant.

Graphically

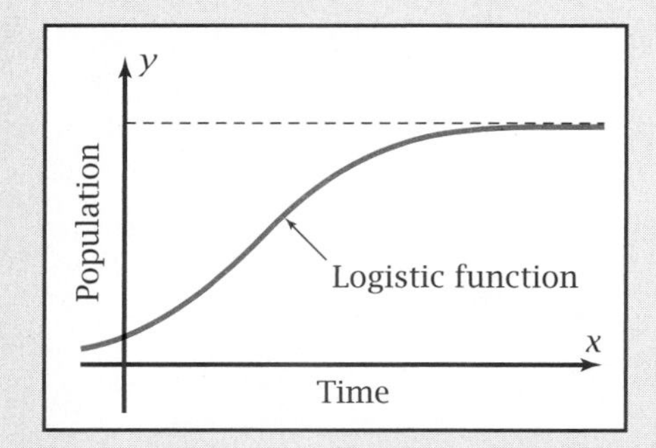

Verbally *I figured out that the icon at the top of each page is the graph of an exponential function. Exponential functions can describe unrestrained population growth, such as that of rabbits if they have no natural enemies. Logistic functions start out like exponential functions but then level off. Logistic functions can model restrained population growth where there is a maximum sustainable population in a certain region.*

7-1 Shapes of Function Graphs

In this chapter you'll learn ways to find a function to fit a real-world situation when the type of function has not been given. You will start by refreshing your memory about graphs of functions you studied in Chapter 1.

OBJECTIVE Discover patterns in linear, quadratic, power, and exponential function graphs.

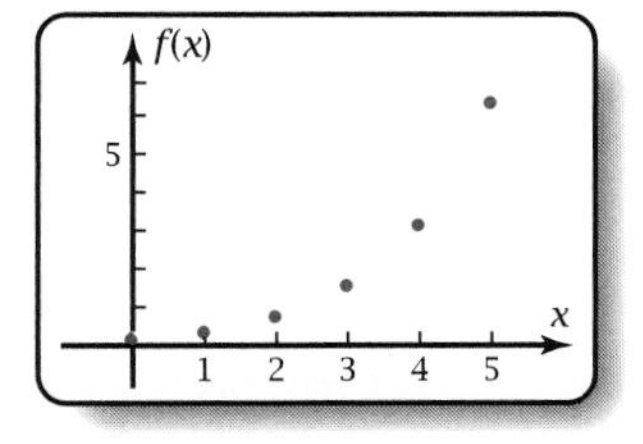

Figure 7-1a

Figure 7-1a shows the plot of points that lie on values of the exponential function $f(x) = 0.2 \times 2^x$. You can make such a plot by storing the x-values in one list and the $f(x)$-values in another and then using the statistics plot feature on your grapher. Figure 7-1b shows that the graph of f contains all of the points in the plot. The **concave** side of the graph faces upward.

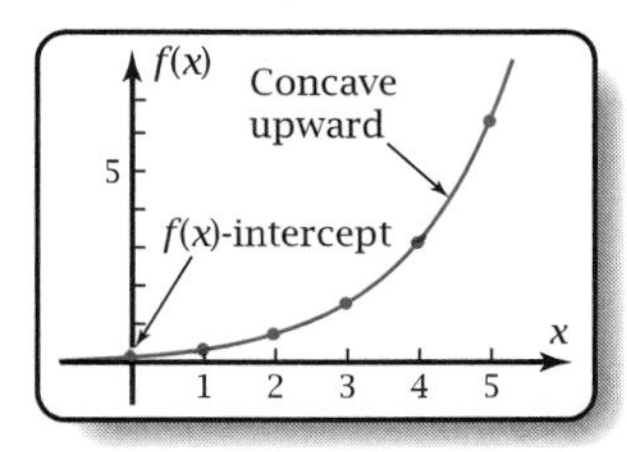

Figure 7-1b

Exploratory Problem Set 7-1

1. *Exponential Function Problem:* In the **exponential function** $f(x) = 0.2 \times 2^x$, $f(x)$ could be the number of thousands of bacteria in a culture as a function of time, x, in hours. Find $f(x)$ for each hour from 0 through 5. Plot the points, and graph the function as in Figure 7-1b. How do you interpret the concavity of the graph in terms of the rate at which the bacteria are growing?

2. *Power Function Problem:* In the **power function** $g(x) = 0.1x^3$, $g(x)$ could be the weight in pounds of a snake that is x feet long. Plot the points for each pound, from $x = 0$ through 6, and graph function g. Because graphs of f in Problem 1 and g in Problem 2 are both increasing and concave up, what graphical evidence could you use to distinguish between the two types of functions? Is the following statement true or false? "The snake's weight increases by the same amount for each foot it increases in length." Give evidence to support your answer.

3. *Quadratic Function Problem 1:* In the **quadratic function** $q(x) = -0.3x^2 + 8x + 7$, $q(x)$ could measure the approximate sales of a new product in the xth week since the product was introduced. Plot the points for every 5 weeks from $x = 0$ through 30, and graph function q. Which way is the concave side of the graph oriented, upward or downward? What feature does the quadratic function graph have that neither the exponential function graph in Problem 1 nor the power function graph in Problem 2 has?

4. *Linear Function Problem:* In the **linear function** $h(x) = 5x + 7$, $h(x)$ could equal the number of cents you pay for a telephone call that is x minutes long. Plot the points for every 3 minutes from $x = 0$ through 18, and graph function h. What does the fact that the graph is neither concave upward nor concave downward tell you about the cents per minute you pay for the call?

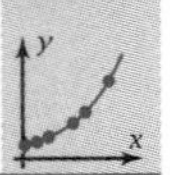

7-2 Identifying Functions from Graphical Patterns

One way to tell what type of function fits a set of points is by recognizing the properties of the graph.

OBJECTIVE Given the graph of a function, know whether the function is exponential, power, quadratic, or linear and find the particular equation algebraically.

Here is a brief review of the basic functions used in modeling. Some of these you have already encountered in Chapter 1.

Linear and Constant Functions

General equation: $y = ax + b$, where a and b are constants and the domain is all real numbers. This equation is in the **slope-intercept form.** (If $a = 0$, the function $y = b$ is a constant function.)

Parent function: $y = x$

Transformed function: $y - y_1 = a(x - x_1)$, called the **point-slope form.** The value y_1 is the vertical translation, a is the vertical dilation, and x_1 is the horizontal translation.

Graphical properties: Figure 7-2a shows that the graph is a straight line with **slope** a (sometimes m).

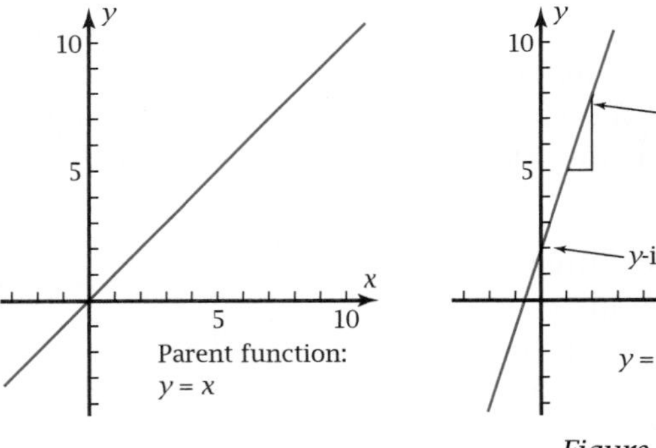

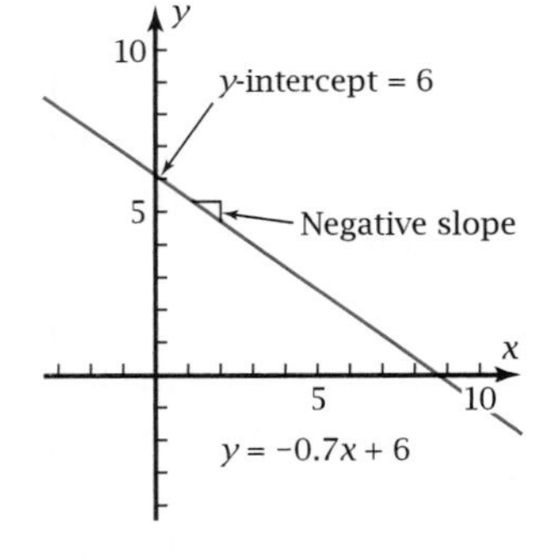

Figure 7-2a
Linear functions

Quadratic Functions

General equation: $y = ax^2 + bx + c$, where $a \neq 0$, a, b, and c are constants, and the domain is all real numbers.

Parent function: $y = x^2$, where the **vertex** is at the origin.

Transformed function: $y - k = a(x - h)^2$, called the **vertex form,** with vertex at (h, k). The value k is the vertical translation, h is the horizontal translation, and a is the vertical dilation.

Graphical properties: The graph is a **parabola** (Greek for "along the path of a ball"), as shown in Figure 7-2b. The graph is concave upward if $a > 0$ and concave downward if $a < 0$.

The eruption of Arenal, an active volcano in Costa Rica. The lava particles follow a parabolic path due to gravitational force.

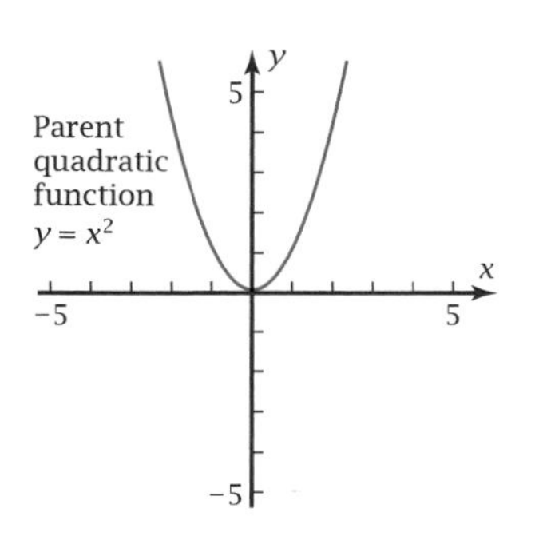

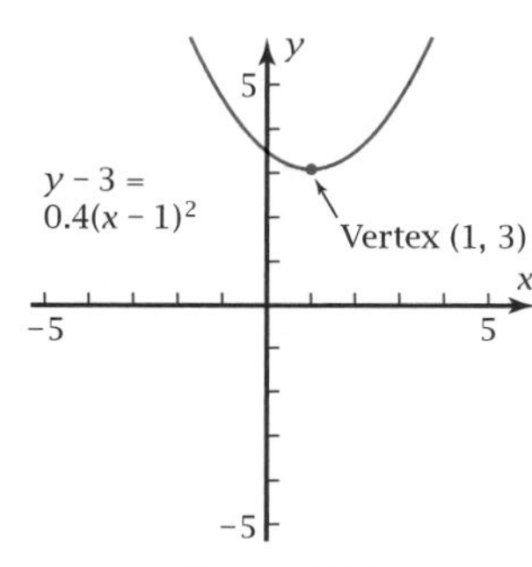

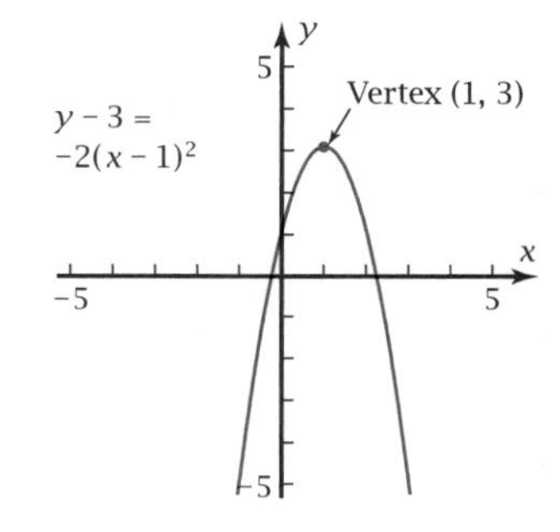

Figure 7-2b
Quadratic functions

Power Functions

General equation, untranslated form: $y = ax^b$, where a and b are nonzero constants and the domain depends on whether b is positive or negative. If b is positive, then the domain is all nonnegative real numbers; if b is negative, then the domain is all positive real numbers.

Parent function: $y = x^b$

Transformed function: $y = a(x - c)^b + d$. The value a is a vertical dilation; c and d are horizontal and vertical translations, respectively. The coefficient a is called the **proportionality constant.**

Verbally: In the equation $y = ax^b$, "y varies with the b power of x" or "y is proportional to the b power of x." If $b > 0$, the relationship is called **direct variation.** If $b < 0$, the variation is called **inverse variation.**

Graphical properties: Figure 7-2c shows three power function graphs for different signs and values of the exponent b. In all three cases, $a > 0$. The shape and concavity of the graph depend on the sign and value of b. The

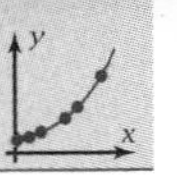

graph contains the origin if $b > 0$; it has the axes as asymptotes if $b < 0$. The function is increasing if $b > 0$; it is decreasing if $b < 0$. The graph is concave upward if $b > 1$ or if $b < 0$ and is concave downward if $0 < b < 1$.

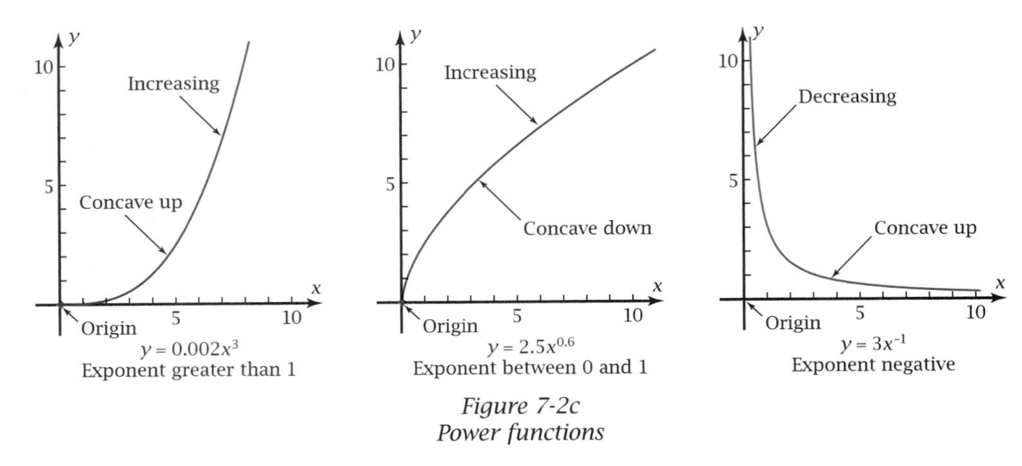

Figure 7-2c
Power functions

Exponential Functions

General equation: $y = ab^x$, where a and b are constants, $a \neq 0$, $b > 0$, $b \neq 1$, and the domain is all of the real numbers.

Parent function: $y = b^x$, where the asymptote is the x-axis.

Transformed function: $y = ab^x + c$, where the asymptote is the line $y = c$.

Verbally: In the equation $y = ab^x$, "y varies exponentially with x."

Graphical properties: Figure 7-2d shows some possible exponential functions that differ according to the values of a and b. The constant a is the y-intercept. The function is increasing if $b > 1$ and decreasing if $0 < b < 1$ (provided $a > 0$). If $a < 0$, the opposite is true. The graph is concave upward if $a > 0$ and concave downward if $a < 0$.

Marie Curie was awarded the Nobel Prize in Chemistry for the discovery of radioactive elements (polonium and radium) in 1911. The breakdown of radioactive elements follows an exponential function.

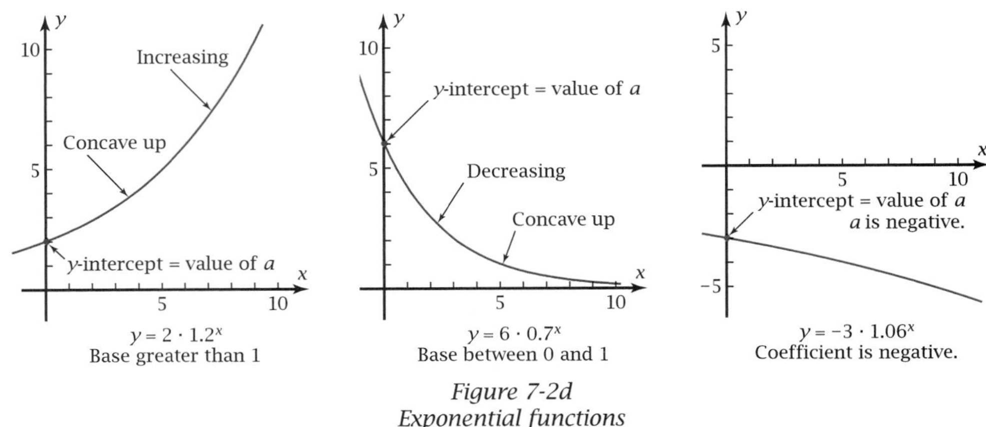

Figure 7-2d
Exponential functions

Mathematicians usually use one of two particular constants for the base of an exponential function. The bases picked are usually either 10, which is the base of the decimal system, or the naturally occuring number e, which equals

2.78128.... To make the equation more general, multiply the variable in the exponent by a constant. The (untranslated) general equations are given in this box.

DEFINITION: Special Exponential Functions

$y = a \cdot 10^{bx}$ **base-10 exponential function**

$y = a \cdot e^{bx}$ **natural (base-*e*) exponential function**

where a and b are constants and the domain is all real numbers.

Note: The equations of these two functions can be generalized by incorporating translations in the x- and y-directions. You'll get: $y = a \cdot 10^{b(x-c)} + d$ and $y = a \cdot e^{b(x-c)} + d$.

Base-e exponential functions have an advantage when you study calculus because the rate of change of e^x is equal to e^x.

▶ EXAMPLE 1 For the function graphed in Figure 7-2e,

a. Identify the kind of function it is.

b. On what interval or intervals is the function increasing or decreasing? Which way is the graph concave?

c. From your experience, describe something in the real world that a function with this shape of graph could model.

d. Find the particular equation for the function, given that points (5, 19) and (10, 6) are on the graph.

e. Confirm that your equation gives the graph in Figure 7-2e.

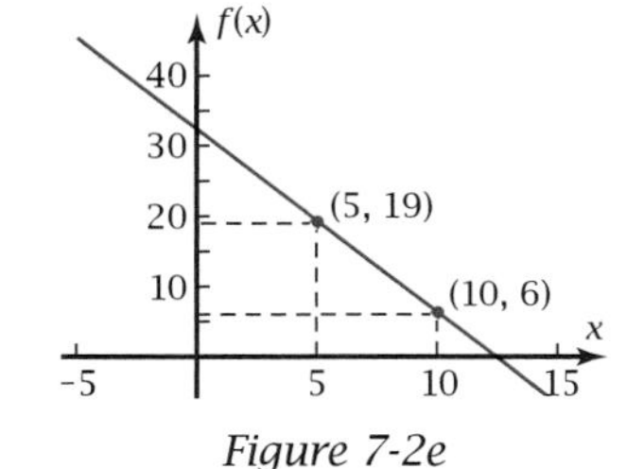

Figure 7-2e

Solution

a. Because the graph is a straight line, the function is linear.

b. The function is decreasing on its entire domain, and the graph is not concave in *either* direction.

c. The function could model anything that decreases at a constant rate. The first-quadrant part of the function could model the number of pages of history text you have left to read as a function of the number of minutes you have been reading.

d.

$f(x) = ax + b$ — Write the general equation. Use $f(x)$ as shown on the graph and a for the slope.

$\begin{cases} 19 = 5a + b \\ 6 = 10a + b \end{cases}$ — Substitute the given values of x and y into the equation of f.

$-13 = 5a \Rightarrow a = -2.6$ — Subtract the first equation from the second.

$6 = 10(-2.6) + b \Rightarrow b = 32$ — Substitute -2.6 for a in one of the equations.

$\therefore f(x) = -2.6x + 32$ — Write the particular equation.

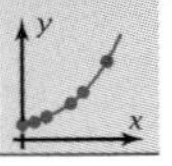

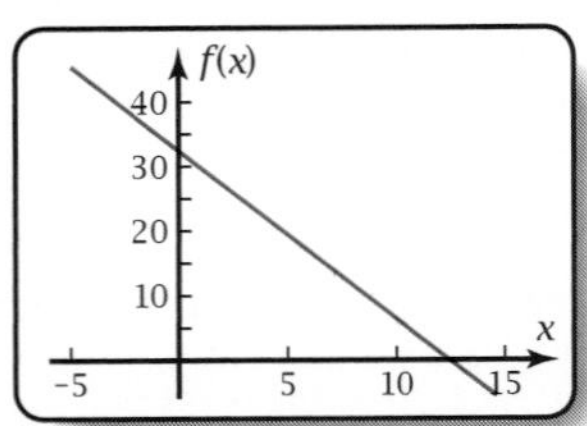

Figure 7-2f

e. Figure 7-2f shows the graph of f, which agrees with the given graph. Note that the calculated slope, −2.6, is *negative,* which corresponds to the fact that $f(x)$ *decreases* as x *increases.* ◀

Note that you could have solved the system of equations in Example 1 using matrices.

$$\begin{cases} 5a + b = 19 \\ 10a + b = 6 \end{cases}$$ The given system.

$$\begin{bmatrix} 5 & 1 \\ 10 & 1 \end{bmatrix}\begin{bmatrix} a \\ b \end{bmatrix} = \begin{bmatrix} 19 \\ 6 \end{bmatrix}$$ Write the system in matrix form.

$$\begin{bmatrix} a \\ b \end{bmatrix} = \begin{bmatrix} 5 & 1 \\ 10 & 1 \end{bmatrix}^{-1}\begin{bmatrix} 19 \\ 6 \end{bmatrix}$$ Multiply both sides by the inverse matrix.

$$= \begin{bmatrix} -2.6 \\ 32 \end{bmatrix}$$ Complete the matrix multiplication.

$a = -2.6$ and $b = 32$

You'll study the matrix solution of linear systems more fully in Section 11-2.

▶ **EXAMPLE 2** For the function graphed in Figure 7-2g,

a. Identify the kind of function it could be.

b. On what interval or intervals is the function increasing or decreasing, and is the graph concave upward or downward?

c. Describe something in the real world that a function with this shape of graph could model.

d. Find the particular equation for the function, given that points (1, 76), (2, 89), and (3, 94) are on the graph.

e. Confirm by plotting that your equation gives the graph in Figure 7-2g.

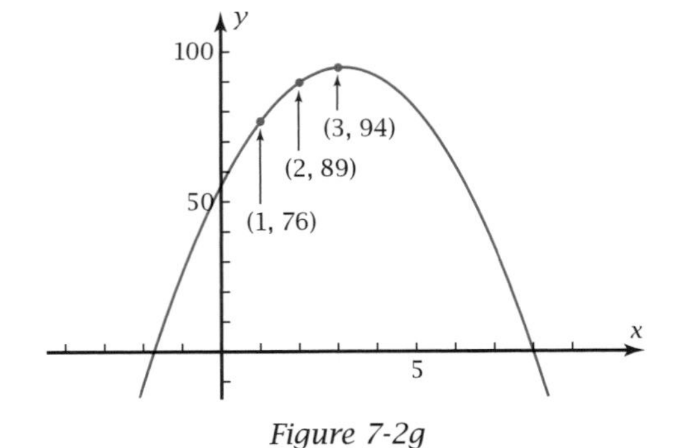

Figure 7-2g

Solution

a. The function could be quadratic because it has a vertex.

b. The function is increasing if $x < 3$ and decreasing if $x > 3$, and it is concave downward.

c. The function could model anything that rises to a maximum and then falls back again, such as the height of a ball as a function of time or the grade you could earn on a final exam as a function of how long you study for it. (Cramming too long might lower your score because of your being sleepy from staying up late!)

d. $y = ax^2 + bx + c$ Write the general equation.

$$\begin{cases} 76 = a + b + c \\ 89 = 4a + 2b + c \\ 94 = 9a + 3b + c \end{cases}$$ Substitute the given x- and y-values.

$$\begin{bmatrix} 1 & 1 & 1 \\ 4 & 2 & 1 \\ 9 & 3 & 1 \end{bmatrix}^{-1} \begin{bmatrix} 76 \\ 89 \\ 94 \end{bmatrix} = \begin{bmatrix} -4 \\ 25 \\ 55 \end{bmatrix}$$ Solve by matrices.

$y = -4x^2 + 25x + 55$ Write the equation.

e. Plotting the graph confirms that the equation is correct. Note that the value of a is negative, which corresponds to the fact that the graph is concave downward. ◀

▶ **EXAMPLE 3** For the function graphed in Figure 7-2h,

a. Identify the kind of function it could be.

b. On what interval or intervals is the function increasing or decreasing, and is the graph concave upward or downward?

c. Describe something in the real world that a function with this shape of graph could model.

d. Find the particular equation for the function you identified in part a, given that points (4, 44.8) and (6, 151.2) are on the graph.

e. Confirm that your equation gives the graph in Figure 7-2h.

Figure 7-2h

Solution

a. The function could be a power function or an exponential function, but a power function is chosen because the graph appears to contain the origin, which exponential functions don't do unless they are translated in the y-direction.

b. The function is increasing on its entire domain, and the graph is concave upward everywhere.

c. The function could model anything that starts at zero and increases at an increasing rate, such as the distance it takes a car to stop as a function of its speed when the driver applies the brakes, or the volumes of geometrically similar objects as a function of their lengths.

d. $y = ax^b$ — Write the untranslated general equation.

$\begin{cases} 44.8 = a \cdot 4^b \\ 151.2 = a \cdot 6^b \end{cases}$ — Substitute the x- and y-values into the equation.

$\frac{151.2}{44.8} = \frac{a \cdot 6^b}{a \cdot 4^b}$ — Divide the second equation by the first to eliminate a.

$3.375 = 1.5^b$ — The a cancels, and $\frac{6^b}{4^b} = \left(\frac{6}{4}\right)^b = 1.5^b$

$\log 3.375 = \log 1.5^b$ — Take the logarithm of both sides to get the variable out of the exponent.

$\log 3.375 = b \log 1.5$

$b = \frac{\log 3.375}{\log 1.5} = 3$

$44.8 = a \cdot 4^3$ — Substitute 3 for b in one of the equations.

$a = \frac{44.8}{4^3} = 0.7$

$\therefore y = 0.7x^3$ — Write the particular equation.

e. Plotting the graph confirms that the equation is correct. Note that the value of b is greater than 1, which corresponds to the fact that the graph is concave upward. ◀

▶ **EXAMPLE 4** For the function graphed in Figure 7-2i,

a. Identify the kind of function it could be.

b. On what interval or intervals is the function increasing or decreasing, and is the graph concave upward or downward?

c. Describe something in the real world that a function with this shape of graph could model.

d. Find the particular equation for the function, given that points (2, 10) and (5, 6) are on the graph.

e. Confirm that your equation gives the graph in Figure 7-2i.

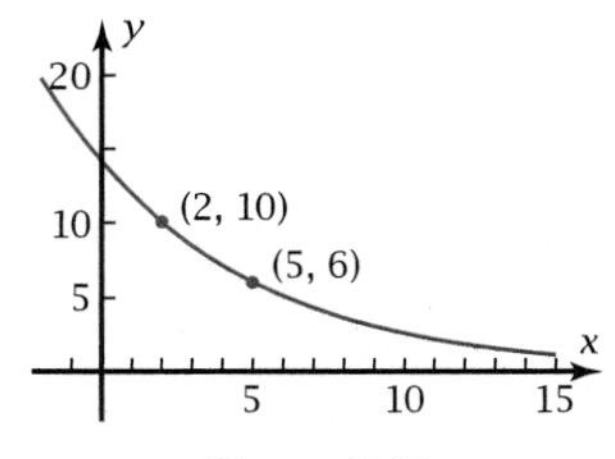

Figure 7-2i

Solution

a. The function could be exponential or quadratic, but exponential is picked because the graph appears to approach the x-axis asymptotically.

b. The function is decreasing and concave upward on its entire domain.

c. The function could model any situation in which a variable quantity starts at some nonzero value and coasts downward, gradually approaching zero, such as the number of degrees a cup of coffee is above room temperature as a function of time since it started cooling.

d. $y = ab^x$ — Write the untranslated general equation.

$\begin{cases} 10 = ab^2 \\ 6 = ab^5 \end{cases}$ — Substitute the given x- and y-values into the equation.

$\dfrac{6}{10} = \dfrac{ab^5}{ab^2}$ — Divide the second equation by the first to eliminate a.

$0.6 = b^3$

$0.6^{1/3} = b$ — Raise both sides to the 1/3 power to eliminate the exponent of b.

$b = 0.8434\ldots$

$10 = a(0.8434\ldots)^2$ — Substitute 0.8434... for b in one of the equations.

$a = 14.0572\ldots$

$\therefore y = 14.0572\ldots(0.8434\ldots)^x$ — Write the equation.

e. Plotting the graph confirms that the equation is correct. Note that the value of b is between 0 and 1, which corresponds to the fact that the function is decreasing. ◀

Problem Set 7-2

Do These Quickly

Q1. If $f(x) = x^2$, find $f(3)$.

Q2. If $f(x) = x^2$, find $f(0)$.

Q3. If $f(x) = x^2$, find $f(-3)$.

Q4. If $g(x) = 2^x$, find $g(3)$.

Q5. If $g(x) = 2^x$, find $g(0)$.

Q6. If $g(x) = 2^x$, find $g(-3)$.

Q7. If $h(x) = x^{1/2}$, find $h(25)$.

Q8. If $h(x) = x^{1/2}$, find $h(0)$.

Q9. If $h(x) = x^{1/2}$, find $h(-9)$.

Q10. What axiom for real numbers is illustrated by $3(x + 5) = 3(5 + x)$?

A. Associative axiom for multiplication

B. Commutative axiom for multiplication

C. Associative axiom for addition

D. Commutative axiom for addition

E. Distributive axiom for multiplication over addition

1. Power functions and exponential functions both have exponents. What major algebraic difference distinguishes these two types of functions?

2. What geometrical feature do quadratic function graphs have that linear, exponential, and power function graphs do not have?

3. Write a sentence or two giving the origin of the word *concave* and how the word applies to graphs of functions.

4. Explain why direct-variation power functions contain the origin but inverse-variation power functions do not.

5. Explain why the **reciprocal function** $f(x) = \frac{1}{x}$ is also a power function.

6. In the definition of quadratic function, what is the reason for the restriction $a \neq 0$?

7. The definition of exponential function, $y = ab^x$, includes the restriction $b > 0$. Suppose that $y = (-64)^x$. What would y equal if $x = \frac{1}{2}$? If $x = \frac{1}{3}$? Why do you think there is the restriction $b > 0$ for exponential functions?

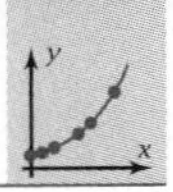

8. *Reading Problem:* Clara has been reading her history assignment for 20 minutes and is now on page 56 in the text. She reads at a (relatively) constant rate of 0.6 pages per minute.
   a. Find the particular equation for the page number she is on as a function of minutes using the point-slope form. Transform your answer to the slope-intercept form.
   b. Which page was Clara on when she started reading the assignment?
   c. The assignment ends on page 63. When would you expect Clara to finish?

9. *Baseball Problem:* Ruth hits a high fly ball to right field. The ball was 4 ft above the ground when she hit it. Three seconds later it reaches its maximum height, 148 ft.

   a. Write an equation in vertex form for the quadratic function expressing the relationship between the height of the ball and time. Transform it so that height is expressed as a function of time.
   b. How high was the ball 5 sec after it was hit?
   c. If nobody catches the ball, how many seconds after it was hit will it reach the ground?

For Problems 10–19, the first-quadrant part of a function graph is shown.

   a. Identify the type of function it could be.
   b. On what interval or intervals is the function increasing or decreasing, and which way is the graph concave?
   c. From your experience, what relationship in the real world could be modeled by a function with this shape of graph?
   d. Find the particular equation for the function if the given points are on the graph.
   e. Confirm that your equation gives the graph shown.

10. 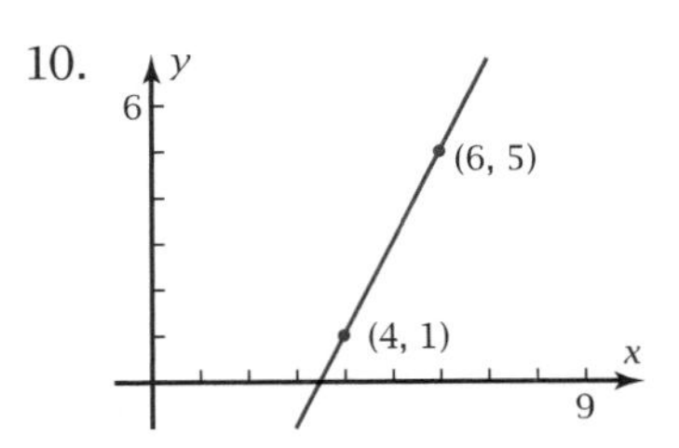

11. 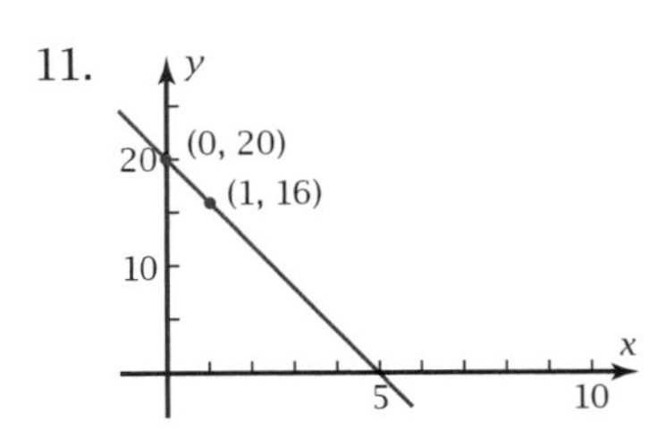

12. 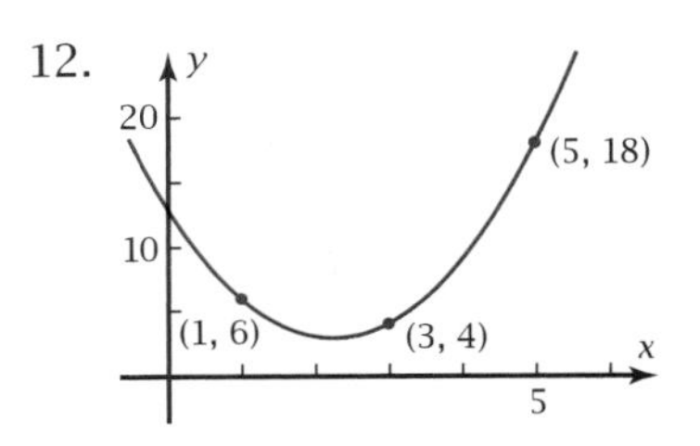

13. 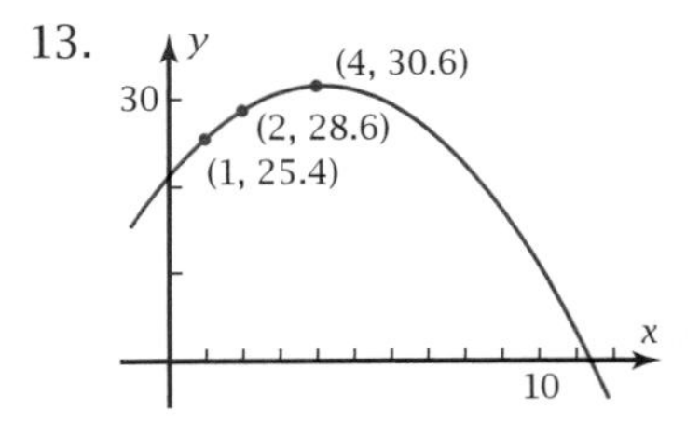

14. 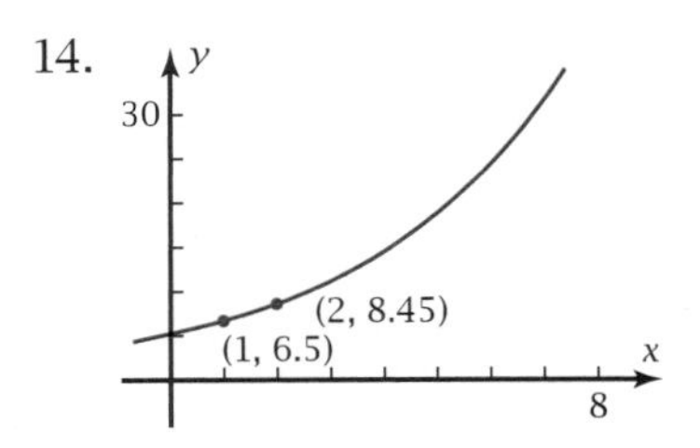

15. 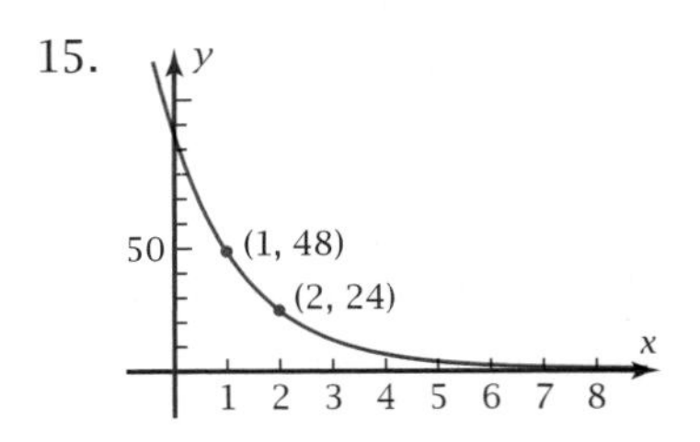

16.

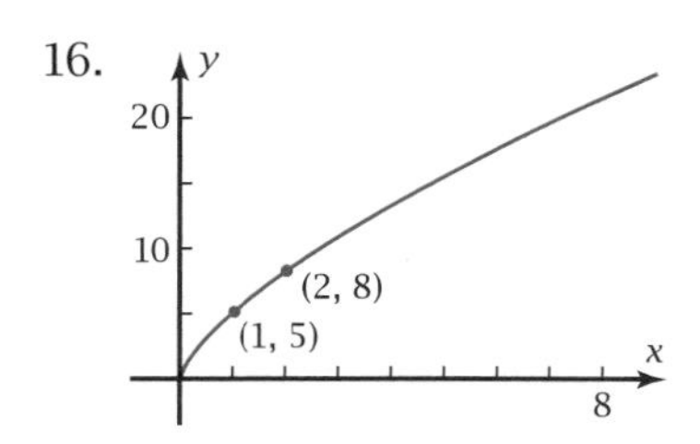

17.

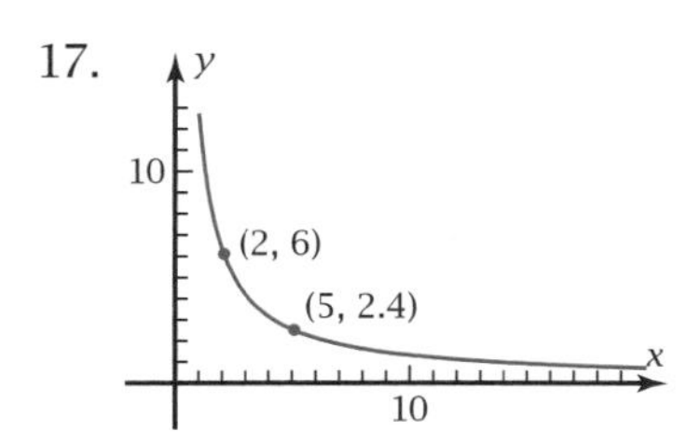

18.

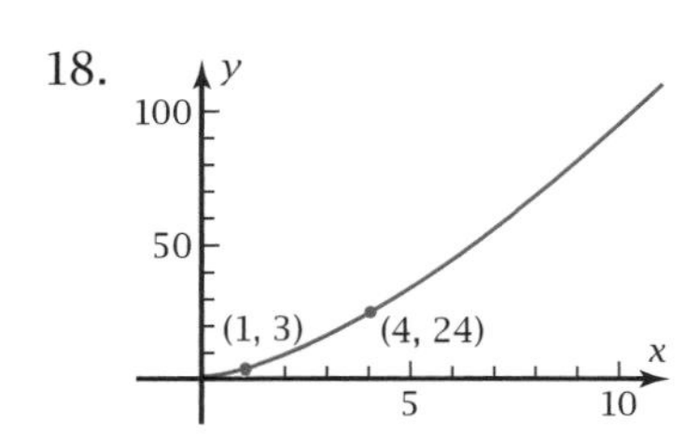

19.

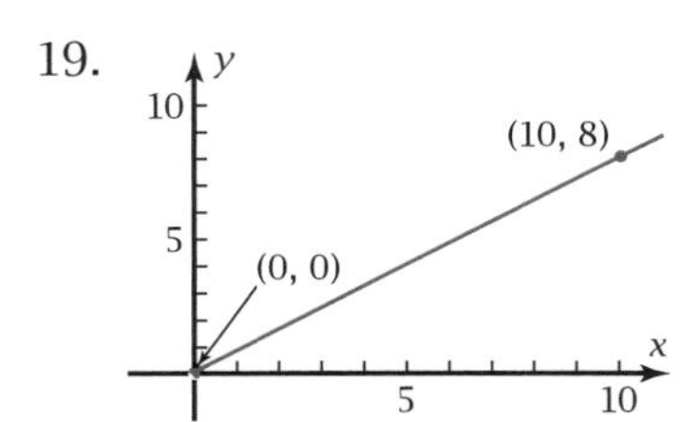

20. Suppose that y increases exponentially with x and that z is directly proportional to the square of x. Sketch the graph of each type of function. In what ways are the two graphs similar to one another? What major graphical difference would allow you to tell which graph is which if they were not marked?

21. Suppose that y decreases exponentially with x and that z varies inversely with x. Sketch the graph of each type of function. Give at least three ways in which the two graphs are similar to one another. What major graphical difference would allow you to tell which graph is which if they were not marked?

22. Suppose that y varies directly with x and that z increases linearly with x. Explain why any direct-variation function is a linear function but a linear function is not necessarily a direct-variation function.

23. Suppose that y varies directly with the square of x and that z is a quadratic function of x. Explain why the direct-square-variation function is a quadratic function but the quadratic function is not necessarily a direct-square-variation function.

24. *Natural Exponential Function Problem:* Figure 7-2j shows the graph of the natural exponential function $f(x) = 3e^{0.8x}$. Let $g(x) = 3b^x$. Find the value of b for which $g(x) = f(x)$. Show graphically that the two functions are equivalent.

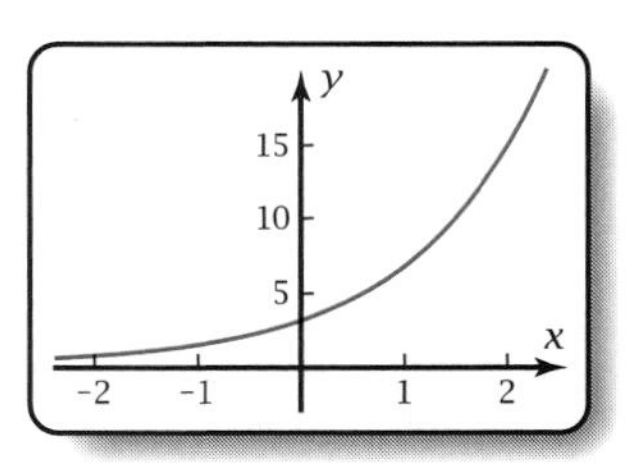

Figure 7-2j

7-3 Identifying Functions from Numerical Patterns

A 16-inch pizza has four times as much area as an 8-inch pizza. A grapefruit whose diameter is 10 cm has eight times the volume of a grapefruit with a 5-cm diameter. In general, when you double the linear dimensions of a three-dimensional object, you multiply the surface area by 4 and the volume by 8.

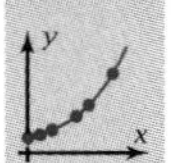

This is an example of the multiply–multiply property of power functions. It is similar to the add–add property for linear functions. Every time you add 1000 miles to the distance you have driven your car, you add a constant amount, say, \$300, to the cost of operating that car.

In this section you will use such patterns to identify the type of function that fits a given set of function values. Then you will find more function values, either by following the pattern or by finding and using the particular equation of the function.

OBJECTIVES

- Given a set of regularly spaced *x*-values and the corresponding *y*-values, identify which type of function they fit (linear, quadratic, power, or exponential).
- Find other function values without necessarily finding the particular equation.

The Add–Add Pattern of Linear Functions

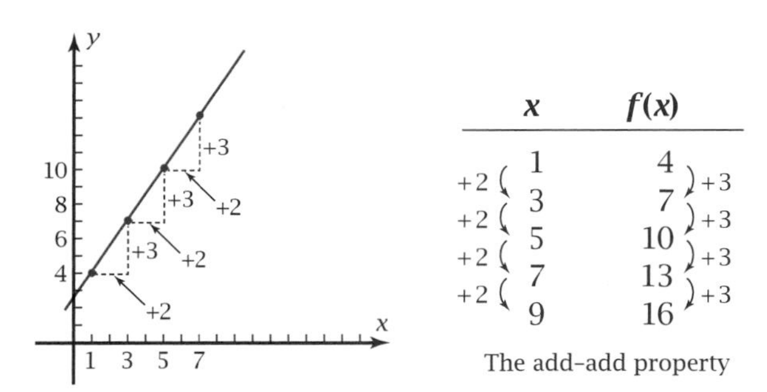

	x	*f*(*x*)	
	1	4	
+2	3	7	+3
+2	5	10	+3
+2	7	13	+3
+2	9	16	+3

The add–add property

Figure 7-3a

Figure 7-3a shows the graph of the linear function $f(x) = 1.5x + 2.5$. As you can see from the graph and the adjacent table, each time you add 2 to x, y increases by 3. This pattern emerges because a linear function has constant slope. Verbally, you can express this property by saying that every time you add a constant to x, you add a constant (not necessarily the same) to y. This property is called the **add–add property** of linear functions.

The Add–Multiply Pattern of Exponential Functions

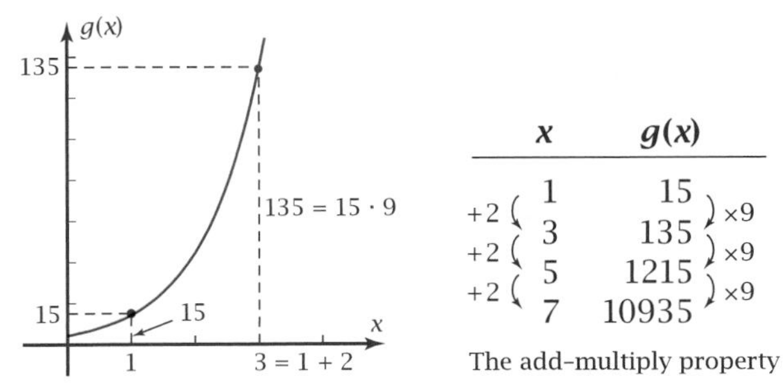

Figure 7-3b

Figure 7-3b shows the exponential function $g(x) = 5 \cdot 3^x$. This time, adding 2 to x results in the corresponding $g(x)$-values being *multiplied* by the constant, 9. This is not coincidental, and here's why the pattern is true.

$g(1) = 5 \cdot 3^1 = 15$

$g(3) = 5 \cdot 3^3 = 135$ (which equals 9 times 15)

You can see algebraically why this is true.

$g(3) = 5 \cdot 3^3$

$= 5 \cdot 3^{1+2}$ Write the exponent as 1 increased by 2.

$= 5 \cdot 3^1 \cdot 3^2$ Product of powers with equal bases property.

$= (5 \cdot 3^1) \cdot 3^2$ Associate 5 and 3^1 to bring $g(1)$ into the expression.

$= 9\, g(1)$

The conclusion is that if you add a constant to x, the corresponding y-value is multiplied by the base raised to that constant. This is called the **add–multiply property** of exponential functions.

The Multiply–Multiply Pattern of Power Functions

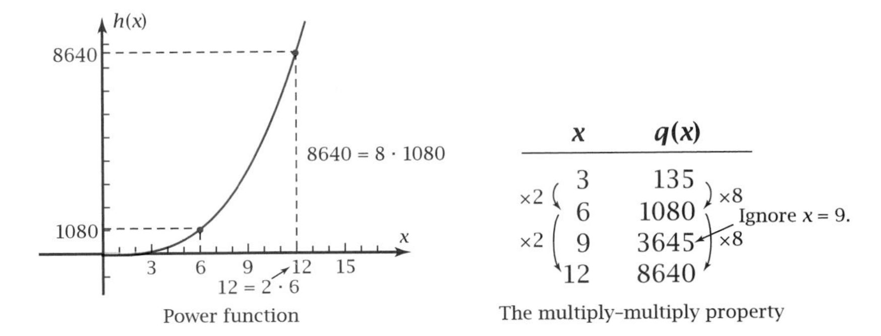

Figure 7-3c

Figure 7-3c shows the graph of the power function $h(x) = 5x^3$. As shown in the table, adding a constant, 3, to x does *not* create a corresponding pattern.

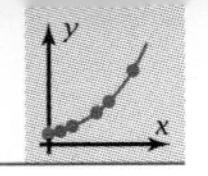

However, a pattern *does* emerge if you pick values of x that change by being *multiplied* by a constant.

$$h(3) = 5 \cdot 3^3 = 135$$

$$h(6) = 5 \cdot 6^3 = 1080 \qquad \text{(which equals } 8 \cdot 135\text{)}$$

$$h(12) = 5 \cdot 12^3 = 8640 \qquad \text{(which equals } 8 \cdot 1080\text{)}$$

If you double the x-value from 3 to 6 or from 6 to 12, the corresponding y-values get multiplied by 8, or 2^3. You can see algebraically why this is true.

$$h(6) = 5 \cdot 6^3$$

$= 5(2 \cdot 3)^3$ — Write the $x = 6$ as twice 3.

$= (5 \cdot 3^3) \cdot 2^3$ — Distribute the power over multiplication and then commute and associate.

$= 8\, h(3)$

In conclusion, if you multiply the x-values by 2, the corresponding y-values are multiplied by 8. This is called the **multiply–multiply property** of power functions. Note that extra points may appear in the table, such as (9, 3645) in Figure 7-3c. They do belong to the function, but the x-values do not fit the "multiply" pattern.

The Constant-Second-Differences Pattern for Quadratic Functions

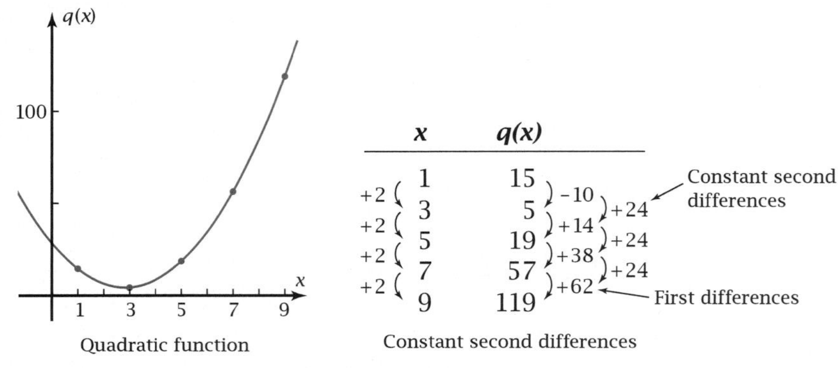

Figure 7-3d

Figure 7-3d shows the quadratic function $q(x) = 3x^2 - 17x + 29$. Because the function decreases and then increases, none of the add and multiply properties already described will apply. However, an extension of the add–add property for linear functions does apply to quadratics. For equally spaced x-values, the *differences* between the corresponding y-values are equally spaced. Thus the differences between these differences (the **second differences**) are constant. This constant is equal to $2ad^2$, twice the coefficient of the quadratic term times the square of the difference between the x-values.

These four properties are summarized in this box.

PROPERTIES: Patterns for Function Values

Add–Add Property of Linear Functions

If f is a linear function, adding a constant to x results in adding a constant to the corresponding $f(x)$-value. That is,

$$\text{if } f(x) = ax + b \text{ and } x_2 = c + x_1, \text{ then } f(x_2) = ac + f(x_1)$$

Multiply–Multiply Property of Power Functions

If f is a power function, multiplying x by a constant results in multiplying the corresponding $f(x)$-value by a constant. That is,

$$\text{if } f(x) = ax^b \text{ and } x_2 = cx_1, \text{ then } f(x_2) = c^b \cdot f(x_1)$$

Add–Multiply Property of Exponential Functions

If f is an exponential function, adding a constant to x results in multiplying the corresponding $f(x)$-value by a constant. That is,

$$\text{if } f(x) = ab^x \text{ and } x_2 = c + x_1, \text{ then } f(x_2) = b^c \cdot f(x_1)$$

Constant-Second-Differences Property of Quadratic Functions

If f is a quadratic function, $f(x) = ax^2 + bx + c$, and the x-values are spaced d units apart, then the second differences between the $f(x)$-values are constant and equal to $2ad^2$.

▶ EXAMPLE 1 Identify the pattern in the following function values and the kind of function that has this pattern.

x	$f(x)$
4	5
5	7
6	11
7	17
8	25

Solution The values have neither the add–add, add–multiply, nor multiply–multiply property. They do exhibit the constant-second-differences property, as shown next. Therefore, a quadratic function fits the data.

	x	$f(x)$	First differences	Second differences
	4	5		
+1			+2	
	5	7		+2
+1			+4	
	6	11		+2
+1			+6	
	7	17		+3
+1			+8	
	8	25		

◀

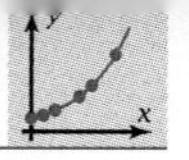

▶ EXAMPLE 2 If function f has values $f(5) = 12$ and $f(10) = 18$, find $f(20)$ if f is

a. A linear function

b. A power function

c. An exponential function

Solution

a. Linear functions have the add–add property. Notice that you add 5 to the first x-value to get the second one and that you add 6 to the first $f(x)$-value to get the second one. Make a table of values ending at $x = 20$. The answer is $f(20) = 30$.

	x	$f(x)$	
	5	12	
+5	10	18	+6
+5	15	24	+6
+5	20	30	+6

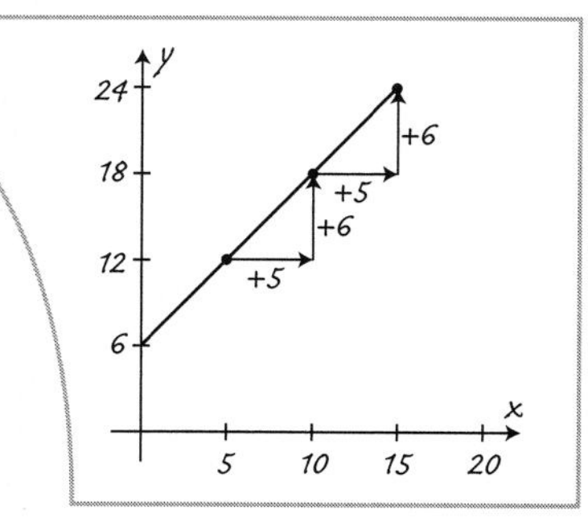

b. Power functions have the multiply–multiply property. When going from the first to the second x- and y-values, notice that you multiply 5 by 2 to get 10 and that you multiply 12 by 1.5 to get 18. Make a table of values ending at $x = 20$. The answer is $f(20) = 27$.

	x	$f(x)$	
	5	12	
×2	10	18	×1.5
×2	20	27	×1.5

c. Exponential functions have the add–multiply property. Notice that adding 5 to x results in the corresponding y-value being multiplied by 1.5. Make a table of values ending at $x = 20$. The answer is $g(20) = 40.5$.

	x	$f(x)$	
	5	12	
+5	10	18	×1.5
+5	15	27	×1.5
+5	20	40.5	×1.5

◀

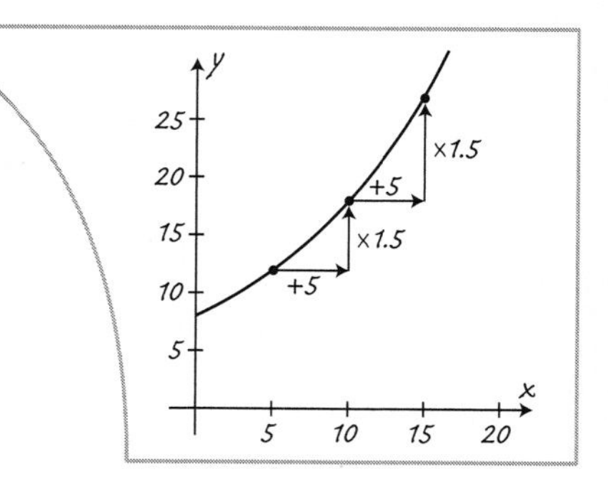

▶ EXAMPLE 3 Describe the effect on y of doubling x if

a. y varies directly with x.

b. y varies inversely with the square of x.

c. y varies directly with the cube of x.

Solution

a. y is doubled (that is, multiplied by 2^1).

b. y is multiplied by $\frac{1}{4}$ (that is, multiplied by 2^{-2}).

c. y is multiplied by 8 (that is, multiplied by 2^3). ◀

▶ **EXAMPLE 4** Suppose that f is a direct square power function and that $f(5) = 1000$. Find $f(20)$.

Solution Since f is a power function, it has the multiply–multiply property. Express $x = 20$ as $4 \cdot 5$. Multiplying x by 4 will multiply the corresponding y by 4^2, so

$$f(20) = f(4 \cdot 5) = 4^2 \cdot f(5) = 16 \cdot 1000 = 16{,}000$$ ◀

▶ **EXAMPLE 5** *Radioactive Tracer Problem:* The compound 18-fluorodeoxyglucose (18-FDG) is composed of radioactive fluorine (18-F) and a sugar (deoxyglucose). It is used to trace glucose metabolism in the heart. 18-F has a half-life of about 2 hours, which means that at the end of each 2-hour time period, only half of the 18-F that was there at the beginning of the time period remains. Suppose a dose of 18-FDG was injected into a patient. Let $f(x)$ be the number of *microcuries* (mCi) of 18-FDG that remain over time, x, in hours, as shown in this table.

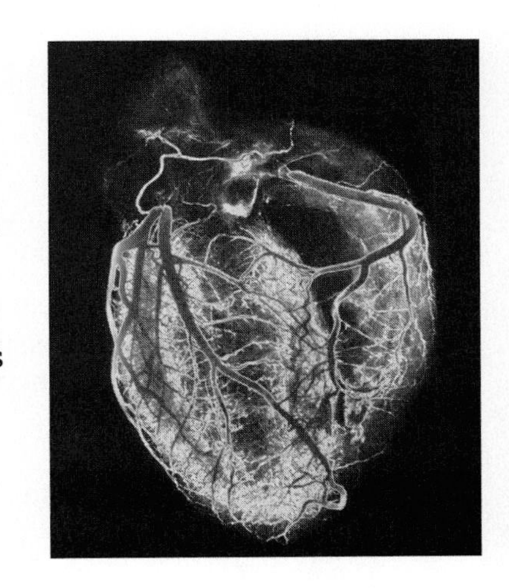

x (hr)	$f(x)$ (mCi)
2	5
4	2.5
6	1.25

a. Find the number of microcuries that remain after 12 hours.

b. Identify the pattern these data points follow. What type of function shows this pattern?

c. Why can't you use the pattern to find $f(25)$?

d. Find a particular equation for $f(x)$. Make a plot to show that all of the $f(x)$-values in the table satisfy the equation.

e. Use the equation to calculate $f(25)$. Interpret the answer.

Solution

a. 8, 10, 12 — Follow the add pattern in the x-values until you reach 12.

0.625, 0.3125, 0.15625 — Follow the multiply pattern in the corresponding y-values.

$f(12) = 0.15625$

b. The data points have the add–multiply property of exponential functions.

c. 22, 24, 26 — Extend the add pattern in the x-values.

The x-values skip over 25, so $f(25)$ cannot be found using the pattern.

d. $f(x) = ab^x$ — General equation of an exponential function.

$\begin{cases} 5 = ab^2 \\ 2.5 = ab^4 \end{cases}$ — Substitute any two of the ordered pairs.

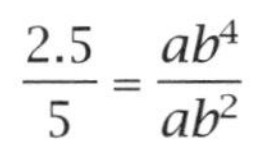

$$\frac{2.5}{5} = \frac{ab^4}{ab^2}$$ Divide the equations. Have the larger exponent in the numerator.

$0.5 = b^2$ Simplify.

$0.5^{1/2} = b$ Raise both sides to the $\frac{1}{2}$ power.

$b = 0.7071\ldots$ Store as b, without round-off.

$5 = a(0.7071\ldots)^2$ Substitute the value for b in one of the equations.

$$a = \frac{5}{0.7071\ldots^2} = 10$$

$\therefore f(x) = 10(0.7071\ldots)^x$ Write the particular equation.

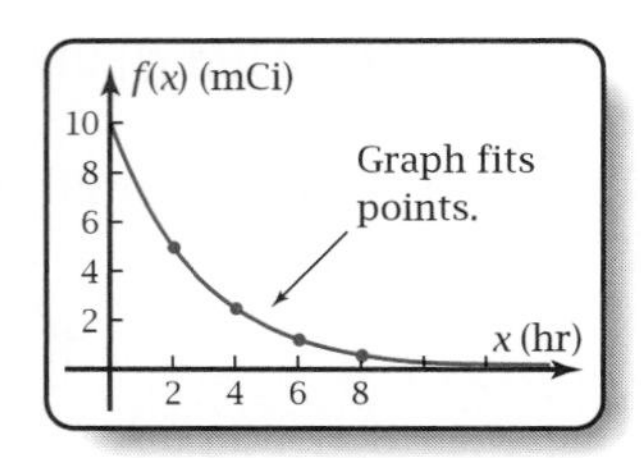

Figure 7-3e

Figure 7-3e shows the graph of f passing through all four given points.

e. $f(25) = 10(0.7071\ldots)^{25} = 0.0017\ldots$

This means that there were about 0.0017 microcuries of 18-FDG after 25 hours. ◀

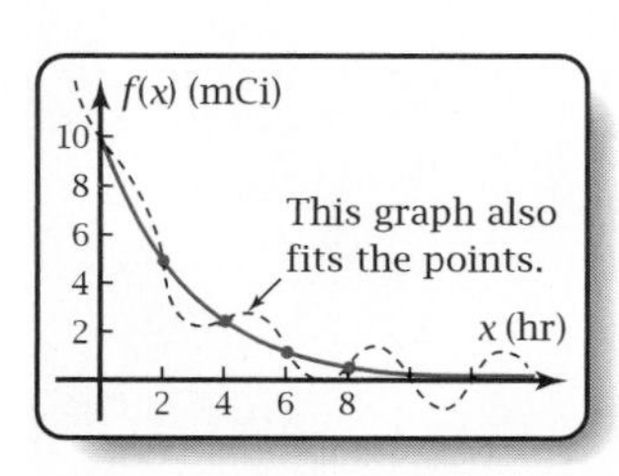

Figure 7-3f

Note that in Example 5, part d calls for "a" function that fits the points. It is possible for other functions to fit the same set of points. For instance, the function

$$g(x) = 10(0.7071\ldots)^x + \sin \tfrac{\pi}{2}x$$

also fits the given points, as shown in Figure 7-3f.

Deciding which function fits better will depend on the situation you model and its theoretical background. Also, you can further test to see whether your model is supported by data. For example, to test whether the second model is correct, you could collect measurements on shorter time intervals and see if there is a wavy pattern in the data.

Problem Set 7-3

Do These Quickly

Q1. Write the general equation for a linear function.

Q2. Write the general equation for a power function.

Q3. Write the general equation for an exponential function.

Q4. Write the general equation for a quadratic function.

Q5. $f(x) = 3 \cdot x^5$ is the equation of a particular —?— function.

Q6. $f(x) = 3 \cdot 5^x$ is the equation of a particular —?— function.

Q7. Name the transformation of $f(x)$ that gives $g(x) = 4 \cdot f(x)$.

Q8. The period of the trigonometric function $y = 3 + 4 \cos \frac{\pi}{5}(x - 6)$ is

A. 10 B. 6 C. 5 D. 4 E. 3

Q9. Sketch the graph of a linear function with negative slope and positive y-intercept.

Q10. Sketch the graph of an exponential function with base greater than 1.

For Problems 1–12, determine whether the data has the add-add, add-multiply, multiply-multiply, or constant-second-differences pattern. Identify the type of function that has the pattern.

1.

x	$f(x)$
2	2700
4	2400
6	2100
8	1800
10	1500

2.

x	$f(x)$
2	1500
4	750
6	500
8	375
10	300

3.

x	$f(x)$
2	12
4	48
6	108
8	192
10	300

4.

x	$f(x)$
2	12
4	48
6	192
8	768
10	3072

5.

x	$f(x)$
2	26
4	52
6	78
8	104
10	130

6.

x	$f(x)$
2	4.6
4	6.0
6	7.4
8	8.8
10	10.2

7.

x	$f(x)$
2	1800
4	450
6	200
8	112.5
10	72

8.

x	$f(x)$
2	400
4	100
6	−200
8	−500
10	−800

9.

x	$f(x)$
2	900
4	100
6	11.1111...
8	1.2345...
10	0.1371...

10.

x	$f(x)$
2	5.6
4	44.8
6	151.2
8	358.4
10	700.0

11.

x	$f(x)$
1	352
3	136
5	64
7	136
9	352

12.

x	$f(x)$
1	25
5	85
9	113
13	109
17	73

For Problems 13–16, find the indicated function value if f is

a. A linear function

b. A power function

c. An exponential function

13. Given $f(2) = 5$ and $f(6) = 20$, find $f(18)$.

14. Given $f(3) = 80$ and $f(6) = 120$, find $f(24)$.

15. Given $f(10) = 100$ and $f(20) = 90$, find $f(40)$.

16. Given $f(1) = 1000$ and $f(3) = 100$, find $f(9)$.

For Problems 17–20, use the given values to calculate the other values specified.

17. Given f is a linear function with $f(2) = 1$ and $f(5) = 7$, find $f(8)$, $f(11)$, and $f(14)$.

18. Given f is a direct cube power function with $f(3) = 0.7$, find $f(6)$ and $f(12)$.

19. Given that $f(x)$ varies inversely with the square of x and that $f(5) = 1296$, find $f(10)$ and $f(20)$.

20. Given that $f(x)$ varies exponentially with x and that $f(1) = 100$ and $f(4) = 90$, find $f(7)$, $f(10)$, and $f(16)$.

For Problems 21–24, describe the effect on y if you double the value of x.

21. Direct square power function

22. Direct fourth power function

23. Inverse-variation power function

24. Inverse-square-variation power function

25. *Volume Problem:* The volumes of similarly shaped objects are directly proportional to the cube of a linear dimension.

Baseball

Volleyball

Figure 7-3g

a. Recall from geometry that the volume, V, of a sphere equals $\frac{4}{3}\pi r^3$, where r is the radius. Explain how the $V = \frac{4}{3}\pi r^3$ formula shows that the volume of a sphere varies directly with the cube of the radius. If a baseball has

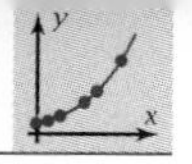

a volume of 100 cm^3, what is the volume of a volleyball that has twice the radius (Figure 7-3g)?

b. King Kong is depicted as having the same proportions as a normal gorilla but as being 10 times as tall. How would his volume (and thus weight) compare to that of a normal gorilla? If a normal gorilla weighs 500 pounds, what would you expect King Kong to weigh? Surprising?!

c. A great white shark 20 ft long weighs about 2000 lb. Fossilized sharks' teeth from millions of years ago suggest that there were once great whites 100 ft long. How much would you expect such a shark to weigh?

d. Gulliver traveled to Lilliput, where people were $\frac{1}{10}$ as tall as normal people. If Gulliver weighed 200 lb, how much would you expect a Lilliputian to weigh?

Iris Weddell White's illustration, The Emperor Visits Gulliver, *in Jonathan Swift's* Gulliver's Travels. *(The Granger Collection, New York)*

26. *Area Problem:* The areas of similarly shaped objects are directly proportional to the square of a linear dimension.

a. Give the formula for the area of a circle. Explain why the area varies directly with the square of the radius.

b. If a grapefruit has twice the diameter of an orange, how do the areas of their rinds compare?

c. When Gutzon Borglum designed the reliefs he carved into Mount Rushmore in South Dakota, he started with models $\frac{1}{12}$ the lengths of the actual reliefs. How does the area of each model compare to the area of each of the final reliefs? Explain why a relatively small decrease in the linear dimension results in a relatively large decrease in the surface areas to be carved?

d. Gulliver traveled to Brobdingnag, where people were 10 times as tall as normal people. If Gulliver had 2 m^2 of skin, how much skin surface would you expect a Brobdingnagian to have had?

27. *Airplane Weight and Area Problem:* In 1896, Samuel Langley successfully flew a model of an airplane he was designing. In 1903, he tried unsuccessfully to fly the full-sized airplane. Assume that the full-sized plane was 4 times the length of the model (Figure 7-3h).

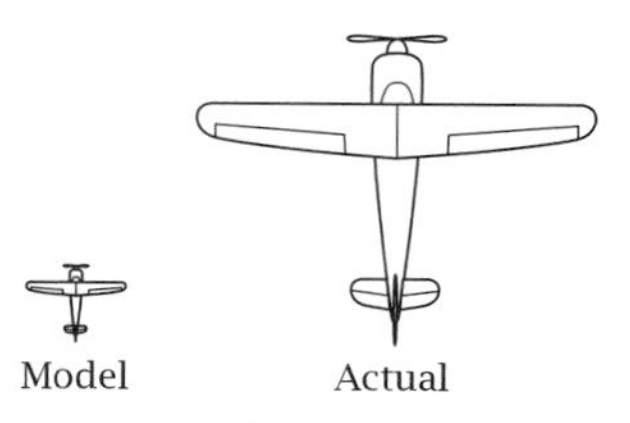

Figure 7-3h

a. The wing area, and thus the lift, of similarly shaped airplanes is directly proportional to the square of the length of each plane. How many times more wing area did the full-sized plane have than the model?

b. The volume, and thus the weight, of similarly shaped airplanes is directly proportional to the cube of the length. How many times heavier was the full-sized plane than the model?

c. Why do you think the model was able to fly but the full-sized plane was not?

28. *Compound Interest Problem 1:* Money left in a savings account grows exponentially with time. Suppose that you invest \$1000 and find that a year later you have \$1100.

a. How much will you have after 2 years? 3 years? 4 years?

b. In how many years will your investment double?

29. *Archery Problem 2:* Ann Archer shoots an arrow into the air. This table lists its heights at various times after she shoots it.

Time (sec)	Height (ft)
1	79
2	121
3	131
4	109
5	55

a. Show that the second differences between consecutive height values in the table are constant.

b. Use the first three points to find the particular equation of the quadratic function that fits these points. Show that the function contains all of the points.

c. Based on the graph you fit to the points, how high was the arrow at 2.3 sec? Was it going up or going down? How do you tell?

d. At what two times was the arrow 100 ft high? How do you explain the fact that there were two times?

e. When was the arrow at its highest? How high was that?

f. At what time did the arrow hit the ground?

30. *The Other Function Fit Problem:* It is possible for different functions to fit the same set of **discrete** data points. Suppose that the data in this table have been given.

x	$f(x)$
2	12
4	48
6	108
8	192
10	300

Figure 7-3i

a. Show that the function $f(x) = 3x^2$ fits the data, as shown in Figure 7-3i.

b. Show that the function $g(x) = 3x^2 + 100 \sin \frac{\pi}{2}x$ also fits the data exactly. Plot the graph, and sketch the result.

c. Let h be a sinusoid of increasing amplitude that touches but does not cross the graph of f at each marked point in Figure 7-3i. Write the particular equation of this sinusoid. Plot the graph, and sketch the results.

31. *Incorrect Point Problem:* By considering second differences, show that a quadratic function does *not* fit the points in this table.

x	y
4	5
5	7
6	11
7	17
8	27

What would the last y-value have to be in order for a quadratic function to fit exactly?

32. *Cubic Function Problem:* Figure 7-3j shows the cubic function

$$f(x) = x^3 - 6x^2 + 5x + 20$$

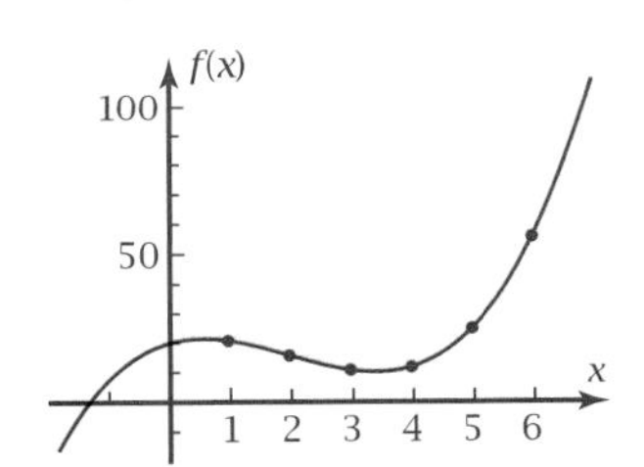

Figure 7-3j

a. Make a table of values of $f(x)$ for each integer value of x from 1 to 6.

b. Show that the **third differences** between the values of $f(x)$ are constant. You can calculate the third differences in a time-efficient way using the list and delta list features of your grapher. If you do it by pencil and paper, be sure to subtract from a function value, the previous value in each case.

c. Make a conjecture about how you could determine whether a **quartic function** (fourth degree) fits a set of points.

33. *The Add–Add Property Proof Problem:* Prove that for a linear function, adding a constant to x adds a constant to the corresponding value of $f(x)$. Do this by showing that if $x_2 = c + x_1$, then $f(x_2)$ equals a constant plus $f(x_1)$. Start by writing the equations for $f(x_1)$ and for $f(x_2)$, and then do the appropriate substitutions and algebra.

34. *The Multiply-Multiply Property Proof Problem:* Prove that for a power function, multiplying x by a constant multiplies the corresponding value of $f(x)$ by a constant as well. Do this by showing that if $x_2 = cx_1$, then $f(x_2)$ equals a constant times $f(x_1)$. Start by writing the equations for $f(x_1)$ and for $f(x_2)$, and then do the appropriate substitutions and algebra.

35. *The Add-Multiply Property Proof Problem 1:* Prove that for an exponential function, adding a constant to x multiplies the corresponding value of $f(x)$ by a constant. Do this by showing that if $x_2 = c + x_1$, then $f(x_2)$ equals a constant times $f(x_1)$. Start by writing the equations for $f(x_1)$ and for $f(x_2)$, and then do the appropriate substitutions and algebra.

36. *The Second-Difference Property Proof Problem:* Let $f(x) = ax^2 + bx + c$. Let d be the constant difference between successive x-values. Find $f(x + d)$, $f(x + 2d)$, and $f(x + 3d)$. Simplify. By subtracting consecutive y-values, find the three first differences. By subtracting consecutive first differences, show that the two second differences equal the constant $2ad^2$.

7-4 Logarithms: Definition, Properties, and Equations

In earlier chapters you learned that the inverse of a function is the relation found by interchanging two variables. For instance,

$y = 2^x$ An exponential function.

$x = 2^y$ The inverse of that exponential function.

You'll see that the inverse of an exponential function is a function, called a **logarithmic function.** In this section you'll learn about **logarithms,** which allow you to solve exponential equations. In the next section you'll see that logarithmic functions have the *multiply-add* property, similar to the add-multiply property for exponential functions.

OBJECTIVE Learn the definition and properties of logarithms, and use logarithms to find algebraic solutions of exponential and logarithmic equations.

Introduction to Logarithmic Functions and the Definition of Logarithm

You can plot the inverse of a function with the help of parametric equations. To plot $y = 2^x$ and its inverse, enter

$$x_1 = t \qquad \qquad x_2 = 2^t$$
$$y_1 = 2^t \quad \text{and} \quad y_2 = t$$

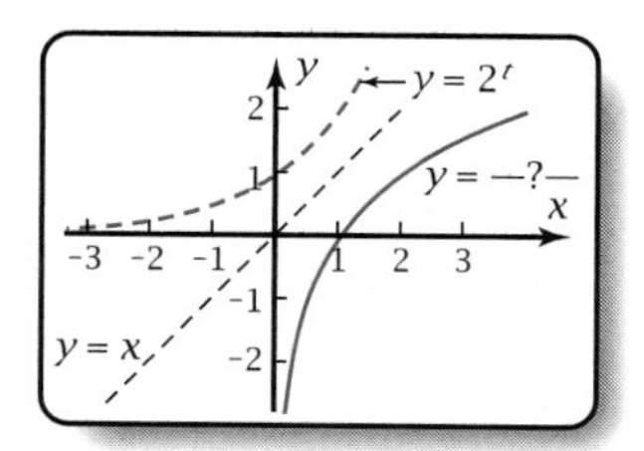

Figure 7-4a

The graphs are shown in Figure 7-4a. Because $y = 2^x$ is a one-to-one function, its inverse relation (solid graph) is a function. The graphs of the function and its inverse are reflections across the line $y = x$.

Function: $y = 2^x$		Inverse: $x = 2^y$	
x	y	x	y
−2	$\frac{1}{4}$	$\frac{1}{4}$	−2
−1	$\frac{1}{2}$	$\frac{1}{2}$	−1
0	1	1	0
1	2	2	1
2	4	4	2

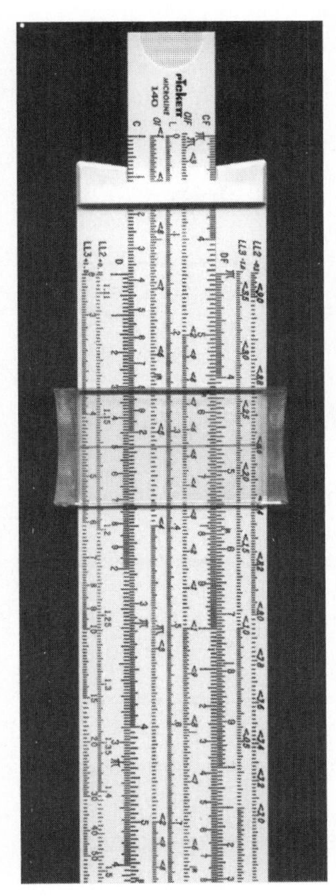

Slide rules, used by engineers in the 19th and early 20th centuries, employ the principle of logarithms for doing complicated calculations.

You can tell that the numerical table corresponds to the two graphs. When you look at the points on the $y = 2^x$ graph, you find that the coordinates of the points $(-2, \frac{1}{4})$, $(-1, \frac{1}{2})$, (0, 1), (1, 2), (2, 4) will all satisfy the $y = 2^x$ equation. Notice that the points $(\frac{1}{4}, -2)$, $(\frac{1}{2}, -1)$, (1, 0), (2, 1), (4, 2) are on the graph of the inverse function and satisfy the equation $x = 2^y$.

The name **logarithm** is used for the exponent of 2 in the expression $2^y = x$. It is written

$y = \log_2 x$ The exponent is called a *logarithm.*

The equation is read "y equals log to the base 2 of x." Your grapher calculates logarithms by means of infinite series because there is no way to calculate them using only the operations of algebra (namely, +, −, ×, ÷, and roots).

The reason for the name is historical. Before there were calculators, base-10 logarithms, calculated approximately using infinite series, were recorded in tables. Then products with many factors, such as

(357)(4.367)(22.4)(3.142)

could be calculated by adding their logarithms (exponents) columnwise in one step rather than by doing several tedious long multiplications on pairs of numbers at a time. Englishman Henry Briggs (1561–1630) and Scotsman John Napier (1550–1616) are credited with inventing this "**log**ical way to do **arithm**etic," from which the name comes.

Here is the definition of logarithm for any base.

DEFINITION: Logarithm

$\log_b x = y$ if and only if $b^y = x$ where $b > 0$, $b \neq 1$, and $x > 0$

Verbally: $\log_b x = y$ means that y is the exponent of b that gives x for the answer. The number x in $\log_b x$ is called the *argument* of the logarithm.

The expression $\log x$ (with base not written) is called the **common logarithm,** with base 10. Another prevalent form of logarithm is $\ln x$, the **natural logarithm,** with base $e = 2.7182\ldots$ (see the later section titled "Logarithms with Any Base: The Change-of-Base Property").

How you read a logarithm gives you a way to remember the definition. Examples 1 and 2 show you how to do this.

▶ **EXAMPLE 1** Write $\log_5 c = a$ in exponential form.

Solution Think this:

- "$\log_5$..." is read "log to the *base* 5...," so 5 is the base.
- A logarithm is an *exponent.* Because the log equals a, a must be the exponent.
- The "answer" I get for 5^a is the argument of the logarithm, c.

Write only this:

$$5^a = c$$

◀

▶ **EXAMPLE 2** Write $z^4 = m$ in logarithmic form.

Solution $\log_z m = 4$

◀

The most important thing to remember about logarithms is that

> A logarithm is an exponent.

Properties of Logarithms

The properties of logarithms are analogous to the properties of exponents. For instance, when you raise a power to a power, you *multiply* the exponents.

$$3^5 = (10^{0.4771...})^5 = 10^{(5)(0.4771...)} = 10^{2.3856...}$$

$$\log_{10} (3^5) = 5 \cdot \log_{10} 3 = (5)(0.4771...) = 2.3856...$$

The logarithm equals the exponent.

When you multiply powers with equal bases, you *add* their exponents. For instance,

$$4 \times 7 = 10^{0.6020...} \times 10^{0.8450...} = 10^{(0.6020... + 0.8450...)} = 10^{1.4771...}$$

$$\log_{10} (4 \times 7) = \log_{10} 4 + \log_{10} 7 = 0.6020... + 0.8450... = 1.4771...$$

When you divide powers with equal bases, you *subtract* the exponent of the denominator from the exponent of the numerator. For instance,

$$5 \div 9 = 10^{0.6989...} \div 10^{0.9542...} = 10^{(0.6989... - 0.9542...)} = 10^{-0.2552...}$$

$$\log_{10} (5 \div 9) = \log_{10} 5 - \log_{10} 9 = 0.6989... - 0.9542... = -0.2552...$$

These three properties are generalized in this box.

Properties of Logarithms

The Logarithm of a Power:

$$\log_b x^y = y \cdot \log_b x$$

Verbally: "The logarithm of a power equals the product of the exponent and the logarithm of the base."

The Logarithm of a Product:

$$\log_b (xy) = \log_b x + \log_b y$$

Verbally: "The logarithm of a product equals the sum of the logarithms of the factors."

The Logarithm of a Quotient:

$$\log_b \frac{x}{y} = \log_b x - \log_b y$$

Verbally: "The logarithm of a quotient equals the logarithm of the numerator minus the logarithm of the denominator."

▶ ***EXAMPLE 3*** Show numerically that $\log 3 + \log 5 = \log (3 \cdot 5)$.

Solution

$\log 3 + \log 5 = 0.4771... + 0.6989... = 1.1760...$ — Find the logarithms with a calculator.

$\log (3 \cdot 5) = \log 15 = 1.1760...$ — Complete the operations in the parentheses first.

$\therefore \log 3 + \log 5 = \log (3 \cdot 5)$ ◀

▶ ***EXAMPLE 4*** Prove algebraically that $\log_b (xy) = \log_b x + \log_b y$.

Proof Let $c = \log_b x$ and $d = \log_b y$. Then

$b^c = x$ and $b^d = y$ — Definition of logarithm.

$xy = b^c b^d$ — Multiply x by y.

$xy = b^{c+d}$ — Product of two powers with equal bases.

$\log_b xy = c + d$ — Use the definition of logarithm. Notice that $c + d$ is the exponent.

$\therefore \log_b xy = \log_b x + \log_b y$, Q.E.D. — Substitution. ◀

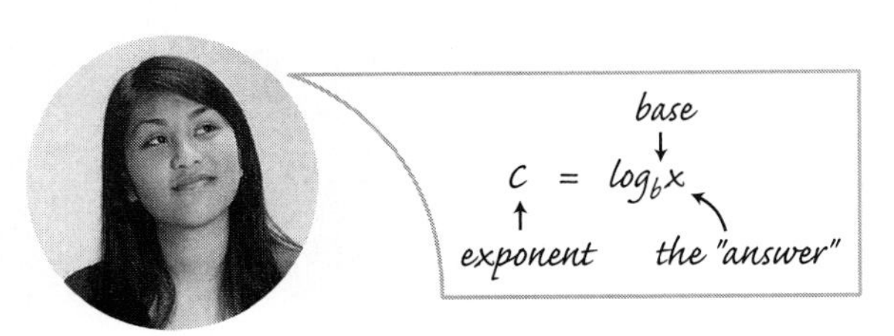

▶ **EXAMPLE 5** If $\log 3 + \log 7 - \log 5 = \log$ —?—, what number goes in the blank?

Solution

$$\log 3 + \log 7 - \log 5 = \log\left(\frac{3 \cdot 7}{5}\right) = \log 4.2$$

So 4.2 goes in the blank.

Check

$$\log 3 + \log 7 - \log 5 = 0.6232\ldots$$

$$\log 4.2 = 0.6232\ldots$$

which agrees. ◀

Solving Logarithmic Equations

You can use the logarithmic properties to solve logarithmic equations. The key to solving these equations involves two main steps. First, you'll rearrange the equation to isolate the logarithmic expression. Then, using the definition of logarithm, you'll rewrite the logarithmic equation as an exponential equation.

▶ **EXAMPLE 6** Solve the equation

$$4 - 3 \ln (x + 5) = 1$$

Solution

$$4 - 3 \ln (x + 5) = 1$$

$$-3 \ln (x + 5) = -3$$ Rearrange the equation to isolate $\ln (x + 5)$.

$$\ln (x + 5) = 1$$

$$e^1 = x + 5$$ Use the definition of logarithm.

$$x = e - 5 = -2.2817\ldots$$ ◀

▶ **EXAMPLE 7** Solve the equation

$$\log_2 (x - 1) + \log_2 (x - 3) = 3$$

Solution

$$\log_2 (x - 1) + \log_2 (x - 3) = 3$$

$$\log_2 (x - 1)(x - 3) = 3$$

Apply the logarithm of a product property.

$$2^3 = (x - 1)(x - 3)$$

Use the definition of logarithm.

$$8 = x^2 - 4x + 3$$

Expand the product.

$$x^2 - 4x - 5 = 0$$

Reduce one side to zero. Use the symmetric property of equality.

A decibel, which measures the relative intensity of sounds, has a logarithmic scale. Prolonged exposure to noise intensity exceeding 85 decibels can lead to hearing loss.

$(x-5)(x+1)=0$ — Solve by factoring.

$x_1 = 5$ or $x_2 = -1$ — Solutions of the quadratic equation.

You have to be cautious here because the solutions in the last step are the solutions of the quadratic equation, and you must make sure they are also solutions of the original logarithmic equation. So check by substituting your solutions into the original equation.

If $x_1 = 5$, then:

$\log_2(5-1) + \log_2(5-3)$

$= \log_2 4 + \log_2 2$

$= 2 + 1 = 3$

If $x_2 = -1$, then:

$\log_2(-1-1) + \log_2(-1-3)$

$= \log_2(-2) + \log_2(-4)$

which is undefined.

$\therefore\ x_1 = 5$ is a solution, but $x_2 = -1$ is not. ◀

▶ EXAMPLE 8 Solve the equation

$$\log(x-1) - \log(x+2) = -1$$

Solution

$\log(x-1) - \log(x+2) = -1$

$\log\dfrac{x-1}{x+2} = -1$ — Use the quotient property of logarithms.

$10^{-1} = \dfrac{x-1}{x+2}$ — Rewrite the equation in exponential form.

$x - 1 = 0.1(x+2)$ — $10^{-1} = \frac{1}{10} = 0.1$. Multiply both sides of the equation by $(x+2)$.

$0.99x = 1.2$ — Distribute the 0.1. Isolate the term containing x.

$x = 1.2121\ldots = 1.\overline{21}$ — Solve for x. ◀

Logarithms with Any Base: The Change-of-Base Property

Calculators usually have only base-10 logarithms or base-e logarithms available. The number e is a constant equal to 2.7812.... It occurs naturally, as does $\pi = 3.1415\ldots$, so logarithms with e as the base are called *natural logarithms.* The two-letter symbol "ln" is used for natural logarithms. It is read "el, en." The three-letter symbol "log" is used for base-10 *common logarithms.*

$\log 5 = \log_{10} 5 = 0.6989\ldots$ meaning that $10^{0.6989\ldots} = 5$

$\ln 5 = \log_e 5 = 1.6094\ldots$ meaning that $e^{1.6094\ldots} = 5$

In calculus you will learn why the seemingly awkward number $e = 2.7182\ldots$ is important enough to deserve a place on calculators and computers.

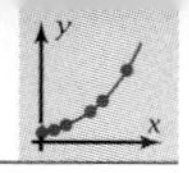

Nautilus shells have a logarithmic spiral pattern.

DEFINITION: Common Logarithm and Natural Logarithm

$\log x$ means $\log_{10} x$.

$\ln x$ means $\log_e x$, where e is a constant equal to 2.71828182846....

In order to calculate or plot logarithms with bases other than e or 10, you'll turn the logarithms you're looking for into logarithms with bases 10 or e. Suppose you want to find $\log_5 17$. This is what you'll do.

Let $x = \log_5 17$

$5^x = 17$ — Use the definition of logarithm.

$\log_{10} 5^x = \log_{10} 17$ — Take $\log_{10}$ of both sides.

$x \log_{10} 5 = \log_{10} 17$ — Use the logarithm of a power property.

$$x = \frac{\log_{10} 17}{\log_{10} 5} = \frac{1}{\log_{10} 5}(\log_{10} 17) = 1.4306\ldots(\log_{10} 17)$$

So to find the base-5 logarithm of a number, you find its base-10 logarithm on your calculator and divide it by the base-10 logarithm of 5. This result is generalized here.

PROPERTY: The Change-of-Base Property for Logarithms

$$\log_a x = \frac{\log_b x}{\log_b a} \quad \text{or, equivalently,} \quad \log_a x = \frac{1}{\log_b a}(\log_b x)$$

▶ EXAMPLE 9 Find ln 29 using the change-of-base property with base-10 logarithm. Check your answer directly by pressing ln 29.

Solution

$$\ln 29 = \frac{\log 29}{\log e} = \frac{1.4623\ldots}{0.4342\ldots} = 3.3672\ldots$$ Press log (29)/log (e).

Directly: $\ln 29 = 3.3672\ldots$

which agrees with the previous value you got using the change-of-base property. ◀

Solving Exponential Equations

Logarithms provide an algebraic way to solve equations that have a variable in the exponent.

▶ EXAMPLE 10 Solve the exponential equation $7^{3x} = 983$ algebraically, using logarithms.

Solution

$7^{3x} = 983$

$\log 7^{3x} = \log 983$ Take the base-10 logarithm of both sides.

$3x \log 7 = \log 983$ Use the logarithm-of-a-power property.

$x = \dfrac{\log 983}{3 \log 7}$ Divide by the coefficient of x.

$x = 1.1803\ldots$ ◀

▶ EXAMPLE 11 Solve the equation and check your answer.

$3^{x-2} = 9^{x+1}$

Solution

$3^{x-2} = 9^{x+1}$

$3^{x-2} = (3^2)^{x+1}$ Create equal bases on both sides of the equation.

$3^{x-2} = 3^{2(x+1)}$ Apply the properties of exponents.

$x - 2 = 2(x + 1)$ If the bases are equal, the exponents have to be equal as well.

$x - 2 = 2x + 2$

$x = -4$

Check $3^{-4-2} = 3^{-6}$ and $9^{-4+1} = 9^{-3} = (3^2)^{-3} = 3^{-6}$

The two sides are equal, so the solution is $x = -4$. ◀

▶ EXAMPLE 12 Solve the equation and check your answer.

$e^{2x} - 3e^x + 2 = 0$

Solution

$e^{2x} - 3e^x + 2 = 0$

$(e^x)^2 - 3e^x + 2 = 0$ Apply the properties of exponents.

You can realize that this is a quadratic equation in the variable e^x. Using the quadratic formula,

$$e^x = \frac{+3 \pm \sqrt{9 - 4(2)}}{2} = \frac{3 \pm 1}{2} = 2 \text{ or } 1$$

You now have to solve these equations.

$e^x = 2$ $e^x = 1$

$x_1 = \ln 2 = 0.6931\ldots$ $x_2 = 0$

Check $e^{2\ln 2} - 3e^{\ln 2} + 2$ $\qquad$ $(e^0)^2 - 3e^0 + 2$

$= (e^{\ln 2})^2 - 3e^{\ln 2} + 2$ $\qquad$ $= 1^2 - 3(1) + 2 = 0$

$= 2^2 - 3(2) + 2 = 0$

Both solutions are correct. ◀

Problem Set 7-4

Do These Quickly 

Q1. In the expression 7^5, the number 7 is called the —?—.

Q2. In the expression 7^5, the number 5 is called the —?—.

Q3. The entire expression 7^5 is called a —?—.

Q4. Write $x^5 \cdot x^7$ as a single exponential expression.

Q5. Write $\dfrac{x^5}{x^7}$ as a single exponential expression.

Q6. Write the expression $(x^5)^7$ without the parentheses.

Q7. For the expression $(xy)^7$, you —?— the exponent 7 to get x^7y^7.

Q8. Expand the square $(x + y)^2$.

Q9. If $y = x^5$, then the inverse function has the equation $y =$ —?—.

Q10. The function $y = 5^x$ is called a(n)

A. Power function

B. Exponential function

C. Polynomial function

D. Linear function

E. Inverse of a power function

For Problems 1–4, find the logarithm on your calculator. Then show numerically that 10 raised to the answer, as a power, gives the argument of the logarithm.

1. log 1066
2. log 2001
3. log 0.0596
4. log 0.314

For Problems 5–8, evaluate the power of 10. Then show that the logarithm of the answer is equal to the original exponent of 10.

5. $10^{-2.7}$
6. $10^{3.5}$
7. $10^{15.2}$
8. 10^{-4}

Problems 9–12 test your knowledge of the definition of logarithm.

9. Write in exponential form: $x = \log_c p$
10. Write in exponential form: $\log_w g = h$
11. Write in logarithmic form: $m = r^k$
12. Write in logarithmic form: $d^z = k$

For Problems 13–20, demonstrate numerically the properties of logarithms.

13. $\log(0.3 \times 0.7) = \log 0.3 + \log 0.7$
14. $\ln(7 \cdot 8) = \ln 7 + \ln 8$
15. $\ln(30 \div 5) = \ln 30 - \ln 5$
16. $\log \frac{2}{8} = \log 2 - \log 8$
17. $\log 2^5 = 5 \log 2$
18. $\ln 5^3 = 3 \ln 5$
19. $\ln \frac{1}{7} = -\ln 7$
20. $\log \frac{1}{1000} = -\log 1000$

Problems 21–36 are short exercises to test your knowledge of the definition and properties of logarithms. Find the missing values.

21. log 7 + log 3 = log —?—
22. log 5 + log 8 = log —?—
23. ln 48 − ln 12 = ln —?—
24. ln 4 − ln 20 = ln —?—
25. log 8 − log 5 + log 35 = log —?—
26. log 2000 − log 40 − log 2 = log —?—
27. 5 ln 2 = ln —?—
28. 4 ln 3 = ln —?—
29. log 125 = (—?—) log 5
30. log 64 = (—?—) log 2

31. $\log_7 33 = \dfrac{\log_{10} 33}{\text{—?—}}$

32. $\log_{0.07} 53 = \dfrac{\ln 53}{\text{—?—}}$

33. $\dfrac{\log_{0.6} x}{\log_{0.6} 3} = \log_{\text{—?—}} x$

34. $\dfrac{\log_{13} n}{\log_{13} 0.5} = \log_{\text{—?—}} n$

35. $\ln \sqrt{x} = (\text{—?—}) \ln x$

36. $\log \sqrt[5]{x} = (\text{—?—}) \log x$

The spiral arms of galaxies follow a logarithmic pattern.

For Problems 37–42, solve the equations and check your answers.

37. $\log (3x + 7) = 0$

38. $2 \log (x - 3) + 1 = 5$

39. $\log_2 (x + 3) + \log_2 (x - 4) = 3$

40. $\log_2 (2x - 1) - \log_2 (x + 2) = -1$

41. $\ln (x - 9)^4 = 8$

42. $\ln (x + 2) + \ln (x - 2) = 0$

For Problems 43–46, solve the exponential equation algebraically, using logarithms.

43. $5^{3x} = 786$

44. $8^{0.2x} = 98.6$

45. $0.8^{0.4x} = 2001$

46. $6^{-5x} = 0.007$

For Problems 47–50, solve the equations and check your answer.

47. $3e^{x-4} + 5 = 10$

48. $4 - e^{2x-3} = 7$

49. $2e^{2x} + 5e^x - 3 = 0$

50. $5 \cdot 2^{2x} - 3 \cdot 2^x - 2 = 0$

51. *Compound Interest Problem 2:* If you invest \$10,000 in a savings account that pays interest at the rate of 7% APR (annual percentage rate), then the amount M in the account after x years is given by the exponential function

$$M = 10{,}000 \times 1.07^x$$

a. Make a table of values of M for each year from 0 to 6 years.

b. How can you conclude that the values in the table have the add-multiply property?

c. Suppose that you want to cash in the savings account when the amount M reaches \$27,000. Set $M = 27{,}000$ and solve the resulting exponential equation algebraically using logarithms. Convert the answer to months, and round appropriately to find how many whole months must elapse before M first exceeds \$27,000.

52. *Population of the United States Problem:* Based on the 1980 and 1990 U.S. censuses, the population increased by an average of 0.94% per year for that time period. That is, the population at the end of any one year was 1.0094 times the population at the beginning of that year.

a. How do you tell that the population function has the add-multiply property?

b. The population in 1980 was about 226.5 million. Write the particular equation expressing population, P, as a function of the n years that have elapsed since 1980.

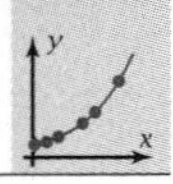

c. Assume that the population continues to grow at the rate of 0.94% per year. Find algebraically the year in which the population first reaches 300 million. In finding the real-world answer, use the fact that the 1980 census was taken as of April 1.

For Problems 53–56, use the change-of-base property to find the indicated logarithm. Show that your answer is correct by raising the base to the appropriate power.

53. $\log_7 29$

54. $\log_8 352$

55. $\log_3 729$ (Surprising?!)

56. $\log_{32} 2$ (Surprising?!)

For Problems 57–60, find the logarithm by applying the definition of logarithm.

57. $x = \log_2 32$

58. $x = \log_5 125$

59. $x = \log_7 49$

60. $x = \log_3 81$

61. *Logarithm of a Power Property Problem:* Prove that $\log_b x^n = n \log_b x$.

62. *Logarithm of a Quotient Property Problem:* Prove that $\log_b \dfrac{x}{y} = \log_b x - \log_b y$.

7-5 Logarithmic Functions

You have already learned about identifying properties for several types of functions.

- Add–add: linear functions
- Add–multiply: exponential functions
- Multiply–multiply: power functions

In this section you'll learn that logarithmic functions have the **multiply–add property.**

OBJECTIVE Show that logarithmic functions have the multiply-add property, and find particular equations by algebra.

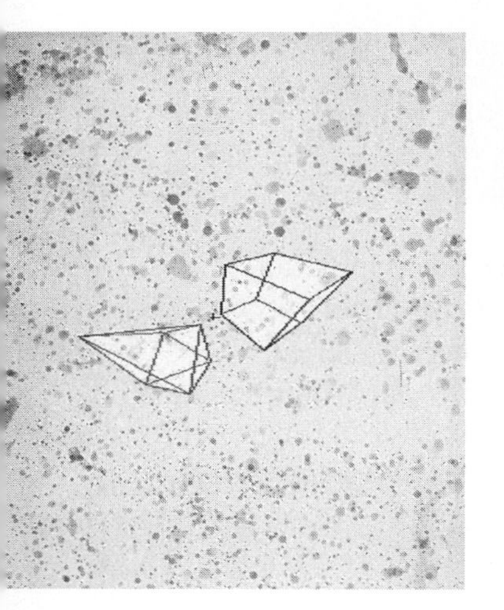

Logarithmic Functions

Figures 7-5a and 7-5b show the natural logarithmic function $y = \ln x$ and the common logarithmic function $y = \log x$ (solid graphs). These functions are inverses of the corresponding exponential functions (dashed graphs), as shown by the fact that the graphs are reflections of $y = e^x$ and $y = 10^x$ across the line $y = x$. Both logarithmic graphs are concave downward. Notice also that the y-values are increasing at a slower and slower rate as x increases. In both cases the y-axis is a vertical asymptote for the logarithmic graph. In addition, you can tell that the domain of these basic logarithmic functions is the set of positive real numbers.

Ronald Davis, Pyramid and Cube. *An acid-resistant substance was used to create this print. Ph, which measures the strength of acids, has a logarithmic scale. (© Gemini G.E.L., Los Angeles, CA, 1983)*

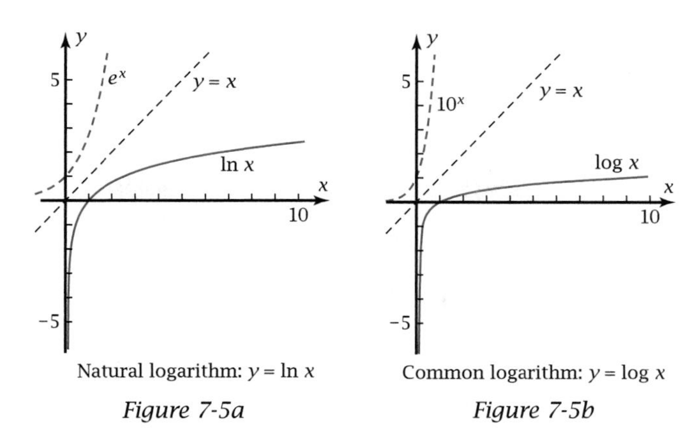

Natural logarithm: $y = \ln x$

Figure 7-5a

Common logarithm: $y = \log x$

Figure 7-5b

The general equation of a logarithmic function on most graphers has constants to allow for vertical translation and dilation.

DEFINITION: General Equation of a Logarithmic Function

Untranslated form: $y = a + b \log_c x$ Base-c logarithmic function.

where a, b, and c are constants, with $b \neq 0$, $c > 0$, and $c \neq 1$. The domain is all positive real numbers.

Transformed function: $y = a + b \log_c (x - d)$

where a is the vertical translation, b is the vertical dilation, and d is the horizontal translation.

Note: Remember that log stands for base-10 logarithm and ln stands for base-e logarithm.

Multiply–Add Property of Logarithmic Functions

The following x- and y-values have the **multiply–add** property. Multiplying x by 3 results in adding 1 to the corresponding y.

	x	y	
	6	1	
×3	18	2	+1
×3	54	3	+1
×3	162	4	+1

By interchanging the variables, you can notice that x is an exponential function of y. You can find its particular equation by algebraic calculations.

$$x = 2 \cdot 3^y$$

This equation can be solved for y as a function of x with the help of logarithms.

$\ln x = \ln (2 \cdot 3^y)$ Take the natural logarithm (ln) of both sides.

$\ln x = \ln 2 + y \ln 3$ Use the product and power properties of logarithms.

$y \ln 3 = \ln x - \ln 2$ Solve for y.

$y = \dfrac{1}{\ln 3} \ln x - \dfrac{\ln 2}{\ln 3}$ Calculate constants.

$y = 0.9102\ldots \ln x - 0.6309\ldots$ By calculator.

This equation is a logarithmic function with $a = -0.6309...$ and $b = 0.9102....$ So if a set of points has the multiply–add property, the points lie on a logarithmic function. Reversing the steps lets you conclude that logarithmic functions have this property in general.

PROPERTY: Multiply–Add Property of Logarithmic Functions

If f is a logarithmic function, then multiplying x by a constant results in adding a constant to the value of $f(x)$. That is,

For $f(x) = a + b \log_c x$, if $x_2 = k \cdot x_1$, then $f(x_2) = \log_c k + f(x_1)$

Particular Equations for Logarithmic Functions

You can find the particular equation for a logarithmic function algebraically by substituting two points that are on the graph of the function and evaluating the two unknown constants. Example 1 shows you how to execute this plan.

▶ EXAMPLE 1 Suppose that f is a logarithmic function with values $f(3) = 7$ and $f(6) = 10$.

a. Without finding the particular equation, find $f(12)$ and $f(24)$.

b. Find the particular equation algebraically using natural logarithms.

c. Confirm that your equation gives the value of $f(24)$ found in part a.

Solution a. Make a table of values using the multiply–add property.

	x	$f(x)$	
	3	7	
×2	6	10	+3
×2	12	13	+3
×2	24	16	+3

$f(12) = 13$ and $f(24) = 16$

b.

$f(x) = a + b \ln x$ — Write the general equation.

$$\begin{cases} 7 = a + b \ln 3 \\ 10 = a + b \ln 6 \end{cases}$$ Substitute the given points.

$3 = b \ln 6 - b \ln 3$ — Subtract the equations to eliminate a.

$b = \dfrac{3}{\ln 6 - \ln 3} = 4.3280...$ — Factor out b and then divide by $\ln 6 - \ln 3$.

$7 = a + 4.3280... \ln 3$ — Substitute 4.3280... for b.

$a = 7 - 4.3280... \ln 3 = 2.2451...$ — Save a and b without round-off.

$f(x) = 2.2451... + 4.3280... \ln x$ — Write the particular equation. Paste it in the y= menu.

c. By calculator, $f(24) = 16$, Q.E.D. ◀

▶ EXAMPLE 2 Plot the graphs of these functions and identify their domains.

a. $f(x) = 3 \log (x - 1)$

b. $f(x) = -\ln (x + 3)$

c. $f(x) = \log_2 (x^2 - 1)$

Solution

a. You can get the graph of the function $f(x) = 3 \log (x - 1)$ through transformations of the parent logarithmic function: a horizontal translation by 1 unit and a vertical dilation by 3 units. Figure 7-5c shows the resulting graph.

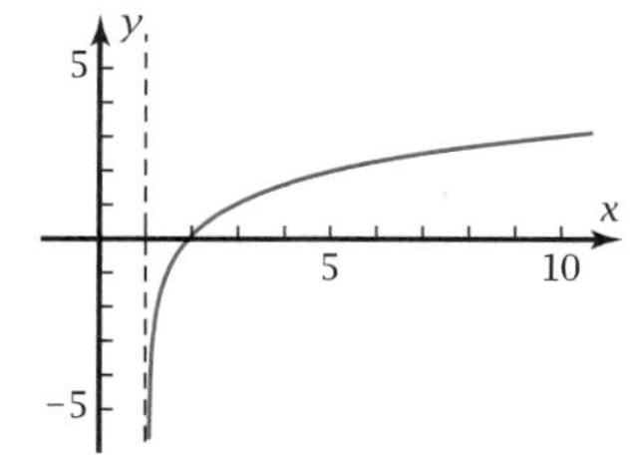

Figure 7-5c

You know that the domain of a logarithmic function is positive real numbers, so the argument of a logarithmic function has to be positive.

$$x - 1 > 0$$

So the domain of the function is $x > 1$. Add 1 to both sides of the inequality.

b. Figure 7-5d shows the graph of the function $f(x) = -\ln (x + 3)$. You can get this graph by reflecting the graph of the $y = \ln x$ function across the x-axis and translating it by -3 units horizontally.

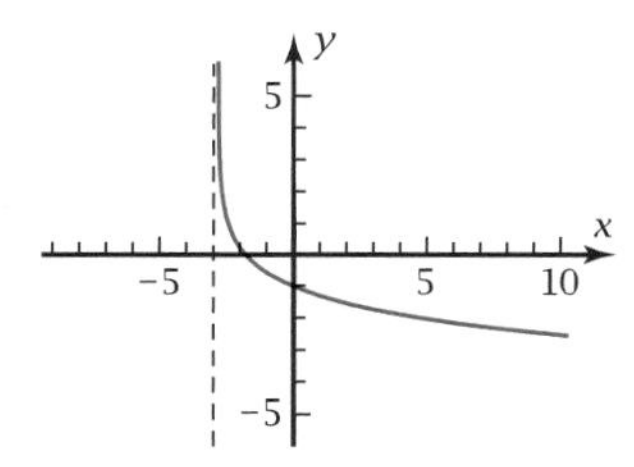

Figure 7-5d

Domain:

$x + 3 > 0$ Argument of a logarithmic function is positive.

So the domain of the function is $x > -3$.

c. In order to graph this function on your grapher, use the change-of-base property.

$$f(x) = \log_2 (x^2 - 1) = \frac{\log (x^2 - 1)}{\log 2}$$

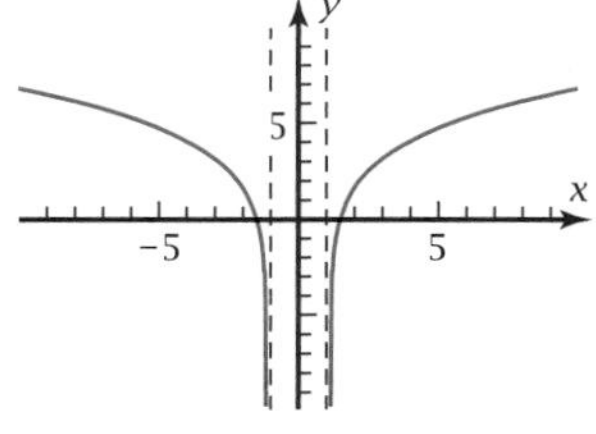

Figure 7-5e

Figure 7-5e shows the resulting graph.

Domain:

$x^2 - 1 > 0$ Argument of a logarithmic function is positive.

You can solve this inequality graphically. Graph the quadratic function and look for the x-values where the function value is greater than zero or where the graph is above the x-axis (see Figure 7-5f).

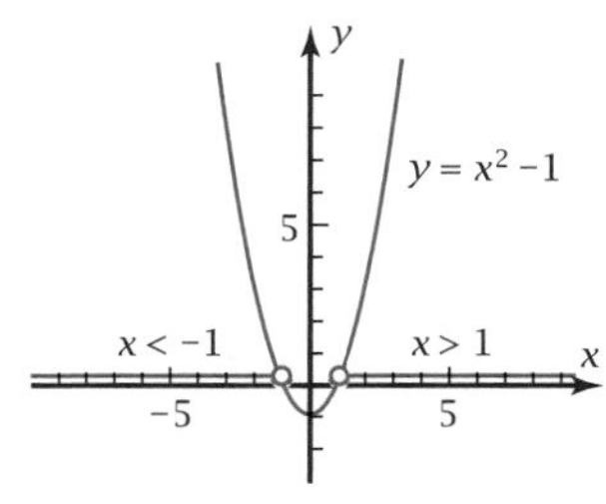

Figure 7-5f

So the domain is $x < -1$ or $x > 1$. ◀

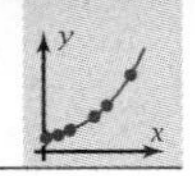

Problem Set 7-5

Do These Quickly 

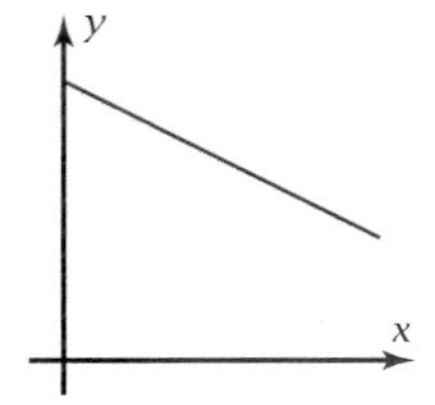

Figure 7-5g

Figure 7-5h

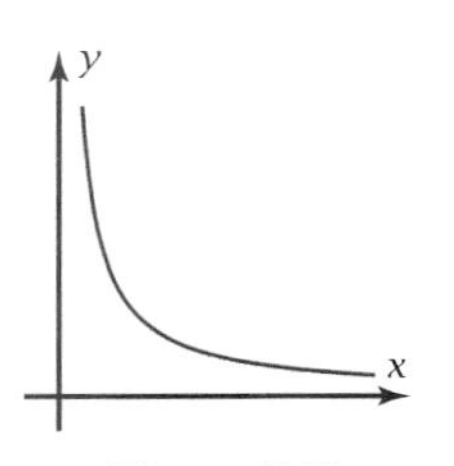

Figure 7-5i

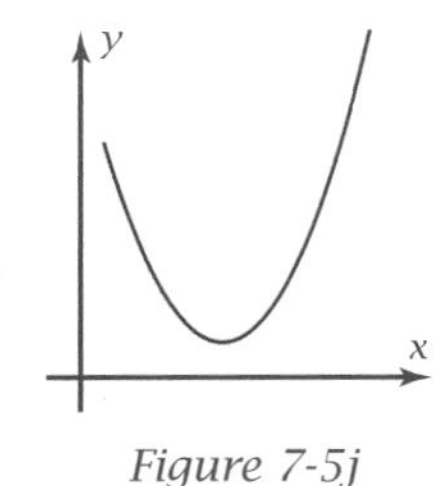

Figure 7-5j

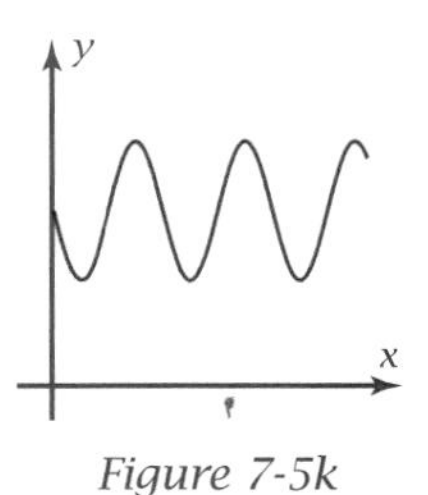

Figure 7-5k

Q1. Name the kind of function in Figure 7-5g.

Q2. Name the kind of function in Figure 7-5h.

Q3. Name the kind of function in Figure 7-5i.

Q4. Name the kind of function in Figure 7-5j.

Q5. Name the kind of function in Figure 7-5k.

Q6. Sketch a reasonable graph: The population of a city depends on time.

Q7. The graph of a quadratic function is called a —?—.

Q8. Expand the square: $(3x - 7)^2$

Q9. Write the next three terms in this sequence: 3, 6, 12, 24, . . .

Q10. The add-multiply property is a characteristic of —?— functions.

For Problems 1 and 2,

a. Show that the values in the table have the multiply-add property.

b. Use the first and last points to find algebraically the particular equation of the natural logarithmic function that fits the points.

c. Show that the equation in part b gives the other points in the table.

1.

x	y
3.6	1
14.4	2
57.6	3
230.4	4
921.6	5

2.

x	y
1	2
10	3
100	4
1000	5

3. *Carbon-14 Dating Problem:* The ages of things, such as wood, bone, and cloth, that are made from materials that had been living can be determined by measuring the percentage of the original radioactive carbon-14 that remains in them. This table contains data on the age as a function of the remaining percentage of carbon-14.

Percentage Remaining	Years Old
100	0
90	874
80	1851
70	2959
60	4238
50	5750

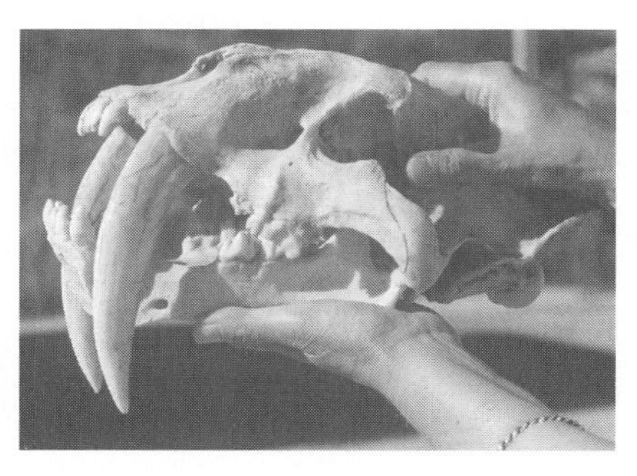

The skull of the saber-toothed cat, which lived in the Pleistocene more than 11,000 years ago.

a. Based on theoretical considerations, it is known that the percentage of carbon-14 remaining is an exponential function of the age. How does this fact indicate that the age should be a logarithmic function of percentage?

b. Using the first and last points, find the particular equation of the logarithmic function that goes through the points. Show that the equation gives values for other points close to the ones in the table.

c. You can use your mathematical model to interpolate between the given data points to find fairly precise ages. Suppose that a piece of human bone were found to have a carbon-14 content of 73.9%. What would you predict its age to be?

d. How old would you predict a piece of wood to be if its carbon-14 content were only 20%?

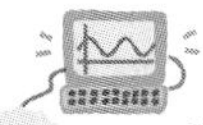

e. Search on the Internet or in some other resource to find out about early hominid fossil remains and carbon-14 dating.

4. *Earthquake Problem 2:* You can gauge the amount of energy released by an earthquake by its **Richter magnitude,** devised by seismologist Charles F. Richter in 1935. The Richter magnitude is a base-10 logarithmic function of the energy released by the earthquake. The following data show Richter magnitude m for earthquakes that release energy equivalent to the explosion of x tons of TNT (*t*ri-*n*itro-*t*oluene).

x (tons)	m (Richter magnitude)
1,000	4.0
1,000,000	6.0

a. Find the particular equation of the *common* logarithmic function $m = a + b \log x$ that fits the two points.

The director of the National Earthquake Service in Golden, Colorado, studies the seismograph display of a magnitude 7.5 earthquake.

b. Use the equation to predict the Richter magnitude for

- The 1964 Alaska earthquake, one of the strongest on record, that released the energy of 5 billion tons of TNT
- An earthquake that would release the amount of solar energy Earth receives every day, 160 trillion tons of TNT
- Blasting done at a construction site that releases the energy of about 30 pounds of TNT

c. The Chilean earthquake of 1960 had a Richter magnitude of about 9.0. How many tons of TNT would it take to produce a shock of this magnitude?

d. True or false? "Doubling the energy released by an earthquake doubles the Richter magnitude." Give evidence to support your answer.

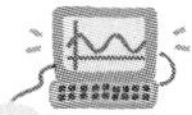

e. Look up Richter magnitude on the Internet or in some other resource. Name one thing you learned that is not mentioned in this problem.

5. *Logarithmic Function Vertical Dilation and Translation Problem:*

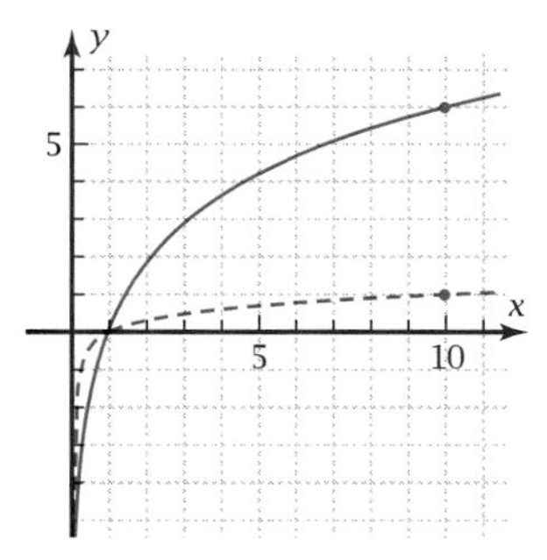

Figure 7-5l

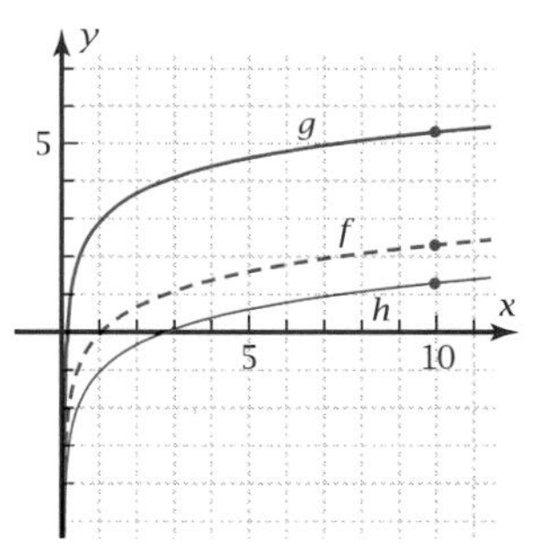

Figure 7-5m

a. Figure 7-5l shows the graph of the common logarithm function $f(x) = \log_{10} x$ (dashed) and a vertical dilation of this graph by a factor of 6, $y = g(x)$ (solid). Write an equation for $g(x)$, considering it as a vertical dilation. Write another equation for $g(x) = \log_b x$, where b is a number other than 10. Find the base.

b. Figure 7-5m shows $f(x) = \ln x$ (dashed). Two vertical translations are shown, $g(x) = 3 + \ln x$ and $h(x) = -1 + \ln x$. Find

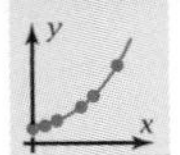

algebraically or numerically the x-intercepts of g and h.

6. *Logarithmic and Exponential Function Graphs Problem:* Figure 7-5n shows the graph of an exponential function $y = f(x)$ and its inverse function $y = g(x)$.

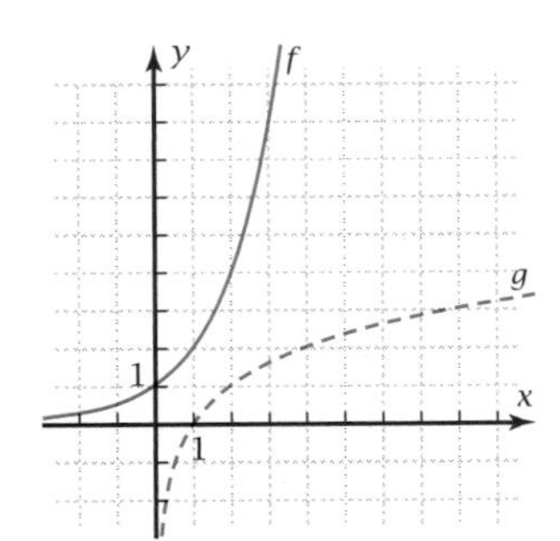

Figure 7-5n

a. The base of the exponential function is an integer. Which integer?

b. Write the particular equation for the inverse function $y = g(x) = f^{-1}(x)$.

c. Confirm that your answers to parts a and b are correct by plotting on your grapher.

d. With your grapher in parametric mode, plot these parametric functions:

$x(t) = f(t)$

$y(t) = t$

What do you notice about the resulting graph?

e. From your answer to part d, explain how you could plot on your grapher the inverse of any given function. Show that your method works by plotting the inverse of $y = x^3 - 9x^2 + 23x - 15$.

For Problems 7–12, graph the functions and identify their domains.

7. $f(x) = -2 \log (x + 3)$

8. $f(x) = \log (3 - 2x)$

9. $f(x) = \log_3 x^2$

10. $f(x) = \ln (x^2 - 4)$

11. $f(x) = \ln 3x$

12. $f(x) = 4 \log_2 (3x + 5)$

13. *The Definition of e Problem:* Figure 7-5o shows the graph of $y = (1 + x)^{1/x}$. If $x = 0$, then y is undefined because of division by zero. If x is close to 0, then $\frac{1}{x}$ is very large. For instance,

$$(1 + 0.0001)^{1/0.0001} = 1.0001^{10000} = 2.71814592\ldots$$

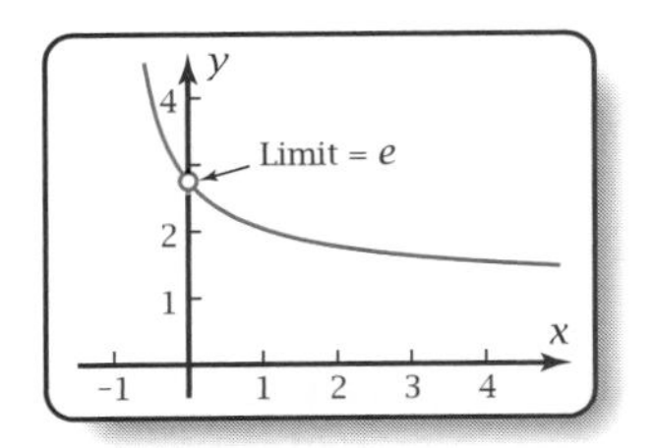

Figure 7-5o

a. Reproduce the graph in Figure 7-5o on your grapher. Use a friendly window that has $x = 0$ as a grid point. Trace to values close to zero, and record the corresponding values of y.

b. Two competing properties influence the expression $(1 + x)^{1/x}$ as x approaches 0. A number greater than 1 raised to a large power is very large. But 1 raised to any power is still 1. Which of these competing properties "wins"? Or is there a "compromise" at some number larger than 1?

c. Call up the number e on your grapher. If it does not have an e key, calculate e^1. What do you notice about the answer to part b and the number e?

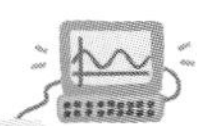

14. *Research Project 1:* On the Internet or via some other reference source, find out about Henry Briggs and John Napier and their contributions to logarithms. See if you can find out why natural logarithms are sometimes called **Napierian logarithms.**

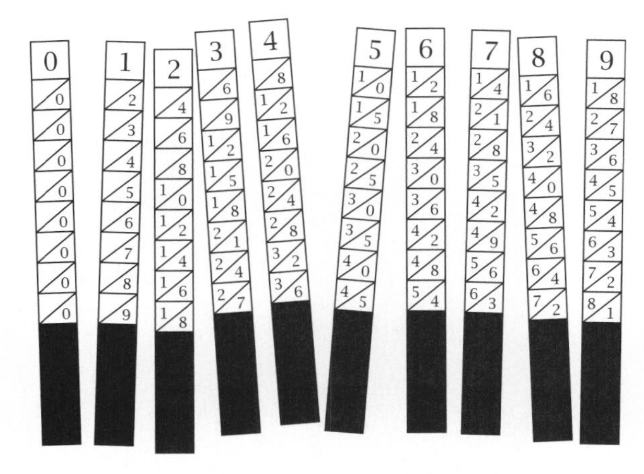

These rods are called "Napier's bones." Invented in the early 1600s, they made multiplication, division, and the extraction of square roots easier.

7-6 Logistic Functions for Restrained Growth

Suppose that the population of a new subdivision is growing rapidly. This table shows monthly population figures.

x (months)	y (houses)
2	103
4	117
6	132
8	148
10	167

Figure 7-6a shows the plot of points and the graph that goes through them. You can tell that it is increasing, concave upward, and has a positive y-intercept, suggesting that an exponential function fits the points. Using the first and last points gives $y = 91.2782\ldots\,(1.0622\ldots)^x$, the curve shown in the figure, which fits the points almost exactly. Suppose that there are only 1000 lots in the subdivision. The actual number of houses will level off, approaching 1000 gradually, as shown in Figure 7-6b.

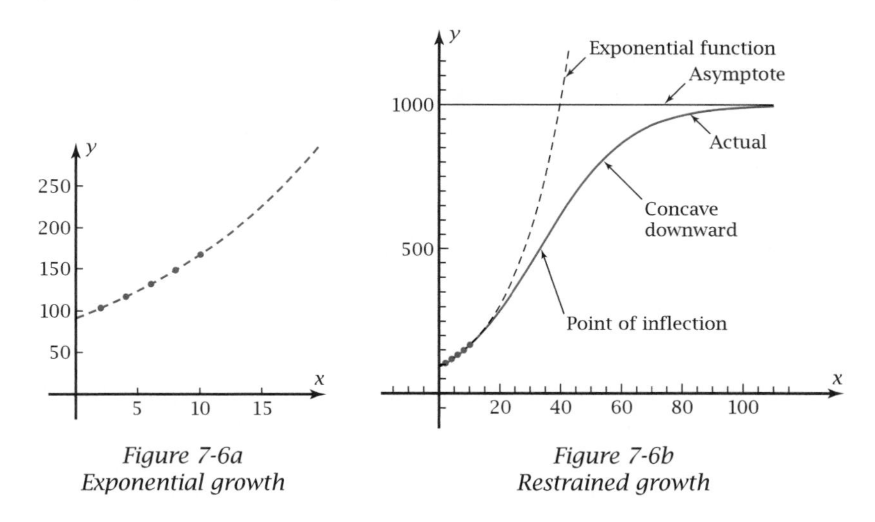

Figure 7-6a
Exponential growth

Figure 7-6b
Restrained growth

In this section you will learn about **logistic functions** that are useful as mathematical models of restrained growth.

OBJECTIVE Fit a logistic function to data for restrained growth.

Figure 7-6c shows the graphs of

$$f(x) = 2^x \qquad \text{and} \qquad g(x) = \frac{2^x}{2^x + 1}$$

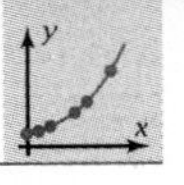

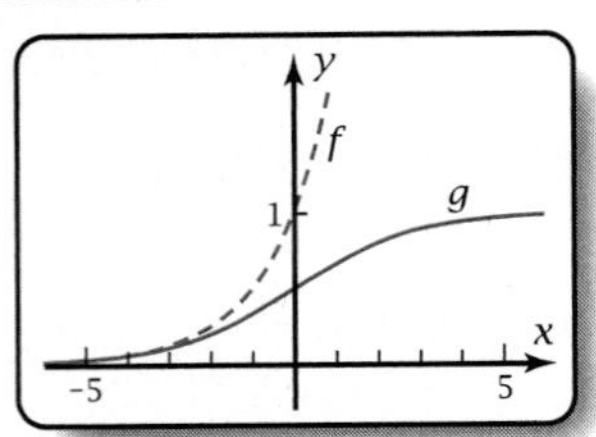

Figure 7-6c

Function f is an exponential function, and function g is a logistic function. For large positive values of x, the graph of g levels off to 1. This is because, for large values of x, 2^x is large compared to the 1 in the denominator. So the denominator is not much different from 2^x, the numerator, and the fraction for $g(x)$ approaches 1.

$$g(x) \rightarrow \frac{2^x}{2^x} = 1$$

For large negative values of x, the 2^x in the denominator is close to zero. So the denominator is close to 1. Thus the fraction for $g(x)$ approaches 2^x.

$$g(x) \rightarrow \frac{2^x}{0+1} = 2^x$$

As you can see in Figure 7-6c, the logistic function is almost indistinguishable from the exponential function for large negative values of x. But for large positive values, the logistic function levels off, as did the number of occupied houses represented by Figure 7-6b. You can fit logistic functions to data sets by the same dilations and translations you have used for other kinds of functions. You'll see how in Example 1.

General Logistic Function

You can transform the equation for function g in Figure 7-6c so that only one exponential term appears.

$$g(x) = \frac{2^x}{2^x + 1}$$

$$= \frac{2^x}{2^x + 1} \cdot \frac{2^{-x}}{2^{-x}} \qquad \text{Multiply by a clever form of 1.}$$

$$= \frac{1}{1 + 2^{-x}}$$

To get a general function of this form, replace the 1 in the numerator with a constant c to give the function a vertical dilation by a factor of c. Replace the exponential term 2^{-x} with ab^{-x} or, equivalently, ae^{-bx} if you want to use the natural exponential function. The result is shown in the box.

DEFINITION: *Logistic Function General Equation*

$$f(x) = \frac{c}{1 + ae^{-bx}} \qquad \text{or} \qquad f(x) = \frac{c}{1 + ab^{-x}}$$

where a, b, and c are constants and the domain is all real numbers.

▶ EXAMPLE 1 Use the information on the occupied houses from the beginning of the section.

x (months)	y (houses)
2	103
4	117
6	132
8	148
10	167

a. Given that there are 1000 lots in the subdivision, use the points for 2 months and 10 months to find the particular equation of the logistic function that satisfies these constraints.

b. Plot the graph of the logistic function from 0 through 100 months. Sketch the result.

c. Make a table showing that the logistic function fits all the points closely.

d. Use the logistic function to predict the number of houses that will be occupied at the value of x corresponding to two years. Which process do you use, extrapolation or interpolation?

e. Find the value of x at the point of inflection. What is the real-world meaning of the fact that the graph is concave upward for times before the point of inflection and concave downward thereafter?

Solution a.

$$y = \frac{1000}{1 + ab^{-x}}$$ The vertical dilation is 1000.

$$\begin{cases} 103 = \dfrac{1000}{1 + ab^{-2}} \\ 167 = \dfrac{1000}{1 + ab^{-10}} \end{cases} \text{ and}$$ Substitute points (2, 103) and (10, 167).

$$\begin{cases} 103 + 103ab^{-2} = 1000 \\ 167 + 167ab^{-10} = 1000 \end{cases} \text{ and}$$ Eliminate the fractions.

$$\begin{cases} 103ab^{-2} = 897 \\ 167ab^{-10} = 833 \end{cases} \text{ and}$$

$$\frac{167ab^{-10}}{103ab^{-2}} = \frac{833}{897}$$ Divide and simplify.

$$b^{-8} = \frac{833}{897} \cdot \frac{103}{167}$$

$$b = \left(\frac{833}{897} \cdot \frac{103}{167}\right)^{-1/8} = 1.0721...$$ Store without rounding.

$$103a(1.0721...)^{-2} = 897$$ Substitute for b.

$$a = 10.0106...$$ Store without rounding.

$$y = \frac{1000}{1 + 10.0106...\,(1.0721...)^{-x}}$$ Write the particular equation.

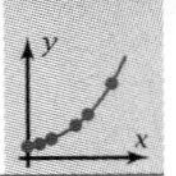

b. Figure 7-6b shows the graph.

c. Make a table of values.

x (months)	y (houses)	Logistic Function	
2	103	103	(exact)
4	117	116.60...	(close)
6	132	131.73...	(close)
8	148	148.50...	(close)
10	167	167	(exact)

d. Trace the function to $x = 24$ for 2 years.

$y = 347.1047... \approx 347$ houses

The process is extrapolation because $x = 24$ is beyond the range of the given points.

e. The point of inflection is halfway between the x-axis and the asymptote at $y = 1000$. Trace the function to a value that is close to $y = 500$. The value of x is approximately 33, so the point of inflection occurs at about 33 months. Before 33 months, the number of houses is increasing at an increasing rate. After 33 months, the number is still increasing but at a decreasing rate. ◄

Note that if a, b, and c are all positive, the logistic function will have two horizontal asymptotes, one at the x-axis and one at $y = c$. The point of inflection occurs halfway between these two asymptotes.

Properties of logistic function graphs are shown in this box.

PROPERTIES: Logistic Functions

The logistic function is $y = \dfrac{c}{1 + ae^{-bx}}$, where $a > 0$, $b \neq 0$, $c > 0$, and a, b, c are constants. The domain is all real numbers. The logistic function has:

- Two horizontal asymptotes: one at $y = 0$ and another at $y = c$
- A point of inflection at $y = \dfrac{c}{2}$

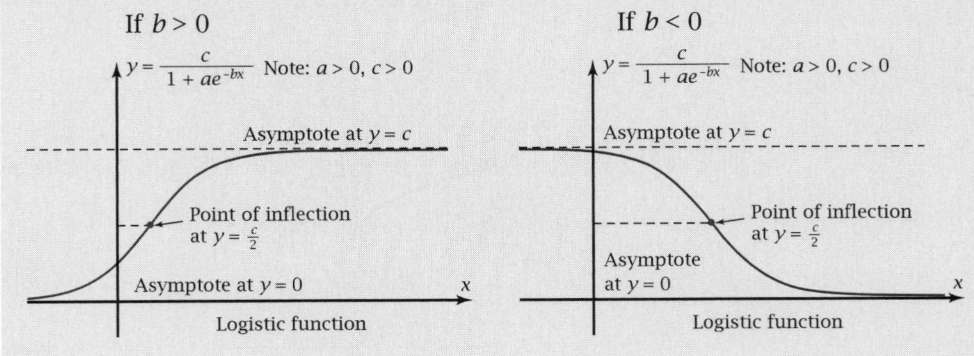

Problem Set 7-6

Q1. An exponential function has the —?— - —?— property.

Q2. A power function has the —?— - —?— property.

Q3. The equation $y = 3 + 5 \ln x$ defines a —?— function.

Q4. The function in Q3 has the —?— - —?— property.

Q5. The expression $\ln x$ is a logarithm with the number —?— as its base.

Q6. Write in exponential form: $h = \log_p m$

Q7. Write in logarithmic form: $c = 5^j$

Q8. If an object rotates at 100 revolutions per minute, how many radians per minute is this?

Q9. Write the general equation for a quadratic function.

Q10. $\cos \pi$ is

A. -1 B. 0 C. 1 D. $\frac{1}{2}$ E. Undefined

---

1. Given the exponential function $f(x) = 1.2^x$ and the logistic function $g(x) = \dfrac{1.2^x}{1.2^x + 1}$,
   a. Plot both graphs on the same screen. Use a domain of $x \in [-10, 10]$. Sketch the result.
   b. How do the two graphs compare for large positive values of x? How do they compare for large negative values of x?
   c. Find the approximate x-value of the point of inflection for function g. For what values of x is the graph of function g concave up? Concave down?
   d. Explain algebraically why the logistic function has a horizontal asymptote at $y = 1$.
   e. Transform the equation of the logistic function so that an exponential term appears only *once*. Show numerically that the resulting equation is equivalent to $g(x)$ as given.

2. Figure 7-6d shows the logistic function

$$f(x) = \frac{3e^{0.2x}}{e^{0.2x} + 4}$$

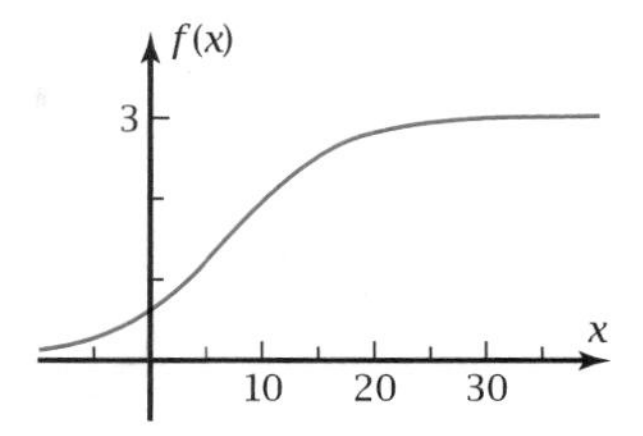

Figure 7-6d

   a. Explain algebraically why the graph has a horizontal asymptote at $y = 3$.
   b. Read the point of inflection from the graph. Find the x-coordinate algebraically.
   c. For what values of x is the graph concave up? Concave down?
   d. Transform the equation so that there is only *one* exponential term. Confirm by graphing that the resulting equation is equivalent to $f(x)$ as given.

3. *Spreading the News Problem:* You arrive at school and meet your mathematics teacher, who tells you today's test has been cancelled! You and your friend spread the good news. The table shows the number of students, y, who have heard the news after x minutes have passed since you and your friend heard the news.

x (min)	y (students)
0	2
10	5
20	13
30	35
40	90

   a. Plot the points. Imagine a function fit to the points. Is the graph of this function concave up or concave down or both?
   b. There are 1220 students in the school. Use the numbers of students at 0 minutes and at 40 minutes to find the equation of the logistic function that meets these constraints.

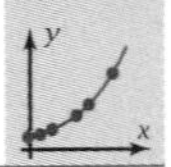

c. Plot the graph of the logistic function for the first three hours.

d. Based on the logistic model, how many students have heard the news at 9:00 a.m. if you heard it at 8:00 a.m.? How long will it be until all but 10 students have heard the news?

4. *Spreading the News Simulation Experiment:* In this experiment you will simulate the spread of the news in Problem 3. Number each student in your class starting at 1. Person 1 stands up and then selects two people at random to "tell" the news to. Do this by selecting two random integers between 1 and the number of students in your class, inclusive. (It is not actually necessary to tell any news!) The random number generator on one student's calculator will help make the random selection. The two people with the chosen numbers stand. Thus after the first iteration there will probably be three students standing (unless a duplicate random number came up). Each of these (three) people selects two more people to "tell" the news to by selecting a total of 6 (or 4?) more random integers. Do this for a total of 10 iterations or until the entire class is standing. At each iteration, record the number of iterations and the total number of people who have heard the news. Describe the results of the experiment. Include such things as

   - The plot of the data points.
   - A function that fits the data, and a graph of this function on the plot. Explain why you chose the function you did.
   - A statement of how well the logistic model fits the data.
   - The iteration number at which the good news was spreading most rapidly.

5. *Ebola Outbreak Epidemic Problem:* In the fall of 2000, an epidemic of the ebola virus broke out in the Gulu district of Uganda. The table shows the total number of people infected from the day the cases were diagnosed as ebola virus infections. The final number of people who got infected during this epidemic is 396. (Ebola is a virus that causes internal bleeding and is fatal in most cases.)

x (days)	y (total infections)
1	71
10	182
15	239
21	281
30	321
50	370
74	394

a. Make a plot of the data points. Imagine a function that fits the data. Is the graph of this function concave upward or downward?

b. Use the second and last points to find the particular equation of a logistic function that fits the data.

c. Plot on the same screen as the plot in part a the logistic function from part b. Sketch the results.

d. Where does the point of inflection occur in the logistic model? What is the real-world meaning of this point?

e. Based on the logistic model, how many people were infected after 40 days?

f. Consult a reference on the Internet or elsewhere to find data about other epidemics. Try to model the spread of the epidemic for which you found data.

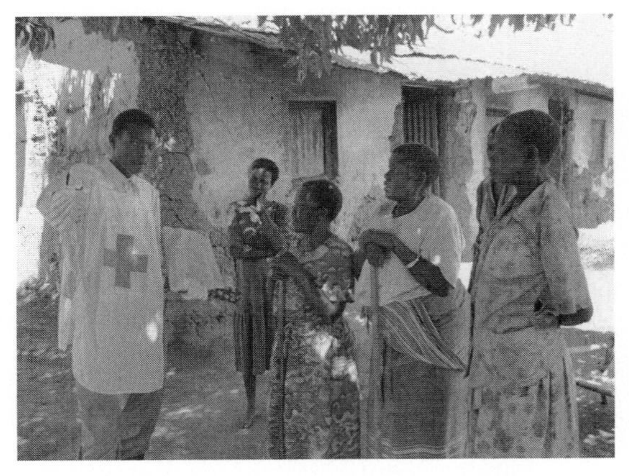

A Red Cross medical officer instructs villagers about the ebola virus in Kabede Opong, Uganda.

6. *Rabbit Overpopulation Problem:* Figure 7-6e shows two logistic functions

$$y = \frac{1000}{1 + ae^{-x}}$$

Both represent the population of rabbits in a particular woods as a function of time x in years. The value of the constant a is to be determined under two different initial conditions.

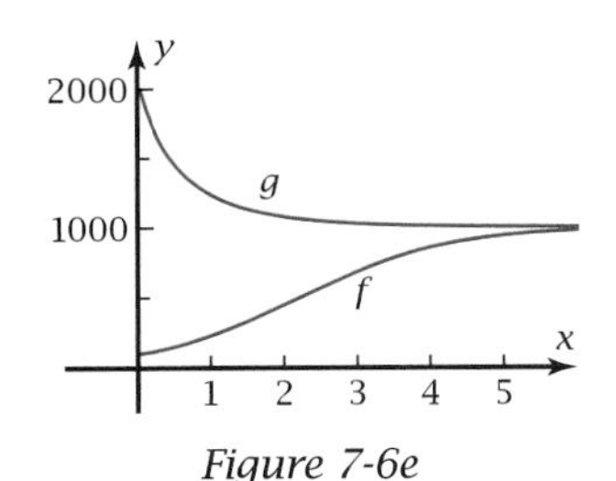

Figure 7-6e

a. For $y = f(x)$ in Figure 7-6e, 100 rabbits were introduced into the woods at time $x = 0$. Find the value of the constant a under this condition. Show that your answer is correct by plotting the graph of f on your grapher.

b. How do you interpret this mathematical model with regard to what happens to the rabbit population under the conditions in part a?

c. For $y = g(x)$ in Figure 7-6e, 2000 rabbits were introduced into the woods at time $x = 0$. Find the value of a under this condition. Show that the graph agrees with Figure 7-6e.

d. How do you interpret the mathematical model under the condition of part c? What seems to be the implication of trying to stock a region with a greater number of a particular species than the region can support?

7. Given the logistic function

$$f(x) = \frac{c}{1 + ae^{-0.4x}}$$

a. Let $a = 2$. Plot on the same screen the graphs of f for $c = 1, 2,$ and 3. Use a domain of $x \in [-10, 10]$. Sketch the results. True or false? "c is a vertical dilation factor."

b. Figure 7-6f shows the graph of f with $c = 2$ and with $a = 0.2, 1,$ and 5. Which graph is which? What transformation does a do on the graph?

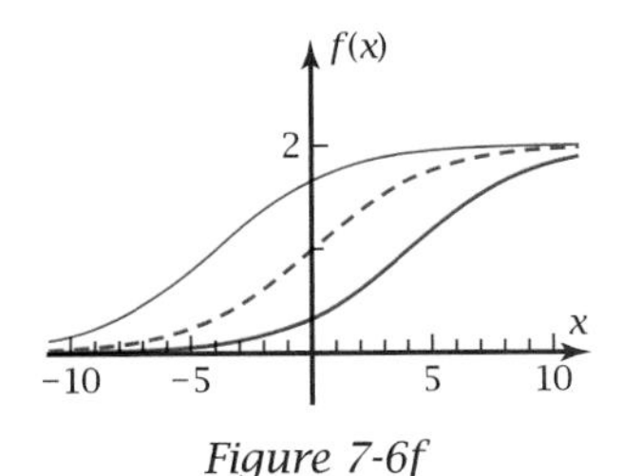

Figure 7-6f

c. Let $g(x) = \dfrac{c}{1 + ae^{-0.4(x-3)}}$.

What transformation of f does this represent? Confirm that your answer is correct by plotting f and g on the same screen using $c = 2$ and $a = 1$.

d. What value of a in the equation for $f(x)$ would produce the same transformation as in part c?

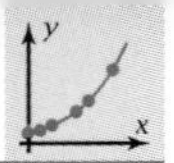

7-7 Chapter Review and Test

In this chapter you have learned graphical and numerical patterns for various types of functions:

- Linear
- Quadratic
- Power
- Exponential
- Logarithmic (the inverse of exponential)
- Logistic (for restrained growth)

These patterns allow you to tell which type of function might fit for a given real-world situation. Once you have selected a function that has appropriate concavity, increasing–decreasing behavior, and numerical behavior, you can find the particular equation by calculating values of the constants. You can check your work by seeing whether the function fits other given points. Once you have the correct equation, you can use it to interpolate between given values or extrapolate beyond given values to calculate y when you know x, or the other way around.

Review Problems

R0. Update your journal with what you have learned in this chapter. Include such things as the definitions, properties, and graphs of the functions just listed. Show typical graphs of the various functions, give their domains, and make connections between, for example, the add–multiply property of the exponential functions and the multiply–add property of the logarithmic functions. Show how you can use logarithms and their properties to solve for unknowns in exponential or logarithmic equations, and explain how these equations arise in finding the constants in the particular equation of certain functions. Tell what you have learned about the constant e and where it is used.

R1. This problem concerns these five function values:

x	$f(x)$
2	1.2
4	4.8
6	10.8
8	19.2
10	30.0

a. On the same screen, plot the points and the graph of $f(x) = 0.3x^2$.

b. Is the function increasing or decreasing? Is the graph concave up or concave down?

c. Name the function in part a. Give an example in the real world that this function might model. Is the y-intercept of f reasonable for this real-world example?

R2. a. Find the particular equation of a linear function containing the points (7, 9) and (10, 11). Give an example in the real world that this function could model.

b. Sketch two graphs showing a decreasing exponential function and an inverse-variation power function. Give two ways in which the graphs are alike. Give one way in which they differ.

c. How do you tell that the function graphed in Figure 7-7a is an exponential function, not a power function? Find the particular equation of the exponential function. Give an example in the real world that this exponential function could model.

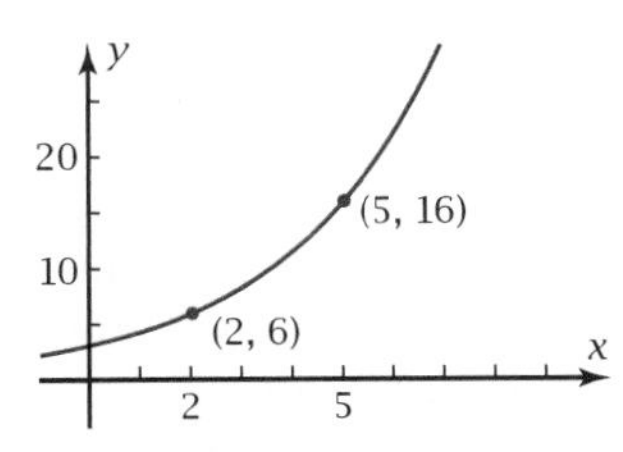

Figure 7-7a

d. Find the particular equation of the quadratic function graphed in Figure 7-7b. How does the equation you get show that the graph is concave downward? Give an example in the real world that this function could model.

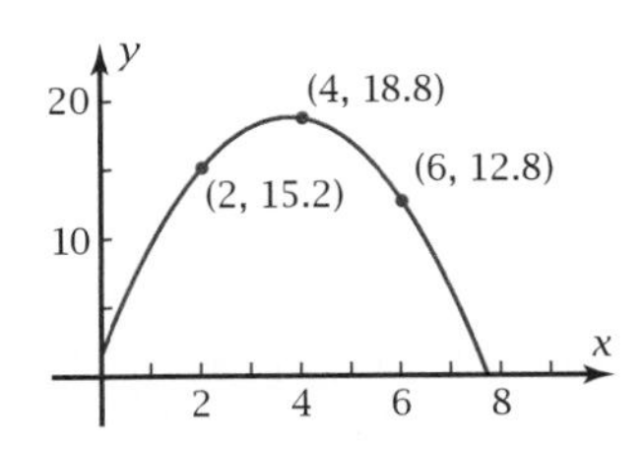

Figure 7-7b

e. A quadratic function has the equation $y - 3 = 2(x - 5)^2$. Where is the vertex of the graph? What does the y-intercept equal?

R3. For each table of values, tell from the pattern whether the function that fits the points is linear, quadratic, exponential, or power.

a.

x	$f(x)$
3	24
6	12
9	6
12	3

b.

x	$g(x)$
3	24
6	12
9	8
12	6

c.

x	$h(x)$
3	24
6	30
9	36
12	42

d.

x	$q(x)$
3	24
6	12
9	18
12	42

e. Suppose that $f(3) = 90$ and $f(6) = 120$. Find $f(12)$ if the function is

i. An exponential function

ii. A power function

iii. A linear function

f. Demonstrate that the add-multiply property for exponential functions is true for $f(x) = 53 \cdot 1.3^x$ by showing algebraically that adding the constant c to x multiplies the corresponding $f(x)$-value by a constant.

R4. a. The most important thing to remember about logarithms is that a logarithm is —?—.

b. Write in exponential form: $p = \log_c m$

c. Write in logarithmic form: $z = 10^p$

d. What does it mean to say that $\log 30 = 1.4771...$?

e. Find $\log_7 30$.

f. Give numerical examples to illustrate these logarithmic properties:

i. $\log (xy) = \log x + \log y$

ii. $\log \dfrac{x}{y} = \log x - \log y$

iii. $\log x^y = y \log x$

g. $\log 48 - \log 4 + \log 5 = \log$ —?—

h. $\ln 7 + 2 \ln 3 = \ln$ —?—

i. Solve the equation: $\log (x + 1) + \log (x - 2) = 1$

j. Solve the equation: $3^{2x-1} = 7^x$

R5. a. On the same screen, plot the graphs of $y_1 = \ln x$ and $y_2 = e^x$. Use the same scales on both axes. Sketch the results. How are the two graphs related to each other and to the 45° line $y = x$?

b. For the natural exponential function $f(x) = 5e^{-0.4x}$, write the equation in the form $f(x) = ab^x$. For the exponential function $g(x) = 4.3 \cdot 7.4^x$, write the equation as a natural exponential function.

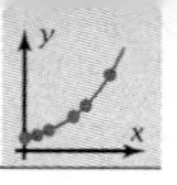

Sunlight Under the Water Problem (R5c–R5e): The intensity of sunlight under water decreases with depth. The table shows the depth, y, in feet, below the surface of the ocean you must go to reduce the intensity of light to the given percentage, x, of what it was at the surface.

x (%)	Depth y (ft)
100	0
50	13
25	26
12.5	39

c. What numerical pattern tells you that a logarithmic function fits the data? Find the particular equation of the function.

d. On the same screen, plot the data and the logarithmic function. Sketch the result.

e. Based on this mathematical model, how deep do you have to go for the light to be reduced to 1% of its intensity at the surface? Do you find this by interpolation or by extrapolation?

R6. a. Plot the graphs on the same screen and sketch the results.

Logistic function: $f(x) = \dfrac{10 \cdot 2^x}{2^x + 10}$

Exponential function: $g(x) = 2^x$

b. Explain why, when x is a large negative number, $f(x)$ is very close to $g(x)$. Explain why, when x is a large positive number, $f(x)$ is close to 10 and $g(x)$ is very large.

c. Transform the equation for $f(x)$ in part a so that it has only one exponential term.

d. Transform the equation for $g(x)$ in part a so that it is expressed in the form $g(x) = e^{kx}$.

e. *Population Problem 1:* A retirement community is built on an island in the Gulf of Mexico. The population grows steadily, as shown in the table.

x (months)	y (people)
6	75
12	153
18	260
24	355

Explain why a logistic function would be a reasonable mathematical model for population as a function of time. If the community has room for 460 residents, find the particular equation of the logistic function that contains the points for 6 months and for 24 months. Show that the equation gives approximately the correct answers for 12 months and 18 months. Plot the graph, and sketch the result. When is the population predicted to reach 95% of the capacity?

Concept Problems

C1. *Rise and Run Property for Quadratic Functions Problem:* The sum of consecutive odd counting numbers is always a perfect square. For instance,

$$1 = 1^2$$
$$1 + 3 = 4 = 2^2$$
$$1 + 3 + 5 = 9 = 3^2$$
$$1 + 3 + 5 + 7 = 16 = 4^2$$

This fact can be used to sketch the graph of a quadratic function by a "rise-run" technique similar to that used for linear functions. Figure 7-7c shows that for $y = x^2$, you can start at the vertex and use the pattern "over 1, up 1; over 1, up 3; over 1, up 5; . . . ."

a. On graph paper, plot the graph of $y = x^2$ by using this rise-run technique. Use integer values of x from 0 to 4. Then repeat the pattern for values of x from 0 to −4.

b. $y = -5 + (x - 2)^2$ is a translation of the graph of $y = x^2$. Locate the vertex, and then plot the graph on graph paper using the rise-run pattern.

c. $y = -5 + 0.3(x - 2)^2$ is a vertical dilation of the graph in part b. Use the rise-run technique for this function, and then plot its graph on the same axes as in part b.

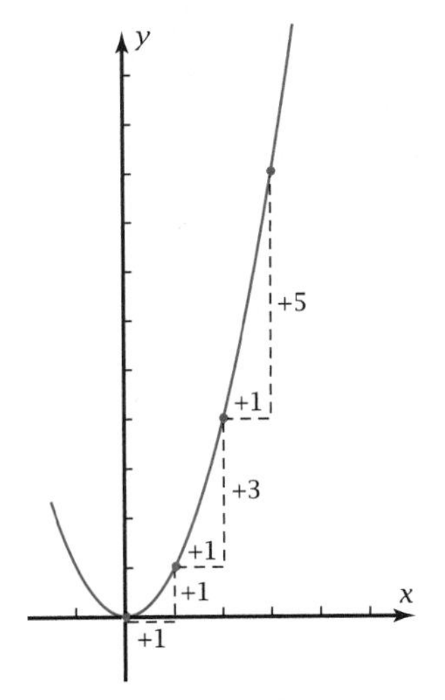

Figure 7-7c

C2. *Log-log and Semilog Graph Paper Problem:* Let $f(x) = 1000 \cdot 0.65^x$ be the number of bacteria remaining in a culture over time, x, measured in hours. Let $g(x) = 0.09x^2$ be the area of skin measured in square centimeters on a snake of length x, measured in centimeters. Figure 7-7d shows the graph of the exponential function f plotted on **semilog** graph paper. Figure 7-7e shows the graph of the power function g plotted on **log-log** graph paper. On these graphs, one or both of the axes have scales proportional to the logarithm of the variable's value. Thus the scales are compressed so that a wide range of values can fit on the same sheet of graph paper. For these two functions, the graphs are straight lines.

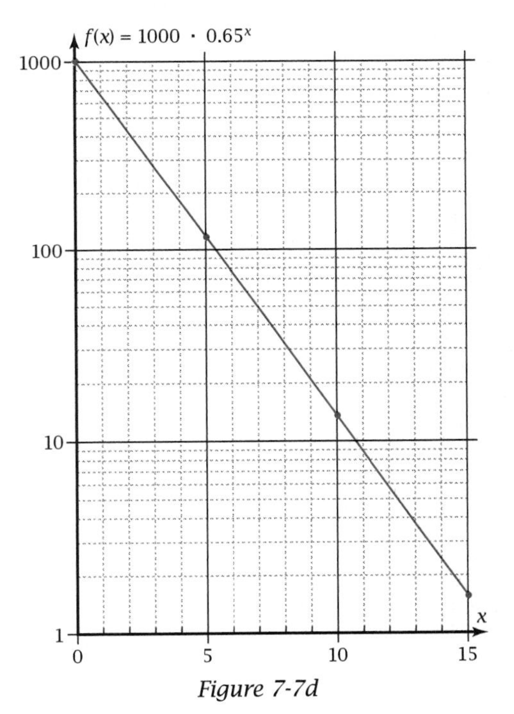

Figure 7-7d

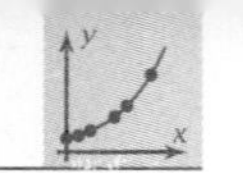

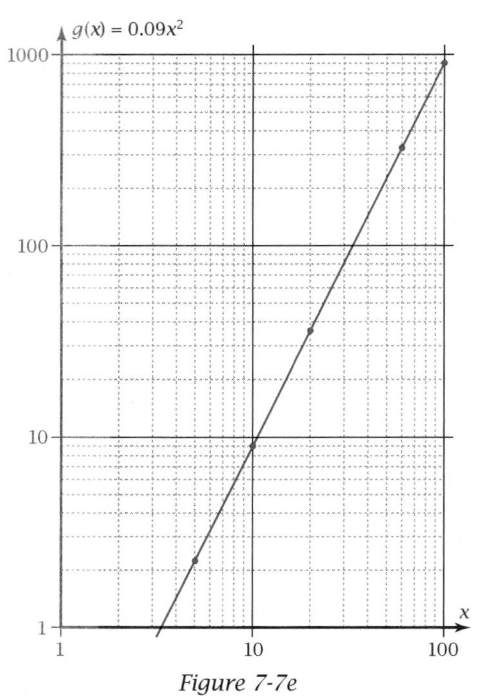

Figure 7-7e

a. Read the values of $f(9)$ and $g(60)$ from the graphs. Then calculate these numbers using the given equations. If your graphical answers are different from the calculated answers, explain what mistakes you made in reading the graphs.

b. You'll need a sheet of semilog graph paper and a sheet of log-log graph paper for graphing. On the semilog paper, plot the function $h(x) = 2 \cdot 1.5^x$ using several values of x in the domain [0, 15]. On the log-log paper, plot the function $p(x) = 700x^{-1.3}$ using several values of x in the domain [1, 100]. What do the graphs of the functions look like?

c. Take the logarithm of both sides of the equation $f(x) = 1000 \cdot 0.65^x$. Use the properties of logarithms to show that log $f(x)$ is a linear function of x. Explain how this is connected to the shape of the graph.

d. Take the logarithm of both sides of the equation $g(x) = 0.09x^2$. Use the properties of logarithms to show that log $g(x)$ is a linear function of log x. How does this relate to the graph in Figure 7-7e?

C3. *Slope Field Logistic Function Problem:* The logistic functions you have studied in this chapter model populations that start at a relatively low value and then rise asymptotically to a **maximum sustainable population.** There may also be a **minimum sustainable population.** Suppose that a new variety of tree is planted on a relatively small island. Research indicates that the minimum sustainable population is 300 trees and that the maximum sustainable population is 1000 trees. A logistic function modeling this situation is

$$y = \frac{300C + 1000e^{0.7x}}{C + e^{0.7x}}$$

where y is the number of trees alive x years after the trees were planted. The 300 and 1000 are the minimum and maximum sustainable populations, respectively, and C is a constant determined by the **initial condition,** the number of trees planted at time $x = 0$.

a. Determine the value of C and write the particular equation if, at time $x = 0$,

   i. 400 trees are planted.

   ii. 1300 trees are planted.

   iii. 299 trees are planted.

b. Plot the graph of each function in part a. Use a window with an x-range of about [0, 10] and a suitable y-range. What are the major differences among the three graphs?

c. Figure 7-7f shows a **slope field** representing functions with the given equation. The line segment through each grid point indicates the slope the graph would have if it passed through that point. On a copy of Figure 7-7f, plot the three equations from part a. How are the graphs related to the line segments on the slope field?

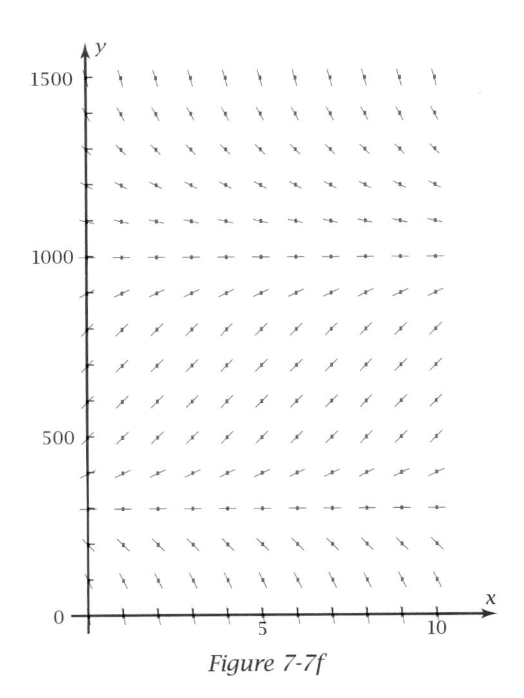

Figure 7-7f

d. Describe the behavior of the tree population for each of the three initial conditions in part a. In particular, explain what happens if too few trees are planted and also if too many trees are planted.

e. Without doing any more computations, sketch on the slope field the graph of the tree population if, at time $y = 0$,

i. 500 trees had been planted.

ii. 1500 trees had been planted.

iii. 200 trees had been planted.

f. How does the slope field allow you to analyze graphically the behavior of many related logistic functions without doing any computations?

Chapter Test

PART 1: No calculators allowed (T1–T7)

T1. Write the general equation of

a. A linear function

b. A quadratic function

c. A power function

d. An exponential function

e. A logarithmic function

f. A logistic function

T2. What type of function could have the graph shown?

a. 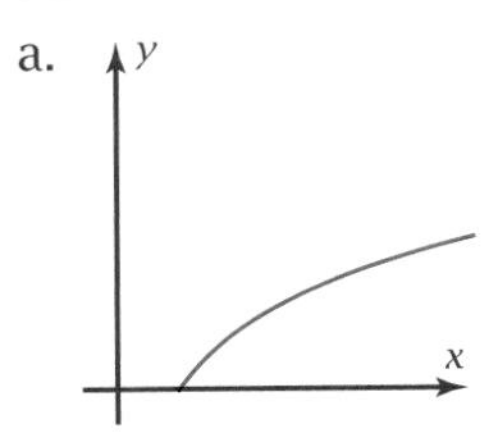

b. 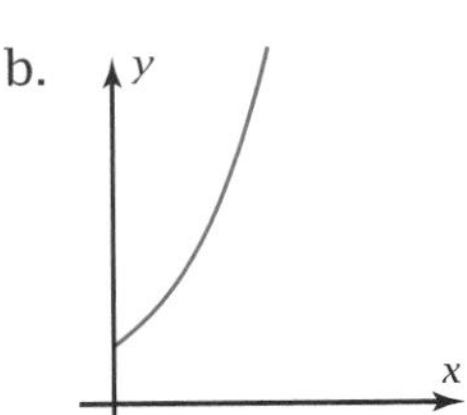

c. 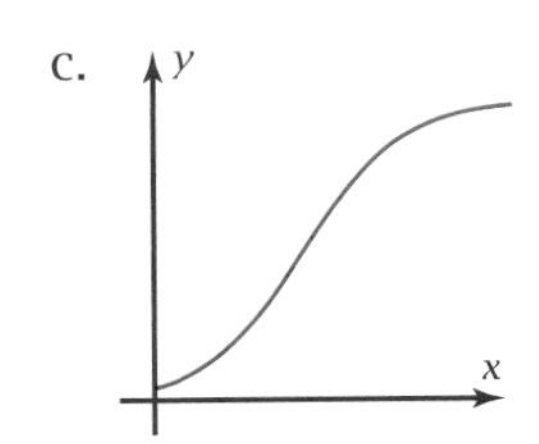

d. 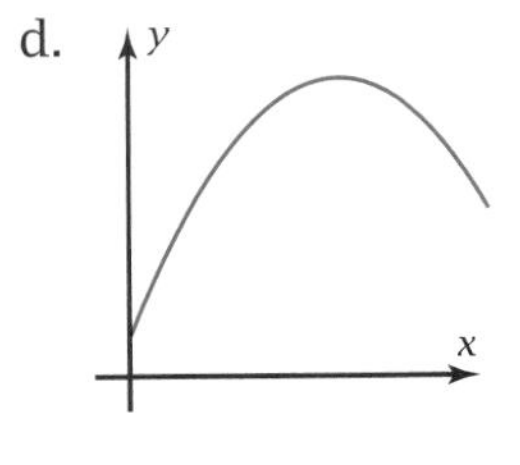

e. 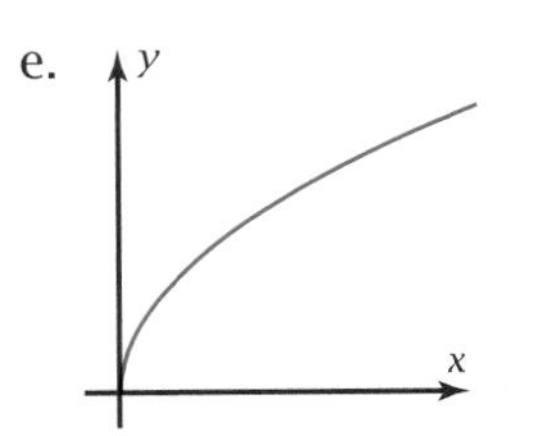

f. 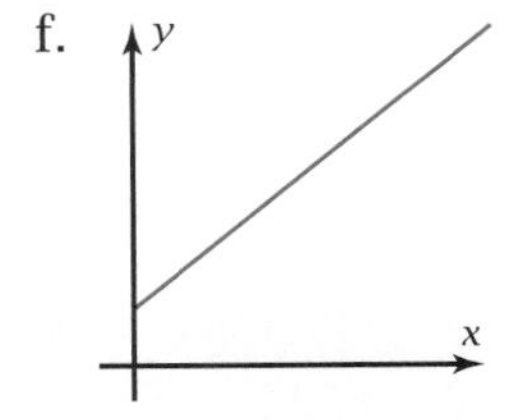

T3. What numerical pattern is followed by regularly spaced data for

a. A linear function

b. A quadratic function

c. A power function of the form $y = ax^b$

d. An exponential function of the form $y = ab^x$

e. A logarithmic function of the form $y = a \log_b x$

T4. Write the equation $\log_a b = c$ in exponential form.

T5. Show how to use the logarithm of a power property to simplify $\log 5^x$.

T6. $\ln 80 + \ln 2 - \ln 20 =$ —?—

T7. $\log 5 + 2 \log 3 = \log$ —?—

T8. Solve the equation: $4^x - 3 \cdot 2^x - 4 = 0$

T9. Solve the equation: $\log_2 (x - 4) - \log_2 (x + 3) = 8$

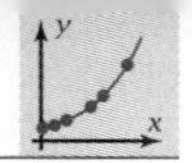

PART 2: Graphing calculators allowed (T10–T28)

Shark Problem: Suppose that from great white sharks caught in the past, fishermen find these weights and lengths. Use this set of data for Problems T10–T14.

x (ft)	$f(x)$ (lb)
5	75
10	600
15	2025
20	4800

T10. Show that the data set in the table has the multiply-multiply property of power functions.

T11. Write the general equation of a power function. Then use the points (5, 75) and (10, 600) to calculate algebraically the two constants in the equation. Store these values without round-off. Write the particular equation.

T12. Confirm that your equation in Problem T11 is correct by showing that it gives the other two points in the table.

T13. From fossilized shark teeth, naturalists think there were once great white sharks 100 feet long. Based on your mathematical model, how heavy would such a shark be? Surprising?!

T14. A newspaper report shows a great white shark that weighed 3000 pounds. Based on your mathematical model, about how long was the shark? Show the method you use.

Coffee Cup Problem: You pour a cup of coffee. Three minutes after you pour it, you find that it is 94.8 F° above room temperature. You record its temperature every 2 minutes thereafter, creating this table of data. Use the data for Problems T15–T18.

x (min)	$g(x)$ (F° above room temperature)
3	94.8
5	76.8
7	62.2
9	50.4
11	40.8

T15. Make a plot of the information. From the plot, tell whether the graph of the function you can fit to the points is concave up or concave down. Explain why an exponential function would be reasonable for this function but a linear and a power function would not.

T16. Find the particular equation of the exponential function that fits the points for $x = 3$ and $x = 11$. Show that the equation gives approximately the correct values for the other three times.

T17. Extrapolate the exponential function backward to estimate the temperature of the coffee when it was poured.

T18. Use your equation to predict the temperature of the coffee half an hour after it was poured.

T19. *The Add-Multiply Property Proof Problem 2:* Prove that if $y = 7(13^x)$, then $\log y$ is a linear function of x.

Model Rocket Problem: A precalculus class launches a model rocket out on the football field. The rocket fires for two seconds. Each second thereafter the class measures its height, finding the values in this table. Use the data for Problems T20–T22.

t (sec)	h (ft)
2	166
3	216
4	234
5	220
6	174

T20. Make a plot of the data points. Imagine fitting a function to the data. Is the graph of this function concave up or concave down? What kind of function would be a reasonable mathematical model for this function?

T21. Show numerically that a quadratic function would fit by showing that the second differences in the h data are constant.

T22. Use any three of the points to find the particular equation of the quadratic function that fits the points. Show that the equation gives the correct values for the other two points.

T23. *Logarithmic Function Problem 1:* A logarithmic function f has $f(2) = 4.1$ and $f(6) = 4.8$. Use the multiply-add property of logarithmic functions to find two more values of $f(x)$. Use the given points to find the particular equation of the form $f(x) = a + b \ln x$.

Population Problem 2: Problems T24–T27 concern a new subdivision that opens in a small town. The population of the town increases as new families move in. The table lists the population of the town various months after the opening of the subdivision.

Months	People
2	363
5	481
7	579
11	830

T24. Find the particular equation of the (untranslated) exponential function f that fits the first and last points. Show that the values of $f(5)$ and $f(7)$ are fairly close to the ones in the table.

T25. Show that the logistic function g gives values for the population that are also fairly close to the values in the table.

$$g(x) = \frac{3500}{1 + 10.8e^{-0.11x}}$$

T26. On the same screen, plot the four given points, the graph of f, and the graph of g. Use a window with an x-range of [0, 70] and a y-range of [0, 5000]. Sketch the result.

T27. Tell why the logistic function g gives more reasonable values for the population than the exponential function f when you extrapolate to large numbers of months.

T28. What did you learn from this test that you did not know before?

CHAPTER

8

Fitting Functions to Data

As a child grows, his or her height is a function of age. But the growth rate is not uniform. There are times when the growth spurts and times when it slows down. However, there might be some mathematical pattern that is true for all children's growth. For example, spurts or slowdowns might occur generally at the same age for boys and for girls. If such a pattern exists, the actual heights will be scattered around some mathematical function that fits the data approximately. The regression techniques you will learn in this chapter can be used to find various types of functions to fit such data and to analyze how good the fit is.

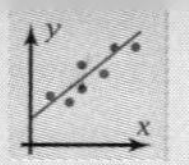

Mathematical Overview

In this chapter you'll learn how to use your grapher to find the best-fitting linear function for a given set of data and how it adapts this technique to fit other types of functions. You'll see how you can calculate the correlation coefficient, and you'll learn ways to find the type of function that is most appropriate for a given set of data. You'll apply the techniques you learn to problems such as predicting the increase of carbon dioxide in the atmosphere, a phenomenon that leads to global warming. You'll gain this knowledge in four ways.

Graphically Figure 8-0a shows a set of points and the linear function that fits the points the best.

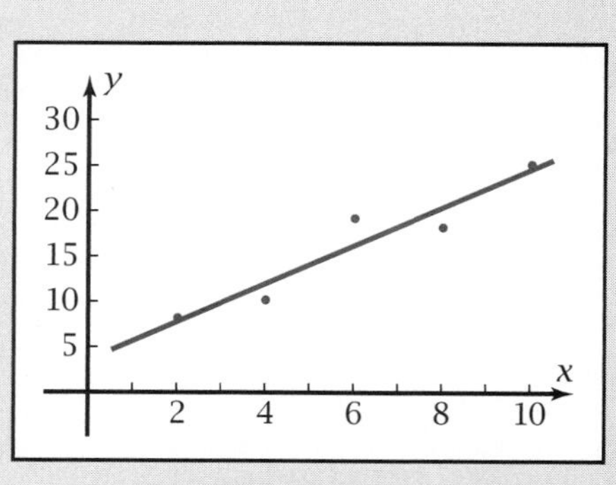

Figure 8-0a

Algebraically The equation of the best-fitting linear function is

$$\hat{y} = 2.1x + 3.4$$

where $\hat{y}$ is the predicted value of y.

Numerically In the last column, 17.60 is the sum of the squares of the residuals. For the best-fitting linear function, this number is a minimum.

	x	y	$\hat{y}$	$y - \hat{y}$	$(y - \hat{y})^2$
	2	8	7.6	0.4	0.16
	4	10	11.8	−1.8	3.24
	6	19	16.0	3.0	9.00
	8	18	20.2	−2.2	4.84
	10	25	24.4	0.6	0.36
Sums:	30	80		0.0	17.60

Verbally *I know how the calculator gets the regression equation and the correlation coefficient. I have learned that even if the correlation coefficient is close to 1, I must look at endpoint behavior and residual plots to decide which function is best. I also know it is risky to extrapolate the function too far beyond the range of the given data.*

8-1 Introduction to Regression for Linear Data

Moe is recovering from surgery. The table and Figure 8-1a show the number of sit-ups he has been able to do on various days afterward. Figure 8-1b shows the best-fitting linear function

$$\hat{y} = 2.1x + 3.4$$

where $\hat{y}$ (pronounced "y hat") is used to distinguish points on this line from the actual y data points. The difference $y - \hat{y}$ is called the residual deviation or, more briefly, the **residual.** In this section you'll learn how to use your grapher to find this equation.

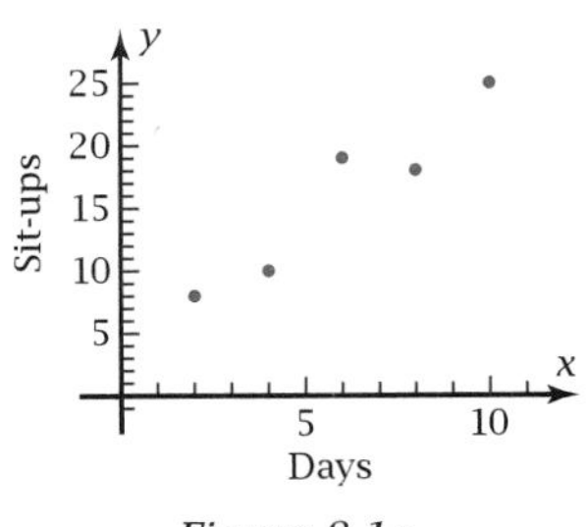

Figure 8-1a

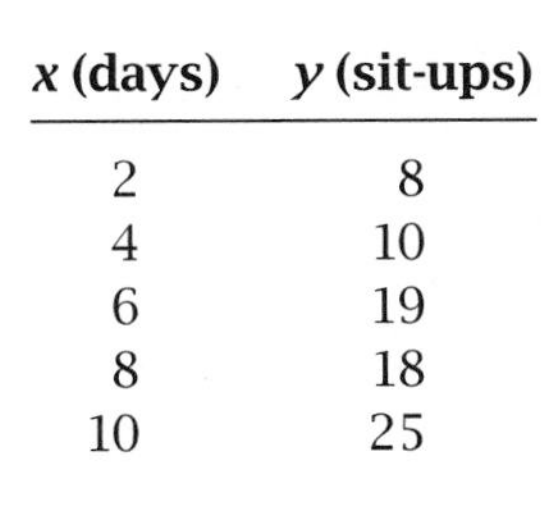

x (days)	y (sit-ups)
2	8
4	10
6	19
8	18
10	25

Figure 8-1b

OBJECTIVE Find the equation of the best-fitting linear function for a set of points by running a linear regression on your grapher, and calculate the sum of the squares of the residuals.

Exploratory Problem Set 8-1

1. Enter the x- and y-values in two lists on your grapher. Then from the statistics menu run a **linear regression** in the form of $\hat{y} = ax + b$. Did you find that the equation is $\hat{y} = 2.1x + 3.4$?

2. Plot the given points and the linear equation from Problem 1 on the same screen. Does the result agree with Figure 8-1b?

3. Show how to use the linear function to predict the number of sit-ups Moe could do two weeks after surgery. What real-world reasons could explain why the actual number of sit-ups might be different from the predicted number?

4. Copy the table. Put in three new columns, one each for
   - The values of $\hat{y}$ calculated by the equation $\hat{y} = 2.1x + 3.4$
   - The residuals, calculated by subtraction, $y - \hat{y}$
   - The squares of the residuals, $(y - \hat{y})^2$

5. Find the **sum of the squares of the residuals.** This number is abbreviated SS_{res}.

6. The **regression line** is the line that makes SS_{res} a *minimum.* Because SS_{res} is a minimum, the regression line is the best-fitting linear function. Show that SS_{res} would be greater for the function $y_2 = 2.1x + 3.5$ that has y-intercept 3.5 instead of 3.4 and also greater for the function $y_3 = 2.2x + 3.4$ that has slope 2.2 instead of 2.1.

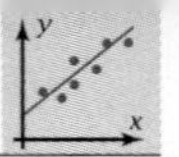

8-2 Deviations, Residuals, and the Correlation Coefficient

In Chapter 7 you found equations for functions that fit given sets of points. If the points are data measured from the real world, they are usually scattered. For instance, Figure 8-2a shows the scatter plot of the weights of 43 fish plotted against their lengths. The scatter of this cloud of points occurs because fish of the same length can have different weights.

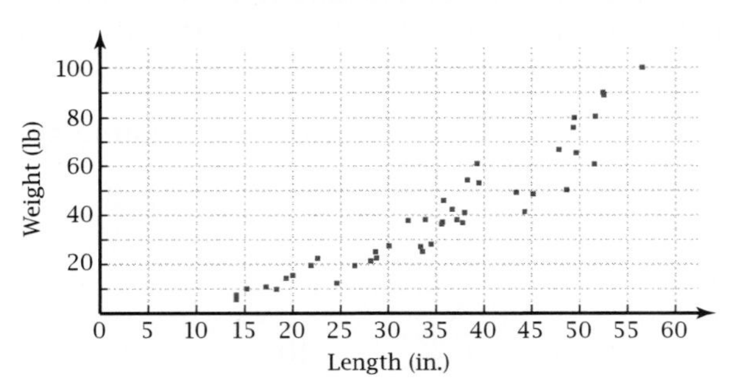

Figure 8-2a

In this section you will learn the basis behind doing regression as you saw in Section 8-1 for points that follow a linear pattern. In the next section you will see how to extend regression to fit data that follow a curved pattern, as in Figure 8-2a.

OBJECTIVE Calculate SS_{res}, the sum of the squares of the residuals, and find out how to determine the equation of the linear function that minimizes SS_{res}.

In Section 8-1 you analyzed data for y sit-ups that Moe could do x days after surgery, as shown in this table. Figure 8-2b shows a scatter plot of the data.

	x (days)	y (sit-ups)
	2	8
	4	10
	6	19
	8	18
	10	25
Sums:	30	80

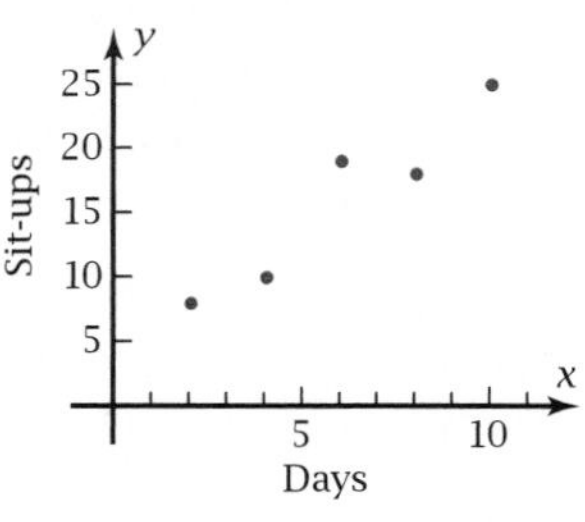

Figure 8-2b

Suppose that you are asked to estimate the number of sit-ups Moe could do, but you are not told the number of days. Your best estimate would be $\bar{y}$ (pronounced "y bar"), the average of the y-values:

$$\bar{y} = \frac{80}{5} = 16$$

Figure 8-2c shows the graph of the constant function $y = \bar{y}$ and the **deviation,** $y - \bar{y}$, of each point from this line.

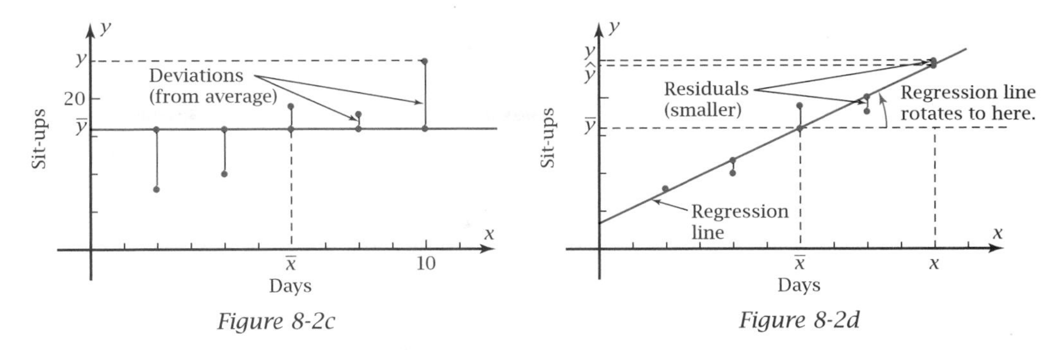

Figure 8-2c *Figure 8-2d*

However, if you *do* know for which day to find an estimate, you can find a better estimate of the number of sit-ups for a particular day by assuming that y is a linear function of x. Figure 8-2d shows that much of each deviation is removed by making the line slanted instead of horizontal. The part of each deviation that remains is called the residual deviation, or residual. If $\hat{y}$ is the value of y for a point on the line, then the residual equals $y - \hat{y}$.

A measure of how well the line fits the point is obtained by squaring each residual, thus making each value positive or zero, and then summing the results. The answer, SS_{res}, is called the **sum of the squares of the residuals.** Doing the same thing for the deviations from the mean gives a quantity called SS_{dev}, the **sum of the squares of the deviations.** Figures 8-2e and 8-2f show that the squares of the residuals are much smaller than the squares of the deviations from the mean.

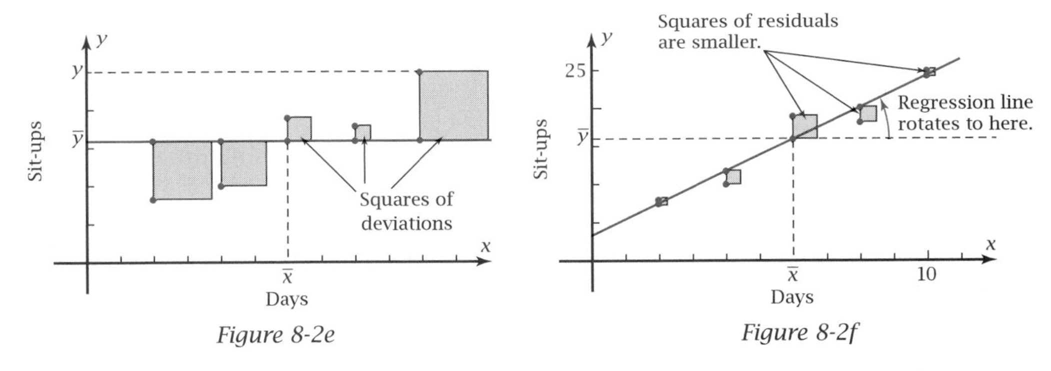

Figure 8-2e *Figure 8-2f*

The **regression equation** $\hat{y} = 2.1x + 3.4$ that you found on your grapher in Section 8-1 is the equation of the linear function for which SS_{res} is the minimum of all possible values. The fraction of the original SS_{dev} that has been removed by using the slanted linear function instead of the constant function is

$$r^2 = \frac{SS_{\text{dev}} - SS_{\text{res}}}{SS_{\text{dev}}}$$

The quantity r^2 is called the **coefficient of determination.** The symbol r^2 is used because its units are the squares of the y-units. The number r is called the **correlation coefficient.** The correlation coefficient indicates how *well* the best-fitting linear function fits the values. The sign of r shows whether the

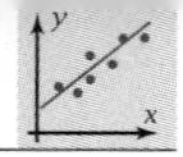

two variables, x and y, are positively or negatively **associated** with each other. If the y-variable increases as x increases, then the correlation coefficient is positive; if the y-variable decreases as x increases, then the correlation coefficient is negative.

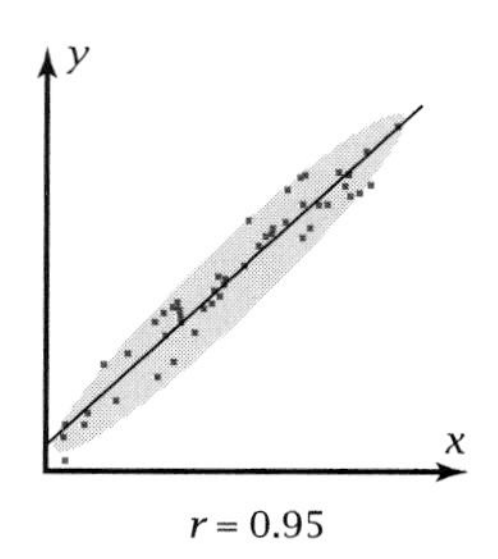

$r = 0.95$

The correlation coefficient also measures the strength of the association. If $r = 1$ or $r = -1$, the function fits the values exactly. This could be the case for a theoretical law in physics or chemistry, where the function that identifies the relationship between the variables follows an exact pattern. Of course, when you collect data, even if there is an underlying law, the data will not fit perfectly, due to measurement errors (from equipment or human inexactness).

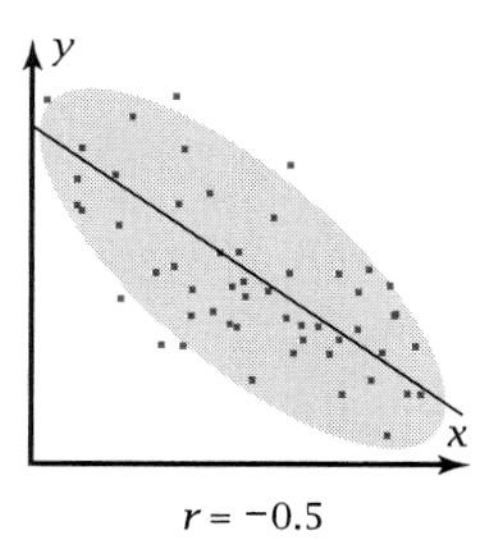

$r = -0.5$

The closer the correlation coefficient is to 1 or −1, the closer the points cluster around the line and the stronger the association is between the variables. If $r = 0$, there is no relationship between x and y. It is helpful to sketch an ellipse around the cloud of points in a scatter plot. If the ellipse is narrow, the correlation is **strong**; if the ellipse is wide, the correlation is **weak.** These scatter plots will help you visualize the strength of the correlation and connect it to the values of the correlation coefficient.

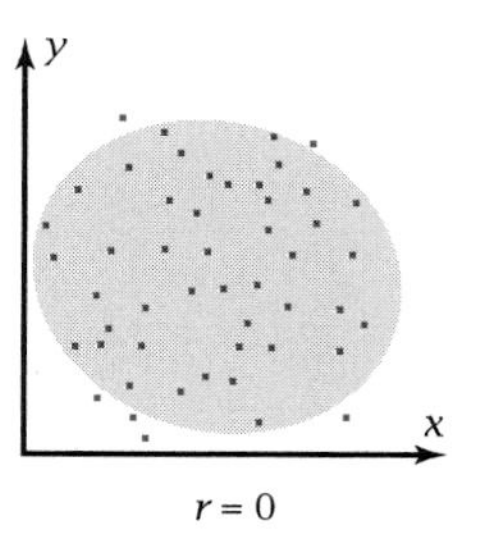

$r = 0$

These examples show how to find SS_{dev} and SS_{res} by operations on lists on your grapher.

▶ EXAMPLE 1 Find SS_{dev} and SS_{res} for the sit-ups data in this section. Use the results to calculate the coefficient of determination and the correlation coefficient. How do you interpret the value of the coefficient of determination?

Solution Enter the values of x and y in columns on a spreadsheet or in two lists on your grapher, say, L_1 and L_2. In a third column or list, compute the squares of the deviations, $(y - 16)^2$. In a fourth column or list, compute the squares of the residuals, $(y - \hat{y})^2$. To do this on your grapher, enter $\hat{y} = 2.1x + 3.4$ as y_1, and then enter $(L_2 - y_1(L_1))$^2 in L_4. The expression $y_1(L_1)$ means the values of $\hat{y}$ (the function in y_1) at the values of x in L_1.

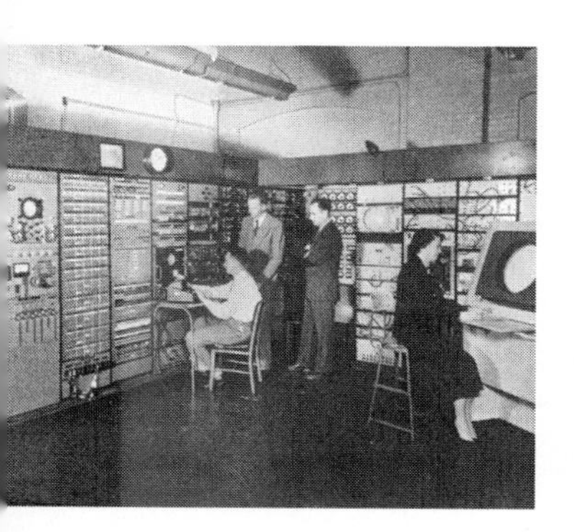
Computer from 1956. Modern computers made statistical data analysis much easier and stimulated the development of different statistical techniques.

x	y	$(y-16)^2$	$(y-\hat{y})^2$	
2	8	64	0.16	$[8 - (2.1(2) + 3.4)]^2 = [8 - 7.6]^2 = 0.4^2 = 0.16$
4	10	36	3.24	
6	19	9	9.00	
8	18	4	4.84	
10	25	81	0.36	
		$SS_{\text{dev}} = 194$	$17.60 = SS_{\text{res}}$	Use the sum command on your grapher.

$$r^2 = \frac{194 - 17.60}{194} = 0.909278\ldots$$

$$r = \sqrt{0.909278\ldots} = 0.9535\ldots$$

Positive square root because the function is increasing.

The coefficient of determination, $r^2 = 0.909...$, indicates that 90.9...% of the original deviation has been accounted for by the linear relationship between x and y, and the remaining 9.0...% is due to other fluctuations in the data. The correlation between the variables is fairly strong. ◀

Notes:

- The values of r^2 and r are found when your grapher calculates linear regression, and you can set it so these values are displayed.
- Using any other linear equation besides $\hat{y} = 2.1x + 3.4$ gives a value of SS_{res} larger than 17.60. For instance, $y = 2.2x + 3.4$ gives $SS_{res} = 19.80$.

Here is a summary of the quantities associated with linear regression.

DEFINITIONS: Deviations, Residuals, the Regression Line, and Correlation

The **deviation** of a data point (x, y) is $y - \bar{y}$, the directed distance of its y-value from $\bar{y}$, where $\bar{y}$ is the average of the y-values.

The **residual** (or residual deviation) of a data point from the line $\hat{y} = mx + b$ is $y - \hat{y}$, the vertical directed distance of its y-value from the line (Figure 8-2g).

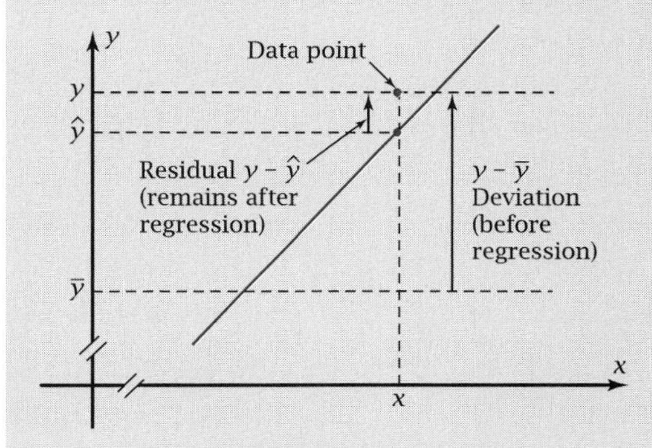

Figure 8-2g

The **sum of the squares of the deviations,** $SS_{dev} = \Sigma(y - \bar{y})^2$, where Σ, the capital Greek letter sigma, means "the sum of the values following the Σ sign."

The **sum of the squares of the residuals** is $SS_{res} = \Sigma(y - \hat{y})^2$.

The **linear regression line** for a set of data is the line for which SS_{res} is a minimum. The **linear regression equation** is the equation of this line, $\hat{y} = mx + b$.

The **coefficient of determination** is $r^2 = \dfrac{SS_{dev} - SS_{res}}{SS_{dev}}$. It is the fraction of SS_{dev} that has been removed by the linear regression.

The **correlation coefficient,** r, is the positive or negative square root of the coefficient of determination. Use the positive square root if the slope of the line is positive, and use the negative square root if the slope is negative.

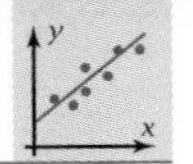

Problem Set 8-2

Do These Quickly

Q1. If $\hat{y} = 3x + 5$ and $x = 4$, by how much does $y = 19$ deviate from $\hat{y}$ on the line?

Q2. If $\hat{y} = 3x + 5$ and $x = 6$, by how much does $y = 20$ deviate from $\hat{y}$ on the line?

Q3. Find the sum of the squares of the residuals if the residuals are 3, −1, −4, and 2.

Q4. How can you tell from the correlation coefficient that a function fits the data perfectly?

Q5. What kind of function has the multiply-multiply property?

Q6. For what kind of function could $f(x + 3) = 8\,f(x)$?

Q7. log 5 + log 7 = log —?—

Q8. What is the least common denominator for $\frac{2}{7} + \frac{3}{8}$?

Q9. In a right triangle, $\dfrac{\text{opposite leg}}{\text{adjacent leg}}$ is the —?— of the angle.

Q10. Expand the square: $(mx + b)^2$

---

1. *Residuals Problem:* Suppose that these data have been measured for two related variables x and y.

x	y
5	11
8	16
11	19
14	27
17	25
20	29
23	33
26	42
29	44
32	51

a. Enter the data in two lists on your grapher. Show by linear regression that the best-fitting linear function is $\hat{y} = 1.4x + 3.8$. Record the correlation coefficient.

b. Make a scatter plot of the data on your grapher. On the same screen, plot $\hat{y}$. How well does the linear function fit the data?

c. Calculate $\bar{x}$ and $\bar{y}$, the averages of x and y. Show algebraically that the average-average point $(\bar{x}, \bar{y})$ is on the regression line.

d. Define new lists to help you calculate the squares of the deviations, $(y - \bar{y})^2$, and the squares of the residuals, $(y - \hat{y})^2$. By summing these lists, calculate SS_{dev} and SS_{res}. Use the results to calculate the coefficient of determination and the correlation coefficient. Does the correlation coefficient agree with the one you recorded in part a?

e. The line $y_2 = 1.5x + 1.95$ also contains the average-average point $(\bar{x}, \bar{y})$, but it has a slope of 1.5 instead of 1.4. Plot the line on the same screen as in part b. Can you tell from the graphs which line fits the data better? Explain. Show that SS_{res} for this line is greater than SS_{res} for the regression line.

2. *New Subdivision Problem:* The data represent actual prices of various garden homes in a new subdivision. The data have been rounded to the nearest 100 square feet and to the nearest 1000 dollars.

Square Feet	Dollars
1900	155,000
2100	168,000
2400	190,000
2500	189,000
2500	207,000
2600	195,000
2600	199,000
2600	199,000
2700	210,000
2800	220,000

a. Run a linear regression on the data. Record the correlation coefficient. Plot the regression equation and the data on the same screen. Use a window with an x-range of [0, 3000]. How can you tell that a linear function fits the data reasonably well?

b. Based on the linear model, how much would you expect to pay for a 5000-ft^2 house in this subdivision? How big a house could you buy for a million dollars? What do you call the process of calculating an x- or y-value outside the given data? What do you call the process of estimating an x- or y-value within the given data?

c. What real-world meaning can you give to the slope and the y-intercept?

d. Find $\bar{x}$ and $\bar{y}$. Show that the average-average point $(\bar{x}, \bar{y})$ is on the regression line.

e. Find SS_{dev} and SS_{res}. Use the results to calculate the coefficient of determination and the correlation coefficient. Do your answers agree with the results from part a?

f. Why is it reasonable for there to be more than one data point with the same x-value?

3. *Gas Tank Problem:* Lisa Carr fills up her car's gas tank and drives off. This table shows the numbers of gallons of gas left in the tank at various numbers of miles she has driven.

Miles	Gallons
6	16.7
22	15.9
44	14.8
50	14.5
60	14.0

a. Run a linear regression on the data. Write down the linear regression equation, and r^2 and r. How do these numbers tell you that the regression line fits the data perfectly? Why is r negative?

b. Calculate the coefficient of determination and the correlation coefficient again, directly from the definition, by calculating SS_{dev} and SS_{res}. Do the answers agree with part a?

c. Plot the data and the regression equation on the same screen. How can you tell graphically that the regression line fits the data perfectly?

d. According to your mathematical model, how much gas does the gas tank hold? How many miles per gallon does Lisa's car get?

e. Show that your mathematical model predicts that the tank is empty after 340 miles. Because this number is found by extrapolation, how confident are you that the car will actually run out of gas after 340 miles?

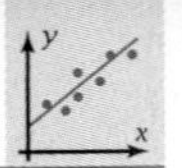

4. *Standardized Test Scores Problem:* Figure 8-2h shows a scatter plot of the scores of 1000 12th-graders on the mathematics part of the SAT (Scholastic Aptitude Test) and their high school grade point averages. By regression, the best-fitting linear function is $\hat{y} = 59.0x + 355$, but the coefficient of determination is only $r^2 \approx 0.14$.

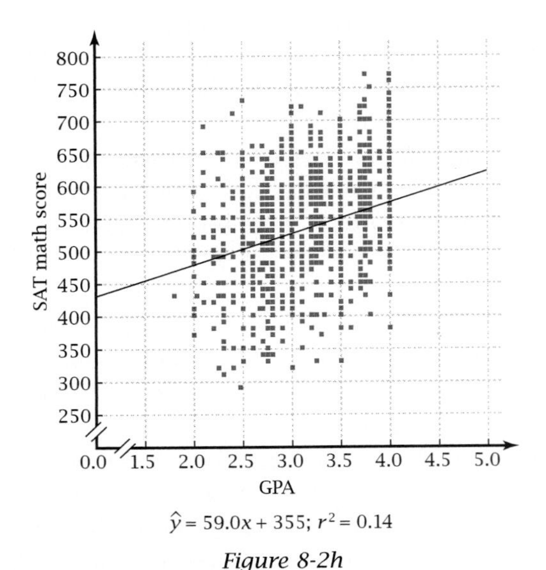

$\hat{y} = 59.0x + 355$; $r^2 = 0.14$

Figure 8-2h

a. Suppose that you have a 4.0 grade point average. According to the regression equation, what score would you be predicted to get on the SAT? How reliable do you think this prediction is?

b. Use the given information to find the correlation coefficient. Explain why you would use the positive square root rather than the negative one.

5. *Data Cloud Problem:* Figure 8-2i shows an elliptical region in which the "cloud" of data points is expected to lie if the correlation coefficient is $r = -0.95$. For each of the following values of r, sketch the cloud you would expect for data with the given correlation coefficient.

   a. $r = 0.95$

   b. $r = 0.8$

   c. $r = -0.7$

   d. $r = 0$

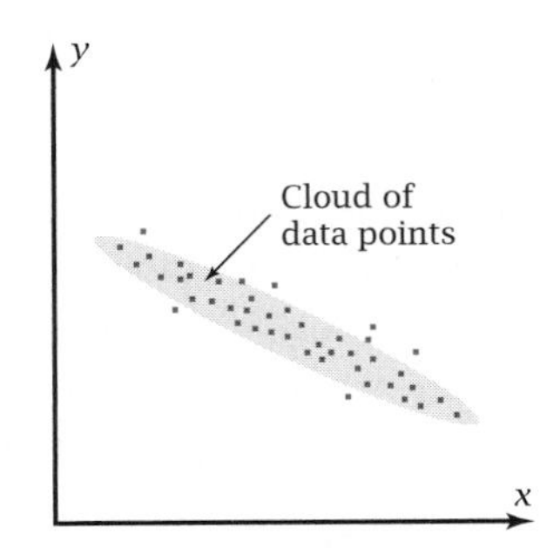

Figure 8-2i

8-3 Regression for Nonlinear Data

In Section 8-2 you learned the basis of linear regression. In this section you will learn how to use regression on your grapher to fit other types of functions to data for which the scatter plot follows a nonlinear pattern.

OBJECTIVE Given a set of data, make a scatter plot, identify the type of function that could model the relationship between the variables, and use regression to find the particular equation that best fits the data.

Figure 8-3a shows the scatter plot of fish weight, y, versus length, x, that you saw in Section 8-2. The table shows the data for these 43 fish.

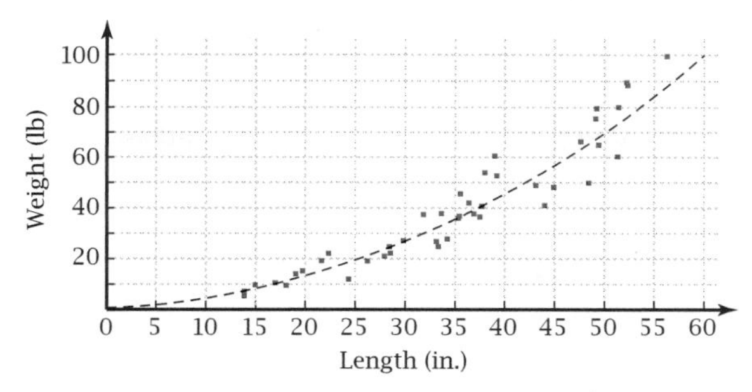

Figure 8-3a

x	y	x	y	x	y	x	y	x	y	x	y
14.0	5.5	22.5	22.3	33.3	27.0	37.1	37.9	45.1	48.4	52.4	89.6
14.0	7.4	24.5	12.2	33.5	25.0	37.7	36.7	47.8	66.4	52.5	88.5
15.1	9.9	26.4	19.3	33.8	38.0	37.9	40.9	48.6	50.1	56.5	99.8
17.1	10.7	28.1	21.2	34.4	28.0	38.2	54.1	49.3	75.4		
18.2	9.7	28.6	24.9	35.5	36.2	39.2	60.7	49.4	79.4		
19.2	14.2	28.7	22.4	35.6	37.0	39.4	52.9	49.6	65.1		
19.9	15.4	30.0	27.3	35.7	45.8	43.3	49.1	51.5	60.5		
21.8	19.4	32.0	37.6	36.6	42.2	44.2	41.2	51.6	79.9		

Because the points follow a curved path that is concave upward, an untranslated power or an exponential function might be a reasonable mathematical model for weight of fish as a function of their lengths. To decide which of the two is more reasonable, consider the **endpoint behavior.** At the left end of the domain, the graph would contain the origin because a fish of zero length would have zero weight. Because an untranslated power function contains the origin, it would be more reasonable than an untranslated exponential function. Press the **power regression** keys to get

$$y = 0.0606\ldots\, x^{1.7990\ldots}$$

Figure 8-3b shows that the graph of this function fits the points reasonably well.

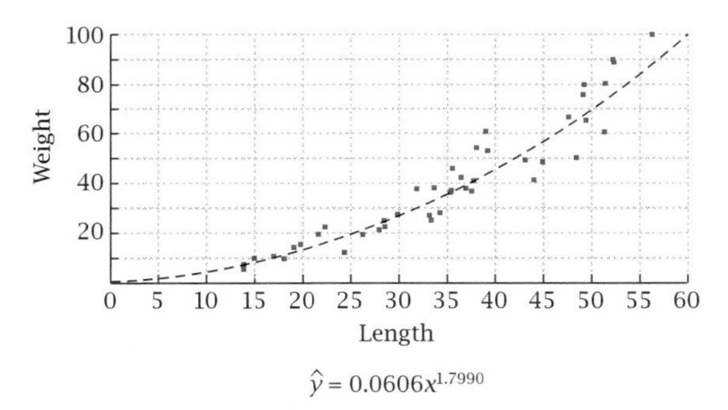

$\hat{y} = 0.0606x^{1.7990}$

Figure 8-3b

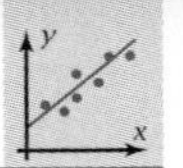

Figure 8-3c shows the result of running **exponential regression** (curved graph) and **linear regression** (straight graph). Neither function has the correct endpoint behavior at $x = 0$.

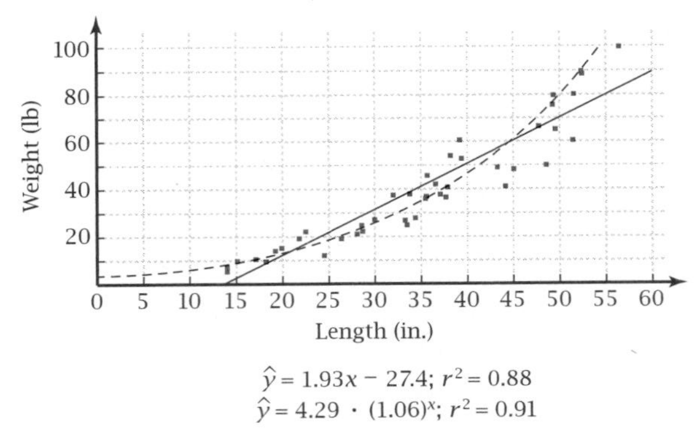

$\hat{y} = 1.93x - 27.4$; $r^2 = 0.88$
$\hat{y} = 4.29 \cdot (1.06)^x$; $r^2 = 0.91$

Figure 8-3c

Your grapher will display the correlation coefficient $r = 0.9669...$ for power regression, $r = 0.9557...$ for exponential regression, and $r = 0.9354...$ for linear regression. A correlation coefficient closer to 1 or -1 indicates a better fit. So in addition to having the wrong endpoint behavior, the other two types of functions have a weaker correlation to the data.

Problem Set 8-3

Do These Quickly

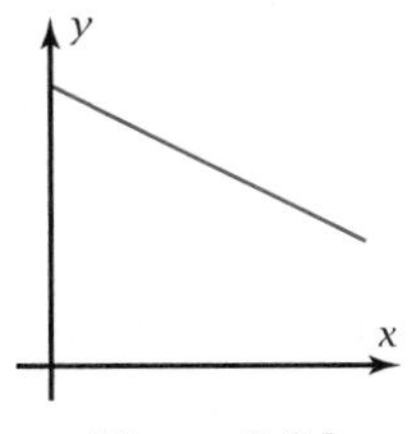

Figure 8-3d

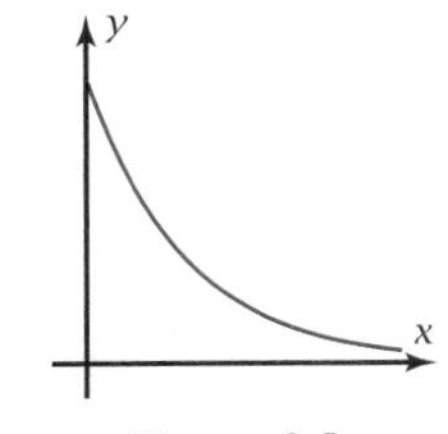

Figure 8-3e

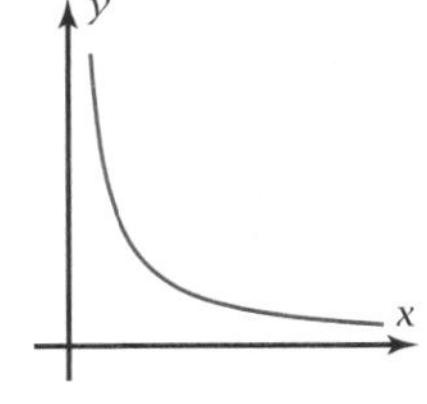

Figure 8-3f

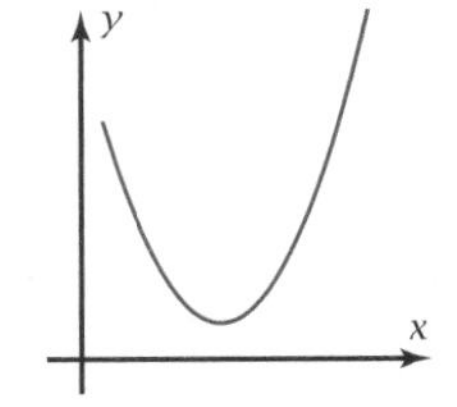

Figure 8-3g

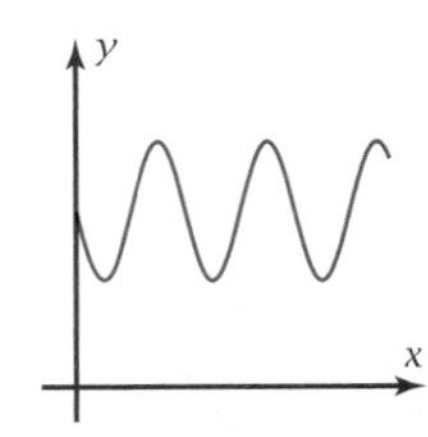

Figure 8-3h

Q1. What type of function is the graph in Figure 8-3d?

Q2. What type of function is the graph in Figure 8-3e?

Q3. What type of function is the graph in Figure 8-3f?

Q4. What type of function is the graph in Figure 8-3g?

Q5. What type of function is the graph in Figure 8-3h?

Q6. Sketch a reasonable graph: The population of a city depends on time.

Q7. The graph of a quadratic function is called a —?—.

Q8. Expand the square: $(3x - 7)^2$

Q9. Write the next three terms in this sequence: 3, 6, 12, 24, . . .

Q10. The "add-multiply" property is a characteristic of —?— functions.

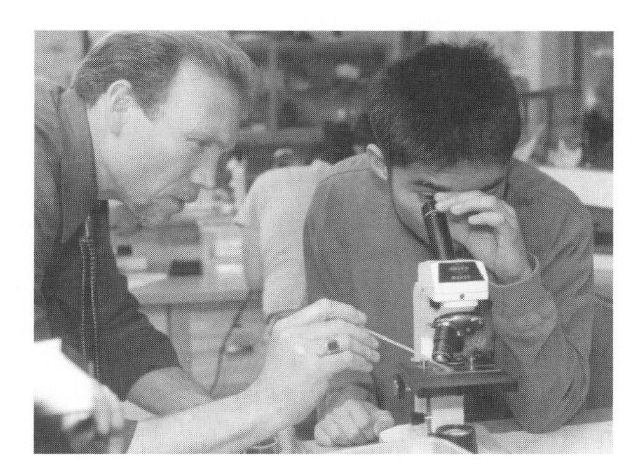

1. *Bacteria Problem:* Thirty-six college biology students start bacterial cultures by taking equal-volume samples from a flask in the laboratory. Each student comes back x hours later in the day and measures the number of bacteria, y, in his or her culture. The results are shown in the table and the scatter plot in Figure 8-3i.

   a. Explain why either an untranslated power function or an untranslated exponential function could be a reasonable mathematical model for the number of bacteria as a function of time. Explain why the exponential function would have a more reasonable left endpoint behavior than a power function.

   b. Run an exponential regression on the data. Write down the equation and the correlation coefficient. Plot the function and the scatter plot on the same screen. Sketch the result on a copy of Figure 8-3i.

   c. According to the exponential model, how many bacteria were in each equal sample when the students took them? What do you predict the number of bacteria will be on the next day, 24 hours after the cultures were started?

   d. After how many hours do you expect the number of bacteria to reach 100,000?

x	y	x	y
0.6	450	4.6	1963
0.8	446	4.7	1774
1.3	588	4.7	2611
1.5	645	4.9	2853
1.6	718	5.0	1848
1.9	729	5.0	3266
2.0	855	5.1	2229
2.3	1008	5.2	2827
3.1	962	5.3	3776
3.5	1570	5.5	2611
3.7	1620	5.7	3126
3.8	1378	5.7	2928
3.8	1561	5.8	3776
3.8	1580	6.3	4067
3.9	1919	6.3	5393
4.2	2210	6.4	5288
4.5	2212	7.3	7542
4.6	2125	8.6	11,042

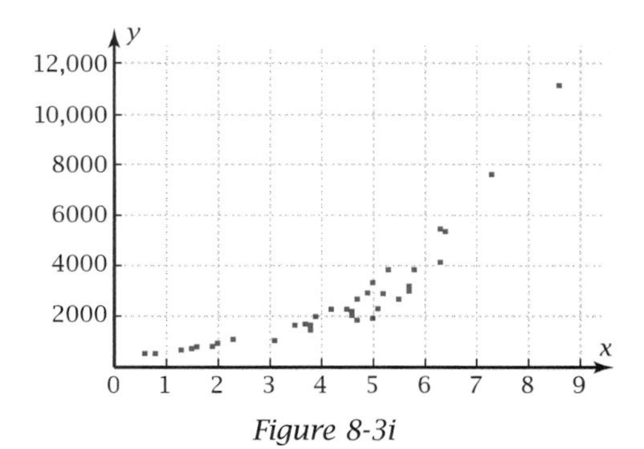

Figure 8-3i

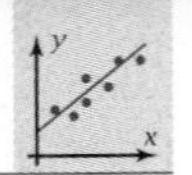

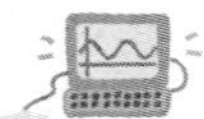

2. *Printed Paragraph Problem:* Ann A. Student types a paragraph on her word processor. Then she adjusts the width of the paragraph by changing the margins. Figure 8-3j and the table show the widths and number of lines the paragraph takes.

x (inches wide)	y (lines long)
6.5	5
5.5	6
5.0	7
3.75	9
3.0	11
1.5	24
1.0	38

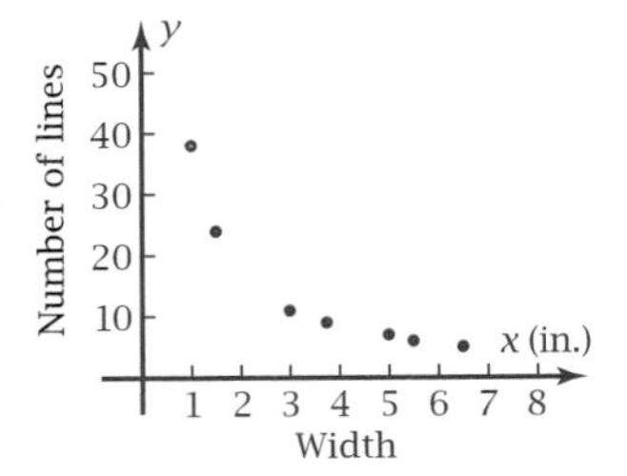

Figure 8-3j

a. Would it be possible to make the width equal zero? Why, then, would a decreasing power function be more reasonable than an exponential function for number of lines as a function of paragraph width? Confirm that a power function fits better by running both power and exponential regressions and comparing the correlation coefficients.

b. Plot the power function and the exponential function from part a on the same screen as a scatter plot of the data. How does the result confirm that the power function fits better?

c. Because each paragraph contains the same words, you might expect the area of the page taken by the paragraph to be constant. In a list on your grapher, calculate the area of each paragraph. Use the fact that there are 7 lines per inch. Make a scatter plot of the areas as a function of paragraph width. By linear regression, show that there is a downward trend in the areas but that the correlation is not very strong. Sketch the graph and points.

3. *Bank Interest Problem:* An ad for a bank lists the numbers of years it will take for the balance in your account to reach a certain level if you invest $1000 in one of their accounts.

Dollars	Years to reach
1000.00	0.00
1100.00	1.91
1500.00	8.11
2000.00	13.86
3000.00	21.97

a. Make a scatter plot of the data. Which way is the graph concave? How can you tell from the shape of the graph that a logarithmic function would fit the data?

b. By **logarithmic regression,** find the particular equation of the best-fitting logarithmic function. How does the correlation coefficient confirm that a logarithmic function fits well?

c. Plot the equation from part b on the scatter plot of part a. Sketch the result.

d. Interpolate using your mathematical model to find out how long it takes for $2500 (the average of $2000 and $3000) to be in the account. Is this number of years equal to the average of 13.86 and 21.97?

e. If you wanted to leave your money in the account until it had reached $5000, how many years would you have to leave it there? Which method do you use, interpolation or extrapolation? Explain.

4. *Planetary Period Problem:* Figure 8-3k and the table show the periods of each planet in years, the distance from the Sun in millions of kilometers, and the planets' masses in relation to Earth's mass, as provided by *The World Almanac and Book of Facts 2001.*

Name	Period (yr)	Orbit (millions of km)	Relative Mass
Mercury	0.24	57.9	0.06
Venus	0.61	108.2	0.82
Earth	1	149.6	1
Mars	1.88	228.0	0.11
Jupiter	11.86	778.5	317.8
Saturn	29.46	1433.5	95.16
Uranus	84.01	2872.6	14.5
Neptune	164.79	4495.6	17.15
Pluto	247.68	5870.5	0.002

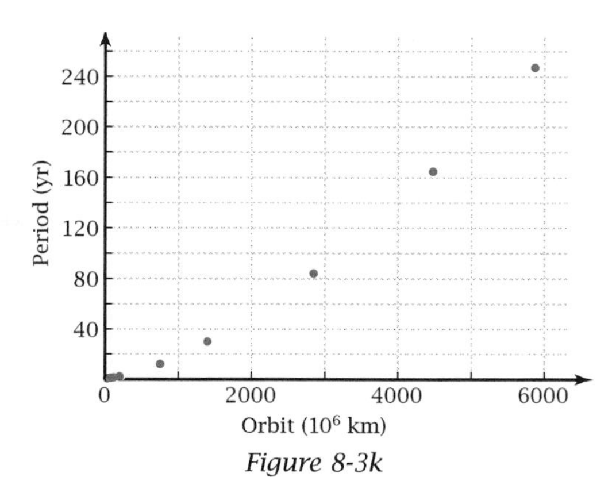

Figure 8-3k

a. If you assume that a power function fits the points, what endpoint behavior are you assuming that the period approaches as the distance from the Sun approaches zero? Run both power and exponential regressions for years as a function of distance in millions of kilometers. Give numerical evidence that the untranslated power function fits better.

b. Plot the power function of part a and the scatter plot on the same screen. Does the power function seem to fit the data well?

c. Make a scatter plot of period as a function of the relative mass of the planet. Show that there is little or no correlation between these two variables.

d. Most asteroids are located in the "asteroid belt" about 430 million km from the Sun. Some scientists believe that they originated by the breakup of a planet or from material that never coalesced to form a planet. If they were originally one planet, what would have been the period of that planet?

This view of the asteroid Ida is a composite of five images taken by the Galileo spacecraft in 1993.

e. Kepler derived his three laws of planetary motion by careful analysis of data such as those in the table. Look up Kepler's third law on the Internet or in a physics text. How well does the result of your power regression agree with that law?

5. *Roadrunners Problem, Part 1:* Naturalists place 30 roadrunners in a game preserve that formerly had no roadrunners. Over the years, the population grows as shown in the table and in Figure 8-3l.

x (years)	y (roadrunners)
0	30
1	44
2	58
3	81
4	110
5	138
6	175
7	203
8	234
9	260
10	276
11	293

y
300
200
100
Number of roadrunners
x
5 10 15
Years

Figure 8-3l

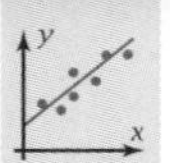

a. Give a physical reason and a graphical reason why a logistic function would be a reasonable mathematical model for the roadrunner population as a function of time. By **logistic regression,** find the particular equation of the best-fitting logistic function. Plot the graph and the data on the same screen. Use a window with an x-range of 0 to 20 years. Sketch the result.

b. What does your mathematical model predict for the roadrunner population at $x = 20$ years? What does the model predict for the maximum sustainable population? At approximately what value of x is the point of inflection, where the rate of population growth is a maximum?

c. Find $\bar{y}$, the average of the y-values. Use $\bar{y}$ to calculate SS_{dev}, the sum of the squares of the deviations. Using the regression function in part a, calculate SS_{res}, the sum of the squares of the residuals. Then find the coefficient of determination. How does this coefficient show that the logistic function fits the points quite well?

6. *Roadrunners Problem, Part 2:* The table and the scatter plot in Figure 8-3m show the change in roadrunner population for each year versus the population at the beginning of that year. For instance, when the population was 81 at the beginning of the third year, it increased by 29 to 110 by the beginning of the fourth year. Thus the rate of increase was 29 roadrunners/year.

Year	x (roadrunners)	y (roadrunners/year)
0	30	14
1	44	14
2	58	23
3	81	29
4	110	28
5	138	37
6	175	28
7	203	31
8	234	26
9	260	16
10	276	17

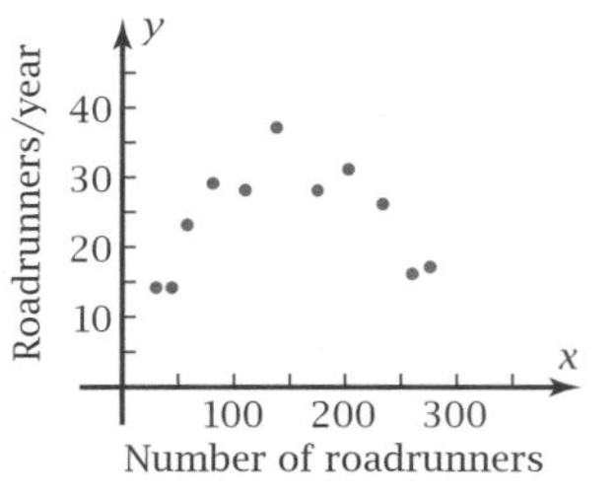

Figure 8-3m

a. The rate of increase of roadrunners is expected to be higher when the population is higher because there are more parents having roadrunner chicks. How, then, can you explain the fact that the population increases at a slower rate when the population is above 150?

b. Assume that the rate of increase is a quadratic function of the population. By **quadratic regression,** find the particular equation of the best-fitting quadratic function. Record the coefficient of determination, R^2. (The capital R is used for regressions other than linear.)

c. Suppose that the naturalists brought in enough roadrunners to increase the population to 400. What does your mathematical model predict for the growth rate? What real-world reason can you think of to explain why this rate is negative?

d. Calculate $\bar{y}$, the average number of roadrunners per year. Put lists in your grapher for the squares of the deviations, $(y - \bar{y})^2$, and for the squares of the residuals, $(y - \hat{y})^2$. By summing these lists, calculate SS_{dev} and SS_{res}. Use the results to calculate the coefficient of determination, R^2. Show that the answer is equal to the value you found by regression in part b.

7. *Linearizing Exponential Data—The Punctured Tire Problem:* Figure 8-3n and the table show pressure, y, measured in pounds per square inch (psi), of air in a car's tire at different times, x, measured in seconds after the tire was punctured. Figure 8-3o shows that log y appears to be a linear function of x.

x (sec)	y (psi)
5	27
10	21
15	16
20	13
25	9
30	7
35	6
40	4
45	3
50	3

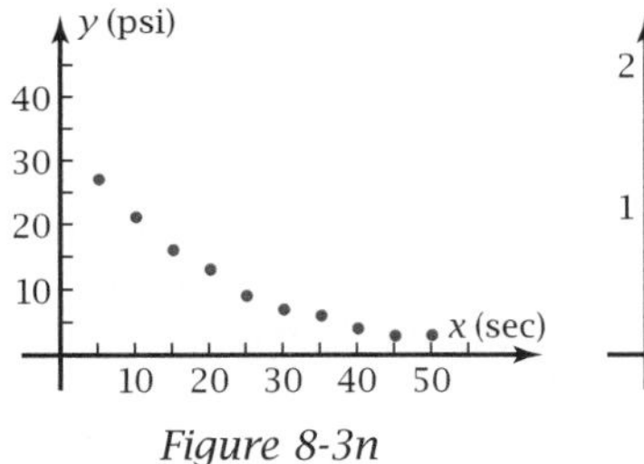

Figure 8-3n

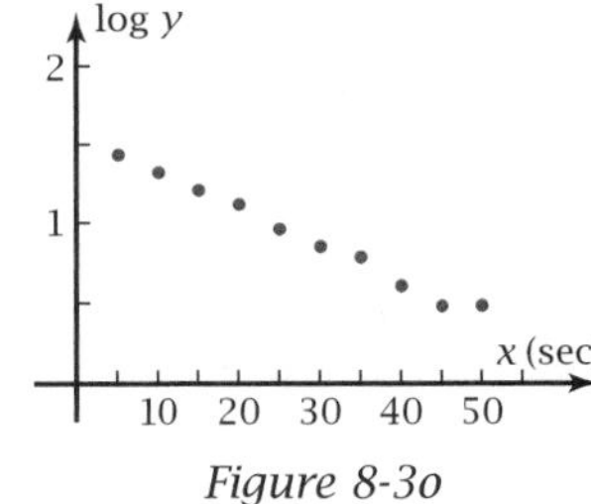

Figure 8-3o

a. By exponential regression, find the equation of the exponential function of the form $y = ab^x$ that best fits the data. Write down the equation and the correlation coefficient r. Plot the equation and the scatter plot on the same screen. Does the exponential function seem to fit the points well?

b. Compute the correlation coefficient directly from the definition by calculating SS_{dev} and SS_{res}, finding the coefficient of determination R^2, and then taking the (negative) square root. Show that the answer does not equal the correlation coefficient r you found in part a.

c. Calculate log y in another list on your grapher. Run a linear regression on log y as a function of x. Show that the correlation coefficient for this regression equals the correlation coefficient for the exponential regression in part a. (Your grapher calculates exponential regression by running linear regression on x and log y.) By appropriate plotting, show that the linear function fits the points in the scatter plot of Figure 8-3o.

d. The exponential regression in part a should have given you the equation

$$y = 34.7990\ldots(0.9493\ldots^x)$$

By taking the logarithm of both sides of this equation, show that log y is the same linear function of x you found in part c.

8. *Linearizing Power Data—The Hose Problem:* The rate water flows from a garden hose depends on the water pressure at the faucet. Figure 8-3p and the table show flow rates y, in gallons per minute (gal/min), for various pressures x, in pounds per square inch (psi). Figure 8-3q shows that log y seems to be a linear function of log x.

x (psi)	y (gal/min)
1	0.9
5	2.0
10	2.8
15	3.5
25	4.5
40	5.7
50	6.4
70	7.5

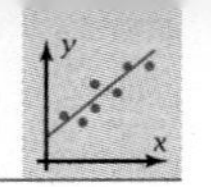

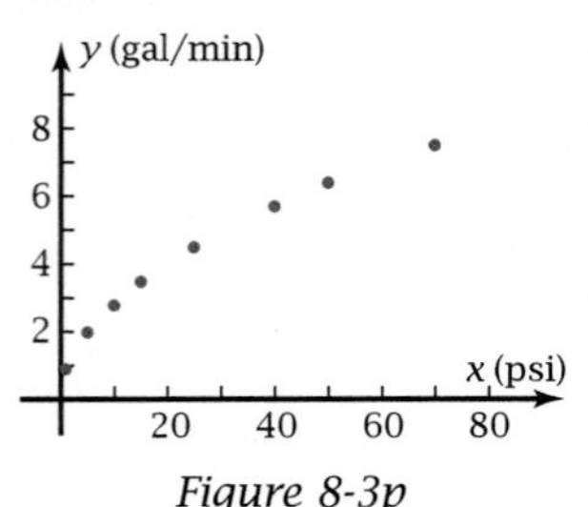

Figure 8-3p

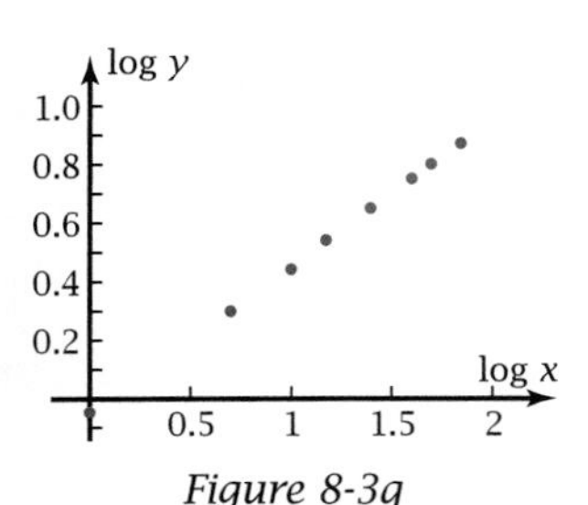

Figure 8-3q

a. By power regression, find the equation of the power function that best fits the data. Write down the equation and the correlation coefficient *r*. Plot the equation and the scatter plot on the same screen. Does the power function seem to fit the points well?

b. Compute the correlation coefficient directly from the definition by calculating SS_{dev} and SS_{res}, finding the coefficient of determination R^2. Show that the square root of R^2 does *not* equal the correlation coefficient *r* you found in part a.

c. Calculate log *x* and log *y* in two other lists on your grapher. Run a linear regression on log *y* as a function of log *x*. Show that the correlation coefficient for this regression equals the correlation coefficient for the power regression in part a. (Your grapher does power regression by running linear regression on log *x* and log *y*.) Plot the regression line and the scatter plot of log *x* and log *y*. How well does the regression line fit the points in the scatter plot?

d. The power regression in part a should have given you the equation

$$y = 0.8955\ldots x^{0.5011\ldots}$$

By taking the logarithm of both sides of this equation, show that log *y* is the same linear function of log *x* as the one you found in part c.

8-4 Residual Plots and Mathematical Models

In Section 8-3 you learned how to do regression analysis to fit functions to data by choosing the function that had the shape of the graph and the correct endpoint behavior. In this section you will obtain further evidence of whether a function fits well by making sure that a plot of the residuals follows no particular pattern.

OBJECTIVE Find graphical evidence for how well a given function fits a set of data by plotting and analyzing the residuals.

Weeks Old	Inches
3	5
4	4
5	7
6	6
7	11
8	11
9	15
10	20
11	21
12	24

A biology class plants one bean each week for 10 weeks. Three weeks after the last bean is planted, the plants have heights shown in the table and in Figure 8-4a. As you can tell, there is a definite upward trend, but it is not absolutely clear whether the best-fitting function is curved or straight.

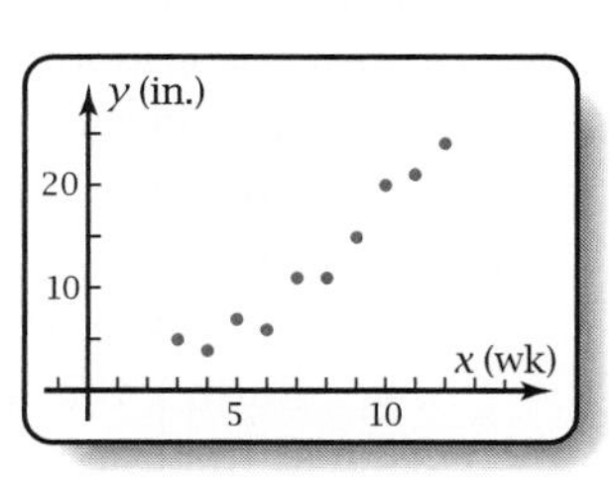

Figure 8-4a

Regression shows roughly the same correlation coefficient for linear and exponential functions.

Linear:
$\hat{y} = 2.1351...x - 4.9636...$
$r = 0.9675...$

Exponential:
$\hat{y} = 2.2538...(1.2267...)^x$
$r = 0.9690...$

Both functions fit reasonably well, as shown in Figures 8-4b and 8-4c.

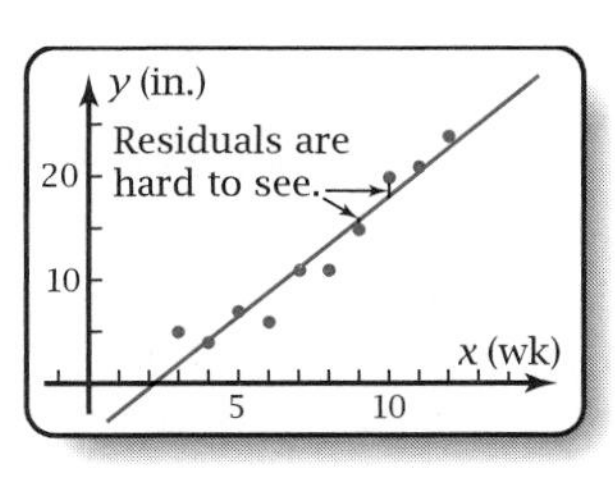

Figure 8-4b
Linear function

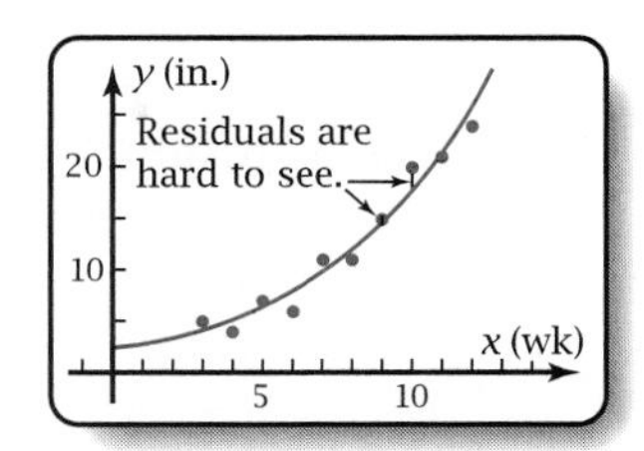

Figure 8-4c
Exponential function

The residuals, $y - \hat{y}$, in these figures are hard to see because they are relatively small. You can see the residuals more easily by making a **residual plot,** which is a scatter plot of the residuals. Figures 8-4d and 8-4e show residual plots for the linear function and for the exponential function, respectively. To make these plots, enter the regression equation for $\hat{y}$ into your grapher, say, in the y= menu as y_1. Assuming that the data are in lists L_1 and L_2, you can calculate the residuals like this:

$L_2 - y_1(L_1)$ The expression $y_1(L_1)$ means the values of the function in y_1 at the values of x in L_1.

Most graphers have a zoom feature that will set the window automatically to fit the statistical data.

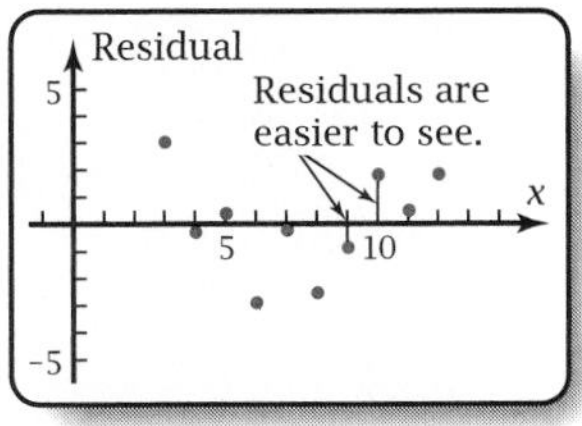

Figure 8-4d
Linear function residuals

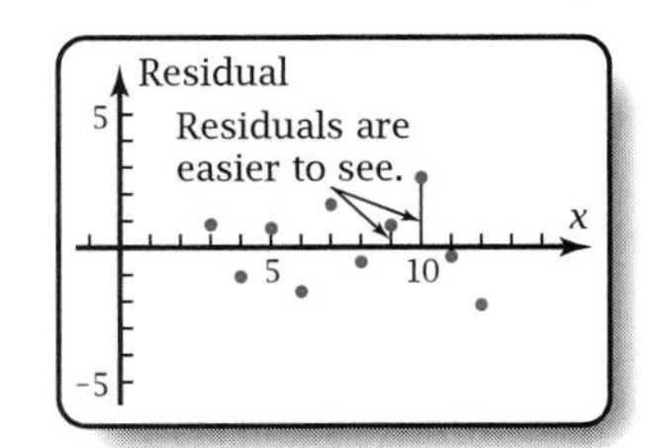

Figure 8-4e
Exponential function residuals

The residuals for the linear function seem to form a pattern, high at both ends and low in the middle. A pattern in the residuals suggests that the data may be nonlinear. The residuals for the exponential function are more random, following no discernible pattern. So the residuals for the exponential function are more likely caused by random variations in the data, such as different growth rates for different bean plants.

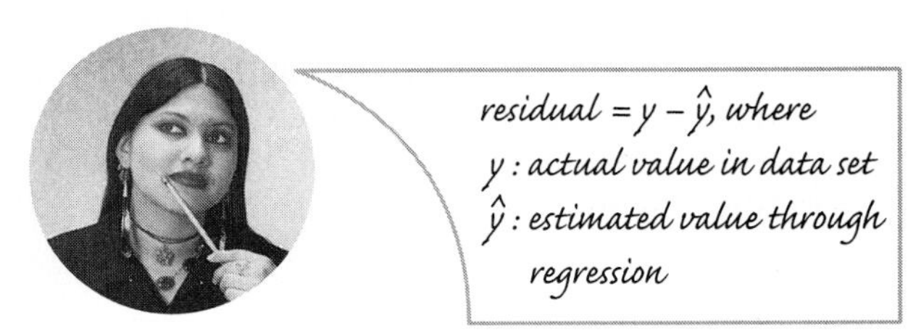

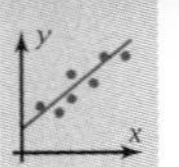

CONCLUSION: Residual Plot Interpretation

If the residual plot follows a regular pattern, then there is a behavior that is not accounted for by the kind of function chosen.

If the residual plot has no identifiable pattern, then the regression equation is likely to account for all but random fluctuations in the data.

Note that the residual plot and the endpoint behavior might give you conflicting information. The residual plot suggests that an exponential function is more reasonable, but it indicates that the bean plants were already sprouted at time $x = 0$ when they were planted.

Problem Set 8-4

Do These Quickly

Q1. Find r^2 if $SS_{res} = 2$ and $SS_{dev} = 10$.

Q2. Find the correlation coefficient if the coefficient of determination is 0.9.

Q3. Find $\hat{y}$ for the point (3, 9) if $\hat{y} = 2x + 7$.

Q4. Find the residual for the point (3, 9) if $\hat{y} = 2x + 7$.

Q5. Find the deviation for the point (3, 9) if $\bar{x} = 5$ and $\bar{y} = 6$.

Q6. Find the y-intercept if $\hat{y} = 2x + 7$.

Q7. Find the x-coordinate of the vertex of this parabola: $y = 3x^2 + 24x + 71$

Q8. 2 log 6 = log —?—.

Q9. What transformation of $y = \sin x$ is the function $y = \sin 2(x - 3)$?

Q10. Find the *exact* value (no decimals) of $\cos \frac{\pi}{6}$.

1. *Radiosonde Air Pressure Problem:* Meteorologists release weather balloons called radiosondes each day to measure data about the atmosphere at various heights. Figure 8-4f and the table show height in meters, pressure in millibars, and temperature in degrees Celsius measured in December 1991 by a radiosonde in Coffeeville, Kansas.

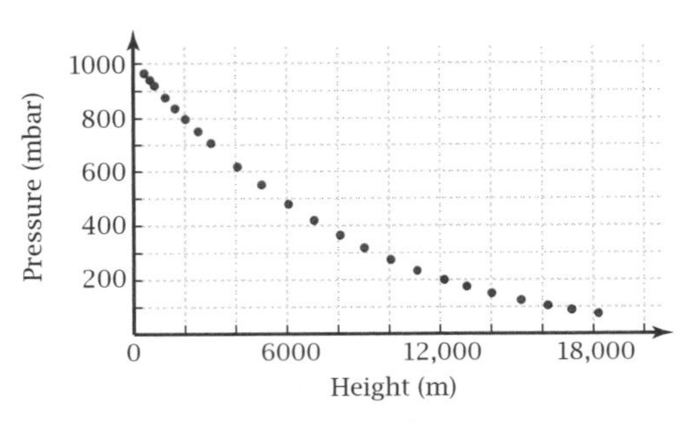

Figure 8-4f

Height	Pressure	Temperature (°C)
400	965	11.4
620	940	12
800	920	11.9
1220	875	11.7
1611	835	11
2018	795	8.6
2500	750	9
3009	705	6.8
4051	620	0.9
5000	550	−6.8
6048	480	−13
7052	420	−19.8
8075	365	−28.5
9005	320	−35.2
10042	275	−43
11086	235	−48.6
12151	200	−46.8
13024	175	−52.6
14004	150	−58.3
15134	125	−63.1
16197	105	−66.5
17120	90	−71.1
18195	75	−70.7

a. Explain why an exponential function would be expected to fit the data well. By regression, find the best-fitting exponential function. Record the correlation coefficient and paste the equation into your grapher's y= menu. Plot the equation and data on the same screen. Do the data points seem to lie along the graph?

b. Put another list in your grapher for the residual of each point. Then make a residual plot and sketch the result. Do the residuals follow a definite pattern, or are they randomly scattered? What real-world phenomenon could account for the deviations?

c. Based on your residual plot in part b, could the exponential function be used to make predictions of pressure correct to the nearest millibar, if this accuracy were necessary?

2. *Hot Water Problem:* Tim put some water in a saucepan and then turned the heat on high. Figure 8-4g and the table show the temperature of the water in degrees Celsius at various times in seconds since he turned on the heat.

Time	Temperature
49	35
62	40
76	45
89	50
103	55
117	60
131	65
145	70
161	75
176	80
190	85
205	90

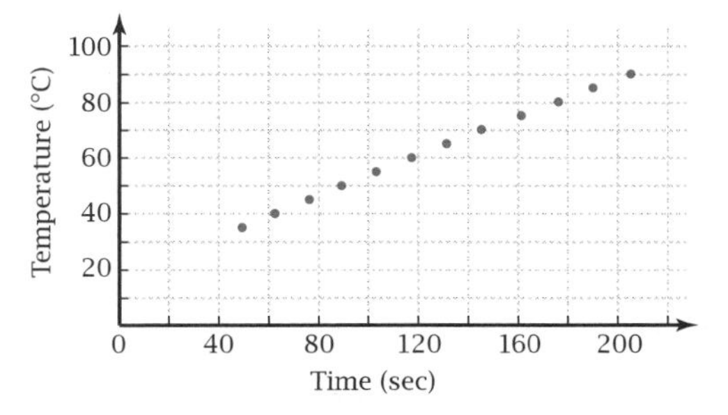

Figure 8-4g

a. Run a linear regression on temperature as a function of time. Record the correlation coefficient. Plot the regression equation and a scatter plot of the data on the same screen. Does the fit of the line to the data confirm the fact that the correlation coefficient is so close to 1?

b. Make a residual plot. Sketch the result. How does the residual plot tell you that there is something in the heating of the water that the linear function does not take into account? What real-world reason do you suppose causes this slight nonlinearity?

c. Based on the linear model, at what time would you expect the water to reach 100°C and boil? Based on your observations in part b, would you expect the water to boil sooner than this or later than this? Explain.

3. *Gas Mileage Problem:* Figure 8-4h and the table show the gasoline consumption rate (mi/gal) of 16 cars of various weights (pounds).

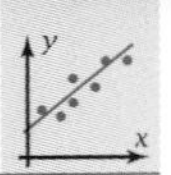

Model	Weight	Mi/gal
Ford Aspire	2140	43
Honda Civic del Sol	2410	36
Honda Civic	2540	34
Ford Escort	2565	34
Honda Prelude	2865	30
Ford Probe	2900	28
Honda Accord	3050	31
BMW 3-series	3250	28
Ford Taurus	3345	25
Ford Mustang	3450	22
Ford Taurus SHO	3545	24
BMW 5-series	3675	23
Lincoln Mark VIII	3810	22
Cadillac Eldorado	3840	19
Cadillac Seville	3935	20
Ford Crown Victoria	4010	22

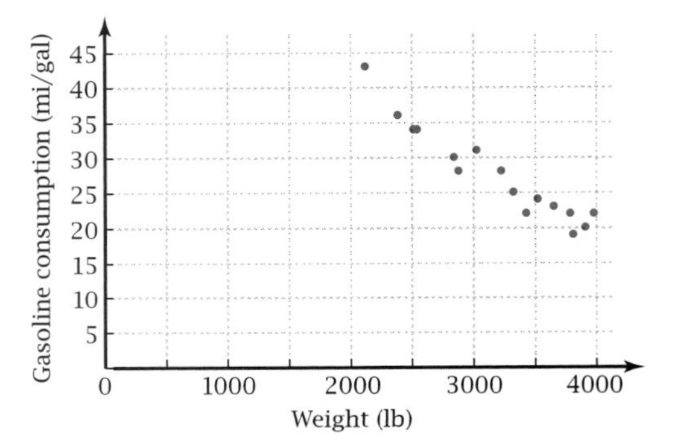

Figure 8-4h

a. Run an exponential regression and a power regression on the data. Show that the correlation coefficient for each function is about the same. Plot each function on the same screen as a scatter plot of the data. Do both functions seem to fit the data well?

b. Make two residual plots, one for the exponential function and one for the power function. Does either residual plot seem to follow any pattern? How do you interpret this answer in terms of which function fits the data more closely?

c. Extrapolate using both the exponential function and the power function to predict the gas mileage for a super-compact car weighing only 500 pounds. Based on your answers, which model—exponential or power regression—has a more reasonable endpoint behavior for light cars? Is there a significant difference in endpoint behavior if you extrapolate for very heavy cars?

4. *Weed Competition Problem:* In the report *Applications of Mathematics: A Nationwide Survey,* submitted to the Ministry of Education, Ontario (1976), A. C. Madgett reports on competition of wild oats (a weed) with various crops. The more wild oats that are growing with a crop, the greater the loss in yield of that crop. Figure 8-4i and the table show information gleaned from pages II-8 to II-10 of that report.

Wild Oat Plants/m^2	Percent Loss of Wheat Crop
1	3
5	8
10	10
20	17
50	25
100	34
150	40
200	48

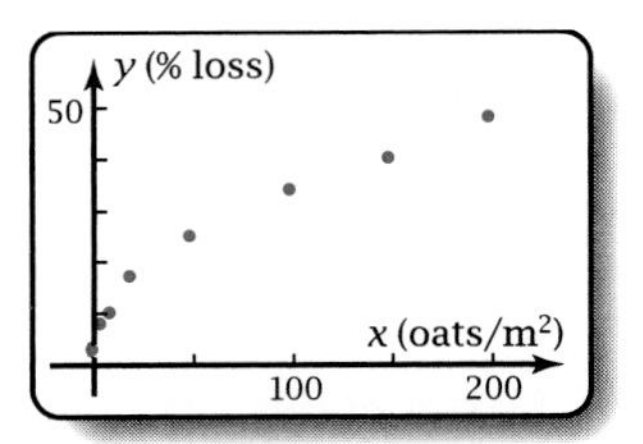

Figure 8-4i

a. From the scatter plot, tell what kind of function is most likely to fit the data. By regression, find the particular equation of this kind of function. Plot it and the scatter plot on the same screen. Sketch the result.

b. Make a residual plot. Sketch the result. What information do you get about the way your selected function fits the data from the residual plot? From the correlation coefficient?

c. Based on your mathematical model, what percent of crop loss would be expected for 500 wild oat plants per square meter? Do you find this number by extrapolation or by interpolation?

d. How many wild oat plants per square meter do you predict it would take to choke out the wheat crop completely? Do you find this number by extrapolation or by interpolation?

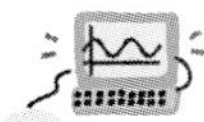

5. *Population Problem 3:* Figure 8-4j and the table below show the population of the United States for various years. The information was obtained from the U.S. Census Bureau at *www.census.gov*. The variable x is the number of years elapsed since 1930.

Year	x (years)	Population (millions)
1940	10	132.1
1950	20	152.3
1960	30	180.7
1970	40	205.1
1980	50	227.2
1990	60	249.5

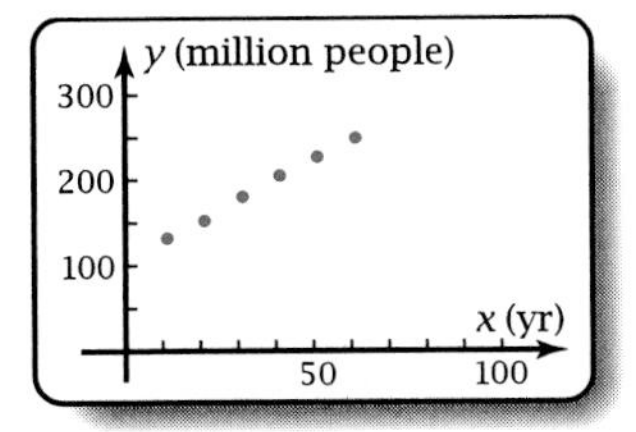

Figure 8-4j

a. Explain why a linear function and an untranslated power function would not have the correct endpoint behavior for dates long before 1930. Explain why an exponential function of the form $y = ab^x$ would not have a reasonable endpoint behavior for dates long after the present.

b. Find the particular equation of the best-fitting logistic function. Plot it and the data on the same screen. Use a window that includes dates back to 1830 and forward to 2130. Sketch the result.

c. What does the logistic model predict for the outcome of the 2000 census? (You can check the Web site to see how close this comes to the actual outcome.) What does the logistic model predict for the maximum sustainable population of the United States?

d. Suppose that the result of the 1990 census had been 250.5 million, one million higher. Run another logistic regression with the modified data. What does the new model predict for the maximum sustainable population?

6. *Wind Chill Problem:* When the wind is blowing on a cold day, the temperature seems to be colder than it really is. For any given actual temperature with no wind, the equivalent temperature due to wind chill is a function of wind speed. Figure 8-4k and the table show data published by NOAA (the National Oceanographic and Atmospheric Administration) (see *www.ncdc.gov/ol/climate/conversion/windchillchart.html*).

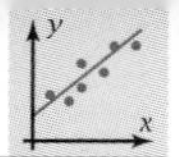

Wind (mi/hr)	Equivalent Temperature (°F)
0	35
5	32
10	22
15	16
20	12
25	8
30	6
35	4
40	3

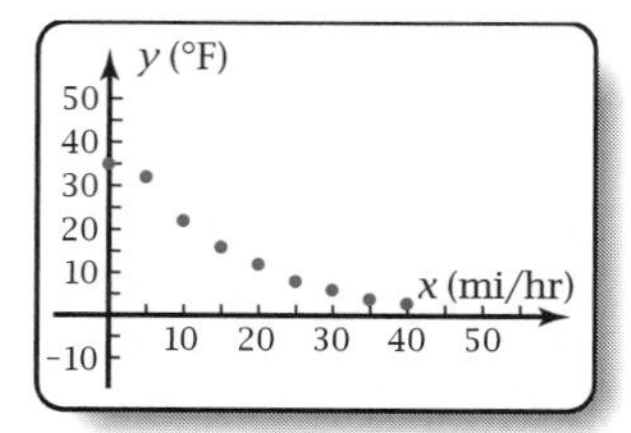

Figure 8-4k

a. Ignoring the data point at 0 mi/hr, run a logarithmic regression to find the best-fitting logarithmic function. Plot it and the data on the same screen. Sketch the result.

b. At what wind speed does the logarithmic function predict that the equivalent temperature would be 0°F? Do you find this number by extrapolation or by interpolation? Explain why an untranslated exponential function would predict that the equivalent temperature would never be zero. Explain why the logarithmic function does not have the correct endpoint behavior at the left end of the domain.

c. Make a residual plot for the logarithmic function. Sketch the result. Do the residuals follow a pattern? What does this fact tell you about how well the logarithmic function fits the data?

d. Check the NOAA Web site mentioned previously to find out the mathematical model NOAA actually uses to compute wind-chill-equivalent temperatures.

7. *Calorie Consumption Problem:* The number of calories an animal must consume per day increases with its body mass. However, the number of calories per kilogram of body mass decreases with mass because larger animals have a lower surface-to-volume ratio. On page 130 of *Studies in Mathematics, Volume XX,* published by Yale University Press (1972), Max Bell reports these data for various mammals:

Animal	Mass (kg)	Cal/kg
Guinea pig	0.7	223
Rabbit	2	58
Human	70	33
Horse	600	22
Elephant	4000	13

a. Make a scatter plot of the data. Explain why the scatter plot does not tell you very much about the relationship between the variables. Then transform the data by taking the logarithm of the mass and the logarithm of the calorie consumption per kilogram. Run a linear regression on the transformed data and plot it on the same screen as a scatter plot of log (cal/kg) versus log (mass).

b. By transforming the equation in part a, show that the calorie consumption is a power function of mass. By running power regression on the original data, show that you get the same power function.

c. The smallest mammal is the shrew. Predict the calorie consumption per kilogram for a 2-g shrew. (The need to eat so much compared to body mass is probably why shrews are so mean!)

d. In the referenced article, Bell reports that a 150,000-kg whale consumes about 1.7 cal/kg. What does the power function predict for the calories per kilogram if you extrapolate it to the mass of a whale? By what percentage does the answer differ from the given 1.7 cal/kg? Think of a reason why the predicted value is so far from the reported value.

8. *Mile Run Record Times:* This table shows that the world record time for the mile run has been decreasing from 1913 through 1985.

Year	Time (s)	Year	Time (s)
1913	254.4	1958	235.5
1915	252.6	1962	234.4
1923	250.4	1964	234.1
1931	249.2	1965	233.6
1933	247.6	1966	231.3
1934	246.8	1967	231.1
1937	246.4	1975	231.0
1942	246.2	1975	229.4
1942	244.6	1979	228.95
1943	242.6	1980	228.80
1944	241.6	1981	228.53
1945	241.4	1981	228.40
1954	239.4	1981	227.33
1954	238.0	1985	226.32
1957	237.2		

The 1500-meter semifinal race at the 2000 Summer Olympics in Sydney, Australia.

a. Make a scatter plot of the data. Use the last two digits of the year for x and the number of seconds for y. A window with a range for y of [220, 260] will allow you to view the entire data set on the screen. Plot the best-fitting linear function on the same screen. Sketch the result.

b. Find the best-fitting exponential function for the data. Plot it on the same screen as in part a. Can you see any difference between the exponential and linear graphs?

c. In 1999, Hicham El Guerrouj of Morocco set a world record of 223.13 seconds for the mile run. Which function comes closer to predicting this result, the linear or the exponential?

d. Make a residual plot for the linear function. Describe any patterns you see in the points.

e. In 1954, Roger Bannister of Great Britain "broke" the 4-minute mile. Until that time, it had been thought that 4 minutes (240 seconds) was the quickest a human being could run a mile. What do the data and residual plot suggest happened in the years just before and just after Bannister's feat?

9. *Meatball Problem:* In *Applications of Mathematics: A Nationwide Survey* (page III-6), A. C. Madgett reports the moisture content of deep-fried meatballs as a function of how long they have been cooked. The following data were gathered to determine the effectiveness of adding whey to hamburger meat to make it retain more moisture during cooking, thus improving its quality.

Cooking Time (min)	Percent Moisture Content	
	No whey	Whey
6	46.4	48.8
8	45.3	—
10	40.6	44.2
12	36.9	41.8
14	33.5	39.2

a. On the same screen, make a scatter plot of both sets of data. Use different symbols for the different data sets. Find the best-fitting linear function for each set of data, and plot the two functions on the same screen as the scatter plots. Sketch the results.

b. Use the linear model to predict the moisture content for the missing data point in the whey data. Which process do you use to do this, extrapolation or interpolation? Explain.

c. Use the linear model to predict how long the meatballs could be cooked without dropping below 30% moisture if they have

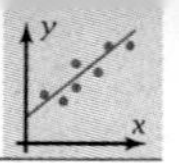

whey and then if they have no whey. Which process do you use to do this, extrapolation or interpolation? Explain.

d. According to your mathematical models, do the two kinds of meatballs have the same moisture content before they are cooked? Give numbers to support your answer.

10. *Television Set Problem:* Here are prices of a popular brand of television set for various sizes of screen.

Diagonal (in.)	Price ($)
5	220
12	190
17	230
27	350
36	500

a. Make a scatter plot of the data. Based on the graph, explain why a quadratic function is a more reasonable mathematical model than either a linear, a logarithmic, an untranslated power, or an exponential function. Confirm your graphical analysis numerically by finding R^2 or r^2, the coefficient of determination, for each of the five kinds of functions.

b. Plot the best-fitting quadratic function and the scatter plot on the same screen. Sketch the result.

c. If the manufacturer made a 21-in. model and a 50-in. model, how much would you expect to pay for each? For which prediction did you use extrapolation, and for which did you use interpolation?

d. Make a residual plot for the quadratic function. Sketch the result. On the residual plot, indicate which sizes of television are slightly overpriced and which are underpriced.

e. Why do prices go *up* for very small television sets?

11. *Journal Problem:* Make an entry explaining how two different types of regression functions can have the same correlation coefficient although one function is preferred over the other. Show how a residual plot and endpoint behavior may sometimes give conflicting information about the function to use.

8-5 Chapter Review and Test

In this chapter you've learned how to fit various types of functions to data. In choosing the type of function, you considered

- The shape in the scatter plot
- The behavior of the chosen type of function at the endpoints
- Whether or not the residual plot has a pattern
- The correlation coefficient

You learned how to calculate SS_{dev}, the sum of the squares of the deviations of the y-values from the horizontal line $y = \bar{y}$, the average of the y-values. Rotating the line to fit the pattern followed by the data reduces the deviations. The linear regression line reduces the deviations enough so that SS_{res}, the sum of the squares of the residuals (residual deviations), is a minimum. The coefficient of determination and the correlation coefficient are calculated from SS_{dev} and SS_{res}. You can use other kinds of regression to fit different types of functions to nonlinear data.

Review Problems

R0. Update your journal with what you have learned in this chapter. Include how your knowledge of the shapes of various function graphs guides you in selecting the type of function you'll choose as a model and how regression analysis lets you find the particular equation of the selected type of function. Mention the ways you have of deciding how well the selected function fits, both within the data and possibly beyond the data. Also, explain how the correlation coefficient and the coefficient of determination are calculated from SS_{dev} and SS_{res}.

R1. Figure 8-5a shows this set of points and the graph of $\hat{y} = 1.6x + 0.9$.

x	y
3	6
5	10
7	9
9	17

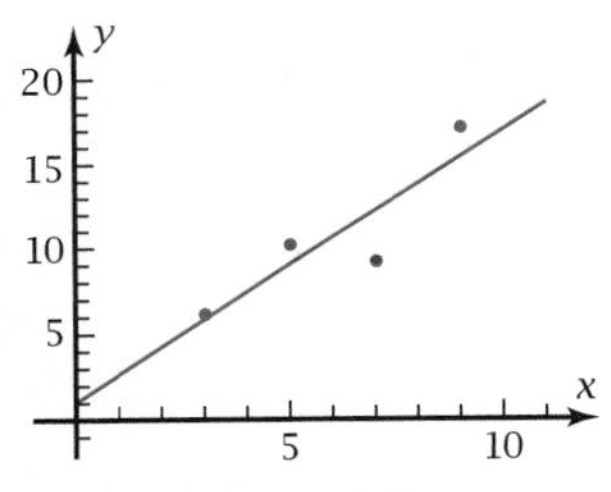

Figure 8-5a

a. By linear regression, confirm that $\hat{y} = 1.6x + 0.9$ is the correct regression equation.

b. Copy the table and put in three new columns, one for the values of $\hat{y}$, one for the values of the residuals, $(y - \hat{y})$, and one for the squares of the residuals, $(y - \hat{y})^2$.

c. By calculation, show that the sum of the squares of the residuals is $SS_{res} = 13.8$.

d. Show that a slight change in the 1.6 slope or the 0.9 y-intercept leads to a higher value of SS_{res}. For instance, you might try $\hat{y} = 1.5x + 1.0$.

R2. a. Figure 8-5b shows the points from Figure 8-5a with a dashed line at $y = \bar{y}$, the average of the y-values of the data. Calculate $\bar{y}$ to show that the figure is correct.

b. Calculate $\bar{x}$. Show algebraically that the point $(\bar{x}, \bar{y})$ is on the regression line $\hat{y} = 1.6x + 0.9$ from Problem R1.

c. On a copy of Figure 8-5b, sketch both the deviation from the average for the point where $x = 9$ and the residual (that is, the residual deviation) for this point. Why do you suppose the word *residual,* which is an adjective, is used as a noun in this instance?

d. Calculate SS_{dev}, the sum of the squares of the deviations. Why do you suppose that SS_{dev} is so much larger than SS_{res}?

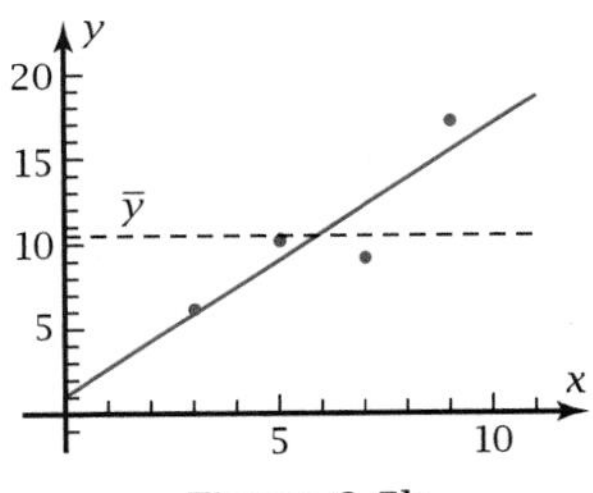

Figure 8-5b

e. Calculate the coefficient of determination,

$$\frac{SS_{dev} - SS_{res}}{SS_{dev}}$$

Then run a linear regression on the points in Problem R1, thus showing that the fraction equals r^2.

f. Calculate the correlation coefficient from r^2. Show that it agrees with the value from the regression on your grapher. Why must you choose the *positive* square root?

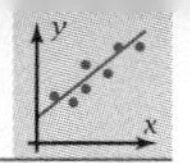

g. Figures 8-5c through 8-5f show the ellipses in which the clouds of data points lie. For each graph, tell whether the correlation coefficient is positive or negative and whether it is closer to 0 or closer to 1 or −1.

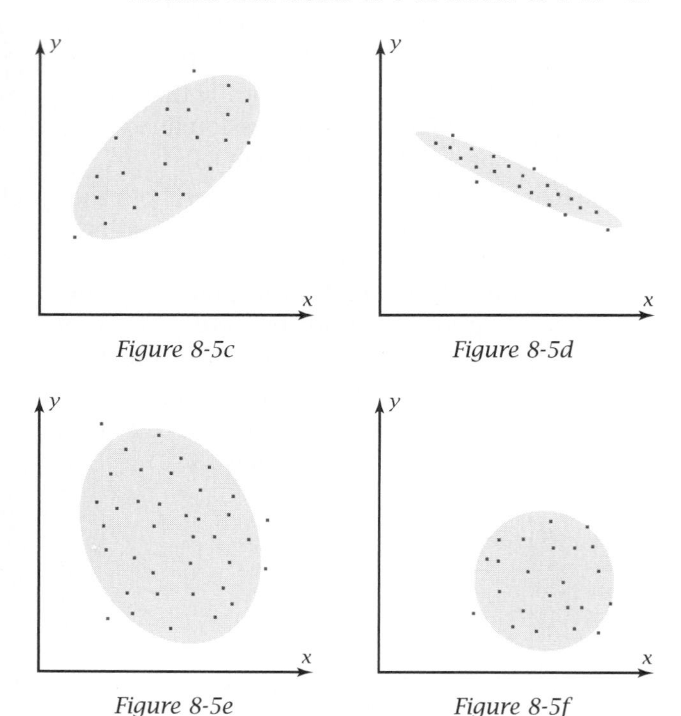

Figure 8-5c

Figure 8-5d

Figure 8-5e

Figure 8-5f

R3. *Learning Curve Problem:* When an article is first manufactured, the cost of making each item is relatively high. As the manufacturer gains experience, the cost per item decreases. The function describing how the cost per item varies with the number of items produced is called the *learning curve.* Suppose that a shoe manufacturer finds the costs per pair of shoes shown in the table and in Figure 8-5g.

Pairs Produced	$/pair
100	60
200	46
500	31
700	26
1000	21

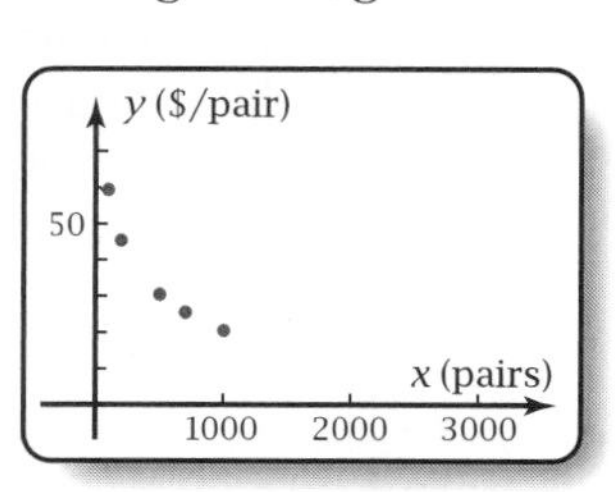

Figure 8-5g

a. Show, using regression analysis, that a logarithmic function and a power function each fit the data about equally well.

b. It is desired to predict the cost per pair of shoes beyond the upper end of the data. Explain why the power function has a more reasonable endpoint behavior than the logarithmic function for large values of pairs produced.

c. Use the power function to predict how many pairs of shoes the shoe manufacturer must produce before the cost per pair drops below $10.00. Do you find this number by extrapolation or by interpolation? How do you know?

d. According to the power function model, what was the cost of manufacturing the first pair of shoes of this style?

e. The learning curve is sometimes described by saying, “Doubling the number manufactured reduces the cost by —?— percent.” What is the percentage for this kind of shoe? What property of power functions does this fact illustrate?

R4. *Carbon Dioxide Problem:* Global warming is caused in part by the increase in the concentration of carbon dioxide in the atmosphere. Suppose that the concentrations in the table were measured monthly over a period of two years. The concentration of carbon dioxide, *y,* is measured in parts per million (ppm), and *x* is the number of months.

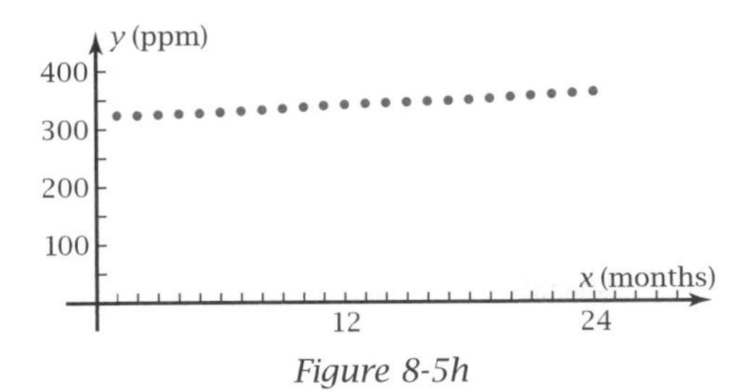

Figure 8-5h

Car exhaust contains carbon dioxide, which contributes to global warming.

a. Explain why a linear or an exponential function of the form $y = ab^x$ might fit the data shown in Figure 8-5h reasonably well, but why an untranslated power function or a logarithmic function would not.

b. Run linear and exponential regressions. Plot both functions on a scatter plot of the data. Show that both functions give close to the actual concentration for $x = 13$ months but yield a significantly different concentration if you extrapolate to 20 years.

c. Make a residual plot using the exponential function. Sketch the result. How do you interpret the residual plot in terms of trends in the real world that the exponential function does not account for?

	x (months)	**y (ppm)**		**x (months)**	**y (ppm)**
January	1	323.8	**January**	13	343.5
February	2	324.8	**February**	14	344.9
March	3	325.9	**March**	15	346.1
April	4	327.2	**April**	16	347.4
May	5	328.1	**May**	17	348.4
June	6	329.6	**June**	18	349.9
July	7	331.5	**July**	19	352.0
August	8	333.5	**August**	20	354.5
September	9	335.5	**September**	21	356.5
October	10	338.0	**October**	22	358.9
November	11	340.2	**November**	23	360.8
December	12	342.1	**December**	24	362.6

Concept Problems

C1. *Sinusoidal Regression Problem:* Some graphers are programmed to calculate **sinusoidal regression.** Suppose these data have been measured.

a. Figure 8-5i shows a plot of the data and the best-fitting sinusoidal function. Run a sinusoidal regression on the data. Find the particular equation and enter it into the y= menu.

b. Confirm that your equation goes through the data points, as shown in Figure 8-5i.

x	y
2	7.7
4	5.4
6	2.7
8	2.1
10	4.2
12	7.0
14	8.0
16	6.2
18	3.4
20	2.0

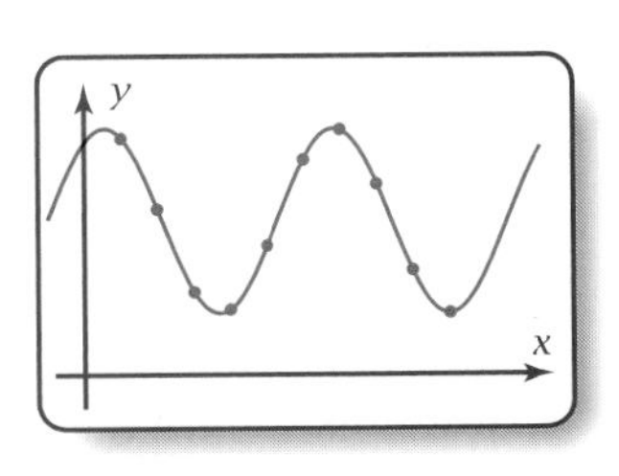

Figure 8-5i

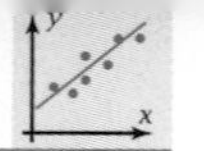

c. Find the approximate values of the period, the amplitude, the phase displacement, and the location of the vertical axis for the sinusoidal function.

C2. *R^2 vs. r^2 Problem:* Figure 8-5j shows the best-fitting exponential function for the points in the table. In this problem you will calculate the coefficient of determination directly from its definition and then compare it with the value found by exponential regression.

x	y
2	3
5	10
17	16
19	23

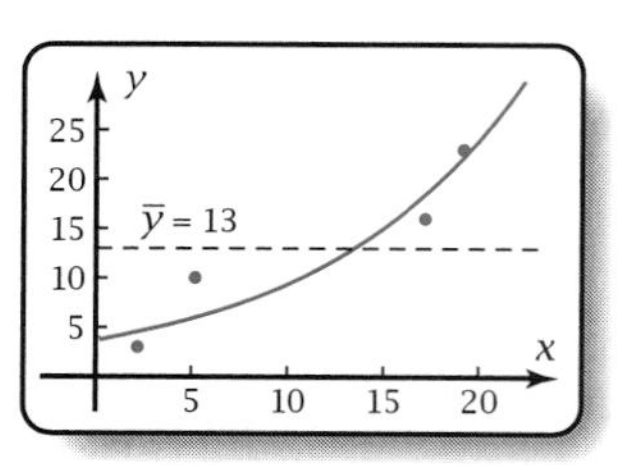

Figure 8-5j

a. By exponential regression, find the equation of the exponential function.

b. Calculate SS_{dev} and SS_{res} by calculating and summing the squares of the deviations and the residuals. Use the results to show that the coefficient of determination equals 0.885496....

c. The coefficient of determination, r^2, from part a equals 0.808484.... Show that this number is the same as the coefficient of determination found by running a *linear* regression on the *logarithm* of the y-values as a function of the x-values.

d. You use the symbol R^2 for the actual coefficient of determination, as in part b. The symbol r^2 is used only for the results of *linear* regression. Why do you think your grapher displays r^2 for exponential and power regression but R^2 for quadratic regression?

Chapter Test

PART 1: No calculators allowed (T1–T8)

For Problems T1–T7, assume that linear regression on a set of data has given $\hat{y} = -2x + 31$.

T1. The average-average point is $(\bar{x}, \bar{y}) = (5, 21)$. Find the deviation of the data point (7, 15).

T2. Find the residual of the data point (7, 15).

T3. Linear regression gives a correlation coefficient of −0.95, and exponential regression gives a correlation coefficient of −0.94. Based on this information, which of these two types of functions fits the data better?

T4. Interpret what it means that the correlation coefficient is negative in Problem T3.

T5. A residual plot shows the patterns for linear regression (Figure 8-5k) and exponential regression (Figure 8-5l). Based on these plots, which kind of function fits better? Explain how you decided.

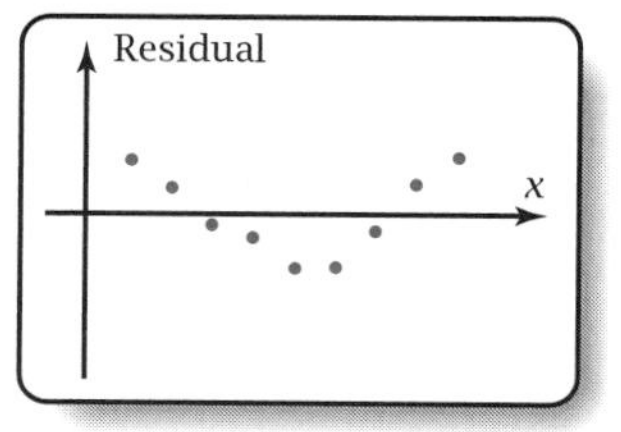

Linear function

Figure 8-5k

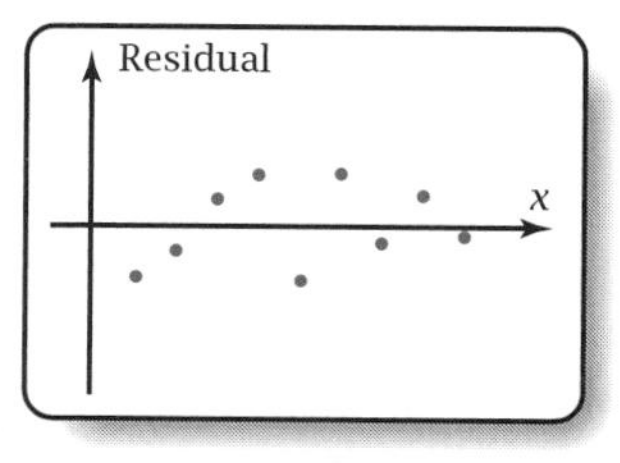

Exponential function

Figure 8-5l

T6. In a real-world context, you expect to have a positive y-intercept. Which function—the linear, the exponential, both, or neither—has this endpoint behavior?

T7. If the function is extrapolated to large x-values, the y-values are expected to approach the x-axis asymptotically. Which function—the linear, the exponential, both, or neither—has this endpoint behavior?

T8. Given SS_{res} and SS_{dev}, how do you calculate the coefficient of determination? How do you calculate the correlation coefficient?

PART 2: Graphing calculators allowed (T9–T17)

The Snake Problem: Herpetologist Herbie Tol raises sidewinder rattlesnakes. He measures the length in centimeters for a baby sidewinder at various numbers of days after it hatches. For Problems T9–T16, use these data.

x (days)	y (cm)
2	5
4	9
6	10
8	14

He runs a linear regression on his grapher and gets

$\hat{y} = 1.4x + 2.5$

He plots a scatter plot and then plots the regression line as shown in Figure 8-5m.

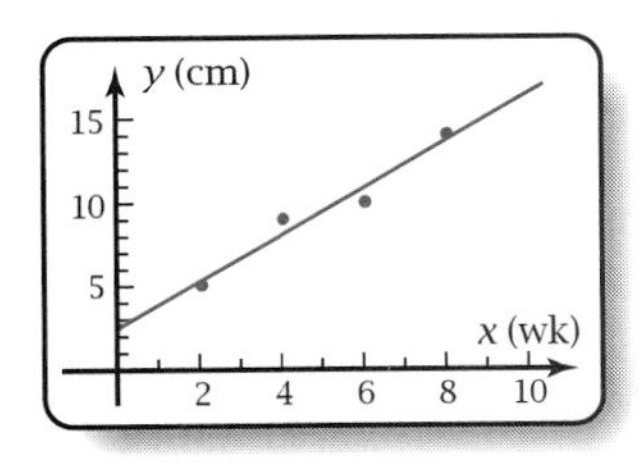

Figure 8-5m

T9. Sketch Figure 8-5m. Demonstrate that you know the meaning of the word *residual* by showing on the sketch the residual for the point where $x = 6$.

T10. Put three lists into the data table, one for the value of $\hat{y}$ using the regression equation, one for the residual deviations, and one for the squares of the residuals.

T11. Calculate SS_{res}, the sum of the squares of the residuals.

T12. Calculate the average-average point, $(\bar{x}, \bar{y})$. Show algebraically that this point satisfies the regression equation.

T13. Herbie's partner, Peter Doubt, notices that the regression line just misses the points (2, 5) and (8, 14). He calculates the equation containing these points and gets

$\hat{y}_2 = 1.5x + 2$

Show that Peter's equation contains the points (2, 5) and (8, 14).

T14. Calculate SS_{res} for Peter's equation. Explain how the answer shows that Peter's equation does not fit the data as well as Herbie's regression equation.

T15. Herbie wants to predict the length of his sidewinder 3 months (91 days) after it hatches. Use the linear regression equation $\hat{y} = 1.4x + 2.5$ to make this prediction. Does this prediction involve extrapolation, or does it involve interpolation?

T16. How much different would the predicted length after 3 months be if Herbie uses Peter's equation, $\hat{y}_2 = 1.5x + 2$?

T17. What did you learn from taking this test that you did not know before?

CHAPTER

9

Probability, and Functions of a Random Variable

To protect themselves from losing their life savings in a fire, homeowners purchase fire insurance. Each homeowner pays an insurance company a relatively small premium each year. From that money, the insurance company pays to replace the few homes that burn. Actuaries at the insurance company use the probability that a given house will burn to calculate the premiums to charge each year so that the company will have enough money to cover insurance claims and to pay employee salaries and other expenses. In this chapter you will learn some of the mathematics involved in these calculations.

Mathematical Overview

In this chapter you'll learn about random experiments and events. Results of random experiments constitute events. You will also learn the concepts of probability and mathematical expectation. The probabilities of various events are values that give you the likelihood of a particular event among all possible events in an experiment. You will gain this knowledge in four ways.

Numerically

$$\text{Probability} = \frac{\text{number of favorable outcomes}}{\text{total number of possible outcomes}}$$

Algebraically

If the probability that a thumbtack will land point up on any one flip is 0.4, then $P(x)$, the probability it will land point up x times in five flips, is this function of a random variable:

$$P(x) = {}_5C_x \cdot 0.6^{5-x} \cdot 0.4^x$$

Graphically

This is the graph of $P(x)$. The graph shows the probability that the thumbtack will land point up, 0, 1, 2, 3, 4, or 5 times in five flips.

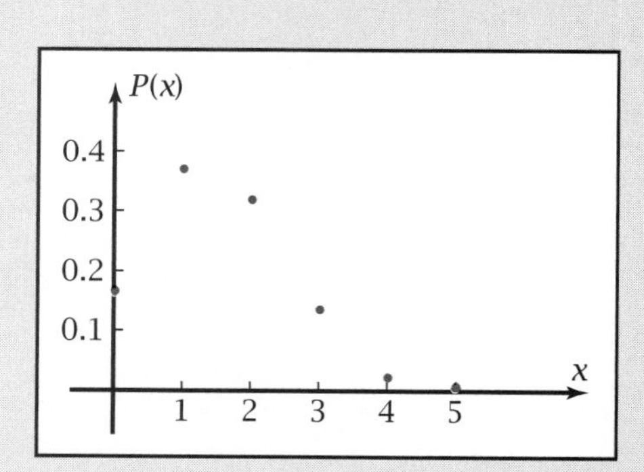

Verbally

I finally understand what they mean by a function of a random variable. You find the probability of each event in some random experiment, like getting "point up" x times when you flip a thumbtack five times. Then you plot the graph of the probability as a function of x. You don't connect the dots, since the number of point-ups has to be a whole number. You can also find your average winnings (called mathematical expectation) if the thumbtack experiment is done for money.

9-1 Introduction to Probability

Suppose two dice are rolled, one white and one blue. Figure 9-1a shows the 36 possible outcomes.

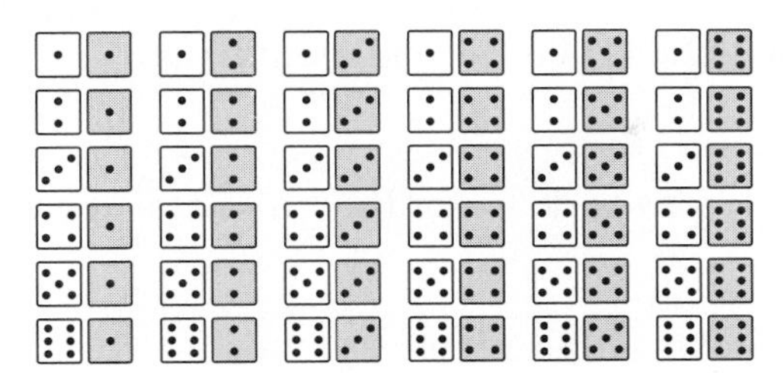

Figure 9-1a

There are five outcomes for which the total on the dice is 6:

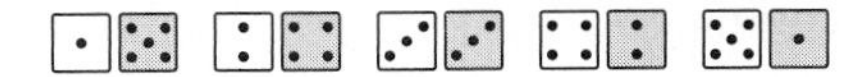

Because each outcome is equally likely, you would expect that in many rolls of the dice, the total would be 6 roughly $\frac{5}{36}$ of the time. This number, $\frac{5}{36}$, is called the **probability** of rolling a 6. In this section you will find the probabilities of other events in the dice-rolling experiment.

OBJECTIVE Find the probability of various events in a dice-rolling experiment.

Exploratory Problem Set 9-1

Two dice are rolled, one white and one blue. Find the probability of each of these events.

1. The total is 10.
2. The total is at least 10.
3. The total is less than 10.
4. The total is at most 10.
5. The total is 7.
6. The total is 2.
7. The total is between 3 and 7, inclusive.
8. The total is between, but does not include, 3 and 7.
9. The total is between 2 and 12, inclusive.
10. The total is 13.
11. The numbers are 2 and 5.
12. The blue die shows 2 and the white die shows 5.
13. The blue die shows 2 or the white die shows 5.

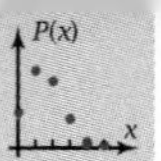

9-2 Words Associated with Probability

You have heard statements such as "It will probably rain today." Mathematicians give the word *probably* a precise meaning by attaching *numbers* to it, such as "The probability of rain is 30%." To understand and use probability, you need to learn the definitions of a few important terms.

OBJECTIVE Distinguish among various words used to describe probability.

For the dice-rolling experiment of Section 9-1, the act of rolling the dice is called a **random experiment.** Each time you roll the dice is called a **trial.** The word *random* lets you know that there is no way of telling beforehand how any roll is going to come out.

Each way the dice could come up, such as

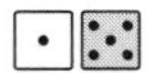

is called an **outcome** or a **simple event.** Outcomes are results of a random experiment. In this experiment, the outcomes are *equally likely.* That is, each has the same chance of occurring.

An **event** is a set of outcomes. For example, the event "the total on two dice is 6" is the five-element set

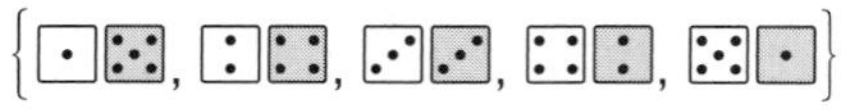

The set of *all* outcomes of an experiment is called the **sample space.** The sample space for the dice-rolling experiment of Section 9-1 is the set of all 36 outcomes shown in Figure 9-1a.

The **probability** of an event may now be defined numerically.

DEFINITION: Probability

If the outcomes of a random experiment are equally likely, then the **probability** that a particular event will occur is

Verbally: $\text{Probability} = \dfrac{\text{number of outcomes in the event}}{\text{number of outcomes in the sample space}}$

Symbolically: $P(E) = \dfrac{n(E)}{n(S)}$

where $n(E)$ is the number of outcomes in event E and $n(S)$ is the number of outcomes in the sample space S.

The symbols in the definition of probability are variations of $f(x)$ notation. The symbols make sense because the probability of an event *depends* on the event, and so forth.

Note that all probabilities are between 0 and 1, inclusive. An event that is *certain* to occur has a probability of 1 because $n(E) = n(S)$. An event that cannot possibly occur has a probability of 0 because $n(E) = 0$.

The chance of rain is rarely 0 percent or 100 percent.

Problem Set 9-2

Do These Quickly

Q1. If you flip a coin, what is the probability that the result will be "heads"?

Q2. If you flip the coin again, what is the probability that the second flip will be "heads"?

Q3. Does the result of the second flip depend on the result of the first flip?

Q4. What is $\frac{3}{5}$ expressed as a percent?

Q5. What is the variance of a set of data if the standard deviation is 9?

Q6. What does the coefficient of determination measure?

Q7. In the expression ab, the numbers a and b are called —?—.

Q8. In the expression $a + b$, the numbers a and b are called —?—.

Q9. True or false: $(ab)^2 = a^2b^2$

Q10. True or false: $(a + b)^2 = a^2 + b^2$

---

1. A card is drawn at random from a standard 52-card deck.
   a. What term is used in probability for the act of drawing the card?
   b. How many outcomes are in the sample space?
   c. How many outcomes are in the event "the card is a face card"?
   d. Calculate P(the card is a face card).
   e. Calculate P(the card is black).
   f. Calculate P(the card is an ace).
   g. Calculate P(the card is between 3 and 7, inclusive).
   h. Calculate P(the card is the ace of clubs).
   i. Calculate P(the card belongs to the deck).
   j. Calculate P(the card is a joker).

2. A penny, a nickel, and a dime are flipped at the same time. Each coin can land either heads up (H) or tails up (T).
   a. What term is used in probability for the act of flipping the coin?
   b. One possible outcome is THT. List all eight outcomes in the sample space.
   c. How many outcomes are in the event "exactly two of the coins show heads"?
   d. Calculate $P(HHT)$.
   e. Calculate P(exactly two heads).
   f. Calculate P(at least two heads).
   g. Calculate P(penny and nickel are tails).
   h. Calculate P(penny or nickel is tails).
   i. Calculate P(none are tails).
   j. Calculate P(zero, one, two, or three heads).
   k. Calculate P(four heads).

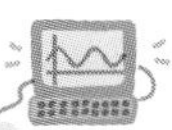

3. *Historical Search Project:* Check the Internet or other sources for information about early contributors to the field of mathematical probability. See if you can find out about the dice problem investigated by Blaise Pascal and Pierre de Fermat that led to the foundations of probability theory.

9-3 Two Counting Principles

Counting the outcomes in an event or sample space is difficult. For example, suppose a CD player is programmed to play eight songs in random order and you want to find the probability that your two favorite songs will play in a row. The sample space for this experiment contains over 40,000 outcomes! In this section you will learn ways of computing numbers of outcomes without actually counting them.

OBJECTIVE Calculate the number of outcomes in an event or sample space.

Independent and Mutually Exclusive Events

Counting outcomes sometimes involves considering two or more events. To find the number of outcomes in a situation involving two events, you need to consider whether *both* events occur or whether *either* one event *or* the other occurs, but not both.

For example, suppose a summer camp offers four outdoor activities and three indoor activities:

Outdoor	Indoor
swimming	pottery
canoeing	computers
volleyball	music
archery	

Morning

Afternoon

On Monday, each camper is assigned an outdoor activity in the morning and an indoor activity in the afternoon. In how many ways can the two activities be chosen?

In this situation, the two events are "an outdoor activity is chosen" and "an indoor activity is chosen." You want to count the number of ways *both* events could occur. You could find the answer by making an organized list of all the possible pairs:

swimming-pottery	swimming-computers	swimming-music
canoeing-pottery	canoeing-computers	canoeing-music
volleyball-pottery	volleyball-computers	volleyball-music
archery-pottery	archery-computers	archery-music

You could also reason like this: There are four choices for the outdoor activity. For each of these choices, there are three choices for the indoor activity. So there are $4 \cdot 3$ or 12 ways of choosing both activities. Note that the events in this situation are said to be **independent** because the way one occurs does not affect the ways the other could occur.

On Tuesday, the outdoor activities are canceled because of a thunderstorm, so each camper is assigned an indoor activity in the morning and a different indoor activity in the afternoon. In how many ways can the two activities be chosen?

In this case, the events are "an indoor activity is chosen" and "a different indoor activity is chosen." Again, you want to count the ways *both* events occur. There are three choices for the first event. After a selection is made for that event, only two choices remain for the second event. So there are $3 \cdot 2$ or 6 ways of choosing both activities. Note that the events in this situation are *not independent* because the way the first event is chosen affects the ways the second could be chosen.

On the day of the camp talent show, there is time for only one activity. Each camper is assigned *either* an outdoor activity *or* an indoor activity. In how many ways can the activity be assigned?

In this situation, the events are **mutually exclusive,** meaning that the occurrence of one of them excludes the possibility that the other will occur. If a camper is assigned an outdoor activity, he or she cannot also be assigned an indoor activity, and vice-versa. Because there are four ways of choosing an outdoor activity and three ways of choosing an indoor activity, there are $4 + 3$ or 7 ways of choosing one type of activity or the other.

These examples illustrate two counting principles.

PROPERTIES: Two Counting Principles

1. Let A and B be two events that occur in sequence. Then

   $$n(A \text{ and } B) = n(A) \cdot n(B \mid A)$$

   where $n(B \mid A)$ is the number of ways B can occur after A has occurred.

2. Let A and B be mutually exclusive events. Then

   $$n(A \text{ or } B) = n(A) + n(B)$$

Notes:

1. If A and B are independent, then $n(B \mid A) = n(B)$, and the first counting principle becomes $n(A \text{ and } B) = n(A) \cdot n(B)$.
2. Events are not mutually exclusive if both of them could happen at the same time. For instance, picking a red card and picking a jack from a deck of cards are not mutually exclusive because two cards in the deck are both red and jacks. Next you'll see an explanation of how to find the number of outcomes in this case.

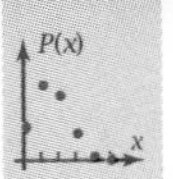

Overlapping Events

Suppose that you draw one card from a standard deck of 52 playing cards. In how many different ways could it be a heart *or* a face card? As shown in Figure 9-3a, $n(\text{heart}) = 13$ and $n(\text{face}) = 12$. But simply adding 13 and 12 gives the *wrong* answer. The cards in the intersection (those that are hearts *and* face cards) have been counted *twice*. An easy way to get the correct answer is to subtract the number that are both hearts and face cards from the 13 + 12. That is,

$$n(\text{heart } or \text{ face}) = n(\text{heart}) + n(\text{face}) - n(\text{heart } and \text{ face})$$

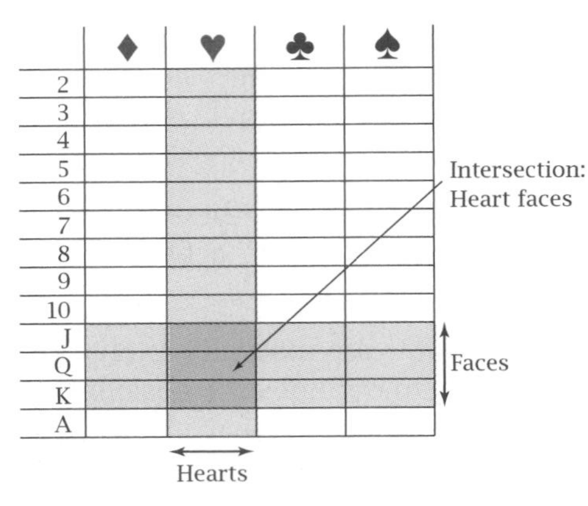

Figure 9-3a

In general, the number of ways event A or B could occur when the two events are *not* mutually exclusive is given by this generalization of the property.

PROPERTY: Non-Mutually Exclusive Events

If A and B are events that are not necessarily mutually exclusive, then

$$n(A \text{ or } B) = n(A) + n(B) - n(A \cap B)$$

Special case: If A and B *are* mutually exclusive, then $n(A \cap B) = 0$ and

$$n(A \text{ or } B) = n(A) + n(B)$$

Problem Set 9-3

Do These Quickly

Q1. Simplify the fraction $\frac{12}{36}$.

Q2. Evaluate: $1 \cdot 2 \cdot 3 \cdot 4 \cdot 5$

Q3. If $n(A) = 71$ and there are 300 elements in the sample space, then $P(A) =$ —?—.

Q4. Multiply: $\left(\frac{2}{7}\right)\left(\frac{3}{4}\right)$

Q5. Add: $\frac{3}{8} + \frac{1}{4}$

Q6. The exact value (no decimals) of $\cos \frac{\pi}{6}$ is —?—.

Q7. How well does a regression equation fit the data if the correlation coefficient is −1?

Q8. $x = 3t^2$ and $y = \cos t$ are equations of a —?— function.

Q9. Factor: $x^2 - 3x - 4$

Q10. The deviation of the number 10 from the mean of 10, 13, 18, and 19 is

A. 9 B. 8 C. 5 D. 3 E. −5

---

1. A salesperson has 7 customers in Denver and 13 customers in Reno. In how many different ways could she telephone
   a. A customer in Denver and then a customer in Reno?
   b. A customer in Denver or a customer in Reno, but not both?
2. A pizza establishment offers 12 vegetable toppings and 5 meat toppings. Find the number of different ways you could select
   a. A meat topping or a vegetable topping
   b. A meat topping and a vegetable topping
3. A reading list consists of 11 novels and 5 biographies. Find the number of different ways a student could select
   a. A novel or a biography
   b. A novel and then a biography
   c. A biography and then another biography
4. A convoy of 20 cargo ships and 5 escort vessels approaches the Suez Canal. In each scenario, how many different ways are there that these vessels could begin to go through the canal?

   a. A cargo ship and then an escort vessel
   b. A cargo ship or an escort vessel
   c. A cargo ship and then another cargo ship
5. The menu at Paesano's lists 7 salads, 11 entrees, and 9 desserts. How many different salad-entree-dessert meals could you select? (Meals are considered to be different if any one thing is different.)
6. Admiral Motors manufactures cars with 5 different body styles, 11 different exterior colors, and 6 different interior colors. A dealership wants to display one of each possible variety of car in its showroom. Explain to the manager of the dealership why the plan would be impractical.
7. Consider the letters in the word LOGARITHM.
   a. In how many different ways could you select a vowel or a consonant?
   b. In how many different ways could you select a vowel and then a consonant?
   c. How many different three-letter "words" (for example, "ORL," "HLG," and "AOI") could you make using each letter no more than once in any one word? (There are three events: "select the first letter," "select the second letter," and "select the third letter." Find the number of ways each event can occur, and then figure out what to do with the three results.)
8. Lee brought two jazz CDs and five rap CDs to play at the class picnic.
   a. How many different ways could he choose a jazz CD and then a rap CD?
   b. How many different ways could he choose a jazz CD or a rap CD?
   c. Lee's CD player allows him to load four CDs at once. The CDs will play in the order he loads them. How many different orderings of four CDs are possible? (See Problem 7 for a hint.)
9. There are 20 girls on the basketball team. Of these, 17 are over 16 years old, 12 are taller than 170 cm, and 9 are both older than 16 and taller than 170 cm. How many of the girls are older than 16 or taller than 170 cm?
10. Lyle's DVD collection includes 37 classic films and 29 comedies. Of these, 21 are classic comedies. How many DVDs does Lyle have that are classics or comedies?
11. The library has 463 books dealing with science and 592 books of fiction. Of these, 37 are science fiction books. How many books are either science or fiction?

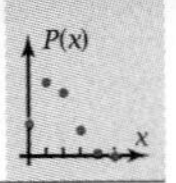

12. The senior class has 367 girls and 425 students with brown hair. Of the girls, 296 have brown hair. In how many different ways could you select a girl or a brown-haired student from the senior class?

13. *Seating Problem:* There are 10 students in a class and 10 chairs, numbered 1 through 10.
    a. In how many different ways could a student be selected to occupy chair 1?
    b. After someone is seated in chair 1, how many different ways are there of seating someone in chair 2?
    c. In how many different ways could chairs 1 and 2 be filled?
    d. If two of the students are sitting in chairs 1 and 2, in how many different ways could chair 3 be filled?
    e. In how many different ways could chairs 1, 2, and 3 be filled?
    f. In how many different ways could all 10 chairs be filled? Surprising?!

14. *Baseball Team Problem 1:* Nine people on a baseball team are trying to decide who will play each position.
    a. In how many different ways could they select a person to be pitcher?
    b. After someone has been selected as pitcher, in how many different ways could they select someone to be catcher?
    c. In how many different ways could they select a pitcher and a catcher?
    d. After the pitcher and catcher have been selected, in how many different ways could they select a first-base player?
    e. In how many different ways could they select a pitcher, catcher, and first-base player?
    f. In how many different ways could all nine positions be filled? Surprising?!

15. *License Plate Problem:* Many states use car license plates that have six characters. Some states use two letters followed by a number from 1 to 9999. Others use three letters followed by a number from 1 to 999.
    a. Which of these two plans allows there to be more possible license plates? How many more?
    b. How many different license plates could there be if the state allowed either two letters and four digits or three letters and three digits?

    c. Assuming there are about 200,000,000 motor vehicles in the United States, would it be possible to have a national license plate program using the plan in part b? Explain.

16. *Telephone Number Problem:* When 10-digit telephone numbers were introduced into the United States and Canada in the 1960s, certain restrictions were placed on the groups of numbers:

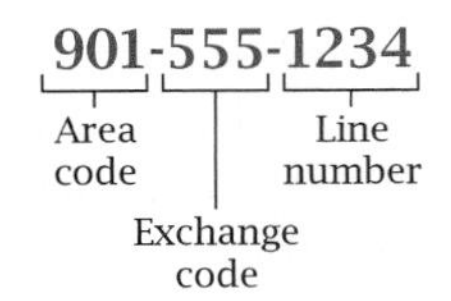

Area Code: 3 digits; the first must not be 0 or 1, and the second must be 0 or 1.

Exchange Code: 3 digits; the first and second must not be 1.

Line Number: 4 digits; at least one must not be 0.

    a. Find the possible numbers of area codes, exchange codes, and line numbers.
    b. How many valid numbers could there be under this numbering scheme?
    c. How many 10-digit numbers could be made if there were no restrictions on the three groups of numbers?

d. What is the probability that a 10-digit number dialed at random would be a valid number under the original restrictions?

e. The total population of the United States and Canada is currently about 300 million. In view of the fact that there are now area codes and exchange codes that do not conform to the original restrictions, what assumption can you make about the number of telephones per person in the United States and Canada?

17. *Journal Problem:* Update your journal with things you have learned about probability and about counting outcomes.

9-4 Probabilities of Various Permutations

Many counting problems involve finding the number of different ways to *arrange,* or *order,* things. For example, the three letters *ABC* can be arranged in six different ways:

ABC *ACB* *BAC* *BCA* *CAB* *CBA*

But the 10 letters *ABCDEFGHIJ* can be arranged in more than 3 *million* different ways! In this section you will learn a time-efficient way to calculate the number of arrangements, or **permutations,** of a set of objects. As a result you will be able to calculate relatively quickly the probability that a permutation selected at random will have certain characteristics.

OBJECTIVE Given a description of a permutation, find the probability of getting that permutation if an arrangement is selected at random.

Here is a formal definition of permutation.

DEFINITION: Permutation

A **permutation** of a set of objects is an arrangement in a definite order of some or all of the elements in that set.

▶ EXAMPLE 1 In how many different ways could you arrange three books on a shelf if you have seven books from which to choose?

Solution The process of selecting an arrangement (permutation) can be divided into three events:

A—Choose a book to go in the first position (Figure 9-4a).

B—Choose a book to go in the second position.

C—Choose a book to go in the third position.

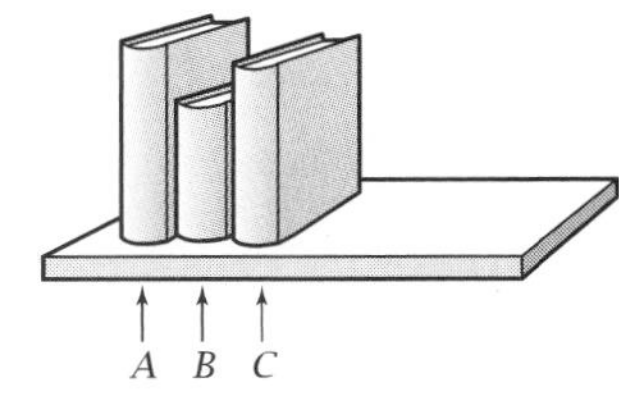

Figure 9-4a

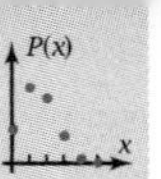

Let n be the number of permutations.

$n = \underline{\ }\ \underline{\ }\ \underline{\ }$	Mark three spaces for the three events.
$n = \underline{7}\ \ \underline{6}\ \ \underline{5}$	7 ways to select the first book; 6 ways to select the second; 5 ways to select the third.
$n = \underline{7} \cdot \underline{6} \cdot \underline{5} = 210$ ways	Apply the counting principle for sequential events. ◀

In Example 1 the answer 210 is "the number of permutations of seven elements taken three at a time."

▶ EXAMPLE 2 A permutation is selected at random from letters in the word SEQUOIA. What is the probability that it has Q in the fourth position and ends with a vowel?

Solution Let E be the set of all favorable outcomes. First, find the number of outcomes (permutations) in E.

$n(E) = \underline{\ }\ \underline{\ }\ \underline{\ }\ \underline{\ }\ \underline{\ }\ \underline{\ }\ \underline{\ }$	Mark seven spaces on which to record the number of ways of selecting each letter.
$n(E) = \underline{\ }\ \underline{\ }\ \underline{\ }\ \underline{1}\ \underline{\ }\ \underline{\ }\ \underline{\ }$	Write 1 in the fourth space, since there is only one Q to go there.
$n(E) = \underline{\ }\ \underline{\ }\ \underline{\ }\ \underline{1}\ \underline{\ }\ \underline{\ }\ \underline{5}$	Write 5 in the last space, since there are 5 ways to select a vowel.
$n(E) = \underline{5}\ \ \underline{4}\ \ \underline{3}\ \ \underline{1}\ \ \underline{2}\ \ \underline{1}\ \ \underline{5}$	There are 5 letters left, so there are 5, 4, 3, 2, and 1 ways to select the remaining letters.
$n(E) = \underline{5} \cdot \underline{4} \cdot \underline{3} \cdot \underline{1} \cdot \underline{2} \cdot \underline{1} \cdot \underline{5} = 600$	Apply the counting principle for sequential events.

Next, find the number of outcomes in the sample space.

$$n(S) = \underline{7} \cdot \underline{6} \cdot \underline{5} \cdot \underline{4} \cdot \underline{3} \cdot \underline{2} \cdot \underline{1} = 5040$$

Finally, find the probability, using the definition.

$$P(E) = \frac{n(E)}{n(S)} = \frac{600}{5040} = \frac{5}{42} = 0.1190\ldots \approx 12\%$$ ◀

Note that in Example 2 the fourth position is a **fixed position,** since there is only one way it can be filled. The last position is a **restricted position**; more than one letter can go there, but the choices are limited to vowels.

Note also that the number of outcomes in the sample space,

$$7 \cdot 6 \cdot 5 \cdot 4 \cdot 3 \cdot 2 \cdot 1$$

is the product of consecutive positive integers ending with 1. This is called a **factorial.** The factorial symbol is the exclamation mark, "!." So

$$7! = 7 \cdot 6 \cdot 5 \cdot 4 \cdot 3 \cdot 2 \cdot 1 = 5040$$ Pronounced "7 factorial."

Most calculators have a built-in factorial function.

DEFINITION: Factorial

For any positive integer n, **n factorial** ($n!$) is given by

$$n! = 1 \cdot 2 \cdot 3 \cdot \cdots \cdot n$$

or, equivalently,

$$n! = n \cdot (n-1) \cdot (n-2) \cdot \cdots \cdot 2 \cdot 1$$

0! is defined to be equal to 1.

Problem Set 9-4

Do These Quickly

Q1. If the outcomes of a random experiment are equally likely, then the probability of an event is defined to be —?—.

Q2. For events A and B, $n(A \text{ and then } B)$ = —?—.

Q3. If events A and B are mutually exclusive, then $n(A \text{ or } B)$ = —?—.

Q4. If events A and B are *not* mutually exclusive, then $n(A \text{ or } B)$ = —?—.

Q5. The set of all possible outcomes of a random experiment is called the —?—.

Q6. —?— functions have the multiply-multiply property.

Q7. The slope of the linear function $4x + 5y = 40$ is —?—.

Q8. The "If" part of a theorem is called the —?—.

Q9. An equation that is true for *all* values of the variable is called a(n) —?—.

Q10. 4% of 700 is —?—.

---

1. The Hawaiian alphabet has 12 letters. How many permutations could be made using
   a. Two different letters
   b. Four different letters
   c. Twelve different letters

2. Fran Tick takes a 10-problem precalculus test. The problems may be worked in any order.
   a. In how many different orders could she work all 10 of the problems?
   b. In how many different orders could she work any 7 of the 10 problems?

3. Triangles are often labeled by placing a different letter at each vertex. In how many different ways could a given triangle be labeled?

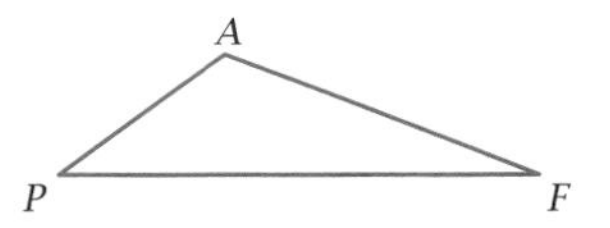

4. Tom, Dick, and Harry each draw two cards from a standard 52-card deck and place them face up in a row. The cards are not replaced. Tom goes first. Find the number of different orders in which
   a. Tom could draw his two cards
   b. Dick could draw his two cards *after* Tom has already drawn
   c. Harry could draw his two cards *after* Tom and Dick have drawn theirs

5. Frost Bank has seven vice presidents, but only three spaces in the parking lot are labeled "Vice President." In how many different ways could these spaces be occupied by the vice presidents' cars?

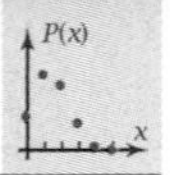

6. A professor says to her class, "You may work these six problems in any order you choose." There are 100 students in the class. Is it possible for each student to work the problems in a different order? Explain.

7. A six-letter permutation is selected at random from the letters in the word NIMBLE.
   a. How many permutations are possible?
   b. How many of these permutations begin with M?
   c. What is the probability that the permutation begins with M?
   d. Express the probability in part c as a percent.
   e. What is the probability that the permutation is NIMBLE?

8. A six-letter permutation is selected at random from the letters in the word NIMBLE. Find the probability of each event.
   a. The third letter is I and the last letter is B.
   b. The second letter is a vowel and the third letter is a consonant.
   c. The second and third letters are both vowels.
   d. The second letter is a consonant and the last letter is E.
   e. The second letter is a consonant and the last letter is L.

9. *Baseball Team Problem 2:* Nine people try out for the nine positions on a baseball team.
   a. In how many different ways could the positions be filled if there are no restrictions on who plays which position?
   b. In how many different ways could the positions be filled if Fred must be the pitcher but the other eight people can take any of the remaining eight positions?
   c. If the positions are selected at random, what is the probability that Fred will be the pitcher?
   d. What is the probability in part c expressed as a percent?

10. *Soccer Team Problem 1:* Eleven girls try out for the 11 positions on a soccer team.

   a. In how many different ways could the 11 positions be filled if there are no restrictions on who plays which position?
   b. In how many different ways could the positions be filled if Mabel must be the goalkeeper?
   c. If the positions are selected at random, what is the probability that Mabel will be the goalkeeper?
   d. What is the probability in part c expressed as a percent?

11. *Baseball Team Problem 3:* Nine people try out for the nine positions on a baseball team. If the players are selected at random for the positions, find the probability of each event.
   a. Fred, Mike, or Jason is the pitcher.
   b. Fred, Mike, or Jason is the pitcher, and Sam or Paul plays first base.
   c. Fred, Mike, or Jason is the pitcher, Sam or Paul plays first base, and Bob is the catcher.

12. *Soccer Team Problem 2:* Eleven girls try out for the 11 positions on the varsity soccer team. If the players are selected at random, find the probability of each event.
   a. Mabel, Keisha, or Diedra is the goalkeeper.
   b. Mabel, Keisha, or Diedra is the goalkeeper, and Alice or Phyllis is the center forward.
   c. Mabel, Keisha, or Diedra is the goalkeeper, Alice or Phyllis is the center forward, and Bea is the left fullback.

13. Eight first-graders line up for a fire drill (Figure 9-4b).

Calvin and Phoebe

Figure 9-4b

a. How many possible arrangements are there?

b. In how many of these arrangements are Calvin and Phoebe next to each other? (*Clue:* Arrange *seven* things—the Calvin and Phoebe pair and the other six children. Then arrange Calvin and Phoebe.)

c. If the eight students line up at random, what is the probability that Calvin and Phoebe will be next to each other?

14. The ten digits, 0, 1, 2, 3, . . . , 9, are arranged at random with no repeats. Find the probability that the numeral formed represents

a. A number greater than 6 billion

b. An even number greater than 6 billion (There are two cases to consider, "first digit is odd" and "first digit is even.")

Permutations with Repeated Elements—Problems 15 and 16: The word CARRIER has seven letters. But there are fewer than 7! permutations, because in any arrangement of these seven letters the three *R*'s are interchangeable. If these *R*'s were distinguishable, there would be 3!, or 6, ways of arranging them. This implies that only $\frac{1}{6}$ (that is, $\frac{1}{3!}$) of the 7! permutations are actually different. So the number of permutations is

$$\frac{7!}{3!} = 840$$

There are four *I*'s, four *S*'s, and two *P*'s in the word MISSISSIPPI, so the number of different permutations of its letters is

$$\frac{11!}{4!\ 4!\ 2!} = 34{,}650$$

15. Find the number of different permutations of the letters in each word.

a. FREELY

b. BUBBLES

c. LILLY

d. MISSISSAUGA

e. HONOLULU

f. HAWAIIAN

16. Nine pennies are lying on a table. Five show heads and four show tails. In how many different ways, such as "HHTHTTHHT," could the coins be lined up if you consider all the heads to be identical and all the tails to be identical?

Circular Permutations—Problems 17–20: In Figure 9-4c, the letters *A, B, C,* and *D* are arranged in a circle. Though these may seem to be different permutations, they are considered the same permutation because the letters have the same position *with respect to each other.* That is, each of the four letters has the same letter to its left and the same letter to its right. An easy way to calculate the number of different **circular permutations** of *n* elements is to fix the position of one element and then arrange the other $(n - 1)$ elements with respect to it (Figure 9-4d). So, for the letters *A, B, C,* and *D,* the number of circular permutations is $n = \underline{1} \cdot \underline{3} \cdot \underline{2} \cdot \underline{1} = \underline{6}$.

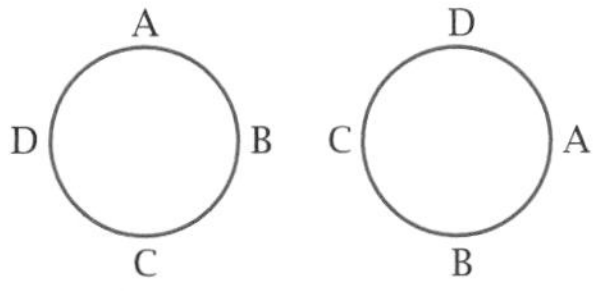

Same circular permutation

Figure 9-4c

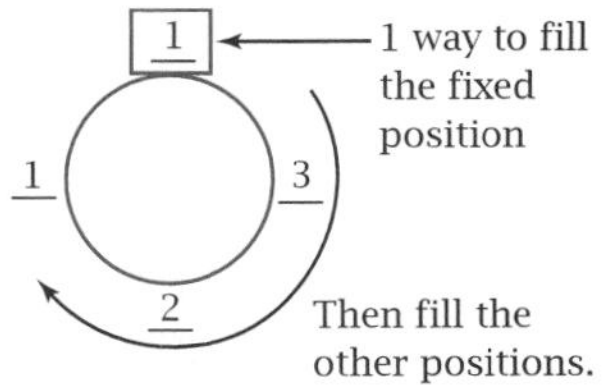

Figure 9-4d

17. How many different circular permutations could be made with these letters?

a. ABCDE

b. QLMXTN

c. LOGARITHM

18. In how many different ways could King Arthur's 12 knights be seated around the Round Table?

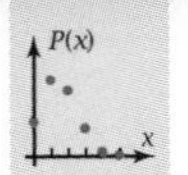

19. Four girls and four boys sit around a merry-go-round.

   a. In how many different ways can they be arranged with respect to each other so that boys and girls alternate?
   b. If they seat themselves at random, what is the probability that boys and girls will alternate?

20. Suppose that you are concerned only with which elements come *between* other elements in a circular permutation, not with which elements are to the left and to the right. Then there are two circular permutations that would be considered the same, a clockwise one and a counterclockwise one (Figure 9-4e). In this case, there would be only *half* the number of circular permutations as calculated earlier.

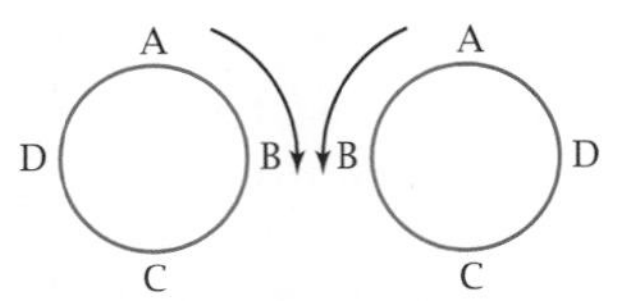

The same "betweenness" property

Figure 9-4e

Find the number of different ways you could arrange

   a. Seven different beads to form a bracelet if you consider only which bead is *between* which other beads
   b. Five keys on a key ring if you consider only which key is *between* which other keys

9-5 Probabilities of Various Combinations

There are 24 different three-letter "words" that can be made from the four letters *A, B, C,* and *D.* These are

ABC	*ACB*	*BAC*	*BCA*	*CAB*	*CBA*	One combination.
ABD	*ADB*	*BAD*	*BDA*	*DAB*	*DBA*	A second combination.
ACD	*ADC*	*CAD*	*CDA*	*DAC*	*DCA*	A third combination.
BCD	*BDC*	*CBD*	*CDB*	*DBC*	*DCB*	A fourth combination.

Because these words are *arrangements* of the letters in a definite order, each one is a *permutation* of four elements taken three at a time.

Suppose you are concerned only with *which* letters appear in the word, not with the order in which they appear. For instance, you would consider *ADC* and *DAC* to be the same because they have the same three letters. Each different group of three letters is called a **combination** of the letters *A, B, C,* and *D.* As shown above, there are 24 different permutations but only four different combinations of the letters *A, B, C,* and *D* taken three at a time. In this section you will learn a time-efficient way to calculate the number of combinations of the elements in a set.

OBJECTIVE Calculate the number of different combinations containing r elements taken from a set containing n elements.

Here is the formal definition of combination.

DEFINITION: Combination

A **combination** of elements in a set is a *subset* of those elements, without regard to the order in which the elements are arranged.

In the example given at the opening of this section, you can see that for every *one* combination, there are *six* possible permutations. So the total number of combinations is equal to the total number of permutations *divided by 6.* That is,

$$\text{Number of combinations} = \frac{24}{6} = 4$$

This idea allows you to calculate a number of combinations by dividing two numbers of permutations.

PROPERTY: Computation of Number of Combinations

$$\text{Number of combinations} = \frac{\text{total number of permutations}}{\text{number of permutations of each } one \text{ combination}}$$

Before proceeding with examples, it helps to define some symbols.

DEFINITIONS: Symbols for Numbers of Combinations and Permutations

$_nC_r$ = number of different combinations of n elements taken r at a time

Pronounced: "*n, C, r*"

$_nP_r$ = number of different permutations of n elements taken r at a time

Pronounced: "*n, P, r*"

Example 1 shows you how to calculate a number of combinations with the help of these symbols.

▶ **EXAMPLE 1** Calculate $_4C_3$.

Solution

$$_4C_3 = \frac{_4P_3}{_3P_3} = \frac{4 \cdot 3 \cdot 2}{3 \cdot 2 \cdot 1} = 4$$ ◀

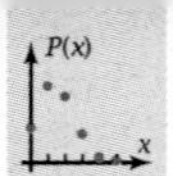

▶ EXAMPLE 2 Write $_9P_4$ as a ratio of factorials. Interpret the answer in terms of the number of elements in the set and the number of elements selected for the permutation.

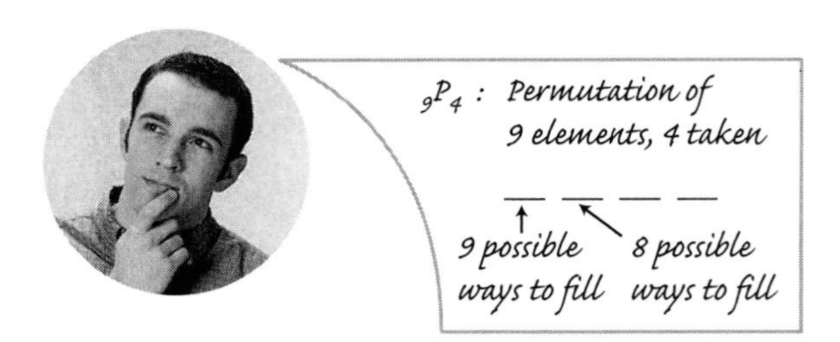

Solution

$$_9P_4 = \underline{9} \cdot \underline{8} \cdot \underline{7} \cdot \underline{6}$$

$$= 9 \cdot 8 \cdot 7 \cdot 6 \cdot \frac{5 \cdot 4 \cdot 3 \cdot 2 \cdot 1}{5 \cdot 4 \cdot 3 \cdot 2 \cdot 1}$$ Multiply by a clever form of 1.

$$= \frac{9 \cdot 8 \cdot 7 \cdot 6 \cdot 5 \cdot 4 \cdot 3 \cdot 2 \cdot 1}{5 \cdot 4 \cdot 3 \cdot 2 \cdot 1}$$

$$= \frac{9!}{5!}$$

The 9 in the numerator is the total number of elements in the set from which the permutation is being made. The 5 in the denominator is the number of elements *not* selected for the permutation. ◀

▶ EXAMPLE 3 Write $_9C_4$ as a ratio of a factorial to a product of factorials. Interpret the answer in terms of the number of elements in the set and the number of elements selected for the combination.

Solution

$$_9C_4 = \frac{_9P_4}{_4P_4}$$ Use the preceding property.

$$= \frac{9!/5!}{4!}$$ From the answer to Example 2.

$$= \frac{9!}{4!\ 5!}$$

The 9 in the numerator is the total number of elements from which the combination is being made. The 4 and 5 in the denominator are the number of elements *selected* for the combination and the number of elements *not* selected for the combination, respectively. ◀

From Examples 2 and 3 you can find relatively simple patterns to use for calculating a number of combinations or permutations.

TECHNIQUE: Calculation of Number of Permutations or Combinations

EXAMPLE: PERMUTATIONS

$${}_9P_4 = \frac{9!}{5!}$$

Same (the 9 in ${}_9P_4$ and the 9 in 9!)

Number of selected elements (the 4)

Number *not* selected (the 5 in 5!)

EXAMPLE: COMBINATIONS

$${}_9C_4 = \frac{9!}{4!\,5!}$$

Same (the 9 in ${}_9C_4$ and the 9 in 9!)

Same (the 4 in ${}_9C_4$ and the 4 in 4!)

Add up to numeral in numerator. (4 + 5 = 9)

With these techniques in mind, you are ready to solve problems in which you must find the probability that a specified kind of combination occurs.

▶ EXAMPLE 4 In how many different ways could you form a committee of three people from a group of seven people? Explain how you know a number of combinations is being asked for, not a number of permutations.

Solution Committees with the same members are different only if the people on the committees have special roles. If there are no special roles, it does not matter how the committee members are arranged. So the answer is a number of combinations, not a number of permutations. Let n(3 people) stand for the number of different three-person committees.

$$n(3 \text{ people}) = {}_7C_3 = \frac{7!}{3!\,4!} = 35 \text{ committees}$$

Use the pattern for combinations. ◀

▶ EXAMPLE 5 If a committee of five is selected at random from a group of six women and three men, find the probability that it will include

a. Eileen and Ben (two of the nine people)

b. Exactly three women and two men

c. At least three women

Solution a. $$P(\text{Eileen and Ben}) = \frac{n(\text{Eileen and Ben})}{n(\text{sample space})}$$ Definition of probability.

The sample space is the set of all possible five-member committees.

$$n(\text{sample space}) = {}_9C_5 = \frac{9!}{5!\,4!} = 126 \text{ committees}$$

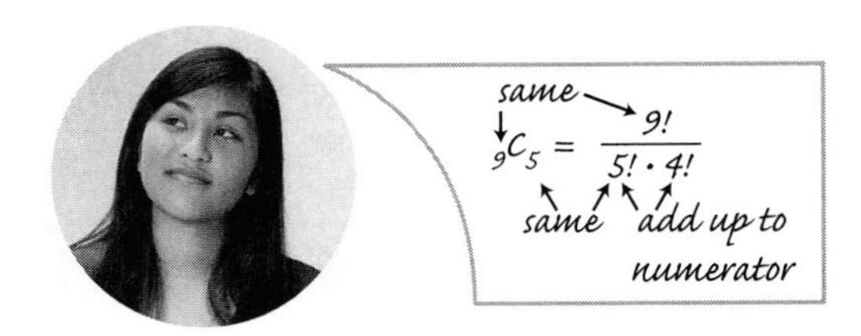

To find n(Eileen and Ben), first select Eileen and Ben (1 way), then select the other three committee members from the seven who remain. There are ${}_7C_3$ ways to select the three committee members.

$$n(\text{Eileen and Ben}) = 1 \cdot {}_7C_3 = \frac{7!}{3!\,4!} = 35$$

By the counting principle for sequential events.

$$\therefore\ P(\text{Eileen and Ben}) = \frac{35}{126} = 0.2777\ldots \approx 28\%$$

The definition of probability.

b. To find the number of three-woman, two-man committees, notice that people are being selected from two different groups. You can divide the "hard" problem of selecting the committee into two "easy" problems, selecting the women and selecting the men. So by the counting principle for sequential events,

$$n(\text{3 women and 2 men}) = n(\text{3 women}) \cdot n(\text{2 men})$$
$$= {}_6C_3 \cdot {}_3C_2$$
$$= \frac{6!}{3!\,3!} \cdot \frac{3!}{2!\,1!}$$
$$= 20 \cdot 3 = 60$$

$$\therefore\ P(\text{3 women and 2 men}) = \frac{60}{126} = 0.4761\ldots \approx 48\%$$

c. If the committee has at least three women, it could have three women *or* four women *or* five women. In each case, the remainder of the committee consists of men. Turn this problem into three easier problems.

$n(3W, 2M) = 60$ From part b.

$n(4W, 1M) = {}_6C_4 \cdot {}_3C_1 = \frac{6!}{4!\,2!} \cdot \frac{3!}{1!\,2!} = 15 \cdot 3 = 45$

$n(5W, 0M) = {}_6C_5 \cdot {}_3C_0 = \frac{6!}{5!\,1!} \cdot \frac{3!}{0!\,3!} = 6 \cdot 1 = 6$ Recall that $0! = 1$.

Because these are mutually exclusive events, you can add the numbers of ways.

$n(\text{at least 3 women}) = 60 + 45 + 6 = 111$

$\therefore P(\text{at least 3 women}) = \frac{111}{126} = 0.8809... \approx 88\%$ ◀

Note that most graphing calculators have built-in functions to calculate numbers of permutations and combinations directly. For instance, to calculate ${}_9C_4$, you might enter 9 nCr 4. The answer would be 126, the same as 9!/(4! 5!).

Problem Set 9-5

Do These Quickly

Q1. 4! = —?—

Q2. $\frac{4!}{4}$ = —?—!

Q3. $\frac{3!}{3}$ = —?—!

Q4. $\frac{2!}{2}$ = —?—!

Q5. $\frac{1!}{1}$ = —?—!

Q6. Why does 0! equal 1, not 0?

Q7. Write ${}_5P_5$ as a factorial.

Q8. Write ${}_nP_n$ as a factorial.

Q9. Express 0.43856... as a percent rounded to the nearest integer.

Q10. The *exact* value (no decimals) of $\tan\frac{\pi}{3}$ is —?—.

For Problems 1–12, evaluate the number of combinations or permutations two ways:

a. Using factorials, as in the examples of this section

b. Directly, using the built-in features of your grapher

1. ${}_5C_3$
2. ${}_6C_4$
3. ${}_{27}C_{19}$
4. ${}_{44}C_{24}$
5. ${}_{10}C_{10}$
6. ${}_{100}C_{100}$
7. ${}_{10}C_0$
8. ${}_{100}C_0$
9. ${}_6P_4$
10. ${}_{11}P_5$
11. ${}_{47}P_{30}$
12. ${}_{50}P_{20}$
13. Twelve people apply to go on a biology field trip, but there is room in the car for only five of them. In how many different ways could the group of five making the trip be chosen? How can you tell that a number of combinations is being asked for, not a number of permutations?
14. Seven people come to an evening bridge party. Only four people can play bridge at any one time, so they decide to play as many games as it takes to use every possible foursome once. How many games would have to be played? Could all of these games be played in one evening?
15. A donut franchise sells 34 varieties of donuts. Suppose one of the stores decides to make sample boxes with six different donuts in each box. How many different sample boxes could be made? Would it be practical to stock one of each kind of box?

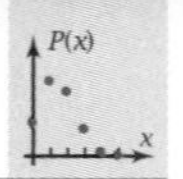

16. Just before each Supreme Court session, each of the nine justices shakes hands with every other justice. How many handshakes take place?

17. Horace Holmsley bought blueberries, strawberries, watermelon, grapes, plums, and peaches. Find the number of different fruit salads he could make if he uses
    a. Three ingredients
    b. Four ingredients
    c. Three ingredients or four ingredients
    d. All six ingredients

18. A pizzeria offers 11 different toppings. Find the number of different kinds of pizza they could make using
    a. Three toppings
    b. Five toppings
    c. Three toppings or five toppings
    d. All 11 toppings

19. A standard deck of playing cards has 52 cards.
    a. How many different 5-card poker hands could be formed from a standard deck?
    b. How many different 13-card bridge hands could be formed?
    c. How can you tell that numbers of combinations are being asked for, not numbers of permutations?

20. The diagonals of a convex polygon are made by connecting the vertices two at a time. However, some of the combinations are *sides,* not diagonals (Figure 9-5a). How many diagonals are there in each convex figure?

Side is *not* a diagonal.

Diagonal

Figure 9-5a

    a. Pentagon (5 sides)
    b. Decagon (10 sides)
    c. n-gon (n sides). From the answer, get a simple formula for the number of diagonals.

21. A set has ten elements. Find the number of subsets that contain exactly
    a. Two elements
    b. Five elements
    c. Eight elements (Explain the relationship between this answer and the answer to part a.)

22. A set has five elements, {❀, ❁, ✌, ✛, ⚯}.
    a. Find the number of different subsets that contain
        i. One element
        ii. Two elements
        iii. Three elements
        iv. Four elements
        v. All five elements
        vi. No elements
    b. How many subsets are there altogether? What relationship does this number have to the number of elements in the set?
    c. Based on your answer to part b, how many subsets would a 10-element set have? A 100-element set?

23. *Review Problem 1:* You draw a 5-card hand from a standard 52-card deck and then arrange the cards from left to right.
    a. After the cards have been selected, in how many different ways could you arrange them?
    b. How many different 5-card hands could be formed without considering arrangement?
    c. How many different 5-card arrangements could be formed from the deck?
    d. Which part(s) of this problem involve permutations and which involve combinations?

24. *Review Problem 2:* At South High School, 55 students entered an essay contest. From these students, 10 are selected as finalists.
    a. After the finalists have been selected, in how many different ways could they be ranked from 1st to 10th?
    b. In how many different ways could the 10 finalists be selected?
    c. How many different 10-student rankings could be made from the 55 entrants?
    d. Which part(s) of this problem involve permutations and which involve combinations?

25. Charlie has 13 socks in his drawer, 7 blue and 6 green. He selects 5 socks at random. Find the probability that he gets
    a. Two blue socks and three green socks
    b. Three blue socks and two green socks
    c. Two blue socks and three green socks, or three blue socks and two green socks
    d. The one sock that has a hole in it

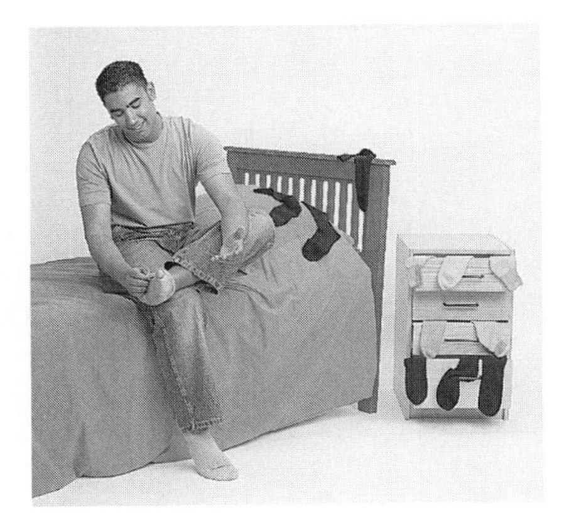

26. In a group of 15 people, 6 are left-handed and the rest are right-handed. If 7 people are selected at random from this group, find the probability that
    a. Three are left-handed and four are right-handed
    b. All are right-handed
    c. All are left-handed
    d. Harry and Peg, two of the left-handers, are selected

27. Three baseball cards are selected at random from a group of seven cards. Two of the cards are rookie cards.
    a. What is the probability that exactly one of the three selected cards is a rookie card?
    b. What is the probability that at least one of the three cards is a rookie card?
    c. What is the probability that none of the three cards is a rookie card?
    d. What is the relationship between the answers to parts b and c?

28. Emma, who is three years old, tears the labels off all ten cans of soup on her mother's shelf (Figure 9-5b). Her mother knows that there are two cans of tomato soup and eight cans of vegetable soup. She selects four cans at random.

Figure 9-5b

    a. What is the probability that exactly one of the four cans contains tomato soup?
    b. What is the probability that at least one of the cans contains tomato soup?

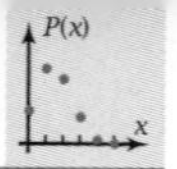

c. What is the probability that none of the four cans contains tomato soup?

d. What is the relationship between the answers to parts b and c?

29. *Light Bulb Problem:* Light bulb manufacturers like to be assured that their bulbs will work. Because testing every bulb is impractical, a random sample of bulbs is tested. Suppose that the quality control department at a light bulb factory decides to test a random sample of 5 bulbs for every 100 that are made.

a. In how many different ways could a 5-bulb sample be taken from 100 bulbs?

b. To check the quality control process, the manager of the factory puts 2 defective bulbs in with 98 working bulbs before the sample is taken. How many different ways could a sample of 5 of these 100 bulbs be selected that contains at least 1 of the defective bulbs?

c. If the 100 bulbs include 2 defective bulbs, what is the probability that the sampling process will reveal at least 1 defective bulb?

d. Based on your answers, do you think that the 5-bulb-in-100 sampling plan is sufficiently effective?

30. A standard 52-card deck of playing cards has 4 suits, with 13 cards in each suit. In a particular game, each of the four players is dealt 13 cards at random.

a. Find the probability that such a 13-card hand has

i. Exactly five spades

ii. Exactly three clubs

iii. Exactly five spades and three clubs

iv. Exactly five spades, three clubs, and two diamonds

b. Which is more probable, getting 4 aces or getting 13 cards of the same suit? Give numbers to support your answer.

31. *Journal Problem:* Update your journal with things you have learned since the last entry. In particular, tell how large numbers of outcomes can be calculated, rather than counted, using the concepts of factorials, combinations, and permutations.

9-6 Properties of Probability

In the preceding sections you learned how to calculate the probability of an event using the definition

$$P(E) = \frac{n(E)}{n(S)}$$

That is, you divided the number of outcomes in an event by the number of outcomes in the sample space. In this section you will learn some properties of probability that will allow you to calculate probabilities *without* having to go all the way back to the definition.

OBJECTIVE Given events A and B, calculate

- $P(A \text{ and } B)$, the probability of the *intersection* of A and B
- $P(A \text{ or } B)$, the probability of the *union* of A and B
- $P(\text{not } A)$ and $P(\text{not } B)$, the probabilities of the *complement* of A and the *complement* of B

Intersection of Events

If A and B are two events, then the **intersection** of A and B, $A \cap B$, is the set of all outcomes in event A *and* event B (Figure 9-6a).

The shaded region is $A \cap B$.

Figure 9-6a

For example, suppose you draw 2 cards from a standard deck of 52 playing cards without replacing the first card before you draw the second. What is the probability that both cards will be black? Here, you are looking for the probability of the intersection of the events "the first card is black" and "the second card is black."

There are 52 ways to choose the first card and 51 ways to choose the second card after the first has already been chosen. So the number of outcomes in the sample space is

$$n(S) = 52 \cdot 51 = 2652$$

There are 26 ways the first card could be black. After the first black card has been drawn, there are only 25 ways the second card could be black. So

$$n(\text{both black}) = 26 \cdot 25 = 650$$ By the counting principle for sequential events.

$$\therefore P(\text{both black}) = \frac{650}{2652} = \frac{25}{102} = 0.2450\ldots$$

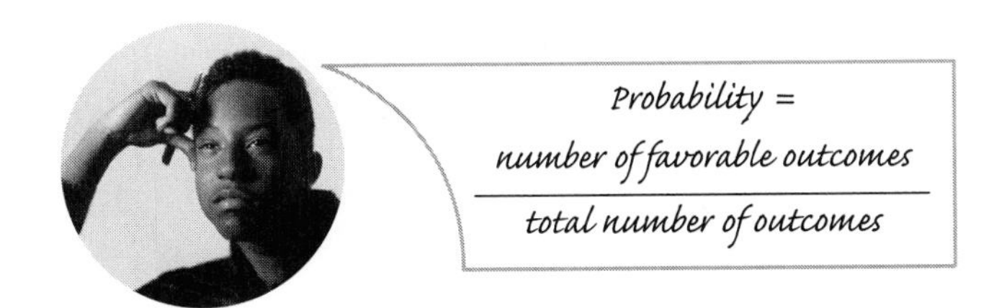

A pattern appears if you do *not* simplify the numbers of outcomes:

$$P(\text{both black}) = \frac{26 \cdot 25}{52 \cdot 51}$$

$$= \frac{26}{52} \cdot \frac{25}{51}$$ Multiplication property of fractions.

$= P(\text{first card is black}) \cdot P(\text{second card is black after the first card is black})$

In this example, the events are *not* independent because the result of the first draw affects the choices for the second draw. However, if the first card were *replaced* before the second card was drawn, then the events would be independent. In this case

$$P(\text{2nd is black after 1st is black}) = P(\text{2nd is black}) = \frac{26}{52}$$

So when the two cards are drawn with replacement

$$\begin{aligned} P(\text{both are black}) &= P(\text{1st is black}) \cdot P(\text{2nd is black}) \\ &= \frac{26}{52} \cdot \frac{26}{52} \\ &= \frac{676}{2704} = 0.25 \end{aligned}$$

PROPERTY: Probability of the Intersection of Two Events

If $P(B \mid A)$ is the probability that B occurs after event A has already occurred, then

$$P(A \text{ and } B) = P(A \cap B) = P(A) \cdot P(B \mid A)$$

If A and B are independent events, then

$$P(A \text{ and } B) = P(A \cap B) = P(A) \cdot P(B)$$

Note that this property corresponds to the counting principle for sequential events, $n(A \text{ and } B) = n(A) \cdot n(B \mid A)$.

Union of Events

If A and B are two events, then the **union** of A and B, $A \cup B$, is the set of all outcomes in event A *or* event B (Figure 9-6b).

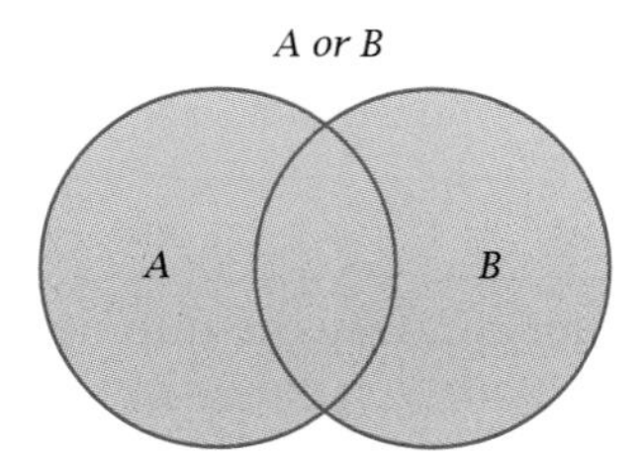

The shaded region is $A \cup B$.

Figure 9-6b

For example, suppose a bag contains 7 chocolate chip cookies, 11 macadamia nut cookies, 12 oatmeal cookies, 4 ginger snaps, and 9 oatmeal-chocolate cookies. If you select 1 cookie at random, what is the probability that it will contain oatmeal or chocolate?

Here, you are looking for the probability of the union of the events "the cookie contains chocolate" and "the cookie contains oatmeal."

The sample space is all the cookies, so

$$n(S) = 7 + 11 + 12 + 4 + 9 = 43$$

The events overlap—that is, there are cookies that contain both oatmeal and chocolate. Use the counting principle for non-mutually exclusive events to count the favorable outcomes:

$$\begin{aligned} n(\text{chocolate or oatmeal}) &= n(\text{chocolate}) + n(\text{oatmeal}) \\ &\quad - n(\text{chocolate} \cap \text{oatmeal}) \\ &= 16 + 21 - 9 \\ &= 28 \end{aligned}$$

$$\therefore P(\text{chocolate or oatmeal}) = \frac{28}{43} = 0.6511\ldots$$

Here again, a pattern appears if you resist the temptation to simplify first:

$$\begin{aligned} P(\text{chocolate or oatmeal}) &= \frac{16 + 21 - 9}{43} \\ &= \frac{16}{43} + \frac{21}{43} - \frac{9}{43} \\ &= P(\text{chocolate}) + P(\text{oatmeal}) \\ &\quad - P(\text{chocolate} \cap \text{oatmeal}) \end{aligned}$$

If two events are mutually exclusive, then their intersection is empty. In this case, you can find the probability of the union of the events simply by adding the probabilities of the two events. For example, there are no cookies that contain both ginger and macadamia nuts, so

$$\begin{aligned} P(\text{ginger or macadamia}) &= P(\text{ginger}) + P(\text{macadamia}) \\ &= \frac{4}{43} + \frac{11}{43} \\ &= \frac{15}{43} = 0.3488\ldots \end{aligned}$$

PROPERTY: Probability of the Union of Two Events

If events A and B are not mutually exclusive, then

$$P(A \text{ or } B) = P(A \cup B) = P(A) + P(B) - P(A \cap B)$$

If events A and B are mutually exclusive, then

$$P(A \text{ or } B) = P(A \cup B) = P(A) + P(B)$$

Note that the first form of the property reduces to the second form when the events are mutually exclusive.

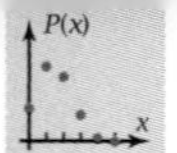

Complementary Events

You can accomplish the third section objective using the property for the union of mutually exclusive events:

Let $P(A)$ be the probability that event A occurs.

Let $P(\text{not } A)$ be the probability that event A does *not* occur.

The events A and "not A" are mutually exclusive, and one or the other is certain to occur. Thus,

$$P(A \text{ or not } A) = P(A) + P(\text{not } A) = 1$$ Probability is 1 if the event is certain to occur.

$$\therefore\ P(\text{not } A) = 1 - P(A)$$

Together, the events A and "not A" complete all the possibilities. Therefore, they are called **complementary events.**

PROPERTY: Complementary Events

The probability that event A will not occur is

$$P(\text{not } A) = 1 - P(A)$$

EXAMPLE 1 Calvin and Phoebe volunteer in the children's ward of a hospital. The probability that Calvin gets mumps as the result of a visit to the ward is $P(C) = 13\%$, and the probability that Phoebe gets mumps is $P(Ph) = 7\%$. Find the probability of each event.

a. Both catch mumps.

b. Calvin does not catch mumps.

c. Phoebe does not catch mumps.

d. Neither Calvin nor Phoebe catches mumps.

e. At least one of them catches mumps.

Solution

a. $$P(C \text{ and } Ph) = P(C) \cdot P(Ph)$$ Assuming C and Ph are independent events.

$$= 0.13 \cdot 0.07$$

$$= 0.0091, \text{ or } 0.91\%$$

b. $$P(\text{not } C) = 1 - P(C)$$ C and "not C" are complementary events.

$$= 1 - 0.13$$

$$= 0.87, \text{ or } 87\%$$

c. $$P(\text{not } Ph) = 1 - P(Ph) = 1 - 0.07$$

$$= 0.93, \text{ or } 93\%$$

d. $P(\text{not } C \text{ and not } Ph) = P(\text{not } C) \cdot P(\text{not } Ph)$

$= 0.87 \cdot 0.93$ — From parts b and c.

$= 0.8091$, or 80.91%

e. $P(\text{at least 1}) = 1 - P(\text{not } C \text{ and not } Ph)$ — Complementary events.

$= 1 - 0.8091$ — From part d.

$= 0.1909$, or 19.09%

Alternate solution for part e:

$P(\text{at least 1}) = P(C) + P(Ph) - P(C \text{ and } Ph)$ — C and Ph are *not* mutually exclusive events.

$= 0.13 + 0.07 - 0.0091$ — From the given probabilities and part a.

$= 0.1909$, or 19.09% ◀

Problem Set 9-6

Do These Quickly

Q1. If A and B are mutually exclusive, then $n(A \text{ or } B) =$ —?—.

Q2. If A and B are *not* mutually exclusive, then $n(A \text{ or } B) =$ —?—.

Q3. $n(A \text{ and } B) = n(A) \cdot n(B \mid A)$, where $n(B \mid A)$ is the number of ways B can happen —?—.

Q4. The number of combinations of five objects taken three at a time is —?—.

Q5. The number of permutations of five objects taken three at a time is —?—.

Q6. Why is a number of permutations greater than the corresponding number of combinations?

Q7. What is the definition of residual deviation?

Q8. Write the exact value (no decimals) of $\sin^{-1} 0.5$.

Q9. $y = 5 \cdot 3^x$ is the particular equation of a(n) —?— function.

Q10. The area of a triangle with sides 5 cm and 3 cm and included angle 30° is —?—.

1. *Calculator Components Problem:* The "heart" of a calculator is one or more chips on which thousands of components are etched. Chips are mass produced and have a fairly high probability of being defective. Suppose that a particular brand of calculator uses two kinds of chip. Chip A has a probability of 70% of being defective, and chip B has a probability of 80% of being defective. If one chip of each kind is randomly selected, find the probability that
   a. Both chips are defective
   b. A is not defective
   c. B is not defective
   d. Neither chip is defective
   e. At least one chip is defective

2. *Car Breakdown Problem:* Suppose you plan to drive on a long trip. The probability your car will have a flat tire is 0.1, and the probability it will have engine trouble is 0.05. What is the probability of each of these events?
   a. No flat tire
   b. No engine trouble
   c. Neither flat tire nor engine trouble

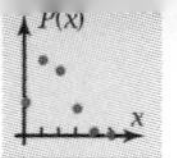

d. Both a flat tire and engine trouble

e. At least one, either a flat tire or engine trouble

3. *Traffic Light Problem 1:* Two traffic lights on Broadway operate independently. Your probability of being stopped at the first light is 40%. Your probability of being stopped at the second one is 70%. Find the probability of being stopped at
   a. Both lights
   b. Neither light
   c. The first light but not the second
   d. The second light but not the first
   e. Exactly one of the lights
4. *Visiting Problem:* Eileen and Ben are away at college. They visit home on random weekends. The probability that Eileen will visit on any given weekend is 20%. The probability that Ben will visit is 25%. On a given weekend, find the probability that
   a. Both children will visit
   b. Neither will visit
   c. Eileen will visit but Ben will not
   d. Ben will visit but Eileen will not
   e. Exactly one child will visit
5. *Backup System Problem:* Vital systems such as electric power generating systems have "backup" components in case one component fails. Suppose that two generators each have a 98% probability of working. The system will continue to operate as long as at least one of the generators is working. What is the probability that the system will continue to operate?

Backup generator

6. *Hide-and-Seek Problem:* The Katz brothers, Bob and Tom, are hiding in the cellar. If either one sneezes, he will reveal their hiding place. Bob's probability of sneezing is 0.6, and Tom's probability is 0.7. What is the probability that at least one brother will sneeze?
7. *Basketball Problem:* Three basketball teams from Lowe High each play on Friday night. The probabilities that the teams win are 70% for varsity, 60% for junior varsity, and 80% for freshman. Find the probability that
   a. All three teams win
   b. All three lose
   c. At least one team wins
   d. The varsity wins and the other two lose
8. *Grade Problem:* Terry Tory has these probabilities of passing various courses: Humanities, 90%; Speech, 80%; and Latin, 95%. Find the probability of each event.
   a. Passing all three
   b. Failing all three
   c. Passing at least one
   d. Passing exactly one
9. *Spaceship Problem 2:* Complex systems such as spaceships have many components. Unless the system has backup components, the failure of any one component could cause the entire system to fail. Suppose a spaceship has 1000 such vital components and is designed without backups.

a. If each component is 99.9% reliable, what is the probability that all 1000 components work and the spaceship does not fail? Does the result surprise you?

b. What is the minimum reliability needed for each component to ensure that there is a 90% probability that all 1000 components will work?

10. *Silversword Problem:* The silversword is a rare relative of the sunflower that grows only atop the 10,000-foot-high Haleakala volcano in Maui, Hawaii. The seeds have only a small probability of germinating, but if enough are planted there is a fairly good chance of getting a new plant. Suppose that the probability that any one seed will germinate is 0.004.

a. What is the probability that any one seed will *not* germinate?

b. If 100 seeds are planted, find the probability that

i. None will germinate

ii. At least one will germinate

c. What is the fewest number of seeds that would need to be planted to ensure a 99% probability that at least one germinates?

11. *Football Plays Problem:* Backbay Polytechnic Institute's quarterback selects passing and running plays at random. By analyzing previous records, an opposing team finds these probabilities:

- The probability that he will pass on first down is 0.4.
- If he passes on first down, the probability that he will pass on second down is 0.3.
- If he selects a running play on first down, the probability that he will pass on second down is 0.8.

a. Find the probability he will pass on

i. First down and second down

ii. First down but not second down

iii. Second down but not first down

iv. Neither first down nor second down

b. Add the four probabilities you have calculated. How do you explain the answer?

12. *Measles and Chicken Pox Problem:* Suppose that in any one year a child has a 0.12 probability of catching measles and a 0.2 probability of catching chicken pox.

a. If these events are *independent* of each other, what is the probability that a child will get *both* diseases in a given year?

b. Suppose statistics show that the probability for getting measles and then chicken pox in the same year is 0.006.

i. Calculate the probability of getting both diseases.

ii. What is the probability of getting chicken pox and then measles in the same year?

c. Based on the given probabilities and your answers to part b, what could you conclude about the effects of the two diseases on each other?

13. *Airplane Engine Problem 1:* One reason airplanes are designed with more than one engine is to increase the planes' reliability. Usually a twin-engine plane can make it to an airport on just one engine should the other engine fail during flight. Suppose that for a twin-engine plane, the probability that any one engine will fail during a given flight is 3%.

a. If the engines operate independently, what is the probability that *both* engines fail during a flight?

b. Suppose flight records indicate that the probability that both engines will fail during a given flight is actually 0.6%. What is the probability that the second engine fails after the first has already failed?

c. Based on your answer to part b, do the engines actually operate independently? Explain.

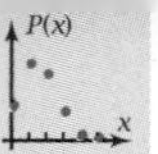

9-7 Functions of a Random Variable

Suppose you conduct the random experiment of flipping a coin five times. The coin is bent so that the probability of heads on any one flip is only 0.4. What is the probability of the event that exactly two of the outcomes are heads and the other three are tails? In this section you'll learn how to calculate the probabilities of all possible events for a random experiment. In the coin-flipping experiment, each probability depends on the number of heads in the event. Thus, the probability is a *function* of the number of heads. Such a function is called the **function of a random variable.**

OBJECTIVE Given a random experiment, find and graph the corresponding probability distribution.

To find $P(3T, 2H)$ for the random experiment just described, it helps to start by looking at a simpler event, $P(TTTHH)$, the probability of three tails and two heads *in that order:*

$$P(TTTHH) = P(T) \cdot P(T) \cdot P(T) \cdot P(H) \cdot P(H)$$

$$= (0.6)(0.6)(0.6)(0.4)(0.4) \qquad P(T) = 1 - P(H) = 1 - 0.4 = 0.6$$

$$= 0.6^3 \cdot 0.4^2$$

There are ten possible outcomes that have exactly two heads:

HHTTT	*HTHTT*	*HTTHT*	*HTTTH*	*THHTT*
THTHT	*THTTH*	*TTHHT*	*TTHTH*	*TTTHH*

Ten is the number of ways of selecting a *group* of two of the five flips to be heads. But this is also the number of *combinations* of five elements taken two at a time, or ${}_5C_2$. So

$$P(3T, 2H) = 10 \cdot 0.6^3 \cdot 0.4^2$$

$$= {}_5C_2 \cdot 0.6^3 \cdot 0.4^2 \qquad \text{By calculator or by computing factorials, } {}_5C_2 = 10.$$

The expression ${}_5C_2 \cdot 0.6^3 \cdot 0.4^2$ is a term in the **binomial series** that comes from expanding

$$(0.6 + 0.4)^5$$

If x stands for the number of times *tails* appears in five flips, then

$$P(x) = {}_5C_x \cdot 0.6^{5-x} \cdot 0.4^x$$

As you can see, the probability $P(x)$ is a *function* of the random variable x. Because this function tells how the 100% probability is *distributed* among the various possible events, it is called a **probability distribution.** In this case, because the probabilities are terms in a binomial series, it is called a **binomial distribution.** A random experiment, such as the coin-flipping experiment, in which the trials are repeated a number of times and in which there are only two possible outcomes for each trial is called a **binomial experiment.** The icon at the top of each page in this chapter is the graph of a binomial distribution.

PROPERTY: Binomial Probability Distribution

Suppose a random experiment consists of repetitions of the same action and that the action has only two possible results. Let E be one of the two possible results.

Let b be the probability that event E occurs in any one repetition.

Let a be the probability that event E does *not* occur in any one repetition.

Let x be the number of times event E occurs in n repetitions.

Then

$$P(x) = {}_nC_x \cdot a^{n-x} \cdot b^x$$

That is, $P(x)$ has the value of the term with b^x as a factor in the binomial series $(a + b)^n$.

► EXAMPLE 1 A bent coin is flipped five times (as described previously). The probability of getting heads on any one toss is 40%.

a. Find all terms in the probability distribution.

b. Show that the total of the probabilities equals 1, and explain the significance of this fact.

c. Plot the graph. Sketch the result.

d. Calculate the probability that the coin lands heads up at least two of the five times.

Solution

a. Because the probability of getting heads on any one flip is 40%, or 0.4, the probability of *not* getting heads (that is, of getting tails) is 1 – 0.4, or 0.6. Let $P(x)$ be the probability that there are exactly x heads in five flips.

$$P(0) = {}_5C_0 \cdot 0.6^5 \cdot 0.4^0 = 1 \cdot 0.6^5 \cdot 0.4^0 = 0.07776$$

$$P(1) = {}_5C_1 \cdot 0.6^4 \cdot 0.4^1 = 5 \cdot 0.6^4 \cdot 0.4^1 = 0.2592$$

$$P(2) = {}_5C_2 \cdot 0.6^3 \cdot 0.4^2 = 10 \cdot 0.6^3 \cdot 0.4^2 = 0.3456$$

$$P(3) = {}_5C_3 \cdot 0.6^2 \cdot 0.4^3 = 10 \cdot 0.6^2 \cdot 0.4^3 = 0.2304$$

$$P(4) = {}_5C_4 \cdot 0.6^1 \cdot 0.4^4 = 5 \cdot 0.6^1 \cdot 0.4^4 = 0.0768$$

$$P(5) = {}_5C_5 \cdot 0.6^0 \cdot 0.4^5 = 1 \cdot 0.6^0 \cdot 0.4^5 = 0.01024$$

A time-efficient way to compute all the probabilities is to put the six values of x, namely, the integers 0 though 5, in one list and then put the formula for $P(x)$ in a second list.

b. The sum of the six probabilities in part a is exactly 1. This indicates that one of the six events listed is certain to happen and that there are no other possible events in this random experiment.

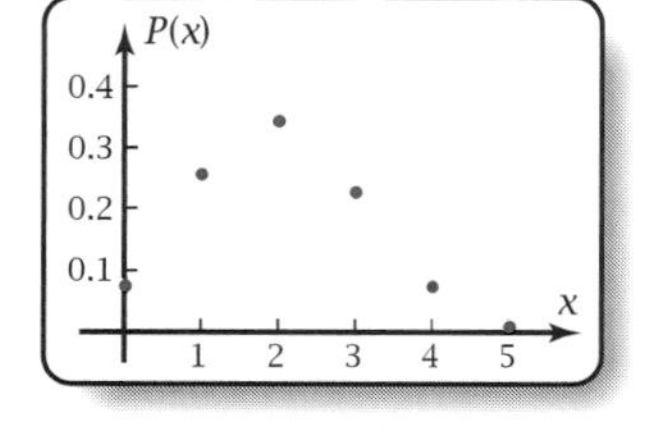

Figure 9-7a

c. Use the stored values in the two lists to make a scatter plot (Figure 9-7a). Note that only integers are in the domain of this function.

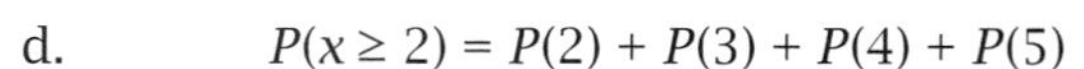

d. $P(x \geq 2) = P(2) + P(3) + P(4) + P(5)$ — The results are mutually exclusive.

$= 0.3456 + 0.2304 + 0.0768 + 0.01024$ — From part a.

$= 0.66304 \approx 66\%$

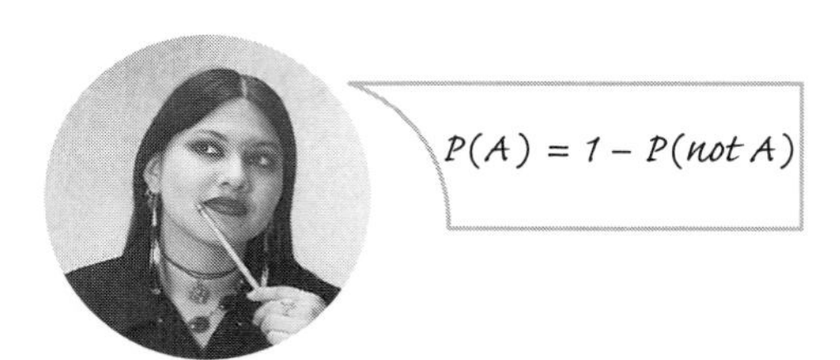

Alternate solution to part d:

$$P(x \geq 2) = 1 - P(0) - P(1)$$
$$= 1 - 0.07776 - 0.2592$$
$$= 0.66304 \approx 66\%$$

◀

Problem Set 9-7

Do These Quickly

Q1. If A and B are mutually exclusive, then $P(A \text{ or } B) =$ —?—.

Q2. If A and B are *not* mutually exclusive, then $P(A \text{ or } B) =$ —?—.

Q3. If A and B happen in that order, then $P(A \text{ and } B) = P(A) \cdot P(B \mid A)$, where $P(B \mid A)$ is —?—.

Q4. If a fair coin is flipped twice, $P(TT) =$ —?—.

Q5. If a fair coin is flipped twice, $P(\text{at least one is heads}) =$ —?—.

Q6. If a fair coin is flipped three times, $P(TTT) =$ —?—.

Q7. If a fair coin is flipped three times, $P(THT \text{ in that order}) =$ —?—. Surprising?!

Q8. If $g(x) = f(x - 2)$, then g is a —?— transformation of f.

Q9. If $g(x) = f(2x)$, then g is a —?— transformation of f.

Q10. Sketch the graph of a logistic function.

---

1. *Heredity Problem:* If a dark-haired mother and father have a particular combination of genes, they have a $\frac{1}{4}$ probability of having a light-haired baby.

   a. What is their probability of having a dark-haired baby?

b. If they have three babies, calculate $P(0)$, $P(1)$, $P(2)$, and $P(3)$, the probabilities of having 0, 1, 2, and 3 dark-haired babies, respectively.

c. Find the sum of the probabilities in part b. How do you interpret the answer?

d. Plot the graph of this probability distribution.

e. What special name is given to this kind of probability distribution?

2. *Multiple-Choice Test Problem 1:* A short multiple-choice test has four questions. Each question has five choices, exactly one of which is right. Willie Passitt has not studied for the test, so he guesses answers at random.

   a. What is the probability that his answer on a particular question is right? What is the probability that it is wrong?

   b. Calculate his probabilities of guessing 0, 1, 2, 3, and 4 answers right.

   c. Perform a calculation that shows that your answers to part b are reasonable.

   d. Plot the graph of this probability distribution.

   e. Willie will pass the test if he gets at least three of the four questions right. What is his probability of passing?

   f. This binomial probability distribution is an example of a function of —?—.

3. If you flip a thumbtack, it can land either point up or point down (Figure 9-7b). Suppose the probability that any one flip will land point up is 0.7. Suppose that the tack is flipped ten times.

Up *Down*

Figure 9-7b

   a. Show how $P(3)$ is calculated.

   b. Calculate $P(x)$ for each of the 11 possible values of x. Make a scatter plot of the probability distribution, and sketch the result.

   c. Which is more probable, that the thumbtack will land point up more than five times or that the thumbtack will land point up at most five times? Show results that support your answer.

4. *Traffic Light Problem 2:* Three widely separated traffic lights on U.S. Route 1 operate independently of each other. The probability that you will be stopped at any one of the lights is 40%.

   a. Show how to calculate the probability of being stopped at exactly two of the three lights.

   b. Which is more probable, being stopped at more than one light or being stopped at one or fewer lights? Show results that support your answer.

   c. Suppose that you make the trip four times, encountering a total of 12 lights. Make a scatter plot of the probability distribution. Sketch the result.

5. *Colorblindness Problem:* Statistics show that about 8% of all males are colorblind. Interestingly, women are less likely to have this condition. Suppose that 20 males are selected at random. Let $P(x)$ be the probability that x of the 20 men are colorblind.

   a. Compute the probability distribution and plot its graph. Use a window that makes the graph fill most of the screen. Sketch the pattern followed by the points on the graph.

   b. From your output in part a, find $P(0)$, $P(1)$, $P(2)$, and $P(3)$.

   c. In a time-efficient way, calculate the probability that at least 4 of the 20 males are colorblind. Show the method you used in the computation.

6. *Eighteen-Wheeler Problem:* Large tractor-trailer trucks usually have 18 tires. Suppose that the probability that any one tire will blow out on a given cross-country trip is 0.03.

   a. What is the probability that any one tire does *not* blow out?

   b. Find the probability that

      i. None of the 18 tires blows out

      ii. Exactly one of the tires blows out

      iii. Exactly two of the tires blow out

      iv. More than two tires blow out

c. If a trucker wants to have a 95% probability of making the trip without a blowout, what must be the reliability of each tire? That is, what would the probability have to be that any one tire blows out?

7. *Perfect Solo Problem:* Clara Nett plays a musical solo. She is quite good and guesses that her probability of playing any one note right is 99%. The solo has 60 notes.

   a. Find the probability that

      i. She plays every note right

      ii. She makes exactly one mistake

      iii. She makes exactly two mistakes

      iv. She makes at least two mistakes

      v. She makes more than two mistakes

   b. What must be Clara's probability of getting any one note right if she wants to have a 95% probability of getting all 60 notes right?

8. *Airplane Engine Problem 2:* One reason commercial airplanes have more than one engine is to reduce the consequences should an engine fail during flight (Figure 9-7c). Under certain circumstances, some counterintuitive things happen when the number of engines is increased. Assume that the probability that any one engine will fail on a given flight is 0.1 (this is high, but assume it anyway).

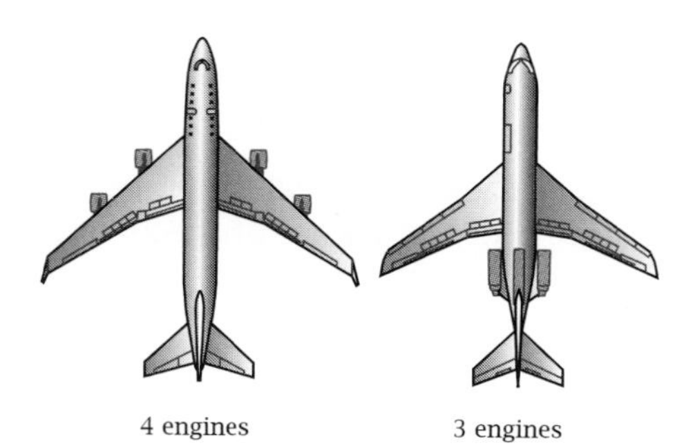

Figure 9-7c

   a. For a plane that has four engines, calculate the probabilities that zero, one, two, three, and all four engines fail during the given flight. Show that the probabilities add to 1, and explain the significance of this fact.

   b. If the plane will keep flying as long as no more than one engine fails, what is the probability that the four-engine plane keeps flying?

   c. Suppose a different kind of plane has three engines of the same reliability and that it, too, will keep flying if no more than one engine fails. What is the probability that the three-engine plane keeps flying?

   d. Based on your computations in this problem, which is safer, the four-engine plane or the three-engine plane?

9. *World Series Project:* Suppose the Dodgers and the Yankees are in the World Series of baseball. A team must win four games to win the World Series. From their season records, you predict that the Dodgers have a probability of 0.6 of beating the Yankees in any one game. Assume this probability is independent of which team has won a preceding game in this World Series.

   a. Find the probability that the Dodgers win the series by winning the first four games.

   b. Find the probability that the Yankees win the series by winning all of the first four games.

   c. For a team to win the series in exactly five games, they must win exactly three of the first four games, then win the fifth game. Find the probability that the Dodgers win the series in five games.

   d. Find the probability that the Yankees win the series in five games.

   e. Find the probability of each of these events:

      i. The Dodgers win the series in six games.

      ii. The Yankees win the series in six games.

      iii. The Dodgers win the series in seven games.

      iv. The Yankees win the series in seven games.

f. Find the probability that the Yankees win the series.

g. What is the most probable length of the series—four, five, six, or seven games?

Problems 10–13 involve probability distributions other than binomial distributions.

10. *Another Dice Problem:* Suppose that a random experiment consists of rolling two dice, one blue and one white, as in Section 9-1.

    a. Plot the probability distribution for each of the following random variables. You may count the outcomes on Figure 9-1a.

        i. x is the sum of the numbers on the two dice.

        ii. x is the number on the blue die minus the number on the white die.

        iii. x is the absolute value of the difference between the number on the blue die and the number on the white die.

    b. For each probability distribution in part a, find the most probable value of x.

11. *Proper Divisors Problem:* An integer from 1 through 10 is selected at random. Let x be the number of proper divisors the integer has. (**A proper divisor** of an integer n is an integer less than n that divides n exactly. For example, 12 has five proper divisors: 1, 2, 3, 4, and 6.)

    a. List the proper divisors and the number of proper divisors for each integer from 1 through 10.

    b. For each possible value of x, identify how many of the integers from 1 through 10 have that number of proper divisors.

    c. Let $P(x)$ be the probability that an integer from 1 through 10 has x proper divisors. Calculate $P(x)$ for each value of x in the domain.

    d. Plot the graph of the probability distribution in part c. Do you see any pattern followed by the points on the graph?

12. *First Girl Problem:* Eva and Paul want to have a baby girl. They know that the probability of having a girl on any single birth is 0.5.

    a. Let x be the number of babies they have, and let $P(x)$ be the probability that the xth baby is the *first* girl. Then $P(1) = 0.5$. $P(2)$ is the probability that the first baby is *not* a girl and that the second baby *is* a girl. Calculate $P(2)$, $P(3)$, and $P(4)$.

    b. Plot the graph of P. Sketch the graph, showing what happens as x becomes large.

    c. Besides being called a probability distribution, what other special kind of function is this?

    d. Show that the sum of the values of $P(x)$ approaches 1 as x becomes very large.

13. *Same Birthday Project:* A group of students compare their birthdays.

    a. What is the probability that Shawn's birthday is *not* the same as Mark's?

    b. If Shawn and Mark have different birthdays, what is the probability that Frieda's birthday is not the same as Shawn's and not the same as Mark's?

    c. What is the probability that Shawn and Mark have different birthdays *and* that Frieda has a birthday different from both of them?

    d. Using the pattern you observe in part c, find the probability that a group of ten students will all have different birthdays. Give a decimal approximation for the result.

    e. What is the probability that in a group of ten students, at least two have the same birthday (that is, *not* all ten have different birthdays)?

    f. Write a program to compute a list of probabilities that at least two people have the same birthday in a group of x people. Store the output in lists of x and $P(x)$ for use in subsequent graphing. Use the program to make a list of $P(x)$ for 2 through 60 people.

    g. Plot the graph of the probability distribution in part f. Use the data in the lists output by the program without doing further computations. Sketch the pattern followed by the points.

h. From the graphical or numerical data, find out how many people must be in a group to have the probability that at least two people will have the same birthday equal to

   i. 50%

   ii. 99%

   Surprising?

14. *Journal Problem:* Update your journal with things you have learned since the last entry. In particular, explain how the properties of probability and the concept of function lead to functions of a random variable.

9-8 Mathematical Expectation

One of the main uses of probability is in calculating the expected value by conducting a random experiment. Insurance companies, for example, can use the expected value to calculate the insurance costs and the expected profit from a particular policy. In this section you will study **mathematical expectation,** a value you can calculate based on the outcomes for each event in a random experiment.

OBJECTIVE Calculate the mathematical expectation of a given random experiment.

At a school carnival, students are awarded points for winning games. At the end of the evening, they may trade in their points for prizes. For a particular game, students start out with 50 points and roll a single die. The payoffs for the game are:

- Roll a 6: Win 100 points (and get your 50 points back)
- Roll a 2 or a 4: Win 10 points (and get your 50 points back)
- Roll an odd number: Win nothing (and lose your 50 points)

Because each outcome is equally likely, you would "expect" to get each number *once* in six rolls of the die. (You probably *won't,* but that is what you expect to happen on the average if you roll the die many times.) If you did roll each number exactly once, your winnings would be

Number	Points Won
1	−50
2	10
3	−50
4	10
5	−50
6	100
Total:	−30

Because you would *lose* 30 points in 6 rolls, your average winnings would be

$$\text{Average winnings} = \frac{-30}{6} = -5 \text{ points per roll}$$

This average winnings is your mathematical expectation. If you play the game thousands of times, you would expect to lose about 5 points per roll, on average.

A pattern shows up if you do *not* carry out the addition when calculating the mathematical expectation in the preceding random experiment. Let E stand for the mathematical expectation:

$$E = \frac{-50 + 10 - 50 + 10 - 50 + 100}{6}$$ Add the six values, one for each outcome, and divide by 6.

$$= \frac{3(-50) + 2(10) + 1(100)}{6}$$ Combine "like terms."

$$= \frac{3(-50)}{6} + \frac{2(10)}{6} + \frac{1(100)}{6}$$ Division distributes over addition.

$$= \frac{3}{6}(-50) + \frac{2}{6}(10) + \frac{1}{6}(100)$$ Properties of fractions.

The $\frac{3}{6}$ is the probability of getting an odd number, and -50 is the value associated with getting an odd number. Similarly, $\frac{2}{6}$ and $\frac{1}{6}$ are the probabilities associated with 10 and 100, which are the values for the other two events. So you can calculate the mathematical expectation of a random experiment by multiplying the probability and the value for each mutually exclusive event and then adding the results. This fact leads to the algebraic definition of mathematical expectation.

DEFINITION: Mathematical Expectation

Algebraic:

The **mathematical expectation,** E, of a random experiment is the sum

$$E = P(A_1)a_1 + P(A_2)a_2 + P(A_3)a_3 + \cdots + P(A_n)a_n$$

or

$$E = \sum_{k=1}^{n} P(A_k)a_k$$

for the n mutually exclusive events $A_1, A_2, A_3, \ldots, A_n$ in the experiment. The values $a_1, a_2, a_3, \ldots, a_n$ correspond to the outcomes of $A_1, A_2, A_3, \ldots, A_n$.

Verbally:

The mathematical expectation is the *weighted average* for a random experiment each time it is run.

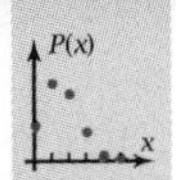

▶ EXAMPLE 1

The basketball toss at an amusement park costs 50¢ to play. To play, you shoot three balls. If you make no baskets, you win nothing (and lose your 50¢). If you make just one basket, you win a key chain worth 5¢. If you make two of the three baskets, you win a stuffed animal worth 60¢. If you make all three baskets, you win a doll worth \$2.50. The basket hoop is small, so your probability of making any one basket is only 30%. What is your mathematical expectation for the game? How do you interpret the answer?

Solution

Let $P(x)$ be your probability of making x baskets. Your probability of missing any one basket is 100 − 30, or 70% (0.7 as a decimal). Therefore,

$$P(0) = {}_3C_0 \cdot 0.7^3 \cdot 0.3^0 = 0.343$$

$$P(1) = {}_3C_1 \cdot 0.7^2 \cdot 0.3^1 = 0.441$$

$$P(2) = {}_3C_2 \cdot 0.7^1 \cdot 0.3^2 = 0.189$$

$$P(3) = {}_3C_3 \cdot 0.7^0 \cdot 0.3^3 = \underline{0.027}$$

Check: Total is 1.000

The payoff for each event is found by subtracting the 50¢ "admission fee" from the amount you win:

x	**Payoff (in cents)**
0	0 − 50 = −50
1	5 − 50 = −45
2	60 − 50 = 10
3	250 − 50 = 200

$\therefore\ E = (0.343)(-50) + (0.441)(-45) + (0.189)(10) + (0.027)(200)$

Definition of expectation.

$= -29.705$

So on average you would expect to *lose* about 30¢ per game if you played many times. (This is the way amusement parks make money on such games!)

Once you understand how mathematical expectation is calculated, you can use list operations on your grapher to do the computations.

x	**$P(x)$**	**Payoff**	**$P(x) \cdot$ Payoff**
0	0.343	−50	−17.15
1	0.441	−45	−19.845
2	0.189	10	1.89
3	0.027	200	5.4
Totals:	1.000		−29.705

$\therefore\ E = -29.705$, or a loss of about 30¢ each time you play the game ◀

Problem Set 9-8

Do These Quickly 

Q1. Expand: $(a + b)^2$

Q2. Expand: $(a + b)^3$

Q3. Expand: $(a + b)^4$

Q4. Evaluate ${}_4C_1$ and ${}_4C_3$.

Q5. How are the answers to Problem Q4 related to the answer to Problem Q3?

Q6. If $a = 0.3$ and $b = 0.7$, then ${}_4C_1\ a^1b^3 =$ —?—.

Q7. How is the answer to Problem Q6 related to a binomial probability distribution?

Q8. Which equals 1, $\cos^2 x + \sin^2 x$ or $\cos^2 x - \sin^2 x$?

Q9. If the mean of a set of data is 37, the data point 35 has a deviation of —?— from the mean.

Q10. Whose name is associated with the normal distribution curve?

---

1. *Uranium Fission Problem:* When a uranium atom splits ("fissions"), it releases 0, 1, 2, 3, or 4 neutrons. Let $P(x)$ be the probability that x neutrons are released. Assume the probability distribution is

x	$P(x)$
0	0.05
1	0.2
2	0.25
3	0.4
4	0.1

   a. What is the mathematically expected number of neutrons released per fission?

   b. The number of neutrons released in any one fission must be an integer. How do you explain the fact that the mathematically expected number in part a is not an integer?

2. *Archery Problem 3:* An expert archer has the probabilities of hitting various rings shown on the target (Figure 9-8a).

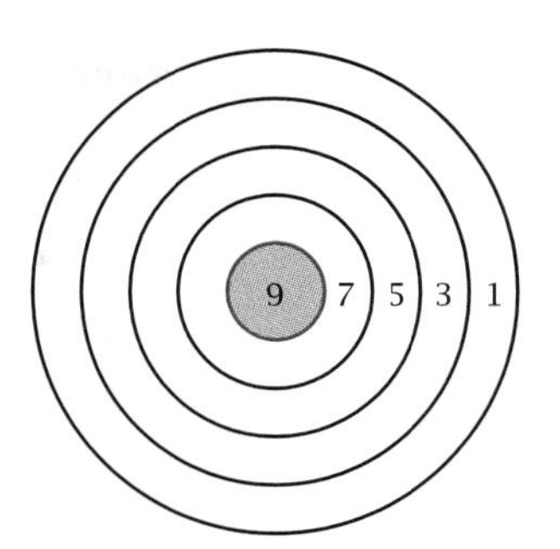

Figure 9-8a

Color	Probability	Points
Gold	0.20	9
Red	0.36	7
Blue	0.23	5
Black	0.14	3
White	0.07	1

   a. What is her mathematically expected number of points on any one shot?

   b. If she shoots 48 arrows, what would her expected score be?

3. *Sales Incentive Problem:* Calvin is a salesperson at a car dealership. At the beginning of the year, the dealership wants to sell the previous year's car models off the lot to make space for new cars. The dealership manager wants minivans to clear out fastest, then station wagons, and other models last, so she puts together two choices of selling incentives for the sales staff during the first week of the new year.

   Option A: Receive a $100 bonus for each vehicle sold of the previous year's models.

   Option B: Receive a $2000 bonus for selling four minivans, two station wagons, one hybrid car, and one sedan of the previous year's models.

   Calvin estimates that his probability of selling four minivans is 50%, selling two station wagons is 70%, selling a hybrid car is 80%, and selling a sedan is 90%.

   a. Calculate the mathematical expectation for Calvin's bonus if he chooses Option A and sells the required four minivans, two station wagons, one hybrid car, and one sedan.

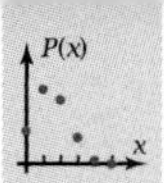

b. Calculate Calvin's probability of being able to sell four minivans, two station wagons, one hybrid car, and one sedan.

c. Calculate the mathematical expectation for Calvin's bonus if he chooses Option B.

d. Based on your answers in parts a–c, which option should he choose?

4. *Seed Germination Problem:* A package of seeds for an exotic tropical plant states that the probability that any one seed germinates is 80%. Suppose you plant four of the seeds.

   a. Find the probabilities that exactly 0, 1, 2, 3, and 4 of the seeds germinate.

   b. Find the mathematically expected number of seeds that will germinate.

5. *Batting Average Problem:* Jackie Robinson's highest batting average was .342, which means that his probability of getting a hit at any one time at bat is 0.342. Suppose that Robinson came to bat five times during a game.

   a. Calculate the probabilities that he got 0, 1, 2, 3, 4, and 5 hits.

   b. What was Robinson's mathematically expected number of hits for this game?

6. *Expectation of a Binomial Experiment:* Suppose you conduct a random experiment that has a binomial probability distribution. Suppose the probability that outcome C occurs on any one repetition is 0.4. Let $P(x)$ be the probability that outcome C occurs x times in five repetitions.

   a. Calculate $P(x)$ for each value of x in the domain.

   b. Find the mathematically expected value of x. (*Hint:* The value if C occurs x times is x.)

   c. Show that the mathematically expected value of x is equal to 0.4 (the probability C occurs on *one* repetition) times 5 (the total number of repetitions).

   d. If the probability that C occurs on any one repetition is b and the probability that C does not occur on one repetition is $a = 1 - b$, prove that in five trials, the expected value of x is $5b$.

   e. From what you have observed in this problem, make a conjecture about the mathematically expected value of x in n repetitions, if the probability that C occurs on any one repetition is b.

   f. If you plant 100 seeds, each of which has a probability of 0.71 of germinating, how many seeds would you expect to germinate?

7. *Multiple-Choice Test Problem 2:* Suppose that you are taking your College Board tests. You answer all the questions you know, and have some time left over. So you decide to guess the answers to the rest of the questions.

   a. Each question is multiple choice with five choices. If you guess at random, what is the probability of getting an answer right? Of getting an answer wrong?

   b. When the testing service grades your paper, they give you 1 point if the answer is right and subtract $\frac{1}{4}$ point if the answer is wrong. What is your mathematically expected score on any question for which you guess at random?

   c. Suppose that, on one question, you can eliminate one choice you know is wrong, and then randomly guess among the other four. What is your mathematically expected score on this question? Surprisingly low, isn't it?

   d. Calculate your mathematically expected score on a question for which you can eliminate two of the choices and then for which you can eliminate three of the choices.

   e. Based on your answers, do you think it is worthwhile guessing answers on a multiple-choice test?

8. *Accident/Illness Insurance Problem:* Some of the highest-paid mathematicians are the *actuaries,* who figure out what you should pay for various types of insurance. Suppose an insurance company has an accident/illness policy that pays \$500 if you get ill during any one year, \$1000 if you have an accident, and \$6000 if you both get ill and have an accident. The premium, or payment, for this policy is \$100 per year. One of your friends, who has studied actuarial science, tells you that your probability of becoming ill in any one year is 0.05 and that your probability of having an accident is 0.03.

   a. Find the probabilities of each event.

      i. Becoming ill and having an accident

      ii. Becoming ill and not having an accident

      iii. Not becoming ill but having an accident

      iv. Not becoming ill and not having an accident

   b. What is the mathematical expectation for this policy?

   c. An insurance policy is *actuarially sound* if the insurance company is expected to make a profit from it. Based on the probabilities assumed, is this policy actuarially sound?

9. *Life Insurance Project, Part 1:* Functions of random variables are used as mathematical models in the insurance business. The following numbers were taken from a *mortality table.* The table shows the probability, $P(x)$, that a person who is alive on his or her xth birthday will die before he or she reaches age $x + 1$.

Age, x	**$P(x)$**
15	0.00146
16	0.00154
17	0.00162
18	0.00169
19	0.00174
20	0.00179

A group of 10,000 15-year-olds get together to form their own life insurance company. For a premium of \$40 per year, they agree to pay \$20,000 to the family of anyone in the group who dies while he or she is 15 through 20 years old.

   a. Calculate $D(15)$, the number out of 10,000 expected to die while they are 15. Round to an integer.

   b. Calculate $A(16)$, the number out of 10,000 expected to be alive on their 16th birthday.

   c. Calculate $D(16)$. Round to an integer.

   d. Make a table of x, $P(x)$, $A(x)$, and $D(x)$ for each value of x from 15 through 20.

   e. Put columns in the table of part d for $I(x)$ and $O(x)$, the income from the \$40 premiums and the amount paid out from the \$20,000 death benefits. Take into account that a person who dies no longer pays premiums the following years.

   f. Calculate $NI(x) = I(x) - O(x)$, the net income of the company each year. Explain why $NI(x)$ *decreases* each year. (There are reasons!)

   g. On average, how much would the company expect to make per year? Would this be enough to pay a full-time employee to operate the company?

10. *Life Insurance Project, Part 2:* A group of 10,000 people, each now 55 years old, is to be insured as described in Problem 9. Upon the death of the insured person at any age from 55 through 59, his or her survivors receive \$20,000. Your job is to calculate the annual premium that should be charged for this policy. Here is the portion of the mortality table that applies to this age group.

Age, x	**$P(x)$**
55	0.01300
56	0.01421
57	0.01554
58	0.01700
59	0.01859

   a. Make a table showing $D(x)$, $A(x)$, and $O(x)$, the numbers of deaths, the number still alive, and the amount paid out in death benefits, respectively.

   b. An administrator is to be paid \$30,000 a year to operate the program. Calculate the total paid out by the company over the five-year period, including the administrator's salary and death benefits.

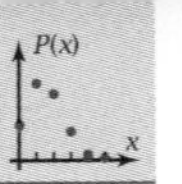

c. Calculate the total income from premiums for the five-year period, taking into consideration the fact that the number of premiums received each year decreases after the first year as the insured persons die. From the result, calculate the premium that must be charged per person per year for the company to break even.

d. Why is the premium in part c so much higher than the $40 per year in Problem 9?

9-9 Chapter Review and Test

In this chapter you analyzed functions in which the independent variable takes on random values. The dependent variable is the probability that a particular value of the random variable occurs. You used the definition of probability as a ratio of numbers of outcomes of a random experiment to derive properties that allow you to calculate probabilities algebraically. You learned that the binomial probability distribution has many real-world applications. Such functions are useful in finding mathematical expectation, which is the potential payoff of a random experiment.

Review Problems

R0. Update your journal with what you have learned in this chapter. Include such things as the definitions of random variable, probability, outcome, event, sample space, permutations, combinations, functions of a random variable, binomial probability distributions, and mathematical expectation. Show how what you have learned allows you to compute numbers of outcomes algebraically rather than by actually counting.

R1. *Quarter, Dime, and Nickel Problem:* A quarter, a dime, and a nickel are marked with 1 on the tails side and 2 on the heads side. All three coins are flipped. The eight possible outcomes are shown in Figure 9-9a. Find the probability of each event.

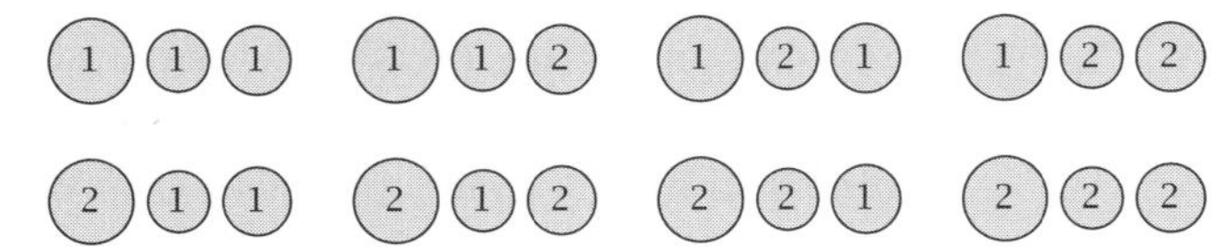

Figure 9-9a

a. The total is 4.

b. The total is 5.

c. The total is 6.

d. The total is 7.

e. The total is odd.

f. The total is between 4 and 6, inclusive.

g. The total is between 3 and 6, inclusive.

h. The quarter shows 2 and the nickel shows 1.

i. The quarter shows 2 or the nickel shows 1.

R2. *Numbered Index Card Problem:* Twenty-five index cards are numbered from 1 through 25. The cards are placed number side down on the table, and one card is drawn at random.

a. How many outcomes are in the sample space?

b. How many outcomes are in the event "the number is odd"?

c. What is the difference between an outcome and an event?

d. Find the probability that

i. The number is odd

ii. The number is divisible by 3

iii. The number has two digits

iv. The number is less than 30

v. The number is at least 30

R3. a. An ice cream shop has 20 flavors of ice cream and 11 flavors of sherbet. Find the number of different ways you could select

i. A scoop of ice cream and a scoop of sherbet

ii. A scoop of ice cream or a scoop of sherbet

b. Using the letters in EXACTING, find the number of different ways you could select

i. A consonant and then a vowel

ii. A consonant and then another consonant

R4. a. The Russian alphabet has 34 characters. Find the number of different permutations that can be made

i. Using 3 different characters

ii. Using 34 different characters

b. How many different 3-letter "words" can be made from the letters in PRECAL?

c. Find the probability that a permutation of all the letters in the word EXACTLY begins with a consonant and ends with a consonant.

d. Find the number of different permutations of all the letters in

HUMUHUMUNUKUNUKU

which are the first 16 letters of the name of Hawaii's state fish.

R5. a. Evaluate ${}_7C_3$ using factorials.

b. What is the difference between a permutation and a combination?

c. The 12th-grade class at Scorpion Gulch High School has 100 students: 53 girls and 47 boys. In how many different ways could they select the following?

i. A group of four students to be class officers

ii. A president, a vice president, a secretary, and a treasurer

iii. A seven-member debate team consisting of four boys and three girls

d. If a seven-member debate team is selected at random as in part c, what is the probability that it will have four boys and three girls?

R6. a. *Car Trouble Problem:* Mr. Rhee's car has a 70% probability of starting, and Ms. Rhee's car has a 80% probability of starting. Find the probability of each event.

i. Neither car will start.

ii. Both cars will start.

iii. Either both cars will start or neither car will start.

iv. Exactly one of the cars will start.

b. *Basketball Game Problem:* High school basketball teams often play each other twice during the season. Suppose Central High has a 60% probability of winning their first game against Tech. If Central wins the first game, they have an 85% probability of winning the second game. If Central loses the first game, they have a 45% probability of winning the second game. Find the probability of each of the following events.

    i. Central wins both games.

    ii. Central wins the first game and loses the second game.

    iii. Central loses the first game and wins the second game.

    iv. Central loses both games.

    v. Show by calculation that the four answers above are reasonable.

R7. *Candle Lighter Problem:* A butane candle lighter does not always light when you pull the trigger. Suppose that a lighter has a 60% probability of lighting on any one pull. You pull the trigger six times. Let $P(x)$ be the probability it lights exactly x of those times.

a. Show the method used to calculate $P(4)$.

b. In a time-efficient way, calculate $P(x)$ for each value of x in the domain.

c. Plot the graph of P as a scatter plot on your grapher. Sketch the result.

d. Find the probability that the lighter lights at least half the time.

e. Why is this random experiment called a *binomial* experiment?

R8. *Airline Overbooking Problem:* A small commuter airline charges $100 for tickets on a particular flight. The plane holds 20 people, so the total revenue for a full flight is $2000. The airline expects a higher revenue by booking 21 passengers and taking their chances that one or more passengers will not show up (the tickets are nonrefundable). Company records indicate that, on average, there is a 10% probability that any one passenger will not show up. However, if everyone does show up, one passenger must be "bumped" and given a $300 payment. (The $100 is not refunded because the ticket can be used on a later flight.)

a. What is the probability that all 21 passengers show up and the airline makes only $1800 ($2100 − $300)? What is the probability that 20 or fewer passengers show up and the airline makes the full $2100 on that flight? What is the airline's mathematically expected revenue if it books 21 passengers?

b. If the airline books 22 passengers, the revenue is $1600 for zero no-shows, $1900 for one no-show, and $2200 for two or more no-shows. What is the airline's mathematically expected revenue if it books 22 passengers?

c. Calculate the mathematically expected revenue if the airline books 23 passengers. Is this more or less than the $2000 it would make if it did no overbooking?

d. What other things besides a possible loss of money might make the airline limit the amount of overbooking it does?

Weighted Average Problem: For parts e and f, a college professor gives students a **weighted average.** Test 1 counts as 10% of the grade. Tests 2, 3, and 4 count as 20% each. The final exam counts as 30% of the grade.

e. To receive a grade of B or above, a student must have a weighted average of at least 80. Suppose that Nita B. Topaz gets scores of 72, 86, 93, 77, and 98 on the five tests, in that order. Will she get at least a B? Show numbers to support your answer.

f. Explain why the mathematics involved in finding a weighted average is the same as that used to find the mathematical expectation of a random experiment.

Concept Problems

C1. *Nuclear Reactor Project:* When a uranium atom inside the reactor of a nuclear power plant is hit by a neutron, it splits (fissions), releasing energy and some new neutrons (Figure 9-9b). The mathematically expected number of new neutrons per fission is 2.3.

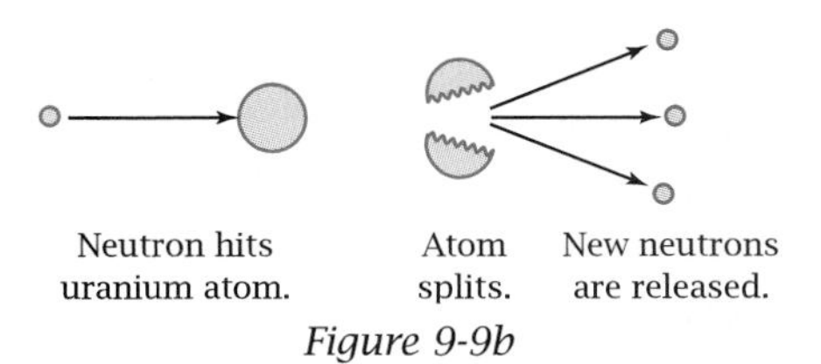

Figure 9-9b

a. Suppose there are 100 neutrons in the reactor initially. If all of these neutrons cause fissions, how many neutrons would you expect there to be after this first "generation" of fissions?

b. If all of the neutrons from the first generation cause fissions, how many would you expect after two generations? After three generations? After four generations? What kind of sequence do these numbers form?

c. If each generation takes 0.001 sec, how many neutrons would you expect there to be after 1 sec? Does this answer surprise you? This is what makes atomic bombs explode!

d. Not all of the neutrons from one generation actually do cause fissions in the next generation. Some leak out of the reactor, some are captured by atoms other than uranium, and some that are captured by uranium atoms do not cause fission.

Assume that

$P(\text{leaking}) = 0.36$

$P(\text{capture by other atom}) = 0.2$

$P(\text{nonfission capture}) = 0.15$

Calculate the probability that *none* of these things happens and thus that the neutron *does* cause a fission in the next generation.

e. Use the probability in part d and the fact that there are 2.3 new neutrons per fission to calculate k, the expected number of new neutrons in the second generation caused by one neutron in the first generation.

f. How many neutrons would you expect there to be after 1 sec under the conditions in part e if each generation still takes 0.001 sec as in part c? Would the reactor explode like a bomb?

g. Why can you say that the number of neutrons is *increasing exponentially* with time?

h. The constant k in part e is called the *multiplication factor.* The *chain reaction* in a nuclear reactor is controlled by moving control rods out or in to absorb fewer or more neutrons. If k is slightly more than 1, the power level increases. If k is less than 1, the power level decreases. If $k = 1$, the power level remains constant and the reactor is said to be *critical.* What would $P(\text{capture by other atom})$, mentioned in part d, have to equal to make the reactor critical?

Chapter Test

PART 1: No calculators allowed (T1–T8)

T1. Calculate the number of permutations of seven objects taken three at a time. Show your method.

T2. Calculate the number of combinations of six objects taken four at a time. Show how this number is calculated using factorials.

T3. What is the difference between a permutation and a combination?

T4. If A and B are independent events and $P(A) = 0.8$ and $P(B) = 0.9$, find $P(A \text{ and } B)$.

T5. If A and B are independent events and $P(A) = 0.8$ and $P(B) = 0.9$, find $P(A \text{ or } B)$.

T6. Suppose that, in each repetition of a random experiment, the probability that A occurs is 0.8. Find the probability that A occurs in exactly two out of three repetitions.

T7. Explain why the random experiment in Problem T6 is called a *binomial* experiment.

T8. Suppose that C, D, and E are three mutually exclusive events of a random experiment and that $P(C)$, $P(D)$, and $P(E)$ are 0.5, 0.3, and 0.2, respectively. If the payoffs are \$10, \$6, and −\$100 for C, D, and E, respectively, find the mathematical expectation of the random experiment.

PART 2: Graphing calculators allowed (T9–T28)

Pick-Three Problem: The 11th-grade class decides to run a lottery to help them finance the prom. A person pays \$1 and picks three different digits. If all three digits match the winning digits, the class pays the person \$100 (but keeps the \$1). If not, the class keeps the \$1.

T9. The sample space for this random experiment contains ${}_{10}C_3$ outcomes. What is the probability that any one pick is the winning combination?

T10. What is the probability that any one pick is *not* the winning combination?

T11. What is the 11th-grade class's payoff if the pick is the winning combination? What is the class's payoff if the pick is not the winning combination?

T12. What is the class's mathematical expectation for any one pick?

T13. How much would the class expect to make from the sale of 1000 picks?

Multiple-Choice Test Problem 3: On a college board test with five different choices for each problem, a test-taker receives one point for each correct answer. For each wrong answer, $\frac{1}{4}$ point is subtracted.

T14. If you guess at random in a question, what is your probability of getting the right answer? Of getting a wrong answer?

T15. What is your mathematically expected number of points for any problem for which you guess the answer?

T16. Suppose you know that three of the five choices are incorrect. What is your probability of guessing the right answer from the remaining two choices? What is your probability of guessing the wrong answer?

T17. What is your mathematically expected number of points for a question for which you can eliminate three of the five choices?

Hezzy's Punctuality Problem: Hezzy Tate has a 30% probability of being late to class on any one day.

T18. What is his probability of *not* being late on any one day?

T19. Show how to calculate Hezzy's probability of being late on exactly two of the five days in a week.

T20. Make a list of Hezzy's probabilities of being late on 0 through 5 days. Write the result.

T21. Perform a calculation that shows that your answers to Problem T20 are reasonable.

T22. Tell the special name of the probability distribution in Problem T20.

T23. Plot a graph of the probability distribution in Problem T20. Sketch the graph.

Cup and Saucer Problem: Wanda washes dishes at a restaurant. Her probability of breaking a cup on any one shift is 8%, and her probability of breaking a saucer is 6%. Calculate her probabilities for the following events.

T24. P(cup and saucer)

T25. P(cup and not saucer)

T26. P(saucer and not cup)

T27. P(not cup and not saucer)

T28. What did you learn as a result of doing this chapter test that you did not know before?

9-10 Cumulative Review, Chapters 7–9

The following Problem Set comprises a comprehensive review of these topics:

- Graphical and numeric properties of elementary functions
- Fitting functions to data
- Linear and other types of regression
- Combinatorics
- Probability
- Functions of a random variable

If you are thoroughly familiar with these topics, you should be able to finish the Problem Set in about two hours.

Problem Set 9-10

PART 1: No calculators (1–13)

1. Write the general equation for
   a. Linear function
   b. Quadratic function
   c. Logarithmic function
   d. Exponential function
   e. Power function
2. Sketch the graphs of these functions.
   a. $f(x) = -2(x - 3)^2 + 5$
   b. $g(x) = e^{x+2}$
   c. $h(x) = \log_{10} x - 1$
3. Name the pattern followed by the y-values of functions with regularly spaced x-values for
   a. Logarithmic functions
   b. Power functions
   c. Quadratic functions
4. Tell what type of function each graph shows.

a.
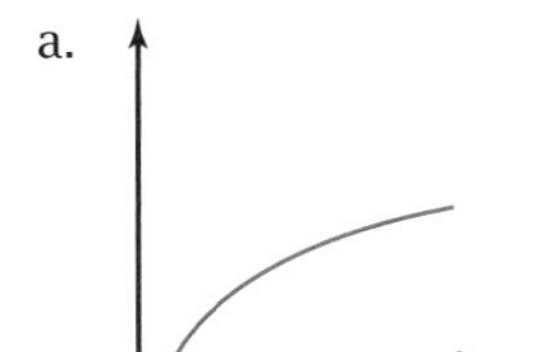

b.
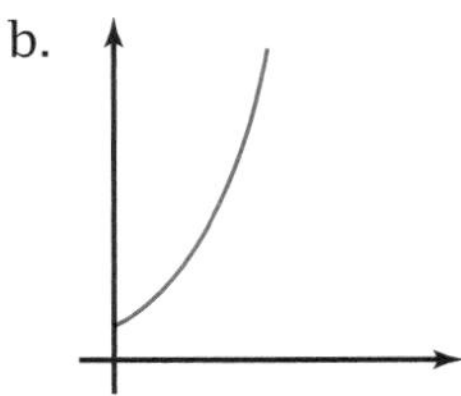

c.
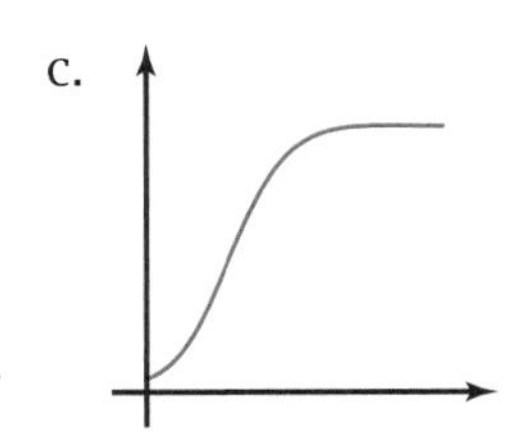

d.
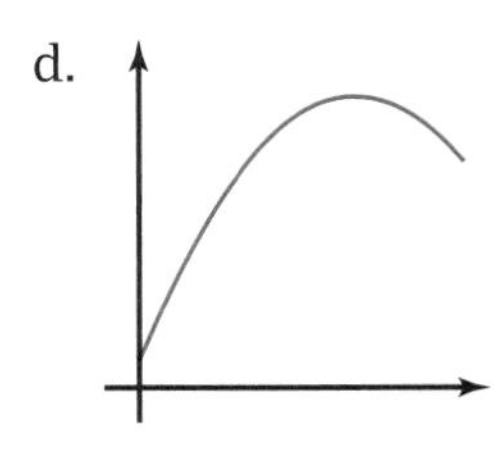

e.
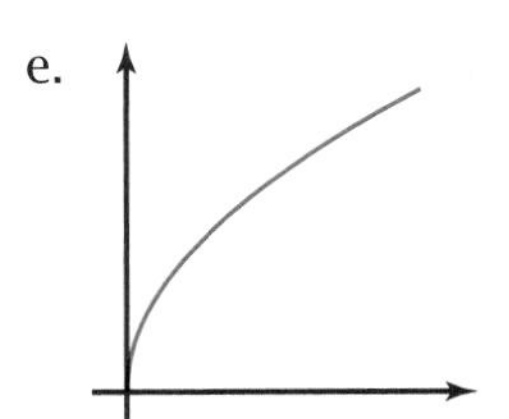

f.
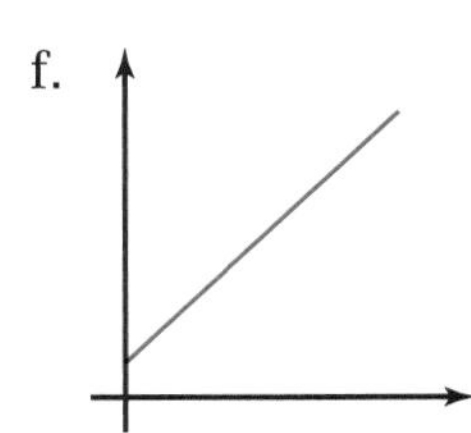

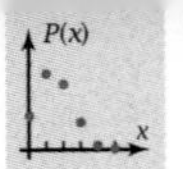

For Problems 5–7, Figure 9-10a shows a set of points with a dashed line drawn across at $y = \overline{y}$, where $\overline{y}$ is the average of the y-values.

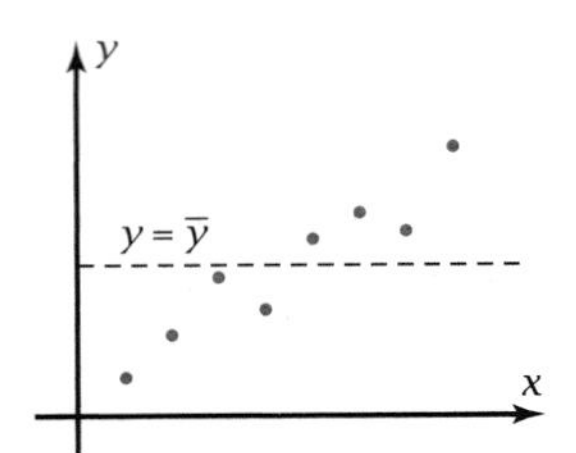

Figure 9-10a

5. On a copy of the figure, sketch the deviation from the average for the rightmost point.

6. Sketch what you think is the best-fitting linear function. Show the residual for the rightmost point. How does the size of the residual compare with the size of the deviation?

7. Based on what you know about the average-average point, $(\overline{x}, \overline{y})$, show where $\overline{x}$ would be on the graph in Problem 5.

8. Suppose that the regression equation for a set of data is $\hat{y} = 3x + 5$. What does the residual equal for the data point (4, 15)?

9. For a regression line, what is true about the sum of the squares of the residuals, SS_{res}?

10. Suppose that the sum of the squares of the deviations is $SS_{dev} = 100$, and that $SS_{res} = 36$. What does the coefficient of determination, r^2, equal?

11. Suppose that a precalculus test has 20 questions. Use factorials to show the number of different ways you could select a group of 7 of these questions to answer. Without actually calculating the number, give the number of different orders a student could work on 7 of these 20 questions.

12. If A and B are independent events with probabilities $P(A) = 0.6$ and $P(B) = 0.8$, find the probability that A or B occurs.

13. If C and D are mutually exclusive events and $P(C) = 0.1$ and $P(D) = 0.2$, find

    a. The probability that C and D both occur

    b. The probability that D does not occur

    c. The mathematical expectation if the payoff for C is \$6.00, the payoff for D is −\$2.00, and the payoff if neither C nor D occurs is \$1.00

PART 2: Graphing calculators are allowed (14–19)

14. *Light Intensity Problem:* The table shows the intensity, y, of light beneath the water surface as a function of distance, x meters, below the surface.

x	y
0	100
3	50
6	25
8	16
11	8
17	2

    a. What pattern do the first three data points follow? What type of function has this pattern?

    b. Find the particular equation for the function in part a algebraically by substituting the second and third points into the general equation. Show that the particular equation gives values for the last three points that are close to the values in the table.

    c. Use the appropriate kind of regression to find the function of the type in part a that best fits all seven of the data points. Write the correlation coefficient, and explain how it indicates that the function fits the data quite well.

    d. Use the regression equation from part c to predict the light intensity at a depth of 14 feet. Which do you use, interpolation or extrapolation, to find this intensity? How do you decide?

15. *Spindletop Problem:* On January 10, 1901, the first oil-well gusher in Texas happened at Spindletop near Beaumont. In the following months, many more wells were drilled. Assume that the data in the table on the next page are number of wells, y, as a function of number of months, x, after January 10, 1901.

x (months)	y (wells)
0	1
1	3
2	10
3	27
4	75
5	150

a. By regression, find the particular equation of the best-fitting logistic function for these data.

b. Plot the logistic function on your grapher. Sketch the result on a copy of Figure 9-10b. Show the point of inflection.

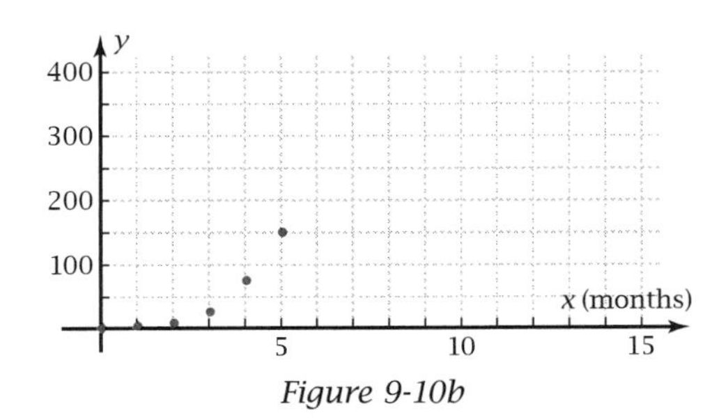

Figure 9-10b

c. How many wells does your model predict were ultimately drilled at Spindletop?

d. Why is a logistic model more reasonable than an exponential model for this problem?

16. *Cricket Problem:* The frequency at which crickets chirp increases as the temperature increases. Let x be the number of degrees Fahrenheit, and let y be the number of chirps per second. Suppose that the following data have been measured and graphed in Figure 9-10c.

x	y
50	35
55	55
60	74
65	93
70	112
75	130
80	147
85	165
90	182
95	200

Figure 9-10c

a. A linear function appears to fit the data. Write the linear regression equation, and give numerical evidence from the regression result that a linear function fits very well.

b. If you extrapolate to temperatures below 50°, what does the linear function indicate will eventually happen to the crickets? What, then, would be a reasonable lower bound for the domain of the linear function?

c. If you extrapolate to temperatures above 95°, what does the linear function indicate will eventually happen to the crickets? Is this a reasonable endpoint behavior?

d. Put a list in your grapher to calculate the residuals, $y - \hat{y}$. Use the results to make a residual plot. Sketch the residual plot. What information do you get from the residual plot concerning how well the linear function fits the data?

17. *Logarithmic Function Problem 2:*

a. Use the definition of logarithm to evaluate $y = \log_9 53$.

b. Use the log of a power property to solve this exponential equation: $3^{4x} = 93$

c. Use the change of base property to evaluate $\log_5 47$ using natural logarithms.

d. Plot the graph of $f(x) = \ln x$. Sketch the result. Explain why $f(1) = 0$. Give numerical evidence to show how $f(5)$ and $f(7)$ are related to $f(35)$.

18. *Probability Distribution Problems:* You flip a thumbtack several times. Suppose that the probability that it lands "point up" on any one flip is 0.4.

a. What is the probability that the thumbtack lands point down?

b. The probability that the first time the thumbtack lands point up is on Flip 2 is the probability that it is *not* point up on Flip 1 and *is* point up on Flip 2. Let $P(x)$ be the probability that the first point up is on Flip x. Find $P(0)$, $P(1)$, $P(2)$, $P(3)$, $P(4)$, and $P(5)$. Plot these values on a copy of Figure 9-10d. What special name refers to this type of probability distribution?

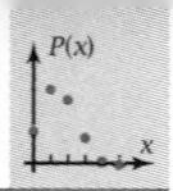

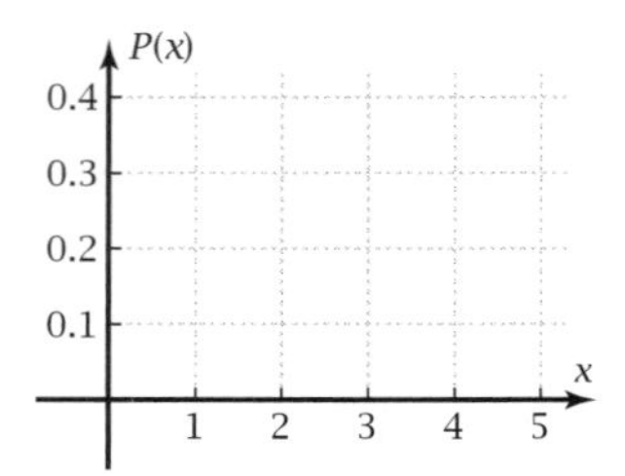

Figure 9-10d

c. You flip the thumbtack six times. What is the probability it lands point up exactly two of those times?

d. Show how the number of combinations of six objects taken two at a time is calculated using factorials. How does this number relate to part c?

e. Calculate the number of permutations of six objects taken two at a time.

19. *Indy 500 Problem:* Hezzy Tate has a 70% probability of finishing in the top ten places in the Indianapolis 500-car race. His wife, Aggie, has an 80% probability of finishing in the top ten.

a. What is the probability that Hezzy and Aggie both finish in the top ten?

b. The probability that Hezzy finishes in the top ten but Aggie does not is 0.14. Show how you could calculate this number.

c. What is the probability that Aggie finishes in the top ten but Hezzy does not?

d. What is the probability that neither of them finishes in the top ten?

e. The Tates decide to go on a vacation to spend their winnings. They will spend the following amounts, depending on which one or ones win.

H and A:	Hawaii, \$8000
H, not A:	California, \$3000
A, not H:	Florida, \$4000
Neither:	Stay home, \$0

Calculate the mathematically expected number of dollars they will spend.

20. What did you learn as a result of taking this test that you did not know before?

CHAPTER

10

Three-Dimensional Vectors

When a building is constructed, it is important for the builders to know relationships between various planes, such as walls, ceilings, and roofs. Three-dimensional vectors let you calculate distances, angles, and intersections of lines and planes in space—quantities that would be hard to find graphically.

Mathematical Overview

In this chapter you will extend what you learned about two-dimensional vectors to vectors in space. You can do some operations, such as vector addition and subtraction, simply by adding a third component to the representation of a vector. Other operations require new techniques—there are two different kinds of multiplication for vectors. The payoff is the ability to calculate such things as the place where a line in space intersects a plane in space so that parts of three-dimensional objects will fit together properly when they are constructed. You will gain this knowledge in four ways.

Numerically

Dot product:

$$(4\vec{i} + 3\vec{j} + 5\vec{k}) \cdot (2\vec{i} + 6\vec{j} + 8\vec{k})$$

$$= (4)(2) + (3)(6) + (5)(8) = 66$$

Cross product:

$$(4\vec{i} + 3\vec{j} + 5\vec{k}) \times (2\vec{i} + 6\vec{j} + 8\vec{k})$$

$$\begin{vmatrix} i & j & k \\ 4 & 3 & 5 \\ 2 & 6 & 8 \end{vmatrix} = -6\vec{i} - 22\vec{j} + 18\vec{k}$$

Graphically

A line intersecting a plane

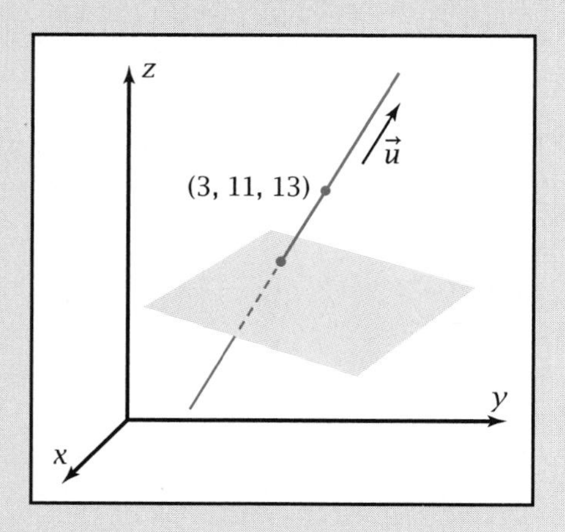

Algebraically

Equation of a line in a three-dimensional space:

$$\vec{r} = \left(5 - \tfrac{2}{3}d\right)\vec{i} + \left(7 + \tfrac{1}{3}d\right)\vec{j} + \left(-1 + \tfrac{2}{3}d\right)\vec{k}$$

Verbally

I finally understand about dot product and cross product. The first is just a number (called a scalar). The second is a vector. It was the other names, "scalar product" and "vector product," that made me realize the difference.

10-1 Review of Two-Dimensional Vectors

In Chapter 6 you learned that vectors are directed line segments. You used them as mathematical models of vector quantities, such as displacement, velocity, and force, that have direction as well as magnitude. In this chapter you will extend your knowledge to vectors in space. First, you will refresh your memory about two-dimensional vectors.

OBJECTIVE Given two vectors, find the resultant vector by adding or subtracting them.

Exploratory Problem Set 10-1

1. Figure 10-1a shows two vectors, $\vec{a}$ and $\vec{b}$. They have magnitude and direction but no fixed location. You find the vector sum $\vec{a} + \vec{b}$ by translating $\vec{b}$ so that its beginning (or tail) is at the end (or head) of $\vec{a}$. The resultant vector goes from the beginning of $\vec{a}$ to the end of $\vec{b}$. Show that you understand how to add two vectors by drawing a sketch showing $\vec{a} + \vec{b}$.

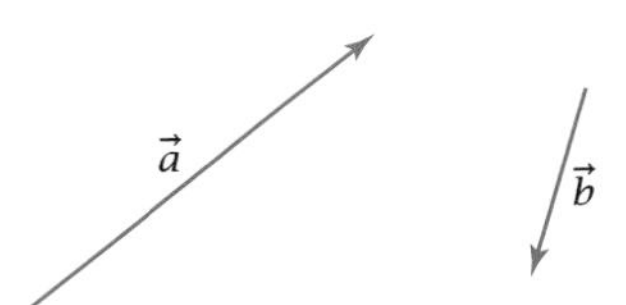

Figure 10-1a

2. You find the vector difference $\vec{a} - \vec{b}$ by adding $-\vec{b}$, the opposite of $\vec{b}$, to $\vec{a}$. The opposite of $\vec{b}$ is a vector of the same length pointing in the opposite direction. Show that you understand vector subtraction by sketching $\vec{a} - \vec{b}$.

3. The **position vector** for a point (x, y) in a coordinate system has a fixed location. It goes from the origin to the point. Figure 10-1b shows the position vector $\vec{p}$ for the point (4, 3). You can think of this vector as the sum of a vector in the x-direction and a vector in the y-direction. On a copy of Figure 10-1b, sketch these two components of $\vec{p}$.

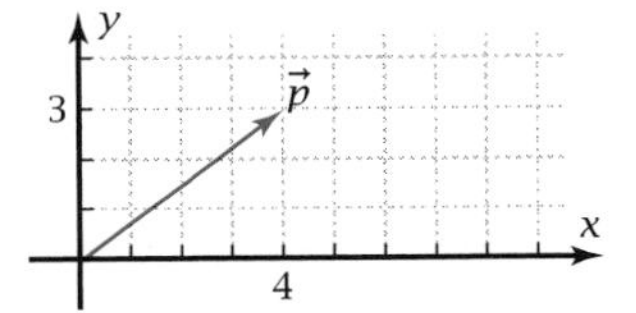

Figure 10-1b

4. In Problem 3, the vector in the x-direction is 4 units long. Vectors $\vec{i}$ and $\vec{j}$ are **unit vectors** in the x- and y-directions, respectively. A unit vector is a vector that is one unit long. So the x-component of $\vec{p}$ is the vector $4\vec{i}$, a product of a **scalar** and a vector. A scalar is a quantity that has only magnitude, not direction. Write the y-component of $\vec{p}$ in Figure 10-1b.

5. Suppose that $\vec{v} = 6\vec{i} + 8\vec{j}$. Sketch $\vec{v}$ as a position vector. Find the length of $\vec{v}$ (the **magnitude**) by means of the Pythagorean theorem. Find the angle $\vec{v}$ makes with the x-axis.

6. Suppose that $\vec{a} = 6\vec{i} + 2\vec{j}$ and $\vec{b} = 3\vec{i} + 5\vec{j}$. Draw $\vec{a}$ on graph paper as a position vector. Draw $\vec{b}$ with its tail at the head of $\vec{a}$. Draw the resultant vector, $\vec{a} + \vec{b}$. Write this sum vector as a position vector. What simple method can you think of to add two vectors if you know their components? How long is vector $\vec{a} + \vec{b}$?

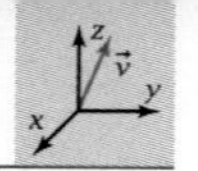

7. Sketch $\vec{c} = -9\vec{i} + 4\vec{j}$ as a position vector. Find its angle in standard position. Is the value given by your calculator the answer?

8. Make a list of all the important words in this section. Put a checkmark by the ones you understand and a question mark by the ones you don't quite understand.

10-2 Two-Dimensional Vector Practice

In this section you will consolidate your knowledge of two-dimensional vectors from Chapter 6 and from Section 10-1. In the next section you will extend these concepts to vectors in three-dimensional space.

OBJECTIVES

- Given the components of a two-dimensional position vector, find its length, a unit vector in its direction, a scalar multiple of it, and its direction angle.
- Given two two-dimensional position vectors, find their sum and their difference.

The box on the next page summarizes the definitions and properties of vectors. Figure 10-2a illustrates some of these definitions. On the top left is a vector $\vec{v}$. On the top right are two equal vectors, $\vec{v}$ and $\vec{a}$, that are translations of one another. On the bottom left is a vector $\vec{v}$ and its opposite, $\vec{a} = -\vec{v}$. On the bottom right is vector $\vec{v}$ that is 5 units long and unit vector $\vec{u}$ in the same direction.

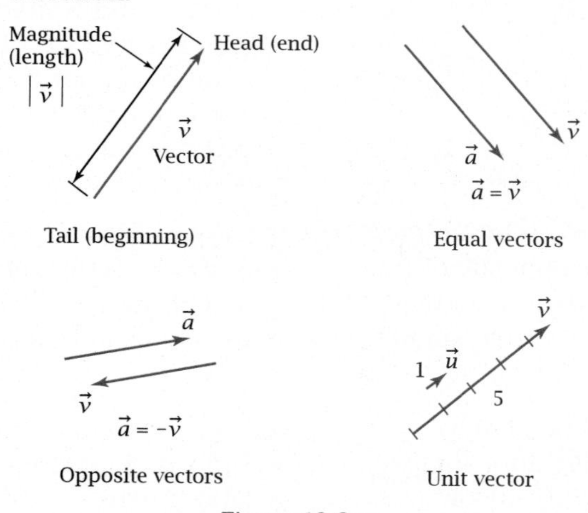

Figure 10-2a

Vectors play an important role in aerodynamics. Here a reproduction of a Wright Brothers' glider is undergoing wind tunnel testing at NASA's Langley Research Center in Hampton, Virginia.

Definitions and Properties Relating to Vectors

- A **vector quantity** is a quantity, such as force, velocity, or displacement, that has both magnitude (size) and direction.
- A **scalar** is a quantity, such as time, speed, or volume, that has only magnitude but no direction.
- A **vector** is a directed line segment that represents a vector quantity. Symbol: $\vec{v}$
- The **tail** of a vector is the point where it begins. The **head** of a vector is the point where it ends. An arrowhead is drawn at the head of a vector.
- The **magnitude,** or **absolute value,** of a vector is its **length.** Symbol: $|\vec{v}|$ If $\vec{v} = x\vec{i} + y\vec{j}$, then $|\vec{v}| = \sqrt{x^2 + y^2}$ by the Pythagorean theorem.
- A **unit vector,** $\vec{u}$, in the direction of $\vec{v}$ is a vector that is one unit long in the same direction as $\vec{v}$. So

$$\vec{u} = \frac{\vec{v}}{|\vec{v}|}$$ Divide the vector by its length.

- Two vectors are **equal** if they have the same magnitude and the same direction. So you may **translate** a vector without changing it, but you can't rotate or dilate it.
- The **opposite** of a vector is a vector of the same length in the opposite direction. Symbol: $-\vec{v}$
- A **position vector,** $\vec{v} = x\vec{i} + y\vec{j}$, starts at the origin and ends at the point (x, y).
- A **displacement vector** is the difference between an object's initial and final positions.

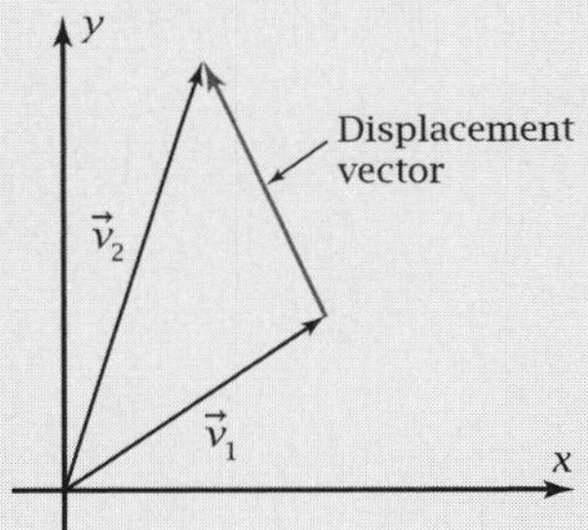

Operations on Vectors

Recall from Chapter 6 the geometrical interpretations of vector sum and vector difference, as shown in Figures 10-2b and 10-2c, respectively. If two vectors are placed head-to-tail, the **vector sum** goes from the beginning of the first vector to the end of the second. If two vectors are placed tail-to-tail, the **vector difference** goes from the head of the second vector to the head of the first. This is the way, for example, you would determine how far you have gone on a trip traveling on a straight line by subtracting odometer readings, "where you ended minus where you started."

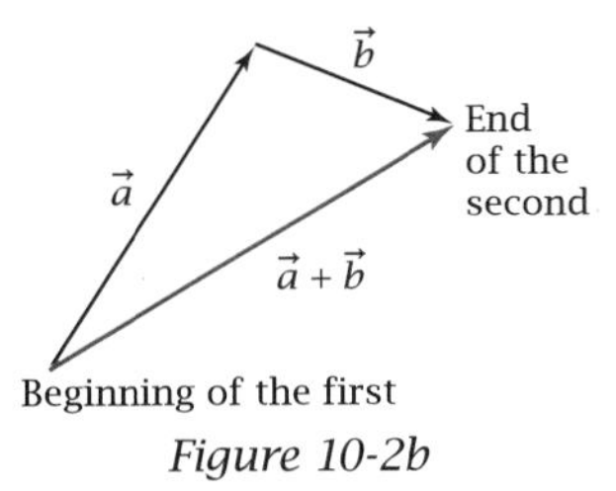

Figure 10-2b

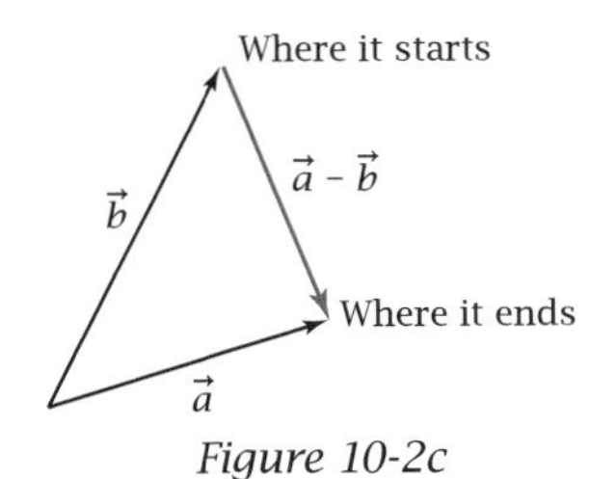

Figure 10-2c

You can add or subtract two vectors easily if you write them in terms of their components in the x- and y-directions. As shown in Figure 10-2d, $\vec{v}$ is the position vector to the point $(x, y) = (-5, 3)$. Vectors $\vec{i}$ and $\vec{j}$ are unit vectors in the x- and y-directions, respectively. So $\vec{v}$ can be written

$$\vec{v} = -5\vec{i} + 3\vec{j}$$

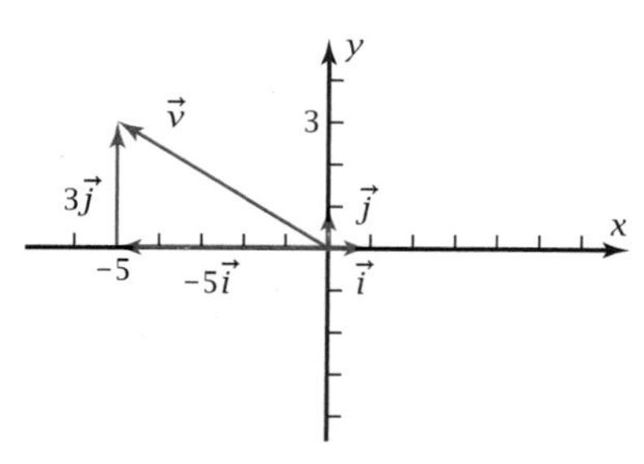

Figure 10-2d

These examples show how to operate with vectors that are written in terms of their components.

► EXAMPLE 1 If $\vec{a} = 3\vec{i} + 8\vec{j}$ and $\vec{b} = 4\vec{i} - 2\vec{j}$,

a. Find $\vec{a} + \vec{b}$.

b. Find $\vec{a} - \vec{b}$.

c. Find $-\vec{a}$.

d. Find $5\vec{a} + 9\vec{b}$.

e. Find $|\vec{a}| + |\vec{b}|$.

f. Find $|\vec{a} + \vec{b}|$.

g. Does $|\vec{a} + \vec{b}| = |\vec{a}| + |\vec{b}|$?

h. Find a unit vector in the direction of $\vec{a}$.

i. Find a vector of length 10 in the direction of $\vec{a}$.

j. Sketch $\vec{b}$ as a position vector.

k. Find the angle in standard position for $\vec{b}$.

Solution

a. $\vec{a} + \vec{b} = (3\vec{i} + 8\vec{j}) + (4\vec{i} - 2\vec{j}) = 7\vec{i} + 6\vec{j}$ — Add the respective components.

b. $\vec{a} - \vec{b} = (3\vec{i} + 8\vec{j}) - (4\vec{i} - 2\vec{j}) = -\vec{i} + 10\vec{j}$ — Subtract the respective components.

c. $-\vec{a} = -(3\vec{i} + 8\vec{j}) = -3\vec{i} - 8\vec{j}$ — Take the opposite of each component.

d. $5\vec{a} + 9\vec{b} = 5(3\vec{i} + 8\vec{j}) + 9(4\vec{i} - 2\vec{j})$

$= 15\vec{i} + 40\vec{j} + 36\vec{i} - 18\vec{j}$

$= 51\vec{i} + 22\vec{j}$ — Combine like terms.

e. $|\vec{a}| + |\vec{b}| = \sqrt{3^2 + 8^2} + \sqrt{4^2 + 2^2}$ — Use the Pythagorean theorem.

$= \sqrt{73} + \sqrt{20}$

$= 13.0161\ldots$

f. $|\vec{a} + \vec{b}| = |7\vec{i} + 6\vec{j}|$ Do what's inside the absolute value sign *first*.

$= \sqrt{7^2 + 6^2}$

$= \sqrt{85}$

$= 9.2195\ldots$

g. No, $|\vec{a} + \vec{b}| \neq |\vec{a}| + |\vec{b}|$, as shown in parts e and f.

h. $\vec{u} = \dfrac{\vec{a}}{|\vec{a}|} = \dfrac{3}{\sqrt{73}}\vec{i} + \dfrac{8}{\sqrt{73}}\vec{j} = 0.3511\ldots\vec{i} + 0.9363\ldots\vec{j}$ Divide the vector by its length.

i. $\vec{v} = 10\vec{u} = \dfrac{30}{\sqrt{73}}\vec{i} + \dfrac{80}{\sqrt{73}}\vec{j} = 3.511\ldots\vec{i} + 9.363\ldots\vec{j}$ Multiply the unit vector by 10.

j. Figure 10-2e shows $\vec{b}$ as a position vector.

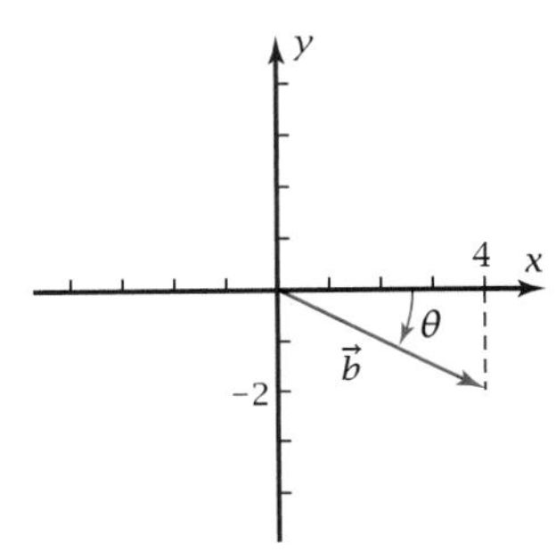

Figure 10-2e

k. $\theta = \tan^{-1}\left(-\frac{2}{4}\right) = -26.5650\ldots° + 180n° \approx -26.6°$ Use $n = 0$. ◀

▶ **EXAMPLE 2** Given point $C(8, 25)$ and $D(17, 3)$,

a. Find vector $\overrightarrow{CD}$, the vector pointing from C to D.

b. Find the position vector of the point $\frac{3}{4}$ of the way from C to D.

Solution a. Sketch points C and D and position vectors $\vec{c}$ and $\vec{d}$ to these points, as in Figure 10-2f.

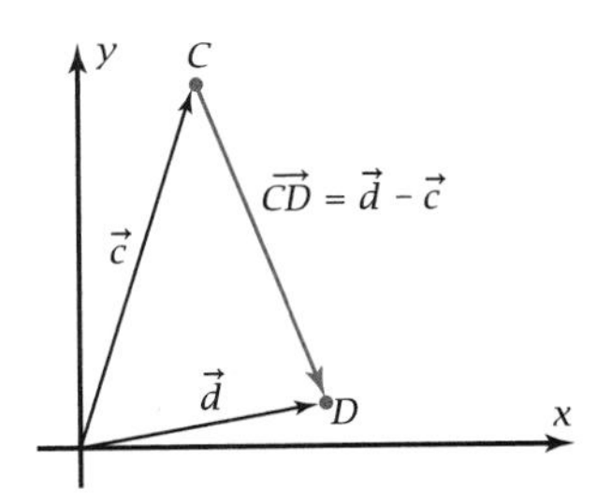

Figure 10-2f

Vector $\overrightarrow{CD}$ starts at C and ends at D. Therefore,

$\overrightarrow{CD} = \vec{d} - \vec{c}$

$= (17\vec{i} + 3\vec{j}) - (8\vec{i} + 25\vec{j})$ Write position vectors to the two points.

$= 9\vec{i} - 22\vec{j}$

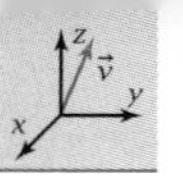

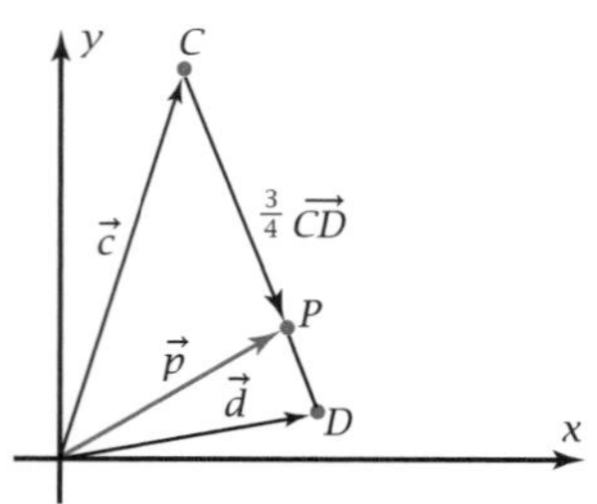

Figure 10-2g

b. Sketch a vector starting at C and going $\frac{3}{4}$ of the way to D, as in Figure 10-2g. This vector will be $\frac{3}{4}$ of $\overrightarrow{CD}$. Because this vector and $\vec{c}$ are head-to-tail, the position vector $\vec{p}$ will be the vector sum.

$$\vec{p} = \vec{c} + \tfrac{3}{4}\overrightarrow{CD}$$
$$= (8\vec{i} + 25\vec{j}) + \tfrac{3}{4}(9\vec{i} - 22\vec{j})$$
$$= 14.75\vec{i} + 8.5\vec{j}$$

◀

Problem Set 10-2

Do These Quickly

For Problems Q1–Q6, express the following values for right triangle ABC in Figure 10-2h.

Q1. $\sin A$

Q2. $\cos C$

Q3. $\tan C$

Q4. $a^2 + c^2$

Q5. The area

Q6. Angle A as an inverse tangent

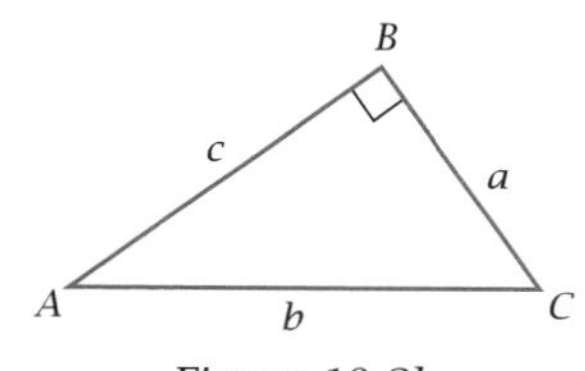

Figure 10-2h

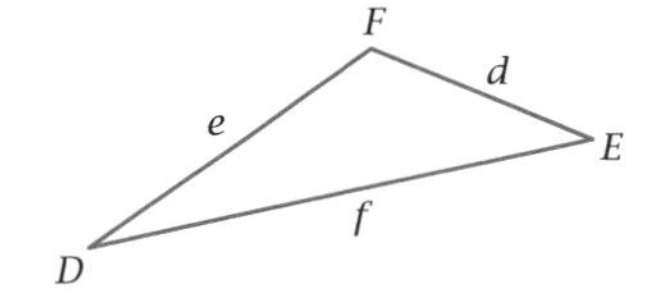

Figure 10-2i

Problems Q7–Q9 refer to oblique triangle DEF in Figure 10-2i.

Q7. Find f^2 by the law of cosines.

Q8. $\dfrac{a}{\sin A}$ = —?— by the law of sines.

Q9. Find the area in terms of two sides and angle D.

Q10. $\cos 180° =$

A. 1 B. $\frac{1}{2}$ C. 0 D. $-\frac{1}{2}$ E. -1

1. For $\vec{a}$ and $\vec{b}$ in Figure 10-2j, draw a sketch showing the geometric meaning of

   a. $\vec{a} + \vec{b}$ b. $\vec{a} - \vec{b}$
   c. $\vec{b} - \vec{a}$ d. $3\vec{a}$
   e. $-2\vec{b}$ f. $|\vec{a}|$

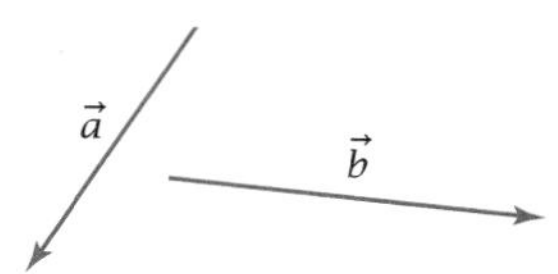

Figure 10-2j

2. For $\vec{c}$ and $\vec{d}$ in Figure 10-2k, draw a sketch showing the geometric meaning of

   a. $\vec{c} + \vec{d}$ b. $\vec{c} - \vec{d}$
   c. $\vec{d} - \vec{c}$ d. $4\vec{d}$
   e. $-3\vec{c}$ f. $|\vec{d}|$

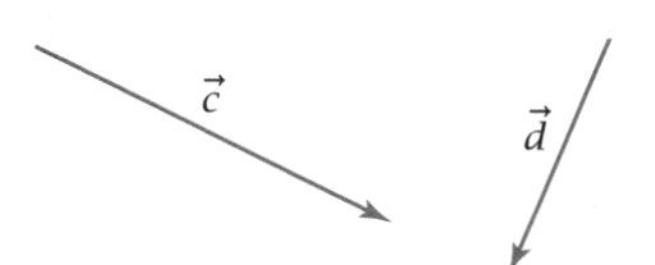

Figure 10-2k

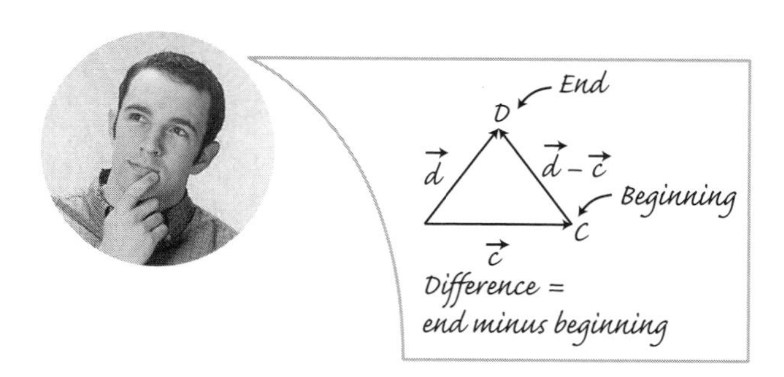

3. If $\vec{a} = 2\vec{i} + 5\vec{j}$ and $\vec{b} = 7\vec{i} - 3\vec{j}$,
   a. Find $\vec{a} + \vec{b}$, $\vec{a} - \vec{b}$, $-\vec{a}$, and $2\vec{a} + 3\vec{b}$.
   b. Find $|\vec{a}| + |\vec{b}|$ and $|\vec{a} + \vec{b}|$. Does $|\vec{a} + \vec{b}| = |\vec{a}| + |\vec{b}|$?
   c. Find a unit vector in the direction of $\vec{a}$. Find a vector of length 10 in the direction of $\vec{a}$.
   d. Sketch $\vec{b}$ as a position vector. Find the angle in standard position for $\vec{b}$.

4. If $\vec{c} = -4\vec{i} + 6\vec{j}$ and $\vec{d} = 9\vec{i} + 8\vec{j}$,
   a. Find $\vec{c} + \vec{d}$, $\vec{c} - \vec{d}$, $\vec{d} - \vec{c}$, and $3\vec{c} - 4\vec{d}$.
   b. Find $|\vec{c}| + |\vec{d}|$ and $|\vec{c} + \vec{d}|$. Does $|\vec{c} + \vec{d}| = |\vec{c}| + |\vec{d}|$?
   c. Find a unit vector in the direction of $\vec{d}$. Find a vector of length 7 in the direction of $\vec{d}$.
   d. Sketch $\vec{c}$ as a position vector. Find the angle in standard position for $\vec{c}$.

For Problems 5–8, find the vector.

5. $\overrightarrow{AB}$ for $A(3, 4)$ and $B(2, 7)$
6. $\overrightarrow{CD}$ for $C(4, 1)$ and $D(3, 5)$
7. $\overrightarrow{BA}$ for $A(7, 3)$ and $B(5, -1)$
8. $\overrightarrow{DC}$ for $C(-2, 3)$ and $D(4, -3)$
9. *Highway Rest Stop Problem:* A highway rest stop will be built 40% of the way from the town of Artesia, A, to the town of Brooks, B, in Figure 10-21. The two towns are located at $A(20, 73)$ and $B(45, 10)$. Big City is located at the origin. The coordinates are in kilometers.

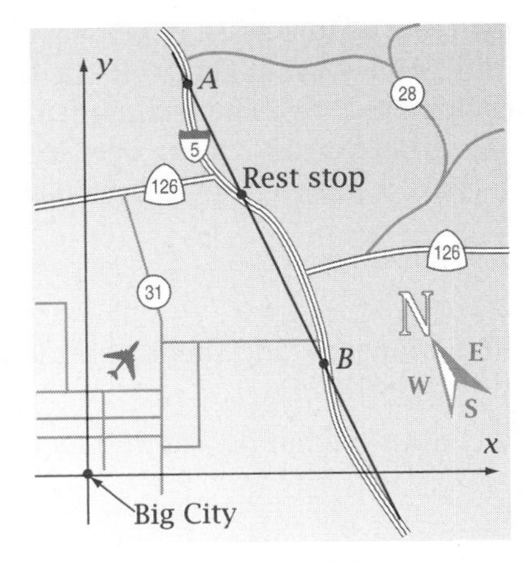

Figure 10-21

   a. Write the position vectors of Artesia and Brooks.
   b. Write the displacement vector from Artesia to Brooks.
   c. Write the displacement vector from Artesia to the rest stop.
   d. Write the position vector to the rest stop.
   e. It is proposed to supply electricity to the rest stop directly from Big City. How long would the electric lines be? At what angle to the x-axis would the lines have to run?
   f. How long would the electric lines have to be if they came from the closer of the two towns, Artesia or Brooks?

10. *Archaeology Problem:* Archaeologists often cut a trial trench through an archaeological site to reveal the different layers under the topsoil. This stratigraphy helps with dating the artifacts they unearth and identifying any geological movements that might have disturbed the original position of objects. They usually lay a grid on the site, much like a coordinate system. Assume that they dig the trial trench from point $C(200, -300)$ to point $D(400, 500)$, where the distances are in yards.

   a. Make a sketch showing the given information.
   b. Write the position vectors for C and D and the vector from C to D.
   c. The crew finds the remnants of a wall 65% of the way from C to D. Write the vector from C to this point.
   d. How long is the trench from point C to the wall?
   e. How far is the wall from the origin?

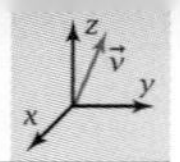

For Problems 11–14, use the techniques of Problems 9 and 10 to find the vector or point.

11. Find the position vector of the point $\frac{1}{3}$ of the way from $A(2, 7)$ to $B(14, 5)$.

12. Find the position vector of the point $\frac{2}{3}$ of the way from $C(11, 5)$ to $D(2, 17)$.

13. Find the midpoint of the segment connecting $E(6, 2)$ and $F(10, -4)$. From the result, give a quick way to find the midpoint of the segment connecting two given points.

14. Find the midpoint of the segment connecting $G(5, 7)$ and $H(-3, 13)$. From the result, give a quick way to find the midpoint of the segment connecting two given points.

15. *Vector Properties Problem:* Prove the properties of vectors that follow. A sketch may help. Express the vectors as the sum of their components, and prove the properties algebraically.

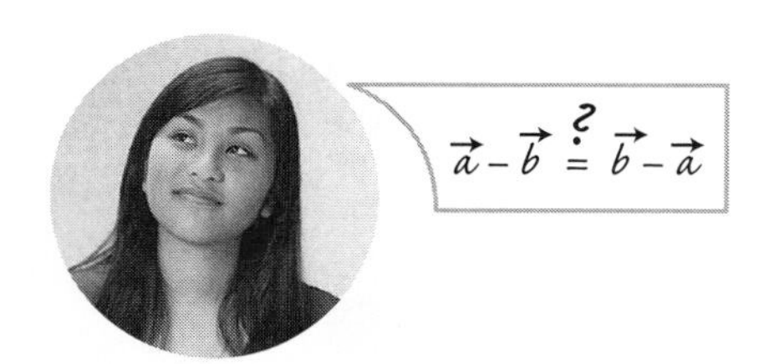

a. Vector addition is commutative.

b. Vector addition is associative.

c. Vector subtraction is *not* commutative.

d. Multiplication by a scalar distributes over vector addition.

e. The set of vectors is closed under addition. (Why is it necessary for there to be a zero vector in order for this closure property to be true?)

16. *Triangle Inequality Problem:*

a. Sketch two nonparallel vectors $\vec{a}$ and $\vec{b}$ head-to-tail. Then draw the sum $\vec{a} + \vec{b}$. What can you say about $|\vec{a} + \vec{b}|$ compared to $|\vec{a}| + |\vec{b}|$?

b. Sketch two parallel vectors $\vec{a}$ and $\vec{b}$ head-to-tail, pointing in the same direction. Then draw the sum $\vec{a} + \vec{b}$. What can you say about $|\vec{a} + \vec{b}|$ compared to $|\vec{a}| + |\vec{b}|$?

c. Use the appropriate theorems and postulates from geometry to prove that the **triangle inequality** shown in the box is true.

Property: Triangle Inequality for Vectors

$$|\vec{a} + \vec{b}| \le |\vec{a}| + |\vec{b}|$$

10-3 Vectors in Space

Suppose that a helicopter rises 700 ft and then moves to a point 300 ft east and 400 ft north of its original ground position. As shown in Figure 10-3a, the position vector has three components rather than just two. The techniques you learned for analyzing two-dimensional vectors in Section 10-2 carry over to three-dimensional vectors. You simply add a third component perpendicular to the other two.

OBJECTIVES

- Given two three-dimensional vectors, find their lengths, add them, subtract them, and use the results to analyze real-world problems.
- If a position vector terminates in the first octant, sketch it on graph paper.

Figure 10-3b shows a two-dimensional representation of a three-dimensional coordinate system. In 3-D, the z-axis points upward, so you draw it going up on your paper. The x-axis is drawn obliquely down to the left, and the y-axis is drawn horizontally to the right. Imagine looking down from somewhere above the first quadrant of the xy-plane. Notice that the x- and y-axes have the same orientation with respect to each other as they do in two dimensions. To assist you visually, the tick marks on the x-axis are drawn horizontally and unit intervals are drawn shorter than on the other two axes.

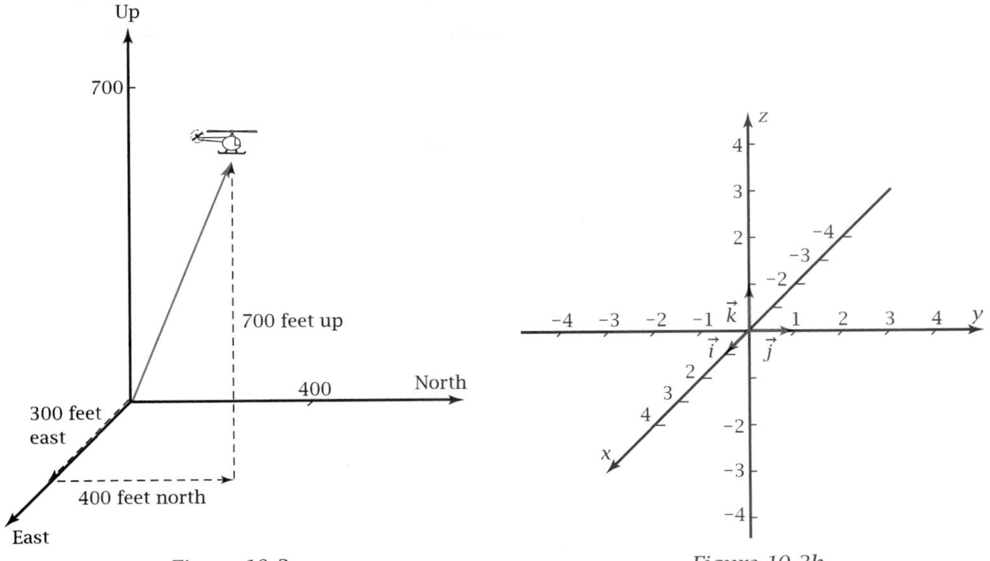

Figure 10-3a *Figure 10-3b*

The xy-plane, yz-plane, and xz-plane divide space into eight regions called **octants.** The region in which all three variables are positive is called the **first octant.** The other octants are not usually named.

The symbols $\vec{i}$, $\vec{j}$, and $\vec{k}$ are used for **unit vectors** in the x-, y-, and z-directions, respectively.

▶ EXAMPLE 1 On graph paper, draw a sketch of the position vector $\vec{p} = 3\vec{i} + 5\vec{j} + 7\vec{k}$. Write the coordinates of the point P at the end (the head) of $\vec{p}$. Find the length of $\vec{p}$.

Solution Draw the three axes, with the x-axis along the diagonal as shown in Figure 10-3c. To get the desired perspective on the x-axis, mark two units on the diagonal for each grid line. Because $\vec{p}$ is a position vector, it starts at the origin. Starting there, draw a vector 3 units long in the x-direction. From its head, draw a second vector 5 units long in the y-direction by counting spaces. From the head of the second vector, draw a third vector 7 units long in the z-direction. Because the vectors are head-to-tail, the sum goes from the beginning of the first vector to the end of the last vector.

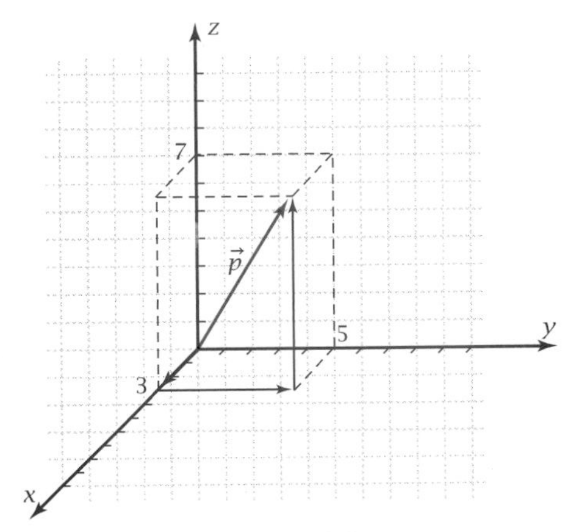

Figure 10-3c

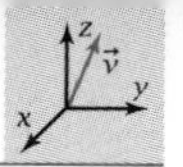

If you draw in the dashed lines, they help make the drawing look three-dimensional. They form a box for which $\vec{p}$ is the main diagonal.

Because $\vec{p}$ is the position vector of an (x, y, z) point, the point has coordinates (3, 5, 7).

You can find the length of the vector using the three-dimensional Pythagorean theorem. It is an extension of the two-dimensional formula, with the square of the third coordinate appearing under the radical sign as well. You'll prove it in Problem 18, in Problem Set 10-3.

$$|\vec{p}| = \sqrt{3^2 + 5^2 + 7^2} = \sqrt{83} = 9.1104\ldots$$ ◀

▶ EXAMPLE 2 Find the displacement vector from point $A(8, 2, 13)$ to point $B(3, 10, 4)$. Use the result to find the distance between the two points.

Solution Sketch two points in three-dimensional coordinates as in Figure 10-3d. It is not necessary to draw them to scale. Draw position vectors $\vec{a}$ and $\vec{b}$ to the two points. The vector from A to B is equal to $\vec{b} - \vec{a}$.

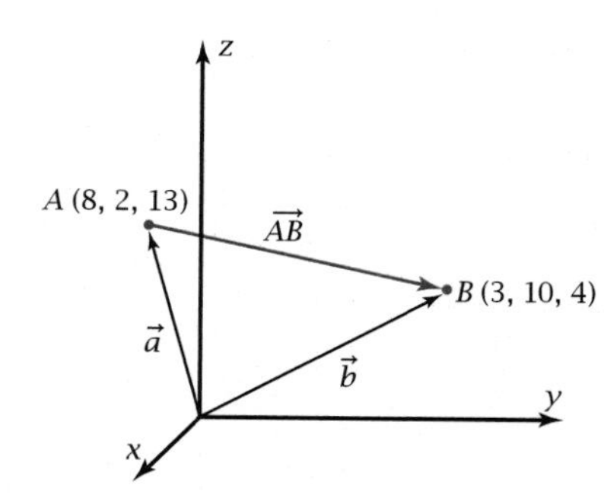

Figure 10-3d

$$\overrightarrow{AB} = \vec{b} - \vec{a} = (3\vec{i} + 10\vec{j} + 4\vec{k}) - (8\vec{i} + 2\vec{j} + 13\vec{k})$$

$$= -5\vec{i} + 8\vec{j} - 9\vec{k}$$ Displacement vector from A to B.

The distance between A and B is the length of $\overrightarrow{AB}$.

$$|\overrightarrow{AB}| = \sqrt{(-5)^2 + 8^2 + (-9^2)} = \sqrt{170} = 13.0384\ldots \text{ units.}$$ ◀

▶ EXAMPLE 3 Find the position vector to the point 70% of the way from point $A(8, 2, 13)$ to point $B(3, 10, 4)$ in Example 2. Write the coordinates of the point.

Solution On the diagram you drew for Example 2, sketch a vector starting at A and ending at a point 70% of the way from A to B, as shown in Figure 10-3e.

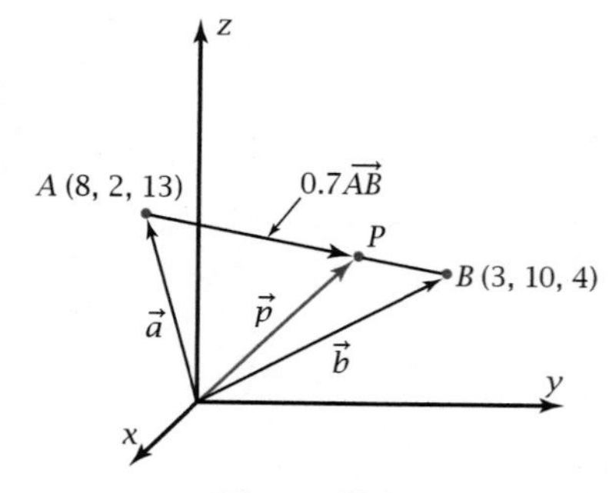

Figure 10-3e

This vector, $0.7\overrightarrow{AB}$, added to $\vec{a}$ is the position vector of the point P, as you can see from the fact that the two vectors are in position for adding (that is, head-to-tail). Using $\vec{p}$ for the position vector,

$$\vec{p} = \vec{a} + 0.7\overrightarrow{AB} = 8\vec{i} + 2\vec{j} + 13\vec{k} + 0.7(-5\vec{i} + 8\vec{j} - 9\vec{k})$$

$$= 8\vec{i} + 2\vec{j} + 13\vec{k} + (-3.5\vec{i} + 5.6\vec{j} - 6.3\vec{k})$$

$$= 4.5\vec{i} + 7.6\vec{j} + 6.7\vec{k}$$ Position vector to point P.

The coordinates of P are (4.5, 7.6, 6.7). ◀

Problem Set 10-3

Do These Quickly (5 min)

Q1. Where a vector starts is called its —?— and where it ends is called its —?—.

Q2. A vector may be translated to another position without changing its —?— or —?—.

Q3. How do you translate two vectors to add them geometrically?

Q4. When you have translated two vectors as in Problem Q3, the sum goes from —?— to —?—.

Q5. How do you translate two vectors to subtract them geometrically?

Q6. When you have translated two vectors as in Problem Q5, the difference goes from —?— to —?—.

Q7. Where does a position vector always start?

Q8. The absolute value of a vector, $|\vec{v}|$, is the same as its —?—.

Q9. True or false? "The difference of two vectors is shorter than their sum."

Q10. What makes a vector a *unit* vector?

For Problems 1–4, draw the position vector on graph paper. Show the circumscribed "box" that makes the vector look three-dimensional. Write the coordinates of the point at the head of the vector.

1. $\vec{p} = 5\vec{i} + 9\vec{j} + 6\vec{k}$
2. $\vec{p} = 8\vec{i} + 2\vec{j} + 7\vec{k}$
3. $\vec{p} = 3\vec{i} + 8\vec{j} + 4\vec{k}$
4. $\vec{p} = 10\vec{i} + 7\vec{j} + 3\vec{k}$
5. Let $\vec{a} = 4\vec{i} + 2\vec{j} - 3\vec{k}$ and $\vec{b} = 7\vec{i} - 5\vec{j} + \vec{k}$.
   a. Find $\vec{a} + \vec{b}$, $\vec{a} - \vec{b}$, and $\vec{b} - \vec{a}$.
   b. Find $3\vec{a}$ and $6\vec{a} - 5\vec{b}$.
   c. Find $|\vec{a} + \vec{b}|$ and $|\vec{a}| + |\vec{b}|$. Does $|\vec{a} + \vec{b}| = |\vec{a}| + |\vec{b}|$?
   d. Find a unit vector in the direction of $\vec{b}$. Find a vector 20 units long in the direction of $\vec{b}$.
6. Let $\vec{c} = -4\vec{i} + 6\vec{j} + 3\vec{k}$ and $\vec{d} = 9\vec{i} + 8\vec{j} - 2\vec{k}$.
   a. Find $\vec{c} + \vec{d}$, $\vec{c} - \vec{d}$, and $\vec{d} - \vec{c}$.
   b. Find $-(\vec{c} + \vec{d})$ and $3\vec{c} - 4\vec{d}$.
   c. Find $|\vec{c}| + |\vec{d}|$ and $|\vec{c} + \vec{d}|$. Does $|\vec{c} + \vec{d}| = |\vec{c}| + |\vec{d}|$?
   d. Find a unit vector in the direction of $\vec{d}$. Find a vector of length 10 in the direction of the opposite of $\vec{d}$.

For Problems 7–10, find the indicated displacement vector. Use the answer to find the distance between the two points.

7. $\overrightarrow{RS}$ for $R(5, 6, 12)$ and $S(8, 13, 6)$
8. $\overrightarrow{PQ}$ for $P(6, 8, 14)$ and $Q(10, 16, 9)$
9. $\overrightarrow{BA}$ for $A(9, 13, -4)$ and $B(3, 6, -10)$
10. $\overrightarrow{DC}$ for $C(2, 9, 0)$ and $D(1, 4, 8)$
11. *Tree House Problem:* Elmer is going to build a tree house in his backyard for the children to play in. The yard is level. He uses one corner of the yard as the origin of a three-dimensional coordinate system. The x- and y-axes run along the ground, and the z-axis is vertical. He finds that the tree house will be at the point $(x, y, z) = (30, 55, 17)$, where the dimensions are in feet. Answer parts a–f.

   a. Sketch the coordinate axes and the point (30, 55, 17).
   b. Write the position vector $\vec{h}$ to the tree house. How high is the tree house above the ground? How far is the tree house from the origin?

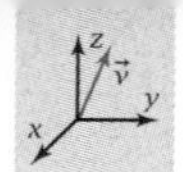

c. A wire is to be stretched from the tree house to the point (10, 0, 8) at the top corner of the back door so that the children can slide messages down it. Write a vector representing the displacement from the tree house to the point on the back door.

d. How long will the wire in part c need to be?

e. The children slide a message down the wire. It gets stuck when it is only 30% of the way from the tree house to the back door. Write a vector representing the displacement from the tree house to the stuck message. How far along the wire did the message go before it got stuck?

f. Write the position vector of the stuck message. How high above the ground is the stuck message?

12. *Space Station Problem 2:* Two communications satellites are in geosynchronous orbit around Earth. (Geosynchronous satellites orbit Earth with a period of 1 day, so they don't appear to move with respect to an observer on the ground.) From a point on the ground, the position vectors to the two satellites are

Satellite 1: $\vec{p}_1 = 18\vec{i} + 5\vec{j} + 12\vec{k}$
Satellite 2: $\vec{p}_2 = 15\vec{i} + 9\vec{j} + 14\vec{k}$

The distances are in thousands of miles. These vectors are shown schematically (not to scale) in Figure 10-3f. A space station is to be located at point P on the line between the two satellites, 40% of the way from Satellite 1 to Satellite 2.

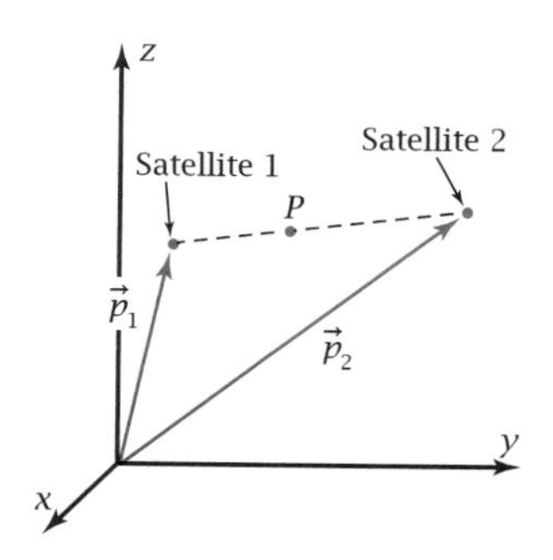

Figure 10-3f

a. Find the displacement vector from Satellite 1 to Satellite 2.

b. Find the position vector of point P, where the space station will be located.

The International Space Station in September, 2000

c. How far will the space station be from Satellite 1? How far will the space station be from the point on the ground?

For Problems 13–16, find the position vector of the indicated point.

13. $\frac{2}{3}$ of the way from (7, 8, 11) to (34, 32, 14)

14. $\frac{1}{3}$ of the way from (5, 1, 23) to (26, 13, 14)

15. 130% of the way from (2, 9, 7) to (4, −3, 1)

16. 270% of the way from (3, 8, 5) to (7, 1, −10)

17. *Perspective Problem:* Prove that if the x-axis is drawn obliquely on a piece of graph paper, as shown in Figure 10-3b, and two units are marked off for each grid line crossed, then the distances along the x-axis are about 70% of the distances along the y- and z-axes.

18. *Three-Dimensional Distances Problem:* Prove the **three-dimensional Pythagorean theorem.** That is, prove that

$$\left|x\vec{i} + y\vec{j} + z\vec{k}\right| = \sqrt{x^2 + y^2 + z^2}$$

This may be done by associating $(x\vec{i} + y\vec{j}) + z\vec{k}$ and then applying the ordinary two-dimensional Pythagorean theorem twice. A sketch may help.

19. *Four-Dimensional Vector Problem:* In Einstein's theory of time and space, time is a fourth dimension. Although it is impossible to draw a vector with more than three dimensions, the techniques you have learned make it possible

to analyze them algebraically. The definitions and techniques for adding, subtracting, and finding lengths can be extended for higher-dimensional vectors. It is convenient to drop the $\vec{i}$, $\vec{j}$, and $\vec{k}$ and to use ordered quadruples, ordered quintuples, and so on to represent the vectors. Let

$$\vec{a} = (3, 5, 2, 7)$$

and

$$\vec{b} = (5, 11, 7, 1)$$

a. Find $|\vec{a}|$ and $|\vec{b}|$.
b. Find $\vec{a} + \vec{b}$.
c. Find $\vec{a} - \vec{b}$.
d. If $\vec{a}$ and $\vec{b}$ are considered to be position vectors, write the displacement vector from the head of $\vec{a}$ to the head of $\vec{b}$.
e. Write the position vector of the point 40% of the way from the head of $\vec{a}$ to the head of $\vec{b}$ in part d.

Albert Einstein (1879–1955) published the theory of special relativity in 1905 and the theory of general relativity in 1916. He received the Nobel Prize in Physics in 1921. Hermann Minkowski (1864–1909) laid the mathematical foundations for Einstein's theory of relativity.

10-4 Scalar Products and Projections of Vectors

You have learned how to add and subtract two vectors and how to multiply a vector by a scalar. In this section you will learn about the **dot product** of two vectors, one way of multiplying vectors. The dot product is also called the **scalar product** or **inner product,** for reasons you will see in this section.

OBJECTIVE Given two vectors, find their dot product. Use the result to find the angle between the vectors and the projection of one vector on the other.

Dot Products (Scalar or Inner Products)

The symbol for the dot product of $\vec{a}$ and $\vec{b}$ is $\vec{a} \cdot \vec{b}$. It is read "$\vec{a}$ dot $\vec{b}$." If you translate the two vectors so that they are tail-to-tail, as in Figure 10-4a, you find the dot product by multiplying the magnitudes of the vectors and the cosine of the angle between them.

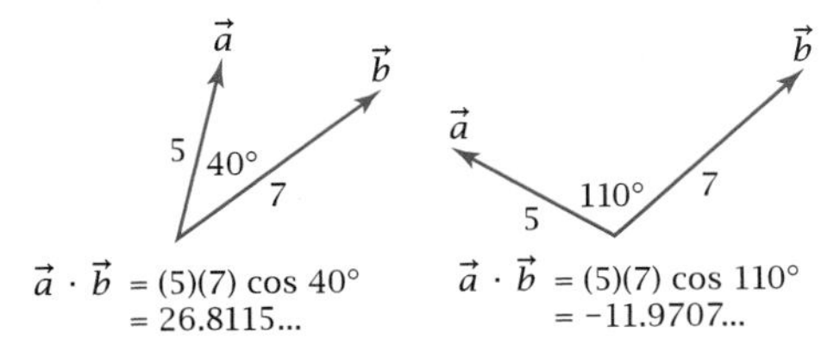

Figure 10-4a

Note that the answer is not a vector. It is a scalar. This is why the dot product is called the scalar product. Note also that if the angle between the vectors is acute, the dot product is positive. If the angle is obtuse, the dot product is negative. Three special cases are shown in Figure 10-4b. If the vectors point in the same direction, the angle between them is zero. Because $\cos 0° = 1$, the dot product is the product of the two magnitudes. Similarly, if the vectors point in opposite directions, the angle is 180°. The dot product is the opposite of the product of the magnitudes because $\cos 180° = -1$. Finally, if the vectors are perpendicular, the dot product is zero because $\cos 90° = 0$.

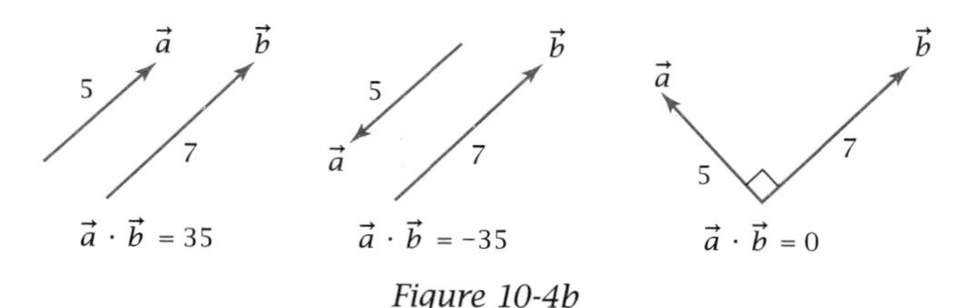

Figure 10-4b

Here is the formal definition of dot product. Memorize it, because it will arise in some unexpected places.

DEFINITION: Dot Product (or Scalar Product or Inner Product)

$$\vec{a} \cdot \vec{b} = |\vec{a}||\vec{b}| \cos \theta$$

where θ is the angle between the two vectors when they are translated tail-to-tail (Figure 10-4c).

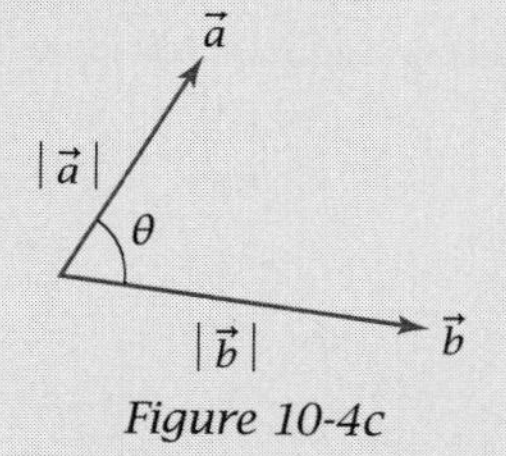

Figure 10-4c

Unfortunately, the definition is not very useful for finding dot products of three-dimensional vectors given by components because you don't know the angle between the vectors. So you seek another way to do the calculation.

Let $\vec{a} = 2\vec{i} + 5\vec{j} + 7\vec{k}$

Let $\vec{b} = 9\vec{i} + 3\vec{j} + 4\vec{k}$

$\vec{a} \cdot \vec{b} = (2\vec{i} + 5\vec{j} + 7\vec{k}) \cdot (9\vec{i} + 3\vec{j} + 4\vec{k})$ Substitute for the two vectors.

$= 18\vec{i} \cdot \vec{i} + 6\vec{i} \cdot \vec{j} + 8\vec{i} \cdot \vec{k}$ Distribute each term in the first vector to each term in the second.

$+ 45\vec{j} \cdot \vec{i} + 15\vec{j} \cdot \vec{j} + 20\vec{j} \cdot \vec{k}$

$+ 63\vec{k} \cdot \vec{i} + 21\vec{k} \cdot \vec{j} + 28\vec{k} \cdot \vec{k}$

Note that $\vec{i}$ is a unit vector, so $\vec{i} \cdot \vec{i} = (1)(1)\cos 0° = 1$. The same is true for $\vec{j} \cdot \vec{j}$ and $\vec{k} \cdot \vec{k}$. However, $\vec{i} \cdot \vec{j} = (1)(1)\cos 90° = 0$. Each of the preceding dot products with perpendicular unit vectors is equal to zero. Therefore,

$$\begin{aligned} \vec{a} \cdot \vec{b} &= 18 + 0 + 0 \\ &\quad + 0 + 15 + 0 \\ &\quad + 0 + 0 + 28 \\ &= 61 \end{aligned}$$

The answer is 61, a scalar.

This calculation reveals a reason for calling the dot product the **inner product.** The numbers that contribute to the dot product are "inside" in the array shown.

The calculation also reveals a quick way to find a dot product from its components.

$$\vec{a} \cdot \vec{b} = (2)(9) + (5)(3) + (7)(4) = 61$$

You multiply the *x*-coefficients, the *y*-coefficients, and the *z*-coefficients and then add.

TECHNIQUE: Computation of Dot Product

If

$$\vec{a} = x_1\vec{i} + y_1\vec{j} + z_1\vec{k}$$

and $\vec{b} = x_2\vec{i} + y_2\vec{j} + z_2\vec{k}$

then

$$\vec{a} \cdot \vec{b} = x_1x_2 + y_1y_2 + z_1z_2$$

Verbally: The dot product of two three-dimensional vectors equals the sum of the respective products of the coefficients for the $\vec{i}$, $\vec{j}$, and $\vec{k}$ unit vectors.

► EXAMPLE 1 Find the dot product $\vec{c} \cdot \vec{d}$ if

$$\vec{c} = 4\vec{i} - 6\vec{j} + 9\vec{k}$$

$$\vec{d} = 2\vec{i} + 5\vec{j} - 3\vec{k}$$

Solution

$$\vec{c} \cdot \vec{d} = (4)(2) + (-6)(5) + (9)(-3) = -49$$

◄

To calculate the dot product, multiply x_1 by x_2, y_1 by y_2, z_1 by z_2, then add.

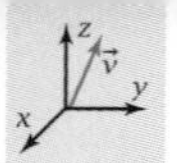

EXAMPLE 2 Use the dot product in Example 1 to find the angle θ between vectors $\vec{c}$ and $\vec{d}$.

Solution

$\vec{c} \cdot \vec{d} = -49$ — From Example 1.

$|\vec{c}||\vec{d}| \cos \theta = -49$ — Use the definition of dot product.

$|\vec{c}| = \sqrt{4^2 + (-6)^2 + 9^2} = \sqrt{133}$ — Find the lengths of $\vec{c}$ and $\vec{d}$.

$|\vec{d}| = \sqrt{2^2 + 5^2 + (-3)^2} = \sqrt{38}$

$\sqrt{133}\sqrt{38} \cos \theta = -49$ — Substitute for the magnitudes of the vectors.

$$\cos \theta = \frac{-49}{\sqrt{133}\sqrt{38}} = -0.6892\ldots$$

$\theta = 133.5709\ldots°$ ◀

Projections of Vectors

Figure 10-4d shows vectors $\vec{a}$ and $\vec{b}$ placed tail-to-tail. Suppose that an object is moving in the direction of $\vec{b}$ and that $\vec{a}$ is a force acting on the object. The component of $\vec{a}$ in the direction of $\vec{b}$ influences such things as the change in the speed of the object and the amount of work done by the force on the moving object. This component is called the **vector projection of $\vec{a}$ on $\vec{b}$.** Light rays shining perpendicular to $\vec{b}$ in the same plane as the two vectors would "project" a shadow on $\vec{b}$ corresponding to this component of $\vec{a}$.

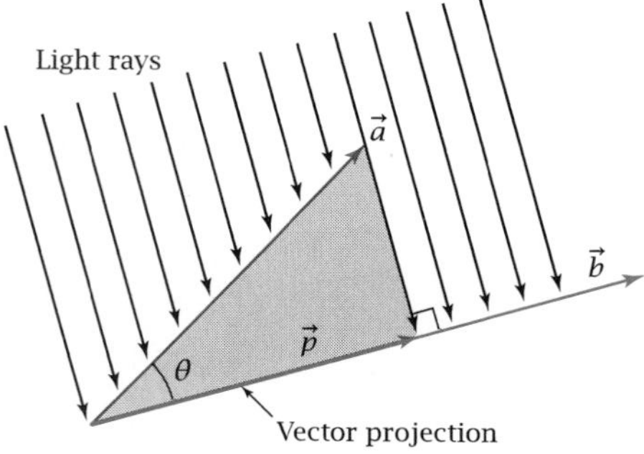

Figure 10-4d

Let $\vec{p}$ be the vector projection of $\vec{a}$ on $\vec{b}$. From trigonometry, you know that the length of $\vec{p}$ is the length of $\vec{a}$ times the cosine of θ, where θ is the angle between the vectors. You can calculate $\vec{p}$ shown in Figure 10-4d by multiplying a unit vector in the direction of $\vec{b}$ by the length of $\vec{p}$. That is,

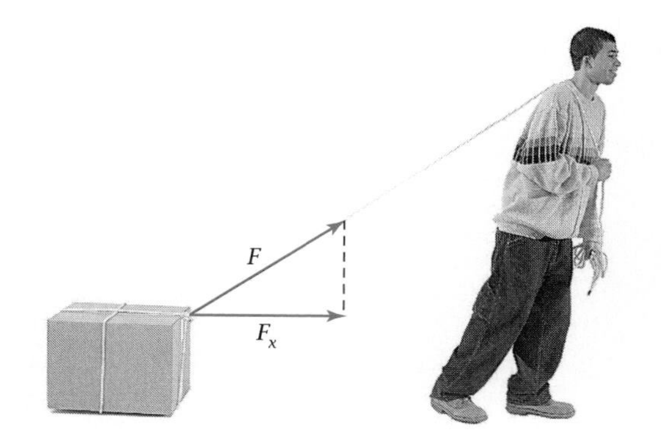

$$\vec{p} = \left(|\vec{a}| \cos \theta\right) \frac{\vec{b}}{|\vec{b}|}$$

▶ ***EXAMPLE 3*** If $\vec{b} = 8\vec{i} - 5\vec{j} + 3\vec{k}$ and $\vec{a}$ is 10 units long at an angle of $\theta = 70°$ to $\vec{b}$, find $\vec{p}$, the vector projection of $\vec{a}$ on $\vec{b}$.

Solution

$|\vec{b}| = \sqrt{8^2 + (-5)^2 + 3^2} = \sqrt{98}$ Find the length of $\vec{b}$.

Unit vector $\vec{u} = \dfrac{\vec{b}}{|\vec{b}|} = \dfrac{8}{\sqrt{98}}\vec{i} - \dfrac{5}{\sqrt{98}}\vec{j} + \dfrac{3}{\sqrt{98}}\vec{k}$ Find a unit vector in the direction of $\vec{b}$.

Length $= |\vec{a}| \cos\theta = 10 \cos 70° = 3.4202\ldots$

$\vec{p} = 3.4202\ldots\left(\dfrac{8}{\sqrt{98}}\vec{i} - \dfrac{5}{\sqrt{98}}\vec{j} + \dfrac{3}{\sqrt{98}}\vec{k}\right)$

$\vec{p} = 2.76\ldots\vec{i} - 1.72\ldots\vec{j} + 1.03\ldots\vec{k}$ ◀

The quantity $|\vec{a}| \cos\theta$ in Example 3 is called the **scalar projection of $\vec{a}$ on $\vec{b}$.** The letter p (without the vector symbol) will be used for the scalar projection.

$p = |\vec{a}| \cos\theta$ Scalar projection of $\vec{a}$ on $\vec{b}$.

If θ is an acute angle, the scalar projection equals the magnitude of $\vec{p}$. If θ is obtuse, the scalar projection is negative and is thus the opposite of the magnitude of $\vec{p}$. Figure 10-4e shows the two cases. If θ is obtuse, the vector projection points in the direction opposite $\vec{b}$.

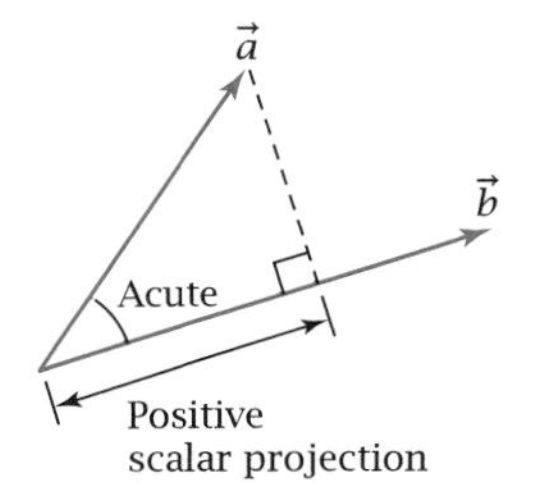

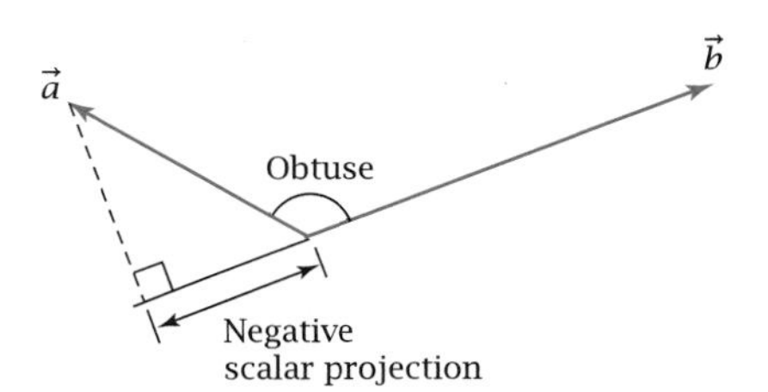

Figure 10-4e

DEFINITIONS: Projections of Vectors

If θ is the angle between $\vec{a}$ and $\vec{b}$ when they are placed tail-to-tail, then the **scalar projection of $\vec{a}$ on $\vec{b}$** is

$p = |\vec{a}| \cos\theta$

If $\vec{u}$ is a unit vector in the direction of $\vec{b}$, then the **vector projection of $\vec{a}$ on $\vec{b}$** is

$\vec{p} = p\vec{u}$

▶ ***EXAMPLE 4*** If $\vec{a} = -4\vec{i} + 5\vec{j} + 9\vec{k}$ and $\vec{b} = 6\vec{i} - 8\vec{j} + \vec{k}$, find

a. The scalar projection of $\vec{a}$ on $\vec{b}$

b. The vector projection of $\vec{a}$ on $\vec{b}$

Solution

a. $\vec{a} \cdot \vec{b} = (-4)(6) + (5)(-8) + (9)(1) = -55$

$|\vec{a}| = \sqrt{(-4)^2 + 5^2 + 9^2} = \sqrt{122}$

$|\vec{b}| = \sqrt{6^2 + (-8)^2 + 1^2} = \sqrt{101}$ Find the dot product and the two lengths.

$\sqrt{122}\sqrt{101}\cos\theta = -55$ Use the definition of dot product.

$$\cos\theta = \frac{-55}{\sqrt{122}\sqrt{101}} = -0.4954...$$

$$\theta = 119.7011...^\circ$$

$\therefore p = \sqrt{122}\cos 119.7011...^\circ = -5.4727...$ Use the definition of scalar projection.

b. $\vec{u} = \dfrac{\vec{b}}{|\vec{b}|} = \dfrac{6\vec{i} - 8\vec{j} + k}{\sqrt{101}}$ Find a unit vector in the direction of $\vec{b}$.

$\therefore \vec{p} = p\vec{u} = -5.4727...\vec{u} = -3.26...\vec{i} + 4.35...\vec{j} - 0.54...\vec{k}$ ◀

Problem Set 10-4

Do These Quickly

Q1. Give two names for the symbol $|\vec{a}|$.

Q2. What is the y-component of $\vec{v} = 3\vec{i} - 5\vec{j} + 2\vec{k}$?

Q3. Find the length of $\vec{v}$ in Problem Q2.

Q4. True or false? "$1\vec{i} + 1\vec{j} + 1\vec{k}$ is a unit vector."

Q5. Find the position vector to the point (5, 8, 6).

Q6. Find the displacement vector from (5, 8, 6) to (11, 3, 7).

Q7. Find the coefficient of determination if $SS_{\text{res}} = 5$ and $SS_{\text{dev}} = 100$.

Q8. Find the number of permutations of five objects taken three at a time.

Q9. Which functions have constant second differences in y-values for equally spaced x-values?

Q10. Without using your calculator, what does $\cos\pi$ equal?

For Problems 1–6, use the definition of dot product to find $\vec{a} \cdot \vec{b}$, where θ is the angle between $\vec{a}$ and $\vec{b}$ when they are placed tail-to-tail.

1. $|\vec{a}| = 30$, $|\vec{b}| = 25$, and $\theta = 37^\circ$
2. $|\vec{a}| = 17$, $|\vec{b}| = 8$, and $\theta = 23^\circ$
3. $|\vec{a}| = 29$, $|\vec{b}| = 50$, and $\theta = 127^\circ$
4. $|\vec{a}| = 40$, $|\vec{b}| = 53$, and $\theta = 126^\circ$
5. $|\vec{a}| = 51$, $|\vec{b}| = 27$, and $\theta = 90^\circ$
6. $|\vec{a}| = 43$, $|\vec{b}| = 29$, and $\theta = 180^\circ$

For Problems 7–12, use the definition of dot product to find the angle between $\vec{a}$ and $\vec{b}$ if the two vectors are placed tail-to-tail.

7. $|\vec{a}| = 20$, $|\vec{b}| = 30$, and $\vec{a} \cdot \vec{b} = 100$
8. $|\vec{a}| = 8$, $|\vec{b}| = 9$, and $\vec{a} \cdot \vec{b} = 24$
9. $|\vec{a}| = 11$, $|\vec{b}| = 17$, and $\vec{a} \cdot \vec{b} = -123$
10. $|\vec{a}| = 300$, $|\vec{b}| = 500$, and $\vec{a} \cdot \vec{b} = -100{,}000$
11. $|\vec{a}| = 60$, $|\vec{b}| = 80$, and $\vec{a} \cdot \vec{b} = 4800$
12. $|\vec{a}| = 29$, $|\vec{b}| = 31$, and $\vec{a} \cdot \vec{b} = 0$

For Problems 13–18, find $\vec{a} \cdot \vec{b}$ and the angle between $\vec{a}$ and $\vec{b}$ when they are tail-to-tail.

13. $\vec{a} = 2\vec{i} + 5\vec{j} + 3\vec{k}$
    $\vec{b} = 7\vec{i} - \vec{j} + 4\vec{k}$
14. $\vec{a} = 3\vec{i} + 2\vec{j} - 4\vec{k}$
    $\vec{b} = 8\vec{i} + 5\vec{j} - 2\vec{k}$

15. $\vec{a} = -3\vec{i} + 5\vec{j} + 2\vec{k}$
    $\vec{b} = 6\vec{i} - 3\vec{j} + \vec{k}$

16. $\vec{a} = 4\vec{i} - 3\vec{j} - 7\vec{k}$
    $\vec{b} = \vec{i} + 5\vec{j} + 3\vec{k}$

17. $\vec{a} = 8\vec{i} + 9\vec{j} - 2\vec{k}$
    $\vec{b} = 3\vec{i} - 4\vec{j} - 6\vec{k}$

18. $\vec{a} = \vec{i} + 3\vec{j} - 5\vec{k}$
    $\vec{b} = -7\vec{i} + 4\vec{j} + \vec{k}$

19. *Sailboat Force Problem:* Two ropes from the sail of a sailboat are both attached to the same cleat on the deck. The force vectors created by the ropes are

$$\vec{F}_1 = 15\vec{i} + 70\vec{j} + 10\vec{k}$$
$$\vec{F}_2 = 30\vec{i} + 50\vec{j} + 5\vec{k}$$

where the forces are in pounds. The vectors are shown (not to scale) in Figure 10-4f.

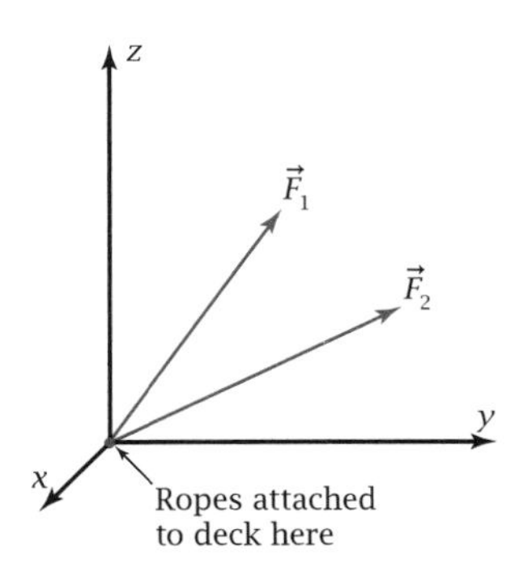

Figure 10-4f

a. Find the resultant force vector.

b. The y-axis runs along the length of the sailboat. The force in the y-direction is what makes the sailboat move forward. How many pounds does the resultant force exert in the y-direction?

c. The x-axis runs across the sailboat. The force in the x-direction makes the ship heel over. How many pounds does the resultant force exert in the x-direction?

d. The z-axis goes perpendicular to the deck. The force in the z-direction tends to pull the cleat out of the deck! How many pounds does the resultant exert in the z-direction?

e. How many pounds, total, does the resultant force exert on the cleat?

f. What is the magnitude of each of the two forces? Do these magnitudes add up to the magnitude of the resultant force in part e?

g. Find the dot product of $\vec{F}_1$ and $\vec{F}_2$. Use the answer to find the angle the two forces make with each other.

20. *Hip Roof Problem:* A house is to be built with a hip roof. The triangular end of the roof is shown in Figure 10-4g. An xyz-coordinate system is set up with its origin at a bottom corner of the roof at the back of the house. The position vector $\vec{h}$ to the front bottom corner, where the angle is marked, and the position vector $\vec{v}$ to the peak of the roof are

$$\vec{h} = 20\vec{i} + 45\vec{j} + 0\vec{k}$$
$$\vec{v} = 10\vec{i} + 35\vec{j} + 8\vec{k}$$

The dimensions are in feet.

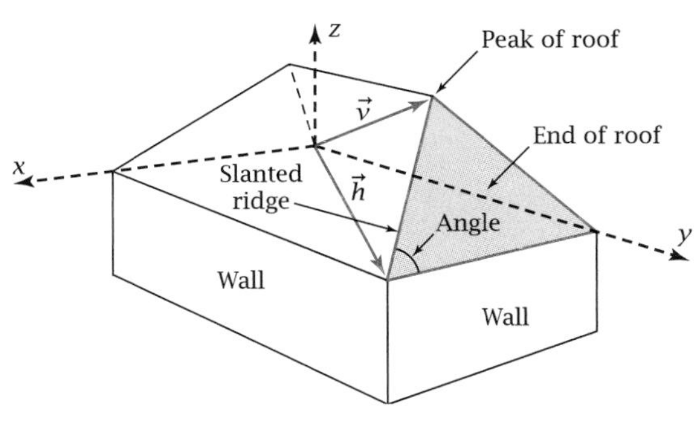

Figure 10-4g

a. The floor of the attic is along the xy-plane. How far will the peak of the roof be above this floor? How can you tell?

b. How long and how wide will the house be? How can you tell?

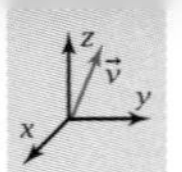

c. Builders need to know how long to make the rafter that goes from the corner with the marked angle to the peak of the roof along the slanted ridge. How long will the rafter be?

d. The roof tiles that come next to the slanted ridge will have to be cut at an angle, as shown in the figure. What angle will this be?

e. Is the triangular end of the roof an isosceles triangle? How can you tell?

21. In Figure 10-4h, vector $\vec{v}$, 10 units long, makes an angle of 28° with $\vec{a} = 7\vec{i} + 3\vec{j} + 4\vec{k}$. Find the vector projection of $\vec{v}$ on $\vec{a}$.

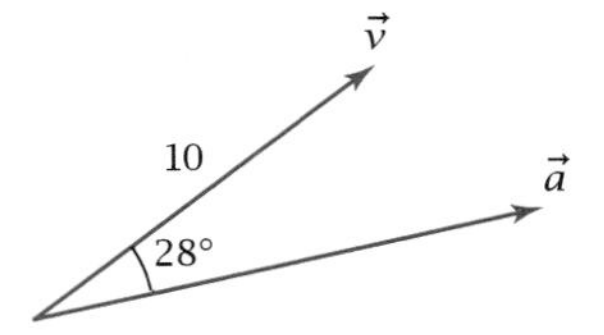

Figure 10-4h

22. In Figure 10-4i, vector $\vec{v}$, 100 units long, makes an angle of 145° with $\vec{b} = 50\vec{i} - 60\vec{j} + 40\vec{k}$. Find the vector projection of $\vec{v}$ on $\vec{b}$.

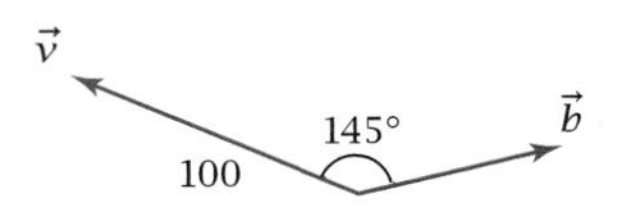

Figure 10-4i

23. *Shortcuts for Projections Problem:* Show that these formulas give the scalar and vector projections of $\vec{a}$ on $\vec{b}$.

Techniques: Formulas for Scalar and Vector Projection

The scalar projection of $\vec{a}$ on $\vec{b}$ is given by

$$p = \frac{\vec{a} \cdot \vec{b}}{|\vec{b}|}$$

The vector projection of $\vec{a}$ on $\vec{b}$ is given by

$$\vec{p} = \frac{\vec{a} \cdot \vec{b}}{|\vec{b}|^2}\vec{b}$$

24. *Vocabulary Problem:* Give three names commonly used for $\vec{a} \cdot \vec{b}$.

For Problems 25–28, find

a. The scalar projection of $\vec{r}$ on $\vec{s}$

b. The vector projection of $\vec{r}$ on $\vec{s}$

25. $\vec{r} = 3\vec{i} + 2\vec{j} + 5\vec{k}$
$\vec{s} = 7\vec{i} - \vec{j} - 3\vec{k}$

26. $\vec{r} = \vec{i} + 4\vec{j} - 7\vec{k}$
$\vec{s} = 5\vec{i} - 2\vec{j} - 3\vec{k}$

27. $\vec{r} = 4\vec{i} - 3\vec{j} + 3\vec{k}$
$\vec{s} = -2\vec{i} + 5\vec{j} + \vec{k}$

28. $\vec{r} = 6\vec{i} - \vec{j} - 7\vec{k}$
$s = \vec{i} - 5\vec{j} + 3\vec{k}$

29. *Cube Problem:* Figure 10-4j shows a cube with one corner at the origin of a three-dimensional coordinate system.

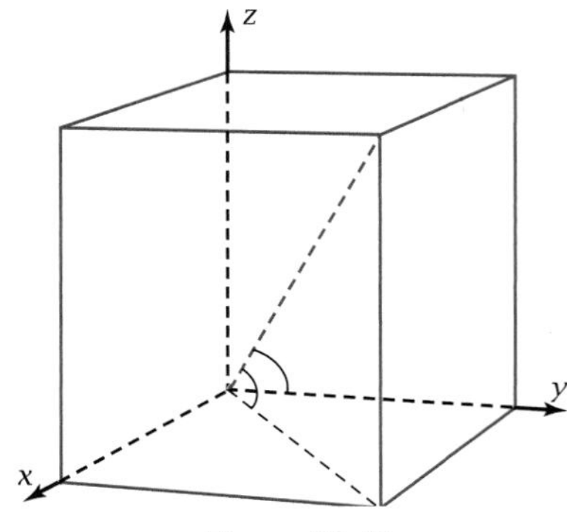

Figure 10-4j

a. Find the angle the main diagonal of the cube makes with one of the edges of the cube.

b. Find the angle the main diagonal makes with a diagonal in a face of the cube.

30. *Journal Problem*: Update your journal with what you have learned since the last entry. Include the definition, computational technique, and uses of the dot product, along with its two other names.

10-5 Planes in Space

Figure 10-5a shows the plane surface of a tilted underground rock formation. By finding an equation relating the x-, y-, and z-coordinates of points on this plane you can calculate how far you would have to drill down to reach the plane. From algebra you may remember that the equation of a plane in space has the form

$$5x + 7y - 4z = 19 \qquad \text{or in general} \qquad Ax + By + Cz = D$$

where x, y, and z are coordinates of a point on the plane and A, B, C, and D stand for constants. In this section you will see how to derive the equation of a plane from a vector normal (perpendicular) to the plane.

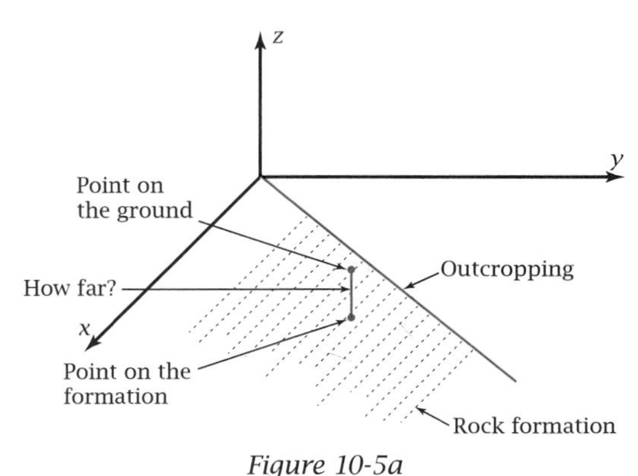

Figure 10-5a

OBJECTIVE Given a point on a plane and a vector perpendicular to the plane, find the particular equation of the plane and use it to find other points on the plane.

Equation of a Plane

Figure 10-5b shows a plane in space with a vector $\vec{n}$ normal to it. Suppose that

$$\vec{n} = 11\vec{i} + 2\vec{j} + 13\vec{k}$$

and that point P_0 on the plane is (3, 5, 7). Point $P(x, y, z)$ is a variable point in the plane. You need to find an equation of the plane relating x, y, and z, where the vector $\overrightarrow{P_0P}$ is perpendicular to $\vec{n}$ for all points P.

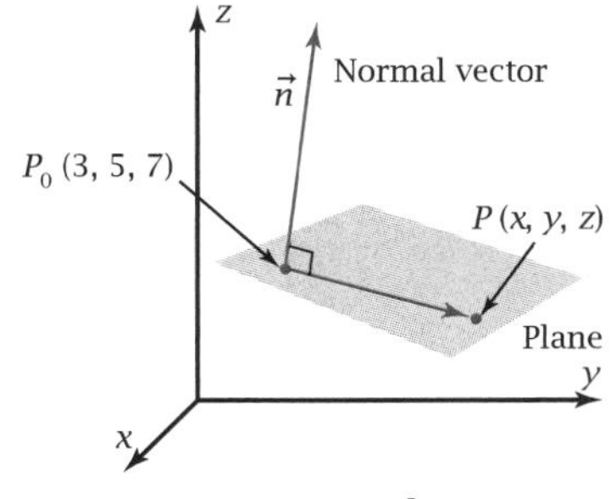

Figure 10-5b

The displacement vector $\overrightarrow{P_0P}$ goes from P_0 to P. By subtracting coordinates, this vector is

$$\overrightarrow{P_0P} = (x - 3)\vec{i} + (y - 5)\vec{j} + (z - 7)\vec{k}$$

Because $\overrightarrow{P_0P}$ is on the plane and $\vec{n}$ is perpendicular to the plane, the dot product of these two vectors equals zero. Some calculations lead to the equation of the plane.

$$\vec{n} \cdot \overrightarrow{P_0P} = 0$$

$$(11\vec{i} + 2\vec{j} + 13\vec{k}) \cdot ((x-3)\vec{i} + (y-5)\vec{j} + (z-7)\vec{k}) = 0$$ Substitute for the two vectors.

$$11(x-3) + 2(y-5) + 13(z-7) = 0$$ Evaluate the dot product.

$$11x + 2y + 13z - 134 = 0$$

$$11x + 2y + 13z = 134$$ This is the equation of the plane in space.

By comparing the answer with the given information, you see that the coefficients in the equation of the plane are the same as the coefficients of the normal vector. This is true in general, and you can use this property to find the equation quickly, as shown in Example 1. (In Problems 15 and 16 of Problem Set 10-5 you will prove the property.)

▶ EXAMPLE 1 Find the equation of the plane containing (3, 5, 7) with normal vector $\vec{n} = 11\vec{i} + 2\vec{j} + 13\vec{k}$.

Solution

$$11x + 2y + 13z = D$$ Substitute the coefficients of the components of $\vec{n}$ into $Ax + By + Cz = D$.

$$11(3) + 2(5) + 13(7) = D$$ Substitute the given point for (x, y, z).

$$134 = D$$

$\therefore$ The equation is $11x + 2y + 13z = 134$. ◀

▶ EXAMPLE 2 Find a vector $\vec{n}$ normal to the plane $7x - 3y + 8z = -51$.

Solution

$$\vec{n} = 7\vec{i} - 3\vec{j} + 8\vec{k}$$ ◀

Note that any multiple of $\vec{n}$ is also an answer to Example 2. In particular, the opposite of $\vec{n}$, $-7\vec{i} + 3\vec{j} - 8\vec{k}$, is also a normal vector to the plane $7x - 3y + 8z = -51$.

▶ EXAMPLE 3 Find an equation of the plane perpendicular to the segment connecting $P_1(3, 8, -2)$ and $P_2(7, -1, 6)$ and passing through the point 30% of the way from P_1 to P_2. Figure 10-5c illustrates the problem in general.

Solution The displacement vector $\overrightarrow{P_1P_2}$ is normal to the plane, so you can write:

$$\vec{n} = (7-3)\vec{i} + (-1-8)\vec{j} + (6-(-2))\vec{k}$$

$$\vec{n} = 4\vec{i} - 9\vec{j} + 8\vec{k}$$

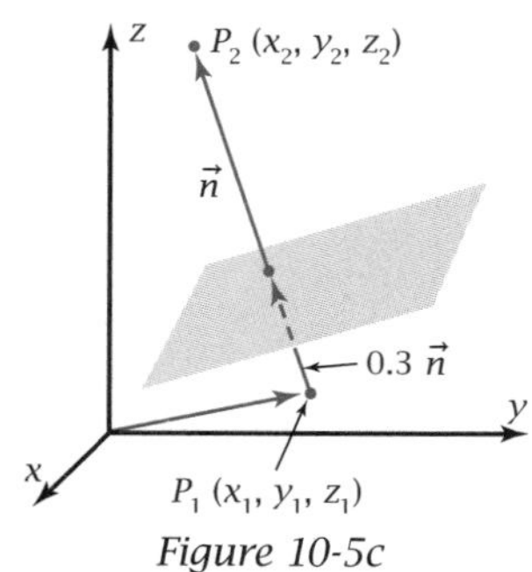

Figure 10-5c

The position vector $\vec{p}$ to the point on the plane equals the position vector to P_1 plus 0.3 times the normal vector $\vec{n}$.

$$\vec{p} = (3\vec{i} + 8\vec{j} - 2\vec{k}) + 0.3(4\vec{i} - 9\vec{j} + 8\vec{k})$$

$$= 4.2\vec{i} + 5.3\vec{j} + 0.4\vec{k}$$

Thus, a point on the plane is (4.2, 5.3, 0.4).

$$4x - 9y + 8z = D$$ The coefficients in the plane's equation are the coefficients of the components of the normal vector.

$$4(4.2) - 9(5.3) + 8(0.4) = D$$

$$-27.7 = D$$

$\therefore$ The equation is $4x - 9y + 8z = -27.7$. ◀

This box summarizes the technique for finding the equation of a plane in space.

TECHNIQUE: Equation of a Plane in Space

1. Use the given information to find a normal vector and a point on the plane.
2. Substitute the coefficients of the components of the normal vector for A, B, and C in the general equation

$$Ax + By + Cz = D$$

3. Substitute the coordinates of the given point for (x, y, z) to calculate the value of D.
4. Write the equation.

▶ EXAMPLE 4 If the equation of a plane is $-7x + 8y + 4z = 200$, find the z-coordinate for point $P(3, 5, z)$ on the plane.

Solution

$$-7(3) + 8(5) + 4z = 200$$ Substitute 3 for x and 5 for y into the equation.

$$4z = 181$$

$$z = 45.25$$ ◀

Problem Set 10-5

Do These Quickly

Q1. Give a name for $\vec{a} \cdot \vec{b}$.

Q2. Give a second name for $\vec{a} \cdot \vec{b}$.

Q3. Give the third name for $\vec{a} \cdot \vec{b}$.

Q4. How can you find the scalar projection of $\vec{a}$ on $\vec{b}$?

Q5. If p is the scalar projection of $\vec{a}$ on $\vec{b}$ and $\vec{u}$ is a unit vector in the direction of $\vec{b}$, what does the vector projection of $\vec{a}$ on $\vec{b}$ equal?

Q6. How do you tell from the dot product whether or not two vectors are perpendicular?

Q7. If $\vec{a} = 3\vec{i} + 2\vec{j} + 1\vec{k}$ and $\vec{b} = 4\vec{i} - 3\vec{j} - 5\vec{k}$, find $\vec{a} \cdot \vec{b}$.

Q8. Find the supplementary angle of 84°.

Q9. Find the supplementary angle of an angle of 1 radian.

Q10. What does $\cos^2 A$ equal in terms of $\sin A$?

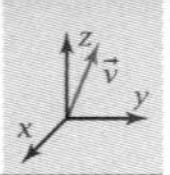

For Problems 1 and 2, find two normal vectors to the plane, pointing in opposite directions.

1. $3x + 5y - 7z = -13$
2. $4x - 7y + 2z = 9$

For Problems 3–8, find a particular equation of the plane described.

3. Perpendicular to $\vec{n} = 3\vec{i} - 5\vec{j} + 4\vec{k}$, containing the point (6, −7, −2)
4. Perpendicular to $\vec{n} = -\vec{i} + 3\vec{j} - 2\vec{k}$, containing the point (4, 7, 5)
5. Perpendicular to the line segment connecting (3, 8, 5) and (11, 2, −3) and passing through the midpoint of the segment
6. Parallel to the plane $3x - 7y + 2z = 11$ and containing the point (8, 11, −3)

Geometrico, Azules, Rojos, Negros Y Blancos, *geometric art by Mario Carreno*

7. Parallel to the plane $5x - 3y - z = -4$ and containing the point (4, −6, 1)
8. Perpendicular to $\vec{n} = 4\vec{i} + 3\vec{j} - 2\vec{k}$ and having an x-intercept of 5 (The **x-intercept of a plane** is the value of x when the other two variables are zero.)
9. A plane has the equation $3x - 7y + 5z = 54$. Points $P_1(6, 2, z_1)$ and $P_2(4, -3, z_2)$ are on the plane. Find the z-coordinates of the two points. How far apart are the points? What is the y-intercept of the plane (the value of y when the other two variables equal zero)?
10. A plane has the equation $4x + 2y - 10z = 300$. Points $P_1(x_1, 4, 5)$ and $P_2(7, y_2, 8)$ are on the plane. Find x_1 and y_2. Calculate the distance between the two points. What is the z-intercept of the plane (the value of z when the other two variables equal zero)?
11. *Geology Problem:* Figure 10-5d shows an underground rock formation that slants up and outcrops at ground level along a line in a field. The x- and y-axes run along perpendicular fence lines. A point on the outcropping is (200, 300, 0), where distances are in meters. A vector normal to the plane of the underground formation is $\vec{n} = 30\vec{i} - 17\vec{j} + 11\vec{k}$.

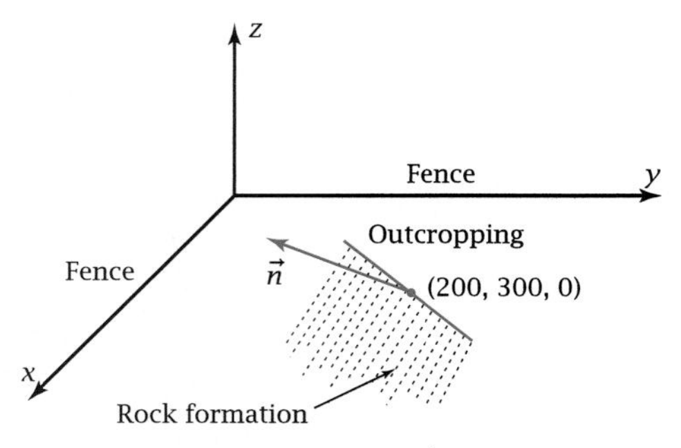

Figure 10-5d

   a. Find an equation of the plane surface of the underground rock formation.
   b. If you follow the outcropping line to the fences, which axis will it cross first, the x-axis or the y-axis? At what point will it meet the fence?
   c. If a well is drilled vertically starting at the point (50, 70, 0), how deep will it be when it first encounters the rock formation?
   d. The angle between the plane of the rock formation and the plane of the ground is called a **dihedral angle.** Geologists call this angle the *dip* of the formation. It is equal to the angle between the normal vectors to the two planes or to the supplement of this angle. Find the acute dip angle.
12. *Roof Valley Problem:* Figure 10-5e shows an L-shaped house that is to be built. Roof 1 and Roof 2 will have normal vectors

$$\vec{n}_1 = 0\vec{i} + 6\vec{j} + 12\vec{k}$$

$$\vec{n}_2 = 6\vec{i} + 0\vec{j} + 12\vec{k}$$

The two roofs will meet at a "valley." Point (30, 30, 10) is at the lower end of the valley. The dimensions are in feet.

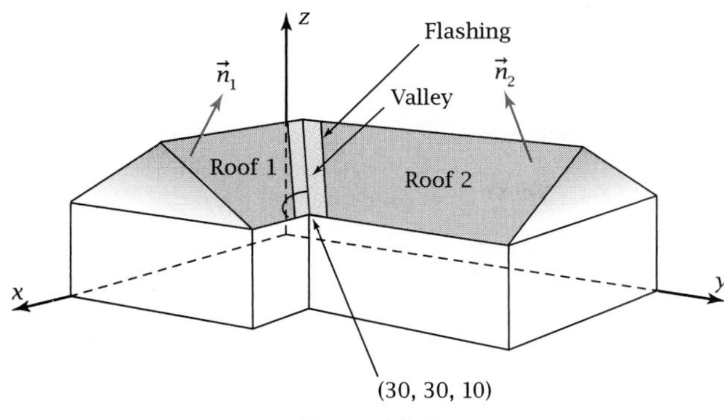

Figure 10-5e

a. Find particular equations of the two plane roofs.

b. The top end of the valley is at (15, 15, z). Use the equation of Roof 1 to calculate the value of z. Show that the point satisfies the equation of Roof 2, and give the real-world meaning of this fact.

c. How high will the ridge of the roof rise above the top of the walls?

d. Write the displacement vector from the bottom of the valley to the top.

e. Find the obtuse angle the valley makes with the bottom edge of Roof 1.

f. A piece of sheet metal flashing is to be fitted into the valley to go underneath the shingles. How long will it be?

g. The two roof sections form a **dihedral angle** equal to the angle between the two normal vectors or to the supplement of this angle. The flashing must be bent to fit this angle. Calculate the obtuse dihedral angle between the two roof sections.

h. Why do you think builders put flashing in roof valleys?

13. Prove that these two planes are perpendicular.

$$2x - 5y + 3z = 10$$
$$7x + 4y + 2z = 17$$

14. Find the value of A that makes these two planes perpendicular.

$$Ax + 3y - 2z = -8$$
$$4x - 5y + z = 7$$

15. *Plane's Equation Proof Problem:* Prove that if $\vec{n} = A\vec{i} + B\vec{j} + C\vec{k}$ is a normal vector to a plane, then a particular equation of the plane is $Ax + By + Cz = D$, where D stands for a constant.

16. *Normal Vector Proof Problem:* Prove the converse of the property in Problem 15. Specifically, prove that if $Ax + By + Cz = D$, where D stands for a constant, then a normal vector to the plane is $\vec{n} = A\vec{i} + B\vec{j} + C\vec{k}$.

10-6 Vector Product of Two Vectors

The dot product of two vectors is a scalar. In this section you will learn about the **cross product** of $\vec{a}$ and $\vec{b}$, written $\vec{a} \times \vec{b}$ and read "$\vec{a}$ cross $\vec{b}$." The main uses of cross products are in fields such as alternating electric current theory and accelerated rotary motion. You will see some geometrical uses of cross products in this section as well.

OBJECTIVE Be able to calculate cross products of two vectors and use cross products for geometrical computations.

Here is the formal definition of cross product.

DEFINITION: Cross Product (Vector Product, Outer Product)

The **cross product** of two vectors, $\vec{a} \times \vec{b}$, is a vector with these properties:

1. $\vec{a} \times \vec{b}$ is perpendicular to the plane containing $\vec{a}$ and $\vec{b}$.
2. The magnitude of $\vec{a} \times \vec{b}$ is

   $$|\vec{a} \times \vec{b}| = |\vec{a}||\vec{b}| \sin \theta$$

   where θ is the angle between the two vectors when they are placed tail-to-tail.
3. The direction of $\vec{a} \times \vec{b}$ is determined by the "right-hand rule." Put the fingers of your right hand so that they curl in the shortest direction *from* the *first* vector *to* the *second* vector. The cross product is in the same direction your thumb points (Figure 10-6a).

Note that the right-hand rule leads you to conclude that $\vec{b} \times \vec{a}$ is the *opposite* of $\vec{a} \times \vec{b}$. As shown in Figure 10-6b, curling your fingers from $\vec{b}$ to $\vec{a}$ makes your thumb point in the opposite direction. So cross multiplication of vectors isn't commutative. The cross product $\vec{b} \times \vec{a}$ equals $-\vec{a} \times \vec{b}$, not $\vec{a} \times \vec{b}$.

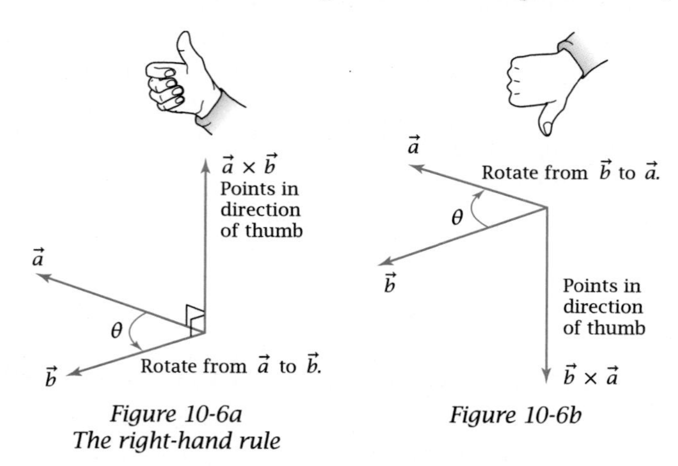

Figure 10-6a
The right-hand rule

Figure 10-6b

Computation of Cross Products from the Definition

The unit vectors $\vec{i}$, $\vec{j}$, and $\vec{k}$ have special properties when they are cross multiplied. Because the angle between a vector and itself is 0° and because $\sin 0° = 0$, the cross product of a vector by itself is zero. For instance, $\vec{i} \times \vec{i} = 0$. Also, $\vec{j} \times \vec{j} = 0$ and $\vec{k} \times \vec{k} = 0$.

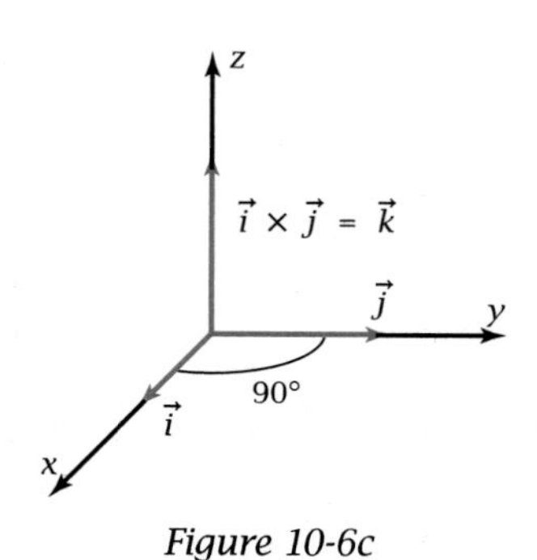

Figure 10-6c

Figure 10-6c shows that $\vec{i} \times \vec{j} = \vec{k}$. Because $\vec{i}$ and $\vec{j}$ are perpendicular,

$$|\vec{i} \times \vec{j}| = |\vec{i}||\vec{j}| \sin 90° = (1)(1)(1) = 1$$

By the right-hand rule, as $\vec{i}$ rotates toward $\vec{j}$, your thumb points in the z-direction. So the answer is $\vec{k}$, a unit vector in the z-direction. Similarly, $\vec{j} \times \vec{k} = \vec{i}$ and $\vec{k} \times \vec{i} = \vec{j}$. These special cases are summarized in the box.

PROPERTIES: Cross Products of the Unit Coordinate Vectors

$$\vec{i} \times \vec{i} = 0 \qquad \vec{i} \times \vec{j} = \vec{k} \qquad \vec{j} \times \vec{i} = -\vec{k}$$

$$\vec{j} \times \vec{j} = 0 \qquad \vec{j} \times \vec{k} = \vec{i} \qquad \vec{k} \times \vec{j} = -\vec{i}$$

$$\vec{k} \times \vec{k} = 0 \qquad \vec{k} \times \vec{i} = \vec{j} \qquad \vec{i} \times \vec{k} = -\vec{j}$$

Note that you can remember the cross products in the middle column because they involve $\vec{i}$ to $\vec{j}$ to $\vec{k}$ and $\vec{k}$ back to $\vec{i}$, in alphabetical order. This fortunate memory aid occurs because of the **right-handed coordinate system** that is being used. It is for this reason that the *x*-axis is shown going to the left and the *y*-axis going to the right, rather than the more intuitive but less useful way of drawing the *x*-axis to the right. Note also that if you reverse the order, as in the rightmost column in the box, the cross products are opposites of the unit vectors.

With these special cases in mind, you can use algebra to compute a cross product.

► ***EXAMPLE 1*** Find $\vec{a} \times \vec{b}$ if

$$\vec{a} = 3\vec{i} + 5\vec{j} + 7\vec{k} \quad \text{and} \quad \vec{b} = 11\vec{i} + 2\vec{j} + 13\vec{k}$$

Solution

$\vec{a} \times \vec{b} = (3\vec{i} + 5\vec{j} + 7\vec{k}) \times (11\vec{i} + 2\vec{j} + 13\vec{k})$ — Substitute for $\vec{a}$ and $\vec{b}$.

$= 33\vec{i} \times \vec{i} + 6\vec{i} \times \vec{j} + 39\vec{i} \times \vec{k}$
$+ 55\vec{j} \times \vec{i} + 10\vec{j} \times \vec{j} + 65\vec{j} \times \vec{k}$
$+ 77\vec{k} \times \vec{i} + 14\vec{k} \times \vec{j} + 91\vec{k} \times \vec{k}$ — Cross each term in the first vector with each term in the second vector.

$= 0 + 6\vec{k} - 39\vec{j}$
$- 55\vec{k} + 0 + 65\vec{i}$
$+ 77\vec{j} - 14\vec{i} + 0$ — Use the special cross products.

$= 51\vec{i} + 38\vec{j} - 49\vec{k}$ — Combine like terms. ◄

Notice that there are zeros down the main diagonal in the next-to-last step of Example 1. The only terms that contribute to the cross product are the “outer” terms in the array, leading to the name **outer product** for cross product. The name **vector product** is also used because the answer is a vector.

Computation of Cross Products by Means of Determinants

The method for computing cross products in Example 1 can seem tedious because you must make sure you have the correct unit vector when you calculate the cross product of unit vectors. Fortunately, there is a more easily remembered technique.

This square array of numbers is called a third-order **determinant.**

$$\begin{vmatrix} \vec{i} & \vec{j} & \vec{k} \\ 3 & 5 & 7 \\ 11 & 2 & 13 \end{vmatrix}$$

You may have encountered determinants in algebra in conjunction with inverting a matrix or solving a system of linear equations.

You form the determinant by writing the three unit vectors along the top row, the coefficients of the first vector in the middle row, and the coefficients of the second vector in the bottom row. Expand this determinant along the top row.

$$\vec{i}\begin{vmatrix} 5 & 7 \\ 2 & 13 \end{vmatrix} - \vec{j}\begin{vmatrix} 3 & 7 \\ 11 & 13 \end{vmatrix} + \vec{k}\begin{vmatrix} 3 & 5 \\ 11 & 2 \end{vmatrix}$$ Remember the − sign for the middle term!

You find the first term by writing the top left element, $\vec{i}$, mentally crossing out its row and its column, and multiplying by the second-order determinant that remains. You do the same for the $\vec{j}$ and the $\vec{k}$. The signs of the expanded determinant alternate, so the second term has a − sign and the third term has a + sign.

To expand the second-order determinants, multiply the top left number by the bottom right, and then subtract the product of the top right number and the bottom left.

$$\vec{i}((5)(13) - (7)(2)) - \vec{j}((3)(13) - (7)(11)) + \vec{k}((3)(2) - (5)(11))$$
$$= 51\vec{i} + 38\vec{j} - 49\vec{k}$$

This is the same as the cross product in Example 1.

Geometrical Applications of Cross Products

You can use the cross product to find a vector normal to two other vectors. You can use the result to find the equation of the plane containing the two vectors, as you did in Section 10-5. Example 2 shows how.

▶ ***EXAMPLE 2*** Find a particular equation of the plane containing the points $P_1(-5, 5, 5)$, $P_2(-3, 2, 7)$, and $P_3(1, 12, 6)$.

Solution Sketch a plane and the three points as in Figure 10-6d. It helps to show the coordinates of the points on your sketch.

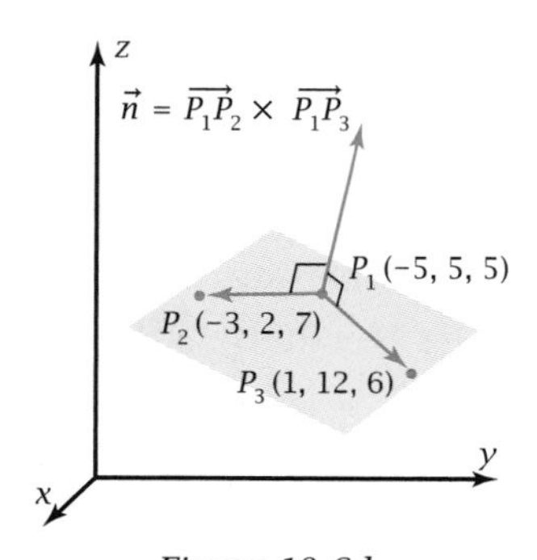

Figure 10-6d

Write the displacement vectors from one point to the other two.

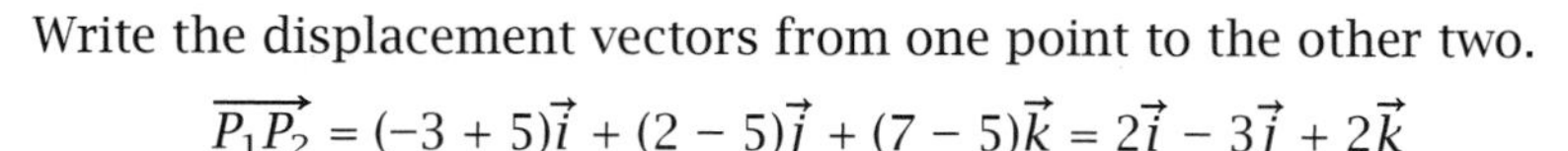

$$\overrightarrow{P_1P_2} = (-3 + 5)\vec{i} + (2 - 5)\vec{j} + (7 - 5)\vec{k} = 2\vec{i} - 3\vec{j} + 2\vec{k}$$
$$\overrightarrow{P_1P_3} = (1 + 5)\vec{i} + (12 - 5)\vec{j} + (6 - 5)\vec{k} = 6\vec{i} + 7\vec{j} + \vec{k}$$

Find the cross product to find a normal vector, $\vec{n}$.

$$\vec{n} = \overrightarrow{P_1P_2} \times \overrightarrow{P_1P_3} = \begin{vmatrix} \vec{i} & \vec{j} & \vec{k} \\ 2 & -3 & 2 \\ 6 & 7 & 1 \end{vmatrix}$$

$$= \vec{i}\begin{vmatrix} -3 & 2 \\ 7 & 1 \end{vmatrix} - \vec{j}\begin{vmatrix} 2 & 2 \\ 6 & 1 \end{vmatrix} + \vec{k}\begin{vmatrix} 2 & -3 \\ 6 & 7 \end{vmatrix}$$

$$= \vec{i}(-3 - 14) - \vec{j}(2 - 12) + \vec{k}(14 + 18)$$ Be careful of double negatives!

$$\vec{n} = -17\vec{i} + 10\vec{j} + 32\vec{k}$$

$$\therefore -17x + 10y + 32z = D$$ The coefficients in the equation equal the coefficients of the components of the normal vector.

$$-17(-5) + 10(5) + 32(5) = D$$ Substitute one of the given points.

$$295 = D$$

The equation is $-17x + 10y + 32z = 295$. ◀

The cross product of two vectors is perpendicular to the original vectors.

Note that Example 2 is the same kind of problem as in Section 10-5. The only difference is the way you calculate the normal vector.

Geometrical Meaning of $|\vec{a} \times \vec{b}|$

The magnitude of the cross product of two vectors has a geometric meaning. Figure 10-6e shows a parallelogram with vectors $\vec{a}$ and $\vec{b}$ as two adjacent sides.

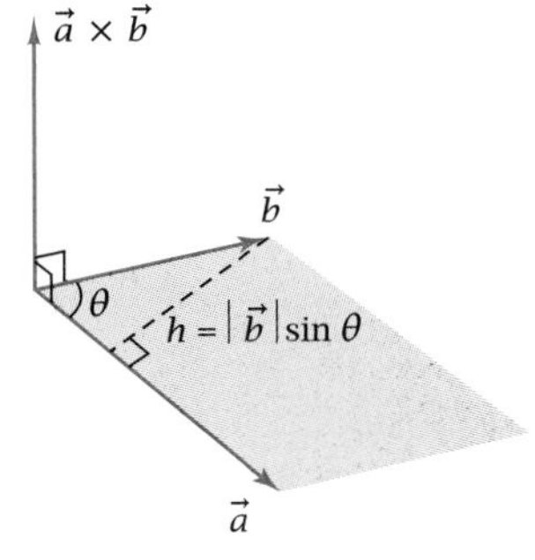

Figure 10-6e

The altitude of the parallelogram is

$$h = |\vec{b}| \sin\theta$$

The base of the parallelogram is $|\vec{a}|$. The area of a parallelogram is base times altitude, so

$$\text{Area} = |\vec{a}||\vec{b}| \sin\theta$$

You should recognize that $|\vec{a}||\vec{b}| \sin\theta$ is defined to be the magnitude of the cross product.

You can find the area of a triangle with two vectors as sides by the same technique because its area is half the area of the corresponding parallelogram. Example 3 shows you how to do this.

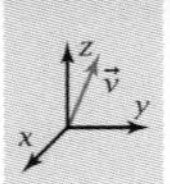

▶ **EXAMPLE 3** Find the area of the triangle with vertices at the points $P_1(-5, 5, 5)$, $P_2(-3, 2, 7)$, and $P_3(1, 12, 6)$.

Solution Figure 10-6f shows the three points (the same as in Example 2), along with the triangle formed by the vectors $\overrightarrow{P_1P_2}$ and $\overrightarrow{P_1P_3}$.

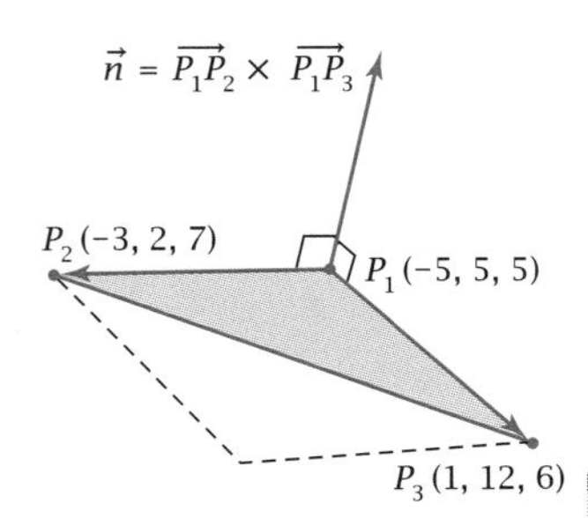

Figure 10-6f

From Example 2, $\overrightarrow{P_1P_2} \times \overrightarrow{P_1P_3} = -17\vec{i} + 10\vec{j} + 32\vec{k}$.

$$\text{Area} = \frac{1}{2}\left|-17\vec{i} + 10\vec{j} + 32\vec{k}\right| = \frac{1}{2}\sqrt{(-17)^2 + 10^2 + 32^2} = \frac{1}{2}\sqrt{1413} = 37.5898\ldots$$ ◀

The box summarizes these geometric properties.

PROPERTY: Area of a Parallelogram or Triangle from the Cross Product

The area of the parallelogram having $\vec{a}$ and $\vec{b}$ as adjacent sides is

$$\text{Area of parallelogram} = |\vec{a} \times \vec{b}|$$

The area of the triangle having $\vec{a}$ and $\vec{b}$ as adjacent sides is

$$\text{Area of triangle} = \tfrac{1}{2}|\vec{a} \times \vec{b}|$$

Problem Set 10-6

Do These Quickly

Q1. If $\vec{a}$ is 7 units long and $\vec{b}$ is 3 units long, what can you say about $|\vec{a} + \vec{b}|$?

Q2. If $\vec{a} \cdot \vec{b} = 25$, $|\vec{a}| = 5$, and $|\vec{b}| = 10$, find the angle between the vectors.

Q3. Find the length of the vector $\vec{i} + \vec{j} + \vec{k}$.

Q4. Find the dot product $(3\vec{i} + 2\vec{j} - \vec{k}) \cdot (\vec{i} + 2\vec{j} + 10\vec{k})$.

Q5. Is the angle between the vectors in Problem Q4 obtuse or acute?

Q6. Sketch two nonzero vectors whose sum is zero.

Q7. Sketch two nonzero vectors whose dot product is zero.

Q8. In ΔABC, if $\sin A = 0.6$, $\sin B = 0.2$, and side $a = 60$, then how long is side b?

Q9. Sketch the graph of a logistic function.

Q10. Find the amplitude of the sinusoid $y = 3 + 4\cos 5\pi(x - 6)$.

For Problems 1–4, find the cross product using determinants.

1. $(3\vec{i} + 4\vec{j} + 2\vec{k}) \times (5\vec{i} + 6\vec{j} + \vec{k})$
2. $(7\vec{i} + 2\vec{j} + 3\vec{k}) \times (6\vec{i} + \vec{j} + 5\vec{k})$
3. $(4\vec{i} - 3\vec{j} - \vec{k}) \times (2\vec{i} - \vec{j} + \vec{k})$
4. $(-3\vec{i} + 8\vec{j} + 2\vec{k}) \times (\vec{i} + 7\vec{j} + 6\vec{k})$

For Problems 5 and 6, find the dot product.

5. $(2\vec{i} + 7\vec{j} - 5\vec{k}) \cdot (9\vec{i} + 3\vec{j} + \vec{k})$
6. $(8\vec{i} - 4\vec{j} - 2\vec{k}) \cdot (5\vec{i} + 6\vec{j} - 7\vec{k})$
7. *Program for Cross Products Problem:* Write a program to calculate the cross product of two vectors. The program should prompt you to enter the three coefficients of each vector. Then it should calculate and display the coefficients of the cross product. Test your program by using it on the vectors in Problem 1. The correct answer is $-8\vec{i} + 7\vec{j} - 2\vec{k}$.

8. *Multipliers of Zero Problem 1:* The multiplication property of zero states that for real numbers x and y, "If $x = 0$ or $y = 0$, then $xy = 0$." Its converse is, "If $xy = 0$, then $x = 0$ or $y = 0$."
   a. Show that there is a multiplication property of zero for cross multiplication.
   b. Show by counterexample that the converse of the property in part a is *false.* That is, find two nonzero vectors whose cross product equals zero.

For Problems 9–11, find a particular equation of the plane containing the given points.

9. (3, 5, 8), (−2, 4, 1), and (−4, 7, 3)
10. (5, 7, 3), (4, −2, 6), and (2, −6, 1)
11. (0, 3, −7), (5, 0, −1), and (4, 3, 9)
12. The cross product of the normal vectors to two planes is a vector that points in the direction of the line of intersection of the planes. Find a particular equation of the plane containing (−3, 6, 5) and normal to the line of intersection of the planes $3x + 5y + 4z = -13$ and $6x - 2y + 7z = 8$.

For Problems 13–16, find the area.

13. The parallelogram determined by $\vec{a} = 2\vec{i} + 3\vec{j} + 6\vec{k}$ and $\vec{b} = 3\vec{i} - 4\vec{j} + 12\vec{k}$
14. The parallelogram determined by $\vec{c} = 4\vec{i} + 4\vec{j} - 7\vec{k}$ and $\vec{d} = -2\vec{i} + 5\vec{j} - 14\vec{k}$
15. The triangle with vertices (3, 7, 5), (2, −1, 7), and (−4, 6, 10)
16. The triangle with vertices (7, 8, 11), (−4, 2, 1), and (3, 8, 2)
17. *Awning Problem 1:* Figure 10-6g shows a triangular awning for the corner of a building. The vertices of the awning are to be at (10, 0, 8), (0, 15, 8), and (0, 0, 13), where the dimensions are in feet.
   a. Find the displacement vectors from the vertex on the z-axis to the other two vertices.
   b. Find a vector normal to the plane.
   c. Find the area of canvas that will be in the awning when it is completed.
   d. Find the lengths of the three sides of the awning and the three vertex angles so that the people who make the awning will know how to cut the canvas.
   e. Find an equation of the plane. Use the equation to find out whether the point (5, 6, 9) is above the awning or below it.

Figure 10-6g

18. *Torque Problem:* Figure 10-6h shows a wrench on a nut and bolt. As you tighten the nut, you exert a force of $\vec{F} = 5\vec{i} + 2\vec{j} + 0\vec{k}$, where the magnitude of the force vector is in pounds. The displacement vector from the center of the bolt to where your hand applies the force is $\vec{d} = 7\vec{i} + 10\vec{j} + 0\vec{k}$, where the magnitude of the displacement is in inches. The force exerts a **torque** on the nut that twists it tight. The torque is defined to be the cross product of the force vector and the displacement vector. Find the torque vector. In which direction does the torque vector act?

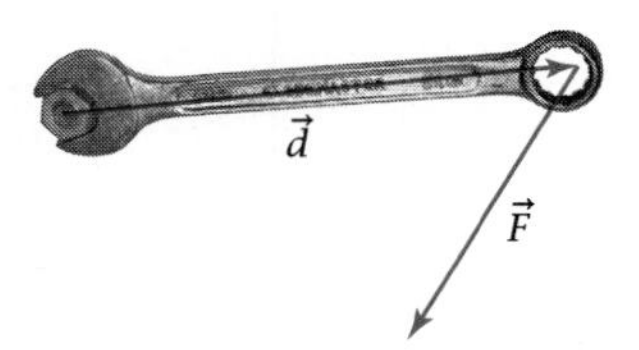

Figure 10-6h

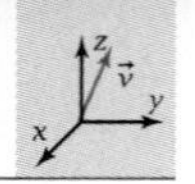

19. Given $\vec{a} = 5\vec{i} - 2\vec{j} + 3\vec{k}$ and $\vec{b} = 4\vec{i} + 7\vec{j} - 6\vec{k}$,
    a. Find $\vec{a} \times \vec{b}$.
    b. Use dot products to show that $\vec{a} \times \vec{b}$ really is normal to both $\vec{a}$ and $\vec{b}$.
    c. Calculate $\vec{a} \cdot \vec{b}$, and use it to find the angle θ between $\vec{a}$ and $\vec{b}$.
    d. Use θ from part c to show that $|\vec{a} \times \vec{b}| = |\vec{a}||\vec{b}| \sin \theta$.
20. Given $\vec{e} = 2\vec{i} + 5\vec{j} - 3\vec{k}$ and $\vec{f} = 7\vec{i} - 4\vec{j} - 2\vec{k}$,
    a. Prove that $\vec{e}$ and $\vec{f}$ are perpendicular.
    b. Show that $|\vec{e} \times \vec{f}| = |\vec{e}||\vec{f}|$.
    c. Explain why the result of part b is consistent with the definition of cross product.
21. Given $\vec{g} = -3\vec{i} + 6\vec{j} - 12\vec{k}$ and $\vec{h} = 5\vec{i} - 10\vec{j} + 20\vec{k}$,
    a. Prove that $\vec{g}$ and $\vec{h}$ are parallel.
    b. Show that $|\vec{g} \times \vec{h}| = 0$.
    c. Explain why the result of part b is consistent with the definition of cross product.
22. A plane is determined by the points (2, 1, 7), (3, 4, 9), and (6, −4, 5). Find its x-, y-, and z-intercepts.
23. If $\vec{u} = 3\vec{i} + 4\vec{j} + 6\vec{k}$ and $\vec{v} = x\vec{i} + \vec{j} - z\vec{k}$, find the values of x and z that make the cross product $\vec{u} \times \vec{v} = 2\vec{i} + 24\vec{j} - 17\vec{k}$.
24. *Pythagorean Quadruples Problem:* You may have observed that the length of a vector sometimes turns out to be an integer. For example,

$$|8\vec{i} + 9\vec{j} + 12\vec{k}| = \sqrt{8^2 + 9^2 + 12^2}$$
$$= \sqrt{289} = 17$$

Four positive integers a, b, c, and d, for which

$$a^2 + b^2 + c^2 = d^2$$

form what is called a **Pythagorean quadruple.** Write a program for your grapher or other computer to find all Pythagorean quadruples for values of a, b, and c up to 20. See if you can get the computer to give only the *primitive* Pythagorean quadruples by eliminating those that are multiples of others, such as 2, 4, 4, 6, which is 2 times 1, 2, 2, 3.

10-7 Direction Angles and Direction Cosines

A beam for a building under construction is to be held in place by three triangular gussets, as shown in Figure 10-7a. So that the beam will point in the correct direction, the gussets must make the correct angles with the three coordinate axes. In this section you will learn how to calculate these **direction angles.** The cosines of these angles are called the **direction cosines.**

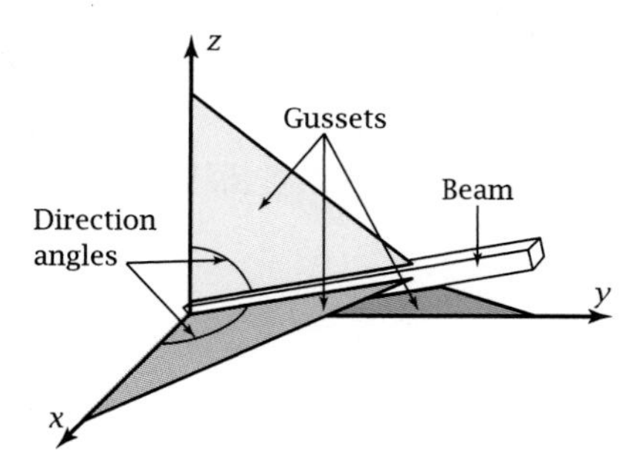

Figure 10-7a

OBJECTIVE Given a vector, find its direction angles and direction cosines, and vice versa.

Figure 10-7b shows a vector $\vec{v}$ and its direction angles. The first three letters of the Greek alphabet, α (alpha), β (beta), and γ (gamma), are commonly used for the three direction angles. Like x, y, and z, the letters come in alphabetical order, corresponding to the three axes.

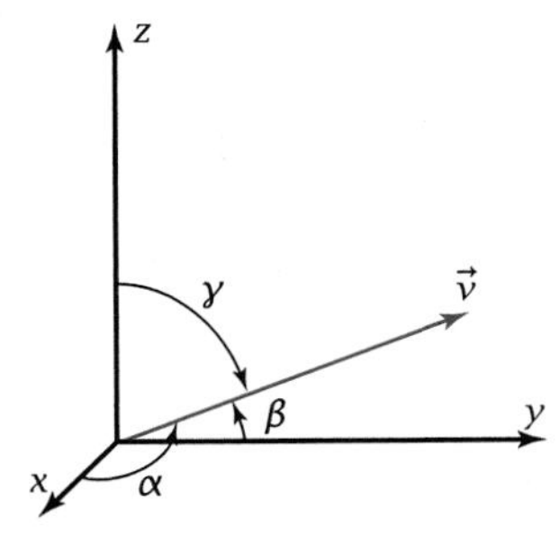

Figure 10-7b

DEFINITIONS: Direction Angles and Direction Cosines

The **direction angles** of a position vector are

α, from the x-axis to the vector

β, from the y-axis to the vector

γ, from the z-axis to the vector

The **direction cosines** of a position vector are the cosines of the direction angles.

$c_1 = \cos \alpha$

$c_2 = \cos \beta$

$c_3 = \cos \gamma$

▶ **EXAMPLE 1** Find the direction cosines and the direction angles for $\vec{v} = 3\vec{i} + 7\vec{j} + 5\vec{k}$.

Solution Find the dot products $\vec{v} \cdot \vec{i}$, $\vec{v} \cdot \vec{j}$, and $\vec{v} \cdot \vec{k}$. Then use these to find the angles.

$\vec{v} \cdot \vec{i} = (3\vec{i} + 7\vec{j} + 5\vec{k}) \cdot (1\vec{i} + 0\vec{j} + 0\vec{k}) = 3$ Equal to the coefficient of $\vec{i}$ in $\vec{v}$.

$\vec{v} \cdot \vec{j} = (3\vec{i} + 7\vec{j} + 5\vec{k}) \cdot (0\vec{i} + 1\vec{j} + 0\vec{k}) = 7$ Equal to the coefficient of $\vec{j}$ in $\vec{v}$.

$\vec{v} \cdot \vec{k} = (3\vec{i} + 7\vec{j} + 5\vec{k}) \cdot (0\vec{i} + 0\vec{j} + 1\vec{k}) = 5$ Equal to the coefficient of $\vec{k}$ in $\vec{v}$.

$|\vec{v}| = \sqrt{3^2 + 7^2 + 5^2} = \sqrt{83}$

$|\vec{i}| = |\vec{j}| = |\vec{k}| = 1$ They are *unit* vectors.

$\sqrt{83}(1) \cos \alpha = 3 \Rightarrow \cos \alpha = \dfrac{3}{\sqrt{83}} \Rightarrow \alpha = 70.7741...^\circ$ Use the definition of dot product.

$\sqrt{83}(1) \cos \beta = 7 \Rightarrow \cos \beta = \dfrac{7}{\sqrt{83}} \Rightarrow \beta = 39.7940...^\circ$

$\sqrt{83}(1) \cos \gamma = 5 \Rightarrow \cos \gamma = \dfrac{5}{\sqrt{83}} \Rightarrow \gamma = 56.7138...^\circ$ ◀

There is nothing special about the sum of α, β, and γ. From the results of Example 1, you can see that in this case $\alpha + \beta + \gamma = 167.282...°$. However, there is a remarkable property about the sum of the squares of the direction cosines.

$$\cos^2\alpha + \cos^2\beta + \cos^2\gamma = \frac{9}{83} + \frac{49}{83} + \frac{25}{83} = \frac{83}{83} = 1$$

The sum of the squares of the direction cosines of a position vector is always 1. This Pythagorean property happens because the three dot products are equal to the three coefficients of the components of $\vec{v}$. When you square the numerators and add them you get 83, the same as the radicand in the magnitude of $\vec{v}$. The ratio will always be 1.

The Pythagorean property explains another property relating to direction cosines. Find a unit vector in the direction of $\vec{v}$ in Example 1.

$$\vec{u} = \frac{\vec{v}}{|\vec{v}|} = \frac{3\vec{i} + 7\vec{j} + 5\vec{k}}{\sqrt{83}} = \frac{3}{\sqrt{83}}\vec{i} + \frac{7}{\sqrt{83}}\vec{j} + \frac{5}{\sqrt{83}}\vec{k}$$

Notice that the coefficients of the components of the unit vector are the direction cosines. These two properties are summarized in this box.

PROPERTIES: Direction Cosines

Pythagorean Property of Direction Cosines

If α, β, and γ are the direction angles of a position vector and $c_1 = \cos\alpha$, $c_2 = \cos\beta$, and $c_3 = \cos\gamma$ are its direction cosines, then

$$\cos^2\alpha + \cos^2\beta + \cos^2\gamma = 1 \qquad \text{or} \qquad c_1^2 + c_2^2 + c_3^2 = 1$$

Unit Vector Property of Direction Cosines

$\vec{u} = c_1\vec{i} + c_2\vec{j} + c_3\vec{k}$ is a unit vector in the direction of the given vector.

▶ ***EXAMPLE 2*** Find the direction angles and the direction cosines of $\vec{v} = 13\vec{i} - 6\vec{j} + 18\vec{k}$ quickly. Use the result to write a unit vector in the direction of $\vec{v}$.

Solution

$$|\vec{v}| = \sqrt{13^2 + 6^2 + 18^2} = \sqrt{529} = 23$$

$$\cos\alpha = \frac{13}{23} \Rightarrow \alpha = 55.5826...°$$

$$\cos\beta = \frac{-6}{23} \Rightarrow \beta = 105.1216...°$$

$$\cos\gamma = \frac{18}{23} \Rightarrow \gamma = 38.4999...°$$

The unit vector is $\vec{u} = \frac{13}{23}\vec{i} - \frac{6}{23}\vec{j} + \frac{18}{23}\vec{k}$

The direction cosines are the coefficients of $\vec{i}$, $\vec{j}$, and $\vec{k}$. ◀

▶ **EXAMPLE 3** If $\alpha = 152°$ and $\beta = 73°$, find γ.

Solution

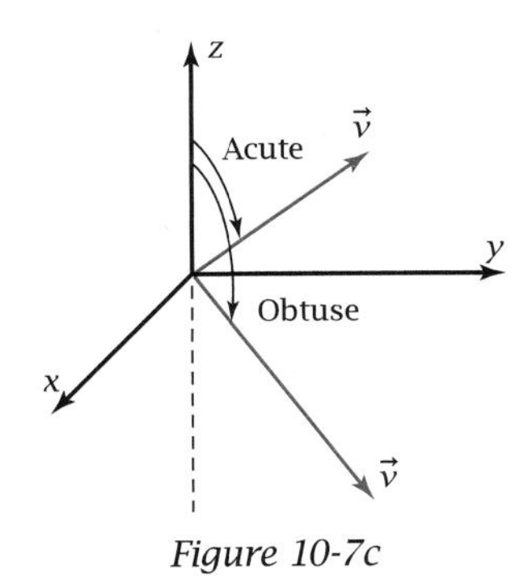

Figure 10-7c

$\cos^2 152° + \cos^2 73° + c_3^2 = 1$ Pythagorean property for direction cosines.

$c_3^2 = 1 - \cos^2 152° - \cos^2 73° = 0.1349...$

$c_3 = \pm 0.3673...$ Don't forget the ±!

$\gamma = 68.4497...°$ or $111.5502...°$ There are two possible values of γ, one acute, one obtuse. ◀

Example 3 shows algebraically that there are two possibilities for the third direction angle if the other two are given. Figure 10-7c illustrates this fact geometrically. One value is acute. The other is its supplementary angle, which is obtuse.

Problem Set 10-7

Do These Quickly

Q1. Write the definition of $\vec{a} \cdot \vec{b}$.

Q2. In the definition of cross product, what does $|\vec{a} \times \vec{b}|$ equal?

Q3. How are the directions of $\vec{a}$ and $\vec{b}$ related to the direction of $\vec{a} \times \vec{b}$?

Q4. $\vec{a} \cdot (\vec{a} \times \vec{b}) =$ —?—.

Q5. How is $\vec{b} \times \vec{a}$ related to $\vec{a} \times \vec{b}$?

Q6. If $|\vec{a} \times \vec{b}| = 50$, find the area of the triangle determined by $\vec{a}$ and $\vec{b}$.

Q7. Find a normal vector for the plane $3x + 4y - 5z = 37$.

Q8. Find another normal vector for the plane in Problem Q7.

Q9. The linear function that best fits a set of data is called the —?— function.

Q10. $a^2 = b^2 + c^2 - 2bc \cos A$ is a statement of the —?—.

For Problems 1 and 2, sketch the vector and show its direction angles.

1. $\vec{v} = 4\vec{i} + 10\vec{j} + 3\vec{k}$
2. $\vec{v} = 5\vec{i} + 4\vec{j} + 9\vec{k}$

For Problems 3–6, find the direction cosines and direction angles for the position vector to the given point.

3. (2, −5, 3)
4. (5, 7, −1)
5. (−4, 8, 19)
6. (10, −15, 6)

For Problems 7–10, find the direction cosines of the vector from the first point to the second.

7. (−3, 7, 1) to (4, 8, −2)
8. (6, 9, 4) to (−2, 10, 1)
9. (2, 9, 4) to (11, 1, 16)
10. (4, 2, −9) to (−7, 10, 7)

For Problems 11 and 12, find a unit vector in the direction from the first point to the second point, and write its direction cosines.

11. (3, 7, −2) to (11, 23, −9)
12. (−5, 3, 2) to (3, 2, 6)

For Problems 13 and 14, prove that the vector is a unit vector, and find its direction cosines and direction angles.

13. $\frac{7}{9}\vec{i} + \frac{4}{9}\vec{j} - \frac{4}{9}\vec{k}$
14. $\frac{1}{9}\vec{i} - \frac{4}{9}\vec{j} + \frac{8}{9}\vec{k}$

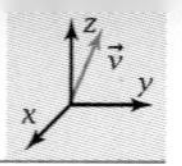

For Problems 15 and 16, find the third direction cosine and the two possible values of the third direction angle.

15. $c_1 = \frac{18}{23}$, $c_2 = -\frac{13}{23}$

16. $c_1 = -\frac{12}{17}$, $c_3 = \frac{8}{17}$

For Problems 17–20, α, β, and γ are direction angles for different position vectors. Find the possible values of γ.

17. $\alpha = 120°$, $\beta = 60°$

18. $\alpha = 110°$, $\beta = 70°$

19. $\alpha = 17°$, $\beta = 12°$

20. $\alpha = 173°$, $\beta = 168°$

21. *Circus Cannon Problem:* In the circus, a dummy clown is shot from a cannon. Its velocity vector as it leaves the cannon is

$$\vec{v} = 5\vec{i} + 11\vec{j} + 7\vec{k}$$

where the 5, 11, and 7 are speeds in the x-, y-, and z-directions, respectively, measured in feet per second. Figure 10-7d shows the *azimuth angle* and the angle of elevation for the cannon.

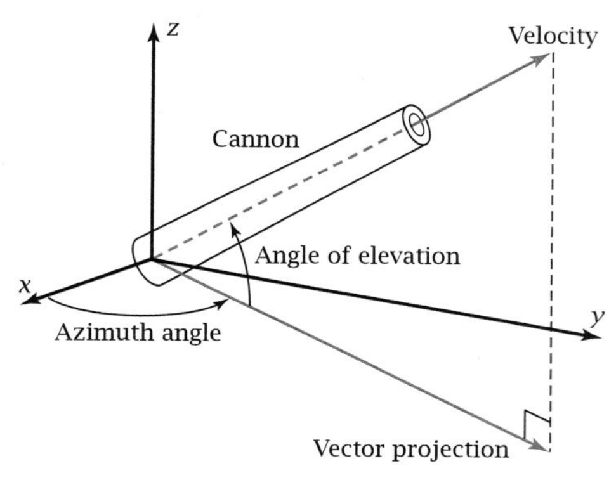

Figure 10-7d

a. At what speed will the dummy leave the cannon?

b. Find the measures of the three direction angles for $\vec{v}$.

c. Use the direction angles to find quickly the measure of the angle of elevation.

d. Find $\vec{p}$, the vector projection of $\vec{v}$ on the (horizontal) xy-plane.

e. The azimuth angle is the angle in the xy-plane from the x-axis to the vector projection $\vec{p}$. Find the measure of this angle.

f. If the angle of elevation were increased by 5°, would this change affect the other two direction angles? Would this change affect the azimuth angle?

22. *Shoe Box Construction Project:* As shown in Figure 10-7e, run a stiff wire (such as coat hanger wire) or a thin stick on the main diagonal of the shoe box. (The front face of the shoe box is not shown so that you can see into the interior.) Let $\vec{v}$ be the vector running along the wire from the bottom corner to the diagonally opposite top corner. Measure the three sides of the shoe box, and use the results to write $\vec{v}$ in terms of its coordinates. Then calculate the three direction angles. Cut out three triangular pieces of cardboard, each with one angle equal to a direction angle. Do the three triangles fit the direction angles? If so, tape the cardboard pieces in place and check your project with your instructor. If not, redo your computations and measurements until the pieces of cardboard do fit. Write up this project in your journal, describing what you have learned as a result of doing the calculations and construction.

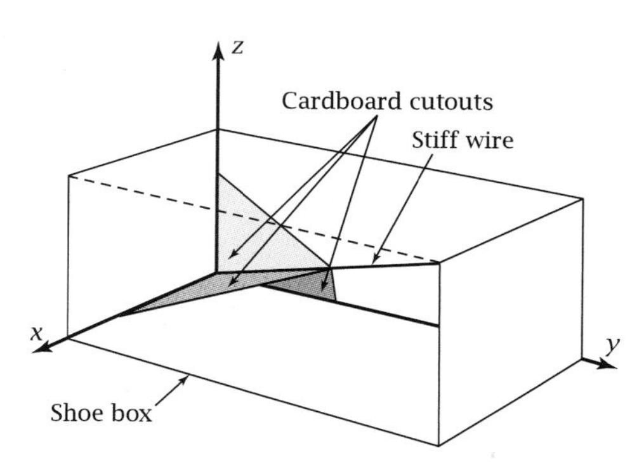

Figure 10-7e

23. *Proof of the Pythagorean Property of Direction Cosines:* Prove that if $\vec{v} = A\vec{i} + B\vec{j} + C\vec{k}$ and c_1, c_2, and c_3 are the direction cosines of $\vec{v}$, then $c_1^2 + c_2^2 + c_3^2 = 1$.

24. *Journal Problem:* Update your journal with what you have learned about vectors since the last entry. In particular, explain the difference between the definitions of cross product and dot product, and describe how you calculate these products. Then give some uses of dot product and cross product.

10-8 Vector Equations of Lines in Space

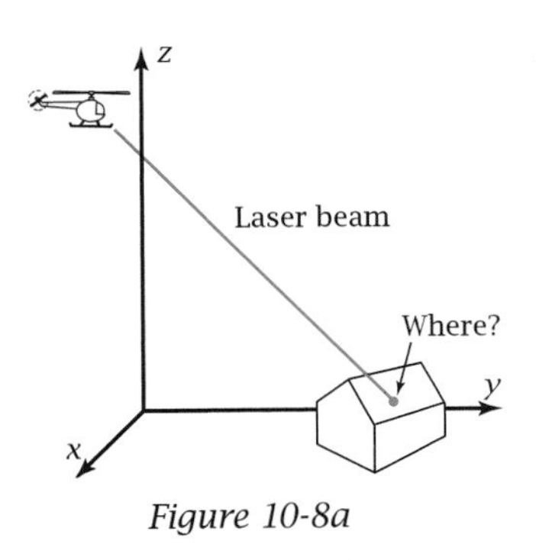

Figure 10-8a

A rescue helicopter tries to locate people stranded on the roof of a house. It shines a light beam on the roof of a house, as shown in Figure 10-8a. By finding an equation relating the x-, y-, and z-coordinates of points on the beam, you can predict where the spot of light will appear on the roof and other information to help with the task. Unfortunately, there is no single Cartesian equation for a line in space as there is for a plane. In this section you will derive a *vector* equation that gives position vectors for points on a line.

OBJECTIVE Given information about a line in space, find a vector equation for the line and use it to calculate coordinates of points on the line.

Figure 10-8b shows a line in space. Vector $\vec{r}$ is the position vector to a variable point $P(x, y, z)$ on the line. Point $P_0(5, 11, 13)$ is a fixed point on the line. Unit vector $\vec{u}$ points along the line. Let d be the directed distance from P_0 to P. Thus, the displacement vector from P_0 to P is

$$\overrightarrow{P_0P} = d\vec{u}$$

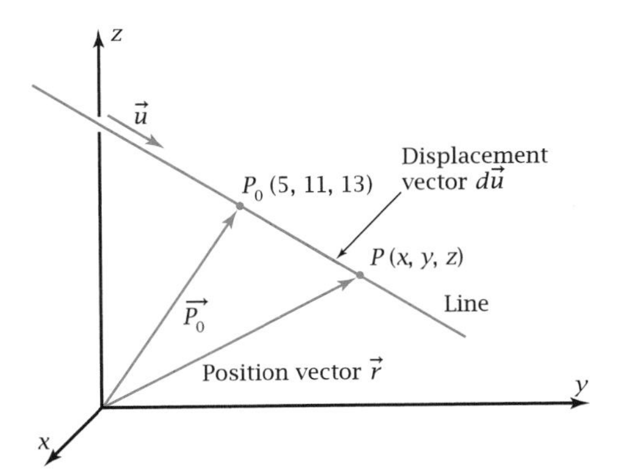

Figure 10-8b

Note that the position vector $\vec{r}$ to the variable point is the sum of $d\vec{u}$ and the position vector $\vec{P_0}$ to the fixed point.

$$\vec{r} = \vec{P_0} + d\vec{u}$$ General vector equation of a line in space.

To calculate points (x, y, z) on the line, you must get the particular equation by substituting for the fixed point and for the unit vector.

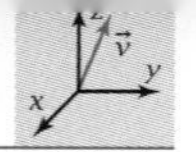

▶ EXAMPLE 1 Find the particular equation of the line that contains the fixed point $P_0(5, 11, 13)$ and that is parallel to the unit vector

$$\vec{u} = \tfrac{3}{7}\vec{i} + \tfrac{6}{7}\vec{j} + \tfrac{2}{7}\vec{k}$$

Solution Make a general sketch of the line as shown in Figure 10-8c.

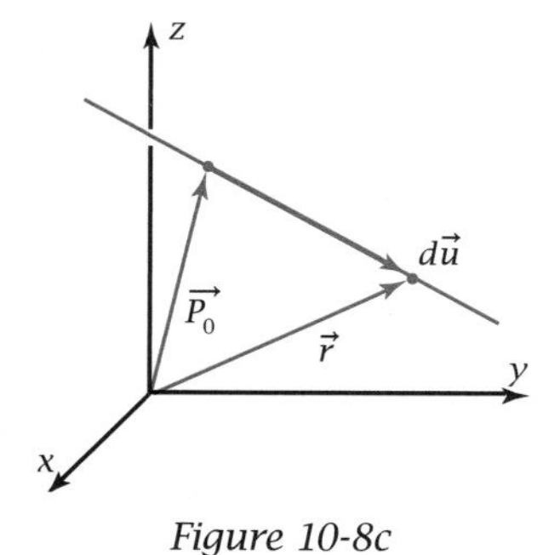

Figure 10-8c

The general equation is $\vec{r} = \vec{P_0} + d\vec{u}$ — $\vec{r}$ is the resultant of $\vec{P_0}$ and $d\vec{u}$.

$\vec{P_0} = 5\vec{i} + 11\vec{j} + 13\vec{k}$ — $\vec{P_0}$ is the position vector to point P_0.

Therefore, the particular equation is

$\vec{r} = (5\vec{i} + 11\vec{j} + 13\vec{k}) + d\left(\tfrac{3}{7}\vec{i} + \tfrac{6}{7}\vec{j} + \tfrac{2}{7}\vec{k}\right)$ — Substitute for $\vec{P_0}$ and $\vec{u}$.

$\vec{r} = \left(5 + \tfrac{3}{7}d\right)\vec{i} + \left(11 + \tfrac{6}{7}d\right)\vec{j} + \left(13 + \tfrac{2}{7}d\right)\vec{k}$ — Combine like terms. ◀

The variable d that represents the directed distance from the fixed point to the variable point is a parameter similar to those you encountered previously.

PROPERTY: General Vector Equation of a Line in Space

$$\vec{r} = \vec{P_0} + d\vec{u}$$

where $\vec{r}$ is the position vector to variable point $P(x, y, z)$ on the line.

$\vec{P_0}$ is the position vector to a fixed point P_0 on the line.

d is a parameter equal to the directed distance from fixed point $P_0(x_0, y_0, z_0)$ to P.

$\vec{u}$ is a unit vector in the direction of the line.

If unit vector $\vec{u} = c_1\vec{i} + c_2\vec{j} + c_3\vec{k}$, then the equation is

$$\vec{r} = (x_0 + c_1 d)\vec{i} + (y_0 + c_2 d)\vec{j} + (z_0 + c_3 d)\vec{k}$$

Once you have the particular equation, you can use it in various ways to find points on the line. For instance, if you want to find the coordinates of a point at a particular distance from P_0, substitute the appropriate value of d and complete the calculations.

▶ EXAMPLE 2 Find the point on the line in Example 1 that is at a directed distance of −21 units from P_0.

Solution

$\vec{r} = (5 + \tfrac{3}{7}(-21))\vec{i} + (11 + \tfrac{6}{7}(-21))\vec{j} + (13 + \tfrac{2}{7}(-21))\vec{k}$ — Substitute −21 for d.

$\vec{r} = (5 - 9)\vec{i} + (11 - 18)\vec{j} + (13 - 6)\vec{k}$

$\vec{r} = -4\vec{i} - 7\vec{j} + 7\vec{k}$

The point is (−4, −7, 7). ◀

To find a point that has a particular value of x, y, or z, you must first calculate the value of d. Then proceed as in Example 2. Example 3 shows how to do this.

▶ EXAMPLE 3 Find the point where the line in Example 1 intersects with the xy-plane.

Solution For any point on the xy-plane, $z = 0$. So you set the z-coordinate of the line equal to zero.

$$13 + \tfrac{2}{7}d = 0 \Rightarrow d = -\tfrac{91}{2}$$ Calculate the value of d.

$$\therefore \vec{r} = \left[5 + \tfrac{3}{7}\left(-\tfrac{91}{2}\right)\right]\vec{i} + \left[11 + \tfrac{6}{7}\left(-\tfrac{91}{2}\right)\right]\vec{j} + \left[13 + \tfrac{2}{7}\left(-\tfrac{91}{2}\right)\right]\vec{k}$$ Substitute $-\tfrac{91}{2}$ for d.

$$\vec{r} = (5 - 19.5)\vec{i} + (11 - 39)\vec{j} + (13 - 13)\vec{k}$$

$$\vec{r} = -14.5\vec{i} - 28\vec{j} + 0\vec{k}$$

The point is $(-14.5, -28, 0)$. As a check, the z-coordinate really does equal 0. ◀

By extending the technique of Example 3, you can calculate the point where a given line intersects a given plane.

▶ EXAMPLE 4 A line containing the point $(5, 3, -1)$ has direction cosines

$$c_1 = \tfrac{6}{11} \qquad c_2 = -\tfrac{2}{11} \qquad \text{and} \qquad c_3 = \tfrac{9}{11}$$

a. Write the particular equation of the line.

b. Find the point where the line intersects the plane $7x + 4y - 2z = 39$.

Solution a. Make a general sketch of the line and plane as shown in Figure 10-8d.

The general equation of the line is $\vec{r} = \vec{P_0} + d\vec{u}$.

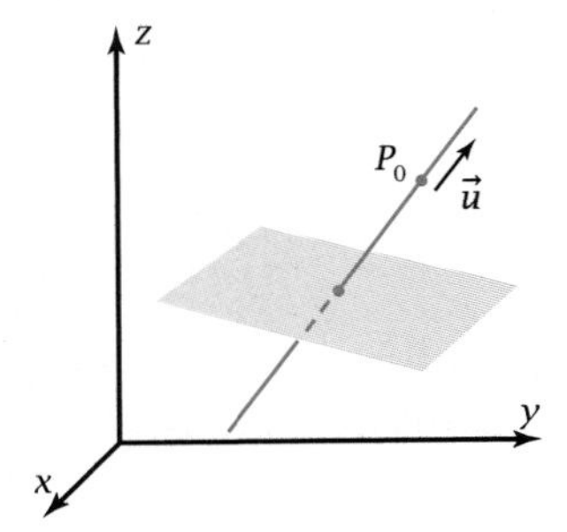

Figure 10-8d

$$\vec{P_0} = 5\vec{i} + 3\vec{j} - \vec{k}$$ The coefficients of the position vector are the coordinates of the fixed point.

$$\vec{u} = \tfrac{6}{11}\vec{i} - \tfrac{2}{11}\vec{j} + \tfrac{9}{11}\vec{k}$$ The coefficients of the components of the unit vector are the direction cosines.

$$\vec{r} = (5\vec{i} + 3\vec{j} - \vec{k}) + d\left(\tfrac{6}{11}\vec{i} - \tfrac{2}{11}\vec{j} + \tfrac{9}{11}\vec{k}\right)$$ Substitute into the general equation.

$$\therefore \vec{r} = \left(5 + \tfrac{6}{11}d\right)\vec{i} + \left(3 - \tfrac{2}{11}d\right)\vec{j} + \left(-1 + \tfrac{9}{11}d\right)\vec{k}$$ The particular equation of the line.

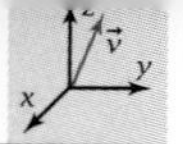

b. Assume that the line intersects the plane at the point (x, y, z). So the point is both on the line and on the plane. This means that the coordinates x, y, and z will satisfy both the equation of the line and the equation of the plane. So you can substitute the x, y, and z for the line from part a into the given equation for the plane and use the result to calculate d.

$$7\left(5 + \tfrac{6}{11}d\right) + 4\left(3 - \tfrac{2}{11}d\right) - 2\left(-1 + \tfrac{9}{11}d\right) = 39$$
$$35 + \tfrac{42}{11}d + 12 - \tfrac{8}{11}d + 2 - \tfrac{18}{11}d = 39$$
$$\tfrac{16}{11}d = -10$$
$$d = -\tfrac{55}{8}$$

$$\therefore \vec{r} = \left[5 + \tfrac{6}{11}\left(-\tfrac{55}{8}\right)\right]\vec{i} + \left[3 - \tfrac{2}{11}\left(-\tfrac{55}{8}\right)\right]\vec{j} + \left[-1 + \tfrac{9}{11}\left(-\tfrac{55}{8}\right)\right]\vec{k}$$ Substitute $-\tfrac{55}{8}$ for d.

$$\vec{r} = 1.25\vec{i} + 4.25\vec{j} - 6.625\vec{k}$$

The point is (1.25, 4.25, −6.625). ◀

Problem Set 10-8

Do These Quickly

Q1. If $\cos\alpha = \frac{1}{2}$, find α.

Q2. For direction angles α, β, and γ, if $\cos^2\alpha = 0.3$ and $\cos^2\beta = 0.2$, find $\cos^2\gamma$.

Q3. Find the magnitude of $\vec{v} = 1\vec{i} - 2\vec{j} + 2\vec{k}$.

Q4. Find the direction cosine c_2 for $\vec{v}$ in Problem Q3.

Q5. Give one major difference between a dot product and a cross product of two vectors.

Q6. If p is the scalar projection of $\vec{a}$ on $\vec{b}$, then the vector projection is $\vec{p}$ = —?—.

Q7. Find $\vec{i} \cdot \vec{j}$.

Q8. Find $\vec{i} \times \vec{j}$.

Q9. If $|\vec{a} \times \vec{b}| = 70$, find the area of the parallelogram with sides $\vec{a}$ and $\vec{b}$.

Q10. What transformation on $f(x)$ is represented by $y = f(3x)$?

For Problems 1 and 2, given the equation of a line, read off

a. The coordinates of the fixed point on the given line.

b. A unit vector, $\vec{u}$, in the direction of the line. Confirm that $\vec{u}$ really is a *unit* vector.

1. $\vec{r} = (5 + \frac{9}{17}d)\vec{i} + (-3 + \frac{12}{17}d)\vec{j} + (4 + \frac{8}{17}d)\vec{k}$
2. $\vec{r} = (6 + \frac{1}{9}d)\vec{i} + (7 + \frac{8}{9}d)\vec{j} + (-5 + \frac{4}{9}d)\vec{k}$
3. Find the point on the line in Problem 1 for which $d = 34$.
4. Find the point on the line in Problem 2 for which $d = 27$.
5. Find the point where the line in Problem 1 intersects the xy-plane.
6. Find the point where the line in Problem 2 intersects the yz-plane.

For Problems 7 and 8, show that $\vec{u}$ is a unit vector. Then write a vector equation of the line parallel to $\vec{u}$ containing the given point.

7. $\vec{u} = \frac{2}{7}\vec{i} + \frac{6}{7}\vec{j} - \frac{3}{7}\vec{k}$, $P_0 = (5, -1, 4)$
8. $\vec{u} = \frac{11}{15}\vec{i} - \frac{2}{15}\vec{j} + \frac{2}{3}\vec{k}$, $P_0 = (-3, 4, 7)$

For Problems 9 and 10, find the direction cosines of $\vec{v}$. Then find a vector equation of the line parallel to $\vec{v}$ containing the given point.

9. $\vec{v} = 2\vec{i} - 3\vec{j} + 4\vec{k}$, $P_0 = (1, -8, -5)$
10. $\vec{v} = \vec{i} + 2\vec{j} - 5\vec{k}$, $P_0 = (-6, 3, -4)$

For Problems 11 and 12, find a vector equation of the line from the first point to the second.

11. (5, 1, −4) to (14, 21, 8)

12. (6, −2, 7) to (10, 6, 26)

For Problems 13 and 14, find the point where the given line intersects the given plane.

13. Line: $\vec{r} = (3 + \frac{2}{3}d)\vec{i} + (4 + \frac{2}{3}d)\vec{j} + (3 - \frac{1}{3}d)\vec{k}$

Plane: $7x - 3y + 5z = -20$

14. Line: $\vec{r} = (4 + \frac{1}{3}d)\vec{i} + (1 + \frac{2}{3}d)\vec{j} + (7 + \frac{2}{3}d)\vec{k}$

Plane: $x + 4y - 3z = 35$

15. *Forensic Bullet Path Problem:* A bullet has pierced the wall and ceiling of a small house (Figure 10-8e), and it may have lodged in the roof. You have studied vectors, so the investigators call on you to calculate the point in the roof where the bullet is expected to be found. The dimensions of the house are in feet.

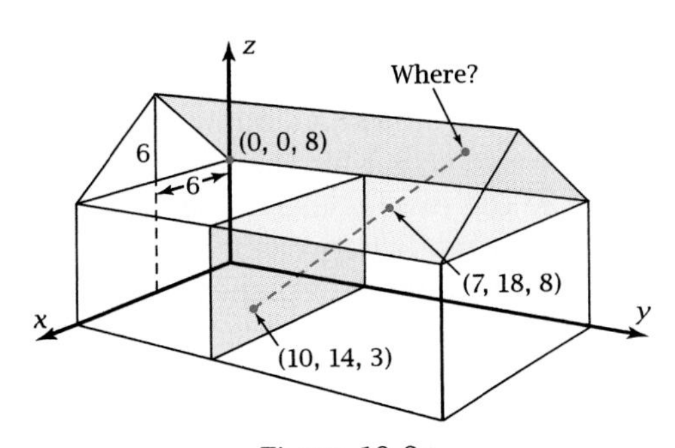

Figure 10-8e

a. You set up a three-dimensional coordinate system as shown in Figure 10-8e, with the origin at the floor in the back corner of the house. The bullet pierced the wall at the point (10, 14, 3) and then pierced the ceiling at (7, 18, 8). Find a unit vector in the direction of the bullet's path.

b. Using (10, 14, 3) as the fixed point, write a vector equation of the line followed by the bullet.

c. How tall is the interior of the house, from floor to ceiling? How can you tell?

d. Figure 10-8e shows that the point (0, 0, 8) is at the back corner of the slanted roof. If you run horizontally in the *x*-direction 6 ft from this point and then rise vertically in the *z*-direction 6 ft, you reach the crest of the roof. Explain why $\vec{n} = -6\vec{i} + 0\vec{j} + 6\vec{k}$ is a vector normal to the plane of the roof that is shaded in Figure 10-8e.

e. Find a Cartesian equation of the plane of the roof in part d.

f. Find the point in the roof at which police may expect to find the bullet.

g. What is the meaning of the word *forensic*, and why is the word appropriate to be used in the title of this problem?

16. *Flood Control Tunnel Problem:* Suppose that you work for a construction company that has been hired to dig a drainage tunnel under a city. The tunnel will carry excess water to the other side of the city during heavy rains, thus preventing flooding (Figure 10-8f). The tunnel is to start at ground level and then slant down until it reaches a point 100 ft below the surface. Then it will go horizontally, far enough to reach the other side of the city (not shown). Your job is to analyze the slanted part of the tunnel.

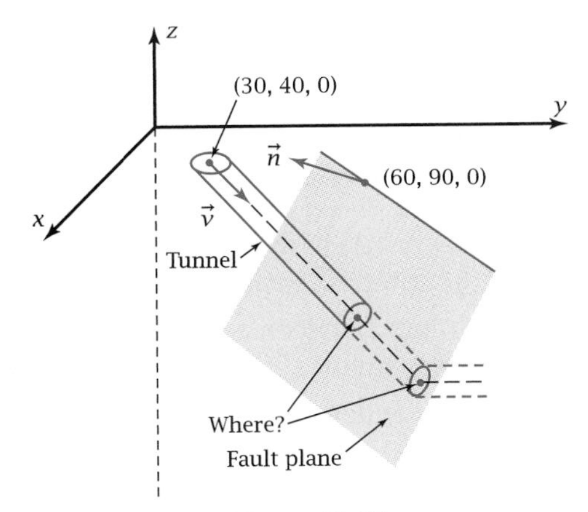

Figure 10-8f

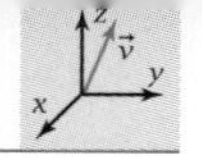

a. The Engineering Department has determined that the tunnel will slant downward in the direction of the vector $\vec{v} = 9\vec{i} + 12\vec{j} - 20\vec{k}$. The centerline of the tunnel starts at the point (30, 40, 0) on the surface. The measurements are in feet. Write the particular vector equation of the centerline.

b. How far along the centerline must the construction crews dig to reach the end of the slanted part of the tunnel, 100 ft below ground? What are the coordinates of this endpoint?

c. Construction crews must be careful when they reach a fault plane that is in the path of the slanted part of the tunnel. The Geology Department has determined that the point (60, 90, 0) is on the fault plane where it outcrops at ground level and that the vector $\vec{n} = 2\vec{i} - 4\vec{j} + \vec{k}$ is normal to the plane (Figure 10-8f). Find the particular equation of the plane.

d. How far along the centerline of the tunnel must the construction crews dig in order to reach the fault plane? What are the coordinates of the point at which the centerline intersects the plane? How far beneath ground level is this point?

10-9 Chapter Review and Test

In this chapter you have extended your knowledge of vectors to three-dimensional space. You extended the operations of addition, subtraction, and multiplication by a scalar to three dimensions simply by giving a third component to the vectors. Two new operations, dot product and cross product, allow you to "multiply" two vectors. Dot multiplication gives a scalar for the answer. Cross multiplication gives a vector perpendicular to the two factor vectors. You have seen how these techniques allow you to find the angle between two vectors, to project one vector onto another vector, to find equations of planes and lines in space, and to find areas of triangles and parallelograms in space.

Review Problems

R0. Update your journal with what you have learned in Chapter 10. Include such things as

- The difference between a position vector and a displacement vector
- The difference between a dot product and a cross product
- The difference between a scalar projection and a vector projection
- The difference between the way you *calculate* a dot product or cross product and what dot and cross products *mean*
- How dot products are used to find equations of planes in space
- How cross products are used to find areas of triangles and parallelograms in space

R1. Figure 10-9a shows two-dimensional vectors $\vec{a} = 3\vec{i} + 4\vec{j}$ and $\vec{b} = 7\vec{i} + 2\vec{j}$.

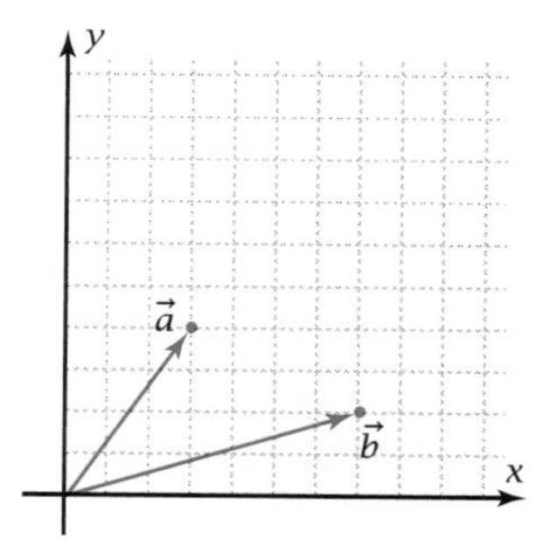

Figure 10-9a

a. Explain why these vectors, as shown, are position vectors.

b. On a copy of Figure 10-9a, sketch $\vec{r} = \vec{a} + \vec{b}$, $\vec{d} = \vec{a} - \vec{b}$, and $\vec{v} = 2\vec{a}$.

c. Write the displacement vector $\vec{d} = \vec{a} - \vec{b}$ in terms of its components.

d. Find the length of resultant vector $\vec{r} = \vec{a} + \vec{b}$.

e. Find the angle $\vec{a}$ makes with the x-axis.

R2. a. For position vectors $\vec{a}$ and $\vec{b}$ in Problem R1, find a displacement vector $\vec{d}$ from the head of $\vec{a}$ to the head of $\vec{b}$.

b. Find a displacement vector from the head of $\vec{a}$ to a point 40% of the way from the head of $\vec{a}$ to the head of $\vec{b}$.

c. Find the position vector for the point 40% of the way from the head of $\vec{a}$ to the head of $\vec{b}$.

d. Write the coordinates of the point 40% of the way from the head of $\vec{a}$ to the head of $\vec{b}$.

R3. a. Draw $\vec{v} = 5\vec{i} + 9\vec{j} + 4\vec{k}$ as a position vector. Show the "box" that makes it look three-dimensional. Indicate the unit vectors $\vec{i}$, $\vec{j}$, and $\vec{k}$ on the drawing.

b. If $\vec{a} = 6\vec{i} - 5\vec{j} + 2\vec{k}$ and $\vec{b} = 3\vec{i} + 4\vec{j} - 7\vec{k}$, find $3\vec{a} - 2\vec{b}$.

c. Find $|\vec{a}|$ for $\vec{a}$ in part b.

d. Find a unit vector in the direction of $\vec{a}$ in part b.

e. If $\vec{a}$ and $\vec{b}$ in part b are position vectors, find the displacement vector from the head of $\vec{a}$ to the head of $\vec{b}$.

f. For $\vec{a}$ and $\vec{b}$ in part b, find the position vector to the point 70% of the way from the head of $\vec{a}$ to the head of $\vec{b}$.

R4. a. Write the definition of dot product. Give two other names for dot product.

b. If $|\vec{a}| = 7$, $|\vec{b}| = 8$, and $\theta = 155°$, find $\vec{a} \cdot \vec{b}$.

c. If $|\vec{a}| = 10$, $|\vec{b}| = 20$, and $\vec{a} \cdot \vec{b} = -35$, find θ.

For Problems R4d–R4h, let $\vec{a} = 6\vec{i} - 5\vec{j} + 2\vec{k}$ and $\vec{b} = 3\vec{i} + 4\vec{j} - 7\vec{k}$.

d. Find $|\vec{a}|$, $|\vec{b}|$, and $\vec{a} \cdot \vec{b}$.

e. Find the angle between $\vec{a}$ and $\vec{b}$ when they are placed tail-to-tail.

f. Find a unit vector in the direction of $\vec{b}$.

g. Find the scalar projection of $\vec{a}$ on $\vec{b}$.

h. Find the vector projection of $\vec{a}$ on $\vec{b}$.

R5. a. Use a dot product to prove that if $\vec{n} = A\vec{i} + B\vec{j} + C\vec{k}$ is normal to a plane, then the equation of the plane is $Ax + By + Cz = D$, where D stands for a constant.

b. Write two normal vectors for the plane $3x - 7y + z = 5$, pointing in opposite directions.

c. Find the particular equation of the plane containing $(6, 2, -1)$, with normal vector $\vec{n} = 2\vec{i} - 7\vec{j} - 3\vec{k}$. Use the equation to find z for the point $P(10, 20, z)$ on the plane.

d. Find the particular equation of the plane perpendicular to the segment with endpoints $(5, 7, 2)$ and $(8, 13, 11)$ if the x-intercept of the plane is $x = 15$.

R6. a. Write the definition of $\vec{a} \times \vec{b}$. Give three names for $\vec{a} \times \vec{b}$.

b. If $|\vec{a}| = 7$, $|\vec{b}| = 8$, and $\theta = 155°$, find $|\vec{a} \times \vec{b}|$.

For Problems R6c–R6e, let $\vec{a} = 3\vec{i} + 2\vec{j} - \vec{k}$ and $\vec{b} = -4\vec{i} + 3\vec{j} + 5\vec{k}$.

c. Find $\vec{a} \times \vec{b}$ and $\vec{b} \times \vec{a}$.

d. Find $\vec{a} \cdot \vec{b}$ and $\vec{b} \cdot \vec{a}$.

e. Find the area of the triangle determined by $\vec{a}$ and $\vec{b}$.

f. Find the particular equation of the plane containing $(2, 5, 8)$, $(3, 7, 4)$, and $(-1, 9, 6)$.

R7. a. Sketch a position vector and show its three direction angles.

b. Find the direction cosines and the direction angles of $\vec{v} = 6\vec{i} - 8\vec{j} + 5\vec{k}$.

c. Vector $\vec{a}$ has direction cosines $c_1 = 0.2$ and $c_2 = -0.3$. Find the two possible values of c_3 and the two possible values of the third direction angle, γ.

d. Show algebraically, using the Pythagorean property, that there is no vector with direction angles $\alpha = 30°$ and $\beta = 40°$. Explain geometrically why such a vector cannot exist.

R8. For Problems R8a–R8d, the position vector $\vec{r}$ to a point on a line is given by the vector equation

$$\vec{r} = (6 + \tfrac{7}{9}d)\vec{i} + (3 + \tfrac{4}{9}d)\vec{j} + (2 - \tfrac{4}{9}d)\vec{k}$$

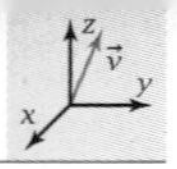

a. Write the fixed point and the fixed vector that appear in the equation. Show that the vector is a unit vector.

b. Show that you understand the meaning of the independent variable d in the equation by finding the coordinates of the point on the line that is a directed distance of -18 units from the fixed point. Explain the significance of the fact that d is negative in this case.

c. Find the coordinates of the point where the line intersects the xz-plane.

d. Find the coordinates of the point where the line intersects the plane $3x - 7y + z = 5$ from part b.

e. Find the vector equation of the line containing the points (2, 8, 4) and (11, 13, 7).

Concept Problems

C1. *Distance Between a Point and a Line Problem:* Figure 10-9b shows a line and a point P_1 not on the line. Vector $\vec{v}$ is parallel to the line, and d is the perpendicular distance between P_1 and the line.

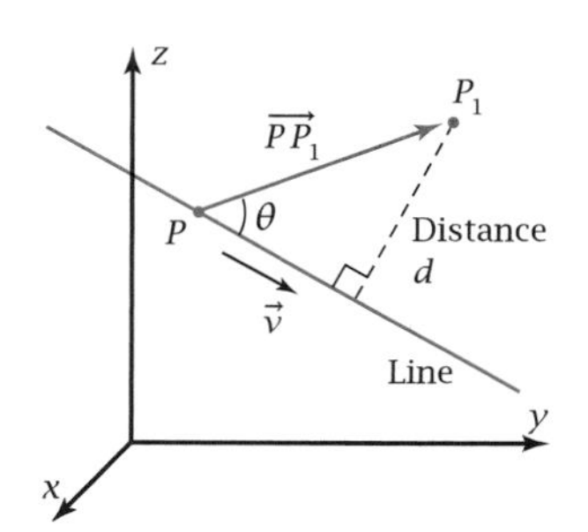

Figure 10-9b

Now suppose that the equation of the line is

$$\vec{r} = (5 + \tfrac{6}{11}t)\vec{i} + (3 - \tfrac{2}{11}t)\vec{j} + (-1 + \tfrac{9}{11}t)\vec{k}$$

and the outside point is $P_1(4, 7, 6)$.

a. By trigonometry, $d = |\overrightarrow{PP_1}| \cdot \sin\theta$. Multiply the right-hand side of this equation by 1 in the form $\frac{|\vec{v}|}{|\vec{v}|}$. Take advantage of the fact that the numerator on the right now equals $|\overrightarrow{PP_1} \times \vec{v}|$ to find d without first finding θ.

b. *Ladder Problem 2:* Figure 10-9c shows a 25-ft ladder leaning against a wall to reach a high window. To miss the flower bed, the ladder is moved over so that its left foot is at (7, 9, 0). The top of the ladder is 24 ft up the wall. Find a vector equation of the line along the left side of the ladder. Given that the rungs on the ladder are 1 ft apart, use the equation to find the rung that is closest to the upper left corner of the window, at the point (0, 5, 18). Find the *perpendicular* distance from the left side of the ladder to (0, 5, 18), taking advantage of the results of part a,

$$d = \frac{|\overrightarrow{PP_1} \times \vec{v}|}{|\vec{v}|}$$

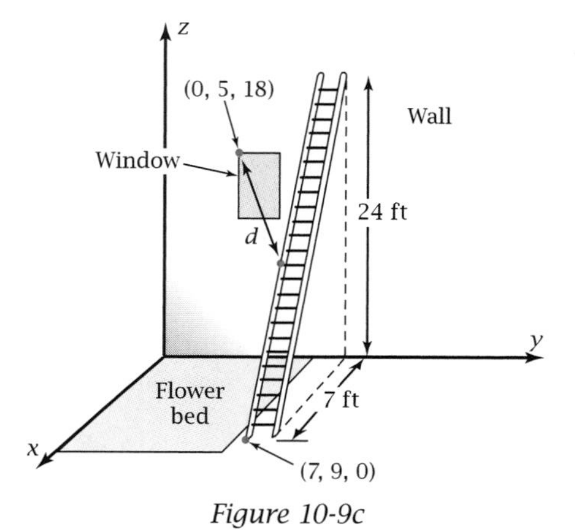

Figure 10-9c

C2. a. *Distance Between Skew Lines Problem:* Figure 10-9d shows Line 1 containing $P_1(3, 8, 5)$ and parallel to $\vec{v}_1 = 6\vec{i} + 3\vec{j} + 5\vec{k}$, and Line 2 containing $P_2(5, 2, 7)$ and parallel to $\vec{v}_2 = 9\vec{i} + 7\vec{j} + 1\vec{k}$. Lines 1 and 2 are **skew** lines because they are not parallel yet do not intersect. The cross product $\vec{v}_1 \times \vec{v}_2$ is perpendicular to both lines. The perpendicular distance d between the two lines is the absolute value of the scalar projection of $\overrightarrow{P_1P_2}$ on this cross product. Find this distance.

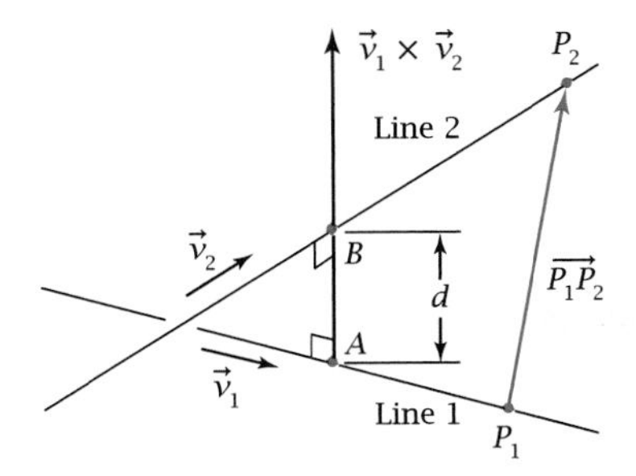

Figure 10-9d

b. *Airplane Near-Miss Velocity Vector Problem:* Flight 007 took off from the point $P_1(3000, 2000, 0)$ on one runway at an airport, where distances are in feet. At the same instant, Flight 1776 was at $P_2(1000, 500, 300)$ preparing to land on another runway that crosses the first one, as shown in Figure 10-9e. Computers in the control tower find that the velocity vectors for the two flights were

Flight 007: $\vec{v}_1 = -100\vec{i} + 50\vec{j} + 20\vec{k}$

Flight 1776: $\vec{v}_2 = 40\vec{i} + 200\vec{j} - 15\vec{k}$

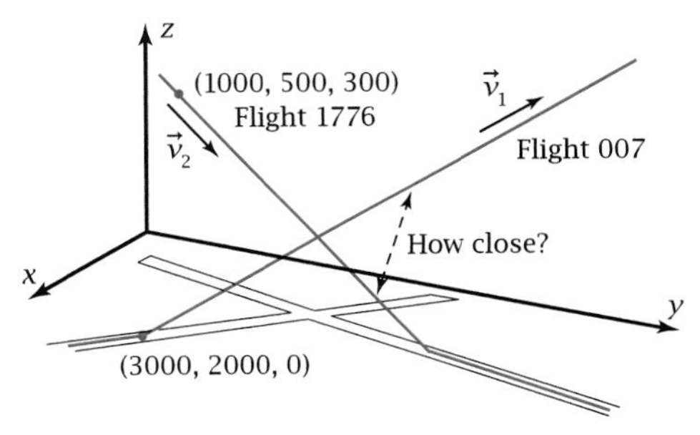

Figure 10-9e

Find the closest the two planes' paths came to each other using the result of part a, namely,

$$d = \frac{|(\vec{v}_1 \times \vec{v}_2) \cdot \overrightarrow{P_1P_2}|}{|\vec{v}_1 \times \vec{v}_2|}$$

c. The speeds in part b are in feet per second. Thus, the position vectors of the planes are

Flight 007: $\vec{r}_1 = (3000 - 100t)\vec{i} + (2000 + 50t)\vec{j} + (0 + 20t)\vec{k}$

Flight 1776: $\vec{r}_2 = (1000 + 40t)\vec{i} + (500 + 200t)\vec{j} + (300 - 15t)\vec{k}$

where the parameter t is time in seconds. Write the displacement vector from Flight 1776 to Flight 007 as a function of time. Using appropriate algebraic or numerical techniques, find the time t at which the flights were closest together by finding the value of t at which the length of the displacement vector was a minimum. Explain why the closest the flights came to each other is *not* the same as the closest the two paths came to each other. Is there cause for concern about how close the flights came to each other?

Chapter Test

PART 1: No calculators allowed (T1–T9)

T1. On a copy of Figure 10-9f,

a. Show the direction angles α, β, and γ for $\vec{v}$.

b. Mark the three unit vectors $\vec{i}$, $\vec{j}$, and $\vec{k}$.

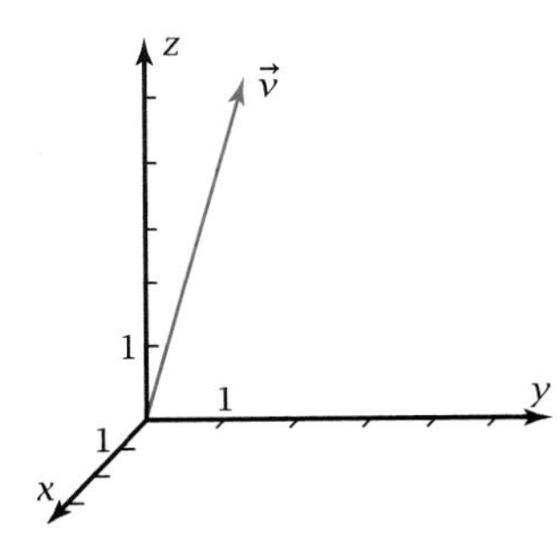

Figure 10-9f

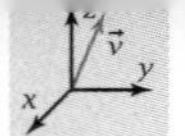

T2. On a copy of Figure 10-9g,

a. Sketch the cross product vector $\vec{a} \times \vec{b}$.

b. Sketch $\vec{p}$, the vector projection of $\vec{a}$ on $\vec{b}$.

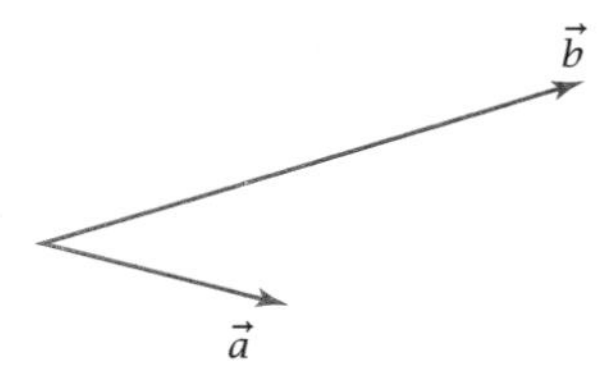

Figure 10-9g

T3. What is the major difference in meaning between $\vec{a} \cdot \vec{b}$ and $\vec{a} \times \vec{b}$?

T4. Write the definition of $\vec{a} \cdot \vec{b}$.

T5. Given: $\vec{a} = 3\vec{i} + 2\vec{j} + 4\vec{k}$

$\vec{b} = 1\vec{i} + 5\vec{j} + 2\vec{k}$

a. Find $\vec{a} \cdot \vec{b}$.

b. Find $\vec{a} \times \vec{b}$.

T6. How can you tell quickly whether or not one three-dimensional vector is perpendicular to another three-dimensional vector?

T7. What does it mean to say that one vector is normal to another vector?

T8. Write a vector normal to the plane $-13x + 10y - 5z = 22$.

T9. What makes a vector a unit vector?

PART 2: Graphing calculators allowed (T10–T31)

Problems T10–T17 refer to the vectors

$\vec{a} = 5\vec{i} + 2\vec{j} + 9\vec{k}$

$\vec{b} = 3\vec{i} + 8\vec{j} + 4\vec{k}$

T10. Find the resultant of $\vec{a}$ and $\vec{b}$.

T11. If $\vec{a}$ and $\vec{b}$ are placed tail-to-tail, find the displacement vector from the head of $\vec{b}$ to the head of $\vec{a}$.

T12. Find $|\vec{a}|$ and $|\vec{b}|$.

T13. Find a unit vector in the direction of $\vec{b}$.

T14. Find the angle between $\vec{a}$ and $\vec{b}$ if they are placed tail-to-tail.

T15. Find a vector perpendicular to both $\vec{a}$ and $\vec{b}$.

T16. Find the area of the triangle formed with $\vec{a}$ and $\vec{b}$ as two of its sides.

T17. Find the scalar projection of $\vec{a}$ on $\vec{b}$.

Problems T18–T22 refer to the line with vector equation

$$\vec{r} = (3 + \tfrac{8}{9}d)\vec{i} + (5 + \tfrac{1}{9}d)\vec{j} + (8 + \tfrac{4}{9}d)\vec{k}$$

T18. Write the coordinates of the fixed point that appears in the equation.

T19. Prove that vector $\vec{u} = \frac{8}{9}\vec{i} + \frac{1}{9}\vec{j} + \frac{4}{9}\vec{k}$ that appears in the equation is a *unit* vector.

T20. Find the point on the line that is at a directed distance of 27 units from the fixed point.

T21. Find the directed distance from the fixed point on the line to the point where the line pierces the *xy*-plane.

T22. Find γ, the direction angle the line makes with the *z*-axis.

Awning Problem 2: For Problems T23–T26, an awning is to be built in the corner of a building, as shown in Figure 10-9h. A vertical column on the left of the awning starts on the *x*-axis and ends at the point (10, 0, 7) on the awning. The dimensions are in feet. A normal vector to the plane of the awning is

$\vec{n} = 7\vec{i} + 5\vec{j} + 10\vec{k}$

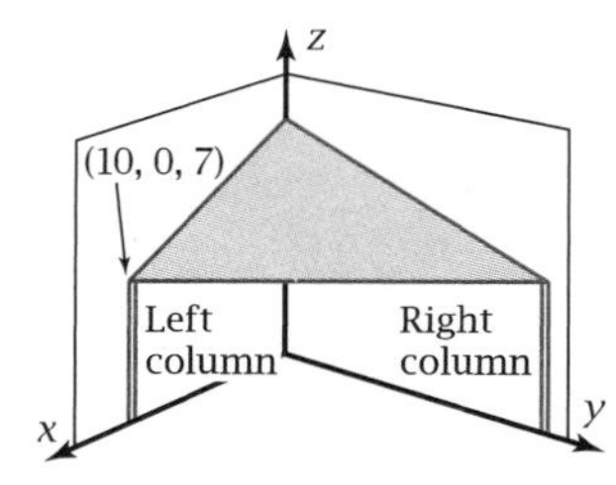

Figure 10-9h

T23. Find the particular equation of the plane.

T24. How long will the column on the right be, where $x = 0$ and $y = 12$ feet?

T25. How high will the awning be at the back corner, where the walls meet?

T26. A light fixture is to be located at the point (4, 6, 9). Find the vertical distance between the light and the awning. Is the light above the awning or below it? How can you tell?

T27. What did you learn as a result of taking this test that you did not know before?

CHAPTER 11

Matrix Transformations and Fractal Figures

The genes of living things contain instructions for the formations of their parts. Each part of the chambered nautilus shell is a transformation—a duplicate of the original shape, with a different size and location. Performing several simple operations iteratively (over and over, starting with the previous result) can produce an amazingly complicated object with self-similar parts.

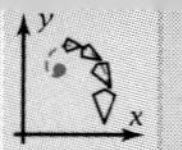

Mathematical Overview

In this chapter you will learn how to use matrices (the plural of *matrix*) to transform two-dimensional figures into complex images. A single transformation, iterated many times, transforms a simple trapezoid into the snail-like figure shown in Figure 11-0a. Iterating several matrix transformations can produce figures so complex that they have fractional dimensions. You will study such *fractal* figures in four ways.

Numerically

$$\begin{bmatrix} 0.9\cos 30^\circ & 0.9\cos 120^\circ & 6 \\ 0.9\sin 30^\circ & 0.9\sin 120^\circ & 2 \\ 0 & 0 & 1 \end{bmatrix}\begin{bmatrix} 2 & -2 & -5 & 5 \\ 5 & 5 & -5 & -5 \\ 1 & 1 & 1 & 1 \end{bmatrix}$$

$$= \begin{bmatrix} 5.30\ldots & 2.19\ldots & 4.35\ldots & 12.11\ldots \\ 6.79\ldots & 4.99\ldots & -4.14\ldots & 0.35\ldots \\ 1 & 1 & 1 & 1 \end{bmatrix}$$

Graphically Spiraling trapezoids

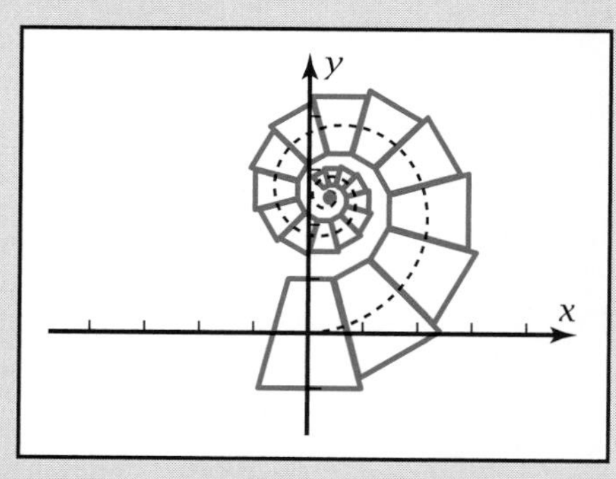

Figure 11-0a

Algebraically 50th iteration: [Image] = $[\text{Transformation}]^{50}$ [Preimage]

Verbally *Images can have fractional dimensions! Wow! I can't imagine how this can be true, but I see how to calculate it mathematically. I'm still not sure how these images can be done one point at a time by Barnsley's method.*

11-1 Introduction to Iterated Transformations

In this section you will explore what happens to a two-dimensional figure when you perform the same transformation over and over, each time applying the transformation to the result of the previous transformation. This process is called **iteration.** The result of each transformation is also called an iteration.

OBJECTIVE See what happens to the perimeter and area of a square when you perform the same set of transformations repeatedly (iteratively).

Exploratory Problem Set 11-1

The left diagram in Figure 11-1a shows a 10-cm by 10-cm square. To create the middle diagram, the original, or *pre-image,* square was transformed into four similar squares, each with sides that are 40% of the original side length. These image squares were then translated so that each has a corner at one of the corners of the pre-image. The right diagram shows the result of applying the same transformation to each of the four squares from the first iteration. In this problem set you will explore the perimeter and area of various iterations.

1. Find the perimeter and area of the pre-image square. Find the *total* perimeter and area of the four squares in the first iteration. Find the *total* perimeter and area of the 16 squares in the second iteration. Display the answers in a table with these column headings: Iteration number, Side length, Total perimeter, and Total area.
2. What pattern do you notice that relates the total perimeter to the iteration number? What pattern relates the total area to the iteration number? What pattern relates the total area to the total perimeter?
3. Using the patterns you observed in Problem 2, find the total perimeter and total area of the third and fourth iterations.
4. Calculate the total perimeter and the total area of the 20th iteration.
5. If the iterations could be carried on infinitely many times, the images would approach a figure called **Sierpiǹski's carpet** or Sierpiǹski's square. What would be the total perimeter of this figure? What would be the total area? Does the answer surprise you? (In this chapter you will encounter other surprises, such as the fact that this figure is less than two-dimensional but more than one-dimensional!)

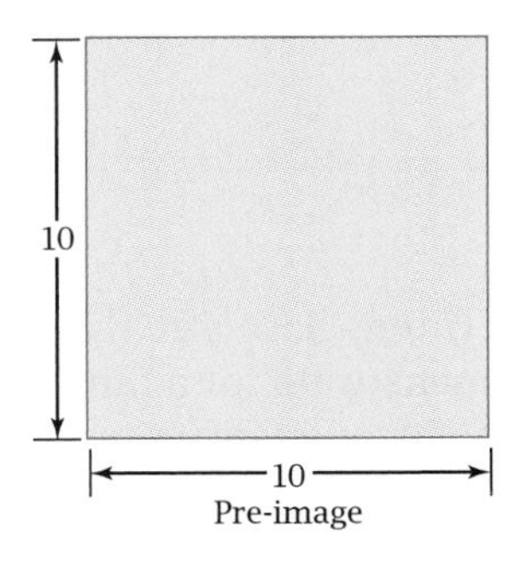

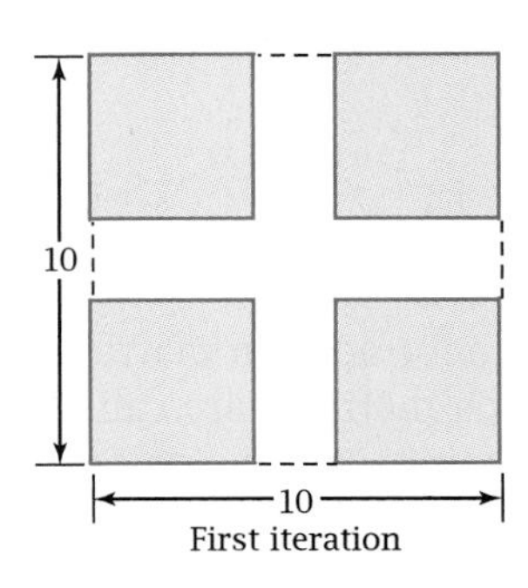

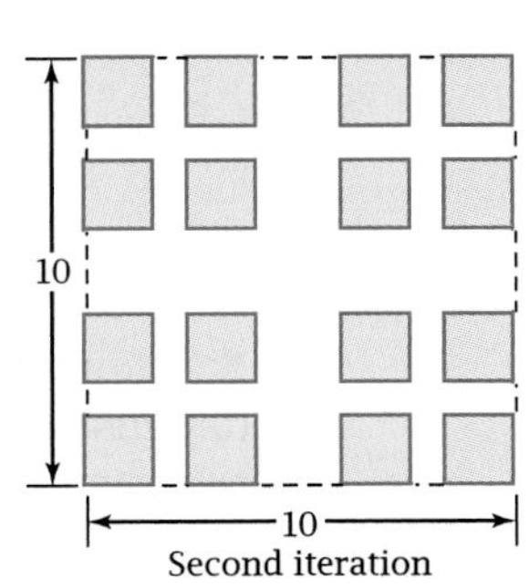

Figure 11-1a

11-2 Matrix Operations and Solutions of Linear Systems

In Section 11-1 a pre-image square was duplicated, dilated, and translated to create four new squares. Iterating these transformations many times creates images that approach a figure with an infinite number of pieces, zero area, and an infinite perimeter. In this section you will refresh your memory about matrices, which you may have studied in earlier courses. In the rest of the chapter you will see how you can use matrices to perform these geometric transformations algebraically.

OBJECTIVES

- Given two matrices, find their sum and product.
- Given a square matrix, find its multiplicative inverse.
- Use matrices to solve a system of linear equations.

A **matrix** is a rectangular array of numbers.

$$\begin{bmatrix} 2 & 5 & 3 \\ -1 & 4 & -2 \end{bmatrix} \quad \begin{bmatrix} 5 & 1 \\ 7 & 3 \\ 2 & -4 \end{bmatrix} \quad \begin{bmatrix} 9 \\ 7 \\ 1 \\ 3 \end{bmatrix} \quad \begin{bmatrix} 2 \\ -5 \end{bmatrix} \quad \begin{bmatrix} 2 & 3 & -4 \\ -1 & 5 & 7 \\ 9 & -8 & -6 \end{bmatrix}$$

2 × 3 matrix 3 × 2 matrix 4 × 1 matrix 2 × 1 matrix 3 × 3 (square) matrix

The numbers in a matrix are called **elements.** When the dimensions of a matrix are stated, the number of rows is always given first. So the first matrix above is called a 2 × 3 (read "two by three") matrix because it has 2 *rows* and 3 *columns.* A *square matrix* has the same number of rows and columns. In this section you will see how to perform operations on matrices.

Addition and Subtraction

You add or subtract two matrices with the same dimensions by adding or subtracting their corresponding elements. For instance,

$$\begin{bmatrix} 5 & 2 & 7 \\ 1 & 3 & 9 \end{bmatrix} + \begin{bmatrix} 4 & 6 & 8 \\ 5 & 1 & 3 \end{bmatrix} = \begin{bmatrix} 9 & 8 & 15 \\ 6 & 4 & 12 \end{bmatrix}$$

Likewise,

$$\begin{bmatrix} 6 & 5 \\ 2 & 3 \end{bmatrix} - \begin{bmatrix} 4 & 7 \\ 5 & 2 \end{bmatrix} = \begin{bmatrix} 2 & -2 \\ -3 & 1 \end{bmatrix}$$

You can add or subtract two matrices only if they have exactly the same dimensions. Such matrices are called **commensurate** for addition or subtraction. If you try adding or subtracting incommensurate matrices on your calculator, you will get an error message.

Multiplication by a Scalar

To multiply a matrix by a scalar (a number), multiply each element of the matrix by that scalar. For instance,

$$5\begin{bmatrix} 2 & 3 & 4 \\ 6 & 1 & -7 \end{bmatrix} = \begin{bmatrix} 10 & 15 & 20 \\ 30 & 5 & -35 \end{bmatrix}$$

Multiplication by a scalar is equivalent to repeated addition. For example, multiplying a matrix by 5 is equivalent to adding together five of the matrices.

Multiplication of Two Matrices

Traffic controllers often use matrices to analyze traffic flow.

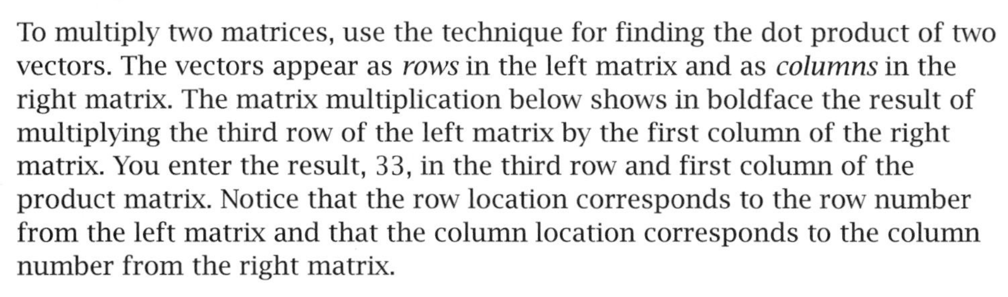

To multiply two matrices, use the technique for finding the dot product of two vectors. The vectors appear as *rows* in the left matrix and as *columns* in the right matrix. The matrix multiplication below shows in boldface the result of multiplying the third row of the left matrix by the first column of the right matrix. You enter the result, 33, in the third row and first column of the product matrix. Notice that the row location corresponds to the row number from the left matrix and that the column location corresponds to the column number from the right matrix.

$$\begin{bmatrix} 2 & 4 & 1 & 3 \\ 1 & 3 & 2 & -1 \\ \mathbf{5} & \mathbf{-2} & \mathbf{3} & \mathbf{0} \end{bmatrix} \begin{bmatrix} \mathbf{6} & 5 \\ \mathbf{0} & 2 \\ \mathbf{1} & -3 \\ \mathbf{2} & 1 \end{bmatrix} = \begin{bmatrix} 19 & 18 \\ 6 & 4 \\ \mathbf{33} & 12 \end{bmatrix}$$

$$(\text{row } 3) \cdot (\text{column } 1) = (5)(6) + (-2)(0) + (3)(1) + (0)(2) = 33$$

For practice, see if you can calculate the other five elements of the product matrix in your head. To keep your place, slide your left index finger along the row of the left matrix while sliding your right index finger down the column of the right matrix.

Make note of these important statements about matrix multiplication.

- To be commensurate for multiplication, the *rows* of the left matrix must have the same number of elements as the *columns* of the right matrix.
- The product matrix has the same number of *rows* as the left matrix and the same number of *columns* as the right matrix.
- Commuting two matrices [A] and [B] might make them **incommensurate** for multiplication. Even when the commuted matrices are commensurate, the product [B][A] can be different from the product [A][B]. Thus matrix multiplication is *not commutative.*

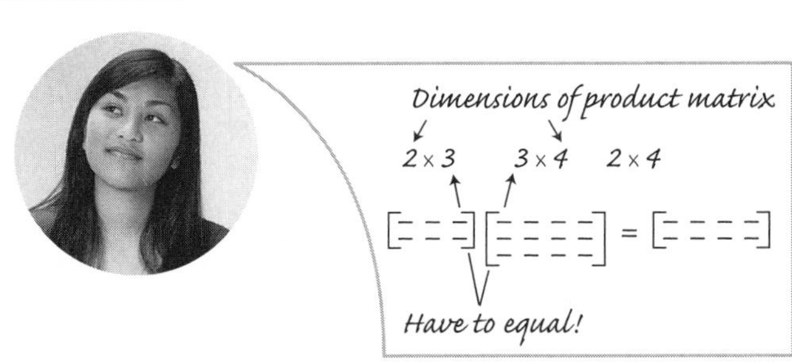

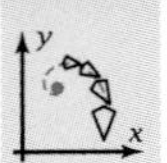

Identities and Inverses for Multiplication of Square Matrices

The 3 × 3 **identity matrix** is the square matrix

$$[\mathrm{I}] = \begin{bmatrix} 1 & 0 & 0 \\ 0 & 1 & 0 \\ 0 & 0 & 1 \end{bmatrix}$$

An identity matrix is a square matrix with 1s along the *main diagonal* (the diagonal from the upper left corner to the lower right corner) and 0s everywhere else. Multiplying a square matrix [A] by the identity matrix in either order leaves [A] unchanged, as you can check either manually or by calculator.

$$\begin{bmatrix} 1 & 0 & 0 \\ 0 & 1 & 0 \\ 0 & 0 & 1 \end{bmatrix}\begin{bmatrix} 4 & -7 & 5 \\ 3 & 6 & 9 \\ 2 & 8 & -1 \end{bmatrix} = \begin{bmatrix} 4 & -7 & 5 \\ 3 & 6 & 9 \\ 2 & 8 & -1 \end{bmatrix} \quad \text{and}$$

$$\begin{bmatrix} 4 & -7 & 5 \\ 3 & 6 & 9 \\ 2 & 8 & -1 \end{bmatrix}\begin{bmatrix} 1 & 0 & 0 \\ 0 & 1 & 0 \\ 0 & 0 & 1 \end{bmatrix} = \begin{bmatrix} 4 & -7 & 5 \\ 3 & 6 & 9 \\ 2 & 8 & -1 \end{bmatrix}$$

If the product of two square matrices is the identity matrix, then those two matrices are **inverses** of each other. The inverse of a matrix [M] is denoted $[\mathrm{M}]^{-1}$. For instance,

$$\begin{bmatrix} 3 & 2 \\ 8 & 7 \end{bmatrix}\begin{bmatrix} \frac{7}{5} & -\frac{2}{5} \\ -\frac{8}{5} & \frac{3}{5} \end{bmatrix} = \begin{bmatrix} 1 & 0 \\ 0 & 1 \end{bmatrix} \quad \text{so} \quad \begin{bmatrix} 3 & 2 \\ 8 & 7 \end{bmatrix}^{-1} = \begin{bmatrix} \frac{7}{5} & -\frac{2}{5} \\ -\frac{8}{5} & \frac{3}{5} \end{bmatrix}$$

Notice that the numerators in the inverse matrix are the elements of the original matrix but rearranged and changed in sign. The denominators are all 5, which is the **determinant** of the first matrix, written det [M]. You can find the determinant of a matrix by using the built-in features on your grapher. For a 2 × 2 matrix, you can calculate the determinant as

det [M] = (upper left times lower right) − (upper right times lower left)

$$\det\begin{bmatrix} 3 & 2 \\ 8 & 7 \end{bmatrix} = (3)(7) - (2)(8) = 5$$

An easy way to find the inverse of a 2 × 2 matrix is to interchange the top left and bottom right elements, change the signs of the other two elements, and multiply by the reciprocal of the determinant of the matrix. That is,

$$\text{If } [\mathrm{M}] = \begin{bmatrix} a & b \\ c & d \end{bmatrix} \quad \text{then} \quad [\mathrm{M}]^{-1} = \frac{1}{\det\,[\mathrm{M}]}\begin{bmatrix} d & -b \\ -c & a \end{bmatrix}$$

The matrix $\begin{bmatrix} d & -b \\ -c & a \end{bmatrix}$ is called the **adjoint** of [M], abbreviated adj [M]. You can calculate the inverse of any square matrix as

$$[\mathrm{M}]^{-1} = \frac{1}{\det\,[\mathrm{M}]} \cdot \text{adj}\,[\mathrm{M}]$$

For square matrices of higher dimension, the adjoint is harder to compute. In this text you will use the built-in feature of your grapher to find the inverse of a matrix.

▶ EXAMPLE 1 Consider the matrix $[M] = \begin{bmatrix} 2 & 3 & 4 \\ 5 & 1 & 2 \\ 6 & 8 & 7 \end{bmatrix}$.

a. Find $[M]^{-1}$, and show that $[M]^{-1}[M]$ equals the identity matrix.

b. Find det [M] and adj [M].

Solution a. Enter [M] in the matrix menu of your grapher. Then press $[M]^{-1}$.

$$[M]^{-1} = \begin{bmatrix} -0.1836... & 0.2244... & 0.0408... \\ -0.4693... & -0.2040... & 0.2365... \\ 0.6938... & 0.0408... & -0.2653... \end{bmatrix}$$

To multiply the answer by [M], press ans*[M] or $[M]^{-1}[M]$:

$$[M]^{-1}[M] = \begin{bmatrix} 1 & 0 & 0 \\ 0 & 1 & 0 \\ 0 & 0 & 1 \end{bmatrix}$$ This is the 3×3 identity matrix.

b. det [M] = 49 Press det [M] on your grapher.

Most graphers cannot calculate the adjoint of a matrix directly. However,

$$[M]^{-1} = \frac{1}{\det [M]} \cdot \text{adj } [M]$$

so you can find adj [M] this way:

$$\text{adj } [M] = 49\ [M]^{-1} = \begin{bmatrix} -9 & 11 & 2 \\ -23 & -10 & 16 \\ 34 & 2 & -13 \end{bmatrix}$$ ◀

Note that if the elements of [M] are integers, then the elements of adj [M] are also integers.

Matrix Solution of a Linear System

You can use the inverse of a matrix to solve a system of linear equations, such as

$$2x + 3y + 4z = 14$$
$$5x + y + 2z = 24$$
$$6x + 8y + 7z = 29$$

You can write the left sides of these three equations as the product of two matrices:

$$\begin{bmatrix} 2 & 3 & 4 \\ 5 & 1 & 2 \\ 6 & 8 & 7 \end{bmatrix} \begin{bmatrix} x \\ y \\ z \end{bmatrix} = \begin{bmatrix} 2x + 3y + 4z \\ 5x + y + 2z \\ 6x + 8y + 7z \end{bmatrix}$$

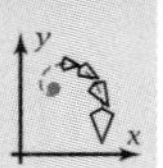

So you can write the system as

$$\begin{bmatrix} 2 & 3 & 4 \\ 5 & 1 & 2 \\ 6 & 8 & 7 \end{bmatrix}\begin{bmatrix} x \\ y \\ z \end{bmatrix} = \begin{bmatrix} 14 \\ 24 \\ 29 \end{bmatrix}$$

Let [C] stand for the coefficient matrix on the far left, [V] for the 3×1 matrix containing the variables, and [A] for the 3×1 matrix on the right containing the "answers." You can write this system (or any other system of n linear equations in n variables) in matrix form:

$$[C][V] = [A]$$

To find the values of the variables in matrix [V], you can eliminate [C] by left-multiplying both sides of the equation by $[C]^{-1}$:

$$[C]^{-1}([C][V]) = [C]^{-1}[A]$$

Associating the $[C]^{-1}$ and [C] gives the identity matrix [I].

$$([C]^{-1}[C])[V] = [C]^{-1}[A]$$

$$[I][V] = [C]^{-1}[A]$$

$$[V] = [C]^{-1}[A] \qquad \text{or} \qquad [C]^{-1}[A] = [V]$$

For these equations, the calculation is

$$\begin{bmatrix} x \\ y \\ z \end{bmatrix} = \begin{bmatrix} 2 & 3 & 4 \\ 5 & 1 & 2 \\ 6 & 8 & 7 \end{bmatrix}^{-1}\begin{bmatrix} 14 \\ 24 \\ 29 \end{bmatrix} = \begin{bmatrix} 4 \\ -2 \\ 3 \end{bmatrix}$$

The solution is $x = 4$, $y = -2$, and $z = 3$.

Problem Set 11-2

Do These Quickly

5 min

Q1. A square dilated to 40% of its original length has —?— percent of the original area.

Q2. After two iterations, Sierpiński's square has a total area that is —?— percent of the pre-image area.

Q3. After two iterations, Sierpiński's square has a total perimeter that is —?— percent of the pre-image perimeter.

Q4. What kind of function has the add-multiply property for regularly spaced x-values?

Q5. Write the general equation for a power function.

Q6. Find the slope of the line perpendicular to the graph of $3x + 7y = 41$.

Q7. How many degrees are there in an angle of $\frac{\pi}{3}$ radians?

Q8. Expand the square $(3x - 5)^2$.

Q9. Find 2% of 3000.

Q10. Find $(3\vec{i} - 2\vec{j} - 7\vec{k}) \cdot (8\vec{i} + 6\vec{j} - \vec{k})$.

For Problems 1–10, perform the given operations by hand. Use your grapher to confirm that your answers are correct.

1. $\begin{bmatrix} 3 & 5 \\ -2 & 4 \\ 7 & 1 \end{bmatrix} + \begin{bmatrix} -5 & 8 \\ 2 & 6 \\ -7 & 10 \end{bmatrix}$

2. $\begin{bmatrix} 5 & 7 & -4 \\ 10 & 0 & -2 \\ 11 & -3 & 12 \end{bmatrix} - \begin{bmatrix} 4 & 5 & -7 \\ 6 & -5 & -8 \\ 4 & -11 & 3 \end{bmatrix}$

3. $4[-8 \quad 5 \quad 3] - 2[-5 \quad -1 \quad 7]$

4. $7\begin{bmatrix} 2 & 8 \\ -4 & 1 \end{bmatrix} + 3\begin{bmatrix} -5 & 1 \\ 2 & -6 \end{bmatrix}$

5. $[-2 \quad 3 \quad 5]\begin{bmatrix} 1 & 4 \\ 7 & -3 \\ -1 & -5 \end{bmatrix}$

6. $\begin{bmatrix} 2 & 4 & -3 \\ 5 & 1 & 2 \\ -1 & 3 & 4 \end{bmatrix}\begin{bmatrix} -1 & 3 & 1 \\ 2 & 4 & 3 \\ 1 & 0 & 2 \end{bmatrix}$

7. $\begin{bmatrix} 4 & 7 & 5 \\ 3 & 2 & -1 \end{bmatrix}\begin{bmatrix} 6 & 8 \\ 3 & -6 \end{bmatrix}$

8. $\begin{bmatrix} 1 & 0 & 0 \\ 0 & 1 & 0 \\ 0 & 0 & 1 \end{bmatrix}\begin{bmatrix} 1 & 4 & 7 \\ 2 & 5 & 8 \\ 3 & 6 & 9 \end{bmatrix}$

9. $\begin{bmatrix} 4 & 4 \\ 5 & 3 \end{bmatrix}\begin{bmatrix} 1 & 0 \\ 0 & 1 \end{bmatrix}$

10. $\begin{bmatrix} 2 & 3 & 4 \\ 5 & 1 & 2 \\ 6 & 5 & 7 \end{bmatrix}\begin{bmatrix} -3 & -1 & 2 \\ -23 & -10 & 16 \\ 19 & 8 & -13 \end{bmatrix}$

11. *Investment Income Problem:* A brokerage company has investments in four states: California, Arkansas, Texas, and South Dakota. The investments are bonds, mortgages, and loans. Matrix [M] shows the numbers of millions of dollars in each investment in each state.

$$[M] = \begin{array}{c} \text{CA} \quad \text{AR} \quad \text{TX} \quad \text{SD} \\ \begin{bmatrix} 32 & 8 & 15 & 2 \\ 15 & 20 & 17 & 9 \\ 14 & 22 & 23 & 7 \end{bmatrix} \end{array} \begin{array}{l} \text{Bonds} \\ \text{Mortgages} \\ \text{Loans} \end{array}$$

The percentages of annual income that the investments yield are bonds, 6%; mortgages, 9%; loans, 11%. These numbers are shown in the yield matrix [Y].

$$[Y] = [0.06 \quad 0.09 \quad 0.11]$$

a. Find the product [Y][M]. Use the product matrix to find the annual income the company gets from investments in Texas. How much of this comes from mortgages?

b. Explain why you cannot find the real-world product [M][Y].

c. Explain why it is impossible in the mathematical world to find the product [M][Y].

12. *Virus Problem:* A virus sweeps through a high school, infecting 30% of the 11th graders and 20% of the 12th graders, as represented by matrix [P].

$$[P] = \begin{array}{c} \text{11th} \quad \text{12th} \\ \begin{bmatrix} 0.3 & 0.2 \\ 0.7 & 0.8 \end{bmatrix} \end{array} \begin{array}{l} \text{Ill} \\ \text{Well} \end{array}$$

There are 100 11th grade boys, 110 11th grade girls, 120 12th grade boys, and 130 12th grade girls, as represented by matrix [S].

$$[S] = \begin{array}{c} \text{Boys} \quad \text{Girls} \\ \begin{bmatrix} 100 & 110 \\ 120 & 130 \end{bmatrix} \end{array} \begin{array}{l} \text{11th Grade} \\ \text{12th Grade} \end{array}$$

a. Show that [P][S] does not equal [S][P].

b. Identify the real-world quantities that the elements of [P][S] represent.

c. Identify the real-world quantities that the elements of [S][P] represent.

For Problems 13 and 14,

a. Find $[M]^{-1}$. Show that $[M]^{-1}[M] = [I]$ and $[M][M]^{-1} = [I]$.

b. Find det [M]. Find adj [M] and show that all the elements of adj [M] are integers.

13. $[M] = \begin{bmatrix} 3 & 5 & 2 \\ 4 & 7 & 7 \\ 5 & 8 & 9 \end{bmatrix}$

14. $[M] = \begin{bmatrix} 3 & 7 & 1 & -2 \\ 4 & 5 & -1 & 6 \\ 2 & 3 & 8 & 1 \\ -5 & 4 & 9 & 7 \end{bmatrix}$

For Problems 15 and 16, find det [M]. Explain why your grapher gives you an error message when you try to find $[M]^{-1}$. Then state what you think a determinant "determines."

15. $[M] = \begin{bmatrix} 6 & 3 \\ 8 & 4 \end{bmatrix}$

16. $[M] = \begin{bmatrix} 1 & 2 & 3 \\ 4 & 5 & 6 \\ 7 & 8 & 9 \end{bmatrix}$

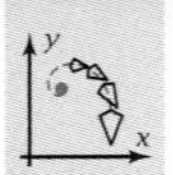

For Problems 17 and 18, solve the system using the inverse of a matrix.

17. $$\begin{aligned} 5x + 3y - 7z &= 3 \\ 10x - 4y + 6z &= 5 \\ 15x + y - 8z &= -2 \end{aligned}$$

18. $$\begin{aligned} w - 5x + 2y - z &= -18 \\ 3w + x - 3y + 2z &= 17 \\ 4w - 2x + y - z &= -1 \\ -2w + 3x - y + 4z &= 11 \end{aligned}$$

19. *Quadratic Function Problem 2:* Recall that a quadratic function has the general equation $y = ax^2 + bx + c$. To find the equation of the particular function that contains the points (4, 13), (6, 29), and (8, 49), you can substitute each pair of x- and y-values into the general equation to get three linear equations with the three unknown constants a, b, and c. Solve them as a system to find the particular quadratic function that contains these three points. Use the equation to predict the value of y when $x = 20$.

20. *Quartic Function Problem:* The general equation of a quartic (fourth-degree) function is $y = ax^4 + bx^3 + cx^2 + dx + e$, where a, b, c, d, and e stand for constants. Find the particular equation of the quartic function that contains the points (1, 15), (2, 19), (3, 75), (4, 273), and (5, 751). Use the equation to predict the value of y when $x = -3$.

21. Show that matrix multiplication is *not* commutative by showing that

$$\begin{bmatrix} 2 & 3 \\ 4 & 5 \end{bmatrix}\begin{bmatrix} 6 & 7 \\ 8 & 9 \end{bmatrix} \neq \begin{bmatrix} 6 & 7 \\ 8 & 9 \end{bmatrix}\begin{bmatrix} 2 & 3 \\ 4 & 5 \end{bmatrix}$$

22. Show that matrix multiplication is not a **well-defined** operation, because not all ordered pairs of matrices can be multiplied.

23. *Multipliers of Zero Problem 2:* For real numbers, the zero-product property states that if the product of two factors is zero, then at least one of the factors is zero. Show that this property is *false* for matrix multiplication by finding two 2×2 matrices whose product is the zero matrix (the matrix in which each element is 0) but for which no element of either matrix is 0. (Find a matrix whose determinant is 0, and multiply it by its adjoint matrix.) The two matrices you find are called **multipliers of zero.**

11-3 Rotation and Dilation Matrices

Matrices have many uses in the real world and in the mathematical world. One use is transforming geometric figures. In this section you will see how to write a matrix that will rotate or dilate a figure.

OBJECTIVE Given a desired dilation and rotation, write a matrix that will perform the transformations when it is multiplied by a matrix representing a geometric figure.

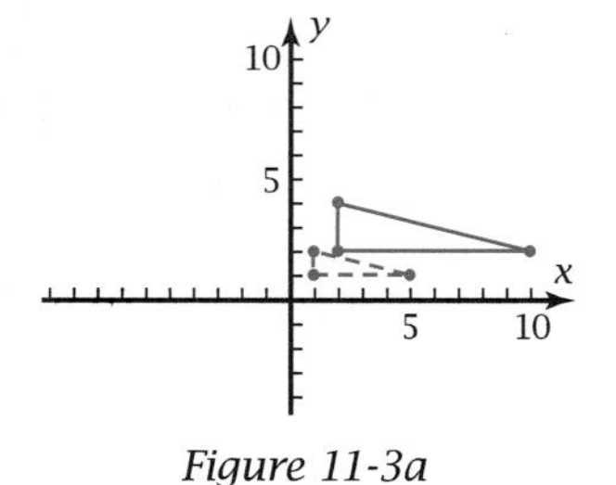

Figure 11-3a

Dilations

You can represent a figure in the plane by a matrix with two rows. For instance, you can represent the smaller triangle shown in Figure 11-3a by this matrix:

$$[M] = \begin{bmatrix} 1 & 5 & 1 \\ 1 & 1 & 2 \end{bmatrix}$$

Each column represents an ordered pair corresponding to one of the vertices. The top element is the x-coordinate and the bottom element is the y-coordinate.

If you multiply the identity matrix by 2, you get

$$[T] = 2[I] = 2\begin{bmatrix} 1 & 0 \\ 0 & 1 \end{bmatrix} = \begin{bmatrix} 2 & 0 \\ 0 & 2 \end{bmatrix}$$

You can use matrix [T] as a **transformation matrix.** The product [T][M] is the **image matrix.**

$$[T][M] = \begin{bmatrix} 2 & 10 & 2 \\ 2 & 2 & 4 \end{bmatrix}$$

If you plot the ordered pairs represented by this matrix, you get the vertices of the larger (solid) triangle in Figure 11-3a. The sides of this **image** triangle are twice as long as those of the original triangle (the **pre-image**). Each point on the image is twice as far from the origin as the corresponding point on the pre-image. The transformation represented by [T] dilates the entire Cartesian plane by a factor of 2, doubling the sides of any figure on the plane.

In general, you dilate by a factor of k by multiplying by the general dilation matrix.

PROPERTY: General Dilation Matrix

Matrix [T] dilates a figure by a factor of k with respect to the origin:

$$[T] = \begin{bmatrix} k & 0 \\ 0 & k \end{bmatrix}$$

Rotations

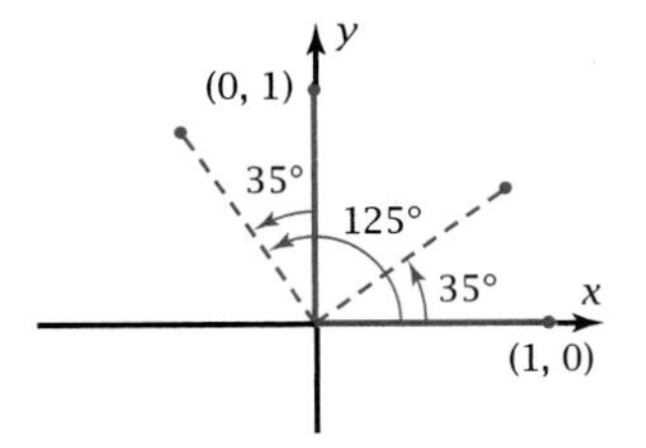

Figure 11-3b

Three-dimensional video images are created using iterated transformations of a pre-image.

You can write the identity matrix this way:

$$\begin{bmatrix} 1 & 0 \\ 0 & 1 \end{bmatrix} = \begin{bmatrix} \cos 0^\circ & \cos 90^\circ \\ \sin 0^\circ & \sin 90^\circ \end{bmatrix}$$

The 1 and 0 in the first column are the coordinates of the endpoint of a unit vector along the positive x-axis. Similarly, the 0 and 1 in the second column are the coordinates of the endpoint of a unit vector pointing in the positive direction on the y-axis. A rotation of 35° counterclockwise moves the endpoints of these unit vectors as shown in Figure 11-3b. The coordinates of the new endpoints are (cos 35°, sin 35°) and (cos 125°, sin 125°), respectively. Replacing the 0° and 90° in the identity matrix with, respectively, 35° and 125° gives a matrix [T] that rotates a figure 35° counterclockwise with respect to the origin.

$$[T] = \begin{bmatrix} \cos 35^\circ & \cos 125^\circ \\ \sin 35^\circ & \sin 125^\circ \end{bmatrix}$$

The angle in the first column is the amount by which the figure is rotated. The angle in the second column is 90° more than the rotation angle. The general rotation matrix for an angle of θ is shown in the following box.

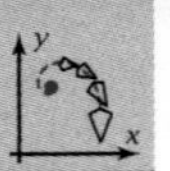

PROPERTY: General Rotation Matrix

Matrix [T] rotates a figure in the plane counterclockwise by an angle of θ.

$$[T] = \begin{bmatrix} \cos\theta & \cos(\theta + 90°) \\ \sin\theta & \sin(\theta + 90°) \end{bmatrix}$$

Verbally: "The *x*-axis rotates to the position of angle θ. The *y*-axis rotates to the position of angle $\theta + 90°$."

▶ EXAMPLE 1 Write a transformation matrix that will rotate a figure clockwise 70° and dilate it by a factor of 1.6. Use it to transform the small pre-image triangle in Figure 11-3a. Plot the pre-image and the image on graph paper.

Solution The desired transformation matrix is the product of the dilation matrix and the rotation matrix. Note that the rotation is clockwise, so the rotation angle is −70°.

$$[T] = \begin{bmatrix} 1.6 & 0 \\ 0 & 1.6 \end{bmatrix}\begin{bmatrix} \cos(-70°) & \cos 20° \\ \sin(-70°) & \sin 20° \end{bmatrix} \qquad \text{The } 20° \text{ is } 90° + (-70°).$$

$$= \begin{bmatrix} 1.6\cos(-70°) & 1.6\cos 20° \\ 1.6\sin(-70°) & 1.6\sin 20° \end{bmatrix}$$

To find the image matrix, multiply [T] by the earlier pre-image matrix [M]:

$$[T] = \begin{bmatrix} 1.6\cos(-70°) & 1.6\cos 20° \\ 1.6\sin(-70°) & 1.6\sin 20° \end{bmatrix}\begin{bmatrix} 1 & 5 & 1 \\ 1 & 1 & 2 \end{bmatrix}$$

$$= \begin{bmatrix} 2.0507\ldots & 4.2396\ldots & 3.5542\ldots \\ -0.9562\ldots & -6.9703\ldots & -0.4090\ldots \end{bmatrix}$$

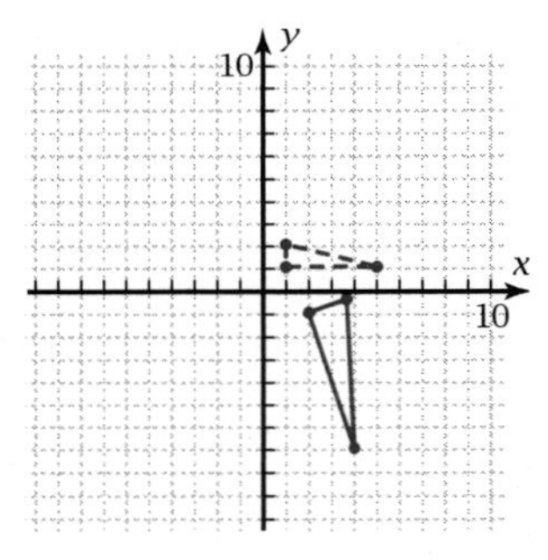

Figure 11-3c

Figure 11-3c shows the pre-image (dashed) and the image (solid) plotted on graph paper. You can confirm with a protractor that the image has been rotated 70° clockwise. You can confirm with a ruler that each point on the image is 1.6 times as far from the origin as the corresponding point on the pre-image. Thus, the lengths of sides in the image have been multiplied by 1.6. ◀

Iterated Transformations

Iterating a transformation means performing the transformation over and over again, each time operating on the image that resulted from the previous transformation.

▶ EXAMPLE 2 Write a transformation matrix [A] that will rotate a figure counterclockwise 60° and dilate it by a factor of 0.9. Starting with the pre-image triangle [M] from Example 1, apply [A] iteratively for four iterations. Plot the pre-image and each of the four images on graph paper. Describe the pattern formed by the triangles.

Solution

$$[A] = \begin{bmatrix} 0.9\cos 60° & 0.9\cos 150° \\ 0.9\sin 60° & 0.9\sin 150° \end{bmatrix}$$

$$[A][M] \approx \begin{bmatrix} -0.3 & 1.5 & -1.1 \\ 1.2 & 4.3 & 1.7 \end{bmatrix}$$

Display only one decimal place to make plotting easier.

$$[A]\ \text{Ans} \approx \begin{bmatrix} -1.1 & -2.7 & -1.8 \\ 0.3 & 3.1 & -0.1 \end{bmatrix}$$

Multiply [A] by the unrounded answer to the previous iteration.

$$[A]\ \text{Ans} \approx \begin{bmatrix} -0.7 & -3.6 & -0.7 \\ -0.7 & -0.7 & -1.5 \end{bmatrix}$$

$$[A]\ \text{Ans} \approx \begin{bmatrix} 0.2 & -1.1 & -1.2 \\ -0.9 & -3.2 & -1.2 \end{bmatrix}$$

Plot the images on graph paper.

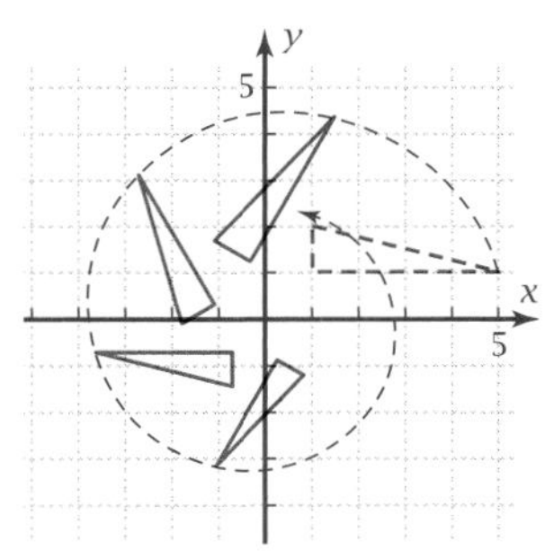

Figure 11-3d

As shown in Figure 11-3d, the images get smaller, and they spiral toward the origin as if they were "attracted" to it. ◀

In Problem 15 of Problem Set 11-3, you will write a program to calculate and plot image matrices iteratively on your grapher.

Problem Set 11-3

Do These Quickly

Q1. Find $\begin{bmatrix} 2 & 5 \\ 3 & 1 \end{bmatrix} + \begin{bmatrix} 4 & 7 \\ 1 & 6 \end{bmatrix}$.

Q2. Find $\begin{bmatrix} 4 & 7 \\ 1 & 6 \end{bmatrix} + \begin{bmatrix} 2 & 5 \\ 3 & 1 \end{bmatrix}$.

Q3. Does [A] + [B] always equal [B] + [A]?

Q4. Based on your answer to Problem Q3, matrix addition is a(n) —?— operation.

Q5. Find $\begin{bmatrix} 2 & 5 \\ 3 & 1 \end{bmatrix}\begin{bmatrix} 4 & 7 \\ 1 & 6 \end{bmatrix}$.

Q6. Find $\begin{bmatrix} 4 & 7 \\ 1 & 6 \end{bmatrix}\begin{bmatrix} 2 & 5 \\ 3 & 1 \end{bmatrix}$.

Q7. Does [A][B] always equal [B][A]?

Q8. Find det $\begin{bmatrix} 4 & 7 \\ 1 & 6 \end{bmatrix}$.

Q9. Find $\begin{bmatrix} 4 & 7 \\ 1 & 6 \end{bmatrix}^{-1}$.

Q10. The position vector for point (3, −7, 5) is —?—.

During World War II, Navajo Code Talkers created a code based on the Navajo language—a code that the Germans and Japanese could not break. Matrices are often used to create and break secret codes.

For Problems 1–6, draw the pre-image represented by the matrix on the right. Assume that the points are connected in the order they appear to form a closed figure. Then carry out the multiplication and plot the image. Describe the transformation.

1. $\begin{bmatrix} 2 & 0 \\ 0 & 2 \end{bmatrix}\begin{bmatrix} 2 & 3 & 1 \\ 1 & 2 & 3 \end{bmatrix}$

2. $\begin{bmatrix} 3 & 0 \\ 0 & 3 \end{bmatrix}\begin{bmatrix} 1 & 3 & 2 \\ 1 & 1 & 4 \end{bmatrix}$

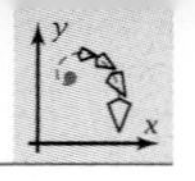

3. $\begin{bmatrix} \frac{1}{2} & 0 \\ 0 & \frac{1}{2} \end{bmatrix} \begin{bmatrix} 2 & 2 & -4 & -4 \\ 6 & -3 & -3 & 6 \end{bmatrix}$

4. $\begin{bmatrix} \frac{1}{3} & 0 \\ 0 & \frac{1}{3} \end{bmatrix} \begin{bmatrix} 9 & -3 & -3 & 9 \\ 0 & 0 & 12 & 12 \end{bmatrix}$

5. $\begin{bmatrix} 0.8 & -0.6 \\ 0.6 & 0.8 \end{bmatrix} \begin{bmatrix} 3 & 3 & 6 & 6 \\ 1 & 2 & 2 & 1 \end{bmatrix}$

6. $\begin{bmatrix} 0.8 & 0.6 \\ -0.6 & 0.8 \end{bmatrix} \begin{bmatrix} 1 & 1 & 2 \\ 2 & 5 & 5 \end{bmatrix}$

For Problems 7–12, write a transformation matrix, and then use it to transform the given figure. Plot the pre-image and the image, confirming that your transformation matrix is correct.

7. Dilate this triangle by a factor of 3.

$$\begin{bmatrix} 1 & 3 & 4 \\ 1 & 1 & 5 \end{bmatrix}$$

8. Dilate this dart by a factor of 2.

$$\begin{bmatrix} 1 & 3 & 5 & 3 \\ 1 & 2 & 1 & 6 \end{bmatrix}$$

9. Rotate the pre-image triangle in Problem 7 clockwise 50°.

10. Rotate the pre-image dart in Problem 8 counterclockwise 70°.

11. Dilate the pre-image triangle in Problem 7 by a factor of 3 and rotate it clockwise 50°.

12. Dilate the pre-image dart in Problem 8 by a factor of 2 and rotate it counterclockwise 70°.

For Problems 13 and 14, write a matrix for the given pre-image, describe the effect the transformation matrix [A] will have, and then iterate four times using the same transformation. Plot each image on graph paper or on a copy of the figure.

13. $[A] = \begin{bmatrix} 0.8 \cos 20° & 0.8 \cos 110° \\ 0.8 \sin 20° & 0.8 \sin 110° \end{bmatrix}$ on Figure 11-3e

14. $[A] = \begin{bmatrix} 0.7 \cos(-40°) & 0.7 \cos 50° \\ 0.7 \sin(-40°) & 0.7 \sin 50° \end{bmatrix}$ on Figure 11-3f

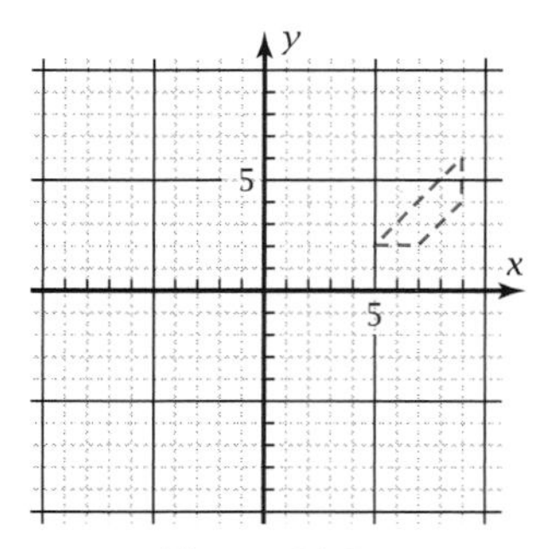

Figure 11-3e

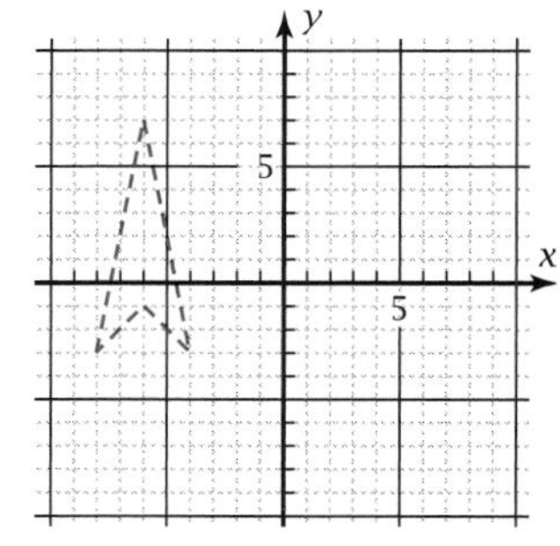

Figure 11-3f

15. *Grapher Program for Iterative Transformations:* Write or download a program to perform iterative transformations. The program should allow you to store a transformation matrix as [A] and a pre-image matrix as [D]. When you run the program, the grapher should first store a copy of [D] as [E] and plot the pre-image on the screen. When you press ENTER, the grapher should multiply [A][E], store the result back in [E], plot the image, and then pause until the ENTER key is pressed again. Check your program using the transformation and pre-image in Example 2.

16. *Grapher Program Test:* Run your program from Problem 15 using the transformation and pre-image in Problem 14. Sketch the path followed by the uppermost point in the pre-image. To what fixed point do the images seem to be attracted?

For Problems 17–20, write a transformation matrix [M] for the transformation described.

17. Counterclockwise rotation of 90°

18. Rotation of 180°

19. Dilation by a factor of 5 with respect to the origin

20. Dilation by a factor of 0.9 with respect to the origin

21. *Journal Problem:* Write in your journal the most important thing you have learned as a result of studying matrix transformations.

11-4 Translation with Rotation and Dilation Matrices

In the last section you saw how to use a transformation matrix to rotate and dilate a figure in the plane. In this section you will learn how to use a matrix to translate a figure to a new position without changing its size or orientation. Then you will explore the combined effects of rotating, dilating, and translating a figure iteratively. Surprisingly, the iterated images are attracted to a fixed point if the dilation reduces the images in size.

OBJECTIVE Given a desired dilation, rotation, and translation, write a matrix that will perform the transformation when it is multiplied by a pre-image matrix, and find the fixed point to which the images are attracted.

Translations

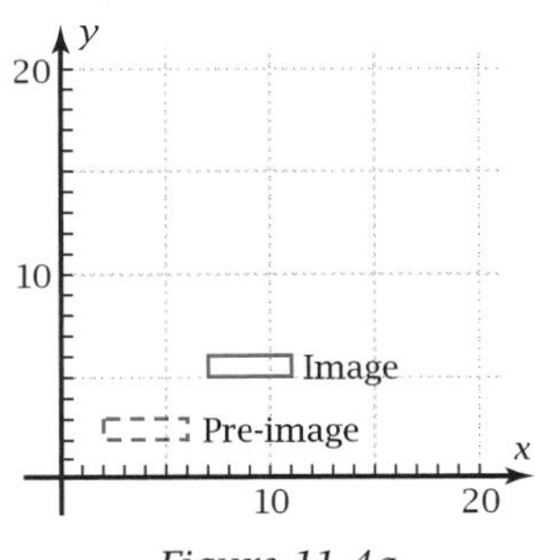

Figure 11-4a

Figure 11-4a shows the (dashed) rectangle represented by this pre-image matrix:

$$[M] = \begin{bmatrix} 2 & 6 & 6 & 3 \\ 2 & 2 & 3 & 2 \end{bmatrix}$$

The (solid) image shows the rectangle **translated 5** units in the x-direction and 3 units in the y-direction. You accomplish this translation algebraically by adding 5 to each x-coordinate in the top row of [M] and adding 3 to each y-coordinate in the bottom row. You could perform this translation using matrix addition:

$$\begin{bmatrix} 5 & 5 & 5 & 5 \\ 3 & 3 & 3 & 3 \end{bmatrix} + \begin{bmatrix} 2 & 6 & 6 & 2 \\ 2 & 2 & 3 & 3 \end{bmatrix} = \begin{bmatrix} 7 & 11 & 11 & 7 \\ 5 & 5 & 6 & 6 \end{bmatrix}$$

However, there is a way to accomplish the translation by multiplying by a transformation matrix rather than by adding. First, insert a third row into [M] containing all 1s. Then write a 3×3 transformation matrix [T] containing the identity matrix in the upper left corner, the translations 5 and 3 in the third column, and the row 0, 0, 1 across the bottom.

$$[T][M] = \begin{bmatrix} 1 & 0 & 5 \\ 0 & 1 & 3 \\ 0 & 0 & 1 \end{bmatrix} \begin{bmatrix} 2 & 6 & 6 & 2 \\ 2 & 2 & 3 & 3 \\ 1 & 1 & 1 & 1 \end{bmatrix}$$

If you multiply these matrices, the image matrix is

$$[T][M] = \begin{bmatrix} 7 & 11 & 11 & 7 \\ 5 & 5 & 6 & 6 \\ 1 & 1 & 1 & 1 \end{bmatrix}$$

The translated figure's coordinates appear along the top two rows, and the bottom row of the image matrix contains all 1s. If you multiply the matrices by hand, you will see what has happened. The 1s in the bottom row of the

pre-image matrix, together with the third column of the translation matrix, cause the 5 and 3 translations to take place. The 0, 0, 1 row in the transformation matrix causes 1, 1, 1, 1 to appear in the bottom row of the image matrix.

Combined Translations, Dilations, and Rotations

The preceding translation matrix [T], as mentioned previously, has the identity matrix in its upper left corner:

$$[T][M] = \begin{bmatrix} 1 & 0 & 5 \\ 0 & 1 & 3 \\ 0 & 0 & 1 \end{bmatrix}$$

If you replace this embedded identity matrix with a rotation and dilation matrix, then multiplying a pre-image matrix by [T] performs all the transformations.

▶ ***EXAMPLE 1*** Write a transformation matrix [T] to rotate a figure counterclockwise 30°, dilate it by a factor of 0.8, and translate it 5 units in the positive x-direction and 3 units in the positive y-direction. Perform the transformation on the kite specified by matrix [M]. Plot the pre-image and the image.

$$[M] = \begin{bmatrix} 8 & 10 & 12 & 10 \\ 7 & 2 & 7 & 8 \\ 1 & 1 & 1 & 1 \end{bmatrix}$$

Solution The transformation matrix has the rotation and dilation matrix as the upper left four elements. The translations appear in the third column.

$$[T] = \begin{bmatrix} 0.8\cos 30^\circ & 0.8\cos 120^\circ & 5 \\ 0.8\sin 30^\circ & 0.8\sin 120^\circ & 3 \\ 0 & 0 & 1 \end{bmatrix}$$

The image is

$$[T][M] = \begin{bmatrix} 7.7425\ldots & 11.1282\ldots & 10.5138\ldots & 8.7282\ldots \\ 11.0497\ldots & 8.3856\ldots & 12.6497\ldots & 12.5425\ldots \\ 1 & 1 & 1 & 1 \end{bmatrix}$$

$$\approx \begin{bmatrix} 7.7 & 11.1 & 10.5 & 8.7 \\ 11.0 & 8.4 & 12.6 & 12.5 \\ 1 & 1 & 1 & 1 \end{bmatrix}$$

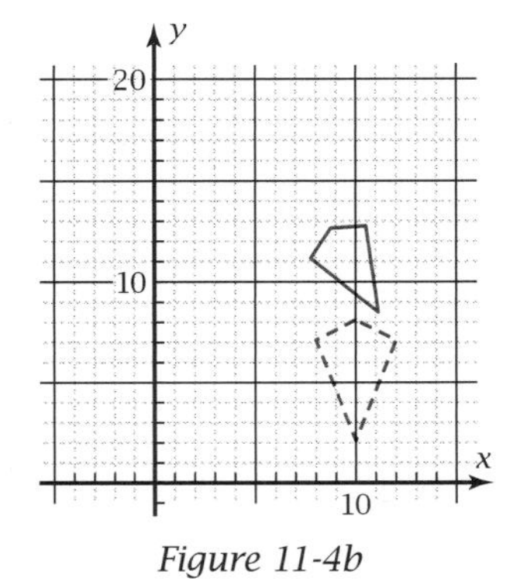

Figure 11-4b

The pre-image and image are shown in Figure 11-4b. If you are plotting on paper, you will find it easier if you set your grapher to round to one decimal place. ◀

If the transformations of Example 1 are performed iteratively, the images spiral around and are attracted to a fixed point. Unlike the pure dilation and rotation transformations of the previous section, however, the attractor is not the origin. You can use the program of Problem 15 in Problem Set 11-3 to plot the iterated images and then find the approximate coordinates of this fixed point.

▶ EXAMPLE 2 Perform the transformations of Example 1 iteratively. Sketch the resulting images. Estimate the coordinates of the point to which the images are attracted.

Solution Figure 11-4c shows the images from the first few iterations and the spiral path followed by the images from subsequent iterations.

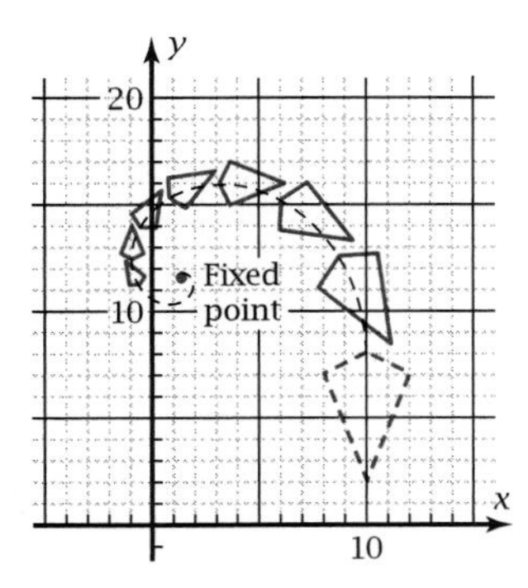

Figure 11-4c

Graphically, the fixed point seems to be attracted to the point at about (1.3, 11.5). To estimate the coordinates of the point numerically, you can display the matrix where the program stores the images (matrix [E]). After 30 iterations, the image matrix is

$$[\text{Image}] = \begin{bmatrix} 1.3122\ldots & 1.3098\ldots & 1.3073\ldots & 1.3098\ldots \\ 11.4914\ldots & 11.4976\ldots & 11.4914\ldots & 11.4901\ldots \\ 1 & 1 & 1 & 1 \end{bmatrix}$$

As you can see, each point in the image has been attracted to a point close to (1.31, 11.49). This point is called a **fixed point attractor** or simply a **fixed point** because if you apply the transformation to it, it does not move—that is, it remains fixed. ◀

Fixed Point Attractors and Limits

The fixed point attractor is the **limit** of the image points as the number of iterations approaches infinity. The fixed point depends only on the transformation matrix, not on the pre-image. Using (x_0, y_0) for a pre-image point, (x_1, y_1) for the image after the first iteration, (x_2, y_2) for the image after the second iteration, and so forth, and using (X, Y) for the fixed point, you can write the limit as shown in the box.

DEFINITION: Fixed Point Limit

If the images approach a **fixed point** (X, Y) when a transformation [T] is performed iteratively, then

$$(X, Y) = \lim_{n \to \infty} (x_n, y_n)$$

where (x_n, y_n) is the image of a point (x_0, y_0) after n iterations.

Verbally: (X, Y) is the limit of (x_n, y_n) as n approaches infinity.

It is possible to calculate a fixed point algebraically and numerically.

▶ EXAMPLE 3 Calculate algebraically the fixed point in Example 2.

Solution If the fixed point is (X, Y), then performing the transformation [T] on (X, Y) will give (X, Y) as the image. Write the fixed point, (X, Y), as a 3×1 matrix and multiply it by the transformation matrix.

$$\begin{bmatrix} 0.8\cos 30° & 0.8\cos 120° & 5 \\ 0.8\sin 30° & 0.8\sin 120° & 3 \\ 0 & 0 & 1 \end{bmatrix}\begin{bmatrix} X \\ Y \\ 1 \end{bmatrix} = \begin{bmatrix} X \\ Y \\ 1 \end{bmatrix}$$

The image is the same as the pre-image.

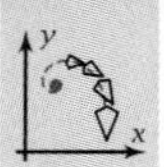

$$\begin{bmatrix} (0.8\cos 30°)X + (0.8\cos 120°)Y + 5 \\ (0.8\sin 30°)X + (0.8\sin 120°)Y + 3 \\ 1 \end{bmatrix} = \begin{bmatrix} X \\ Y \\ 1 \end{bmatrix}$$

Multiply the matrices on the left.

$$(0.8\cos 30°)X + (0.8\cos 120°)Y + 5 = X$$

Equate the top two rows of the matrices.

$$(0.8\sin 30°)X + (0.8\sin 120°)Y + 3 = Y$$

$$(0.8\cos 30° - 1)X + (0.8\cos 120°)Y = -5$$

Get X and Y on the left and the constants on the right.

$$(0.8\sin 30°)X + (0.8\sin 120° - 1)Y = -3$$

$$\begin{bmatrix} 0.8\cos 30° - 1 & 0.8\cos 120° \\ 0.8\sin 30° & 0.8\sin 120° - 1 \end{bmatrix}\begin{bmatrix} X \\ Y \end{bmatrix} = \begin{bmatrix} -5 \\ -3 \end{bmatrix}$$

Write the system in matrix form.

$$\begin{bmatrix} X \\ Y \end{bmatrix} = \begin{bmatrix} 0.8\cos 30° - 1 & 0.8\cos 120° \\ 0.8\sin 30° & 0.8\sin 120° - 1 \end{bmatrix}^{-1}\begin{bmatrix} -5 \\ -3 \end{bmatrix} = \begin{bmatrix} 1.3205\ldots \\ 11.4858\ldots \end{bmatrix}$$

Solve.

So the fixed point is (1.3205..., 11.4858...), which confirms the approximate values found graphically and numerically. ◀

If you perform transformation [T] iteratively on matrix [M], the images are

Iteration 1: $[T][M]$

Iteration 2: $[T]([T][M]) = ([T][T])[M] = [T]^2[M]$

Iteration 3: $[T]([T]([T][M])) = ([T][T][T])[M] = [T]^3[M]$

These equations are true because matrix multiplication is associative. To find the transformation matrix for the 30th iteration, you would calculate

$$[T]^{30} = \begin{bmatrix} -0.0012\ldots & 0.0000\ldots & 1.3222\ldots \\ 0.0000\ldots & -0.0012\ldots & 11.5000\ldots \\ 0 & 0 & 1 \end{bmatrix}$$

The four elements in the rotation and dilation part of the matrix are close to zero because the dilation, 0.8, is less than 1 and is being raised to a high power. The two elements in the translation part of the matrix are close to the coordinates of the fixed point. If the rotation and translation elements were equal to zero, the resulting matrix would translate any point (a, b) to the fixed point.

$$\begin{bmatrix} 0 & 0 & X \\ 0 & 0 & Y \\ 0 & 0 & 1 \end{bmatrix}\begin{bmatrix} a \\ b \\ 1 \end{bmatrix} = \begin{bmatrix} X \\ Y \\ 1 \end{bmatrix}$$

Example 4 shows how to take advantage of this fact to find the fixed point numerically.

▶ **EXAMPLE 4** Calculate the fixed point in Example 3 numerically in a time-efficient way.

Solution

$$[T]^{100} = \begin{bmatrix} 0.0000\ldots & 0.0000\ldots & 1.3205\ldots \\ 0.0000\ldots & 0.0000\ldots & 11.4858\ldots \\ 0 & 0 & 1 \end{bmatrix}$$

Raise [T] to a higher power. Be sure the rotation-dilation is close to 0.

The fixed point is $(X, Y) \approx (1.3205\ldots, 11.4858\ldots)$.

Fixed point is in the translation part. ◀

The information in Examples 1 through 4 is summarized in this box.

PROPERTY: General Rotation, Dilation, and Translation Matrix

When applied to a matrix of the form

$$[M] = \begin{bmatrix} x_1 & x_2 & \ldots \\ y_1 & y_2 & \ldots \\ 1 & 1 & \ldots \end{bmatrix}$$

the transformation matrix

$$[T] = \begin{bmatrix} d\cos A & d\cos(A + 90°) & h \\ d\sin A & d\sin(A + 90°) & k \\ 0 & 0 & 1 \end{bmatrix}$$

- Dilates by a factor of d
- Rotates counterclockwise by A degrees
- Translates by h units in the x-direction and k units in the y-direction

The techniques for finding the fixed point are summarized in the box at the bottom of the next page.

Problem Set 11-4

Do These Quickly

Q1. What is the dilation factor for an 80% reduction?

Q2. Find the image of (1, 0) under a counterclockwise rotation of 30°.

Q3. Find the dimensions of [A][B] if [A] is 3 × 5 and [B] is 5 × 2.

Q4. If $[M] = \begin{bmatrix} 7 & 2 \\ 8 & 4 \end{bmatrix}$, find det [M].

Q5. Find $[M]^{-1}$ for [M] in Problem Q4.

Q6. Find $[M]^{-1}[M]$ for [M] in Problem Q4.

Q7. Find $[M]^2$ for [M] in Problem Q4.

James Maxwell (1831–1879), a Scottish physicist, developed the unifying theory of quantum mechanics expressed in a set of matrix equations.

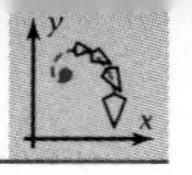

Q8. Explain why $[D] = \begin{bmatrix} 3 & 5 \\ 3 & 5 \end{bmatrix}$ has no multiplicative inverse.

Q9. Find the dot product $(3\vec{i} + 7\vec{j} - 2\vec{k}) \cdot (4\vec{i} - \vec{j} + 5\vec{k})$.

Q10. Is the angle between the two vectors in Problem Q9 acute or obtuse?

---

1. Consider the rectangle with vertices (3, 2), (7, 2), (7, 4), and (3, 4).
   a. Plot the rectangle on graph paper. Write a transformation matrix [A] to rotate this rectangle counterclockwise 20°, dilate it by a factor of 0.9, and translate it 6 units in the x-direction and −1 unit in the y-direction. Write a matrix [M] for the pre-image rectangle, apply the transformation, and plot the image on graph paper.
   b. Enter the matrices on your grapher and perform the transformation in part a iteratively, plotting the images using the program from Problem 15 in Section 11-3. Sketch the path followed by the images. To what fixed point do the images seem to be attracted?
   c. Find the approximate location of the fixed point numerically by finding $[A]^{100}[M]$. Does it agree with the answer you found graphically in part b?
   d. Find the location of the fixed point algebraically. Show that your answer agrees with the answers you found graphically and numerically in parts b and c.
2. Consider the dart with vertices (7, 1), (9, 2), (11, 1), and (9, 5).
   a. Plot the dart on graph paper. Write a transformation matrix [A] to rotate the dart counterclockwise 40°, dilate it by a factor of 0.8, and translate it −3 units in the x-direction and 4 units in the y-direction. Write a matrix [M] for the pre-image figure, apply the transformation, and plot the image on the graph paper.

(Problem Set 11-4 continued)

PROCEDURE: Fixed Point of a Linear Transformation

If a linear transformation has a fixed point (X, Y), then

$$(X, Y) = \lim_{n\to\infty} (x_n, y_n)$$

where (x_n, y_n) is the image of a point (x_0, y_0) after n iterations.

To find the fixed point (X, Y) algebraically:

1. Write the equation $[T]\begin{bmatrix} X \\ Y \\ 1 \end{bmatrix} = \begin{bmatrix} X \\ Y \\ 1 \end{bmatrix}$.
2. Multiply the matrices on the left side of the equation.
3. Equate the first and second elements from the resulting matrix on the left to the X and Y in the matrix on the right.
4. Solve the resulting system of equations for X and Y.

To find the fixed point (X, Y) numerically,

1. Raise [T] to a high power.
2. Make sure the rotation and dilation elements are close to zero.
3. Write the fixed point from the translation part of the resulting matrix.

b. Enter the matrices on your grapher and perform the transformation in part a iteratively, plotting the images using the program from Problem 15 in Section 11-3. Sketch the path followed by the images. To what fixed point do the images seem to be attracted?

c. Find the approximate location of the fixed point numerically by finding $[A]^{100}[M]$. Does it agree with the answer you found graphically in part b?

d. Find the location of the fixed point algebraically. Show that your answer agrees with the answers you found graphically and numerically in parts b and c.

3. *Fixed Point Problem:* Figure 11-4d shows a rectangle that is to be the pre-image for a set of linear transformations. In this problem you will find out which matrix determines the fixed point, the transformation matrix or the pre-image matrix.

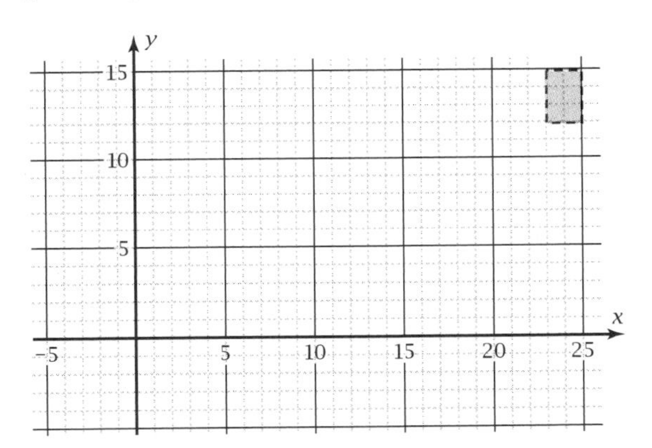

Figure 11-4d

a. Describe the transformations accomplished by matrix $[T_1]$.

$$[T_1] = \begin{bmatrix} 0.8\cos(-20°) & 0.8\cos 70° & 2 \\ 0.8\sin(-20°) & 0.8\sin 70° & 6 \\ 0 & 0 & 1 \end{bmatrix}$$

b. Write a matrix $[M_1]$ for the rectangle in Figure 11-4d. Apply transformation $[T_1]$ iteratively to $[M_1]$ using your grapher program. To what fixed point do the images converge? Show this fixed point on a copy of Figure 11-4d, along with the path the images follow to reach this point.

c. Apply transformation $[T_1]$ iteratively to another rectangle, $[M_2]$ below. Are the images attracted to the same fixed point as for $[M_1]$? Sketch the pre-image and the path the images follow.

$$[M_2] = \begin{bmatrix} 2 & -2 & -2 & 2 \\ 10 & 10 & 0 & 0 \\ 1 & 1 & 1 & 1 \end{bmatrix}$$

d. Write a transformation matrix $[T_2]$ that performs this set of transformations.

- A 70% reduction; that is a dilation by a factor of $k = 0.7$.
- A 35° counterclockwise rotation
- A translation of 7 units in the x-direction and −3 units in the y-direction

Apply $[T_2]$ iteratively to the rectangle $[M_1]$. Are the images attracted to the same fixed point as for $[T_1]$? On a copy of Figure 11-4d, sketch the path the images follow to reach the fixed point.

e. Based on your results for parts b–d, which determines the location of the fixed point attractor, the transformation matrix or the pre-image matrix? Does applying $[T_2]$ to $[M_2]$ support your conclusion?

4. *Third Row Problem:* Multiply the given matrices "by hand." From the results, explain the effect of the 1, 1 in the third row of the pre-image matrix [M]. Explain the effect of the 0, 0, 1 in the third row of the transformation matrix [T] and why it is important for this effect to happen.

$$[T][M] = \begin{bmatrix} 2 & 0 & 7 \\ 0 & 2 & 3 \\ 0 & 0 & 1 \end{bmatrix}\begin{bmatrix} 8 & 5 \\ 4 & 9 \\ 1 & 1 \end{bmatrix}$$

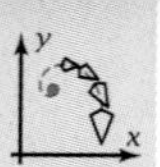

11-5 Strange Attractors for Several Iterated Transformations

In the last section you saw how an iterated transformation can cause images to be attracted to a fixed point. In this section you will see what happens when several different transformations are performed iteratively. Instead of being attracted to a single point, the images are attracted to a figure of remarkable complexity. Sometimes these **strange attractors** have shapes that look like trees, ferns, snowflakes, or islands.

OBJECTIVE Given several different transformations, perform them iteratively, starting with a pre-image, and plot the resulting images.

Strange Attractors Geometrically

Figure 11-5a shows a rectangular pre-image whose matrix is

$$[M] = \begin{bmatrix} 2 & 2 & -2 & -2 \\ 0 & 10 & 10 & 0 \\ 1 & 1 & 1 & 1 \end{bmatrix}$$

Figure 11-5b shows the four images that result from applying four transformations—[A], [B], [C], and [D]—to the pre-image in Figure 11-5a.

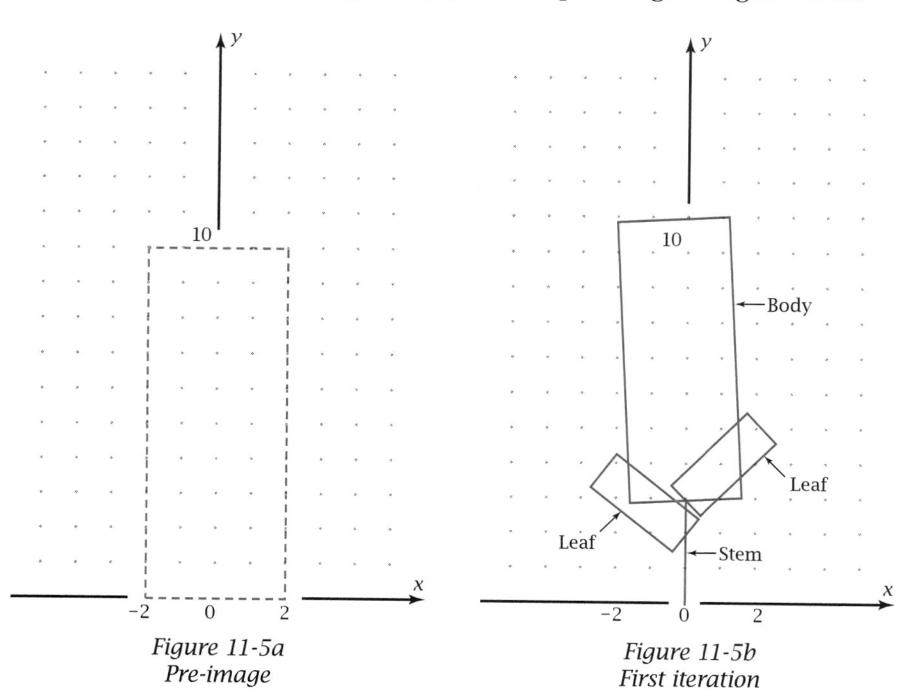

Figure 11-5a
Pre-image

Figure 11-5b
First iteration

The image matrices for these transformations are

$$[A][M] = \begin{bmatrix} 0.8\cos 3° & 0.8\cos 93° & 0 \\ 0.8\sin 3° & 0.8\sin 93° & 3 \\ 0 & 0 & 1 \end{bmatrix}[M] \approx \begin{bmatrix} 1.6 & 1.2 & -2.0 & -1.6 \\ 3.1 & 11.1 & 10.9 & 2.9 \\ 1 & 1 & 1 & 1 \end{bmatrix}$$

$$[B][M] = \begin{bmatrix} 0.3\cos 52° & 0.3\cos 142° & 0 \\ 0.3\sin 52° & 0.3\sin 142° & 2 \\ 0 & 0 & 1 \end{bmatrix}[M] \approx \begin{bmatrix} 0.4 & -2.0 & -2.7 & -0.4 \\ 2.6 & 4.3 & 3.4 & 1.5 \\ 1 & 1 & 1 & 1 \end{bmatrix}$$

$$[C][M] = \begin{bmatrix} 0.3\cos(-46°) & 0.3\cos 44° & 0 \\ 0.3\sin(-46°) & 0.3\sin 44° & 3 \\ 0 & 0 & 1 \end{bmatrix}[M] \approx \begin{bmatrix} 0.4 & 2.6 & 1.7 & -0.4 \\ 2.6 & 4.7 & 5.5 & 3.4 \\ 1 & 1 & 1 & 1 \end{bmatrix}$$

$$[D][M] = \begin{bmatrix} 0 & 0 & 0 \\ 0 & 0.3 & 0 \\ 0 & 0 & 0 \end{bmatrix}[M] = \begin{bmatrix} 0 & 0 & 0 & 0 \\ 0 & 3 & 3 & 0 \\ 1 & 1 & 1 & 1 \end{bmatrix}$$

The first three transformations are dilations and rotations. Matrix [D] dilates by 0.3 in the y-direction and by 0 in the x-direction, shrinking the rectangle to a line segment along the y-axis.

Figure 11-5c shows the 16 images that result from applying each of the four transformations to each of the four images from the first iteration. Note that the figure created by this second iteration has three pieces that are similar to each other and to the image from the first iteration. Each piece has a linear "stem," a large rectangle coming out of the stem (body), and two small rectangles at the base of the large rectangle (leaves). A fourth piece is a long stem, extending down to the origin.

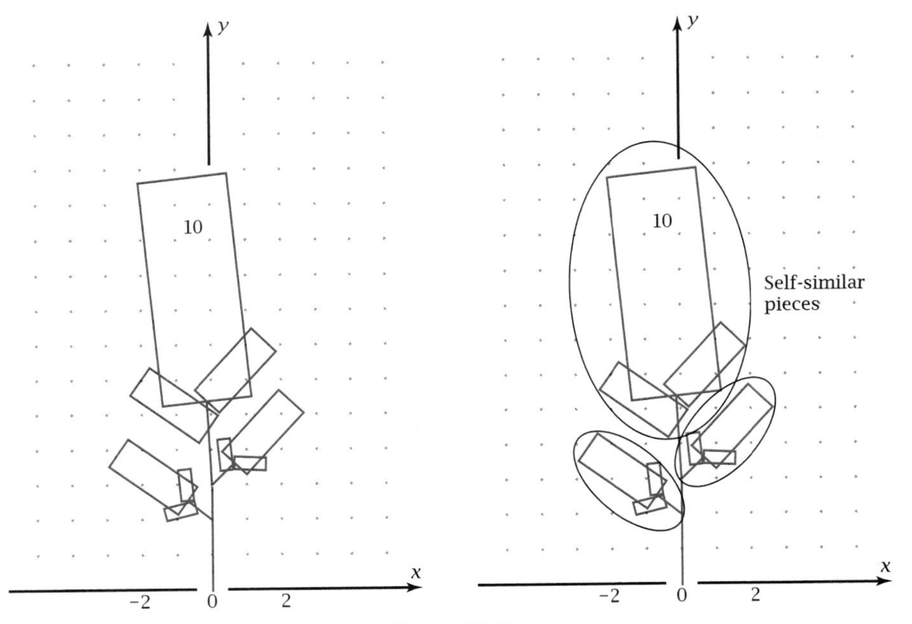

Figure 11-5c
Second iteration

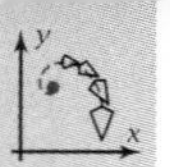

If a third iteration is done, each of the four transformations will be performed on the 16 images from the second iteration, giving 64 new images. The 20th iteration has 4^{20} or 1,099,511,627,776 images and resembles the fern leaf shown in Figure 11-5d.

Strange Attractors Numerically—Barnsley's Method

As you have seen, plotting images created by applying several transformations iteratively can be tedious. In 1988, Michael Barnsley published a method for plotting such images more efficiently. Instead of starting with a pre-image figure, you start with *one point*. You then select *one* of the four transformations at random, apply it to that point, and plot the image point. Then you again select one of the four transformations at random, apply it to the first image, and plot the new point. As more and more points are plotted, there are regions to which the points are attracted and regions they avoid. The resulting image is an approximation of the image you would get by carrying out the iterations in Figure 11-5c an infinite number of times. The more points you plot, the better the approximation. Figure 11-5d shows this **strange attractor** plotted with 200 points, 1000 points, and 5000 points.

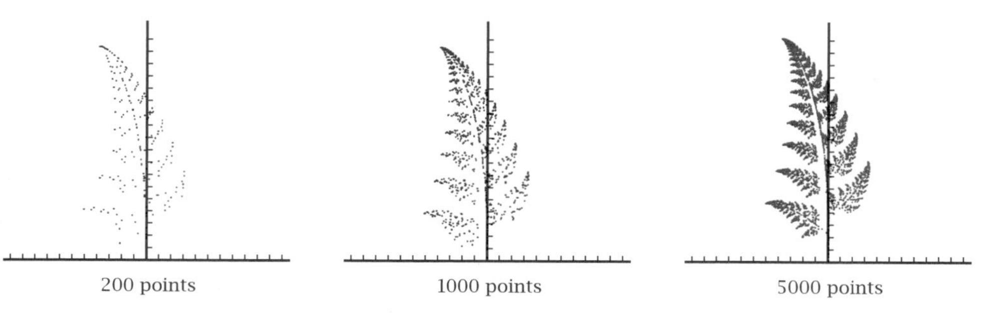

Figure 11-5d

If you were to magnify any one of the branches of the fern leaf by the proper amount, it would look exactly like the whole leaf. Figures with this quality are said to be **self-similar.** In Section 11-6 you will learn about such images, called **fractals.** They are so "fractured" that their dimensions turn out to be fractions.

The technique of plotting a strange attractor pointwise is called *Barnsley's method.* To give the images reasonable point densities, you must select a probability for each transformation. For Figure 11-5d there was an 80% probability of picking transformation [A], which produces the main form of the leaf; 9% each for [B] and [C], which produce the two side branches; and 2% for [D], which draws the stem of the fern leaf.

Self-similarity is readily apparent in cauliflower.

PROCEDURE: Barnsley's Method

To plot a strange attractor pointwise:

1. Select several transformation matrices.
2. Select a probability for each matrix with a dilation factor that is less than 1.
3. Select any point to be the pre-image.
4. Pick a transformation at random using the appropriate probability, apply the transformation to the point, and plot the image.
5. Repeat the previous step on the image from the step before. Do this for as many iterations as you choose.

Problem Set 11-5

Do These Quickly

5 min

Q1. Write a 2 × 2 matrix to dilate a figure by a factor of 2.

Q2. Write a 2 × 2 matrix to rotate a figure clockwise 12°.

Q3. Write a 3 × 3 matrix for a 60% reduction, a 23° counterclockwise rotation, a 4-unit x-translation, and a −3-unit y-translation.

Q4. After a 60% reduction, a segment 10 cm long becomes —?— cm long.

Q5. A 60% reduction transforms a 100-cm^2 rectangle to one with area —?—.

The Ba-ila settlement in southern Zambia has a fractal design.

Q6. If a rotation takes the x-axis to $\theta = 40°$, then it takes the y-axis to $\theta =$ —?—.

Q7. Write cos 30° exactly, in radical form.

Q8. By the cofunction property, what does $\sin\left(\frac{\pi}{2} - x\right)$ equal?

Q9. How many degrees are there in $\frac{\pi}{3}$ radians?

Q10. $\vec{a}$ is 7 units long, $\vec{b}$ is 8 units long, and the angle between them when they are placed tail-to-tail is 38°. How long is $\vec{a} \times \vec{b}$?

---

1. *Sierpiński's Triangle Problem:* Figure 11-5e shows a triangular pre-image whose matrix is

$$[M] = \begin{bmatrix} 15 & 0 & -15 \\ -10 & 20 & -10 \\ 1 & 1 & 1 \end{bmatrix}$$

Figure 11-5f shows the first iteration of three transformations. Each transformation reduces the triangle by 50%. Transformation [A] translates the dilated image so that its lower right vertex coincides with the lower right vertex of the pre-image. Transformation [B] translates the dilated image so that its lower left vertex coincides with the lower left vertex of the pre-image. Transformation [C] translates

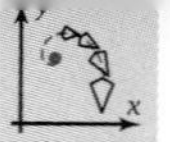

the dilated image so that its upper vertex coincides with the upper vertex of the pre-image.

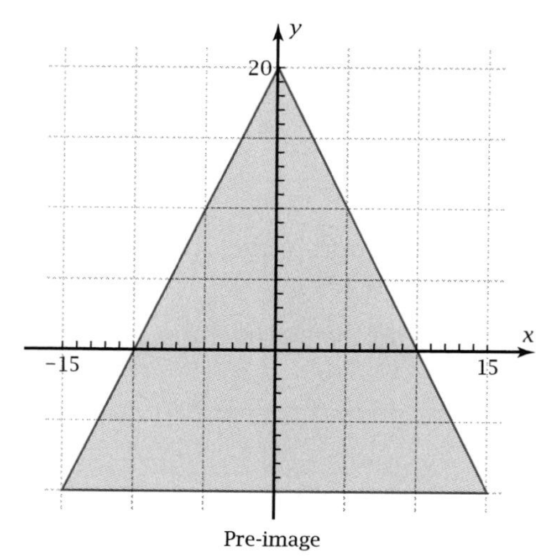

Pre-image

Figure 11-5e

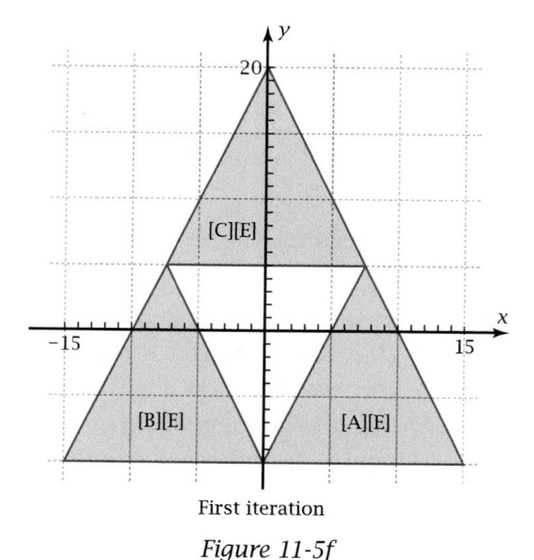

First iteration

Figure 11-5f

a. Write transformation matrices [A], [B], and [C].

b. Apply the nine transformations required for the second iteration. That is, do [A][A][M], [A][B][M], [A][C][M], [B][A][M], [B][B][M], [B][C][M], [C][A][M], [C][B][M], and [C][C][M].

c. On graph paper, plot the nine images from the second iteration in part b.

d. How many images will be in the third iteration? The 20th iteration?

e. Find the area of the pre-image triangle. Find the total area of the three triangles in the first iteration. Use what you observe in these calculations to find a formula for the total area of the nth image.

f. If the iterations are carried on forever, the figure is called **Sierpiǹski's triangle** or Sierpiǹski's gasket. What is the area of Sierpiǹski's triangle? Does the answer surprise you?

2. *Sierpiǹski's Square Problem:* Figure 11-5g shows the square pre-image whose matrix is

$$[M] = \begin{bmatrix} 20 & 20 & 0 & 0 \\ 20 & 0 & 0 & 20 \\ 1 & 1 & 1 & 1 \end{bmatrix}$$

Figure 11-5h shows the first iteration of four transformations. Each transformation gives the square a 40% reduction. The four transformations then translate the reduced images to the four corners of the original pre-image.

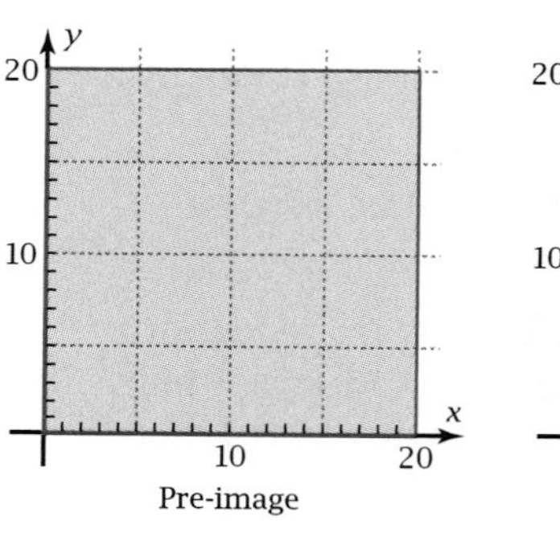

Pre-image

Figure 11-5g

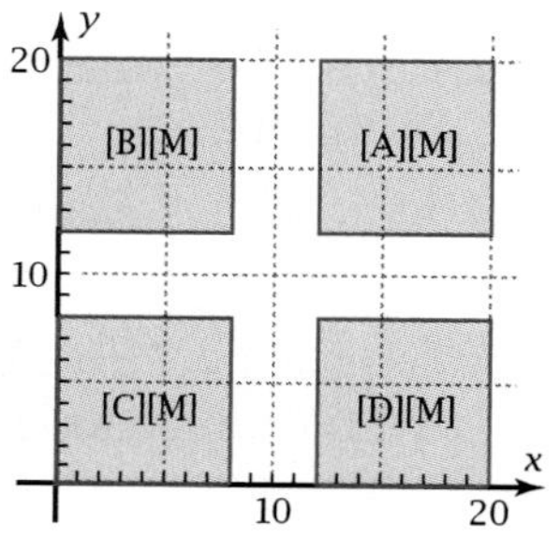

First iteration

Figure 11-5h

a. Write a transformation matrix for each of the four transformations.

b. Perform the 16 transformations required for the second iteration. Plot the 16 images on graph paper.

c. How many images will there be in the third iteration? The 20th iteration?

d. Find the perimeter of the pre-image. Find the total perimeter of the squares in the first iteration. Find the total perimeter of the squares in the third iteration. By extending the pattern you observe in your answers, find the total perimeter of the squares in the 20th iteration. What happens to the total perimeter as the number of iterations becomes very large?

e. As a research project, find out about Waclaw Sierpiński, the man for whom Sierpiński's triangle and square are named.

3. *Barnsley's Method Program:* Write or download a program that will plot a strange attractor using Barnsley's method. Before running the program you should store up to four transformation matrices and the probability associated with each matrix. The program should allow you to input a starting pre-image point and the number of points to plot. Then the program should iteratively select a transformation at random, apply it to the preceding image, and plot the new image.

4. *Barnsley's Method Program Debugging:* Test your program for Barnsley's method by plotting the fern-shaped strange attractor shown in Figure 11-5d. Use the transformation matrices [A], [B], [C], and [D] shown on page 477, with probabilities of 0.9, 0.09, 0.09, and 0.02, respectively. Use (1, 1) as the initial pre-image point, and plot a sufficient number of points to get a reasonably good image. When your program is working, run it again using a different pre-image point. Does the pre-image you select seem to change the final image? Run the program again using five times as many points. Describe the similarities and differences in the final image created by using more points.

5. Use your Barnsley's method program to plot Sierpiński's triangle from Problem 1. Use a probability of $\frac{1}{3}$ for each of the three transformations. Only three transformations are involved, so use a probability of 0 for the fourth transformation. You should get an image similar to that shown in Figure 11-5i.

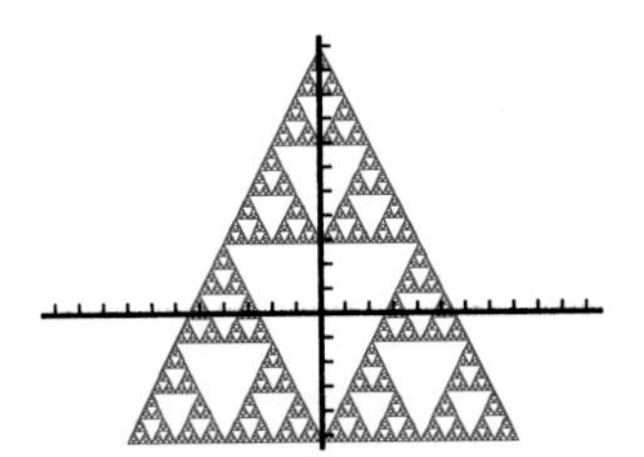

Figure 11-5i

6. Use your Barnsley's method program to plot Sierpiński's square from Problem 2. Use a probability of $\frac{1}{4}$ for each of the four transformations. You should get an image similar to that shown in Figure 11-5j.

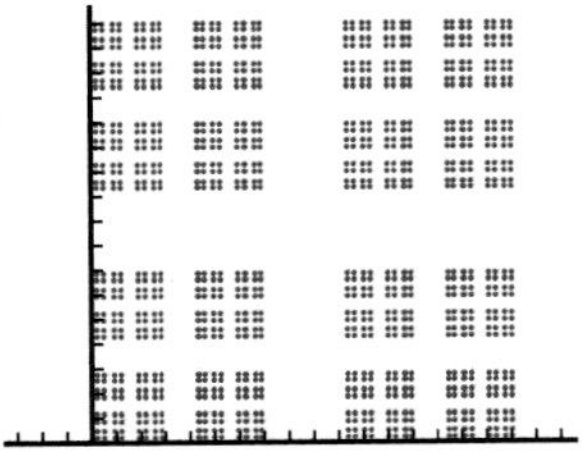

Figure 11-5j

7. Change the matrices for Sierpiński's square in Problem 6 so that the dilation is 0.5 instead of 0.4. You must also change the translations so that the upper right square's upper right corner still goes to the point (20, 20), and so forth. Explain why the pattern of the points in Figure 11-5j disappears when the dilation is changed to 0.5.

8. Change the matrices for Sierpiński's square in Problem 6 so that the dilation is 0.6 instead of 0.4. You must also change the translations so that the upper right square's upper right corner still goes to the point (20, 20), and so forth. Does any pattern seem to appear in the points?

9. *Foerster's Tree Problem:* Figure 11-5k shows a vertical segment 10 units high, starting at the origin. This pre-image is to be transformed into a "tree" with three pieces, each 6 units long, as shown in Figure 11-5l. The three pieces satisfy these conditions:
   - The left branch is rotated +30° from the trunk and starts at $y = 5$.
   - The trunk starts at the origin.
   - The right branch is rotated −30° from the trunk and starts at $y = 4$.

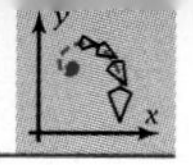

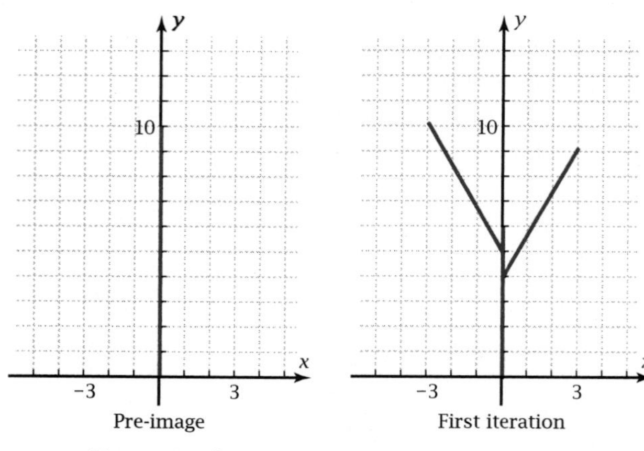

Figure 11-5k *Figure 11-5l*

a. Write the pre-image on the left as a 3 × 2 matrix, [D], with two 1s in the bottom row.

b. Write three 3 × 3 transformation matrices to do the following:

[A] should transform the pre-image to the left branch.

[B] should transform the pre-image to the trunk.

[C] should transform the pre-image to the right branch.

c. Figure 11-5l shows the three images in the first iteration. To get the nine images in the second iteration, you multiply each image by [A], [B], and [C]. Calculate the nine images [A][A][D], [A][B][D], [A][C][D], [B][A][D], [B][B][D], [B][C][D], [C][A][D], [C][B][D], and [C][C][D]. Round the entries of the image matrices to one decimal place. Plot the nine images on graph paper.

d. Use your Barnsley's method program to see what the tree would look like if the iterations were done infinitely many times. Use a probability of $\frac{1}{3}$ for each transformation.

e. The tree in this problem "attracts" the points. What special name is given to such an attractor?

f. Calculate the sum of the lengths of the images in the first, second, third, and 100th iterations of this tree. If the iterations were done infinitely many times, what would the sum of the lengths of the images approach? Does the answer surprise you?

g. Calculate the fixed points for each of the three transformations in this problem. How do these points relate to points on the graph in part d?

10. *Koch's Snowflake Problem:* Figure 11-5m shows the line segment represented by the pre-image matrix

$$[M] = \begin{bmatrix} 12 & 12 \\ 6 & -6 \\ 1 & 1 \end{bmatrix}$$

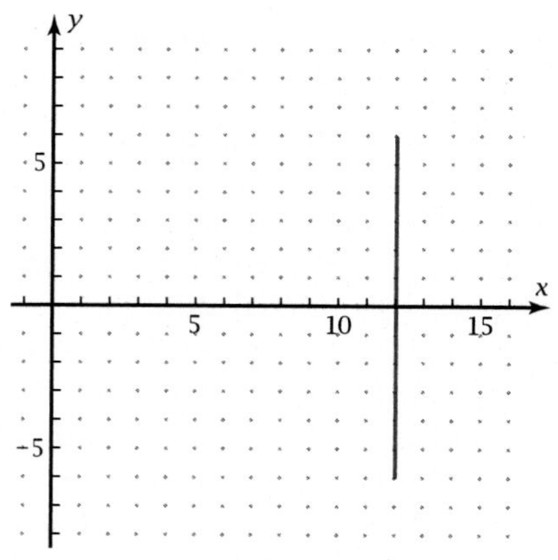

Figure 11-5m

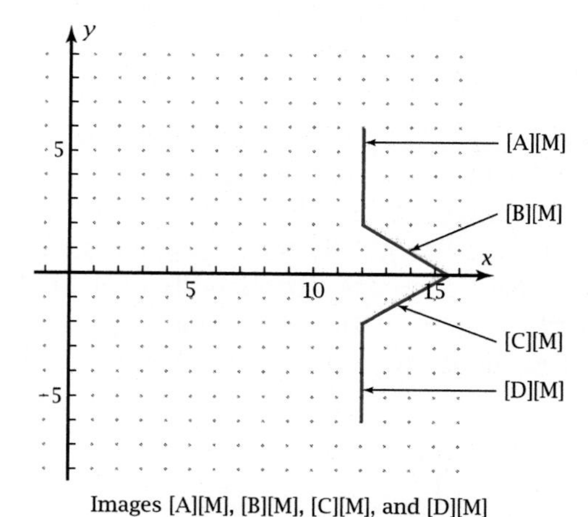

Figure 11-5n

Figure 11-5n shows the images of these four transformations applied to the pre-image matrix.

$$[A] = \begin{bmatrix} \frac{1}{3} & 0 & 8 \\ 0 & \frac{1}{3} & 4 \\ 0 & 0 & 1 \end{bmatrix}$$

$$[B] = \begin{bmatrix} \frac{1}{3}\cos 60° & \frac{1}{3}\sin 150° & 11.7320... \\ \frac{1}{3}\sin 60° & \frac{1}{3}\sin 150° & -2.4641... \\ 0 & 0 & 1 \end{bmatrix}$$

$$[C] = \begin{bmatrix} \frac{1}{3}\cos(-60°) & \frac{1}{3}\cos 30° & 11.7320... \\ \frac{1}{3}\sin(-60°) & \frac{1}{3}\sin 30° & -2.4641... \\ 0 & 0 & 1 \end{bmatrix}$$

$$[D] = \begin{bmatrix} \frac{1}{3} & 0 & 8 \\ 0 & \frac{1}{3} & -4 \\ 0 & 0 & 1 \end{bmatrix}$$

a. Draw a sketch showing how [A] dilates [M] by a factor of $\frac{1}{3}$ and then translates the dilated image so that its top point is at the top of the pre-image.

b. Show algebraically that the rotation and dilation part of [B] moves the point (12, 6) at the top of the pre-image to the point ($4\cos 60° + 2\cos 150°$, $4\sin 60° + 2\sin 150°$). Then show how the translation part of [B] moves this to the point (12, 2) at the bottom of image [A][M].

c. Based on your answers to parts a and b, tell what effects transformations [C] and [D] have on the pre-image segment.

d. In the second iteration, the four transformations are applied to each of the four images. Write matrices for the 16 images in the second iteration, with elements rounded to one decimal place. Plot these images on dot paper or graph paper.

e. If the transformations are applied infinitely many times, the result is part of Helge von Koch's **snowflake curve.** The first and second iterations of that curve are shown, respectively, in Figure 11-5o and in Figure 11-5p. The result of many iterations is shown in Figure 11-5q. On a copy of Figure 11-5q, circle two parts of the snowflake curve that are of different sizes that show that the snowflake curve is self-similar.

f. The length of the pre-image in Figure 11-5m is 12 units. The first iteration in Figure 11-5n has total length 16 units because it has four segments that are each four units long. What is the total length of the second iteration? The third iteration? The fourth iteration? Find the length of the 100th iteration. What would be the total length of this part of the final snowflake curve? Does the answer surprise you?

g. Use Barnsley's method to show that you get the same strange attractor when you start with *one* point as a pre-image and perform the four transformations iteratively, at random, on the resulting images. Use a probability of $\frac{1}{4}$ for each transformation.

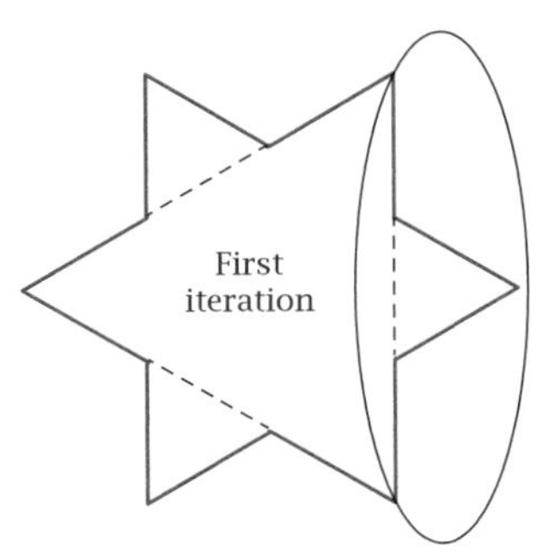

Figure 11-5o

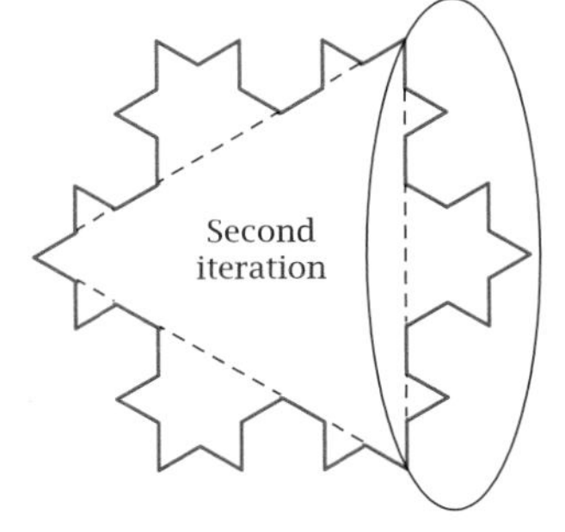

Figure 11-5p

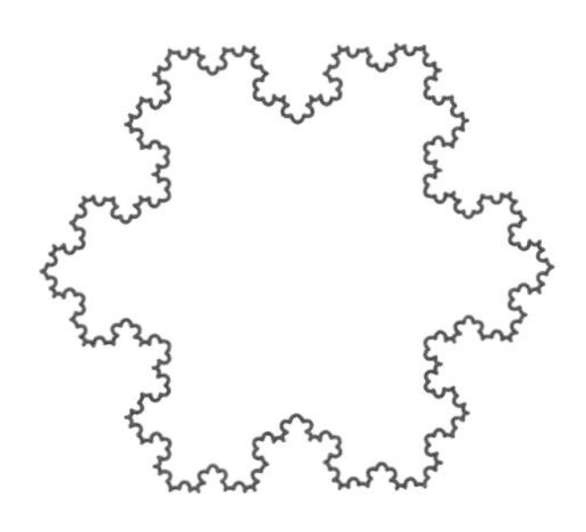

Figure 11-5q

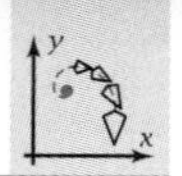

11. *Fixed Points in a Strange Attractor:* Transformation [A] for the fern image shown in Figure 11-5d is

$$[A] = \begin{bmatrix} 0.8\cos 3^\circ & 0.8\cos 93^\circ & 0 \\ 0.8\sin 3^\circ & 0.8\sin 93^\circ & 3 \\ 0 & 0 & 1 \end{bmatrix}$$

a. Find the fixed point for this transformation. To what part of the fern image does this fixed point correspond?

b. Make a conjecture about the approximate locations of the fixed points for transformations [B], [C], and [D] for the fern image in Figure 11-5d.

c. Compute the fixed points in part b numerically by raising the transformation matrices to a high power. Do the computations confirm or refute your conjecture?

11-6 Fractal Dimensions

In Section 11-5 you had the chance to explore Koch's snowflake curve and Sierpiǹski's triangle. These are shown in Figure 11-6a. You may have found in your explorations that the lengths of successive iterations of the snowflake curve follow an increasing geometric sequence: 12, 16, 21.3333..., 28.4444..., ... The areas of successive iterations of Sierpiǹski's triangle form a decreasing geometric sequence: 337.5, 253.125, 189.84375, ... So the length of the snowflake curve approaches infinity and the area of Sierpiǹski's triangle approaches zero.

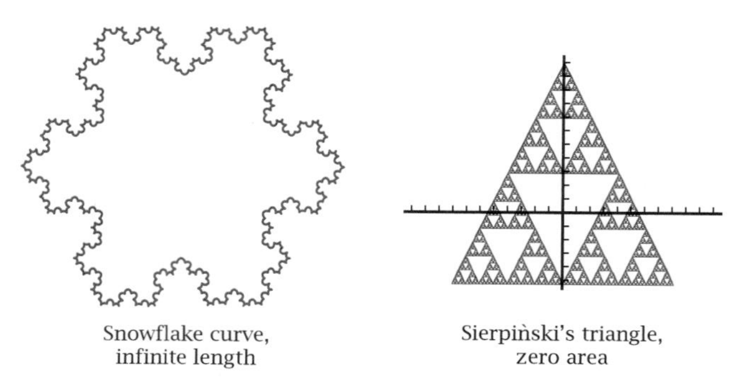

Figure 11-6a

In this section you will make sense out of these seeming contradictions by learning a precise definition of the dimension of a figure. Both figures in Figure 11-6a are called **fractals** because their dimensions are fractions. As you'll see, the dimensions of both these fractals are between 1 and 2.

OBJECTIVE Given a figure formed by iteration of several transformation matrices, determine its fractal dimension.

A solid (three-dimensional) cube is a *self-similar* object because it can be broken into smaller cubes that are similar to one another and to the original cube.

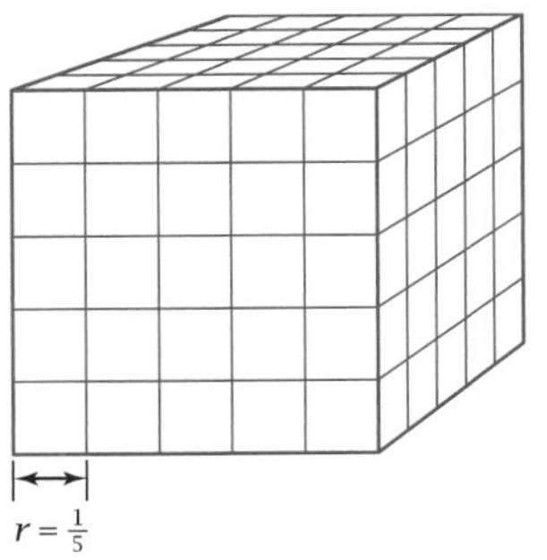

Figure 11-6b

Figure 11-6b shows a cube with edge length 1 unit divided into smaller cubes, each with edge length $\frac{1}{5}$ unit. Note that each of the smaller cubes is also self-similar.

There are $N = 5^3$, or 125, small cubes. The exponent 3 in this equation is the dimension of the cube. With the help of logarithms, you can isolate the exponent 3.

$$\log 5^3 = \log N$$

$$3 \log 5 = \log N$$

$$3 = \frac{\log N}{\log 5}$$

The 5 in the equation is equal to $\frac{1}{r}$, where r, in this case $\frac{1}{5}$, is the ratio of the edge length of one small cube to the length of the original pre-image cube. Substituting this information into the last equation gives

$$3 = \frac{\log N}{\log \frac{1}{r}}$$

This equation is the basis for the definition of dimension credited to Felix Hausdorff, a German mathematician who lived from 1868 to 1942.

DEFINITION: Hausdorff Dimension

If an object is transformed into N self-similar pieces, and the ratio of the length of each piece to the length of the original object is r, and the subdivisions can be done infinitely many times, then the dimension D of the object is

$$D = \frac{\log N}{\log \frac{1}{r}}$$

This is called the **Hausdorff dimension.**

To see how this definition applies to the snowflake curve, consider the pre-image and first iteration of any one segment in the curve. As Figure 11-6c shows, the segment is transformed to 4 self-similar segments, each of which is $\frac{1}{3}$ as long as the original segment, so $N = 4$ and $r = \frac{1}{3}$.

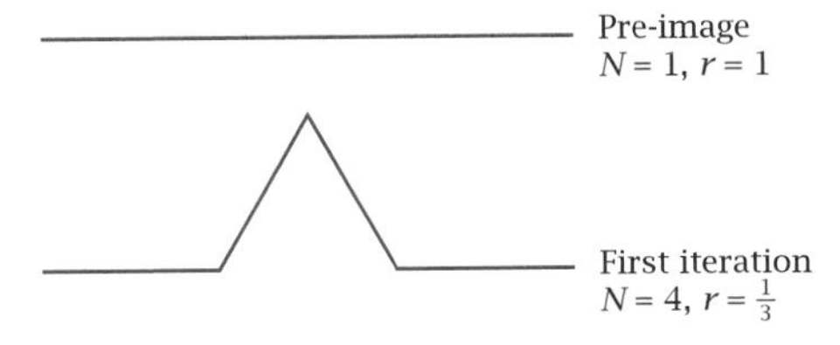

Figure 11-6c

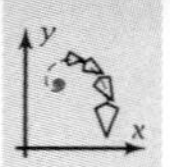

The iterations are done infinitely many times in the same pattern, so Hausdorff's definition of dimension applies. Thus the dimension of the snowflake curve is

$$D = \frac{\log 4}{\log \frac{1}{\frac{1}{3}}} = \frac{\log 4}{\log 3} = 1.2618\ldots$$

So the snowflake curve is 1.2618... dimensional! The dimension of a fractal measures the fractal's "space-filling ability." The curve's infinite length helps explain why it has a fractional dimension. If it had finite length, it could be "straightened out" into line segments and would therefore be one-dimensional. However, because the snowflake curve is infinite, it can never be straightened out all the way; there will always be "spikes" that extend into the second dimension.

▶ EXAMPLE 1 At each iteration in the generation of the snowflake curve, any one segment in the preceding iteration is divided into four self-similar segments, each of which is one-third as long as the previous segment.

a. For iterations 0 through 4, make a table of values showing the iteration number (n), the number of segments (N), the ratio of the length of each segment to the length of the pre-image (r), and $\frac{1}{r}$.

b. Calculate the dimension D of the snowflake using N and r from iteration 1. Show that you get the *same* value of D using N and r from iteration 4. How does the name used for this sort of figure reflect the fact that the Hausdorff dimension is not an integer?

c. Perform linear regression for $\log N$ as a function of $\log\left(\frac{1}{r}\right)$. Plot the points and the equation on the same screen. Show numerically that the slope of the line equals the dimension of the snowflake.

d. If the pre-image is 12 units long, calculate the total length of the images at each iteration, 1 through 4. Use the pattern you observe to calculate the total length of the images at the 50th iteration. Explain why the total length of the snowflake approaches infinity as the iterations continue.

Solution a. The total-length values are computed in part d.

Iteration	N	r	$\frac{1}{r}$	Total Length, L
0	1	1	1	$12(1)(1) = 12$
1	4	$\frac{1}{3}$	3	$12(4)\left(\frac{1}{3}\right) = 16$
2	16	$\frac{1}{9}$	9	$12(16)\left(\frac{1}{9}\right) = 21.3333\ldots$
3	64	$\frac{1}{27}$	27	$12(64)\left(\frac{1}{27}\right) = 28.4444\ldots$
4	256	$\frac{1}{81}$	81	$12(256)\left(\frac{1}{81}\right) = 37.9259\ldots$

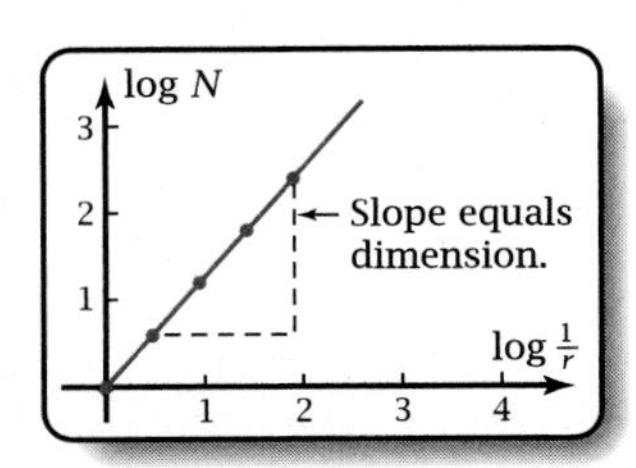

Figure 11-6d

The coastline of Cape Cod, Massachusetts. The length of a coastline can be estimated using the same calculation methods as for the length of fractals.

b. $$D = \frac{\log 4}{\log \frac{1}{\frac{1}{3}}} = \frac{\log 4}{\log 3} = 1.2618\ldots$$

$$D = \frac{\log 256}{\log \frac{1}{\frac{1}{81}}} = \frac{\log 256}{\log 81} = 1.2618\ldots,$$ which is the same value.

It's called a *fractal,* indicating that its Hausdorff dimension is a fraction.

c. $$\log N = (1.2618\ldots) \log \frac{1}{r} + 0$$ Fit is exact because $r = 1$.

The 1.2618... slope equals the dimension from part b, as in Figure 11-6d.

d. The total lengths are shown in the table in part a. The total length at each iteration is the original length, 12, multiplied by the number of segments N times the ratio of the length of each segment to the original length. From this pattern, the equation for the length L as a function of the iteration number n is

$$L = 12\left(\frac{4}{3}\right)^n$$

Substituting 50 for n in this equation gives

$$L = 12\left(\frac{4}{3}\right)^{50} = 21{,}189{,}371.5\ldots$$

The values of L form a geometric sequence with common ratio $\frac{4}{3}$. Thus the values of L are unbounded as n increases. The number of iterations to generate the snowflake curve is infinite, so the length is infinite. ◀

Problem Set 11-6

Do These Quickly

Q1. What is the common ratio of this geometric sequence? 100, 90, 81, 72.9, . . .

Q2. If you apply one linear transformation iteratively, the images can be attracted to a —?—.

Q3. If you apply several linear transformations iteratively, the images can be attracted to a —?—.

Q4. What rotation is caused by this matrix?

$$\begin{bmatrix} 0.7\cos(-35°) & 0.7\cos 55° & -4 \\ 0.7\sin(-35°) & 0.7\sin 55° & 2 \\ 0 & 0 & 1 \end{bmatrix}$$

Q5. What dilation is caused by the matrix in Problem Q4?

Q6. What x-translation is caused by the matrix in Problem Q4?

Q7. What y-translation is caused by the matrix in Problem Q4?

Q8. Explain the purpose of the 0, 0, 1 in the bottom row of the matrix in Problem Q4.

Q9. What kind of function has the "multiply-multiply" property?

Q10. If $g(x) = f\left(\frac{1}{3}x\right)$, what transformation is performed on $f(x)$ to get $g(x)$?

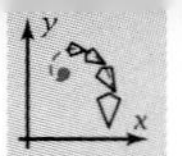

1. *Dimension Definition Applied to a Square Problem:* Figure 11-6e shows a square region divided into 25 self-similar squares, each with side length $\frac{1}{5}$ the side length of the original square.

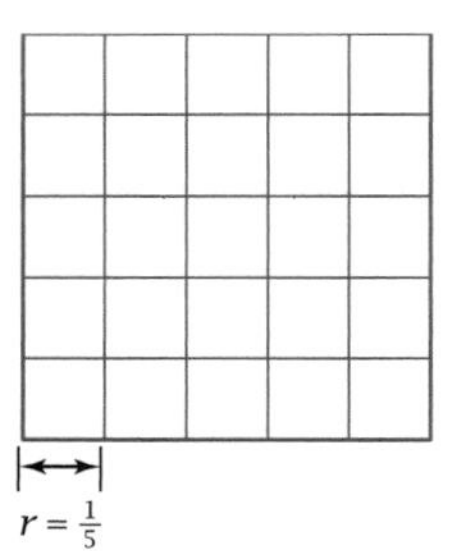

Figure 11-6e

   a. Show that the Hausdorff dimension leads you to conclude that a square is two-dimensional.

   b. If the square were cut into smaller squares with sides $r = 0.01$ times the length of the original sides, show that the Hausdorff dimension would still lead to the conclusion that a square is two-dimensional.

   c. What allows you to conclude that the Hausdorff dimension really does apply to the square when it is cut into smaller self-similar squares?

2. *Dimension of Sierpiński's Triangle Problem:* Figure 11-6f shows the pre-image and the first iteration of Sierpiński's triangle. The original pre-image triangle is transformed into three self-similar triangles, each of which has sides one-half the length of the sides in the pre-image.

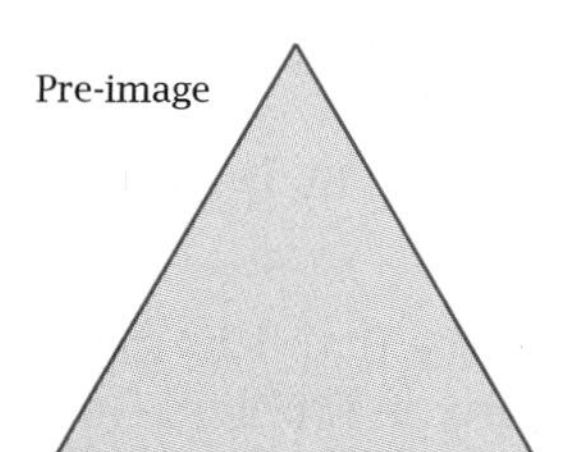

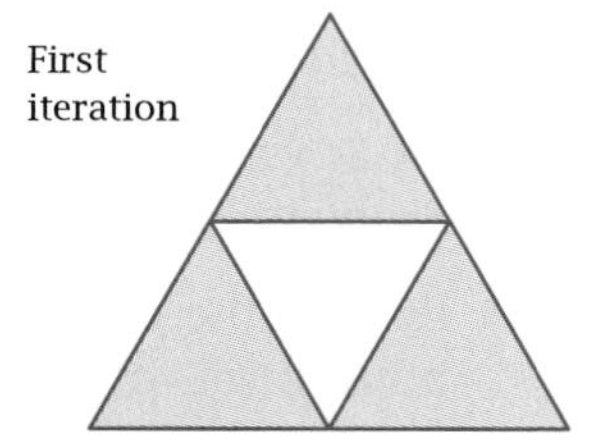

Figure 11-6f

   a. For iterations 0 through 4, make a table of values showing the iteration number (n), the number of triangles (N), the ratio of the side length of each triangle to the side length of the pre-image (r), and $\frac{1}{r}$.

   b. Calculate the dimension D of the Sierpiński's triangle using iterations 1 and 4. Show that you get the same value of D using the values for N and r from either iteration 1 or 4. What word is used to describe the fact that the dimension is not an integer?

   c. Perform a linear regression for log N as a function of log $\left(\frac{1}{r}\right)$. Plot the points and the equation on the same screen. Show numerically that the slope of the line equals the dimension of Sierpiński's triangle.

   d. If the pre-image is an equilateral triangle with sides 16 cm long, calculate the total perimeter of the images in each iteration, 1 through 4. Use the pattern you observe to calculate the total perimeter of the images in the 50th iteration. Explain why the total perimeter of Sierpiński's triangle approaches infinity as the iterations continue.

3. *Dimension of Sierpiński's Square Problem:* Figure 11-6g shows the pre-image and the first iteration of Sierpiński's square. The original pre-image is divided into four self-similar squares, each of which has side lengths that are 40% of the side lengths in the pre-image.

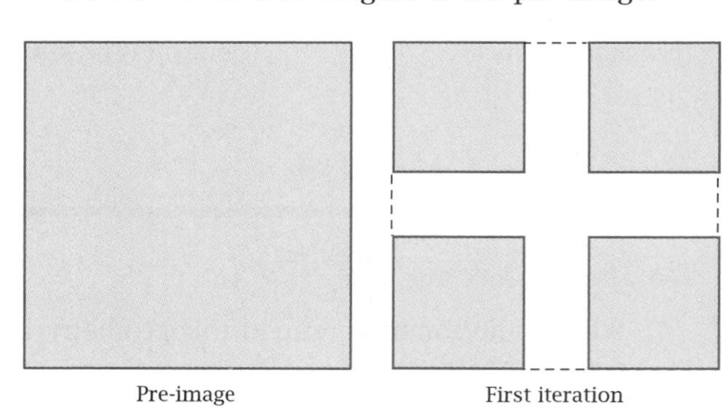

Figure 11-6g

   a. The complete Sierpiński's square is formed by iterating infinitely many times. What, then, is the dimension of the final Sierpiński's square?

   b. Calculate the total area of the final Sierpiński's square. How does the result correspond to the dimension of the square?

   c. Calculate the total perimeter of the final Sierpiński's square. How does the result correspond to the dimension of the square?

d. Suppose that the self-similar squares at each iteration had side lengths that were 50% of the side lengths of the preceding image. How would this change the dimension of the square? Why is it *not* correct to call Sierpiǹski's square a fractal in this case?

e. Suppose that the self-similar squares at each iteration had side lengths that were 60% of the side lengths of the preceding image. How would this change the dimension of the square? How would it affect the total area of the square?

4. *Conclusions Problem:* At the beginning of this section you read that the snowflake curve has infinite length and that Sierpiǹski's triangle has zero area. How is the fractal dimension of a figure related to its total length and area?

5. *Journal Problem:* Write an entry in your journal telling the most significant things you have learned about iterated transformations and fractals by studying this chapter.

11-7 Chapter Review and Test

In this chapter you have seen how you can use the concepts of *matrix* and *determinant* to solve systems of linear equations and to apply linear transformations to geometric figures. Performed iteratively, a linear transformation can cause the images to be attracted to a fixed point. If several such transformations are done iteratively, the images can be attracted to a figure of great complexity, sometimes resembling an object in nature, such as a fern or a tree. These strange attractors are called *fractals* because they have fractional Hausdorff dimensions rather than integer dimensions.

Applying linear transformations to a simple starting shape can produce images of startling realism—evident in this computer-generated landscape.

Review Problems

R0. Update your journal with what you have learned in this chapter. Include such things as

- The one most important thing you have learned as a result of studying this chapter
- The new terms you have learned and what they mean
- The ways in which matrix transformations can change an image
- How Barnsley's method for generating fractal images differs from repeating transformations on all images in the previous iteration
- Hausdorff's definition of dimension

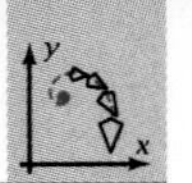

R1. Figure 11-7a shows a line segment 1 unit long. In the first iteration, an image is formed by removing the middle third of the segment. In the second iteration, an image is formed by removing the middle third of each segment in the first iteration. If the iterations are done infinitely many times, the image is called the **Cantor set,** after Georg Cantor, a German mathematician who lived from 1845 to 1918. How many segments are there in the 10th iteration? What is the total length of the segments in the 10th iteration? What does the total length approach as the number of iterations increases without bound?

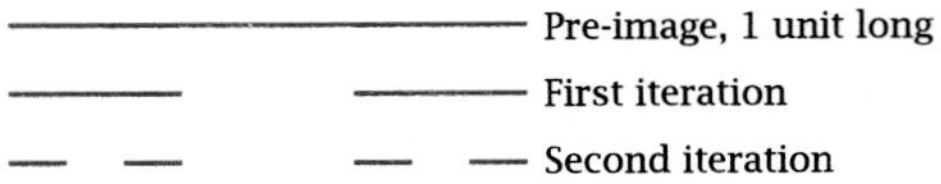

Pre-image, 1 unit long

First iteration

Second iteration

Figure 11-7a

R2. a. Evaluate: $9\begin{bmatrix} 5 & 2 \\ 7 & -1 \end{bmatrix} - 6\begin{bmatrix} 3 & 8 \\ 5 & 4 \end{bmatrix}$

b. Evaluate: $\begin{bmatrix} 3 & -5 & 2 \\ -1 & 4 & 3 \end{bmatrix}\begin{bmatrix} 6 \\ 2 \\ -3 \end{bmatrix}$

c. Evaluate: $\det\begin{bmatrix} 3 & 8 \\ 5 & 4 \end{bmatrix}$

d. Solve this system by using matrices.

$$3x - 5y + 2z = -7$$
$$4x + y - 6z = 33$$
$$9x - 8y - 7z = 38$$

R3. a. Describe the transformations produced by matrix [T].

$$[T] = \begin{bmatrix} 0.6\cos 30° & 0.6\cos 120° \\ 0.6\sin 30° & 0.6\sin 120° \end{bmatrix}$$

b. Plot the pre-image triangle specified by

$$[M] = \begin{bmatrix} 5 & 8 & 7 \\ 1 & 2 & 3 \end{bmatrix}$$

c. Plot the image from the third iteration, [T][T][T][M], on the same grid as the pre-image triangle.

d. By how many degrees has the image in part c been rotated from the original pre-image?

e. Show that the image vertex closest to the origin is exactly 0.6^3 times as far from the origin as the corresponding vertex in the pre-image.

R4. a. Describe the transformations produced by matrix [A].

$$[A] = \begin{bmatrix} 0.6\cos 30° & 0.6\cos 120° & 5 \\ 0.6\sin 30° & 0.6\sin 120° & 2 \\ 0 & 0 & 1 \end{bmatrix}$$

b. Plot the pre-image rectangle specified by

$$[M] = \begin{bmatrix} 0 & 10 & 10 & 0 \\ 0 & 0 & 3 & 3 \\ 1 & 1 & 1 & 1 \end{bmatrix}$$

c. Plot the image at the third iteration, [A][A][A][M].

d. To what fixed point are the images attracted when transformation [A] is performed iteratively many times? How do you calculate this point numerically in a time-efficient way?

e. Tell the purpose of the row 1, 1, 1, 1 in [M] and of the row 0, 0, 1 in [A].

R5. Figure 11-7b shows the rectangular pre-image

$$[M] = \begin{bmatrix} -3 & 3 & 3 & -3 \\ -1 & -1 & 1 & 1 \\ 1 & 1 & 1 & 1 \end{bmatrix}$$

It also shows the images of [A][M] and [B][M], where [A] is the transformation given in Review Problem R4, part a, and [B] rotate and dilates the same as [A] but translates in the opposite direction.

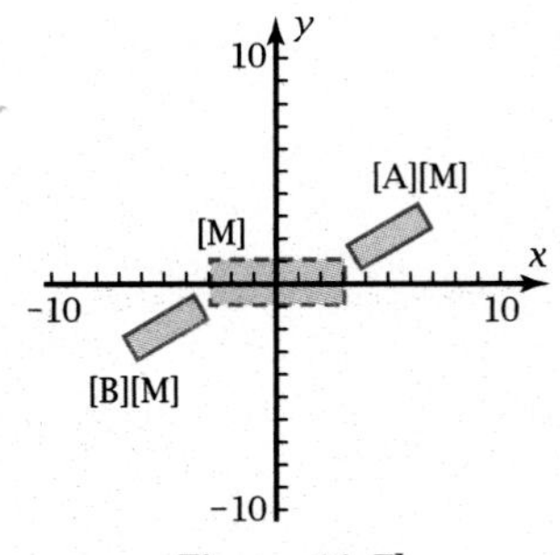

Figure 11-7b

a. Write a matrix for transformation [B].

b. The two rectangles in Figure 11-7b form the first iteration for a more complicated geometric figure. In the second iteration, [A] and [B] are applied to both images in the first iteration. Perform these transformations and plot the resulting four images on graph paper.

c. The figure that results from applying [A] and [B] iteratively infinitely many times can be plotted approximately using Barnsley's method. The result is shown in Figure 11-7c. Confirm on your grapher that the attractor in Figure 11-7c is correct. Write a paragraph describing the procedure used in Barnsley's method.

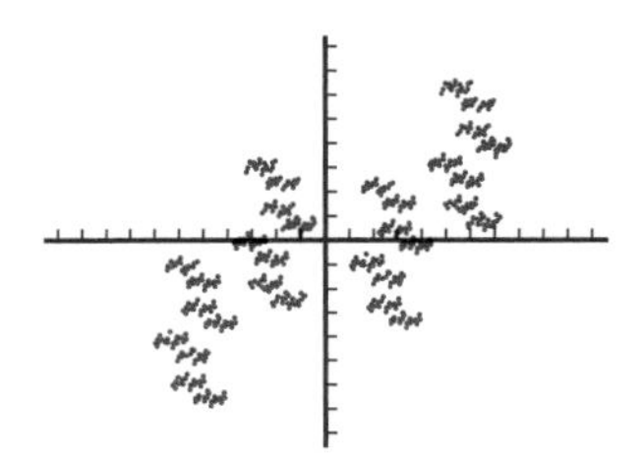

Figure 11-7c

d. On a copy of Figure 11-7c, circle two parts of the figure of different sizes, each of which is similar to the entire figure.

e. On the copy of Figure 11-7c, plot the fixed point you found for transformation [A] in Review Problem R4, part d. How does this fixed point seem to correspond to the fractal image?

f. For the pre-image and the first three iterations, make a table showing the iteration number, the number of rectangles, the perimeter of each rectangle, and the total perimeter of the figure. From the pattern in the table, find the total perimeter of the 50th iteration. If the iterations were done infinitely many times, the result would be the strange attractor in Figure 11-7c. What is the total perimeter of the strange attractor?

R6. a. State the definition of the Hausdorff dimension.

b. Show that the fractal in Figure 11-7c is more than one-dimensional but less than two-dimensional.

c. How does the result from part b agree with the total perimeter of the fractal in Figure 11-7c?

d. Find the area of the pre-image rectangle and the total areas of the rectangles in the first, second, and third iterations. What number does the area of the fractal in Figure 11-7c approach as the number of iterations increases without bound? How does this answer agree with the dimension you calculated in part b?

e. Suppose that the dilations for matrices [A] and [B] from Review Problems R4 and R5 were changed from 0.6 to 0.5. Use Barnsley's method to plot the resulting figure. Use probabilities of 50% for each matrix.

f. Calculate the dimension of the figure in part e. Calculate the number that the total perimeter of the rectangles approaches as the number of iterations approaches infinity. How do the results of these calculations explain the change in the figure caused by reducing the dilation to 0.5?

g. If the dilation in transformations [A] and [B] were reduced to 0.4, what would be the dimension of the resulting figure? What number would the total perimeter of the figure approach? How does this number correspond to the dimension of the figure?

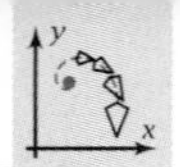

Concept Problem

C1. *Research Problem 2:* Explore literature about fractals. From the information you find, choose a topic and write a detailed report. Mention how your topic corresponds to what you have been learning in this chapter and how it has extended your knowledge of iterated transformations and fractals. Here are some suggested sources.

- Barnsley, Michael F. *Fractals Everywhere.* Boston: Academic Press, 1988.
- Mandelbrot, Benoit B. *The Fractal Geometry of Nature.* New York: W. H. Freeman & Company, 1983.
- Peitgen, Heinz-Otto, Hartmut Jürgens, and Dietmar Saupe. *Fractals for the Classroom.* New York: Springer-Verlag, 1992.
- Prusinkiewicz, Przemyslaw, and Aristid Lindenmayer. *The Algorithmic Beauty of Plants.* New York: Springer-Verlag, 1990.

The set of points in the center, dark region of the image is called the Mandelbrot set.

Chapter Test

PART 1: No calculators allowed (T1–T6)

T1. Multiply: $\begin{bmatrix} 3 & 5 & -2 \\ 4 & 1 & 7 \end{bmatrix}\begin{bmatrix} 6 & 8 \\ -1 & 9 \\ 0 & 3 \end{bmatrix}$

T2. Explain why the matrices in Problem T1 are commensurate for multiplication.

T3. In what way is matrix multiplication similar to the multiplication of two vectors?

T4. Find det $\begin{bmatrix} 3 & 8 \\ 2 & 7 \end{bmatrix}$. Use the result to find $\begin{bmatrix} 3 & 8 \\ 2 & 7 \end{bmatrix}^{-1}$. Show that the product of the matrix and its inverse is equal to the identity matrix.

T5. What transformation is represented by this matrix?

$$\begin{bmatrix} 0.9\cos 15^\circ & 0.9\cos 105^\circ & 3 \\ 0.9\sin 15^\circ & 0.9\sin 105^\circ & 2 \\ 0 & 0 & 1 \end{bmatrix}$$

T6. A line segment is transformed by iterative matrix multiplication. The matrices for the pre-image and the images of the first and second iterations are

$$\begin{bmatrix} 5 & 5 \\ 2 & 6 \\ 1 & 1 \end{bmatrix} \qquad \begin{bmatrix} 6.7 & 5.1 \\ 6.4 & 9.2 \\ 1 & 1 \end{bmatrix} \qquad \begin{bmatrix} 6.1 & 3.8 \\ 10.1 & 11.4 \\ 1 & 1 \end{bmatrix}$$

Pre-image   First iteration   Second iteration

Plot these three images on graph paper or dot paper. Tell what will happen to the images as more and more iterations are performed.

PART 2: Graphing calculators allowed (T7–T19)

For Problems T7–T18, the graph in Figure 11-7d shows a vertical segment 10 units high, starting at the origin. This pre-image is to be transformed into a "tree" made from three segments, each 5 units long. The three segments satisfy these conditions:

- The left branch starts at (0, 5) and is rotated +20° from the trunk.
- The trunk extends from (0, 0) to (0, 5).
- The right branch starts at (0, 5) and is rotated −30° from the trunk.

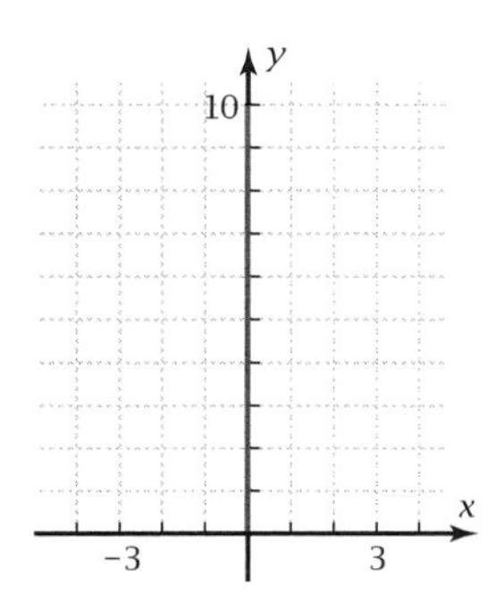

Figure 11-7d

T7. Write three 3 × 3 transformation matrices to do the following:

[A] should transform the pre-image to the left branch.

[B] should transform the pre-image to the trunk.

[C] should transform the pre-image to the right branch.

T8. Multiply each of [A], [B], and [C] by the pre-image matrix. Write the three image matrices, rounding the entries to one decimal place. Plot the three images on graph paper.

T9. The three images in Problem T8 are the results of the first iteration. If the three transformations are done to each of these three images, the nine resulting images form the second iteration. Calculate the image [A][C][D] of the second iteration, and plot it on the same axes as the images from Problem T8.

T10. If the iterations are done infinitely many times, the resulting tree is a fractal. If the transformations are performed at random on a single point, each time using the image from the time before, the points are attracted to the *same* fractal figure. Use Barnsley's method with 1000 points to plot this strange attractor.

T11. Figure 11-7e shows the strange attractor from Problem T10 plotted with 1000 points. On a copy of Figure 11-7e, illustrate self-similarity by circling two parts of the figure of different size, each of which is similar to the whole tree.

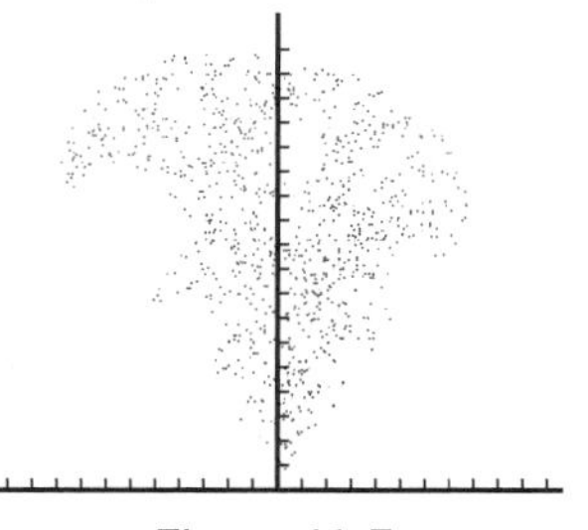
Figure 11-7e

T12. The first iteration has three images, each 5 units long. The second iteration has nine images, each 2.5 units long. Calculate the sum of the lengths of the images at iterations 0, 1, 2, 3, and 100.

T13. If the iterations were performed infinitely many times, what would the sum of the lengths of the images approach?

T14. Each iteration divides each previous segment into three self-similar pieces, each 0.5 times as long as the previous segment. Let N be the number of pieces, and let r be the ratio of the length of each piece to the length of the original pre-image. Complete a table for 0–5 iterations giving values for n, r, $\frac{1}{r}$, and N.

T15. Complete the statement: "Each time $\frac{1}{r}$ is multiplied by 2, N is multiplied by —?—."

T16. Write Hausdorff's definition of dimension.

T17. Calculate the dimension of the tree that would result if the iterations were done infinitely many times.

T18. Strange attractors such as the one in Figure 11-7e result from iterating several different transformations. If just *one* transformation is iterated, the images are attracted to a *single* fixed point. Find the fixed point to which the images are attracted if [A] is performed iteratively on the pre-image [D]. Show the fixed point and the pattern followed by the images on your graph from Problem T8.

T19. What did you learn as a result of taking this test that you did not know before?

Analytic Geometry of Conic Sections and Quadric Surfaces

CHAPTER 12

Three-dimensional figures that model objects such as the Kobe Port Tower in Japan and the parabolic antenna are generated by rotating conic sections—hyperbolas, parabolas, or ellipses—about an axis of symmetry. These figures have rich geometrical and algebraic properties. The general name *conic sections* comes from the fact that they can be generated geometrically by a plane slicing through (sectioning) a cone.

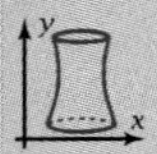

Mathematical Overview

In this chapter you will learn how to apply algebraic techniques to analyze ellipses, hyperbolas, parabolas, and circles, all of which are formed by planes sectioning a cone. Fixed focal points and directrix lines help define these *conic sections.* The shapes of spaceship and comet paths are conic sections, and reflective surfaces are formed by rotating the conic section shapes about their axes. You will study conic sections in four ways.

Graphically Ellipse (Figure 12-0a)

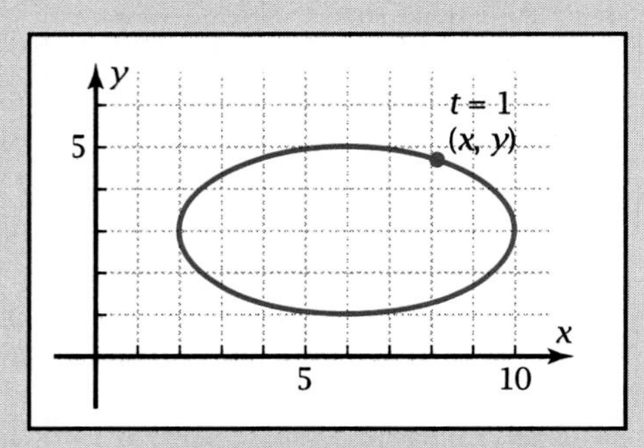

Figure 12-0a

Algebraically Cartesian equation of an ellipse:

$x^2 + 4y^2 - 12x - 24y + 56 = 0$

Parametric equations of an ellipse:

$x = 6 + 4\cos t$

$y = 3 + 2\sin t$

Numerically $t = 1$: $x = 6 + 4\cos 1 = 8.1612\ldots$

$y = 3 + 2\sin 1 = 4.6829\ldots$

Verbally *I learned how to plot an ellipse on my grapher either by using the parametric equations or directly from the Cartesian equation $x^2 + 4y^2 - 12x - 24y + 56 = 0$. Both graphs come out the same. I also learned how to transform the parametric equations into Cartesian form, and vice versa. I had to remember the Pythagorean properties of cosine and sine.*

12-1 Introduction to Conic Sections

Figures 12-1a to 12-1d show the graphs of a circle, an ellipse, a hyperbola, and a parabola. These graphs are called **conic sections**—or, in short, conics—because they are formed by a plane cutting, or sectioning, a cone. They are relations whose equations are quadratics with two variables.

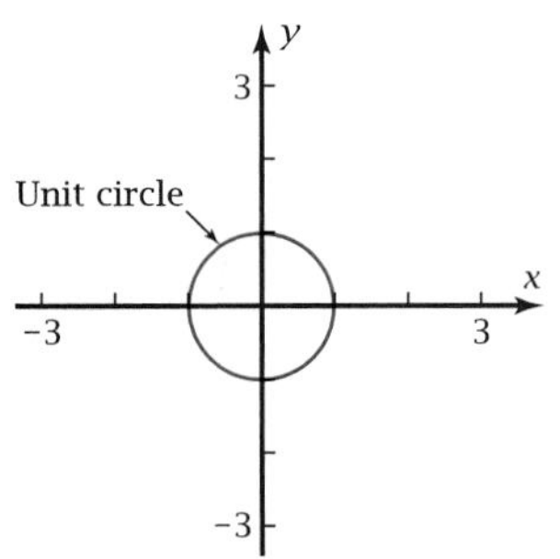

Figure 12-1a

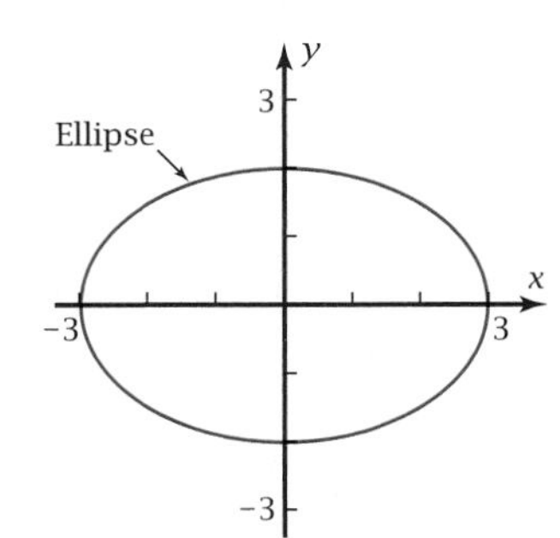

Figure 12-1b

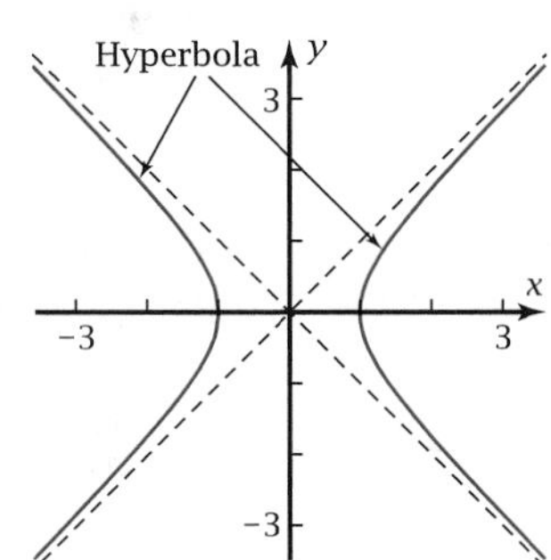

Figure 12-1c

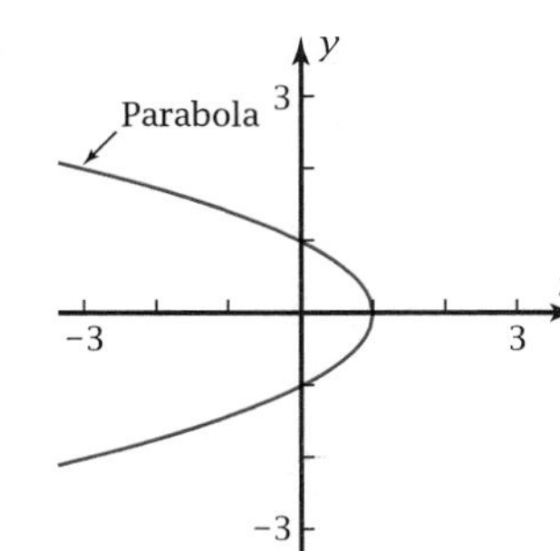

Figure 12-1d

OBJECTIVE Given a quadratic equation with two variables, plot its graph and formulate conclusions.

Exploratory Problem Set 12-1

1. Plot $x^2 + y^2 = 1$ by solving for y in terms of x. Enter the two solutions as y_1 and y_2. (One is a positive square root and the other is a negative square root.) Use a friendly window that includes the integers as grid points and has equal scales on the two axes. Based on the Pythagorean theorem, explain why the graph is the unit circle in Figure 12-1a.

2. Plot $4x^2 + 9y^2 = 36$ by first solving for y in terms of x. Show that the result is the ellipse in Figure 12-1b.

3. The ellipse in Problem 2 is a dilation of the unit circle by a factor of 3 in the x-direction and by a factor of 2 in the y-direction. By making the right side equal 1, transform the given equation to this equivalent form:

$$\left(\frac{x}{3}\right)^2 + \left(\frac{y}{2}\right)^2 = 1$$

This form is sometimes called the **standard form** of the equation of the ellipse. Where do the two dilations show up in the transformed equation?

4. Plot $x^2 - y^2 = 1$ by solving for y in terms of x. Show that the result is the hyperbola in Figure 12-1c. Plot the two lines $y = x$ and $y = -x$. How are these lines related to the graph?

5. Plot the hyperbola $4x^2 - 9y^2 = 36$. Use a friendly window with an x-range of about $[-10, 10]$, and use equal scales on the two axes. Show that the asymptotes now have slopes of $\pm\frac{2}{3}$ instead of ± 1.

6. Transform the equation in Problem 5 to make the right side equal 1, as in Problem 3. Show that the hyperbola in Problem 5 is a dilation of the hyperbola in Figure 12-1c with an x-dilation of 3 and a y-dilation of 2. Tell where these dilation factors appear in the transformed equation.

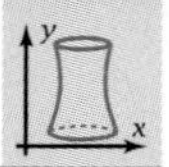

7. The equation $x + y^2 = 1$ has only one squared term. Solve the equation for y in terms of x and plot the two solutions as y_1 and y_2. Show that the graph is the parabola in Figure 12-1d.

8. How could you tell from the equation before it is transformed whether its graph will be a circle, an ellipse, a hyperbola, or a parabola?

12-2 Parametric and Cartesian Equations of the Conic Sections

The paths of satellites or comets traveling in space under the action of gravity are circles, ellipses, parabolas, or hyperbolas. Figure 12-2a shows how these conic sections are formed by a plane sectioning one or both **nappes** of a cone at various angles. The graphs of quadratic equations in two variables are conic sections, as you saw in Section 12-1.

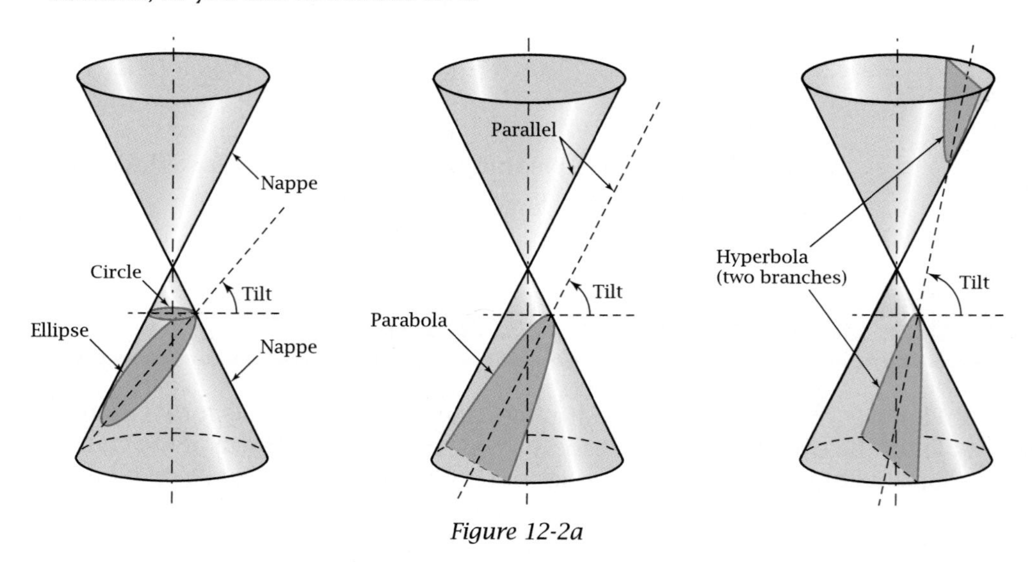

Figure 12-2a

OBJECTIVE Given a Cartesian or parametric equation of a conic section, sketch or plot the graph, and given the graph, find an equation.

Cartesian Equations

Figure 12-2b shows five parent graphs. The equation for each conic section is shown below its graph.

Hyperbolas approach **asymptotes** as x approaches positive or negative infinity. For the unit hyperbolas in Figure 12-2b the asymptotes have equations

$$y = x \qquad \text{and} \qquad y = -x$$

You can tell the type of graph the equation has by looking at the signs of the squared terms.

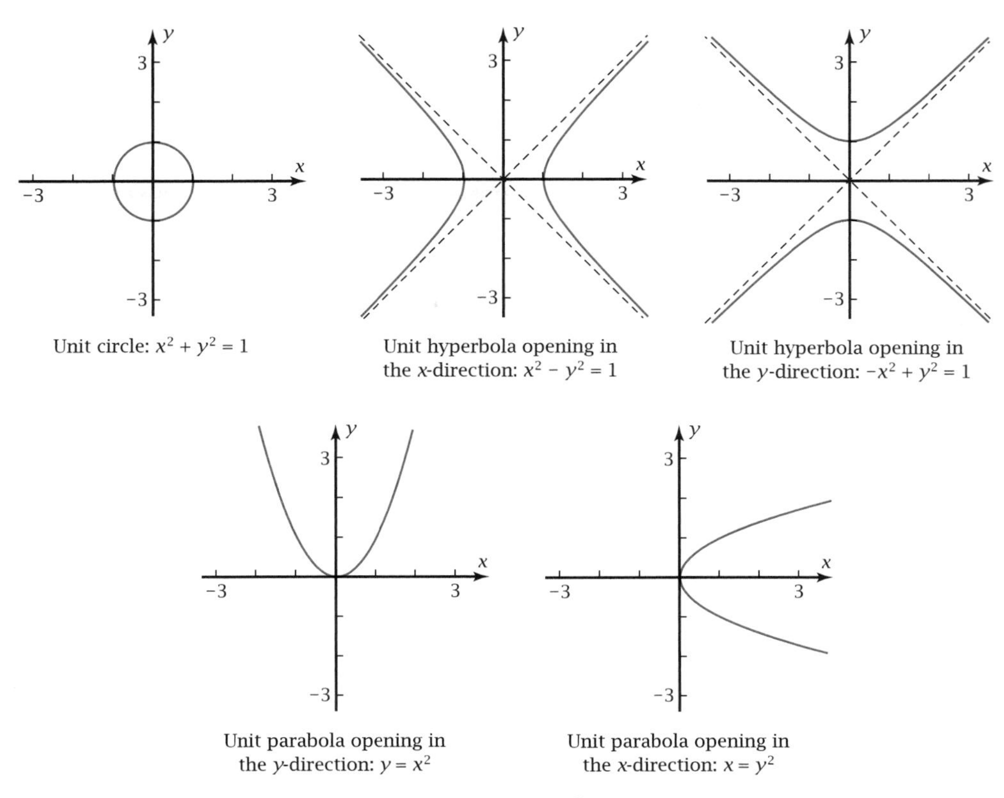

Figure 12-2b

PROPERTIES: Recognition of Conic Sections from Equations

If a quadratic equation in two variables has no *xy*-term, then the graph is:

- A circle if x^2 and y^2 have equal coefficients
- An ellipse if x^2 and y^2 have unequal coefficients but the same sign
- A hyperbola if x^2 and y^2 have opposite signs
- A parabola if only one of the two variables is squared

Any conic section with its axes of symmetry parallel to the *x*- or *y*-axis can be formed by dilating and translating its parent graph. For instance, the ellipse in Figure 12-2c is formed by dilating the unit circle by a factor of 2 in the *x*-direction and by a factor of 3 in the *y*-direction.

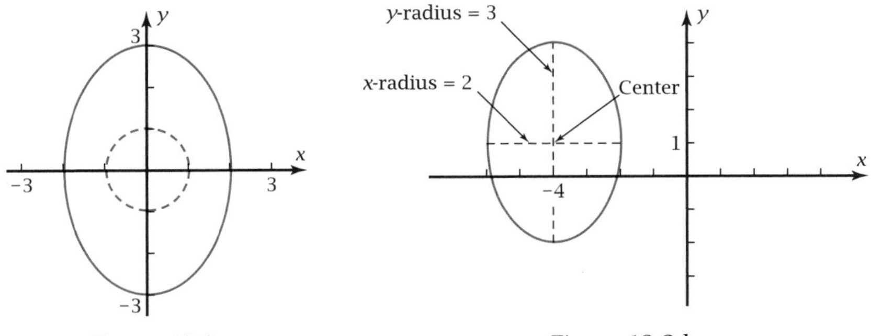

Figure 12-2c

Figure 12-2d

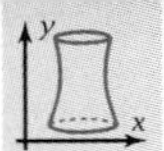

As you recall from previous work, you dilate in a particular direction by *dividing* the respective variable by the dilation factor. So the equation of the ellipse in Figure 12-2c is

$$\left(\frac{x}{2}\right)^2 + \left(\frac{y}{3}\right)^2 = 1$$

Here you can call the dilation factors 2 and 3 the **x-radius** and the **y-radius,** respectively. You translate a graph in either direction by *subtracting* a constant from the respective variable. Figure 12-2d shows the ellipse of Figure 12-2c translated so that its center is at (–4, 1) instead of (0, 0). In this case, the particular equation of the ellipse is

$$\left(\frac{x+4}{2}\right)^2 + \left(\frac{y-1}{3}\right)^2 = 1$$

EXAMPLE 1 Sketch the graph of $\left(\frac{x-2}{5}\right)^2 + \left(\frac{y-4}{3}\right)^2 = 1.$

Solution Analysis:

- The graph will be an ellipse. — x^2 and y^2 have coefficients with the same sign.
- The center is at (2, 4). — The x- and y-translations are 2 and 4, respectively.
- The x-radius is 5 and the y-radius is 3. — The same as the dilation factors.

First plot the center point (2, 4). Then plot points ±5 from the center in the x-direction and ±3 units from the center in the y-direction. These points are called the **vertices** of the ellipse and give the endpoints of its **major axis** and **minor axis,** respectively. Sketch the ellipse by connecting the four points you have plotted with a smooth curve, as shown in the right graph of Figure 12-2e.

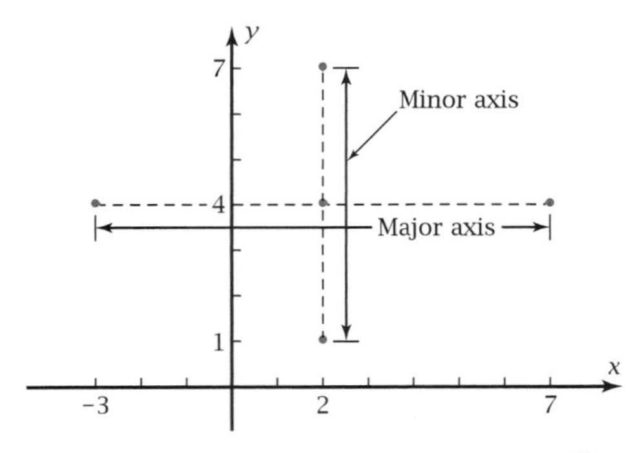

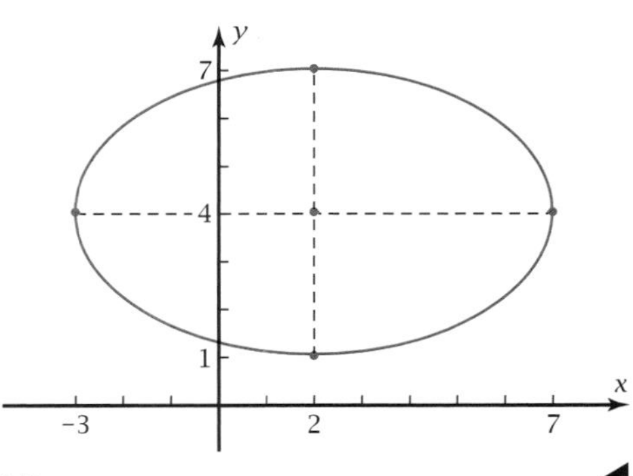

Figure 12-2e

EXAMPLE 2 Transform the equation in Example 1 to the form

$$Ax^2 + Bxy + Cy^2 + Dx + Ey + F = 0$$

where A, B, C, D, E, and F are constants.

Solution

$$\left(\frac{x-2}{5}\right)^2 + \left(\frac{y-4}{3}\right)^2 = 1$$

$$\frac{(x-2)^2}{25} + \frac{(y-4)^2}{9} = 1$$

$$9(x-2)^2 + 25(y-4)^2 = 225$$ Multiply both sides by (25)(9) to eliminate the fractions.

$$9(x^2 - 4x + 4) + 25(y^2 - 8y + 16) = 225$$

$$9x^2 - 36x + 36 + 25y^2 - 200y + 400 = 225$$

$$9x^2 + 25y^2 - 36x - 200y + 211 = 0$$ Commute and associate the terms and make the right side equal zero. ◀

Note that the x^2- and y^2-terms have the same sign but unequal coefficients, indicating that the graph is an ellipse. Note also that there is no xy-term. In Section 12-5 you will learn that an xy-term rotates the graph.

▶ **EXAMPLE 3** Sketch the graph of $-\left(\frac{x+5}{2}\right)^2 + \left(\frac{y-4}{3}\right)^2 = 1$.

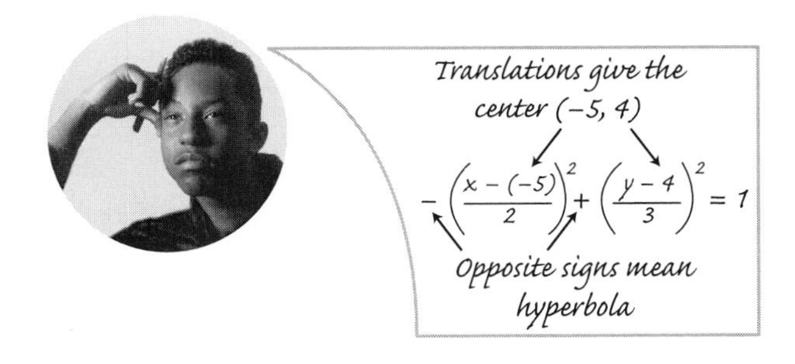

Solution Analysis:

- The graph is a hyperbola. — The squared terms have opposite signs.
- It opens in the y-direction. — The y-containing term is positive.
- The center is at point $(-5, 4)$. — These values make the x- and y-terms zero, respectively.
- The asymptotes have slopes $\pm\frac{3}{2}$. — Slope is $\pm\frac{y\text{-dilation}}{x\text{-dilation}}$.

Sketch the center, vertices, and asymptotes (Figure 12-2f, left). The asymptotes cross at the center and have slopes given by

$$\text{Slope} = \pm\frac{y\text{-dilation}}{x\text{-dilation}}$$

Then sketch the graph (Figure 12-2f, right). Be sure the branches of the graph get closer and closer to the asymptotes and do not curve away.

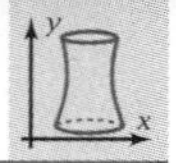

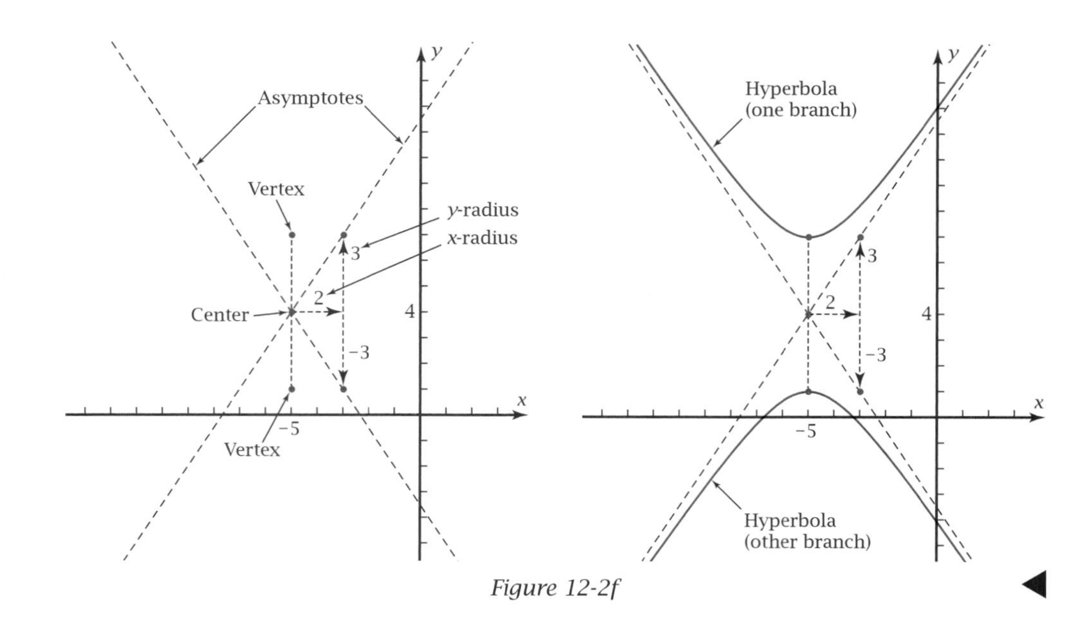

Figure 12-2f

Conic Sections by Grapher from the Cartesian Equation

The general Cartesian equation of a conic section is

$$Ax^2 + Bxy + Cy^2 + Dx + Ey + F = 0$$

where A, B, C, D, E, and F are constants. You can plot the graph by first writing the equation as a quadratic function of y and then using the quadratic formula:

$$Cy^2 + (Bx + E)y + (Ax^2 + Dx + F) = 0$$

$$y = \frac{-(Bx + E) \pm \sqrt{(Bx + E)^2 - 4(C)(Ax^2 + Dx + F)}}{2C}$$ By the quadratic formula.

You can write or download a program to use this result. The input would be the six coefficients, A, B, C, D, E, and F. The program should paste the two equations for y, one with the + sign and one with the − sign, into the y= menu. Then the grapher should plot the two functions.

▶ EXAMPLE 4 Plot the graph of $9x^2 + 25y^2 - 36x + 200y + 211 = 0$ from Example 2.

Solution Run the program. Input 9 for A, 0 for B, 25 for C, −36 for D, 200 for E, and 211 for F.

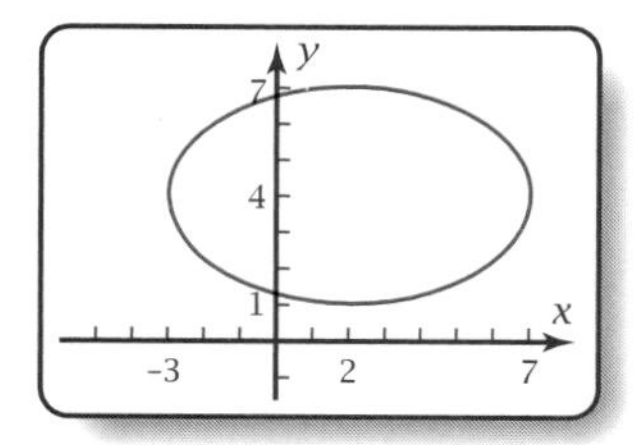

Figure 12-2g

Set a friendly window with an x-range of at least [−3, 7] that has the integers as grid points. Set a window with a y-range so that both axes have equal scales.

The result is the ellipse shown in Figure 12-2g.

Zoom square to make equal scales on both axes.

Parametric Equations of Ellipses, Circles, and Hyperbolas

You can also plot an ellipse, a circle, or a hyperbola parametrically using the center and the two dilations. The parametric equations for the unit circle and two unit hyperbolas are shown in the box.

PROPERTIES: Parametric Equations of the Unit Circle and Unit Hyperbola

For the unit circle centered at the origin:

$$x = \cos t$$
$$y = \sin t$$

For the unit hyperbola centered at the origin:

$$x = \sec t \qquad \text{Opening in the } x\text{-direction.}$$
$$y = \tan t$$

$$x = \tan t \qquad \text{Opening in the } y\text{-direction.}$$
$$y = \sec t$$

where $0 \le t \le 2\pi$.

▶ EXAMPLE 5 Use parametric equations to plot the hyperbola in Example 3:

$$-\left(\frac{x+5}{2}\right)^2 + \left(\frac{y-4}{3}\right)^2 = 1$$

Solution The hyperbola has x- and y-dilations of 2 and 3 and x- and y-translations of -5 and 4, respectively. The secant goes with y because the hyperbola opens in the y-direction.

$x = -5 + 2 \tan t$ — The value added indicates translation; the multiplication coefficient indicates dilation.

$y = 4 + 3 \sec t$ — Enter this as $y = 4 + 3/\cos t$.

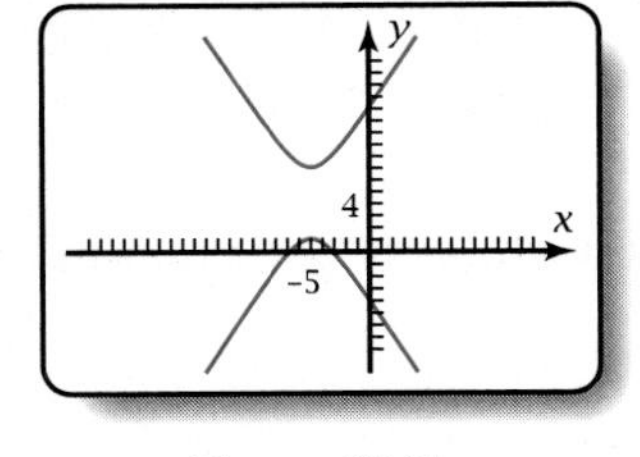

Figure 12-2h

Graph these equations with your grapher in parametric mode. Use radian mode and a t-range of 0 to 2π (one revolution). Use a window centered on $(-5, 4)$, the center of the hyperbola. Use equal scales on the two axes. The graph is shown in Figure 12-2h. ◀

Cartesian Equations of Parabolas

Parabolas have equations in which only one of the variables is squared. Example 6 shows a parabola that opens in the x-direction.

▶ EXAMPLE 6 Sketch the graph of $(x - 5) + 0.5(y - 2)^2 = 0$. Confirm the result by grapher.

Solution Analysis:

- The graph is a parabola opening in the x-direction because the x-term is not squared.
- The x- and y-translations are 5 and 2, respectively. Thus the vertex is at point $(5, 2)$.
- The axis of symmetry is the horizontal line $y = 2$.

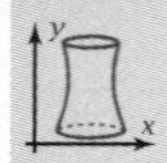

Find the x-intercept to get another point on the graph:

$$(x - 5) + 0.5(0 - 2)^2 = 0 \qquad \text{Substitute 0 for } y \text{ and solve for } x.$$

$$x - 5 + 2 = 0 \quad \Rightarrow \quad x = 3$$

Graph the parabola showing its vertex, x-intercept, and a point symmetric to the x-intercept. The graph is shown in Figure 12-2i.

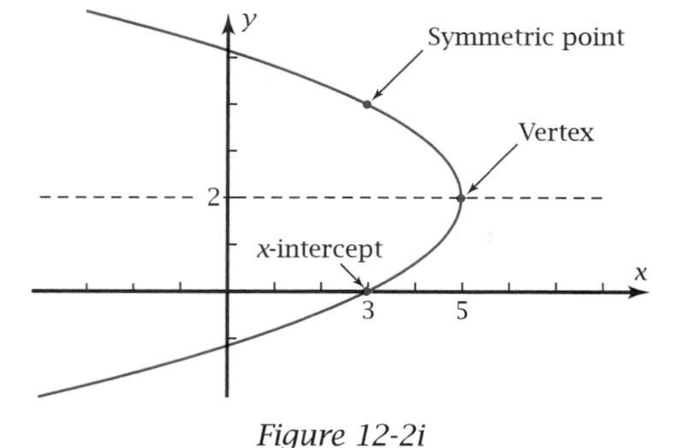

Figure 12-2i

To plot the graph on your grapher, transform the given equation and use the program of Example 4.

$$x - 5 + 0.5(y^2 - 4y + 4) = 0$$

$$0.5y^2 + x - 2y - 3 = 0$$

Enter $A = 0$, $B = 0$, $C = 0.5$, $D = 1$, $E = -2$, and $F = -3$. Your graph should resemble the one in Figure 12-2i. ◀

Problem Set 12-2

Do These Quickly

For Problems Q1–Q4, identify the transformations applied to the parent cosine function to get the circular function $y = 3 + 4 \cos 5(x - 6)$.

Q1. What is the horizontal translation?

Q2. What is the vertical translation?

Q3. What is the horizontal dilation?

Q4. What is the vertical dilation?

Q5. What is the distinguishing characteristic of fractal figures?

Q6. If $\vec{a} = 3\vec{i} + 5\vec{j}$ and $\vec{b} = 4\vec{i} - 6\vec{j}$, find $\vec{a} + \vec{b}$.

Q7. Find $\vec{a} \cdot \vec{b}$ for the vectors in Problem Q6.

Q8. If the probability that Event A happens is 0.8, what is the probability that Event A does not happen?

Q9. How many permutations are possible of the letters in the word PIANO?

Q10. Write the Pythagorean property involving sine and cosine.

For Problems 1–4, sketch the graph.

1. $x^2 + y^2 = 1$
2. $x^2 - y^2 = 1$
3. $-x^2 + y^2 = 1$
4. $x = y^2$

For Problems 5–12,

a. Name the conic section simply by looking at the Cartesian equation.

b. Sketch the graph.

c. Transform the given equation to an equation of the form

$$Ax^2 + Bxy + Cy^2 + Dx + Ey + F = 0$$

d. Plot the Cartesian equation using the result of part c. Does it agree with part b?

5. $\left(\frac{x-3}{2}\right)^2 + \left(\frac{y-1}{4}\right)^2 = 1$
6. $\left(\frac{x+2}{7}\right)^2 + \left(\frac{y-4}{3}\right)^2 = 1$
7. $-\left(\frac{x-2}{5}\right)^2 + \left(\frac{y+1}{3}\right)^2 = 1$
8. $\left(\frac{x+3}{4}\right)^2 - \left(\frac{y+3}{2}\right)^2 = 1$
9. $\left(\frac{x+1}{6}\right)^2 + \left(\frac{y-2}{6}\right)^2 = 1$
10. $\left(\frac{x-4}{10}\right)^2 + \left(\frac{y-2}{10}\right)^2 = 1$

11. $0.2(x-1)^2 + (y-6) = 0$

12. $(x+6) - 1.5(y-3)^2 = 0$

For Problems 13–20,

a. Name the conic section simply by looking at the parametric equations.

b. Sketch the graph.

c. Plot the parametric equations. Does your sketch in part b agree with the graph?

13. $x = \cos t$
 $y = \sin t$

14. $x = \sec t$
 $y = \tan t$

15. $x = 3 \tan t$
 $y = 2 \sec t$

16. $x = 3 \cos t$
 $y = 2 \sin t$

17. $x = 4 + 5 \sec t$
 $y = 3 + 2 \tan t$

18. $x = -2 + 3 \tan t$
 $y = 1 + 4 \sec t$

19. $x = -6 + 5 \cos t$
 $y = -2 + 5 \sin t$

20. $x = 3 + 4 \cos t$
 $y = -2 + 4 \cos t$

For Problems 21–24,

a. Write a Cartesian equation for the conic section.

b. Write parametric equations for the conic section.

c. Confirm that your answer to part b is correct by plotting the parametric equations.

21.

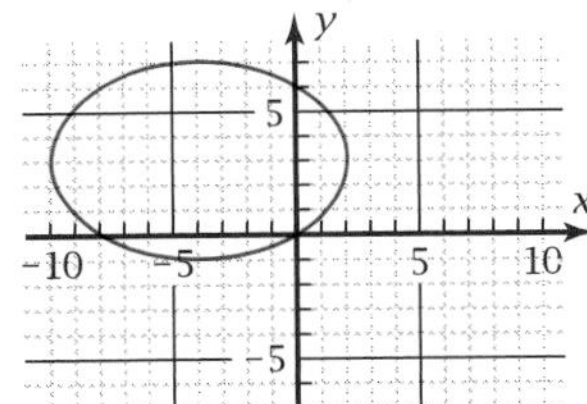

22.

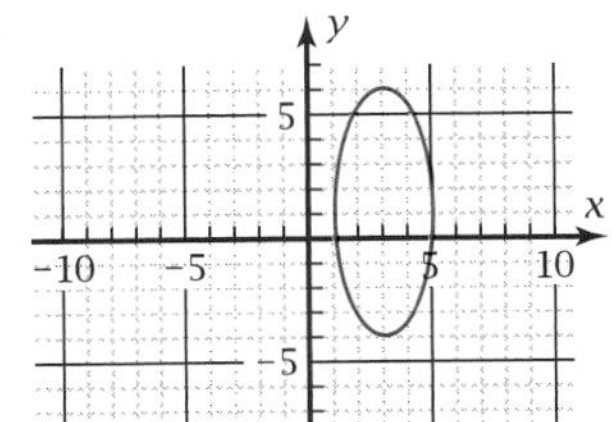

23.

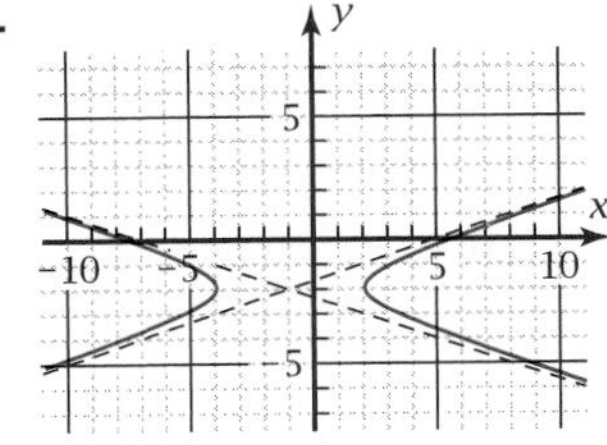

24.

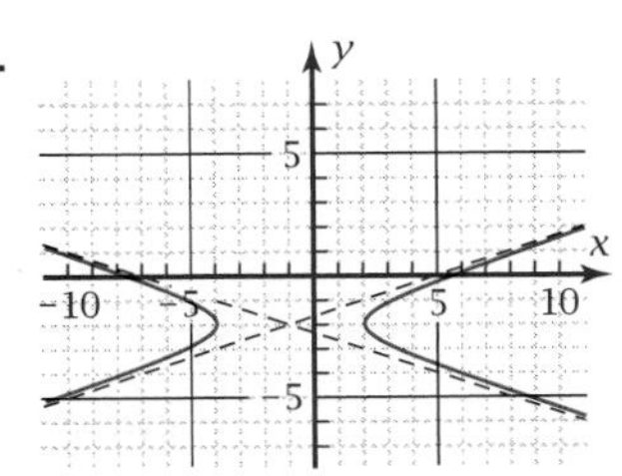

25. *Spaceship Problem 3:* A spaceship orbits in an elliptical path close to the Sun, which is at the origin. If x and y are in millions of miles, the equation of the orbit is

$$\left(\frac{x-12}{13}\right)^2 + \left(\frac{y}{5}\right)^2 = 1$$

a. Sketch the elliptical orbit.

b. What is the closest the spaceship is to the Sun? The farthest?

c. If $x = 20$, what are the two possible values of y?

d. At the points in part c, how far is the spaceship from the Sun?

26. *Meteor Problem:* Astronomers detect a meteor approaching Earth. They determine that its path is the branch of the hyperbola

$x = -50 + 40 \sec t$

$y = 30 \tan t$

as t ranges from -0.5π to 0.5π. The center of Earth is at the origin, and x and y are in thousands of miles.

a. Plot the branch of the hyperbola. Sketch the result.

The Willamete meteorite in the American Museum of Natural History in New York. For the Clackamas tribe, this meteorite represented a union of sky, earth, and water.

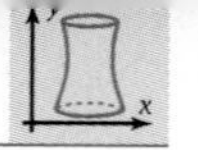

b. When $t = -1$ radian, what are the x- and y-coordinates of the meteor? How far is it from the center of Earth at this time?

c. At what value of t is the meteor closest to Earth? At that time, how far is the meteor from the surface of Earth? (Earth's diameter is about 7920 miles.)

d. Before Earth's gravity deflected the meteor into its curved path, it was traveling straight along one asymptote of the hyperbola. What is the Cartesian equation of this asymptote?

e. What do you suppose is the physical significance of the other asymptote of the hyperbola?

27. *Hyperbola Proof Problem:* For the hyperbola in Problem 17,

$$x = 4 + 5 \sec t$$
$$y = 3 + 2 \tan t$$

a. Transform the first equation so that $\sec t$ is expressed in terms of x, 4, and 5. Transform the second equation so that $\tan t$ is expressed in terms of y, 3, and 2.

b. Square both sides of both transformed equations. Don't expand the squares on the side that involves x or y.

c. Subtract the squared equations.

d. Based on the Pythagorean property for secant and tangent, explain why the result of part c is equivalent to the Cartesian equation of a hyperbola.

28. *Ellipse Proof Problem:* Transform these parametric equations to Cartesian form, as in Problem 27, taking advantage of the Pythagorean property for cosine and sine:

$$x = 3 + 2 \cos t$$
$$y = 1 + 5 \sin t$$

29. *Completing the Square Problem:* It is possible to go from the Cartesian equation back to the form showing translations and dilations by reversing the process of squaring the binomials. The process, called **completing the square,** is illustrated here. Suppose that

$$16x^2 - 25y^2 - 128x - 100y + 556 = 0$$

a. Show the steps in transforming the equation to

$$16(x^2 - 8x \qquad) - 25(y^2 + 4y \qquad) = -556$$

b. You can transform the expressions inside the parentheses to trinomial squares by taking half the coefficient of the linear term, squaring it, and adding it inside the parentheses. Show the steps in transforming the equation to

$$16(x^2 - 8x + 16) - 25(y^2 + 4y + 4)$$
$$= -556 + 256 - 100 = -400$$

c. Show the transformations that give

$$-\left(\frac{x-4}{5}\right)^2 + \left(\frac{y+2}{4}\right)^2 = 1$$

d. Plot the original equation. Show that the graph is the same hyperbola you would get by sketching the equation you got in part c.

30. *Ellipse from the Cartesian Equation Problem:* By completing the square, transform this equation of an ellipse to make the dilations and transformations visible.

$$25x^2 + 4y^2 - 150x + 8y + 129 = 0$$

Confirm, by plotting the given equation, that the ellipse actually does have the features your transformed equation indicates.

31. *xy-Term Problem:* Figure 12-2j shows the graphs of

$$9x^2 + 25y^2 = 225 \qquad \text{and}$$
$$9x^2 - 20xy + 25y^2 = 225$$

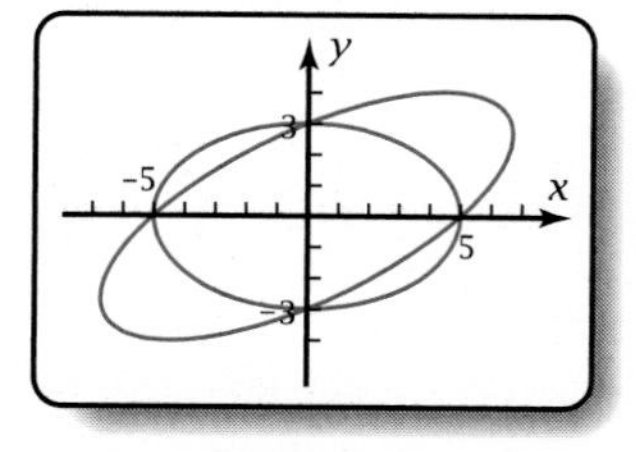

Figure 12-2j

a. Plot these graphs using the program of Example 4. (With low-resolution graphics, the ends of the second graph may not close.)

b. What changes and what does not change when an xy-term is added to the equation?

12-3 Quadric Surfaces and Inscribed Figures

Figure 12-3a shows a parabola in the yz-plane in a three-dimensional coordinate system. The axis of symmetry is along the y-axis. Figure 12-3b shows the three-dimensional surface generated by rotating this parabola about its axis of symmetry. The surface is called a **paraboloid.** The suffix *-oid* means "like," so a paraboloid is "parabola-like."

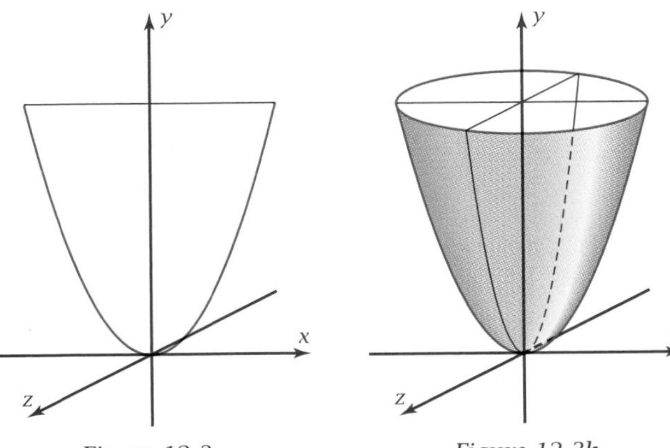

Figure 12-3a *Figure 12-3b*

Real-world objects such as reflectors in flashlights and receivers that pick up quarterbacks' voices at football games have this shape. In general, the three-dimensional analog of a conic section is a **quadric surface.** Quadric surfaces in general are defined by quadratic equations in three variables. In this section you'll encounter some special quadric surfaces that are generated by rotating conic sections about their axes, such as a paraboloid. You'll also study plane and solid figures that can be inscribed inside these surfaces.

A paraboloid surface reflects the parallel electromagnetic rays of television broadcasts into its focal point.

OBJECTIVE Given the equation of a conic section, sketch the surface generated by rotating it about one of its axes, and find the area or volume of a figure inscribed either in the plane region bounded by the graph or in the solid region bounded by the surface.

The parabola in Figure 12-3a was rotated about its axis of symmetry. The axes of symmetry of ellipses and hyperbolas are given different names to distinguish between them. Figure 12-3c shows that for an ellipse the names are *major axis* and *minor axis.* The names refer to the relative sizes of the two axes, not to the directions in which they point. For a hyperbola, the names are **transverse axis** (from vertex to vertex) and **conjugate axis** (perpendicular to the transverse axis). The latter name comes from the conjugate hyperbola, which has the same asymptotes but opens in the other direction.

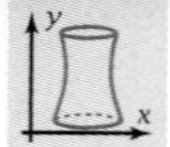

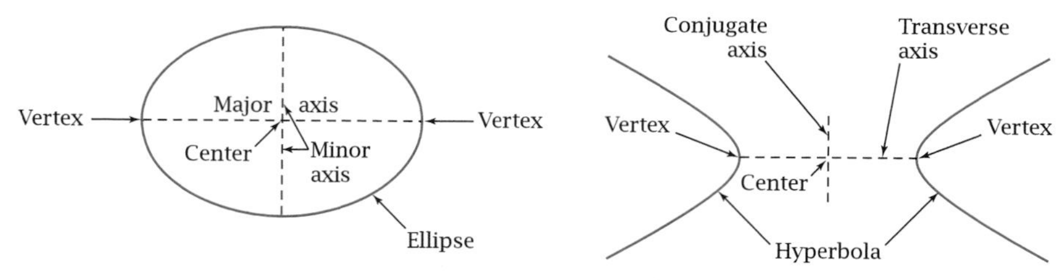

Figure 12-3c

Rotating an ellipse about one of its axes generates an **ellipsoid.** If the rotation is about the major axis, the ellipsoid is called a **prolate spheroid,** reminiscent of a football or an egg. If the rotation is about the minor axis, the ellipsoid is called an **oblate spheroid,** like a round pillow. Figure 12-3d shows these shapes.

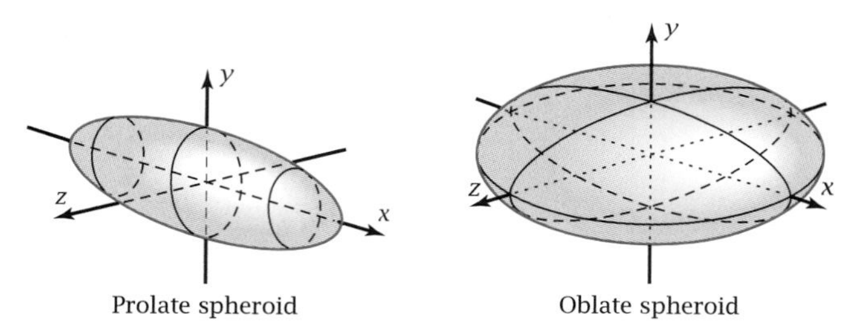

Figure 12-3d

Rotating a hyperbola about one of its axes generates a **hyperboloid.** If the rotation is about the transverse axis, the two branches of the hyperbola form two disconnected surfaces, giving a **hyperboloid of two sheets.** If the rotation is about the conjugate axis, the surface is connected and is called a **hyperboloid of one sheet.** The icon at the top of each even-numbered page in this chapter is a hyperboloid of one sheet. Figure 12-3e shows these shapes.

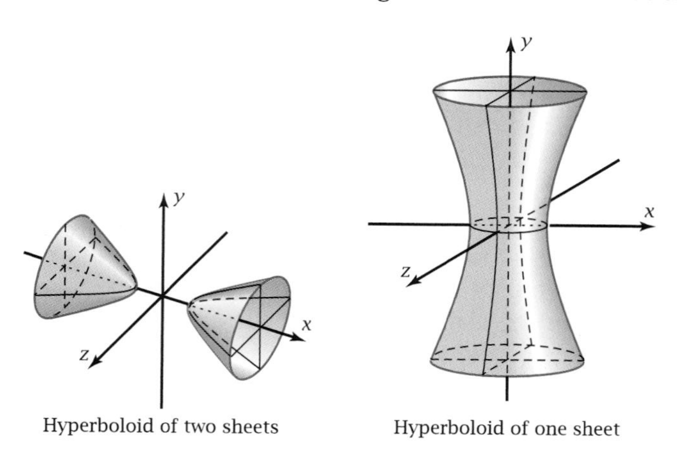

Figure 12-3e

Hyperboloids of one sheet have the remarkable property that they can be generated by rotating a line around an axis skew to it. The decorative table in Figure 12-3f shows this property. The power plant cooling towers shown in the figure take advantage of the fact that they can be built with straight reinforcing materials, without any internal support structure.

Figure 12-3f

▶ **EXAMPLE 1** Sketch the hyperboloid formed by rotating about the y-axis the hyperbola $-9x^2 + y^2 = 9$ from $x = 0$ to $x = 2$.

Solution

$$-9x^2 + y^2 = 9$$

$$-\left(\frac{x}{1}\right)^2 + \left(\frac{y}{3}\right)^2 = 1$$

Figure 12-3g shows the hyperbola centered at the origin, opening in the y-direction with asymptotes having slopes ± 3. Show the circular cross sections in perspective, using dashed lines where the cross section is hidden.

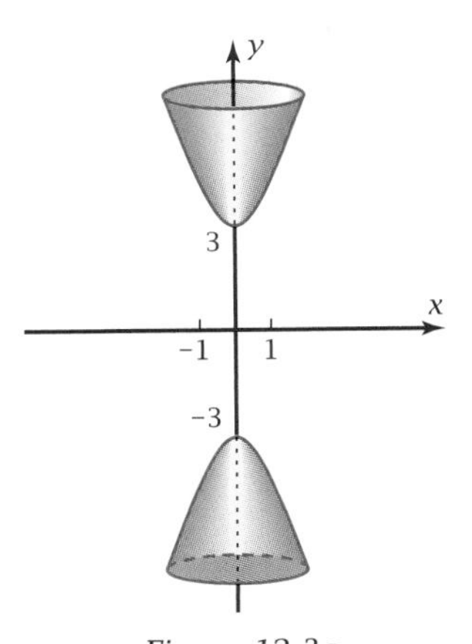

Figure 12-3g

◀ *Le Corbusier used hyperbolic-paraboloid surfaces in the architectural design of this 1958 World Expo building, the Phillips Pavilion.*

▶ **EXAMPLE 2** Rectangles of various proportions are inscribed in the region under the half-ellipse $64x^2 + 25y^2 = 1600$, $y \geq 0$, as shown in Figure 12-3h. A vertex of the rectangle is the **sample point** (x, y) that lies on the ellipse. The rectangle can be tall and skinny, short and wide, or somewhere in between, depending on where the sample point is placed on the ellipse. The area of the rectangle depends on the location of the sample point. Place the sample point in Quadrant I.

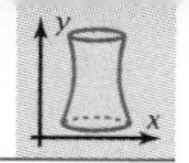

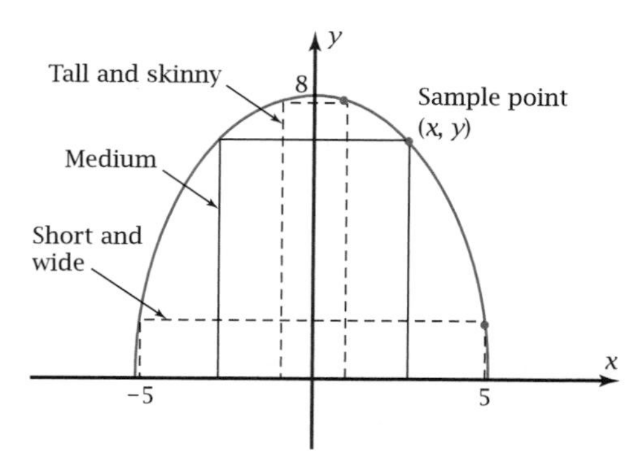

Figure 12-3h

a. Write the area of a representative rectangle in terms of x and y at the sample point.

b. Transform the equation in part a so that the area is a function of x alone.

c. Make a table of values of area as a function of x for each 1 unit from $x = 0$ to $x = 5$. Based on your table, approximately what value of x in this interval seems to give the maximum area?

d. Plot the graph of area as a function of x. Use the maximum feature of your grapher to find the value of x that produced the maximum area and to find that maximum.

Solution

a. Let A represent the area.

$$A = 2xy$$

The width of the rectangle is $2x$, not x.

b.

$$64x^2 + 25y^2 = 1600, \quad y \geq 0$$

Transform the ellipse equation so that y is in terms of x.

$$25y^2 = 1600 - 64x^2$$

$$y^2 = \frac{64}{25}(25 - x^2)$$

$$y = \frac{8}{5}\sqrt{25 - x^2}$$

Why just the *positive* square root?

$$\therefore A = 3.2x\sqrt{25 - x^2}$$

$A = 2xy$.

c.

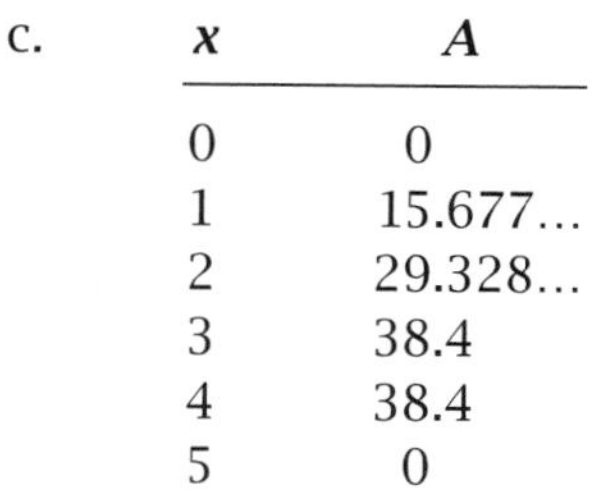

x	A
0	0
1	15.677...
2	29.328...
3	38.4
4	38.4
5	0

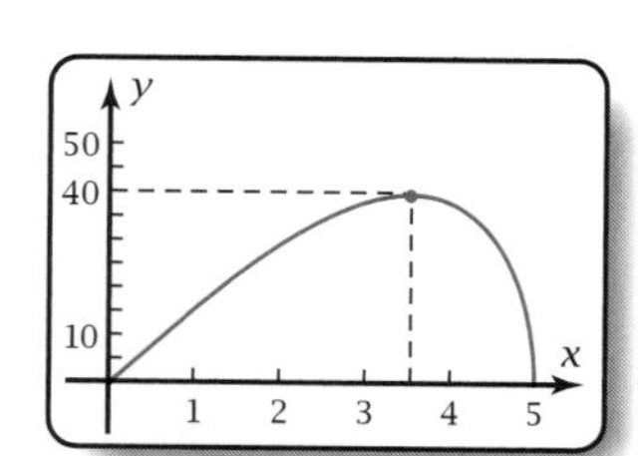

Figure 12-3i

The maximum area seems to occur at a value of x between 3 and 4.

d. Figure 12-3i shows the graph of area as a function of x. The maximum of 40 occurs at $x = 3.5355....$ ◀

▶ EXAMPLE 3 The part of the parabola $y = 4 - x^2$ in the first quadrant is rotated about the y-axis to form a paraboloid. A cylinder is inscribed in the region under this surface with the center of its lower base at the origin and the points on the circumference of its upper base on the parabola.

a. Sketch the paraboloid and the cylinder.

b. Find the volume of the cylinder as a function of its radius.

c. Find the maximum volume the cylinder can have, and then find the radius of this maximal cylinder.

Solution

a. Sketch the parabola as shown in Figure 12-3j. Pick a sample point (x, y) on the parabola in Quadrant I. Then draw the paraboloid by considering what happens as the parabola rotates about its axis of symmetry. The upper base of the cylinder will be traced by the sample point. The two circular bases will appear as ellipses in the figure because they are being viewed in perspective.

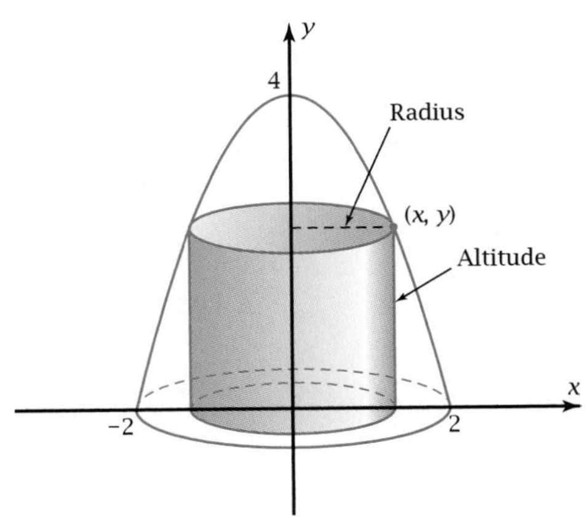

Figure 12-3j

b. Let V = volume of the cylinder.

$V = \pi x^2 y$ Volume = π(radius)^2(altitude).

$V = \pi x^2(4 - x^2)$ Substitute for y.

c. Sketch the volume function as shown in Figure 12-3k.

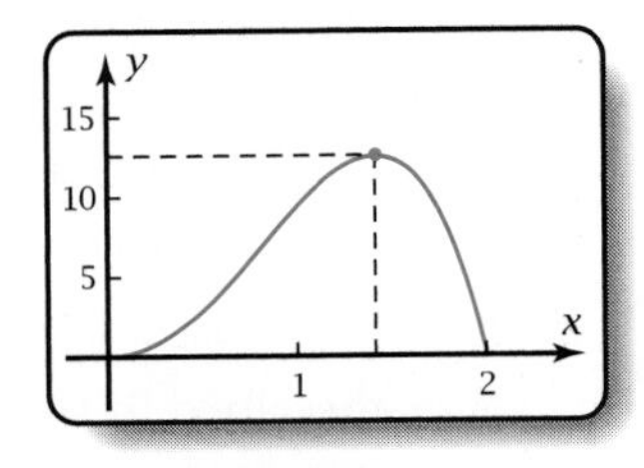

Figure 12-3k

Using the maximum feature of the grapher, the maximum volume is 12.5663... cubic units at $x = 1.4142...$ (which is the square root of 2). ◀

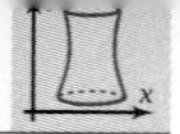

Recall that the lateral area of a solid is its surface area, excluding the area of the bases. These formulas from geometry will help you with the problems in Problem Set 12-3.

PROPERTIES: Geometry Formulas

Figure	Volume	Surface Area	
Cylinder	$V = \pi r^2 h$	$L = 2\pi rh$	Lateral area
		$S = 2\pi rh + 2\pi r^2$	Total area
Cone	$V = \frac{1}{3}\pi r^2 h$	$L = \pi rl$	Lateral area
		(l = slant height)	
		$S = \pi rl + \pi r^2$	Total area
Sphere	$V = \frac{4}{3}\pi r^3$	$S = 4\pi r^2$	

Problem Set 12-3

Do These Quickly

For Problems Q1–Q7, name the conic section for each equation.

Q1. $x^2 + 4y^2 + 5x + 6y = 100$

Q2. $x^2 - 4y^2 + 5x + 6y = 100$

Q3. $-x^2 + 4y^2 + 5x + 6y = 100$

Q4. $4x^2 + 4y^2 + 5x + 6y = 100$

Q5. $4x^2 + 5x + 6y = 100$

Q6. $4y^2 + 5x + 6y = 100$

Q7. $x = 3 + 5 \cos t$
$y = 4 + 2 \sin t$

Q8. Complete the square: $5x^2 + 30x + 58$

Q9. Complete the square: $y^2 + 10y + 10$

Q10. Which operation causes dilation of a figure, multiplication or addition?

For Problems 1–10, sketch the quadric surface.

1. Paraboloid formed by rotating the part of the graph of $y = x^2$ from $x = 0$ to $x = 3$ about the y-axis
2. Paraboloid formed by rotating the part of the graph of $y = 9 - x^2$ that lies in the first quadrant about the y-axis
3. Ellipsoid formed by rotating the graph of $4x^2 + y^2 = 16$ about the x-axis
4. Ellipsoid formed by rotating the graph of $4x^2 + y^2 = 16$ about the y-axis
5. Hyperboloid formed by rotating the part of $4x^2 - y^2 = 4$ from $x = 1$ to $x = 2$ about the y-axis
6. Hyperboloid formed by rotating the part of $-x^2 + y^2 = 9$ from $y = 3$ to $y = 6$ about the x-axis
7. Hyperboloid formed by rotating the part of $x^2 - 4y^2 = 4$ from $x = -5$ to $x = 5$ about the x-axis
8. Hyperboloid formed by rotating the part of $-x^2 + y^2 = 9$ from $x = -6$ to $x = 6$ about the y-axis
9. Cone formed by rotating the part of the line $y = 3x$ from $x = -2$ to $x = 2$ about the y-axis
10. Cone formed by rotating the part of the line $y = 0.5x$ from $x = -6$ to $x = 6$ about the x-axis
11. *Triangle in Parabola Problem:* A triangle is inscribed in the region bounded by the parabola $y = 9 - x^2$ and the x-axis (Figure 12-3l). A vertex of the triangle is at the origin, and the opposite side is parallel to the x-axis. Another vertex touches the parabola at the sample point (x, y) in the first quadrant. Plot the area

of the triangle as a function of x, and sketch the result. Find the value of x that maximizes the area of the triangle, and find the area of this maximal triangle.

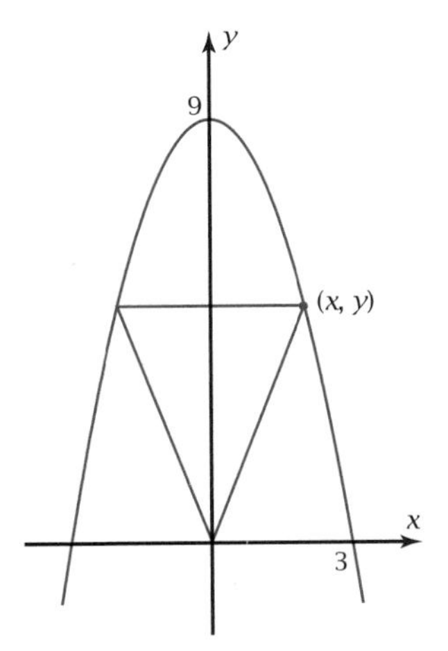

Figure 12-3l

12. *Rectangle in Ellipse Problem:* A rectangle is inscribed in the ellipse $9x^2 + 25y^2 = 225$. The sides of the rectangle are parallel to the coordinate axes. Sketch the ellipse and the rectangle. Then find the area of the rectangle in terms of a sample point at which a vertex of the rectangle touches the ellipse in the first quadrant. Plot the area of the rectangle as a function of x, and sketch the result. Find the value of x that maximizes the area of the rectangle, and find this maximum area.

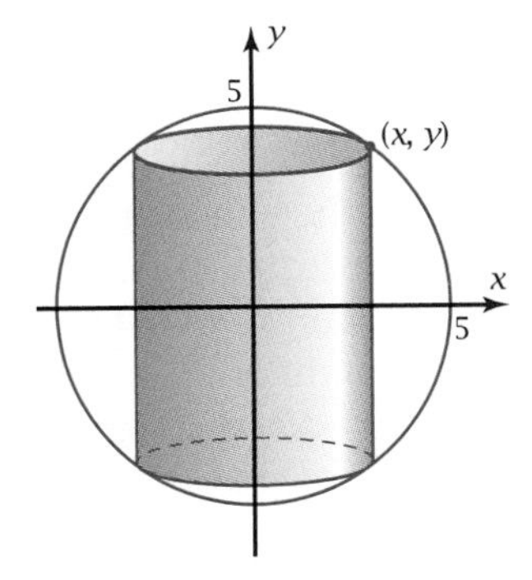

Figure 12-3m

13. *Cylinder in Sphere Volume Problem:* The circle $x^2 + y^2 = 25$ is rotated about the y-axis to form a sphere (Figure 12-3m). A cylinder is inscribed in the sphere, with its axis along the y-axis. Find an equation for the volume of the cylinder in terms of a sample point (x, y) in the first quadrant where the upper base of the cylinder touches the circle. Plot the volume of the cylinder as a function of x, and sketch the result. Find the value of x that gives the maximum volume, and find this maximum volume.

14. *Cylinder in Ellipsoid Problem:* The ellipse $x^2 + 4y^2 = 4$ is rotated about the x-axis to form an ellipsoid. A cylinder is inscribed in the ellipsoid, with its axis along the x-axis. The right base of the cylinder touches the ellipse at the sample point (x, y) in the first quadrant. Sketch the ellipsoid and the cylinder. Then plot the volume of the cylinder as a function of x. Sketch the graph. Find the value of x that maximizes the volume of the cylinder, and find this maximum volume.

15. *Cylinder in Sphere Area Problem:* A cylinder is inscribed in a sphere of radius 5 units, as in Figure 12-3m.
    a. Find the radius and altitude of the cylinder of maximum lateral area.
    b. Find the radius and altitude of the cylinder of maximum total area.
    c. Does the cylinder of maximum lateral area also have the maximum total area?
    d. Does the cylinder of maximum total area also have maximum volume, as in Problem 13?

16. *Cylinder in Ellipsoid Area Problem:* A cylinder is inscribed in the ellipsoid of Problem 14.
    a. Find the radius and altitude of the cylinder of maximum lateral area.
    b. Find the radius and altitude of the cylinder of maximum total area.
    c. Does the cylinder of maximum lateral area also have the maximum total area?
    d. Does the cylinder of maximum total area also have maximum volume, as in Problem 14?

17. *Submarine Problem 2:* The bow of a submarine has the shape of the half-ellipsoid formed by rotating about the x-axis the right half of the ellipse

$$225x^2 + 900y^2 = 202{,}500$$

where x and y are in feet. The ellipsoid (a doubly curved surface) is to be shaped from thin metal that is relatively easy to mold. The pressure hull, made of thick metal, is in the form of a cylinder (a singly curved surface) inscribed in the ellipsoid (Figure 12-3n). How should the cylinder be constructed to give it the maximum volume? How much of the heavy steel plate will be needed to form the curved walls of the cylinder?

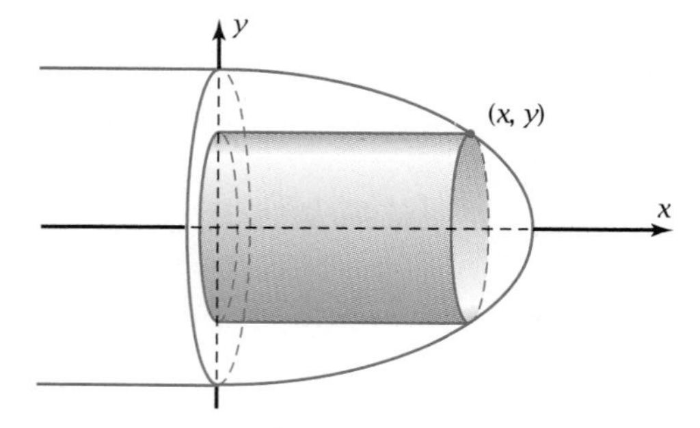

Figure 12-3n

12-4 Analytic Geometry of the Conic Sections

The Old Senate Chamber in the U.S. Capitol in Washington, D.C., is a famous whispering chamber.

For each vertex of a conic section, there is a special fixed point on the concave side called the **focus.** For ellipses, light or sound rays starting at one focus are reflected toward the other focus, making the ellipse a useful shape for auditoriums and "whispering chambers." For parabolas, rays starting at the focus are reflected parallel to the axis of symmetry, making the parabola a useful shape for headlight reflectors and dish TV antennas. For hyperbolas, the reflected rays diverge, but when extended backwards they pass through the other focus.

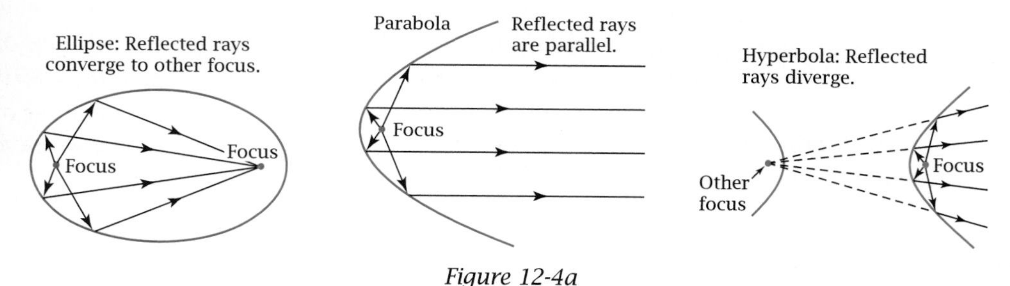

Figure 12-4a

For each focus of any conic section, there is a fixed line on the convex side, called the **directrix,** perpendicular to the axis of symmetry. For each point on the graph, its distance from the focus is directly proportional to its distance from the corresponding directrix. The proportionality constant is called the **eccentricity,** written e. As shown in Figure 12-4b, if d_1 is the distance from a point on the graph to the directrix and d_2 is the distance from that point to the focus, then

$$d_2 = ed_1$$

The name *eccentricity* is given because the closer e is to zero, the rounder an ellipse is. The closer e is to 1, the longer and more "eccentric" the ellipse is. If the eccentricity is 1 or greater, the graph is a parabola or hyperbola, respectively.

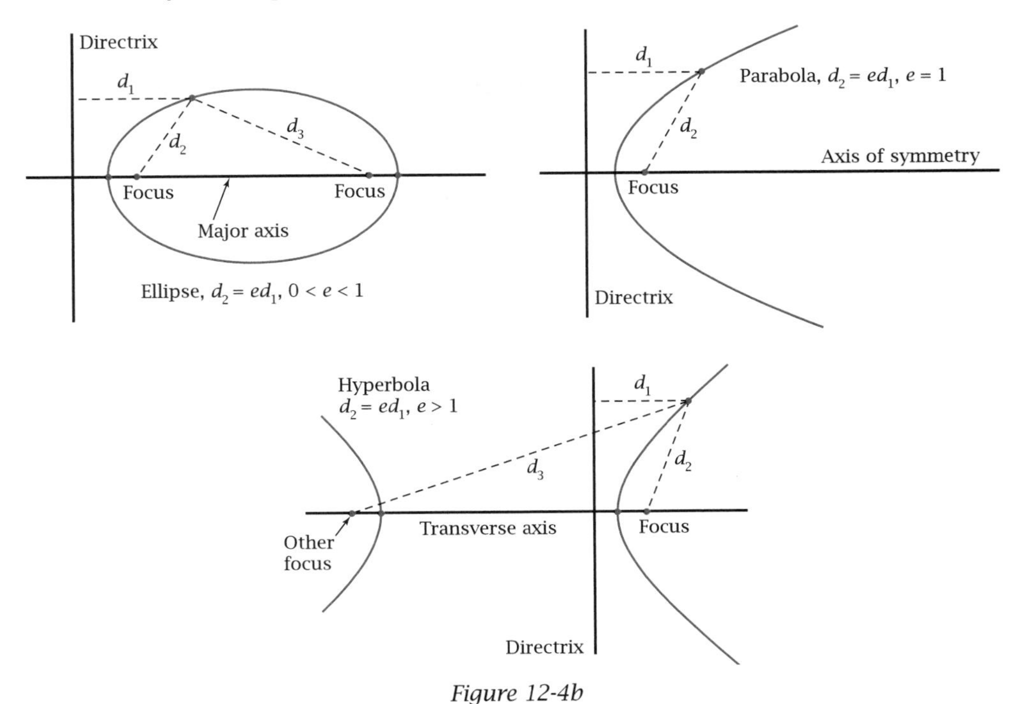

Figure 12-4b

OBJECTIVE Given the equation of a conic section, find the foci, the directrix, and the eccentricity, and vice versa.

The focus, directrix, and eccentricity properties of conic sections are summarized in the box.

PROPERTY: Focus, Directrix, and Eccentricity of a Conic Section

If d_1 is the distance from the point (x, y) on a conic section to its directrix and d_2 is the distance from (x, y) to the corresponding focus, then

$$d_2 = ed_1 \qquad \text{or, equivalently,} \qquad e = \frac{d_2}{d_1}$$

Verbally: "The distance to the focus is e times the distance to the directrix."

"The eccentricity is the ratio $\dfrac{\text{distance from point to focus}}{\text{distance from point to directrix}}$."

$e > 1$	Hyperbola	
$e = 1$	Parabola	
$0 < e < 1$	Ellipse	
$e = 0$	Circle	The directrix is infinitely far away.

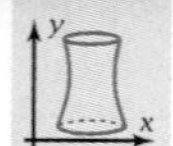

Radii of an Ellipse and of a Hyperbola

An ellipse has four constant "radii," identified with letters in Figure 12-4c.

- a, **major radius,** from the center to a vertex along the major axis
- b, **minor radius,** from the center to one end of the minor axis
- c, **focal radius,** from the center to a focus along the major axis
- d, **directrix radius,** from the center to a directrix along the major axis

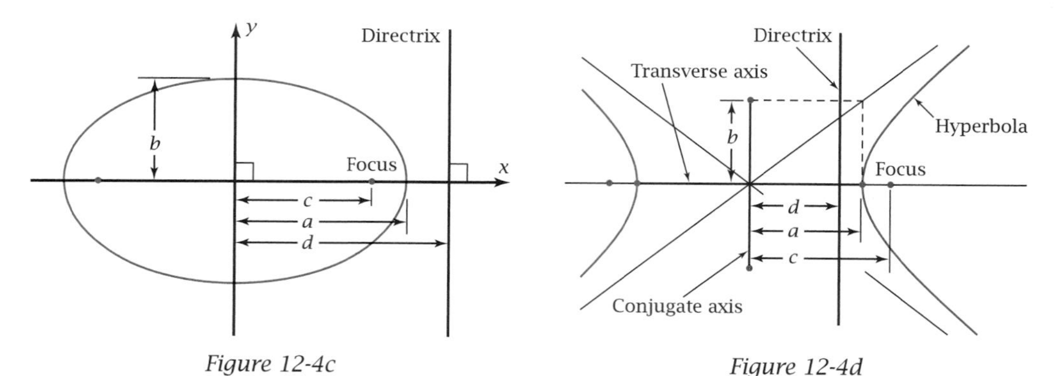

Figure 12-4c *Figure 12-4d*

The radii for a hyperbola are labeled with the same letters, as shown in Figure 12-4d. Each focus is on the concave side of the corresponding vertex, and the directrix is on the convex side. Recall that the transverse axis goes from vertex to vertex and that the conjugate axis goes through the center, perpendicular to the transverse axis. You'll learn how the transverse, conjugate, and focal radii are related later in this section.

- a, **transverse radius,** from the center to a vertex along the transverse axis
- b, **conjugate radius,** from the center to one endpoint of the conjugate axis (length of the tangent segment from the vertex to an asymptote)
- c, **focal radius,** from the center to a focus along the transverse axis
- d, **directrix radius,** from the center to a directrix along the transverse axis

Focal Distances for an Ellipse or a Hyperbola

Figure 12-4b (on the previous page) shows distances d_2 and d_3 from a point on an ellipse or hyperbola to the two foci. An ellipse has the property that the sum of these distances is constant. For a hyperbola, the difference between these distances is constant. These properties allow you to define the ellipse and the hyperbola geometrically.

DEFINITIONS: Ellipse and Hyperbola

An **ellipse** is the set of all points *P* in a plane for which the sum of the distances from point *P* to two fixed points (the foci) is a constant.

A **hyperbola** is the set of all points *P* in a plane for which the absolute value of the difference of the distances from point *P* to two fixed points (the foci) is constant.

Figure 12-4e shows how you can demonstrate this property for ellipses. Tie two pins 10 cm apart, and stick them at the foci points (4, 0) and (−4, 0) in a coordinate system drawn on centimeter graph paper. Place the pencil as shown and draw a curve, keeping the string taut. For any point (*x*, *y*) on the resulting ellipse, the sum of its distances from the two foci equals the constant length of the string, 10 cm in this case. This distance is the length of the major axis or major diameter, $2a$, twice the major radius.

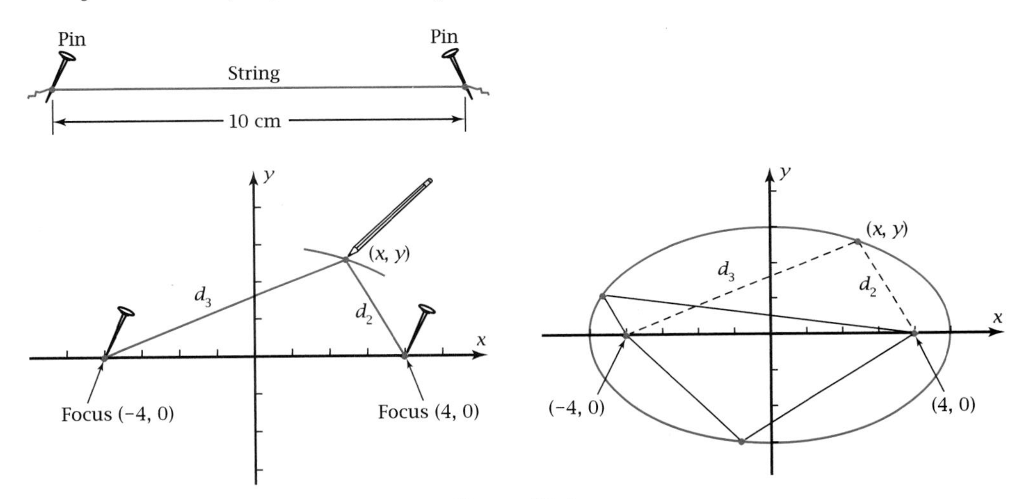

Figure 12-4e

PROPERTY: Two-Foci Properties for Ellipses and Hyperbolas

If d_2 and d_3 are the distances from point (*x*, *y*) on an ellipse or a hyperbola to the two foci and *a* is the major radius or transverse radius, then

$d_2 + d_3 = 2a$ for ellipses

$|d_2 - d_3| = 2a$ for hyperbolas

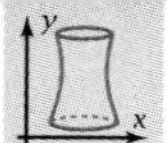

Pythagorean Property for an Ellipse or a Hyperbola

The major, minor, and focal radii of an ellipse are related by a Pythagorean property. Placing the pencil of Figure 12-4e at the end of the minor axis (Figure 12-4f) shows that the major radius, a (half the length of the string), equals the hypotenuse of a right triangle whose legs are the minor radius, b, and the focal radius, c. By the Pythagorean theorem, $a^2 = b^2 + c^2$.

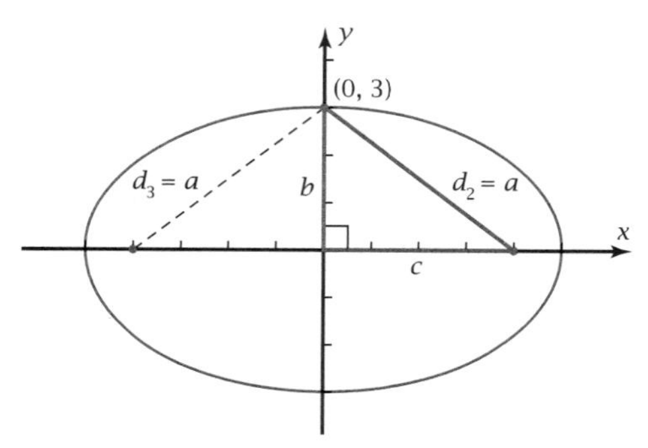

Figure 12-4f

Figure 12-4g shows that for hyperbolas the focal radius equals the hypotenuse of a right triangle whose legs are the transverse radius, a, and the conjugate radius, b. Thus $c^2 = a^2 + b^2$.

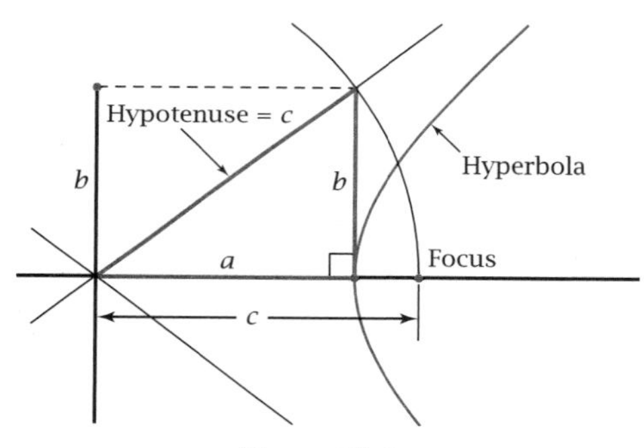

Figure 12-4g

PROPERTIES: Pythagorean Properties of Ellipses and Hyperbolas

For an ellipse, if a is the major radius, b is the minor radius, and c is the focal radius, then

$$a^2 = b^2 + c^2$$

For a hyperbola, if a is the transverse radius, b is the conjugate radius, and c is the focal radius, then

$$c^2 = a^2 + b^2$$

Radii and Eccentricity Properties for an Ellipse or a Hyperbola

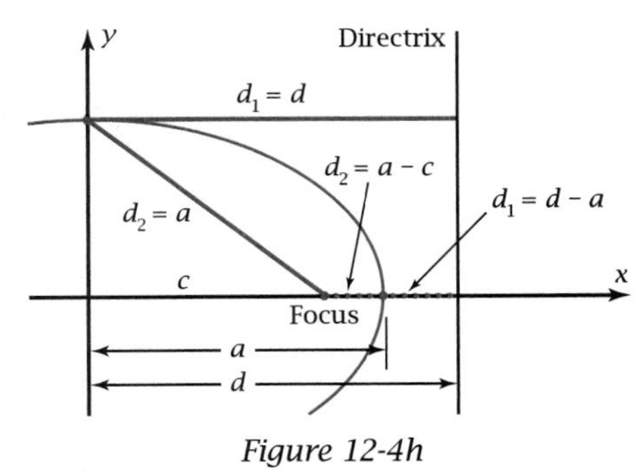

Figure 12-4h

The major (or transverse) radius, focal radius, and directrix radius for an ellipse or a hyperbola are related by the eccentricity, *e*. Figure 12-4h shows a point on the ellipse at the end of the minor axis. Here, the distance d_1 to the directrix equals *d*, the directrix radius, and the distance d_2 to the focus equals *a*, the major radius. Substituting *a* for d_2 and *d* for d_1 in $d_2 = ed_1$, you get

$a = ed$ Major radius equals the product of *e* and the directrix radius.

If the point is at the vertex of the ellipse, then $d_1 = d - a$ and $d_2 = a - c$. Substituting $a - c$ for d_2 and $d - a$ for d_1 in $d_2 = ed_1$, you get

$$a - c = e(d - a)$$

Knowing that $a = ed$, you can transform the last equation to

$c = ea$ Focal radius equals the product of *e* and the major radius.

Figure 12-4i shows a similar relationship for a hyperbola. The equations $a = ed$ and $c = ea$ are also true for hyperbolas.

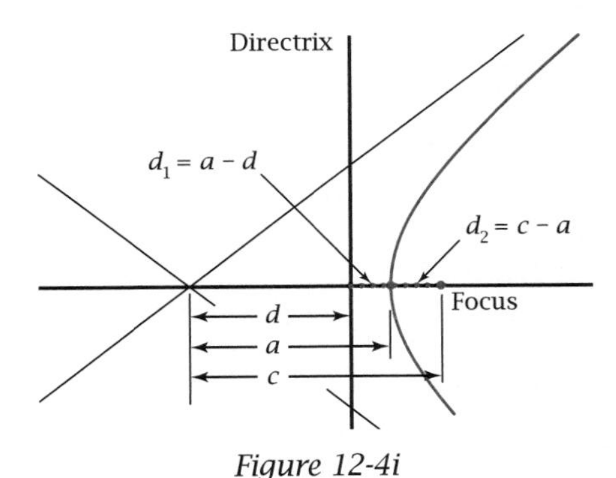

Figure 12-4i

PROPERTY: Radii and Eccentricity of an Ellipse or a Hyperbola

If *e* is the eccentricity of an ellipse or a hyperbola, then

$c = ea$ Focal radius equals the product of the eccentricity and the major (transverse) radius.

$a = ed$ Major (transverse) radius equals the product of eccentricity and the directrix radius.

$e = \frac{c}{a}$ and $e = \frac{a}{d}$ Eccentricity $= \frac{\text{focal radius}}{\text{major radius}} = \frac{\text{major radius}}{\text{directrix radius}}$

The geometric properties of ellipses and hyperbolas are summarized in this box.

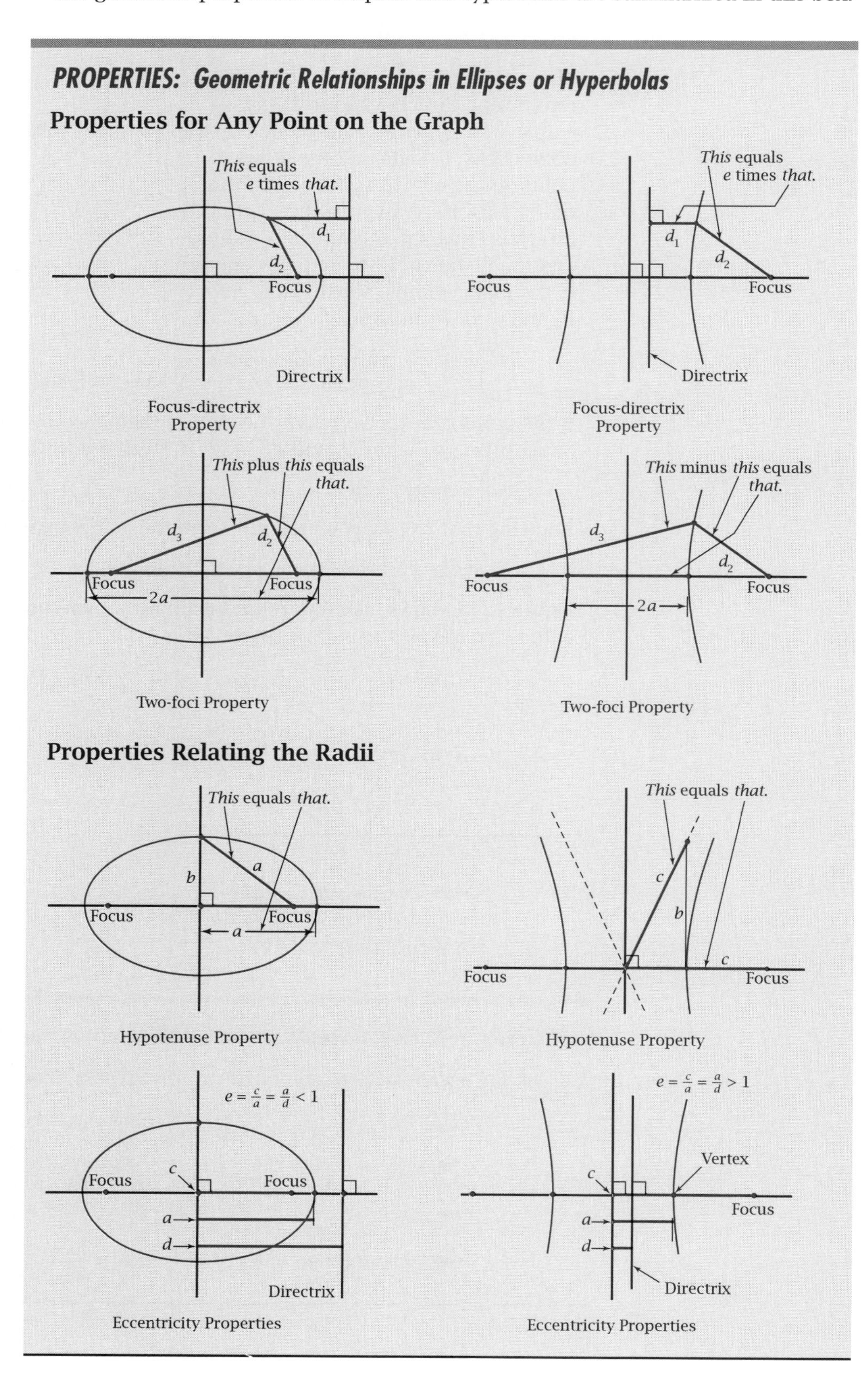

Focal Distance of a Parabola

The radius properties just given do not apply to parabolas because a parabola has only one vertex and no center. For a parabola, $d_2 = d_1$ because the eccentricity equals 1. This property lets you define parabolas geometrically.

The Guggenheim Museum in Bilbao, Spain, designed by Frank Gehry. Modern buildings, such as this one, incorporate quadric surfaces.

DEFINITION: Parabola

A **parabola** is the set of all points P in a plane for which point P's distance to a fixed point (the focus) is equal to its distance to a fixed line (the directrix).

Using this property you can find the equation of a parabola with vertex at the origin in terms of the distance p from the vertex to the focus or directrix (Figure 12-4j). This distance, p, is called the **focal distance** of a parabola.

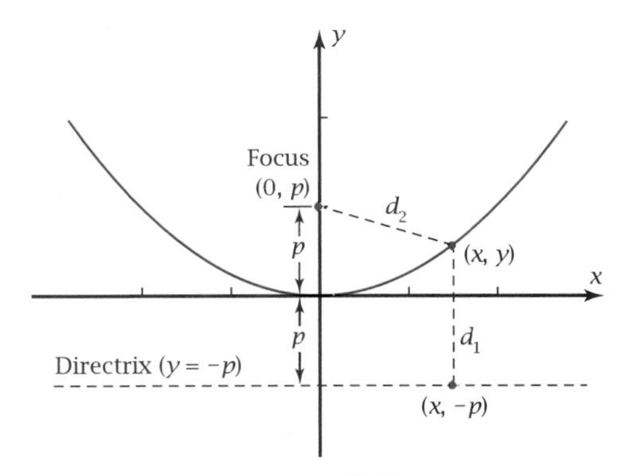

Figure 12-4j

Pick a point (x, y) on the parabola (Figure 12-4j). By the focus-directrix property,

$$d_2 = d_1 \qquad \text{The eccentricity of a parabola is 1.}$$

$$\sqrt{x^2 + (y - p)^2} = |y + p| \qquad \text{By the distance formula.}$$

$$x^2 + y^2 - 2py + p^2 = y^2 + 2py + p^2 \qquad \text{Square both sides and expand.}$$

$$x^2 = 4py$$

Or, $y = \frac{1}{4p}x^2$

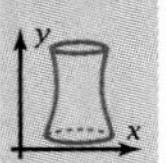

PROPERTY: Focus-Directrix Equation of Parabolas

The equation of a parabola with vertex at the origin and axis of symmetry along a coordinate axis is

$$y = \frac{1}{4p}x^2 \qquad \text{or} \qquad x = \frac{1}{4p}y^2$$

where p is the distance from the vertex to the focus or to the directrix.

► **EXAMPLE 1** For the ellipse $49x^2 + 16y^2 = 784$,

a. Sketch the graph. Show the two foci and the two directrices (plural of *directrix*).

b. Find the major, minor, and focal radii, the eccentricity, and the directrix radius.

c. Calculate y if $x = 3$. Show that the distance from (x, y) to one focus is e times its distance to the corresponding directrix.

Solution First, transform the equation to find the two dilation factors.

a. $49x^2 + 16y^2 = 784$ — Write the given equation.

$\left(\frac{x}{4}\right)^2 + \left(\frac{y}{7}\right)^2 = 1$ — Make the right side equal 1 to find the two dilation factors.

Sketch the graph (Figure 12-4k, left) with x-radius = 4, y-radius = 7. Show the foci on the concave side of the corresponding vertices and the directrices on the convex side.

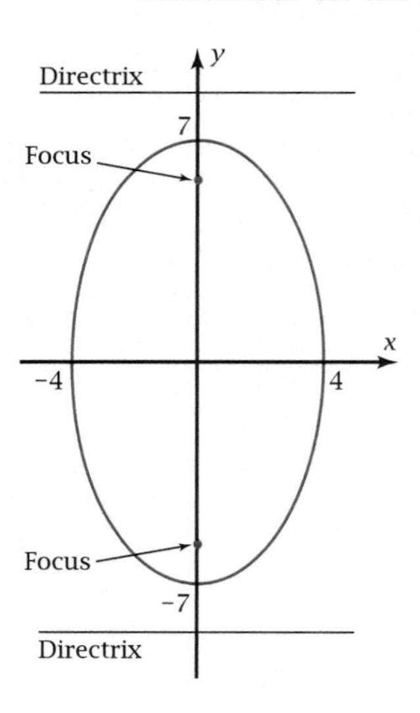

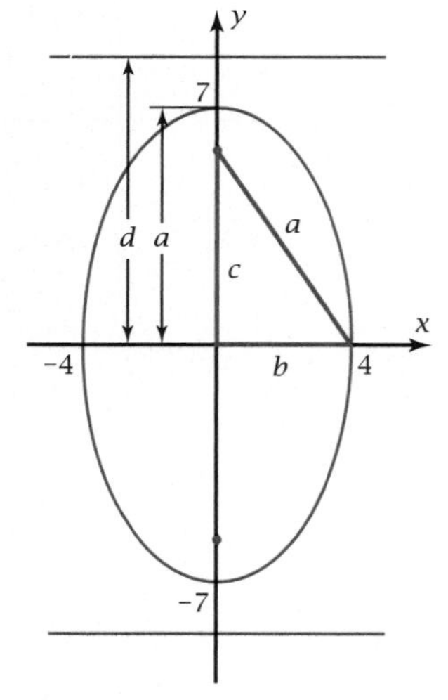

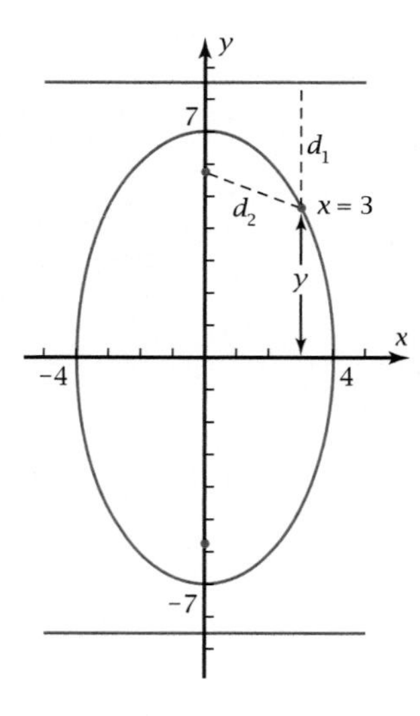

Figure 12-4k

b. Major radius: $a = 7$ — The major radius is the longer radius (Figure 12-4k, middle).

Minor radius: $b = 4$

$c^2 = a^2 - b^2 = 7^2 - 4^2 = 33$ — Use the Pythagorean property: Hypotenuse = major radius.

Focal radius: $c = \sqrt{33} = 5.7445...$ — Store as c in your grapher.

Eccentricity: $e = \frac{c}{a} = \frac{\sqrt{33}}{7} = 0.8206...$ — e is between 0 and 1, as is true for ellipses. Store as e.

$e = \frac{a}{d} \Rightarrow d = \frac{a}{e} = \frac{7}{\frac{\sqrt{33}}{7}} = \frac{49}{\sqrt{33}} = 8.5298...$ — d is from center to directrix.

Note that $d = 8.5298... > 7$, which means that the directrix is on the convex side of the vertex, and that $c = 5.7445... < 7$, which means that the focus is on the concave side.

c. $49(3^2) + 16y^2 = 784$ — Substitute 3 for x in the given equation.

$$y = \pm\sqrt{\frac{343}{16}} = \pm 4.6300...$$

The right ellipse in Figure 12-4k shows the point (3, 4.6300...) and distances d_1 and d_2.

$$d_1 = d - y = 8.5298... - 4.6300... = 3.8997...$$

$$d_2 = \sqrt{3^2 + (c - y)^2} = \sqrt{3^2 + 1.1144...^2} = 3.2003...$$

Use the Pythagorean theorem.

$ed_1 = (0.8206...)(3.8997...) = 3.2003...$, which equals d_2. ◀

▶ EXAMPLE 2 Consider a conic section with given eccentricity and foci.

a. Find the particular equation of the conic with eccentricity 1.25 and foci (6, 2) and (−4, 2). Identify the conic.

b. Sketch the graph, showing the foci and directrices. You may first plot the conic parametrically or by using a program like the one in Section 12-2.

Solution a. The conic is a hyperbola because $e > 1$. Sketch the foci (Figure 12-4l). The hyperbola opens in the x-direction because the transverse axis (through the foci) is horizontal.

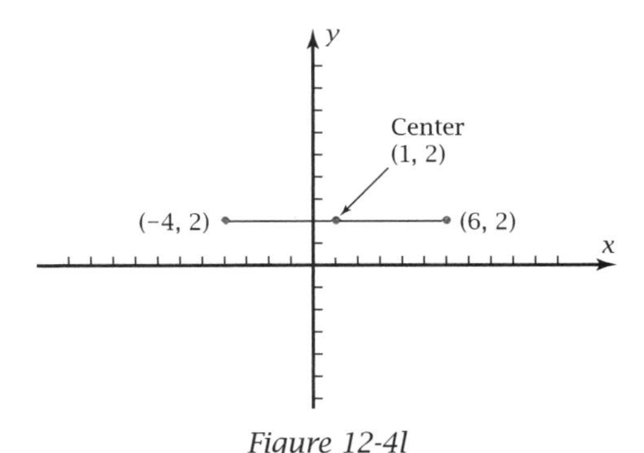

Figure 12-4l

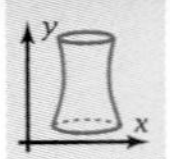

The center is at (1, 2).	Average −4 and 6 to find the x-coordinate.
$c = 6 - 1 = 5$	Focal radius goes from center to focus.
$e = \dfrac{c}{a} \Rightarrow a = \dfrac{c}{e} = \dfrac{5}{1.25} = 4$	Use the eccentricity to find the transverse radius.
$b^2 = c^2 - a^2 = 25 - 16 = 9 \Rightarrow b = 3$	Use the Pythagorean property to find the conjugate radius.
$e = \dfrac{a}{d} \Rightarrow d = \dfrac{a}{e} = \dfrac{4}{1.25} = 3.2$	Use the eccentricity to find the directrix radius.
Equation is $\left(\dfrac{x-1}{4}\right)^2 - \left(\dfrac{y-2}{3}\right)^2 = 1$	a goes under the positive term; b, under the negative.

b. See the graph in Figure 12-4m. Each focus is on the concave side of the vertex, and each directrix is on the convex side.

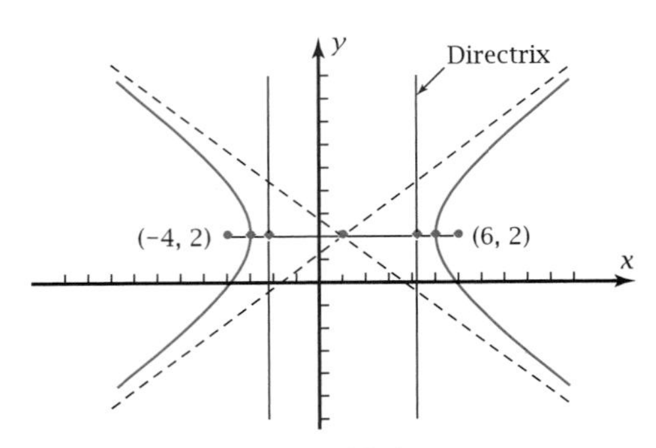

Figure 12-4m

Parametrically,

$x = 1 + 4 \sec t$ Secant goes with the direction the hyperbola opens.

$y = 2 + 3 \tan t$

To use the program of Section 12-2, first transform the Cartesian equation to

$$9x^2 - 16y^2 - 18x + 64y - 199 = 0$$

◀

▶ **EXAMPLE 3** Find the particular Cartesian equation of the conic with focus (−1, 2), directrix $x = 3$, and eccentricity $e = \frac{3}{4}$. How is the result consistent with the eccentricity property of conics? Plot the graph on your grapher using the program of Section 12-2. Sketch the result.

Solution Draw a sketch showing the given directrix and focus and a point (x, y) on the graph (Figure 12-4n). Show d_1 from point (x, y) to the directrix and d_2 from point (x, y) to the focus.

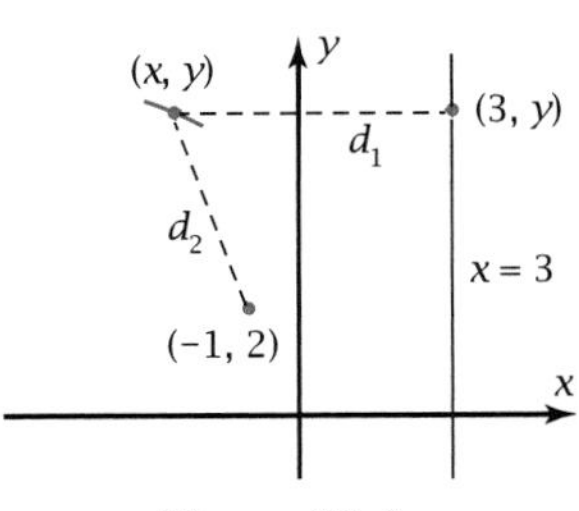

Figure 12-4n

$$d_2 = ed_1$$ Focus-directrix property of conics.

$$\sqrt{(x+1)^2 + (y-2)^2} = \tfrac{3}{4}|x-3|$$ By the distance formula.

$$16(x^2 + 2x + 1 + y^2 - 4y + 4) = 9(x^2 - 6x + 9)$$ Square both sides and simplify.

$$7x^2 + 16y^2 + 86x - 64y - 1 = 0$$ Make the right side equal 0.

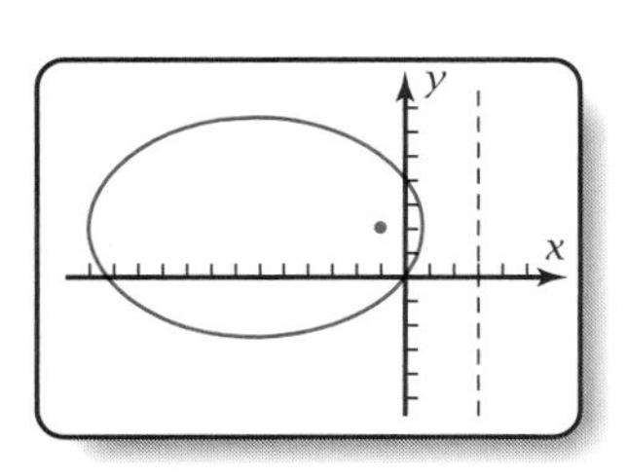

Figure 12-4o

The squared terms have the same sign but unequal coefficients, indicating an ellipse. This is consistent with the fact that $e = \frac{3}{4}$ is between 0 and 1.

Figure 12-4o shows the ellipse with the given focus and directrix. ◀

Problem Set 12-4

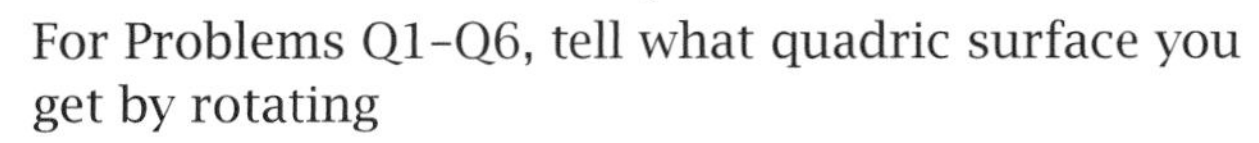

Do These Quickly

For Problems Q1–Q6, tell what quadric surface you get by rotating

Q1. An ellipse about its major axis

Q2. An ellipse about its minor axis

Q3. A hyperbola about its transverse axis

Q4. A hyperbola about its conjugate axis

Q5. A circle about its diameter

Q6. A parabola about its axis of symmetry

Q7. How do you recognize a hyperbola from its Cartesian equation?

Q8. How do you distinguish between a circle and an ellipse from the Cartesian equation?

Q9. How do you tell which way a parabola opens from its Cartesian equation?

Q10. What is the origin of the word *ellipse*?

---

1. *Hyperbola Problem 1:* Figure 12-4p shows the hyperbola $16x^2 - 9y^2 = 144$. Its foci are at $(-5, 0)$ and $(5, 0)$, and a directrix is the line $x = 1.8$. Its eccentricity is $e = \frac{5}{3}$.

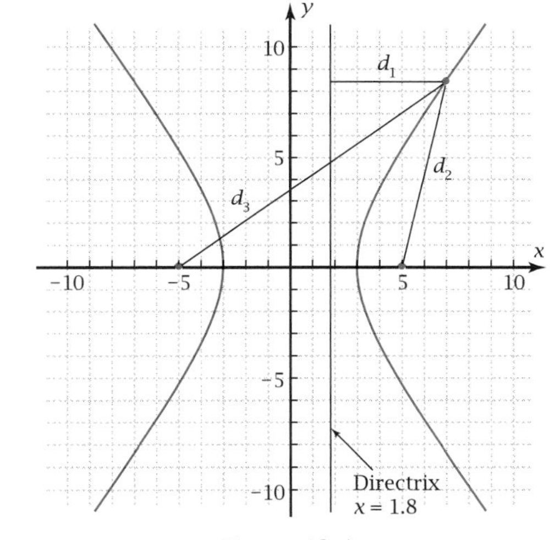

Figure 12-4p

a. A point on the hyperbola in Quadrant I has x-coordinate 7. Calculate y for this point. Does it agree with the graph? Store the answer as y in your grapher.

b. Use the Pythagorean theorem and the result of part a to calculate these distances:
   - d_1 from the point $(7, y)$ to the directrix
   - d_2 from the point $(7, y)$ to the focus $(5, 0)$
   - d_3 from the point $(7, y)$ to the focus $(-5, 0)$

c. Show that $d_2 = ed_1$.

d. Show that $|d_2 - d_3| = 6$, the length of the transverse axis (between the vertices).

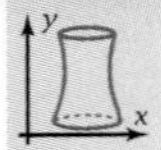

e. Find the x- and y-dilations. Which of these is the transverse radius, a, and which is the conjugate radius, b?

f. As shown in Figure 12-4p, the focal radius is $c = 5$. Show that $c^2 = a^2 + b^2$, the Pythagorean property for hyperbolas.

g. Show that the directrix radius, $d = 1.8$, satisfies the equation $a = ed$ and that the focal radius $c = 5$ satisfies $c = ea$.

2. *Ellipse Problem 1:* Figure 12-4q shows the ellipse $25x^2 + 9y^2 = 225$. Its foci are at $(0, -4)$ and $(0, 4)$, and a directrix is the line $y = 6.25$. Its eccentricity is $e = 0.8$.

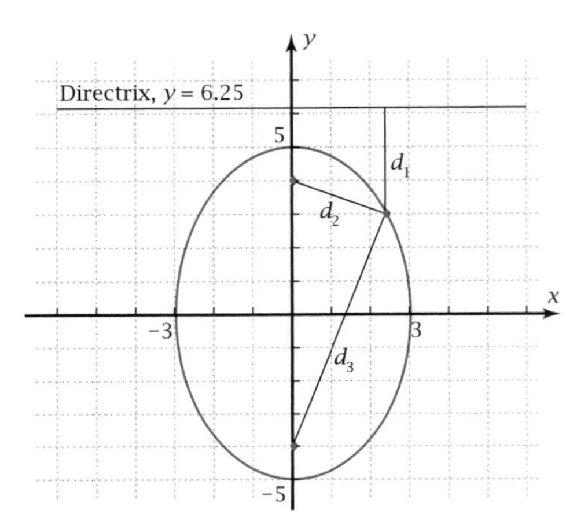

Figure 12-4q

The Coliseum in Rome, Italy, is an elliptical amphitheater that once seated 50,000 people.

a. On the ellipse a point in Quadrant I has y-coordinate 3. Calculate x for this point. Does your answer agree with the graph? Store the answer as x in your grapher.

b. Use the Pythagorean theorem and the result of part a to calculate these distances:
   - d_1 from the point $(x, 3)$ to the directrix
   - d_2 from the point $(x, 3)$ to the focus $(0, 4)$
   - d_3 from the point $(x, 3)$ to the focus $(0, -4)$

c. Show that $d_2 = ed_1$.

d. Show that $d_2 + d_3 = 10$, the length of the major axis.

e. Find the x- and y-dilations. Which of these is the major radius, a, and which is the minor radius, b?

f. As shown in Figure 12-4q, the focal radius is $c = 4$. Show that $a^2 = b^2 + c^2$, the Pythagorean property for ellipses.

g. Show that the directrix radius, $d = 6.25$, satisfies the equation $a = ed$ and that the focal radius, $c = 4$, satisfies $c = ea$.

3. *Parabola Problem 1:* Figure 12-4r shows the parabola $y = \frac{1}{8}x^2$. Its focus is at $(0, 2)$, and its directrix is the line $y = -2$.

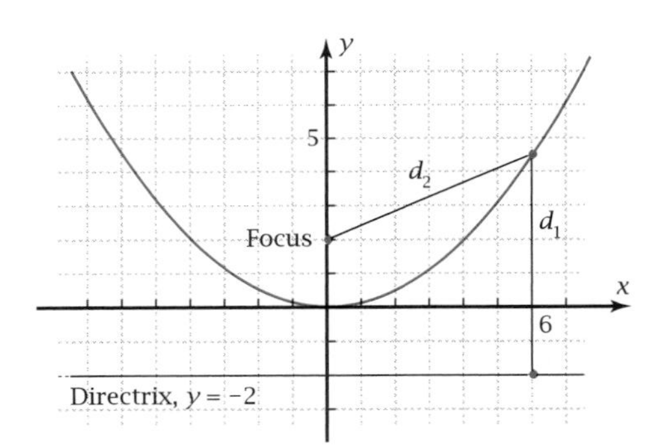

Figure 12-4r

a. The vertex is at $(0, 0)$. Explain how this fact confirms that the parabola's eccentricity is 1.

b. The point shown in the first quadrant has $x = 6$. Calculate y for this point. Does your answer agree with the graph?

The paraboloid mirror behind the lightbulb in a headlight's focus reflects its light in parallel rays.

c. Calculate the following distances:
   - d_1 from the point $(6, y)$ to the directrix
   - d_2 from the point $(6, y)$ to the focus

d. Show that $d_2 = d_1$ and that this fact is consistent with the eccentricity of 1 for a parabola.

4. *Circle Problem:* The circle $x^2 + y^2 = 25$ can be considered an ellipse with major and minor radii equal to each other.

   a. Find the major and minor radii.

   b. Based on the Pythagorean property for ellipses, explain why the focal radius of a circle is zero. Where, then, are the foci of a circle?

   c. The eccentricity of an ellipse is $e = \frac{c}{a}$, where c and a are the focal radius and major radius, respectively. Explain why the eccentricity of a circle is zero. Why is the name *eccentricity* appropriate in this case?

   d. The eccentricity of an ellipse is also equal to $\frac{a}{d}$, where d is the directrix radius. Based on the answer to part c, explain why the directrix of a circle is infinitely far from the center.

5. *Conic Construction Problem 1:* Plot on graph paper the conic with focus (0, 0), directrix $x = -6$, and eccentricity $e = 2$. Put the x-axis near the middle of the graph paper and the y-axis just far enough from the left side to fit the directrix on the paper. Plot the points for which d_1 from the directrix equals 2, 4, 6, 8, and 10. Connect the points with a smooth curve. Which conic section have you graphed?

6. *Conic Construction Problem 2:* Plot on graph paper the conic with focus (0, 0), directrix $x = -6$, and eccentricity $e = 1$. Plot points for which the distance d_1 from the directrix equals 3, 6, 10, and 20. Connect the points with a smooth curve. Which conic section have you graphed?

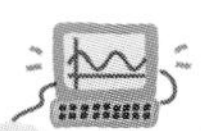

7. *Computer Graphics Project 1:* Use a graphing utility such as The Geometer's Sketchpad to plot the conics in Problems 5 and 6. Sketch the results. How do the graphs confirm your conclusion about the kind of conic section plotted?

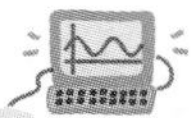

8. *Computer Graphics Project 2:* Figure 12-4s shows a fixed directrix and a fixed focus, with conics of varying eccentricity. Create a custom tool in The Geometer's Sketchpad or other graphing utility to reproduce these graphs. Draw a vertical line for the directrix and a point 6 units to its right for the focus. Define a set of points to be e times as far from the focus as they are from the directrix. Make a slider for e so that e can be varied from 0 through 2.

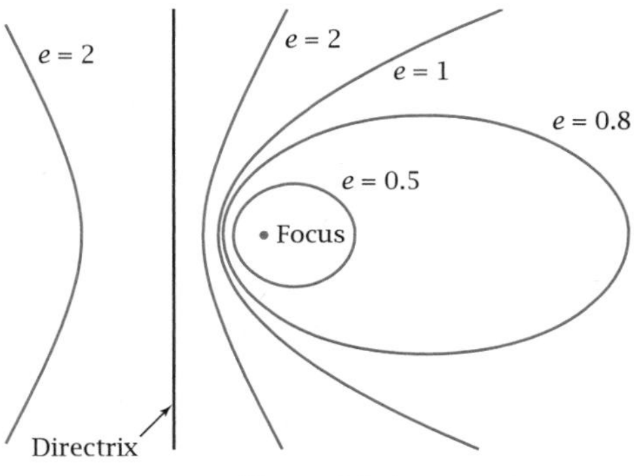

Figure 12-4s

9. *Mars Orbit Problem:* Mars is in an elliptical orbit around the Sun, with the Sun at one focus. The *aphelion* (the point farthest from the Sun) and the *perihelion* (the point closest to the Sun) are 155 million miles and 128 million miles, respectively, as shown in Figure 12-4t (not to scale).

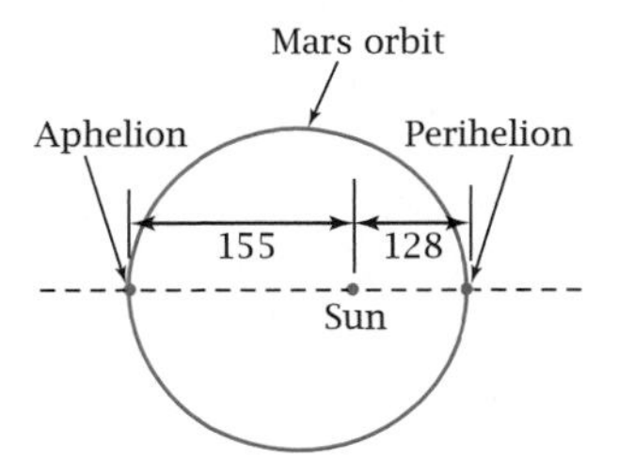

Figure 12-4t

   a. How long is the major axis of the ellipse? What is the major radius?

   b. Find the focal radius and the minor radius of the ellipse.

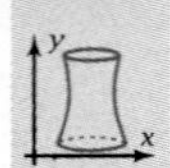

c. Write a Cartesian equation for the ellipse, with the center at the origin and the major axis along the x-axis.

d. At the two equinoxes (times of equal day and night), the angle at the Sun between the major axis and Mars is 90°. At these times, what is the value of x? How far is Mars from the Sun?

e. Find the eccentricity of the ellipse.

f. How far from the Sun is the closer directrix of the ellipse?

g. Write parametric equations for the ellipse. Plot the graph using parametric mode. Zoom appropriately to make equal scales on the two axes.

h. The ellipse you plotted in part g looks almost circular. How do the major and minor radii confirm this? How does the eccentricity confirm this?

10. *Comet Path Problem:* Figure 12-4u shows the path of a comet approaching Earth. The path is a conic section with eccentricity $e = 1.1$ and directrix radius $d = 100$ thousand miles.

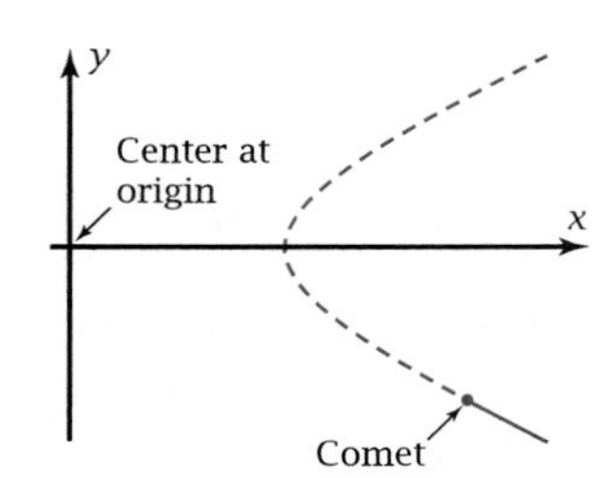

Figure 12-4u

a. How can you tell from the given information that the path is a hyperbola?

b. The center of the hyperbola is at the origin in Figure 12-4u and the transverse axis is along the x-axis. On a sketch of the figure, show the focus and the directrix.

c. The center of Earth is at the focus of the hyperbola. Find the coordinates of the focus.

d. The comet is closest to Earth when it is at the vertex. How close does it come to the center of Earth? How close does it come to the surface of Earth, 4000 miles from the center?

e. Write parametric equations of the hyperbola. What range of t-values will generate the branch shown?

f. When the comet is at the point shown, $x = 200$ thousand miles. At this time, what does the parameter t equal? What does y equal? How far is the comet from the center of Earth?

For Problems 11–20,

a. Identify the conic section.

b. Calculate four radii and the eccentricity.

c. Plot the graph. Sketch the result.

11. $\dfrac{x^2}{9} + \dfrac{y^2}{25} = 1$

12. $\dfrac{x^2}{289} + \dfrac{y^2}{64} = 1$

13. $-\dfrac{x^2}{36} + \dfrac{y^2}{9} = 1$

14. $\dfrac{x^2}{9} - \dfrac{y^2}{16} = 1$

15. $\left(\dfrac{x-1}{4}\right)^2 + \left(\dfrac{y+2}{3}\right)^2 = 1$

16. $-\left(\dfrac{x+1}{3}\right)^2 + \left(\dfrac{y-2}{16}\right)^2 = 1$

17. $5x^2 - 3y^2 = -30$

18. $16x^2 + 25y^2 = 1600$

19. $x = -\dfrac{1}{4}y^2 + 3$

20. $x = \dfrac{1}{8}y^2 + 1$

For Problems 21–32,

a. Draw a sketch showing the given information. Sketch the conic section.

b. Find the particular equation (Cartesian or parametric).

c. Plot the graph on your grapher. Does your sketch in part a agree with the plot?

21. Focus (0, 0), directrix $y = 3$, eccentricity $e = 2$. Identify the conic section.

22. Focus (0, 0), directrix $x = 5$, eccentricity $e = \frac{3}{4}$. Identify the conic section.

23. Focus (0, 0), directrix $y = -4$, eccentricity $e = 1$. Identify the conic section.

24. Focus (0, 0), directrix $x = \frac{1}{2}$, eccentricity $e = 1$. Identify the conic section.

25. Focus (2, −3), directrix $y = 0$, eccentricity $e = \frac{1}{2}$. Identify the conic section.

26. Focus (3, 1), directrix $x = 2$, eccentricity $e = 4$. Identify the conic section.

27. Ellipse with foci (12, 0) and (−12, 0) and constant sum of distances equal to 26

28. Hyperbola with foci (0, 5) and (0, −5) and constant difference of distances equal to 8

29. Hyperbola with vertices (−1, 3) and (5, 3) and slope of asymptotes $\pm\frac{2}{3}$

30. Ellipse with vertices (4, −2) and (4, 8) and minor radius 3

31. Parabola with focus (2, 3) and directrix $y = 5$

32. Parabola with focus (4, 5) and vertex (4, 2)

33. *Latus Rectum Problem:* The **latus rectum** of a conic section is the chord through a focus parallel to the directrix (Figure 12-4v). By appropriate substitution into the equations, find the length of the latus rectum for the conics in Problems 9, 11, and 17.

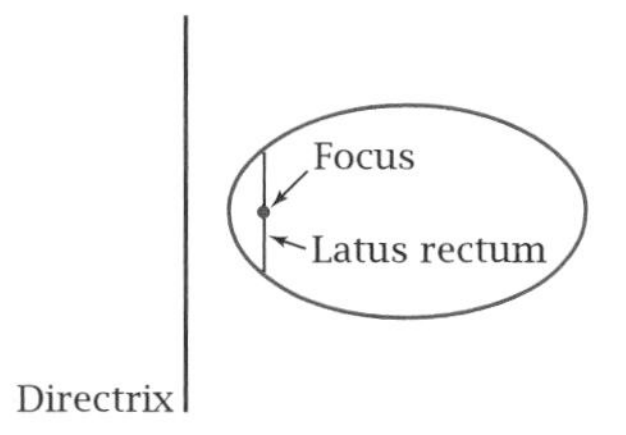

Figure 12-4v

12-5 Parametric and Cartesian Equations for Rotated Conics

Previously in this chapter you saw how to write parametric equations for circles, ellipses, and hyperbolas. In this section you will see how you can transform the parametric equations to rotate the conic through a given angle using a rotation matrix as in Section 11-3. The Cartesian equations for these rotated conics turn out to contain an xy-term as well as x^2- and y^2-terms.

OBJECTIVES

- Plot a conic section rotated by a specified angle to the coordinate axes.
- Identify a rotated conic from its Cartesian equation.
- Plot a rotated conic using its Cartesian or parametric equations.

Parametric Equations

The parametric equations of the conic sections centered at the origin and with major or transverse axis along the x- or y-axis are shown in the box.

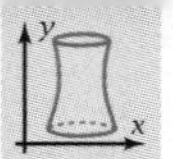

PARAMETRIC EQUATIONS OF CONIC SECTIONS: Center at Origin

Where a is the major radius or transverse radius
and b is the minor radius or conjugate radius

	x-axis	y-axis	
Circle or ellipse:	$x = a\cos t$ $y = b\sin t$	$x = b\cos t$ $y = a\sin t$	$a \geq b$
Hyperbola:	$x = a\sec t$ $y = b\tan t$	$x = b\tan t$ $y = a\sec t$	
Parabola:	$x = at^2$ $y = t$	$x = t$ $y = at^2$	Vertex at origin.

▶ ***EXAMPLE 1*** Plot the hyperbola centered at (1, 3) that has transverse radius 5 and conjugate radius 2 if the transverse axis makes an angle of 25° with the x-axis (Figure 12-5a).

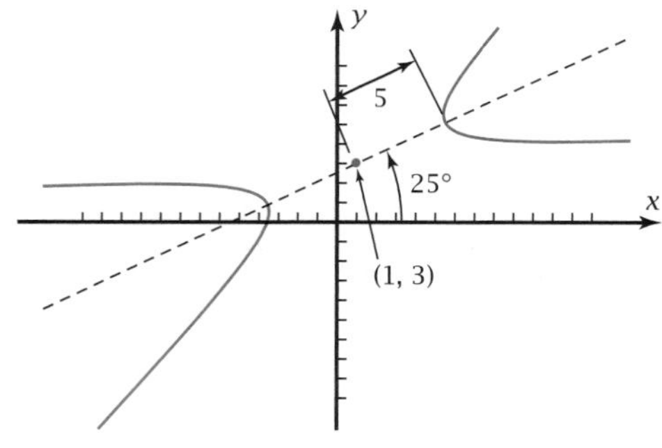

Figure 12-5a

Solution The parent equations for the unrotated hyperbola centered at the origin with transverse axis along the x-axis are

$$x = 5\sec t$$

$$y = 2\tan t$$

See the preceding box; a and b are also equal to dilations.

$$\begin{bmatrix} \cos 25° & \cos 115° \\ \sin 25° & \sin 115° \end{bmatrix}\begin{bmatrix} 5\sec t \\ 2\tan t \end{bmatrix}$$

Multiply by the rotation matrix (Section 11-3): 115° = 90° + 25°.

$$= \begin{bmatrix} 5\cos 25° \sec t + 2\cos 115° \tan t \\ 5\sin 25° \sec t + 2\sin 115° \tan t \end{bmatrix}$$

$$= \begin{bmatrix} 4.5315\ldots \sec t - 0.8452\ldots \tan t \\ 2.1130\ldots \sec t + 1.8126\ldots \tan t \end{bmatrix}$$

$$x = 1 + 4.5315\ldots \sec t - 0.8452\ldots \tan t$$

$$y = 3 + 2.1130\ldots \sec t + 1.8126\ldots \tan t$$

Write x and y from the matrix and add the translations.

With your grapher in parametric mode and degree mode, plot these two parametric equations. Use a range for t of [0°, 360°]. The graph should look like that in Figure 12-5a. Be sure to use equal scales on the two axes. ◀

Cartesian Equation with xy-Term

Beginning in Section 12-2, you have been using a program to plot conics in function mode if the equation has the form

$$Ax^2 + Bxy + Cy^2 + Dx + Ey + F = 0$$

This program can be used to explore the graphs of conics for which the xy-term is not zero.

▶ EXAMPLE 2 Plot the graph of $9x^2 - 40xy + 25y^2 - 8x + 2y - 28 = 0$. Which conic section does the graph appear to be? Does this conclusion agree with what you have learned about the coefficients of the x^2- and y^2-terms?

Solution Graph the hyperbola using the program of Section 12-2. The graph is shown in Figure 12-5b. The result appears to be a hyperbola whose axes are tilted at an angle to the coordinate axes. Based on the coefficients of x^2 and y^2, the graph would be expected to be an ellipse, not a hyperbola. ◀

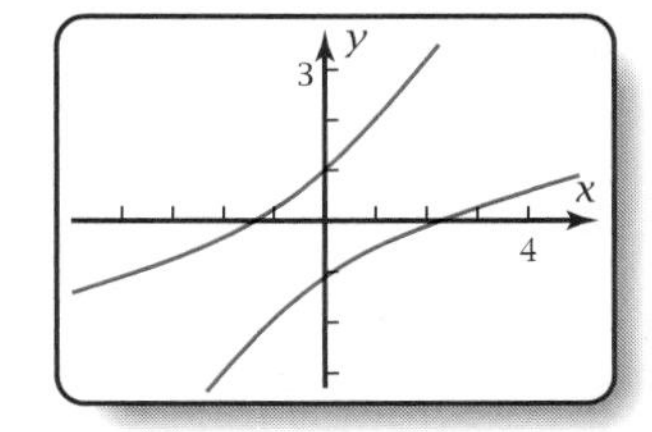

Figure 12-5b

▶ EXAMPLE 3 Plot the graph of $9x^2 - 20xy + 25y^2 - 8x + 2y - 28 = 0$ on the same screen as the hyperbola in Figure 12-5b. Describe the similarities and differences.

Solution The graph shown in Figure 12-5b is the thin curve in Figure 12-5c. With $-20xy$ instead of $-40xy$, the graph is now an ellipse. The rotated ellipse has the same x- and y-intercepts as the hyperbola in Example 2, as you can tell by the fact that the xy-term equals zero if either x or y is zero. ◀

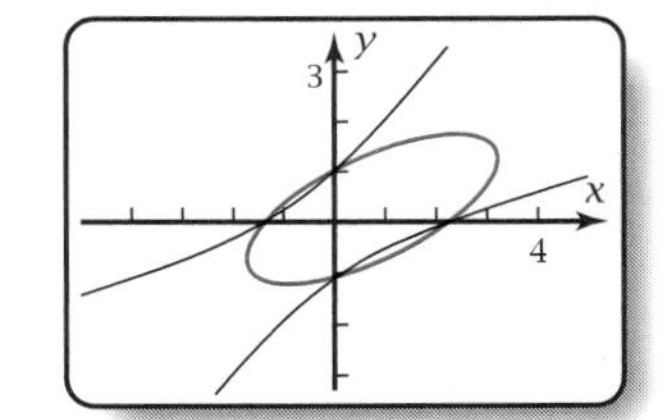

Figure 12-5c

▶ EXAMPLE 4 Plot the graph of $9x^2 - 30xy + 25y^2 - 8x + 2y - 28 = 0$ on the same screen as the hyperbola and ellipse in Figure 12-5c. Describe the similarities and differences.

Solution The new graph is a parabola, shown in orange in Figure 12-5d. An xy-term with the right coefficient makes the rotated conic a parabola. As before, the x- and y-intercepts of the graphs are equal. ◀

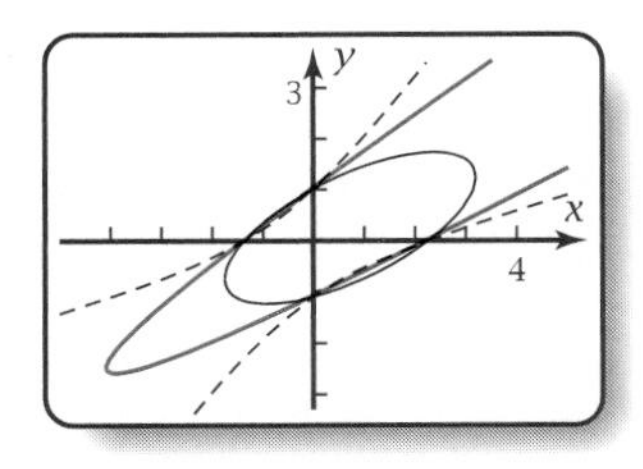

Figure 12-5d

Identifying a Conic by the Discriminant

In Examples 2, 3, and 4, the xy-coefficient takes on different values. As the absolute value of the coefficient increases, the graph changes from an ellipse to a parabola to a hyperbola. It is possible to tell which figure the graph will be from the coefficients of the three quadratic terms, x^2, xy, and y^2. To see what that relationship is, start with the equation for the hyperbola in Example 2, and solve it for y in terms of x using the quadratic formula.

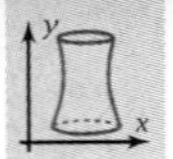

$$9x^2 - 40xy + 25y^2 - 8x + 2y - 28 = 0$$ Write the equation from Example 2.

$$25y^2 + (-40x + 2)y + (9x^2 - 8x - 28) = 0$$ Write the equation as a quadratic in y.

$$y = \frac{-(-40x + 2) \pm \sqrt{(-40x + 2)^2 - 4(25)(9x^2 - 8x - 28)}}{2(25)}$$ Use the quadratic formula.

$$y = \frac{20x - 1 \pm \sqrt{175x^2 + 160x + 701}}{25}$$ Simplify.

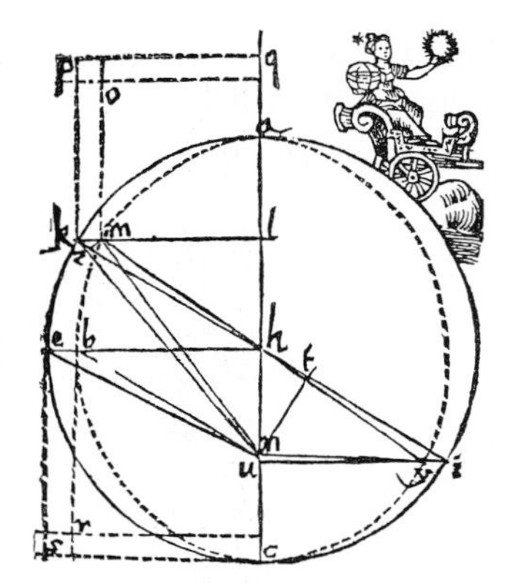

Johannes Kepler's diagram of the elliptical path of planet m around the Sun at point n—from his book Astronomia Nova *(1609).*

This transformation is the basis for the program you have been using since Section 12-2. The grapher plots two equations for y, one with the + sign and one with the − sign. Repeating the calculation for the equations in Examples 3 and 4 gives the following results.

Ellipse, $-20xy$: $$y = \frac{10x - 1 \pm \sqrt{-125x^2 + 180x + 701}}{25}$$

Parabola, $-30xy$: $$y = \frac{15x - 1 \pm \sqrt{0x^2 + 170x + 701}}{25}$$

Hyperbola, $-40xy$: $$y = \frac{20x - 1 \pm \sqrt{175x^2 + 160x + 701}}{25}$$

The graph will be an ellipse if the coefficient of x^2 under the radical sign is negative, a parabola if the x^2-coefficient is zero, and a hyperbola if the x^2-coefficient is positive. To find out what this coefficient is in general, repeat the steps just given for the hyperbola in Example 2 using A, B, C, D, E, and F for the six coefficients.

$$Ax^2 + Bxy + Cy^2 + Dx + Ey + F = 0$$ Write the general equation.

$$Cy^2 + (Bx + E)y + (Ax^2 + Dx + F) = 0$$ Write the equation as a quadratic in y.

$$y = \frac{-(Bx + E) \pm \sqrt{(Bx + E)^2 - 4(C)(Ax^2 + Dx + F)}}{2C}$$ Use the quadratic formula.

Expanding the expression under the radical sign gives

$$B^2x^2 + 2BEx + E^2 - 4ACx^2 - 4CDx - 4CF =$$
$$(B^2 - 4AC)x^2 + (2BE - 4CD)x + (E^2 - 4CF)$$

The x^2-coefficient, $B^2 - 4AC$, is called the **discriminant,** like $b^2 - 4ac$ in the familiar quadratic formula. This box summarizes the results so far.

PROPERTY: Discriminant of a Conic Section

For $Ax^2 + Bxy + Cy^2 + Dx + Ey + F = 0$, the **discriminant** is $B^2 - 4AC$.

$B^2 - 4AC < 0 \Rightarrow$ Circle or ellipse (not a circle if $B \neq 0$)

$B^2 - 4AC = 0 \Rightarrow$ Parabola

$B^2 - 4AC > 0 \Rightarrow$ Hyperbola

▶ EXAMPLE 5 For $2x^2 - 5xy + 8y^2 + 5x - 56y + 120 = 0$, use the discriminant to identify which conic the graph is. Confirm by plotting on your grapher.

Solution $B^2 - 4AC = 25 - 4(2)(8) = -39 < 0 \Rightarrow$ Graph will be an ellipse.

The graph in Figure 12-5e confirms that it is an ellipse. ◀

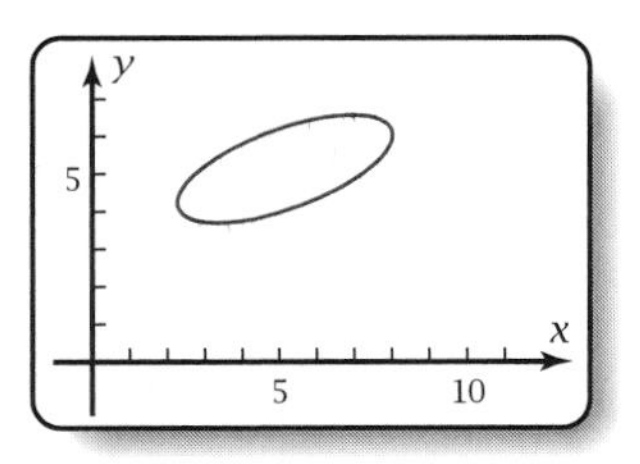

Figure 12-5e

Note that if you are using a low-resolution grapher, such as a handheld graphing calculator, the two branches of the ellipse might not meet. Figure 12-5f shows you what the graph might look like. When you sketch the graph, you should show it closing, as in Figure 12-5e.

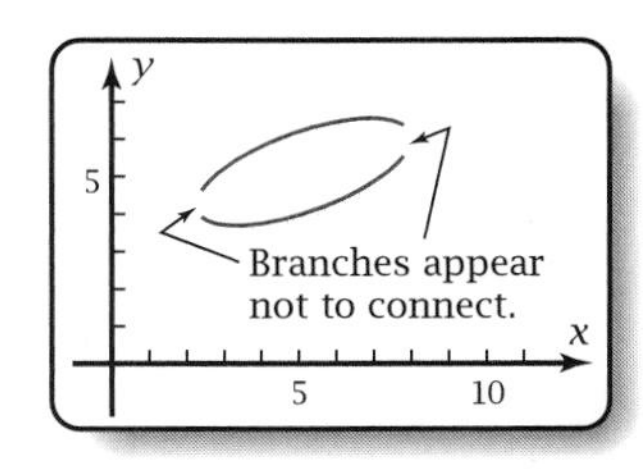

Figure 12-5f

Problem Set 12-5

Do These Quickly

Q1. Find the major radius of this ellipse:

$$\left(\frac{x}{8}\right)^2 + \left(\frac{y}{17}\right)^2 = 1$$

Q2. Find the focal radius of the ellipse in Problem Q1.

Q3. Find the eccentricity of the ellipse in Problem Q1.

Q4. Find the directrix radius for the ellipse in Problem Q1.

Q5. Find the conjugate radius of this hyperbola:

$$\left(\frac{x}{3}\right)^2 - \left(\frac{y}{4}\right)^2 = 1$$

Q6. Find the focal radius of the hyperbola in Problem Q5.

Q7. Find the eccentricity of the hyperbola in Problem Q5.

Q8. Find the directrix radius for the hyperbola in Problem Q5.

Q9. How far apart are the vertices of the hyperbola in Problem Q5?

Q10. What is the eccentricity of the parabola $y = 0.1x^2$?

For Problems 1–6, find the parametric equations of the conic section shown, and confirm that your answer is correct by plotting the graph on your grapher.

1. Ellipse

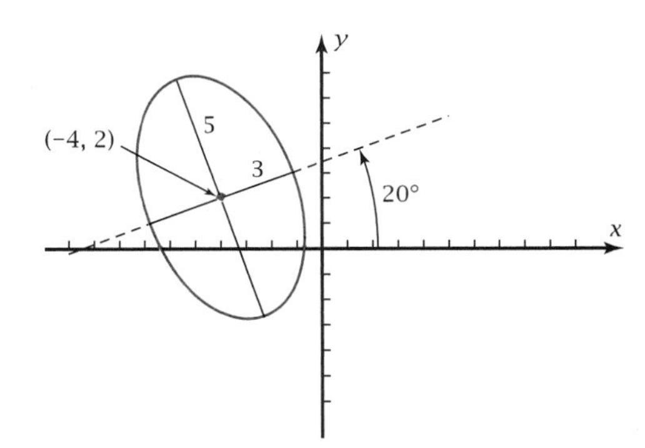

2. Ellipse

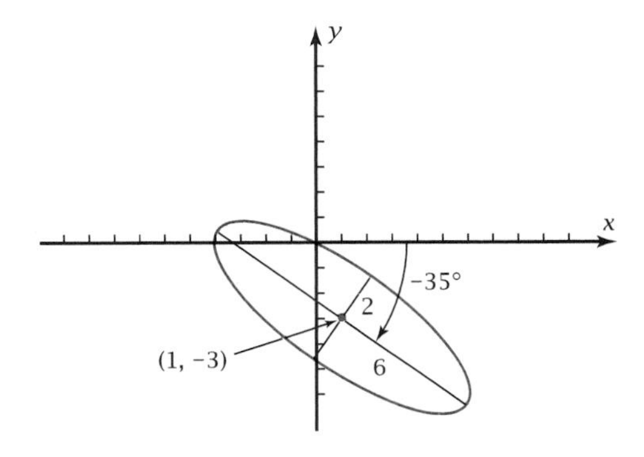

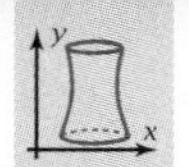

3. Hyperbola

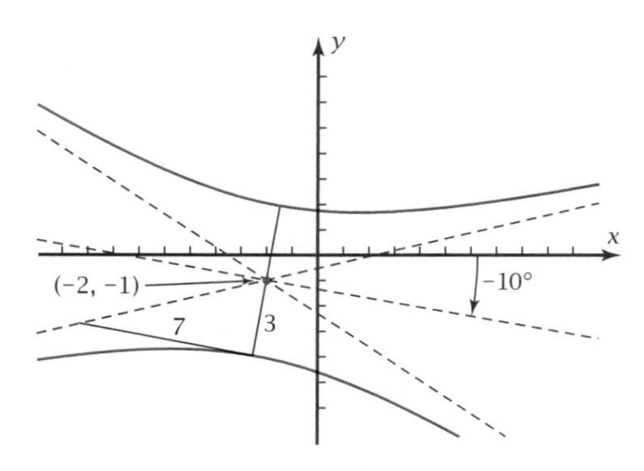

4. Hyperbola

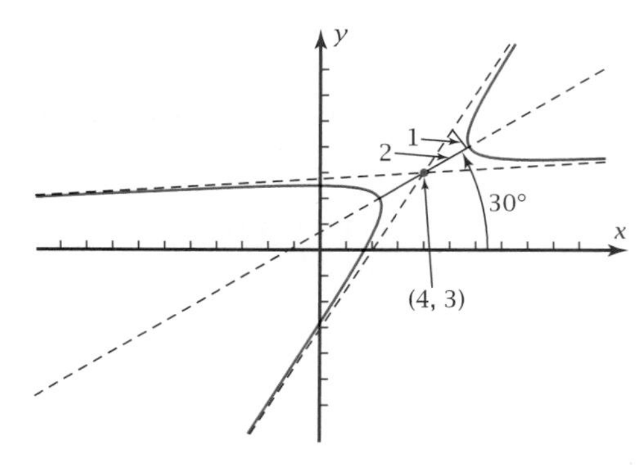

5. Parabola
(Use a t-range of $[-10, 10]$)

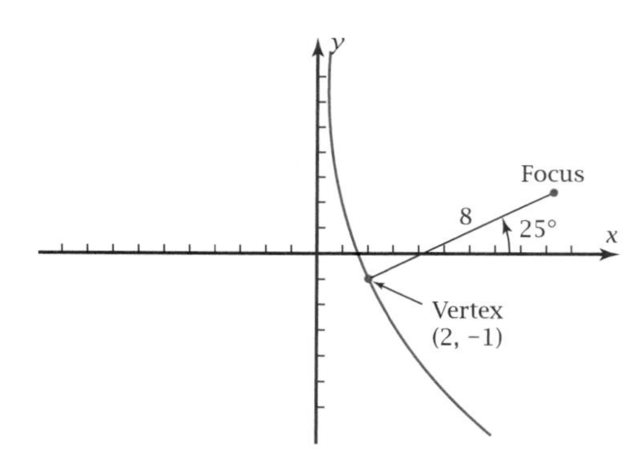

6. Parabola
(Use a t-range of $[-10, 10]$)

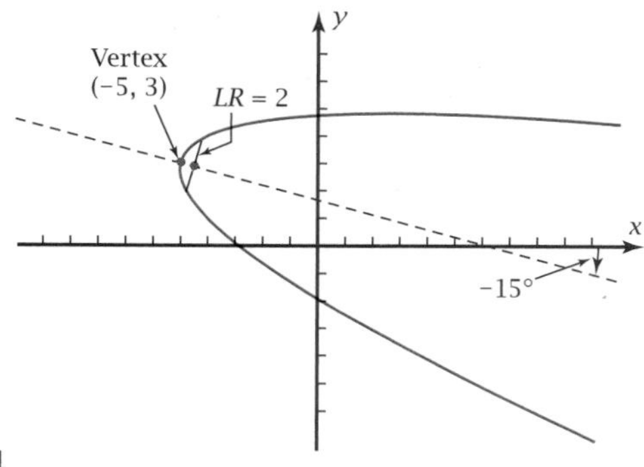

For Problems 7–12, find the parametric equations of the conic section described. Plot the graph on your grapher and sketch the result.

7. Ellipse with center (8, 5), eccentricity 0.96, and major radius 25 at an angle of −30° to the x-axis. Use a window with an x-range of $[-30, 30]$ and equal scales on the two axes.

8. Ellipse with center (6, −2), eccentricity 0.8, and major radius 5 at an angle of 70° to the x-axis. Use a window with an x-range of $[-10, 10]$ and equal scales on the two axes.

9. Hyperbola with center (0, 0), eccentricity 1.25, and transverse radius 4 at an angle of 15° to the x-axis. Use a window with an x-range of $[-10, 10]$ and equal scales on the two axes.

10. Hyperbola with center (−5, 10), eccentricity $\frac{25}{24}$, and transverse radius 24 at an angle of −20° to the x-axis. Use a window with an x-range of $[-50, 50]$ and equal scales on the two axes.

11. Parabola with vertex at point (−8, 5), focus $\frac{1}{4}$ unit from the vertex, and axis of symmetry at −30° to the x-axis, opening to the lower right. Use a t-range of $[-5, 5]$, a window with an x-range of $[-10, 10]$, and equal scales on the two axes.

12. Parabola with vertex at the origin, focus in the first quadrant 10 units from the vertex, and axis of symmetry at 45° to the x-axis. Use a t-range of $[-10, 10]$, a window with an x-range of $[-10, 10]$, and equal scales on the two axes.

For Problems 13–18, use the discriminant to determine which conic section the graph will be. Confirm your conclusion by plotting the graph. Sketch the result. (In your sketch, connect any gaps left by low-resolution graphers.)

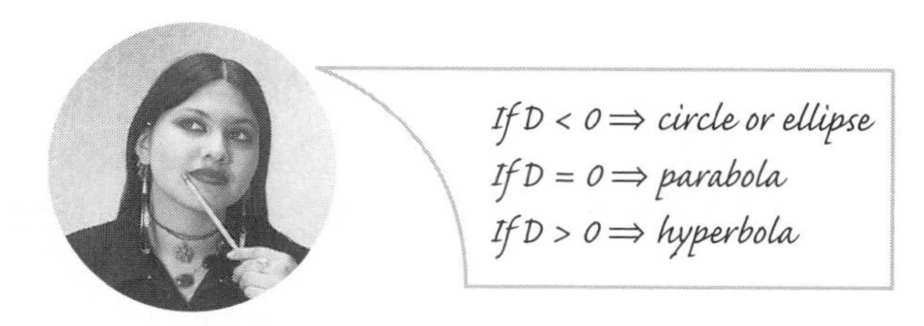

13. $3x^2 - 5xy + 9y^2 - 21x + 35y - 50 = 0$

14. $10x^2 - 4xy + 2y^2 + 87x - 13y + 100 = 0$

15. $3x^2 - 10xy + 6y^2 - 12x + 4y + 10 = 0$

16. $8x^2 + 40xy + 2y^2 - 20x - 10y - 31 = 0$

17. $x^2 + 6xy + 9y^2 - 3x - 4y - 10 = 0$

18. $16x^2 - 40xy + 25y^2 + 80x - 150y - 200 = 0$

19. *Discriminant of Unrotated Conics Problem:* If there is no xy-term, you can identify a conic section from the signs of the squared terms. For instance, the conic is an ellipse if x^2 and y^2 have the same sign. Show that the property you have learned about identifying a conic from the sign of the discriminant is consistent with the properties you have learned for identifying an unrotated conic from the signs of the squared terms.

20. *Inverse Variation Function Problem:* An inverse variation power function has the particular equation $y = \frac{12}{x}$. Prove that the graph is a hyperbola. What angle does its transverse axis make with the x-axis?

21. *Rotation and Dilation from Parametric Equations Project:* Here are the general parametric equations for an ellipse centered at the origin with x- and y-radii a and b, respectively, that has been rotated counterclockwise through an angle of α. Assume that t is in the range of $[0, 2\pi]$.

$$x = (a \cos \alpha) \cos t + (b \cos (\alpha + 90°)) \sin t$$
$$y = (a \sin \alpha) \cos t + (b \sin (\alpha + 90°)) \sin t$$

a. Use the rotation matrix to show how these equations follow from the unrotated ellipse equations

$$x = a \cos t$$
$$y = b \sin t$$

b. A particular ellipse has parametric equations

$$x = 3 \cos t - 2 \sin t$$
$$y = 5 \cos t + 1.2 \sin t$$

Calculate algebraically the major and minor radii and the angle α that one of the axes makes with the x-axis. Show by graphing that your answers are correct.

22. *Parabolic Lamp Reflector Project:* Figure 12-5g, left, shows in perspective a paraboloid that forms the reflector for a table lamp. The reflector is 12 in. long in the x-direction and has a radius of 5 in. The circular lip with 5-in. radius is shown in perspective as an ellipse with major radius 5 in. and minor radius 2 in.

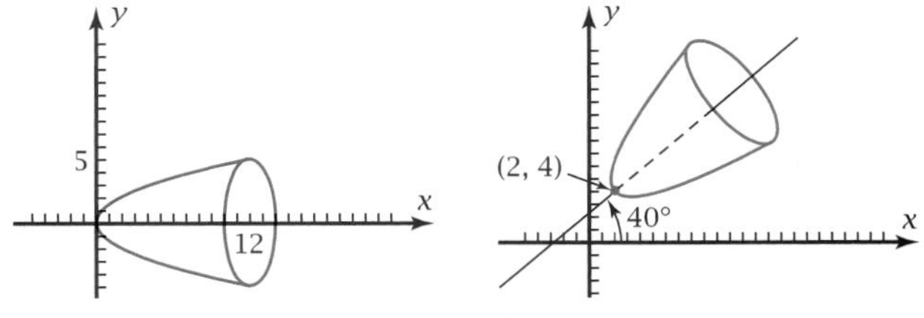

Figure 12-5g

a. Write parametric equations for the parabola shown. Pick a suitable t-range and plot the graph. Does the parabola start and end at the points shown in the figure?

b. Write parametric equations for the ellipse. Plot the ellipse on the same screen as the parabola. If your grapher does not allow you to pick different t-ranges for the two curves, you will have to be clever about setting the period for the cosine and sine in the parametric equations so that the entire ellipse will be plotted.

c. The figure on the right in Figure 12-5g shows the same lamp reflector rotated so that its axis is at 40° to the x-axis. Find parametric equations of the rotated parabola and rotated ellipse. Plot the two graphs on the same screen. If the result does not look like the given figure, keep working on your equations until the figure is correct.

23. *Increasing xy-Term Problem:* Here are four conic section equations that differ only in the coefficient of the xy-term. Identify each conic with the help of the discriminant. Plot all four graphs on the same screen. What graphical feature do you notice is the same for all four? How can you tell algebraically that your graphical observation is correct?

a. $x^2 + y^2 - 10x - 8y + 16 = 0$

b. $x^2 + xy + y^2 - 10x - 8y + 16 = 0$

c. $x^2 + 2xy + y^2 - 10x - 8y + 16 = 0$

d. $x^2 + 4xy + y^2 - 10x - 8y + 16 = 0$

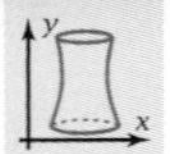

12-6 Applications of Conic Sections

This giant parabolic mirror is a solar furnace on the side of a building harvesting solar power in Odeillo, France.

You have learned how to find equations of ellipses, parabolas, hyperbolas, and circles from their analytic properties. These figures appear in the real world as paths of thrown objects, orbits of planets and comets, and shapes of bridges. They even appear in solutions of business problems. In this section you will consolidate your knowledge of conic sections by applying them to some real-world problems.

OBJECTIVE Given a situation from the real world in which conic sections appear, create a mathematical model and use it to make predictions and interpretations.

Problem Set 12-6

Do These Quickly

Q1. Why are parabolas, ellipses, hyperbolas, and circles called *conic sections*?

Q2. Why does a hyperbola have two branches, while a parabola has only one branch?

Q3. Write a Cartesian equation for a unit hyperbola opening in the y-direction.

Q4. Write parametric equations for a unit hyperbola opening in the y-direction.

Q5. Complete the square: $x^2 - 10x +$ —?—

Q6. Sketch a paraboloid.

Q7. Write the definition of *eccentricity* in terms of d_1 from the directrix and d_2 from the focus.

Q8. Sketch an ellipse, showing the approximate location of the two foci.

Q9. Tell what is special about the distances from the two foci to a point on an ellipse.

Q10. If an ellipse has semimajor axis 7 and semiminor axis 3, what is the focal radius?

1. *Coffee Table Problem:* A furniture manufacturer wishes to make elliptical tops for coffee tables 52 in. long and 26 in. wide, as shown in Figure 12-6a. A pattern is to be cut from plywood so that the outline of the tabletops can easily be marked on the tabletop's surface. Give detailed instructions for a rapid way to mark the ellipse on the plywood. What will be the eccentricity of the tabletops?

Figure 12-6a

2. *Stadium Problem:* The plan for a new football stadium calls for the stands to be in a region defined by two concentric ellipses (Figure 12-6b). The outer ellipse is to be 240 yd long and 200 yd wide. The inner ellipse is to be 200 yd long and 100 yd wide. A football field of standard dimensions, 120 yd by 160 ft, will be laid out in the center of the inner ellipse.

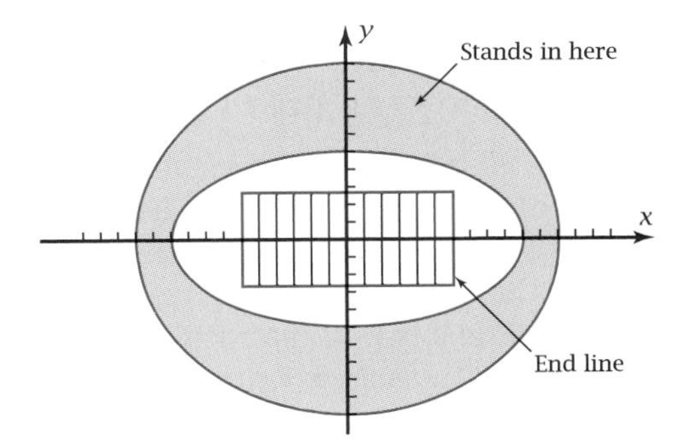

Figure 12-6b

   a. Find particular equations of the two ellipses.
   b. Find the eccentricities of the two ellipses.
   c. How much clearance will there be between the corner of the field and the inner ellipse in the direction of the end line?
   d. The area of an ellipse is πab, where a and b are the major and minor radii, respectively. To the nearest square yard, what is the area of the stands? If each seat takes about 0.8 square yards, what will be the approximate seating capacity of the stadium?
   e. Show that the familiar formula for the area of a circle is a special case of the formula for the area of an ellipse.

3. *Bridge Problem:* Figure 12-6c is a photograph of Bixby Bridge in Big Sur, California. A similar bridge under construction is shown in Figure 12-6d. The span of the bridge is to be 1000 ft, and it is to rise 250 ft at the vertex of the parabola. The roadway is horizontal and will pass 20 ft above the vertex. Vertical columns extend between the parabola and the roadway, spaced every 50 ft horizontally.

Figure 12-6c

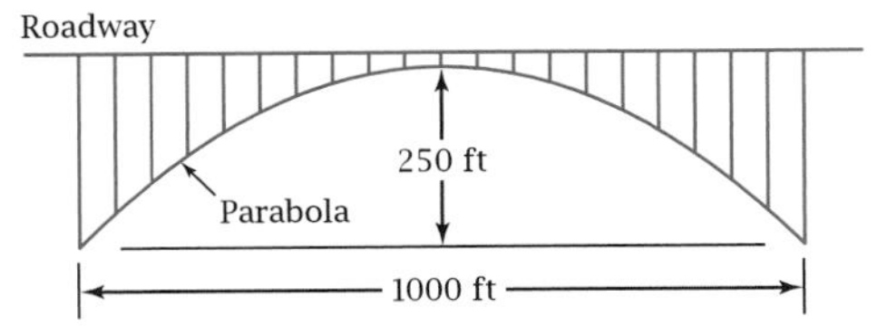

Figure 12-6d

   a. Using convenient axes, find the particular equation of the parabola.
   b. The construction company that builds the bridge must know how long to make each vertical column. Make a table of values showing these lengths.
   c. To order enough steel to make the vertical columns, the construction company must know the total length. By appropriate operations on the values in the table in part b, calculate this total length. Observe that there is a row of columns on both sides of the bridge.

4. *Halley's Comet Problem:* Halley's comet moves in an elliptical orbit around the Sun, with the Sun at one focus. It passes within about 50 million miles of the Sun once every 76 years. The other end of its orbit is about 5000 million miles from the Sun, beyond the orbit of Uranus.

   a. Find the eccentricity of the comet's orbit.
   b. Find the particular equation of the orbit. Put the origin at the Sun and the x-axis along the major axis.

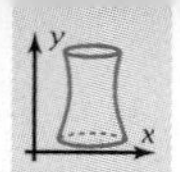

c. The orbit's major axis is 5050 million miles long. How wide is it?

d. How far is the comet from the Sun when the line from its position to the Sun is perpendicular to the major axis?

e. Where is the directrix of the ellipse corresponding to the Sun?

5. *Meteor Tracking Problem 1:* Suppose that you have been hired by Palomar Observatory near San Diego. Your mission is to track incoming meteors to predict whether or not they will strike Earth. Since Earth has a circular cross section, you decide to set up a coordinate system with its origin at Earth's center (Figure 12-6e). The equation of Earth's surface is

$$x^2 + y^2 = 40$$

where x and y are distances in *thousands* of kilometers.

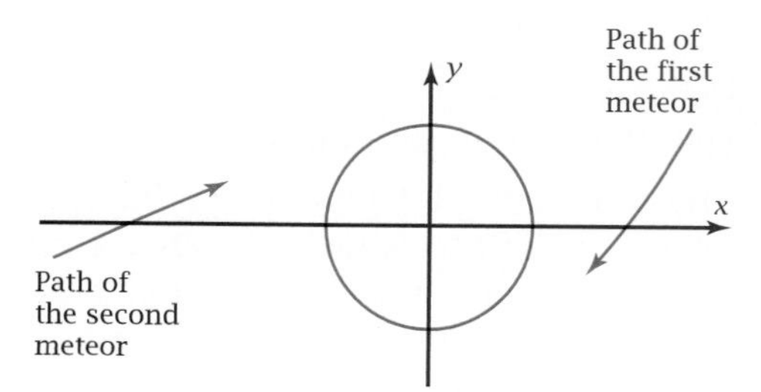

Figure 12-6e

a. The first meteor you observe is moving along a path whose equation is

$$x^2 - 18y = 144$$

What geometrical figure is the path? Find out graphically whether or not the meteor's path intersects Earth's surface.

b. Confirm your conclusion to part a algebraically by solving the system of equations

$$x^2 + y^2 = 40$$
$$x^2 - 18y = 144$$

To do this, try eliminating x algebraically and solving the resulting equation for y.

c. The second meteor you observe is moving along a path whose equation is

$$x^2 - 4y^2 + 80y = 340$$

What geometrical figure is the path? Confirm graphically that the path *does* intersect Earth's surface. Find numerically the point at which the meteor will strike Earth.

d. Find algebraically the point at which the meteor in part c will strike Earth.

6. *Meteor Tracking Problem 2:* A meteor originally moving along a straight path will be deflected by Earth's gravity into a path that is a conic section. If the meteor is moving fast enough, the path will be a hyperbola with the original straight-line path as an asymptote. Assume that a meteor is approaching the vicinity of Earth along a hyperbola with general equations

$$x = a \sec t$$
$$y = b \tan t$$

a. Suppose that at $t = -1.4$, the position of the meteor is $(x, y) = (29.418, -11.596)$, where x and y are displacements in millions of miles. Find the particular values of a and b.

b. Plot the branch of the hyperbola in the range for t of $\left(-\frac{\pi}{2}, \frac{\pi}{2}\right)$. Sketch the result.

c. Earth is at the focus of the hyperbola that is closest to the vertex in part b. What is the focal radius? How far is the meteor from Earth when it is at the vertex of the hyperbola?

d. Assuming that the meteor does not hit Earth, its path as it leaves Earth's vicinity will approach the other asymptote of the hyperbola. What is the equation of this other asymptote? Show both asymptotes on your sketch in part b.

7. *Marketing Problem 1:* A customer located at point (x, y) can purchase goods from Supplier 1 located at $(0, 0)$ or from Supplier 2 located at $(6, 0)$, where x and y are in miles (Figure 12-6f). The charge for delivery of the goods is

\$10.00 per mile from Supplier 1 and \$20.00 per mile from Supplier 2.

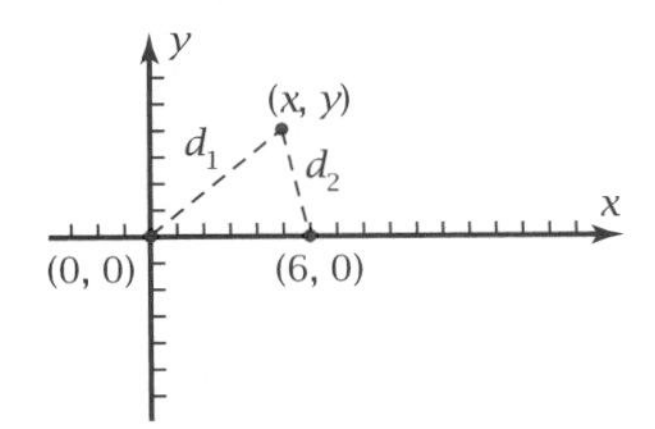

Figure 12-6f

a. To decide which supplier to use for a given point (x, y), it would help the customer to know all points for which the total cost of shipping is the same for either supplier. These points will satisfy

$$10d_1 = 20d_2$$

where d_1 and d_2 are the distances from Supplier 1 and Supplier 2, respectively, to the point (x, y). Find the particular Cartesian equation for this set of points. Show that the graph is a circle.

b. If point (x, y) is close enough to Supplier 2, it is cheaper to pay the higher cost per mile. Shade the region in which it is cheaper to receive goods from Supplier 2.

c. If a customer is located at point (15, 0), which supplier is closer? Which supplier's shipping charges would be less? Is this surprising?

8. *Marketing Problem 2:* Suppose that the shipping charge from both suppliers in Problem 7 is \$10.00 but that the purchase price of the goods is \$980 from Supplier 1 and \$1000 from Supplier 2.

a. Find the particular equation of the set of points (x, y) for which the total cost (shipping plus goods) is the same for both suppliers.

b. Explain why the graph of the set of points in part a is a hyperbola.

c. Explain why one branch of the hyperbola in part a is not relevant to this application. Sketch the other branch.

d. From which supplier should you purchase the goods if you are located at point (x, y) = (7, 20). Justify your answer.

9. *Hyperboloid Project:* The icon at the top of the left-hand pages in this chapter is a hyperboloid of one sheet. An enlarged version of the hyperboloid is given in Figure 12-6g. The top and bottom circles appear as ellipses in the figure because they are seen in perspective. Find parametric equations of the ellipses and hyperbola, and use the results to plot the hyperboloid. If your grapher does not allow you to set different *t*-ranges for different graphs, find a clever way to make the figure come out correctly.

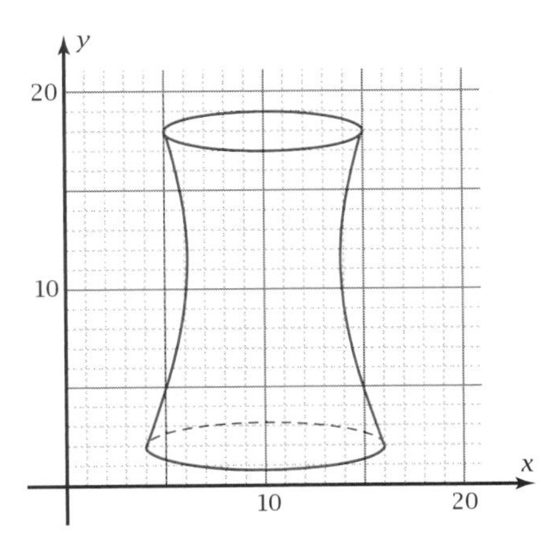

Figure 12-6g

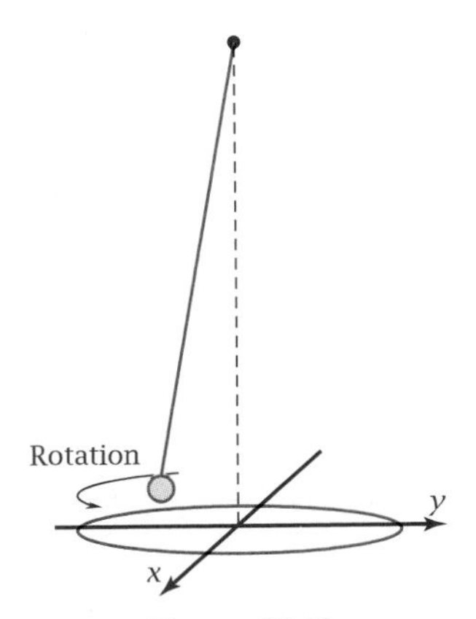

Figure 12-6h

10. *Elliptical Pendulum Project:* Make a pendulum by tying a small weight to the end of a string, as shown in Figure 12-6h. Tie the other end of the string to the ceiling or an overhead doorway so that it clears the floor by a centimeter or two. Place two metersticks perpendicular to each other to represent the x- and y-axes. The origin should be under the rest position of the pendulum. Measure the

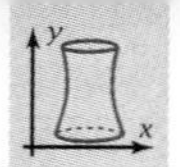

period of the pendulum by pulling it out to $x = 30$ cm and counting the time for ten swings. Confirm that the period is the same if you start the pendulum from $y = 20$ cm. Then set the pendulum in motion by holding it at $x = 30$ on the x-axis and pushing it in the y-direction just hard enough so that it crosses the y-axis at $y = 20$. Figure out parametric equations for the elliptical path that the pendulum traces in terms of the parameter t seconds since the pendulum was started. Using the equations, calculate the expected position of the pendulum at time $t = 7$ seconds. Then set the pendulum in motion again. Start timing $t = 0$ when the pendulum crosses the positive x-axis the second time. Does the pendulum pass through the point you calculated when $t = 7$?

12-7 Chapter Review and Test

In this chapter you have studied ellipses, circles, parabolas, and hyperbolas. These conic sections get their names from plane slices of cones. Each one has geometrical or analytic properties, such as foci, directrices, eccentricity, and axes of symmetry. You have analyzed the graphs algebraically, both in Cartesian form and in parametric form. Rotation of conic section graphs about their axes of symmetry produces three-dimensional quadric surfaces. You have also seen how conic sections appear in the real world, such as in the shapes of bridges and the paths of celestial objects.

Review Problems

R0. Update your journal with what you have learned in this chapter. Include such things as:

- The way conic sections are formed by slicing a cone with a plane
- The origins of the names *ellipse, parabola,* and *circle*
- Cartesian equations of the conic sections
- Parametric equations of the conic sections
- Dilations, translations, and rotations of the conic sections
- Quadric surfaces formed by rotating conic sections about axes of symmetry
- Focus, directrix, and eccentricity properties of conic sections
- Some real-world places where conic sections appear

R1. Without plotting, tell whether the graph will be a circle, an ellipse, a parabola, or a hyperbola.

a. $x^2 + y^2 = 36$ b. $x^2 + 9y^2 = 36$

c. $4x^2 - y^2 = 36$ d. $4x^2 - 4y^2 = 36$

e. $4x^2 + y = 36$

R2. a. Write the equation for the specified parent function.

i. Cartesian equation for a unit circle

ii. Parametric equations for a unit circle

iii. Parametric equations for a unit hyperbola opening in the x-direction

iv. Cartesian equation for a unit hyperbola opening in the y-direction

v. Cartesian equation for a unit parabola opening in the x-direction

vi. Cartesian equation for a unit parabola opening in the y-direction

b. For the ellipse with equation

$$\left(\frac{x-2}{7}\right)^2 + \left(\frac{y+3}{4}\right)^2 = 1$$

i. Sketch the graph.

ii. Transform to the form

$$Ax^2 + Cy^2 + Dx + Ey + F = 0$$

iii. Plot the graph using the program of Section 12-2. Does it agree with your sketch?

iv. Write the parametric equations and use them to plot the graph.

c. For the hyperbola in Figure 12-7a,

i. Write a Cartesian equation.

ii. Write parametric equations.

iii. Plot the parametric equations. Does the graph agree with Figure 12-7a?

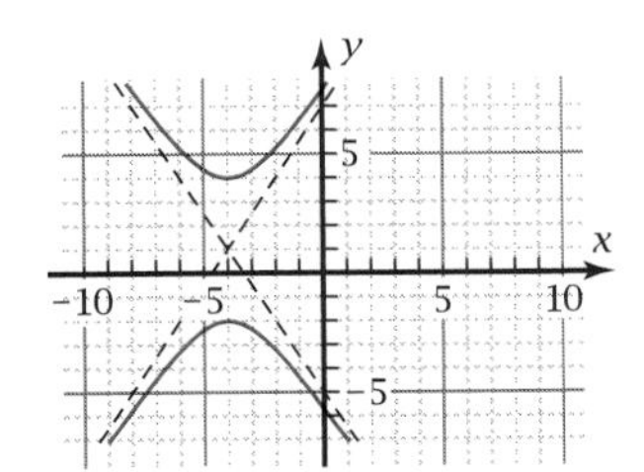

Figure 12-7a

R3. a. Sketch the quadric surface.

i. Ellipsoid formed by rotating the graph of $4x^2 + y^2 = 16$ about the y-axis

ii. Hyperboloid formed by rotating the graph of $x^2 - 9y^2 = 9$ about the x-axis

iii. Hyperboloid formed by rotating the graph of $4x^2 - y^2 = 4$ about the y-axis

b. A paraboloid is formed by rotating about the x-axis the part of the parabola $x = 4 - y^2$ that lies in the first quadrant. A cylinder is inscribed in the paraboloid, with its axis along the x-axis, its left base containing the origin, and its right base touching the paraboloid (Figure 12-7b). Find the radius and altitude of the cylinder of maximum volume. Find this maximum volume.

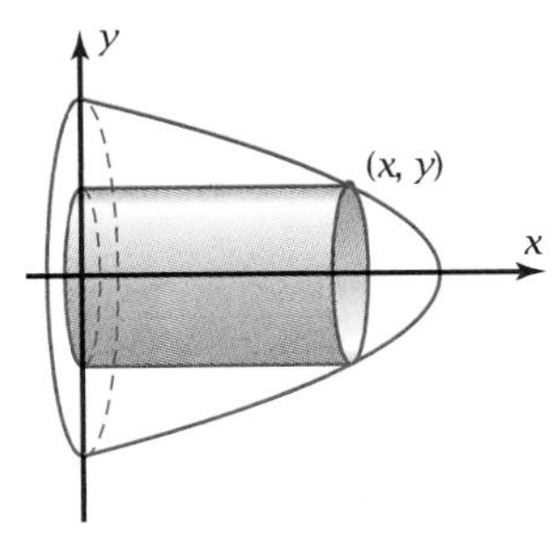

Figure 12-7b

R4. For parts a and b, Figure 12-7c shows an ellipse of eccentricity 0.8, with foci at the points (−6.4, 0) and (6.4, 0) and one directrix at $x = 10$. The point shown on the ellipse has $x = 3$.

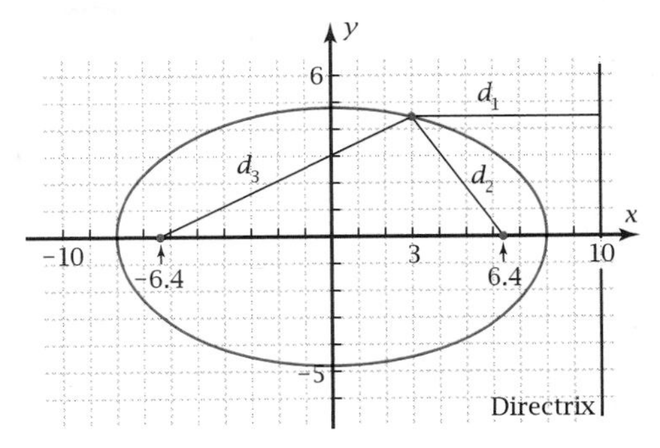

Figure 12-7c

a. Measure the three distances d_1, d_2, and d_3 using the scales shown. Confirm that $d_2 = ed_1$. Confirm that $d_2 + d_3 = 16$, the major diameter.

b. Write a Cartesian equation for the ellipse. Use the equation to calculate y for the point where $x = 3$. Use the value of y and the Pythagorean theorem to calculate d_2 and d_3 exactly. Do your measurements in part a agree with the calculated values? Does $d_2 + d_3$ equal 16 exactly?

c. *Billiards Table Problem:* An elliptical billiards table is to be built with eccentricity 0.9 and foci at the points (0, −81) and (0, 81), where x and y are in centimeters. Find the major and minor radii. Write parametric equations for the ellipse, plot them on your grapher, and sketch the result. If a ball at one focus is hit in any direction and bounces at the edge of the table, through what special point will the path of the ball go?

d. Identify the conic section whose equation is shown here. Sketch the graph. Find the four radii and the eccentricity.

$$\left(\frac{x}{4}\right)^2 - \left(\frac{y}{3}\right)^2 = 1$$

e. A parabola has directrix $x = 4$ and focus at point (0, 0). Sketch this information. Find the particular Cartesian equation. Plot the graph and sketch it on your paper.

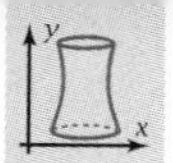

f. Find the parametric equations of the hyperbola with eccentricity $\frac{5}{3}$ and foci at points (2, 1) and (2, 7). Plot the graph and sketch the result.

R5. a. Find parametric equations for the ellipse in Figure 12-7d. Confirm that your answer is correct by plotting on your grapher.

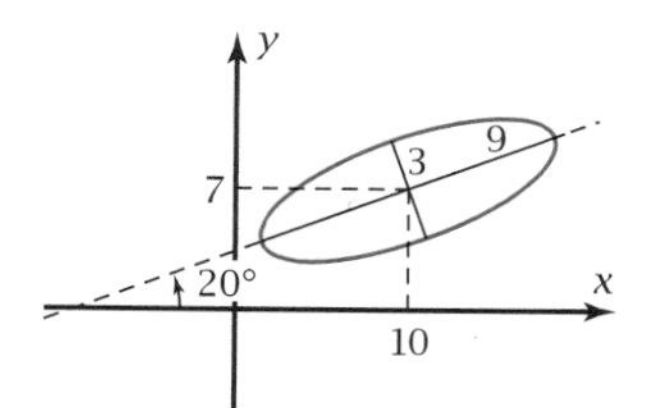

Figure 12-7d

b. Find parametric equations of the hyperbola with eccentricity 2 centered at point (3, −4) if its transverse radius is 6 and its transverse axis makes an angle of 35° with the x-axis. Confirm that your answer is correct by plotting on your grapher.

c. Find parametric equations of the parabola with vertex at point (1, 2) and focus at point (4, 5). Confirm that your answer is correct by plotting on your grapher.

d. Use the discriminant to identify the conic. Confirm by plotting the graph. Sketch the graph.

i. $4x^2 + 2xy + 9y^2 + 15x - 13y - 19 = 0$

ii. $4x^2 + 12xy + 9y^2 + 15x - 13y - 19 = 0$

iii. $4x^2 + 22xy + 9y^2 + 15x - 13y - 19 = 0$

e. What graphical feature do all three of the conics in part d have in common?

R6. *Parabolic Antenna Problem:* A satellite dish antenna is to be constructed in the shape of a paraboloid (Figure 12-7e). The paraboloid is formed by rotating about the x-axis the parabola with focus at point (25, 0) and directrix $x = -25$, where x and y are in inches. The diameter of the antenna is to be 80 in.

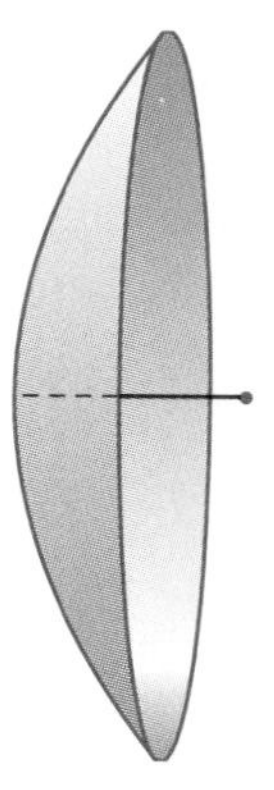

Figure 12-7e

a. Find the equation of the parabola and the domain of x.

b. Sketch the graph of the parabola, showing the focus and the directrix.

c. A receiver is to be placed at the focus. Figure 12-7f (on the next page) suggests that the receiver would touch the ground if the antenna were placed "face down." Determine algebraically whether or not this geometrical observation is correct.

Concept Problems

C1. *Reflecting Property of a Parabola Problem:* Figure 12-7f shows the parabolic cross section of the television dish antenna shown in Figure 12-7e. The equation of the parabola is $x = 0.01y^2$.

a. Confirm that the focus of the parabola is at point (0, 25), as shown in Figure 12-7f.

b. A television signal ray comes in parallel to the x-axis at $y = 20$. Find the coordinates of the point at which the ray strikes the parabola.

c. Calculate the angle A between the incoming ray and a line to the focus from the point where the incoming ray strikes the parabola.

d. An incoming ray and its reflected ray make equal angles with a line tangent to the curved surface. Use your answer to part c to calculate angles B and C if these angles have equal measure. With a protractor, measure A, B, and C in Figure 12-7f, thus confirming your calculations.

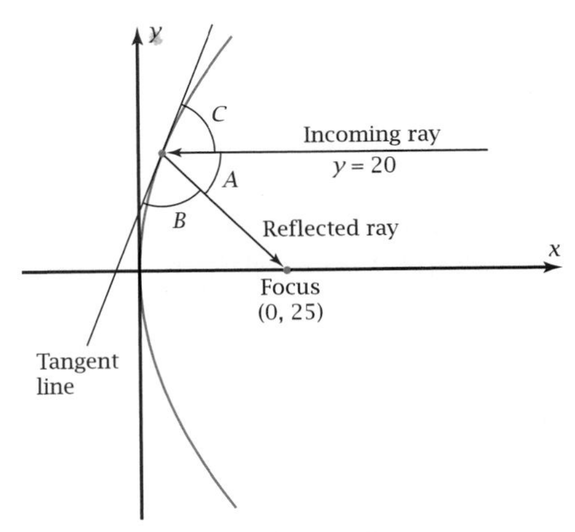

Figure 12-7f

e. Find the particular equation of the tangent line. Use angle C to find its slope. Plot the parabola and the tangent line on your grapher. Zoom in on the point of tangency. What do you notice about the tangent line and the graph as you zoom in?

f. Pick another incoming ray. Calculate the measure of angle A. Let C be half of the supplement of A, as it was in part d. Is the line at angle C with the incoming ray tangent to the graph at the point where the incoming ray strikes the graph? How do you know?

g. Write a conjecture about the direction the reflected ray takes whenever the incoming ray is parallel to the axis of the parabola. Your conjecture should give you insight into why the name *focus* is used and why television satellite antennas and other listening devices are made in the shape of a paraboloid.

C2. *Systems of Quadratic Equations Problem:* Figure 12-7g shows an ellipse, a hyperbola, and a line. The equations for each pair of these graphs form a system of equations. In this problem you will solve the systems by finding the points at which these graphs intersect each other.

a. Solve the ellipse and hyperbola system graphically, to one decimal place.

b. Solve the ellipse and line system graphically, to one decimal place.

c. Solve the hyperbola and line system graphically, to one decimal place.

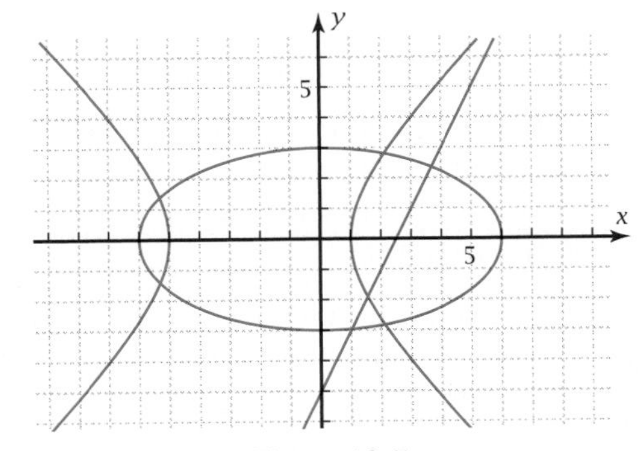

Figure 12-7g

d. The equations of the ellipse and hyperbola in Figure 12-7g are

$$x^2 - y^2 + 4x - 5 = 0$$

$$x^2 + 4y^2 - 36 = 0$$

Quick! Tell which is which. How do you know?

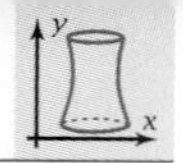

e. Solve the ellipse and hyperbola system algebraically. To do this, first eliminate y by adding a multiple of the first equation to the second. Solve the resulting quadratic equation for x. Finally, substitute the two resulting x-values into one of the original equations and calculate y. What do you notice about two of the four solutions? How does this observation agree with the graphs?

f. Solve the ellipse and hyperbola system numerically. Do this by plotting the two graphs on your grapher and then using the intersect feature. Do the answers agree with parts a and e?

g. The equation of the line in Figure 12-7g is

$$2x - y = 5$$

Solve the ellipse–line system algebraically. To do this, first solve the linear equation for y in terms of x and then substitute the result for y in the ellipse equation. After you solve the resulting quadratic equation, substitute the two values of x into the *linear* equation to find y. Do the answers agree with part b?

h. Solve the hyperbola–line system algebraically. Which solution does *not* appear in the graphical solution of part c?

Chapter Test

PART 1: No calculators allowed (T1–T10)

For Problems T1–T6, identify the conic section.

T1. $4x^2 + 9y^2 + 24x + 36y - 72 = 0$

T2. $9x^2 - 25y^2 + 36x + 200y - 589 = 0$

T3. $x^2 - 14x - 36y + 13 = 0$

T4. $x^2 + 3xy + 4y^2 - 400 = 0$

T5. $x^2 + 4xy + 4y^2 - 400 = 0$

T6. $x^2 + 5xy + 4y^2 - 400 = 0$

T7. Sketch a hyperboloid of one sheet.

T8. Give another name for an ellipsoid formed by rotating an ellipse about its major axis. Name something in the real world that has (approximately) this shape.

T9. Sketch an ellipse. Show the approximate locations of the foci and the directrices. Pick a point on the ellipse and draw its distances to a directrix and to the corresponding focus. How are the two distances related to each other?

T10. Draw a sketch showing an ellipse as a section of a cone.

PART 2: Graphing calculators allowed (T11–T18)

T11. Sketch the graph of

$$-\left(\frac{x-3}{6}\right)^2 + \left(\frac{y-1}{2}\right)^2 = 1$$

T12. Write parametric equations for the graph in Problem T11. Plot the graph. Does your sketch in Problem T11 agree with the plotted graph?

T13. Transform the equation in Problem T11 to the form $Ax^2 + Cy^2 + Dx + Ey + F = 0$.

T14. An ellipsoid is formed by rotating the ellipse $x^2 + 4y^2 = 16$ about the y-axis. A cylinder is inscribed in the ellipsoid with its axis along the y-axis and its two bases touching the ellipsoid (Figure 12-7h). Plot the graph of the volume of the cylinder as a function of its radius, x. Find numerically the radius and altitude of the cylinder of maximum volume, and approximate the value of this maximum volume.

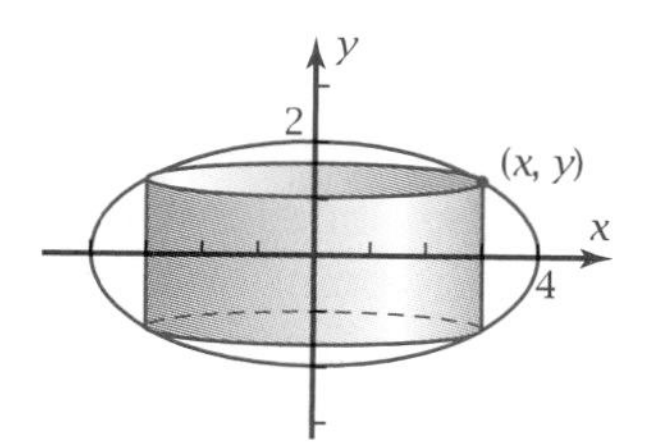

Figure 12-7h

T15. *Satellite Problem 2:* A satellite is in elliptical orbit around Earth, as shown in Figure 12-7i. The major radius of the ellipse is 51 thousand miles, and the focal radius is 45 thousand miles. The center of Earth is at one focus of the ellipse.

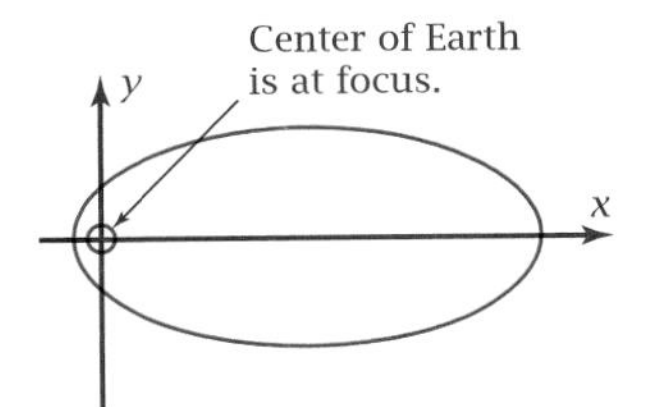

Figure 12-7i

a. Where is the center of the ellipse? What is the minor radius? What is the eccentricity?

b. Write the Cartesian equation of the ellipse. Use the equation to find the y-intercepts of the ellipse.

c. The satellite is closest to Earth when it is at a vertex of the ellipse. The radius of Earth is about 4000 miles. What is the closest the satellite comes to the surface of Earth?

T16. Find the particular equation of the parabola with focus at point $(-4, -5)$ and directrix $y = 2$.

T17. Figure 12-7j shows an ellipse centered at point $(1, -2)$, with major radius 7 units long making an angle of $-25°$ to the x-axis and minor radius 3 units. Find parametric equations of the ellipse. Confirm your answer by plotting the graph.

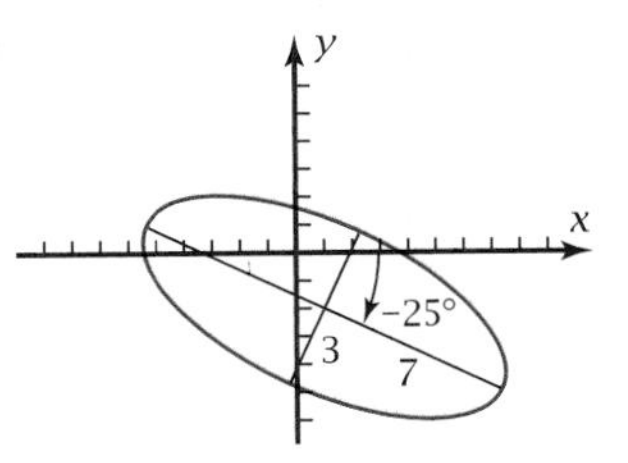

Figure 12-7j

T18. What did you learn as a result of taking this test that you did not know before?

CHAPTER

13

Polar Coordinates, Complex Numbers, and Moving Objects

Paths traced by rotating objects can be modeled by coordinates in which the independent variable is an angle and the dependent variable is a directed distance from the origin. These *polar coordinates,* along with the parametric functions of Chapter 4, give a way to find equations for such things as the *involute of a circle,* which is the shape of the surfaces of the gear teeth shown in the photograph. This shape allows one gear to transmit its motion to the other in a smooth manner.

Mathematical Overview

In this chapter you will learn about polar coordinates, where r and θ are used instead of x and y. Polar coordinates allow you to plot complicated graphs such as the five-leaved rose in the icon at the top of each even-numbered page and in Figure 13-0a. These coordinates have surprising connections to imaginary and complex numbers and to the parametric functions you studied in earlier chapters. You will study polar coordinates in four ways.

Graphically Five-leaved rose

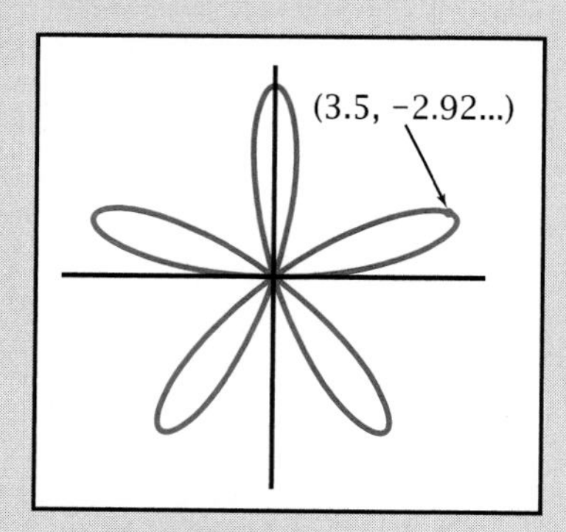

Figure 13-0a

Algebraically $r = 3 \sin 5\theta$ Polar equation of a rose.

Numerically $\theta = 3.5$ radians: $r = -2.9268...$ units The value of r can be negative!

Verbally *I was surprised to find that a point in the first quadrant could have a negative r-value. Now I realize that if r is negative, you just go around to the angle and then plot the point in the opposite direction.*

13-1 Introduction to Polar Coordinates

The graphs of the trigonometric functions you have plotted so far have been in the familiar Cartesian coordinate system, where points are located by x- and y-coordinates. A more natural way to plot such graphs is to locate points by an angle θ in standard position and a distance r from the origin. Such graphs are said to be plotted in **polar coordinates.**

OBJECTIVE Given an equation in polar coordinates, plot the graph on polar coordinate paper.

Exploratory Problem Set 13-1

1. On polar coordinate paper (Figure 13-1a), plot the point $(r, \theta) = (7, 30°)$ by going around to an angle of 30° and then going out 7 units from the **pole** (the origin).

2. Plot the point $(r, \theta) = (-7, 210°)$ by going around to 210° and then going *back* 7 units from the pole. What do you notice about this point and the point in Problem 1?

3. Plot the points shown in the table. Connect the points in order with a smooth curve.

θ	r	θ	r
0°	10.0	195°	−5.7
15°	9.7	210°	−4.9
30°	8.9	225°	−3.7
45°	7.7	240°	−2.0
60°	6.0	255°	0.0
75°	4.1	270°	2.0
90°	2.0	285°	4.1
105°	0.0	300°	6.0
120°	−2.0	315°	7.7
135°	−3.7	330°	8.9
150°	−4.9	345°	9.7
165°	−5.7	360°	10.0
180°	−6.0		

4. Put your grapher in polar mode and degree mode. Set the window so the range for θ is from 0° to 360° and the θ-step is 5°. Use a range for x of at least [−10, 10] and a range for y of at least [−7, 7]. Enter the **polar equation** $r = 2 + 8 \cos \theta$ in the y= menu and plot the graph. Press zoom square to make the scales equal on the two axes. What do you notice about the graph?

5. From the format menu, select polar grid coordinates. Then trace to $\theta = 150°$. Does the point on the graph agree with the point in the table?

6. What did you learn as a result of doing this problem set that you did not know before?

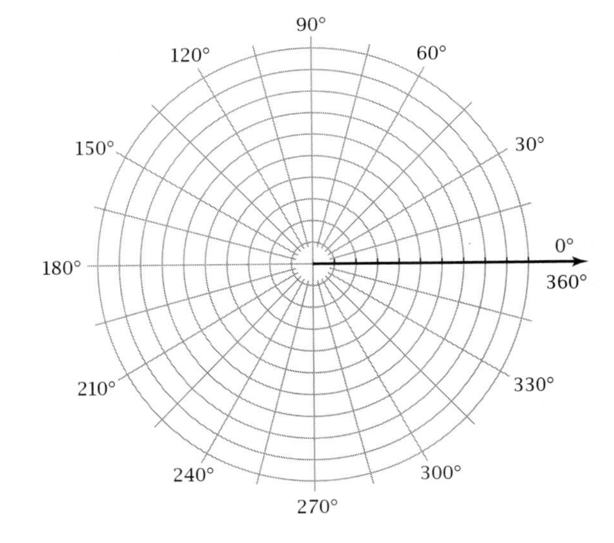

Figure 13-1a

13-2 Polar Equations of Conics and Other Curves

Figure 13-2a shows the graph of $r = 2 + 8\cos\theta$ that you plotted in Problem Set 13-1. The figure is called a **limaçon of Pascal.** The *ç* in *limaçon* is pronounced like an *s*. *Limaçon* is a French word for "snail." Graphs of polar functions may also be familiar conic sections, such as the ellipse $r = \frac{9}{5 - 4\cos\theta}$ shown in Figure 13-2b. In this section you will use algebra to see why the graphs of some polar functions are conic sections.

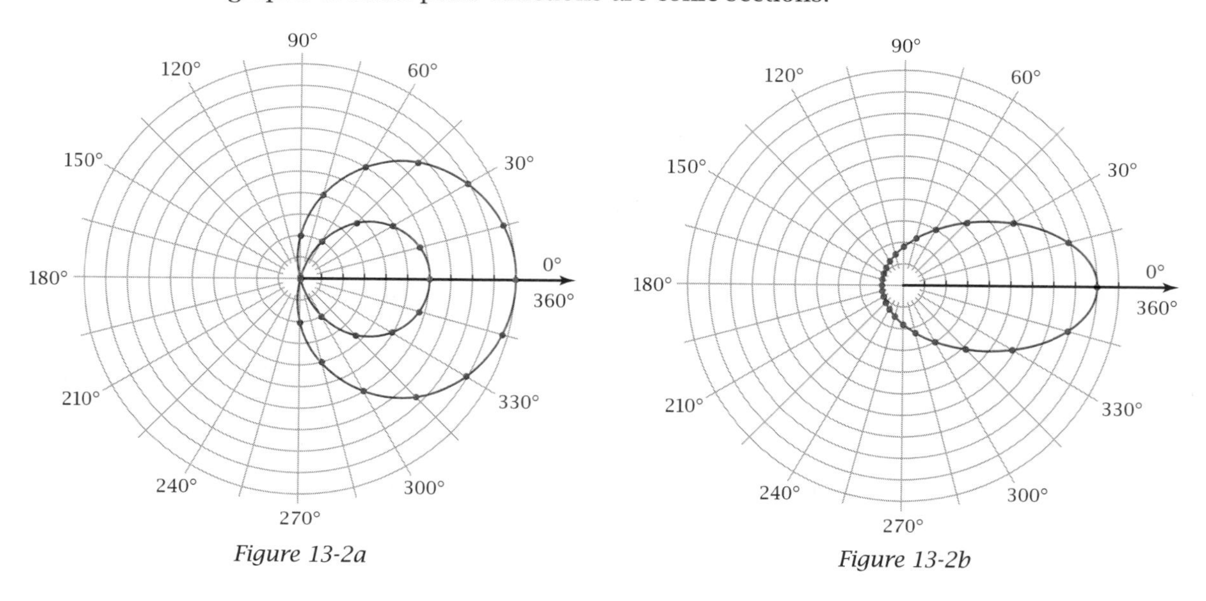

Figure 13-2a

Figure 13-2b

OBJECTIVES

- Given a polar equation, plot the graph.
- Given the polar equation of a conic section, transform it to Cartesian coordinates.

Background on Polar Coordinates

On the left in Figure 13-2c, a point is labeled with its Cartesian coordinates (x, y). In the middle, the same point is labeled with its polar coordinates (r, θ). In the polar coordinate system, the origin is called the **pole** and the positive horizontal axis is called the **polar axis.** The figure on the right shows the relationships among x, y, r, and θ.

Airplanes use the azimuth as the polar "axis" for collecting navigation information.

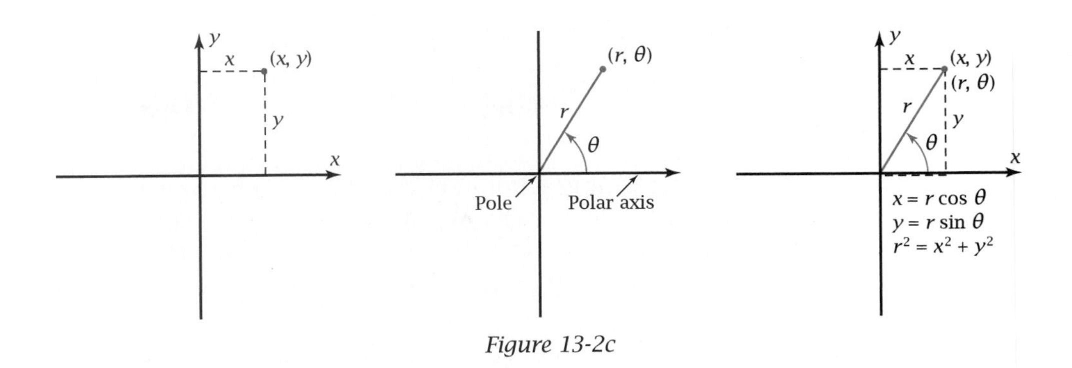

Figure 13-2c

DEFINITIONS: Polar Coordinates

The pole in the polar coordinate system is the same as the origin in the Cartesian coordinate system. The **polar axis** is shown in the same position as the positive x-axis in the Cartesian coordinate system.

A point in polar coordinates is written as an ordered pair (r, θ), where

- θ is the measure of an angle in standard position whose terminal side contains the point
- r (for "radius") is the directed distance from the pole to the point in the direction of the terminal ray of θ

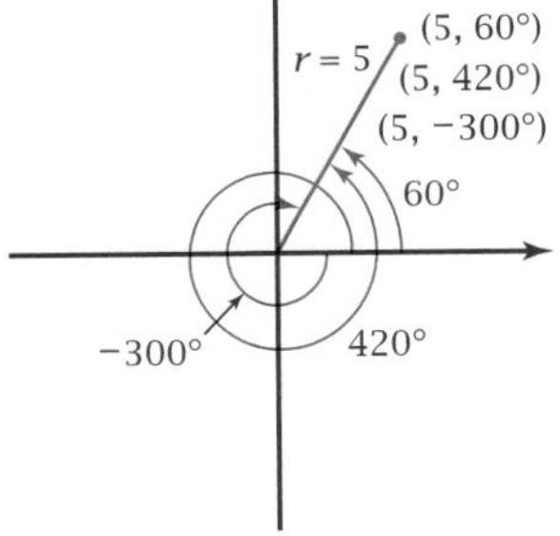

Many polar coordinates for the same point

Figure 13-2d

Note: Although the angle θ is the independent variable, it is customary to write ordered pairs as (r, θ) instead of (θ, r). Also, it is customary to use θ for the angle whether it is in degrees or radians.

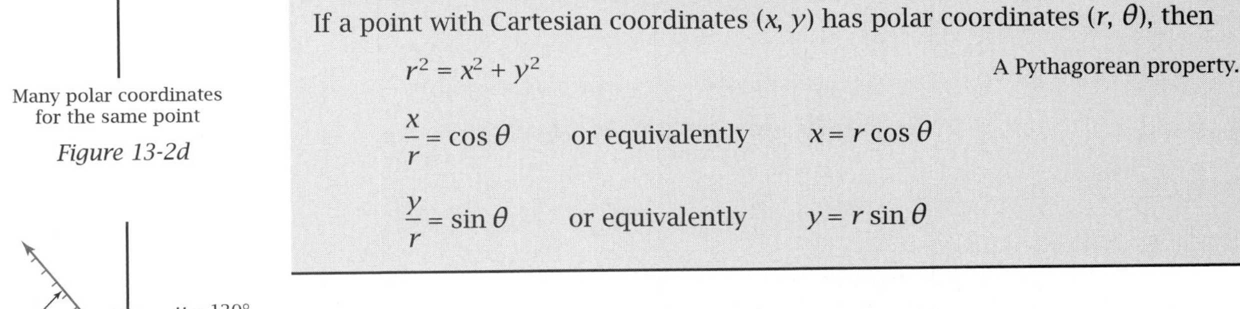

PROPERTIES: Polar and Cartesian Coordinates

If a point with Cartesian coordinates (x, y) has polar coordinates (r, θ), then

$r^2 = x^2 + y^2$ A Pythagorean property.

$\frac{x}{r} = \cos\theta$ or equivalently $x = r\cos\theta$

$\frac{y}{r} = \sin\theta$ or equivalently $y = r\sin\theta$

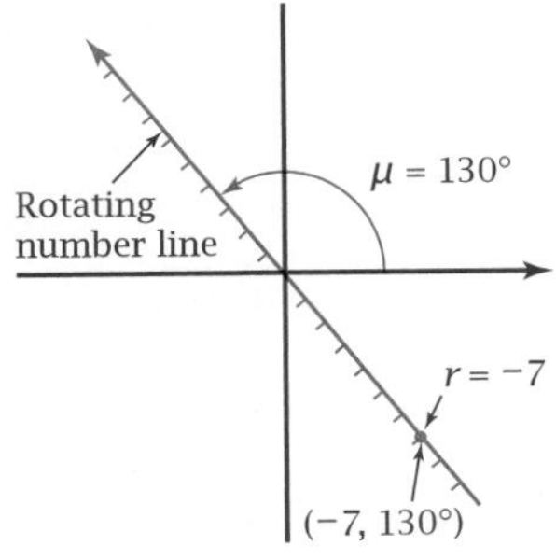

Figure 13-2e

Because an infinite number of coterminal angles pass through any given point, a point in polar coordinates may be represented by an infinite number of ordered pairs. Figure 13-2d shows three ordered pairs for the same point.

Points in polar coordinates can have negative values of r. To plot the point $(-7, 130°)$, imagine a number line rotating about the pole to an angle of 130° (Figure 13-2e). Because $r = -7$, measure 7 units in the negative direction along the rotated number line.

Graphs of Polar Equations

To plot the graph of a polar equation such as

$$r = 1 + 2\cos\theta$$

put your grapher in polar mode. Set the window so that the x- and y-axes have equal scales. You must also select a range for the independent variable θ. Usually a range of $[0°, 360°]$ with a θ-step of 5° or $[0, 2\pi]$ with a θ-step of 0.1 radian or $\frac{\pi}{48}$ radian will give a fairly smooth graph in a reasonable length of time.

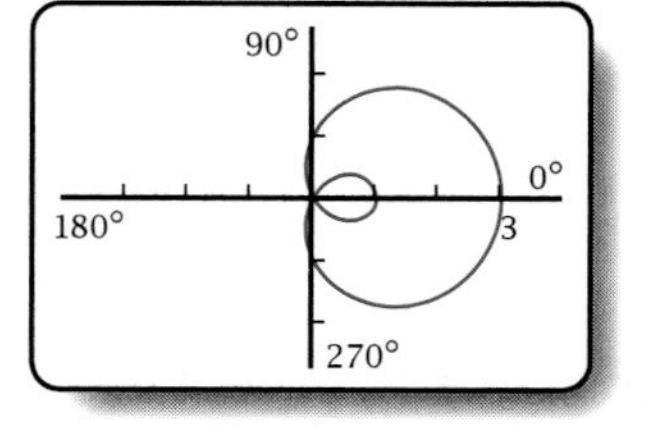

Figure 13-2f

Like the graph you plotted in Section 13-1, the graph of $r = 1 + 2\cos\theta$ in Figure 13-2f is a limaçon of Pascal. If you trace the graph, you'll find that the inner loop corresponds to θ-values between 120° and 240°. To see why, it helps to plot an **auxiliary Cartesian graph** (Figure 13-2g) of the equation

$$y = 1 + 2\cos\theta$$

The Cartesian graph reveals that when θ is between 120° and 240°, r is negative. So the points are plotted in the opposite direction (Figure 13-2h) for θ in this range. Note also that where the graph goes through the pole, it is tangent to the lines $\theta = 120°$ and $\theta = 240°$.

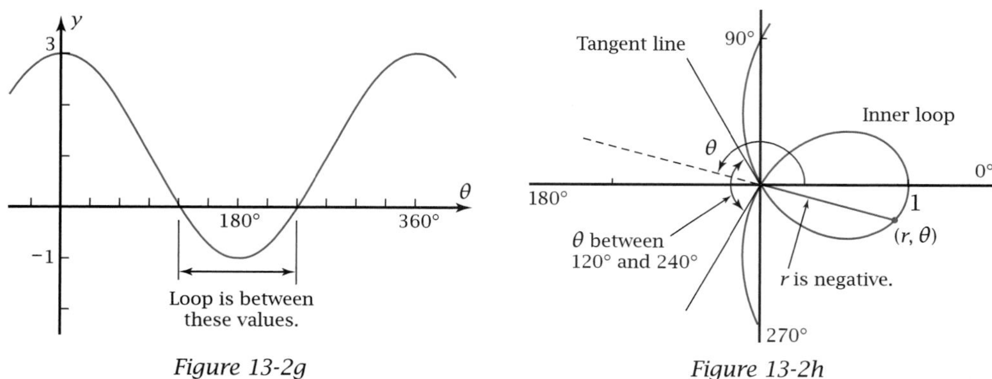

Figure 13-2g

Figure 13-2h

DEFINITION: Limaçon in Polar Coordinates

A **limaçon** is a figure with polar equation

$$r = a + b\cos\theta \qquad \text{or} \qquad r = a + b\sin\theta$$

If $|a| < |b|$, then the limaçon has an inner loop.

If $|a| > |b|$, then the limaçon has no inner loop.

If $|a| = |b|$, then the limaçon has a cusp at the pole and is called a **cardioid** (which means "heartlike").

A **cusp** is a point at which the graph of a relation changes direction abruptly.

▶ EXAMPLE 1 Plot the graph of the five-leaved rose $r = 6 \sin 5\theta$. Find numerically the first range of positive θ-values for which r is negative. Confirm your answer by plotting this part of the graph.

Solution The graph is shown in Figure 13-2i. Notice the five "leaves."

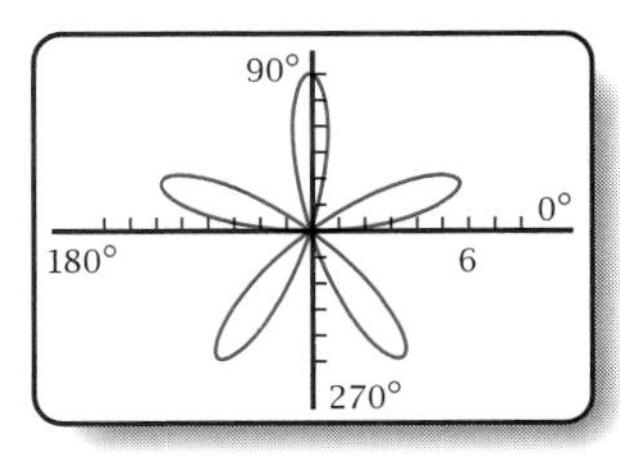

Figure 13-2i

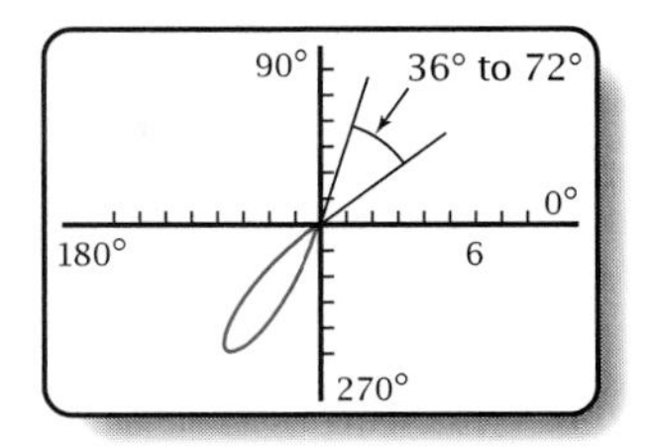

Figure 13-2j

θ	r
0°	0
15°	5.7955...
30°	3
45°	−4.24326...
60°	−5.1961...
75°	1.5529...

Make a table of values of θ and r starting at 0° and stepping by 15°. Negative values first occur at 45° and 60°. By narrowing your search you will find that r changes from positive to negative at 36° and from negative back to positive at 72°. So $36° < \theta < 72°$ is the first range of positive θ-values for which r is negative.

You can confirm the answer graphically as shown in Figure 13-2j or algebraically by solving $6 \sin 5\theta = 0$ to get $\theta = 0°, 36°, 72°, 108°, \ldots$ ◀

Note that the icon at the top of even-numbered pages in this chapter is a five-leaved rose.

Conics in Polar Coordinates

If you graph the reciprocal of the polar equation of a limaçon, the result is, surprisingly, a conic section! Figure 13-2k shows two examples.

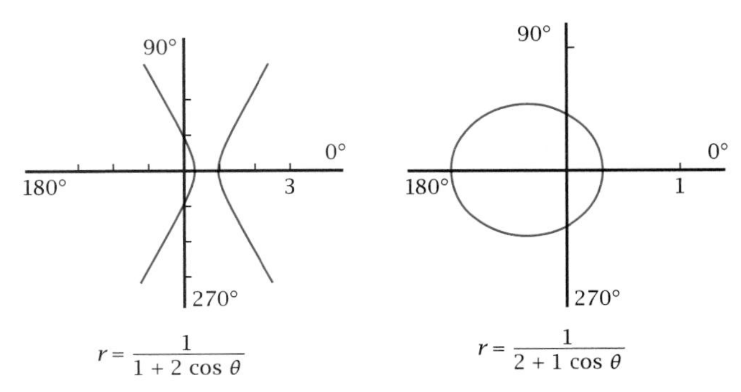

Figure 13-2k

▶ EXAMPLE 2 Plot the graph of $r = \dfrac{9}{5 - 4\cos\theta}$.

Show algebraically that the graph is an ellipse.

Solution The graph (Figure 13-2l) has an elliptical shape.

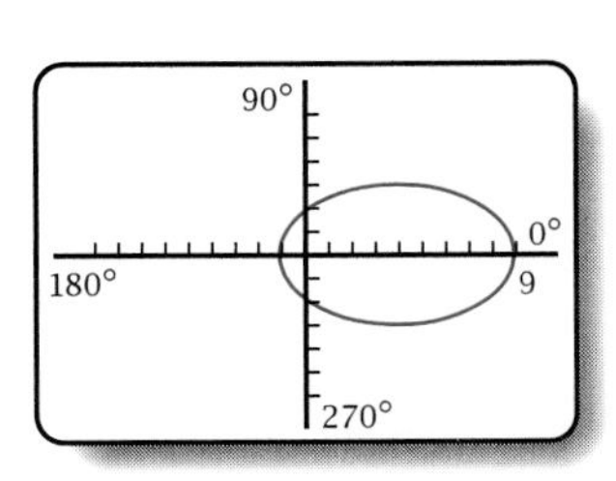

Figure 13-2l

To verify that the graph is an ellipse, write the equation in Cartesian form.

$5r - 4r\cos\theta = 9$	Eliminate the fraction by multiplying by $5 - 4\cos\theta$.
$5\sqrt{x^2 + y^2} - 4x = 9$	Substitute $\sqrt{x^2 + y^2}$ for r and x for $r\cos\theta$.
$5\sqrt{x^2 + y^2} = 4x + 9$	Isolate the radical term on one side of the equation.
$25x^2 + 25y^2 = 16x^2 + 72x + 81$	Square both sides to eliminate the radical term.
$9x^2 + 25y^2 - 72x - 81 = 0$	
Therefore, the graph is an ellipse.	x^2 and y^2 have the same sign but different coefficients.

One focus of the ellipse in Example 2 is at the pole. The eccentricity is $\left|\frac{-4}{5}\right|$, or 0.8, the absolute value of the ratio of the coefficients in the denominator of the polar equation. The general properties of conics with one focus at the pole are summarized in the box.

PROPERTY: Conic Sections in Polar Coordinates

The general polar equation of a conic section with one focus at the pole is

$$r = \frac{k}{a + b\cos\theta} \quad \text{or} \quad r = \frac{k}{a + b\sin\theta}$$

The eccentricity of the conic is $e = \left|\frac{b}{a}\right|$.

$|k| = |aep|$, where p is the distance between the focus and the directrix.

- If $|b| < |a|$, then the graph is an ellipse ($e < 1$).
- If $|b| > |a|$, then the graph is a hyperbola ($e > 1$).
- If $|b| = |a|$, then the graph is a parabola ($e = 1$).

This box summarizes the relationship between conics and limaçons.

PROPERTY: Relationship Between Conics and Limaçons

The r-values for a conic section are reciprocals of the r-values for a limaçon.

- If the limaçon has a loop, then the conic is a hyperbola.
- If the limaçon has no loop, then the conic is an ellipse.
- If the limaçon is a cardioid, then the conic is a parabola.

Polar Equations of Special Circles and Lines

Figure 13-2m shows a circle of radius 9 units centered on the polar axis passing through the pole. Point (r, θ) is on the circle. The triangle shown is a right triangle because the angle at (r, θ) is inscribed in the semicircle. By the definition of cosine, $\frac{r}{9} = \cos\theta$. So a polar equation for the circle is

$$r = 9\cos\theta$$

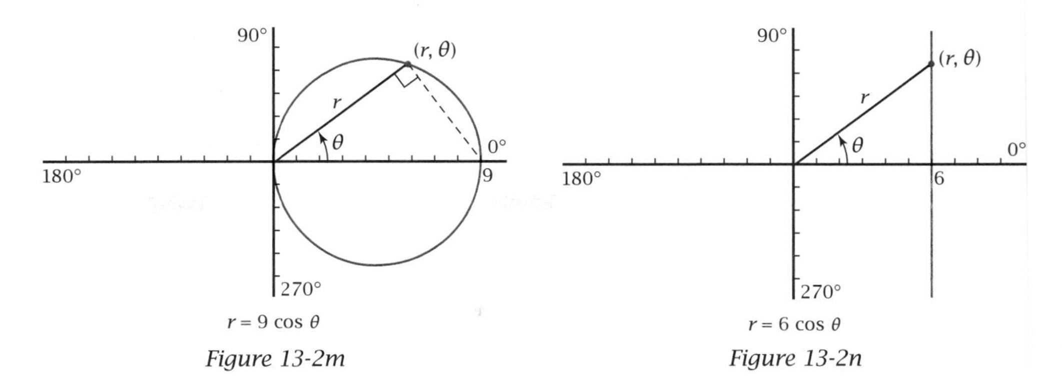

Figure 13-2m *Figure 13-2n*

Similarly, the graph of $r = 9 \sin \theta$ is a circle passing through the pole centered on the line $\theta = 90°$.

Figure 13-2n shows a point (r, θ) on the line perpendicular to the polar axis and 6 units from the pole. In the right triangle formed by r, the line, and the polar axis, $\frac{r}{6} = \sec \theta$. So a polar equation for the line is

$$r = 6 \sec \theta$$

The graph of $r = 6 \csc \theta$ is a line parallel to the polar axis and 6 units from the pole. This box contains a summary of the equations for these special circles and lines.

PROPERTIES: Polar Equations of Special Circles and Lines

The equations are for the circles and lines shown.

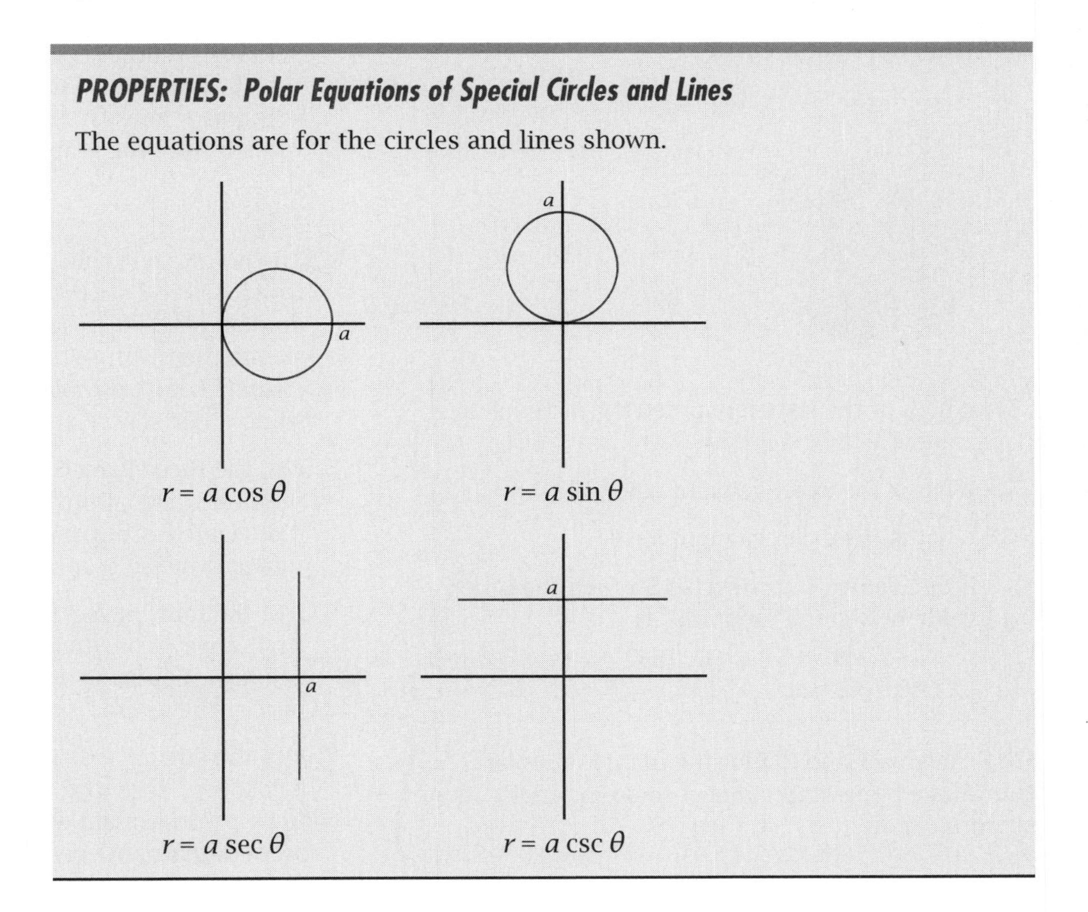

Problem Set 13-2

Do These Quickly

For Q1–Q7, refer to Figure 13-2o.

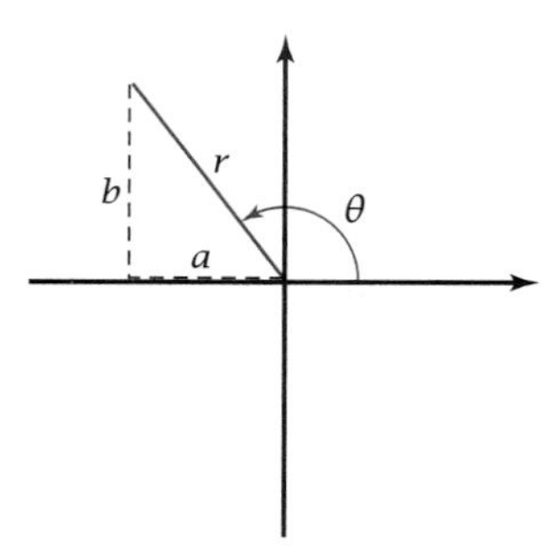

Figure 13-2o

Q1. What does sin θ equal?

Q2. What does cos θ equal?

Q3. What does tan θ equal?

Q4. What does cot θ equal?

Q5. What does sec θ equal?

Q6. What does csc θ equal?

Q7. Which of the six trigonometric functions of angle θ will be negative?

Q8. What is the exact value of cos 150°?

Q9. What is the exact value of $\sin \frac{\pi}{4}$?

Q10. If one value of arcsin x is 15°, then the other value between 0° and 360° is

A. 75° B. 105° C. 165° D. 195°
E. None of these

For Problems 1 and 2, plot the points on polar coordinate paper and connect them in order with a smooth curve.

1.

θ	r	θ	r
0°	0	105°	−4.8
15°	1.3	120°	−7.5
30°	4.3	135°	−7.1
45°	7.1	150°	−4.3
60°	7.5	165°	−1.3
75°	4.8	180°	0
90°	0		

2.

θ	r	θ	r
0°	0	105°	−7.2
15°	0.1	120°	−3.0
30°	0.6	135°	−1.4
45°	1.4	150°	−0.6
60°	3.0	165°	−0.1
75°	7.2	180°	0
90°	(infinite)		

3. The points in Problem 1 are for the **bifolium** $r = 20 \cos \theta \sin^2 \theta$. Plot the graph in the domain [0°, 360°]. Trace to $\theta = 225°$ and sketch the result. What is happening to the graph as θ goes from 180° to 360°? Why do you suppose the graph is called a *bifolium*?

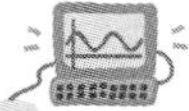

4. The points in Problem 2 are for the **cissoid of Dioclese** $r = 2 \sin \theta \tan \theta$. Plot the graph in the domain [0°, 360°]. Trace to $\theta = 240°$ and sketch the result. What is the root of the word *cissoid*? Look up *Dioclese* on the Internet or some other source and tell what you find.

5. Plot the **three-leaved rose** $r = 10 \sin 3\theta$. Find the first range of positive θ-values for which r is negative. Confirm your answer by plotting this part of the graph.

6. Plot the **four-leaved rose** $r = 5 \cos 2\theta$. Find the first range of positive θ-values for which r is negative. Confirm your answer by plotting this part of the graph.

7. Plot the circle $r = 6 \sin \theta$. Use a domain of [0°, 360°]. Trace to $\theta = 300°$. Draw a sketch to help you explain why the point is in the second quadrant even though 300° is a fourth-quadrant angle.

8. Plot the line $r = 2 \csc \theta$. Use a domain of $[0°, 360°]$. Trace to $\theta = 240°$. Draw a sketch to help you explain why the point is in the first quadrant, even though 240° is a third-quadrant angle.

9. Show algebraically that the graph of $r = 6 \sin \theta$ in Problem 7 is a circle through the pole.

10. Show algebraically that the graph of $r = 2 \csc \theta$ in Problem 8 is the line $y = 2$.

11. Figure 13-2p shows the graph of $r = 9 \cos \frac{\theta}{2}$. Plot the graph using two revolutions. Find the first range of positive values of θ for which r is negative. Sketch this part of the graph.

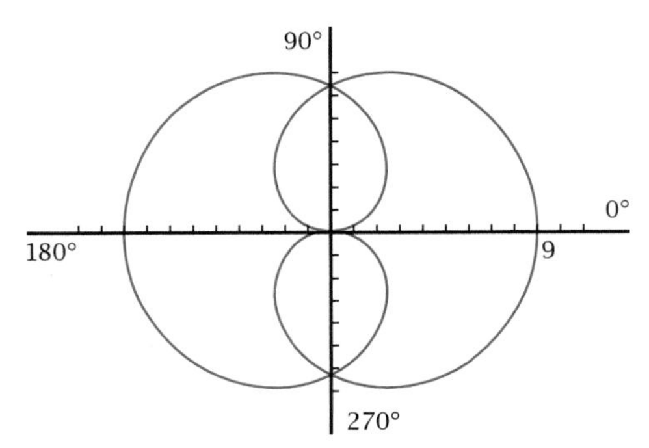

Figure 13-2p

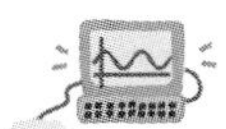

12. Figure 13-2q shows a **lemniscate of Bernoulli,** $r = \sqrt{9 \cos 2\theta}$. Plot the graph on your grapher. Describe the behavior of the graph as it goes to the pole. Explain why there are values of θ for which there is no graph. Explain why there are no negative values of r. Look up the name *Bernoulli* on the Internet or in some other source. See how many different members of the Bernoulli family you can find. Reveal approximately when each person lived and in what country.

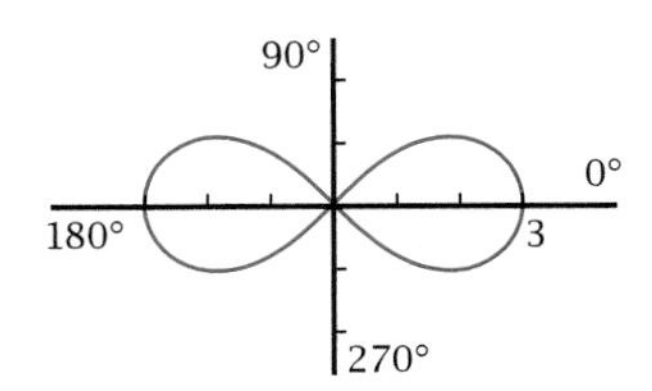

Figure 13-2q

13. Figure 13-2r shows the first three revolutions of the **spiral** $r = \frac{2\theta}{\pi}$, where θ is measured in radians. Plot the graph on your grapher in the domain $[0, 6\pi]$. At what three positive values of r does the graph cross the polar axis? Extend the domain to $[-6\pi, 6\pi]$. Describe what you see on the grapher.

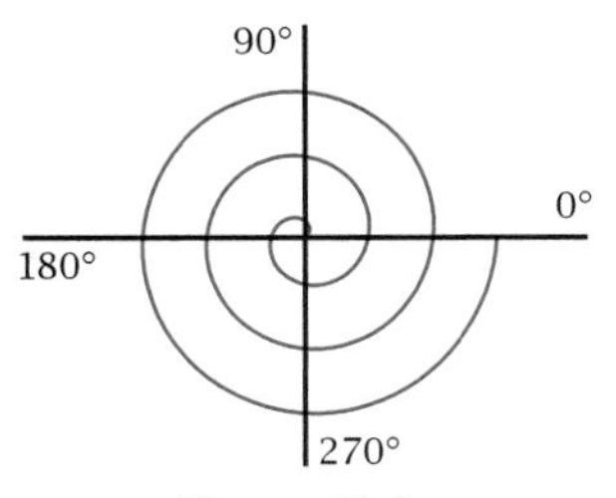

Figure 13-2r

14. Figure 13-2s shows a **conchoid of Nicomedes,** $r = 8 + 3 \csc \theta$. Plot the graph on your grapher. Use a range for θ of $[0°, 360°]$ with a θ-step of 5°. What happens to the graph as θ approaches 180° and 360°? At the point P shown in the figure, is r positive or negative? What range of θ-values generates the loop below the horizontal axis? Look up the name *Nicomedes* on the Internet or in some other source. Describe what you find out.

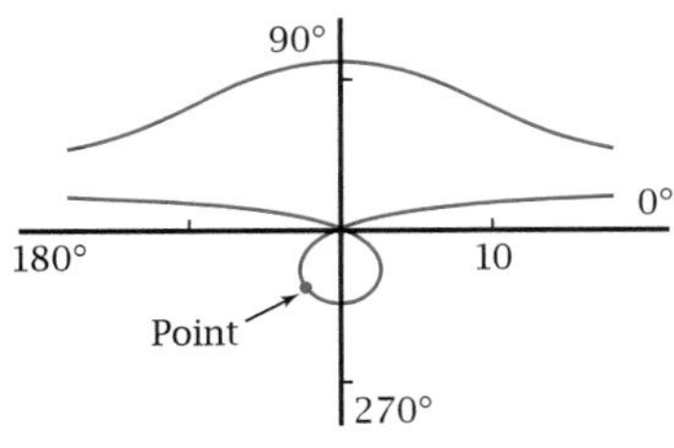

Figure 13-2s

15. *Circles Problem:* Each of the graphs in Figure 13-2t is a circle of diameter 1 unit passing through the pole. Write the polar equation of each circle.

a.

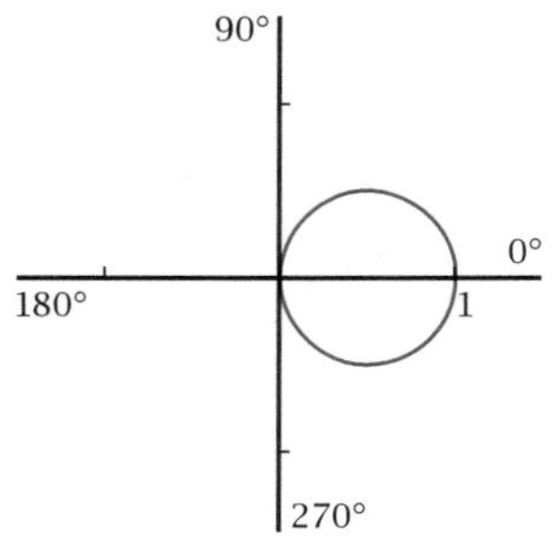

b.

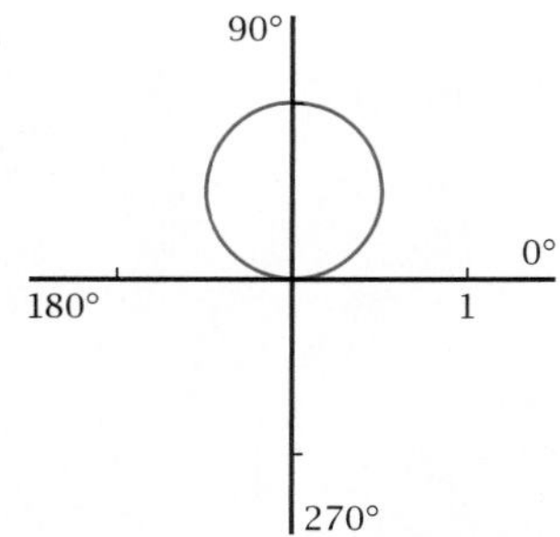

c.

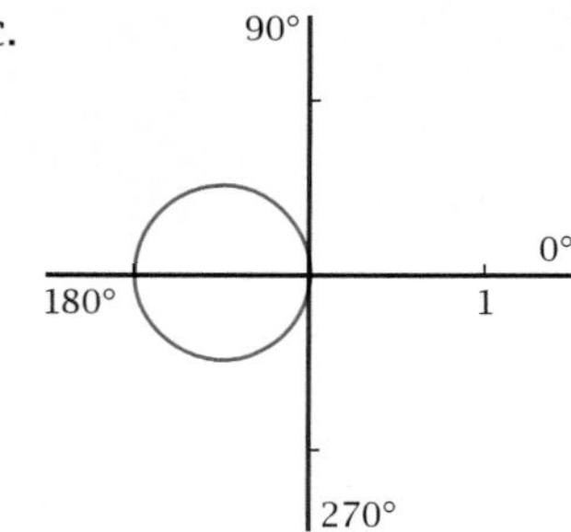

d.

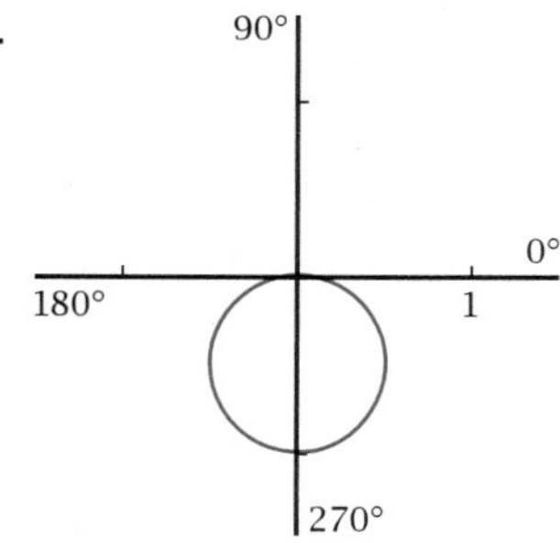

Figure 13-2t

16. *Lines Problem:* Each of the graphs in Figure 13-2u is a line 1 unit from the pole. Write the polar equation of each line.

a.

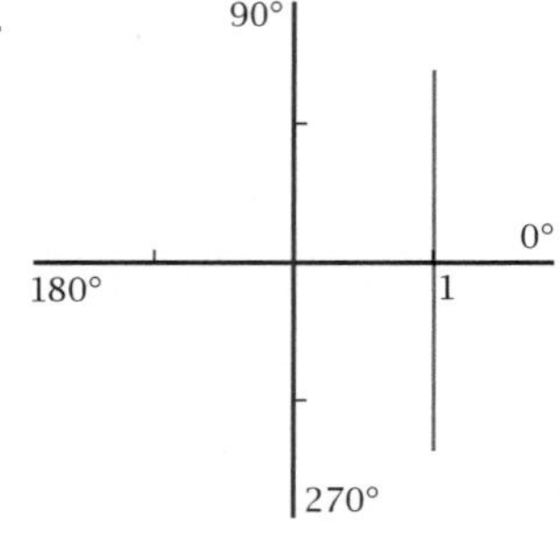

b.

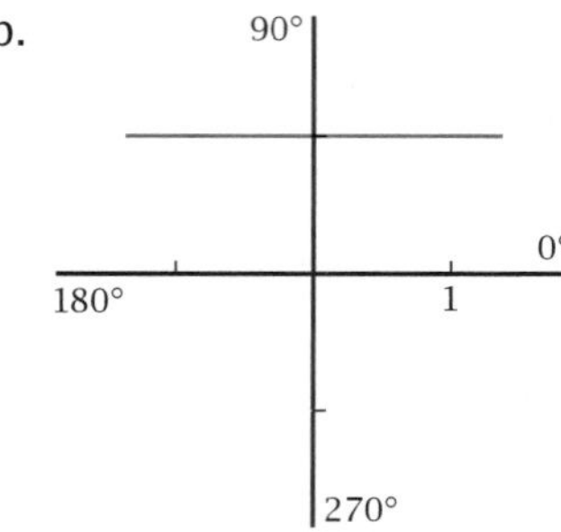

c.

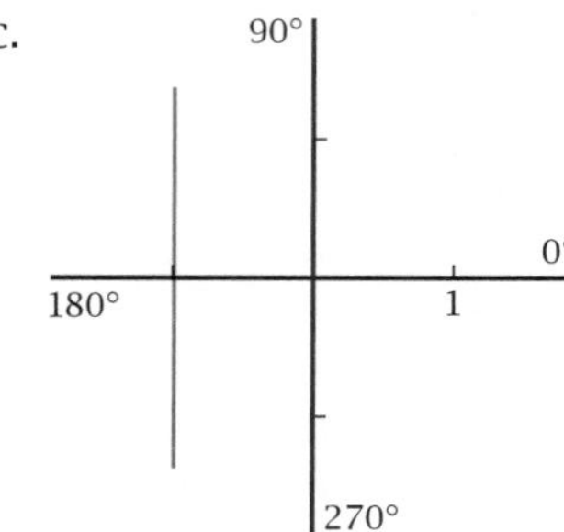

d.

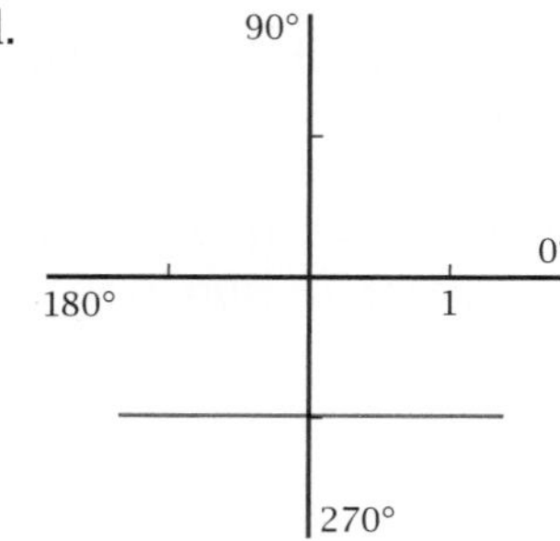

Figure 13-2u

17. *Hyperbola Problem 2:* Consider the polar equation $r = \frac{8}{3 + 5 \cos \theta}$.

    a. Plot the graph. Sketch the result.

    b. Show algebraically that the graph is a hyperbola by transforming the equation to Cartesian form.

    c. Where is one focus of the hyperbola? What is the eccentricity?

18. *Ellipse Problem 2:* Consider the polar equation $r = \frac{10}{3 + 2 \sin \theta}$.

    a. Plot the graph. Sketch the result.

    b. Show algebraically that the graph is an ellipse by transforming the equation to Cartesian form.

    c. Where is one focus of the ellipse? What is the eccentricity?

19. *Parabola Problem 2:* Consider the polar equation $r = \frac{6}{1 + \cos \theta}$.

    a. Plot the graph. Sketch the result.

    b. Show algebraically that the graph is a parabola by transforming the equation to Cartesian form.

    c. Where is the focus? How can you tell from the equation that the eccentricity equals 1?

20. *Rotated Polar Graphs Problem:* Figure 13-2v shows two ellipses,

$$r = \frac{19}{10 - 9 \cos \theta} \quad \text{and}$$

$$r = \frac{19}{10 - 9 \cos (\theta - 30°)}$$

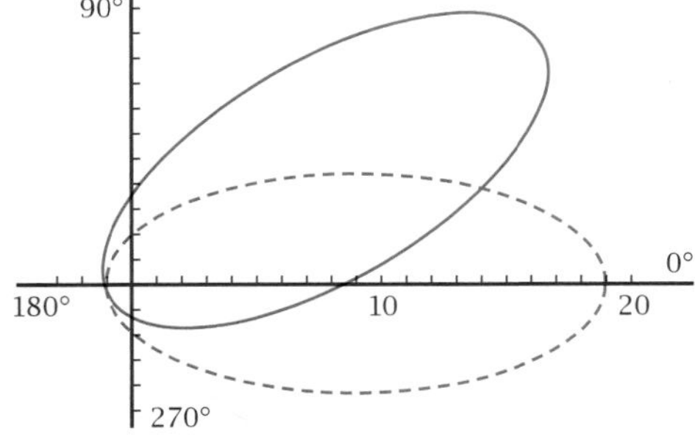

Figure 13-2v

    a. Which graph is which? What is the effect of subtracting 30° from θ in the second equation?

    b. Write an equation that would rotate the dashed graph by 90° counterclockwise. Confirm that your answer is correct by plotting the graph.

    c. Explain the difference in the effect of subtracting a constant from θ in polar coordinates and subtracting a constant from θ in Cartesian coordinates.

    d. The graph of $r = 3 \sec \theta$ is a line perpendicular to the polar axis 3 units from the pole. Write an equation that would rotate the line 60° clockwise. Plot the graph. Where does the line cross the polar axis? Is the line still a perpendicular distance of 3 units from the pole? How can you tell?

21. *Roller Skating Problem:* Figure 13-2w shows a roller skating loop as it appears in a manual of the Roller Skating Rink Operations of America. The figure is composed of arcs of circles that are easy to mark on the rink floor. The finished figure resembles a limaçon with a loop. Find a polar equation for the limaçon. Confirm your answer by plotting the limaçon on your grapher.

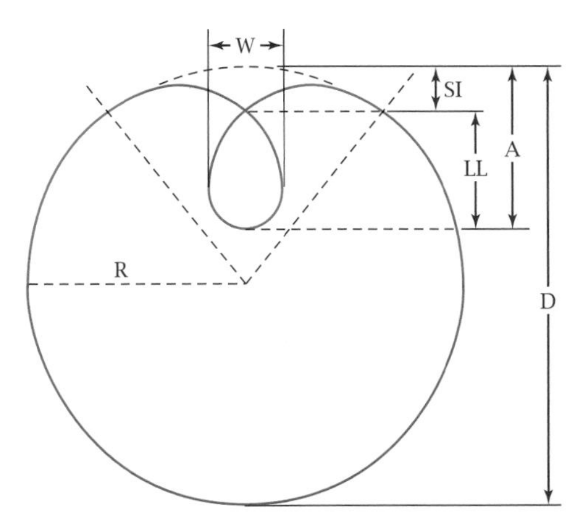

Figure 13-2w

Loop Dimensions

D	240 cm
SI ($\frac{1}{8}$ D)	30 cm
LL ($\frac{1}{4}$ D)	60 cm
A	90 cm
W ($\frac{1}{6}$ D)	40 cm

22. *Rose Problem:* The general equation of a rose in polar coordinates is

$$r = k \cos n\theta$$

where n is an integer.

a. Plot the four-leaved rose $r = 9 \cos 2\theta$.

b. Plot the three-leaved rose $r = 9 \cos 3\theta$.

c. What is the relationship between n and the number of leaves in the rose? You may want to plot other roses to confirm your conclusion.

d. Plot the five-leaved rose that appears in the icon at the top of even-numbered pages in this chapter. Make the distance from the pole to the tip of a leaf equal to 7 units.

13-3 Intersections of Polar Curves

The limaçon $r_1 = 3 + 2 \cos \theta$ and the four-leaved rose $r_2 = 5 \sin 2\theta$ (Figure 13-3a) appear to intersect at eight points, for instance, P_1 and P_2. In this section you will discover that P_1 is an intersection point but that P_2 is not. In fact, only four of the eight points are really intersections. The other four are points where the graphs cross but for different values of θ.

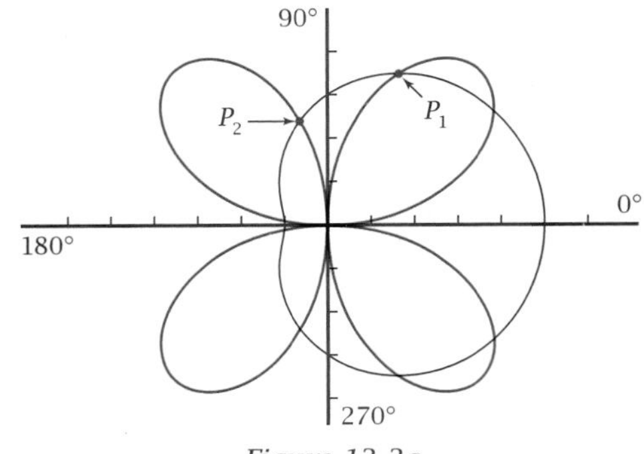

Figure 13-3a

OBJECTIVE Given two polar curves, find the intersection points.

If you plot the equations $r_1 = 3 + 2 \cos \theta$ and $r_2 = 5 \sin 2\theta$ with your grapher in simultaneous mode, the two graphs will be drawn at the same time.

Figure 13-3b shows the effect of pausing at about 65°. Both graphs are at point P_1 for a value of θ close to 65°. But if you continue plotting to about 105° (Figure 13-3c), the limaçon is at point P_2 and the rose is at point P_3. Because the rose has a negative value of r and the limaçon has a positive value of r at this angle, point P_2 is not an intersection point.

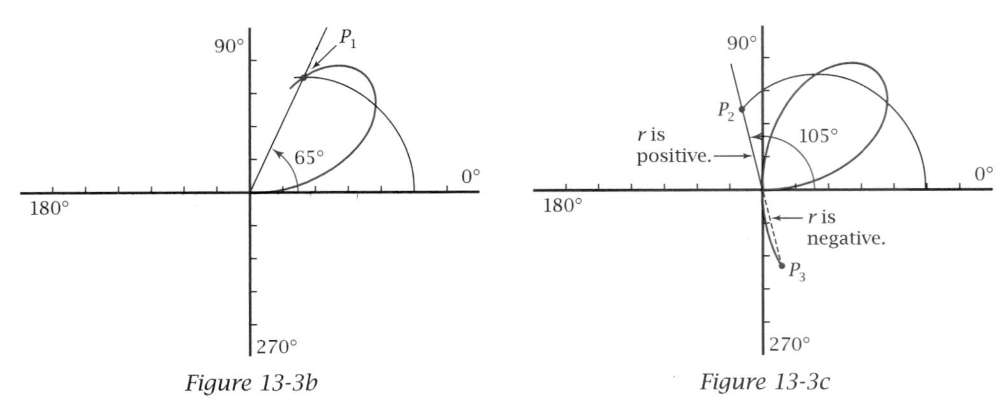

Figure 13-3b

Figure 13-3c

▶ EXAMPLE 1 Find the intersection points of the limaçon $r_1 = 3 + 2\cos\theta$ and the four-leaved rose $r_2 = 5\sin 2\theta$.

Solution To see which points are actually intersections, it helps to plot auxiliary Cartesian graphs. Figure 13-3d shows graphs of $y_1 = 3 + 2\cos\theta$ and $y_2 = 5\sin 2\theta$. The true intersections occur where the auxiliary graphs intersect. You can find the precise values using the intersect feature.

$$(4.6529..., 34.2630...°), (3.8511..., 64.8126...°), (1.0103..., 185.8289...°), (2.4855..., 255.0953...°)$$

These true intersection points are marked in Figure 13-3e.

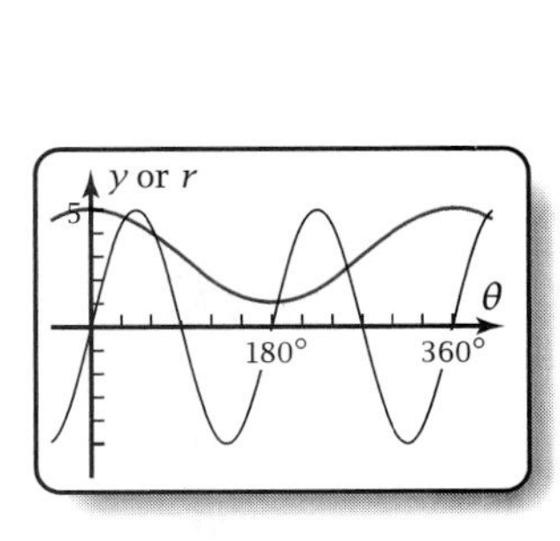

Figure 13-3d

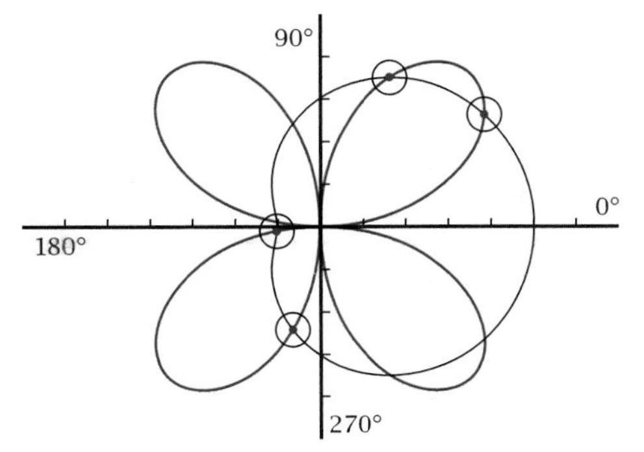

Figure 13-3e

◀

Problem Set 13-3

Do These Quickly

Q1. Find the Cartesian coordinates of the polar point $(r, \theta) = (6, 30°)$.

Q2. Find polar coordinates of the Cartesian point $(x, y) = (3, 7)$.

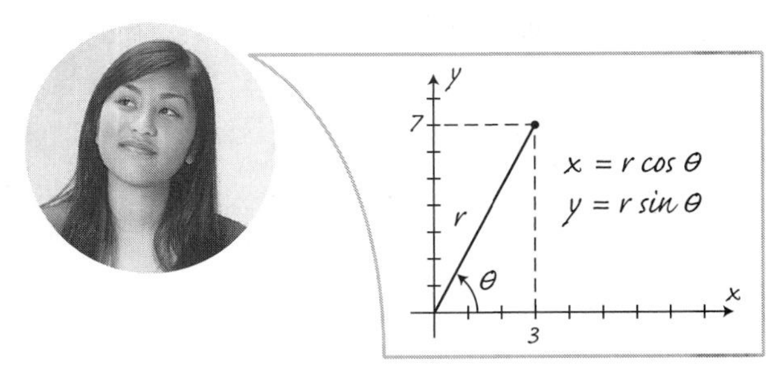

For Q3–Q6, write another ordered pair for the polar point (6, 30°) with the given specifications.

Q3. r and θ both positive.

Q4. r positive and θ negative.

Q5. r negative and θ positive.

Q6. r and θ both negative.

Q7. What geometric figure has the polar equation $r = 5(3 + 4\cos\theta)$?

Q8. What geometric figure has the polar equation $r = \frac{5}{3 + 4\cos\theta}$?

Q9. Write the polar equation of the circle in Figure 13-3f.

Q10. Write parametric equations for the circle in Figure 13-3f.

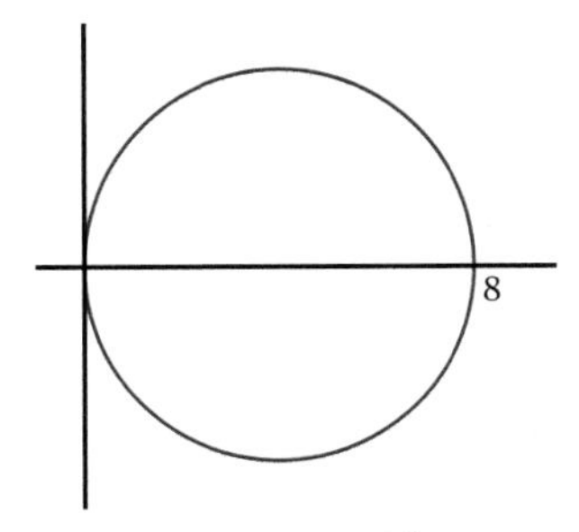

Figure 13-3f

For Problems 1–8, find the coordinates of the intersection points and mark them on a sketch of the graphs.

1. Ellipse $r_1 = \frac{5}{3 - 2\cos\theta}$ and hyperbola $r_2 = \frac{5}{2 + 3\cos\theta}$ (Figure 13-3g)

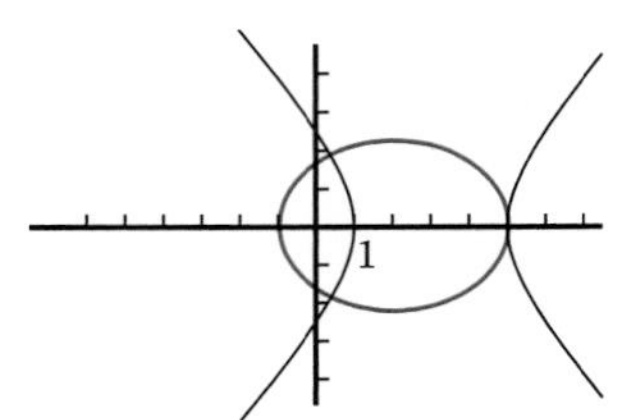

Figure 13-3g

2. Limaçon $r_1 = 2 + 3\cos\theta$ and ellipse $r_2 = \frac{5}{3 + 2\cos\theta}$ (Figure 13-3h)

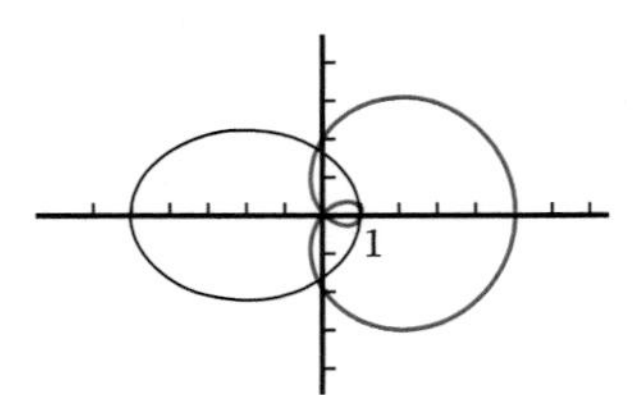

Figure 13-3h

3. Limaçon $r_1 = -1 - 5\cos(\theta - 180°)$ and limaçon $r_2 = 3 + 2\cos\theta$ (Figure 13-3i)

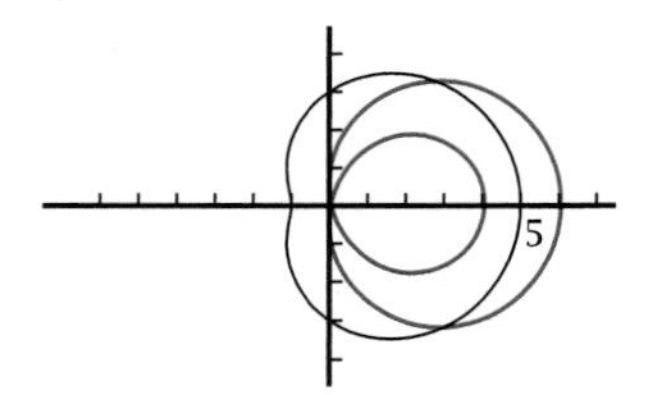

Figure 13-3i

4. Limaçon $r_1 = 1 + 5\cos\theta$ and line $r_2 = 3\sec(\theta - 30°)$ (Figure 13-3j)

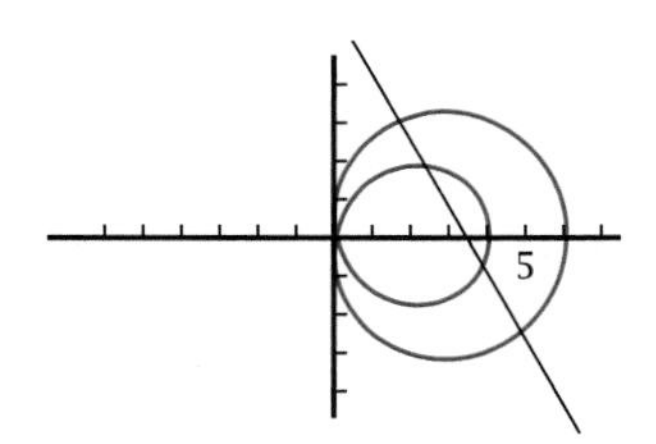

Figure 13-3j

5. Conchoid $r_1 = 3 + \csc\theta$ and ellipse $r_2 = \frac{5}{3 - 2\sin\theta}$ (Figure 13-3k)

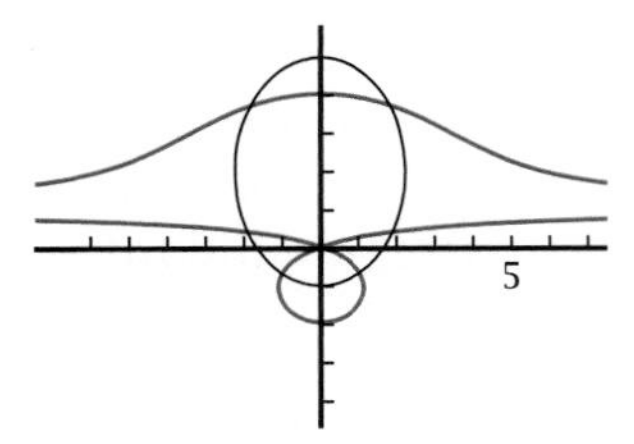

Figure 13-3k

6. Limaçon $r_1 = 3 - 2\sin\theta$ and rose $r_2 = 5\cos 2\theta$ (Figure 13-3l)

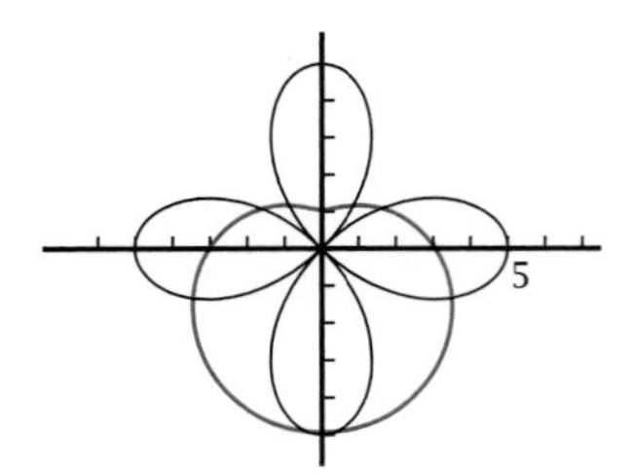

Figure 13-3l

7. Spiral $r_1 = 0.5\theta$ and rose $r_2 = 5\cos 2\theta$ (Figure 13-3m). Use radian mode and positive values of θ.

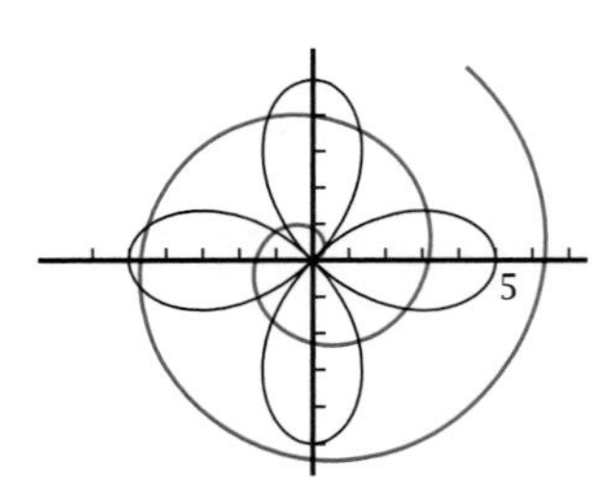

Figure 13-3m

8. Spiral $r_1 = 0.5\theta$ and circle $r_2 = 5\cos\theta$ (Figure 13-3n). Use radian mode and positive values of θ.

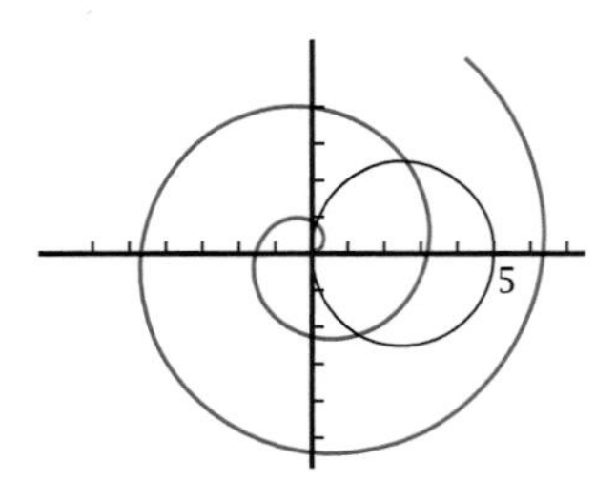

Figure 13-3n

9. *False Intersection Point Problem:* Plot on the same screen the limaçons

$r_1 = 3 + 2\sin\theta$

$r_2 = -3 - 2\sin(\theta - 180°)$

Use simultaneous mode and path style. Describe what you observe about the point at which each graph is being plotted and the final graphs. Tracing to a particular value of θ on the final graphs might help. Then plot the circle $r_3 = 4$. Show that the false intersection of r_3 with r_2 is the true intersection of r_3 with r_1. From what you have observed, write instructions that could be used to find the false intersections of two polar curves.

13-4 Complex Numbers in Polar Form

An **imaginary number** is the square root of a negative number. A **complex number** is the sum of a real number and an imaginary number. For instance, the complex number $z = 4 + \sqrt{-9}$ is the sum of the real number 4 and the imaginary number $\sqrt{-9}$. Using i for the **unit imaginary number** $\sqrt{-1}$, you can write

$$z = 4 + i\sqrt{9} = 4 + 3i$$

You can represent complex numbers by points in a Cartesian coordinate system called the **complex plane,** as shown in Figure 13-4a. The horizontal coordinate of a point represents the real part of the number, and the vertical coordinate represents the imaginary part.

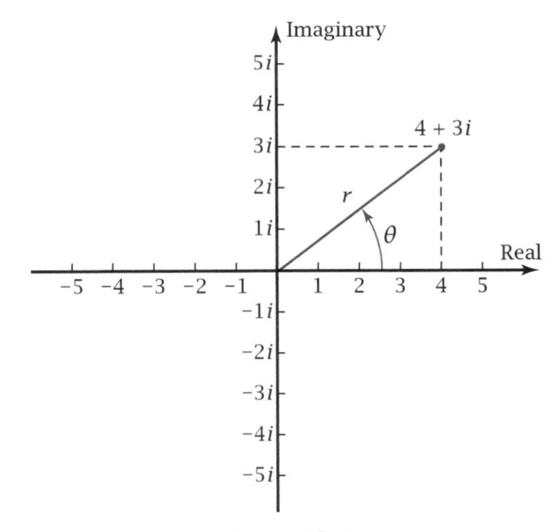

Figure 13-4a

You can write a complex number in **polar form** by writing the real and imaginary parts in terms of the polar coordinates of the point. In this section you will make remarkable discoveries about products and roots of complex numbers by writing the numbers in polar form.

OBJECTIVE Operate with complex numbers in polar form.

These definitions apply to complex numbers in Cartesian form.

DEFINITIONS: Imaginary and Complex Numbers

$i = \sqrt{-1}$	The **unit imaginary number.**
$i^2 = -1$	Squaring a square root removes the radical sign.
$z = a + bi$	General form of a complex number in Cartesian form. The real number a is called the **real part** of z. The real number b is called the **imaginary part** of z.
$a - bi$	The **complex conjugate** of $a + bi$.

You operate with complex numbers in the same way you operate with other binomials. For example, to add or subtract complex numbers, combine like terms:

$$(4 + 3i) - (5 - 2i) = 4 + 3i - 5 + 2i$$
$$= -1 + 5i$$

To multiply complex numbers, expand and combine like terms:

$$(4 + 3i)(5 - 2i) = 20 - 8i + 15i - 6i^2$$
$$= 20 + 7i - 6(-1)$$
$$= 26 + 7i$$

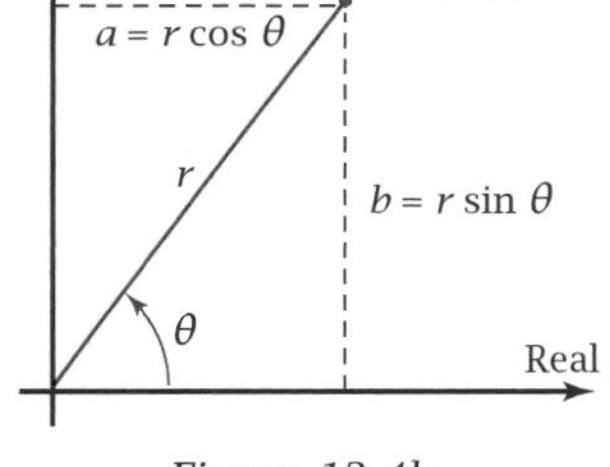

Figure 13-4b

Figure 13-4b shows the number $z = a + bi$ in the complex plane. The polar coordinates of z are (r, θ).

By the definitions of cosine and sine,

$$a = r\cos\theta \qquad \text{and} \qquad b = r\sin\theta$$

Therefore, z can be written

$$z = (r\cos\theta) + i(r\sin\theta) \qquad \text{or} \qquad z = r(\cos\theta + i\sin\theta) \qquad \text{Factor out the } r.$$

The expression $(\cos\theta + i\sin\theta)$ is written "cis θ," pronounced "sis" as in "sister." The *c* comes from *cosine,* the *i* from the unit imaginary number, and the *s* from *sine.* Thus, any complex number can be written

$$z = r \text{ cis } \theta$$

DEFINITION: Polar Form of a Complex Number

$$z = r \text{ cis } \theta = r(\cos\theta + i\sin\theta)$$

where r is called the **modulus** (or magnitude) of z and θ is called the **argument** of z (either degrees or radians).

The **absolute value** of a complex number is the modulus of that number:

$$|z| = r$$

These relationships let you convert between the polar and Cartesian forms of a complex number.

PROPERTY: Relationships Between Polar and Cartesian Form

If $z = a + bi = r \text{ cis } \theta$, then

$$a = r\cos\theta$$

$$b = r\sin\theta$$

$$r^2 = a^2 + b^2$$

EXAMPLE 1 Transform $z = -5 + 7i$ to polar form.

Solution

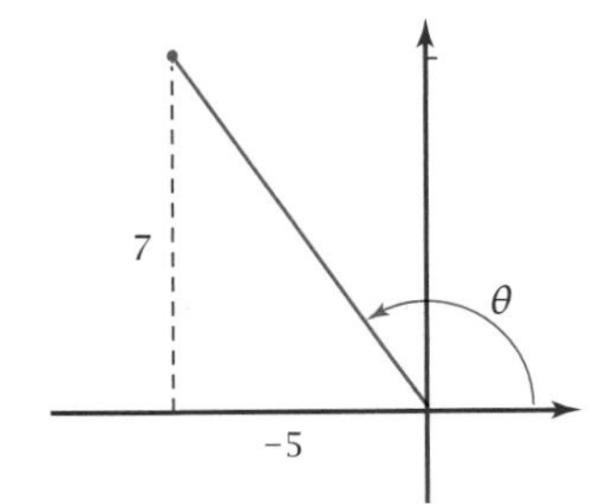

Figure 13-4c

Sketch the number on the complex plane (Figure 13-4c).

$r = \sqrt{(-5)^2 + 7^2} = \sqrt{74}$ Use only the coefficient 7, not $7i$.

$\theta = \arctan \dfrac{7}{-5} = -54.4623...° + 180n° = 125.5376...°$

θ is in Quadrant II.

$\therefore z = \sqrt{74} \text{ cis } (125.5376...°)$

EXAMPLE 2 Transform $z = 5 \text{ cis } 144°$ to Cartesian form.

Solution

$z = 5(\cos 144° + i \sin 144°)$ Definition of cis.

$= -4.0450... + (2.9389...)i$

$\approx -4.05 + 2.94i$

EXAMPLE 3 If $z_1 = 3 \text{ cis } 83°$ and $z_2 = 2 \text{ cis } 41°$, find the product $z_1 z_2$.

Solution

Long Way

$z_1 z_2 = 3(\cos 83° + i \sin 83°) \cdot 2(\cos 41° + i \sin 41°)$ Definition of cis θ.

$= (3)(2)[\cos 83° \cos 41° + i(\sin 83° \cos 41°) + i(\cos 83° \sin 41°)$
$+ i^2(\sin 83° \sin 41°)]$

$$= 6[\cos 83° \cos 41° - \sin 83° \sin 41°)$$
$$+ i(\sin 83° \cos 41° + \cos 83° \sin 41°)]$$

$$= 6[\cos (83° + 41°) + i \sin (83° + 41)]$$

Composite argument properties of Chapter 5.

$$= 6 \text{ cis } 124°$$

Short Way

$$z_1 z_2 = (3 \text{ cis } 83°)(2 \text{ cis } 41°)$$

$$= 6 \text{ cis } 124°$$ Multiply the moduli; add the arguments. ◀

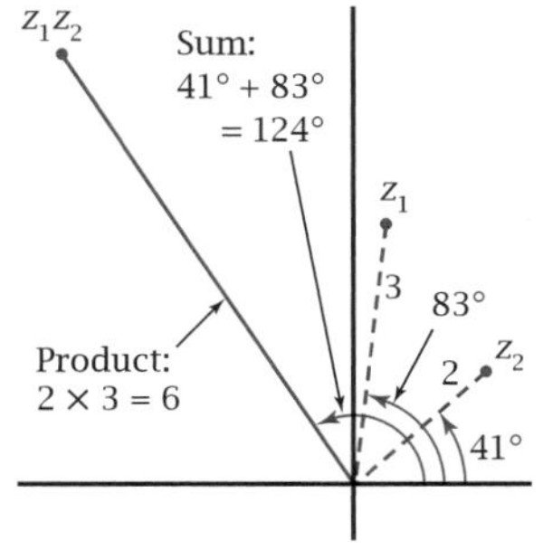

Figure 13-4d

PROPERTY: Product of Two Complex Numbers in Polar Form

If $z_1 = r_1 \text{ cis } \theta_1$ and $z_2 = r_2 \text{ cis } \theta_2$, then

$$z_1 z_2 = r_1 r_2 \text{ cis } (\theta_1 + \theta_2)$$

Verbally: "Multiply the moduli; add the arguments."

Figure 13-4d illustrates this property using the complex numbers in Example 3.

▶ ***EXAMPLE 4*** Find the reciprocal of $z = 2 \text{ cis } 29°$.

Solution **Long Way**

$$\frac{1}{z} = \frac{1}{2\,(\cos 29° + i \sin 29°)}$$ Definition of cis θ.

$$= \frac{1}{2} \cdot \frac{1}{\cos 29° + i \sin 29°} \cdot \frac{\cos 29° - i \sin 29°}{\cos 29° - i \sin 29°}$$ Multiply by a clever form of 1.

$$= \frac{1}{2} \cdot \frac{\cos 29° - i \sin 29°}{\cos^2 29° - i^2 \sin^2 29°}$$ Product of conjugate binomials in the denominator.

$$= \frac{1}{2} \cdot \frac{\cos 29° - i \sin 29°}{\cos^2 29° + \sin^2 29°}$$ $i^2 = -1$.

$$= \frac{1}{2}(\cos 29° - i \sin 29°)$$ Pythagorean property for cosine and sine.

$$= \frac{1}{2}(\cos (-29°) + i \sin (-29°))$$ Odd-even properties for cosine and sine.

$$= \frac{1}{2} \text{ cis } (-29°)$$ Definition of cis θ.

Short Way

$$\frac{1}{z} = \frac{1}{2 \text{ cis } 29°} = \frac{1}{2} \text{ cis } (-29°)$$ Take the reciprocal of the modulus and the opposite of the argument. ◀

PROPERTY: Reciprocal of a Complex Number in Polar Form

If $z = r \text{ cis } \theta$, then

$$\frac{1}{z} = \frac{1}{r} \text{ cis } (-\theta)$$

Verbally: "Take the reciprocal of the modulus and the opposite of the argument."

▶ EXAMPLE 5 If $z_1 = 5 \text{ cis } 71°$ and $z_2 = 2 \text{ cis } 29°$, find $\frac{z_1}{z_2}$.

Solution **Long Way**

$$\frac{z_1}{z_2} = \frac{5 \text{ cis } 71°}{2 \text{ cis } 29°}$$

$$= (5 \text{ cis } 71°)\left(\frac{1}{2} \text{ cis } (-29°)\right)$$ Apply the reciprocal property.

$$= \frac{5}{2} \text{ cis } 42°$$ Apply the multiplication property.

Short Way

$$\frac{z_1}{z_2} = \frac{5 \text{ cis } 71°}{2 \text{ cis } 29°} = \frac{5}{2} \text{ cis } 42$$ Divide the moduli; subtract the arguments. ◀

PROPERTY: Quotient of Two Complex Numbers in Polar Form

If $z_1 = r_1 \text{ cis } \theta_1$ and $z_2 = r_2 \text{ cis } \theta_2$, then

$$\frac{z_1}{z_2} = \frac{r_1}{r_2} \text{ cis } (\theta_1 - \theta_2)$$

Verbally: "Divide the moduli; subtract the arguments."

▶ EXAMPLE 6 If $z = 2 \text{ cis } 29°$, find z^5.

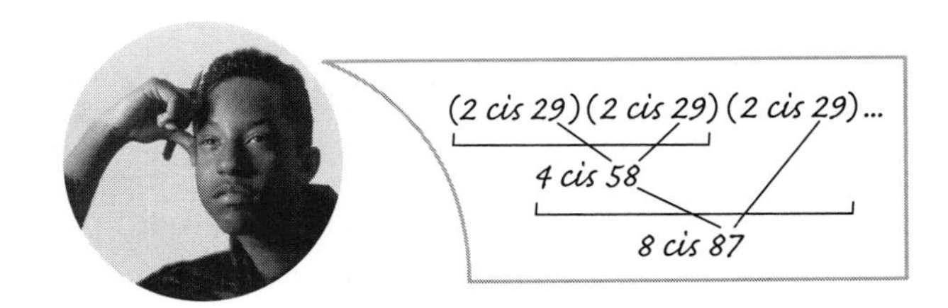

Solution **Long Way**

$$z^5 = (2 \text{ cis } 29°)^5$$
$$= (2 \text{ cis } 29°)(2 \text{ cis } 29°)(2 \text{ cis } 29°)(2 \text{ cis } 29°)(2 \text{ cis } 29°)$$
$$= 2^5 \text{ cis } (5 \cdot 29)$$ Apply the product property repeatedly.
$$= 32 \text{ cis } 145°$$

Short Way

$$z^5 = 2^5 \text{ cis } (5 \cdot 29°) = 32 \text{ cis } 145°$$ Raise the modulus to the power; multiply the argument by the exponent. ◀

The "short way" to raise a complex number to a power is known as **De Moivre's theorem.**

PROPERTY: De Moivre's Theorem

If $z = r \text{ cis } \theta$, then

$$z^n = r^n \text{ cis } n\theta$$

Verbally: "Raise the modulus to the power, and multiply the argument by the exponent."

Because De Moivre's theorem is true for fractional exponents, you can use it to find *roots* of a complex number in polar form. There is a surprise, as you will see in Example 7.

▶ ***EXAMPLE 7*** If $z = 8 \text{ cis } 60°$, find the cube roots of z, $\sqrt[3]{z}$.

Solution There are multiple polar coordinates for a given point. In general, $z = 8 \text{ cis } (60° + 360k°)$, where k stands for an integer. So there are multiple cube roots for a complex number:

$$\sqrt[3]{z} = z^{1/3}$$
$$= (8 \text{ cis } 60° + 360k°)^{1/3}$$
$$= 8^{1/3} \text{ cis } \frac{1}{3}(60° + 360k°)$$ By De Moivre's theorem.
$$= 2 \text{ cis } (20° + 120k°)$$
$$= 2 \text{ cis } 20°, 2 \text{ cis } 140°, 2 \text{ cis } 260°, 2 \text{ cis } 380°, \ldots$$

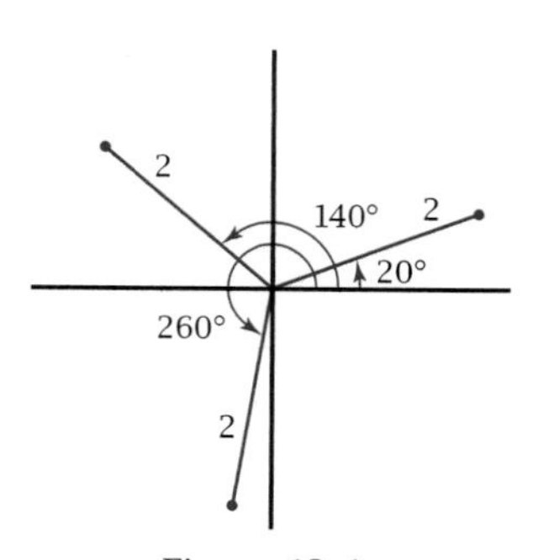

Figure 13-4e

But 2 cis 380° is coterminal with 2 cis 20°, so only the first three results are *distinct* cube roots.

$$\sqrt[3]{z} = 2 \text{ cis } 20°, 2 \text{ cis } 140°, \text{ or } 2 \text{ cis } 260°$$ ◀

In general, a complex number has exactly n distinct nth roots. Figure 13-4e shows that the three cube roots of 8 cis 60° are equally spaced around the pole in the complex plane.

Problem Set 13-4

Do These Quickly (5 min)

Q1. Write Cartesian coordinates for the polar point (3, 90°).

Q2. Write polar coordinates for the Cartesian point (−3, 3).

Q3. What figure is the graph of $r = 10\cos\theta$?

Q4. If the limaçon $r = a + b\cos\theta$ has a loop, how are a and b related?

Q5. If the limaçon $r = a + b\cos\theta$ is a cardioid, how are a and b related?

Q6. How do you tell whether a point where two polar curves cross is really an intersection?

Q7. How are the graphs of $r_1 = 2\cos\theta$ and $r_2 = 10\cos\theta$ related?

Q8. How are the graphs of $r_1 = 2\cos\theta$ and $r_3 = 2\cos(\theta - 50°)$ related?

Q9. Find the discriminant: $3x^2 + 5x + 11 = 0$

Q10. The polar graph of $r = \frac{10}{2 + 3\cos\theta}$ is a(n)

A. Hyperbola B. Ellipse C. Parabola
D. Circle E. None of these

For Problems 1–12, write the complex number in polar form, r cis θ.

1. $-1 + i$
2. $1 - i$
3. $\sqrt{3} - i$
4. $1 + i\sqrt{3}$
5. $-4 - 3i$
6. $-3 + 4i$
7. $5 + 7i$
8. $-11 - 2i$
9. 1
10. i
11. $-i$
12. -8

For Problems 13–22, write the complex number in Cartesian form, $a + bi$.

13. 8 cis 34°
14. 11 cis 247°
15. 6 cis 120°
16. 8 cis 150°
17. $\sqrt{2}$ cis 225°
18. $3\sqrt{2}$ cis 45°
19. 5 cis 180°
20. 9 cis 90°
21. 3 cis 270°
22. 2 cis 0°

For Problems 23–26, find

a. $z_1 z_2$ b. $\frac{z_1}{z_2}$

c. z_1^2 d. z_2^3

23. $z_1 = 3$ cis 47°, $z_2 = 5$ cis 36°
24. $z_1 = 2$ cis 154°, $z_2 = 3$ cis 27°
25. $z_1 = 4$ cis 238°, $z_2 = 2$ cis 51°
26. $z_1 = 6$ cis 19°, $z_2 = 4$ cis 96°

For Problems 27–36, find the indicated roots and sketch the answers on the complex plane.

27. Cube roots of 27 cis 120°
28. Cube roots of 8 cis 15°
29. Fourth roots of 16 cis 80°
30. Fourth roots of 81 cis 64°
31. Square roots of i
32. Square roots of $-i$
33. Cube roots of 8
34. Cube roots of −27
35. Sixth roots of −1
36. Tenth roots of 1

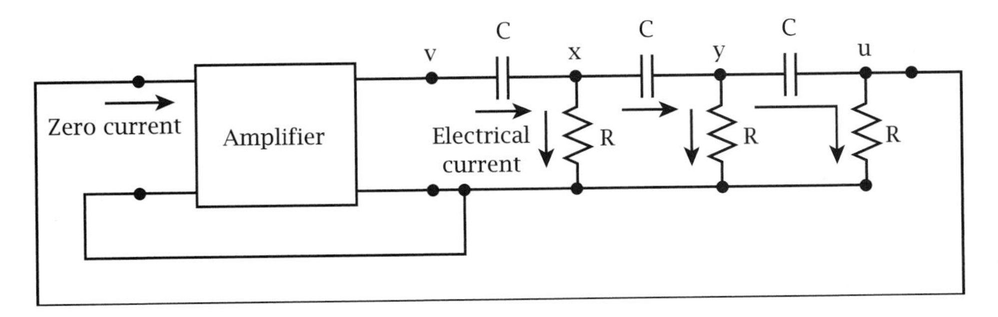

This electrical circuit forms the basis of variable-frequency sound generators and "works" because of $\sqrt{-1}$.

37. *Triple Argument Properties Problem:* By De Moivre's theorem,

$$(\cos\theta + i\sin\theta)^3 = \cos 3\theta + i\sin 3\theta$$

Expand the expression on the left. By equating the real parts and the imaginary parts on the left and right sides of the resulting equation, derive **triple argument properties** expressing $\cos 3\theta$ and $\sin 3\theta$ in terms of sines and cosines of θ.

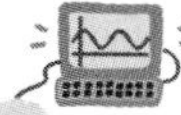

38. *Research Project 2:* From the Internet or some other source, find out about De Moivre. For instance, learn about his major mathematical contributions, the books he wrote, and his life.

39. *Journal Problem:* Update your journal. Include such things as the use of polar coordinates to represent complex numbers and how polar coordinates make it relatively easy to find products, quotients, roots, and powers.

Abraham De Moivre (1667–1754), a French mathematician, pioneered the development of analytic geometry and the theory of probability, and introduced complex numbers in trigonometry. (The Granger Collection, New York)

13-5 Parametric Equations for Moving Objects

So far in this chapter you have seen how you can plot ellipses, limaçons, and other graphs relatively easily in polar coordinates and how polar coordinates lead to a way of analyzing complex numbers. In this section you will return to the polar graphs you studied at the beginning of the chapter and analyze them with the help of parametric functions and vectors. Many of the graphs you will encounter in this section come from moving objects, such as a wheel rolling along a line or around a circle, a string unwinding from a circle, or a projectile moving through the air.

OBJECTIVE Given a geometrical description of the path followed by a moving object, write parametric equations for the path and plot it on your grapher.

▶ ***EXAMPLE 1*** A ship moves with an eastward velocity of 21 km/hr and a northward velocity of 13 km/hr. At time $t = 0$ hr the ship is at the point P_0 (−43, 19), where the distances are in kilometers from a lighthouse (Figure 13-5a).

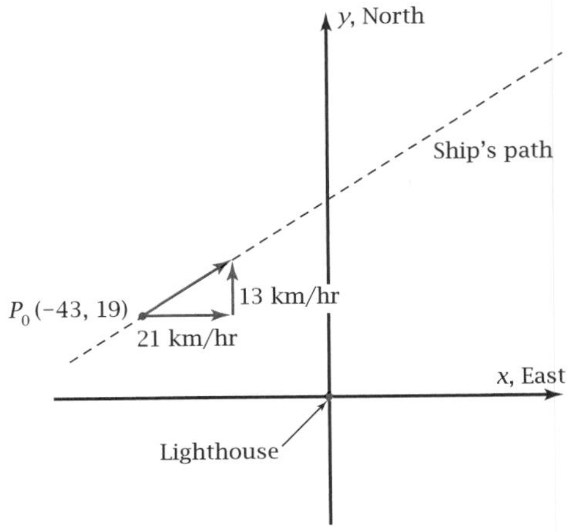

Figure 13-5a

a. Find parametric equations for the ship's path, using t hours as the parameter.

b. Confirm that your answer is correct by plotting it on your grapher.

c. Predict the time when the ship will be 60 km north of the lighthouse. How far east or west of the lighthouse will it be at this time?

Solution a.

$$x = -43 + 21t$$
$$y = 19 + 13t \qquad \text{Distance = rate × time.}$$

b. The graph (Figure 13-5b) confirms that the equations represent the given path. Use equal scales on the two axes.

Figure 13-5b

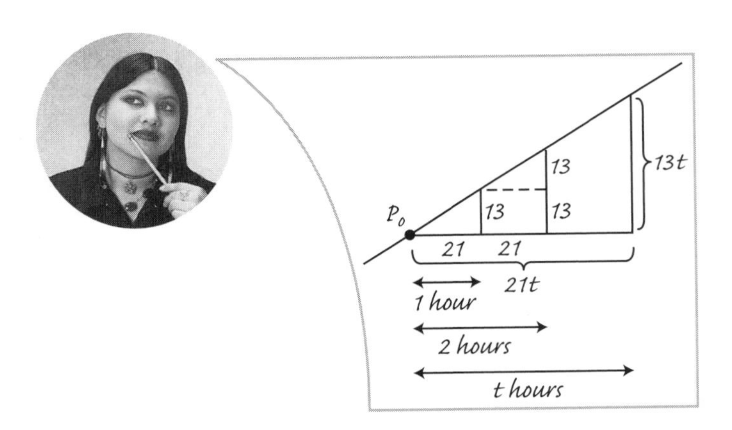

c.

$$60 = 19 + 13t \Rightarrow t = 3.1538... \approx 3.15 \text{ hr} \approx 3 \text{ hr } 9 \text{ min}$$

Set $y = 60$ and solve for t.

$$x = -43 + 21(3.1538...) = 23.2307...$$

Substitute the solution into the x-equation.

The ship will be approximately 23.23 km east of the lighthouse. ◀

▶ ***EXAMPLE 2*** As a wheel rolls along a straight-line path, a fixed point on the rim of the wheel traces a curve called a **cycloid.** The wheel in Figure 13-5c has a radius of 6 cm and rolls in a positive

direction along the x-axis. Let the parameter t represent the number of radians the wheel has rolled since a point $P(x, y)$ on its rim was at the origin. Find the parametric equations for the cycloid traced by point P. Check your equation by graphing.

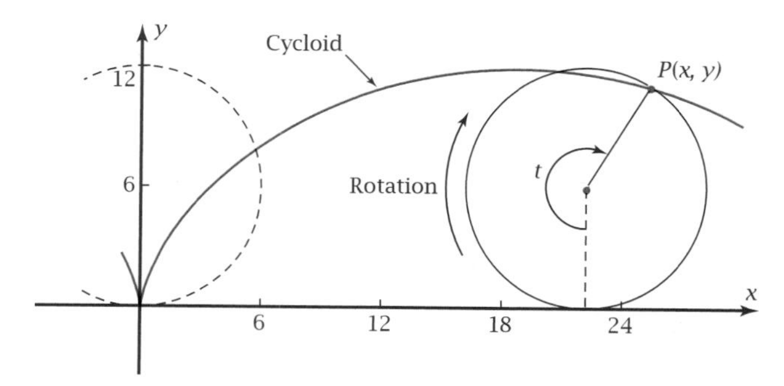

Figure 13-5c

Solution Let $\vec{r}$ be the position vector to the point $P(x, y)$ on the circle after it has rolled t radians.

Figure 13-5d shows that you can write $\vec{r}$ as the sum of three other vectors:

$$\vec{r} = \vec{v}_1 + \vec{v}_2 + \vec{v}_3$$

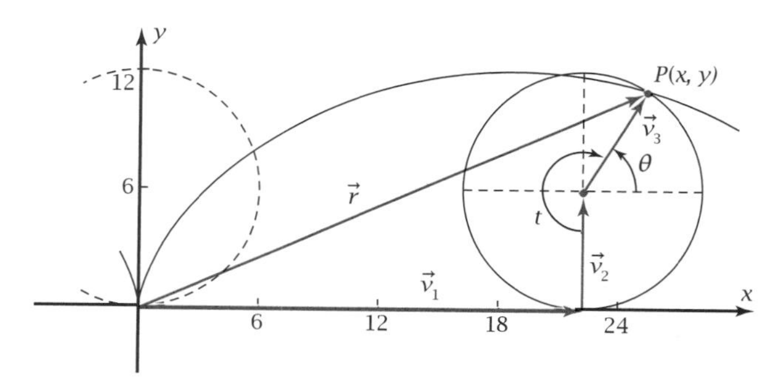

Figure 13-5d

Vector $\vec{v}_1$ is horizontal, and its length is the distance the wheel has rolled.

$$|\vec{v}_1| = 6t \Rightarrow \vec{v}_1 = (6t)\vec{i}$$ Arc length = (radius)(central angle in radians).

Vector $\vec{v}_2$ is vertical, with constant length 6 units.

$$\vec{v}_2 = 6\vec{j}$$

Vector $\vec{v}_3$ extends from the center of the wheel to point $P(x, y)$. It has constant length 6 units, but it rotates clockwise. Let θ be the angle in standard position for $\vec{v}_3$. Then

$$\vec{v}_3 = (6 \cos \theta)\vec{i} + (6 \sin \theta)\vec{j}$$ Definitions of cosine and sine.

To get θ in terms of t, observe that θ starts at 1.5π radians (270°) when $t = 0$ and decreases as t increases. Therefore,

$$\theta = 1.5\pi - t$$

$$\vec{v}_3 = (6 \cos (1.5\pi - t))\vec{i} + (6 \sin (1.5\pi - t))\vec{j}$$

$$\therefore\ \vec{r} = (6t)\vec{i} + 6\vec{j} + (6\cos(1.5\pi - t))\vec{i} + (6\sin(1.5\pi - t))\vec{j}$$

Add the three vectors.

$$\vec{r} = (6t + 6\cos(1.5\pi - t))\vec{i} + (6 + 6\sin(1.5\pi - t))\vec{j}$$

Combine like terms.

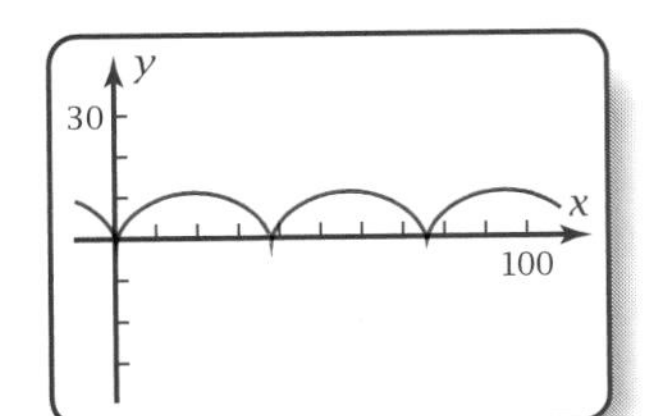

Figure 13-5e

This is called a **vector equation** for the cycloid. The parametric equations are the x- and y-components of the vector equation.

$$x = 6t + 6\cos(1.5\pi - t)$$
$$y = 6 + 6\sin(1.5\pi - t)$$

The graph is shown in Figure 13-5e. Use radian mode and equal scales on both axes. Note that although y is a periodic function of x, it is not a sinusoid. The low points are cusps rather than rounded curves.

You can use the composite argument properties of Chapter 5 to simplify the cosine and sine terms in Example 2.

$$\cos(1.5\pi - t) = \cos 1.5\pi \cos t + \sin 1.5\pi \sin t = -\sin t$$

$$\sin(1.5\pi - t) = \sin 1.5\pi \cos t - \cos 1.5\pi \sin t = -\cos t$$

The parametric equations are thus

$$x = 6t - 6\sin t$$
$$y = 6 - 6\cos t$$

The general parametric equations of a cycloid are listed in this box.

PROPERTY: Parametric Equations of a Cycloid

As a wheel of radius a rolls along the x-axis, a point on the rim of the wheel that starts at the origin traces the path of a cycloid whose parametric equations are

$$x = a(t - \sin t)$$
$$y = a(1 - \cos t)$$

where t is the number of radians the wheel has rolled since the point was at the origin.

Problem Set 13-5

Do These Quickly

Q1. Add: $(3 - 7i) + (5 + 6i)$

Q2. Add: $(3\vec{i} - 7\vec{j}) + (5\vec{i} + 6\vec{j})$

Q3. Add: 20 cis 80° + 10 cis 50°

Q4. Multiply: (20 cis 80°)(10 cis 50°)

Q5. Divide: (20 cis 80°) ÷ (10 cis 50°)

Q6. Cube: $(2\text{ cis } 20°)^3$

Q7. Write 20 cis 80° in the form $a + bi$.

Q8. Write $3 - 7i$ in polar form.

Q9. Write the composite argument property for $\cos(A - B)$.

Q10. Which two of the six trigonometric functions are even functions?

1. *Airplane's Path Problem:* An airplane is flying with a velocity of 300 km/hr west and 100 km/hr north. At time $t = 0$ hours the plane is at the point (473, 155), where the distances are in kilometers from a Federal Aviation Agency station located at the origin (Figure 13-5f).

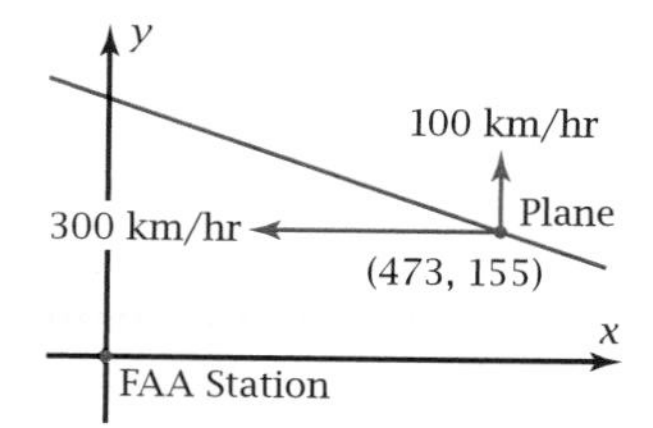

Figure 13-5f

   a. Write parametric equations for the plane's path, using t hours as the parameter.
   b. If the plane continues on this path, when will it be due north of the FAA station? How far north of the station will it be?
   c. What is the plane's actual speed?

2. *Walking Problem 2:* Calvin is walking at a speed of 6 ft/sec along a path that makes an angle of 55° with the x-axis. At time $t = 0$ he is at the point (263, 107), where the distances are in feet from a particular traffic light (Figure 13-5g).

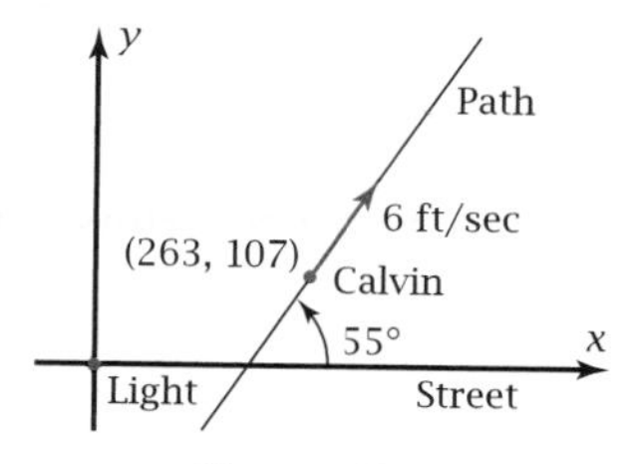

Figure 13-5g

   a. What are Calvin's speeds in the x- and y-directions?
   b. Write parametric equations for his position as a function of the parameter t seconds.
   c. A street goes along the x-axis. Assuming Calvin was walking at his 6 ft/sec pace before $t = 0$, at what time t did he cross the street?
   d. How far from the light does the path cross the street?

3. *Projectile Motion Problem:* Sir Francis Drake's ship fires a cannonball at an enemy galleon. At time $t = 0$ seconds the cannonball has an initial velocity of 200 ft/sec and a 20° angle of elevation (Figure 13-5h). You are to find the cannonball's position $P(x, y)$ as a function of time. Assume the origin represents the point from which the cannonball was fired.

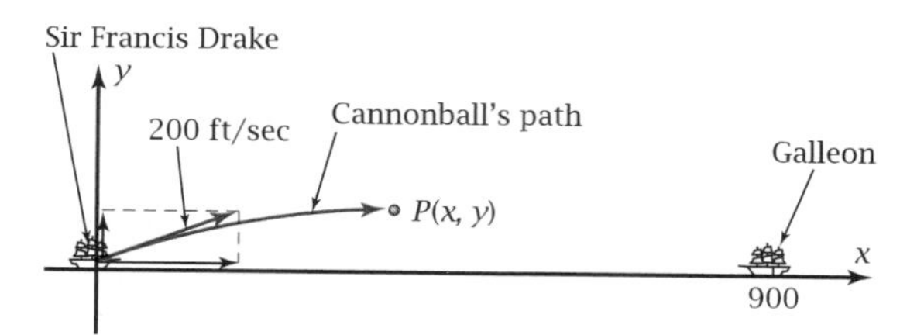

Figure 13-5h

   a. Find parametric equations of the cannonball's path. To do this, assume there is no air friction, so the horizontal velocity remains what it was at time $t = 0$. The vertical velocity is what it was at $t = 0$, minus $16t^2$ due to the action of gravity. Also, recall that distance = (rate)(time).
   b. Plot the parametric equations. Use a window for x large enough to show the point where the cannonball hits the water. Sketch the result.
   c. The galleon is at $x = 900$ ft from Sir Francis' ship. The tops of the sails are 40 ft above the surface of the water. Will the cannonball fall short of the galleon, pass over the galleon, or hit it somewhere between the waterline and the tops of the sails? Show how you get your answer.
   d. To be most effective, the cannonball should hit the galleon right at the waterline ($y = 0$). At what angle of elevation should the cannonball be fired to accomplish this objective?

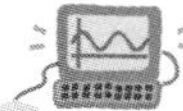

   e. On the Internet or in some other reference source, find out who Sir Francis Drake was, when he lived, and to which country the enemy galleon might have belonged.

4. *Ship Collision Project:* Two ships are steaming through the fog. At time $t = 0$ minutes, their positions and velocities are

   Ship A: Point $(x, y) = (2000, 600)$,
   velocity 500 m/min on an angle of 140°

   Ship B: Point $(x, y) = (200, 300)$,
   velocity 400 m/min on an angle of 80°

   a. Write parametric equations for the position (x, y) of each ship, with x and y in meters and the parameter t in minutes.
   b. Plot the parametric equations for both ships on the same screen. Use simultaneous mode so that you can see where each ship is with respect to the other as the graphs are being plotted. Based on what you observe, do the ships collide, almost collide, or miss each other by a significant amount?
   c. Use the distance formula to help you write a Cartesian equation for the distance between the ships as a function of time. Plot this function on another screen and sketch the result.
   d. By appropriate operations on the function of part c, find numerically the time the ships are the closest and how close they get. Based on your answer, should the ships have changed their courses to avoid a collision, or will they miss each other by a safe distance?

   e. In July 1956, there was a serious collision between the ships *Andrea Doria* and *Stockholm.* On the Internet or in some other reference source, find out what happened.

5. *Ellipse from Geometrical Properties Problem:* Figure 13-5i shows concentric circles of radii 3 and 5 centered at the origin of a Cartesian coordinate system. A ray from the center at an angle of t radians to the x-axis cuts the two circles at points A and B, respectively. From point A a horizontal line is drawn, and from point B a vertical line is drawn. These two lines intersect at point P on the graph of a curve.

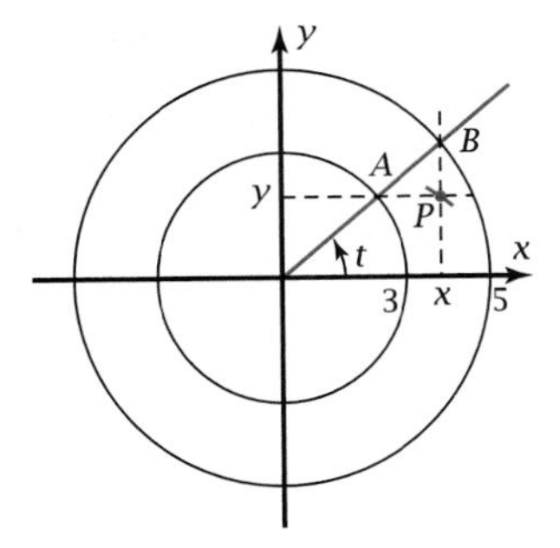

Figure 13-5i

   a. On a copy of Figure 13-5i, pick other values of the angle t and plot more points using the given specifications. Connect the resulting points with a smooth curve. Does the graph seem to be an ellipse?
   b. Write parametric equations for the point $P(x, y)$ in terms of the parameter t using the given geometrical description. Is the result the same as the parametric equations of an ellipse from Section 4-5?
   c. Plot the parametric equations on your grapher. Use equal scales on the two axes. Also, plot the two circles. Do the circles have the same relationship to the curve as in your sketch?
   d. The parameter t can be eliminated from the two parametric equations to give a single equation involving only the variables x and y. Clever use of the Pythagorean properties will allow you to do this. Write a Cartesian equation for this curve. How can you tell the equation represents an ellipse?

6. *Serpentine Curve Problem:* Figure 13-5j shows the **serpentine curve,** so called for its snakelike shape. A fixed circle of radius 5 has its center on the x-axis and passes through the origin. A variable line from the origin makes an angle of t radians with the x-axis. It intersects the circle at point A, and it intersects the fixed line $y = 5$ at point B. A horizontal line from A and a vertical line from B intersect at point $P(x, y)$ on the serpentine curve.

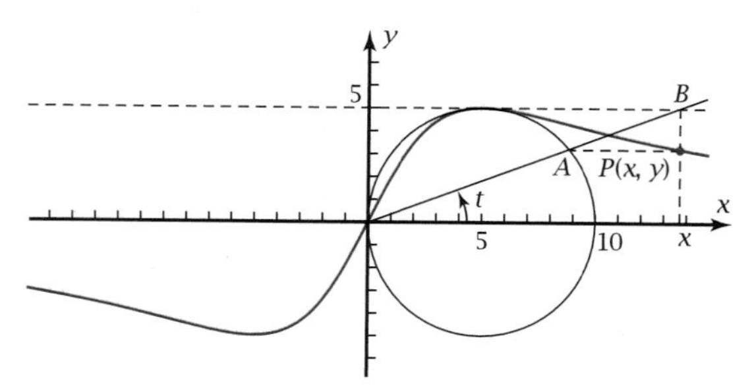

Figure 13-5j

a. On a copy of Figure 13-5j pick a different value of t between 0 and $\frac{\pi}{2}$ and plot the corresponding point P as described. Plot another point for a t-value between $\frac{\pi}{2}$ and π. Show that the resulting points really are on the serpentine curve.

b. Find the parametric equations for x and y in terms of the parameter t. (To find y, first find the distance from the origin to point A. You can do this by recalling the polar equation of a circle or by drawing a right triangle inscribed in the semicircle with right angle at A and hypotenuse 10.)

c. Confirm that your parametric equations are correct by plotting them on your grapher. Use a window with an x-range at least as large as the one shown, and use equal scales on both axes.

d. The point P in Figure 13-5j corresponds to $t = 0.35$ radian. Confirm that this is correct by showing that the values of x and y you get from the equation agree with the values in the figure.

7. *Flanged Wheel Prolate Cycloid Problem:* Train wheels have flanges that project beyond the rims to keep the wheels from slipping off the track. A point P on the flange traces a **prolate cycloid** as the wheel turns. Figure 13-5k shows

an example. The radius of the flange has been exaggerated so you can see more clearly what a prolate cycloid looks like. Assume the wheel has radius 50 cm, the flange has radius 70 cm, and that the wheel has rotated t radians since the point $P(x, y)$ was farthest below the track.

- Let $\vec{r}$ (not shown) be the position vector to point $P(x, y)$ on the flange.
- Let $\vec{v}_1$ be the vector from the origin to the point where the wheel touches the track.
- Let $\vec{v}_2$ be the vector from that point to the center of the wheel.
- Let $\vec{v}_3$ be the vector from the center of the wheel to the point on the flange.

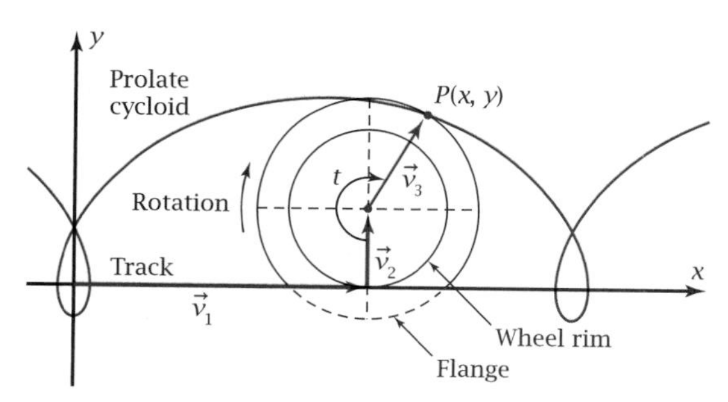

Figure 13-5k

a. Explain why $\vec{r} = \vec{v}_1 + \vec{v}_2 + \vec{v}_3$.

b. The length of $\vec{v}_1$ equals the distance the wheel has rolled. Vector $\vec{v}_2$ is a constant vector in the vertical direction. Write $\vec{v}_1$ and $\vec{v}_2$ in terms of their components.

c. Vector $\vec{v}_3$ goes from the center of the wheel to point $P(x, y)$. Write $\vec{v}_3$ as a function of t. Use the result to write $\vec{r}$ as a vector function of t.

d. Plot the graph of $\vec{r}$ using parametric mode. Does the graph look like Figure 13-5k?

e. How far does P move in the x-direction between $t = 0$ radians and $t = 0.1$ radian? How do you explain the fact that the displacement is *negative*, even though the wheel is going in the positive x-direction?

8. *Epicycloid Problem:* Figure 13-5l (top diagram) shows the **epicycloid** traced by a point on the rim of a wheel of radius 2 cm as it rotates, without slipping, around the outside of a circle of radius 6 cm. The wheel starts with point $P(x, y)$ at (6, 0). The parameter t is the

number of radians from the positive x-axis to a line through the center of the wheel.

- Let $\vec{r}$ (not shown) be the position vector to point $P(x, y)$.
- Let $\vec{v}_1$ be the vector from the origin to the center of the wheel.
- Let $\vec{v}_2$ be the vector from the center of the wheel to point $P(x, y)$.

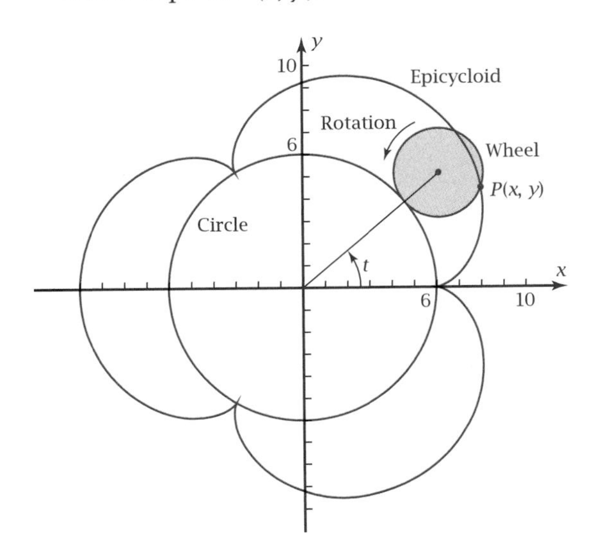

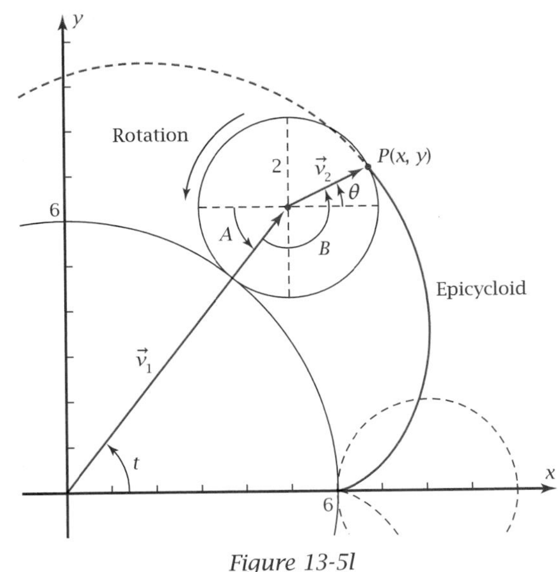

Figure 13-5l

a. Find $\vec{v}_1$ in terms of t and the unit vectors $\vec{i}$ and $\vec{j}$. Find $\vec{v}_2$ in terms of angle θ in Figure 13-5l and the unit vectors $\vec{i}$ and $\vec{j}$.

b. Write θ in terms of t by observing that θ starts at $-\pi$ radians when $t = 0$. Thus, θ is given by the sum $-\pi + A + B$, where A and B are shown in the lower diagram of Figure 13-5l. Express angles A and B in terms of t. Note that the arc of the wheel subtended by angle B equals the arc of the circle subtended by angle t, because the wheel rotates without slipping. Note also that the length of an arc of a circle equals the central angle in radians times the radius.

c. Write a vector equation for $\vec{r}$ as a function of t by observing that $\vec{r} = \vec{v}_1 + \vec{v}_2$. Use the result to plot the epicycloid using parametric mode. Use a t-range large enough to get one complete cycle, and use equal scales on both axes. Does the graph agree with Figure 13-5l? If not, go back and check your work.

d. Plot the 6-cm circle on the same screen as in part c. Does the result agree with the figure?

9. *Involute of a Circle Problem 2:* Figure 13-5m shows an **involute of a circle.** A string is wrapped around a circle of radius 5 units. A pen is tied to the string at the point (5, 0). Then the string is unwound in the counterclockwise direction. The involute is the spiral path followed by the pen as the string unwinds. The curve is interesting because gear teeth made with their surfaces in this shape transmit the rotation smoothly from one gear to the next.

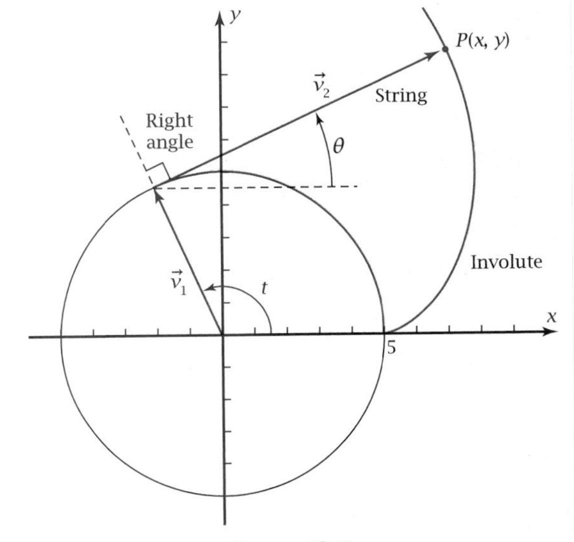

Figure 13-5m

a. The parameter t is the radian measure of the angle between the positive x-axis and the line from the origin to the point of tangency of the string. Vector $\vec{v}_1$ goes from the center of the circle to the point of tangency. Vector $\vec{v}_2$ goes from the point of tangency to point $P(x, y)$ on the involute. Find a vector equation for position vector $\vec{r}$ (not shown) to point P in terms of the parameter t. Note that $\vec{v}_1$ and $\vec{v}_2$ are perpendicular because $\vec{v}_1$ is a radius to the point of tangency. Confirm that your equation is correct by plotting it on your grapher. Use at least three revolutions for the t-range.

b. Construct an involute by wrapping a string around a roll of tape, tying a pencil or pen to the end of the string, and tracing the path as you unwind the string. Does the involute you plotted in part a agree with this actual involute?

10. *Roller Coaster Problem 2:* Figure 13-5n shows part of a roller coaster track. Its shape is a **prolate cycloid** (see Problem 7) traced by a point $P(x, y)$ on the flange of a wheel as the wheel rolls, without slipping, through an angle of t radians underneath the line $y = 20$. Point $P(x, y)$, 12 feet from the center of the wheel, is at its top position, directly above the origin, when angle $t = 0$. The position vector $\vec{r}$ (not shown) to point $P(x, y)$ is the sum of four other vectors starting at the origin, $\vec{v}_1 + \vec{v}_2 + \vec{v}_3 + \vec{v}_4$. Find a vector equation for the track in terms of t. By plotting on your grapher, show that your equation is correct.

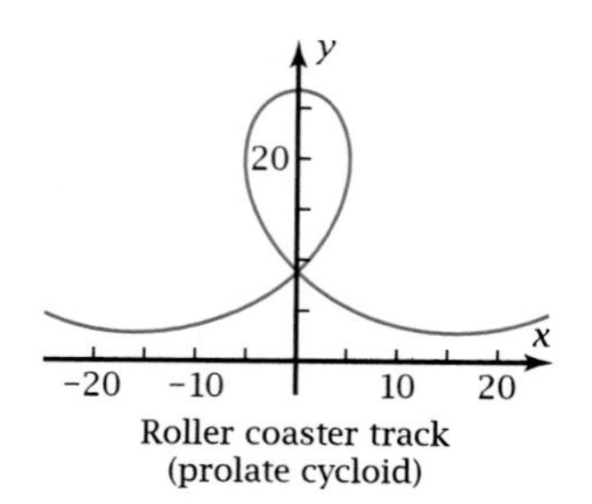

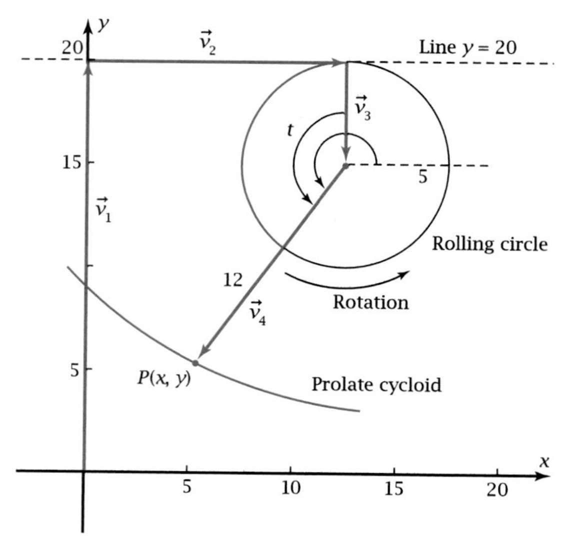

Figure 13-5n

11. *Parametric Equations for Polar Curves Problem:* Figure 13-5o shows the ellipse with the following polar equation, superimposed on a rectangular coordinate grid:

$$r = \frac{9}{5 - 4\cos\theta}$$

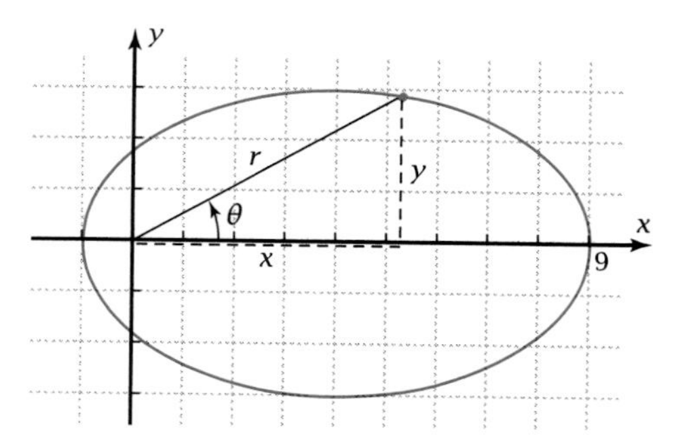

Figure 13-5o

a. Confirm on your grapher that the equation gives the graph in Figure 13-5o. Note that the value of θ shown in the figure is 0.5 radian.

b. Use the definitions of cosine and sine and the value of r from the polar equation to write parametric equations for the ellipse. Confirm that your parametric equations give the same ellipse and that tracing to $\theta = 0.5$ radian gives the same point shown.

c. The ellipse in Figure 13-5o has center at (4, 0), x-radius 5, and y-radius 3. Use this information to write another set of parametric equations for the ellipse in terms of the parameter t.

d. Plot the parametric equations in part c. Does the ellipse coincide with the one shown in Figure 13-5o? Does tracing to $t = 0.5$ radian on this ellipse give the same point as tracing to $\theta = 0.5$ radian on the polar version of the ellipse? Do your findings surprise you?

13-6 Chapter Review and Test

In this chapter you started by plotting graphs in polar coordinates. Polar coordinates allowed you to write complex numbers in polar form. In this form you saw that the product of two complex numbers has a magnitude equal to the product of the two magnitudes and an angle equal to the sum of the two angles. By De Moivre's theorem you were able to find powers and roots of complex numbers. Vectors and parametric functions gave you a way to describe the path followed by a point on a moving object.

Review Problems

R0. Update your journal with what you have learned in this chapter. Include such things as:

- Multiple ways to write polar coordinates for the same point
- Graphs such as limaçons and conic sections in polar coordinates
- The fact that r can be negative in polar coordinates
- False and actual intersections of polar curves
- How a complex number can be written in polar form
- De Moivre's theorem, and products, quotients, and roots of complex numbers
- Parametric equations for cycloids and other paths of moving objects

R1. Plot the following points on polar coordinate paper. Show especially what happens when r is negative. Connect the points with a smooth curve.

θ	r
45°	7.1
60°	5.0
75°	2.6
90°	0
105°	−2.6
120°	−5.0
135°	−7.1

R2. a. Plot the three-leaved rose $r = 10 \cos 3\theta$. Sketch the graph.

b. Plot the limaçon $r = 3 - 5 \cos \theta$. Sketch the graph.

c. Plot the hyperbola $r = \frac{8}{3 - 5 \cos \theta}$. Sketch the graph.

d. Plot the circle $r = 2 \cos (\theta - 60°)$. Sketch the graph.

e. Plot the line $r = 2 \sec (\theta - 60°)$. Sketch the graph.

f. Prove algebraically that the graph in part c is a hyperbola by transforming the equation

to Cartesian form. Where is the focus of the hyperbola? What is its eccentricity?

R3. Find the true intersections of $r_1 = 4 + 6 \cos \theta$ and $r_2 = 5 - 3 \cos \theta$ in Figure 13-6a. At what other point(s) do the graphs cross but not intersect?

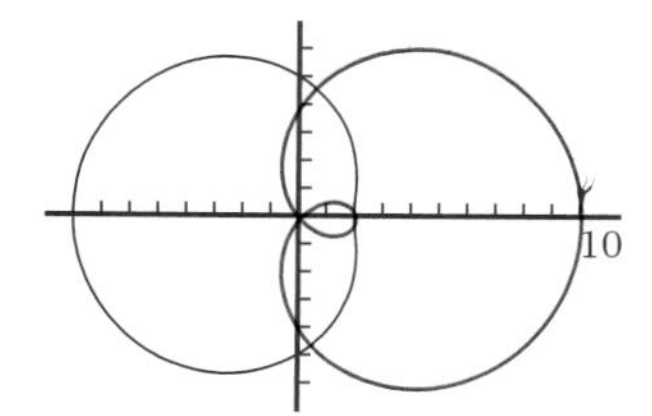

Figure 13-6a

R4. a. Write in polar form: $-5 + 12i$

b. Write in Cartesian form: 7 cis 234°

c. Multiply: (2 cis 52°)(5 cis 38°)

d. Divide: (51 cis 198°) ÷ (17 cis 228°)

e. Raise to the power: $(2 \text{ cis } 27°)^5$

f. Raise to the power: $(8 \text{ cis } 120°)^{1/3}$

g. Sketch the three answers to part f on the complex plane.

R5. a. Express this sum in Cartesian form.

$$10 \text{ cis } 43° + 7 \text{ cis } 130° - 5 \text{ cis } 215°$$

b. Write the answer to part a as a complex number in polar form.

R6. a. *Parametric Line Problem:* As you travel eastward on Highway I-90 from Cleveland to Erie, the highway makes an angle of 24° north of east. Suppose you start at Cleveland at time $t = 0$ hours and drive 60 mi/hr. Write parametric equations for your position $P(x, y)$ as functions of t hours, where x and y are in miles east and north of Cleveland, respectively. The highway passes through Erie, which is 50 miles east of Cleveland. How far north of Cleveland is Erie?

b. *Quarter and Dime Epicycloid Problem:* A quarter is placed with its center at the origin of a coordinate system. A dime is placed with its center on the x-axis, as shown in Figure 13-6b. The quarter is held fixed while the dime rotates counterclockwise around it without slipping. Figure 13-6c shows the dime rotated to the place where a line from the origin through its center makes an angle of t radians with the x-axis. Point P on the dime traces an epicycloid. The quarter has radius 12 mm, and the dime has radius 9 mm.

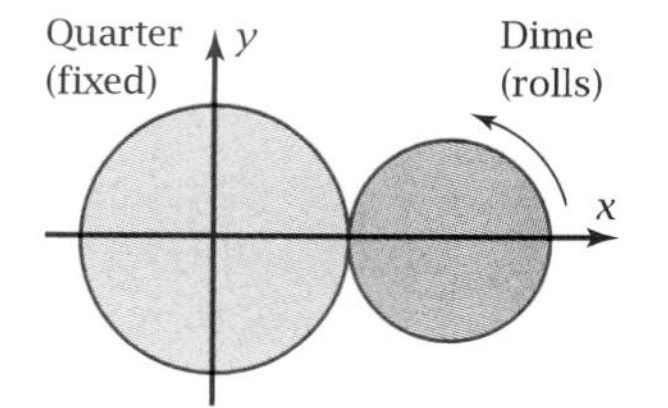

Figure 13-6b

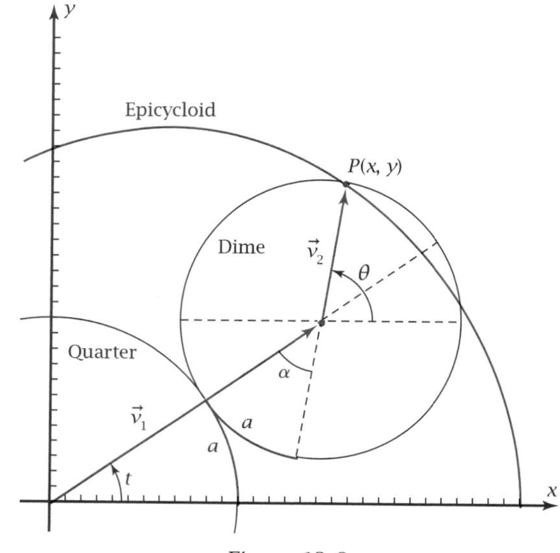

Figure 13-6c

i. Do you agree that the circles shown in Figure 13-6b really are the size of a quarter and dime? Do you agree that the radii are 12 mm and 9 mm, respectively?

ii. Derive a vector equation for point $P(x, y)$ on the edge of the rolling dime in terms of t. To help you do this, notice that the position vector $\vec{r}$ to P (not shown in Figure 13-6c) is equal to $\vec{v}_1 + \vec{v}_2$, where $\vec{v}_1$ extends from the origin to the center of the dime and $\vec{v}_2$ extends from the center of the dime to the point P. Once you find $\vec{v}_2$ in terms of θ in Figure 13-6c, you can get $\theta = \frac{7}{3}t$ by realizing that arc a on the quarter equals arc a on the dime, thus letting you get α in terms of t.

iii. Plot one complete cycle of the epicycloid with your grapher in parametric mode. Note that it takes more than one revolution for the graph to close.

Concept Problems

C1. *Planetary Motion Science Fiction Problem:* Figure 13-6d shows a small planet with a 3-mile radius orbiting a black hole. The orbit is circular with a radius of 10 miles. As it orbits the black hole, the planet rotates counterclockwise. Vector $\vec{v}_1$ goes from the center of the black hole to the center of the planet. Vector $\vec{v}_2$ goes from the center of the planet to point $P(x, y)$ on the surface of the planet. Angles A and B are in standard position at the centers of the black hole and the planet, respectively. Vector $\vec{r}$ (not shown) is the position vector to point $P(x, y)$.

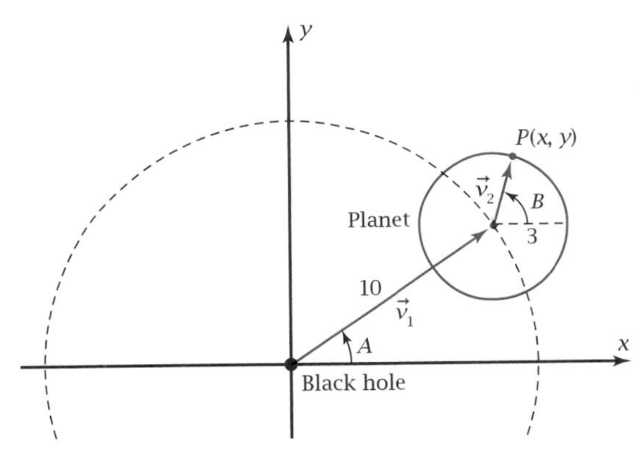

Figure 13-6d

a. Write $\vec{r}$ in terms of the unit vectors $\vec{i}$ and $\vec{j}$ and the angles A and B.

b. At time $t = 0$ both angles A and B equal 0. The planet rotates counterclockwise at 12 radians per hour, and it orbits the black hole counterclockwise at 2 radians per hour. Write equations expressing A and B in terms of t. Use the results to write $\vec{r}$ as a function of t.

c. Plot the path of point P on your grapher. Use radian mode and equal scales on both axes. Sketch the result.

d. Explain why there are only *five* loops in the graph, in spite of the fact that the angular velocity of the planet is *six* times the angular velocity of orbit.

e. Plot the graph of the path of P under the following conditions.

   i. The planet slows from 12 radians per hour to 8 radians per hour.

   ii. The planet rotates clockwise at 12 radians per hour instead of counterclockwise.

   iii. The planet rotates at exactly the right angular velocity to make cusps instead of loops in the path.

C2. *Gear Tooth Problem:* The surfaces of gear teeth are made in the shape of an **involute of a circle** (Figure 13-6e). This form is used because it allows the motion of one gear to be transmitted uniformly to the motion of another. An involute is the path traced by the end of a string as it is unwound from around a circle. In this problem you will see how the polar coordinates of a point on an involute are related to the Cartesian coordinates that can be found by parametric equations. You will do this by using vectors in the form of complex numbers.

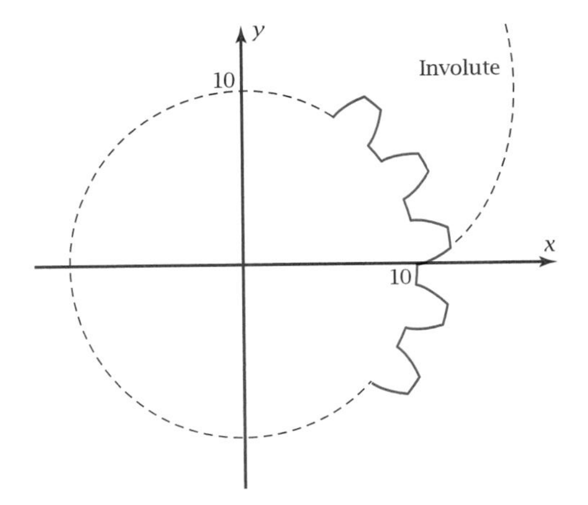

Figure 13-6e

a. Suppose a gear is to have a radius of 10 cm (to the inside of the teeth). The gear tooth surface is to have the shape of the involute formed by unwrapping a string from this

circle. Vector $\vec{v}_1$ (Figure 13-6f) goes from the center of the circle to the point of tangency of the string. Write $\vec{v}_1$ as a complex number in polar form, in terms of the angle t radians.

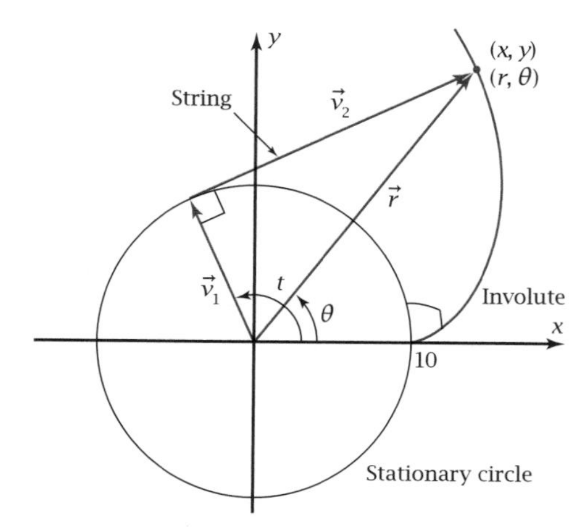

Figure 13-6f

b. Vector $\vec{v}_2$ goes along the string from the point of tangency to the point on the involute. Explain why its length is the same as the arc of the circle subtended by angle t. Write $\vec{v}_2$ as a complex number in polar form. Observe that, with respect to the horizontal axis, the angle for $\vec{v}_2$ is $\frac{\pi}{2}$ less than t because $\vec{v}_2$ is perpendicular to $\vec{v}_1$. Use the appropriate properties to get $\vec{v}_2$ in terms of functions of t.

c. Vector $\vec{r}$ is the position vector to point $P(x, y)$ on the involute. Show that

$$\vec{r} = 10(\cos t + t \sin t) + 10i(\sin t - t \cos t)$$

d. Show that r (the length of $\vec{r}$) is given by $r = 10\sqrt{1 + t^2}$.

e. The gear teeth are to be 2 cm deep, which means that the outer radius of the gear will be 12 cm. If the inside of the tooth shown in Figure 13-6f is at $t = 0$, what will be the value of t at the outside of the tooth? What will be the value of θ for this value of t?

f. Machinists who make the gear need to know the degree measure of angle θ in part e. Find this measure in degrees and minutes, to the nearest minute.

C3. *General Polar Equation of a Circle Problem:* The general polar equation of a circle that does not pass through the pole can be found with the help of the law of cosines. Suppose a circle of radius a is centered at the point with polar coordinates (k, α), as shown in Figure 13-6g.

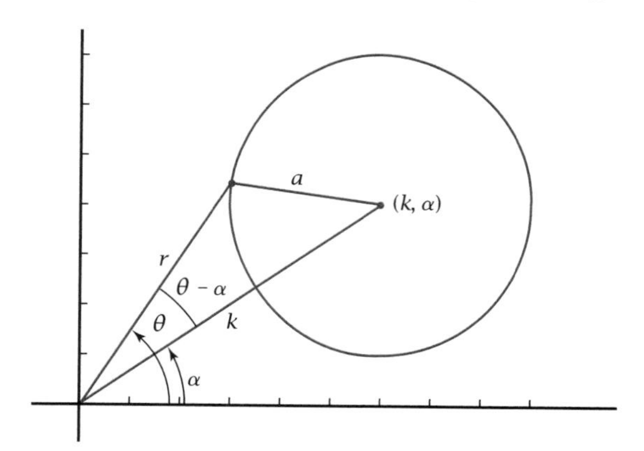

Figure 13-6g

a. Use the law of cosines to write an equation relating r, a, k, and the angle $(\theta - \alpha)$. Solve the equation for r with the help of the quadratic formula.

b. The quadratic formula has an ambiguous sign $\pm$ in it. Use the form of the solution with the + sign to plot the circle with radius 3 centered at $(7, 40°)$. Use a θ-range of 0° to 360°. Does the grapher plot the graph as one continuous circle?

c. Use the equation for r that has the − sign. How does the graph relate to the one in part b?

d. Find the two values of r if $\theta = 50°$.

e. Show algebraically that there are no values of r if $\theta = 90°$.

Chapter Test

PART 1: No calculators allowed (T1–T9)

T1. Plot the points on polar coordinate paper. Connect the points with a smooth curve.

θ	r	θ	r
150°	−8.4	195°	−5.6
165°	−5.6	210°	−8.4
180°	−5.0		

T2. Write the polar equation of the circle in Figure 13-6h.

T3. Write the polar equation of the line in Figure 13-6i.

T4. Write the definition of $r \text{ cis } \theta$.

T5. Multiply: (5 cis 37°)(3 cis 54°)

T6. Divide: (3 cis 100°) ÷ (12 cis 20°)

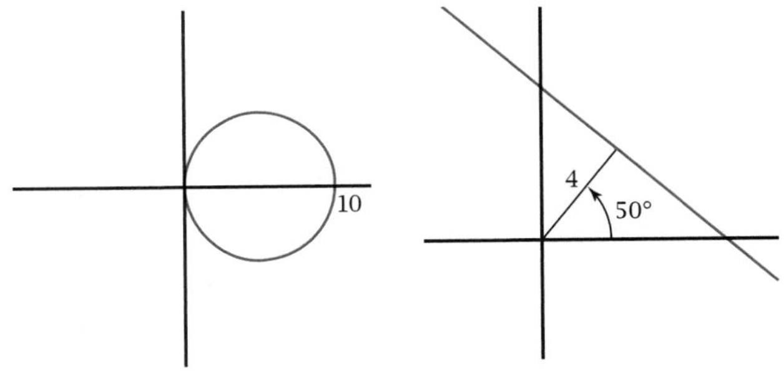

Figure 13-6h *Figure 13-6i*

T7. Raise to the power: $(4 \text{ cis } 50°)^3$

T8. Write i as a complex number in polar form. Use the result and De Moivre's theorem to find the two square roots of i. Sketch the answers on the complex plane.

T9. Write two other polar ordered pairs for the point (7, 30°), one with a positive value of r and a positive value of θ and the other with a negative value of r and a positive value of θ.

PART 2: Graphing calculators allowed (T10–T17)

T10. The polar equation of the graph in Problem T1 is

$$r = \frac{5}{2 + 3\cos\theta}$$

Eliminate the fraction by multiplying both sides by the denominator of the right side. Then derive a Cartesian equation, thus showing that the graph really is a hyperbola.

T11. The graphs of $r = 3 \sin 4\theta$ and $r = 3 \sin 5\theta$ are both roses. Plot each equation on your grapher. How many "leaves" are on each rose? How can you tell from the coefficient of θ how many leaves will be in a rose graph?

T12. Write $24 - 7i$ as a complex number in polar form.

T13. Write 6 cis 300° as a complex number in rectangular form.

T14. Figure 13-6j shows the two polar curves

$r_1 = 5 + 4\cos\theta$ and $r_2 = 1 + 6\sin\theta$

a. Which graph is which? What special name is given to each graph?

b. The graphs cross at $\theta = 90°$. Is this a true intersection? Explain.

c. The graphs cross at a point in the first quadrant. Show that this is a true intersection, and find its polar coordinates.

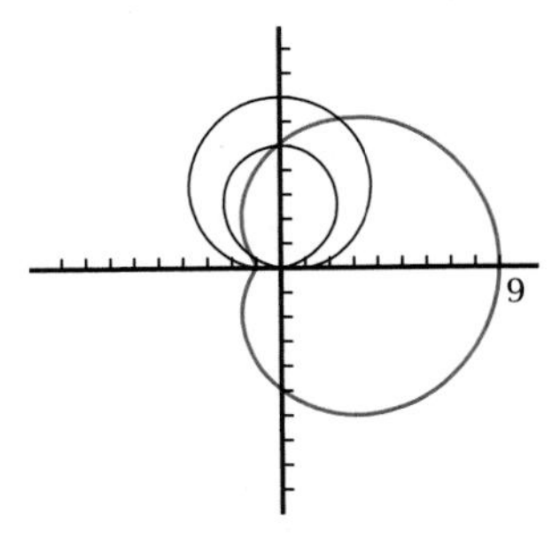

Figure 13-6j

T15. *Airplane Looping Problem:* A stunt pilot is doing a loop with her plane. As shown in Figure 13-6k, there are three forces acting on the plane:

Wing lift: 2500 lb at 127°

Propeller thrust: 700 lb at 37° (perpendicular to the lift)

Gravitational force: 2000 lb straight down

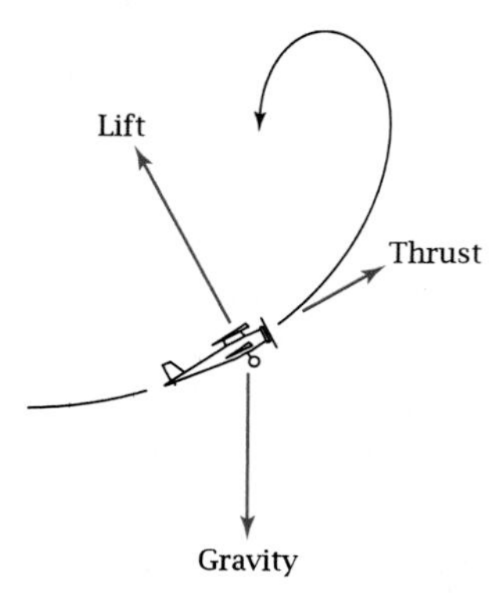

Figure 13-6k

Find the sum of these three force vectors as a complex number in polar form.

T16. *Car Wheel Curtate Cycloid Problem:* Figure 13-6l shows the **curtate cycloid** path traced by the valve stem (where you put in the air) on a car tire as the car moves. *Curtate* comes from the Latin word *curtus,* which means "shortened." The wheel rotates t radians from a point where the valve stem was at its lowest.

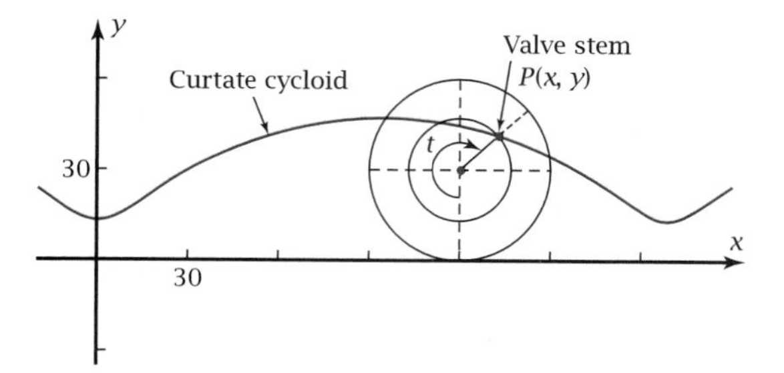

Figure 13-6l

a. The wheel has a radius of 30 cm. The valve stem is 17 cm from the center of the wheel. Write the position vector to point P on the curtate cycloid as a sum of three other vectors, $\vec{v}_1$ from the origin to the point where the tire touches the road, $\vec{v}_2$ from the head of $\vec{v}_1$ to the center of the wheel, and $\vec{v}_3$ from the center of the wheel to the valve stem. Use the result to write parametric equations for x and y in terms of the parameter t.

b. If you have not already done so, simplify your equations in part a using the composite argument properties to get equations that have only t as the argument.

c. Confirm that your equations are correct by plotting three cycles of the curtate cycloid. Use equal scales on the two axes.

d. On the same screen, plot a sinusoid with the same period, amplitude, and high and low points as the curtate cycloid. Sketch the results. Tell how you can distinguish between the two graphs.

T17. What did you learn as a result of taking this practice test that you did not know before?

CHAPTER

14

Sequences and Series

Positioning spaceships requires highly accurate computations because slight errors can make a difference of many miles in the landing point or cause an unacceptable reentry angle. Using a series of powers, you can calculate cosines and sines to as many decimal places as desired, just by using the operations of addition, subtraction, and multiplication many times. In this chapter you will learn about these power series as well as other series that may be used to calculate such things as compound interest on money in a savings account and cumulative effects of repeated doses of medication.

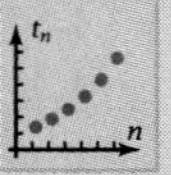

Mathematical Overview

In this chapter you will learn about sequences of numbers and about series, which are sums of terms of sequences. Geometric and arithmetic series make logical mathematical models for functions such as compound interest, where the amount of money in an account increases by jumps each month rather than rising continuously. You will look at sequences and series in several ways.

Numerically

Sequence: 3, 6, 12, 24, . . .

Series: $3 + 6 + 12 + 12 + 24 + \cdots$

Partial sum: $S_{10} = 3 + 6 + 12 + 24 + \cdots + 1536 = 3069$

Algebraically

$$S_{10} = 3 \cdot \frac{1 - 2^{10}}{1 - 2} = 3069$$

Graphically

Figure 14-0a shows the terms of the geometric sequence as a function of n, the term number. The dotted line indicates the continuous function graph that fits the discrete values of the sequence.

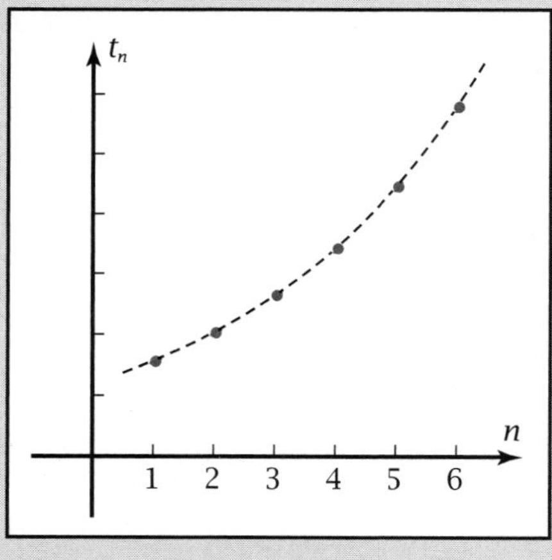

Figure 14-0a

Verbally

I learned that arithmetic and geometric series are similar. In arithmetic series the terms progress by adding a constant, and in geometric series the terms progress by multiplying by a constant, like linear and exponential functions.

14-1 Introduction to Sequences and Series

Most of the functions you have studied up to now have been **continuous.** The graphs have been smooth curves. Where **discrete** data points have been measured, you looked for the continuous function that fits best. In this chapter you will study **sequences** of numbers, such as

5, 7, 9, 11, 13, . . .

and **series** formed by adding the terms of a sequence, such as

$5 + 7 + 9 + 11 + 13 + \cdots$

OBJECTIVES

- Given a few terms in a sequence or series of numbers, find more terms.
- Given a series, find the sum of a specified number of terms.

Exploratory Problem Set 14-1

1. The infinite set of numbers 5, 7, 9, 11, . . . is an **arithmetic sequence.** It progresses by adding 2 to one term to get the next term. What does the tenth term equal? How many 2s would you have to add to the first term, 5, to get the tenth term? How could you get the tenth term quickly? Find the 100th term quickly.
2. Enter the first ten terms of the sequence in Problem 1 in a list in your grapher, and enter the term numbers 1, 2, 3, 4, . . . , 10 in another list. Make a point plot of term value as a function of term number. Sketch.
3. What kind of continuous function contains all the points in the plot of Problem 2?
4. The infinite sum $5 + 7 + 9 + 11 + \cdots$ is an **arithmetic series.** Because the series has an infinite number of terms, you cannot add them all. But you can add *part* of the terms. Find the tenth **partial sum** by adding the first ten terms.
5. Find the average of the first and tenth term in the partial sum of Problem 4. Multiply this number by 10. What do you notice about the answer? Use the pattern you observe to find the 100th partial sum of the series. Show how you did it.
6. Calculate the first ten partial sums of the series in Problem 4 and enter them in a third data list. Make a point plot of partial sum as a function of number of terms, using the term numbers in one of the data lists from Problem 2. Sketch the result.
7. Run regressions to find out which kind of continuous function exactly fits the partial sums in Problem 6. Write its particular equation. Use the result to find quickly the 100th partial sum of the series.
8. The infinite set of numbers 6, 12, 24, 48, . . . is a **geometric sequence.** How do the terms progress from one to the next? Find the tenth term of the sequence.
9. The infinite sum $6 + 12 + 24 + 48 + \cdots$ is a **geometric series.** Find the tenth partial sum of the series.
10. What did you learn as a result of doing this problem set that you did not know before?

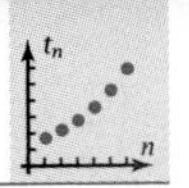

14-2 Arithmetic, Geometric, and Other Sequences

(The Granger Collection, New York)

Suppose you have \$40 in a piggy bank that you are saving to spend on a special project. You take on a part-time job that pays \$13 per day. Each day you put this cash into the piggy bank. The number of dollars in the bank is a function of the number of days you have worked.

Days (term numbers, n):	1	2	3	4	5	...
Dollars (terms, t_n):	53	66	79	92	105	...

But it is a **discrete function** rather than a **continuous function.** After $3\frac{1}{2}$ days you still have the same \$79 as you did after 3 days. A function like this, whose domain is the set of positive integers, is called a **sequence.** In this section you will look for patterns in sequences that allow you to calculate a term from its term number or to find the term number of a given term.

OBJECTIVES

- Represent sequences explicitly and recursively.

Given information about a sequence,

- Find a term when given its term number.
- Find the term number of a given term.

▶ EXAMPLE 1 For the sequence of dollars 53, 66, 79, 92, 105, . . . shown at the beginning of this section,

a. Sketch the graph of the first few terms of the sequence.

b. Find t_{100}, the 100th term of the sequence.

c. Write an equation for t_n, the nth term of the sequence, in terms of n.

Solution

a. The graph in Figure 14-2a shows discrete points. You may connect the points with a dashed line to show the pattern, but don't make it a solid line because sequences are defined on the set of natural numbers. So there are no points in between consecutive terms in a sequence.

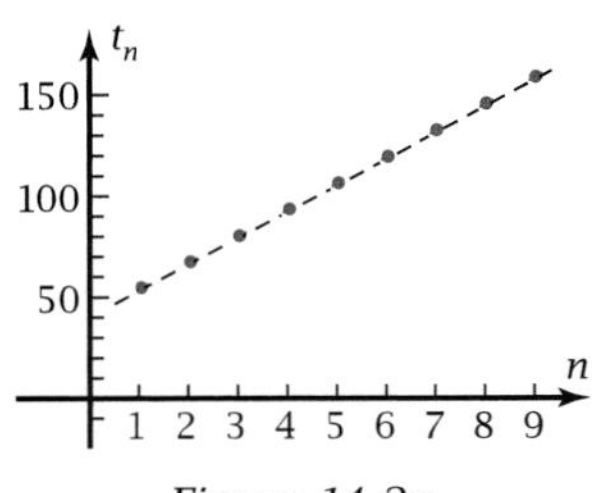

Figure 14-2a

b. To find a pattern, write the term number, n, in one column and the term, t_n, in another column. Then show the 13s being added to the preceding terms to get the next terms.

n	t_n	
1	53	
		+13
2	66	
		+13
3	79	
		+13
4	92	

To get the *fourth* term, you add the common difference of 13 *three* times to 53. So to get the 100th term, you add the common difference 99 times to 53.

$$t_{100} = 53 + 99(13) = 1340$$

c. $t_n = 53 + 13(n - 1)$ or $t_n = 40 + 13n$ ◀

The sequence in Example 1 is called an **arithmetic sequence.** You get each term by adding the same constant to the preceding term. You can also say that the difference of consecutive terms is a constant. This constant is called the **common difference.**

The pattern "add 13 to the previous term to get the next term" in Example 1 is called a *recursive* pattern for the sequence. You can write an algebraic **recursion formula:**

$$t_n = t_{n-1} + 13$$

Of course, in this example it is necessary to specify the value of the first term, $t_1 = 53$. The **sequence mode** on your grapher makes calculating terms recursively easy. Here's how you would enter the equation in the y= menu on a typical grapher:

$n\text{Min} = 1$	Enter the beginning value of the term number, n.
$u(n) = u(n - 1) + 13$	Enter the recursion formula. $u(n)$ stands for t_n.
$u(n\text{Min}) = \{53\}$	Enter the first term. The braces are used in case there is more than one given term.

Pressing table gives

n	$u(n)$
1	53
2	66
3	79
⋮	⋮

The pattern $t_n = 53 + 13(n - 1)$ you saw in part c of Example 1 is called an **explicit formula** for the sequence. It "explains" how to calculate any desired term without finding the terms before it.

▶ EXAMPLE 2

When you leave money in a savings account, the interest is **compounded.** This means that interest is paid on the previously earned interest as well as on the amount originally deposited. If the interest is 6% per year, compounded once a year, the amount at the beginning of any year is 1.06 times the amount at the beginning of the previous year. Suppose that parents invest \$1000 in an account on their baby's first birthday.

a. Find recursively the first four terms, t_1, t_2, t_3, and t_4, in the sequence of amounts.

b. Make a point plot of t_n as a function of n for the first 18 birthdays.

c. Calculate explicitly the value of t_{18}, the amount on the 18th birthday.

d. Write an explicit formula for the amount, t_n, as a function of the birthday number, n.

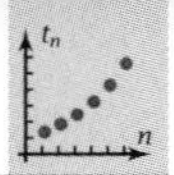

e. If the money is left in the account and the interest rate stays the same, when would the amount first exceed \$11,000?

Solution

a.

Birthday, n	Dollars, t_n
1	1000.00
2	1060.00
3	1123.60
4	1192.02

$\times 1.06$ between consecutive terms.

Calculate *without* rounding; round the *answers.*

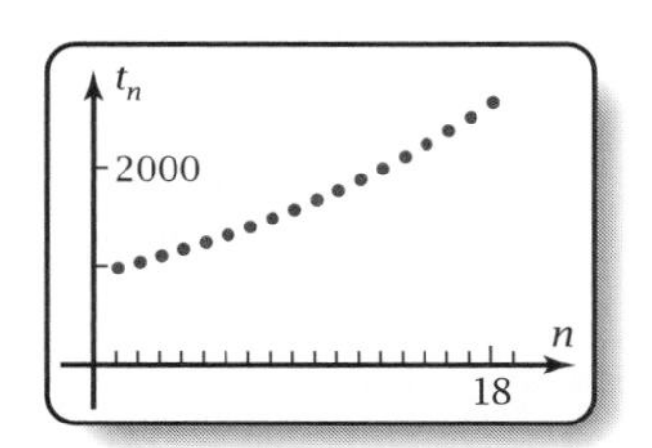

Figure 14-2b

b. With your grapher in sequence mode, enter the recursion formula in the y= menu.

$$n\text{Min} = 1$$

$$u(n) = u(n - 1) * 1.06$$

$$u(n\text{Min}) = \{1000\}$$

Figure 14-2b shows the point plot.

c. For the *fourth* birthday, you multiply 1000 by *three* factors of 1.06. So for the 18th birthday, you multiply 1000 by 17 factors of 1.06:

$$t_{18} = 1000 \times 1.06^{17} = 2692.7727\ldots \approx \$2692.77$$

d. $t_n = 1000 \times 1.06^{n-1}$ Multiply 1000 by $(n - 1)$ factors of 1.06.

e. Algebraic solution:

$$1000 \times 1.06^{n-1} = 11{,}000$$

$$1.06^{n-1} = 11$$

$$(n - 1) \log 1.06 = \log 11$$

$$n - 1 = \frac{\log 11}{\log 1.06} = 41.1522\ldots \Rightarrow n = 42.1522\ldots$$

The amount would first exceed \$11,000 on the 43rd birthday.

Realize that n was rounded *upward* in this case.

Numerical solution: With your grapher in sequence mode, make a table and scroll down to $n = 43$, the first year in which t_n exceeds 11,000. ◀

The sequence in Example 2 is called a **geometric sequence.** You get each term by multiplying the previous term by the same constant. You can also say that the ratio of consecutive terms is a constant. This constant is called the **common ratio.** Notice that this pattern is the same as the add-multiply property of exponential functions, which you learned about in Chapter 7.

▶ **EXAMPLE 3** For the sequence 6, 12, 20, 30, 42, 56, 72, . . .

a. Find a recursion formula for t_n as a function of t_{n-1}. Use it to find the next few terms.

b. Find an explicit formula for t_n as a function of n. Use it to find t_{100}.

Solution

a. Make a table showing term number and corresponding term value.

n	t_n		
1	6		2×3
		$+6$	
2	12		3×4
		$+8$	
3	20		4×5
		$+10$	
4	30		5×6
		$+12$	
5	42		6×7
		$+14$	
6	56		7×8
		$+16$	
7	72		8×9

The terms progress by adding amounts that increase by 2 each time. So

$$t_n = t_{n-1} + 2(n + 1)$$

Enter this recursion formula in the y= menu with nMin = 1 and $t_1 = u(n\text{Min}) = 6$. Set the table to start at 7 so that the last given value appears in the table, thus checking your work.

n	t_n
7	72
8	90
9	110
10	132
⋮	⋮

b. The terms are also products of consecutive integers, term number plus 1 and term number plus 2, as shown in the table in part a. So an explicit formula is

$$t_n = (n + 1)(n + 2)$$

$$t_{100} = (101)(102) = 10{,}302$$ ◀

Example 3 illustrates that the recursion formula is useful for finding the next few terms, but the explicit formula for t_n in terms of n lets you find terms farther along in the sequence without having to find all of the intermediate terms.

These definitions pertain to sequences.

DEFINITION: Sequences

A **sequence** is a function whose domain is the set of positive integers. The independent variable is the term number, n, and the dependent variable is the term value, t_n.

A **recursion formula** for a sequence specifies t_n as a function of the preceding term, t_{n-1}.

An **explicit formula** for a sequence specifies t_n as a function of n.

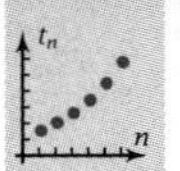

Notes:

- A sequence can have a finite or infinite number of terms, depending on whether its domain is finite or infinite.
- A recursion formula gives an easy way to find the next few terms.
- An explicit formula is useful for calculating terms later in the sequence or for calculating the term number for a given term.

DEFINITIONS: Arithmetic and Geometric Sequences

An **arithmetic sequence** is a sequence in which each term is formed recursively by *adding* a constant to the previous term. The constant added is called the **common difference.**

A **geometric sequence** is a sequence in which each term is formed recursively by *multiplying* the previous term by a constant. The constant multiplier is called the **common ratio.**

Notes:

- An arithmetic sequence is a *linear function* of the term number.
- A geometric sequence is an *exponential function* of the term number.

There are techniques for finding a specified term or for finding the term number of a given term.

TECHNIQUES: Terms, Term Numbers, and Graphs of Sequences

To find more terms in a sequence, make a table of term numbers and terms and then

- Find a recursive pattern and follow the pattern to the desired term.
- Find an explicit formula for t_n in terms of n and substitute a value for n.

To find the value of n for a given term,

- Follow the recursive pattern until you reach the given term.
- Substitute the given term into the explicit formula and solve for n.

To plot the graph of a sequence,

- Make a table of n and t_n on your grapher, then plot the points (n, t_n).
- Set your grapher in sequence mode, enter the formulas for t_n, and then graph.

Problem Set 14-2

Do These Quickly

Q1. What kind of function has the add-multiply property?

Q2. What kind of function has the multiply-add property?

Q3. What kind of function has the add-add property?

Q4. What kind of function has the multiply-multiply property?

Q5. If y is a direct cube power function of x, then what does doubling x do to y?

Q6. What is an integer?

Q7. Solve: $x^2 + 7x + 6 = 0$

Q8. Add the vectors $3\vec{i} + 4\vec{j}$ and $2\vec{i} - 5\vec{j}$.

Q9. Add the complex numbers $3 + 4i$ and $2 - 5i$.

Q10. Write polar coordinates of the point $(-5, 30°)$ using a *positive* value of r.

For Problems 1–10,

a. Tell whether the sequence is arithmetic, geometric, or neither.

b. Write the next two terms.

c. Find t_{100}.

d. Find the term number of the term after the first ellipsis marks.

1. 27, 36, 48, . . . , 849490.02..., . . .
2. 27, 31, 35, . . . , 783, . . .
3. 58, 45, 32, . . . , −579, . . .
4. 100, 90, 81, . . . , 3.0903..., . . .
5. 54.8, 137, 342.5, . . . , 3266334.53..., . . .
6. 67.3, 79, 90.7, . . . , 38490.1, . . .
7. 50, −45, 40.5, . . . , −15.6905..., . . .
8. −1234, −1215.7, −1197.4, . . . , 2426, . . .
9. 0, 3, 8, 15, 24, 35, 48, 63, 80, 99, . . . , 3248, . . .
10. 4, 10, 18, 28, 40, 54, 70, . . . , 178504, . . .
11. *Grain of Rice Problem:* A story is told that the person who invented chess centuries ago was to be rewarded by the king. The inventor gave the king a simple request: "Place one grain of rice on the first square of a chess board, place two grains on the second, then four, eight, and so forth, till all 64 squares are filled." What kind of sequence do the numbers of grains form? On which square would the number of grains first exceed 1000? How many grains would be on the last square? Why do you think the king was upset about having granted the inventor's request?

12. *George Washington's Will Problem:* Suppose you find that when George Washington died in 1799, he left \$1000 in his will to your ancestors. The money has been in a savings account ever since, earning interest. The amounts 1, 2, and 3 years after Washington died were \$1050.00, \$1102.50, and \$1157.63, respectively. Show that these numbers form a

Arithmetic sequences:

$t_1 + d + d + \cdots$ (t_2 = $t_1 + d$; t_3 = $t_1 + d + d$; ...)

Geometric sequences:

$t_1 \cdot r \cdot r \cdot \cdots$ (t_2 = $t_1 \cdot r$; t_3 = $t_1 \cdot r \cdot r$; ...)

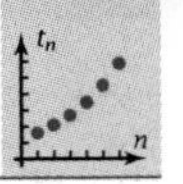

geometric sequence (allowing for round-off, if necessary). When will (or did) the total in the account first exceed $1,000,000? How much would be in the account this year? Why do you think banks have rules limiting the number of years money can be left in a dormant account before they stop paying interest on it?

13. *Depreciation Problem:* The Internal Revenue Service (IRS) assumes that an item that can wear out, such as a house, car, or computer, **depreciates** by a constant number of dollars per year. (If the item is used in a business, the owner is allowed to subtract the amount of the depreciation from the business's income before figuring taxes.) Suppose that an office building is originally valued at $1,300,000.
    a. If the building depreciates by $32,500 per year, write the first few terms of the sequence of values of the building after 1, 2, 3, . . . years. What kind of sequence do these numbers follow? How much will the building be worth after 30 years? How long will it be until the building is *fully depreciated*? Why does the IRS call this *straight-line depreciation*?
    b. Suppose that the IRS allows the business to take *accelerated depreciation,* each year deducting 10% of the building's value at the beginning of the year. Write the first few terms in the sequence of values in each year of its life. How much will the business get to deduct the first, second, and third years of the building's life? How old will the building be when the business gets to deduct less than $32,500, which is the amount using straight-line depreciation?

14. *Piggy Bank Problem:* Suppose that you decide to save money by putting $5 into a piggy bank the first week, $7 the second week, $9 the third week, and so forth.
    a. What kind of sequence do the deposits form? How much will you deposit at the end of the tenth week? In what week will you deposit $99?
    b. Find the total you would have in the bank at the end of the tenth week. Show that you can calculate this total by averaging the first and the tenth deposits and then multiplying this average by the number of weeks.
    c. What is the total amount you would have in the bank at the end of a year? (Do the computation in a time-efficient way.)

15. *Laundry Problem:* An item of clothing loses a certain percentage of its color with each washing. Suppose that a pair of blue jeans loses 9% of its color with each washing. What percentage remains after the first, second, and third washings? What kind of sequence do these numbers form? What percentage of the original color would be left after 20 washings? How many washings would it take until only 10% of the original color remains?

16. *Ancestors Problem:* Your ancestors in the first, second, and third generations back are your biological parents, grandparents, and great-grandparents, respectively.

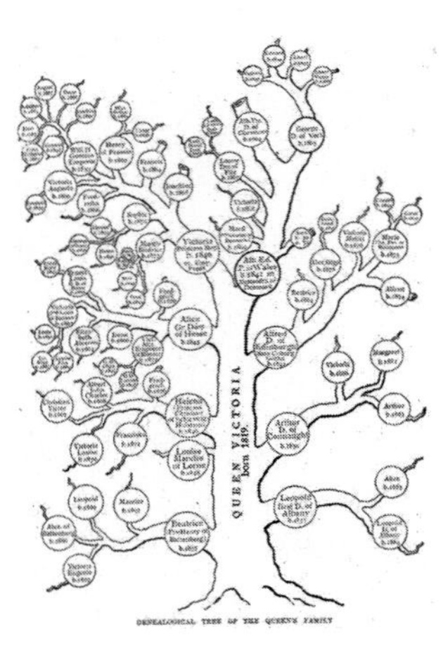

The genealogy tree of Queen Victoria (The Granger Collection, New York)

    a. Write the number of ancestors you have (living or dead) in the first, second, and third generations back. What kind of sequence do these numbers form? How many ancestors do you have in the 10th generation back? In the 20th generation back?
    b. As the number of generations back gets larger, the calculated number of ancestors increases without limit and will eventually exceed the population of the world. What do you conclude must be true to explain this seeming contradiction?

17. *Fibonacci Sequence Problem:* These numbers form the **Fibonacci sequence:**

$$1, 1, 2, 3, 5, 8, 13, 21, 34, 55, \ldots$$

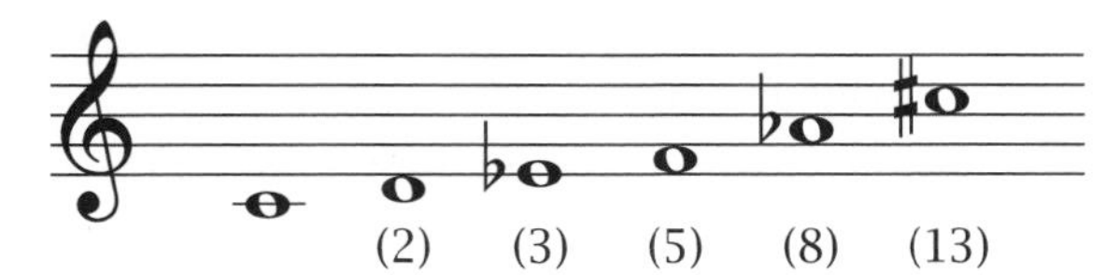

The successive tones of Béla Bartók's musical scale increase in a Fibonacci sequence of halftones.

a. Figure out the recursion pattern followed by these **Fibonacci numbers.** Write the next two terms of the sequence. Enter the recursion formula into your grapher. You will need to enter $u(n\text{Min}) = \{1, 1\}$ to show that the first two terms are given. Make a table of Fibonacci numbers and scroll down to find the 20th term of the sequence.

b. Find the first ten ratios, r_n, of the Fibonacci numbers, where

$$r_n = \frac{t_{n+1}}{t_n}$$

Show that these ratios get closer and closer to the **golden ratio,**

$$r = \frac{\sqrt{5} + 1}{2} = 1.61803398\ldots$$

c. Find a pinecone, a pineapple, or a sunflower, or a picture of one of these. Each has sections formed by intersections of two spirals, one in one direction and another in the opposite direction. Count the number of spirals in each direction. What do you notice about these numbers?

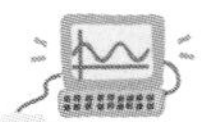

d. Look up Leonardo Fibonacci (also known as Leonardo of Pisa) on the Internet or via another reference source. Find out when and where he lived. See if you can find out how he related the sequence to the growth of a population of rabbits, and why, therefore, his name is attached to the sequence.

18. *Factorial Sequence Problem:* These numbers form the sequence of factorials:

    1, 2, 6, 24, 120, 720, . . .

    a. Figure out a recursive pattern in the sequence and use it to write the next two factorials.

    b. Recall from Chapter 9 that you use the exclamation mark, !, to designate a factorial. For example, 6! = 720. Write a recursion formula and use it to find 10! and 20!. What do you notice about the magnitude of the values? Think of a possible reason the exclamation mark is used for factorials.

19. *Staircase Problem:* Debbie can take the steps of a staircase 1 at a time or 2 at a time. She wants to find out how many different ways she can go up staircases with different numbers of steps. She realizes that there is one way she can go up a staircase of 1 step and two ways she can go up a staircase of 2 steps (1 and 1, or both steps simultaneously).

    a. Explain why the number of ways she can go up 3- and 4-step staircases are 3 and 5, respectively.

    b. If Debbie wants to get to the 14th step of a staircase, she can reach it either by taking 1 step from the 13th step or 2 steps from the 12th step. So the number of ways to get to step 14 is the number of ways to get to step 13 plus the number of ways to get to step 12. Let n be the number of steps in the staircase, and let t_n be the number of different ways she can go up that staircase. Write a recursion formula for t_n as a function of t_{n-1} and t_{n-2}. Use the recursion formula to find the number of ways she could go up a 20-step staircase. On your grapher, you must enter $u(n\text{Min}) = \{2, 1\}$ to show that $t_2 = 2$ and $t_1 = 1$.

    c. How does the number of ways of climbing stairs relate to the Fibonacci sequence in Problem 17?

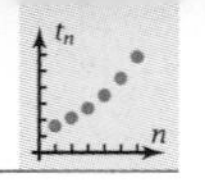

d. In how many different ways could Debbie go up the 91 steps to the top of the pyramid in Chichen Itza, Mexico? Surprising?

20. *Mortgage Payment Problem:* Suppose that a family borrows \$150,000 to purchase a house. They agree to pay back this *mortgage* at \$1074.65 per month. But part of that payment goes to pay the interest for the month on the balance remaining. The interest rate is 6% per year, so they pay 0.5% per month. The balance b_n remaining after month n is given by the recursion formula

$$b_n = b_{n-1} + 0.005b_{n-1} - 1074.65$$

a. Explain the meaning of each of the three terms in the recursion formula.

b. Find b_{12}, the balance remaining at the end of the first year of the mortgage. How much money did the family pay altogether for the year? How much of this amount went to pay interest, and how much went to reducing the balance of the mortgage?

c. After how many months will the balance have dropped to zero and the mortgage be paid off?

21. *Arithmetic and Geometric Means Problem:* **Arithmetic means** and **geometric means** between two numbers are terms between two numbers that form an arithmetic or geometric sequence with the two numbers.

a. Insert three arithmetic means between 47 and 84 so that 47, —?—, —?—, —?—, 84 is part of an arithmetic sequence.

b. Insert three geometric means between 3 and 48 so that 3, —?—, —?—, —?—, 48 is part of a geometric sequence.

c. There are two different sets of geometric means in part b, one of which involves only positive numbers and another that involves both positive and negative numbers. Find the set of means you didn't find in part b.

14-3 Series and Partial Sums

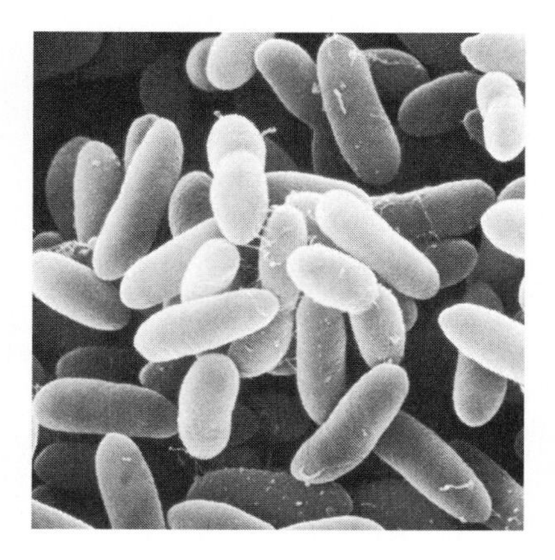

A population of bacteria grows by subdividing. Suppose that the number of new bacteria in any one generation is a term in the sequence

$$5, 12, 21, 32, 45, \ldots$$

The total number of bacteria present at any time n is the sum of the terms in the sequence,

$$5 + 12 + 21 + 32 + 45 + \cdots + t_n$$

The indicated sum of the terms of a sequence is called a series as you saw it defined in Section 14-1. The total number of bacteria at the fifth generation, for instance, is the fifth partial sum of the series,

$$5 + 12 + 21 + 32 + 45 = 115$$

On the fifth day there are 45 new bacteria, for a total of 115 bacteria.

In this section you will learn ways of calculating partial sums of series. You will also encounter **binomial series** that come from raising a binomial to a power, such as $(a + b)^{10}$.

OBJECTIVES

- Given a series, find a specified partial sum, or find the number of terms if the partial sum is given.
- Use sigma notation to write partial sums.
- Given a power of a binomial, expand it as a binomial series.

Numerical Computation of Partial Sums of Series: Sigma Notation

The next example shows you how to calculate a partial sum of a series directly, by adding the terms on your grapher.

▶ EXAMPLE 1 For the series $5 + 12 + 21 + 32 + 45 + \cdots$, above, calculate S_{100}, the 100th partial sum.

Solution The terms in the given series are products of integers, as shown in this table:

n	t_n	Pattern
1	5	5×1
2	12	6×2
3	21	7×3
4	32	8×4
⋮	⋮	⋮
n	$(n + 4)(n)$	

Write or download into your grapher a program to compute partial sums. Call it Series. The program should use a formula stored in the y= menu to calculate the term values. The program should allow you to put in the desired number of terms. Then it should enter a loop that calculates the term values one at a time and accumulates them by adding each term to a variable such as S (for sum). At each iteration the program should have your grapher display the current partial sum. The final output should be the last partial sum. Your program should give

$$S_{100} = 358{,}550$$ ◀

A partial sum can be written compactly using sigma notation. The symbol Σ, the uppercase Greek letter sigma, is often used to indicate a sum.

$$S_{100} = \sum_{n=1}^{100} (n + 4)(n)$$

The expression on the right side of the equation is read "The sum from $n = 1$ to 100 of $(n + 4)(n)$." It means to substitute $n = 1, 2, 3, \ldots, 100$ into the formula, do the computations, and add the results. The variable n is called the **term index.** You may recall sigma notation from your work in Chapter 8.

Finding Partial Sums of Arithmetic Series Algebraically

Suppose that you put \$7 in a piggy bank the first day, \$10 the second day, \$13 the third day, and so forth. The amounts you put in follow the arithmetic sequence

$$7, 10, 13, 16, 19, \ldots$$

The total in the piggy bank on any one day is a partial sum of the arithmetic series that comes from adding the appropriate number of terms in the preceding sequence.

Day 1: 7

Day 2: $7 + 10 = 17$

Day 3: $7 + 10 + 13 = 30$

Day 4: $7 + 10 + 13 + 16 = 46$

$\vdots$

Suppose you want to find S_{10}, the tenth partial sum of the series:

$$S_{10} = 7 + 10 + 13 + 16 + 19 + 22 + 25 + 28 + 31 + 34 = 205$$

A pattern shows up if you add the first and last terms, the second and next-to-last, and so forth.

$$\begin{aligned} S_{10} &= (7 + 34) + (10 + 31) + (13 + 28) + (16 + 25) + (19 + 22) \\ &= 41 + 41 + 41 + 41 + 41 \\ &= 5(41) \\ &= 205 \end{aligned}$$

So a time-efficient way to find the partial sum algebraically is to add the first and last terms and then multiply the result by the number of pairs of terms:

$$S_{10} = \frac{10}{2}(7 + 34) = 5(41) = 205$$

By associating the 2 in the denominator with the 7 + 34, you can see that

$$S_{10} = 10 \cdot \frac{7 + 34}{2} = 10 \cdot 20.5 = 205$$

So the nth partial sum is the same as the sum of n terms, each of which is equal to the *average* of the first and last terms. This fact allows you to see why the pattern works for an odd number of terms as well as for an even number.

▶ ***EXAMPLE 2*** Find algebraically the 100th partial sum of the arithmetic series $53 + 60 + 67 + \cdots$. Check the answer by computing the partial sum numerically, as in Example 1.

Solution

$t_{100} = 50 + 99(7) = 746$ — Add 99 common differences to 53 to get the 100th term.

$S_{100} = \frac{100}{2}(53 + 746) = 39{,}950$ — There are $\frac{100}{2}$ pairs, each equal to the first term plus the last term.

Check: Enter $y_1 = 53 + (x - 1)(7)$. Then run the grapher program to get

$$S_{100} = 39{,}950$$

which agrees with the algebraic solution. ◀

Finding Partial Sums of Geometric Series Algebraically

If you make regular deposits (for example, monthly) into a savings account, the total you have in the account at any given time is a partial sum of a geometric series. It is possible to calculate such partial sums algebraically.

Consider this sixth partial sum of a geometric series:

$$S_6 = 7 + 21 + 63 + 189 + 567 + 1701$$

The first term is 7, and the common ratio is 3. If you multiply both sides of the equation by 3 and subtract the result from the original sum, you get

$$\begin{aligned} S_6 &= 7 + 21 + 63 + 189 + 567 + 1701 \\ 3S_6 &= 21 + 63 + 189 + 567 + 1701 + 5103 \\ \hline S_6 - 3S_6 &= 7 + 0 + 0 + 0 + 0 - 5103 \end{aligned}$$

Subtract, top minus bottom.

The middle terms *telescope,* or cancel out, leaving only the first term of the top equation minus the last term of the bottom equation. So

$$S_6 - 3S_6 = 7 - 7 \cdot 3^6$$ 5103 is $7 \cdot 3^6$.

$$S_6(1 - 3) = 7(1 - 3^6)$$ Factor out S_6 on the left and 7 on the right.

$$S_6 = 7 \cdot \frac{1 - 3^6}{1 - 3}$$ Divide by $(1 - 3)$.

Because 7 is the first term, t_1, 3 is the common ratio, r, and 6 is the number, n, of terms to be added, you can conclude that, in general,

$$S_n = t_1 \cdot \frac{1 - r^n}{1 - r}$$

Verbally, you can remember this result by saying, "First term times a fraction. The fraction is 1 minus the common ratio to the n, divided by 1 minus the common ratio."

▶ EXAMPLE 3 If you deposit \$100 a month into an account that pays 6% per year interest, compounded monthly, then each deposit earns 0.5% per month interest. For instance, after four deposits, the first one has earned three months' interest, the second has earned two months' interest, the third has earned one month's interest, and the last has earned no interest. Thus, the total is

$$S_4 = 100 + 100(1.005) + 100(1.005)^2 + 100(1.005)^3$$

a. How much will be in the account after ten years (120 deposits)? How much of this is interest?

b. How long will it take until the total in the account first exceeds \$50,000?

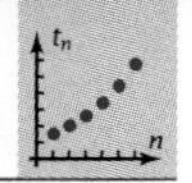

Solution

a. Algebraically: The series is geometric, with first term 100 and common ratio 1.005.

$$S_{120} = 100 \cdot \frac{1 - 1.005^{120}}{1 - 1.005} = 16{,}387.9346\ldots \approx \$16{,}387.93$$

Numerically: Enter $y = 100(1.005)^{x-1}$ into the y= menu and run the series program.

$$S_{120} = 16{,}387.9346\ldots \approx \$16{,}387.93$$

The amount of interest is \$16,387.93 − \$12,000 = \$4,387.93.

b. Algebraically:

$$50{,}000 = 100 \cdot \frac{1 - 1.005^n}{1 - 1.005}$$

$$-2.5 = 1 - 1.005^n$$

$$1.005^n = 3.5$$

$$n \log 1.005 = \log 3.5$$

$$n = \frac{\log 3.5}{\log 1.005} = 251.1784\ldots$$

It will take 252 months for the amount in the account to exceed \$50,000.

Numerically: Store $100(1.005)^{x-1}$ in the y= menu and run the series program until the amount first exceeds 50,000. This will be \$50,287.41 at 252 months. ◀

Convergent and Divergent Geometric Series

This is an example of a classic problem. Assume a person who is 200 cm from another person is allowed to take steps each of which is half the remaining distance. So the steps will be of length

100, 50, 25, 12.5, 6.25, . . .

The total distance traveled will be given by the geometric series

$$100 + 50 + 25 + 12.5 + 6.25 + \cdots$$

The person in motion will never go the entire 200 cm, but the partial sums of the series **converge** to 200 as a limit.

$$S_5 = 193.75$$

$$S_{10} = 199.8046\ldots$$

$$S_{20} = 199.999809\ldots$$

$$S_{30} = 199.9999998\ldots$$

The algebraic formula for S_n shows you why this happens:

$$S_n = 100 \cdot \frac{1 - 0.5^n}{1 - 0.5}$$

The value of 0.5^n approaches zero as n becomes large. So you can write

$$\lim_{n \to \infty} S_n = 100 \cdot \frac{1 - 0}{1 - 0.5} = 100 \cdot 2 = 200$$

The symbol in front of the S_n is read "The limit as n approaches infinity." A geometric series will converge to a limit if the common ratio r satisfies the inequality $|r| < 1$. If $|r| \geq 1$, then the terms of the series do not go to zero, and thus the series **diverges.** The partial sums do not approach a limit.

▶ ***EXAMPLE 4*** To what limit does the geometric series $50 + 45 + 40.5 + \cdots$ converge? How many terms must be added in order for the partial sums to be within one unit of this limit?

Solution

$$r = \frac{45}{50} = 0.9$$

$$\therefore S_n = 50 \cdot \frac{1 - 0.9^n}{1 - 0.9} = 500(1 - 0.9^n)$$ Divide 50 by (1 − 0.9).

$$\lim_{n \to \infty} S_n = 500(1 - 0) = 500$$ 0.9^n approaches 0 as n approaches infinity.

$$499 = 500(1 - 0.9^n)$$ Substitute (500 − 1) for S_n.

$$0.998 = 1 - 0.9^n$$

$$0.9^n = 0.002$$

$$n \log 0.9 = \log 0.002$$ Take the log of both sides.

$$n = \frac{\log 0.002}{\log 0.9} = 58.9842\ldots$$ Solve.

To be within one unit of the limit, 59 terms must be added.

Round to the next higher term. ◀

$0.9^5 = 0.590\ldots$
$0.9^{10} = 0.348\ldots$
$0.9^{50} = 0.0051\ldots$
$0.9^{100} = 0.000026\ldots$
The larger the exponent, the closer the result is to zero.

Binomial Series

If you expand a power of a binomial expression, you get a series with a finite number of terms. For instance,

$$(a+b)^5 = a^5 + 5a^4b + 10a^3b^2 + 10a^2b^3 + 5ab^4 + b^5$$

Such a series is called a **binomial series.** There are several patterns that you can see in the binomial series that comes from expanding $(a+b)^n$. A binomial series is also called a **binomial expansion.**

- There are $(n+1)$ terms.
- Each term has degree n.
- The powers of a start at a^n and decrease by 1 with each term. The powers of b start at b^0 and increase by 1 with each term.
- The sum of the powers in each term is n.
- The coefficients are symmetrical with respect to the ends of the series.
- The coefficients form a row of **Pascal's triangle:**

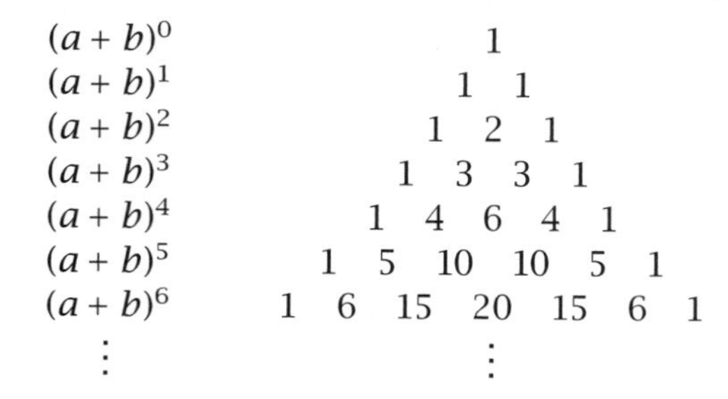

$$\begin{array}{lc}
(a+b)^0 & 1 \\
(a+b)^1 & 1 \quad 1 \\
(a+b)^2 & 1 \quad 2 \quad 1 \\
(a+b)^3 & 1 \quad 3 \quad 3 \quad 1 \\
(a+b)^4 & 1 \quad 4 \quad 6 \quad 4 \quad 1 \\
(a+b)^5 & 1 \quad 5 \quad 10 \quad 10 \quad 5 \quad 1 \\
(a+b)^6 & 1 \quad 6 \quad 15 \quad 20 \quad 15 \quad 6 \quad 1 \\
\vdots & \vdots
\end{array}$$

Each number in the interior of the triangle is the sum of the two numbers to its left and right in the row directly above. The first and last number in each row is 1.

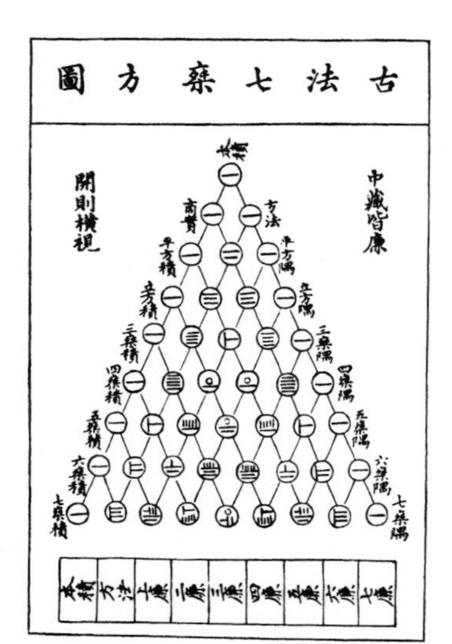

Yang Hui, a Chinese mathematician, discovered the relationship between the numbers in consecutive rows of this triangle approximately 500 years earlier than Pascal did.

- The coefficients can be calculated recursively using parts of the preceding term.

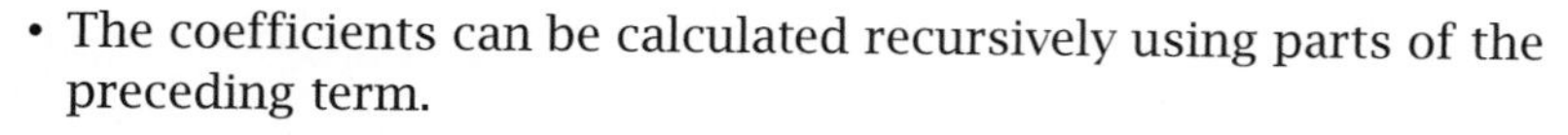

$$\frac{(\text{coefficient of } a)(\text{exponent of } a)}{\text{term number}} = \text{coefficient of next term}$$

For example, $5a^4b$ is the *second* term in the binomial series given earlier. To find the coefficient of the third term, write

$$\frac{(5)(4)}{2} = 10$$

- The coefficients can also be calculated *algebraically.* They are equal to numbers of combinations of n objects taken r at a time, where r is the exponent of b. From Section 9-5 you may recall, for example, that

$${}_5C_2 = \frac{5!}{3!\,2!} = \frac{5 \cdot 4 \cdot 3 \cdot 2 \cdot 1}{(3 \cdot 2 \cdot 1)(2 \cdot 1)} = 10$$

So the term containing b^2 is

$$\frac{5!}{3!\,2!}a^3b^2$$

The factorials in the denominator are the same as the exponents of a and b. The factorial in the numerator is the sum of these exponents, which is equal to the original exponent of the binomial. Because they appear as coefficients of the terms in a binomial series, expressions such as $\frac{5!}{3!\,2!}$ are sometimes called **binomial coefficients.** In general, the coefficient of the term containing b^r is $\frac{n!}{(n-r)!\,r!}$. Using combination notation, you can write this expression as ${}_nC_r$. Another common way to write the same expression is $\binom{n}{r}$, which is read "n choose r." Recall that 0! is defined as 1. So $\binom{n}{0} = 1$ and $\binom{n}{n} = 1$. Using this notation, you can write a general formula for finding the terms of a binomial series (or "expanding a binomial").

Binomial Formula (or Binomial Theorem)

For any positive integer n and any numbers a and b,

$$(a+b)^n = a^n + \binom{n}{1}a^{n-1}b + \binom{n}{2}a^{n-2}b^2 + \cdots + \binom{n}{n-2}a^2b^{n-2} + \binom{n}{n-1}ab^{n-1} + b^n$$

Note: You can also write the binomial formula compactly with sigma notation:

$$(a+b)^n = \sum_{r=0}^{n}\binom{n}{r}a^{n-r}b^r$$

▶ EXAMPLE 5 Use the binomial formula to expand the binomial $(x-2y)^4$.

Solution

$$\begin{aligned}(x-2y)^4 &= x^4 + \binom{4}{1}x^3(-2y) + \binom{4}{2}x^2(-2y)^2 + \binom{4}{3}x(-2y)^3 + (-2y)^4 \\ &= x^4 + 4x^3(-2y) + 6x^2(-2y)^2 + 4x(-2y)^3 + (-2y)^4 \\ &= x^4 - 8x^3y + 24x^2y^2 - 32xy^3 + 16y^4\end{aligned}$$

◀

▶ EXAMPLE 6 Find the eighth term of the binomial series that comes from expanding $(3-2x)^{12}$.

Solution The eighth term contains $(-2x)^7$. Therefore,

$$\text{8th term} = \frac{12!}{5!\,7!}(3)^5\,(-2x)^7 = 792(243)(-128x^7) = -24634368x^7$$

◀

This table summarizes the properties of series from this section.

PROPERTY: Formulas for Arithmetic, Geometric, and Binomial Series

Arithmetic Series

The terms of the arithmetic sequence, t_n, progress by *adding* a common difference d.

$t_n = t_1 + (n - 1)d$ Add $(n - 1)$ common differences to the first term.

The nth partial sum of the arithmetic series is

$S_n = \frac{n}{2}(t_1 + t_n)$ Add $\frac{n}{2}$ terms, each equal to the sum of the first and last terms.

Geometric Series

The terms of the geometric sequence, t_n, progress by *multiplying* by a common ratio r.

$t_n = t_1 \cdot r^{n-1}$ Multiply the first term by $(n - 1)$ common ratios.

The nth partial sum of the geometric series is

$S_n = t_1 \cdot \frac{1 - r^n}{1 - r}$ Multiply the first term by a fraction.

The sum of the infinite geometric series is

$$\lim_{n \to \infty} S_n = t_1 \cdot \frac{1}{1 - r} \quad (\text{if } |r| < 1)$$

Binomial Series

The terms come from expanding a binomial $(a + b)^n$:

$$\text{Term with } b^r \text{ is } \frac{n!}{(n - r)!\, r!} a^{n-r} b^r$$

The sum of the binomial series, or binomial formula, is

$$(a + b)^n = \sum_{r=0}^{n} \binom{n}{r} a^{n-r} b^r$$

Problem Set 14-3

Do These Quickly

Q1. Write the next two terms of this arithmetic sequence: 10, 20, . . .

Q2. Write the next two terms of this geometric sequence: 10, 20, . . .

Q3. Write the next two terms of this **harmonic sequence:** $\frac{1}{3}, \frac{1}{4}, \frac{1}{5}, \frac{1}{6}, \ldots$

Q4. Write the next two terms of this factorial sequence: 1, 2, 6, . . .

Q5. Find the 101st term of the arithmetic sequence with first term 20 and common difference 3.

Q6. Find t_{101} for the geometric sequence with $t_1 = 20$ and $r = 1.1$.

Q7. Multiply the complex numbers $(2 + 3i)(2 - 3i)$ and simplify.

Q8. Find the dot product of these vectors: $(2\vec{i} + 3\vec{j}) \cdot (2\vec{i} - 3\vec{j})$

Q9. Which conic section is the graph of this equation? $x^2 - y^2 + 3x - 5y = 100$

Q10. Multiply these matrices:

$$\begin{bmatrix} 3 & 2 \\ 5 & 1 \end{bmatrix}\begin{bmatrix} 4 & 3 \\ 6 & 2 \end{bmatrix}$$

---

1. *Arithmetic Series Problem:* A series has a partial sum

$$S_{10} = \sum_{n=1}^{10} (3 + (n - 1)(5))$$

   a. Write out the terms of the partial sum. How you can tell that the series is arithmetic?

   b. Evaluate S_{10} three ways: numerically, by adding the 10 terms; algebraically, by averaging the first and last terms and multiplying by the number of terms; and numerically, by storing the formula for t_n in the y= menu and using your grapher program. Are the answers the same?

   c. Evaluate S_{100} for this series. Which method did you use?

2. *Geometric Series Problem:* A series has a partial sum

$$S_6 = \sum_{n=1}^{6} 5 \cdot 3^{n-1}$$

   a. Write out the terms of the partial sum. How you can tell that the series is geometric?

   b. Evaluate S_6 three ways: numerically, by adding the six terms; algebraically, by using the pattern (first term)(fraction involving r); and numerically, by storing the formula for t_n in the y= menu and using your grapher program. Are the answers the same?

   c. Evaluate S_{20} for this series. Which method did you use?

3. *Convergent Geometric Series Pile Driver Problem:* A pile driver pounds a piling (a column) into the ground for a new building that is being constructed (Figure 14-3a). Suppose that on the first impact the piling moves 100 cm into the ground. On the second impact the piling moves another 80 cm into the ground. Assume that the distances the piling moves with each impact form a geometric sequence.

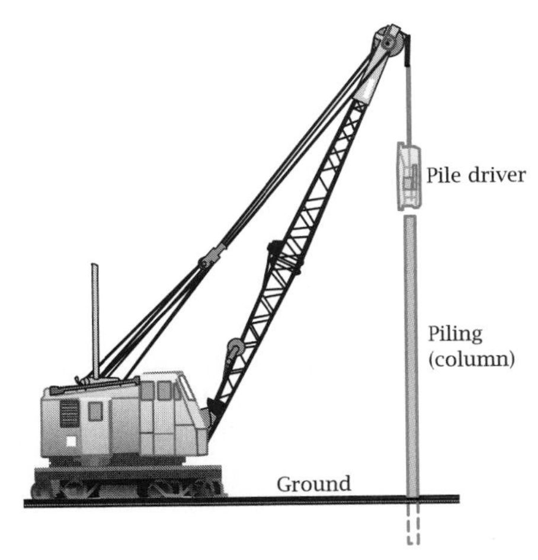

Figure 14-3a

   a. How far will the piling move on the tenth impact? How far into the ground will it be after ten impacts?

   b. Run the partial series program using $n = 100$. What do you notice about the partial sums as the grapher displays each one? What does it mean to say that the partial sums are "converging to 500"? What is the real-world meaning of this limit to which the series converges?

   c. Show how to calculate algebraically that the limit to which the partial sums converge is 500.

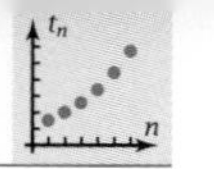

4. *Harmonic Series Divergence Problem:* If you stack a deck of cards so that they just barely balance, the top card overhangs by $\frac{1}{2}$ the deck length, the second card overhangs by $\frac{1}{3}$ the deck length, the third card overhangs by $\frac{1}{4}$ the deck length, and so on (Figure 14-3b).

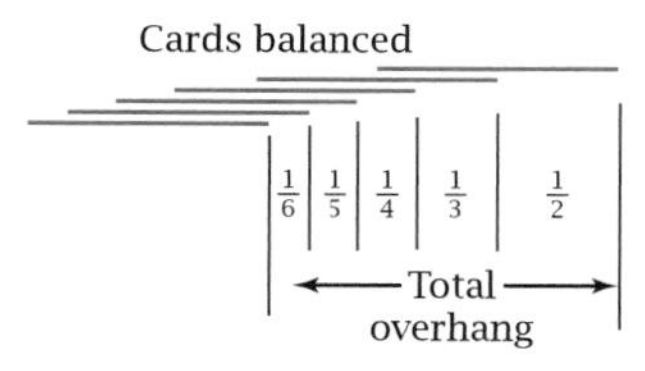

Figure 14-3b

The total overhang for n cards is thus a partial sum of the **harmonic series**

$$\frac{1}{2}+\frac{1}{3}+\frac{1}{4}+\frac{1}{5}+\frac{1}{6}+\frac{1}{7}+\cdots$$

a. The figure indicates that the total overhang for three cards is greater than the length of the deck. Show numerically that this is true.

b. How many cards would you have to stack in order for the total overhang to exceed two deck lengths?

c. What would the total overhang be for a standard 52-card deck? Surprising?

d. Associate the terms of the harmonic series this way:

$$\frac{1}{2}+\left(\frac{1}{3}+\frac{1}{4}\right)+\left(\frac{1}{5}+\frac{1}{6}+\frac{1}{7}+\frac{1}{8}\right)+\cdots$$

The terms are grouped into groups of 1, 2, 4, 8, 16, . . . terms. Show that each group of terms is greater than or equal to $\frac{1}{2}$. How does this fact allow you to conclude that the partial sums of a harmonic series *diverge* and can get larger than any real number?

5. *Geometric Series for Compound Interest Problem:* Money in an Individual Retirement Account (IRA) earns interest at a rate that usually is higher than for other accounts. Suppose that you invest \$100 in an IRA that pays 12% per year interest, compounded monthly. The amount during any one month is 1.01 times the amount the month before. The amounts follow this geometric sequence:

Month:	1	2	3	4	5	6	. . .
Dollars:	100	100(1.01)	$100(1.01)^2$	$100(1.01)^3$	$100(1.01)^4$	$100(1.01)^5$	. . .

If you make regular \$100 deposits each month, there is a geometric sequence for each deposit:

Month:	1	2	3	4	5	6	. . .
Dollars:	100	100(1.01)	$100(1.01)^2$	$100(1.01)^3$	$100(1.01)^4$	$100(1.01)^5$	. . .
		100	100(1.01)	$100(1.01)^2$	$100(1.01)^3$	$100(1.01)^4$	. . .
			100	100(1.01)	$100(1.01)^2$	$100(1.01)^3$	. . .
				100	100(1.01)	$100(1.01)^2$	. . .
					100	100(1.01)	. . .
						100	. . .

a. Explain why the amount in the IRA at the fifth month is the fifth *partial sum* of a geometric *series.* Calculate this amount using a time-efficient method.

b. If you continue the regular \$100 monthly deposits, how much will be in the IRA at the end of ten years? How much of this will be interest?

c. How many months would it take before the total first exceeds \$100,000?

6. *Present Value Compound Interest Problem:* Suppose that money is invested in a savings account at 6% annual interest, compounded monthly. Because the interest rate is 0.5% per month, the amounts in the account each month form a geometric sequence with common ratio 1.005.

   a. Find the amount you would have to invest now to have $10,000 at the end of ten years. This amount is called the *present value* of $10,000.

   b. If you invest x dollars a month into this account, the total at the end of each month is a partial sum of a geometric series with x as the first term and common ratio 1.005. How much would you have to invest each month in order to have $10,000 at the end of ten years?

7. *Geometric Series Mortgage Problem:* Suppose that someone gets a $100,000 *mortgage* (loan) to buy a house. The interest rate is $I = 1\%$ or 0.01 per month (12% per year), and the payments are $P = \$1050.00$ per month. Most of the monthly payment goes to pay the interest for that month, with the rest going to pay on the *principal,* thus reducing the *balance, B,* owed on the loan. The table shows payment, interest, principal, and balance for the first few months.

   a. Show that you understand how the table is constructed by calculating the row for month 4.

   b. The balance, B_1, after one month is given by $B_1 = B_0 + B_0I - P = B_0(1 + I) - P$. Show that the balance after four months can be written

$$B_4 = B_0(1 + I)^4 - P(1 + I)^3 - P(1 + I)^2 - P(1 + I) - P$$

Month, n	Payment, P	Interest	Principal	Balance, B_n
0				100,000.00
1	1050.00	1000.00	50.00	99,950.00
2	1050.00	999.50	50.50	99,899.50
3	1050.00	998.99	51.01	99,848.49

   Then use the formula for the partial sum of a geometric series to show that

$$B_4 = B_0(1 + I)^4 - P\frac{1 - (1 + I)^4}{1 - (1 + I)}$$

$$= B_0(1 + I)^4 + \frac{P}{I}(1 - (1 + I)^4)$$

   c. The formula in part b can be generalized to find B_n by replacing the 4s with n's. Use this information to calculate the number of months it takes to pay off the mortgage; that is, find the value of n for which $B_n = 0$.

   d. Plot the graph of B_n as a function of n from $n = 0$ to the time the mortgage is paid off. Sketch a smooth curve showing the pattern followed by the points. True or false: "Halfway through the duration of the mortgage, half of the mortgage has been paid off." Explain your reasoning.

8. *Geometric Series by Long Division Problem:* The limit, S, of the partial sums for a convergent geometric series is given by

$$S = t_1 \cdot \frac{1}{1 - r}$$

   where t_1 is the first term of the sequence and r is the common ratio.

   a. Use long division to divide $(1 - r)$ into 1. Show that the result is the geometric series

$$S = t_1 + t_1r + t_1r^2 + t_1r^3 + t_1r^4 + t_1r^5 + \cdots$$

   b. The result illustrates the way you can do mathematical problems "backward" as well as "forward." Give another instance in which you can use this forward-and-backward phenomenon.

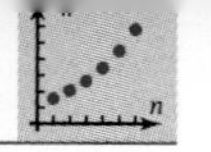

9. *Thumbtack Binomial Series Problem:* If you flip a thumbtack five times, there are six possible numbers of "point-ups" you could get, namely, 0, 1, 2, 3, 4, and 5. If the probability of "point-up" on any one flip is 60% (0.6), then the probabilities of each of the six outcomes are terms in the binomial series that comes from expanding

   $(0.4 + 0.6)^5$

   The probability of exactly three "point-ups" is the term that contains 0.6^3.

   a. Calculate the six terms of the binomial series.

   b. What is the probability of exactly three "point-ups"?

   c. Explain why the probability of no more than three "point-ups" is a partial sum of this binomial series. Calculate this probability.

   d. Calculate the probability of no more than six "point-ups" in ten flips of the thumbtack. Is the answer the same as the probability of no more that three "point-ups" in five flips?

10. *Snowflake Curve Series Problem:* Figure 14-3c shows Koch's snowflake curve, which you may have encountered in Section 11-5.

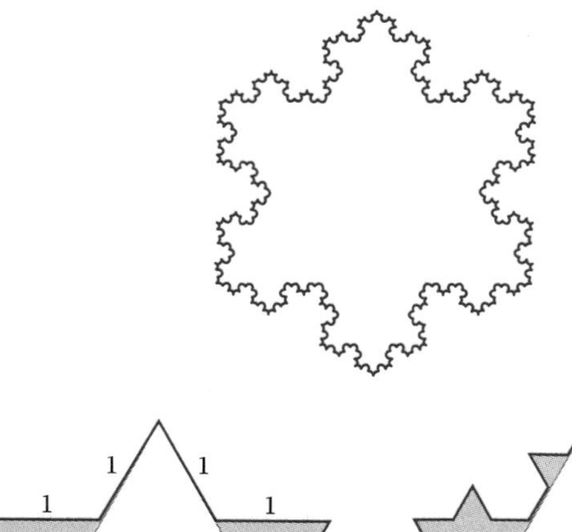

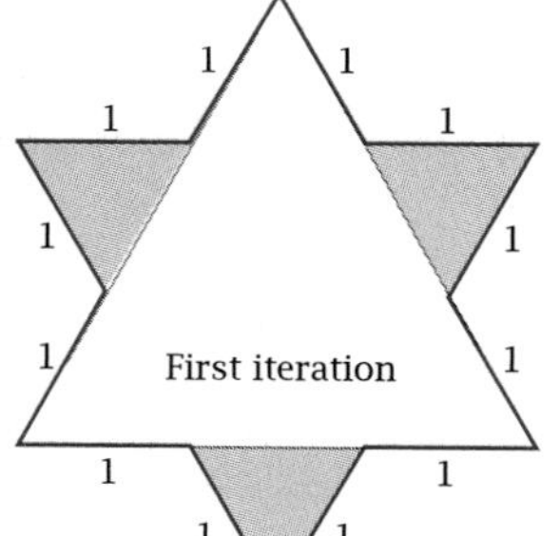

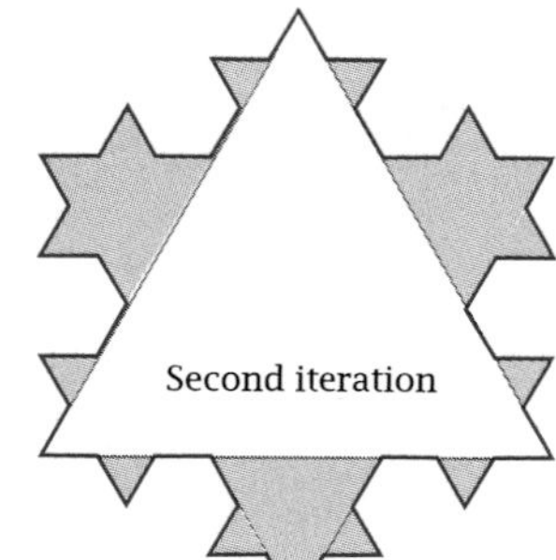

Figure 14-3c

In the first iteration, segments 1 unit long are marked on the sides of an equilateral triangle of side 3, and three equilateral triangles (shaded) are drawn. In the second iteration, equilateral triangles of side $\frac{1}{3}$ unit are constructed on each side of the first iteration. The iterations are carried on this way infinitely. The snowflake curve is the boundary of the resulting figure.

   a. Find the total perimeter of the first iteration. Find the total perimeter of the second iteration. What kind of sequence do the lengths of the iterations form? How do you conclude that the perimeter of the completed snowflake curve is *infinite*?

   b. What is the total area of the shaded triangles in the first iteration? What area is added to this with the shaded triangles in the second iteration? What kind of series do the areas of the iterations form? Does this series converge? If so, to what number? If not, show why not.

For Problems 11–14, write out the terms of the partial sum and add them.

11. $S_5 = \sum_{n=1}^{5} 2n + 7$

12. $S_7 = \sum_{n=1}^{7} n^2$

13. $S_6 = \sum_{n=1}^{6} 3^n$

14. $S_8 = \sum_{n=1}^{6} n!$

For Problems 15–22, each series is either geometric or arithmetic. Find the indicated partial sum.

15. For $2 + 10 + 50 + \cdots$, find S_{11}.

16. For $97 + 131 + 165 + \cdots$, find S_{37}.

17. For $24 + 31.6 + 39.2 + \cdots$, find S_{54}.

18. For $36 + 54 + 81 + \cdots$, find S_{29}.

19. For $1000 + 960 + 920 + \cdots$, find S_{78}.

20. For $1000 + 900 + 810 + \cdots$, find S_{22}.

21. For $50 - 150 + 450 - \cdots$, find S_{10}.

22. For $32.5 - 52 + 83.2 - \cdots$, find S_{41}.

For Problems 23–28, the series is either arithmetic or geometric. Find n for the given partial sum.

23. $32 + 43 + 54 + \cdots$, find n if $S_n = 4407$.

24. $13 + 26 + 52 + \cdots$, find n if $S_n = 425{,}971$.

25. $18 + 30 + 50 + \cdots$, find n if $S_n \approx 443{,}088$.

26. $97 + 101 + 105 + \cdots$, find n if $S_n = 21{,}663$.

27. $97 + 91 + 85 + \cdots$, find n if $S_n = 217$. (Surprising?)

28. $60 + 54 + 48.6 + \cdots$, find n if $S_n \approx 462.74$.

For Problems 29–36, state whether or not the geometric series converges. If it does converge, find the limit to which it converges.

29. $100 + 90 + 81 + \cdots$

30. $25 + 20 + 16 + \cdots$

31. $40 + 50 + 62.5 + \cdots$

32. $200 - 140 + 98 - \cdots$

33. $300 + 90 + 27 + \cdots$

34. $20 + 60 + 180 + \cdots$

35. $1000 - 950 + 902.5 - \cdots$

36. $360 + 240 + 160 + \cdots$

For Problems 37–42, expand as a binomial series and simplify.

37. $(x - y)^3$

38. $(4m - 5n)^2$

39. $(2x - 3)^5$

40. $(3a + 2)^4$

41. $(x^2 + y^3)^6$

42. $(a^3 - b^2)^5$

For Problems 43–52, find the indicated term in the binomial series.

43. $(x + y)^8$, y^5-term

44. $(p + j)^{11}$, j^4-term

45. $(p - j)^{15}$, j^{11}-term

46. $(c - d)^{19}$, d^{15}-term

47. $(x^3 - y^2)^{13}$, x^{18}-term

48. $(x^3 - y^2)^{24}$, x^{30}-term

49. $(3x + 2y)^8$, y^5-term

50. $(3x + 2y)^7$, y^4-term

51. $(r - q)^{15}$, 12th term

52. $(a - b)^{17}$, 8th term

53. *Journal Problem:* Update your journal with things you have learned so far in this chapter. Include the difference between a sequence and a series and what makes a sequence or series arithmetic or geometric. Tell how partial sums of series can be calculated numerically and how calculations for arithmetic, geometric, and binomial series can be done algebraically.

14-4 Chapter Review and Test

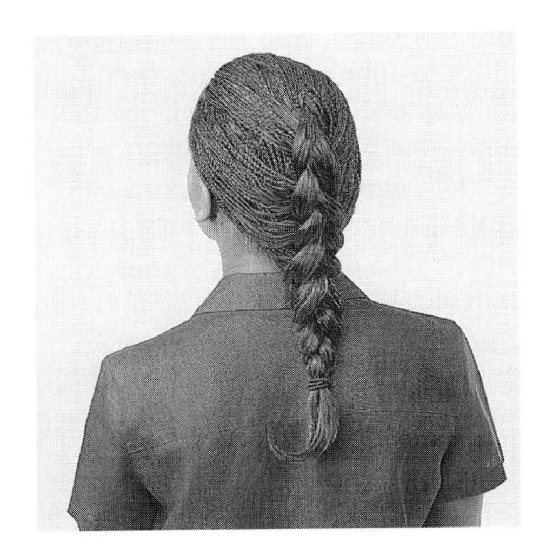

In this chapter you studied sequences of numbers. You can consider these sequences to be functions in which the independent variable is a positive integer (the term number) and the dependent variable is the term itself. You can find the term values either recursively as a function of the preceding term or explicitly as a function of the term number. Then you learned about series, which are indicated sums of the terms of sequences. You can calculate partial sums of series numerically using features available on graphing calculators. For arithmetic and geometric series, it is possible to calculate partial sums algebraically, taking advantage of the add-add property and the add-multiply property, respectively.

The French braid of this woman can be described by a finite arithmetic sequence.

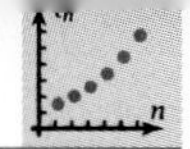

Review Problems

R0. Update your journal with what you have learned in this chapter. Include such things as:

- The difference between a sequence and a series
- The similarities and differences between arithmetic and geometric series
- The meaning and computation of terms in a binomial series
- Numerical ways to compute term values and partial sums for various kinds of series
- Algebraic ways to compute terms and partial sums of arithmetic and geometric series
- An algebraic way to find the sum of an infinite geometric series with $|r| < 1$

R1. a. Find the next two terms of the arithmetic sequence 5, 8, 11, 14, ...

b. Find the sixth partial sum of the arithmetic series $5 + 8 + 11 + 14 + \cdots$.

c. Show that the sixth partial sum in part b equals 6 times the average of the first and last terms.

d. Find the next two terms of the geometric sequence 5, 10, 20, 40, ...

e. Find the sixth partial sum of the geometric series $5 + 10 + 20 + 40 + \cdots$.

R2. a. Tell whether the sequence 23, 30, 38, . . . is arithmetic, geometric, or neither.

b. Find t_{200}, the 200th term of the arithmetic sequence 52, 61, 70, ...

c. 3571 is a term in the sequence in part b. What is its term number?

d. Find t_{100}, the 100th term of the geometric sequence with $t_1 = 200$ and $r = 1.03$.

e. $t_n = 5644.6417...$ is a term of the sequence in part d. What does n equal?

f. Find a recursion formula and write the next three terms of the sequence 0, 3, 8, 15, 24, ...

g. Find an explicit formula and use it to calculate the 100th term of the sequence in part f.

h. *Monthly Interest Problem:* Suppose that you invest \$3000 in an account that pays 6% per year interest (0.5% per month), compounded monthly. At any time during the first month, you have \$3000 in the account. At any time during the second month, you have \$3000 plus the interest for the first month, and so on.

i. Show that the month number and the amount in the account during that month have the add-multiply property of exponential functions. Explain why a sequence is a more appropriate mathematical model than is a continuous exponential function.

ii. In what month would the amount first exceed \$5000?

iii. If you leave the money in the account until you retire 50 years from now, how much would be in the account?

R3. a. Write the terms of this partial sum:

$$\sum_{n=1}^{6}(2 + (n - 1)(3))$$

b. How do you know the series in part a is an arithmetic series?

c. Show that the partial sum in part a equals 6 times the average of the first and last terms.

d. 418,435 is a partial sum of the series in part a. Which partial sum is it?

e. Find the 200th partial sum of the geometric series with $t_1 = 4000$ and $r = 0.95$. Do this numerically by adding the terms using the appropriate features of your grapher.

f. Find the 200th partial sum in part e again, algebraically, using the formula for S_n.

g. 78,377.8762... is a partial sum of the series in part e. Which partial sum is it?

h. To what limit do the partial sums of the series in part e converge?

i. Write the term containing b^7 in the binomial series for $(a - b)^{13}$.

j. *Vincent and Maya's Walking Problem:* Vincent and Maya start walking at the same time from the same point and in the same direction. Maya starts with a 12-inch step and increases her stride by a half inch each step. She goes 21 steps then stops. Vincent starts with a 36-inch step, and each subsequent step is 90% as long as the preceding one.

   i. What kind of sequence do Maya's steps follow? What kind of sequence do Vincent's steps follow?

   ii. How long is Maya's last step? How long is Vincent's 21st step?

   iii. After each has taken 21 steps, who is ahead? Show how you reached your conclusion.

   iv. If Vincent keeps walking in the same manner, will he ever get to where Maya stopped? Explain.

k. *Vitamin C Dosage Problem:* When you take a dose of medication, the amount of medication in your body jumps up immediately to a higher value and then decreases as the medication is used up and expelled. Suppose you take 500-mg doses of vitamin C each 6 hours to fight a cold. Assume that by the time of the second dose only 60% of the first dose remains. How much vitamin C will you have just after you take the second dose? Just after the third dose? Show that the total amount of vitamin C in your system is a partial sum of a geometric series. How much will you have at the end of the fourth day, when you have just taken the 16th dose? What limit does the amount of vitamin C approach as the number of doses approaches infinity?

Concept Problems

C1. *Tree Problem 1:* A treelike figure is drawn in the plane, as shown in Figure 14-4a. The first year, the tree grows a trunk 2 m long. The next year, two branches, each 1 m long, grow at right angles to each other from the top of the trunk, symmetrically to the line of the trunk. In subsequent years, each branch grows two new branches, each half as long as the preceding branch.

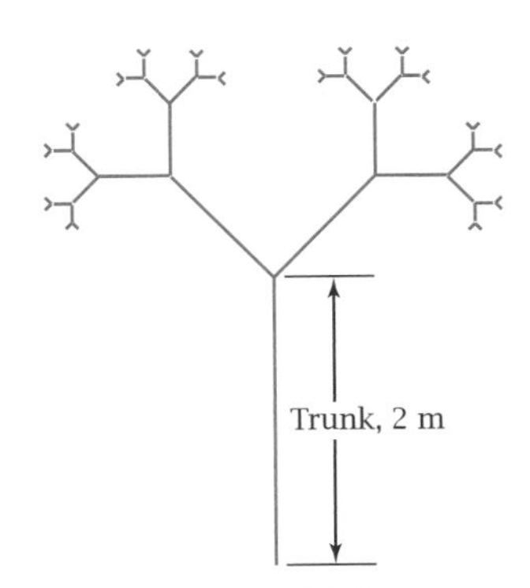

Figure 14-4a

a. Show that the lengths of the branches form a geometric sequence.

b. Find the height of the tree after 2 years, 3 years, and 4 years.

c. Show that the height of the tree is the partial sum of *two* geometric series. What is the common ratio of each?

d. If the tree keeps growing like this forever,

   i. What limit will the height approach?

   ii. What limit will the width approach?

   iii. What limit will the length of each branch approach?

   iv. What limit will the total length of all branches approach?

   v. How close to the ground will the lowest branches come?

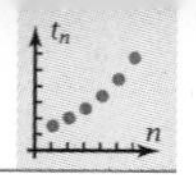

C2. *Bode's Law Problem:* In 1766, Johann Titus, a German astronomer, discovered that the distances of the planets from the Sun are proportional to the terms of a rather simple sequence:

Number	Name	Term
1	Mercury	4
2	Venus	7
3	Earth	10
4	Mars	16
5	Ceres	28
6	Jupiter	52
7	Saturn	100

(Mercury, at 4, does not fit the sequence. Ceres is one of the asteroids in the asteroid belt.)

a. Describe the pattern followed by the terms. Use this pattern to find the term for Uranus, planet 8.

b. Planets 9 and 10, Neptune and Pluto, have distances corresponding to the numbers 305 and 388, respectively. Are these numbers terms in the sequence? Justify your answer.

c. Find a formula for t_n, the term value, in terms of n, the planet number. For what values of n is the formula valid?

d. The asteroid Ceres was found by looking at a distance from the Sun calculated by Bode's law. If you were to look beyond Pluto for planet 11, how far from the Sun would Bode's law suggest that you look? (Earth is 93 million miles from the Sun.)

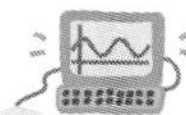

e. Consult the Internet or some other reference (such as *Scientific American* magazine, July 1977, page 128) to see why Titus's discovery is called Bode's law.

C3. *Binomial Series with Noninteger Exponent Problem:* If you raise a binomial to a noninteger power, such as $(a + b)^{1.8}$, you can still find the coefficient of the next term by the pattern

$$\frac{(\text{coefficient})(\text{exponent of } a)}{(\text{term number})}$$

a. Write the first five terms of the binomial series for $(a + b)^{1.8}$.

b. The series in part a has an infinite number of terms. A binomial series for a positive integer exponent may also be considered to have an infinite number of terms. Use the coefficient pattern in this problem to show what happens beyond the b^5-term in the expansion of $(a + b)^5$ and why such series seem to have a finite number of terms.

C4. *Power Series Problem:* The following are three **power series.** Each term involves a nonnegative integer power of x. (They are also called **Maclaurin series** or **Taylor series.**)

$$f(x) = 1 + x + \frac{1}{2!}x^2 + \frac{1}{3!}x^3 + \cdots$$

$$g(x) = x - \frac{1}{3!}x^3 + \frac{1}{5!}x^5 - \frac{1}{7!}x^7 + \cdots$$

$$h(x) = 1 - \frac{1}{2!}x^2 + \frac{1}{4!}x^4 - \frac{1}{6!}x^6 + \cdots$$

a. Evaluate the seventh partial sum of the series for $f(0.6)$ (terms through x^6). Show that the answer is close to the value of $e^{0.6}$, where e is the base of natural logarithms.

b. Evaluate the fourth partial sum of the series for $g(0.6)$. Show that the answer is close to the value of sin 0.6. Show that the fifth partial sum is even closer to the value of sin 0.6.

c. Evaluate the fourth partial sum of the series for $h(0.6)$. Which function on your calculator does $h(x)$ seem to be close to?

d. Figure 14-4b shows the graphs of y_1—a cubic function equal to the second partial sum for $g(x)$—and $y_2 = \sin x$. Graph these functions on your grapher. Then add the next three terms of the series for $g(x)$ to the equation for y_1. You'll get a 9th-degree function. Plot the two graphs again. Sketch the two graphs, and describe what you observe.

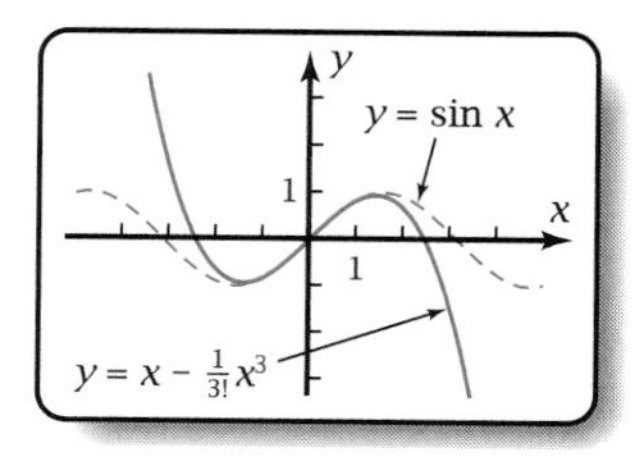

Figure 14-4b

Chapter Test

PART 1: No calculators allowed (T1–T9)

T1. A geometric series has first term $t_1 = 6$ and common ratio $r = 2$. Write the first five terms of the series. Find the fifth partial sum, S_5, numerically by adding up the terms of the series.

T2. An arithmetic series has first term $t_1 = 7$ and common difference $d = 3$. Write the first six terms of the series. Find the sixth partial sum, S_6, numerically by adding up the terms and algebraically using the sum of the first and last terms in an appropriate way.

T3. Write an algebraic formula for S_n, the nth partial sum of an arithmetic series, in terms of n, t_1, and t_n.

T4. Is this series arithmetic, geometric, or neither? Give numerical evidence to support your answer.

$$3 + 7 + 12 + 18 + 25 + \cdots$$

T5. Evaluate the partial sum numerically, by writing out and adding the terms.

$$S_5 = \sum_{k=1}^{4} 7 \cdot 3^{k-1}$$

T6. Is the series in Problem T5 arithmetic, geometric, or neither? Give numerical evidence to support your answer.

T7. Write the term containing b^9 in the binomial series for $(a - b)^{15}$. Leave the answer in factorial form.

T8. Write a recursive formula for t_n for this arithmetic sequence: 17, 21, 25, . . .

T9. Write an explicit formula for t_n for this geometric sequence: 7, 14, 28, . . .

PART 2: Graphing calculators allowed (T10–T24)

T10. In Problem T1, the fifth partial sum is $6 + 12 + 24 + 48 + 96$, which equals 186. Show that the algebraic formula for the sum S_n of a geometric series gives 186 for the fifth partial sum. Show numerically that the fifth partial sum is 186 using your Series program.

T11. The arithmetic series in Problem T2 is $7 + 10 + 13 + \cdots$. Calculate the 200th term, t_{200}. Use the answer to calculate the 200th partial sum, S_{200}, algebraically, with the help of the pattern for partial sums of an arithmetic series. Confirm that your answer is correct by adding the terms numerically using your Series program.

Bouncing Ball Problem: Imagine that you're bouncing a ball. Each time the ball bounces, it comes back to 80% of its previous height. For Problems T12–T14, assume that the starting height of the ball was 5 feet.

T12. What kind of sequence describes the ball's successive heights at each bounce? How long is the *total* path the ball covers in ten bounces?

T13. Find the formula for the ball's height at its nth bounce. Find the formula for the *total* length of the path of the ball during n bounces.

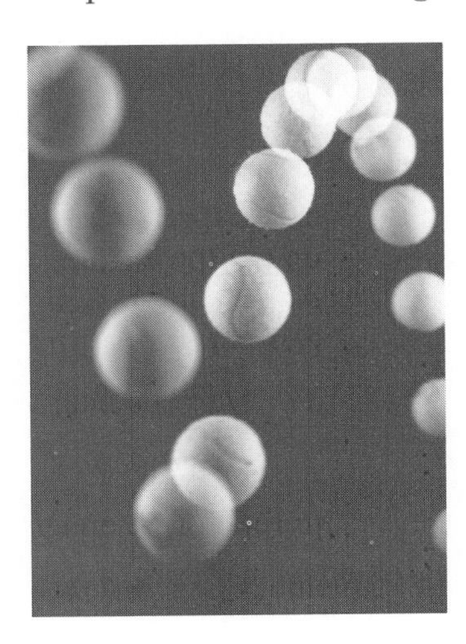

T14. The series in Problem T13 converges to a certain number. Based on your answers, what is the number to which it converges? Show algebraically that this is correct.

Pushups Problem: For Problems T15 and T16, Emma starts an exercise program. On the first workout she does five pushups. The next workout, she does eight pushups. She decides to let the number of pushups in each workout be a term in an arithmetic series.

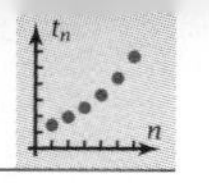

T15. Find algebraically the number of pushups she does on the tenth workout and the total number of pushups she has done after ten workouts.

T16. If Emma were able to keep up the arithmetic series of pushups, one of the terms in the series would be 101. Calculate this term number algebraically.

Medication Problem: For Problems T17–T20, Natalie takes 50 mg of allergy medicine each day. By the next day, some of the medicine has decomposed, but the rest is still in her body. Natalie finds that the amount still in her body after n days is given by this partial sum, where k is an integer:

$$S_n = \sum_{k=1}^{n} 50(0.8^{k-1})$$

T17. Demonstrate that you know what sigma notation means by writing out the first three terms of this series. What does S_3 equal?

T18. Run your Series program to find S_{40} numerically.

T19. After how many days does the amount in her body first exceed 200 mg?

T20. Does the amount of active medicine in Natalie's body seem to be converging to a certain number, or does it just keep getting bigger without limit? How can you tell?

Loan Problem: For Problems T21 and T22, Leonardo borrows $200.00 from his parents to buy a new calculator. They require him to pay back 10% of his unpaid balance at the end of each month.

T21. Find Leonardo's unpaid balances at the end of 0, 1, 2, and 3 months. Is the sequence of unpaid balances arithmetic, geometric, or neither? How do you know?

T22. Leonardo must pay off the rest of the loan when his unpaid balance has dropped below $5.00. After how many months will this have happened? Indicate the method you use. (It is not enough to say, "I used my calculator" or "Guess and check.")

T23. Find the seventh term of the binomial series $(a - b)^{12}$.

T24. What did you learn as a result of taking this test that you did not know before?

Polynomial and Rational Functions, Limits, and Derivatives

CHAPTER 15

A bee flies back and forth past a flower. Its distance from the flower is a function of time. Its speed varies, so you can't find distance simply by multiplying rate by time. You can use the concept of *limit,* which you saw in connection with geometric series, to find the bee's instantaneous speed by taking the limit of its average speed over shorter and shorter time intervals.

Mathematical Overview

In this chapter you'll learn about polynomial functions. You'll find the zeros of these functions, a generalization of the concept of x-intercept. The techniques you learn will allow you to analyze rational functions, in which $f(x)$ equals a ratio of two polynomials. These ratios can represent average rates of change. The limit is an instantaneous rate of change, called a *derivative.* You will do the investigations in four ways.

Graphically The graph of the polynomial function, $f(x) = x^4 - 8x^2 - 8x + 15$, has two x-intercepts (Figure 15-0a).

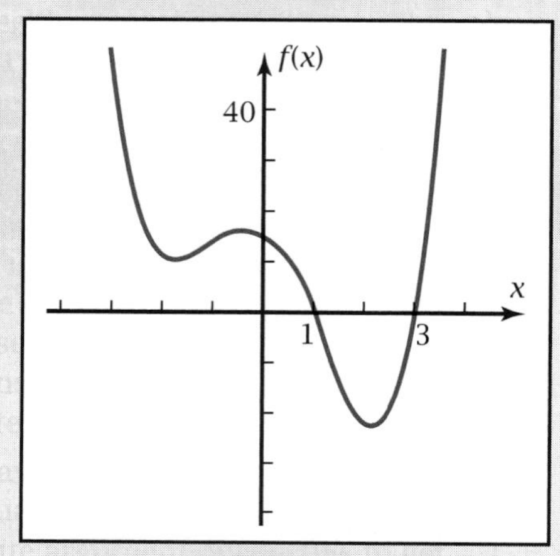

Figure 15-0a

Algebraically

$$f(x) = x^4 + 0x^3 - 8x^2 - 8x + 15$$
$$= (x - 1)(x - 3)(x + 2 + i)(x + 2 - i)$$

Numerically $f(1) = 0$, $f(3) = 0$, $f(-2 - i) = 0$, and $f(-2 + i) = 0$

Zeros are $x = 1$, $x = 3$, $x = -2 - i$, and $x = -2 + i$.

Verbally *I learned that a fourth-degree function has exactly four zeros if the domain of the function includes complex numbers. In other words, you are allowed to have complex numbers for the zeros.*

15-1 Review of Polynomial Functions

Figure 15-1a shows the graphs of three cubic functions, f, g, and h. In this problem set you will duplicate these graphs on your grapher and learn some algebraic properties of cubic functions.

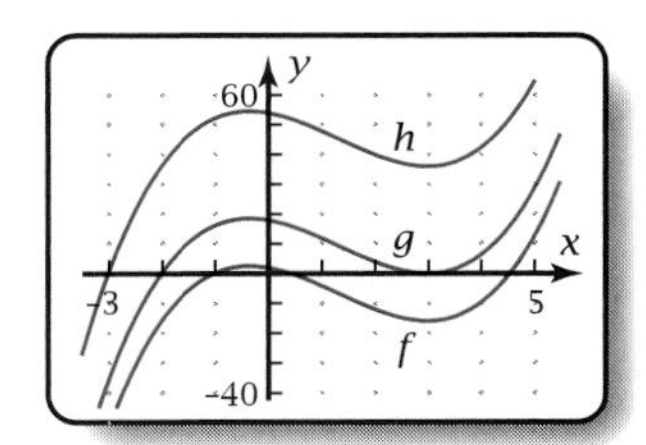

Figure 15-1a

OBJECTIVE Discover some properties of cubic functions and their graphs.

Exploratory Problem Set 15-1

1. Function f has equation $f(x) = x^3 - 4x^2 - 3x + 2$. Confirm on your grapher that function f has the graph shown in Figure 15-1a.
2. The graph of f intersects the x-axis at three places. The x-values at these places are called **zeros** of $f(x)$ because $f(x) = 0$ for these x-values. Find these three zeros graphically.
3. Show algebraically that $x = -1$ is a zero of $f(x)$.
4. Because $f(-1) = 0$, you can write $f(x) = (x + 1)(\text{—?—})$. By appropriate calculations, find the unknown factor. Then find the other two zeros of $f(x)$ algebraically.
5. Function g has the equation $g(x) = x^3 - 4x^2 - 3x + 18$. Confirm on your grapher that function g has the graph shown in Figure 15-1a. What similarities and differences do you notice in the graphs of functions f and g?
6. Show algebraically that -2 is a zero of $g(x)$. Explain why the equation for g can be written $g(x) = (x + 2)\,(\text{—?—})$. Find the other factor. If possible, factor the result further into two linear factors. What do you notice about these two factors?
7. There are three zeros of $g(x)$, one for each linear factor. Explain why $g(x)$ has a **double zero.** What feature does the graph of g have at the double zero?
8. Function h has equation $h(x) = x^3 - 4x^2 - 3x + 54$. Confirm on your grapher that the graph of h in Figure 15-1a is correct. How many zeros does $h(x)$ have?
9. Show algebraically that $h(-3) = 0$ and therefore that $h(x) = (x + 3)(\text{—?—})$. By finding the other factor and setting it equal to zero, find the two **complex zeros** of $h(x)$.
10. What is true about the graph of a cubic function if it has complex zeros?
11. Summarize what you have learned as a result of doing this problem set.

15-2 Graphs and Zeros of Polynomial Functions

In Section 15-1 you encountered cubic functions, polynomial functions of degree three. In this section you'll learn how to recognize the degree of a polynomial function from its graph and how to find *zeros,* values of x at which y equals zero. Some zeros are real numbers equal to the x-intercepts, and others are complex numbers that are not on the graph.

OBJECTIVE Given a polynomial function,

- Tell from the graph what degree it might be, and vice versa
- Find the zeros from the equation or graph

Graphs of Polynomial Functions

Recall the general equation of a polynomial function from Chapter 1. Here are some examples of polynomial functions.

$f(x) = 4x^2 - 7x + 3$	Quadratic function, 2nd degree
$f(x) = 2x^3 - 5x^2 + 4x + 7$	Cubic function, 3rd degree
$f(x) = x^4 + 6x^3 - 3x^2 + 5x - 8$	Quartic function, 4th degree
$f(x) = -6x^5 - x^3 + 2x$	Quintic function, 5th degree
$f(x) = 5x^9 + 4x^8 - 11x^3 + 63$	9th-degree function (no special name)

In each case, $f(x)$ is equal to a **polynomial** expression. The only operations performed on the variable in a polynomial are the **polynomial operations,** namely, +, −, and ×. The **degree** of a one-variable polynomial is the same as the greatest exponent of the variable. Thus, the degree tells the greatest number of variables *multiplied* together. The coefficient of the highest-degree term is called the **leading coefficient.** Notice that each term has a power of a variable, so you can think of a polynomial function as a sum of a finite number of power

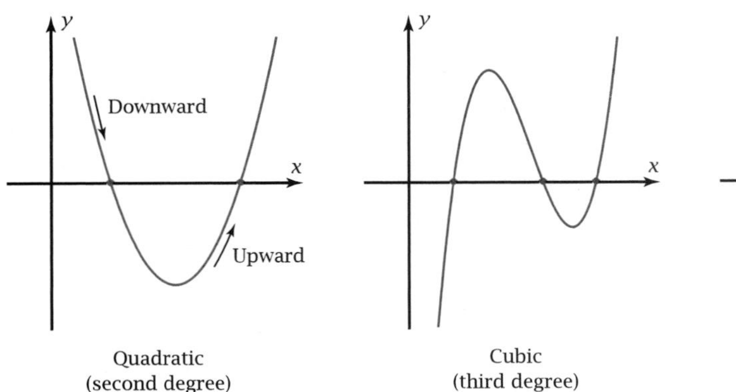

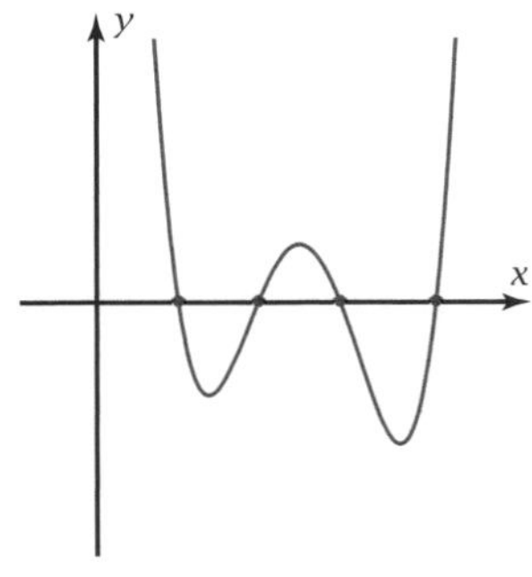

Quadratic (second degree) Two zeros

Cubic (third degree) Three zeros

Quartic (fourth degree) Four zeros

Figure 15-2a

functions. The exponents of a polynomial function must be nonnegative integers so that there is no division by a variable and no root of a variable.

As you can see from Figure 15-2a, a quadratic function has two branches, downward and upward, resulting in at most two zeros and one **extreme point** (or **vertex** or **critical point**). A cubic function can have three branches, resulting in three zeros and two extreme points, and a quartic function can have four branches, giving four zeros and three extreme points. In general, the graph of a polynomial function of degree n can have up to n increasing and decreasing branches, resulting in up to n zeros where these branches cross the x-axis and up to $n - 1$ extreme points.

Here is the formal definition of a zero of a function, which you'll learn how to find next.

DEFINITION: Zero of a Function

A **zero of a function** f is an x-value, c, for which $f(c) = 0$.

Finding Zeros of a Polynomial Function: Synthetic Substitution

Synthetic substitution is a quick pencil-and-paper method to evaluate a polynomial function. Suppose that $f(x) = x^3 - 9x^2 - x + 105$ and that you want to find $f(6)$ by synthetic substitution.

- Write 6 and the coefficients of $f(x)$ like this:

6	1	−9	−1	105	
					Leave space here.
					Leave space here.

- Bring down the leading coefficient, 1, below the line, multiply it by the 6, and write the answer in the next column, under the −9.

6	1	−9	−1	105
		6		
	1			

- Add the −9 and the 6, and write the answer, −3, below the line. Multiply it by the 6, and write the answer, −18, above the line. Repeat the same steps, add and multiply. The final result is

6	1	−9	−1	105
		6	−18	−114
	1	−3	−19	−9

$\therefore f(6) = -9$ The value of the function $f(6)$ is the last number below the line.

Check:

$$f(6) = 6^3 - 9(6^2) - 1(6) + 105 = -9$$

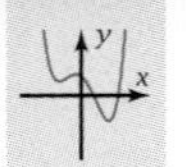

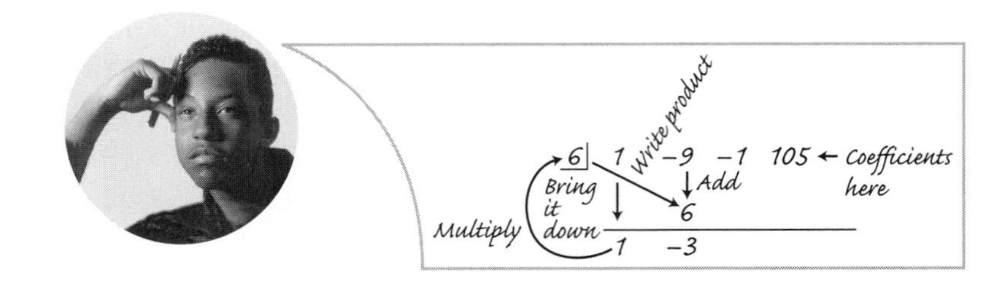

To see why synthetic substitution works, factor the polynomial into **nested form:**

$$f(x) = x^3 - 9x^2 - x + 105$$

$$= (1x - 9)x^2 - x + 105$$

$$= ((1x - 9)x - 1)x + 105 \qquad \text{Nested form.}$$

In this form you can evaluate the polynomial by repeating the steps:

Multiply by x.

Add the next coefficient.

Synthetic substitution is closely related to long division of polynomials. To see why, divide $f(x)$ by $(x - 6)$, a linear binomial that equals 0 when x is 6. First, divide the x from $(x - 6)$ into the x^3 from the polynomial. Write the answer, x^2, above the x^2-term of the polynomial.

$x^2 - 3x - 19$	Quotient
$x - 6\overline{)x^3 - 9x^2 - x + 105}$	
$\underline{x^3 - 6x^2}$	Multiply x^2 and $(x - 6)$.
$-3x^2 - x$	Subtract and bring $-x$ down.
$\underline{-3x^2 + 18x}$	Multiply $-3x$ and $(x - 6)$.
$-19x + 105$	Subtract and bring the next term down.
$\underline{-19x + 114}$	Multiply -19 and $(x - 6)$.
-9	Remainder.

$$\therefore \frac{x^3 - 9x^2 - x + 105}{x - 6} = x^2 - 3x - 19 + \frac{-9}{x - 6} \qquad \text{"Mixed-number" form.}$$

Note: The term *"mixed-number" form* is used here because of the similarity of this form to the result of whole-number division when there is a remainder. For example, when you divide 13 by 4, the quotient is 3 and the remainder is 1, which you can write as $\frac{13}{4} = 3\frac{1}{4}$.

Notice that the coefficients of the quotient, 1, −3, and −19, are the values below the line in the synthetic substitution process. Thus, synthetic substitution gives a way to do long division of a polynomial by a linear binomial expression. Just substitute the value of x for which the linear binomial equals zero.

The fact that the *remainder* after division by $(x - 6)$ equals the *value* of $f(6)$ is an example of the **remainder theorem.**

PROPERTY: The Remainder Theorem

If $p(x)$ is a polynomial, then $p(c)$ equals the remainder when $p(x)$ is divided by the quantity $(x - c)$.

COROLLARY: The Factor Theorem

$(x - c)$ is a factor of polynomial $p(x)$ if and only if $p(c) = 0$.

The corollary is true because if the remainder equals zero, then $(x - c)$ divides $p(x)$ evenly. As a result, $(x - c)$ is a *factor* of $p(x)$.

▶EXAMPLE 1

Let $f(x) = x^3 - 4x^2 - 3x + 2$.

Let $g(x) = x^3 - 4x^2 - 3x + 18$.

Let $h(x) = x^3 - 4x^2 - 3x + 54$.

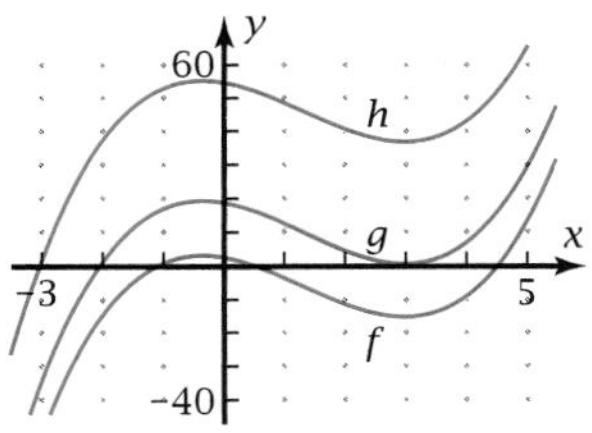

The graphs are shown in Figure 15-2b.

Figure 15-2b

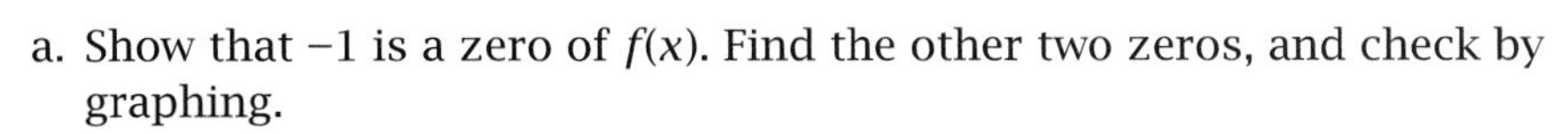

a. Show that -1 is a zero of $f(x)$. Find the other two zeros, and check by graphing.

b. Show that -2 is a zero of $g(x)$. Find the other two zeros, and check by graphing.

c. Show that -3 is a zero of $h(x)$. Find the other two zeros, and check by graphing.

Solution

a.
$$\begin{array}{r|rrrr} -1 & 1 & -4 & -3 & 2 \\ & & -1 & 5 & -2 \\ \hline & 1 & -5 & 2 & \underline{0} \end{array}$$

Synthetically substitute -1 for x.

The remainder is zero.

Therefore, -1 is a zero of $f(x)$. — $f(-1) = 0$ because $f(-1)$ equals the remainder.

$$f(x) = (x + 1)(x^2 - 5x + 2)$$

$(x + 1)$ is a factor of $f(x)$ because it is 0 when $x = -1$; coefficients of the other factor appear in the bottom row of the synthetic substitution.

$$x^2 - 5x + 2 = 0$$

Set the other factor equal to zero.

$x = 4.5615\ldots$ and $0.4384\ldots$ are also zeros of $f(x)$.

Solve by the quadratic formula.

The graph of f in Figure 15-2b crosses the x-axis at about -1, 0.4, and 4.6, which agree with the algebraic solutions.

b. $\begin{array}{r|rrrr} -2 & 1 & -4 & -3 & 18 \\ & & -2 & 12 & -18 \\ \hline & 1 & -6 & 9 & \underline{0} \end{array}$

Synthetically substitute −2 for x.

The remainder is zero.

Therefore, −2 is a zero of $g(x)$. — $g(-2) = 0$ because it equals the remainder.

$g(x) = (x + 2)(x^2 - 6x + 9)$ — $(x + 2)$ is a factor of $g(x)$ because it is 0 when $x = -2$.

$g(x) = (x + 2)(x - 3)(x - 3)$ — The second factor can itself be factored.

−2, 3, and 3 are zeros of $g(x)$. — 3 is called a *double zero* of $g(x)$.

The graph of g in Figure 15-2b crosses the x-axis at −2 and touches the axis at 3, in agreement with the algebraic solutions.

c. $\begin{array}{r|rrrr} -3 & 1 & -4 & -3 & 54 \\ & & -3 & 21 & -54 \\ \hline & 1 & -7 & 18 & \underline{0} \end{array}$

Synthetically substitute −3 for x.

The remainder is zero.

Therefore, −3 is a zero of $h(x)$. — $h(-3) = 0$ because it equals the remainder.

$h(x) = (x + 3)(x^2 - 7x + 18)$ — $(x + 3)$ is a factor of $h(x)$ because it is 0 when $x = -3$.

$x^2 - 7x + 18 = 0$ — Set the other factor equal to zero.

$$x = \frac{7 \pm \sqrt{7^2 - 4(1)(18)}}{2(1)}$$ Use the quadratic formula.

$$= 3.5 \pm 0.5\sqrt{-23}$$

$$= 3.5 + 2.3979\ldots i \quad \text{or} \quad 3.5 - 2.3979\ldots i$$

Complex solutions are a conjugate pair.

The graph of h in Figure 15-2b crosses the x-axis only at −3, which agrees with the algebraic solutions. ◀

Notes:

- In part b of Example 1, $x = 3$ is called a **double zero** of $f(x)$. It appears twice, once for each factor $(x - 3)$. As you can see from Figure 15-2b, the graph of g just touches the x-axis and does not cross it. So there are three zeros, −2, 3, and 3, although there are only two distinct x-intercepts.
- In part c of Example 1, there are three zeros of $h(x)$ but only one x-intercept, as you can see in Figure 15-2b. The other two zeros are nonreal complex numbers. The two complex zeros are *complex conjugates* of each other.

The results of Example 1 illustrate the **fundamental theorem of algebra** and its corollaries.

PROPERTY: The Fundamental Theorem of Algebra and Its Corollaries

A polynomial function has at least one zero in the set of complex numbers.

Corollary

An nth-degree polynomial function has exactly n zeros in the set of complex numbers, counting multiple zeros.

Corollary

If a polynomial has only real coefficients, then any nonreal complex zeros appear in conjugate pairs.

Be sure you understand what the theorem says. The real numbers are a subset of the complex numbers, so the zeros of the function could be nonreal or real complex numbers. The x-intercepts correspond to the real-number zeros of the function. From now on, these real-number zeros will be referred to as *real zeros.*

▶ **EXAMPLE 2** Identify the degree and the number of real and nonreal complex zeros that the polynomial function in Figure 15-2c could have. Tell whether the leading coefficient is positive or negative.

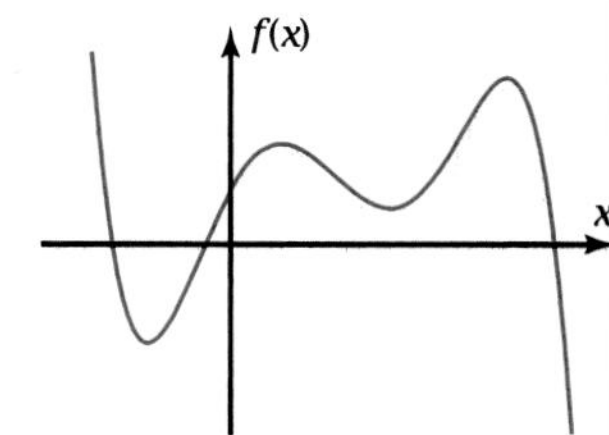

Figure 15-2c

Solution The function could be 5th degree because it has five branches—down, up, down, up, down—resulting in four extreme points in the domain shown.

There are three real zeros because the graph crosses the x-axis at three places. There are two nonreal complex zeros because the total number of zeros must be five.

The leading coefficient is negative because $f(x)$ becomes very large in the negative direction as x becomes large in the positive direction. ◀

Sums and Products of Zeros

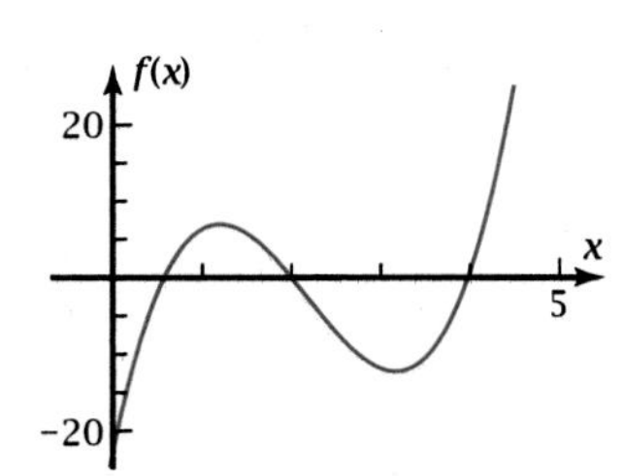

Figure 15-2d

Figure 15-2d shows the graph of the cubic function

$$f(x) = 5x^3 - 33x^2 + 58x - 24$$

The three zeros are $x = z_1 = 0.6$, $x = z_2 = 2$, and $x = z_3 = 4$. By factoring out the leading coefficient, 5, the sum and the product of these zeros appear in the equation.

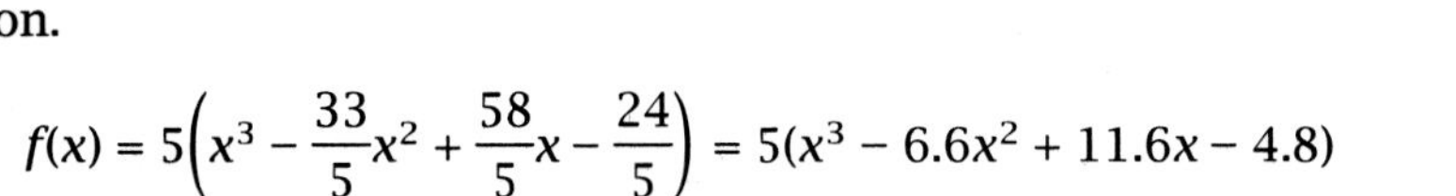

$$f(x) = 5\left(x^3 - \frac{33}{5}x^2 + \frac{58}{5}x - \frac{24}{5}\right) = 5(x^3 - 6.6x^2 + 11.6x - 4.8)$$

$$z_1 + z_2 + z_3 = 0.6 + 2 + 4 = 6.6$$

The opposite of the quadratic coefficient.

$$z_1 z_2 z_3 = (0.6)(2)(4) = 4.8$$

The opposite of the constant term.

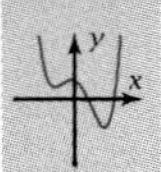

The sum of the products of the zeros taken two at a time equals the linear coefficient.

$$z_1z_2 + z_1z_3 + z_2z_3 = (0.6)(2) + (0.6)(4) + (2)(4) = 11.6$$

Equal to the linear coefficient.

To see why these properties are true, start with $f(x)$ in factored form, expand it, and combine like terms *without* completing the calculations.

$$\begin{aligned} f(x) &= 5(x - 0.6)(x - 2)(x - 4) \\ &= 5(x^3 - 0.6x^2 - 2x^2 - 4x^2 + (0.6)(2)x + (0.6)(4)x + (2)(4)x - (0.6)(2)(4)) \\ &= 5(x^3 - \underbrace{(0.6 + 2 + 4)}_{\text{Opposite of sum}}x^2 + \underbrace{((0.6)(2) + (0.6)(4) + (2)(4))}_{\text{Sum of pairwise products}}x - \underbrace{(0.6)(2)(4)}_{\text{Opposite of product}}) \end{aligned}$$

This property is true, in general, for any cubic function. In Problems 34 and 35 of Problem Set 15-2 you will extend this property to quadratic functions and to higher-degree functions.

PROPERTY: Sums and Products of the Zeros of a Cubic Function

If $p(x) = ax^3 + bx^2 + cx + d$ has zeros z_1, z_2, and z_3, then

$$z_1 + z_2 + z_3 = -\frac{b}{a}$$ Sum of the zeros.

$$z_1z_2 + z_1z_3 + z_2z_3 = \frac{c}{a}$$ Sum of the pairwise products of the zeros.

$$z_1z_2z_3 = -\frac{d}{a}$$ Product of the zeros.

This property allows you to tell something about the zeros of the function without actually calculating them. It also gives you a way to find the particular equation of a cubic function when you know its zeros.

▶ EXAMPLE 3 Find the zeros of the function

$$f(x) = x^3 - 13x^2 + 59x - 87$$

Then show that the sum of the zeros, the sum of their pairwise products, and the product of all three zeros correspond to the coefficients of the equation that defines function f.

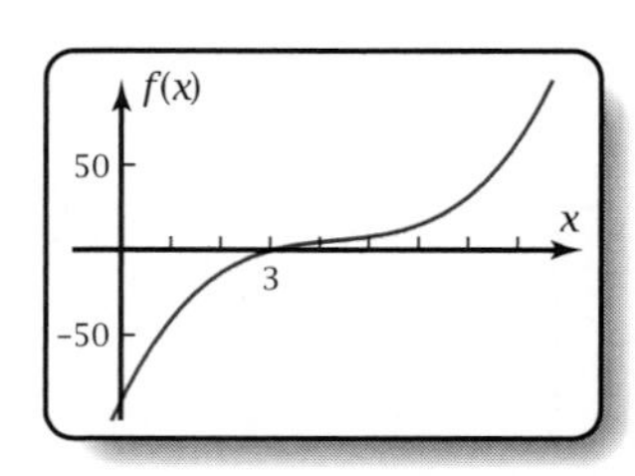

Figure 15-2e

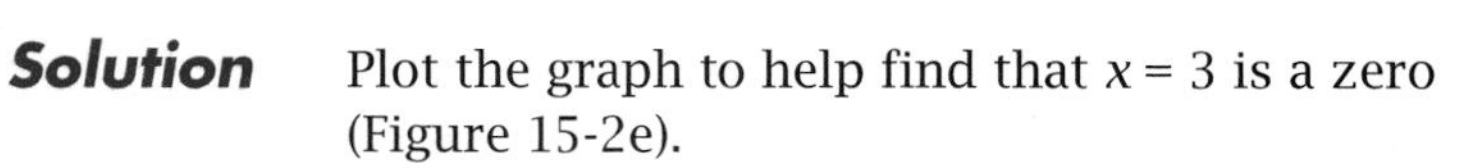

Solution Plot the graph to help find that $x = 3$ is a zero (Figure 15-2e).

Do synthetic substitution to find the other factor.

$$\begin{array}{r|rrrr} 3 & 1 & -13 & 59 & -87 \\ & & 3 & -30 & 87 \\ \hline & 1 & -10 & 29 & \underline{0} \end{array}$$

$$f(x) = (x - 3)(x^2 - 10x + 29)$$

$$x^2 - 10x + 29 = 0 \quad \Leftrightarrow \quad x = \frac{10 \pm \sqrt{100 - 4(1)(29)}}{2(1)}$$

$$= \frac{10 \pm \sqrt{-16}}{2} = 5 \pm 2i$$

Sum: $3 + (5 + 2i) + (5 - 2i) = 13$ — The opposite of the quadratic coefficient.

Pairwise-product sum: $3(5 + 2i) + 3(5 - 2i) + (5 + 2i)(5 - 2i)$

$= 15 + 6i + 15 - 6i + 25 + 4$ — Recall that $i^2 = -1$.

$= 59$ — Equals the linear coefficient.

Product: $3(5 + 2i)(5 - 2i) = 3(25 + 4) = 87$ — The opposite of the constant term. ◀

▶ EXAMPLE 4 Find the particular equation of a cubic function with integer coefficients if the zeros have the given sum, product, and sum of pairwise products. Confirm these properties after finding the zeros of the function.

Sum: $-\frac{5}{3}$ Sum of pairwise products: $-\frac{58}{3}$ Product: $\frac{40}{3}$

Solution The particular equation of one possible function is

$$y = x^3 + \frac{5}{3}x^2 - \frac{58}{3}x - \frac{40}{3}$$

The particular equation of a function with integer coefficients is

$$f(x) = 3x^3 + 5x^2 - 58x - 40$$

By graph (Figure 15-2f), the zeros are $x = -5$, $x = -\frac{2}{3}$, and $x = 4$.

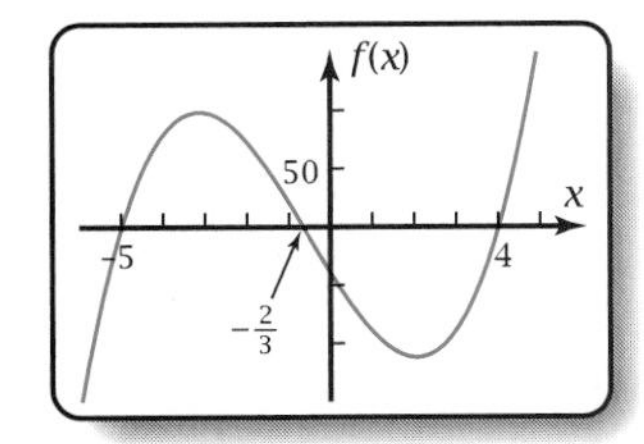

Figure 15-2f

Check:

Sum: $(-5) + \frac{-2}{3} + 4 = \frac{-5}{3}$ (Correct)

Sum of the pairwise products:

$$(-5)\left(\frac{-2}{3}\right) + (-5)(4) + \left(\frac{-2}{3}\right)(4) = \frac{-58}{3} \quad \text{(Correct)}$$

Product: $(-5)\left(\frac{-2}{3}\right)(4) = \frac{40}{3}$ (Correct) ◀

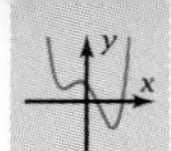

Problem Set 15-2

Do These Quickly 

Q1. Sketch the graph of $y = x^2$.

Q2. Sketch the graph of $y = x^3$.

Q3. Sketch the graph of $y = x^4$.

Q4. Sketch the graph of $y = 2^x$.

Q5. Multiply the complex numbers and simplify: $(3 + 2i)(5 + 6i)$

Q6. Solve the equation using the quadratic formula: $x^2 - 14x + 54 = 0$

Q7. Solve the equation using the quadratic formula: $x^2 - 14x + 58 = 0$

Q8. What transformation of $f(x) = \sin x$ is indicated? $g(x) = \sin 2x$

Q9. What type of function has the add–multiply property?

Q10. Because 37° and 53° add up to 90°, they are called —?— angles.

1. Given $p(x) = x^3 - 5x^2 + 2x + 8$,
   a. Plot the graph using an appropriate domain. How many branches (increasing or decreasing) does the graph have? How is this number related to the degree of $p(x)$?
   b. Find graphically the three real zeros of $p(x)$.
   c. Show by synthetic substitution that $(x + 1)$ is a factor of $p(x)$. Use the result to write the other factor, and factor it further if you can.
   d. Explain the relationship between the zeros of $p(x)$ and the factors of $p(x)$.

2. Given $p(x) = x^3 - 3x^2 + 9x + 13$,
   a. Plot the graph using an appropriate domain. From the graph, find the real zero of $p(x)$.
   b. Show by synthetic substitution that $(x + 1)$ is a factor of $p(x)$. Write the other factor.
   c. Find the other two zeros. Substitute one of the two complex zeros into the equation for $p(x)$, and thus show that it really is a zero.
   d. Explain the relationship between the zeros of $p(x)$ and the graph of $p(x)$.

For the polynomial functions in Problems 3–6, give the degree, the number of real zeros (counting double zeros as two), and the number of nonreal complex zeros the function could have.

3. 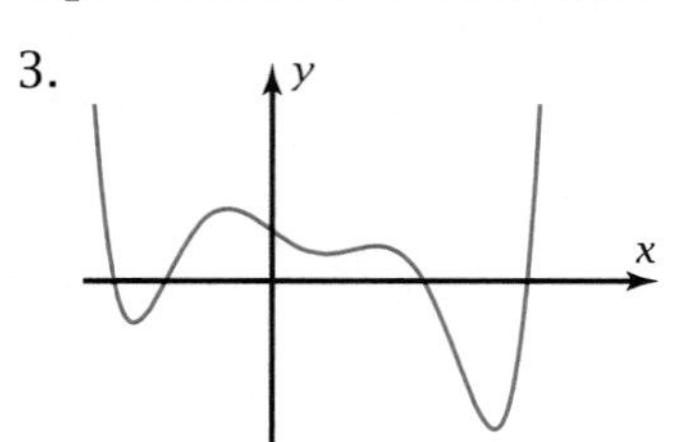

4. 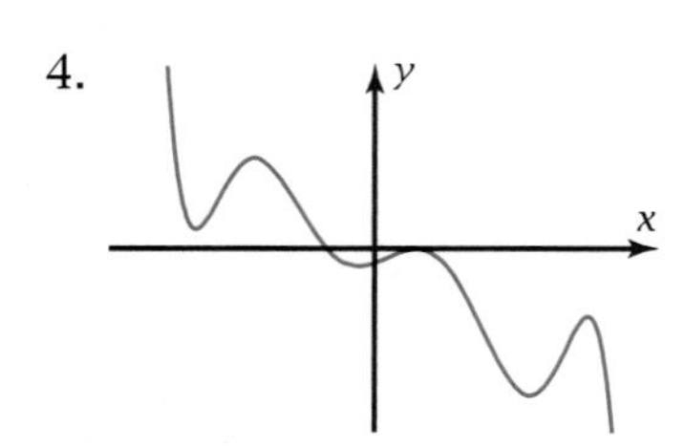

5. 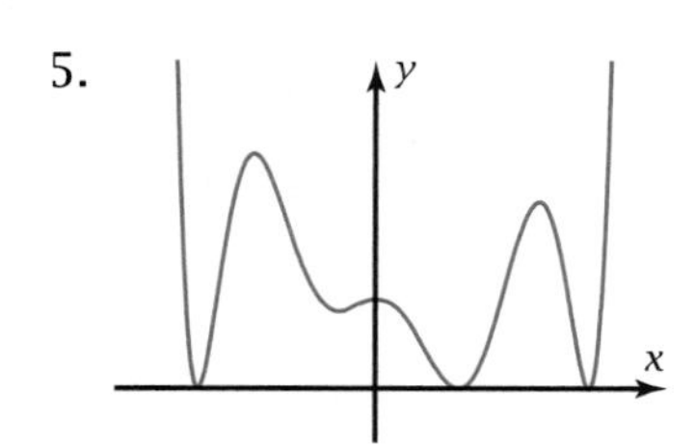

6. 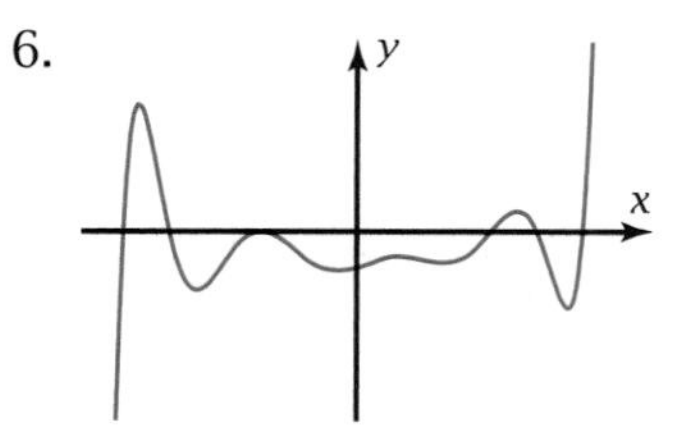

For Problems 7–18, sketch the graph of the polynomial function described, or explain why no such function can exist. The expression *complex zero* will be used to mean a *nonreal complex number.*

7. Cubic function with two distinct negative zeros, one positive zero, and a positive y-intercept

8. Cubic function with a negative double zero and a positive zero, and a negative leading coefficient

9. Cubic function with one real zero, two complex zeros, and a positive leading coefficient

10. Cubic function with no real zeros

11. Cubic function with no extreme points

12. Quartic function with no extreme points

13. Quartic function with no real zeros

14. Quartic function with two distinct positive zeros, two distinct negative zeros, and a negative y-intercept

15. Quartic function with two double zeros

16. Quartic function with two distinct real zeros and two complex zeros

17. Quartic function with five distinct real zeros

18. Quintic function with five distinct real zeros

For Problems 19–22, use the coefficients to find quickly the sum, the product, and the sum of the pairwise products of the zeros, using the properties. Then find the zeros and confirm that your answers satisfy the properties.

19. $f(x) = x^3 - x^2 - 22x + 40$

20. $f(x) = x^3 + x^2 - 7x - 15$

21. $f(x) = -5x^3 - 18x^2 + 7x + 156$

22. $f(x) = 2x^3 - 9x^2 - 8x + 15$

For Problems 23–26, find a particular equation of the cubic function, with zeros as described, if the leading coefficient equals 1. Then find the zeros and confirm that your answers satisfy the given properties.

23. Sum: 4; sum of the pairwise products: −11; product: −30

24. Sum: 9; sum of the pairwise products: 26; product: 24

25. Sum: 8; sum of the pairwise products: 29; product: 52

26. Sum: −5; sum of the pairwise products: 4; product: 10

For Problems 27 and 28,

a. By synthetic substitution, find $p(c)$.

b. Write $\frac{p(x)}{x-c}$ in mixed number form.

27. $p(x) = x^3 - 7x^2 + 5x + 4$, $c = 2$ and $c = -3$

28. $p(x) = x^3 - 9x^2 + 2x - 5$, $c = 3$ and $c = -2$

29. State the remainder theorem.

30. State the factor theorem.

31. State the fundamental theorem of algebra.

32. State the two corollaries of the fundamental theorem of algebra.

33. *Synthetic Substitution Program Problem:* Write a program for your grapher or computer to do synthetic substitution. Store the coefficients of the polynomial, including ones that equal zero, in a list before you run the program. The input should be the degree of the polynomial and the value of x at which you want to evaluate the function. Store the output of the program, the coefficients of the quotient polynomial and the remainder (which equals the value of the polynomial), in another list. Test your program with $p(x) = x^3 - 7x^2 + 5x + 4$ from Problem 27, with $c = 2$. You should get the four numbers 1, −5, −5, −6, which imply that the quotient is $x^2 - 5x - 5$, with a remainder of −6, when $p(x)$ is divided by $(x - 2)$, and that $p(2) = -6$.

34. *Quadratic Function Sum and Product of Zeros Problem:* By the quadratic formula, the zeros of the general quadratic function $f(x) = ax^2 + bx + c$ are

$$z_1 = \frac{-b + \sqrt{b^2 - 4ac}}{2a} \quad \text{and}$$

$$z_2 = \frac{-b - \sqrt{b^2 - 4ac}}{2a}$$

By finding $z_1 + z_2$ and $z_1 z_2$, show that there is a property of the sum and product of zeros of a quadratic function that is similar to the corresponding property for cubic functions.

35. *Quartic Function Sum and Product of Zeros Problem:* By repeated synthetic substitution or with your grapher or computer, find the zeros of

$$f(x) = 2x^4 + 3x^3 - 14x^2 - 9x + 18$$

Then find the following quantities:

- Sum of the zeros ("products" of the zeros taken one at a time)
- Sum of all possible products of zeros taken two at a time

- Sum of all possible products of zeros taken three at a time
- Product of the zeros ("sum" of the products taken all four at a time)

From the results of your calculations, make a conjecture about how the property of the sums and products of the zeros can be extended to functions of degree higher than 3.

36. *Reciprocals of the Zeros Problem:* Prove that if

$$p(x) = ax^3 + bx^2 + cx + d$$

has zeros z_1, z_2, and z_3, then the function

$$q(x) = dx^3 + cx^2 + bx + a$$

has zeros $\frac{1}{z_1}$, $\frac{1}{z_2}$, and $\frac{1}{z_3}$.

37. *Horizontal Translation and Zeros Problem:* Let $f(x) = x^3 - 5x^2 + 7x - 12$. Let $g(x)$ be a horizontal translation of $f(x)$ by 1 unit in the positive direction. Find the particular equation of $g(x)$. Show algebraically that each zero of $g(x)$ is 1 unit larger than the corresponding zero of $f(x)$.

15-3 Fitting Polynomial Functions to Data

Figure 15-3a shows four cubic functions that could model the position of a moving object as a function of time. From left to right in the figure, the object

Olympic gold medalist Apolo Anton Ohno at the 2002 Winter Olympics in Salt Lake City, Utah

- Passes a reference point ($y = 0$) while slowing down (decreasing slope) and then speeds up again (increasing slope)
- Stops momentarily (zero slope) and then continues forward
- Reverses direction (negative slope) and then continues forward
- Approaches the reference point ($y = 0$) going in the negative direction

In all four cases, there is only one x-intercept, indicating that the other two zeros are nonreal complex numbers.

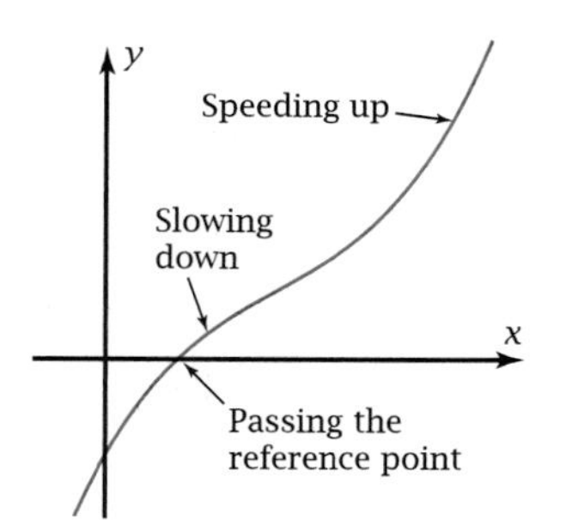

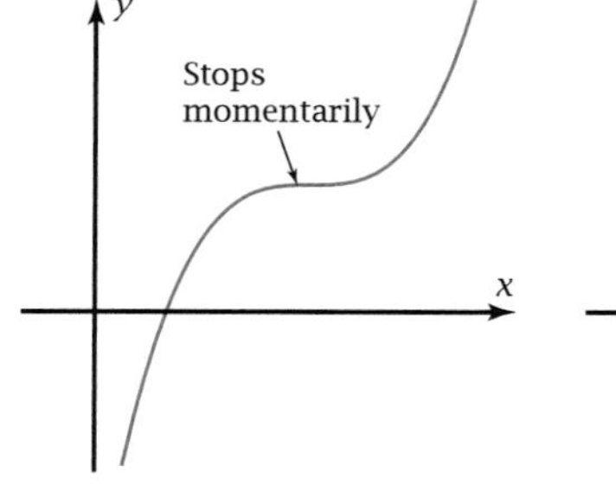

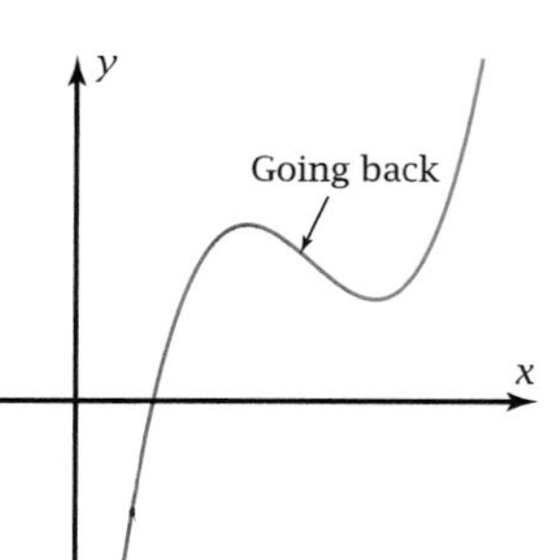

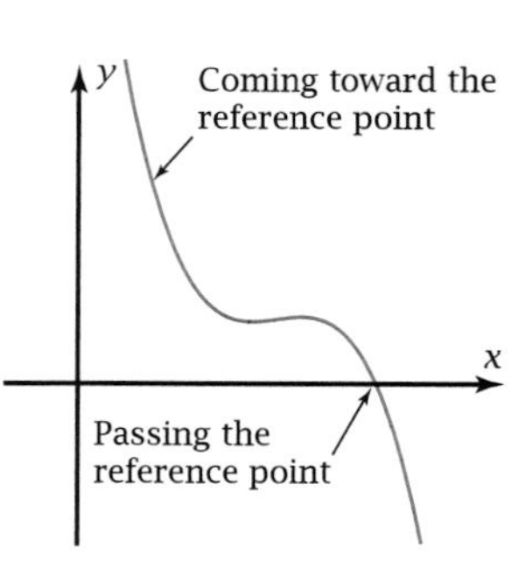

Figure 15-3a

In this section you will apply familiar curve-fitting techniques to polynomial functions.

OBJECTIVE Given a set of points, find the particular equation of the polynomial function that fits the data exactly or fits the best for a given degree.

▶ EXAMPLE 1 A cubic function f contains the points (6, 38), (5, 74), (2, 50), and (−1, 80).

a. Find the particular equation algebraically. Check by cubic regression.

b. Verify the answer by plotting and tracing on the graph.

Solution a.

$$f(x) = ax^3 + bx^2 + cx + d$$

Write the general equation.

$$\begin{cases} 216a + 36b + 6c + d = 38 \\ 125a + 25b + 5c + d = 74 \\ 8a + 4b + 2c + d = 50 \\ -a + b - c + d = 80 \end{cases}$$

Substitute 6 for x and 38 for $f(x)$, and so forth, to get a system of equations.

$$\begin{bmatrix} 216 & 36 & 6 & 1 \\ 125 & 25 & 5 & 1 \\ 8 & 4 & 2 & 1 \\ -1 & 1 & -1 & 1 \end{bmatrix}^{-1} \begin{bmatrix} 38 \\ 74 \\ 50 \\ 80 \end{bmatrix} = \begin{bmatrix} -2 \\ 15 \\ -19 \\ 44 \end{bmatrix}$$

Solve the system using matrices.

$$\therefore f(x) = -2x^3 + 15x^2 - 19x + 44$$

Cubic regression gives the same equation, with $R^2 = 1$.

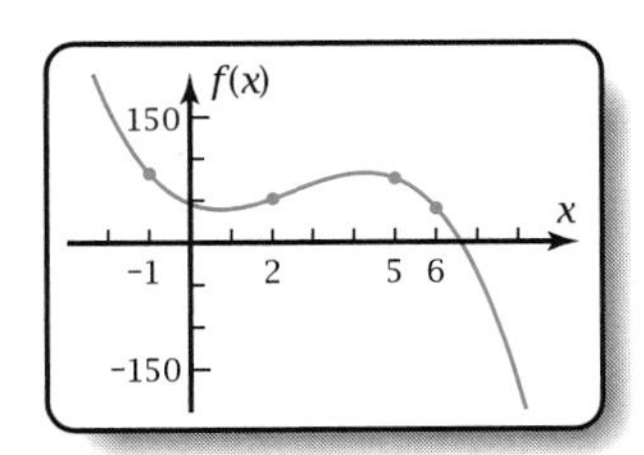

Figure 15-3b

b. Graph, Figure 15-3b. Tracing to $x = 6$, 5, 2, and −1 confirms that the given points are on the graph. Note that for large positive values of x the values of $f(x)$ get larger in the negative direction, consistent with the fact that the leading coefficient is negative. ◀

You recall the constant-second-differences property for quadratic functions. For a cubic function, the *third* differences between the y-values are constant. The table shows the result for $f(x) = -2x^3 + 15x^2 - 19x + 44$ from Example 1.

x	$f(x)$	First differences	Second differences	Third differences
−1	80			
		−36		
0	44		30	
		−6		−12
1	38		18	
		12		−12
2	50		6	
		18		−12
3	68		−6	
		12		−12
4	80		−18	
		−6		−12
5	74		−30	
		−36		
6	38			

Use the ΔList feature of your grapher.

PROPERTY: Constant-nth-Differences Property

For an nth-degree polynomial function, if the x-values are equally spaced, then the $f(x)$-values have constant nth differences.

In Problem 12 of Problem Set 15-3 you will prove this property.

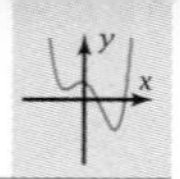

▶ EXAMPLE 2 An object moving in a straight line passes a reference point at time $x = 2$ seconds. It slows down, stops, reverses direction, and passes the reference point going backwards. Then it stops and reverses direction again, passing the reference point a third time. The table shows its displacements, $f(x)$, in meters, at various times.

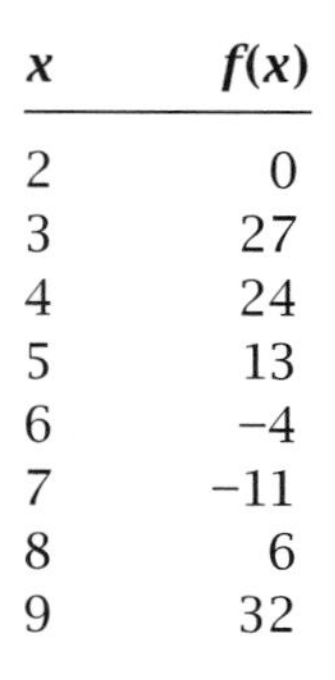

x	$f(x)$
2	0
3	27
4	24
5	13
6	−4
7	−11
8	6
9	32

a. Make a scatter plot of the data. Explain why a cubic function would be a reasonable mathematical model for displacement as a function of time.

b. Find the particular equation of the best-fitting cubic function. Plot the graph of the function on the scatter plot in part a.

c. Use the equation in part b to calculate the approximate time the object passed the reference point going backwards.

d. Show that a quartic function gives a coefficient of determination closer to 1 but that it has the wrong endpoint behavior for the given information.

Solution

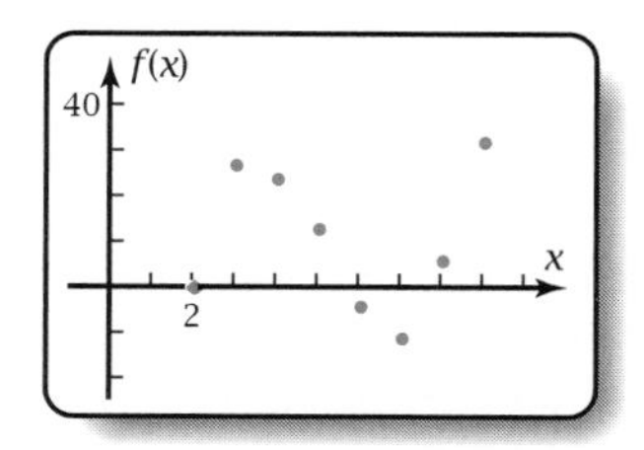

Figure 15-3c

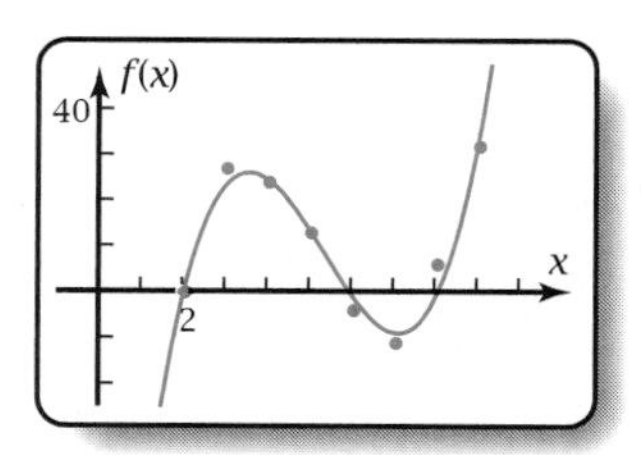

Figure 15-3d

a. Figure 15-3c shows the scatter plot. A cubic function is a reasonable mathematical model because it can reverse direction twice (has two extreme points), as shown by the scatter plot.

b. Using cubic regression, the particular equation is

$$f(x) = 1.5782\ldots x^3 - 25.0119\ldots x^2 + 117.2669\ldots x - 145.4761\ldots$$

$R^2 = 0.9611\ldots$, indicating a reasonably good fit because it is close to 1. The graph of f in Figure 15-3d shows that it is a reasonably good fit.

c. From the graph or the table, the x-value when the object passes the reference point going backwards is close to 6. Use the zeros, intersect, or solver feature on your grapher.

$$x \approx 5.8908\ldots \approx 5.9 \text{ seconds}$$

d. Quartic regression gives a coefficient of determination $R^2 = 0.9861\ldots$. However, the quartic function would have a third extreme point and cross the x-axis a fourth time, which wasn't mentioned in the statement of the problem. ◀

Problem Set 15-3

Do These Quickly

Q1. Multiply: $(x - 3)(x - 5)$

Q2. Multiply and simplify: $(3 + 2i)(5 + 4i)$

Q3. Expand the square: $(x - 7)^2$

Q4. Expand the square and simplify: $(5 + 3i)^2$

Q5. What is the maximum number of extreme points a quintic function graph can have?

Q6. One zero of a particular cubic function with real-number coefficients is $-7 + 4i$. What is another zero?

Q7. Sketch the graph of a cubic function with a positive double zero and an x^3-coefficient of −2.

Q8. Find the sum of the zeros of

$$f(x) = 2x^3 + 7x^2 - 5x + 13$$

Q9. If polynomial $p(x)$ has a remainder of 7 when divided by $(x - 5)$, find $p(5)$.

Q10. If polynomial $p(x)$ has $p(\frac{-3}{2}) = 0$, then a factor of $p(x)$ is

A. $x - 3$ B. $x + 3$ C. $x - 2$
D. $x + 2$ E. $2x + 3$

---

1. Given $P(x) = x^3 - 5x^2 + 2x + 10$,
   a. Plot the graph using an appropriate domain. Sketch the result.
   b. How many zeros does the function have? How many extreme points does the graph have? How are these numbers related to the degree of $p(x)$?
   c. Make a table of values of $p(x)$ for each integer value of x from 4 to 9. Show that the third differences between the $p(x)$-values are constant.
2. Given $P(x) = -x^4 + 6x^3 + 6x^2 - 12x + 11$,
   a. Plot the graph using an appropriate domain. Sketch the result.
   b. How many zeros does the function have? How many extreme points does the graph have? How are these numbers related to the degree of $p(x)$?
   c. Make a table of values of $p(x)$ for each integer value of x from -2 to 4. Show that the fourth differences between the $p(x)$-values are constant.
3. These values give the coordinates of points that are on the graph of function f:

x	$f(x)$
2	25.4
3	13.1
4	−3.8
5	−23.5
6	−44.2
7	−64.1

   a. Make a scatter plot of the points.
   b. Show that the third differences between the $f(x)$-values are constant.
   c. Find algebraically the particular equation of the cubic function that fits the first four points. Show that the cubic regression on all six points gives the same equation. Plot the equation of the function on the same screen as the scatter plot from part a.
4. These values give the coordinates of points that are on the graph of function g:

x	$g(x)$
2	−3
3	−25
4	−31
5	27
6	221
7	647
8	1425

   a. Make a scatter plot of the points.
   b. Show that the fourth differences between the $g(x)$-values are constant.
   c. Find algebraically the particular equation of the quartic function that fits the first five points. Show that quartic regression on all seven points gives the same equation. Plot the equation of the function on the same screen as the scatter plot of part a.
5. *Diving Board Problem:* Figure 15-3e shows a person standing on the end of a diving board. Theoretical results on strength of materials indicate that the deflection of such a *cantilever beam* below its horizontal rest position at any point x from the built-in end of the beam is a cubic function of x. Suppose that the following deflections, $f(x)$, are measured, in thousandths of an inch, when x is measured in feet.

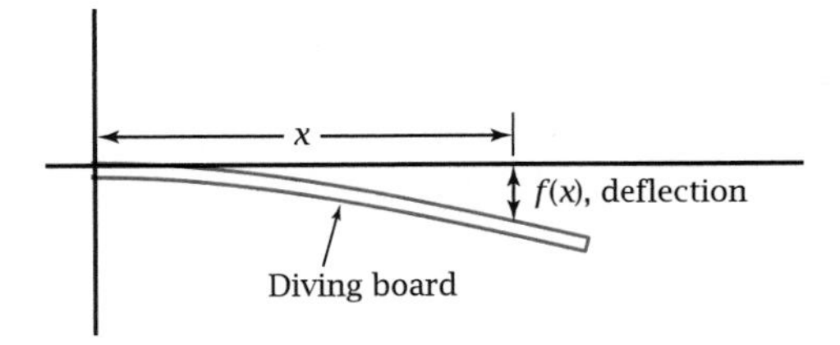

Figure 15-3e

x (ft)	$f(x)$ (thousandths of an inch)
0	0
1	116
2	448
3	972

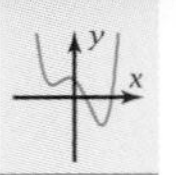

a. Find the particular equation of the cubic function that fits the data. Show that the linear and constant coefficients are zero.

b. How far does the diving board deflect at its end, $x = 10$ feet, where the person is standing?

c. Show that the function has another zero but that it is out of the domain of the function.

d. Sketch the graph of the function, showing both vertices. Darken the part of the graph that is in the domain determined by the 10-foot-long diving board.

6. *Two-Stage Rocket Problem:* A two-stage rocket is fired straight up. After the first stage finishes firing, the rocket slows down until the second stage starts firing. Its altitudes, $h(x)$, in feet above the ground, at each 10 seconds after firing are

x	$h(x)$
10	1750
20	3060
30	3510
40	3700
50	4230
60	5700

a. Show that the $h(x)$-values have constant third differences.

b. What type of function will fit the data exactly? Find its particular equation.

c. Plot the graph of h. Based on the graph, does the rocket start coming back down before the second stage fires? How can you tell?

d. The first stage of the rocket was fired at time $x = 0$. How do you explain the fact that the function in part b has a zero at $x = 3$ seconds?

7. *Television Set Pricing:* A retail store has various sizes of television sets made by the same manufacturer. The price for a set is a function of the screen size, measured in inches along the diagonal. Suppose that the prices are

x (in.)	$p(x)$ (dollars)
2	160
5	100
7	120
12	250
17	220
21	200
27	340
32	680
35	1100

a. Make a scatter plot of the data. Based on the scatter plot, tell why a quartic function is a more appropriate mathematical model than a cubic function.

b. Find the particular equation of the best-fitting quartic function, $p(x)$. Plot function p on the same screen as the scatter plot.

c. Based on the quartic model, which size television set is most overpriced?

d. What real-world reason can you think of to explain why 17- and 21-in. sets are less expensive than the smaller, 12-in. set?

8. *Pilgrim's Bean Crop Problem:* When the Pilgrims arrived in America, they brought along seeds from which to grow crops. If they had planted beans, the number of bean plants would have increased rapidly, leveled off, and then decreased with the approach of winter. The next spring, a new crop would have come up. Suppose that the number of bean plants,

Squanto Teaching Pilgrims
by Charles W. Jefferys
(The Granger Collection, New York)

$B(x)$, as a function of x weeks since the planting is given by this table:

x	$B(x)$
3	59
4	113
5	160
6	203
7	240
8	272

a. Find the particular equation of the best-fitting cubic function. Plot the equation and the data on the same screen. Sketch the result.

b. According to the cubic model, what is the maximum number of bean plants they had in the first year? After how many weeks did the number of bean plants reach this maximum? Did any plants survive through the next winter? If so, what is the smallest number of plants, and when did the number reach this minimum? If not, when did the last plant die, and when did the first plant emerge the next spring?

c. Show that if $B(8)$ had been 273 instead of 272, the conclusions of part b would be much different. (This phenomenon is called *sensitive dependence on initial conditions.*)

d. What year did the first Pilgrims arrive in America? What did they name the place where they landed?

9. *River Bend Problem:* A river meanders back and forth across Route 66. Three crossings are 1.7 mi, 3.8 mi, and 5.5 mi east of the intersection of Route 66 and Farm Road 13, or FM 13 (Figure 15-3f).

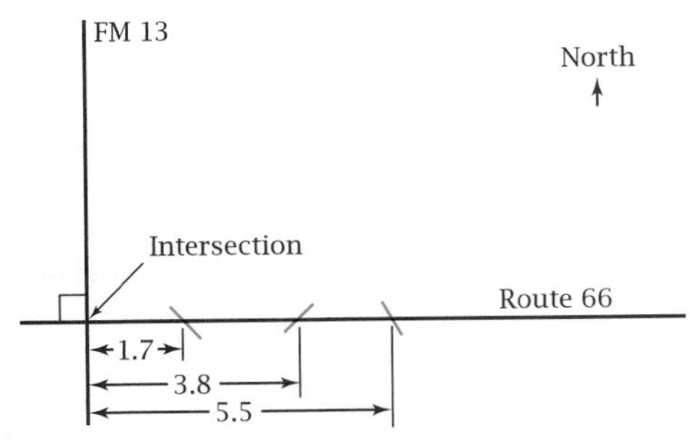

Figure 15-3f

a. Use the sum and product of the zeros properties to find quickly an equation of the cubic function $y = x^3 + bx^2 + cx + d$ that has 1.7, 3.8, and 5.5 as zeros. What is the y-intercept?

b. Let $f(x)$ be the number of miles north of Route 66 for a point on the river that is x miles east of FM 13. Suppose that the river crosses FM 13 at a point 4.1 miles north of the intersection. What *negative* vertical dilation of the equation in part a would give a cubic function with the correct $f(x)$-intercept as well as the correct x-intercepts? Write the particular equation for $f(x)$.

c. Plot function f. Sketch the result. What is the graphical significance of the fact that the leading coefficient is negative?

d. Based on the cubic model, what is the farthest south of Route 66 that the river goes between the 1.7-mile crossing and the 3.8-mile crossing? What is the farthest north of Route 66 that the river goes between the 3.8-mile crossing and the 5.5-mile crossing? How far east of the 5.5-mile crossing would you have to go for the river to be 10 miles south of Route 66?

10. *Airplane Payload Problem:* The number of kilograms of "payload" an airplane can carry equals the number of kilograms the wings can lift minus the mass of the airplane, minus the mass of the crew and their equipment. Use these facts to write an equation of the payload as a function of the airplane's length.

- The plane's mass is directly proportional to the cube of the plane's length.
- The plane's lift is directly proportional to the square of the plane's length.

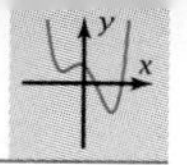

a. Assume that a plane of a particular design and length $L = 20$ m can lift 2000 kg and has a mass of 800 kg. Write an equation for the lift and an equation for the mass as functions of L.

b. Assume that the crew and their equipment have a mass of 400 kg. Write the particular equation for $P(L)$, the payload the plane can carry in kilograms.

c. Make a table of values of $P(L)$ for each 10 m from 0 to 50 m.

d. Function P is cubic and thus has three zeros. Find these three zeros, and explain what each represents in the real world.

11. *Behavior of Polynomial Functions for Large Values of x:* Figure 15-3g shows

$f(x) = x^3 - 7x^2 + 10x + 2$ (solid) and

$g(x) = x^3$ (dashed)

The graph on the right is zoomed out by a factor of 4 in the x-direction and by a factor of 64 (equal to 4^3) in the y-direction.

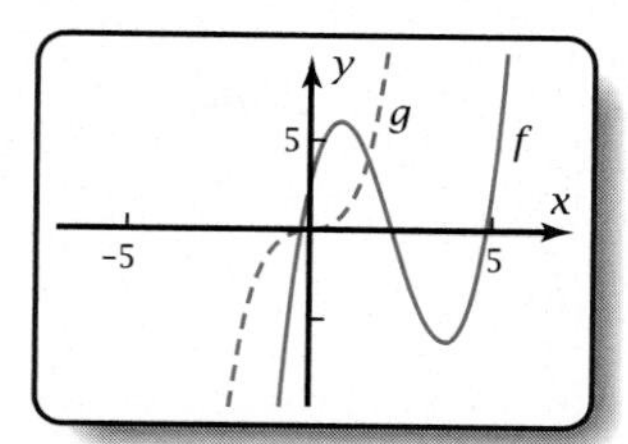

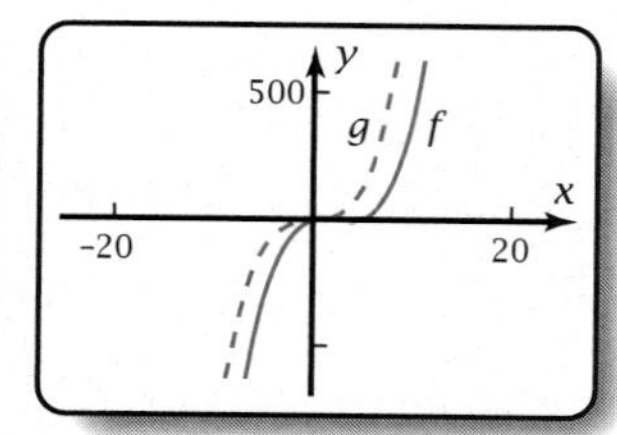

Figure 15-3g

a. Plot the two graphs on your grapher with a window as shown in the graph on the left. Set the zoom factors on your grapher to 4 in the x-direction and 64 in the y-direction. Then zoom out by these factors. Does the result resemble the graph on the right in Figure 15-3g?

b. Zoom out again by the same factors. Sketch the resulting graphs.

c. What do you notice about the shapes of the two graphs as you zoom out farther and farther? Can you still see the intercepts and vertices of the f graph? What do you think is the reason for saying that the highest-degree term **dominates** the function for large values of x?

12. *Constant-nth-Differences Proof Project:* Let $f(x) = ax^3 + bx^2 + cx + d$.

a. Find algebraically four consecutive values of $f(x)$ for which the x-values are k units apart. That is, find $f(x)$, $f(x + k)$, $f(x + 2k)$, and $f(x + 3k)$. Expand the powers.

b. Show that the third differences between the values in part a are independent of x and are equal to $6ak^3$.

c. Let $g(x) = 5x^3 - 11x^2 + 13x - 19$. Find $g(3)$, $g(10)$, $g(17)$, $g(24)$, and $g(31)$. By finding the third differences between consecutive values, show numerically that the conclusion of part b is correct.

13. *Coefficient of Determination Review Problem:*

a. Enter the data from Example 1 into your grapher and perform the cubic regression. Confirm that the coefficient of determination is 0.9611..., as shown in the example.

b. Using the appropriate list features on your grapher, find SS_{res}, the sum of the squares of the residual deviations of each data point from the regression curve.

c. Find the mean of the given $f(x)$-values. Find SS_{dev}, the sum of the squares of the deviations of each data point from this mean value.

d. You recall that the coefficient of determination is defined to be the fraction of SS_{dev} that is removed by the regression. That is,

$$R^2 = \frac{SS_{dev} - SS_{res}}{SS_{dev}}$$

Confirm that this formula gives 0.9611..., the value found by regression.

14. *Journal Problem:* Enter into your journal what you have learned so far about higher-degree polynomial functions. Include such things as shapes of the graphs and constant differences and their relationship to the degree of the polynomial.

15-4 Rational Functions: Discontinuities, Limits, and Partial Fractions

The function

$$y = \frac{x+2}{x^2 - x - 6}$$

is called a **rational function** because y equals a *ratio* of two polynomials. In this section you will learn what happens to graphs of rational functions at values of x at which the denominator equals zero.

OBJECTIVE Find discontinuities in the graphs of rational functions, and identify the kind of discontinuities they are.

Discontinuities and Limits

In Figure 15-4a, rational functions f and g have equations that look almost alike. Both $f(3)$ and $g(3)$ are undefined because division by zero is undefined, making their graphs **discontinuous** at $x = 3$, as shown in Figure 15-4a. However, the effect of the discontinuity is quite different on each graph.

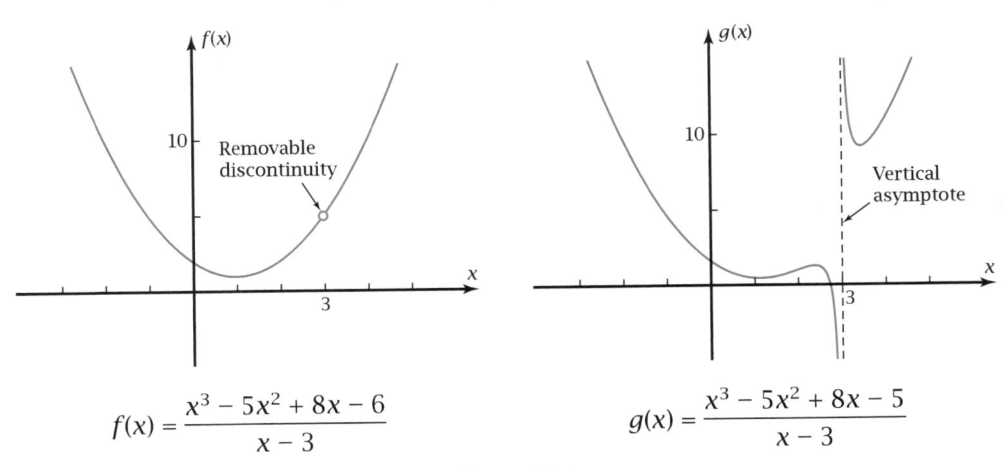

$$f(x) = \frac{x^3 - 5x^2 + 8x - 6}{x - 3} \qquad g(x) = \frac{x^3 - 5x^2 + 8x - 5}{x - 3}$$

Figure 15-4a

This table shows numerically what happens if x is close to 3.

x	$f(x)$	$g(x)$
2.9	4.61	−5.39
2.99	4.9601	−55.0399
2.999	4.996001	−995.003999
3	Undefined	Undefined
3.001	5.004001	1005.004001
3.01	5.0401	105.0401
3.1	5.41	15.41

Each function has a **discontinuity** at $x = 3$.

Function f has a **removable discontinuity** at $x = 3$. A point is missing from the graph. Because $f(x)$ gets closer and closer to 5 as x approaches 3, the discontinuity could be "removed" by defining $f(3) = 5$. The discontinuity in function g at $x = 3$ means that there is a **vertical asymptote.** As x approaches 3, $g(x)$ approaches positive or negative infinity.

You can see algebraically what happens to these functions at $x = 3$ by substituting 3 for x in each equation. The answers have these forms.

$f(3)$: $\frac{0}{0}$ Indeterminate form

$g(3)$: $\frac{1}{0}$ Infinite form

M. C. Escher's wood engraving, Smaller and Smaller, *demonstrates limit—as the areas of these geckos' bodies get closer and closer to zero when they approach the center. (M. C. Escher's* Smaller and Smaller *© 2002 Cordon Art B.V.-Baarn-Holland. All rights reserved.)*

The form $\frac{0}{0}$ is called an **indeterminate form.** You can't "determine" that $f(x)$ is close to 5 just by looking at the form $\frac{0}{0}$. The form $\frac{1}{0}$ is called an **infinite form** because the quotient in $g(x)$ keeps getting larger and larger in absolute value as x gets closer and closer to 3. The form $\frac{1}{0}$ is always infinite.

The number 5 is called the **limit** of $f(x)$ as x approaches 5, written

$\lim_{x \to 3} f(x) = 5$ Pronounced "The limit of $f(x)$ as x approaches 3 is 5."

DEFINITION: Limit

$L = \lim_{x \to c} f(x)$ if and only if you can keep $f(x)$ arbitrarily close to L by keeping x sufficiently close enough to c (but not *equal* to c).

Fortunately, there is an algebraic way to find the limit of a rational function at a removable discontinuity. Example 1 shows you how.

▶ EXAMPLE 1 Find $\lim_{x \to 3} f(x)$ for

$$f(x) = \frac{x^3 - 5x^2 + 8x - 6}{x - 3}$$

Solution First do synthetic substitution of $x = 3$ (which makes the denominator equal 0) into the polynomial in the numerator.

$$\begin{array}{r|rrrr} 3 & 1 & -5 & 8 & -6 \\ & & 3 & -6 & 6 \\ \hline & 1 & -2 & 2 & \underline{0} \end{array}$$

The remainder = 0, so $(x - 3)$ is a factor of the numerator.

$$f(x) = \frac{(x-3)(x^2 - 2x + 2)}{x - 3}$$

Write the numerator in factored form.

$$f(x) = x^2 - 2x + 2, \quad \text{provided } x \neq 3$$

Simplify to "remove" the discontinuity algebraically.

$$\lim_{x \to 3} f(x) = 3^2 - 2(3) + 2 = 5$$

Substitute 3 for x in the quotient polynomial. ◀

When you cancel the $(x - 3)$ factors in the next-to-last line of Example 1, you remove the discontinuity algebraically, and that is why it is called "removable." Although $f(x)$ is undefined at $x = 3$, once you've simplified it, you can evaluate the quotient polynomial at $x = 3$. Substituting 3 into the polynomial gives the exact value of the limit.

▶ **EXAMPLE 2** Find $\lim_{x \to 3} g(x)$ for

$$g(x) = \frac{x^3 - 5x^2 + 8x - 5}{x - 3}$$

Solution

$$\begin{array}{r|rrrr} 3 & 1 & -5 & 8 & -5 \\ & & 3 & -6 & 6 \\ \hline & 1 & -2 & 2 & \underline{1} \end{array}$$

The remainder = 1, so $(x - 3)$ is *not* a factor.

$$g(x) = x^2 - 2x + 2 + \frac{1}{x - 3}$$

Write $g(x)$ in mixed-number form.

$$\lim_{x \to 3} g(x) = \infty$$

The discontinuity cannot be removed. ◀

The symbol "∞" in Example 2 stands for **infinity.** Note that ∞ is not a number. It is used to indicate that the value of the limit is greater in absolute value than any real number.

PROPERTIES: Discontinuities and Indeterminate Forms

For the rational algebraic function $f(x) = \dfrac{p(x)}{q(x)}$

- f has a **discontinuity** at $x = c$ if $q(c) = 0$.
- f has a **vertical asymptote** at $x = c$ if $f(c)$ has the **infinite form** $\dfrac{\text{(nonzero number)}}{0}$.
- f may have a **removable discontinuity** at $x = c$ if $f(c)$ has the **indeterminate form** $\frac{0}{0}$. Cancel the common factor to see if the discontinuity is removed.
- The y-value of a removable discontinuity at $x = c$ is the **limit** of $f(x)$ as x approaches c.
- The limit can be found graphically, numerically, or algebraically.

Partial Fractions

The rational function

$$f(x) = \frac{9x - 7}{x^2 - x - 6} = \frac{9x - 7}{(x + 2)(x - 3)}$$

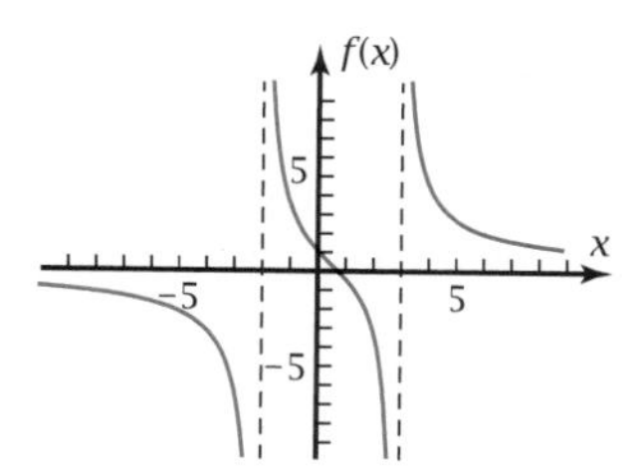

Figure 15-4b

has a quadratic denominator. From the factored form you can tell that there are two discontinuities, one at $x = -2$ and the other at $x = 3$. There are vertical asymptotes at both discontinuities because neither factor in the denominator cancels. The graph is shown in Figure 15-4b.

Rational functions such as this can come from adding two rational expressions with linear denominators. For instance,

$$\frac{5}{x+2} + \frac{4}{x-3} = \frac{5(x-3)}{(x+2)(x-3)} + \frac{4(x+2)}{(x-3)(x+2)} = \frac{9x-7}{(x+2)(x-3)}$$

It is possible to start with the expression on the far right and reverse the process of addition to find the two **partial fractions** that were added to form $f(x)$. Example 3 shows you how.

▶ **EXAMPLE 3** Resolve

$$f(x) = \frac{9x - 7}{x^2 - x - 6}$$

into partial fractions.

Solution

$$f(x) = \frac{9x - 7}{(x + 2)(x - 3)}$$ Factor the denominator.

$$= \frac{A}{x+2} + \frac{B}{x-3}$$ A and B stand for unknown constants.

$$= \frac{A(x-3) + B(x+2)}{(x+2)(x-3)}$$ Find a common denominator and add the fractions.

$$= \frac{(A+B)x + (-3A+2B)}{(x+2)(x-3)}$$ Arrange the numerator as an x-term and a constant term.

$$\therefore (A + B)x = 9x \quad \text{and} \quad (-3A + 2B) = -7$$ Equate the x-terms and constant terms.

$$\begin{cases} A + B = 9 \\ -3A + 2B = -7 \end{cases}$$

$$\begin{bmatrix} 1 & 1 \\ -3 & 2 \end{bmatrix}^{-1} \begin{bmatrix} 9 \\ -7 \end{bmatrix} = \begin{bmatrix} 5 \\ 4 \end{bmatrix}$$ Solve by matrices.

$$\therefore f(x) = \frac{5}{x+2} + \frac{4}{x-3}$$ Write the answer. ◀

A quicker way to find the partial fractions in Example 3, attributed to British mathematician Oliver Heaviside (1850–1925), is:

$$f(x) = \frac{9x - 7}{(x + 2)(x - 3)}$$

Look at the factored form of the denominator and find the values of x at which the factors are zero. So if $x = -2$, then $(x + 2)$ is zero. Then cover up the $(x + 2)$ factor with your finger and substitute -2 for x in what remains.

$$f(x) = \frac{9x - 7}{(\quad)(x - 3)}$$

Calculate: $\frac{9(-2) - 7}{-2 - 3} = 5$

The value you get is 5, the numerator of the factor covered up. Similarly, you can find the numerator of the other factor by covering up the $(x - 3)$ factor and substituting 3 into what is left. The value $x = 3$ makes that factor equal zero.

$$f(x) = \frac{9x - 7}{(x + 2)(\quad)}$$

Calculate: $\frac{9(3) - 7}{3 + 2} = 4$

The value you get is 4; that is the numerator of the other covered-up factor.

$$\therefore f(x) = \frac{5}{x + 2} + \frac{4}{x - 3}$$

Problem Set 15-4

Do These Quickly

5 min

Q1. Sketch the graph of a quadratic function.

Q2. Sketch the graph of a cubic function with three real zeros.

Q3. Sketch the graph of a cubic function with two nonreal complex zeros.

Q4. Sketch the graph of a quartic function with four real zeros.

Q5. How many vertices can the graph of a 7th-degree function have?

Q6. Write the zeros of this function: $y = (3x - 5)(2x + 7)(x - 4)$

Q7. How many times does the graph of $g(x) = (x - 4)(x^2 + 5x + 2)$ cross the x-axis?

Q8. What transformation of f gives g, if $g(x) = f\left(\frac{1}{3}x\right)$?

Q9. If you multiply a 3×2 matrix by a 2×4 matrix, what dimension matrix results?

Q10. The exact value of $\sin \frac{\pi}{3}$ is

A. 0 B. 0.5 C. $\frac{\sqrt{3}}{2}$

D. $\frac{\sqrt{3}}{3}$ E. $\frac{\sqrt{2}}{2}$

For Problems 1 and 2, plot the graphs of the two functions on the same screen. Use a friendly window with an x-range that includes the integers as grid points. Sketch the results, showing any vertical asymptotes or removable discontinuities.

1. $f(x) = \frac{x^3 - 10x^2 + 24x - 16}{x - 2}$ and

$g(x) = \frac{x^3 - 10x^2 + 24x - 17}{x - 2}$

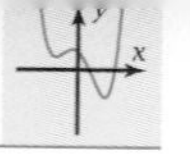

2. $r(x) = \dfrac{x^2 - x - 12}{x - 4}$ and $h(x) = \dfrac{x^2 - x - 11}{x - 4}$

3. For functions f and g in Problem 1, make a table of values for $f(x)$ and $g(x)$ for values of x that get closer and closer to 2 from both sides. How do the values in the table relate to the graphs in Problem 1?

4. For functions r and h in Problem 2, make a table of values for $r(x)$ and $h(x)$ for values of x that get closer and closer to 4 from both sides. How do the values in the table relate to the graphs in Problem 2?

5. For both functions in Problem 1, use synthetic substitution either to remove the removable discontinuity or to write the equation in mixed-number form.

6. For both functions in Problem 2, use synthetic substitution either to remove the removable discontinuity or to write the equation in mixed-number form.

7. For functions f and g in Problem 1, find algebraically $\lim_{x \to 2} f(x)$ and $\lim_{x \to 2} g(x)$.

8. For functions r and h in Problem 2, find algebraically $\lim_{x \to 4} r(x)$ and $\lim_{x \to 4} h(x)$.

9. Function g in Problem 1 is the sum of a quadratic function and a transformation of the reciprocal function $y = \frac{1}{x}$. What transformation was applied to the reciprocal function?

10. Function h in Problem 2 is the sum of a linear function and a transformation of the reciprocal function $y = \frac{1}{x}$. What transformation was applied to the reciprocal function?

For Problems 11 and 12, plot the function using a friendly window with an x-range that includes integers as grid points. Sketch the results showing any vertical asymptotes.

11. $f(x) = \dfrac{2x - 22}{x^2 + 2x - 8}$

12. $g(x) = \dfrac{7x - 2}{x^2 - x - 2}$

13. Resolve $f(x)$ in Problem 11 into partial fractions.

14. Resolve $g(x)$ in Problem 12 into partial fractions.

15. The partial fractions in Problem 13 are transformations of the reciprocal function $y = \frac{1}{x}$. Identify the transformations that were applied to get each fraction.

16. The partial fractions in Problem 14 are transformations of the reciprocal function $y = \frac{1}{x}$. Identify the transformations that were applied to get each fraction.

17. What is the difference in meaning between an *indeterminate* form and an *infinite* form?

18. What is meant by a *removable discontinuity*? What process can you use to remove a removable discontinuity?

19. *Step Discontinuity Problem 2:* The function in Figure 15-4c has a step discontinuity at $x = 2$. The value of $f(x)$ approaches a different number as x approaches 2 from the left side than it does as x approaches 2 from the right side. The function in Figure 15-4d has step discontinuities at each integer value of x. It could represent the cost of postage for a first-class letter as a function of the number of ounces the letter weighs.

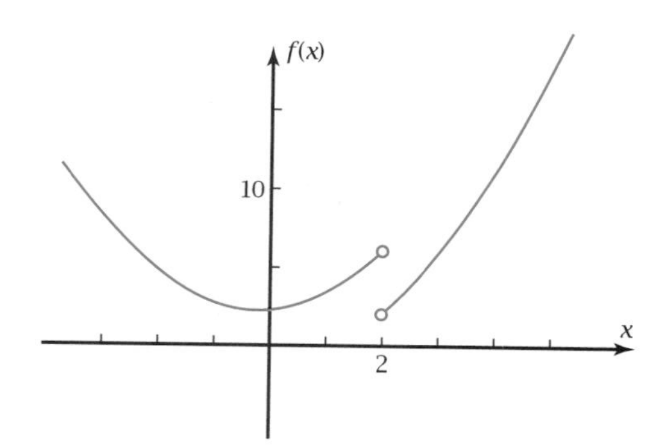

Figure 15-4c

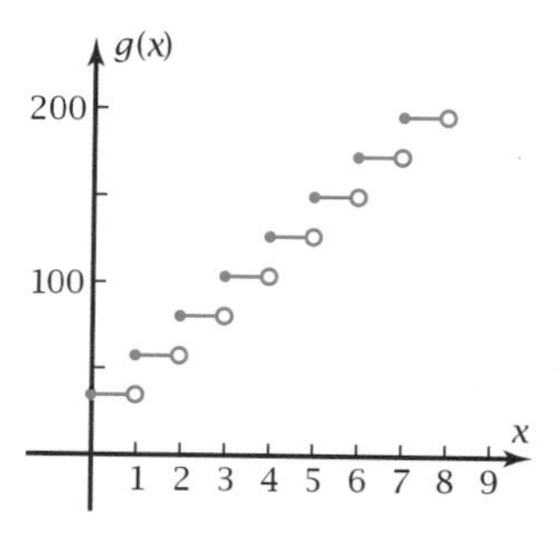

Figure 15-4d

a. The equation in Figure 15-4c is

$$f(x) = x^2 - \frac{2|x-2|}{x-2}$$

Plot this function using dot style and a friendly window with an x-range including $x = 2$. What number does $f(x)$ approach as x approaches 2 from the left side? From the right side?

b. The equation in Figure 15-4d is $g(x) = 34 + 23\,\text{int}(x)$, where $\text{int}(x)$ is the greatest integer less than or equal to x. Plot the graph using dot style. What is $g(2.99)$? $g(3)$?

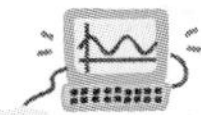

20. *Research Project 3:* Look up Oliver Heaviside on the Internet or via another reference source. Using these sources, try to find why the "cover up" method for partial fractions is valid.

15-5 Instantaneous Rate of Change of a Function: The Derivative

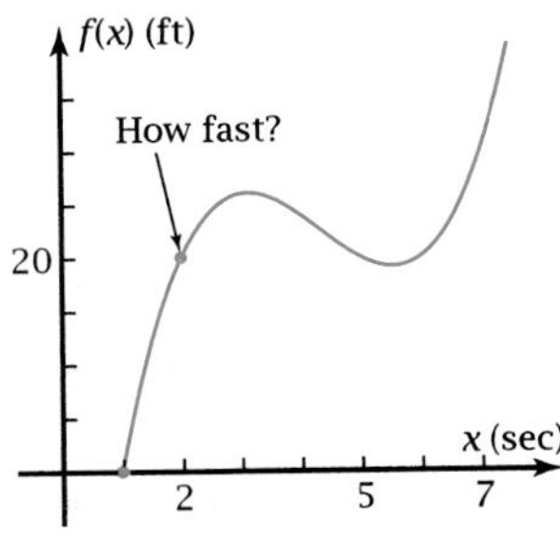

Figure 15-5a

Suppose that a bird takes off from the ground at time $x = 1$ sec. It climbs for a while, then dives for a while, and then swoops back up again. Figure 15-5a shows what its height might be as a function of time.

From the graph you can tell that the bird is still climbing at $x = 2$ sec. The question is, "At what *rate* is the bird climbing at the instant $x = 2$?" In this section you'll learn how to calculate the *derivative* of certain kinds of functions, which tells you the *instantaneous rate.*

OBJECTIVE Given the particular equation of a polynomial function, find the instantaneous rate of change (the derivative) at a given point, and interpret the answer graphically.

Instantaneous Rate (Derivative) Numerically

Rate equals distance divided by time. An "instant" is 0 sec long. In 0 sec, the bird mentioned before would travel 0 ft. So the instantaneous rate takes on the indeterminate form

$$\text{Instantaneous rate} = \frac{0}{0}$$

About 300 years ago, Isaac Newton (1642–1727) in England and Gottfried Wilhelm Leibniz (1646–1716) in Germany solved this problem with the help of the limit concept, which you have seen in connection with asymptotes and

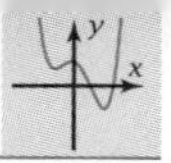

Gottfried Wilhelm Leibniz (1646–1716), a German mathematician, made significant contributions to the development of calculus, including the power rule for differentiation. (The Granger Collection, New York)

removable discontinuities in the last section and with fixed points and geometric series earlier.

Suppose that the bird's height, depicted in Figure 15-5a, is given by the function

$$f(x) = x^3 - 13x^2 + 52x - 40$$

where x is in seconds and $f(x)$ is in feet. To get an approximation for the instantaneous rate that the bird is climbing at $x = 2$, first find the *average* rate over a small time interval. Using your grapher, you will find

$$f(2) = 20 \qquad \text{and} \qquad f(2.1) = 21.131$$

So the bird climbed 1.131 ft in 0.1 sec, for an average rate of

$$\text{Average rate} = \frac{1.131}{0.1} = 11.31 \text{ ft/sec (feet per second)}$$

To get a better estimate for the instantaneous rate, use smaller time intervals.

$$f(2.01) = 20.119301 \qquad \text{Average rate} = \frac{0.119301}{0.01} = 11.9301 \text{ ft/sec}$$

$$f(2.001) = 20.011993001 \qquad \text{Average rate} = \frac{0.011993001}{0.001} = 11.993001 \text{ ft/sec}$$

The average rates seem to be approaching a limit of 12 ft/sec as the length of the time interval gets closer and closer to zero. This limit, 12 ft/sec, is the instantaneous rate, or instantaneous velocity. This instantaneous rate is called the **derivative** of the time-height function for the bird's flight.

Instantaneous Rate (Derivative) Algebraically

In the bird flight example, the distance the bird rose between 2 sec and x sec was

$$\text{distance} = f(x) - f(2) = (x^3 - 13x^2 + 52x - 40) - 20 = x^3 - 13x^2 + 52x - 60$$

The time it took was

$$\text{length of time} = x - 2$$

So the average rate, $r(x)$, is given by the rational function

$$r(x) = \frac{x^3 - 13x^2 + 52x - 60}{x - 2}$$

By graphing or by synthetic substitution you can find that 2 is a zero of the numerator and that the expression for $r(x)$ can be simplified.

$$r(x) = \frac{(x - 2)(x^2 - 11x + 30)}{x - 2}$$

$$r(x) = x^2 - 11x + 30 \qquad \text{provided } x \neq 2$$

The graph of the average rate, $r(x)$, has a removable discontinuity at $x = 2$, as shown in Figure 15-5b. The discontinuity is removable because the $(x - 2)$ factor, which makes the denominator zero when $x = 2$, is canceled by the $(x - 2)$ factor in the numerator.

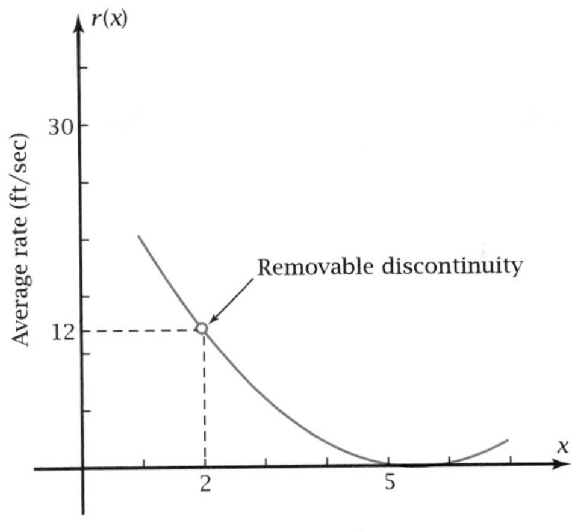

Figure 15-5b

The instantaneous rate at $x = 2$ is the y-value at this discontinuity. It is the *limit* of the average rate as x approaches 2. This instantaneous rate is the derivative of the time-height function at $x = 2$ as you've seen it defined previously.

$$\text{Instantaneous rate}_{x=2} = \text{derivative}_{x=2} = \lim_{x \to 2} (\text{average rate})$$
$$= 2^2 - 11(2) + 30 = 12$$

So the bird was climbing at the rate of 12 ft/sec at the instant $x = 2$ sec.

Instantaneous Rate (Derivative) Graphically

The derivative of a function has a remarkable geometric relationship to the graph of the function. Figure 15-5c shows the height of the bird, as in Figure 15-5a. A line with slope 12 (the derivative at $x = 2$) is drawn at the point on the graph where $x = 2$. As you can see from the figure, the line is **tangent** to the graph.

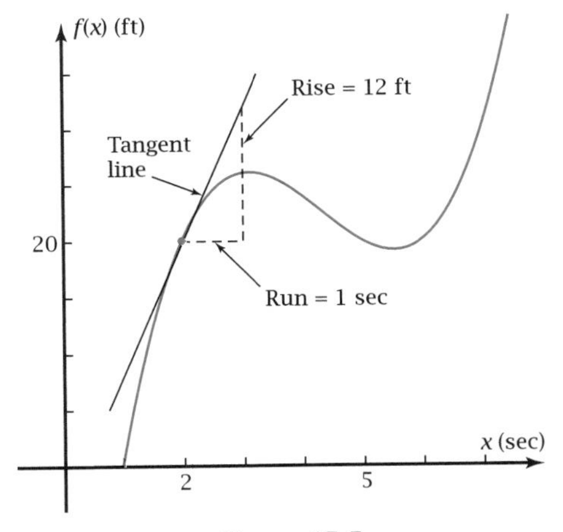

Figure 15-5c

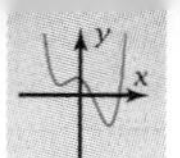

Here is the summary of what you've learned so far.

DEFINITIONS: Average and Instantaneous Rates of Change

- The **average rate of change,** $r(x)$, of function $f(x)$ on an interval starting at $x = c$ is the change in the y-value of the function divided by the corresponding change in the x-value. It is given by the rational function

$$r(x) = \frac{f(x) - f(c)}{x - c} \qquad \text{(Average rate)}$$

- The **instantaneous rate of change** of $f(x)$ at $x = c$ is called the **derivative** and is denoted $f'(x)$, said "f prime of x." It is equal to the **limit** of the average rate as x approaches c.

$$f'(x) = \lim_{x \to c} \frac{f(x) - f(c)}{x - c} \qquad \text{(Instantaneous rate)}$$

- The value of the derivative of $f(x)$ at $x = c$ equals the slope of the **tangent line** to the graph of f at $x = c$.

Equation of a Tangent Line

Using the techniques you learned in algebra, you can find the equation of the line tangent to a graph at a given point and plot it on your grapher.

▶ ***EXAMPLE 1*** Find the particular equation of the line tangent to the graph of $f(x) = x^3 - 13x^2 + 52x - 40$ at the point where $x = 2$. Plot the graph and the tangent line on the same screen.

Solution

$y = mx + b$ — In the general equation of a line, m is the slope, b is the y-intercept.

$y = 12x + b$ — From the previous work, the derivative (slope) is 12.

$20 = 12(2) + b$ — From the previous work, the point (2, 20) is on the graph.

$-4 = b$

$\therefore y = 12x - 4$

Figure 15-5d shows that the line is tangent to the graph.

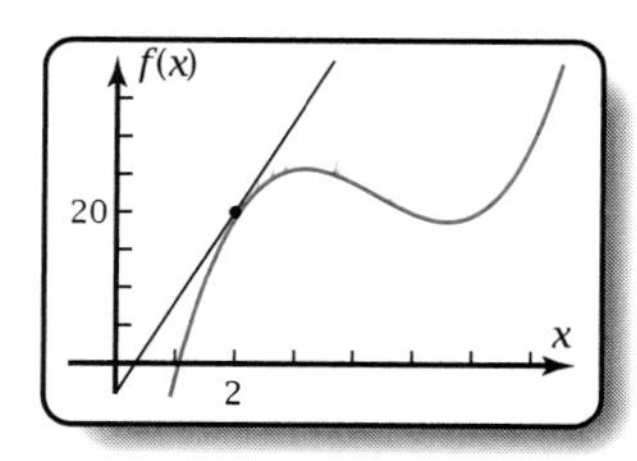

Figure 15-5d

◀

Problem Set 15-5

Do These Quickly 5 min

For Problems Q1–Q4, a rational algebraic function has equation $r(x) = \frac{p(x)}{q(x)}$. What do you know about $p(x)$ and $q(x)$ for each given condition?

Q1. There is a discontinuity at $x = 4$.

Q2. The discontinuity at $x = 4$ is removable.

Q3. The discontinuity at $x = 4$ yields a vertical asymptote.

Q4. $r(5) = 0$

Q5. What is the form $\frac{0}{0}$ called?

Q6. What is the form $\frac{1}{0}$ called?

Q7. What is the slope of the linear function $y = 3x + 5$?

Q8. What is the slope of the linear function for which $f(2) = 7$ and $f(5) = 19$?

Q9. Write a linear factor of polynomial function f if $f(3) = 0$.

Q10. If $(3 - 5i)$ is a zero of a polynomial with real-number coefficients, what is another complex zero?

---

1. Given $f(x) = x^3 - 6x^2 + 8x + 5$,
   a. Estimate numerically the instantaneous rate of change of $f(x)$ at $x = 2$ and at $x = 4$.
   b. Find algebraically the instantaneous rate of change of $f(x)$ at $x = 2$ and at $x = 4$. Do the numerical answers in part a agree with these values?
   c. Find an equation for the line through the point on the graph at $x = 4$ with slope equal to the derivative at that point. Plot the line and $f(x)$ on the same screen. Sketch the result. Is the line really tangent to the graph at this point?
2. Given $g(x) = x^3 - 2x^2 - x + 6$,
   a. Estimate numerically the instantaneous rate of change of $g(x)$ at $x = 1$ and at $x = 3$.
   b. Find algebraically the instantaneous rate of change of $g(x)$ at $x = 1$ and at $x = 3$. Do the numerical answers in part a agree with these values?
   c. Find an equation for the line through the point on the graph at $x = 1$ with slope equal to the derivative at that point. Plot the line and $g(x)$ on the same screen. Sketch the result. Is the line really tangent to the graph at this point?
3. Figure 15-5e shows the position of a moving object in time.

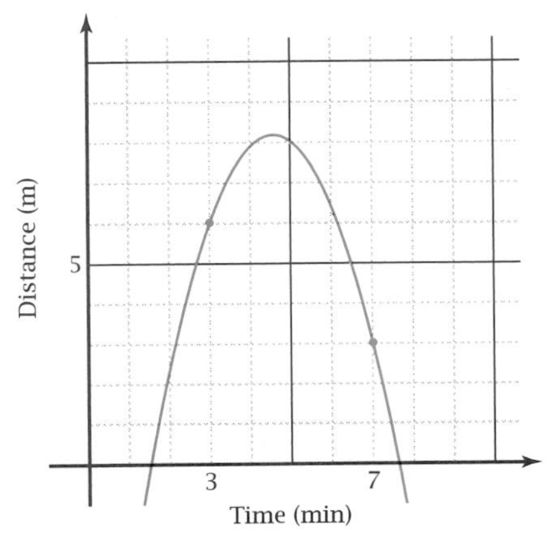

Figure 15-5e

   a. On a copy of the figure, draw lines tangent to the graph at times 3 min and 7 min.
   b. Estimate graphically the derivative of the function at these two times.
   c. Tell whether the object is speeding up or slowing down at these times, and at what rate.
4. Figure 15-5f shows the position of a moving object in time.

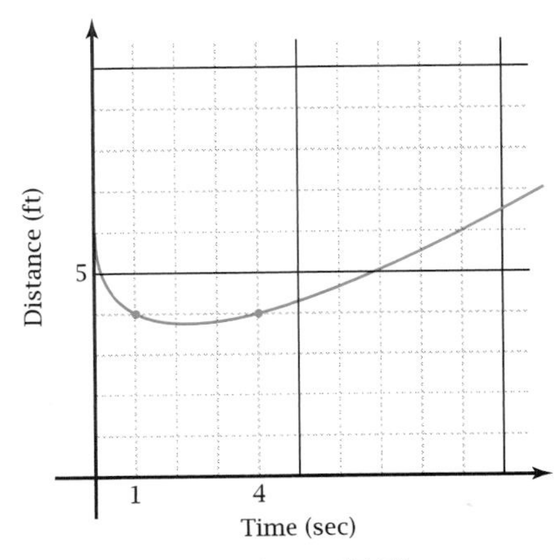

Figure 15-5f

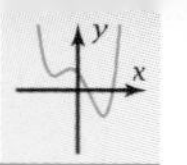

a. On a copy of the figure, draw lines tangent to the graph at times 1 sec and 4 sec.

b. Estimate graphically the derivative of the function at these two times.

c. Tell whether the object is speeding up or slowing down at these times, and at what rate.

5. *Tim and Lum's Board Pricing Problem:* Tim Burr and his brother Lum own a lumber company. Figure 15-5g shows the price, in cents, they charge for boards of varying length. For shorter boards, the price per foot decreases. For longer boards, the price per foot increases, because tall trees are harder to find.

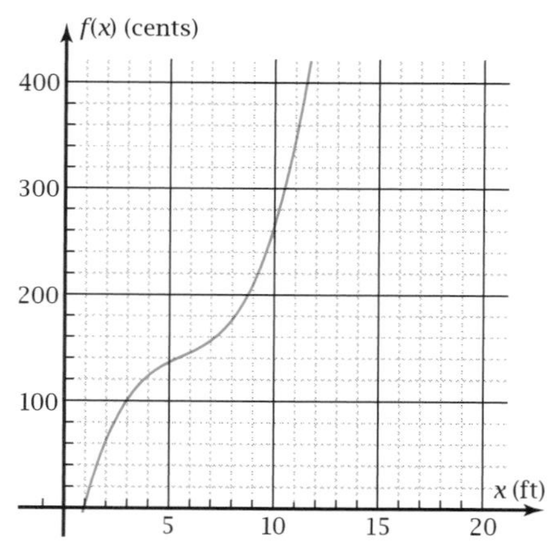

Figure 15-5g

a. The particular equation of this function is the cubic polynomial

$$f(x) = x^3 - 17x^2 + 105x - 89$$

where x is the length of the board, in feet, and $f(x)$ is the price of the board, in cents. Show by synthetic substitution that $x = 1$ is a zero of this function. What is the real-world meaning of this fact? Show algebraically that the other two zeros of this function are nonreal complex numbers.

b. The graph goes off scale. If the vertical scale were extended far enough, what price would you pay for a 20-ft board? How long would a board be that cost $10.00?

c. Write a rational algebraic function for the average rate of change from 8 ft to x ft. By simplifying this fraction, remove the removable discontinuity. Use the result to show that the instantaneous rate of change at $x = 8$ is exactly 25 cents per foot.

d. On a copy of Figure 15-5g, plot a line through the point on the graph at $x = 8$ with slope equal to the instantaneous rate of change you calculated in part c. How is this line related to the graph?

6. *Bumblebee Problem:* A bumblebee flies past a flower. It decides to go back for another look. Just before it reaches the flower, it turns and flies off. Its displacement from the flower, $d(t)$, in feet, is given by

$$d(t) = 0.2t^3 - 1.8t^2 + 3t + 5.6$$

where t is time, in seconds. Figure 15-5h shows the graph of this function.

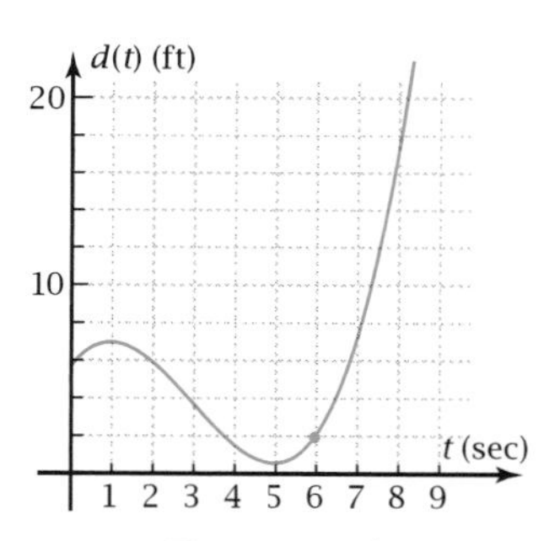

Figure 15-5h

a. What is the bee's average velocity from $t = 6$ to $t = 6.01$ seconds? (Average velocity is the average rate for the time-displacement function.)

b. Write an expression for the bee's average velocity from 6 seconds to t seconds. By simplifying the resulting fraction, calculate algebraically its instantaneous velocity at $t = 6$ seconds.

c. On a copy of Figure 15-5h, plot a line through the point at $t = 6$ with slope equal to the instantaneous velocity at time $t = 6$. How does the line relate to the graph?

d. Figure 15-5h indicates graphically that there are no positive values of t for which the bee is at the flower. Show algebraically that this is true.

e. At what negative value of t was the bee at the flower? At what positive time was it closest to the flower? How close?

7. *Derivative Shortcut for Power Function Problem:* In this problem you will learn how to find a function that gives the instantaneous rate of change of a polynomial function.

a. Let $f(x) = x^3 - 11x^2 + 36x - 26$. Show algebraically that the instantaneous rate of change of f is 4 at $x = 2$ and -3 at $x = 3$.

b. Let $g(x) = 3x^2 - 22x + 36$. Show that $g(2)$ and $g(3)$ equal the instantaneous rates of change of f in part a.

c. What operation could you perform on the terms of the polynomial $f(x)$ to derive the terms of $g(x)$? Explain how these operations apply to the constant term, -26.

d. Figure 15-5i shows the graph of f. Estimate from the graph the x-coordinates of the two extreme points.

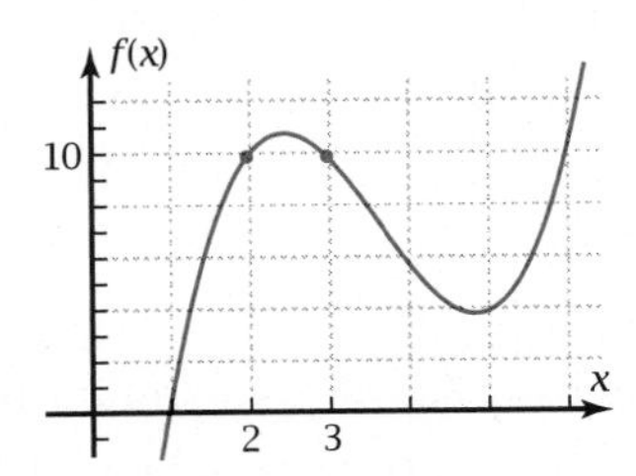

Figure 15-5i

e. At the extreme points in part d the tangent line will be horizontal, and thus the instantaneous rate will equal zero. Find algebraically the values of x at which $g(x) = 0$. Show that the values you estimated graphically agree with these exact values.

f. Calculate quickly the rate at which $f(x)$ is changing at $x = 5$. In what way does your answer agree with the graph?

g. Make a conjecture: "If $f(x) = x^{1776}$, then the instantaneous rate of change of $f(x)$ at any particular point is given by $g(x) =$ —?—." (A revolutionary idea?)

8. *Instantaneous Rate Quickly Problem:* Use the pattern in Problem 7 to find quickly the instantaneous rate of change of $f(x) = 5x^2 - 51x + 17$ at $x = 3$. Is $f(x)$ increasing or decreasing at this value of x? How can you tell?

The Derivative Function: The **derivative function** is the function that gives the instantaneous rate of change of a given function at any x-value. The name "derivative" is used because its equation can be "derived" from the given equation. In Problem 7, g is the derivative function of function f. For polynomial functions, the derivative function can be found as described in the box. When you study **calculus,** you'll learn how to derive this property.

Property: Derivative Function of a Polynomial Function

If $f(x) = x^n$, where n stands for a nonnegative integer, then $f'(x) = nx^{n-1}$.

Verbally: To find the derivative of a power function, multiply by the original exponent and decrease the exponent by 1.

If $f(x) = a_n x^n + \cdots + a_1 x + a_0$, where the coefficients are real numbers and the exponents are nonnegative integers, then $f'(x) = a_n(nx^{n-1}) + \cdots + a_1$.

Verbally: To find the derivative of a polynomial function, take the derivative of each term, multiplying by the coefficient of that term.

For Problems 9–16, use the pattern described in the box to find the equation for the derivative function, $f'(x)$. Remember that you can write a constant, c, as $c \cdot x^0$.

9. $f(x) = x^7$
10. $f(x) = x^9$
11. $f(x) = 8x^6$
12. $f(x) = 12x^{10}$
13. $f(x) = 9x^3 - 5x^2 + 2x - 16$
14. $f(x) = 11x^3 - 3x^2 - 13x + 37$
15. $f(x) = x^6 - 3^6$
16. $f(x) = x^5 + 4^5$

For Problems 17–22, find the derivative function, $f'(x)$. Use the fact that the derivative is zero at an extreme point to find the x-coordinates of all extreme points. Confirm your answer graphically. Save the graphs of Problems 17 and 18 for Problems 23 and 24.

17. $f(x) = x^3 - 12x^2 + 36x + 17$
18. $f(x) = x^3 - 3x^2 - 9x + 7$
19. $f(x) = x^3 - 4x^2 + x + 6$
20. $f(x) = x^3 - 5x + 6$
21. $f(x) = x^3 + x - 2$
22. $f(x) = 2x^3 - 4x^2 + 3x + 5$

For Problems 23 and 24, find the particular equation of the line tangent to the graph of function f at the given value of $x = c$. Plot the function and the line on the same screen. Sketch the result, showing that the line really is tangent to the graph.

23. $f(x) = x^3 - 12x^2 + 36x + 17$ (Problem 17), at $c = 3$
24. $f(x) = x^3 - 3x^2 - 9x + 7$ (Problem 18), at $c = 5$
25. *Derivative of an Exponential Function Problem:* Figure 15-5j shows the graph of the exponential function $f(x) = 2^x$.

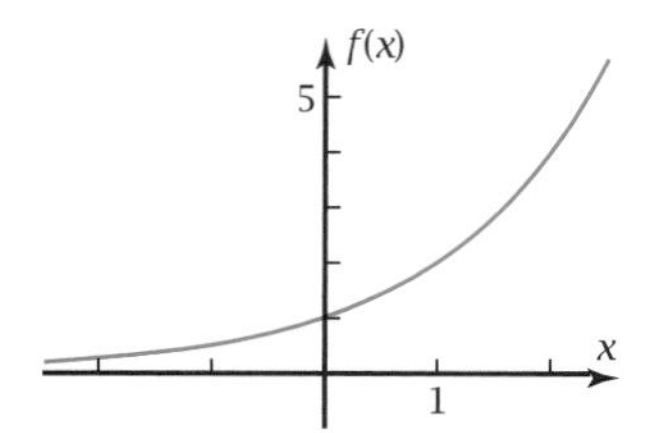

Figure 15-5j

a. If the "multiply by the original exponent, and decrease the exponent by 1" pattern worked for exponential functions, what would $f'(x)$ equal for this function?
b. Find $f'(0)$ using the results of part a. Based on the graph in Figure 15-5j, explain why this number could not possibly be the derivative of $f(x) = 2^x$ at $x = 0$.
c. Explain why the counterexample in part b constitutes a proof that the derivative of an exponential function *cannot* be found using the same shortcut as for the derivative of a power function.

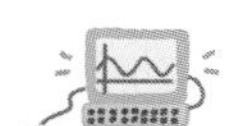

26. *Historical Research Problem:* On the Internet or via another reference source, look up Sir Isaac Newton and Gottfried Wilhelm Leibniz. Write a paragraph or two about each person. Include, if possible, their contributions to the subject of calculus, the branch of mathematics that concerns derivatives and instantaneous rates of change of functions.

15-6 Chapter Review and Test

In this chapter you have extended your knowledge of higher-degree polynomial functions. You have refreshed your memory about regression by using it to find particular equations of cubic and quartic functions. You can use these equations as mathematical models for moving objects that speed up and slow down, perhaps reversing directions more than once. You learned how to write a rational function for the average rate of change of a polynomial function. You found the instantaneous rate by removing the removable discontinuity in the average value function so that you could take the limit as the length of the time interval approached zero.

Review Problems

R0. Update your journal with what you have learned in this chapter. Include such things as those listed here.

- The one most important thing you have learned as a result of studying Chapter 15
- The new terms you have learned and what they mean
- Significant features of polynomial function graphs
- What kind of real-world situations polynomial functions can model
- How rational functions can be used to find instantaneous rates of change

R1. Figure 15-6a shows three cubic functions

$f(x) = -x^3 + 2x^2 + 5x - 6$

$g(x) = -x^3 + 2x^2 + 4x - 8$

$h(x) = -x^3 + 2x^2 + 3x - 10$

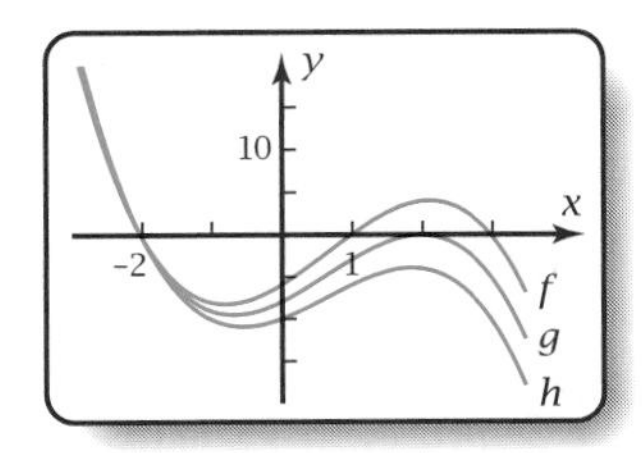

Figure 15-6a

a. Show algebraically that −2 is a zero of each function. That is, show that $f(-2)$, $g(-2)$, and $h(-2)$ all equal zero.

b. Transform the equation for each function to the form $(x + 2)$(other factor). You may use your grapher or computer software.

c. By setting each "other" factor equal to zero, find algebraically the other zeros of each function. Explain how the answers agree with the graphs.

d. What is the meaning of a *double zero* of a function? What is meant by a *complex zero*?

R2. *Stock Market Problem:* The stock for a new biotechnology company goes on the market. It is an immediate success, and the price of the stock rises dramatically. The price per share of stock is shown in the table.

Weeks	Dollars per Share
1	14.00
2	26.00
3	60.00
4	110.00
5	170.00
6	234.00

a. Find algebraically the particular equation of the cubic function that fits the first four data points. Show the steps leading to your answer.

b. Show that the fifth and sixth data points agree with the cubic function in part a.

c. Run a cubic regression on the data. Tell how the result confirms the fact that the cubic function in part a fits all six data points.

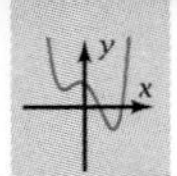

d. Plot the graph of the cubic function in part a. Sketch the result.

e. If you use this mathematical model to extrapolate a few months into the future, what does it predict will eventually happen to the price of the stock? Think of a real-world reason why this behavior might occur.

R3. a. *Train Problem, Part 1:* The locomotive of a train heading north stops to do some switching operations. The locomotive goes back and forth across a railroad crossing.

As it does so, the front end of the locomotive's displacement, $d(x)$, in feet, north of the crossing at time x, in minutes, is

$$d(x) = 120x^3 - 1200x^2 + 3480x - 2400$$

By synthetic substitution, show that 5 is a zero of this function. Find the other zeros. Explain how the number of zeros agrees with the fundamental theorem of algebra and its corollary. If the train is 700 ft long, will the locomotive ever be far enough north of the crossing so that the other end of the train does not block the crossing? How do you arrive at your answer?

b. Use long division to divide $x^3 - 4x^2 + 7x + 11$ by $x - 2$. Write the answer in mixed-number form. Explain how the answer agrees with the remainder theorem.

c. By synthetic substitution, show that 2 is a zero of $f(x) = x^3 - 10x^2 + 57x - 82$. Find the other two zeros. Plot the graph and sketch the result.

d. Figure 15-6b shows the graph of a polynomial function. Give the degree of the function, the number of real zeros, and the number of nonreal complex zeros.

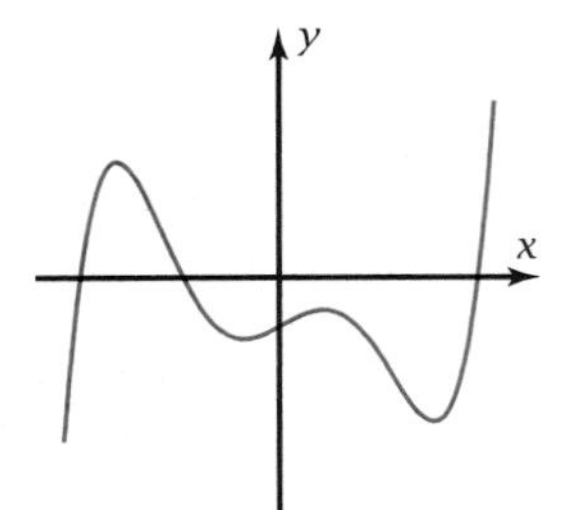

Figure 15-6b

R4. a. Given the rational functions

$$f(x) = \frac{x^3 - 13x^2 + 57x - 81}{x - 3} \quad \text{and}$$

$$g(x) = \frac{x^3 - 13x^2 + 57x - 80}{x - 3}$$

plot both graphs on the same screen. Use a friendly window that includes $x = 3$ as a grid point. Sketch the result, identifying which function is which.

b. Function f in part a has a removable discontinuity at $x = 3$. Remove the discontinuity by simplifying. Find the limit of $f(x)$ as x approaches 3. Explain the graphical significance of the answer.

c. Function g in part a has a vertical asymptote at $x = 3$. Write the equation for $g(x)$ in mixed-number form, and explain why the discontinuity at $x = 3$ cannot be removed.

d. What is the form $\frac{0}{0}$ called? What is the form $\frac{\text{nonzero}}{0}$ called?

e. *Partial Fractions Problem:* Given the rational function

$$h(x) = \frac{9x - 18}{x^2 - 5x + 4}$$

express $h(x)$ as partial fractions. Identify what transformation each fraction is of the reciprocal function $y = \frac{1}{x}$. Show that the graphs of h in partial fraction form and in the original form are identical. Sketch the result. Find algebraically the zero of $h(x)$.

R5. *Train Problem, Part 2:* Figure 15-6c shows the graph of the position of the locomotive in Review Problem R3. The particular equation is

$$d(x) = 120x^3 - 1200x^2 + 3480x - 2400$$

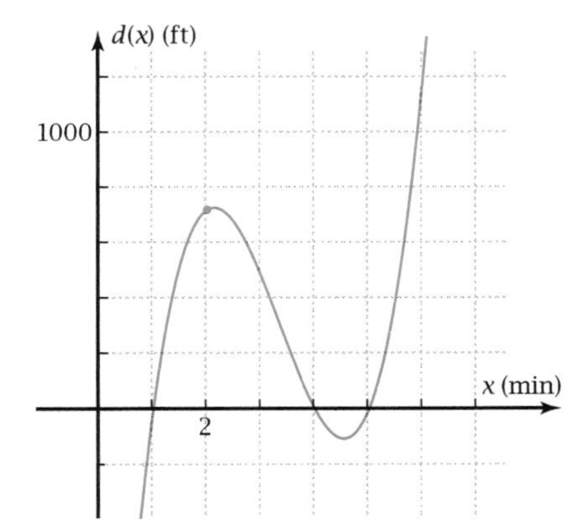

Figure 15-6c

a. What is the average velocity of the locomotive from $x = 2$ to $x = 2.1$ min?

b. Write the particular equation of the rational function that gives the average velocity for the time interval $[2, x]$. Include the units.

c. Plot the graph of the average velocity function in part b using a friendly window with $x = 2$ as a grid point. Sketch the result. Identify the feature that occurs in the graph at $x = 2$.

d. Simplify the equation in part b by removing the removable discontinuity at $x = 2$.

e. Find the instantaneous velocity at $x = 2$ by taking the limit of the average velocity as x approaches 2. What is this instantaneous velocity called?

f. On a photocopy of Figure 15-6c, graph a line with slope equal to the instantaneous velocity at $x = 2$ that contains the point on the graph where $x = 2$. How is the line related to the graph?

g. At $x = 2$ minutes, is the train going north or south? How can you tell?

h. Let $f'(x) = 360x^2 - 2400x + 3480$. Show that $f'(2)$ is equal to the instantaneous velocity that you calculated in part e. What algebraic operations could you perform on the equation of $f(x)$ to get the equation for $f'(x)$?

Concept Problems

C1. *Graphs of Complex Zeros Problem:* Figure 15-6d shows the graph of the cubic function

$$f(x) = x^3 - 18x^2 + 105x - 146$$

In this problem you will find out that you can find the complex zeros of the function geometrically from the graph.

a. Confirm that the graph agrees with the equation by making a table of values of $f(x)$ for each integer value of x that is on the graph.

b. Confirm that 2 is a real zero of $f(x)$. Find the two nonreal complex zeros, $x = a \pm bi$.

c. Through the point (2, 0), draw a line that is tangent to the graph at a point to the right of the low point at $x = 7$. Write the x-coordinate of the point of tangency. Write the slope of the tangent line. Show that the real part of the complex zeros, a, equals the x-coordinate of the point of tangency and that the coefficient of the imaginary part, b,

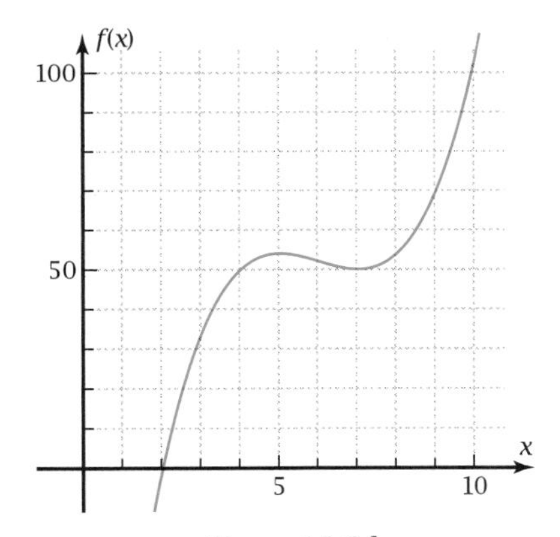

Figure 15-6d

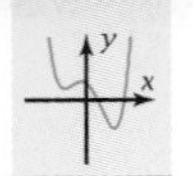

equals the square root of the slope of the tangent line.

d. Use the results of part c to find the complex zeros of the cubic function in Figure 15-6e.

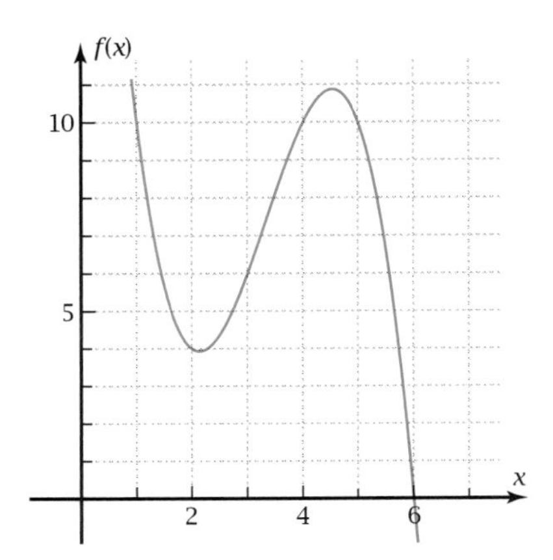

Figure 15-6e

e. For integer values of x, the values of $f(x)$ in Figure 15-6e are also integers. Use four of these points to find the particular equation for $f(x)$. Then calculate the three zeros and thus show algebraically that your graphical answers to part d are correct.

f. Find Paul J. Nahin's book *An Imaginary Tale: The Story of* $\sqrt{-1}$, published by Princeton University Press in 1998. Consult pages 27–30 for an in-depth analysis of the property you have learned in this problem.

C2. *The Rational Root Theorem:*

a. By synthetic substitution show that −1 is a zero of

$$f(x) = 6x^3 + 17x^2 - 24x - 35$$

Use the results to show that $f(x)$ factors into these linear factors:

$$f(x) = (x + 1)(2x + 7)(3x - 5)$$

What rational numbers are the other two zeros of $f(x)$?

b. This box contains a statement of the **rational root theorem** from algebra. Show that the zeros of $f(x)$ in part a agree with the conclusion of this theorem.

> **Property: The Rational Root Theorem**
> If the rational number $\frac{n}{d}$ is a zero of the polynomial function p (or a root of the polynomial equation $P(x) = 0$), then n is a factor of the constant term of the polynomial and d is a factor of the leading coefficient.

c. Show that these two polynomial functions agree with the rational root theorem.

$$g(x) = 3x^3 - 19x^2 + 13x + 35$$

$$h(x) = 6x^3 - 35x^2 - 31x + 280$$

Chapter Test

PART 1: No calculators allowed (T1–T9)

T1. Multiply: $(x - 5)(x - 2)$

T2. Use the answer to Problem T1 to calculate the product $(x - 5)(x - 2)(x + 1)$.

T3. Write the zeros of the function $f(x) = (x - 5)(x - 2)(x + 1)$.

T4. Sketch the graph of function f in Problem T3. Show where the zeros are and what the shape of the graph is.

T5. Sketch the graph of a quartic (4th-degree) function that has four real zeros.

T6. Figure 15-6f shows the function $g(x) = x^3 + x^2 - 7x - 15$. How can you tell from the graph that there are two zeros that are nonreal complex numbers?

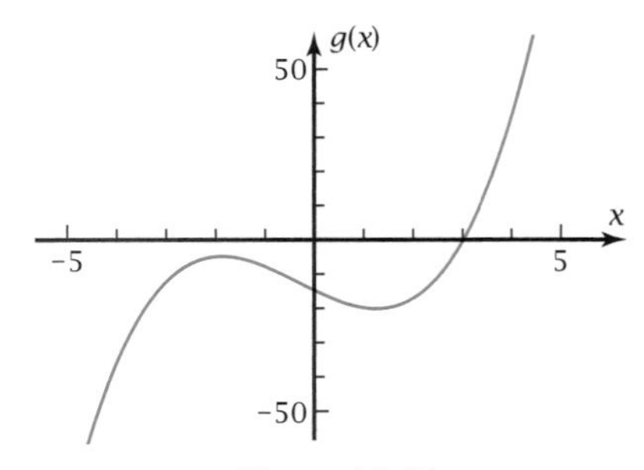

Figure 15-6f

T7. From the graph in Problem T6, $x = 3$ seems to be a zero of $g(x)$. Use synthetic substitution to show algebraically that this is correct.

T8. Write $g(x)$ in Problem T6 as a product of a linear factor and another factor.

T9. Show that the other factor in Problem T8 has no real zeros.

PART 2: Graphing calculators allowed (T10–T21)

T10. A cubic function contains these points. Use the first four points to find algebraically the particular equation of the cubic function. Confirm that the equation is correct by showing that the equation fits all six data points.

x	$f(x)$
2	19.4
3	40.1
4	74.2
5	123.5
6	189.8
7	274.9

Driving Problem: For Problems T11–T20, Hezzy Tate drives through an intersection. At time $t = 2$ sec she crosses the stripe at the beginning of the intersection. She slows down a bit, but does not stop, and then speeds up again. Hezzy is good at mathematics, and she figures that her displacement, $d(t)$, in feet, from the first stripe is given by

$$d(t) = t^3 - 12t^2 + 54t - 68$$

T11. Use synthetic substitution to show that $t = 2$ is a zero of $d(t)$.

T12. Use the results of the synthetic substitution and the quadratic formula to find the other two zeros of $d(t)$.

T13. How do the zeros of $d(t)$ confirm the fact that Hezzy does not stop and go back across the stripe?

T14. What is Hezzy's average velocity from $t = 3$ to $t = 3.01$ sec?

T15. Write the equation for the rational algebraic function equal to Hezzy's average velocity from 3 sec to t sec.

T16. Plot the graph of the rational function in Problem T15. Use a friendly window with a t-range that includes $t = 3$ as a grid point.

T17. What feature does the graph in Problem T16 have at $t = 3$?

T18. By appropriate simplification of the fraction in Problem T15, calculate Hezzy's instantaneous velocity at time $t = 3$.

T19. Figure 15-6g shows $d(t)$ as a function of time. Plot a line though the point on the graph at $t = 3$ with slope equal to the instantaneous velocity. Consider the different scales on the two axes.

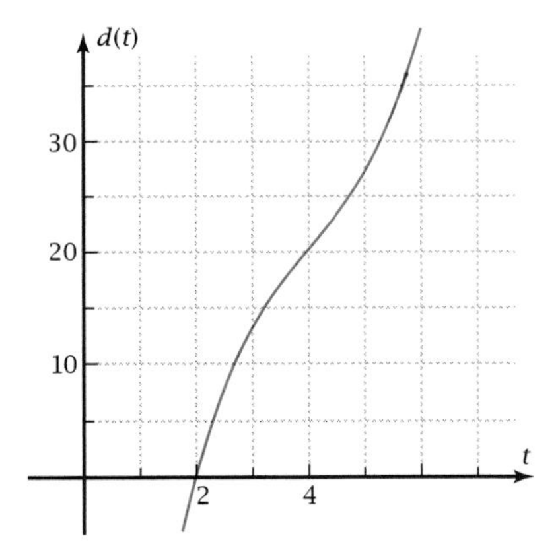

Figure 15-6g

T20. How is the line in Problem T19 related to the graph?

T21. What did you learn as a result of taking this test that you did not know before?

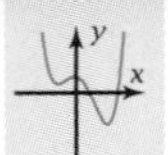

15-7 Cumulative Review, Chapters 10–15

This problem set is a final exam on the topics of

- Three-dimensional vectors
- Matrix transformations
- Conic sections
- Polar coordinates
- Parametric functions
- Complex numbers
- Sequences and series
- Polynomial functions
- Limits and derivatives

If you are thoroughly familiar with these topics, you should be able to finish the problem set in about three hours.

Cumulative Review Problems

Air Show Problem: (Problems 1–8) A pilot is doing stunts with her plane at an air show.

At times $t = 0$ sec and $t = 10$ sec, her position vectors from the control tower are

$\vec{a} = 7\vec{i} + 4\vec{j} + 3\vec{k}$ at $t = 0$

$\vec{b} = 10\vec{i} + 20\vec{j} + 5\vec{k}$ at $t = 10$

where $\vec{i}$ and $\vec{j}$ are unit vectors in the horizontal xy-plane and $\vec{k}$ is a vertical unit vector in the direction of the z-axis. The magnitude of the unit vector is a hundred meters.

1. Sketch vectors $\vec{a}$ and $\vec{b}$ tail-to-tail. Show the displacement vector from the head of $\vec{a}$ to the head of $\vec{b}$.

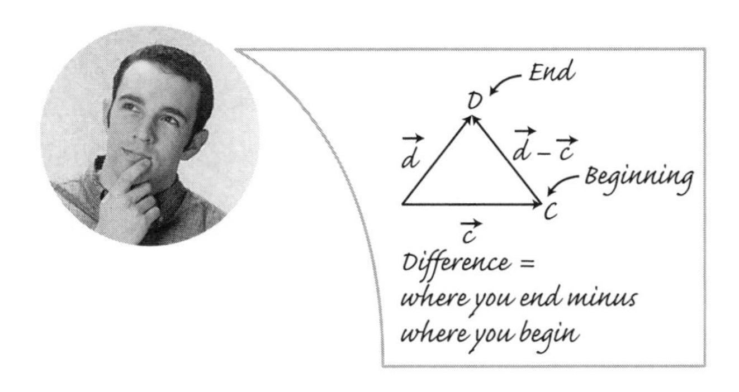

2. Calculate the displacement vector in Problem 1.
3. At 10 sec, how far was the pilot from her position at 0 sec?

4. Was the plane higher up or lower down at 10 sec than it was at 0 sec? By how many meters?

5. How large is the angle between vectors $\vec{a}$ and $\vec{b}$?

6. Find the scalar projection of $\vec{a}$ on $\vec{b}$.

7. The cross product of $\vec{a}$ and $\vec{b}$ is

$$\vec{a} \times \vec{b} = -40\vec{i} - 5\vec{j} + 100\vec{k}$$

Show that $\vec{a} \times \vec{b}$ is perpendicular to $\vec{a}$.

8. Find the area of the triangle formed by $\vec{a}$ and $\vec{b}$.

Doug's Iterative Transformation Problem: (Problems 9–16) Doug is an archaeologist. He unearths a set of paving stones that follow a spiral pattern generated by this matrix:

$$[A] = \begin{bmatrix} 0.45 & -0.78 & 20 \\ 0.78 & 0.45 & 10 \\ 0 & 0 & 1 \end{bmatrix}$$

9. Given that the dilation is 0.9, what is the angle of rotation?

10. The pre-image matrix is

$$[D] = \begin{bmatrix} 2 & -2 & -5 & 5 \\ 5 & 5 & -5 & -5 \\ 1 & 1 & 1 & 1 \end{bmatrix}$$

Use your Itrans program to apply [A] for 40 iterations. A window with an x-range of $[-50, 50]$ and a y-range of $[-20, 50]$ will be reasonable. Find numerically to 1 decimal place the coordinates of the fixed point to which the images have been attracted.

11. Show that you understand how matrix multiplication is done by multiplying

$$\begin{bmatrix} 0.45 & -0.78 & 20 \\ 0.78 & 0.45 & 10 \\ 0 & 0 & 1 \end{bmatrix}\begin{bmatrix} x \\ y \\ 1 \end{bmatrix}$$

12. At the fixed point, the image of (x, y) must equal (x, y). Thus,

$$\begin{bmatrix} 0.45 & -0.78 & 20 \\ 0.78 & 0.45 & 10 \\ 0 & 0 & 1 \end{bmatrix}\begin{bmatrix} x \\ y \\ 1 \end{bmatrix} = \begin{bmatrix} x \\ y \\ 1 \end{bmatrix}$$

Use this information to calculate algebraically the coordinates of the fixed point.

13. Doug finds a pattern on a wall that resembles the fractal image shown in Figure 15-7a.

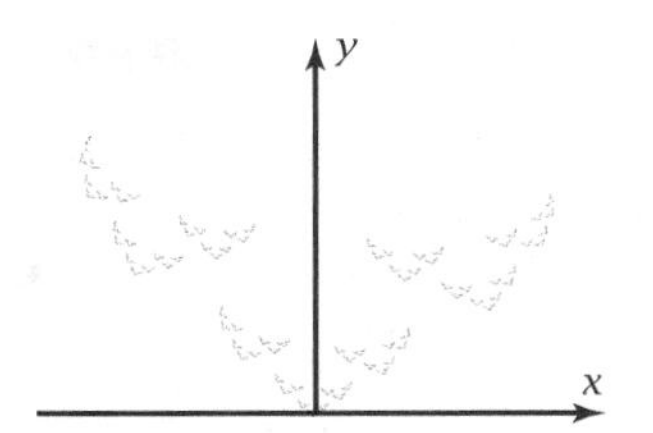

Figure 15-7a

Circle two parts of different size, each of which is similar to the entire figure.

14. One of the three transformations that generated Figure 15-7a is

$$[A] = \begin{bmatrix} 0.4\cos 20^\circ & 0.4\cos 110^\circ & 5 \\ 0.4\sin 20^\circ & 0.4\sin 110^\circ & 3 \\ 0 & 0 & 1 \end{bmatrix}$$

Write another transformation matrix [B] to do all of these things:

- Rotate 20° clockwise
- Dilate by a factor of 0.4
- Translate −5 in the x-direction
- Translate 4 in the y-direction

15. Write a third transformation matrix [C] to dilate by a factor of 0.4 without rotating or translating.

The Great Wave *by Katsushika Hokusai. The waves in this painting have a fractal pattern.*

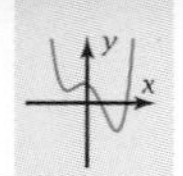

16. Check your answers to Problems 14 and 15 by running your Barnsley program with 1000 points. Use probabilities of $\frac{1}{3}$, $\frac{1}{3}$, $\frac{1}{3}$, and 0 for the four transformations the program is expecting. The window shown is [−9.4, 9.4] for x and [0, 12.4] for y.

Fractional Dimension Problem: (Problems 17–21) Figure 15-7b shows a 9 × 9 square pre-image and the first iteration of four transformations performed on the pre-image. Each transformation dilates the pre-image to $\frac{1}{3}$ of its original length but translates by different amounts.

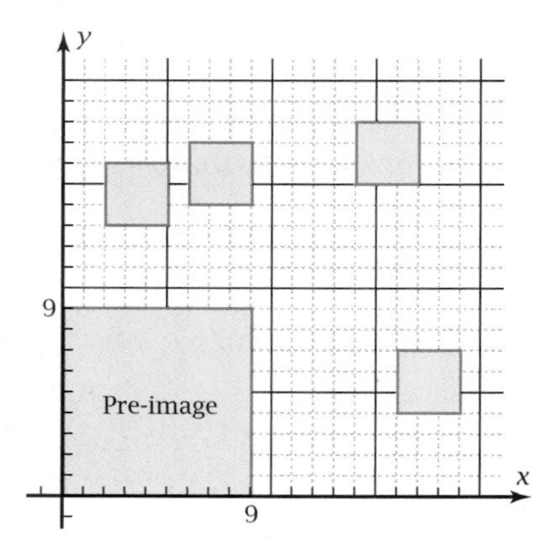

Figure 15-7b

17. How many images, N, will there be in the second iteration and in the third iteration? What is the total perimeter of the images in the first iteration, in the second iteration, and in the third iteration?

Iteration	N	Perimeter of Each	Total Perimeter
0	1	36	36
1	4	12	
2			
3			

18. Following the pattern in the table, what will be the total perimeter of the 50th iteration?

19. What is the total area of the first iteration? The second iteration? The third iteration?

20. If the transformations of this problem are performed using Barnsley's method, what will be the dimension of the resulting fractal image?

21. What limit do the total areas approach as the number of iterations approaches infinity? What limit do the total perimeters approach? Explain why these answers are consistent with the dimension you calculated in Problem 20.

Annie's Conic Section Problems: (Problems 22–26) Annie takes a test on conic sections.

22. How can she tell without plotting that the graph of this conic section will be a hyperbola? Plot the graph and sketch the result.

$$25x^2 - 9y^2 - 200x + 18y = -391$$

23. Figure 15-7c shows the ellipse $9x^2 + 25y^2 = 225$. Find the focal radius, eccentricity, and directrix radius.

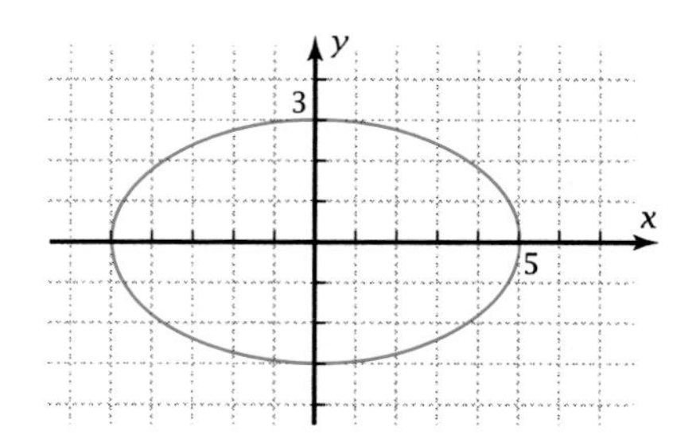

Figure 15-7c

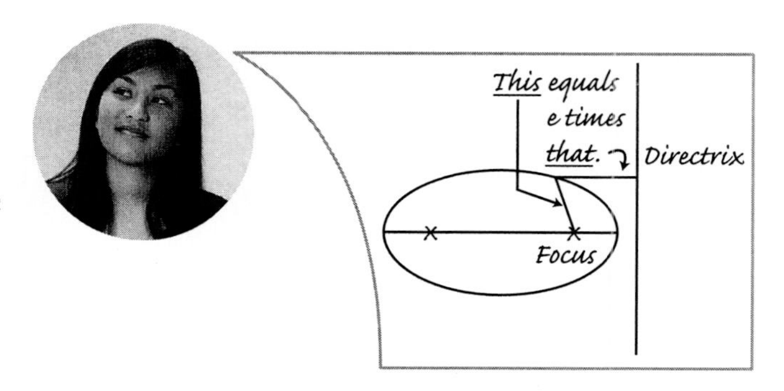

24. On a photocopy of Figure 15-7c, pick a point on the ellipse. Measure the distances from the point to the two foci. How do these distances compare to the length of the major axis? Plot the directrix on the right side of the ellipse. Measure the distance from the point on the graph to the directrix. How is this distance related to the eccentricity and the distance from the point to the focus on the right side?

25. Write parametric equations for the ellipse in Figure 15-7c. Then write parametric equations for the ellipse after it is translated 2 units in the positive x-direction and translated 1 unit in the positive y-direction. Confirm that your equations are correct by plotting on your grapher.

26. Figure 15-7d shows the ellipsoid formed by rotating the ellipse of Figure 15-7c about the x-axis. A cylinder is inscribed in this ellipsoid, with its axis along the x-axis. Find the volume of the cylinder in terms of the sample point (x, y) where the cylinder touches the ellipse. Use the result to find the radius and altitude (length) of the cylinder of maximum volume.

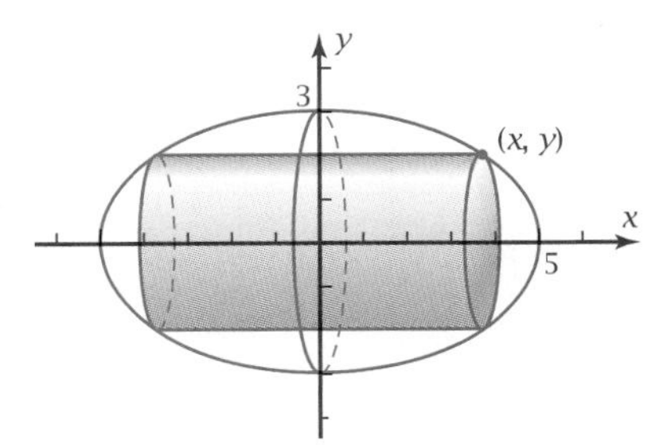

Figure 15-7d

Polar Coordinate Problems (Problems 27–31)

27. Figure 15-7e shows the graph of the circle $r = 3960$ in polar coordinates, representing Earth with radius 3960 miles. It also shows part of the elliptical path of a spaceship with polar equation

$$r = \frac{5600}{1 + 0.5 \cos \theta}$$

If the spaceship continues on its present path, will it hit Earth's surface? If so, what are the polar coordinates of the point where it will hit? If not, explain how you know it will not.

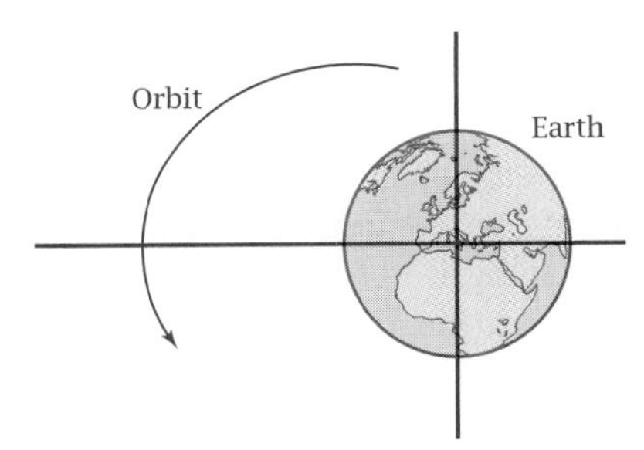

Figure 15-7e

28. Multiply the complex numbers 5 cis 70° and 8 cis 40°. Write the answer as a complex number in the form $r(\cos \theta + i \sin \theta)$. Transform the answer to $a + bi$ form.

29. Write the three cube roots of 64 as complex numbers in polar form.

30. Plot the **cardioid** $r = 24 - 24 \cos \theta$. Use a window with an x-range of $[-50, 50]$ and a y-range that makes equal scales on both axes. Sketch the result. Why do you think the figure is called a *cardioid*?

31. Figure 15-7f shows a quarter (radius 12 mm) centered at the origin. Another quarter rolls around it (without slipping). A point on the moving quarter traces an **epicycloid of one cusp.** Find parametric equations for the path. It will help if you sketch the rolling quarter in another position, draw a vector from the origin to the center of the rolling quarter, and then draw another vector from there to the point on the graph. Confirm that your equations are correct by plotting on your grapher. How does the graph relate to the cardioid in Problem 30?

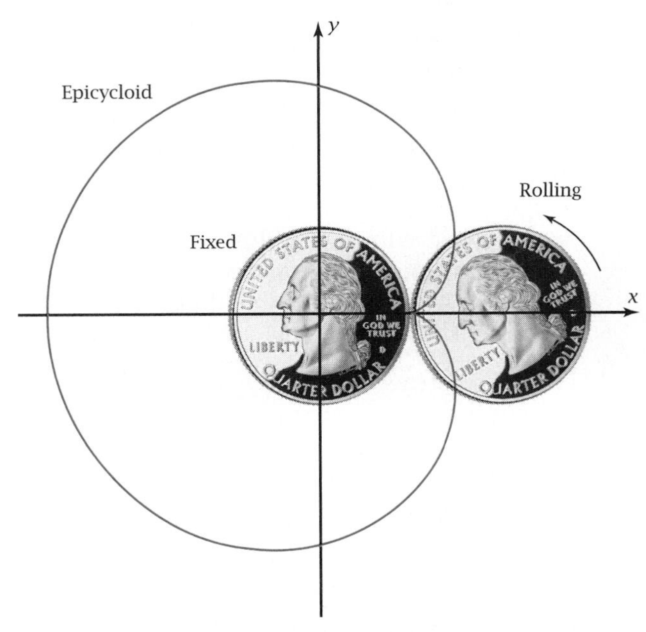

Figure 15-7f

Sequences and Series Problems (Problems 32–39)

32. Give the first four terms of the arithmetic *sequence* with first term 7 and common difference 5.

33. Give the fourth partial sum of the geometric *series* with first term 8 and common ratio 3.

34. You put \$12 into a piggy bank one week, \$15 the next week, \$18 the next week, and so on. How much will you put in on the tenth week?

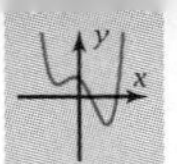

How much, total, will be in the bank during the tenth week (after the tenth deposit)?

35. After a long time, you break open the piggy bank in Problem 34 and invest \$500 of the money in a bank account that pays 9% per year interest, compounded monthly. How much, total, do you have in the bank during the 36th month? How much of this amount is interest? How much more do you have in the account than you would have had if the bank compounded interest only once a year? Why is a sequence a more reasonable mathematical model than a continuous function for this problem?

36. Wildlife conservationists find that the number of catfish in a particular lake is dropping by 10% each year. So the number of catfish after any 1 year is 0.9 times what it was at the beginning of that year. Assume that at present, time $t = 0$ years, there are 100 fish in the lake. The conservationists decide to add 30 more catfish at the end of each year. Write a recursion formula and use it to predict the number of catfish at the end of 1, 2, 3, 4, and 5 years. If the same recursive formula holds for many years, will the catfish population level off and approach a limit, or will it continue to increase without limit? Explain how you can decide.

37. For the sequence 2, 5, 10, 17, 26, 37, 50, . . . , there is a relatively simple pattern relating the term values to the term numbers. By finding this pattern, write an explicit formula for t_n as a function of n. Use the formula to find t_{100}.

38. One of the terms in the sequence of Problem 37 equals 5042. Use the formula to calculate algebraically the term number of this term.

39. The following series is called a ***p*-series** because the term index in the denominator is raised to a "power." Write out the first four terms of the series. Find S_{100}, the 100th partial sum.

$$\sum_{n=1}^{\infty} \frac{1}{n^{1.2}}$$

Tree Problem 2: (Problems 40–46) Ann R. Burr has a tree nursery in which she stocks various sizes of live oak trees. Figure 15-7g shows the price, $f(x)$ dollars, she charges for a tree, of height x feet, planted on the customer's property. For short trees, the price per foot decreases as the height increases. For taller trees, the price per foot increases because the trees are harder to move and plant.

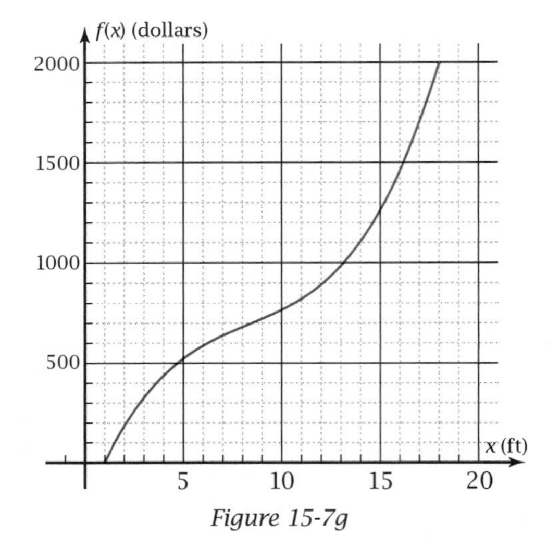

Figure 15-7g

40. The particular equation of this function is the cubic polynomial

$$f(x) = x^3 - 25x^2 + 249x - 225$$

Show by synthetic substitution that $x = 1$ is a zero of this function. What is the real-world meaning of this fact?

41. Show algebraically that the other two zeros of this function are nonreal complex numbers.

42. What is the sum of the zeros of this function?

43. The graph goes off scale. If the vertical scale were extended far enough, how much would a 20-ft tree cost?

44. Write a rational algebraic function for the average rate of change in price from 10 ft to x ft. By simplifying this fraction, remove the removable discontinuity. Use the result to show that the instantaneous rate of change at $x = 10$ is exactly 49 dollars per foot.

45. On a photocopy of Figure 15-7g, plot a line through the point on the graph at $x = 10$ with slope equal to the instantaneous rate of change in Problem 44. How is this line related to the graph?

46. What is the most important thing you learned as a result of studying for this final exam?

APPENDIX

Kinds of Numbers, Axioms, and Other Properties

In algebra you learned names for various kinds of numbers, such as real numbers, imaginary numbers, and rational numbers. You also learned commutative axioms, closure axioms, transitive axioms, and so on, as well as other properties such as the multiplication property of zero that can be proved from these axioms. In this appendix you will refresh your memory about the names of these kinds of numbers, axioms, and other properties, and some definitions. You'll also see, in the examples, how to use axioms to prove properties.

Kinds of Numbers

Complex numbers are numbers of the form $a + bi$, where a and b are **real numbers** and $i = \sqrt{-1}$. The real numbers form a subset of the complex numbers, for which the number $b = 0$. You can group real numbers several ways: as positive or negative, rational or irrational, algebraic or transcendental, and so forth. Figure A-1a shows the set of complex numbers and some of its subsets.

An important thing to realize is that a **rational number** is a number that *can be* expressed as a ratio of two integers. It does not have to be written that way. For instance,

23 is rational because it can be written $\frac{23}{1}$, $\frac{46}{2}$, and so on.

$\sqrt{9}$ is rational because it can be written as 3, which equals $\frac{3}{1}$.

$2\frac{3}{7}$ is rational because it can be written $\frac{17}{7}$.

3.87 is rational because it can be written $\frac{387}{100}$.

5.3333... (repeating) is rational because it can be written $5\frac{1}{3}$, which equals $\frac{16}{3}$.

Irrational numbers, on the other hand, *cannot* be expressed as the ratio of two integers. Irrational numbers include the nth root of any integer that is not an exact nth power. For instance, you can prove by contradiction that $\sqrt{3}$ is irrational. To do this, assume that $\sqrt{3}$ is a rational number, so it could be written as

$$\sqrt{3} = \frac{a}{b}$$

where a and b are relatively prime integers. But then

$$\left(\sqrt{3}\right)^2 = 3 = \frac{a^2}{b^2}$$

which is impossible because a and b have no common factors to cancel. So you've arrived at a contradiction, which could be caused only by the faulty assumption that $\sqrt{3}$ is a rational number.

An **algebraic number** is a number that is a solution to a polynomial equation with rational-number coefficients. You can think of it as a number that can be expressed using a finite number of algebraic operations, including taking the roots.

Transcendental numbers are numbers that are not algebraic. They "transcend," or go beyond, the algebraic operations. The most notable examples you may know of are π and e. Most trigonometric and circular function values, as well as most logarithm values, are also transcendental.

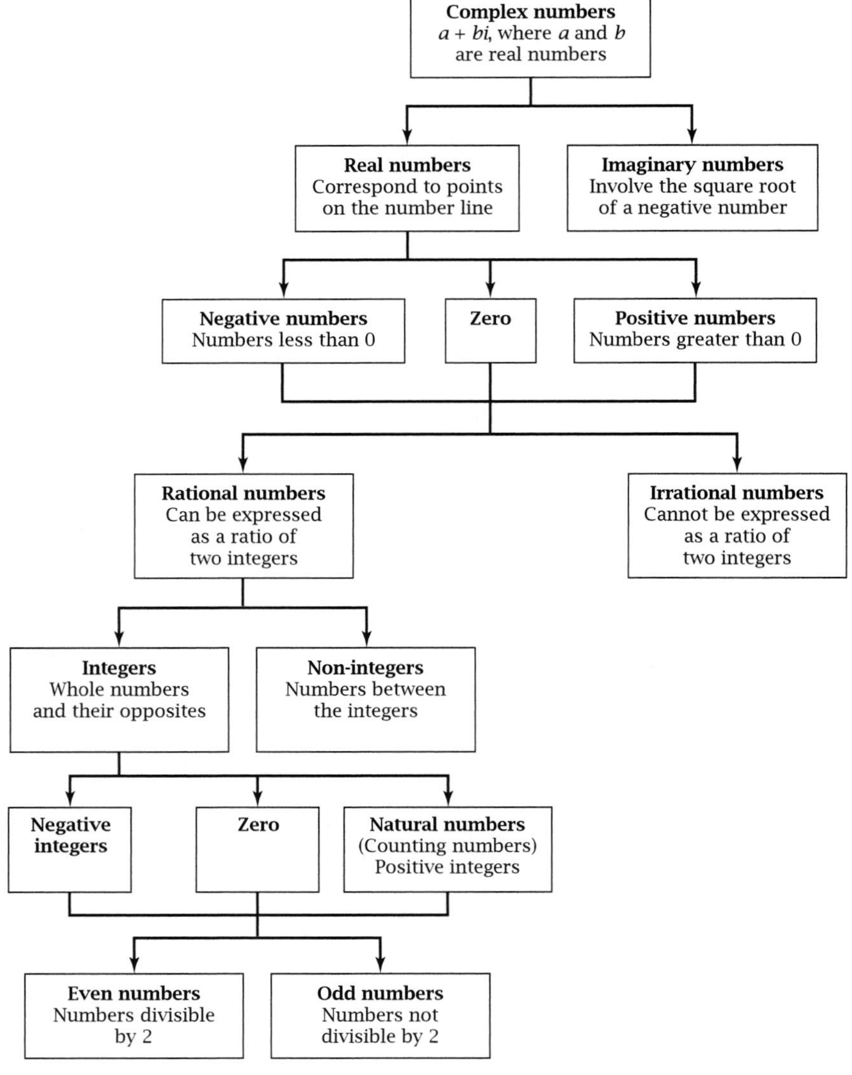

Figure A-1a

Axioms for Addition and Multiplication

There are 11 basic axioms for real numbers that apply to the operations of addition and multiplication. These axioms are called the **field axioms.** An *axiom* is a property that is assumed to be true so that it can be used as the basis for a mathematical system. Generally speaking, mathematicians prefer to have as few axioms as possible and to prove other properties using these axioms. The box lists the field axioms and what they state.

PROPERTIES: The Field Axioms

If x, y, and z are real numbers, then the following statements are true.

1. **Closure Under Addition**

   $x + y$ is a unique real number.

   "You can't get out of the set of real numbers when you add two real numbers."

2. **Closure Under Multiplication**

   xy is a unique real number.

   "You can't get out of the set of real numbers when you multiply two real numbers."

3. **Additive Identity**

   $$0 + x = x + 0 = x$$

   "Adding zero does not change a number."

4. **Multiplicative Identity**

   $$1 \cdot x = x \cdot 1 = x$$

   "Multiplying by 1 does not change a number."

5. **Additive Inverses**

   Every real number x has a unique additive inverse $-x$ such that $x + (-x) = 0$.

   "You can undo the addition of a number by adding its opposite."

6. **Multiplicative Inverses**

   Every real number x (except 0) has a unique multiplicative inverse $\frac{1}{x}$ such that $x \cdot \frac{1}{x} = 1$.

   "You can undo multiplication by a number if you multiply by its reciprocal."

(continued)

Properties: The Field Axioms, continued

7. Commutative Axiom for Addition

$$x + y = y + x$$

"You can commute two terms in a sum without changing the answer."

8. Commutative Axiom for Multiplication

$$xy = yx$$

"You can commute two factors in a product without changing the answer."

9. Associative Axiom for Addition

$$(x + y) + z = x + (y + z)$$

"You can associate terms in a sum differently without changing the answer."

10. Associative Axiom for Multiplication

$$(xy)z = x(yz)$$

"You can associate factors in a product differently without changing the answer."

11. Distributive Axiom for Multiplication Over Addition

$$x(y + z) = xy + xz$$

"You can distribute multiplication over addition without changing the answer."

Subtraction and division can be defined in terms of addition and multiplication with the aid of the inverse axioms.

DEFINITIONS: Subtraction and Division

$$x - y = x + (-y)$$

"Subtracting a number means adding its opposite."

$$x \div y = x \cdot \frac{1}{y}$$

"Dividing by a number means multiplying by its reciprocal."

Axioms for Equality and Order

There are three axioms for equality and three axioms for order (inequality), all of which state facts related to the = sign, the < sign, and the > sign. The term *order* pertains to the order in which numbers appear on the number line.

PROPERTIES: Axioms for Equality and for Order

If x, y, and z stand for real numbers, then the following statements are true.

Reflexive Axiom for Equality

$$x = x$$

"A real number is equal to itself, so a variable stands for the same number wherever it appears in an expression."

Symmetric Axiom for Equality

If $x = y$, then $y = x$.

"You can reverse the sides of an equation without affecting the equality."

Transitive Axioms for Equality and Order

If $x = y$ and $y = z$, then $x = z$.

If $x < y$ and $y < z$, then $x < z$.

If $x > y$ and $y > z$, then $x > z$.

"If the first number equals the second number and the second number equals the third number, then the first number equals the third number." (And so on.)

"Equality *goes through* (hence the name 'transit…') from first to last number."

Trichotomy Axiom (or Comparison Axiom)

For any two given numbers x and y, exactly *one* of these is true:

$$x < y$$

$$x = y$$

$$x > y$$

"A number y cuts the number line into three pieces (hence the name *trichotomy*): numbers less than it, numbers equal to it, and numbers greater than it."

Properties That Can Be Proved from the Axioms

The other familiar properties of real numbers can be proved from the axioms. Four examples are shown here. Other provable properties are listed after Example 4.

▶ EXAMPLE 1 **Substitution into Sums and Products**

$x + y$ and $z + y$ stand for the same number, provided $x = z$.

xy and zy stand for the same number, provided $x = z$.

"You can substitute equal quantities for equal quantities in a sum or a product."

Proof By the closure axioms, $x + y$ and xy each stand for a *unique* real number. Thus, it does not matter what symbol is used for x if $x = z$; both have the same value. Given that $y = y$, from the reflexive axiom, $x + y = z + y$ and $xy = zy$. Q.E.D. ◀

▶ EXAMPLE 2 **Addition Property of Equality**

If $x = y$, then $x + z = y + z$.

"You can add the same number to both sides of an equation without affecting the equality."

Proof

$x + z = x + z$	Reflexive axiom: A number equals itself.
$x = y$	Given.
$\therefore\ x + z = y + z$, Q.E.D.	Substitution into a sum. ◀

The **converse** of a property is the statement you get by interchanging the hypothesis (the "if" part) and the conclusion (the "then" part). The converse of a property may or may not be true. For instance, "If you run marathons, then you are in good shape" is true. But the converse, "If you are in good shape, then you run marathons," is false. So converses must also be proved.

▶ EXAMPLE 3 **Converse of the Addition Property of Equality**

If $x + z = y + z$, then $x = y$.

"You can cancel equal terms on both sides of the equal sign."

Proof

$x + z = y + z$	Given.
$(x + z) + (-z) = (y + z) + (-z)$	Addition property of equality: Add the opposite of z to both sides.
$x + [z + (-z)] = y + [z + (-z)]$	Associative axiom for addition.
$x + 0 = y + 0$	Additive inverse axiom.
$\therefore\ x = y$, Q.E.D.	Additive identity axiom. ◀

Note that the proof may seem to involve an excessive number of steps, especially because you are so familiar with "adding the opposite to both sides." However, to constitute a proof, each step must be justified by an axiom or by a previously proved property.

▶ **EXAMPLE 4** **Combining Like Terms**

Prove that $2x + 3x = 5x$.

Proof

$2x + 3x = (2 + 3)x$	Write one side of the desired equation. Use the distributive axiom (read in reverse).
$= 5x$	Arithmetic.
$\therefore\ 2x + 3x = 5x$, Q.E.D.	Transitive axiom for equality. ◀

The box shows these four properties as well as others. Each one can be proved from the axioms and from those properties that appear before it in the box.

Properties of Real Numbers Provable from the Axioms

If x, y, and z are real numbers, then the following statements are true.

1. **Substitution into Sums and Products**

   $x + y$ and $z + y$ stand for the same number, provided $x = z$.

   xy and zy stand for the same number, provided $x = z$.

   "You can substitute equal quantities for equal quantities in a sum or a product."

2. **Addition Property of Equality**

   If $x = y$, then $x + z = y + z$

   "You can add the same number to both sides of an equation without affecting the equality."

3. **Converse of the Addition Property of Equality**

   If $x + z = y + z$, then $x = y$

   "You can cancel equal terms on both sides of the equal sign."

4. **Combining Like Terms**

   Example: $2x + 3x = 5x$

   "You can combine like terms by adding their coefficients."

5. **Multiplication Property of Equality**

   If $x = y$, then $xz = yz$

   "You can multiply both sides of an equation by the same number without affecting the equality."

(continued)

Properties of Real Numbers Provable from the Axioms, continued

6. Cancellation Property of Equality for Multiplication

If $xz = yz$ and $z \neq 0$, then $x = y$

"You can divide both sides of an equation by the same nonzero number."

7. Opposite of an Opposite

$$-(-x) = x$$

"The opposite of the opposite of a number is the original number."

8. Reciprocal of a Reciprocal

If $x \neq 0$, then $\dfrac{1}{\frac{1}{x}} = x$

"The reciprocal of the reciprocal of a nonzero number is the original number."

9. Reciprocal of a Product

If $x \neq 0$ and $y \neq 0$, then $\dfrac{1}{xy} = \dfrac{1}{x} \cdot \dfrac{1}{y}$

"The reciprocal of a product can be split into the product of the two reciprocals."

10. Multiplication Property of Fractions

If $x \neq 0$ and $y \neq 0$, then $\dfrac{ab}{xy} = \dfrac{a}{x} \cdot \dfrac{b}{y}$

"A quotient of two products can be split into a product of two fractions."

11. Multiplication Property of Zero

For any real number x, $x \cdot 0 = 0$

"Zero times any real number is zero."

12. Converse of the Multiplication Property of Zero

If $xy = 0$, then $x = 0$ or $y = 0$

"The only way a product can equal zero is for one of the factors to equal zero."

13. Multiplication Property of –1

$$-1 \cdot x = -x$$

"–1 times a number equals the opposite of that number."

(continued)

Properties of Real Numbers Provable from the Axioms, continued

14. Product of Two Opposites

For two positive numbers, x and y,

$$(-x)(-y) = xy$$

"Negative times negative is positive."

15. Opposites of Equal Numbers

If $x = y$, then $-x = -y$

"If two real numbers are equal, then their opposites are equal. Therefore, you can take the opposite of both sides of an equation without affecting the equality."

16. Reciprocals of Equal Numbers

If $x = y \neq 0$, then $\frac{1}{x} = \frac{1}{y}$

"If two nonzero numbers are equal, then their reciprocals are equal. Therefore, you can take the reciprocal of both sides of an equation, unless it involves taking the reciprocal of zero, without affecting the equality."

17. The Square of a Real Number

$x^2 \geq 0$ for any real number x

"The square of a real number is never negative."

18. Distributive Property for Subtraction

$$x(y - z) = xy - xz$$

"Multiplication distributes over subtraction."

19. Distributive Property for Division

$$\frac{x + y}{z} = \frac{x}{z} + \frac{y}{z} \qquad z \neq 0$$

"Division distributes over addition."

(Reading this property from right to left explains why you can add fractions that have a common denominator.)

$$\frac{x - y}{z} = \frac{x}{z} - \frac{y}{z} \qquad z \neq 0$$

"Division distributes over subtraction."

(continued)

Properties of Real Numbers Provable from the Axioms, continued

20. Division of a Number by Itself

$$\frac{n}{n} = 1 \qquad n \neq 0$$

"A nonzero number divided by itself equals 1."

21. Reciprocal of 1

$$\frac{1}{1} = 1$$

"1 is its own reciprocal."

22. Dividing by 1

$$\frac{n}{1} = n$$

"Any number divided by 1 equals that number."

23. Dividing Numbers with Opposite Signs

$$\frac{-x}{y} = -\frac{x}{y}$$

"A negative number divided by a positive number is negative."

$$\frac{x}{-y} = -\frac{x}{y}$$

"A positive number divided by a negative number is negative."

24. Opposite of Sum and Difference

$$-(x + y) = -x + (-y)$$

"The opposite of a sum equals the sum of the opposites."

$$-(x - y) = y - x$$

"$x - y$ and $y - x$ are opposites of each other."

APPENDIX

B Mathematical Induction

You recall that the distributive axiom states that

$$a(x_1 + x_2) = ax_1 + ax_2$$

In words, "Multiplication distributes over a sum of *two* terms." It seems reasonable that multiplication distributes over sums of three terms, four terms, and so forth. In general,

$$a(x_1 + x_2 + x_3 + \cdots + x_n) = ax_1 + ax_2 + ax_3 + \cdots + ax_n$$

In this appendix you will learn about **mathematical induction,** a technique by which you can prove that this **extended distributive property** is true for *any* number of terms, no matter how large.

Unfortunately, the field axioms are not sufficient to prove this extended distributive property. Another axiom, the **well-ordering axiom,** allows the proof to be done.

PROPERTY: The Well-Ordering Axiom

Any nonempty set of positive integers has a *least* element.

The truth of this axiom, like that of most axioms, should be obvious to you. The name comes from the fact that a set is said to be "well-ordered" if its elements can be arranged in order, starting with a least element. The set of positive real numbers and the set of all integers do not have this property. There is a restriction to nonempty sets because the empty set has no elements at all and thus cannot have a least element.

You can prove the extended distributive property by contradiction. You assume that it is *false* and then show that this assumption leads to a contradiction. Example 1 shows how to do this.

▶ **EXAMPLE 1** **Extended Distributive Property**

Prove that $a(x_1 + x_2 + x_3 + \cdots + x_n) = ax_1 + ax_2 + ax_3 + \cdots + ax_n$ is true for any integer $n \geq 2$.

Proof (by contradiction)

Assume that the property is false. Then there is a positive integer $n = p$ for which

$$a(x_1 + x_2 + x_3 + \cdots + x_p) \neq ax_1 + ax_2 + ax_3 + \cdots + ax_p$$

By the distributive axiom, the property is true for $n = 2$. That is,

$$a(x_1 + x_2) = ax_1 + ax_2$$

Draw Venn diagrams for two sets of positive integers (Figure B-1a),

T = {positive integers n for which the property is true}

F = {positive integers n for which the property is false}

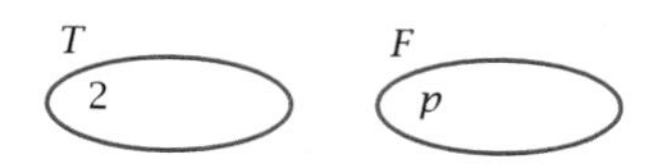

Figure B-1a

The integer 2 is an element of T, and (by assumption) p is an element of F. Write these integers in the Venn diagram of Figure B-1a.

Because F is a nonempty set of positive integers, the well-ordering axiom allows you to conclude that it has a least element, ℓ. Because ℓ is the *least* element of F, $\ell - 1$ is not an element of F. Because 2 is in T, $\ell \geq 3$, and thus both ℓ and $\ell - 1$ are positive integers. Because $\ell - 1$ is a positive integer and is not in F, it must be an element of T. Write $\ell - 1$ and ℓ in the Venn diagram (Figure B-1b).

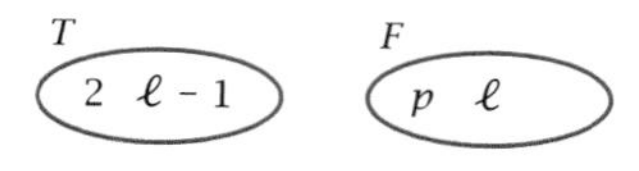

Figure B-1b

Because $\ell - 1$ is in T, the property is true for $n = \ell - 1$. Because ℓ is in F, the property is false for $n = \ell$. Write the (true) statement of the property if $n = \ell - 1$.

$$a(x_1 + x_2 + x_3 + \cdots + x_{\ell-1}) = ax_1 + ax_2 + ax_3 + \cdots + ax_{\ell-1}$$

The distributive property if $n = \ell - 1$.

Now, start with $n = \ell$.

$$a(x_1 + x_2 + x_3 + \cdots + x_{\ell-1} + x_\ell)$$

$= a((x_1 + x_2 + x_3 + \cdots + x_{\ell-1}) + x_\ell)$ — Associate the first $\ell - 1$ terms.

$= a(x_1 + x_2 + x_3 + \cdots + x_{\ell-1}) + ax_\ell$ — Multiplication distributes over a sum of two terms.

$= ax_1 + ax_2 + ax_3 + \cdots + ax_{\ell-1} + ax_\ell$ — Substitute from the statement of the distributive property if $n = \ell - 1$.

Therefore, the property is true for $n = \ell$ terms. But this statement contradicts the statement that the property is false for $n = \ell$ terms. The only place this contradiction could have arisen is the assumption that there is a positive integer $n = p$ for which the property is false. Thus, there is no positive integer n for which the property is false, and it is true for all integers $n \geq 2$. Q.E.D. ◀

Once you understand the process, the proof may be shortened. All you need to do is (1) prove that assuming the property is true for *one* value of n implies that it is true for the *next* value of n, and (2) prove that there is one value of n for which it actually *is* true. These two ideas combine to form the **induction principle.**

PROPERTY: The Induction Principle

If

(1) there is a positive integer n_0 for which a property is true, and

(2) for any integer $k \geq n_0$ assuming the property is true for $n = k$ allows you to conclude it is also true for $n = k + 1$,

then the property is true for *any* integer $n \geq n_0$.

Note that the step where you demonstrate that the property is true for one value of n is called the **anchor.** This step "anchors" the induction. The step in which you assume that the property is true for $n = k$ is called the **induction hypothesis.**

Example 2 shows how the proof of the extended distributive property can be shortened with the help of the induction principle. Proofs done this way are said to be done by **mathematical induction,** which is the topic of this appendix.

▶ EXAMPLE 2 Extended Distributive Property, Again

Prove that $a(x_1 + x_2 + x_3 + \cdots + x_n) = ax_1 + ax_2 + ax_3 + \cdots + ax_n$ is true for any integer $n \geq 2$.

Proof (by induction on n)

Anchor: $a(x_1 + x_2) = ax_1 + ax_2$ by the distributive axiom. Therefore, the property is true for $n = 2$.

Induction Hypothesis: Assume the property is true for some positive integer $n = k$, where $k \geq 2$. That is, assume that

$$a(x_1 + x_2 + x_3 + \cdots + x_k) = ax_1 + ax_2 + ax_3 + \cdots + ax_k$$

Demonstration for $n = k + 1$:

For $n = k + 1$,

$$a(x_1 + x_2 + x_3 + \cdots + x_k + x_{k+1})$$

$$= a((x_1 + x_2 + x_3 + \cdots + x_k) + x_{k+1})$$ Associate the first k terms.

$$= a(x_1 + x_2 + x_3 + \cdots + x_k) + ax_{k+1}$$ By the anchor (distribute over *two* terms).

$$= ax_1 + ax_2 + ax_3 + \cdots + ax_k + ax_{k+1}$$ Substitute, using the induction hypothesis.

Conclusion: Because (1) the property is true for one value of n, namely, $n = 2$, and because (2) assuming it is true for $n = k$ implies that it is true for $n = k + 1$, you can conclude that

$$a(x_1 + x_2 + x_3 + \cdots + x_n) = ax_1 + ax_2 + ax_3 + \cdots + ax_n$$

is true for any integer $n \geq 2$. Q.E.D. ◀

You may have detected one weakness in the proof of the extended distributive property. Without ever stating it, you have assumed an extended *associative* property and an extended *transitive* property. The proof of the extended associative property is a bit tricky; it is presented as Example 3.

▶ ***EXAMPLE 3*** **Extended Associative Property for Addition**

Prove that

$$x_1 + x_2 + x_3 + \cdots + x_{n-1} + x_n = (x_1 + x_2 + x_3 + \cdots + x_{n-1}) + x_n$$

for any integer $n \geq 3$.

Proof The associative property, $(x_1 + x_2) + x_3 = x_1 + (x_2 + x_3)$, guarantees a unique sum of three numbers. So you can write $x_1 + x_2 + x_3 = (x_1 + x_2) + x_3$. This alternative form of the associative property is the starting point you'll use.

Anchor: For $n = 3$,

$$x_1 + x_2 + x_3 = (x_1 + x_2) + x_3 \qquad \text{Agreed-upon order of operations.}$$

Induction Hypotheses: Assume that for $n = k$, where $k \geq 3$:

$$x_1 + x_2 + x_3 + \cdots + x_{k-1} + x_k = (x_1 + x_2 + x_3 + \cdots + x_{k-1}) + x_k$$

Demonstration for $n = k + 1$: If there are $k + 1$ terms, then

$$x_1 + x_2 + x_3 + \cdots + x_{k-1} + x_k + x_{k+1}$$

$$= ((\ldots((x_1 + x_2) + x_3) + \cdots + x_{k-1}) + x_k) + x_{k+1}$$

Order of operations, sum of $k + 1$ terms.

$$= ((x_1 + x_2 + x_3 + \cdots + x_{k-1}) + x_k) + x_{k+1}$$

Order of operations, sum of $k - 1$ terms.

$$= (x_1 + x_2 + x_3 + \cdots + x_{k-1} + x_k) + x_{k+1}$$

Induction hypothesis.

Therefore, the property is true for a sum of $k + 1$ terms.

Conclusion:

$$\therefore x_1 + x_2 + x_3 + \cdots + x_{n-1} + x_n = (x_1 + x_2 + x_3 + \cdots + x_{n-1}) + x_n$$

for any integer $n \geq 3$. Q.E.D. ◀

Induction is useful for proving that the formulas for series work for *any* finite number of terms. Example 4 shows you how to prove that the formula for the partial sum S_n of a geometric series is true for any value of n, no matter how large.

▶ **EXAMPLE 4** **Partial Sum of a Geometric Series Property**

Prove that $S_n = a + ar + ar^2 + ar^3 + \cdots + ar^{n-1} = \dfrac{a(1 - r^n)}{1 - r}$ for all integers $n \geq 1$.

(Note that a is used for t_1 for simplicity of writing and that it is written in the numerator of the partial sum formula instead of out in front of the fraction.)

Proof (by induction on n)

Anchor: If $n = 1$, then $S_1 = a$. The formula gives $\dfrac{a(1 - r^1)}{1 - r} = a$, which anchors the induction.

Induction Hypothesis: Assume that the formula works for $n = k$. That is, assume that

$$S_k = a + ar + ar^2 + ar^3 + \cdots + ar^{k-1} = \frac{a(1 - r^k)}{1 - r}$$

Demonstration for n = k + 1:

S_{k+1}

$= a + ar + ar^2 + ar^3 + \cdots + ar^{k-1} + ar^k$ — Definition of geometric series.

$= (a + ar + ar^2 + ar^3 + \cdots + ar^{k-1}) + ar^k$ — Extended associative property.

$= \dfrac{a(1 - r^k)}{1 - r} + ar^k$ — Induction hypothesis.

$= \dfrac{a(1 - r^k) + ar^k(1 - r)}{1 - r}$ — Find a common denominator and add fractions.

$= \dfrac{a - ar^k + ar^k - ar^{k+1}}{1 - r}$ — Distribute.

$= \dfrac{a(1 - r^{k+1})}{1 - r}$ — Combine like terms, then factor out a.

which is the formula with $k + 1$ substituted for n.

Conclusion:

$$\therefore S_n = a + ar + ar^2 + ar^3 + \cdots + ar^{n-1} = \frac{a(1 - r^n)}{1 - r}$$

for all integers $n \geq 1$. Q.E.D. ◀

You should be careful not to read too much into the conclusion of an induction proof. The proof is good only for any *finite* number of terms. If there is an infinite number of terms, the sum may or may not converge to a real number, depending on the value of r.

▶ **EXAMPLE 5** **Partial Sum for the Series of Squares**

Prove that $S_n = 1 + 4 + 9 + 16 + \cdots + n^2 = \dfrac{n(n + 1)(2n + 1)}{6}$ for all integers $n \geq 1$.

Proof *Anchor:* If $n = 1$, then $S_1 = 1$. The formula gives $\frac{1(2)(3)}{6} = 1$ as well, which anchors the induction.

Induction Hypothesis: Assume that the formula is correct for $n = k$. That is, assume that

$$S_k = 1 + 4 + 9 + 16 + \cdots + k^2 = \frac{k(k+1)(2k+1)}{6}$$

Demonstration for $n = k + 1$:

S_{k+1}

$= 1 + 4 + 9 + \cdots + k^2 + (k+1)^2$	Definition of the sum of square series.
$= (1 + 4 + 9 + \cdots + k^2) + (k+1)^2$	Extended associative property.
$= \dfrac{k(k+1)(2k+1)}{6} + (k+1)^2$	Induction hypothesis.
$= \dfrac{k(k+1)(2k+1) + 6(k+1)^2}{6}$	Write the quotient as one fraction with common denominator.
$= \dfrac{(k+1)(k(2k+1) + 6(k+1))}{6}$	Factor the $(k + 1)$ term in the numerator.
$= \dfrac{(k+1)(2k^2 + k + 6k + 6)}{6}$	Multiply the factors in the second parentheses.
$= \dfrac{(k+1)(2k^2 + 7k + 6)}{6}$	Combine like terms in the second parentheses.
$= \dfrac{(k+1)(k+2)(2k+3)}{6}$	Factor the second parentheses in the numerator.
$= \dfrac{(k+1)((k+1)+1)(2(k+1)+1)}{6}$	Set apart the $(k + 1)$ terms inside the parentheses.

which is the formula with $(k + 1)$ substituted for n.

Conclusion:

$$S_n = 1 + 4 + 9 + 16 + \cdots + n^2 = \frac{n(n+1)(2n+1)}{6}$$

is true for all integers $n \geq 1$. Q.E.D. ◀

You can use mathematical induction to prove other extended field, equality, and order axioms, properties of exponentiation and logarithms, formulas for sequences and series, and some interesting properties of numbers.

Answers to Selected Problems

CHAPTER 1

Problem Set 1-1

1. a. 20 m; −17.5 m; it is below the top of the cliff.
   b. ≈0.3 sec, 3.8 sec; ≈5.4 sec
   c. 5 m
   d. There is only one altitude for any given time; some altitudes correspond to more than one time.
   e. Domain: $0 \le x \le 5.4$; range: $-30 \le y \le 25.1$

3. a.

x	y
0	50000
12	49362
24	48642
36	47832
48	46918
60	45889
72	44729
84	43422
96	41949
108	40290
120	38420
132	36313
144	33938
156	31263

   b. Changing ΔTbl to 1, you see that the balance becomes negative at the end of month 241, so the balance will become 0 during month 241.

x	y
235	2862.2
236	2340.5
237	1813.5
238	1281.3
239	743.79
240	200.88
241	−347.4

   c.

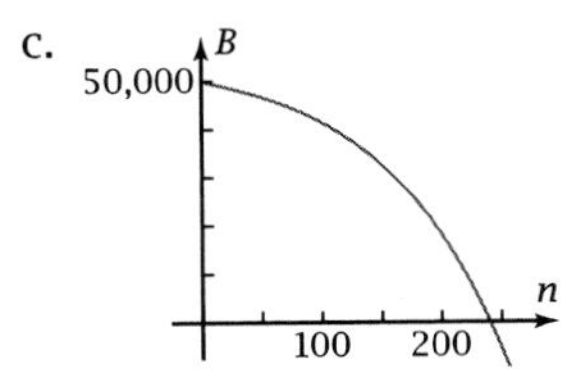

   d. False

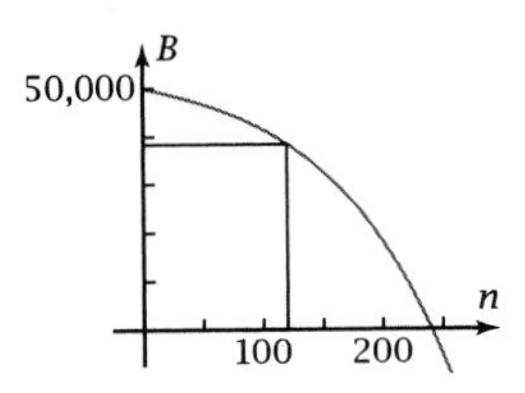

   e. Domain: $0 \le x \le 241$, x an integer; range: $0 \le y \le 50{,}000$. The values are only calculated at whole-month intervals.

5. This graph assumes that the element heats from a room temperature of 72°F to nearly a maximum temperature of 350°F in one minute.

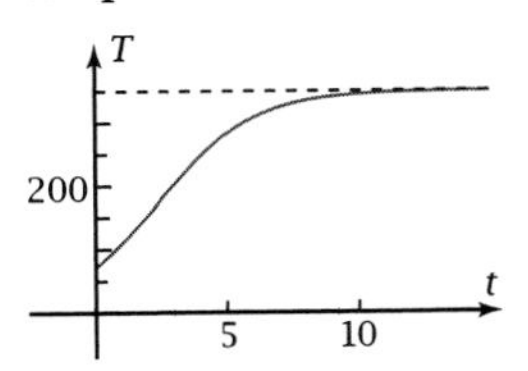

Domain: $x \ge 0$ sec; range: $72°\text{F} \le y < 350°\text{F}$

Problem Set 1-2

1. a.

b. $3 \le f(x) \le 23$

c. Linear

d. Answers may vary; e.g., the cost (in thousands of dollars) of manufacturing x items if each item costs $2000 to manufacture and there is a $3000 start-up cost

3. a.

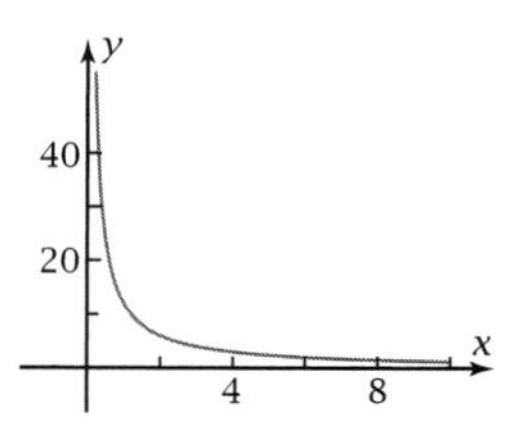

b. $f(x) \ge 1.2$

c. Inverse variation

d. Answers may vary; e.g., the time it takes to go 12 mi at x mi/hr

5. a.

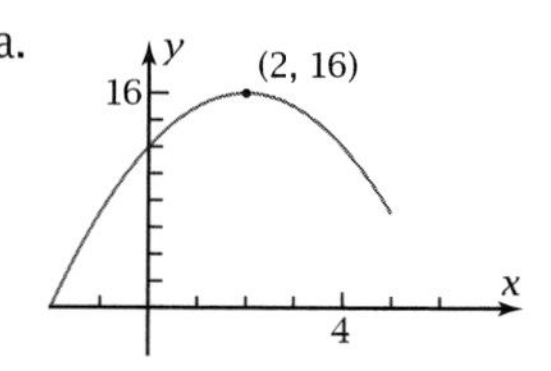

b. y-intercept at $y = 12$; no x-intercepts; no asymptotes

c. $7 \le y \le 16$

7. a.

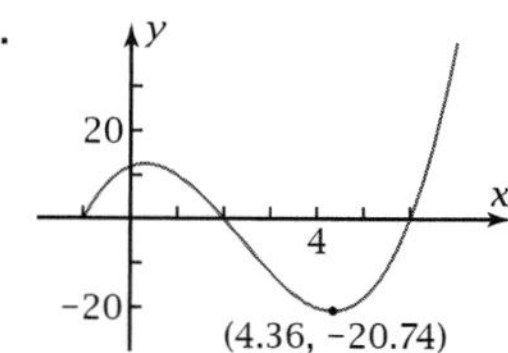

b. y-intercept at $y = 12$; x-intercepts at $x = -1$, $x = 2$, and $x = 6$; no asymptotes

c. $-20.7453\ldots \le y \le 40$

9. a.

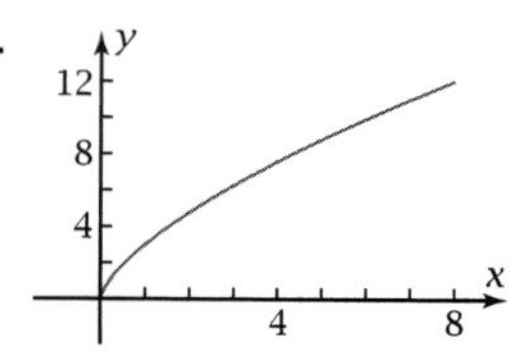

b. y-intercept at $y = 0$; x-intercept at $x = 0$; no asymptotes

c. $0 \le y \le 12$

11. a.

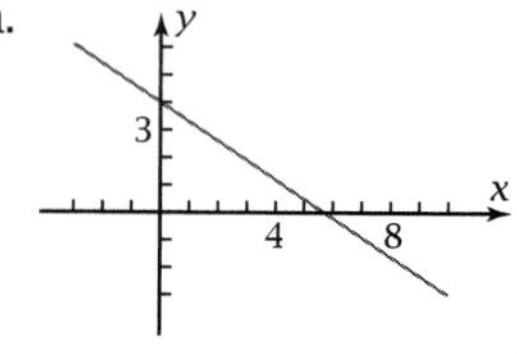

b. y-intercept at $y = 4$; x-intercept at $x = 5\frac{5}{7}$; no asymptotes

c. $-3 \le y \le 6.1$

13. a.

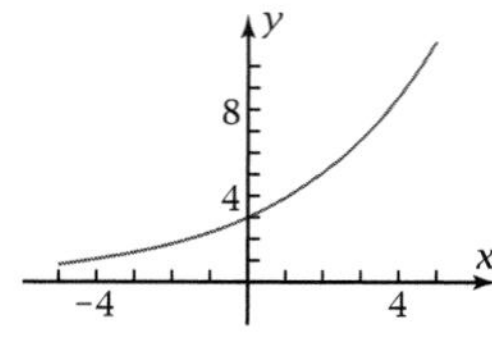

b. y-intercept at $y = 3$; no x-intercepts; asymptote at $y = 0$ (the x-axis)

c. $0.8079\ldots \le y \le 11.1387\ldots$

15. a.

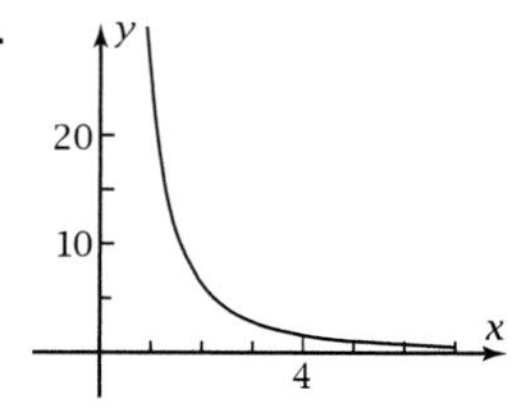

b. No y-intercept; no x-intercept; asymptotes at $y = 0$ (the x-axis) and $x = 0$ (the y-axis)

c. $y > 0$

17. a.

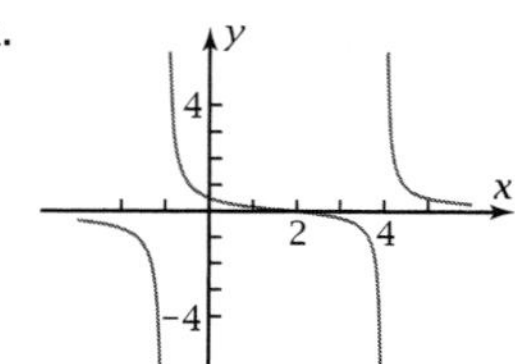

b. y-intercept at $y = \frac{1}{2}$; x-intercept at $x = 2$; asymptotes at $x = -1$ and $x = 4$

c. All real numbers

19. Exponential

21. Linear

23. Quadratic (polynomial)

25. Power

27. Rational

29. a.

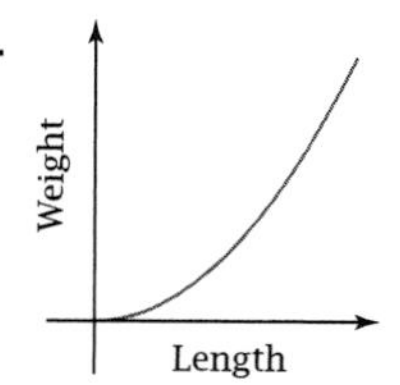

b. Power (cubic)

31. a.

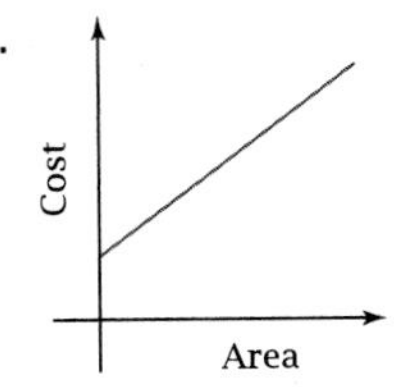

b. Linear

33. Function; no x-value has more than one corresponding y-value.

35. Not a function; there is at least one x-value with more than one corresponding y-value.

37. Not a function; there is at least one x-value with more than one corresponding y-value.

39. a. A vertical line through a given x-value crosses the graph at the y-values that correspond to that x-value. So if a vertical line crosses the graph more than once, it means that that x-value has more than one corresponding y-value.

b. (Sketch not shown.) In Problem 33, any vertical line crosses the graph at most once, but in Problem 35, any vertical line between the two endpoints crosses the graph twice.

41. $x - 2$, that is, the number (or the variable representing it) that is being substituted into f

Problem Set 1-3

1. a. $g(x) = 2\sqrt{9 - x^2}$

b.

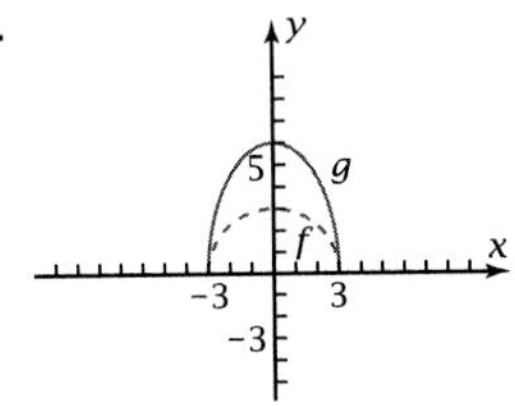

c. y-dilation of 2

3. a. $g(x) = \sqrt{9 - (x - 4)^2}$

b.

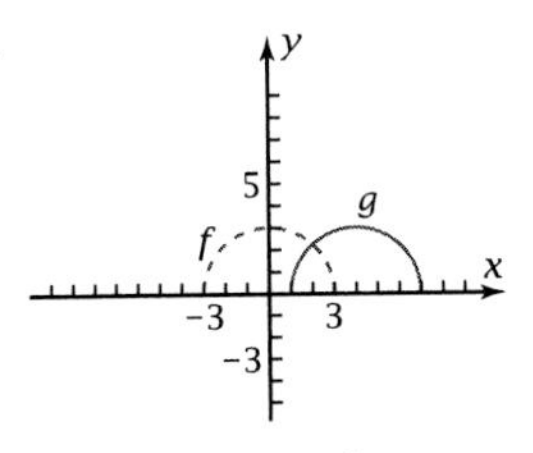

c. x-translation of 4

5. a. $g(x) = 1 + \sqrt{9 - \left(\frac{x}{2}\right)^2}$

b.

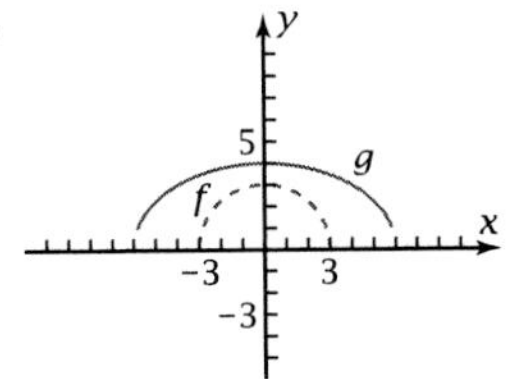

c. x-dilation of 2, y-translation of 1

7. a. y-translation of 7

b. $g(x) = 7 + f(x)$

9. a. x-dilation of 3

b. $g(x) = f\left(\frac{x}{3}\right)$

11. a. x-translation of 6, y-dilation of 3

b. $g(x) = 3 \cdot f(x - 6)$

13. No. The domain of $f(x)$ is $x \leq 1$, but the domain of the graph is $-3 \leq x \leq 1$. That restriction must be added to the definition of $f(x)$.

15. a.

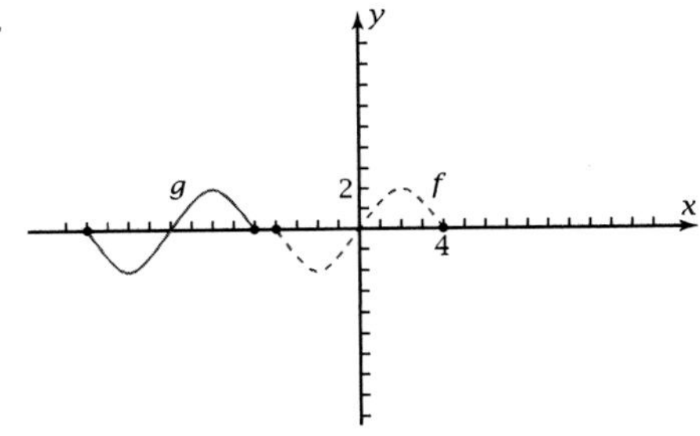

b. x-translation of −9

17. a.

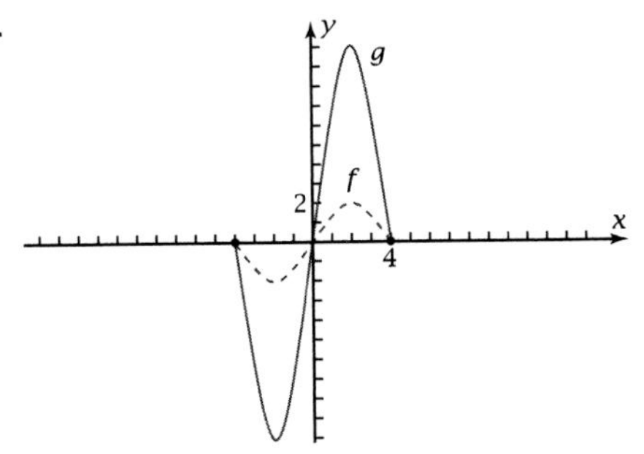

b. y-dilation of 5

19. a.

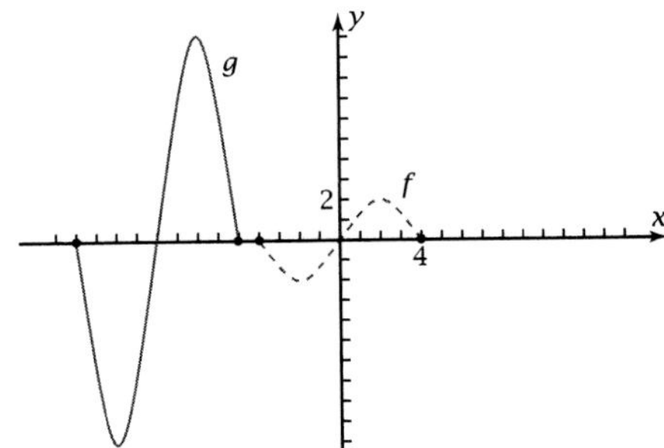

b. y-dilation of 5, x-translation of −9

Problem Set 1-4

1. a. $R(t) = 5 + 7t$

   b. $R(4) = 33$ cm; $R(10) = 75$ cm;
   $A(4) = 3421.1944\ldots\text{ cm}^2$;
   $A(10) = 17{,}671.45686\ldots\text{ cm}^2$

   c. $A(t) = A(R(t))$ or $\pi(49t^2 + 70t + 25)$

   d. $t = (17.3080\ldots)$ sec. The negative answer doesn't fit the context.

3. a.

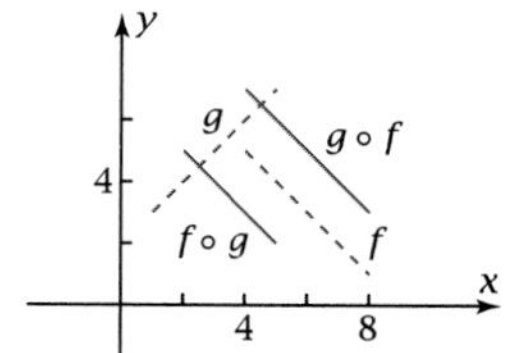

   b. $f(g(3)) = 4$

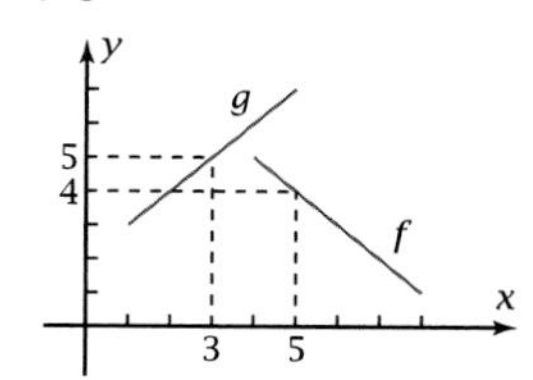

   c. 6 is not in the domain of g, so $g(x)$ is undefined. $g(1) = 3$, but 3 is not in the domain of f.

   d. $f(g(x)) = 7 - x$

   e. Domain of $f \circ g$: $2 \le x \le 5$
   Domain of $g \circ f$: $4 \le x \le 8$
   Both match the graph.

   f. $2 \le f(g(x)) \le 5$. This agrees with the graph.

   g. $f(f(5)) = 5$; $g(5) = 7$, and 7 is not in the domain of g.

5. a. Yes

   b. For $3\sqrt{x-4}$ to be a real number, we must have $x - 4 \ge 0 \Rightarrow x \ge 4$.

   c.

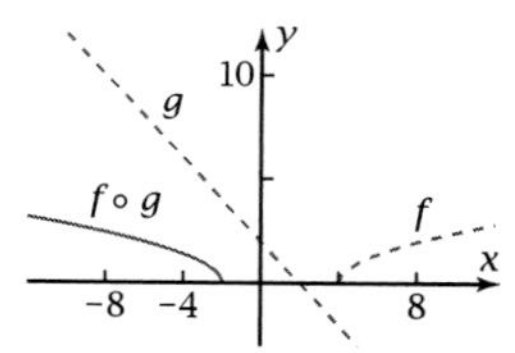

   d. $f(g(x)) = 3\sqrt{-x-2}$; for $3\sqrt{-x-2}$ to be a real number, we must have $-x - 2 \ge 0 \Rightarrow x \le -2$.

   e. $g(f(x)) = 2 - 3\sqrt{x-4}$. Domain is $x \ge 4$.

7. a. $f(g(3)) = 3$; $f(g(7)) = 7$; $g(f(5)) = 5$; $g(f(8)) = 8$.
   Conjecture: For all values of x, $f(g(x)) = g(f(x)) = x$.

   b. $f(g(-9))$ is undefined; $g(f(-9)) = 9 \ne -9$. No.

   c.

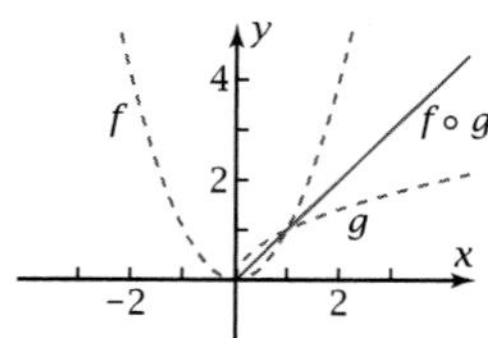

   $f(g(x)) = x$, but g is defined only for nonnegative x, so $f \circ g$ is defined only for nonnegative x.

   d.

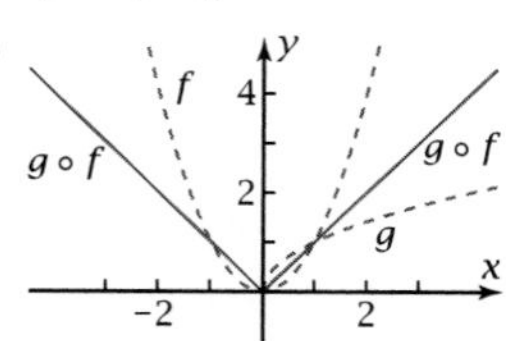

   e. $g(f(x)) = g(x^2) = \sqrt{x^2} = \begin{cases} x & \text{if } x \ge 0 \\ -x & \text{if } x < 0 \end{cases} = |x|$

9. If the dotted graph is $f(x)$, $1 \le x \le 5$, then the solid graph is $f(-x)$, $-5 \le x \le -1$. In terms of composition of functions, the solid graph is $f(g(x))$, where $g(x) = -x$.

11. a. Translation 3 units to the right

    b. Horizontal dilation by a factor of 2

    c.

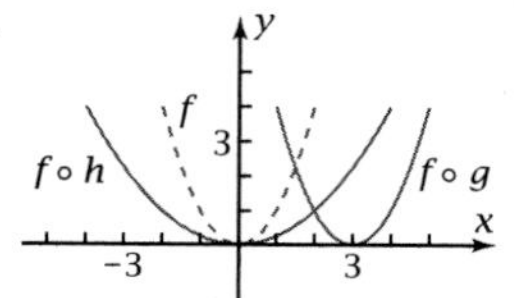

    Yes

Problem Set 1-5

1. Function

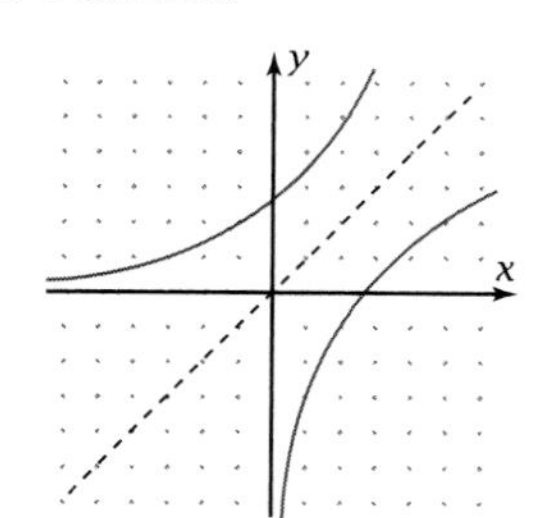

3. Not a function

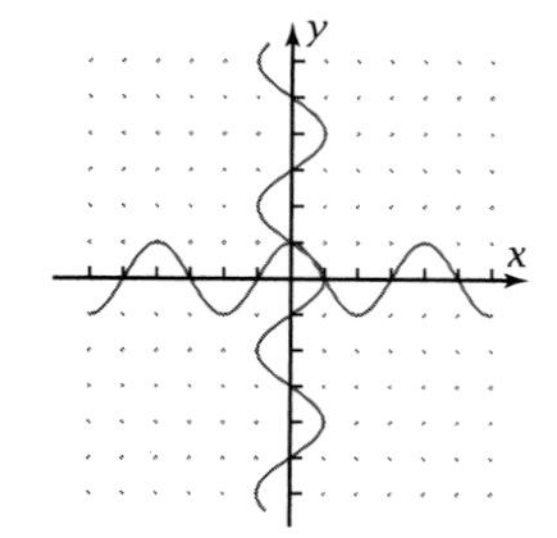

5. Function

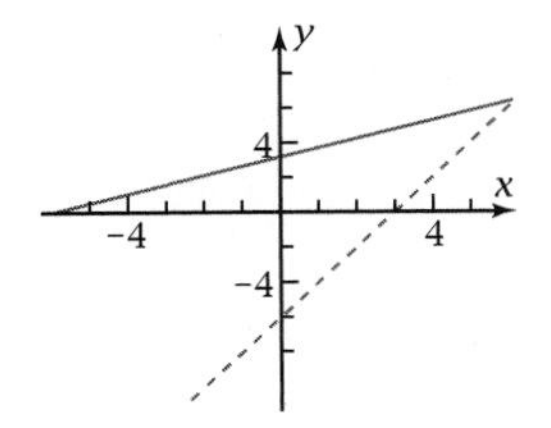

7. Not a function

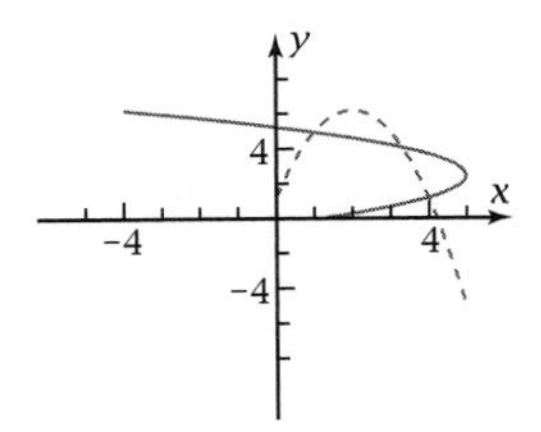

9. Function

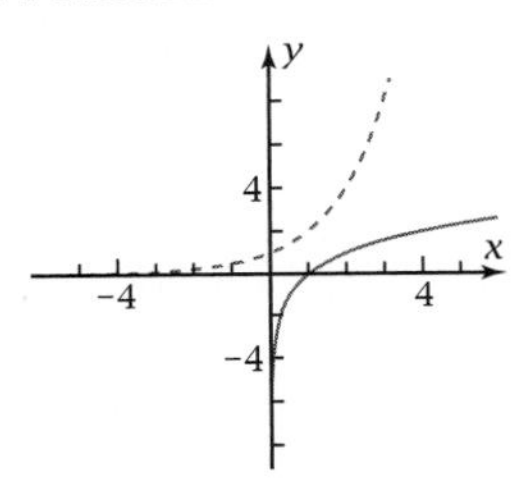

11. Function

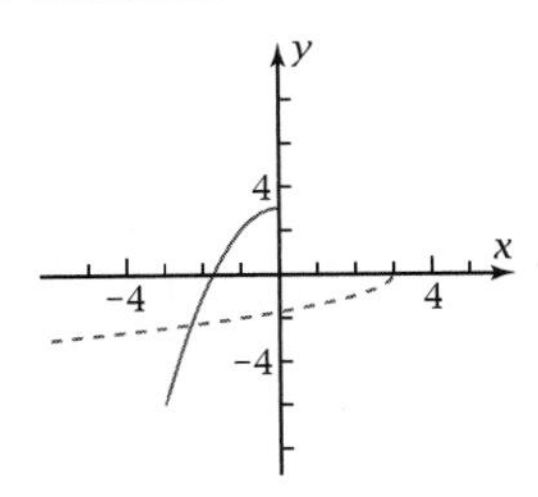

13. Function

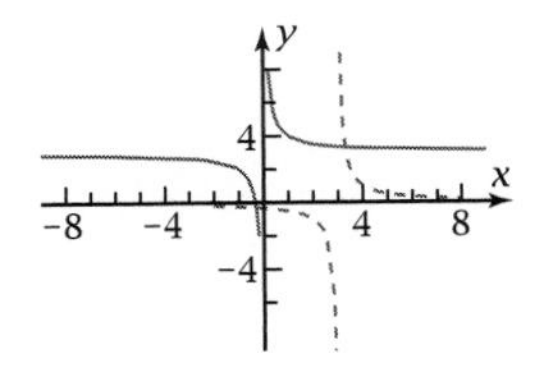

15. Function

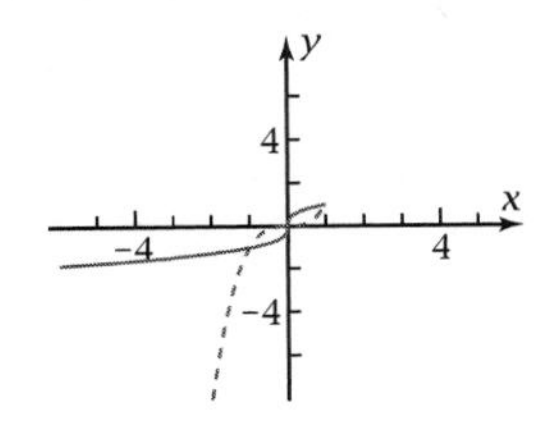

17. $y = f^{-1}(x) = \frac{1}{2}x + 3$

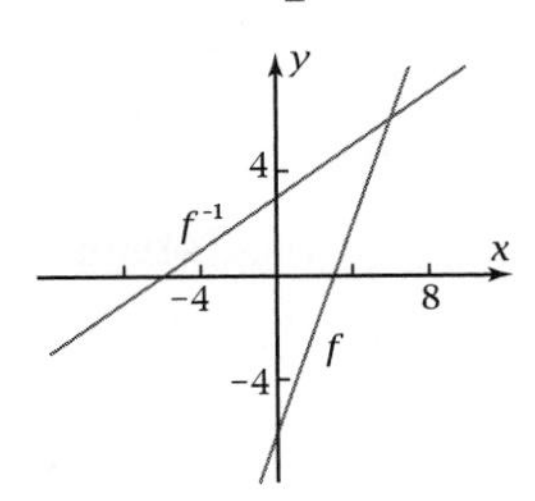

The inverse relation is a function.

19. $y = \pm\sqrt{-2x - 4}$

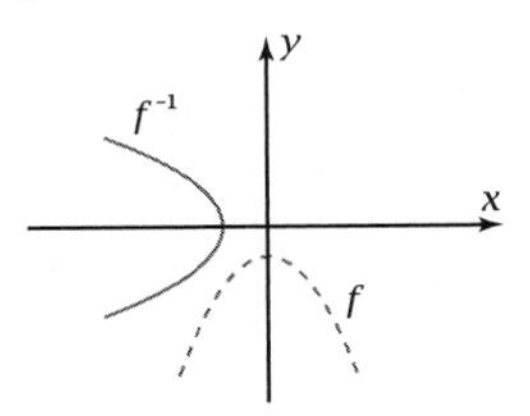

The inverse relation is not a function.

21. $f(f(x)) = \dfrac{1}{f(x)} = \dfrac{1}{\left(\frac{1}{x}\right)} = x,\ x \neq 0$

23. a. $c(1000) = 550$. If you drive 1000 mi in a month, your monthly cost is \$550.

b. $c^{-1}(x) = 4x - 1200$. $c^{-1}(x)$ is a function because no input produces more than one output. $c^{-1}(437) = 4(437) - 1200 = 548$. You would have a monthly cost of \$437 if you drove 548 mi in a month.

c.

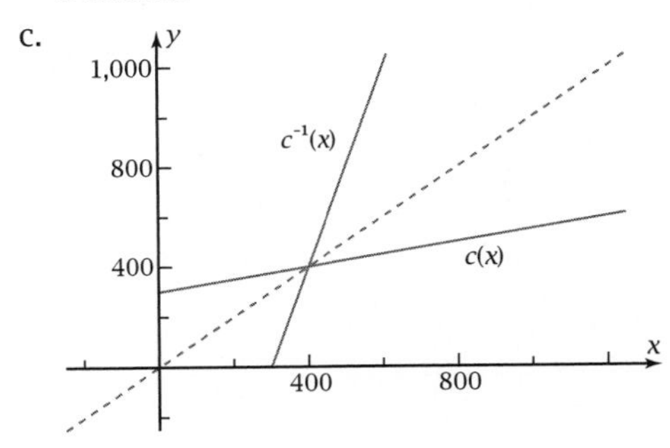

25. a. $y = d^{-1}(x) = \sqrt{\dfrac{x}{0.057}}$; because the domain of d is $x \geq 0$, the range of d^{-1} is $d^{-1}(x) \geq 0$.

b. $d^{-1}(200) = 59.234\ldots$; a 200-ft skid mark is caused by a car moving at a speed of about 59 mi/hr.

c.

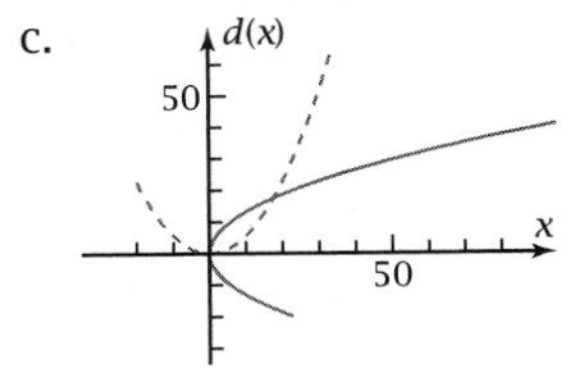

d. Because the domain of d now contains negative numbers, the range of d^{-1} contains negative numbers. Because the range of d^{-1} contains negative numbers, $d^{-1} = \pm\sqrt{\dfrac{x}{0.057}}$, which is not a function.

27. No. The value $y = 2000$ comes from both $x = 60$ and $x = 80$.

Problem Set 1-6

1. a.

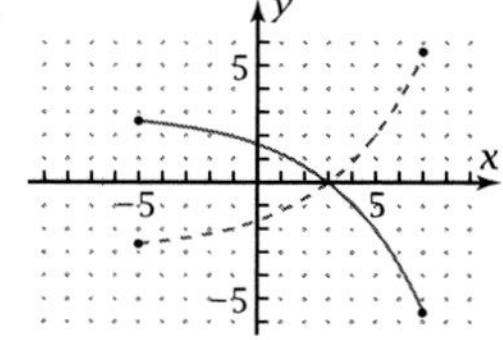

b.

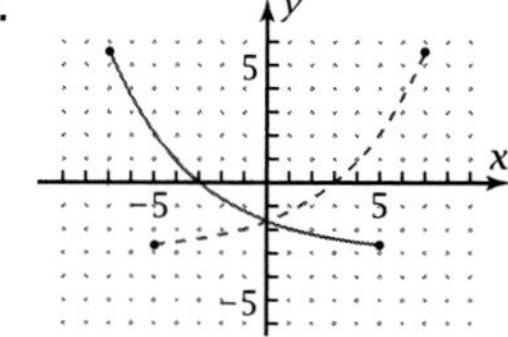

c.

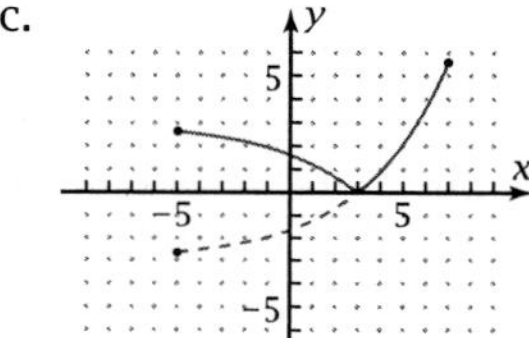

d.

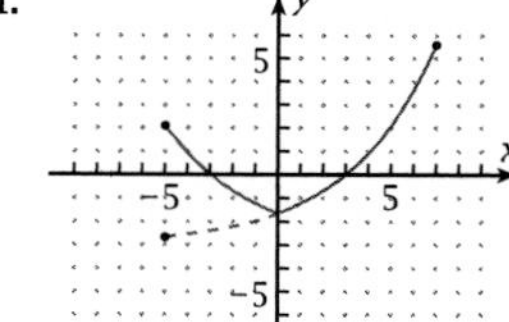

3. a.

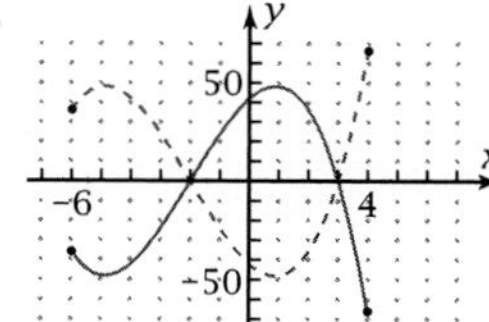

b.

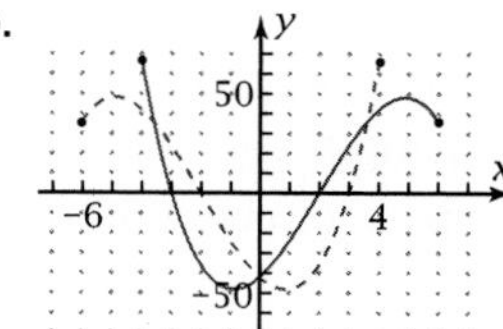

c.

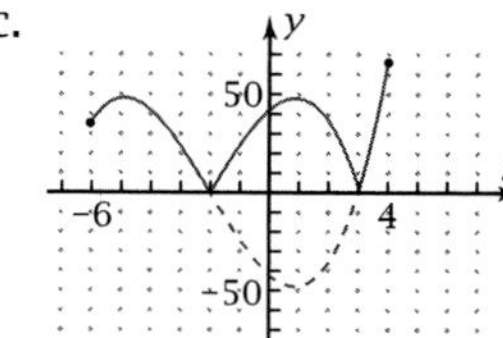

d.

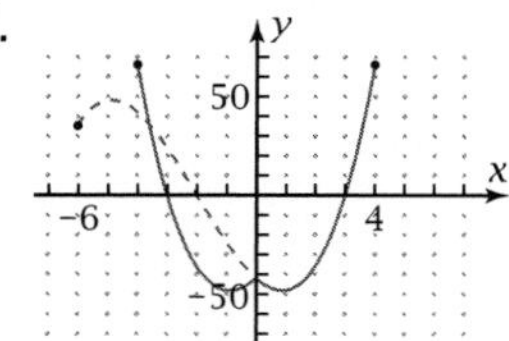

5. The graphs match.

7. a.

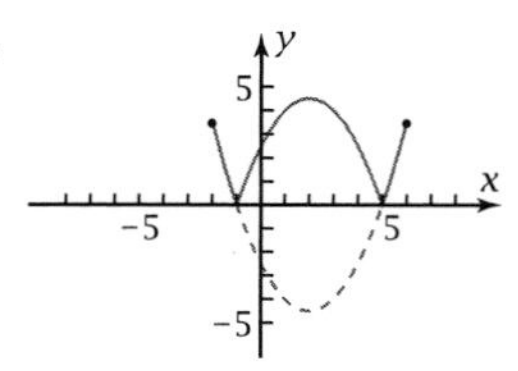

This transfomation reflects all the points on the graph below the x-axis across the x-axis.

b.

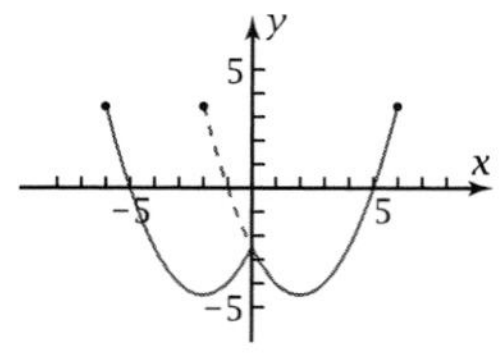

This transformation reflects $f(x)$ for positive values of x across the y-axis.

c. $|f(3)| = 4$; $f(|-3|) = -4$; The domain of $f(|x|)$ includes -3 because the transformation of f reflects all the points containing positive values of x across the y-axis.

d.

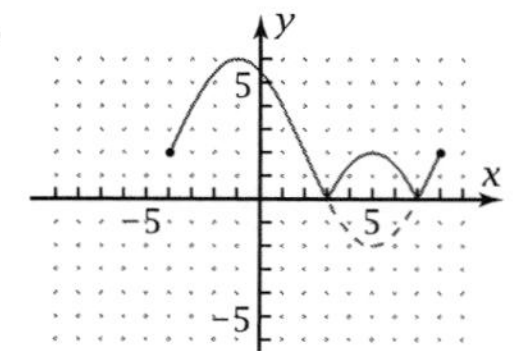

9. a.

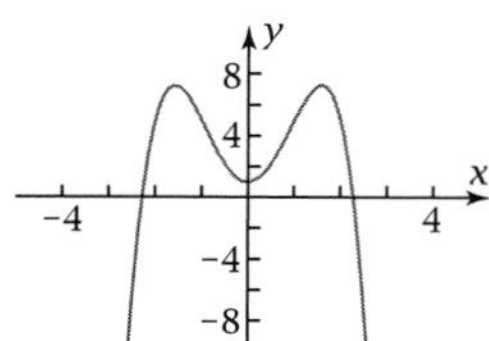

A negative number raised to an even power is equal to the absolute value of that number raised to the same power. So for $\pm x$, the same corresponding y-value occurs, and therefore $f(x) = f(-x)$.

b.

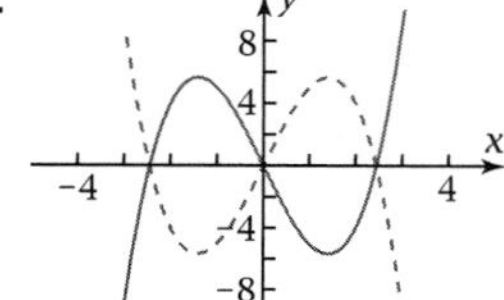

A negative number raised to an odd power is equal to the opposite of the absolute value of that number raised to the same power. Because each term in $g(x)$ has x raised to an odd power, $g(-x)$ transforms g in the same way as $-g(x)$ does.

c. Graph h is odd; graph j is even.

d.

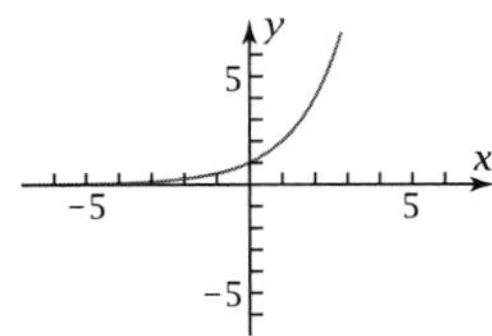

The function $e(x)$ is neither odd nor even. $e(-x) \neq -e(x)$

11. a. The graphs agree.

b. $g(x) = 2\dfrac{|x-1|}{x-1} + 3$

c. $f(x) = (x-3)^2 - 2 \cdot \dfrac{|x-5|}{x-5}$

The graphs agree.

13. a. $a = 0.0375$; $b = 2{,}400{,}000{,}000$

b. $y_1 = 0.0375x/(0 \le x \text{ and } x \le 4000)$;
$y_2 = 2{,}400{,}000{,}000/x^2/(x \ge 4000)$

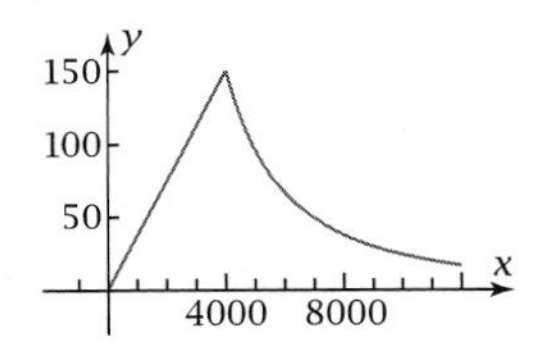

c. $y(3000) = 112.5$ lb; $y(5000) = 96$ lb

d. $x = 1333.\overline{3}$ mi and $x = 6928.2032\ldots$ mi

Problem Set 1-8

R1. a. ≈ 17 psi

b.

x	y
0	35
1	24.5
2	17.15
3	12.005
4	8.4035
5	5.8825

c. Domain: $0 \le x \le \approx 5.5$; range: $5 \le y \le 35$

d. Asymptote

e. Answers will vary.

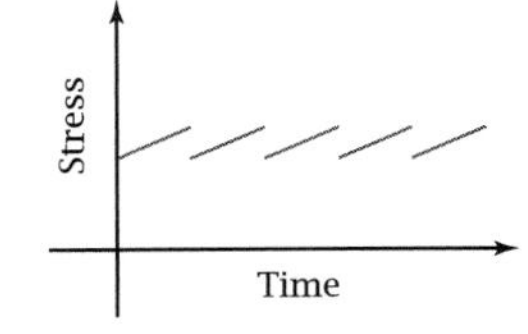

R2. a. Linear

b. Polynomial (cubic)

c. Exponential

d. Power

e. Rational

f. Answers will vary; for example, number of items manufactured and total manufacturing cost

g. $13 \le f(x) \le 37$

h. Answers will vary.

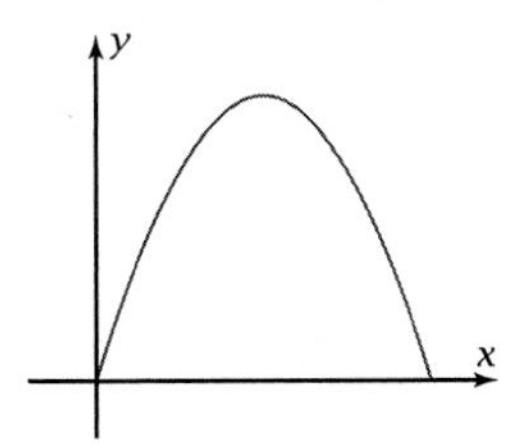

Exponential

i. 1-8b: exponential; 1-8c: polynomial (probably quadratic); 1-8d: power (possibly quadratic)

j. Figure 1-8e passes the vertical line test: No vertical line intersects the graph more than once, so no y-value corresponds to more than one x-value. Figure 1-8f fails the vertical line test: There is at least one vertical line that intersects the graph more than once, so more than one y-value corresponds to the same x-value.

R3. a. Horizontal dilation by a factor of 3, vertical translation by -5; $g(x) = \sqrt{4 - \left(\dfrac{x}{3}\right)^2} - 5$

b. Horizontal translation by $+4$, vertical dilation by a factor of 3.

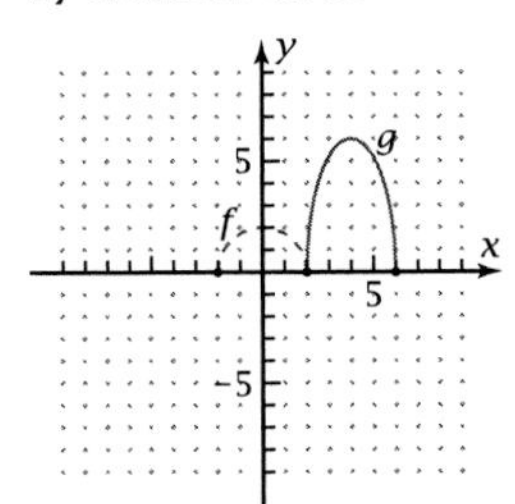

R4. a. $h(t) = 3t + 20$

b. $h(5) = 3(5) + 20 = 35$ in.;
$W(h(5)) = 0.004(35)^{2.5} \approx 29$ lb

c.

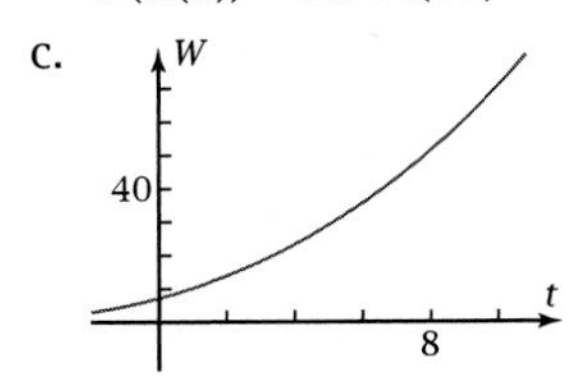

d. No; the graph is curved.

e. $0 \le x \le 6$

f.

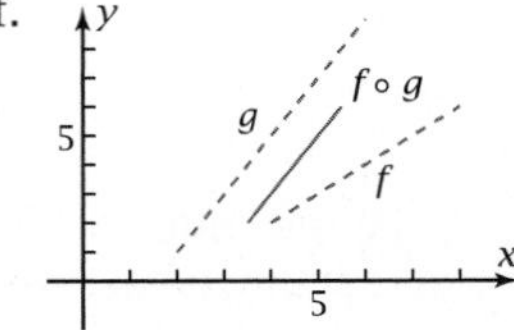

g. $f(g(4)) = 3$

h. $f(g(3)) = f(2(3) - 3) = f(3)$, which is undefined, since 3 is not in the domain of f.

i. $\frac{7}{2} \le x \le \frac{11}{2}$, which agrees with the graph

R5. a. The inverse does not pass the vertical line test.

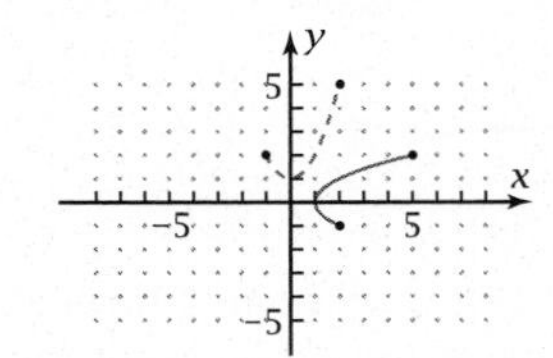

b.

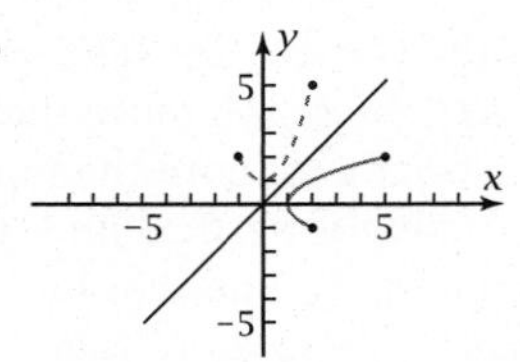

The domain of f corresponds to the range of f^{-1}. The range of f corresponds to the domain of f^{-1}.

c. $x = y^2 + 1 \Rightarrow y = \pm\sqrt{x - 1}$. The $\pm$ reveals that there are two different y-values for some x-values.

d.

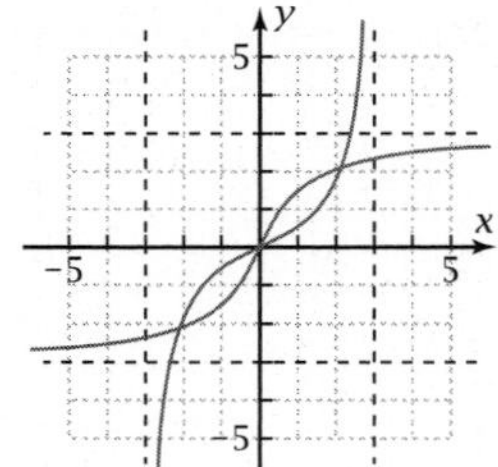

It passes the vertical line test; asymptotes.

e. $V(5) = 523.598\ldots$ in.^{3}; $x = 2.879\ldots$ in.;

$V^{-1}(x) = \sqrt[3]{\frac{3x}{4\pi}}$; $V(x)$ has the third power of x and

$V^{-1}(x) = \left(\frac{3x}{4\pi}\right)^{1/3}$ has the one-third power of x.

V^{-1} would be more useful when you are trying to find the radius for different volumes of spheres.

f. Because no y corresponds to more than one x in the original function, no x corresponds to more than one y in the inverse relation, so the inverse is a function.

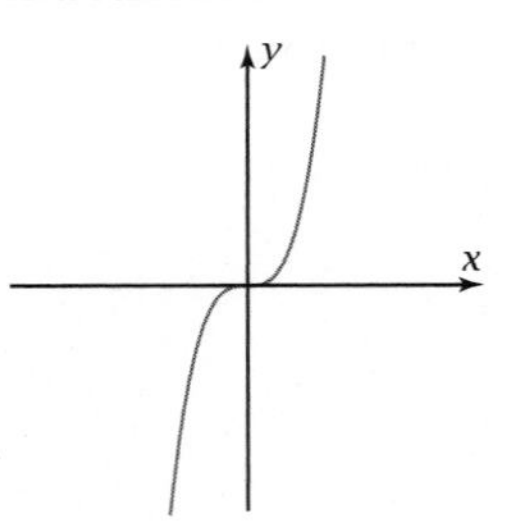

R6. a.

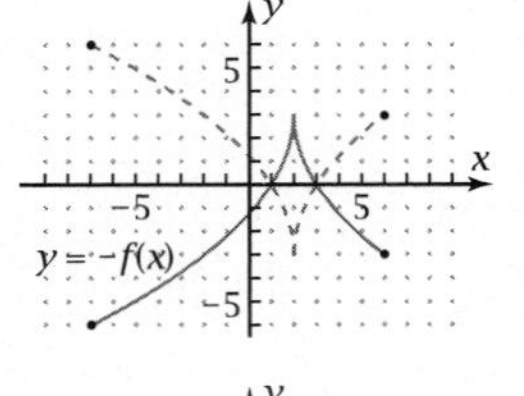

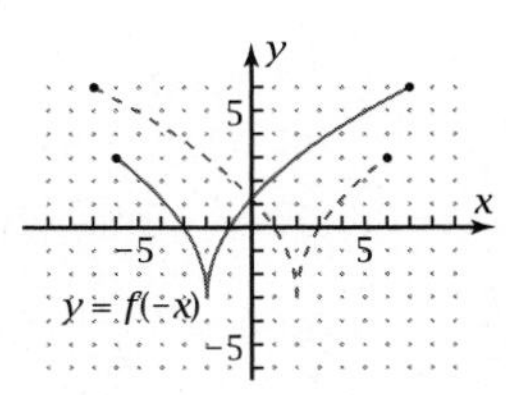

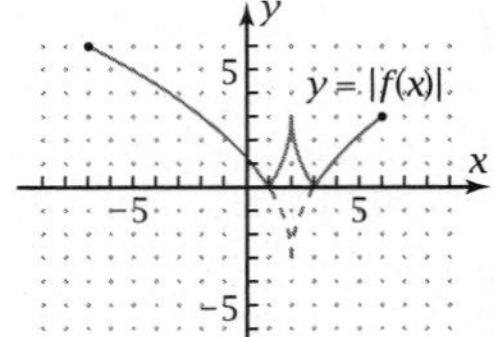

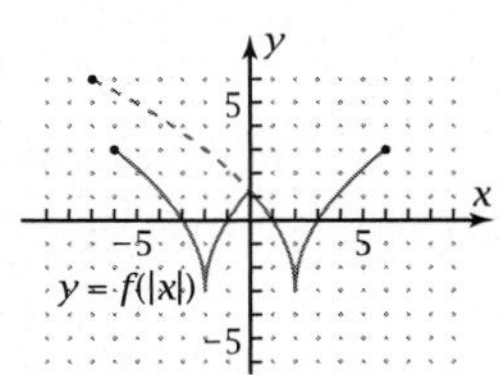

b. The graph agrees with Figure 1-8k; each of the graphs agrees with those in part a.

c. Because power functions with odd powers satisfy the property $f(-x) = -f(x)$ and power functions with even powers satisfy the property $f(-x) = f(x)$

d.

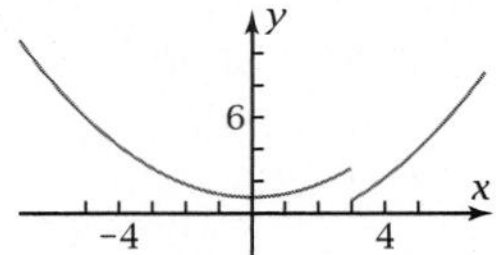

Discontinuity

R7. Answers will vary.

CHAPTER 2

Problem Set 2-1

1.

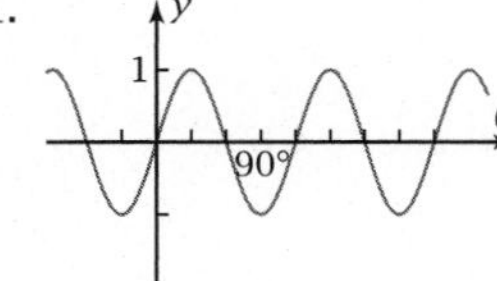

3. $y_2 = 11 + 9 \sin x$, shown with the original graph

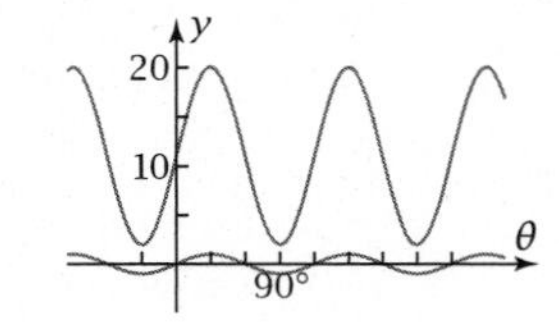

Problem Set 2-2

1. $\theta_{ref} = 50°$

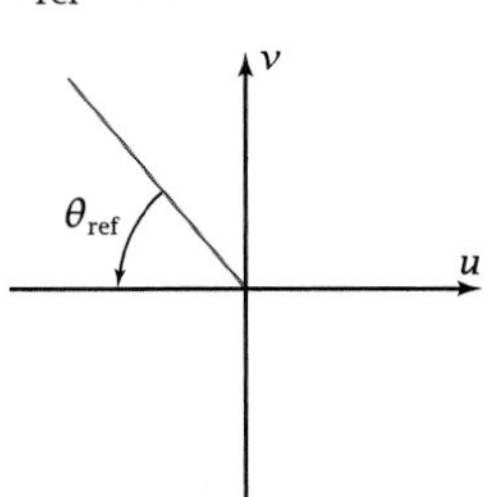

3. $\theta_{ref} = 79°$

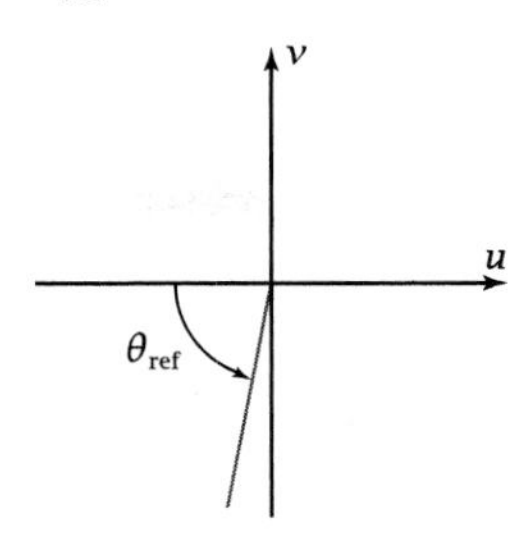

5. $\theta_{ref} = 18°$

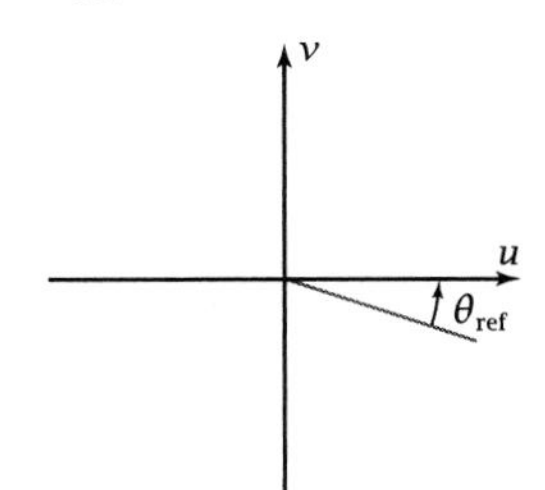

7. $\theta_{ref} = 54°$

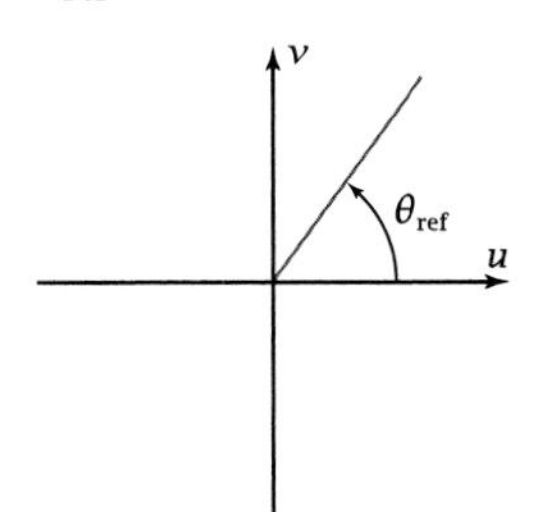

9. $\theta_{ref} = 20°$

11. $\theta_{ref} = 65°$

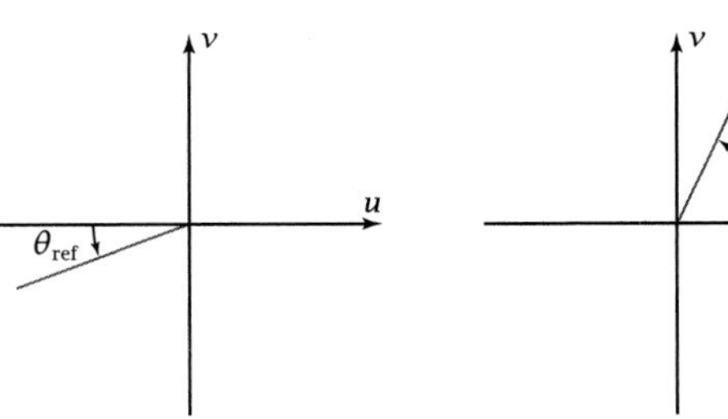

13. $\theta_{ref} = 81.4°$

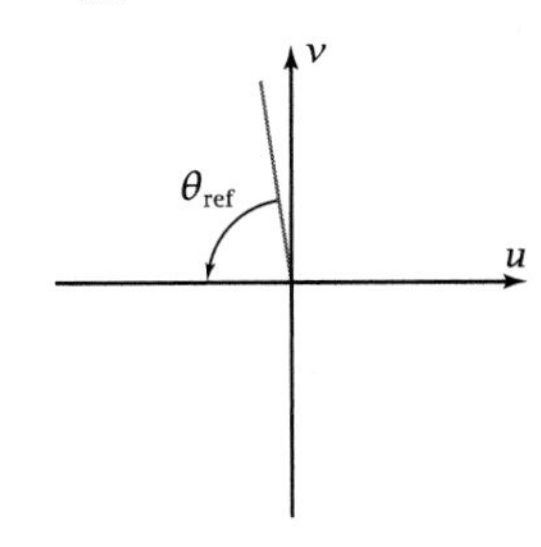

15. $\theta_{ref} = 25.9°$

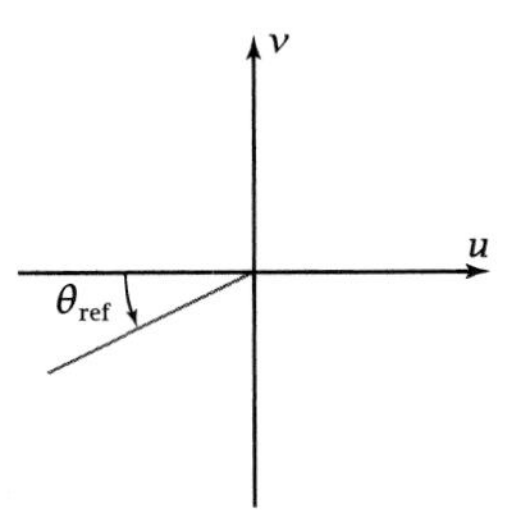

17. $\theta_{ref} = 81°$

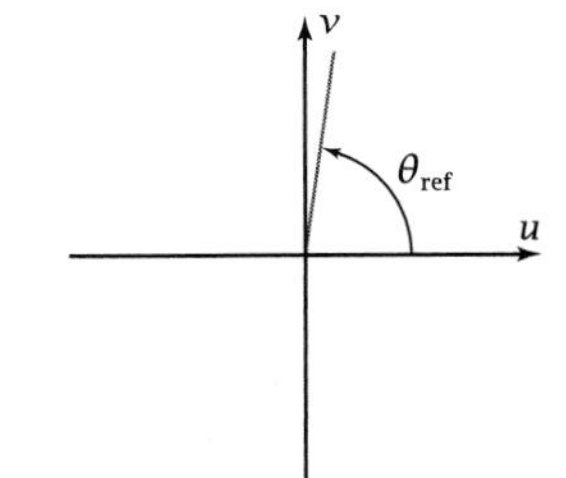

19. $\theta_{ref} = 46°$

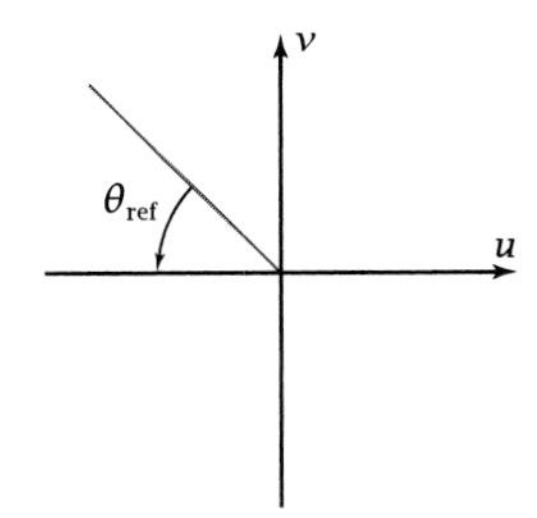

21. $\theta_{ref} = 34°23'$

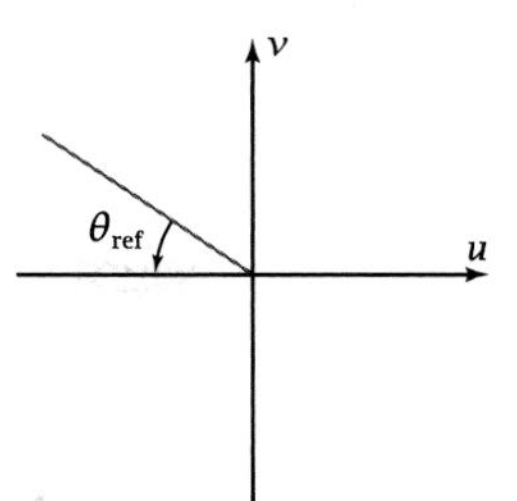

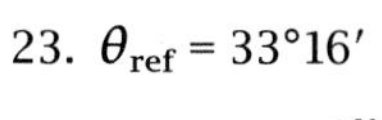

23. $\theta_{ref} = 33°16'$

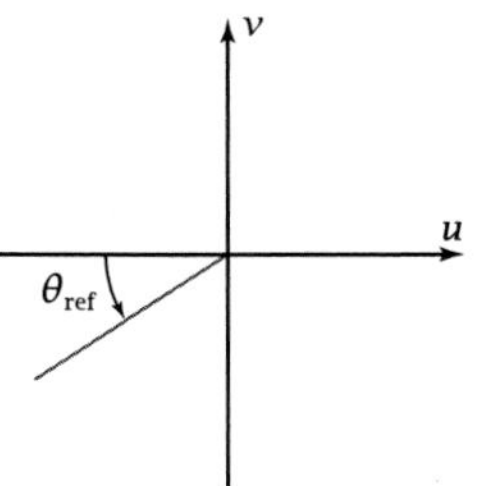

25. $\theta_{ref} = 51°45'9''$

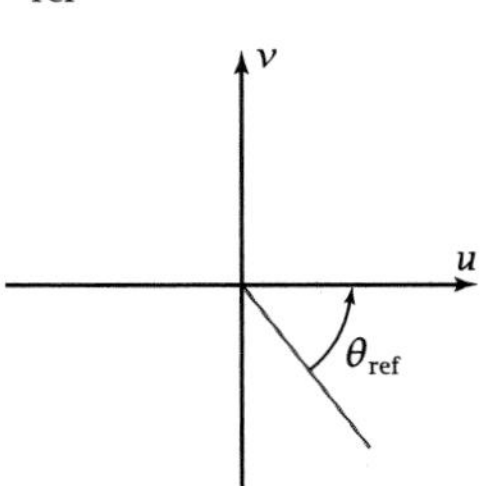

27.

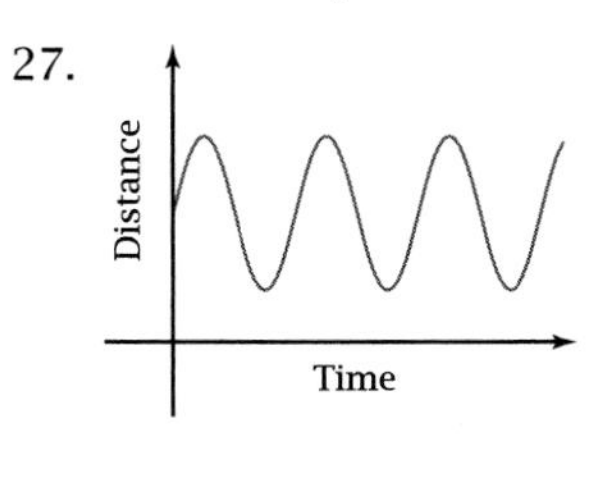

29. $g(x) = 4 + f(x - 1)$

Problem Set 2-3

1. $\theta_{ref} = 70°$

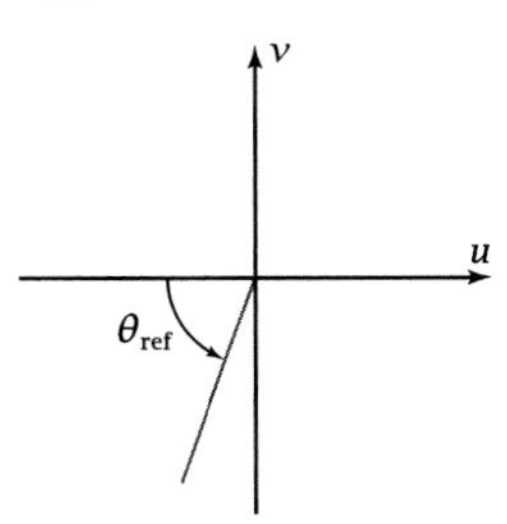

$\sin 250° = -0.9396\ldots$, $\sin 70° = 0.9396\ldots$,
$\sin 250° = -\sin 70°$

3. $\theta_{ref} = 40°$

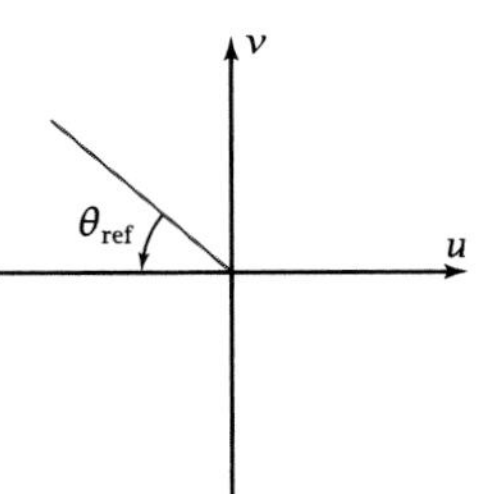

$\cos 140° = -0.7660\ldots$, $\cos 40° = 0.7660\ldots$,
$\cos 140° = -\cos 40°$

5. $\theta_{\text{ref}} = 60°$

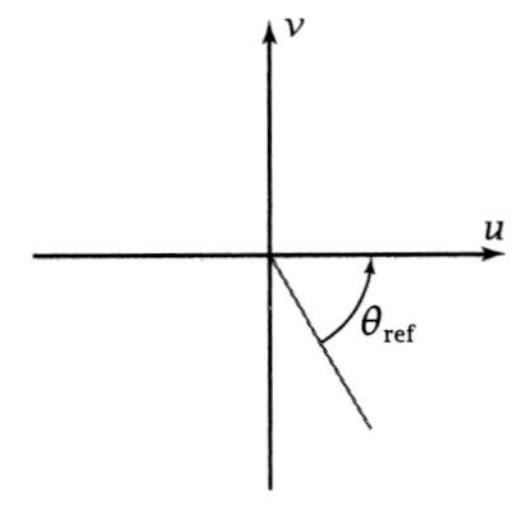

$\cos 300° = 0.5$, $\cos 60° = 0.5$,
$\cos 300° = \cos 60°$

7. $r = \sqrt{170}$; $\sin\theta = \dfrac{11}{\sqrt{170}}$; $\cos\theta = \dfrac{7}{\sqrt{170}}$

9. $r = \sqrt{29}$; $\sin\theta = \dfrac{5}{\sqrt{29}}$; $\cos\theta = \dfrac{-2}{\sqrt{29}}$

11. $r = 4\sqrt{5}$; $\sin\theta = \dfrac{-2}{\sqrt{5}}$; $\cos\theta = \dfrac{1}{\sqrt{5}}$

13. $r = 25$; $\sin\theta = \dfrac{-7}{25}$; $\cos\theta = \dfrac{-24}{25}$

15. θ-translation of $y = \sin\theta$ by $+60°$

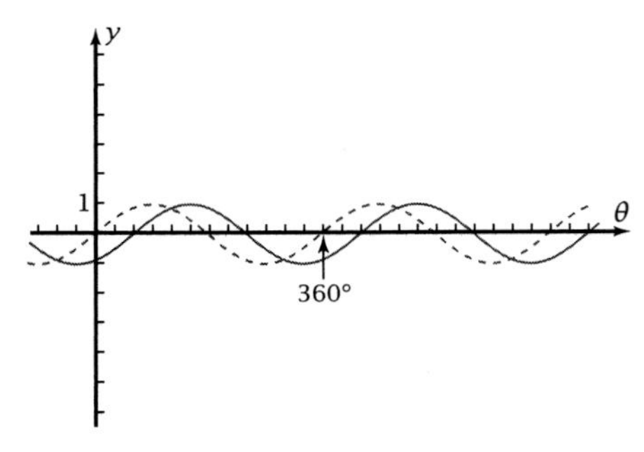

17. y-dilation of $y = \cos\theta$ by 3

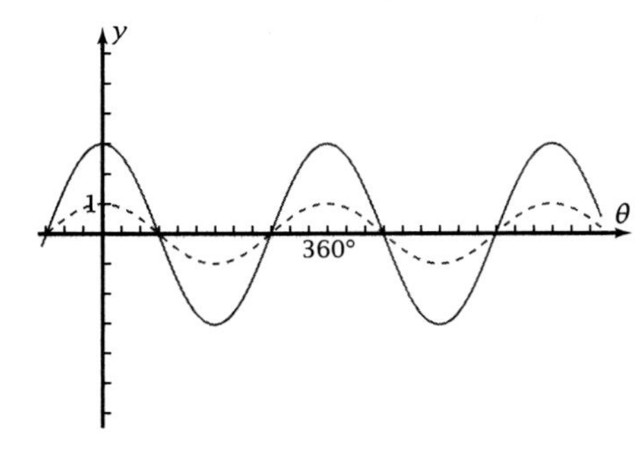

19. θ-dilation of $y = \cos\theta$ by $\frac{1}{2}$, y-translation by $+3$

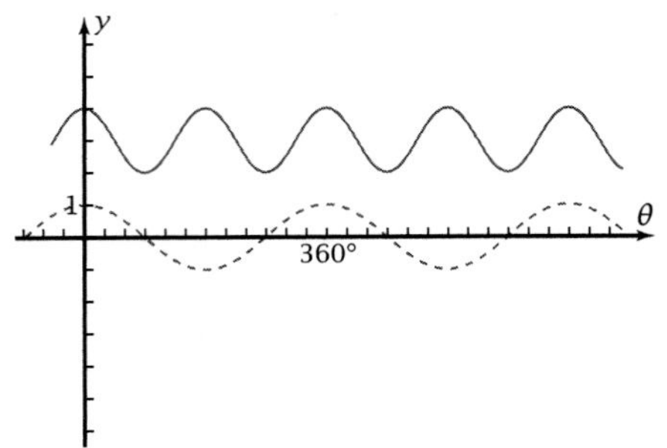

21.

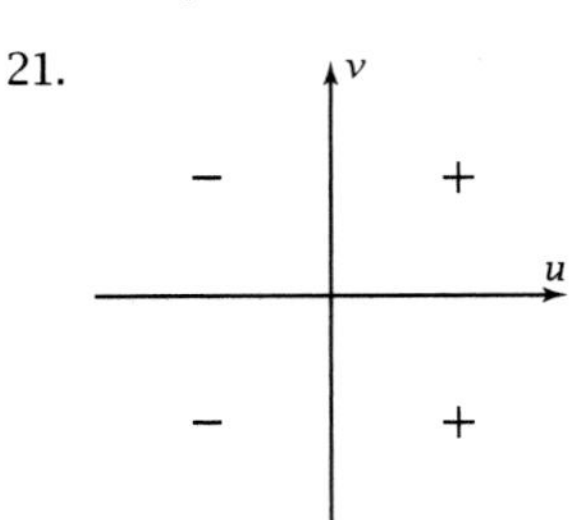

23. Examples will vary.

Problem Set 2-4

1. $\cot 38° \approx 1.2799$
3. $\sec 238° \approx -1.8871$
5. $\csc(-179°) \approx -57.2987$
7. $\sin\theta = -\frac{3}{5}$ $\quad \cos\theta = \frac{4}{5}$

   $\tan\theta = -\frac{3}{4}$ $\quad \cot\theta = -\frac{4}{3}$

   $\sec\theta = \frac{5}{4}$ $\quad \csc\theta = -\frac{5}{3}$

9. $\sin\theta = -\dfrac{7\sqrt{74}}{74}$ $\quad \cos\theta = -\dfrac{5\sqrt{74}}{74}$

   $\tan\theta = \frac{7}{5}$ $\quad \cot\theta = \frac{5}{7}$

   $\sec\theta = -\dfrac{\sqrt{74}}{5}$ $\quad \csc\theta = -\dfrac{\sqrt{74}}{7}$

11. $\sin\theta = \frac{4}{5}$ $\quad \cos\theta = -\frac{3}{5}$

    $\tan\theta = -\frac{4}{3}$ $\quad \cot\theta = -\frac{3}{4}$

    $\sec\theta = -\frac{5}{3}$ $\quad \csc\theta = \frac{5}{4}$

13. $\sin\theta = -\dfrac{\sqrt{15}}{4}$ $\quad \cos\theta = \frac{1}{4}$

    $\tan\theta = -\sqrt{15}$ $\quad \cot\theta = -\dfrac{\sqrt{15}}{15}$

    $\sec\theta = 4$ $\quad \csc\theta = -\dfrac{4\sqrt{15}}{15}$

15. $\sin 60° = \frac{\sqrt{3}}{2}$ $\qquad \cos 60° = \frac{1}{2}$

$\tan 60° = \sqrt{3}$ $\qquad \cot 60° = \frac{\sqrt{3}}{3}$

$\sec 60° = 2$ $\qquad \csc 60° = \frac{2\sqrt{3}}{3}$

17. $\sin(-315°) = \frac{\sqrt{2}}{2}$ $\qquad \cos(-315°) = \frac{\sqrt{2}}{2}$

$\tan(-315°) = 1$ $\qquad \cot(-315°) = 1$

$\sec(-315°) = \sqrt{2}$ $\qquad \csc(-315°) = \sqrt{2}$

19. $\sin 180° = 0$ $\qquad \cos 180° = -1$

$\tan 180° = 0$ $\qquad \cot 180°$ is undefined

$\sec 180° = -1$ $\qquad \csc 180°$ is undefined

21. $\sin 180° = 0$

23. $\cos 240° = -\frac{1}{2}$

25. $\tan 315° = -1$

27. $\cot 0°$ is undefined

29. $\sec 150° = -\frac{2\sqrt{3}}{3}$

31. $\csc 45° = \sqrt{2}$

33. a. $\sin\theta = 0$ at $\theta = 0°, 180°, 360°$
    b. $\cos\theta = 0$ at $\theta = 90°, 270°$
    c. $\tan\theta = 0$ at $\theta = 0°, 180°, 360°$
    d. $\cot\theta = 0$ at $\theta = 90°, 270°$
    e. $\sec\theta \neq 0$ for all θ
    f. $\csc\theta \neq 0$ for all θ

35. $\sin 30° + \cos 60° = 1$

37. $\sec^2 45° = 2$

39. $\sin 240° \csc 240° = 1$

41. $\tan^2 60° - \sec^2 60° = -1$

43. a. 67°
    b. $\cos 23° = 0.9205\ldots$, $\sin 67° = 0.9205\ldots$; they are equal.
    c. "Complement"

Problem Set 2-5

1. $\sin^{-1} 0.3 = 17.4567\ldots°$ because $\sin 17.4567\ldots° = 0.3$.

3. $\tan^{-1} 7 = 81.8658\ldots°$ because $\tan 81.8658\ldots° = 7$.

5. $\cos(\sin^{-1} 0.8) = 0.6$
   $\theta = \sin^{-1} 0.8$ represents an angle of a right triangle with sides 3, 4, and 5.

7. a. They are not one-to-one functions.
   b. Sine: $-90° \le \theta \le 90°$; cosine: $0° \le \theta \le 180°$; they are one-to-one functions.
   c. $\sin^{-1}(-0.9) = -64.1580\ldots°$. On the principal branch only negative angles correspond to negative values of the sine.

9. a. ≈6.0 m $\qquad$ b. ≈76°

11. ≈25.4°

13. ≈1198 ft

15. a. ≈34.4° $\qquad$ b. ≈10.1 cm

17. a. ≈33.5 m $\qquad$ b. ≈17.5 m

19. a. ≈108 m deep; ≈280 m from starting point
    b. ≈2790 m

21. a. ≈1 ft 5 in. $\qquad$ b. ≈8 ft 3 in.
    c. ≈6 ft 10 in.

23. Answers will vary.

Problem Set 2-6

R1. a. The graphs match.
    b. y-dilation by 0.7, y-translation by +2; $y = 2 + 0.7\sin\theta$; the result agrees with the graph.
    c. Sinusoid

R2. a. $\theta_{ref} = 70°$ $\qquad$ b. $\theta_{ref} = 79°$

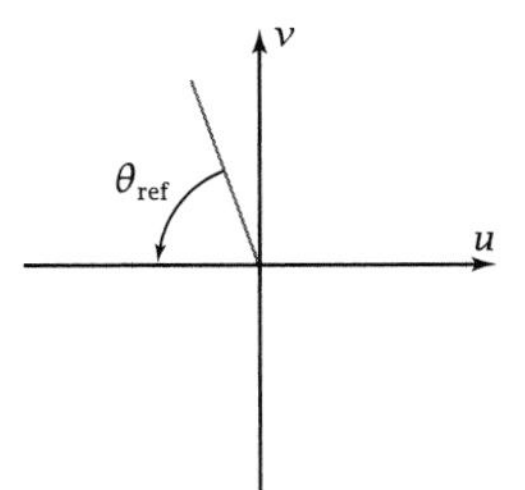

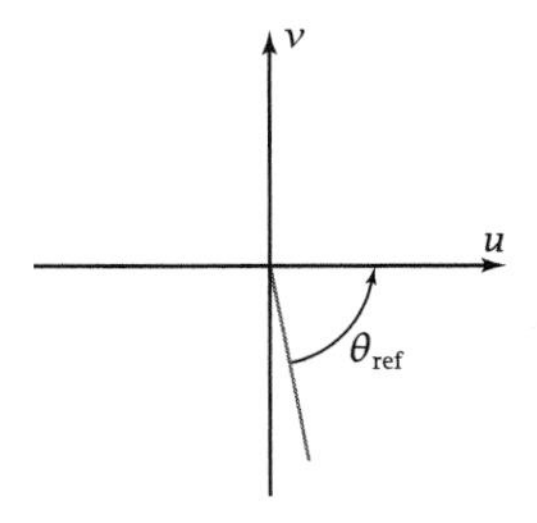

c. $\theta_{ref} = 76°$

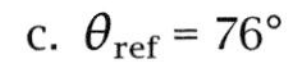

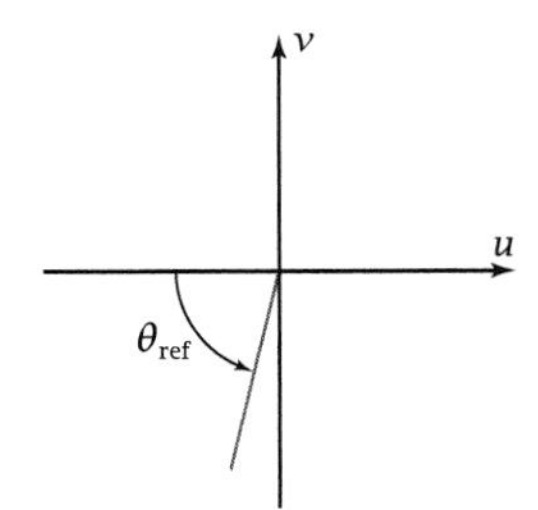

R3. a. $\sin\theta = \frac{7\sqrt{74}}{74}$; $\cos\theta = -\frac{5\sqrt{74}}{74}$
    b. $\sin 160° = 0.3420\ldots$,
    $\cos 160° = -0.9396\ldots$,
    $\theta_{ref} = 20°$

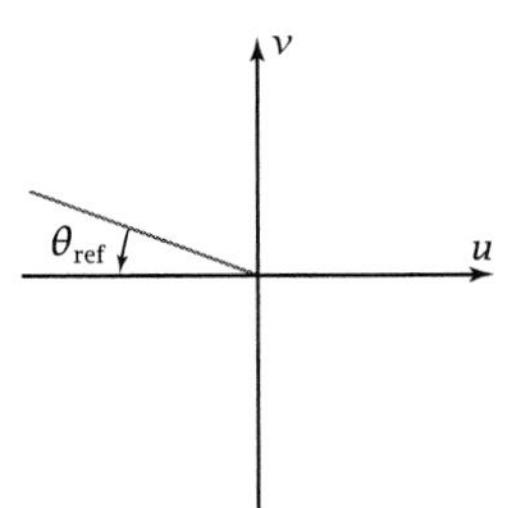

160° terminates in Quadrant II, above the x-axis (so $\sin 160° > 0$) and to the left of the y-axis (so $\cos 160° < 0$).

c. 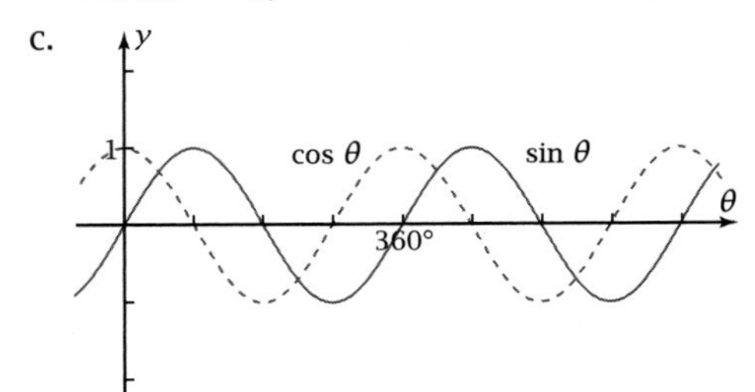

d. Quadrants III and IV

e. y-translation of +4, x-dilation of $\frac{1}{2}$

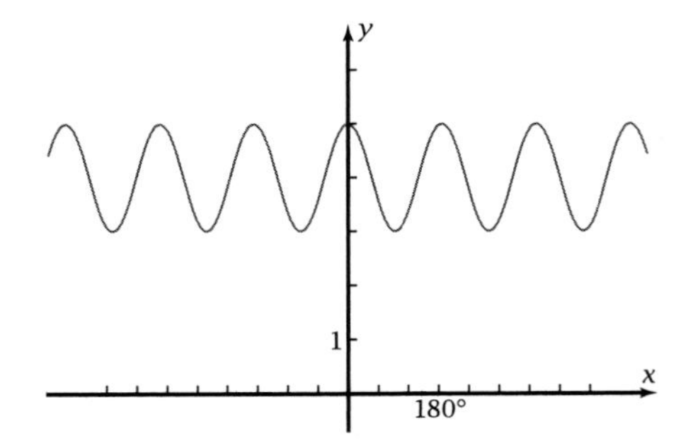

R4. a. $\csc 256° = -1.0306\ldots$

b. $\sin 150° = \frac{1}{2}$ $\quad\cos 150° = -\frac{\sqrt{3}}{2}$

$\tan 150° = -\frac{\sqrt{3}}{3}$ $\quad\cot 150° = -\sqrt{3}$

$\sec 150° = -\frac{2\sqrt{3}}{3}$ $\quad\csc 150° = 2$

c. $\sec\theta = -\sqrt{2}$

d. $\cos\theta = -\frac{3\sqrt{34}}{34}$

e. $\sec(-120°) = -2$

f. $\tan^2 30° - \csc^2 30° = -3\frac{2}{3}$

g. The endpoint (u, v) is (0, 1), so $\tan 90° = \frac{v}{u} = \frac{1}{0}$, which is undefined.

R5. a. $\cos^{-1} 0.6 = 53.1301\ldots°$. This means that $\cos(53.1301\ldots°) = 0.6$.

b. i. ≈321 m

ii. ≈603 m

iii. ≈75°

iv. Answers will vary.

CHAPTER 3

Problem Set 3-1

1. Amplitude = 1

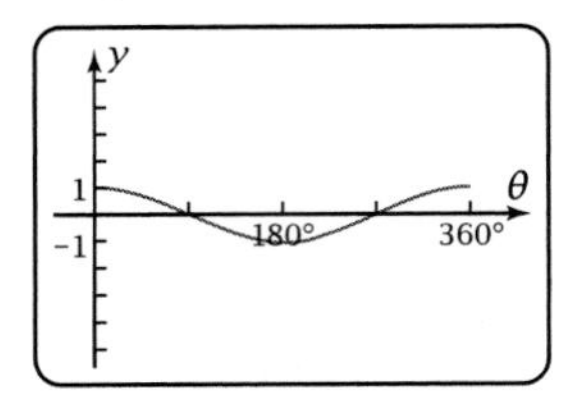

3. Period = 360° for both functions

5. 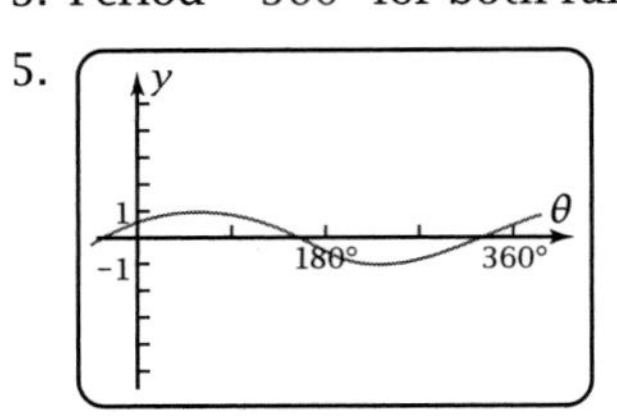

Horizontal translation of +60°

7. 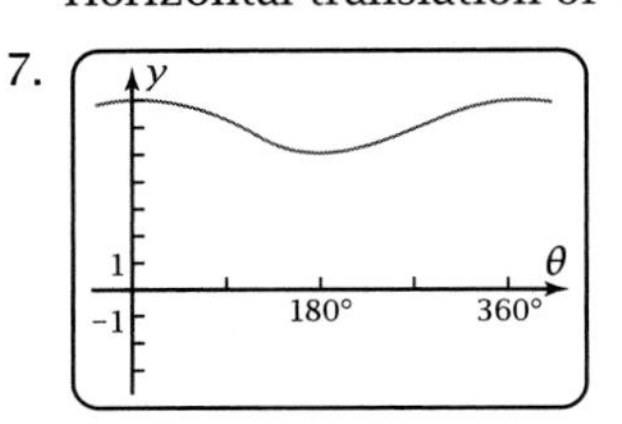

Vertical translation of +6

9. Amplitude = 5
Period = 120°
Phase displacement = 60°
Sinusoidal axis = 6

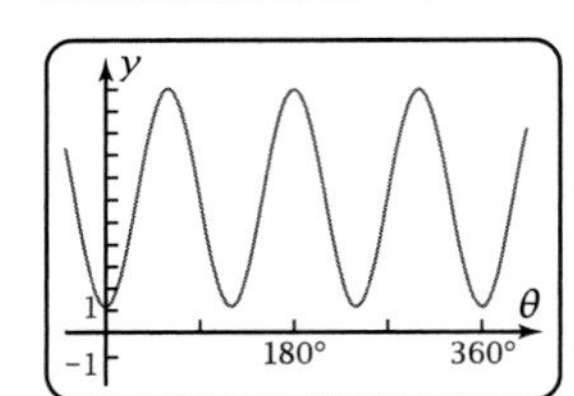

Problem Set 3-2

1. Amplitude = 4
Period = 120°
Phase displacement = −10°
Sinusoidal axis = 7

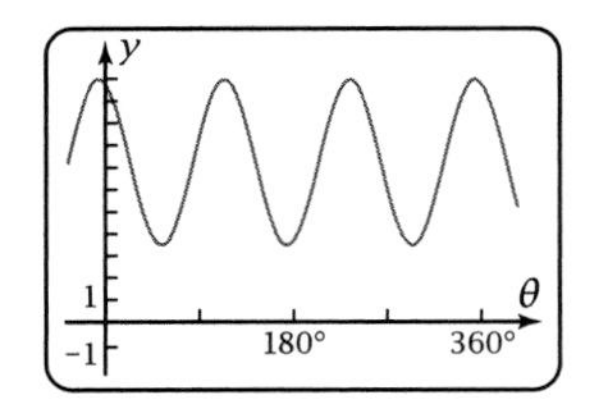

3. Amplitude = 20
Period = 720°
Phase displacement = 120°
Sinusoidal axis = −10

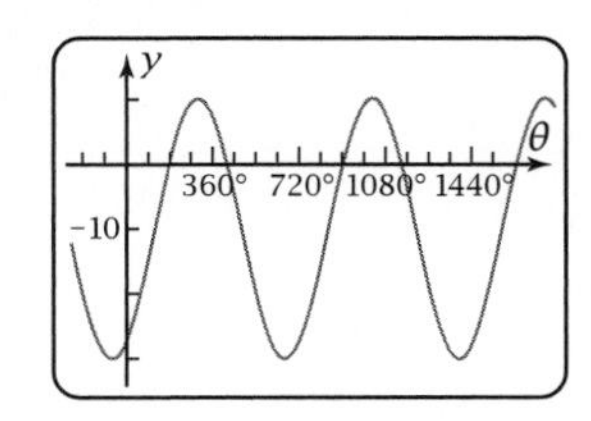

5. a. $y = 9 + 6 \cos 2(\theta - 20°)$
   b. Amplitude = 6
      Period = 180°
      Frequency = $\frac{1}{180}$ cycle/day
      Phase displacement = 20°
      Sinusoidal axis = 9
   c. $y = 10.0418...$ at $\theta = 60°$
      $y = 8.7906...$ at $\theta = 1234°$

7. a. $y = -3 + 5 \cos 3(\theta - 10°)$
   b. Amplitude = 5
      Period = 120°
      Frequency = $\frac{1}{120}$ cycle/day
      Phase displacement = 10°
      Sinusoidal axis = −3
   c. $y = -8$ at $\theta = 70°$
      $y = 1.9931...$ at $\theta = 491°$

9. $y = 1.45 - 1.11 \sin 10(\theta - 2°)$

11. $y = 1.7 \cos (\theta - 30°)$

13. $y = 7 \cos 3\theta$

15. $y = 35 + 15 \sin 90\theta$

17. $y = 4 - 9 \sin \frac{9}{13} (\theta + 60°)$

19. $y = 12.4151...$ at $\theta = 300°$
    $y = 2.0579...$, 1.9420... below the sinusoidal axis, at $\theta = 5678°$

21. $y = 4 + 3 \cos 5(\theta - 6°)$

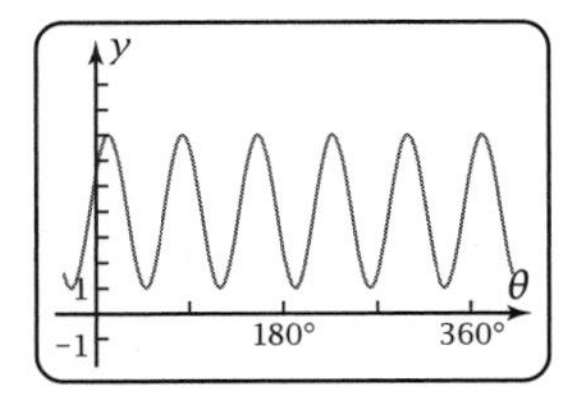

23. a. $y = 6 + 4 \cos 3(\theta + 10°)$
    b. $y = 6 - 4 \cos 3(\theta - 50°)$
    c. $y = 6 + 4 \sin 3(\theta + 40°)$
    d. $y = 6 - 4 \sin 3(\theta - 20°)$

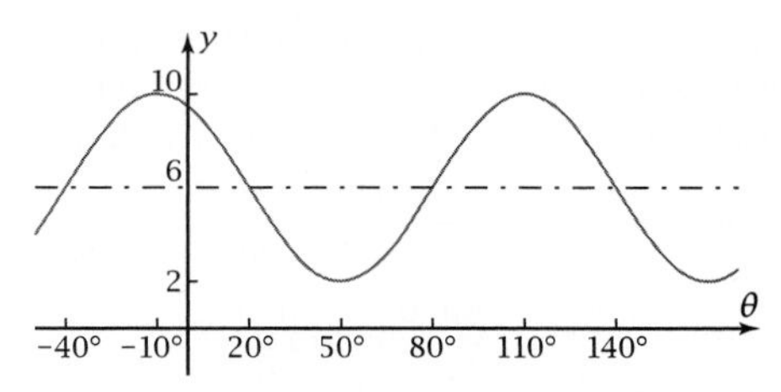

25. a. 60 cycles/deg. Thinking in terms of complete cycles (60 of them) gives a clearer mental picture than thinking in terms of fractions (1/60) of a cycle.
    b. Period = 1.2°/cycle
       Frequency = $\frac{5}{6}$ cycle/deg
       The frequency is 360° divided by 300.

27. a. Subtract 3 from both sides, then divide both sides by 4. Also,
       $$2(\theta - 5°) = \frac{\theta - 5°}{\frac{1}{2}}$$
    b. Now *all* numbers are the opposites for translations and reciprocals for dilations of the actual value of the transformations.
    c. It isolates y as a function of θ, making it easier to calculate y, given θ.

Problem Set 3-3

1. a.

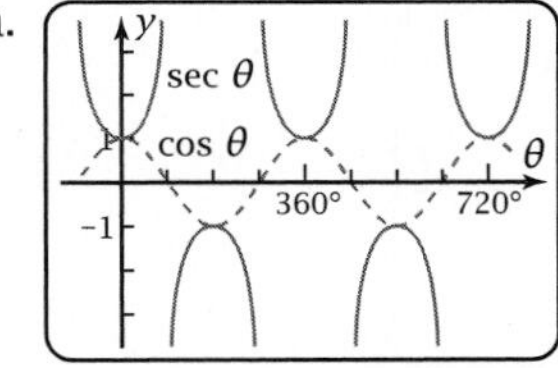

   b. Asymptotes occur where $\cos \theta = 0$, at $\theta = 90° \pm 180n°$.
   c. Yes, where $\cos \theta = \pm 1$, at $\theta = \pm 180n°$.
   d. No, the concavity changes at the asymptotes, not at points on the graph.

3.

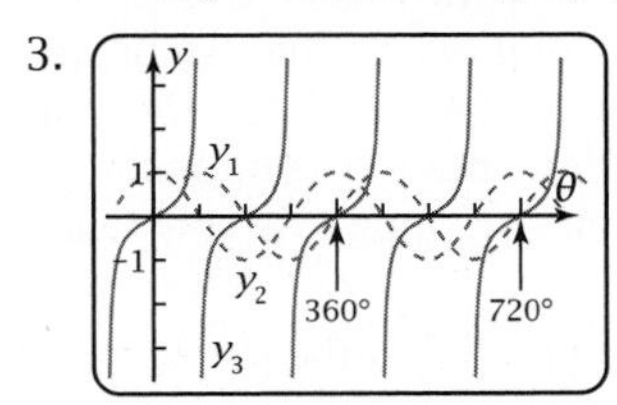

The graph of y_3 is the same as the graph of $\tan \theta$.

5. See Figure 3-3a.

7. $\sin(\theta \pm 180°n) = -\sin\theta$ and $\cos(\theta \pm 180°n) = -\cos\theta$, where n is an odd integer. (If n were even, then you would have $\sin(\theta \pm 180°n) = -\sin\theta$.) Therefore,

$$\tan(\theta \pm 180°n) = \frac{\sin(\theta \pm 180°n)}{\cos(\theta \pm 180°n)}$$
$$= \frac{-\sin\theta}{-\cos\theta} = \tan\theta$$

9. The domain is $\theta \neq 90° \pm 180n°$. The range is $|y| \geq 1$.

11. θ-translation of +5°, θ-dilation of $\frac{1}{3}$; y-dilation of 5, y-translation of +2

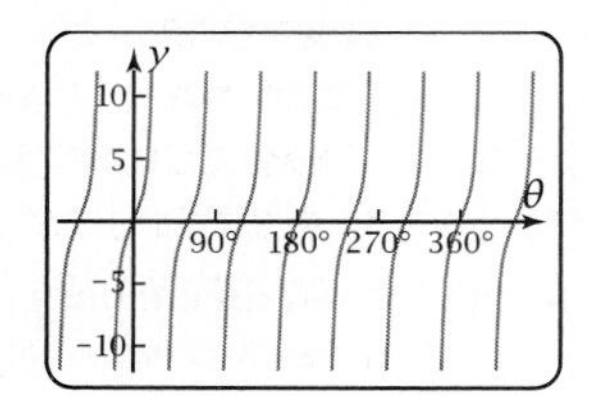

13. θ-translation of −50°, θ-dilation of 2; y-dilation of 6, y-translation of +4

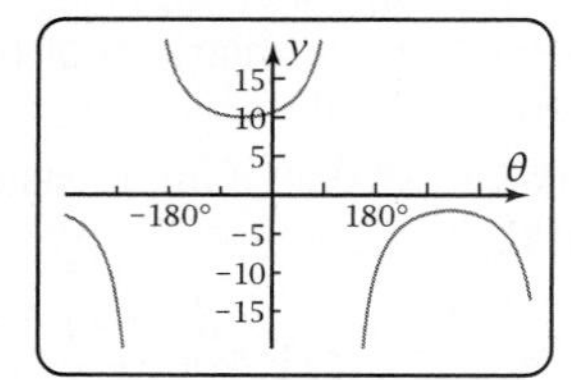

15. a.

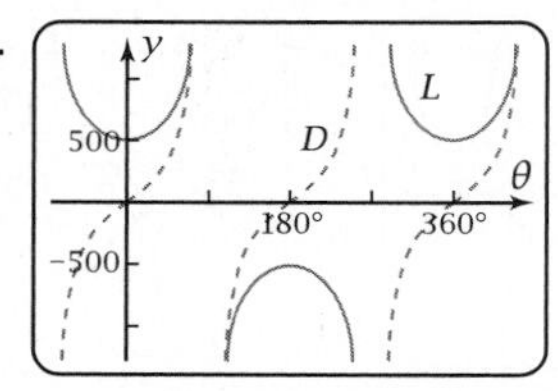

b. $D \approx 714$ m along the shore

$L \approx 872$ m

c. $\theta \approx 75.5°$

d. For $90° < \theta < 180°$, the spot of light is in the opposite direction along the beach. L is defined as being measured along *one* of the two beams. Because the *other* beam, which strikes the beach when $90° < \theta < 270°$, points in the opposite direction, a measurement along it is considered negative.

e. The beam of light is parallel to the shore.

Problem Set 3-4

1. a.

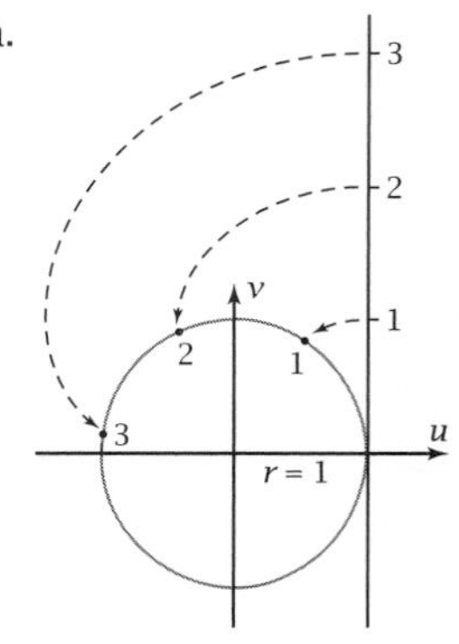

b.

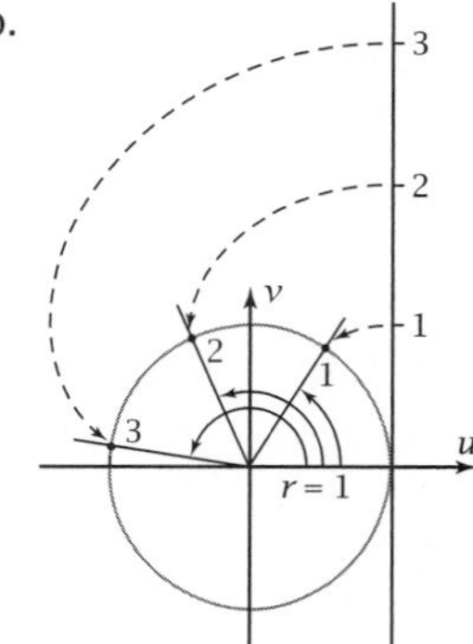

c. The arc length on the unit circle equals the radian measure.

3. $\frac{\pi}{3}$

5. $\frac{\pi}{6}$

7. $\frac{2}{3}\pi$

9. $-\frac{5}{4}\pi$

11. $\frac{37}{180}\pi = 0.6457...$

13. $\frac{123}{180}\pi = 2.1467...$

15. 18°

17. 30°

19. 15°

21. 135°

23. 270°

25. 19.4805...°

27. 72.1926...°

29. 57.2957...°

31. −0.9589...

33. 1.1192...

35. 0.3046...

37. 0.3217...

39. $\frac{\sqrt{3}}{2}$

41. $\frac{\sqrt{3}}{3}$

43. 1

45. 4

47. 1

49. $y = 5 + 7\cos 30(\theta - 2°)$

51. $x = 13.9255...$ cm

53. $\theta = 64.6230...°$

Problem Set 3-5

1. $\frac{\pi}{6}$ units

3. $\frac{\pi}{2}$ units

5. 60°

7. 45°

9. $\frac{\pi}{2}$ units

11. 2 units

13. 1.5574...

15. −1.0101...

17. 1.2661...

19. 0.2013...

21. $\sin\frac{\pi}{3} = \frac{\sqrt{3}}{2}$

23. $\tan\frac{\pi}{6} = \frac{\sqrt{3}}{3}$

25. Period = 10
Amplitude = 2
Phase displacement = +4
Sinusoidal axis = +3

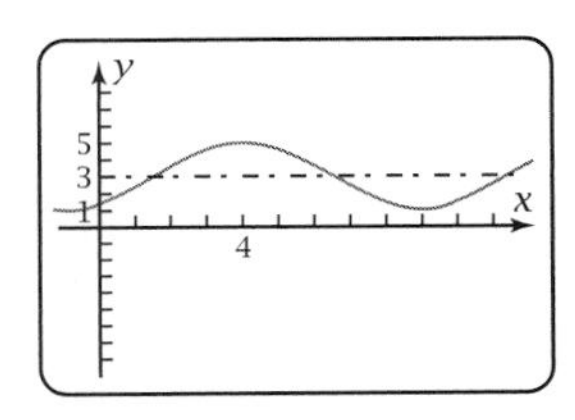

27. Period = 8
Amplitude = 6
Phase displacement = −1
Sinusoidal axis = +2

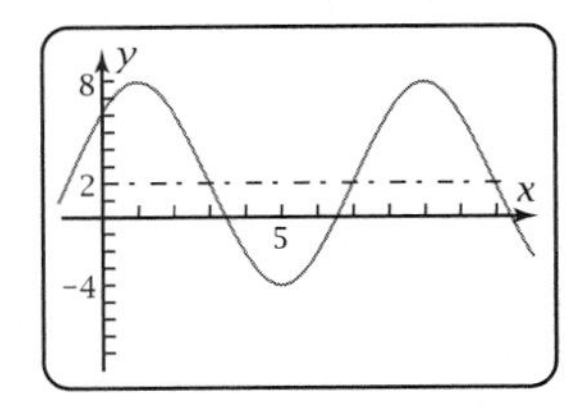

29. Period = 4
Asymptotes at $\pm 4n$
Points of inflection at $2 \pm 4n$

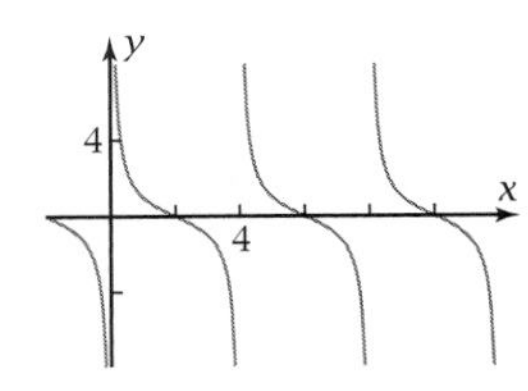

31. Period = 2π
Asymptotes at $\frac{\pi}{2} \pm n\pi$
Critical points at $(\pm 2n\pi, 3)$ and $(\pm[2n + 1]\pi, 1)$

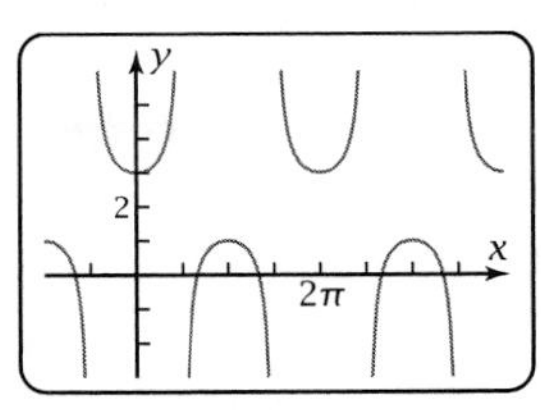

33. $y = 5 + 2\cos\frac{\pi}{3}(x - 1)$

35. $y = -2 + 5\cos\frac{\pi}{15}(x + 5)$

37. $y = \csc\frac{\pi}{6}x$

39. $y = 3\tan x$

41. $z = -8 + 2\sin 5\pi(t - 0.17)$

43. $z(0.4) = -8.9079...$
$z(50)$ is 0.9079... below the sinusoidal axis.

45. a. Horizontal translation of $+\frac{\pi}{2}$;
$$\sin x = \cos\left(x - \frac{\pi}{2}\right)$$
b. The graph would coincide with itself and appear unchanged.
c. $+2\pi$ or -2π or any multiple of $\pm 2\pi$
d. 2π is the period of the sine function. A horizontal translation by a multiple of 2π will result in a graph that coincides with itself.

47. a. This lets u, v, x, and y all be represented on the same diagram—x is now an arc, and y is now either u or v, depending on whether you are talking about $y = \cos x$ or $y = \sin x$.
b. A radian measure corresponds to an angle measure, using $m^R(\theta) = m°(\theta) \cdot \frac{\pi}{180°}$, but because a radian measure is a pure number, it can represent something other than an angle in an application problem.

Problem Set 3-6

1. 0.4510..., 5.8321..., 6.7342..., 12.1153..., 13.0173...

3. 1.7721..., 4.5110..., 8.0553..., 10.7942..., 14.3385...

5. a. $x \approx 1, 5, 21, 25$
b. $y = 2 + 5\cos\frac{\pi}{10}(x - 3)$

c. $x \approx 0.9516\ldots, 5.0483\ldots, 20.9516\ldots, 25.0483\ldots$

d. $x = 3 \pm \frac{10}{\pi}\left(\cos^{-1}\frac{4}{5} + 2\pi n\right)$

7. a. $x \approx -2.9, -0.5, 1.1, 3.5, 5.1$

b. $y = -2 + 4\cos\frac{\pi}{2}(x - 0.3)$

c. $x \approx -2.8608\ldots, -0.5391\ldots, 1.1391\ldots, 3.4608\ldots, 5.1391\ldots$

d. $x = 0.3 \pm \frac{2}{\pi}(\cos^{-1}\frac{1}{4} + 2\pi n)$

9. a. $x \approx -10.6, -3.4, 5.4, 12.6, 21.4$

b. $y = 1 - 3\cos\frac{\pi}{8}(x - 1)$

c. $x \approx -10.5735\ldots, -3.4264\ldots, 5.4264\ldots, 12.5735\ldots, 21.4264\ldots$

d. $x = 1 \pm \frac{8}{\pi}\left[\cos^{-1}\left(-\frac{1}{6}\right) + 2\pi n\right]$

11. a. $\theta \approx 130°, 170°, 310°$

b. $y = 6 + 4\cos 2(\theta - 60°)$

c. $\theta \approx 129.2951\ldots°, 170.7048\ldots°, 309.2951\ldots°$

d. $\theta = 60° \pm \frac{1}{2}\left(\cos^{-1}\frac{3}{4} + 360n°\right)$

13. a. $x = 2.6905\ldots, 3.5926\ldots, 8.9737\ldots, 9.8758\ldots, 15.2569\ldots, 16.1589\ldots$

b. $x = 203.7524\ldots$

Problem Set 3-7

1. a.

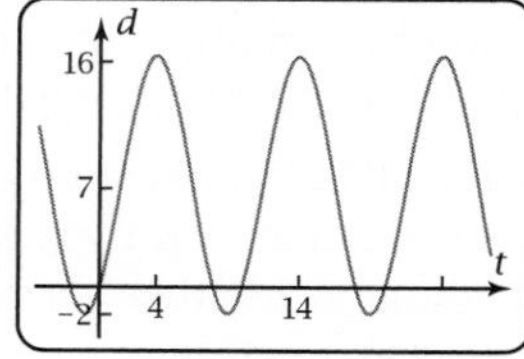

b. −2 ft; this number must be negative because part of the wheel is underwater.

c. $d = 7 + 9\cos\frac{\pi}{5}(t - 4)$

d. $d = 4.2188\ldots$ ft

e. $t = 0.0817\ldots$ sec; the wheel must have been coming out of the water.

f. "Mark Twain" was riverboat terminology indicating that the water was 2 fathoms deep.

3. a.

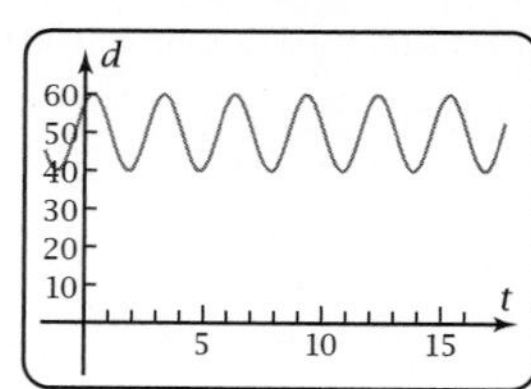

b. $d = 50 + 10\cos\frac{\pi}{1.5}(t - 0.3)$

c. $d = 43.3086\ldots$ cm

d. $d = 58.0901\ldots$ cm

e. $t = 0.0846\ldots$ sec

5. a. $y = 12 + 15\cos\frac{\pi}{50}x$

b.

x	Length
0 m	27 m
2 m	26.8817... m
4 m	26.5287... m
6 m	25.9466... m
8 m	25.1446... m
10 m	24.1352... m
12 m	22.9345... m
14 m	21.5613... m
16 m	20.0374... m
18 m	18.3866... m
20 m	16.6352... m
22 m	14.8107... m
24 m	12.9447... m
26 m	11.0581... m
28 m	9.1892... m
30 m	7.3647... m
32 m	5.6133... m
34 m	3.9625... m
36 m	2.4386... m
38 m	1.0654... m

c. $y = 12 + 15\cos\frac{\pi}{50}x \Rightarrow x = \frac{50}{\pi}\cos^{-1}\frac{y - 12}{15}$

y	Length
0 m	39.7583... m
2 m	36.6139... m
4 m	33.9530... m
6 m	31.5494... m
8 m	29.2961... m
10 m	27.1284... m
12 m	25 m
14 m	22.8715... m
16 m	20.7038... m
18 m	18.4505... m
20 m	16.0469... m
22 m	13.3860... m
24 m	10.2416... m
26 m	5.8442... m

d. The sum of vertical timbers is about 324 m, and the sum of horizontal timbers is about 331 m.

7. a. 11 years

b. $S = 60 + 50\cos\frac{2\pi}{11}(t - 1948)$

c. $S(2020) \approx 12$ sunspots

d. $t = 2021.3333...$; maximum in 2025

9. a.

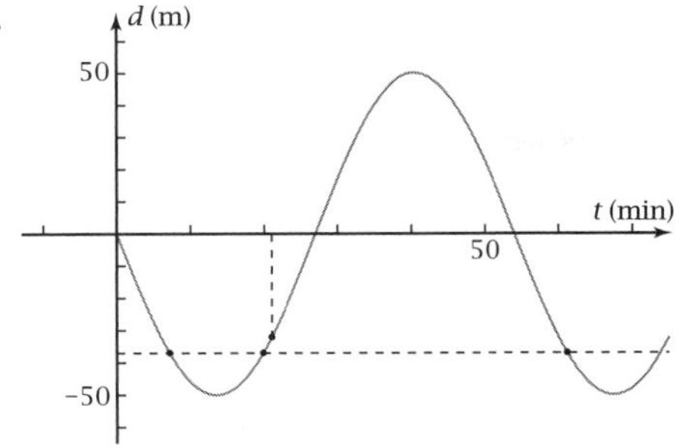

b. $t = 40.5$ min

c. $d = -50 \sin \frac{\pi}{27} t$

d. $d = -32.1393...$ m

e. $t = 7.1597...$ min, $19.8402...$ min, $61.1597...$ min

11. a.

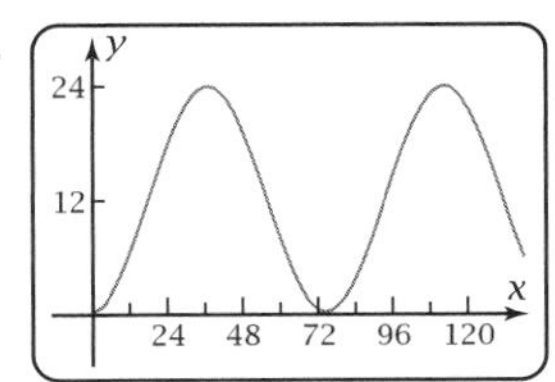

b. $y = 12 - 12 \cos \frac{x}{12}$

c. $y = 8.2161...$ in.

d. $x = 17.8483...$ in. and $57.5498...$ in.

13. a. Frequency = 60 cps; period = $\frac{1}{60}$ sec

b. Wavelength = 220 in.

c. 34 ft 4.5 in.

15. Answers will vary.

Problem Set 3-8

R1. a.

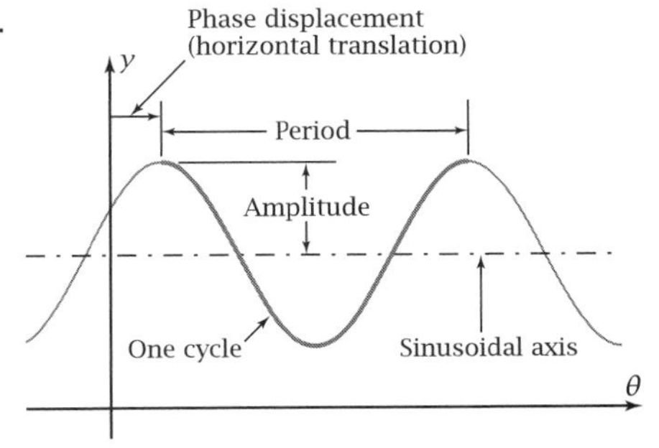

b. Argument

R2. a.

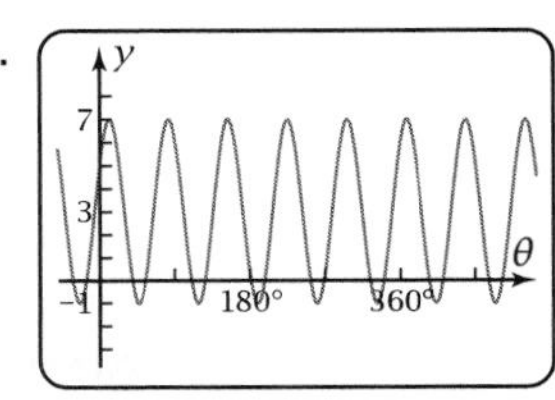

Amplitude = 4
Period = 72°
Sinusoidal axis = 3
Phase displacement = 10°

b. $y = -7 + 3 \cos \frac{360}{28}(\theta - 10°)$

$y = -7 + 3 \sin \frac{360}{28}(\theta - 3°)$

c. $y = 50 + 70 \sin \frac{360}{48}(\theta - 8°)$

d. Point of inflection at $\theta = 8°$, critical point at $\theta = 20°$

e. Frequency = $\frac{1}{7.5°}$, or $\frac{2}{15}$ of a cycle per degree

R3. a.

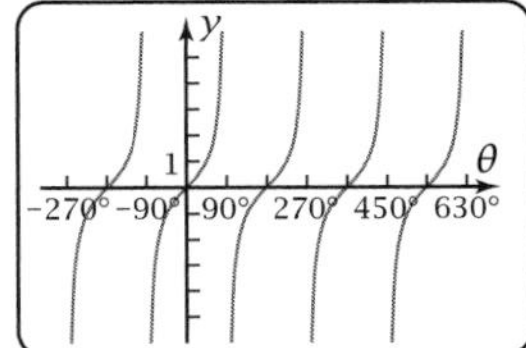

b. $\sin(\theta \pm 180°n)$ and $\cos(\theta \pm 180°n)$ are both opposites of $\sin \theta$ and $\cos \theta$.

Therefore, $\tan(\theta \pm 180°n) = \dfrac{\sin(\theta \pm 180°n)}{\cos(\theta \pm 180°n)}$

$= \dfrac{-\sin \theta}{-\cos \theta} = \tan \theta$

c.

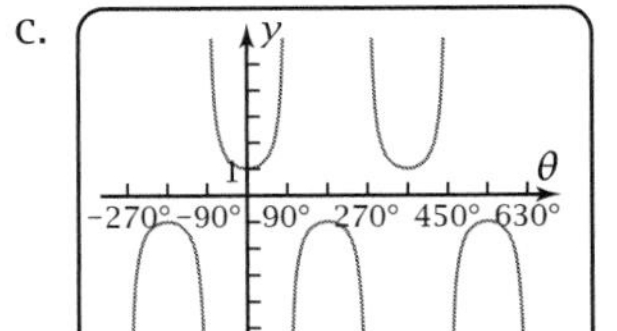

$y = \sec \theta = \dfrac{1}{\cos \theta}$

d. $\sin \theta = 0$ at $\theta = 0°, 180°, 360°$, so $\csc \theta = \dfrac{1}{\sin \theta}$ is undefined at these points.

e. The cosecant graph changes concavity only at the asymptotes, not at any points that are actually on the graph, so it has no points of inflection. The cotangent graph is always decreasing, so it has no critical points. It is concave upward to the left of $90° + 180°n$ and concave downward to the right. So $90° + 180°n$ are the points of inflection.

f. θ-translation of +40°, θ-dilation of 3; y-dilation of 0.4, y-translation of +2

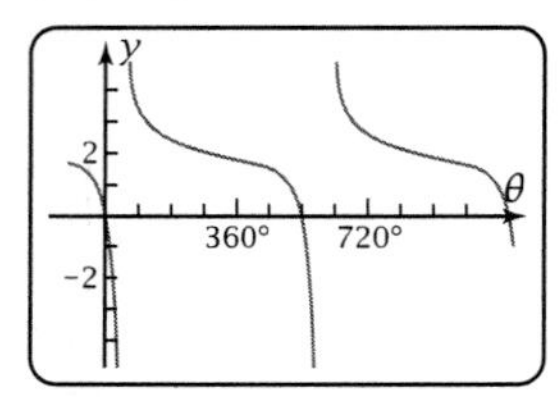

Period = 540°
The value of y is unbounded, so the "amplitude" is infinite.

R4. a. $30° = \dfrac{\pi}{6}$ radians, $45° = \dfrac{\pi}{4}$ radians, $60° = \dfrac{\pi}{3}$ radians

b. 2 radians $= \dfrac{720°}{2\pi} = 114.5915...°$

c. $\cos 3 = -0.9899...$, $\cos 3° = 0.9986...$

d. $\cos^{-1} 0.8 = 0.6435...$, $\csc^{-1} 2 = \dfrac{\pi}{6}$

e. Arc length = 17 units

R5. a.

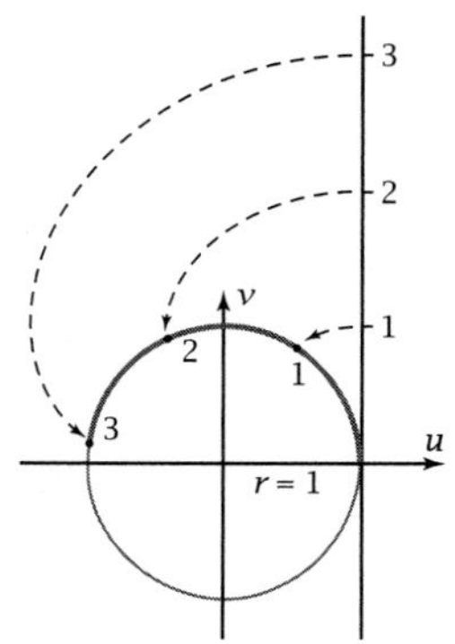

b. $60° \rightarrow \dfrac{\pi}{3}$ units

2.3 radians → 2.3 units

c. $\sin 2° = 0.0348...$, $\sin 2 = 0.9092...$

d. $\cos^{-1} 0.6 = 0.9272...$

e. $\cos \dfrac{\pi}{6} = \dfrac{\sqrt{3}}{2}$, $\sec \dfrac{\pi}{4} = \sqrt{2}$, $\tan \dfrac{\pi}{2}$ is undefined

f.

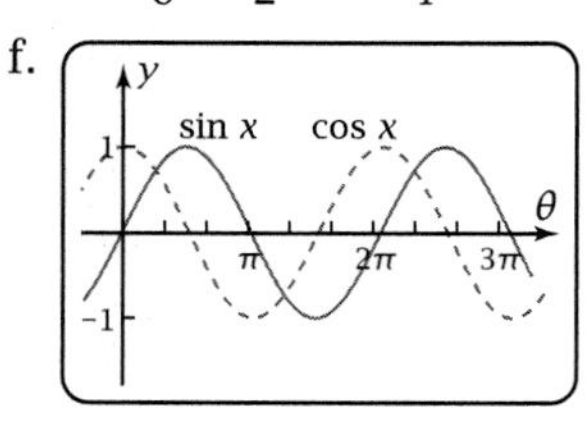

g. Period $= 2\pi \div \dfrac{\pi}{10} = 20$

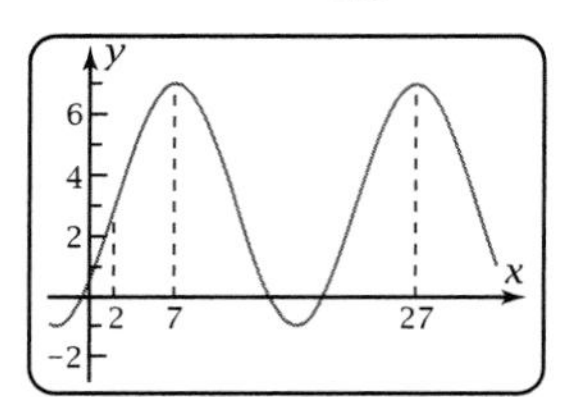

h. $y = -10 - 35 \sin \frac{\pi}{20}(x - 13)$

R6. a. $\arccos 0.8 = \cos^{-1} 0.8 + 2\pi n$

b. 0.6435..., 6.9266..., 13.2098...

c. 102.0015...

d. $x \approx -7.6, -4.4, 8.4, 11.6$;
$y = 6 + 5 \cos \frac{2\pi}{16}(x - 2)$, so
$x = -7.6063..., -4.4148..., 8.3510..., 11.5425...$;
$x = -7.6386..., -4.3613..., 8.3613..., 11.6386...$

e. $x = 24.3613...$

R7. a.

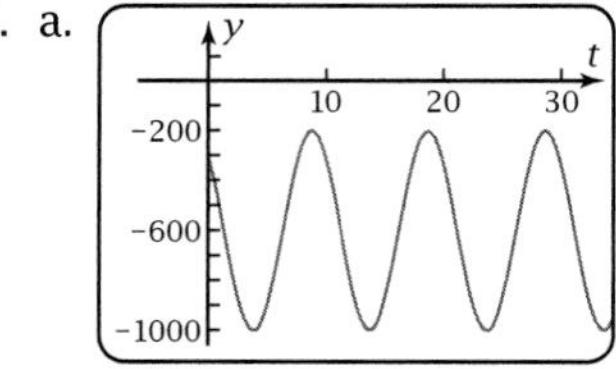

b. $y = -600 + 400 \cos \frac{\pi}{5}(t - 9)$

c. $y = -276.3932...$ m
Submarine could communicate.

d. $0.1502...$ min $< t < 7.8497...$ min

CHAPTER 4

Problem Set 4-1

1. The sum is 1.

3.

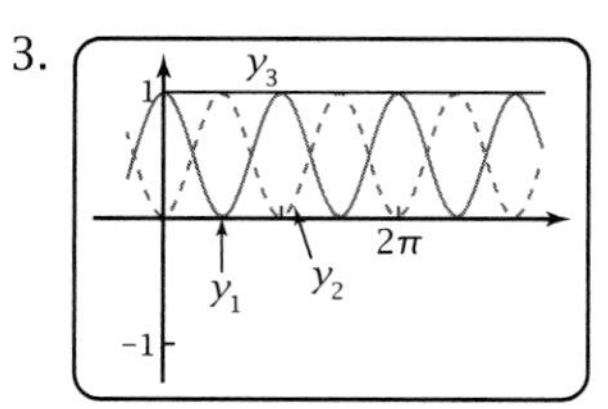

y_1 and y_2 are symmetrical with respect to the $y = \frac{1}{2}$ line, so the amount that one graph is above $\frac{1}{2}$ is the same as the amount the other graph is below $\frac{1}{2}$. When added,

$$\left(\frac{1}{2} - a\right) + \left(\frac{1}{2} + a\right) = 1$$

5. Because $r = 1$, $\sin 50° = \frac{v}{r} = v$, and $\cos 50° = \frac{u}{r} = u$.

Problem Set 4-2

1. $\sec x = \dfrac{1}{\cos x}$

3. $\tan x = \dfrac{\sin x}{\cos x}$

5. Cosine and sine are the lengths of the legs of a right triangle whose hypotenuse is a radius of a unit circle. The Pythagorean theorem then says $(\cos x)^2 + (\sin x)^2 = 1^2$.

7. 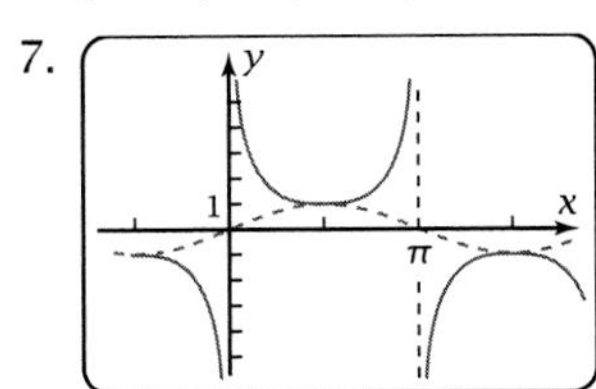

Asymptotes at $\theta = 0°, 180°, 360°, \ldots$

9. $\cos^2 x + \sin^2 x = 1$
$\Rightarrow \cos^2 x + \sin^2 x - \cos^2 x = 1 - \cos^2 x$
$\Rightarrow \sin^2 x = 1 - \cos^2 x$

11. $\cos^2 x + \sin^2 x = 1$ $\quad \cos x \cdot \sec x = 1$
$\csc^2 x - \cot^2 x = 1$ $\quad \sin x \cdot \csc x = 1$
$\sec^2 x - \tan^2 x = 1$ $\quad \tan x \cdot \cot x = 1$

13. $\sec x = \dfrac{1}{\cos x} \Leftrightarrow \csc x = \dfrac{1}{\sin x}$
$\cot x = \dfrac{1}{\tan x} \Leftrightarrow \tan x = \dfrac{1}{\cot x}$
$\tan x = \dfrac{\sin x}{\cos x} = \dfrac{\sec x}{\csc x} \Leftrightarrow \cot x = \dfrac{\cos x}{\sin x} = \dfrac{\csc x}{\sec x}$
$\cos^2 x + \sin^2 x = 1 \Leftrightarrow \sin^2 x + \cos^2 x = 1$
$1 + \tan^2 x = \sec^2 x \Leftrightarrow 1 + \cot^2 x = \csc^2 x$

Problem Set 4-3

1. $\cos x \tan x = \cos x \cdot \dfrac{\sin x}{\cos x}$
$= \left(\cos x \cdot \dfrac{1}{\cos x}\right) \cdot \sin x$
$= \sin x$
$\therefore \cos x \tan x = \sin x$, Q.E.D.

3. $\sec A \cot A \sin A = \cot A \sin A \dfrac{1}{\cos A}$
$= \cot A \dfrac{\sin A}{\cos A}$
$= \cot A \tan A = 1$
$\therefore \sec A \cot A \sin A = 1$, Q.E.D.

5. $\sin^2 \theta \sec \theta \csc \theta = \sin^2 \theta \dfrac{1}{\cos \theta} \dfrac{1}{\sin \theta}$
$= \dfrac{\sin \theta}{\cos \theta}$
$= \tan \theta$
$\therefore \sin^2 \theta \sec \theta \csc \theta = \tan \theta$, Q.E.D.

7. $\cot R + \tan R = \dfrac{\cos R}{\sin R} + \dfrac{\sin R}{\cos R}$
$= \dfrac{\cos R}{\sin R} \cdot \dfrac{\cos R}{\cos R} + \dfrac{\sin R}{\sin R} \cdot \dfrac{\sin R}{\cos R}$
$= \dfrac{\cos^2 R}{\sin R \cos R} + \dfrac{\sin^2 R}{\sin R \cos R}$
$= \dfrac{\cos^2 R + \sin^2 R}{\sin R \cos R}$
$= \dfrac{1}{\sin R \cos R}$
$= \dfrac{1}{\sin R} \cdot \dfrac{1}{\cos R}$
$= \csc R \sec R$
$\therefore \cot R + \tan R = \csc R \sec R$, Q.E.D.

9. $\csc x - \sin x = \dfrac{1}{\sin x} - \dfrac{\sin x}{\sin x} \cdot \sin x$
$= \dfrac{1}{\sin x} - \dfrac{\sin^2 x}{\sin x}$
$= \dfrac{1 - \sin^2 x}{\sin x}$
$= \dfrac{\cos^2 x}{\sin x}$
$= \dfrac{\cos x}{\sin x} \cdot \cos x$
$= \cot x \cos x$
$\therefore \csc x - \sin x = \cot x \cos x$, Q.E.D.

11. $\tan x (\cot x \cos x + \sin x)$
$= \dfrac{\sin x}{\cos x}\left(\dfrac{\cos x}{\sin x} \cos x + \sin x\right)$
$= \dfrac{1}{\cos x}\left(\sin x \cdot \dfrac{\cos x}{\sin x} \cdot \cos x + \sin^2 x\right)$
$= \sec x (\cos^2 x + \sin^2 x)$
$= \sec x$
$\therefore \tan x (\cot x \cos x + \sin x) = \sec x$, Q.E.D.

13. $(1 + \sin B)(1 - \sin B) = 1 - \sin^2 B = \cos^2 B$
$\therefore (1 + \sin B)(1 - \sin B) = \cos x$, Q.E.D.

15. $(\cos \phi - \sin \phi)^2 = \cos^2 \phi - 2 \cos \phi \sin \phi + \sin^2 \phi$
$= \cos^2 \phi + \sin^2 \phi - 2 \cos \phi \sin \phi$
$= 1 - 2 \cos \phi \sin \phi$
$\therefore (\cos \theta - \sin \theta)^2 = 1 - 2\cos \theta \sin \theta$, Q.E.D.

17. $(\tan n + \cot n)^2 = \tan^2 n + 2 \tan n \cot n + \cot^2 n$

$= \tan^2 n + 2 + \cot^2 n$

$= (\tan^2 n + 1) + (1 + \cot^2 n)$

$= \sec^2 n + \csc^2 n$

$\therefore (\tan n + \cot n)^2 = \sec^2 n + \csc^2 n$, Q.E.D.

19. $\dfrac{\csc^2 x - 1}{\cos x} = \dfrac{\cot^2 x}{\cos x}$

$= \cot^2 x \dfrac{1}{\cos x}$

$= \cot x \cdot \dfrac{\cos x}{\sin x} \cdot \dfrac{1}{\cos x}$

$= \cot x \cdot \dfrac{1}{\sin x}$

$= \cot x \csc x$

$\therefore \dfrac{\csc^2 x - 1}{\cos x} = \cot x \csc x$, Q.E.D.

21. $\dfrac{\sec^2 \theta - 1}{\sin \theta} = \dfrac{\tan^2 \theta}{\sin \theta}$

$= \tan^2 \theta \dfrac{1}{\sin \theta}$

$= \tan \theta \dfrac{\sin \theta}{\cos \theta} \dfrac{1}{\sin \theta}$

$= \tan \theta \dfrac{1}{\cos \theta}$

$= \tan \theta \sec \theta$

$\therefore \dfrac{\sec^2 \theta - 1}{\sin \theta} = \tan \theta \sec \theta$, Q.E.D.

23. $\dfrac{\sec A}{\sin A} - \dfrac{\sin A}{\cos A} = \dfrac{1}{\cos A \sin A} - \dfrac{\sin A}{\cos A}$

$= \dfrac{1}{\cos A \sin A} - \dfrac{\sin^2 A}{\cos A \sin A}$

$= \dfrac{1 - \sin^2 A}{\cos A \sin A}$

$= \dfrac{\cos^2 A}{\cos A \sin A}$

$= \dfrac{\cos A}{\sin A}$

$= \cot A$

$\therefore \dfrac{\sec A}{\sin A} - \dfrac{\sin A}{\cos A} = \cot A$, Q.E.D.

25. $\dfrac{1}{1 - \cos x} + \dfrac{1}{1 + \cos x}$

$= \dfrac{1 + \cos x}{(1 + \cos x)(1 - \cos x)} + \dfrac{1 - \cos x}{(1 - \cos x)(1 + \cos x)}$

$= \dfrac{1 + \cos x}{1 - \cos^2 x} + \dfrac{1 - \cos x}{1 - \cos^2 x}$

$= \dfrac{1 + \cos x + 1 - \cos x}{1 - \cos^2 x}$

$= \dfrac{2}{\sin^2 x}$

$= 2 \csc^2 x$

$\therefore \dfrac{1}{1 - \cos x} + \dfrac{1}{1 + \cos x} = 2 \csc^2 x$, Q.E.D.

27. $\sec x (\sec x - \cos x) = \sec^2 x - \sec x \cos x$

$= \sec^2 x - 1$

$= \tan^2 x$

$\therefore \sec x (\sec x - \cos x) = \tan^2 x$, Q.E.D.

29. $\sin x (\csc x - \sin x) = \sin x \csc x - \sin^2 x$

$= 1 - \sin^2 x$

$= \cos^2 x$

$\therefore \sin x (\csc x - \sin x) = \cos^2 x$, Q.E.D.

31. $\csc^2 \theta - \cos^2 \theta \csc^2 \theta = \csc^2 \theta(1 - \cos^2 \theta)$

$= \csc^2 \theta(\sin^2 \theta)$

$= 1$

$\therefore \csc^2 \theta - \cos^2 \theta \csc^2 \theta = 1$, Q.E.D.

33. $(\sec \theta + 1)(\sec \theta - 1) = \sec^2 \theta - 1 = \tan^2 \theta$

$\therefore (\sec \theta + 1)(\sec \theta - 1) = \tan^2 \theta$, Q.E.D.

35. $(2 \cos x + 3 \sin x)^2 + (3 \cos x - 2 \sin x)^2$

$= 4 \cos^2 x + 12 \cos x \sin x + 9 \sin^2 x$

$+ 9 \cos^2 x - 12 \cos x \sin x + 4 \sin^2 x$

$= 4 \cos^2 x + 4 \sin^2 x + 9 \sin^2 x + 9 \cos^2 x$

$= 4 + 9 = 13$

$\therefore (2 \cos x + 3 \sin x)^2 + (3 \cos x - 2 \sin x)^2 = 13$, Q.E.D.

37.

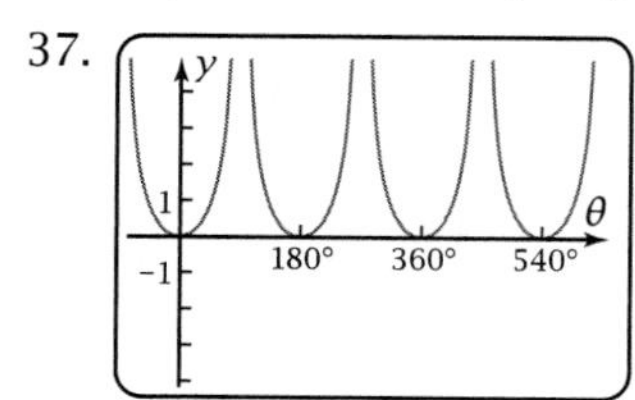

39.

x	$(2 \cos x + 3 \sin x)^2 + (3 \cos x - 2 \sin x)^2$
0	13
1	13
2	13
3	13
4	13
5	13

41. For example, $\cos\frac{\pi}{4} = \frac{\sqrt{2}}{2} = 0.7071...$, but

$1 - \sin\frac{\pi}{4} = 1 - \frac{\sqrt{2}}{2} = 0.2928....$

43. $\sec^2 A + \tan^2 A \sec^2 A = \sec^2 A(1 + \tan^2 A)$
$= \sec^2 A(\sec^2 A) = \sec^4 A$

$\therefore \sec^2 A + \tan^2 A \sec^2 A = \sec^4 A$, Q.E.D.

45. $$\frac{1}{\sin x \cos x} - \frac{\cos x}{\sin x} = \frac{1}{\sin x \cos x} - \frac{\cos x}{\sin x} \cdot \frac{\cos x}{\cos x} = \frac{1 - \cos^2 x}{\sin x \cos x} = \frac{\sin^2 x}{\sin x \cos x} = \frac{\sin x}{\cos x} = \tan x$$

$\therefore \frac{1}{\sin x \cos x} - \frac{\cos x}{\sin x} = \tan x$, Q.E.D.

47. $$\frac{1}{1 + \cos p} = \frac{1 - \cos p}{(1 + \cos p)(1 - \cos p)} = \frac{1 - \cos p}{1 - \cos^2 p} = \frac{1 - \cos p}{\sin^2 p} = \frac{1}{\sin^2 p} - \frac{\cos p}{\sin^2 p} = \csc^2 p - \frac{1}{\sin p} \cdot \frac{\cos p}{\sin p} = \csc^2 p - \csc p \cot p$$

$\therefore \frac{1}{1 + \cos p} = \csc^2 p - \csc p \cot p$, Q.E.D.

49. $$\frac{1 + \sin x}{1 - \sin x} = \frac{(1 + \sin x)(1 + \sin x)}{(1 - \sin x)(1 + \sin x)} = \frac{1 + 2\sin x + \sin^2 x}{1 - \sin^2 x} = \frac{1 + 2\sin x + \sin^2 x}{\cos^2 x} = \frac{1}{\cos^2 x} + \frac{2\sin x}{\cos^2 x} + \frac{\sin^2 x}{\cos^2 x} = \sec^2 x + 2 \cdot \frac{1}{\cos x} \cdot \frac{\sin x}{\cos x} + \tan^2 x = 2\sec^2 x + 2\sec x \tan x - 1$$

$\therefore \frac{1 + \sin x}{1 - \sin x} = 2\sec^2 x + 2\sec x \tan x - 1$, Q.E.D.

51. $$\sec^2\theta + \csc^2\theta = \frac{1}{\cos^2\theta} + \frac{1}{\sin^2\theta} = \frac{\sin^2\theta}{\cos^2\theta\sin^2\theta} + \frac{\cos^2\theta}{\cos^2\theta\sin^2\theta} = \frac{\sin^2\theta + \cos^2\theta}{\cos^2\theta\sin^2\theta} = \frac{1}{\cos^2\theta\sin^2\theta} = \sec^2\theta\csc^2\theta$$

$\therefore \sec^2\theta + \csc^2\theta = \sec^2\theta\csc^2\theta$, Q.E.D.

53. $$\frac{1 - 3\cos x - 4\cos^2 x}{\sin^2 x} = \frac{(1 - 4\cos x)(1 + \cos x)}{(1 - \cos^2 x)} = \frac{(1 - 4\cos x)(1 + \cos x)}{(1 - \cos x)(1 + \cos x)} = \frac{1 - 4\cos x}{1 - \cos x}$$

$\therefore \frac{1 - 3\cos x - 4\cos^2 x}{\sin^2 x} = 1 - \cos x$, Q.E.D.

Problem Set 4-4

1. a. $\theta = 44.4270...° + 360n°$ or $135.5729...° + 360n°$
   b. $\theta = 44.4270...°, 135.5729...°, 404.4270...°, 495.5729...°$

3. a. $x = -0.2013... + 2\pi n$ or $3.3429... + 2\pi n$
   b. $x = 3.3429..., 6.0818..., 9.6261..., 12.3650...$

5. a. $\theta = -75.9637...° + 180n°$
   b. $\theta = 104.0362...°, 284.0362...°, 464.0362...°, 644.0362...°$

7. a. $x = 1.4711... + \pi n$
   b. $x = 1.4711..., 4.6127..., 7.7543..., 10.8959...$

9. a. $\theta = \pm 78.4630...° + 360n°$
   b. $\theta = 78.4630...°, 281.5369...°, 438.4630...°, 641.5369...°$

11. 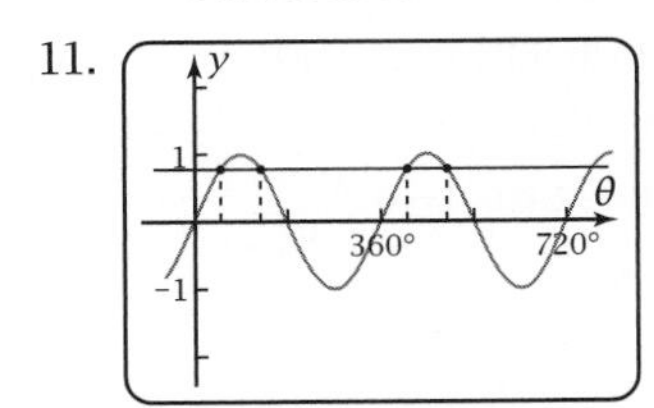

13. No angle has a cosine of 2 ($-1 \le \cos x \le 1$ for all x), but there are infinitely many angles whose tangent is 2; the opposite side can be twice the adjacent side, and $\tan(1.1071... + n\pi) = 2$.

15. $\theta = 120°, 300°, 480°, 660°$

17. $\theta = -257°, -17°, 103°, 343°$

19. $x = 0.3918..., 1.6081..., 2.3918..., 3.6081..., 4.3918..., 5.6081...$

21.

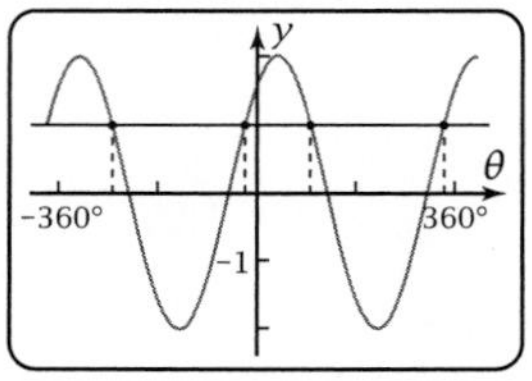

23. $\theta = 0°, 120°, 240°, 360°, 480°, 600°, 720°$

25. $\theta = 210°, 330°, 570°, 690°$

27. $\theta = 48.189...°, 90°, 270°, 311.810...°$

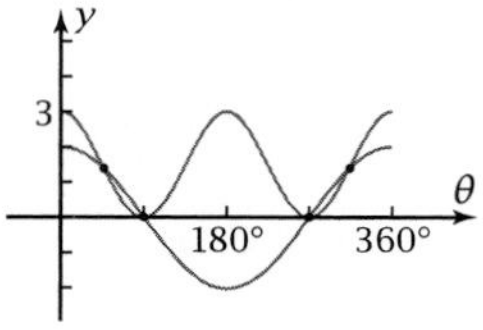

29. $x = \dfrac{3\pi}{10}, \dfrac{7\pi}{10}, \dfrac{11\pi}{10}, \dfrac{19\pi}{10}$

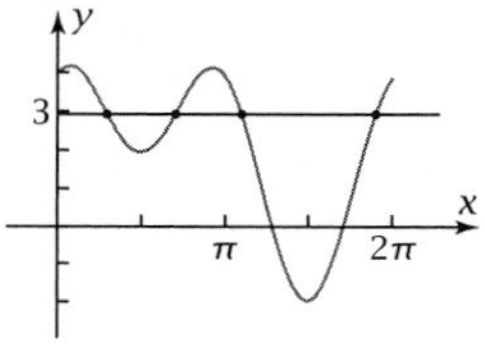

31. a. $y = 500 \tan \theta$
    b. $\theta = 5t°$
    $y = 500 \tan 5t$
    c. $t = 10.0388...$ sec, $46.0388...$ sec, $82.0388...$ sec, $118.0388...$ sec

33. a. $x = 0$, $x \approx 1.3, 2.4$
    b. $x = 0, 1.2901..., 2.3730...$
    c. x appears both algebraically (as x) and transcendentally (in the argument of tangent).

35. a. Graph should match Figure 4-4o.
    b.

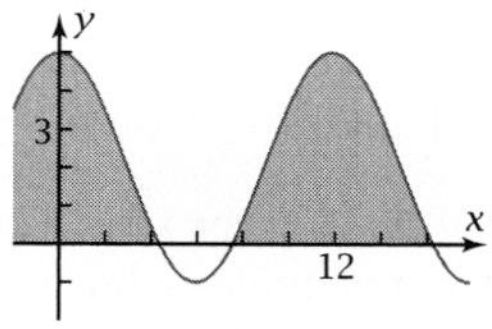

    c. $7.6063... \le x \le 16.3936...$

37. The equation is an identity for all $x \ne \dfrac{\pi}{2} \pm \pi n$. The two graphs coincide.

Problem Set 4-5

1. a.

t	x	y
−2.0	−5	−5
−1.5	−3.5	−4
−1.0	−2	−3
−0.5	−0.5	−2
0	1	−1
0.5	2.5	0
1.0	4	1
1.5	5.5	2
2.0	7	3

   b.

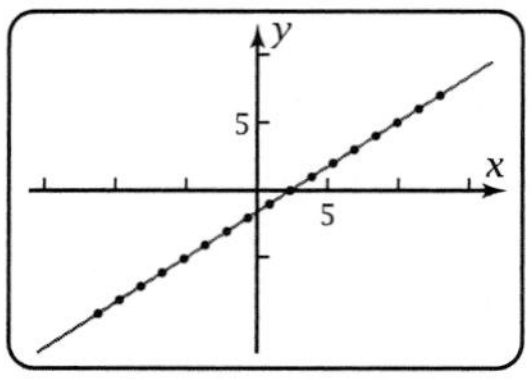

   c. The graphs agree.

3. a.

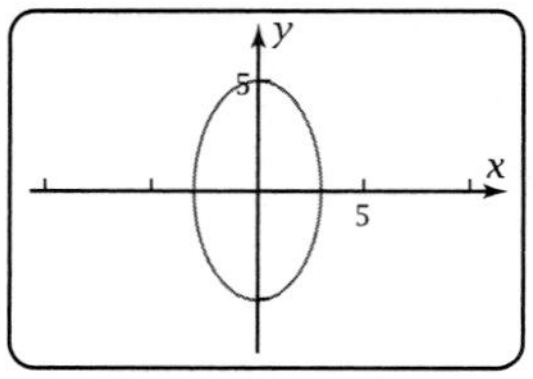

   b. $\left(\dfrac{x}{3}\right)^2 + \left(\dfrac{y}{5}\right)^2 = 1$
   c. The equation in part b is an ellipse because its x-radius is 3 and its y-radius is 5, and they are not the same.

5. a.

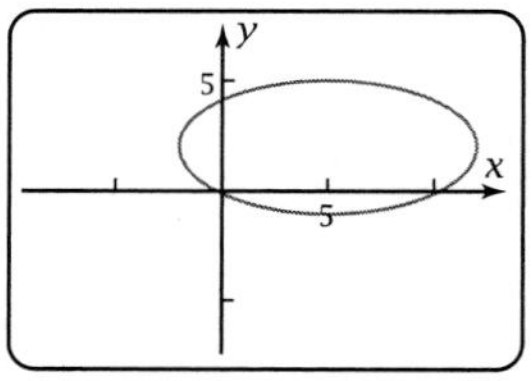

   b. $\left(\dfrac{x-5}{7}\right)^2 + \left(\dfrac{y-2}{3}\right)^2 = 1$
   c. The equation in part b is an ellipse centered at (5, 2) with x-radius 7 and y-radius 3.

7. a. $x = 6 + 5 \cos t$, $y = 9 + \sin t$
   $0° \le t \le 360°$
   b. The graphs should match the figure.

9. a. $x = 1 + 0.4 \cos 0.5t$, $y = 4 + 2 \sin 0.5t$
   $x = 14 + 0.4 \cos t$, $y = 4 + 2 \sin t$
   $180° \le t \le 540°$
   b. The graphs should match the figure.

11. a. $x = 8 + 5\cos 0.5t$, $y = 2 + \sin 0.5t$
$x = 8 + 3\cos t$, $y = 9 + 0.6\sin t$
$360° \le t \le 720°$

b. The graphs should match the figure.

13. a. $x = 5 + 0.8\cos(t + 180°)$,
$y = 6 + 4\sin(t + 180°)$
$x = 5 + 4\cos t$, $y = 6 + 4\sin t$
$-90° \le t \le 90°$

b. The graphs should match the figure.

15. a. (60 m, 75.9 m)

b. $t = 5$ sec
$y(5) = 77.5$ m

c. $t_1 = 0.8355\ldots$ sec, $t_2 = 7.3277\ldots$ sec
$x(t_1) = 16.7103\ldots$ m
$x(t_2) = 146.5549\ldots$ m

d. $x = 160 \text{ m} \Rightarrow t = 8 \text{ sec} \Rightarrow y = 6.4$ m
$y(8 \text{ sec}) > 2$ m, so the ball will go over the fence.

e. $y = 2x - \dfrac{4.9}{400}x^2$

17.

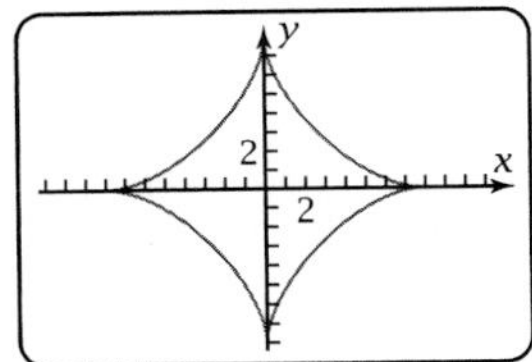

19.

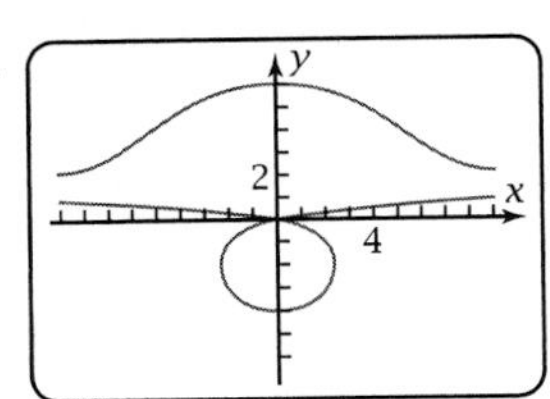

21. a.

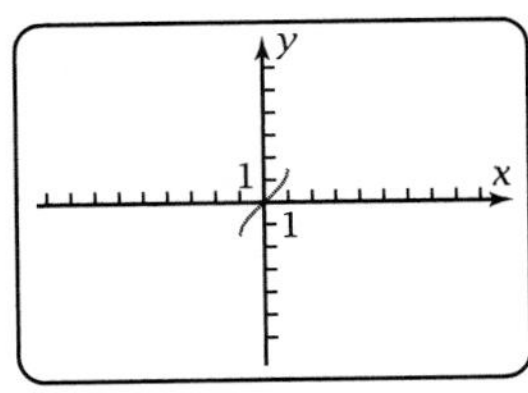

This shows only the principal branch of the relation.

b.

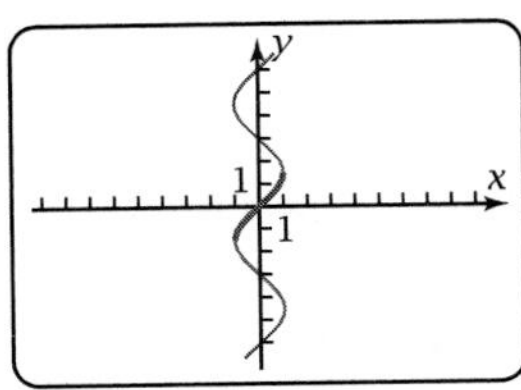

This graph shows every y for which $\sin y = x$.

c.

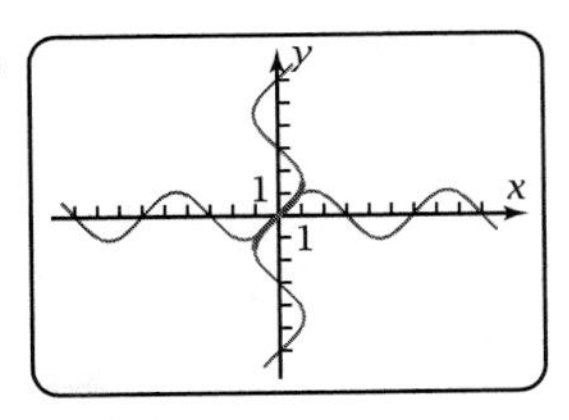

The graphs are reflections of each other across the line $y = x$.

d.

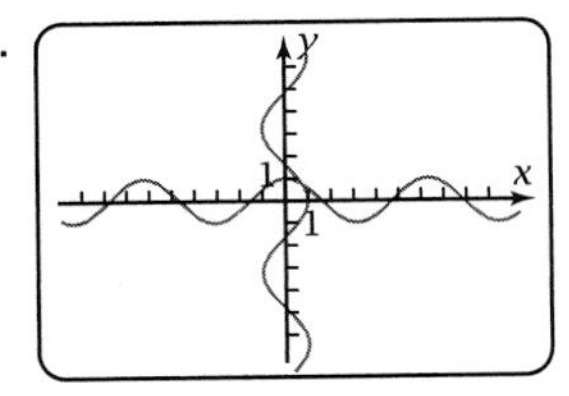

The graphs are reflections of each other across the line $y = x$.

e.

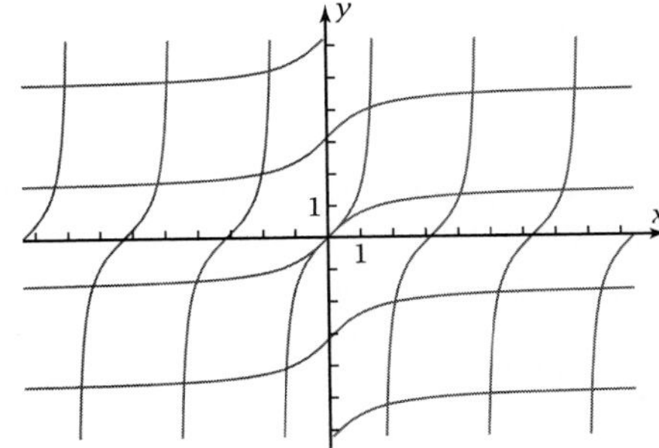

The graphs are reflections of each other across the line $y = x$.

Problem Set 4-6

1. Graphs should match the darker portion of the corresponding graphs in Figure 4-6d.

3.

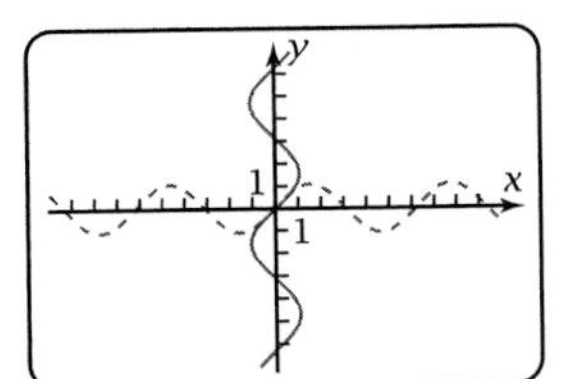

You can make a table of values and show that the (x, y) pairs of one graph are the same as the (y, x) values of the other. Exchanging x and y is equivalent to reflecting across the line $y = x$.

5. Quadrant I: $\tan\left(\cos^{-1}\dfrac{4}{5}\right) = \dfrac{3}{4}$

7. Quadrant I: $\sin\left(\tan^{-1}\dfrac{5}{12}\right) = \dfrac{5}{13}$

9. Quadrant IV: $\cos\left(\sin^{-1}\left(-\dfrac{8}{17}\right)\right) = \dfrac{15}{17}$

11. Quadrant I: $\sec\left(\cos^{-1}\frac{2}{3}\right) = \frac{3}{2}$

13. Undefined

15. 3 cannot equal a cosine of any angle. In other words, 3 is not in the domain of arccosine.

17. $\cos(\sin^{-1}x) = \sqrt{1-x^2}$, $-1 \le x \le 1$. The graphs match.

19. $\sin(\tan^{-1}x) = \dfrac{x}{\sqrt{x^2+1}}$, all real x. The graphs match.

21. $\sin(\sin^{-1}x) = x$, $-1 \le x \le 1$. The graphs match.

23. a. $y = f(x) \Leftrightarrow x = f^{-1}(y)$, so $x = f^{-1}(y) = f^{-1}[f(x)]$

    b. $y = f^{-1}(x) \Leftrightarrow x = f(y)$, so $x = f(y) = f[f^{-1}(x)]$

25. a. $y = 100 + 150\sin\left(\dfrac{\pi}{700}(x - 162.5956...)\right)$

    b. The tunnel is about 1025.2 m long. The bridge is 374.8 m long.

    c. Now the tunnel is about 950.7 m long, and the bridge is 449.3 m long.

Problem Set 4-7

R1. a. u and v are the legs of a right triangle with hypotenuse 1.

b. $\cos\theta = \dfrac{\text{horizontal coordinate}}{\text{radius}} = \dfrac{u}{1} = u$

$\sin\theta = \dfrac{\text{vertical coordinate}}{\text{radius}} = \dfrac{v}{1} = v$

c. $u^2 + v^2 = 1$ and $u = \cos\theta$ and $v = \sin\theta$
$\Rightarrow (\cos\theta)^2 + (\sin\theta)^2 = 1$

d. For $\theta = 30°$, $\sin\theta = \dfrac{1}{2}$ and $\cos\theta = \dfrac{\sqrt{3}}{2}$

$\Rightarrow \sin^2\theta = \dfrac{1}{4}$ and $\cos^2\theta = \dfrac{3}{4}$

$\Rightarrow \sin^2\theta + \cos^2\theta = \dfrac{1}{4} + \dfrac{3}{4} = 1$

e.

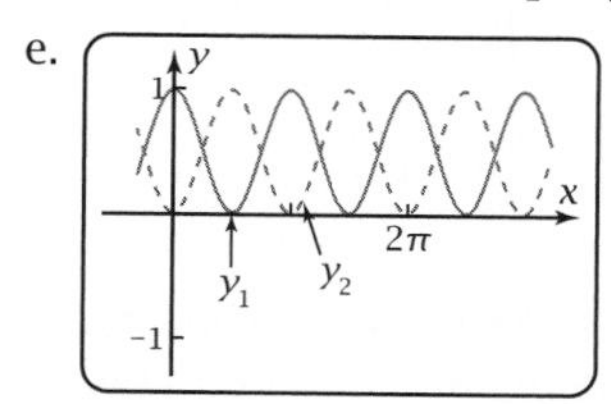

The graphs are symmetrical across $y = \frac{1}{2}$; one graph is above $\frac{1}{2}$ by the same amount as the other is under it.

R2. a. $\tan x = \dfrac{\sin x}{\cos x}$, $\cot x = \dfrac{\cos x}{\sin x}$

b. $\tan x = \dfrac{\sec x}{\csc x}$, $\cot x = \dfrac{\csc x}{\sec x}$

c. $\sin x \cdot \csc x = 1$, $\cos x \cdot \sec x = 1$, $\tan x \cdot \cot x = 1$

d.

x	$\cos^2 x$	$\sin^2 x$	$\cos^2 x + \sin^2 x$
0	1	0	1
1	0.2919...	0.7080...	1
2	0.1731...	0.8268...	1
3	0.9800...	0.0199...	1
4	0.4272...	0.5727...	1
5	0.0804...	0.9195...	1

e. i. $\sin^2 x = 1 - \cos^2 x$

ii. $\tan^2 x = \sec^2 x - 1$

iii. $\csc^2 x = \cot^2 x + 1$

f.

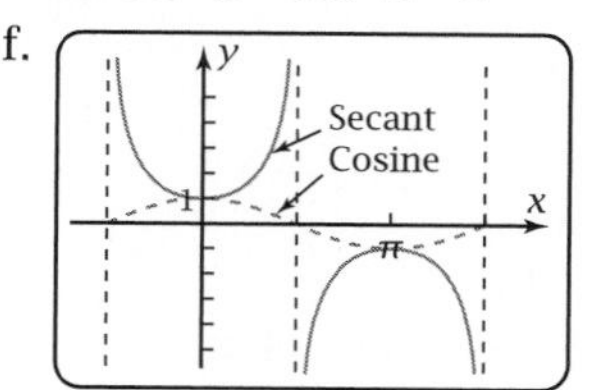

R3. a. $\tan A \sin A + \cos A = \dfrac{\sin A}{\cos A}\sin A + \dfrac{\cos A}{\cos A}\cos A$

$= \dfrac{\sin^2 A + \cos^2 A}{\cos A} = \dfrac{1}{\cos A}$

$= \sec A$

$\therefore \tan A \sin A + \cos A = \sec A$ for $A \ne \dfrac{\pi}{2} + n\pi$, Q.E.D.

b. $(\cos B + \sin B)^2 = \cos^2 B + 2\cos B \sin B + \sin^2 B$

$= (\cos^2 B + \sin^2 B) + 2\cos B \sin B$

$= 1 + 2\cos B \sin B$

$\therefore (\cos B + \sin B)^2 = 1 + 2\cos B \sin B$ for all real B, Q.E.D.

c. $\dfrac{1}{1+\sin C} + \dfrac{1}{1-\sin C}$

$= \dfrac{1-\sin C}{(1+\sin C)(1-\sin C)} + \dfrac{1+\sin C}{(1-\sin C)(1+\sin C)}$

$= \dfrac{1-\sin C}{1-\sin^2 C} + \dfrac{1+\sin C}{1-\sin^2 C} = \dfrac{1-\sin C}{\cos^2 C} + \dfrac{1+\sin C}{\cos^2 C}$

$= \dfrac{2}{\cos^2 C} = 2\sec^2 C$

$\therefore \dfrac{1}{1+\sin C} + \dfrac{1}{1-\sin C} = 2\sec^2 C$ for $C \ne \dfrac{\pi}{2} + n\pi$, Q.E.D.

d. $\csc D(\csc D - \sin D)$

$= \csc^2 D - \csc D \sin D = \csc^2 D - 1$

$= \cot^2 D$

$\therefore \csc D(\csc D - \sin D) = \cot^2 D$ for $D \ne n\pi$, Q.E.D.

e. $(3 \cos E + 5 \sin E)^2 + (5 \cos E - 3 \sin E)^2$

$= 9 \cos^2 E + 30 \cos E \sin E + 25 \sin^2 E$
$\quad + 25 \cos^2 E - 30 \cos E \sin E + 9 \sin^2 E$

$= 9 \cos^2 E + 9 \sin^2 E + 25 \sin^2 E + 25 \cos^2 E$

$= 9 + 25 = 34$

$\therefore (3 \cos E + 5 \sin E)^2 + (5 \cos E - 3 \sin E)^2 = 34,$

Q.E.D.

f.

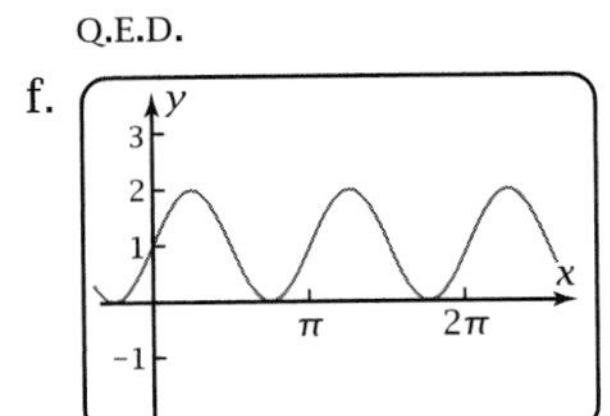

g.

x	$(3 \cos E + 5 \sin E)^2 + (5 \cos E - 3 \sin E)^2$
0	34
1	34
2	34
3	34
4	34
5	34

R4. a. $\theta = 17.4576...° + 360n°$ or $162.5423...° + 360n°$

b. $x = 0.275, 0.775, 1.275, 1.775$

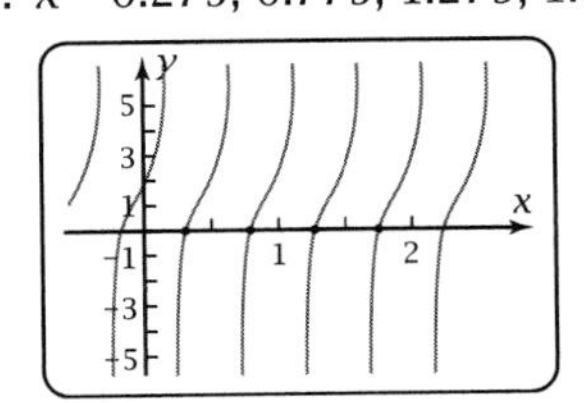

c. $\theta = 60°, 240°, 300°, 420°$

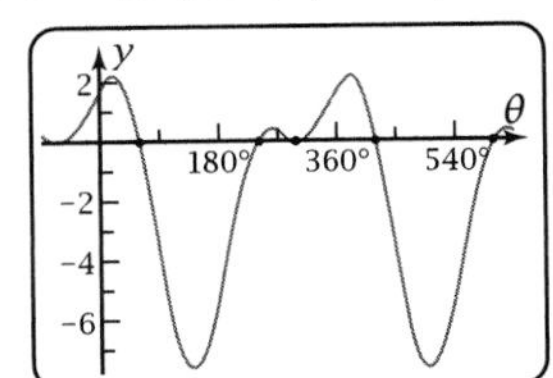

R5. a.

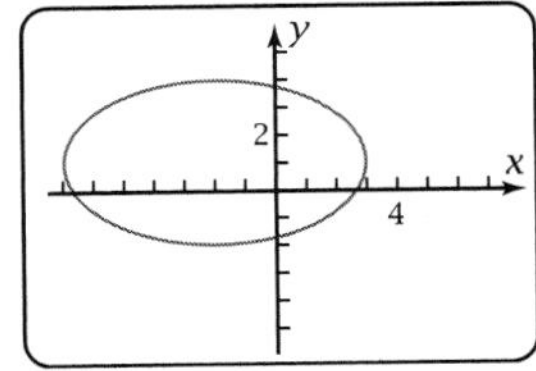

b. $\left(\frac{x+2}{5}\right)^2 + \left(\frac{y-1}{3}\right)^2 = 1$

c. The equation in part b is an ellipse because the x- and y-radii are not equal. The ellipse is centered at $(-2, 1)$ with x-radius 5 and y-radius 3.

d. $x = 7 + 5 \cos t$, $y = 2 + 0.8 \sin t$, $180° \le t \le 360°$

R6. a. Use $x = \cos t$, $y = t$, $-7 \le t \le 7$ (or whatever y-limits you use for your graphing window).

b.

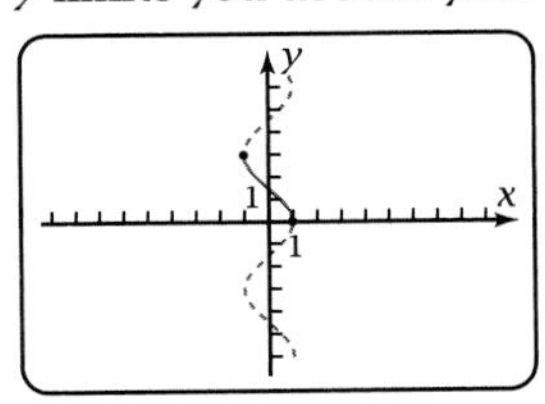

c. It is the inverse.

d. $\sin(\tan^{-1} 2) = \frac{2}{\sqrt{5}}$

e. $y = \tan(\cos^{-1} x) = \frac{\sqrt{1 - x^2}}{x}$

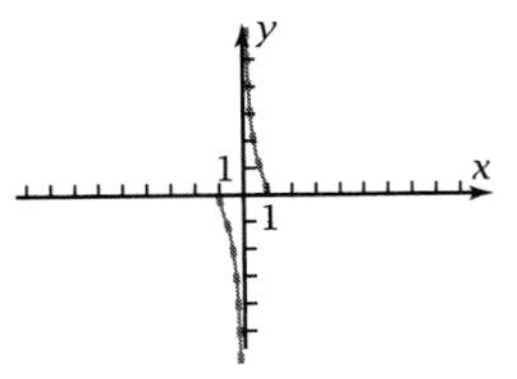

f. In Quadrants I and II, where $\cos^{-1} x$ is defined, you have, $\cos^{-1} x = \theta \Leftrightarrow \cos \theta = x$, so $\cos(\cos^{-1} x) = \cos \theta = x$.

g.

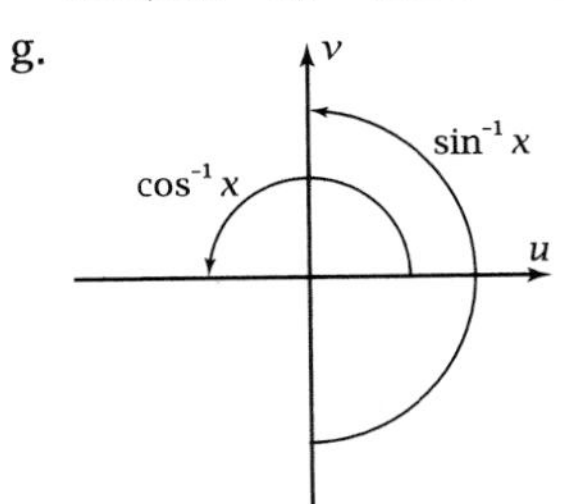

h. Arccos x means the angle (or arc) whose cosine is x.

CHAPTER 5

Problem Set 5-1

1. The graphs agree.

3. Amplitude = 5
   Phase displacement = 53.1301...°
   Answers are reasonably close to estimates.

5. $y_3(\theta) = y_4(\theta)$ for all θ

θ	$y_3 = y_4$
0°	3
60°	4.9641...
120°	1.9641...
180°	−3
240°	−4.9641...
300°	−1.9641...
360°	3

7. Let $\theta = \pi$ and $D = \frac{\pi}{2}$. Then $\cos(\theta - D) = \cos\frac{\pi}{2} = 0$, but $\cos\theta - \cos D = \cos\pi - \cos\frac{\pi}{2} = -1$.

Problem Set 5-2

1. $y = 13\cos(\theta - 22.6198...°)$

3. $y = 25\cos(\theta - 106.9577...°)$

5. $y = \sqrt{185}\cos(\theta + 126.0273...°)$

7. $y = 6\sqrt{2}\cos(\theta + 45°)$

9. $y = 2\cos(\theta - 30°)$

11. $y = 5\cos(x - 4.0688...)$

13. 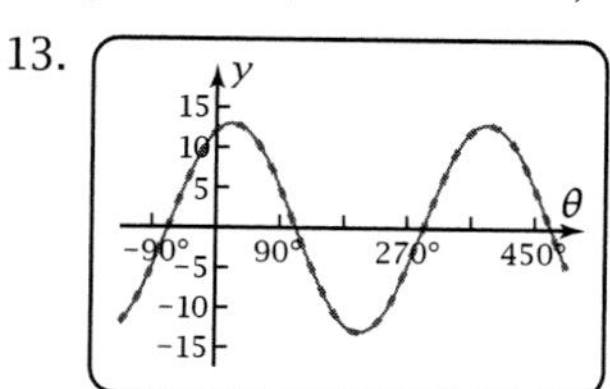

15. $y = \sqrt{2}\cos\left(3x - \frac{\pi}{4}\right)$
Horizontal dilation by $\frac{1}{3}$

17. Consider $A = \pi$ and $B = \frac{\pi}{2}$.
Then $\cos(A - B) = \cos\frac{\pi}{2} = 0$,
but $\cos A - \cos B = \cos\pi - \cos\frac{\pi}{2} = -1$.

19. $y = 5\sqrt{3}\cos\theta + 5\sin\theta$

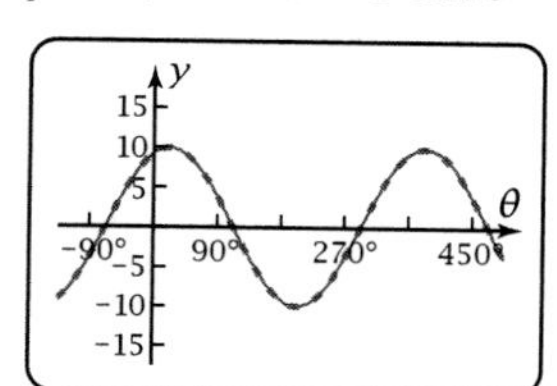

21. $y = -2.5 \cdot \sqrt{3}\cos 3\theta + 2.5\sin 3\theta$

23. $\theta = \cos^{-1}\dfrac{3}{\sqrt{74}} + \cos^{-1}\dfrac{5}{\sqrt{74}} = 124.0518°$

25. $x = \cos^{-1}\dfrac{5}{\sqrt{73}} + \left(2\pi - \cos^{-1}\dfrac{-8}{\sqrt{73}}\right) = 4.4460...$

27. $\cos 2\theta = \cos(5\theta - 3\theta)$
$= \cos 5\theta\cos 3\theta + \sin 5\theta\sin 3\theta$;
$0.3 = \cos 5\theta\cos 3\theta + \sin 5\theta\sin 3\theta = \cos 2\theta$
$\Rightarrow 2\theta = \pm\cos^{-1} 0.3 + 360n° \Rightarrow \theta = 36.2711...°$, $143.7288...°$, $216.2711...°$, $323.7288...°$

29. a. $\cos 70° = 0.3420... = \sin 20°$
b. $\cos(90° - \theta) = \cos 90°\cos\theta + \sin 90°\sin\theta$
$= 0 \cdot \cos\theta + 1 \cdot \sin\theta = \sin\theta$
c. *Co-* means complementary.

31. See the derivation in the text.

Problem Set 5-3

1. Let $A = B = 90°$. Then $\sin(A + B) = \sin 180° = 0$
$\neq 2 = \sin 90° + \sin 90° = \sin A + \sin B$

3. $\tan(60° - 30°) = \dfrac{1}{\sqrt{3}}$ and $\dfrac{\tan 60° - \tan 30°}{1 + \tan 60°\tan 30°} = \dfrac{1}{\sqrt{3}}$

5.

x	$\cos(-x)$	$\cos x$
0°	1	1
30°	0.8660...	0.8660...
60°	0.5	0.5
90°	0	0
120°	−0.5	−0.5
150°	−0.8660...	−0.8660...
180°	−1	−1

7. The graphs are the same.

9. a. The v-coordinates are opposite for θ and $-\theta$.
b. The u-coordinates are the same for θ and $-\theta$.
c. The slopes of the rays for θ and $-\theta$ are opposites.
d. For any function $f(x)$, if $f(-x) = f(x)$, then $\dfrac{1}{f(-x)} = \dfrac{1}{f(x)}$ when defined. If $f(-x) = -f(x)$, then $\dfrac{1}{f(-x)} = -\dfrac{1}{f(x)}$ when defined.

11. $\cos(\theta - 90°) = \cos\theta\cos 90° + \sin\theta\sin 90°$
$= \cos\theta \cdot 0 + \sin\theta \cdot 1 = \sin\theta$

13. $\cos\left(x - \frac{\pi}{2}\right) = \cos x\cos\frac{\pi}{2} + \sin x\sin\frac{\pi}{2}$
$= \cos x \cdot 0 + \sin x \cdot 1 = \sin x$

15. $\sin(\theta + 60°) - \cos(\theta + 30°)$
$= (\sin\theta\cos 60° + \cos\theta\sin 60°)$
$- (\cos\theta\cos 30° - \sin\theta\sin 30°)$
$= \dfrac{1}{2}\sin\theta + \dfrac{\sqrt{3}}{2}\cos\theta - \dfrac{\sqrt{3}}{2}\cos\theta + \dfrac{1}{2}\sin\theta = \sin\theta$

17. $\sqrt{2}\cos\left(x - \dfrac{\pi}{4}\right) = \sqrt{2}\cos x\cos\dfrac{\pi}{4} + \sqrt{2}\sin x\sin\dfrac{\pi}{4}$
$= \sqrt{2}\cos x \cdot \dfrac{\sqrt{2}}{2} + \sqrt{2}\sin x \cdot \dfrac{\sqrt{2}}{2}$
$= \cos x + \sin x$

19. $\sin 3x\cos 4x + \cos 3x\sin 4x = \sin(3x + 4x) = \sin 7x$

21. a. $x = 2\pi - 0.6 \pm 0.451...$
b. $x = 5.232...$ or $6.134...$

23. a. $\theta = 22.5° + 180n°$ or $67.5° + 180n°$
b. $\theta = 22.5°, 67.5°, 202.5°$ or $247.5°$

25. a. $x = \dfrac{\pi}{3} + \pi n$
b. $x = \dfrac{\pi}{3}$ or $\dfrac{4\pi}{3}$

27. $\cos(A - B) = \dfrac{297}{425}$

29. $\tan(A - B) = \dfrac{304}{297}$

31. $\sin(A + B) = \dfrac{416}{425}$

33. $\sin 15° = \sin(45° - 30°)$
$= \sin 45° \cos 30° - \cos 45° \sin 30°$
$= \dfrac{\sqrt{2}}{2} \cdot \dfrac{\sqrt{3}}{2} - \dfrac{\sqrt{2}}{2} \cdot \dfrac{1}{2}$
$= \dfrac{\sqrt{6} - \sqrt{2}}{4}$

35. $\sin 75° = \cos(90° - 75°) = \cos 15° = \dfrac{\sqrt{6} + \sqrt{2}}{4}$

37. $\tan 15° = \dfrac{\sin 15°}{\cos 15°} = \dfrac{\left(\dfrac{\sqrt{6} - \sqrt{2}}{4}\right)}{\left(\dfrac{\sqrt{6} + \sqrt{2}}{4}\right)} = \dfrac{\sqrt{6} - \sqrt{2}}{\sqrt{6} + \sqrt{2}}$

$= \dfrac{\sqrt{6} - \sqrt{2}}{\sqrt{6} + \sqrt{2}} \cdot \dfrac{\sqrt{6} - \sqrt{2}}{\sqrt{6} - \sqrt{2}} = \dfrac{8 - 4\sqrt{3}}{4} = 2 - \sqrt{3}$

39. a. $\theta = 90° - \sin^{-1} x \Rightarrow \sin^{-1} x = 90° - \theta$
$\Rightarrow x = \sin(90° - \theta) = \cos\theta$

b. $-90° \le \sin^{-1} x \le 90° \Rightarrow -90° \le 90° - \theta \le 90°$
$\Rightarrow -180° \le -\theta \le 0° \Rightarrow 0° \le \theta \le 180°$

c. θ is the unique value of arccos x such that $0 \le \theta \le 180°$.

41. $\cos(A + B + C)$
$= \cos A \cos B \cos C - \cos A \sin B \sin C$
$- \sin A \cos B \sin C - \sin A \sin B \cos C$

Problem Set 5-4

1. a. y_1 is the tall single arch and trough, y_2 the short wiggly line.

b, c.

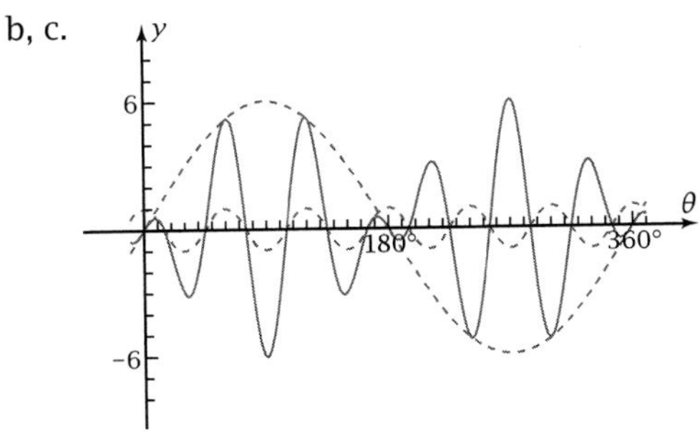

d. Sinusoid with variable amplitude $6 \sin\theta$

3. $y = 3 \cos\theta + 2 \sin 13\theta$

5. $y = 5 \sin\theta \cos 11\theta$

7. $y = 2 \cos \frac{\pi}{3}x + 4 \sin 4\pi x$

9. $y = 4 \sin \frac{\pi}{6}x \sin 3\pi x$

11. $y = 2 \sin 2x \sin 28x$

13. a. $y = 3 \cos 120\pi x + \cos 800\pi x$

b. $\frac{1}{60}$ sec; $\frac{1}{400}$ sec

c. 60 cycles/sec; 400 cycles/sec

d. Yes

Problem Set 5-5

1. $2 \sin 41° \cos 24° = \sin 65° + \sin 17°$

3. $2 \cos 53° \cos 49° = \cos 102° + \cos 4°$

5. $2 \cos 3.8 \sin 4.1 = \sin 7.9 + \sin 0.3$

7. $2 \sin 3x \sin 7.2 = \cos(3x - 7.2) - \cos(3x + 7.2)$

9. $\cos 46° + \cos 12° = 2 \cos 29° \cos 17°$

11. $\sin 2 + \sin 6 = 2 \sin 4 \cos 4$

13. $\cos 2.4 - \cos 4.4 = 2 \sin 3.4 \sin 1$

15. $\sin 3x - \sin 8x = -2 \cos 5.5x \sin 2.5x$

17. $y = 2 \cos\theta \cos 9\theta = \cos 10\theta + \cos 8\theta$

19. $y = \cos x + \cos 15x = 2 \cos 8x \cos 7x$

21. $\sin 3x - \sin x = 2 \cos 2x \sin x = 0 \Rightarrow \sin x = 0$ or
$\cos 2x = 0 \Rightarrow x = 0, \pi, 2\pi, \frac{\pi}{4}, \frac{3\pi}{4}, \frac{5\pi}{4}, \frac{7\pi}{4}$

23. $\cos 5\theta + \cos 3\theta = 2 \cos 4\theta \cos\theta = 0 \Rightarrow \cos 4\theta = 0$ or $\cos\theta = 0 \Rightarrow \theta = 22.5°, 67.5°, 112.5°, 157.5°, 202.5°, 247.5°, 292.5°, 337.5°, 90°, 270°$

25. $\cos x - \cos 5x = 2 \sin 3x \sin 2x$
$= 2 \sin 3x (2 \sin x \cos x)$
$= 4 \sin 3x \sin x \cos x$

27. $\cos x + \cos 2x + \cos 3x = \cos 2x + (\cos x + \cos 3x)$
$= \cos 2x + (2 \cos 2x \cos x)$
$= (1 + 2 \cos x)\cos 2x$

29. $\cos(x + y) \cos(x - y)$
$= \frac{1}{2} \cos[(x + y) + (x - y)] + \frac{1}{2} \cos[(x + y) - (x - y)]$
$= \frac{1}{2} \cos 2x + \frac{1}{2} \cos 2y$
$= \frac{1}{2} (\cos 2x + 1) + \frac{1}{2} (\cos 2y - 1)$
$= \frac{1}{2}[\cos(x + x) + \cos(x - x)]$
$+ \frac{1}{2}[\cos(y + y) - \cos(y - y)]$
$= \frac{1}{2}(2 \cos x \cos x) + \frac{1}{2}(-2 \sin y \sin y)$
$= \cos^2 x - \sin^2 y$

31. a. $f(t) = \cos 442\pi t + \cos 438\pi t$

b. $f(t) = 2 \cos 440\pi t \cos 2\pi t$

c. Combined note will sound like A220 getting louder and softer twice per second.

33. a. Carrier wave: $y = 200 \cos(1200 \cdot 2\pi t)$
Sound wave: $y = \cos(40 \cdot 2\pi t)$

b. $200 \cos(1200 \cdot 2\pi t) \cos(40 \cdot 2\pi t)$
$= 100 \cos(1240 \cdot 2\pi t) + 100 \cos(1160 \cdot 2\pi t)$

c. The graphs are identical.

d. Frequency = 1200 kc/sec

e. Hertz (Hz)

Problem Set 5-6

1. $\cos 2x$ has period π and amplitude 1, but $2 \cos x$ has period 2π and amplitude 2.

3.

x	$\sin 2x$	$2 \sin x \cos x$
0	0	0
0.5	0.8414	0.8414
1	0.9093	0.9093
1.5	0.1411	0.1411
2	−0.7568	−0.7568
2.5	−0.9589	−0.9589
3	−0.2794	−0.2794

5.

x	$\cos 2x$	$2 \cos^2 x - 1$
0	1	1
0.5	0.5403	0.5403
1	−0.4161	−0.4161
1.5	−0.9900	−0.9900
2	−0.6536	−0.6536
2.5	0.28366	0.28366
3	0.96017	0.96017

7.

x	$\sin 0.5x$	$\sqrt{0.5(1 - \cos x)}$
0	0	0
30	0.25882	0.25882
60	0.5	0.5
90	0.70711	0.70711
120	0.86603	0.86603
150	0.96593	0.96593
180	1	1

9.

x	$\cos 0.5x$	$-\sqrt{0.5(1 + \cos x)}$
360	−1	−1
390	−0.9659	−0.9659
420	−0.866	−0.866
450	−0.7071	−0.7071
480	−0.5	−0.5
510	−0.2588	−0.2588
540	0	0

11. a.

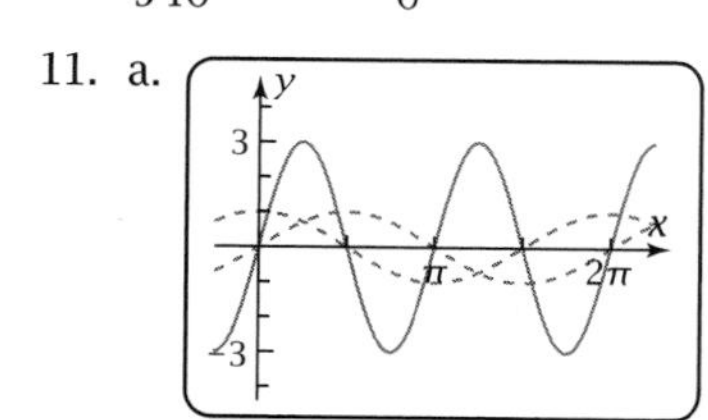

b. $y = 6 \sin x \cos x = 3 \sin 2x$

c. The double argument property is used.

13. a.

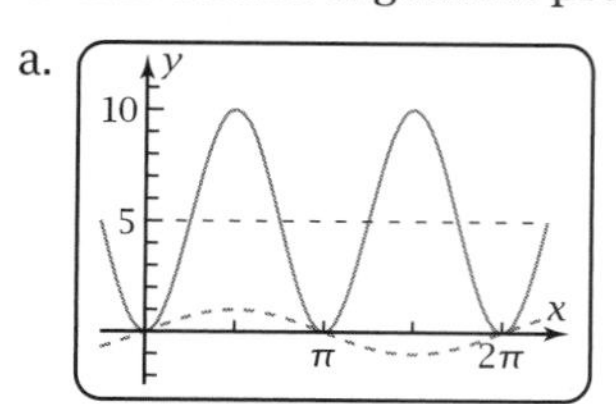

b. $y = 10 \sin^2 x = 5 - 5 \cos 2x$

c. The double argument property is used.

15. a.

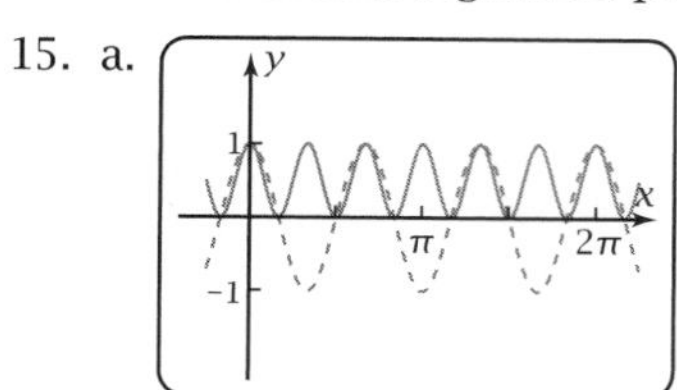

b. $y = \cos^2 3x = \frac{1}{2} + \frac{1}{2} \cos 6x$

c. The double angle formula is used.

17. Not a sinusoid: The graph is not symmetrical around any horizontal central axis.

19. a. The dotted curve represents y_1, and the solid curve represents y_2.

b. The + should be used on $[720n° - 180°, 720n° + 180°]$ and the − on $[720n° + 180°, 720n° + 540°]$.

c. $|y_1(\theta)| = y_2(\theta)$

d. By definition, $\sqrt{x^2}$ means the *positive* value, so $\sqrt{x^2} = \begin{cases} x, & x \geq 0 \\ -x, & x < 0 \end{cases} = |x|$. The derivation of the half argument properties take the square root of the squares of sine and cosine.

21. a. $\sin 2A = \frac{24}{25}$; $\cos \frac{1}{2}A = \sqrt{.8}$

b. $A = 53.1301...°$

23. a. $\sin 2A = \frac{24}{25}$; $\cos \frac{1}{2}A = -\sqrt{0.2}$

b. $A = 233.1301...°$

25. a. $\sin 2A = -\frac{24}{25}$; $\cos \frac{1}{2}A = \sqrt{0.8}$

b. $A = 666.8698...°$

27. $\sin 2x = \sin(x + x) = \sin x \cos x + \cos x \sin x$
$= 2 \sin x \cos x$

29. $\sin x \cos x = \frac{1}{2} \sin 2x$

31. a. The graphs are the same for both expressions.

b. $\tan 2A = \tan(A + A) = \dfrac{\tan A + \tan A}{1 - \tan A \tan A}$

$$= \frac{2 \tan A}{1 - \tan^2 A}$$

$$\therefore \tan 2A = \frac{2 \tan A}{1 - \tan^2 A}, \text{ Q.E.D.}$$

c. $\tan 2A = \dfrac{\sin 2A}{\cos 2A} = \dfrac{2\sin A\cos A}{\cos^2 A - \sin^2 A} \cdot \dfrac{\left(\frac{1}{\cos^2 A}\right)}{\left(\frac{1}{\cos^2 A}\right)}$

$= \dfrac{\left(\frac{2\sin A}{\cos A}\right)}{1 - \frac{\sin^2 A}{\cos^2 A}} = \dfrac{2\tan A}{1 - \tan^2 A}$

$\therefore \tan 2A = \dfrac{2\tan A}{1 - \tan^2 A}$, Q.E.D.

33. $x = \dfrac{\pi}{6}, \dfrac{\pi}{3}, \dfrac{7\pi}{6}, \dfrac{4\pi}{3}$

35. $\theta = 45°, 135°, 225°, 315°$

37. $x = \dfrac{\pi}{3}, \dfrac{5\pi}{3}, \dfrac{7\pi}{3}, \dfrac{11\pi}{3}$

39. $\sin 2x = 2\sin x\cos x \cdot \dfrac{\sec^2 x}{\sec^2 x} = \dfrac{2\sin x\sec x}{\sec^2 x}$

$= \dfrac{2\frac{\sin x}{\cos x}}{\sec^2 x} = \dfrac{2\tan x}{1 + \tan^2 x}$

41. $\sin 2\phi = 2\sin\phi\cos\phi \cdot \dfrac{\sin\phi}{\sin\phi} = 2\sin^2\phi\,\dfrac{\cos\phi}{\sin\phi}$

$= 2\cot\phi\sin^2\phi$

43. $\sin^2 5\theta = \frac{1}{2}[1 - \cos(2 \cdot 5\theta)] = \frac{1}{2}(1 - \cos 10\theta)$

Problem Set 5-7

R1. a. Amplitude = 13

Phase displacement = $\cos^{-1}\frac{5}{13} = 67.3801...°$

$y = 13\cos(\theta - 67.3801...°)$

b. The graphs agree.

R2. a. y_1 is the solid graph, y_2 the dashed one. You can tell that cosine does not distribute because the graphs are different.

b. $\cos(\theta - 60°) = \dfrac{1}{2}\cos\theta + \dfrac{\sqrt{3}}{2}\sin\theta$

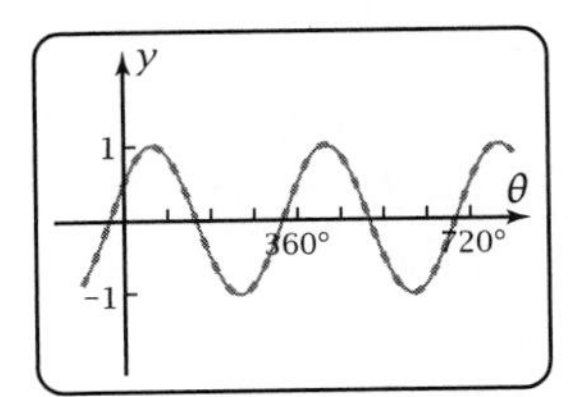

c. $8\cos\theta + 15\sin\theta = 17\cos(\theta - 61.9275...)$

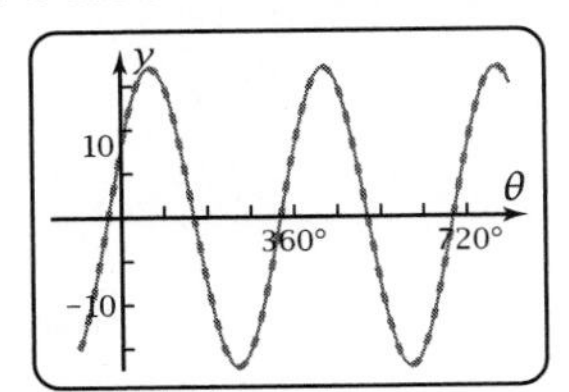

The two graphs are the same.

d. $-9\cos x + 7\sin x = \sqrt{130}\cos(x - 2.4805...)$

x	$-9\cos x + 7\sin x$	$\sqrt{130}\cos(x - 2.4805...)$
0	−9	−9
$\frac{\pi}{4}$	−1.4142	−1.4142
$\frac{\pi}{2}$	7	7
$\frac{3\pi}{4}$	11.31	11.31
π	9	9
$\frac{5\pi}{4}$	1.4142	1.4142
$\frac{3\pi}{2}$	−7	−7

e. $x = 1.8027..., 5.7674...$

These are the x-coordinates of the intersection of the two graphs.

R3. a. $\sin(-x) = -\sin x$, $\cos(-x) = \cos x$, $\tan(-x) = -\tan x$

b. $\sin(x + y) = \sin x\cos y + \cos x\sin y$

c. $\cos(x + y) = \cos x\cos y - \sin x\sin y$

d. $\tan(x - y) = \dfrac{\tan x - \tan y}{1 + \tan x\tan y}$

e. $\cos(90° - \theta) = \sin\theta$

f. $\cot\left(\frac{\pi}{2} - x\right) = \tan x$, $x \neq \frac{\pi}{2} + n\pi$

g. $\csc\left(\frac{\pi}{2} - x\right) = \sec x$, $x \neq \frac{\pi}{2} + n\pi$

h. The graph is symmetrical about the origin.

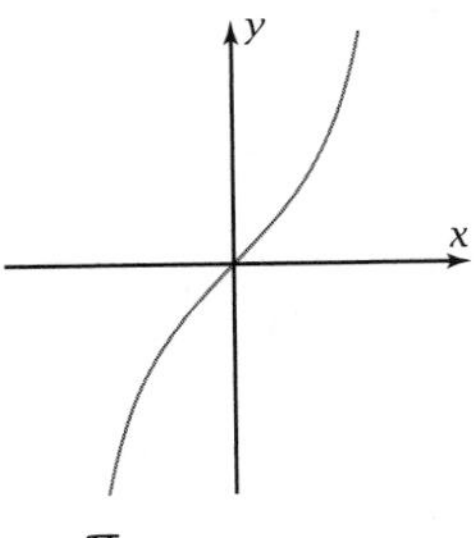

$x \neq \dfrac{\pi}{2} + n\pi$

i. $x = 0.5796..., 2.5619..., 3.7212..., 5.7035...$

These are the x-coordinates of the intersection of the two graphs.

R4. a.

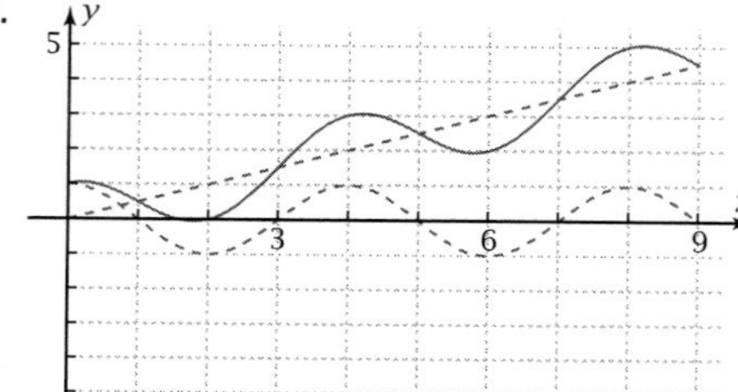

b.

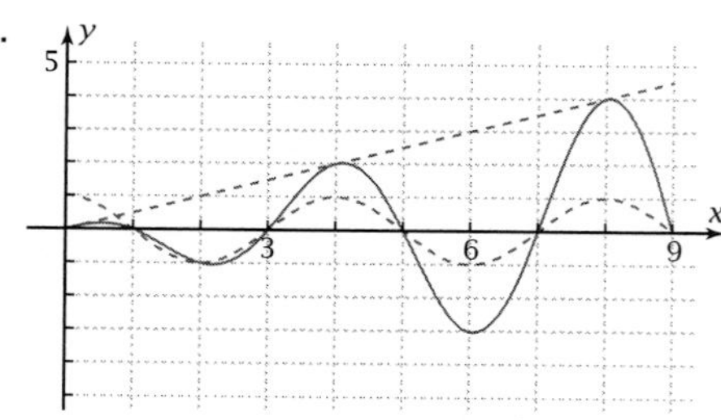

c. First graph: $y = \frac{x}{2} + \cos\frac{\pi x}{2}$

Second graph: $y = \frac{x}{2}\cos\frac{\pi x}{2}$

d. Sinusoid with variable sinusoidal axis $y_0 = \frac{x}{2}$; sinusoid with variable amplitude $A = \frac{x}{2}$

e. $y = 3\cos 3\theta \sin 36\theta$

f. $y = 3\cos 25x + 2\sin 2x$

R5. a. $\cos 13° \cos 28° = \frac{1}{2}\cos 41° + \frac{1}{2}\cos 15°$

b. $\sin 5 - \sin 8 = -2\cos\frac{13}{2}\sin\frac{3}{2}$

c. $4\sin x \sin 11x = -2\cos 12x + 2\cos 10x$

d. $\theta = 0°, 90°, 180°, 270°, 360°$

e. $$\cos\left(x + \frac{\pi}{3}\right)\cos\left(x - \frac{\pi}{3}\right)$$
$$= \frac{1}{2}\cos\left[\left(x + \frac{\pi}{3}\right) + \left(x - \frac{\pi}{3}\right)\right]$$
$$+ \frac{1}{2}\cos\left[\left(x + \frac{\pi}{3}\right) - \left(x - \frac{\pi}{3}\right)\right]$$
$$= \frac{1}{2}\cos 2x + \frac{1}{2}\cos\frac{2\pi}{3}$$
$$= \frac{1}{2}(2\cos^2 x - 1) + \frac{1}{2}\left(-\frac{1}{2}\right) = \cos^2 x - \frac{1}{2} - \frac{1}{4}$$
$$= \cos^2 x - \frac{3}{4}$$

R6. a. $y = \frac{1}{2} - \frac{1}{2}\cos 6x$

b. For $x = 0$, $\cos 2x = 1$ and $2\cos x = 2$

c. $\cos 2x = 2\cos^2 x - 1$

d. $\tan 2x = \dfrac{2\tan x}{1 - \tan^2 x}$, $x \neq \frac{\pi}{4} + \frac{n\pi}{2}$, $x \neq \frac{\pi}{2} + n\pi$

e. $\cos 2A = \frac{527}{625} = -0.8432$; $\cos\frac{1}{2}A = \frac{4}{5} = 0.8$

$A = 73.7397...°$

f. $$\sin 2A = 2\sin A\cos A \cdot \frac{\cos A}{\cos A}$$
$$= 2\frac{\sin A}{\cos A} \cdot \cos^2 A = 2\tan A\cos^2 A$$

$\sin\left(2 \cdot \frac{\pi}{2}\right) = \sin\pi = 0$, but $2\tan\frac{\pi}{2}\cos^2\frac{\pi}{2}$ is undefined because $\tan\frac{\pi}{2}$ is undefined. This discrepancy was introduced when you multiplied by $\frac{\cos A}{\cos A}$, which equals $\frac{0}{0}$ when $A = \frac{\pi}{2}$.

g. $\theta = 60°, 300°, 420°, 660°$. This is an algebraic solution. The graphical method gives the same solution.

CHAPTER 6

Problem Set 6-1

1. All measurements seem correct.

3.

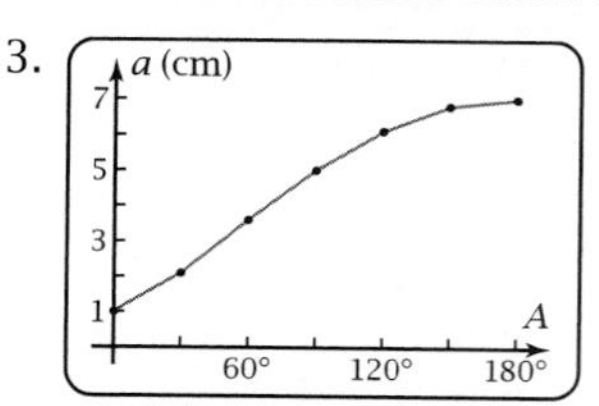

5. $-2 \cdot 3 \cdot 4\cos A$

Problem Set 6-2

1. $r \approx 3.98$ cm

3. $r \approx 4.68$ ft

5. $m\angle U \approx 28.96°$

7. $m\angle T \approx 134.62°$

9. This is not a possible triangle.

11. $m\angle O = 90°$

13. Measurements are equivalent.

15. a. 542.7249... ft

b. \$2035.22

c. \$2747.55

17. $X = (4\cos Z, 4\sin Z)$, $Y = (5, 0)$, so
$$z^2 = (4\cos Z - 5)^2 + (4\sin Z - 0)^2$$
$$= 4^2\cos^2 Z - 2 \cdot 4 \cdot 5\cos Z + 25 + 4^2\sin^2 Z$$
$$= 4^2(\sin^2 Z + \cos^2 Z) + 5^2 - 2 \cdot 4 \cdot 5\cos Z$$
$$= 4^2 + 5^2 - 2 \cdot 4 \cdot 5\cos Z$$

Problem Set 6-3

1. 5.44 ft²

3. 6.06 cm²

5. 26.9814... cm²

7. 4.3906... in.^{2}

9. a. $5 + 6 < 13$, so the triangle inequality shows that no triangle can have these three sides.

   b. $s = \frac{1}{2}(5 + 6 + 13) = 12$ cm
   Area $= \sqrt{-504}$. According to Hero's formula, the triangle would have to have an impossible area.

11. a. $A = \frac{1}{2} \cdot 4 \cdot 3 \sin\theta = 6 \sin\theta$

   b.

θ	A
0°	0.0000
15°	1.5529
30°	3.0000
45°	4.2426
60°	5.1962
75°	5.7956
90°	6.0000
105°	5.7956
120°	5.1962
135°	4.2426
150°	3.0000
165°	1.5529
180°	0.0000

   c. False. The function increases from 0° to 90°, then decreases from 90° to 180°.

   d. $0° \le \theta \le 180°$; the sine function becomes negative outside this domain.

13. $\frac{h}{4} = \sin Z$

   $h = 4 \sin Z$

   $A = \frac{1}{2}bh = \frac{1}{2}(5)(4)\sin Z$

Problem Set 6-4

1. $b \approx 5.23$ cm
   $c \approx 10.08$ cm

3. $h \approx 249.92$ yd
   $s \approx 183.61$ yd

5. $a \approx 9.32$ m
   $p \approx 4.91$ m

7. $a \approx 214.74$ ft
   $l \approx 215.26$ ft

9. a. $Z = 180° - (42° + 58°) = 80°$
   $x = 679.4530...$ m
   $y = 861.1306...$ m

   b. $67,220.70.

   c. A savings of $105,421.05 over y and $38,200.35 over x

11. a. $A = 33.1229...°$

   b. $C = 51.3178...°$

   c. $C = 128.6821...°$

   d. This is the complement of 51.3178...° and is one of the *general* solutions to $\sin^{-1}\frac{10 \sin A}{7}$.

   e. The principal values of $\cos^{-1} x$ go from 0° to 180°; a negative argument will give an obtuse angle, and a positive argument will give an acute angle, always the actual angle in the triangle. But the principal values of $\sin^{-1} x$ go from −90° to 90°; a negative argument will never happen in a triangle problem, but a positive argument will only give an acute angle, while the actual angle in the triangle may be the obtuse complement of the acute angle.

13. $A = \frac{1}{2}xy \sin Z = \frac{1}{2}yz \sin X = \frac{1}{2}zx \sin Y$

   So $\frac{1}{2}xy \sin Z = \frac{1}{2}yz \sin X$

   $x \sin Z = z \sin X$

   $\frac{x}{\sin X} = \frac{z}{\sin Z}$

   and similarly.

Problem Set 6-5

1. $c = 5.3153...$ cm or 1.3169... cm

3. $c \approx 7.79$ cm

5. There is no solution. Side b is too short.

7. $s \approx 5.52$ in.

9. $C \approx 23.00°$ or 157.00°

11. $Z \approx 43.15°$

13. a. $x \approx 38.65$ mi

   b. The other answer is approximately −12.94 mi. This means 12.94 mi to the west of Ocean City.

   c. $K \approx 99.29°$

Problem Set 6-6

1. $|\vec{a} + \vec{b}| \approx 14.66$ cm
   $\alpha \approx 45.84°$

3. $|\vec{a} + \vec{b}| \approx 11.69$ in.
   $\alpha \approx 150.00°$

5. a. Lucy's bearing is 150.9453...°.

   b. The starting point's bearing from Lucy is 330.9453...°.

   c. 205.9126... m

7. $6.0376...\vec{i} - 5.2484...\vec{j}$

9. $-6.1344...\vec{i} + 14.4519...\vec{j}$

11. a. $-12.8175...\vec{i} + 54.3745...\vec{j}$

   b. $|\vec{r}| = 55.8648...$ units
   $\theta = 103.2640...°$

13. $|\vec{r}| = 70$ mi
$\theta = 41.7867...°$

15. $|\vec{r}| = 167.8484...$ mi/hr
$\theta = 304.6074°$

17. $\vec{r} = 6.8478...\vec{i} + 51.2124...\vec{j}$
$|\vec{r}| = 51.6682...$ newtons
$\theta = 82.3838...°$

19.

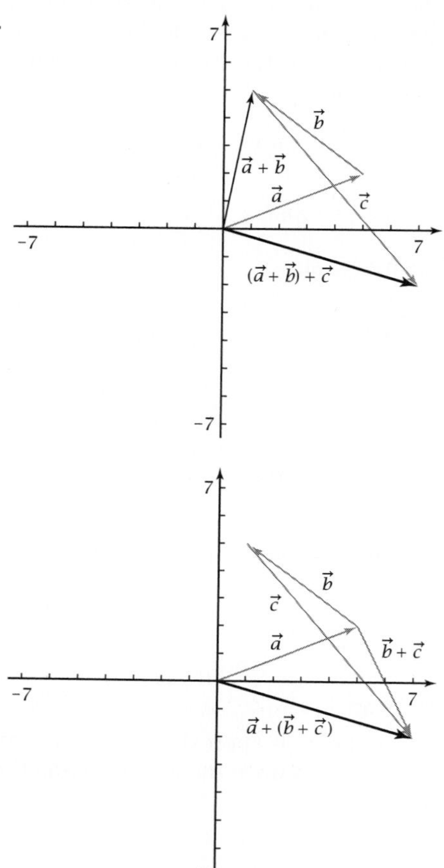

21. If $a\vec{i} + b\vec{j}$ and $c\vec{i} + d\vec{j}$ are any two vectors, then a, b, c, and d are real numbers. So $a + c$ and $b + d$ are also real numbers because the real numbers are closed under addition. Therefore, the sum $(a + c)\vec{i} + (b + d)\vec{j}$ exists and is a vector, so the set of vectors is closed under addition. The zero vector is necessary so that the sum of any vector $a\vec{i} + b\vec{j}$ and its opposite $-a\vec{i} - b\vec{j}$ will exist.

Problem Set 6-7

1. a. Window ≈ 12.1 ft
Roof ≈ 20.1 ft
b. Area ≈ 120.6 ft^2

3. a. The shelter will be able to display about 364 pumpkins.
b. $\theta = 33.1229...°$

5. First find θ, the angle at the peak, then find the desired angle as $180° - (50° + \theta)$. But
$$\sin\theta = \frac{30 \sin 50°}{20} \approx 1.15$$
which is not the sine of any angle. It is impossible to build the truss to the specifications. Either the 20-ft side is too short or the 30-ft side is too large or the 50° angle is too large.

7. Let A be the observer, B the launching pad, C the missile when at 21°, and D the missile when at 35°.
a. $CD = 0.6327...$ km
b. 0.1265... km/sec
c. 58.7692...°

9. a. $|\vec{r}| \approx 537.0$ km/hr
b. $|\vec{r}| \approx 463.4$ km/hr

11. a.

θ	Lift (lb)	Horizontal Component (lb)
0°	500000	0
5°	501910	43744
10°	507713	88163
15°	517638	133975
20°	532089	181985
25°	551689	233154
30°	577350	288675

b. The centripetal force is stronger, so the plane is being forced more strongly away from a straight line into a circle.
c. The horizontal component is 0, so there is no centripetal force to push the plane out of a straight path.
d. $\theta \approx 33.56°$
e. It would start to fall because the vertical component would be less than 500,000 lb and could not support it. Together with the turning caused by the horizontal component, this would result in a spiral downward.

13. $|\vec{r}| \approx 290.7$ km/hr

15. 357.5381... km

17. a. 133.3576...°
b. 6838.2143... m^2

19. a, b. Answers will vary.
c. Label the 95° angle A, and label the rest of the vertices clockwise as B through F.
$$AF = \sqrt{AE^2 + 17^2 - 2 \cdot AE \cdot 17 \cos \angle AEF}$$
$$= 30.6817...\text{ m}$$

d. For a nonconvex polygon, you might not be able to divide it into triangles that fan out radially from a single vertex.

Problem Set 6-8

R1. a. Answers may vary slightly. They should be approximately 2.5, 4.6, 6.4, 7.8, 8.7.

b. 9; 1

c. $\sqrt{41} \approx 6.4$; yes

d. 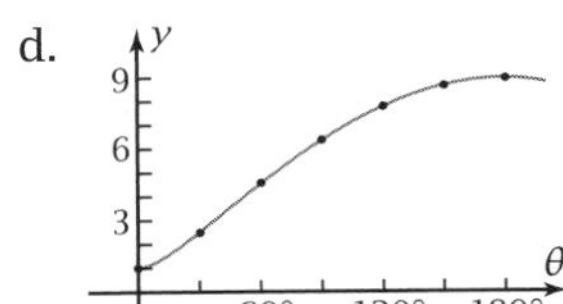

The data seem to fit the law of cosines. The shape is a sinusoid.

R2. a. 77.9295… ft

b. ≈ 113.5781…°

c. $3 + 5 < 10$. Also, whichever angle you try to calculate, you get an impossible cosine:

$$\frac{3^2 + 5^2 - 10^2}{2 \cdot 3 \cdot 5} = -2.2$$

$$\frac{5^2 + 10^2 - 3^2}{2 \cdot 5 \cdot 10} = 1.16$$

$$\frac{10^2 + 3^2 - 5^2}{2 \cdot 10 \cdot 3} = 1.4$$

d. 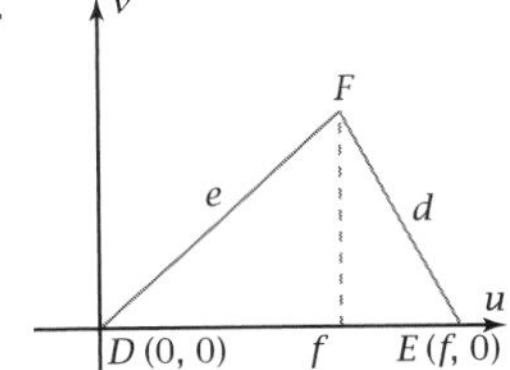

$E = (f, 0)$

$F = (e \cos \theta, e \sin \theta)$

$$\begin{aligned} d^2 &= (e \cos \theta - f)^2 + (e \sin \theta - 0)^2 \\ &= e^2 \cos^2 \theta - 2ef \cos \theta + f^2 + e^2 \sin^2 \theta \\ &= e^2(\cos^2 \theta + \sin^2 \theta) + f^2 - 2ef \cos \theta \\ &= e^2 + f^2 - 2ef \cos \theta \end{aligned}$$

R3. a. 340.4928… ft^2

b. $\theta = 103.1365587…°$
$A = 42.84857057…$ mi^2 using both methods.

c. $\theta = 41.8°$ or $138.2°$

d. Base = d, altitude = $e \sin F$

$$A = \frac{1}{2}bh = \frac{1}{2}de \sin F$$

R4. a. 7.0852… in.

b. 9.2718… m

c. 46.1415…° or 133.8584…°

d. $\frac{1}{2}de \sin F = \frac{1}{2}df \sin E$

$e \sin F = f \sin E$

$$\frac{e}{\sin E} = \frac{f}{\sin F}$$

and similarly.

R5. a. $x \approx 11.4$ cm or 3.4 cm

b. $\sin \phi = \frac{8 \sin 85°}{5} \approx 1.6$, which is not the sine of any angle.

c. $\theta \approx 38.7°$

d. $x \approx 10.5$ cm

R6. a. $|\vec{r}| = 4.0813…$
$\phi = 165.2°$

b. $\vec{a} + \vec{b} = 12\vec{i} - 3\vec{j}$
$|\vec{r}| \approx 12.4$
$\theta \approx 346.0°$

c. $|\vec{r}| \approx 132.8$ mi
$\theta \approx 165.5°$

d. $\vec{r} = (-255.1704…)\vec{i} + (-138.4578…)\vec{j}$
$|\vec{r}| \approx 290.3$ km/hr
$\theta \approx 208.5°$

R7. a. 137.7798… km, so it is out of range.

b. $x = 177.170054…$ km or 325.111375… km

c. $(-520 \cos 40°)^2 - 4(57{,}600) = -71.722.76…$, so x is undefined.

d. $\theta \approx 22.6°$
$x = 240$ km

e. $240^2 = 57{,}600 = (177.17…)(325.11…)$.
The theorem states that if P is a point exterior to circle C, PR cuts C at Q and R, and PS is tangent to C at S, then $PQ \cdot PR = PS^2$.

f. Nagoya Airport is closer by 35.2 km.

g. 7.5946…°

h. 3026.5491… lb

i. The helicopter can tilt so that the thrust vector exactly cancels the wind vector.

Problem Set 6-9

1. $f(x) = ax^2 + bx + c$, where $a \neq 0$
2. Horizontal dilation by $\frac{1}{3}$, vertical dilation by 5
3. Horizontal translation by 3, vertical translation by -3; $h(x) = f(x - 3) - 3$
4. 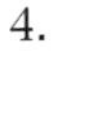 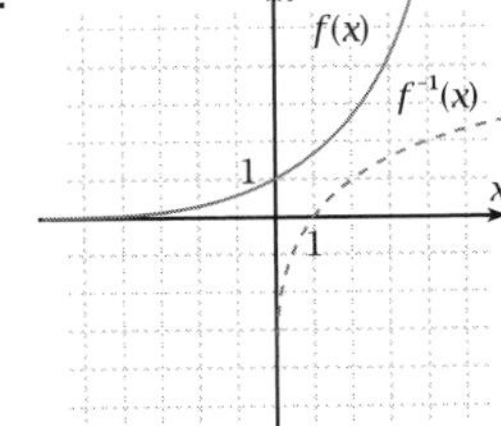

5. Odd

6. $g(x) = 3f\left[\frac{1}{2}(x-4)\right] + 5$

7. 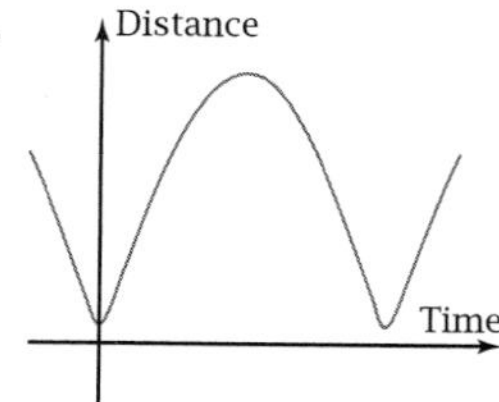

8. 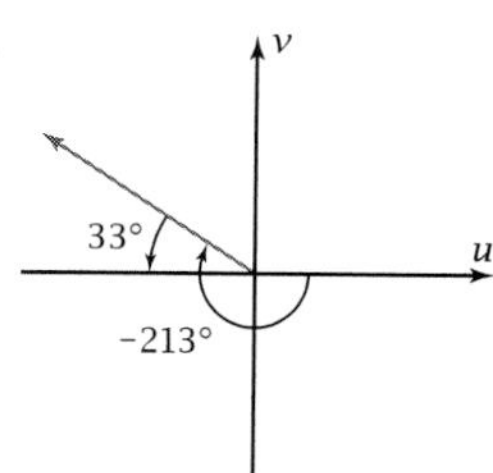

9. $\sin\theta = -\frac{5}{13}$, $\cos\theta = \frac{12}{13}$,
$\tan\theta = -\frac{5}{12}$, $\cot\theta = -\frac{12}{5}$,
$\sec\theta = -\frac{13}{12}$, $\csc\theta = -\frac{13}{5}$

10. $\sin 240° = -\frac{\sqrt{3}}{2}$

11. 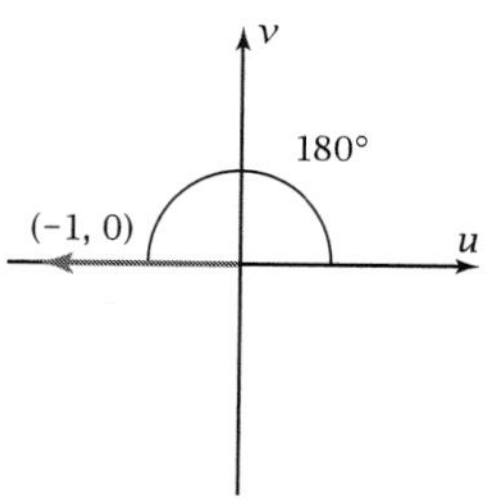

$\cos 180°$ = the u-coordinate on the x-axis = −1

12. θ (degrees)

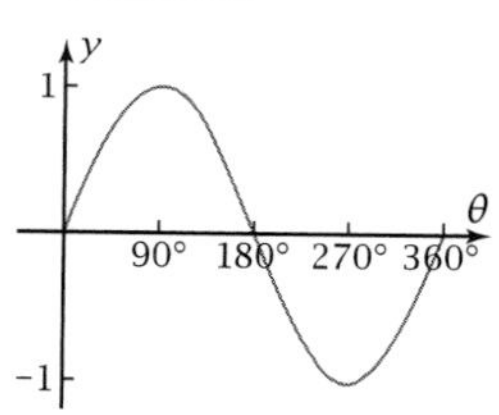

13. Sinusoidal

14. 2π; π; $\frac{\pi}{2}$; $\frac{\pi}{4}$

15. $2 \cdot \frac{180°}{\pi} = 114.5915...°$

16. 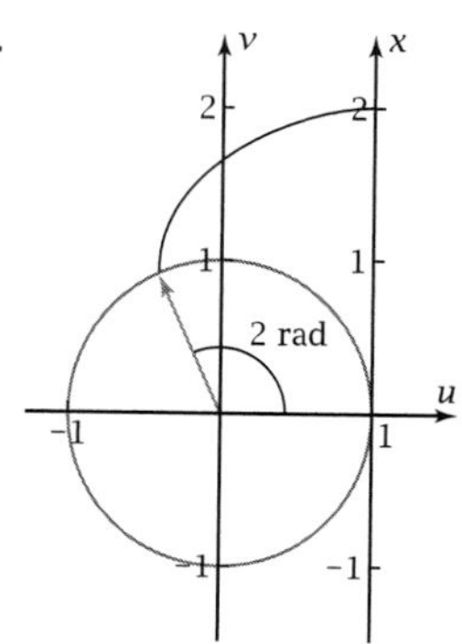

17. 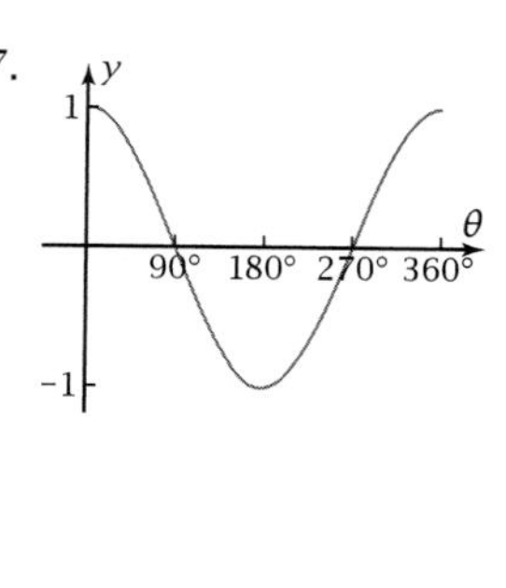

18. a. $\frac{1}{5}$ b. 4 c. −6 d. 3

19. a. 2π times horizontal dilation is the period
 b. Amplitude
 c. Phase displacement
 d. Sinusoidal axis

20. $y = 2 + 3\cos\frac{\pi}{5}(x-1)$

21. $y = -3.4452...$

22. $x \approx 2.3, 9.7, 12.3$

23. 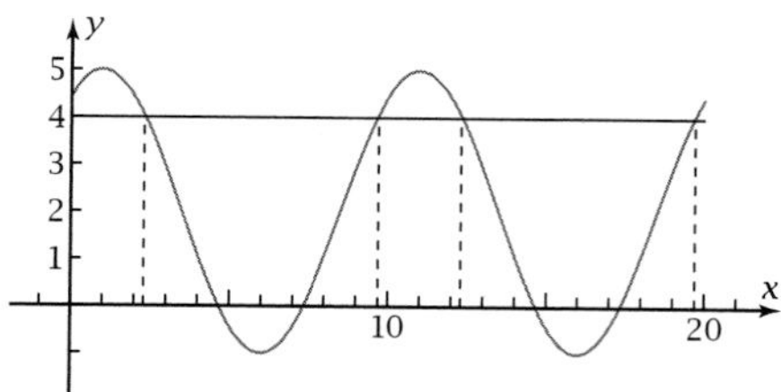

24. $d = 5000 - 4000\cos\frac{\pi}{50}t$

25. Reciprocal properties:

$\sec\theta = \dfrac{1}{\cos\theta}$

Quotient properties:

$\tan\theta = \dfrac{\sin\theta}{\cos\theta}$

Pythagorean properties:

$\sin^2\theta + \cos^2\theta = 1$

26. $\sec^2 x \sin^2 x + \tan^4 x = \dfrac{\sin^2 x}{\cos^2 x} + \dfrac{\sin^4 x}{\cos^4 x}$

$= \dfrac{\sin^2 x\cos^2 x}{\cos^4 x} + \dfrac{\sin^4 x}{\cos^4 x} = \dfrac{\sin^2 x\cos^2 x + \sin^4 x}{\cos^4 x}$

$= \dfrac{\sin^2 x(\cos^2 x + \sin^2 x)}{\cos^4 x} = \dfrac{\sin^2 x}{\cos^4 x}$

$\therefore \sec^2 x\sin^2 x + \tan^4 x = \dfrac{\sin^2 x}{\cos^4 x}$

for $\cos x \neq 0$ ($x \neq 90° + n \cdot 180°$), Q.E.D.

27. $\cos(x - y) = \cos x\cos y + \sin x\sin y$
Cosine of first, cosine of second, plus sine of first, sine of second

28. $\cos 34° = 0.8290... = \sin 56°$

29. $\cos(90° - \theta) = \cos 90°\cos\theta + \sin 90°\sin\theta$
$= 0 \cdot \cos\theta + 1 \cdot \sin\theta = \sin\theta$;
$\cos(34°) = \sin(90° - 34°) = \sin 56°$

30. $A = 5$
$D = 53.1301...°$
$\therefore 3\cos\theta + 4\sin\theta = 5\cos(\theta - 53.1301...°)$

31. $6 \sin 2\theta = 6 \sin(\theta + \theta)$
$= 6(\sin\theta \cos\theta + \cos\theta \sin\theta)$
$= 6 \cdot 2 \sin\theta \cos\theta = 12 \sin\theta \cos\theta$

32. $\cos 2x = 1 - 2\sin^2 x$, so
$\sin^2 x = \frac{1}{2} - \frac{1}{2}\cos 2x$ which is a sinusoid.

33. $y = 3\cos 6\theta + 2\sin 30\theta$

34. $y = 5\sin\theta \cos 12\theta$

35. $y = \cos 21\theta + \cos 19\theta$

36. In the problem, the period of $\cos 20\theta$ is 18°; the period of $\cos\theta$ is 360°. These are much different. In the answer, the period of $\cos 21\theta$ is 17.1428...°; the period of $\cos 19\theta$ is 18.9473...°. These are fairly nearly equal.

37. $\theta = 78.6900...°$

38. $y = 0.4115... + 2n\pi$ rad
or $2.7300... + 2n\pi$ rad

39. See Figure 6-9f.

40. Domain is $-1 \le x \le 1$
Range is $0° \le y \le 180°$

41. $\theta \approx 63.4°, 243.4°, 423.4°, 603.4°$

42. In $\triangle ABC$, $c^2 = a^2 + b^2 - 2ab\cos C$ (and similarly for a^2 and b^2). The square of one side of a triangle is the sum of the squares of the other two sides minus twice their product times the cosine of the angle between them.

43. In $\triangle ABC$,
$$\frac{a}{\sin A} = \frac{b}{\sin B} = \frac{c}{\sin C}$$
The length of one side of a triangle is to the sine of the angle opposite it as the length of any other side is to the sine of the angle opposite that side.

44. For $\triangle ABC$, Area $= \frac{1}{2}ab\sin C = \frac{1}{2}bc\sin A = \frac{1}{2}ca\sin B$.
The area of a triangle is $\frac{1}{2}$ the product of any two sides and the sine of the angle between them.

45. 134.6183...°

46. Area = 14.9478... ft^2

47. a. $= 2\vec{i} + 16\vec{j}$
b. $|\vec{r}| \approx 16.1$
$\theta \approx 82.9°$

c. $\vec{a}, \vec{b}, \vec{a} + \vec{b}$

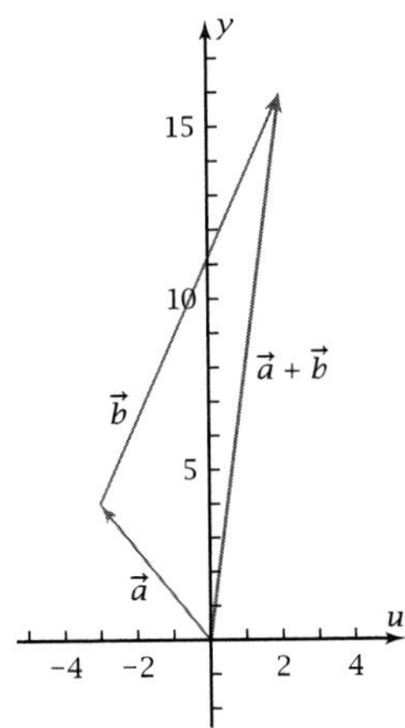

d. False. This is only true if $\vec{a}$ and $\vec{b}$ are at the same angle.

48. In units of 1000 miles:
a. $y = \sqrt{41 - 40\cos\theta}$ b. $y = \sqrt{41 - 40\cos\frac{\pi}{50}t}$
c.

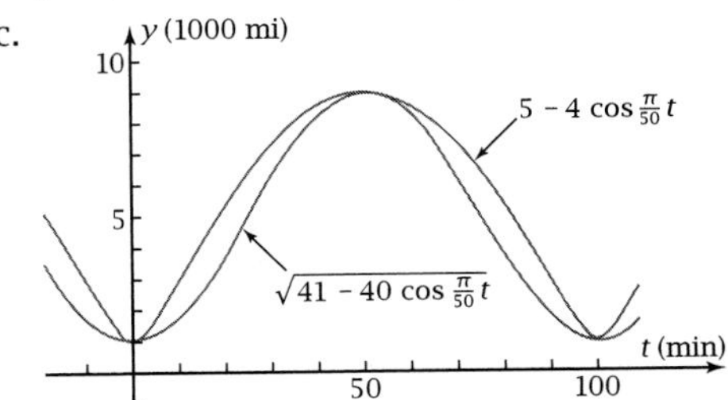

CHAPTER 7

Problem Set 7-1

1.

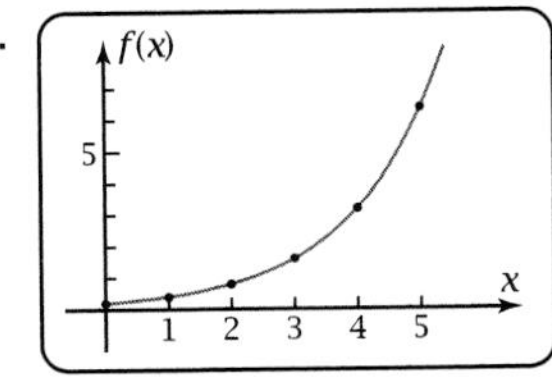

The "hollow" section is upward. The bacteria are growing faster and faster.

3.

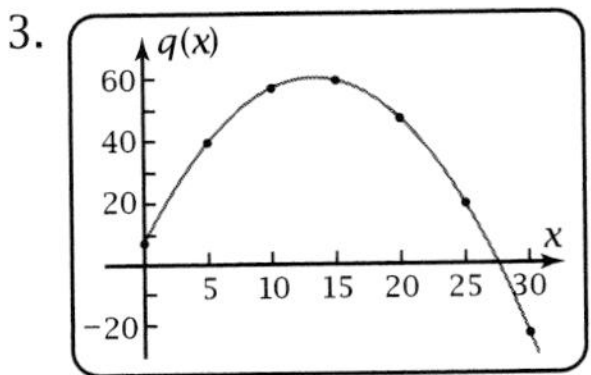

The graph is concave downward. This graph possesses a maximum (high point) at $x = 13\frac{1}{3}$.

Problem Set 7-2

1. In power functions, the exponent is constant and the independent variable is in the base. In exponential functions, the base is constant and the independent variable is in the exponent.

3. Answers will vary. The term *concave* is from the Latin *cavus,* meaning "hollow." The concavity of a curved portion of a graph is the "inside" of that curve.

5. $\frac{1}{x} = x^{-1}$

7. $(-64)^{1/2}$ is undefined, but $(-64)^{1/3} = -4$. The restriction allows the function to be defined for all values of x.

9. a. $y = -1.6625(x-4)^2 + 30.6$
   b. $y(5) = 28.9375$ ft
   c. $x = 8.2902...$ sec

11. a. Linear
    b. Decreasing for all real number values of x, not concave
    c. Answers will vary.
    d. $y = -4x + 20$
    e. The graphs match.

13. a. Quadratic
    b. Increasing for $x < \frac{81}{22}$ and decreasing for $x > \frac{81}{22}$, concave downward
    c. Answers will vary.
    d. $y = -\frac{11}{15}x^2 + \frac{27}{5}x - \frac{311}{15}$
    e. The graphs match.

15. a. Exponential
    b. Decreasing for all real number values of x, concave upward
    c. Answers will vary.
    d. $y = 96 \cdot (0.5)^x$
    e. The graphs match.

17. a. Power (inverse)
    b. Decreasing for $x > 0$, concave upward
    c. Answers will vary.
    d. $y = 12x^{-1}$
    e. The graphs match.

19. a. Linear
    b. Increasing for $x \geq 0$, not concave
    c. Answers will vary.
    d. $y = 0.8x$
    e. The graphs match.

21.

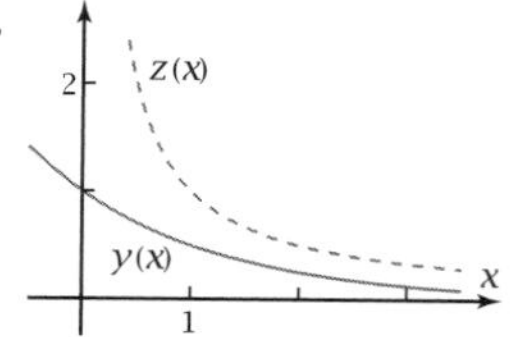

Both graphs are concave, both graphs approach zero as x grows large, and both graphs never intersect the horizontal axis. But the exponential function does intersect the vertical axis, whereas the inverse graph does not.

23. You can write a power function proportional to the square of x in the quadratic form $y = ax^2 + bx + c$ with $b = 0$ and $c = 0$. But no quadratic function $y = ax^2 + bx + c$ with $b \neq 0$ or $c \neq 0$ can be written as a power function $y = ax^2$.

Problem Set 7-3

1. Add-add property: linear

3. Multiply-multiply property: power; and constant-second-differences property: quadratic

5. Add-add property: linear

7. Multiply-multiply property: power

9. Add-multiply property: exponential

11. Constant-second-differences property: quadratic

13. a. 65
    b. 80
    c. 1280

15. a. 70
    b. 81
    c. 72.9

17. $f(8) = 13$, $f(11) = 19$, $f(14) = 25$

19. $f(10) = 324$, $f(20) = 81$

21. Multiply y by 4

23. Divide y by 2

25. a. $V(r)$ has the form $V = ar^3$, where $a = \frac{4}{3}\pi$. The volleyball would have volume 800 cm^3.
    b. His volume would be 1000 times that of a normal gorilla. 500 lb $\cdot$ $(10)^3$ = 500,000 lb
    c. 250,000 lb
    d. 0.2 lb, or 3.2 oz

27. a. 16 times more wing area
    b. 64 times heavier
    c. The full-sized plane had four times as much weight per unit of wing area as the model.

29. a. $[H(3) - H(2)] - [H(2) - H(1)] = -32$ ft
$[H(4) - H(3)] - [H(3) - H(2)] = -32$ ft
$[H(5) - H(4)] - [H(4) - H(3)] = -32$ ft

b. $H(t) = -16t^2 + 90t + 5$; $H(4) = 109$; $H(5) = 55$

c. $H(2.3) = 127.36$ ft, going up. The height seems to peak at 3 sec.

d. $t = 1.4079\ldots$ sec (going up) or $4.2170\ldots$ sec (coming down)

e. The vertex is at $t = 2.8125$ sec; $H(22.8125 \text{ sec}) = 131.5625$ ft

f. $t = 5.6800\ldots$ sec

31. $[y(6) - y(5)] - [y(5) - y(4)] = 2$
$[y(7) - y(6)] - [y(6) - y(5)] = 2$
$[y(8) - y(7)] - [y(7) - y(6)] = 4$
If $y(8)$ were 25, then a quadratic function would fit.

33. If $f(x) = ax + b$, then $f(x_1 + c) = a(x_1 + c) + b$
$= ax_1 + ac + b = (ax_1 + b) + ac = f(x_1) + ac$

35. If $f(x) = ab^x$, then $f(c + x_1)ab^{c+x_1} = a(b^c \cdot b^{x_1})$
$= b^c \cdot ab^{x_1} = b^c \cdot f(x_1)$

Problem Set 7-4

1. $3.0277\ldots$; $10^{3.0277\ldots} = 1066$

3. $-1.2247\ldots$; $10^{-1.2247\ldots} = 0.0596$

5. $0.001995\ldots$; $\log 0.001995\ldots = -2.7$

7. $1.5848\ldots \times 10^{15}$; $\log(1.5848\ldots \times 10^{15}) = 15.2$

9. $e^x = p$

11. $\log_r m = k$

13. $\log 0.21 = -0.6777\ldots = -0.5228\ldots + (-0.1549\ldots)$

15. $\ln 6 = 1.7917\ldots = 3.4011\ldots - 1.6094\ldots$

17. $\log 32 = 1.5051\ldots = 5(0.3010\ldots)$

19. $-1.9459\ldots = -(1.9459\ldots)$

21. 21

23. 4

25. 56

27. 32

29. 3

31. $\log_{10} 7$

33. 3

35. $\frac{1}{2}$

37. $x = -2$
Check: $\log 1 = 0 \Rightarrow 10^0 = 1$

39. $x_1 = 5$ $\quad x_2 = -4$
Check: 1. $\log_2 8 + \log_2 1 = 3 + 0 = 3$
2. $\log_2(-1) + \log_2(-8) \Rightarrow$ no solution

41. $x = 16.3890\ldots$
Check: $\ln(e^2)^4 = \ln e^8 = 8 \ln e = 8(1) = 8$

43. $x = 1.3808\ldots$

45. $x = -85.1626\ldots$

47. $x \ln \frac{5}{3} = 4.5108\ldots$
Check: $3e^{\ln(5/3)} + 5 = 3 \cdot \frac{5}{3} + 5 = 10$

49. $x = \ln \frac{1}{2} = -0.6931$
Check: $2e^{2\ln(1/2)} + 5e^{\ln(1/2)} - 3 = 2\left(\frac{1}{4}\right) + 5\left(\frac{1}{2}\right) - 3 = 0$

51. a.

x	M
0	10000
1	10700
2	11449
3	12250
4	13108
5	14026
6	15007

b. $f(x + c) = 10{,}000 \times 1.07^{x+c}$
$= 10{,}000 \times (1.07^x \cdot 1.07^c)$
$= (10{,}000 \times 1.07^x) \cdot 1.07^c = 1.07^c \cdot f(x)$
Exponential functions always have the add-multiply property.

c. $x = 14.6803\ldots$ yr; 177 mo

53. $\dfrac{\log 29}{\log 7} = 1.7304\ldots$; $7^{1.7304\ldots} = 29$

55. $\dfrac{\log 729}{\log 3} = 6$; $3^6 = 729$

57. $\log_2 32 = 5$

59. $\log_7 49 = 2$

61. Let $c = \log_b x$, so $x = b^c$. Then $x^n = (b^c)^n = b^{cn}$, so $\log_b x^n = cn = nc = n \log_b x$.

Problem Set 7-5

1. a. $\dfrac{14.4}{3.6} = \dfrac{57.6}{14.4} = \dfrac{230.4}{57.6} = \dfrac{921.6}{230.4} = 4$

b. $y = \log_4\left(\dfrac{10}{9}x\right)$

c. Equation fits data.

3. a. The inverse of an exponential function is a logarithmic function.

b. $y = 38{,}203.6483\ldots - 8{,}295.8190\ldots \ln p$; $r = -0.9999\ldots$, which is very close to -1.

c. $y(73.9) = 38{,}203.6483\ldots - 8{,}295.8190\ldots \ln 73.9$
$= 2509.1214 \approx 2509$ years old

d. $y(20) = 13{,}351.5954\ldots \approx 13{,}350$ years old

5. a. $g(x) = 6 \log_{10} x$; $g(x) = \log_{1.4677\ldots} x$

b. For g, $x = e^{-3} = 0.0497\ldots$;
for h, $x = e = 2.7182\ldots$

7. Domain: $x > -3$

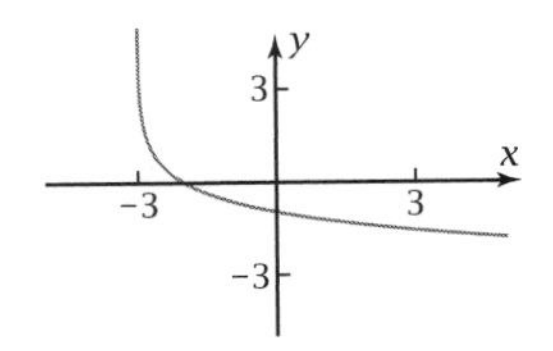

9. Domain: $x \neq 0$

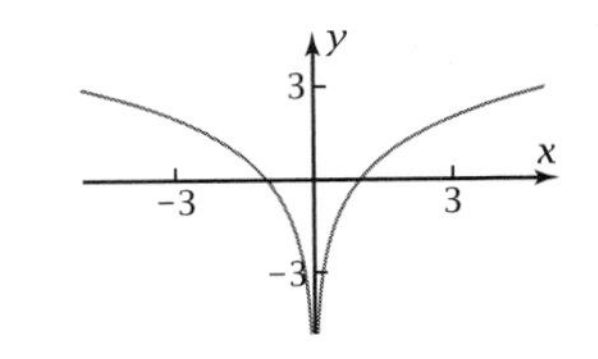

11. Domain: $x > 0$

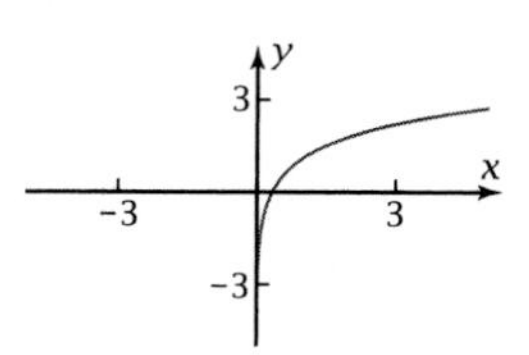

13. a.

x	y
−0.00100	2.7196...
0.00100	2.7169...
−0.00010	2.7184...
0.00010	2.7181...
−0.00001	2.7182...
0.00001	2.7182...

b. The two properties balance out so that as x approaches 0, y approaches 2.7182....

c. $e = 2.7182...$; they are the same.

Problem Set 7-6

1. a.

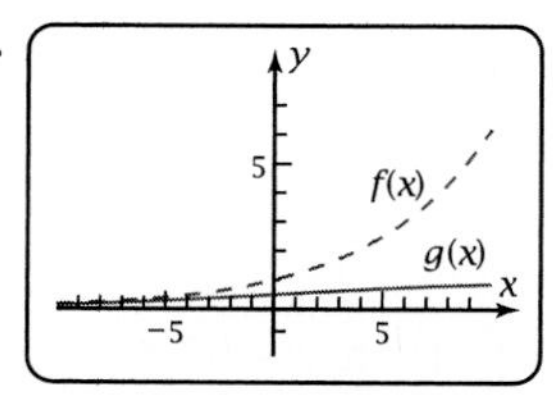

b. The graphs are almost the same for large negative values of x but widely different for large positive values of x.

c. Point of inflection is at $x = 0$; g is concave up for $x < 0$ and concave down for $x > 0$.

d. As x grows very large, the 1 in the denominator is less significant than the 1.2^x, so

$$g(x) = \frac{1.2^x}{1.2^x + 1} \approx \frac{1.2^x}{1.2^x} = 1$$

e. $g(x) = \dfrac{1}{1 + 1.2^{-x}}$, which is equivalent to the original.

3. a. Concave up

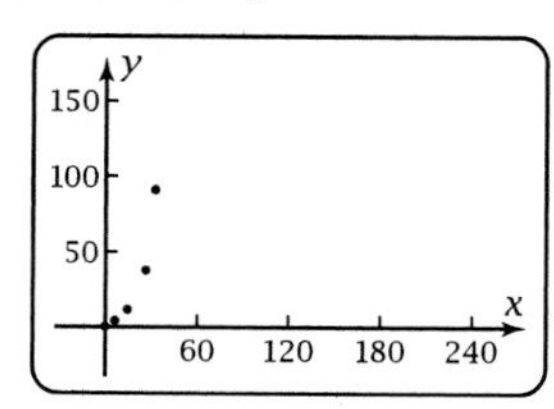

b. $y = \dfrac{1220}{1 + (1218)(1.1211...)^{-x}}$

c.

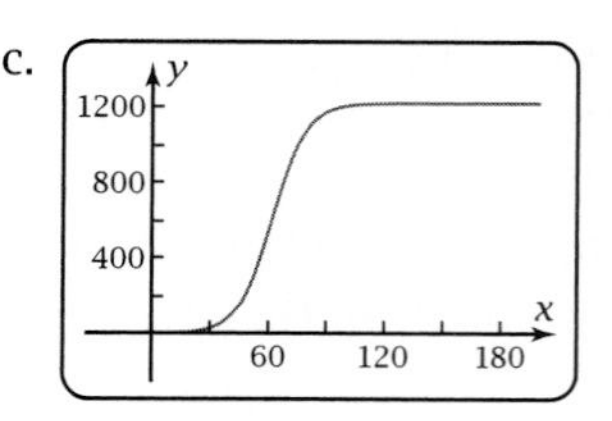

d. $y(60) = 536.2073...$

5. a. Concave down

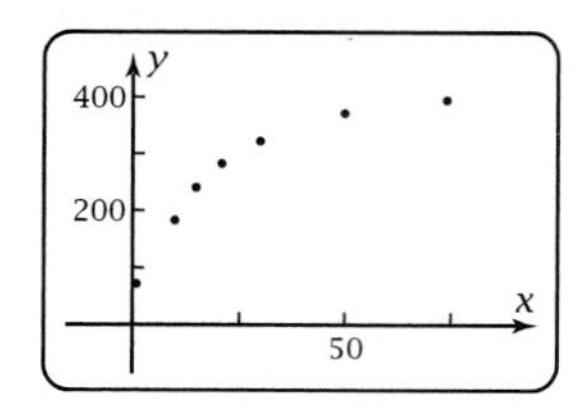

b. $y = \dfrac{396}{1 + (2.752...)(1.0888)^{-x}}$

c.

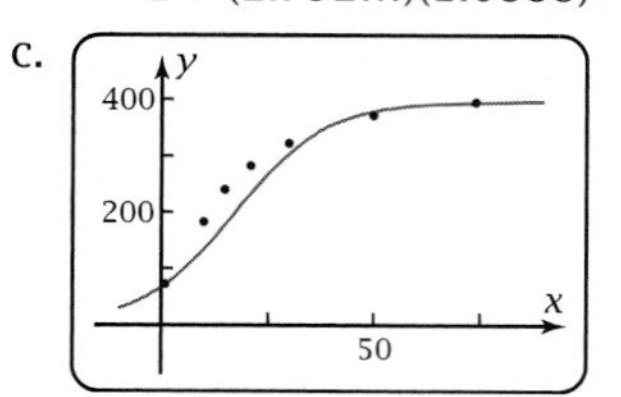

d. The point of inflection occurs at (11.9037..., 198). Before approximately 12 days passed, the rate of new infection was increasing; after that, the rate was decreasing.

e. $y = 362.7742...$; after 40 days, approximately 363 people were infected.

f. Answers will vary.

7. a. True: c is a vertical dilation factor.

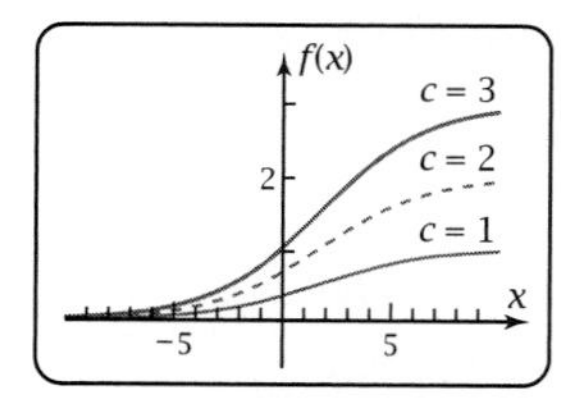

b. Changing a translates the graph horizontally by $\frac{\ln a}{0.4}$.

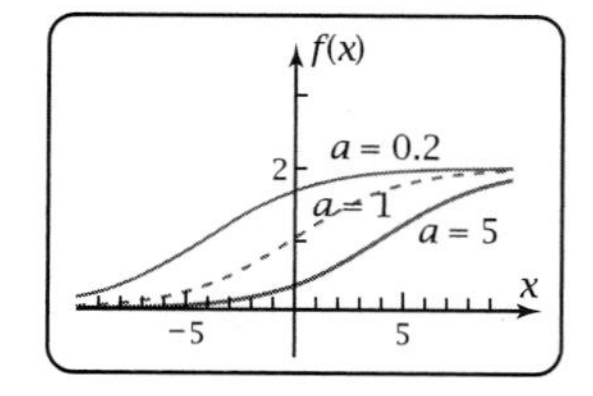

c. Horizontal translation by 3

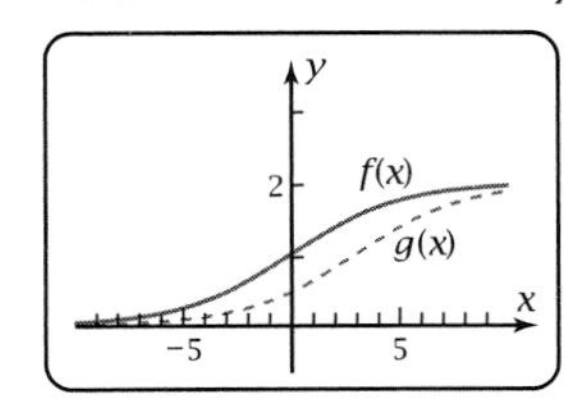

d. $a = e^{1.2}$

Problem Set 7-7

R1. a.

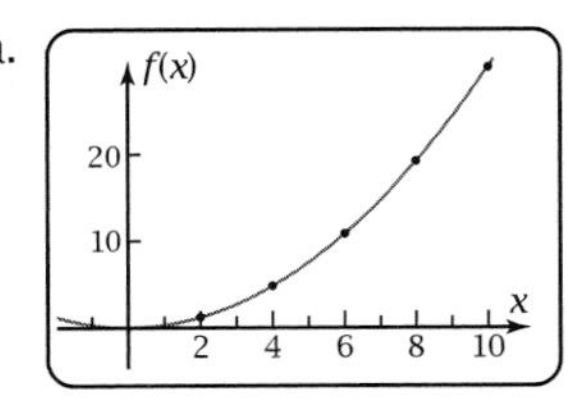

b. Increasing; concave up

c. Quadratic power function; answers will vary.

R2. a. $y = \frac{2}{3}x + \frac{13}{3}$

b.

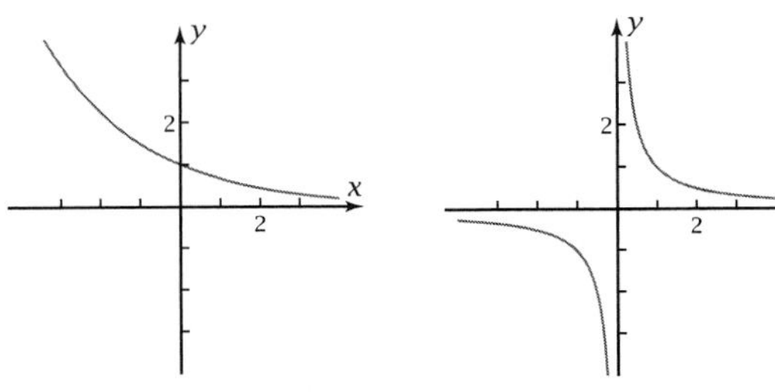

Both are decreasing. Both have the x-intercept as an asymptote. But the exponential function crosses the y-axis, whereas the inverse function has no y-intercept.

c. The y-intercept is nonzero.
$y = (3.1201\ldots)(1.3867\ldots)^x$

d. $y = -1.2x^2 + 9x + 2$; $a = -1.2$. If a is negative, the graph is concave down.

e. Vertex (5, 3); y-intercept at $y = 53$

R3. a. Exponential; $f(x) = 48\left(\sqrt[3]{0.5}\right)^x$

b. Power (inverse variation);
$g(x) = 72x^{-1}$

c. Linear; $h(x) = 2x + 18$

d. Quadratic; $q(x) = x^2 - 13x + 54$

e. i. $f(12) = 213\frac{1}{3}$
ii. $f(12) = 160$
iii. $f(12) = 180$

f. $f(x + c) = 53 \times 1.3^{x+c} = 53 \times 1.3^x \times 1.3^c = 1.3^c \cdot f(x)$

R4. a. An exponent
b. $c^p = m$
c. $p = \log_{10} z$
d. $10^{1.4771\ldots} = 30$
e. $\log_7 30 = 1.7478\ldots$

f. Answers will vary. Sample answers:
i. $\log(100 \cdot 10) = \log 1000 = 3$
$\log 100 + \log 10 = 2 + 1 = 3$
ii. $\log \frac{10{,}000}{1000} = \log 10 = 1$
$\log 10{,}000 - \log 1000 = 4 - 3 = 1$
iii. $\log 10^3 = \log 1000 = 3$
$3 \log 10 = 3 \cdot 1 = 3$

g. 60
h. 63

i. 1) $x = -3$ or 2) $x = 4$
Check: 1) $\log(-2) + \log(-5) \Rightarrow$ undefined
2) $\log 5 + \log 2 = 1 \Rightarrow \log 10 = 1$

j. $x = 4.3714\ldots$
Check: $3^{2(4.3714\ldots)-1} = 4946.7129\ldots$
$7^{4.3714\ldots} = 4946.7129\ldots$

R5. a. y_1 and y_2 are reflections of each other across the line $y = x$.

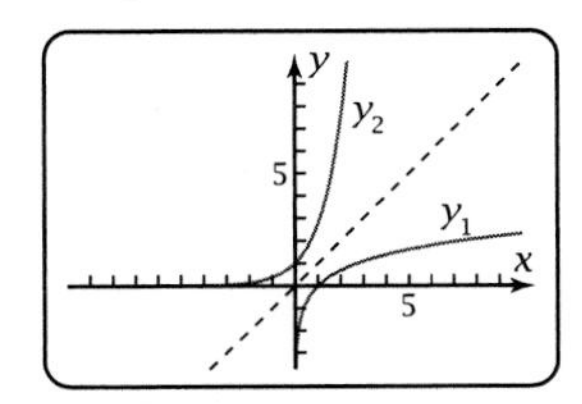

b. $f(x) = 5 \cdot 0.5703\ldots^x$
$g(x) = 4.3e^{2.0014\ldots x}$

c. Multiply–add property
$y = -13 \log_2 \frac{x}{100}$

d.

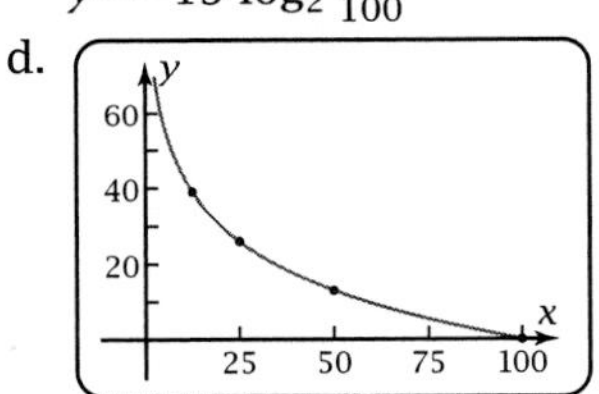

e. $x = 94.8077\ldots$ ft deep

R6. a.

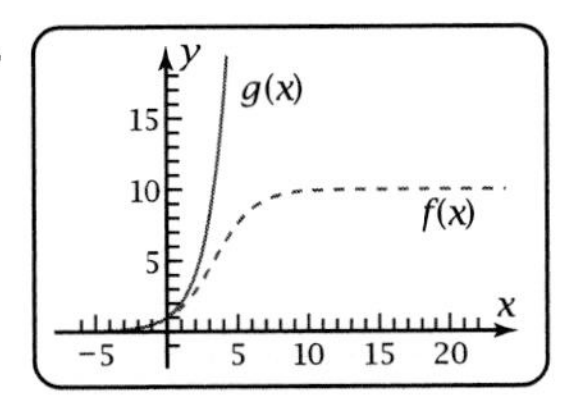

b. When x is a large negative number, the denominator of $f(x)$ is essentially equal to 10, so

$$f(x) = \frac{10 \cdot 2^x}{2^x + 10} \approx \frac{10 \cdot 2^x}{10} = 2^x = g(x)$$

But for large positive x, the 10 in the denominator of $f(x)$ is negligible compared to the 2^x, so

$$f(x) = \frac{10 \cdot 2^x}{2^x + 10} \approx \frac{10 \cdot 2^x}{2^x} = 10$$

c. $f(x) = \dfrac{10}{1 + 10 \cdot 2^{-x}}$

d. $g(x) = e^{(\ln 2)x}$

e. The size of the population would be limited by the capacity of the island.

$f(x) = \dfrac{460}{1 + (13.2906\ldots)(1.1718\ldots)^{-x}}$

$f(12) = 154.233\ldots$

$f(18) = 260.507\ldots$

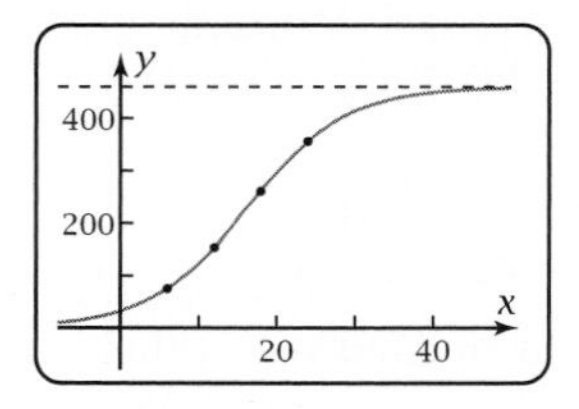

$x = 34.8878\ldots$ months

CHAPTER 8

Problem Set 8-1

1. Yes, $\hat{y} = 2.1x + 3.4$

3. $\hat{y}(14) = 2.1(14) + 3.4 = 32.8$ sit-ups. Explanations may vary. Fourteen days is an extrapolation from the given data, and extrapolating frequently gives incorrect predictions.

5. $SS_{\text{res}} = 17.60$

Problem Set 8-2

1. a. A graphing calculator gives $\hat{y} = 1.4x + 3.8$, with $r = 0.9842\ldots$.

b.

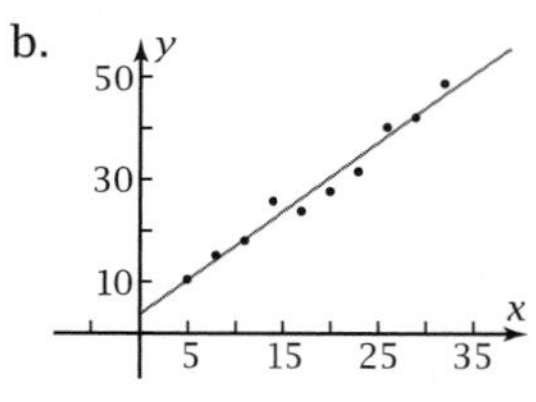

c. $\bar{x} = 18.5$, $\bar{y} = 29.7$; $\hat{y}(\bar{x}) = 1.4(18.5) + 3.8 = 29.7 = \bar{y}$

d.

$y - \bar{y}$	$(y - \bar{y})^2$	$y - \hat{y}$	$(y - \hat{y})^2$
−18.7	349.69	0.2	0.04
−13.7	187.69	1.0	1.00
−10.7	114.49	−0.2	0.04
−2.7	7.29	3.6	12.96
−4.7	22.09	−2.6	6.76
−0.7	0.49	−2.8	7.84
3.3	10.89	−3.0	9.00
12.3	151.29	1.8	3.24
14.3	204.49	−0.4	0.16
21.3	453.69	2.4	5.76
	1502.10		46.80

$SS_{\text{dev}} = 1502.10$, $SS_{\text{res}} = 46.8$,

$r = \sqrt{\dfrac{SS_{\text{dev}} - SS_{\text{res}}}{SS_{\text{dev}}}} = \sqrt{0.9688\ldots} = 0.9842\ldots$,

which agrees with part a.

e.

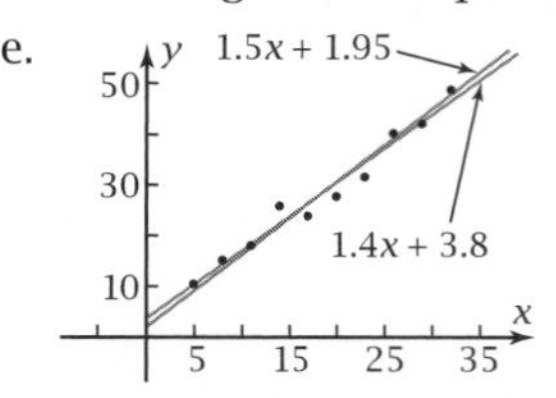

It is hard to tell which line fits better.

$\hat{y} = 1.5x + 1.95$	$y - \hat{y}$	$(y - \hat{y})^2$
9.45	1.55	2.4025
13.95	2.05	4.2025
18.45	0.55	0.3025
22.95	4.05	16.4025
27.45	−2.45	6.0025
31.95	−2.95	8.7025
36.45	−3.45	11.9025
40.95	1.05	1.1025
45.45	−1.45	2.1025
49.95	1.05	1.1025

$SS_{\text{res}} = 54.2250$, which is larger than SS_{res} for the regression line.

3. a. $\hat{y} = -0.05x + 17$, $r^2 = 1$, $r = -1$, which means a perfect fit.

b. $\bar{y} = 15.18$ gal

$y - \bar{y}$	$(y - \bar{y})^2$	$y - \hat{y}$	$(y - \hat{y})^2$
1.52	2.3104	0	0
0.72	0.5184	0	0
−0.38	0.1444	0	0
−0.68	0.4624	0	0
−1.18	1.3924	0	0

$SS_{\text{dev}} = 4.828$, $SS_{\text{res}} = 0$

$r^2 = 1$, $r = -1$

c.

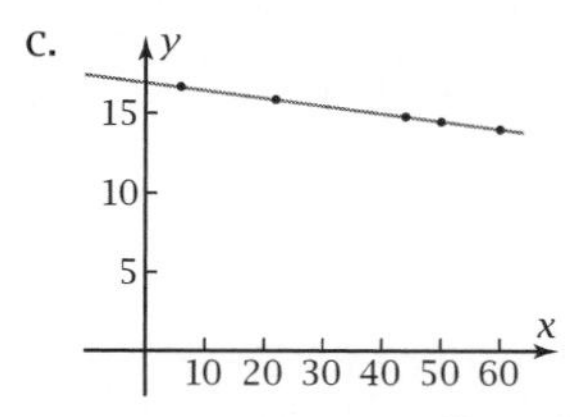

Data points are all on the line.

d. At $x = 0$ miles, the tank holds $\hat{y} = 17$ gal. The car gets 20 mi/gal.

e. $\hat{y}(340) = -0.05 \cdot 340 + 17 = 0$ gal

5. a.

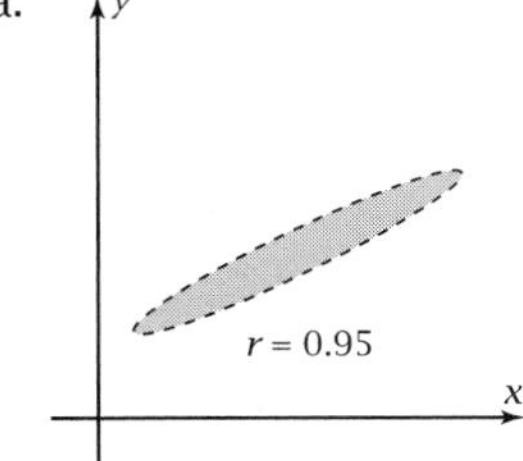

b.

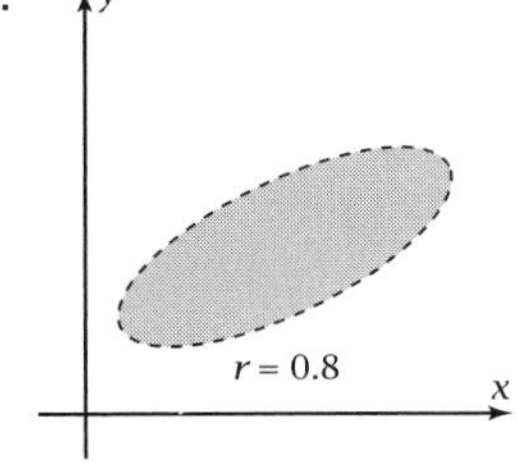

c.

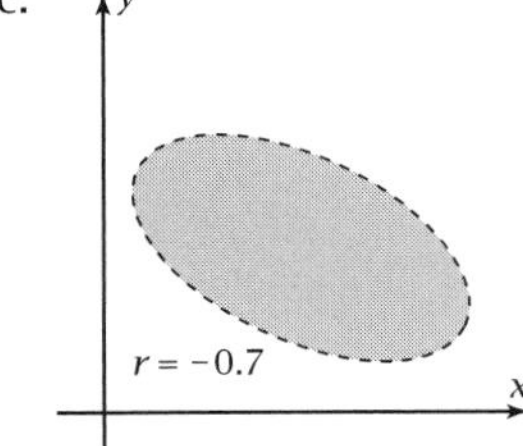

d.

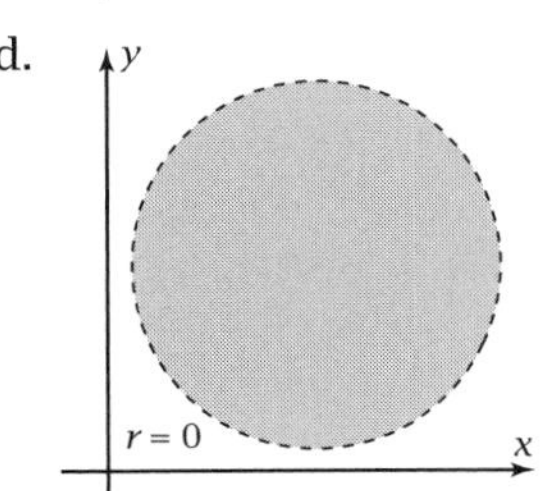

Problem Set 8-3

1. a. Both a power function and an exponential function have the proper right endpoint behavior: increasing to infinity. Only an exponential function has the correct left endpoint behavior: being nonzero.

   b. $\hat{y} = 346.9291... \cdot 1.4972...^x$, with $r = 0.9818...$

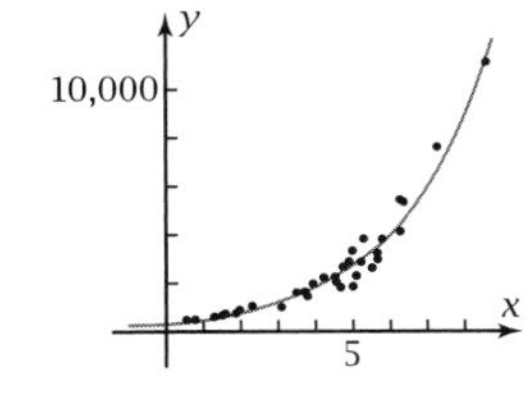

   c. $\hat{y}(0) = 346.9291... \approx 347$ bacteria

   $\hat{y}(24) = 5{,}584{,}729.3315... \approx 5.6$ million bacteria

   d. $x = 14.0331... \approx 14.0$ hr

   Check: $\hat{y}(14.0331...) = 100{,}000$ bacteria

3. a.

   Concave downward. The graph decreases more steeply (presumably to $-\infty$) toward $x = 0$ and increases less steeply as x gets larger.

   b. $\hat{y} = -138.1230... + 19.9956... \ln x$;

   $r = 0.99999999799...$, which is nearly 1

   c.

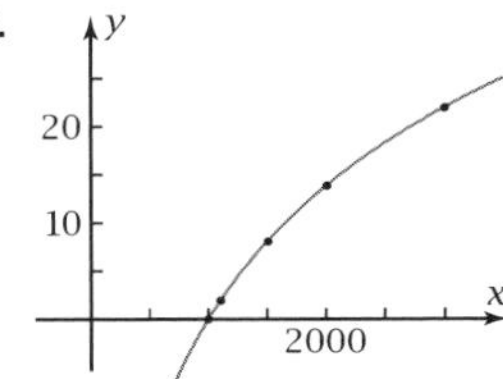

   d. $\hat{y}(2500) = 18.3236... \approx 18.32$ yr

   $$\neq \frac{13.86 + 21.97}{2} = 17.915 \text{ yr}$$

   e. $\hat{y}(5000) = 32.1835... \approx 32.18$ yr

   Extrapolation, because $5000 > 3000$. The bank probably uses a simple formula to calculate interest that would make the regression equation apply for all values of x.

5. a. Growth is basically exponential, but physical limits eventually make the population level off. A logistic function fits data that have asymptotes at both endpoints but are exponential in the middle.

   $$\hat{y} = \frac{327.5140...}{1 + 10.0703...e^{-0.4029...x}}$$

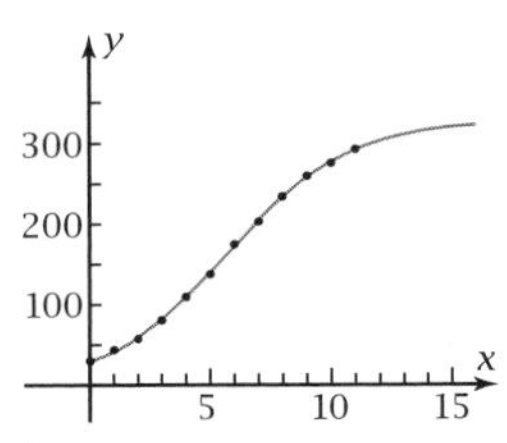

   b. $\hat{y}(20) \approx 326$ roadrunners

   $\hat{y} \rightarrow 327.5140... \approx 328$ roadrunners as $x \rightarrow \infty$

   The inflection point appears to be at $x \approx 5.5$ yr.

c. $\bar{y} = 158.5$ roadrunners

$y - \bar{y}$	$(y - \bar{y})^2$	$y - \hat{y}$	$(y - \hat{y})^2$
−128.5	16,512.25	0.4153...	0.1725...
−114.5	13,110.25	1.6330...	2.6669...
−100.5	10,100.25	−1.5676...	2.4574...
−77.5	6,006.25	−0.7496...	0.5619...
−48.5	2,352.25	1.1633...	1.3533...
−20.5	420.25	−1.7936...	3.2171...
16.5	272.25	2.3942...	5.7325...
44.5	1,980.25	−1.7203...	2.9596...
75.5	5,700.25	0.2077...	0.0431...
101.5	10,302.25	1.6919...	2.8628...
117.5	13,806.25	−1.7752...	3.1515...
134.5	18,090.25	0.4914...	0.2414...
SS_{dev} = 98,653.00		SS_{res} = 25.4205...	

$r^2 = 0.9997... \approx 1$

7. a. $\hat{y} = 34.7990... \cdot 0.9493...^x$, $r = -0.9952...$

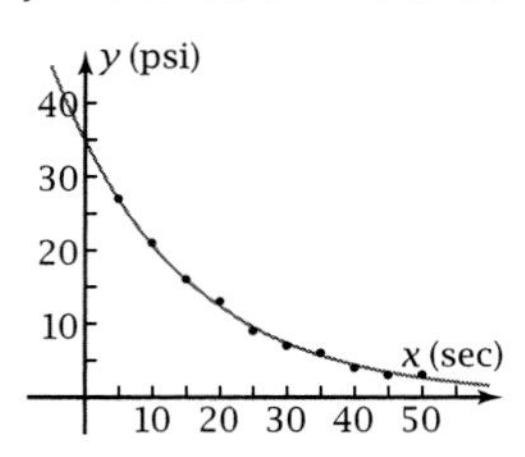

The exponential function is a good fit, both visually and according to r.

b. $\bar{y} = 10.9$ psi

$y - \bar{y}$	$(y - \bar{y})^2$	$y - \hat{y}$	$(y - \hat{y})^2$
16.1	259.21	0.1679...	0.0282...
10.1	102.01	0.3109...	0.0966...
5.1	26.01	0.0475...	0.0022...
2.1	4.41	0.6997...	0.4896...
−1.9	3.61	−0.4842...	0.2344...
−3.9	15.21	−0.3128...	0.0978...
−4.9	24.01	0.3613...	0.1305...
−6.9	47.61	−0.3477...	0.1209...
−7.9	62.41	−0.3523...	0.1241...
−7.9	62.41	0.4151...	0.1723...
SS_{dev} = 606.90		SS_{res} = 1.4971...	

$r^2 = 0.9975...$
$r = -0.9987... \neq -0.9952...$ from part a.

c. $\log \hat{y} = -0.0225...x + 1.5415...$
$r = -0.9952...$, as in part a.

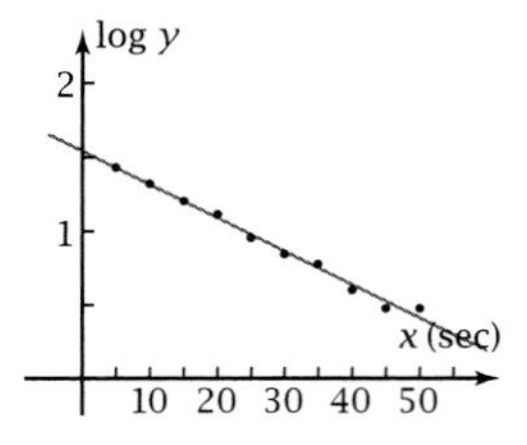

[−5, 60, 5, −0.25, 2, 0.5]

d. $\log \hat{y} = -0.0225...x + 1.5414...$, as in part c.

Problem Set 8-4

1. a. The scatter plot is decreasing, is concave up, has a finite value for $x = 0$, and seems to approach 0 as $x \to \infty$.
$\hat{y} = 1076.9102... \cdot 0.9998...^x$, $r = -0.9985...$

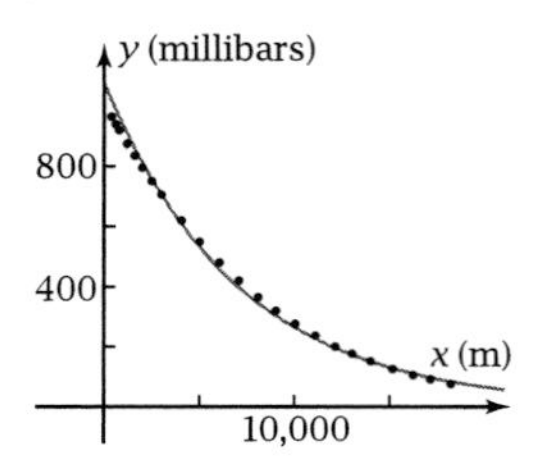

The points seem to lie very near or on the line.

b.

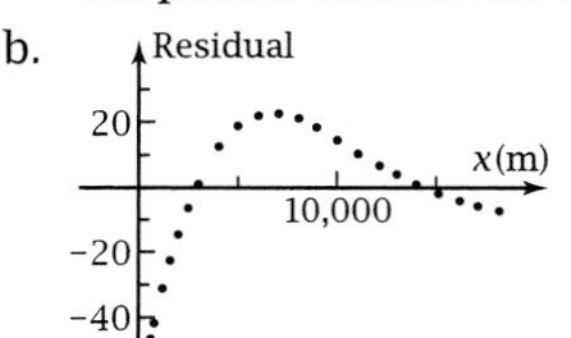

The residuals have a very definite pattern.

c. No, the residuals are as great as ≈ -52.7 millibars.

3. a. Exponential: $\hat{y} = 91.7362... \cdot 0.9996...^x$, $r = -0.9646...$

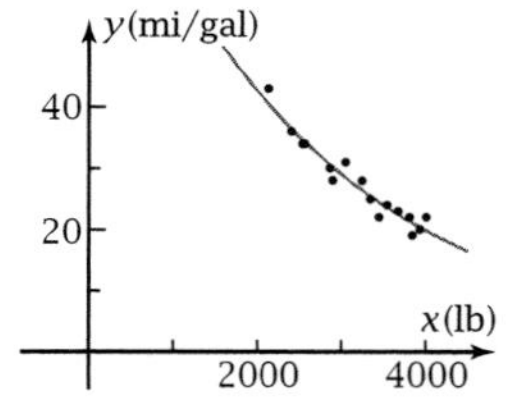

Power: $\hat{y} = 338{,}947.0156\ldots x^{-1.1721\ldots}$,
$r = -0.9672\ldots$

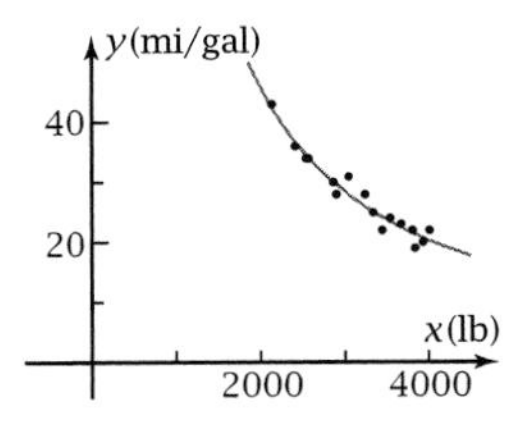

The functions appear to fit about equally well, both graphically and by their r-values.

b. Exponential:

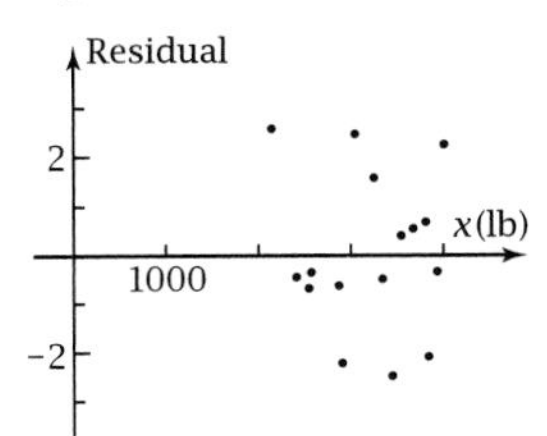

Power:

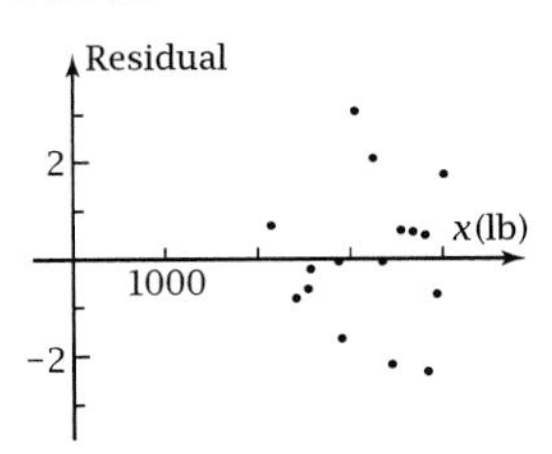

Both residual plots are fairly random. It is unclear which fits better.

c. Exponential: $\hat{y}(500) = 75.7486\ldots$ mi/gal
Power: $\hat{y}(500) = 232.5653\ldots$ mi/gal
The exponential model is much more reasonable. The right endpoint behavior is not significantly different.

5. a. Both would give a population of 0 at some finite time in the past. The exponential function would have even more rapid growth as years go by.

b. $\hat{y} = \dfrac{351.8082\ldots}{1 + 2.2400\ldots e^{0.0282\ldots}}$

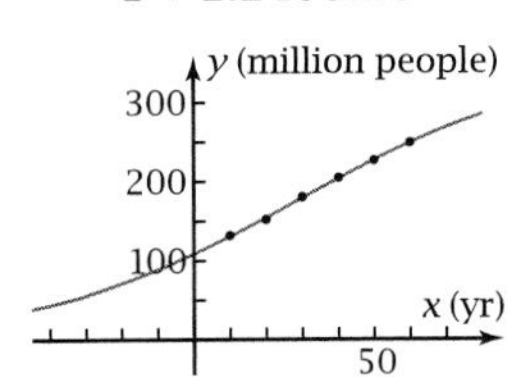

c. $\hat{y}(70) = 268.7380\ldots \approx 268.7$ million
$\hat{y} \rightarrow 351.8082\ldots$ million as $x \rightarrow \infty$

d. $\hat{y} = \dfrac{360.8765\ldots}{1 + 2.3126\ldots e^{0.0275\ldots}}$

$\hat{y} \rightarrow 360.8765\ldots$ million as $x \rightarrow \infty$

7. a.

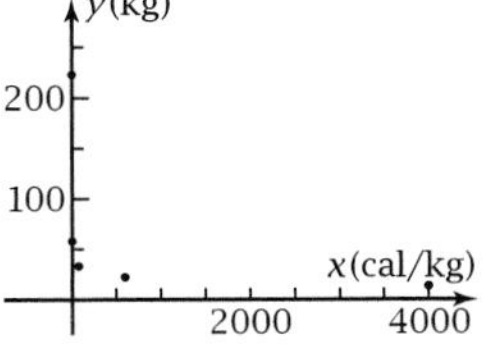

The leftmost data points are squeezed too close to the y-axis, and the rightmost data points too close to the x-axis, to show their relationship clearly.

ln (mass)	**ln (cal/kg)**
−0.1594...	2.3483...
0.3010...	1.7634...
1.8450...	1.5185...
2.7781...	1.3424...
3.6020...	1.1139...

$\log \hat{y} = -0.2774\ldots \log x + 2.0817\ldots$
$r = -0.9366\ldots$

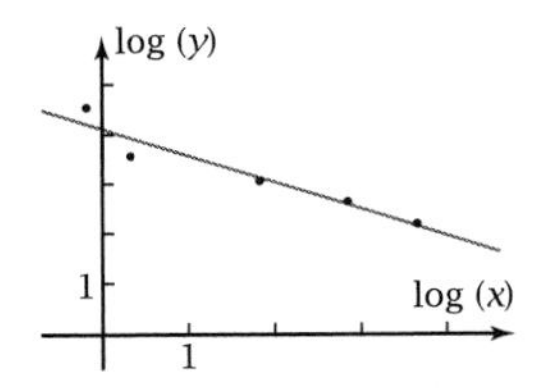

b. $\hat{y} = 120.7251\ldots x^{-0.2774}$, which is the same equation as found by power regression, with $r = -0.9366\ldots$, as in part a.

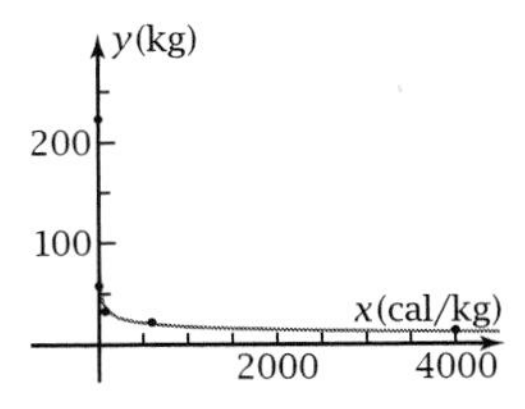

c. $\hat{y}(0.002) = 676.9192\ldots$ cal/kg

d. $\hat{y}(150{,}000) = 4.4244\ldots$ cal/kg, approximately 160% larger than or 260% of the actual value. You are extrapolating to an x-value quite far out of the data range. The logarithmic function is already not a good predictor for very small values of x (the residuals are very large), and it is not good for very large values either.

9. a. $\hat{y}_{\text{no whey}} = -1.71x + 57.64$

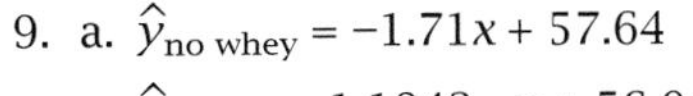

$\hat{y}_{\text{whey}} = -1.1942\ldots x + 56.04$

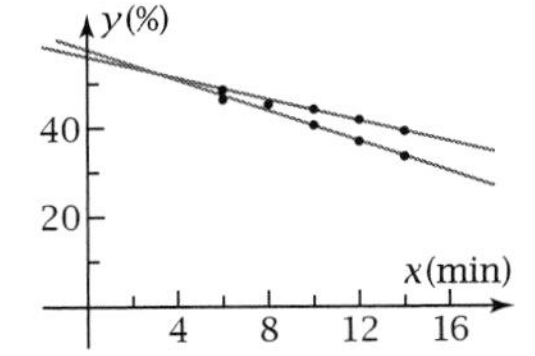

b. $\hat{y}_{whey}(8) = 46.4847...\%$; interpolation

c. If whey is used: $x = 21.8038...$
If no whey is used: $x = 16.1637...$
Extrapolation in both cases

d. $\hat{y}_{no\ whey}(0) \approx 57.6\%$, $\hat{y}_{whey}(0) \approx 56.0\%$.
They are close.

Problem Set 8-5

R1. a. A graphing calculator confirms that $\hat{y} = 1.6x + 0.9$.

b.

$\hat{y}$	$y - \hat{y}$	$(y - \hat{y})^2$
5.7	0.3	0.09
8.9	1.1	1.21
12.1	−3.1	9.61
15.3	1.7	2.89

c. $\Sigma(y - \hat{y})^2 + 13.80$

d. For $\hat{y} = 1.5x + 1.0$:

$\hat{y}$	$y - \hat{y}$	$(y - \hat{y})^2$
5.5	0.5	0.25
8.5	1.5	2.25
11.5	−2.5	6.25
14.5	2.5	6.25

$\Sigma(y - \hat{y})^2 + 15.00$

R2. a. $\bar{y} = 10.5$

b. $\bar{x} = 6$; $\hat{y}(6) = 10.5$

c.

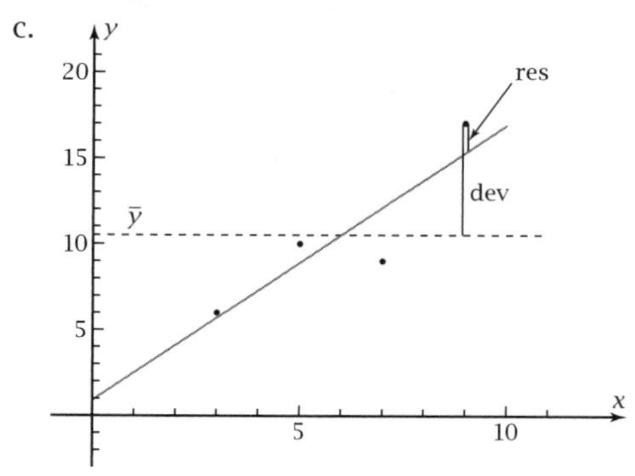

d.

$y - \hat{y}$	$(y - \hat{y})^2$
−4.5	20.25
−0.5	0.25
−1.5	2.25
6.5	42.25

$SS_{dev} = 65.00$. The deviations $y - \bar{y}$ don't take into account the variation $\hat{y}(x)$.

e. $r^2 = 0.7876...$

f. $r = 0.8875...$
The regression line has positive slope, so you choose the positive branch of the square root.

g. 8-5c: positive, closer to 1
8-5d: negative, very close to −1
8-5e: negative, closer to 0
8-5f: very close to 0 (neither very positive nor very negative)

R3. a. Logarithmic:
$\hat{y} = 136.6412... - 16.8782... \ln x$,
$r = -0.9979...$

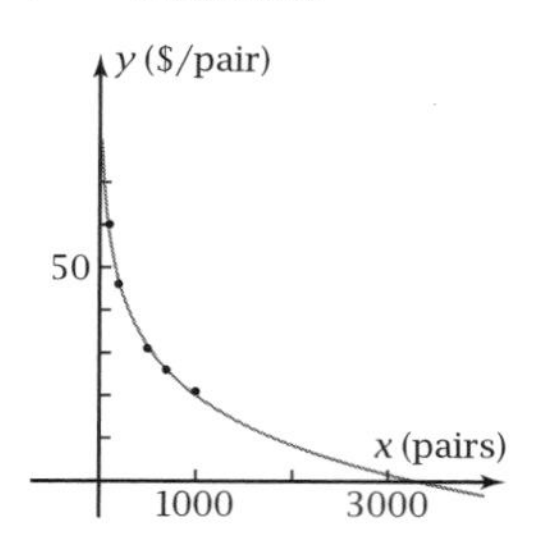

Power: $\hat{y} = 488.0261...x^{-0.4494...}$,
$r = -0.9970...$

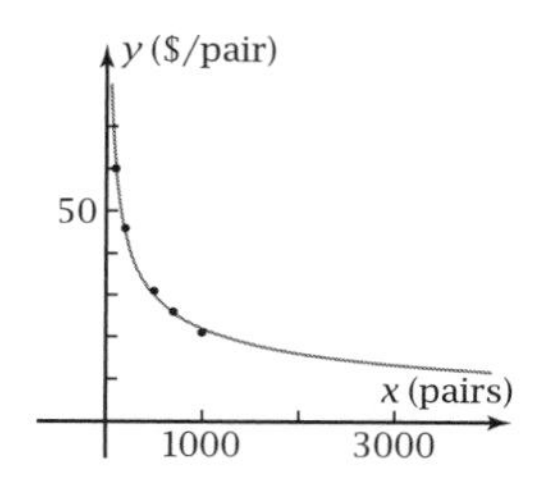

b. The logarithmic function predicts that the price of a pair eventually reaches \$0.00 and then even becomes negative. The power function never predicts a price of \$0.00 or lower.

c. $x = 5714.7585... \rightarrow 5714$ pairs
This is found by extrapolation.

d. $\hat{y}(1) = 488.0261... \approx \$488.03$ per pair

e. $\dfrac{\hat{y}(2x)}{\hat{y}(x)} = 0.7323... \approx 73\%$; multiply–multiply

R4. a. A power function would have $\hat{y}(0) = 0$, and a logarithmic function would have $\hat{y}(0) \rightarrow -\infty$, but both the exponential function and linear function would have $\hat{y}(0) \approx 300$.

b. Linear: $\hat{y} = 1.7115...x + 320.5101...$,
$r = 0.9974...$

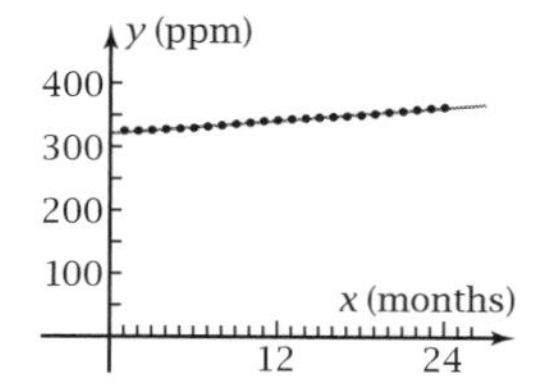

Exponential: $\hat{y} = 320.9749... \cdot 1.0050...^x$,
$r = 0.9977...$

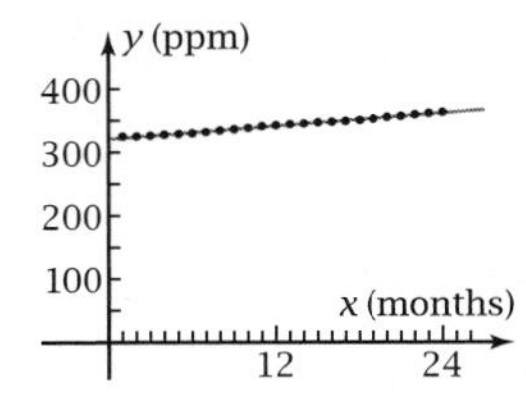

Linear: $\hat{y}(13) = 342.5542... \approx 342.6$ ppm
Exponential: $\hat{y}(13) = 342.7599... \approx 342.8$ ppm
Actual: 343.5 ppm
Linear: $\hat{y}(20 \cdot 12) = 731.2753... \approx 731.3$ ppm
Exponential:
$\hat{y}(20 \cdot 12) = 1066.9906... \approx 1067.0$ ppm

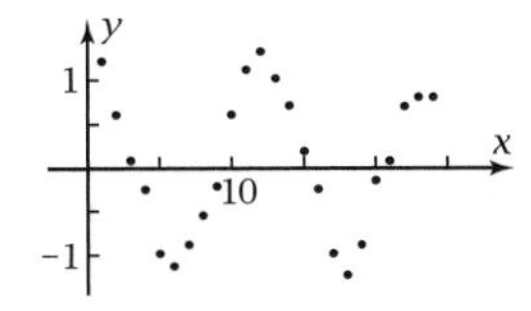

There is a definite sinusoidal pattern, with a period of about one year and a maximum in the winter and a minimum in the summer.

CHAPTER 9

Problem Set 9-1

1. $\frac{1}{12}$
3. $\frac{5}{6}$
5. $\frac{1}{6}$
7. $\frac{5}{9}$
9. $\frac{1}{1}$
11. $\frac{1}{18}$
13. $\frac{1}{12}$

Problem Set 9-2

1. a. A random experiment
   b. 52 c. 12 d. $\frac{3}{13}$ e. $\frac{1}{2}$
   f. $\frac{1}{13}$ g. $\frac{5}{13}$ h. $\frac{1}{52}$ i. 1
   j. 0
3. Answers will vary.

Problem Set 9-3

1. a. 91 b. 20
3. a. 16 b. 55 c. 20
5. 693
7. a. 9 b. 18 c. 504
9. 20
11. 1018
13. a. 10 b. 9 c. 90
    d. 8 e. 720 f. 3,628,800
15. a. The second plan gives 10,799,100 more plates.
    b. 24,317,748
    c. No, there would not be enough plates.

Problem Set 9-4

1. a. 132 b. 11,880 c. 479,001,600
3. 15,600
5. 210
7. a. 720 b. 120 c. $\frac{1}{6}$
   d. $16\frac{2}{3}\%$ e. $\frac{1}{720}$
9. a. 362,880 b. 40,320
   c. $\frac{1}{9}$ d. $11\frac{1}{9}\%$
11. a. $\frac{1}{3}$ b. $\frac{1}{12}$ c. $\frac{1}{84}$
13. a. 40,320 b. 10,080 c. $\frac{1}{4}$
15. a. 360 b. 840 c. 20
    d. 415,800 e. 5040 f. 3360
17. a. 24 b. 120 c. 40,320
19. a. 144 b. $\frac{1}{35}$

Problem Set 9-5

1. 10
3. 2,220,075
5. 1
7. 1
9. 360
11. $7.2710818848902 \times 10^{44}$
13. 792; "group"
15. 1,344,904; no
17. a. 20 b. 15 c. 35 d. 1
19. a. 2,598,960 b. 635,013,559,600
    c. The order of the cards in a "hand" is not important.
21. a. 45 b. 252
    c. 45. Choosing 8 elements to include is the same as choosing 2 elements *not* to include.
23. a. 120 b. 2,598,960 c. 311,875,200
    d. Permutation: parts a and c; combination: part b
25. a. $\frac{140}{429} \approx 33\%$ b. $\frac{175}{429} \approx 41\%$
    c. $\frac{105}{143} \approx 73\%$ d. $\frac{5}{13} \approx 38\%$
27. a. $\frac{4}{7} \approx 57\%$ b. $\frac{5}{7} \approx 71\%$ c. $\frac{2}{7} \approx 29\%$
    d. $P(\text{b}) = 1 - P(\text{c})$. This is because part b is the opposite (complement) of part c.
29. a. 75,287,520 b. 7,376,656 c. $\frac{97}{990} \approx 9.8\%$
    d. No, because it is not likely that the defective bulbs will be found.

Problem Set 9-6

1. a. 56% b. 30% c. 20%
   d. 6% e. 94%
3. a. 28% b. 18% c. 12%
   d. 42% e. 54%
5. 99.96%
7. a. 33.6% b. 2.4% c. 97.6% d. 5.6%
9. a. ≈36.77% b. ≈99.99%
11. a. i. 12%
    ii. 28%
    iii. 48%
    iv. 12%
    b. 100%. These are all the possibilities there are.
13. a. 0.09% b. 20%
    c. They are not independent. An engine is more likely to fail if the other one has already failed.

Problem Set 9-7

1. a. $\frac{3}{4}$
   b. $P(0) = \frac{1}{64} \approx 1.6\%$; $P(1) = \frac{9}{64} \approx 14\%$;
   $P(2) = \frac{27}{64} \approx 42\%$; $P(3) = \frac{27}{64} \approx 42\%$
   c. 100%. These are all the possibilities there are.
   d.

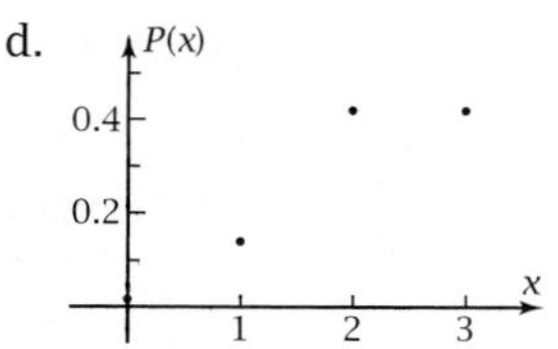

   e. Binomial
3. a. $P(3) = {}_{10}C_3(0.7)^3(0.3)^7 = 0.090\ldots$
   b.

x	P(x)
0	0.000005...
1	0.0001...
2	0.0014...
3	0.0090...
4	0.0367...
5	0.1029...
6	0.2001...
7	0.2668...
8	0.2334...
9	0.1210...
10	0.0282...

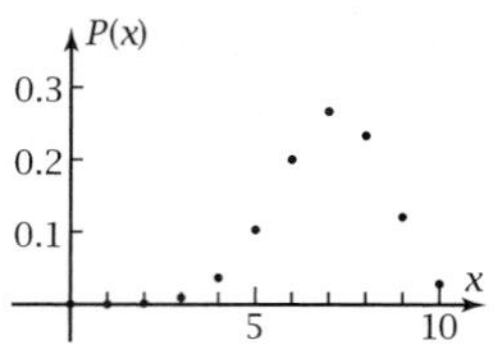

   c. $P(\text{more than } 5) = 0.8497\ldots$
   $P(\text{at most } 5) = 0.1502\ldots$
   "More than 5" is more likely.
5. a. $P(x) = {}_{20}C_x a^{20-x} b^x$

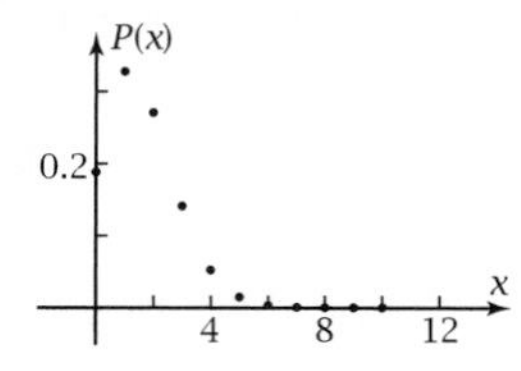

   b.

x	P(x)
0	0.18869...
1	0.32816...
2	0.27109...
3	0.14143...

   c. $1 - [P(0) + P(1) + P(2) + P(3)] = 0.07063\ldots$
7. a. i. 0.5471... ii. 0.3316...
   iii. 0.0988... iv. 0.1212...
   v. 0.0224...
   b. 0.9991... ≈ 99.91%
9. a. 0.1296 b. 0.0256
   c. 0.2073... d. 0.0614...
   e. i. 0.2073... ii. 0.0921...
   iii. 0.1658... iv. 0.1105...
   f. 0.7102... ≈ 71%
   g. A six-game series is most likely, followed by seven, five, and then four games.
11. a.

x	PDs of x	No. of PDs of x
1	–	0
2	1	1
3	1	1
4	1, 2	2
5	1	1
6	1, 2, 3	3
7	1	1
8	1, 2, 4	3
9	1, 3	2
10	1, 2, 5	3

b.

x	Nos. That Have x PDs	No. of Nos. That Have x PDs
0	1	1
1	2, 3, 5, 7	4
2	4, 9	2
3	6, 8, 10	3

c.

x	$P(x)$
0	$\frac{1}{10}$
1	$\frac{4}{10}$
2	$\frac{2}{10}$
3	$\frac{3}{10}$

d.

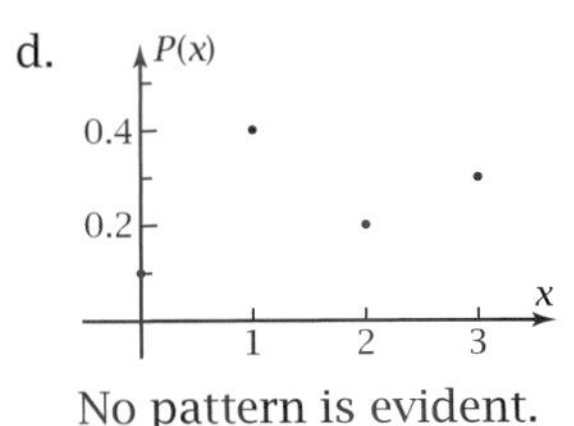

No pattern is evident.

13. a. 0.9972...   b. 0.9945...

c. $\frac{364 \cdot 363}{365^2} = 0.9918...$

d. $\frac{_{365}P_{10}}{365^{10}} = 0.8830...$

e. $1 - \frac{_{365}P_{10}}{365^{10}} = 0.1169...$

f. Here are a few selected values:

x	$P(x)$
10	0.1169...
20	0.4114...
30	0.7063...
40	0.8912...
50	0.9703...
60	0.9941...

Using the sequential function mode on the TI-83, enter in the y= menu:
nMin = 1
u(n) = 1 − (1 − u(n − 1))(366 − n)/365
u(nMin) = {0}

g.

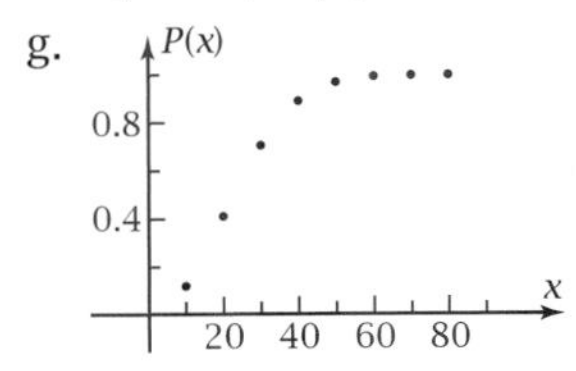

h. i. There must be at least 23 people.
ii. There must be at least 57 people.

Problem Set 9-8

1. a.

Number · P(number)
0.0
0.2
0.5
1.2
0.4

Total: 2.3 neutrons per fission

b. Mathematical expectation is a mathematical abstraction, meaning, for example, that you would expect 10 fissions to produce 23 neutrons, 100 to produce 230, and so on.

3. a.

No. of Cars	$P(A)$	a	$P(A) \cdot a$
4 Minivans	0.5	\$400	\$200
2 Station Wagons	0.7	\$200	\$140
1 Hybrid	0.8	\$100	\$80
1 Sedan	0.9	\$100	\$90

The mathematical expectation of Option A is \$510.

b. P(selling all required cars) = 0.252

c. a = \$2000
E(selling all required cars) = P(selling all required cars) · a = (0.252)(\$2000) = \$504

d. He should choose Option A.

5. a, b. Let X = no. of hits. Then $P(X) = {_5C_x}(0.3)^x(0.7)^{5-x}$

x	$P(x)$	$x \cdot P(x)$
0	0.16807	0.00000
1	0.36015	0.36015
2	0.30870	0.61740
3	0.13230	0.39690
4	0.02835	0.11340
5	0.00243	0.01215

$E(x) = 1.5$

7. a. $\frac{1}{5}; \frac{4}{5}$ b. 0   c. $\frac{1}{16}$ d. $\frac{1}{6}; \frac{3}{8}$

e. If you can eliminate at least one answer, it is worthwhile.

9. a. $D(15) \approx 15$ b. $A(16) = 9985$ c. $D(16) \approx 15$

d.

x	$P(x)$	$A(x)$	$D(x)$
15	0.00146	10,000	15
16	0.00154	9,985	15
17	0.00162	9,970	16
18	0.00169	9,954	17
19	0.00174	9,937	17
20	0.00179	9,920	18

e. $I(x) = 40 \cdot A(x)$ $\quad$ $O(x) = 20{,}000 \cdot D(x)$

$I(x) = 40 \cdot A(x)$	$O(x) = 20{,}000 \cdot D(x)$
\$400,000	\$300,000
\$399,400	\$300,000
\$398,800	\$320,000
\$398,160	\$340,000
\$397,480	\$340,000
\$396,800	\$360,000

f.

$NI(x) = I(x) - O(x)$
\$100,000
\$99,400
\$78,800
\$58,160
\$57,480
\$36,800

The company has less income each year because (1) there are fewer people still alive to pay and (2) it has more outgo because there are more people dying.

g. $\sum_{x=15}^{20} NI(x) = \$430{,}640$ for six years, which averages to \$71,773 per year. This is definitely enough to pay a full-time employee.

Problem Set 9-9

R1. a. $\frac{3}{8}$ b. $\frac{3}{8}$ c. $\frac{1}{8}$
d. 0 e. $\frac{1}{2}$ f. $\frac{7}{8}$
g. 1 h. $\frac{1}{4}$ i. $\frac{3}{4}$

R2. a. 25 b. 13
c. An outcome is one of the equally likely results of a random experiment. An event is a set of outcomes.
d. i. $\frac{13}{25}$ ii. $\frac{8}{25}$
iii. $\frac{16}{25}$ iv. 1
v. 0

R3. a. i. 220 ii. 31
b. i. 15 ii. 20

R4. a. i. 35,904 ii. $34! \approx 2.9523 \times 10^{38}$
b. 120 c. $\frac{10}{21}$ d. 32,432,400

R5. a. $\dfrac{7!}{3! \cdot 4!} = 35$
b. In a permutation, order is important. In a combination, it is not. That is, rearrangements of the same choice of objects are considered different permutations but the same combination.
c. i. 3,921,225
ii. 94,109,400
iii. 4,178,378,490

d. $0.2610\ldots \approx 26.1\%$

R6. a. i. 6% ii. 56%
iii. 62% iv. 38%
b. i. 51% ii. 9%
iii. 18% iv. 22%
v. 51% + 9% + 18% + 22% = 100%

R7. a. $P(4) = {}_6C_4(0.6)^4(0.4)^2 = 0.31104$
b.

x	$P(x)$
0	0.004096
1	0.036864
2	0.138240
3	0.276480
4	0.311040
5	0.186624
6	0.046656

c.

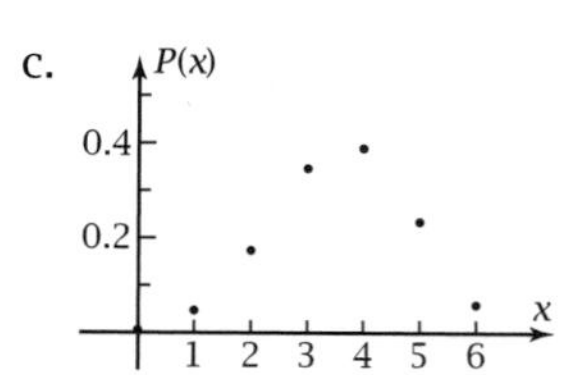

d. 0.54432
e. The probabilities are the terms in the expansion $(0.6 + 0.4)^6$ of the binomial 0.6 + 0.4.

R8. a. $P(21) = {}_{21}C_{21}(0.9)^{21}(0.1)^0 = 0.1094\ldots$
$P(20 \text{ or fewer}) = 1 - P(21) = 0.8905\ldots$
$P(21) \cdot a(21) + P(20 \text{ or fewer}) \cdot a(20 \text{ or fewer})$
$= 0.1094\ldots \cdot \$1800 + 0.8905\ldots \cdot \$2100$
$\approx \$2067.17$
b. $P(22) = 0.0984\ldots$
$P(21) = 0.2407\ldots$
$P(20 \text{ or fewer}) = 0.6608\ldots$
$E(\text{booking 22 passengers}) \approx \$2068.53$
c. $P(23) = 0.0886\ldots$
$P(22) = 0.2264\ldots$
$P(21) = 0.2768\ldots$
$P(20 \text{ or fewer}) = 0.4080\ldots$
$0.0886\ldots \cdot \$1400 + 0.2264\ldots \cdot \$1700$
$+ 0.2768\ldots \cdot \$2000 + 0.4080\ldots \cdot \$2300$
$\approx \$2001.29$
This is about \$1.29 more than they would make without overbooking.
d. Answers may vary. Customer dissatisfaction and federal regulation are two possible reasons.
e. $(0.1)(72) + (0.2)(86) + (0.2)(93) + (0.2)(77) + (0.3)(98) = 87.8$. She will get at least a B.

f. The probability of an outcome may be thought of as weighing the value associated with that outcome proportionately to its chance of occurring.

Problem Set 9-10

1. a. $y = mx + b$
   b. $y = ax^2 + bx + c$
   c. $y = a + b \ln x$ or $y = a + b \log x$
   d. $y = ab^x$
   e. $y = ax^b$

2. 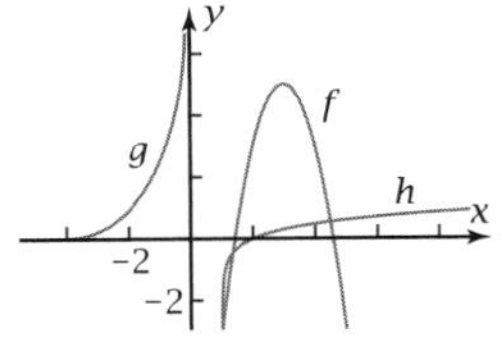

3. a. Multiply-add
   b. Multiply-multiply
   c. Constant second differences

4. a. Logarithmic
   b. Exponential
   c. Logistic
   d. Quadratic
   e. Power
   f. Linear

5. 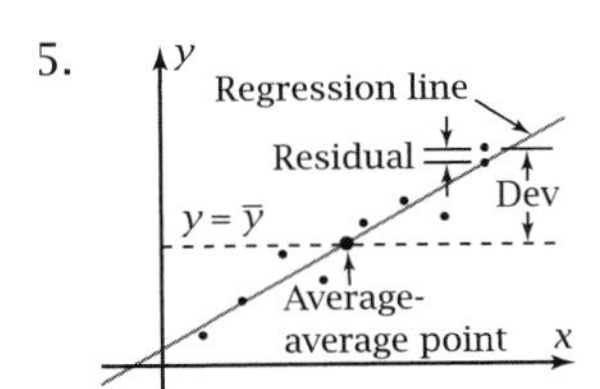

6. $\hat{y} = 0.8766\ldots x + 0.6290\ldots$
   The residual is smaller than the deviation.

7. The average-average point is $(\bar{x}, \bar{y}) = (4.5, 4.5738\ldots)$.

8. $\hat{y} = 3(4) + 5 = 17$
   Residual $= 15 - 17 = -2$

9. SS_{res} is a minimum.

10. $r^2 = \dfrac{100 - 36}{100} = 0.64$

11. Number of selections is ${}_{20}C_7 = \dfrac{20!}{7!\,13!}$

    Number of orders is ${}_{20}P_7 = 7! \cdot {}_{20}C_7 = \dfrac{20!}{13!}$

12. $P(A \text{ or } B) = P(A) + P(B) - P(A \text{ and } B)$
    $= 0.6 + 0.8 - (0.6)(0.8) = 0.92$

13. a. $P(C \text{ and } D) = 0$ (The events are mutually exclusive.)
    b. $P(\text{not } D) = 1 - P(D) = 1 - 0.2 = 0.8$
    c. $E = (0.1)(6.00) + (0.2)(-2.00) + (0.7)(1.00)$
    $= \$0.90$

14. a. Add-multiply (add 3 to x, multiply y by 0.5); Exponential function
    b. $y = 100(0.7937\ldots)^x$

x	y Given	y Calculated	
8	16	15.7490	Close
11	8	7.8745	Close
17	2	1.9686	Close

    c. $\hat{y} = 99.9085\ldots(0.7946\ldots)^x$
    $r = -0.999993537\ldots$, indicating a good fit because it is close to -1.
    d. $\hat{y} = 3.9989\ldots \approx 4.0$ units
    Interpolation, because 14 m is within the given data.

15. a. $\hat{y} = \dfrac{263.8737\ldots}{1 + 314.7597\ldots e^{-1.2056\ldots x}}$
    b. Actual point of inflection is $(4.7707\ldots, 131.9318\ldots)$.

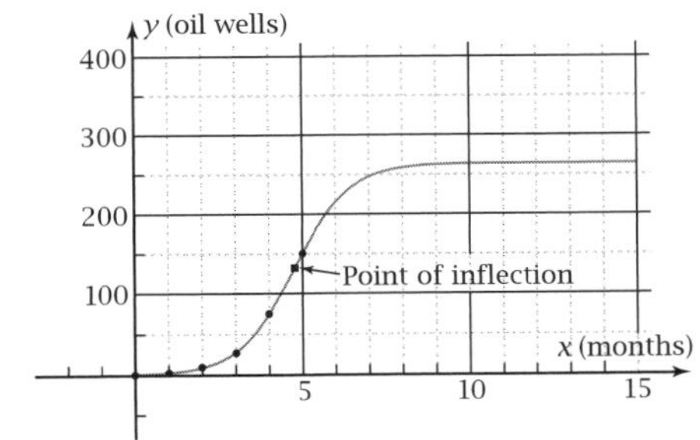

    c. The horizontal asymptote is at $y = 263.8737\ldots$, meaning that about 264 wells were ultimately drilled.
    d. An exponential model predicts that the number of oil wells would grow without bound. The logistic model shows that the number of wells will level off because of overcrowding.

16. a. $\hat{y} = 3.6472\ldots x - 145.1272\ldots$
    $r = 0.999717\ldots$, indicating a good fit.
    b. Extrapolating to lower values of x indicates that the number of chips per minute becomes negative, which is unreasonable. Solving for $\hat{y} = 0$ gives $x = 39.7906\ldots$, indicating the domain should be about $x \geq 40$.

c. Extrapolation to higher values of x indicates that the chirp rate keeps increasing. Actually, the crickets will die if the temperature gets too high.

d.

x	Residual
50	−2.23
55	−0.47
60	0.29
65	1.05
70	1.81
75	1.58
80	0.34
85	0.10
90	−1.12
95	−1.36

The graph shows a definite pattern, so there is something in the data that is not accounted for by the linear function.

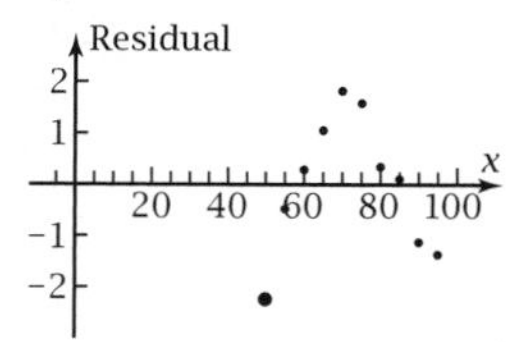

17. a. $y = \dfrac{\log 53}{\log 9} = 1.8069\ldots$

b. $x = 1.0314\ldots$

c. $\log_5 47 = 2.3922\ldots$

d.

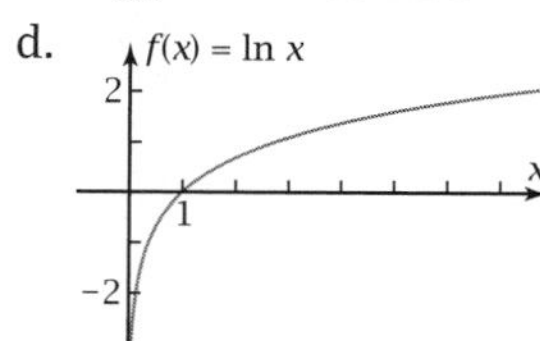

$f(1) = 0$ because $\ln 1 = 0$, because $\ln 1 = \log_e 1$, and $e^0 = 1$.

18. a. $P(\text{point down}) = 1 - 0.4 = 0.6$

b. $P(0) = {}_5C_0 \cdot 0.4^0 \cdot 0.6^5 = 0.07776$
$P(1) = {}_5C_1 \cdot 0.4^1 \cdot 0.6^4 = 0.2592$
$P(2) = {}_5C_2 \cdot 0.4^2 \cdot 0.6^3 = 0.3456$
$P(3) = {}_5C_3 \cdot 0.4^3 \cdot 0.6^2 = 0.2304$
$P(4) = {}_5C_4 \cdot 0.4^4 \cdot 0.6^1 = 0.0768$
$P(5) = {}_5C_1 \cdot 0.4^5 \cdot 0.6^0 = 0.01024$

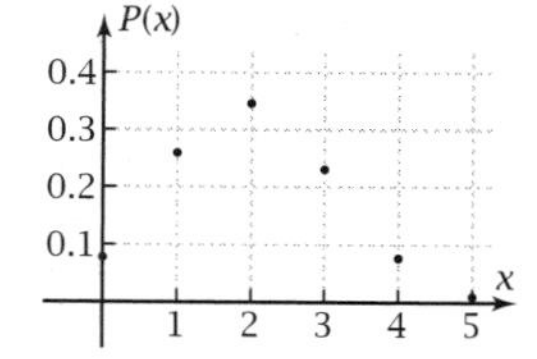

Binomial distribution

c. $P(2) = {}_6C_2 \cdot 0.4^2 \cdot 0.6^4 = 0.31104$

d. ${}_6C_2 = \dfrac{6!}{2!\,4!} = 15$

This is the number of different ways there could be 2 ups and 4 downs, each of which has the probability $0.4^2 \cdot 0.6^4$.

e. ${}_6P_2 = \dfrac{6!}{4!} = 30$

19. a. $P(\text{H and A}) = (0.7)(0.8) = 0.56$

b. $P(\text{not A}) = 1 - 0.8 = 0.2$
$P(\text{H, not A}) = (0.7)(0.2) = 0.14$

c. $P(\text{A, not H}) = (0.8)(0.3) = 0.24$

d. $P(\text{Neither}) = (0.3)(0.2) = 0.06$

e. $ME = (0.56)(8000) + (0.14)(3000) + (0.24)(4000) + (0.06)(0) = \$5680$

20. Answers will vary.

CHAPTER 10

Problem Set 10-1

1.

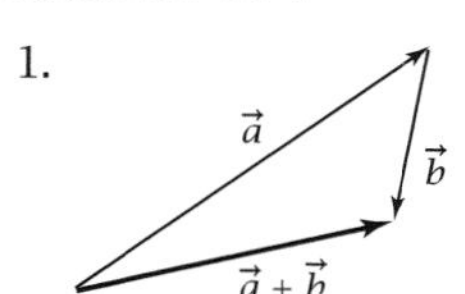

3.

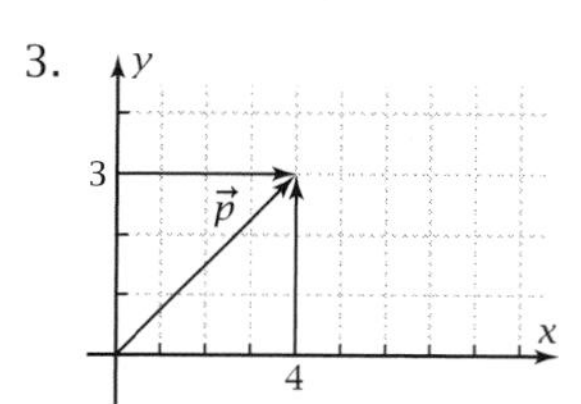

5.

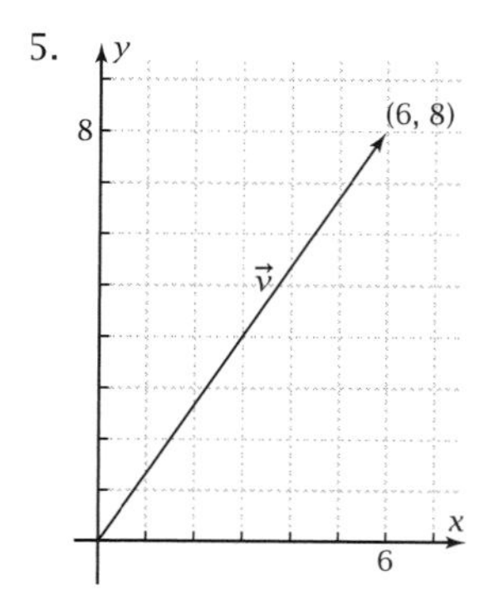

$\sqrt{6^2 + 8^2} = 10;$

$\tan^{-1}\left(\frac{8}{6}\right) = 53.1301\ldots° \approx 53.1°$

7.

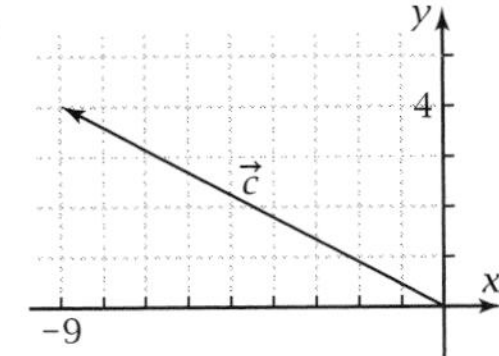

$\tan^{-1}\left(\frac{4}{-9}\right) = 156.0375...° \approx 156.0°$
The calculator gives $-23.962...°$.

Problem Set 10-2

1. a.

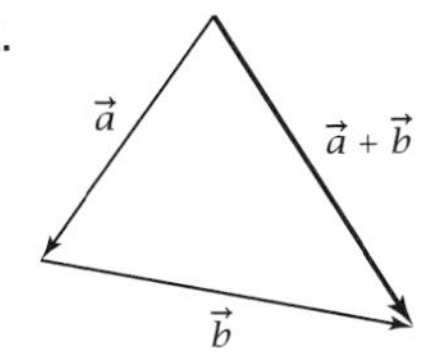

b.

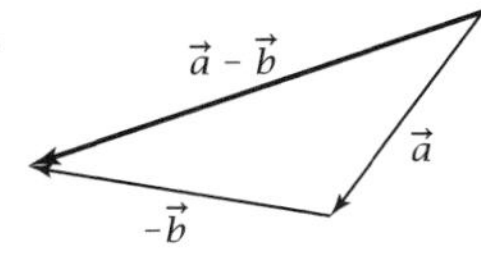

c.

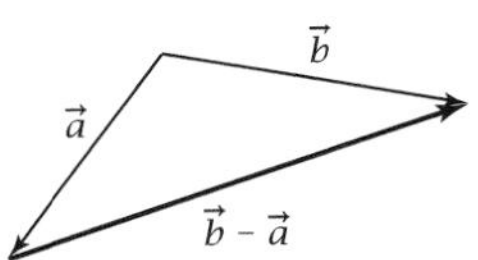

d.

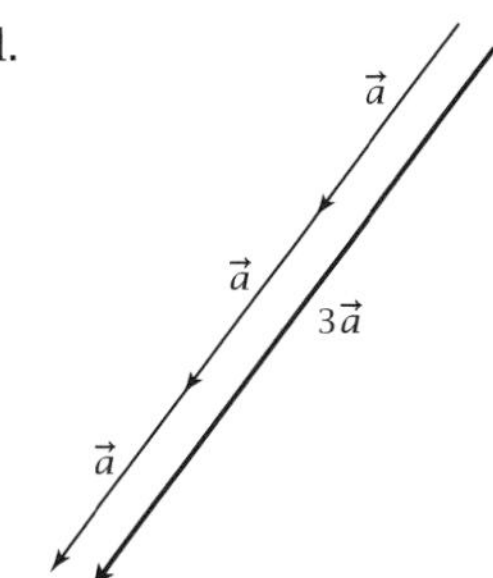

e.

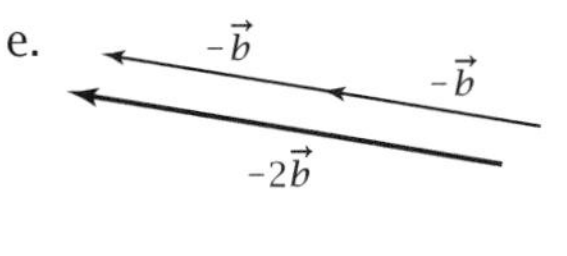

f.

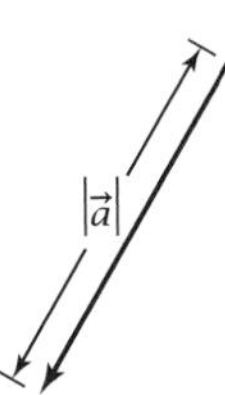

3. a. $\vec{a} + \vec{b} = 9\vec{i} + 2\vec{j}$
$\vec{a} - \vec{b} = -5\vec{i} + 8\vec{j}$
$-\vec{a} = -2\vec{i} - 5\vec{j}$
$2\vec{a} + 3\vec{b} = 25\vec{i} + \vec{j}$

b. $|\vec{a}| + |\vec{b}| = \sqrt{29} + \sqrt{58}$
$|\vec{a} + \vec{b}| = \sqrt{85}$
No

c. $\vec{u} = \dfrac{2}{\sqrt{29}}\vec{i} + \dfrac{5}{\sqrt{29}}\vec{j}$
$= (0.3713...)\vec{i} + (0.9284...)\vec{j}$
$10\vec{u} = \dfrac{20}{\sqrt{29}}\vec{i} + \dfrac{50}{\sqrt{29}}\vec{j}$
$= (3.713...)\vec{i} + (9.284...)\vec{j}$

d.

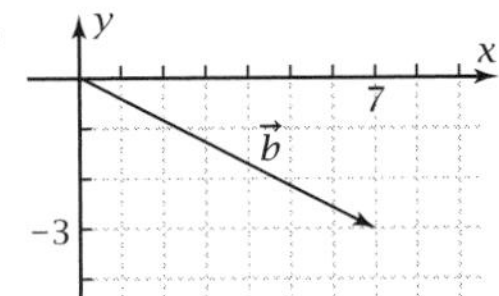

$\tan^{-1}\left(\frac{-3}{7}\right) = 336.8014...° \approx 336.8°$

5. $\overrightarrow{AB} = -\vec{i} + 3\vec{j}$

7. $\overrightarrow{BA} = 2\vec{i} + 4\vec{j}$

9. a. $\vec{A} = 20\vec{i} + 73\vec{j}$, $\vec{B} = 45\vec{i} + 10\vec{j}$
b. $\overrightarrow{AB} = 25\vec{i} - 63\vec{j}$
c. $\overrightarrow{AR} = 10\vec{i} - 25.2\vec{j}$
d. $\vec{R} = 30\vec{i} + 47.8\vec{j}$
e. $|\vec{R}| \approx 56.43$ km; $\theta \approx 57.9°$
f. $|\overrightarrow{AR}| \approx 27.1$ km to Artesia

11. $\vec{P} = \vec{A} + \dfrac{1}{3}\overrightarrow{AB} = 6\vec{i} + \dfrac{19}{3}\vec{j}$

13. $\overrightarrow{EF} = 4\vec{i} - 6\vec{j}$
$\dfrac{1}{2}\overrightarrow{EF} = 2\vec{i} - 3\vec{j}$
$\vec{M} = \vec{E} + \dfrac{1}{2}\overrightarrow{EF} = 8\vec{i} - \vec{j}$
But this is just
$\dfrac{1}{2}(16\vec{i} - 2\vec{j}) = \dfrac{1}{2}(\vec{E} + \vec{F})$
The position vector of the midpoint of two points is just the average of their position vectors.

15. a.

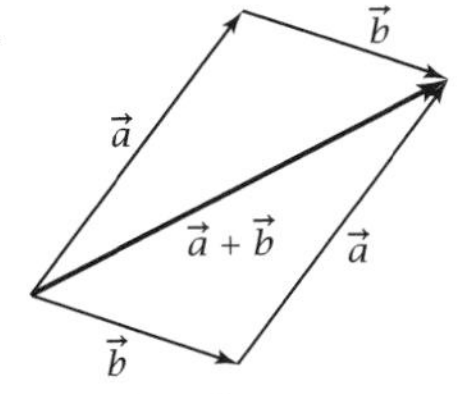

$\vec{a} + \vec{b} = \vec{b} + \vec{a}$

b. Use the definition of vector addition and the associativity of addition for scalars.

c.

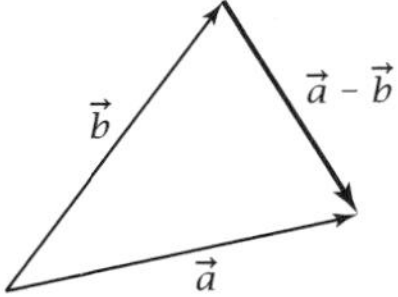

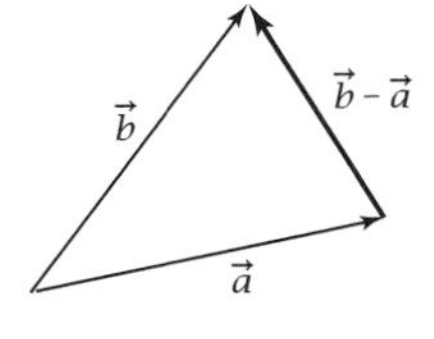

$\vec{a} - \vec{b} \neq \vec{b} - \vec{a}$

d. Use the definition of vector addition, the definition of scalar multiplication, and the distribution property of scalars.

e. Use the definition of vector addition and the property that the real numbers are close under addition.

If there were no zero vector, a sum of the form $(a\vec{i} + b\vec{j}) + [(-a)\vec{i} + (-b)\vec{j}] = [a + (-a)\vec{i} + [b + (-b)\vec{j}]$ $= 0\vec{i} + 0\vec{j}$ would not yield a vector.

Problem Set 10-3

1.

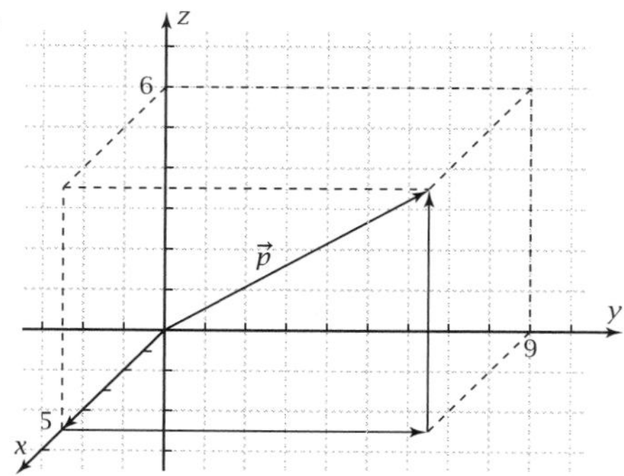

3.

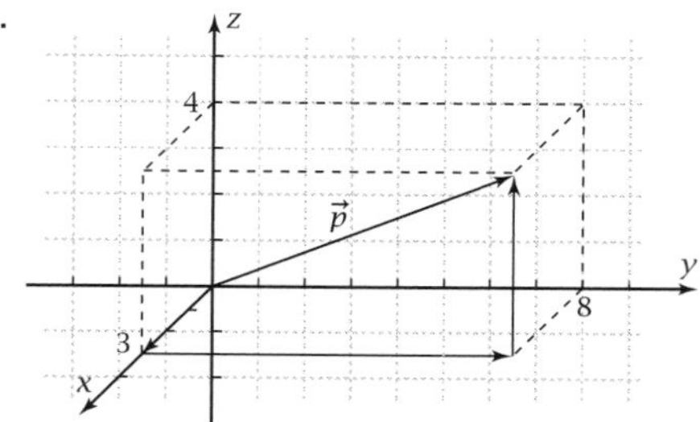

5. a. $\vec{a} + \vec{b} = 11\vec{i} - 3\vec{j} - 2\vec{k}$

$\vec{a} - \vec{b} = -3\vec{i} + 7\vec{j} - 4\vec{k}$

$\vec{b} - \vec{a} = 3\vec{i} - 7\vec{j} + 4\vec{k}$

b. $3\vec{a} = 12\vec{i} + 6\vec{j} - 9\vec{k}$

$6\vec{a} - 5\vec{b} = -11\vec{i} + 37\vec{j} - 23\vec{k}$

c. $|\vec{a} + \vec{b}| = \sqrt{134} = 11.5758...$

$|\vec{a}| + |\vec{b}| = \sqrt{29} + \sqrt{75} = 14.0454...$

No

d. $\vec{u} = \dfrac{7\sqrt{3}}{15}\vec{i} - \dfrac{\sqrt{3}}{3}\vec{j} + \dfrac{\sqrt{3}}{15}\vec{k}$

$= (0.8082...)\vec{i} - (0.5773...)\vec{j} + (0.1154...)\vec{k}$

$20\vec{u} = \dfrac{28\sqrt{3}}{3}\vec{i} - \dfrac{20\sqrt{3}}{3}\vec{j} + \dfrac{4\sqrt{3}}{3}\vec{k}$

$= (16.1658...)\vec{i} - (11.5470...)\vec{j} + (2.3094...)\vec{k}$

7. $\overrightarrow{RS} = 3\vec{i} + 7\vec{j} - 6\vec{k}$

$|\overrightarrow{RS}| = \sqrt{94}$

9. $\overrightarrow{BA} = 6\vec{i} + 7\vec{j} + 6\vec{k}$

$|\overrightarrow{BA}| = 11$

11. a.

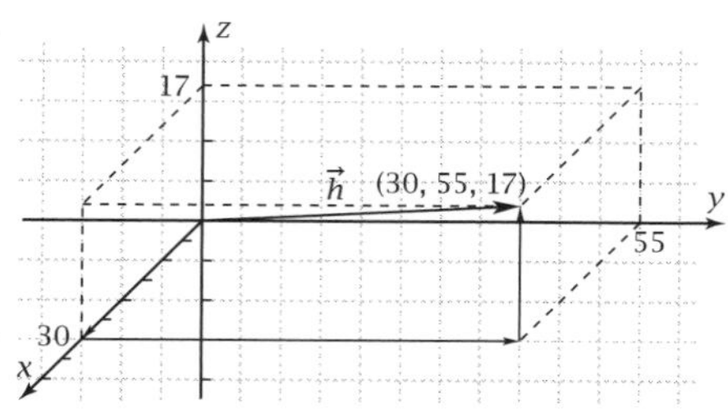

b. $\vec{h} = 30\vec{i} + 55\vec{j} + 17\vec{k}$; 17 ft; 64.9153... ft

c. $\vec{d} - \vec{h} = -20\vec{i} - 55\vec{j} - 9\vec{k}$

d. $\sqrt{3506}$ ft $\approx$ 59.2 ft

e. $0.3(\vec{d} - \vec{h}) = -6\vec{i} - 16.5\vec{j} - 2.7\vec{k}$;

$\sqrt{315.54}$ ft $\approx$ 17.8 ft

f. $\vec{h} + 0.3(\vec{d} - \vec{h}) = 24\vec{i} + 38.5\vec{j} + 14.3\vec{k}$; 14.3 ft

13. $25\vec{i} + 24\vec{j} + 13\vec{k}$

15. $4.6\vec{i} - 6.6\vec{j} - 0.8\vec{k}$

17.

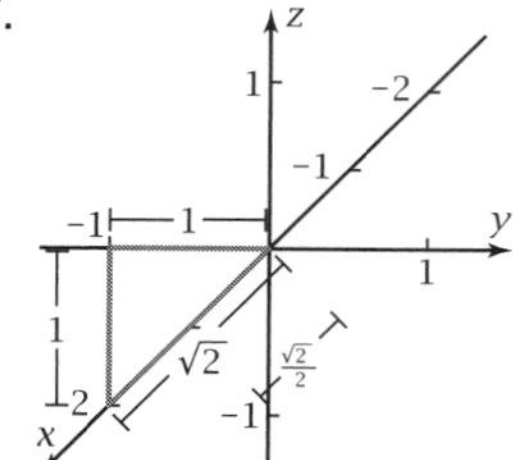

By the Pythagorean theorem, the length of 2 x-units is equal to $\sqrt{2}$ y- or z-units. So each x-unit is equal to $\dfrac{\sqrt{2}}{2}$, or about 70%, of a y- or z-unit.

19. a. $|\vec{a}| = \sqrt{87} = 9.3273...$; $|\vec{b}| = 14$

b. $\vec{a} + \vec{b} = (8, 16, 9, 8)$

c. $\vec{a} - \vec{b} = (-2, -6, -5, 6)$

d. $\overrightarrow{AB} = \vec{b} - \vec{a} = (2, 6, 5, -6)$

e. $\vec{a} + 0.4(\vec{b} - \vec{a}) = (3.8, 7.4, 4, 4.6)$

Problem Set 10-4

1. $\vec{a} \cdot \vec{b} = 30 \cdot 25 \cdot \cos 37° = 598.9766...$

3. $\vec{a} \cdot \vec{b} = 29 \cdot 50 \cdot \cos 127° = -872.6317...$

5. $\vec{a} \cdot \vec{b} = 51 \cdot 27 \cdot \cos 90° = 0$

7. $\theta = \cos^{-1}\dfrac{100}{20 \cdot 30} = 80.4095...°$

9. $\theta = \cos^{-1}\dfrac{-123}{11 \cdot 17} = 131.1288...°$

11. $\theta = \cos^{-1}\dfrac{4800}{60 \cdot 80} = \cos^{-1}(1) = 0°$

13. $\vec{a} \cdot \vec{b} = 21$
$|\vec{a}| = \sqrt{38}$; $|\vec{b}| = \sqrt{66}$;
$\theta = 65.2077...°$

15. $\vec{a} \cdot \vec{b} = -31$
$|\vec{a}| = \sqrt{38}$; $|\vec{b}| = \sqrt{46}$;
$\theta = 137.8564...°$

17. $\vec{a} \cdot \vec{b} = 0$; $\theta = 90°$

19. a. $\vec{F}_{res} = 45\vec{i} + 120\vec{j} + 15\vec{k}$
b. 120 lb
c. 45 lb
d. 15 lb
e. $\sqrt{16{,}650}$ lb $\approx$ 129 lb
f. $|\vec{F}_1| = \sqrt{5225}$ lb $\approx$ 72.3 lb
$|\vec{F}_2| = \sqrt{3425}$ lb $\approx$ 58.5 lb
$|\vec{F}_1| + |\vec{F}_2| \approx 130.8$ lb $\neq 129$ lb $\approx \vec{F}_{res}$; no
g. $\vec{F}_1 \cdot \vec{F}_2 = 4000$; $\theta \approx 19.0°$

21. $|\vec{v}| \cos\theta = 8.8294...$
$\vec{u} = \dfrac{\vec{a}}{|\vec{a}|} = \dfrac{7}{\sqrt{74}}\vec{i} + \dfrac{3}{\sqrt{74}}\vec{j} + \dfrac{4}{\sqrt{74}}\vec{k}$
$(8.8294...)\vec{u} = (7.1848...)\vec{i} + (3.0792...)\vec{j} + (4.1056...)\vec{k}$

23. $\cos\theta = \dfrac{\vec{a} \cdot \vec{b}}{|\vec{a}| \cdot |\vec{b}|} \Rightarrow$
$p = |\vec{a}| \cos\theta = |\vec{a}| \cdot \dfrac{\vec{a} \cdot \vec{b}}{|\vec{a}| \cdot |\vec{b}|} = \dfrac{\vec{a} \cdot \vec{b}}{|\vec{b}|}$
$\vec{u} = \dfrac{\vec{b}}{|\vec{b}|} \Rightarrow \vec{p} = p\vec{u} = \dfrac{\vec{a} \cdot \vec{b}}{|\vec{b}|} \cdot \dfrac{\vec{b}}{|\vec{b}|} = \dfrac{\vec{a} \cdot \vec{b}}{|\vec{b}|^2}\vec{b}$

25. a. $p = |\vec{r}| \cos\theta = 0.5207...$
b. $\vec{p} = (|\vec{r}| \cos\theta)\vec{u} = \dfrac{28}{59}\vec{i} - \dfrac{4}{59}\vec{j} - \dfrac{12}{59}\vec{k}$

27. a. $p = |\vec{r}| \cos\theta = -3.6514...$
b. $\vec{p} = (|\vec{r}| \cos\theta)\vec{u} = \dfrac{4}{3}\vec{i} - \dfrac{10}{3}\vec{j} - \dfrac{20}{3}\vec{k}$

29. a. $\theta = \cos^{-1} \dfrac{1}{\sqrt{3}} \approx 54.7°$
b. $\theta = \cos^{-1} \dfrac{2}{\sqrt{3} \cdot \sqrt{2}} \approx 35.3°$

Problem Set 10-5

1. $3\vec{i} + 5\vec{j} - 7\vec{k}$, $-3\vec{i} - 5\vec{j} + 7\vec{k}$

3. $3x - 5y + 4z = 45$

5. $8x - 6y - 8z = 18$

7. $5x - 3y - z = D$;
$D = 5(4) - 3(-6) - 1(1) = 37$; $5x - 3y - z = 37$

9. $3(6) - 7(2) + 5z_1 = 54 \Rightarrow 5z_1 = 50 \Rightarrow z_1 = 10$
$3(4) - 7(-3) + 5z_2 = 54 \Rightarrow 5z_2 = 21 \Rightarrow z_2 = \dfrac{21}{5}$
$P_1(6, 2, 10)$ $\quad P_2\left(4, -3, \dfrac{21}{5}\right)$
$d = \sqrt{(6-4)^2 + (2-(-3))^2 + \left(10 - \dfrac{21}{5}\right)^2} = 7.914...$
$-7y = 54 \Rightarrow y = \dfrac{-54}{7}$

11. a. $30x - 17y + 11z = 900$
b. It intersects the x-axis first, at $x = 30$ m.
c. $z = 53.636...$ m
d. $\theta = \cos^{-1} \dfrac{11}{\sqrt{1310}} \approx 72.3°$

13. $\vec{n}_1 = 2\vec{i} - 5\vec{j} + 3\vec{k}$; $\vec{n}_2 = 7\vec{i} + 4\vec{j} + 2\vec{k}$;
$\vec{n}_1 \cdot \vec{n}_2 = 2 \cdot 7 - 5 \cdot 4 + 3 \cdot 2 = 0$

15. Let $P_0(x_0, y_0, z_0)$ be a fixed point in the plane, and let $P(x, y, z)$ be an arbitrary different point in the plane so that the displacement vector from P_0 to P is $\vec{d} = (x - x_0)\vec{i} + (y - y_0)\vec{j} + (z - z_0)\vec{k}$. Then $\vec{n} \cdot \vec{d} = 0$
$\Leftrightarrow A(x - x_0) + B(y - y_0) + C(z - z_0) = 0$
$\Leftrightarrow Ax + By + Cz - Ax_0 - By_0 - Cz_0 = 0$
$\Leftrightarrow Ax + By + Cz - D = 0$, where $D = Ax_0 + By_0 + Cz_0$

Problem Set 10-6

1. $-8\vec{i} + 7\vec{j} - 2\vec{k}$

3. $-4\vec{i} - 6\vec{j} + 2\vec{k}$

5. 34

7. Student program

9. $19x + 24y - 17z = 41$

11. $12x + 14y - 3z = -63$

13. $A = |\vec{a} \times \vec{b}|$
$= |(2\vec{i} + 3\vec{j} + 6\vec{k}) \times (3\vec{i} - 4\vec{j} + 12\vec{k})|$
$= |60\vec{i} - 6\vec{j} - 17\vec{k}| = \sqrt{60^2 + 6^2 + 17^2}$
$= \sqrt{3925} = 62.6498...$

15. $A = \frac{5}{2}\sqrt{182} = 33.7268...$

17. a. $\vec{d}_{z,x} = 10\vec{i} - 5\vec{k}$; $\vec{d}_{z,y} = 15\vec{j} - 5\vec{k}$
b. $\vec{n} = 75\vec{i} + 50\vec{j} + 150\vec{k}$ or, dividing by 25, $3\vec{i} + 2\vec{j} + 6\vec{k}$
c. $A = 87.5$ ft^2
d. $|\vec{d}_{z,x}| = 5\sqrt{5}$ ft $\approx$ 11.2 ft
$|\vec{d}_{z,y}| = 5\sqrt{10}$ ft $\approx$ 15.8 ft
$|\vec{d}_{x,y}| = 5\sqrt{13}$ ft $\approx$ 18.0 ft
$\theta_x = \cos^{-1} \dfrac{4}{\sqrt{65}} \approx 60.3°$
$\theta_y = \cos^{-1} \dfrac{9}{\sqrt{130}} \approx 37.9°$
$\theta_z = \cos^{-1} \dfrac{1}{5\sqrt{2}} \approx 81.9°$
e. $3x + 2y + 6z = 78$; $3(5) + 2(6) + 6z = 78 \Rightarrow z = 8.5$
So (5, 6, 9) is above the plane.

19. a. $\vec{a} \times \vec{b} = (5\vec{i} - 2\vec{j} + 3\vec{k}) \times (4\vec{i} + 7\vec{j} - 6\vec{k})$
$= -9\vec{i} + 42\vec{j} + 43\vec{k}$
b. $\vec{a} \cdot (\vec{a} \times \vec{b}) = 0$; $\vec{b} \cdot (\vec{a} \times \vec{b}) = 0$
c. $\vec{a} \cdot \vec{b} = -12$; $\theta = 101.1687\ldots°$
d. $|\vec{a} \times \vec{b}| = 60.7782\ldots$
$|\vec{a}| \cdot |\vec{b}| \cdot \sin\theta = 60.7782\ldots$

21. a. $\vec{h} = 5\vec{i} - 10\vec{j} + 20\vec{k}$
$= -\frac{5}{3}(-3\vec{i} + 6\vec{j} - 12\vec{k}) = -\frac{5}{3}\vec{g}$
b. $|\vec{g} \times \vec{h}| = |0\vec{i} + 0\vec{j} + 0\vec{k}| = 0$
c. $|\vec{g} \times \vec{h}| = |\vec{g}| \cdot |\vec{h}| \sin 0° = |\vec{g}|\,|\vec{h}|\,0 = 0$

23. $z = 2$, $x = 5$

Problem Set 10-7

1.

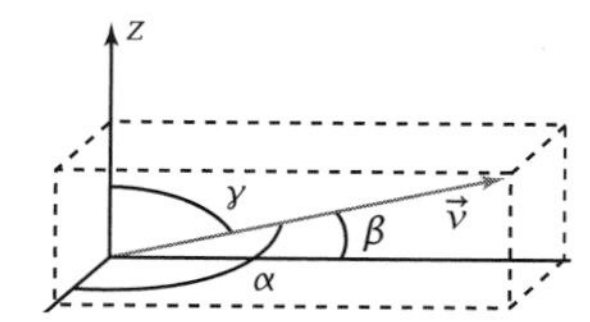

3. $c_1 = \frac{2}{\sqrt{38}}$; $\alpha = \cos^{-1} c_1 = 71.0681\ldots°$
$c_2 = \frac{-5}{\sqrt{38}}$; $\beta = \cos^{-1} c_2 = 144.2042\ldots°$
$c_3 = \frac{3}{\sqrt{38}}$; $\gamma = \cos^{-1} c_3 = 60.8784\ldots°$

5. $c_1 = \frac{-4}{21}$; $\alpha = \cos^{-1} c_1 = 100.9805\ldots°$
$c_2 = \frac{8}{21}$; $\beta = \cos^{-1} c_2 = 67.6073\ldots°$
$c_3 = \frac{19}{21}$; $\gamma = \cos^{-1} c_3 = 25.2087\ldots°$

7. $c_1 = \frac{7}{\sqrt{59}}$; $c_2 = \frac{1}{\sqrt{59}}$; $c_3 = \frac{-3}{\sqrt{59}}$

9. $c_1 = \frac{9}{17}$; $c_2 = \frac{-8}{17}$; $c_3 = \frac{12}{17}$

11. $\vec{u} = \frac{8}{3\sqrt{41}}\vec{i} + \frac{16}{3\sqrt{41}}\vec{j} - \frac{7}{3\sqrt{41}}\vec{k}$
$c_1 = \frac{8}{3\sqrt{41}}$; $c_2 = \frac{16}{3\sqrt{41}}$; $c_3 = \frac{-7}{3\sqrt{41}}$

13. $\left(\frac{7}{9}\right)^2 + \left(\frac{4}{9}\right)^2 + \left(\frac{4}{9}\right)^2 = \frac{49}{81} + \frac{16}{81} + \frac{16}{81} = \frac{81}{81} = 1$
$c_1 = \frac{7}{9}$; $\alpha = \cos^{-1} c_1 = 38.9424\ldots°$
$c_2 = \frac{4}{9}$; $\beta = \cos^{-1} c_2 = 63.6122\ldots°$
$c_3 = \frac{-4}{9}$; $\gamma = \cos^{-1} c_3 = 116.3877\ldots°$

15. $c_3 = \pm\frac{6}{23}$
$\gamma = \cos^{-1} c_3 = 74.8783\ldots°$ or $105.1216\ldots°$

17. $c_3 = \pm\sqrt{0.5}$; $\gamma = \cos^{-1} c_3 = 45°$ or $135°$

19. $c_3 = \pm\sqrt{-0.8712\ldots}$, for which the cosine is undefined.

21. a. $\sqrt{195}$ ft/sec ≈ 14.0 ft/sec
b. $\alpha = \cos^{-1}\frac{5}{\sqrt{195}} = 69.0190\ldots°$
$\beta = \cos^{-1}\frac{11}{\sqrt{195}} = 38.0264\ldots°$
$\gamma = \cos^{-1}\frac{7}{\sqrt{195}} = 59.9152\ldots°$
c. The angle of elevation is the same as $90° - \gamma = 30.0847\ldots°$
d. $\vec{p} = 5\vec{i} + 11\vec{j}$
e. $\alpha = \cos^{-1}\frac{5}{\sqrt{146}} = 65.5560\ldots°$
f. The two direction angles would change. Because $\cos^2\alpha + \cos^2\beta + \cos^2\gamma = 1$, any change in γ has to affect at least one of α and β. The azimuth angle would also change.

23. $c_1^2 + c_2^2 + c_3^2 = \left(\frac{A}{|\vec{v}|}\right)^2 + \left(\frac{B}{|\vec{v}|}\right)^2 + \left(\frac{C}{|\vec{v}|}\right)^2$
$= \left(\frac{A}{\sqrt{A^2 + B^2 + C^2}}\right)^2 + \left(\frac{B}{\sqrt{A^2 + B^2 + C^2}}\right)^2$
$+ \left(\frac{C}{\sqrt{A^2 + B^2 + C^2}}\right)^2$
$= \frac{A^2 + B^2 + C^2}{A^2 + B^2 + C^2} = 1$

Problem Set 10-8

1. a. $(5, -3, 4)$
b. $\frac{9}{17}\vec{i} + \frac{12}{17}\vec{j} + \frac{8}{17}\vec{k}$
$\left(\frac{9}{17}\right)^2 + \left(\frac{12}{17}\right)^2 + \left(\frac{8}{17}\right)^2$
$= \frac{81}{289} + \frac{144}{289} + \frac{64}{289} = \frac{289}{289} = 1$

3. $(23, 21, 20)$

5. $\left(\frac{1}{2}, -9, 0\right)$

7. $\left(\frac{2}{7}\right)^2 + \left(\frac{6}{7}\right)^2 + \left(-\frac{3}{7}\right)^2 = \frac{4}{49} + \frac{36}{49} + \frac{9}{49} = \frac{49}{49} = 1$

9. $c_1 = \frac{2}{\sqrt{29}}$, $c_2 = \frac{-3}{\sqrt{29}}$, $c_3 = \frac{4}{\sqrt{29}}$
$\vec{r} = \left(1 + \frac{2}{\sqrt{29}}d\right)\vec{i} + \left(-8 - \frac{3}{\sqrt{29}}d\right)\vec{j} + \left(-5 + \frac{4}{\sqrt{29}}d\right)\vec{k}$

11. $\vec{r} = \left(5 + \frac{9}{25}d\right)\vec{i} + \left(1 + \frac{4}{5}d\right)\vec{j} + \left(-4 + \frac{12}{25}d\right)\vec{k}$

13. $\left(-\frac{79}{3}, -\frac{76}{3}, \frac{53}{3}\right)$

15. a. $\vec{u} = -\frac{3\sqrt{2}}{10}\vec{i} + \frac{2\sqrt{2}}{5}\vec{j} + \frac{\sqrt{2}}{2}\vec{k}$

b. $\vec{P}_0 + d\vec{u}$

$= \left(10 - \frac{3\sqrt{2}}{10}d\right)\vec{i} + \left(14 + \frac{2\sqrt{2}}{5}d\right)\vec{j} + \left(3 + \frac{\sqrt{2}}{2}d\right)\vec{k}$

c. 8 ft, because the floor is at $z = 0$ and you know a point on the ceiling at $z = 8$.

d. The roof is perpendicular to the xz-plane, so the normal to the roof is parallel to the xz-plane and therefore has $0\vec{j}$ as its y-component. Because the triangular part of the wall is a 45°-45° right triangle, simple geometry shows that the vector through (6, 0, 8) and (0, 0, 14) is a normal. This vector is

$(0 - 6)\vec{i} + (0 - 0)\vec{j} + (14 - 8)\vec{k} = -6\vec{i} + 0\vec{j} + 6\vec{k}$

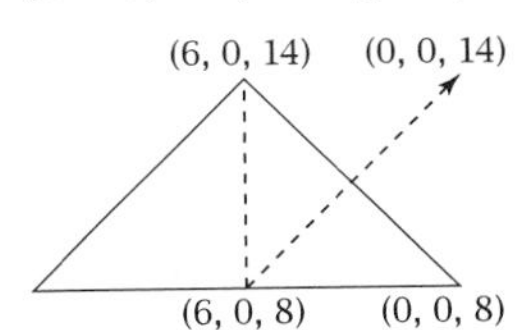

e. $-6x + 6z = 48$

f. (4.375, 21.5, 12.375)

g. *Forensic* means "belonging to, used in, or suitable to public discussion and debate." The evidence about the bullet and its path could be used in a trial.

Problem Set 10-9

R1. a. They start at the origin and go to a point.

b.

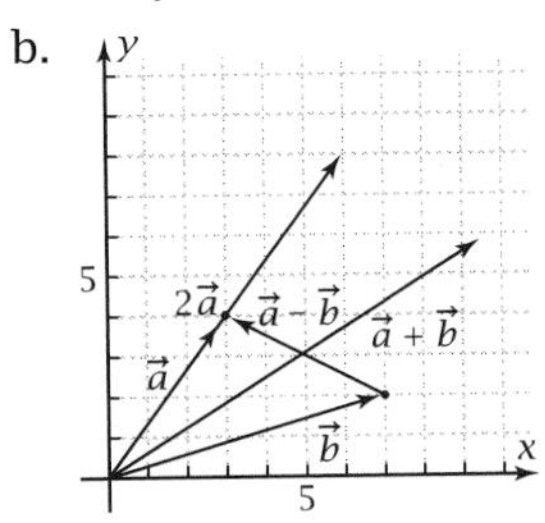

c. $\vec{d} = -4\vec{i} + 2\vec{j}$

d. $|\vec{r}| = |\vec{a} + \vec{b}| = 2\sqrt{34}$ units

e. $\theta = \tan^{-1}\frac{4}{3} = 53.1301...°$

R2. a. $\vec{d} = 4\vec{i} - 2\vec{j}$

b. $\vec{d} = 1.6\vec{i} - 0.8\vec{j}$

c. $\vec{p} = 4.6\vec{i} + 3.2\vec{j}$

d. (4.6, 3.2)

R3. a.

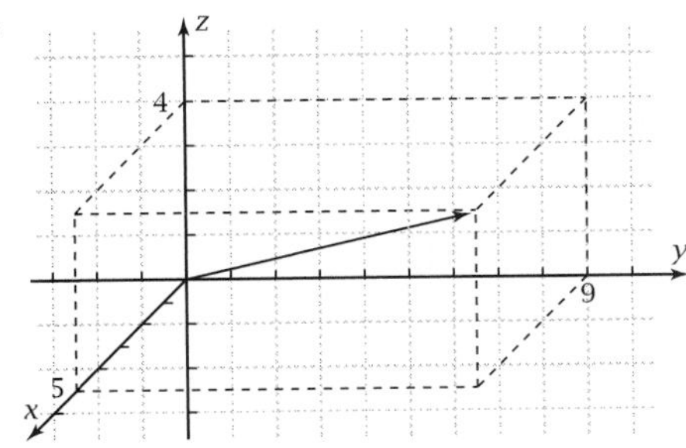

b. $3\vec{a} - 2\vec{b} = 12\vec{i} - 23\vec{j} + 20\vec{k}$

c. $|\vec{a}| = \sqrt{65}$ units

d. $\vec{u} = \frac{\vec{a}}{|\vec{a}|} = \frac{6}{\sqrt{65}}\vec{i} - \frac{5}{\sqrt{65}}\vec{j} + \frac{2}{\sqrt{65}}\vec{k}$

e. $\vec{d} = -3\vec{i} + 9\vec{j} - 9\vec{k}$

f. $\vec{p} = 3.9\vec{i} + 1.3\vec{j} - 4.3\vec{k}$

R4. a. $\vec{a} \cdot \vec{b} = |\vec{a}| \cdot |\vec{b}| \cdot \cos\theta$, where θ is the angle between $\vec{a}$ and $\vec{b}$ placed tail-to-tail. Scalar product and inner product are two other names.

b. $\vec{a} \cdot \vec{b} = -50.7532...$ units

c. $\theta = 100.0786...°$

d. $|\vec{a}| = \sqrt{65}$ units; $|\vec{b}| = \sqrt{74}$ units

$\vec{a} \cdot \vec{b} = -16$

e. $\theta = 103.3382...°$

f. $\vec{u} = \frac{3}{\sqrt{74}}\vec{i} + \frac{4}{\sqrt{74}}\vec{j} - \frac{7}{\sqrt{74}}\vec{k}$

g. $p = \frac{-16}{\sqrt{74}}$ units. (The negative value indicates that the projection points in the opposite direction from $\vec{b}$.)

h. $\vec{p} = -\frac{24}{37}\vec{i} - \frac{32}{37}\vec{j} + \frac{56}{37}\vec{k}$

R5. a. Let $P_0 = (x_0, y_0, z_0)$ be a fixed point on the plane, and let $P = (x, y, z)$ be an arbitrary point on the plane. Then $\overrightarrow{P_0P} = (x - x_0)\vec{i} + (y - y_0)\vec{j} + (z - z_0)\vec{k}$ is contained within the plane and so is normal to $\vec{n}$. Thus:

$\overrightarrow{P_0P} \cdot \vec{n} = 0$

$A(x - x_0) + B(y - y_0) + C(z - z_0) = 0$

$Ax + By + Cz = Ax_0 + By_0 + Cz_0$

$Ax + By + Cz = D$, where $D = Ax_0 + By_0 + Cz_0$

b. $\vec{n} = 3\vec{i} - 7\vec{j} + \vec{k}$, $\vec{n} = -3\vec{i} + 7\vec{j} - \vec{k}$

c. $2x - 7y - 3z = 1$; $z = \frac{-121}{3}$

d. The specific equation is $3x + 6y + 9z = 45$.

R6. a. $\vec{a} \times \vec{b}$ is a vector perpendicular to both $\vec{a}$ and $\vec{b}$, with magnitude $|\vec{a} \times \vec{b}| = |\vec{a}| \cdot |\vec{b}| \sin\theta$, where θ is the angle between $\vec{a}$ and $\vec{b}$ placed tail-to-tail and with direction given by the right-hand rule. Three names for $\vec{a} \times \vec{b}$ are cross product, vector product, and outer product.

b. $|\vec{a} \times \vec{b}| = 23.6666\ldots$ units

c. $\vec{a} \times \vec{b} = 13\vec{i} - 11\vec{j} + 17\vec{k}$

$\vec{b} \times \vec{a} = -13\vec{i} + 11\vec{j} - 17\vec{k}$

d. $\vec{a} \cdot \vec{b} = -11$

$\vec{b} \cdot \vec{a} = -11$

e. $A = \frac{1}{2}\sqrt{579}$ units^2

f. $6x + 7y + 5z = D;$

$D = 6(2) + 7(5) + 5(8) = 87;$

$6x + 7y + 5z = 87$

R7. a.

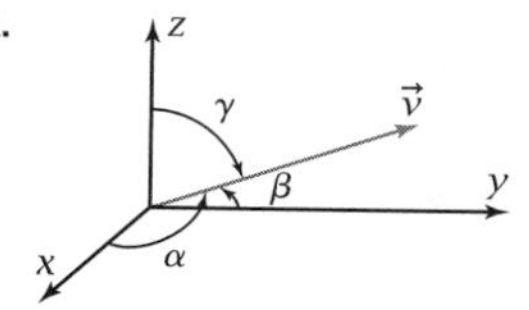

b. $c_1 = \frac{6}{5\sqrt{5}} = 0.5366\ldots;$

$c_2 = \frac{-8}{5\sqrt{5}} = -0.7155\ldots;$

$c_3 = \frac{1}{\sqrt{5}} = 0.4472\ldots;$

$\alpha = 57.5436\ldots°; \beta = 135.6876\ldots°;$

$\gamma = 63.4349\ldots°$

c. $c_3 = \pm\sqrt{0.87}$; $\gamma = \cos^{-1} c_3 = 21.1342\ldots°$ or $158.8657\ldots°$

d. $c_3 = \pm\sqrt{1 - \cos 30° - \cos 40°} = \pm\sqrt{-0.6320\ldots}$, which is imaginary.

R8. a. $\vec{P}_0 = 6\vec{i} + 3\vec{j} + 2\vec{k}; \vec{u} = \frac{7}{9}\vec{i} + \frac{4}{9}\vec{j} - \frac{4}{9}\vec{k};$

$\left(\frac{7}{9}\right)^2 + \left(\frac{4}{9}\right)^2 + \left(-\frac{4}{9}\right)^2 = 1$

b. $\vec{P} = \vec{r}(-18)$

$= \left[6 + \frac{7}{9}(-18)\right]\vec{i} + \left[3 + \frac{4}{9}(-18)\right]\vec{j} + \left[2 - \frac{4}{9}(-18)\right]\vec{k}$

$= -8\vec{i} - 5\vec{j} + 10\vec{k}$; $(-8, -5, 10)$

The point lies in the opposite direction from $(6, 3, 2)$ than the direction of $\vec{u}$.

c. $\left(\frac{19}{2}, 5, 0\right)$

d. $\left(\frac{24}{11}, \frac{9}{11}, \frac{46}{11}\right)$

e. $\vec{r}(t) = (2 + 9t)\vec{i} + (8 + 5t)\vec{j} + (4 + 3t)\vec{k}$

CHAPTER 11

Problem Set 11-1

1.

Iteration Number	Side Length	Total Perimeter	Total Area
0	10 cm	40 cm	100 cm^2
1	4 cm	64 cm	64 cm^2
2	1.6 cm	102.4 cm	40.96 cm^2

3. $P(3) = 163.84$ cm; $A(3) = 26.2144$ cm^2
$P(4) = 262.144$ cm; $A(4) = 16.777216$ cm^2

5. The perimeter approaches infinity, while the area approaches 0.

Problem Set 11-2

1. $\begin{bmatrix} -2 & 13 \\ 0 & 10 \\ 0 & 11 \end{bmatrix}$

3. $[-22 \;\; 22 \;\; -2]$

5. $[14 \;\; -42]$

7. Undefined

9. $\begin{bmatrix} 4 & 4 \\ 5 & 3 \end{bmatrix}$

11. a. $[Y][M] = [4.81 \quad 4.70 \quad 4.96 \quad 1.70]$
The company's total annual income from Texas is \$4.96 million. Of this, \$1.53 million is earned annually from mortgages in Texas.

b. As the matrices are set up, there would be no way to match up the bonds with the bond interest rates, mortgages with mortgage interest rates, and loans with loan interest rates. However, real-world analysts can simply write [M] as a 4×3 matrix and [Y] as a 3×1 matrix.

c. The number of rows of the first matrix does not equal the number of columns of the second.

13. a. $[M]^{-1} = \begin{bmatrix} 0.7 & -2.9 & 2.1 \\ -0.1 & 1.7 & -1.3 \\ -0.3 & 0.1 & 0.1 \end{bmatrix}$

$[M]^{-1}[M] = [M][M]^{-1} = \begin{bmatrix} 1 & 0 & 0 \\ 0 & 1 & 0 \\ 0 & 0 & 1 \end{bmatrix} = [I]$

b. $\det[M] = 10$

$\text{adj}[M] = [M]^{-1} \cdot \det[M] = \begin{bmatrix} 7 & -29 & 21 \\ -1 & 17 & -13 \\ 3 & 1 & 1 \end{bmatrix}$

15. Finding the inverse requires dividing by the determinant, but $\det\begin{bmatrix} 6 & 3 \\ 8 & 4 \end{bmatrix} = 0$.

17. $\begin{bmatrix} 5 & 3 & -7 \\ 10 & -4 & 6 \\ 15 & 1 & -8 \end{bmatrix}^{-1} \begin{bmatrix} 3 \\ 5 \\ -2 \end{bmatrix} = \begin{bmatrix} 1.22 \\ 6.9 \\ 3.4 \end{bmatrix}$;

$x = 1.22$, $y = 6.9$, $z = 3.4$

19. $y = 0.5x^2 + 3x - 7$; $y(20) = 253$

21. $\begin{bmatrix} 2 & 3 \\ 4 & 5 \end{bmatrix} \begin{bmatrix} 6 & 7 \\ 8 & 9 \end{bmatrix} = \begin{bmatrix} 36 & 41 \\ 64 & 73 \end{bmatrix}$

$\begin{bmatrix} 6 & 7 \\ 8 & 9 \end{bmatrix} \begin{bmatrix} 2 & 3 \\ 4 & 5 \end{bmatrix} = \begin{bmatrix} 40 & 53 \\ 52 & 69 \end{bmatrix}$

23. $\begin{bmatrix} 1 & 1 \\ 1 & 1 \end{bmatrix} \begin{bmatrix} 1 & -1 \\ -1 & 1 \end{bmatrix} = \begin{bmatrix} 0 & 0 \\ 0 & 0 \end{bmatrix}$

Problem Set 11-3

1. 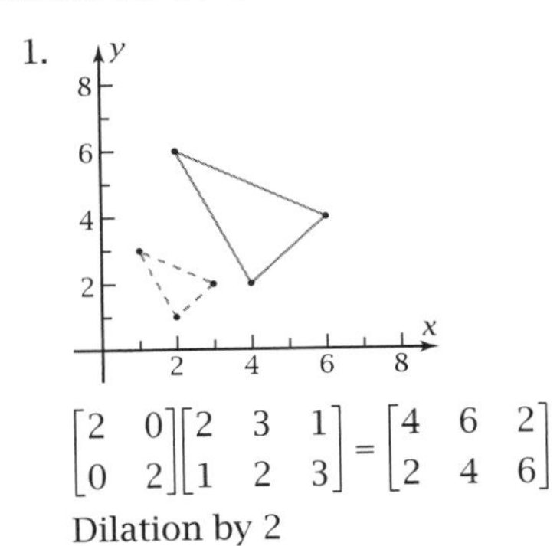

$\begin{bmatrix} 2 & 0 \\ 0 & 2 \end{bmatrix} \begin{bmatrix} 2 & 3 & 1 \\ 1 & 2 & 3 \end{bmatrix} = \begin{bmatrix} 4 & 6 & 2 \\ 2 & 4 & 6 \end{bmatrix}$

Dilation by 2

3. 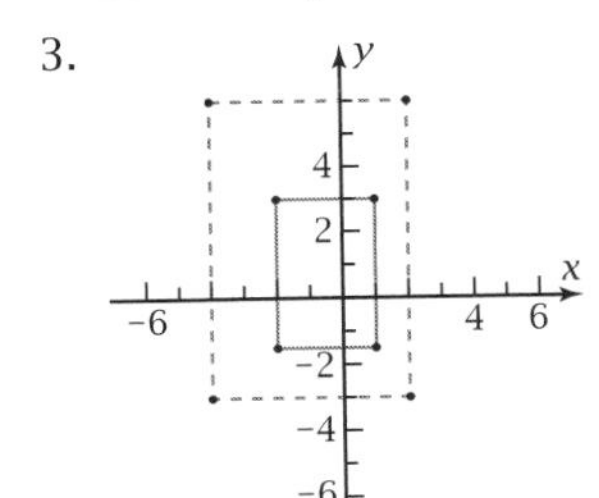

$\begin{bmatrix} \frac{1}{2} & 0 \\ 0 & \frac{1}{2} \end{bmatrix} \begin{bmatrix} 2 & 2 & -4 & -4 \\ 6 & -3 & -3 & 6 \end{bmatrix} = \begin{bmatrix} 1 & 1 & -2 & -2 \\ 3 & -1.5 & -1.5 & 3 \end{bmatrix}$

Dilation by $\frac{1}{2}$

5. 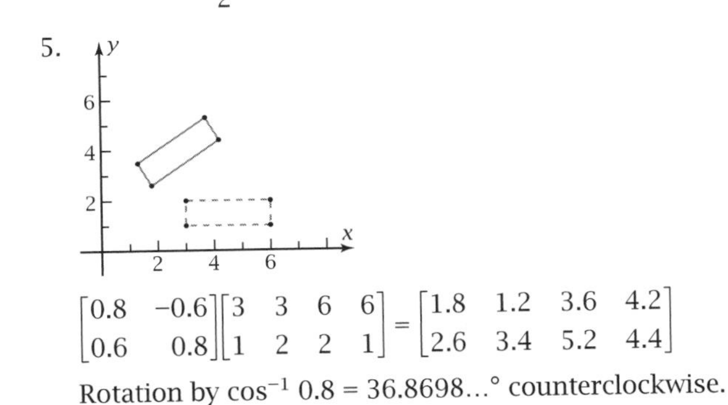

$\begin{bmatrix} 0.8 & -0.6 \\ 0.6 & 0.8 \end{bmatrix} \begin{bmatrix} 3 & 3 & 6 & 6 \\ 1 & 2 & 2 & 1 \end{bmatrix} = \begin{bmatrix} 1.8 & 1.2 & 3.6 & 4.2 \\ 2.6 & 3.4 & 5.2 & 4.4 \end{bmatrix}$

Rotation by $\cos^{-1} 0.8 = 36.8698...°$ counterclockwise.

7. $\begin{bmatrix} 3 & 0 \\ 0 & 3 \end{bmatrix} \begin{bmatrix} 1 & 3 & 4 \\ 1 & 1 & 5 \end{bmatrix} = \begin{bmatrix} 3 & 9 & 12 \\ 3 & 3 & 15 \end{bmatrix}$

9. $\begin{bmatrix} \cos(-50°) & \cos 40° \\ \sin(-50°) & \sin 40° \end{bmatrix} \begin{bmatrix} 1 & 3 & 4 \\ 1 & 1 & 5 \end{bmatrix}$

$= \begin{bmatrix} 1.4088... & 2.6944... & 6.4013... \\ -0.1232... & -1.6553... & 0.1497... \end{bmatrix}$

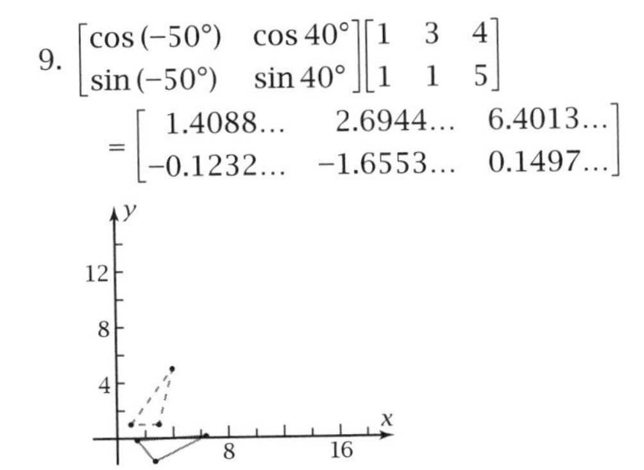

11. $\begin{bmatrix} \cos(-50°) & \cos 40° \\ \sin(-50°) & \sin 40° \end{bmatrix} \begin{bmatrix} 3 & 0 \\ 0 & 3 \end{bmatrix} \begin{bmatrix} 1 & 3 & 4 \\ 1 & 1 & 5 \end{bmatrix}$

$= \begin{bmatrix} 4.2264... & 8.0832... & 19.2041... \\ -0.3697... & -4.9660... & 0.4492... \end{bmatrix}$

13. [A] will dilate the image by 0.8 and rotate it counterclockwise by 20°. The pre-image matrix is

$\begin{bmatrix} 5 & 7 & 9 & 9 \\ 2 & 2 & 4 & 6 \end{bmatrix}$

$[A]\begin{bmatrix} 5 & 7 & 9 & 9 \\ 2 & 2 & 4 & 6 \end{bmatrix}$

$= \begin{bmatrix} 3.2115... & 4.7150... & 5.7613... & 5.1240... \\ 2.8715... & 3.4188... & 5.4695... & 6.9730... \end{bmatrix}$

$[A]^2\begin{bmatrix} 5 & 7 & 9 & 9 \\ 2 & 2 & 4 & 6 \end{bmatrix}$

$= \begin{bmatrix} 1.6285... & 2.6091... & 2.7668... & 1.9441... \\ 3.0374... & 3.8602... & 5.6635... & 6.6440... \end{bmatrix}$

$[A]^3\begin{bmatrix} 5 & 7 & 9 & 9 \\ 2 & 2 & 4 & 6 \end{bmatrix}$

$= \begin{bmatrix} 0.3931... & 0.9051... & 0.5303... & -0.3564... \\ 2.7290... & 3.6158... & 5.0146... & 5.5266... \end{bmatrix}$

$$[A]^4\begin{bmatrix}5 & 7 & 9 & 9\\ 2 & 2 & 4 & 6\end{bmatrix}$$
$$= \begin{bmatrix}-0.4511... & -0.3088... & -0.9733... & -1.7801...\\ 2.1591... & 2.9658... & 3.9149... & 4.0571...\end{bmatrix}$$

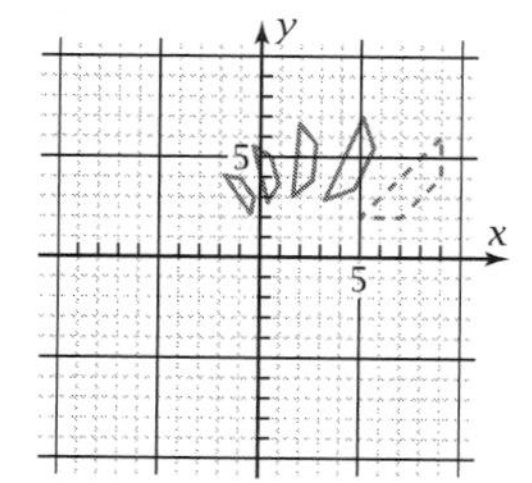

15. Student program

17. $[M] = \begin{bmatrix}\cos 90° & \cos 180°\\ \sin 90° & \sin 180°\end{bmatrix} = \begin{bmatrix}0 & -1\\ 1 & 0\end{bmatrix}$

19. $[M] = \begin{bmatrix}5 & 0\\ 0 & 5\end{bmatrix}$

Problem Set 11-4

1. a. $[A] = \begin{bmatrix}0.8457... & -0.3078... & 6\\ 0.3078... & 0.8457... & -1\\ 0 & 0 & 1\end{bmatrix}$;

$$[M] = \begin{bmatrix}3 & 7 & 7 & 3\\ 2 & 2 & 4 & 4\\ 1 & 1 & 1 & 1\end{bmatrix}$$

[A][M]
$$= \begin{bmatrix}7.9215... & 11.3044... & 10.6887... & 7.3058...\\ 1.6149... & 2.8461... & 4.5376... & 3.3063...\\ 1 & 1 & 1 & 1\end{bmatrix}$$

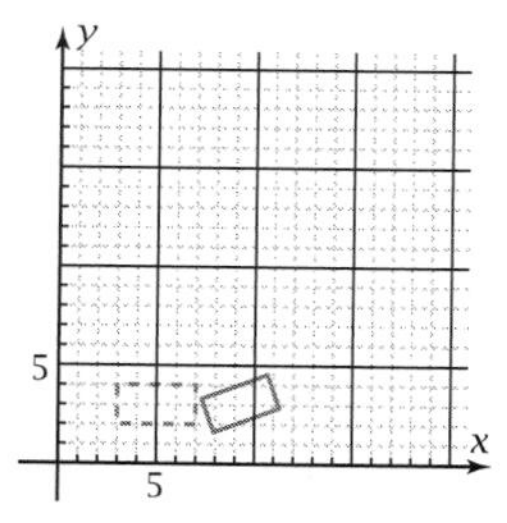

b.

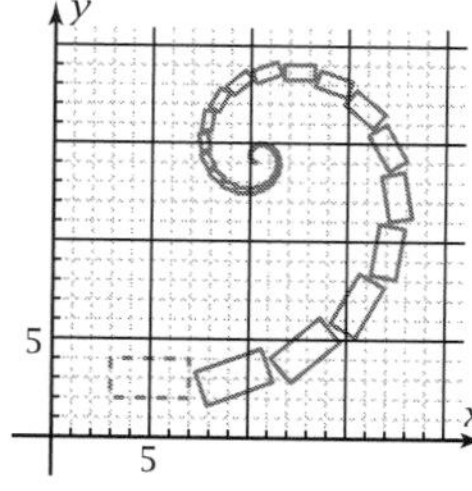

The images seem to be attracted to (10, 14).

c. $[A]^{100}[M]$
$$= \begin{bmatrix}0.0000... & 0.0000... & 10.4045...\\ 0.0000... & 0.0000... & 14.2778...\\ 0 & 0 & 1\end{bmatrix};$$

(10.4045..., 14.2778...) ≈ (10, 14) from part b.

d. $\begin{bmatrix}0.9\cos 20° & 0.9\cos 110° & 6\\ 0.9\sin 20° & 0.9\sin 110° & -1\\ 0 & 0 & 1\end{bmatrix}\begin{bmatrix}X\\ Y\\ 1\end{bmatrix} = \begin{bmatrix}X\\ Y\\ 1\end{bmatrix}$

$$\rightarrow \begin{cases}0.9X\cos 20° + 0.9Y\cos 110° + 6 = X\\ 0.9X\sin 20° + 0.9Y\sin 110° - 1 = Y\end{cases}$$
$$\rightarrow \begin{cases}(0.9\cos 20° - 1)X + (0.9\cos 110°)Y = -6\\ (0.9\sin 20°)X + (0.9\sin 110° - 1)Y = 1\end{cases}$$
$$\rightarrow \begin{bmatrix}0.9\cos 20° - 1 & 0.9\cos 110°\\ 0.9\sin 20° & 0.9\sin 110° - 1\end{bmatrix}\begin{bmatrix}X\\ Y\end{bmatrix} = \begin{bmatrix}-6\\ 1\end{bmatrix}$$
$$\rightarrow \begin{bmatrix}X\\ Y\end{bmatrix} = \begin{bmatrix}0.9\cos 20° - 1 & 0.9\cos 110°\\ 0.9\sin 20° & 0.9\sin 110° - 1\end{bmatrix}^{-1}\begin{bmatrix}-6\\ 1\end{bmatrix}$$
$$= \begin{bmatrix}10.4044...\\ 14.2773...\end{bmatrix} \approx \begin{bmatrix}10.4045...\\ 14.2778...\end{bmatrix} \text{ from part c}$$

3. a. The figure will be rotated 20° clockwise and shrunk by a factor of 0.8, then translated 2 units horizontally and 6 units vertically.

b. $[M_1] = \begin{bmatrix}25 & 23 & 23 & 25\\ 15 & 15 & 12 & 12\\ 1 & 1 & 1 & 1\end{bmatrix}$

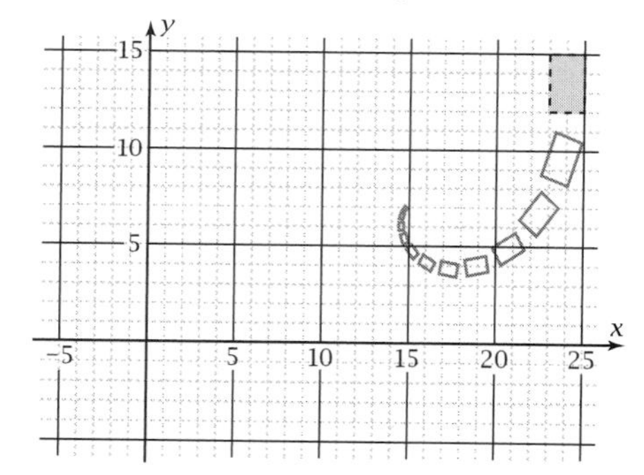

The rectangles converge to ≈(16, 7).

c.

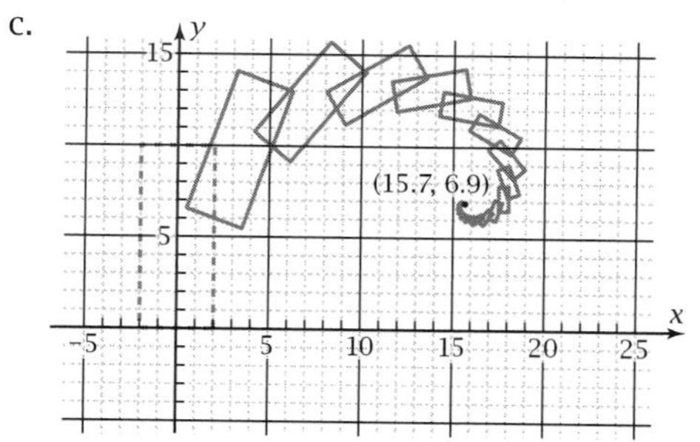

The images seem to be attracted to the same fixed point.

d. $[T_2] = \begin{bmatrix} 0.7\cos 35° & 0.7\cos 125° & 7 \\ 0.7\sin 35° & 0.7\sin 125° & -3 \\ 0 & 0 & 1 \end{bmatrix}$

$= \begin{bmatrix} 0.5734\ldots & -0.4015\ldots & 7 \\ 0.4015\ldots & 0.5734\ldots & -3 \\ 0 & 0 & 1 \end{bmatrix}$

Applying $[T_2]$ iteratively to $[M_1]$:

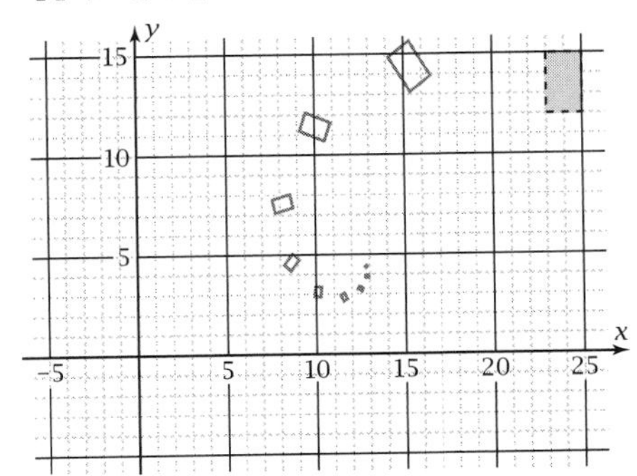

The rectangles converge to $\approx (12, 4.5)$.

e. The transformation matrix determines the fixed point attractor. To demonstrate this, note that applying $[T_1]$ iteratively to either $[M_1]$ or $[M_2]$ gives the same fixed point attractor; but applying $[T_1]$ iteratively to $[M_1]$ gives a different fixed point attractor from the one yielded by applying $[T_2]$ iteratively to $[M_1]$. Applying $[T_2]$ iteratively to $[M_2]$ supports this:

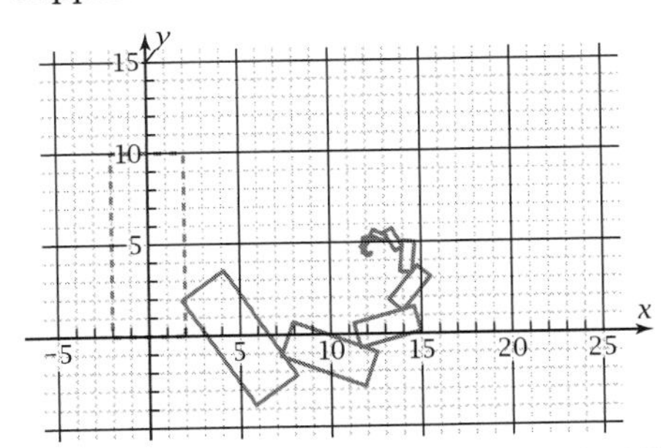

The rectangles still converge to $\approx (12, 4.5)$.

Problem Set 11-5

1. a. $[A] = \begin{bmatrix} 0.5 & 0 & 7.5 \\ 0 & 0.5 & -5 \\ 0 & 0 & 1 \end{bmatrix}$; $[B] = \begin{bmatrix} 0.5 & 0 & -7.5 \\ 0 & 0.5 & -5 \\ 0 & 0 & 1 \end{bmatrix}$;

$[C] = \begin{bmatrix} 0.5 & 0 & 0 \\ 0 & 0.5 & 10 \\ 0 & 0 & 1 \end{bmatrix}$

b. $[A][A][M] = \begin{bmatrix} 15 & 11.25 & 7.5 \\ -10 & -2.5 & -10 \\ 1 & 1 & 1 \end{bmatrix}$

$[A][B][M] = \begin{bmatrix} 7.5 & 3.75 & 0 \\ -10 & -2.5 & -10 \\ 1 & 1 & 1 \end{bmatrix}$

$[A][C][M] = \begin{bmatrix} 11.25 & 7.5 & 3.75 \\ -2.5 & 5 & -2.5 \\ 1 & 1 & 1 \end{bmatrix}$

$[B][A][M] = \begin{bmatrix} 0 & -3.75 & -7.5 \\ -10 & -2.5 & -10 \\ 1 & 1 & 1 \end{bmatrix}$

$[B][B][M] = \begin{bmatrix} -7.5 & -11.25 & -15 \\ -10 & -2.5 & -10 \\ 1 & 1 & 1 \end{bmatrix}$

$[B][C][M] = \begin{bmatrix} -3.75 & -7.5 & -11.25 \\ -2.5 & 5 & -2.5 \\ 1 & 1 & 1 \end{bmatrix}$

$[C][A][M] = \begin{bmatrix} 7.5 & 3.75 & 0 \\ 5 & 12.5 & 5 \\ 1 & 1 & 1 \end{bmatrix}$

$[C][B][M] = \begin{bmatrix} 0 & -3.75 & -7.5 \\ 5 & 12.5 & 5 \\ 1 & 1 & 1 \end{bmatrix}$

$[C][C][M] = \begin{bmatrix} 3.75 & 0 & -3.75 \\ 12.5 & 20 & 12.5 \\ 1 & 1 & 1 \end{bmatrix}$

c.

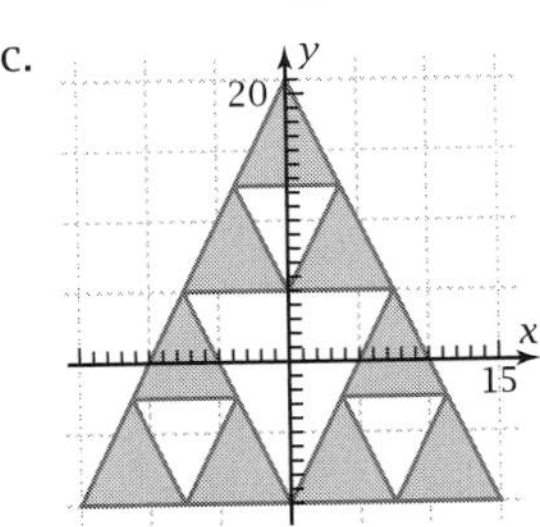

d. 3rd iteration: $3^3 = 27$ images

20th iteration: $3^{20} = 3{,}486{,}784{,}401$ images

e. Pre-image area: $450 = 450 \cdot 1 = 450 \cdot \left(\frac{3}{4}\right)^0$

1st iteration area: $450 \cdot \frac{3}{4} = 337.5 = 450 \cdot \left(\frac{3}{4}\right)^1$

nth iteration area: $450 \cdot \left(\frac{3}{4}\right)^n$

f. As $n \to \infty$, the area approaches zero. Sierpiñski's triangle has zero area!

3. Student program

5. The results should resemble the figure.

7. $[A] = \begin{bmatrix} 0.5 & 0 & 10 \\ 0 & 0.5 & 10 \\ 0 & 0 & 1 \end{bmatrix}$

$[B] = \begin{bmatrix} 0.5 & 0 & 0 \\ 0 & 0.5 & 10 \\ 0 & 0 & 1 \end{bmatrix}$

$[C] = \begin{bmatrix} 0.5 & 0 & 0 \\ 0 & 0.5 & 0 \\ 0 & 0 & 1 \end{bmatrix}$

$[D] = \begin{bmatrix} 0.5 & 0 & 10 \\ 0 & 0.5 & 0 \\ 0 & 0 & 1 \end{bmatrix}$

The combined image space of all four transformations now covers the entire square.

9. a. $[D] = \begin{bmatrix} 0 & 0 \\ 0 & 10 \\ 1 & 1 \end{bmatrix}$

b. $[A] = \begin{bmatrix} 0.5196\ldots & -0.3 & 0 \\ 0.3 & 0.5196\ldots & 5 \\ 0 & 0 & 1 \end{bmatrix}$

$[B] = \begin{bmatrix} 0.6 & 0 & 0 \\ 0 & 0.6 & 0 \\ 0 & 0 & 1 \end{bmatrix}$

$[C] = \begin{bmatrix} 0.5196\ldots & 0.3 & 0 \\ -0.3 & 0.5196\ldots & 4 \\ 0 & 0 & 1 \end{bmatrix}$

c. $[A][A][D] \approx \begin{bmatrix} -1.5 & -4.6 \\ 7.6 & 9.4 \\ 1 & 1 \end{bmatrix}$

$[A][B][D] \approx \begin{bmatrix} 0 & -1.8 \\ 5 & 8.1 \\ 1 & 1 \end{bmatrix}$

$[A][C][D] \approx \begin{bmatrix} -1.2 & -1.2 \\ 7.1 & 10.7 \\ 1 & 1 \end{bmatrix}$

$[B][A][D] \approx \begin{bmatrix} 0 & -1.8 \\ 3 & 6.1 \\ 1 & 1 \end{bmatrix}$

$[B][B][D] = \begin{bmatrix} 0 & 0 \\ 0 & 3.6 \\ 1 & 1 \end{bmatrix}$

$[B][C][D] \approx \begin{bmatrix} 0 & 1.8 \\ 2.4 & 5.5 \\ 1 & 1 \end{bmatrix}$

$[C][A][D] \approx \begin{bmatrix} 1.5 & 1.5 \\ 6.6 & 10.2 \\ 1 & 1 \end{bmatrix}$

$[C][B][D] \approx \begin{bmatrix} 0 & 1.8 \\ 4 & 7.1 \\ 1 & 1 \end{bmatrix}$

$[C][C][D] \approx \begin{bmatrix} 1.2 & 4.3 \\ 6.1 & 7.9 \\ 1 & 1 \end{bmatrix}$

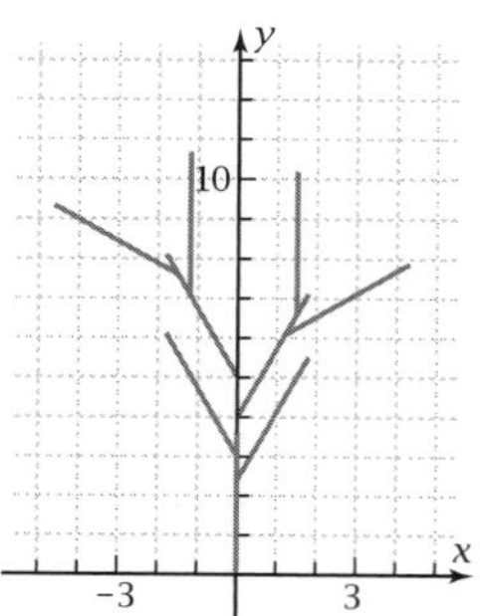

d. 5000 iterations:

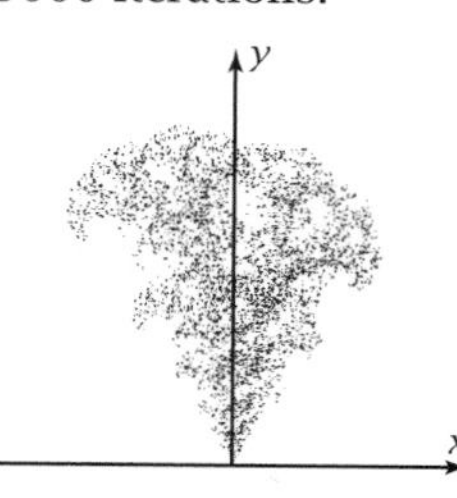

e. Strange attractor

f. Pre-image: 10 units

1st iteration: 18 units

2nd iteration: 32.4 units

3rd iteration: 58.32 units

100th iteration: $3.3670\ldots \times 10^{26}$ units

If the iterations were done forever, the length would become infinite.

g. $(X_A, Y_A) = (-4.6762\ldots, 7.4880\ldots)$

This point appears to be in the "foliage" of the left "branch" of the tree.

$(X_B, Y_B) = (0, 0)$

This is the "root" of the tree, i.e., the base of the "trunk."

$(X_C, Y_C) = (3.7410\ldots, 5.9904\ldots)$

This is in the "foliage" of the right "branch" of the tree.

11. a. $\begin{bmatrix} X_A \\ Y_A \end{bmatrix} = \begin{bmatrix} -2.9769... \\ 14.2984... \end{bmatrix}$

This is the topmost point of the "fern."

b. (X_B, Y_B) is the tip of the lower left "frond," (X_C, Y_C) is the tip of the lower right "frond," and (X_D, Y_D) is the base of the "stem."

c. $(X_B, Y_B) = (-0.6561..., 2.2628...)$
$(X_C, Y_C) = (0.9616..., 3.5276...)$
$(X_D, Y_D) = (0, 0)$
These results confirm the conjecture.

Problem Set 11-6

1. a. $r = \frac{1}{5}$, $N = 25 = 5^2$

$$D = \frac{\log N}{\log \frac{1}{r}} = \frac{\log 25}{\log 5} = \frac{2 \log 5}{\log 5} = 2$$

b. $r = 0.01 = \frac{1}{100}$, $N = 10{,}000 = 100^2$

$$D = \frac{\log N}{\log \frac{1}{r}} = \frac{\log 10{,}000}{\log 100} = \frac{2 \log 100}{\log 100} = 2$$

c. The smaller squares are identical and self-similar, and you can carry out the division process infinitely.

3. a. $r = 0.4 = \frac{1}{2.5}$, $N = 4$

$$D = \frac{\log N}{\log \frac{1}{r}} = \frac{\log 4}{\log 2.5} = 1.5129...$$

b. As n grows infinite, $A_n = A_0 \cdot (4 \cdot 0.4^2)^n = A_0 \cdot 0.64^n$ approaches zero. This is consistent with the dimension being less than 2.

c. As n grows infinite, $L_n = L_0 \cdot (4 \cdot 0.4)^n = L_0 \cdot 1.6^n$ also grows infinite. This is consistent with the dimension being greater than 1.

d. $D = 2$
The dimension is a whole number, not a fraction.

e. $D = 2.7138...$

As n grows infinite, the sum of the areas of the smaller squares $A = A_0 \cdot (4 \cdot 0.6^2)^n = A_0 \cdot 1.44^n$ also grows infinite. This is consistent with the dimension being greater than 2. However, the actual total area of the figure is the same as the area of the original square because the smaller squares overlap each other.

Problem Set 11-7

R1. 1024 segments; 0.0173... units;
$\left(\frac{2}{3}\right)^n \rightarrow 0$ units as $n \rightarrow \infty$

R2. a. $9\begin{bmatrix} 5 & 2 \\ 7 & -1 \end{bmatrix} - 6\begin{bmatrix} 3 & 8 \\ 5 & 4 \end{bmatrix} = \begin{bmatrix} 27 & -30 \\ 33 & -33 \end{bmatrix}$

b. $\begin{bmatrix} 3 & -5 & 2 \\ -1 & 4 & 3 \end{bmatrix}\begin{bmatrix} 6 \\ 2 \\ -3 \end{bmatrix} = \begin{bmatrix} 2 \\ -7 \end{bmatrix}$

c. $\begin{vmatrix} 3 & 8 \\ 5 & 4 \end{vmatrix} = -28$

d. $x = 2$, $y = 1$, $z = -4$

R3. a. Dilation by 0.6 and rotation about the origin by 30° counterclockwise.

b.

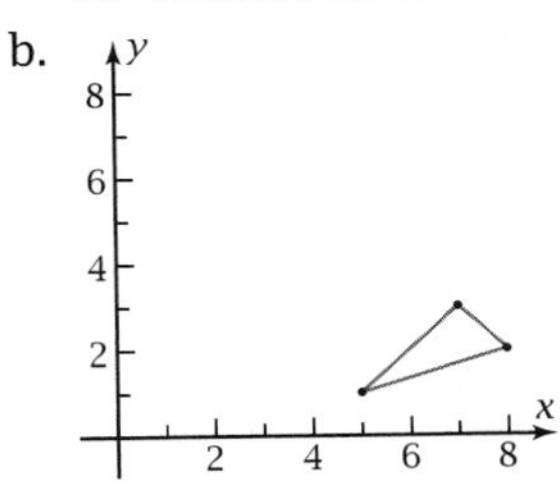

c. $[T]^3\begin{bmatrix} 5 & 8 & 7 \\ 1 & 2 & 3 \end{bmatrix} = \begin{bmatrix} -0.216 & -0.432 & -0.648 \\ 1.08 & 1.728 & 1.512 \end{bmatrix}$

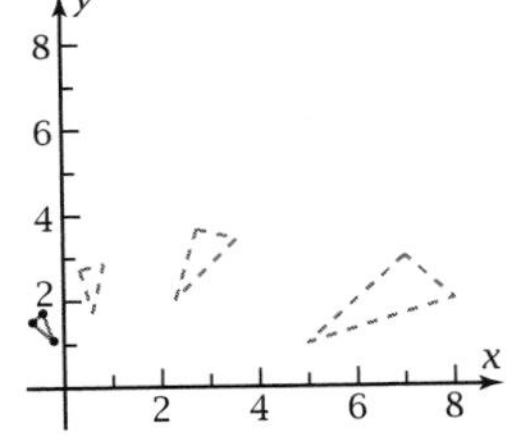

d. The graph has been rotated 90°.

e. Distance of pre-image vertex:
$d = \sqrt{5^2 + 1^2} = \sqrt{26}$
Distance of image vertex:
$d = \sqrt{(-0.216)^2 + 1.08^2} = \sqrt{1.213056} = \sqrt{26 \cdot 0.6^6}$
$= 0.6^3 \cdot \sqrt{26}$

R4. a. Dilation by 0.6, rotation by 30° counterclockwise about the origin, horizontal translation by 5, vertical translation by 2

b.

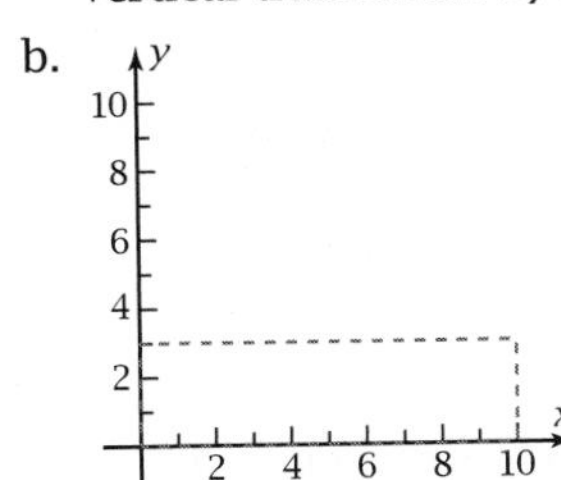

c.

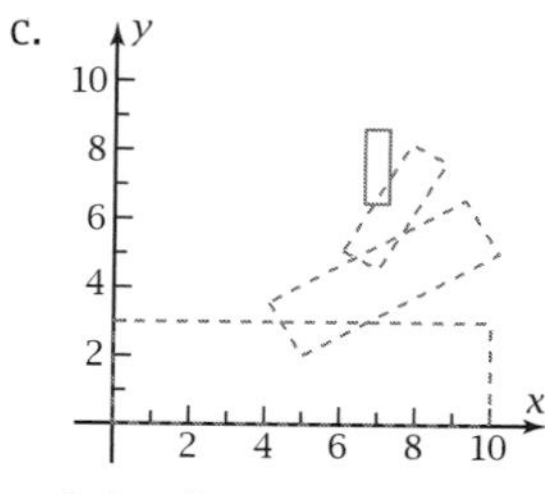

d. $\begin{bmatrix} X \\ Y \end{bmatrix} = \begin{bmatrix} 0.6\cos 30° - 1 & 0.6\cos 120° \\ 0.6\sin 30° & 0.6\sin 120° - 1 \end{bmatrix}^{-1} \begin{bmatrix} -5 \\ -2 \end{bmatrix}$

$= \begin{bmatrix} 5.6175... \\ 7.6714... \end{bmatrix}$

e. The third row in [M] adds the appropriate translation factor to each row of the image, and the third row of [A] ensures that the third row of the image will be the same as the third row of the pre-image.

R5. a. $[B] = \begin{bmatrix} 0.6\cos 30° & 0.6\cos 120° & -5 \\ 0.6\sin 30° & 0.6\sin 120° & -2 \\ 0 & 0 & 1 \end{bmatrix}$

b. [A][A][M]

$= \begin{bmatrix} 7.2263... & 6.1463... & 6.7698... & 7.8498... \\ 5.6545... & 3.7839... & 3.4239... & 5.2945... \\ 1 & 1 & 1 & 1 \end{bmatrix}$

[A][B][M]

$= \begin{bmatrix} 3.2301... & 2.1501... & 2.7736... & 3.8536... \\ 0.5760... & -1.2945... & -1.6545... & 0.2160... \\ 1 & 1 & 1 & 1 \end{bmatrix}$

[B][A][M]

$= \begin{bmatrix} -2.7736... & -3.8536... & -3.2301... & -2.1501... \\ 1.6545... & -0.2160... & -0.5760... & 1.2945... \\ 1 & 1 & 1 & 1 \end{bmatrix}$

[B][B][M]

$= \begin{bmatrix} -6.7698... & -7.8498... & -7.2263... & -6.1463... \\ -3.4239... & -5.2945... & -5.6545... & -3.7839... \\ 1 & 1 & 1 & 1 \end{bmatrix}$

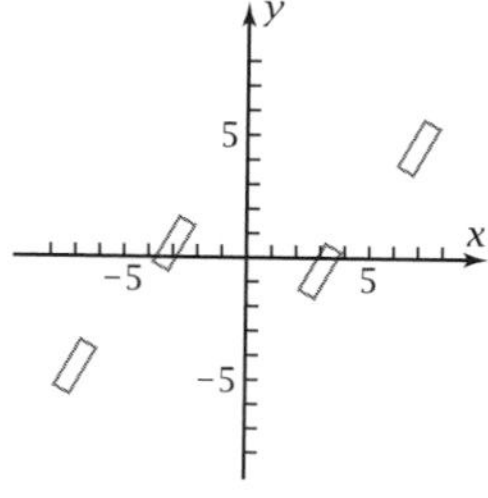

c. The attractor is correct. Probabilities are assigned to the transformations (in this case, equal probabilities of 0.25 each). An initial point is chosen. Then a random value between 0 and 1 determines which transformation is performed on that point. Then the procedure is repeated on the resulting point. As many iterations as desired are performed.

d.

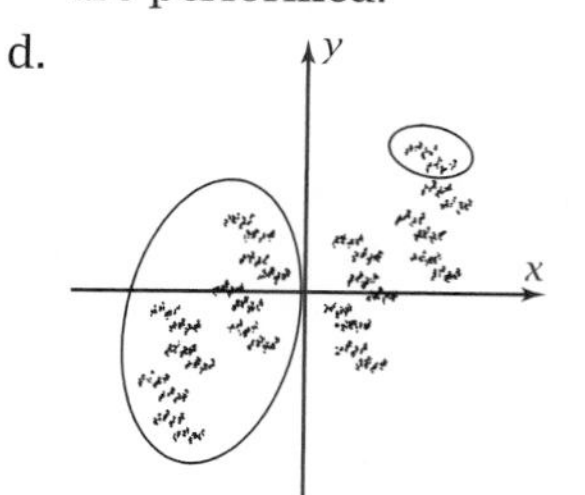

e. The fixed point, (5.6175..., 7.6714...), attracts the upper right end of the fractal image.

f.

n	N	Perimeter of One Rectangle	Total Perimeter
0	1	16	16
1	2	9.6	19.2
2	4	5.76	23.04
3	8	3.456	27.648
⋮	⋮	⋮	⋮
50	$1.1258... \cdot 10^{15}$	$1.2932... \cdot 10^{-10}$	145,607.0104

As the number of iterations approaches infinity, the total perimeter also approaches infinity.

R6. a. If an object is cut into N identical self-similar pieces, the ratio of the length of each piece to the length of the original object is r, and the subdivisions can be carried on infinitely, then the dimension D of the object is

$$D = \frac{\log N}{\log \frac{1}{r}}$$

b. $r = 0.6 = \frac{3}{5}$, $N = 2$

$$D = \frac{\log N}{\log \frac{1}{r}} = \frac{\log 2}{\log \frac{5}{3}} = 1.3569...$$

c. As n grows infinite, the total perimeter becomes infinite. This is consistent with the dimension being greater than 1.

d.

n	N	Area of One Rectangle	Total Area
0	1	12	12
1	2	4.32	8.64
2	4	1.5552	6.2208
3	8	0.559872	4.478976

As n grows infinite, the total area $A = 12 \cdot 0.72^n$ approaches zero. This is consistent with the dimension being less than 2.

e. The attractor is still quite similar but with more blank space:

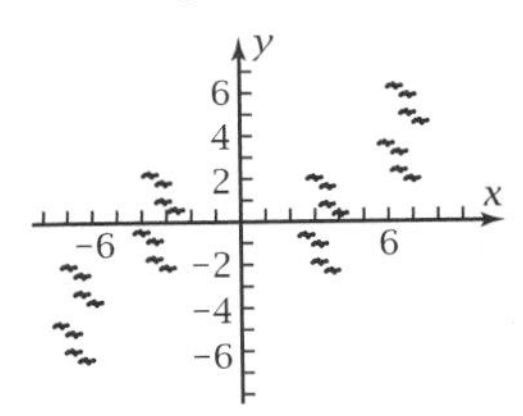

f. $r = 0.5 = \frac{1}{2},\ N = 2$

$$D = \frac{\log N}{\log \frac{1}{r}} = \frac{\log 2}{\log 2} = 1$$

$P_n = 2^n(0.5)^n \cdot 16 = 16$, so the perimeter remains 16 as n approaches infinity. This is consistent with the dimension being equal to 1.

g. $r = 0.4 = \frac{1}{2.5},\ N = 2$

$$D = \frac{\log N}{\log \frac{1}{r}} = \frac{\log 2}{\log 2.5} = 0.7564\ldots$$

$P_n = 2^n(0.4)^n \cdot 16 = (0.8)^n \cdot 16$

As n grows infinite, the total perimeter approaches zero. This is consistent with the dimension being less than 1.

CHAPTER 12

Problem Set 12-1

1.

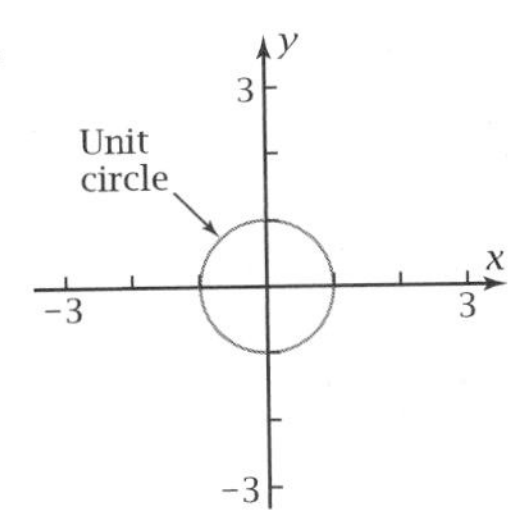

$y = \pm\sqrt{1 - x^2}$

The graph consists of all points whose distance to the origin is 1:

$\sqrt{(x-0)^2 + (y-0)^2} = \sqrt{x^2 + y^2} = 1$

3. $\frac{4}{36}x^2 + \frac{9}{36}y^2 = 1$

$\frac{x^2}{9} + \frac{y^2}{4} = 1$

$\left(\frac{x}{3}\right)^2 + \left(\frac{y}{2}\right)^2 = 1$

The dilations 3 and 2 appear as the denominators of x and y, respectively.

5.

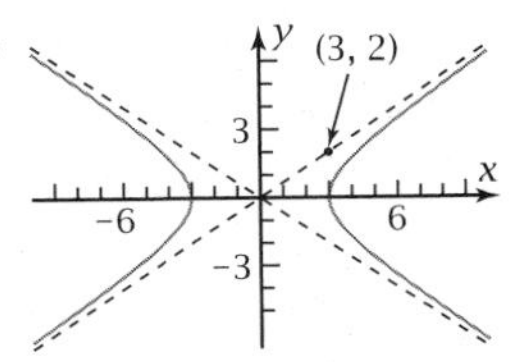

The asymptotes are $y = \frac{2x}{3}$ and $y = -\frac{2x}{3}$, which have slopes $\pm\frac{2}{3}$.

7. $y = \pm\sqrt{1 - x}$. The graph is the parabola in Figure 12-1d.

Problem Set 12-2

1. The unit circle shown in Figure 12-2b
3. The unit hyperbola shown in Figure 12-2b
5. a. Ellipse

b.

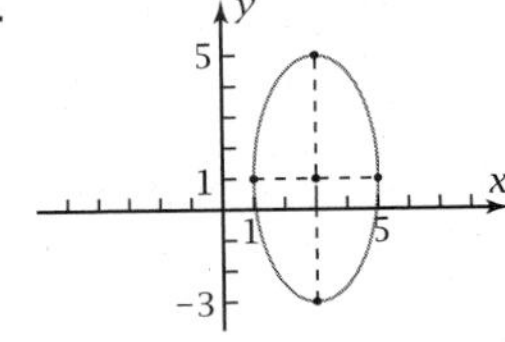

c. $4x^2 + y^2 - 24x - 2y + 21 = 0$

d. The graphs agree.

7. a. Hyperbola that opens in y-direction

b.

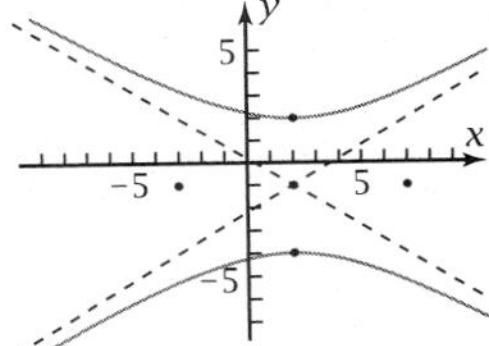

c. $-9x^2 + 25y^2 + 36x + 50y - 236 = 0$

d. The graphs agree.

9. a. Circle

b.

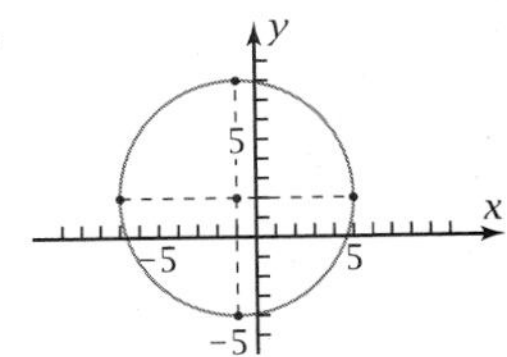

c. $x^2 + y^2 + 2x - 4y - 31 = 0$

d. The graphs agree.

11. a. Parabola opening vertically

b.

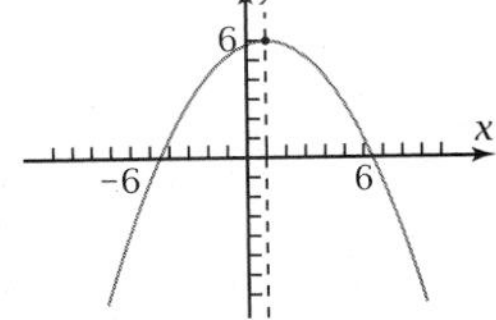

c. $0.2x^2 - 0.4x + y - 5.8 = 0$

d. The graphs agree.

13. a. Circle

b.

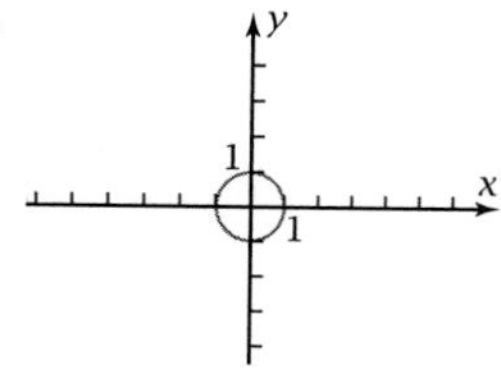

c. The graphs agree.

15. a. Hyperbola opening in the y-direction

b.

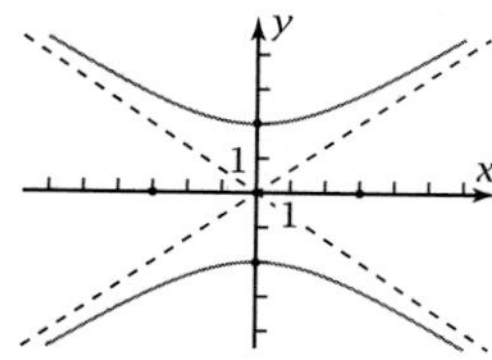

c. The graphs agree.

17. a. Hyperbola opening in the x-direction

b.

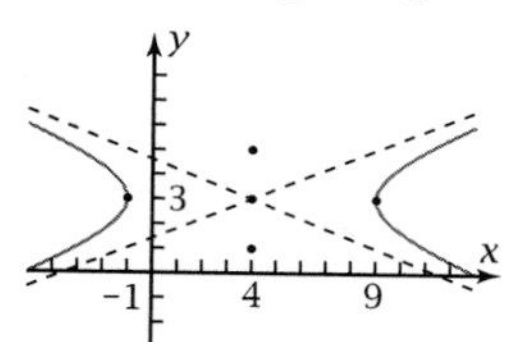

c. The graphs agree.

19. a. Circle

b.

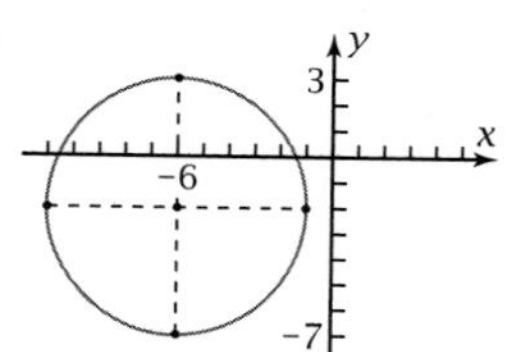

c. The graphs agree.

21. a. $\left(\frac{x+4}{6}\right)^2 + \left(\frac{y-3}{4}\right)^2 = 1$

b. $x = -4 + 6 \cos t$, $y = 3 + 4 \sin t$

c. The graphs match.

23. a. $-\left(\frac{x-4}{4}\right)^2 + \left(\frac{y+2}{3}\right)^2 = 1$

b. $x = 4 + 4 \tan t$, $y = -2 + 3 \sec t$

c. The graphs match.

25. a.

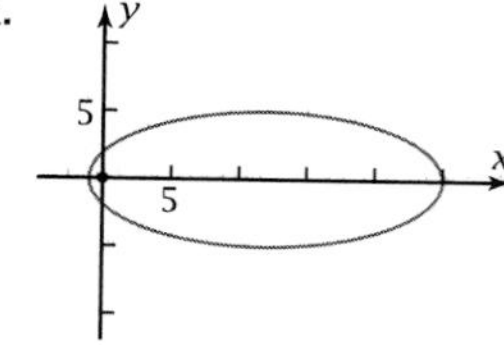

b. The closest the spaceship is to the Sun is 1 million miles away. The farthest the spaceship is from the Sun is 25 million miles away.

c. $y = \pm\frac{5}{13}\sqrt{105} = \pm 3.9411...$ million miles

d. The distance of $\left(20, \frac{5}{13}\sqrt{105}\right)$ from the Sun is 20.3846... million miles.

27. a. $\sec t = \frac{x-4}{5}$

$\tan t = \frac{y-3}{2}$

b. $\sec^2 t = \left(\frac{x-4}{5}\right)^2$

$\tan^2 t = \left(\frac{y-3}{2}\right)^2$

c. $\sec^2 t - \tan^2 t = \left(\frac{x-4}{5}\right)^2 - \left(\frac{y-3}{2}\right)^2$

d. $\sec^2 t - \tan^2 t = 1$, so the equation in part c is equivalent to $\left(\frac{x-4}{5}\right)^2 - \left(\frac{y-3}{2}\right)^2 = 1$, which is the equation of a hyperbola.

29. a. $16x^2 - 25y^2 - 128x - 100y + 556 = 0$

$$16x^2 - 128x - 25y^2 - 100y = -556$$
$$16x^2 - 16 \cdot 8x - 25y^2 + (-25) \cdot 4y = -556$$
$$16(x^2 - 8x) - 25(y^2 + 4y) = -556$$

b. $16(x^2 - 8x) + 256 - 25(y^2 + 4y) - 100$

$$= -556 + 256 - 100$$
$$16(x^2 - 8x) + 16 \cdot 16 - 25(y^2 + 4y) + (-25) \cdot 4$$
$$= -400$$
$$16(x^2 - 8x + 16) - 25(y^2 + 4y + 4) = -400$$

c. $\frac{16(x^2 - 8x + 16)}{-400} - \frac{25(y^2 + 4y + 4)}{-400} = \frac{-400}{-400} = 1$

$$-\frac{x^2 - 8x + 16}{25} + \frac{y^2 + 4y + 4}{16} = 1$$
$$-\frac{(x-4)^2}{25} + \frac{(y+2)^2}{16} = 1$$
$$-\left(\frac{x-4}{5}\right)^2 + \left(\frac{y+2}{4}\right)^2 = 1$$

d.

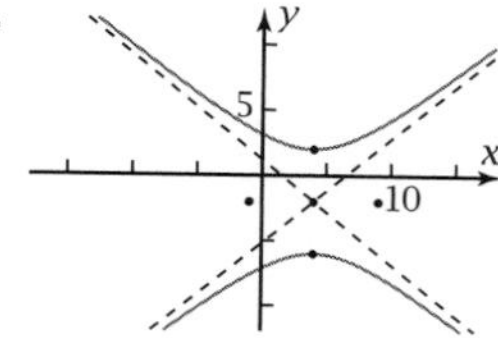

31. a. See Figure 12-2j.

b. The type of figure remains the same, but the figure may be rotated and the shape may be distorted.

Problem Set 12-3

1.

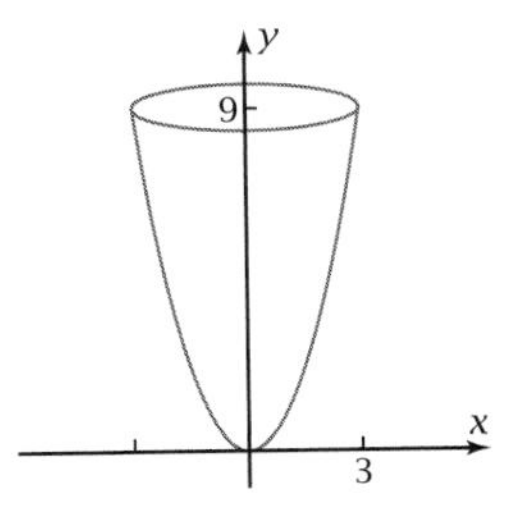

3.

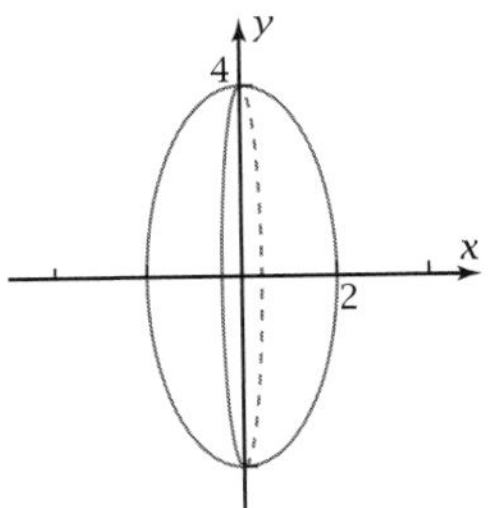

5.

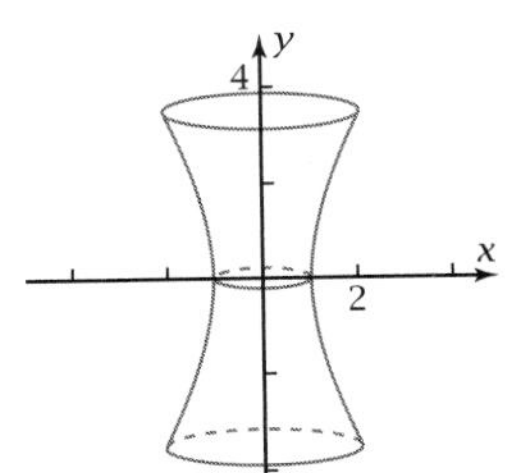

7.

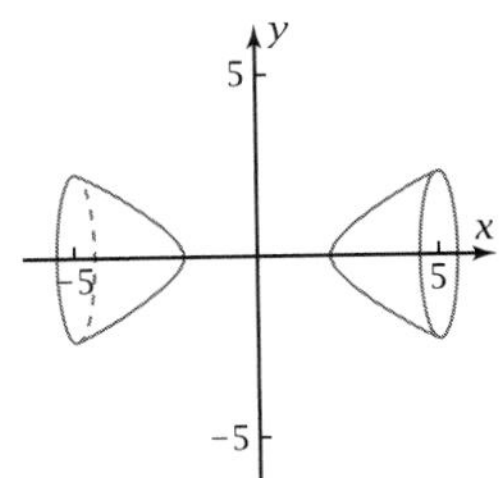

9.

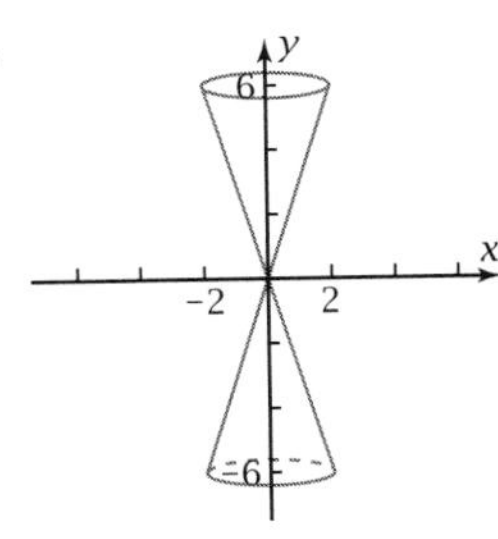

11. $A = 9x - x^3$

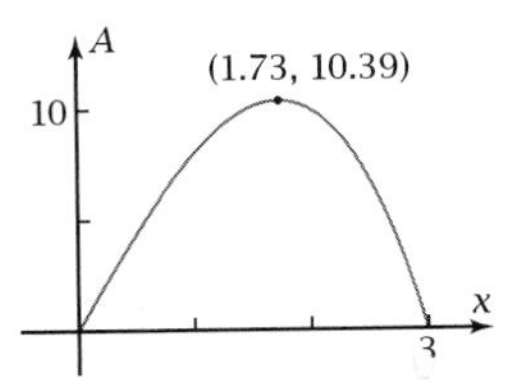

Maximum area at $x = 1.7320\ldots$

$A = 10.3923\ldots$

13. $x^2 + y^2 = 25 \Rightarrow y = \sqrt{25 - x^2}$ (you only need the positive value)

$V = \pi r^2 h = \pi x^2 \cdot 2y = 2\pi x^2\sqrt{25 - x^2}$

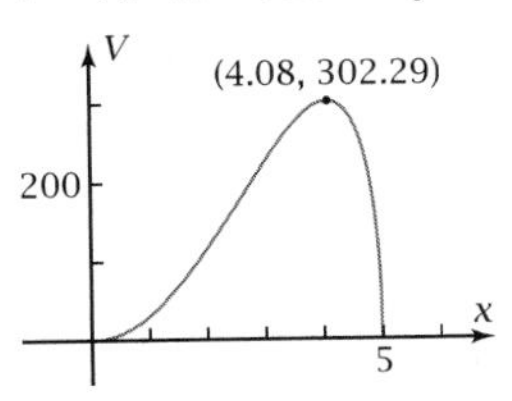

Maximum volume at $x = 4.0824\ldots$

$V = 302.2998\ldots$

15. $x^2 + y^2 = 25 \Rightarrow y = \sqrt{25 - x^2}$ (you only need the positive value)

a. $A = 2\pi rh = 2\pi x \cdot 2y = 4\pi x\sqrt{25 - x^2}$

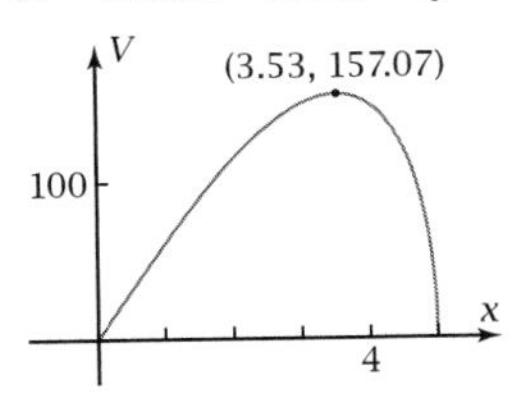

Maximum area at $x = 3.5355\ldots$

$A = 157.0796\ldots$

$r = 3.5355\ldots$; $h = 12.5$

b. $A = 2\pi r^2 + 2\pi rh = 2\pi x^2 + 2\pi x \cdot 2y$

$= 2\pi x\,(x + 2\sqrt{25 - x^2})$

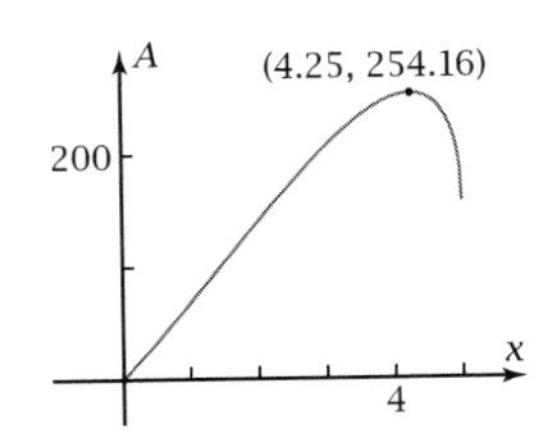

Maximum area at $x = 4.2532\ldots$

$A = 254.1601\ldots$

$r = 4.2532\ldots$; $h = 2.6286\ldots$

c. No

d. No

17. $225x^2 + 900y^2 = 202{,}500$

$\Rightarrow \left(\frac{x}{30}\right)^2 + \left(\frac{y}{15}\right)^2 = 1$, so 0 ft $< x <$ 30 ft

Also, $225x^2 + 900y^2 = 202{,}500$

$\Rightarrow y^2 = \frac{1}{4}(900 - x^2)$, so

$V = \pi r^2 h = \pi y^2 x$

$= \pi \cdot \frac{1}{4}(900 - x^2) \cdot x = \frac{\pi}{4}(900x - x^3)$

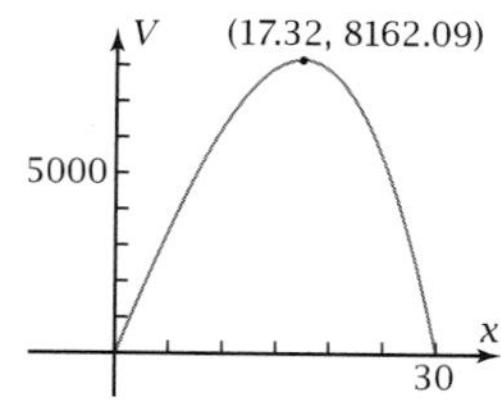

Maximum volume at $x = 17.3205\ldots$ ft

$V = 12.2474\ldots$ ft^3; $h = 17.3205\ldots$ ft

$A_{\text{lateral}} = 2\pi rh = 2\pi yx$

$= 2\pi \cdot \frac{1}{2}\sqrt{900 - x^2} \cdot x = \pi x\sqrt{900 - x^2}$

$A(x_{\max}) = A(17.3205\ldots) = 1332.8648\ldots$ ft^2

Problem Set 12-4

1. a. $y = \frac{1}{3}\sqrt{640} = 8.4327\ldots$ This agrees with the graph.

b. $d_1 = 5.2 \quad d_2 = \frac{26}{3} \quad d_3 = \frac{44}{3}$

c. $d_2 = \frac{26}{3} = \frac{5}{3} \cdot 5.2 = ed_1$

d. $|d_2 - d_3| = \left|\frac{26}{3} - \frac{44}{3}\right| = \left|\frac{-18}{3}\right| = 6$

e. x-dilation is $3 = a$, the transverse radius; y-dilation is $4 = b$, the conjugate radius.

f. $c^2 = 5^2 = 3^2 + 4^2 = a^2 + b^2$

g. $a = 3 = \frac{5}{3} \cdot 1.8 = ed$

$c = 5 = \frac{5}{3} \cdot 3 = ea$

3. a. The vertex is equidistant from the focus and the directrix. The eccentricity is the ratio of the distances from a point on the curve to the focus and to the directrix, so $e = 1$.

b. $y = \frac{1}{8}(6^2) = \frac{36}{8} = 4.5$ agrees with the graph.

c. $d_1 = |4.5 - (-2)| = 6.5$

$d_2 = \sqrt{(6 - 0)^2 + (4.5 - 2)^2} = \sqrt{36 + 6.25} = 6.5$

d. $d_1 = 6.5 = d_2$

The eccentricity is $e = \frac{d_2}{d_1} = \frac{6.5}{6.5} = 1$.

5. 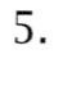

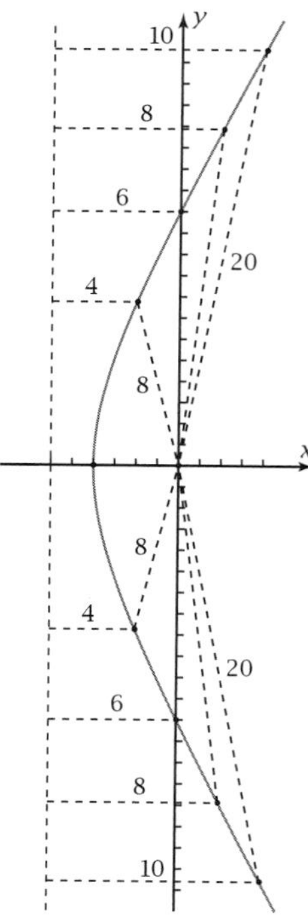

7. The graph looks the same as in Problems 5 and 6.

9. a. The major axis is 283 million miles long. The major radius, a, is 141.5 million miles long.

b. $c = 13.5$ million miles

$b = \sqrt{19{,}840} = 140.8545\ldots$ million miles

c. $\frac{x^2}{20{,}022.25} + \frac{y^2}{19{,}840} = 1$

d. $x = c = 13.5$ million miles

Distance to Sun is

$y = 140.2120\ldots$ million miles.

e. $e = 0.0954\ldots$

f. The distance is 1469.6296 million miles.

g. $x = 141.5 \cos t$

$y = 140.8545\ldots \sin t$

The graph looks similar to Figure 12-4u.

h. The major and minor radii are nearly equal, and the eccentricity is close to zero.

11. a. Ellipse

b. $a = 5$, $b = 3$, $c = 4$, $d = \frac{25}{4} = 6\frac{1}{4}$, $e = \frac{4}{5}$

c.

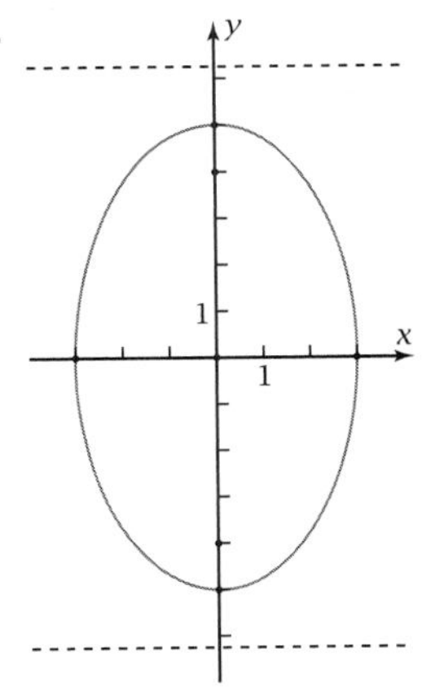

13. a. Hyperbola opening vertically
 b. $a = 3$, $b = 6$, $c = \sqrt{45} = 6.7082\ldots$,
 $d = \sqrt{1.8} = 1.3416\ldots$, $e = \sqrt{5} = 2.2360\ldots$
 c.

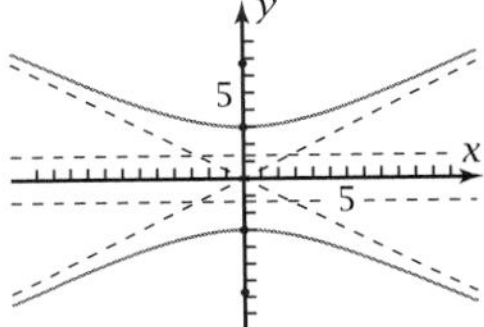

15. a. Ellipse
 b. $a = 4$, $b = 3$, $c = \sqrt{7} = 2.6457\ldots$, $d = \dfrac{16}{\sqrt{7}} = 6.0474\ldots$,
 $e = \dfrac{\sqrt{7}}{4} = 0.6614\ldots$
 c.

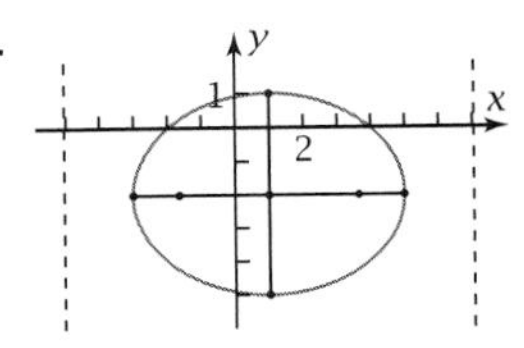

17. a. Hyperbola opening vertically
 b. $a = \sqrt{10} = 3.1622\ldots$, $b = \sqrt{6} = 2.4494\ldots$,
 $c = 4$, $d = 2.5$, $e = \dfrac{2\sqrt{10}}{5} = 1.2649\ldots$
 c.

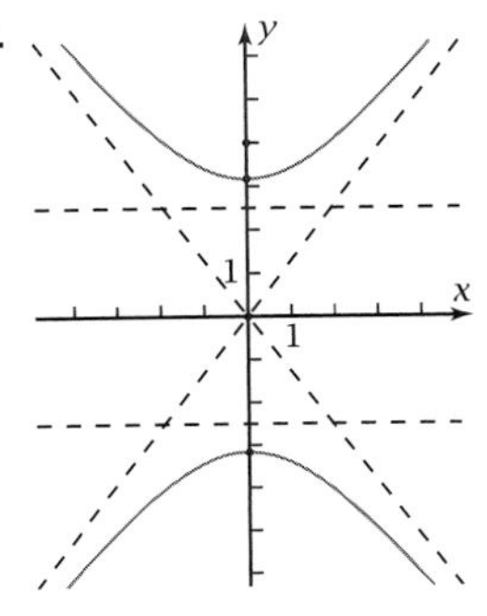

19. a. Parabola opening left
 b. $p = -1$, $c = -1$, $d = 1$, $e = 1$
 c.

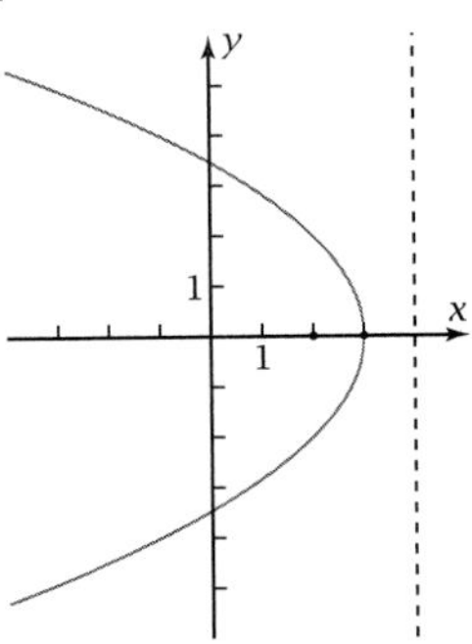

21. a. Hyperbola opening vertically

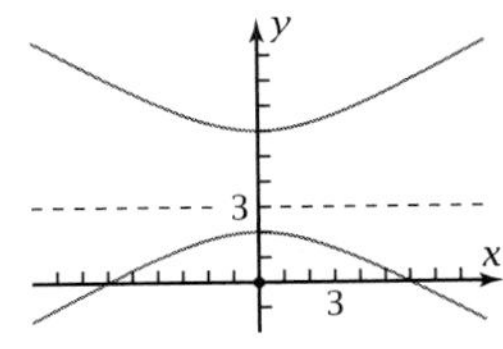

 b. $-\left(\dfrac{x}{2\sqrt{3}}\right)^2 + \left(\dfrac{y-4}{2}\right)^2 = 1$
 c. The graphs agree.

23. a. Parabola opening vertically

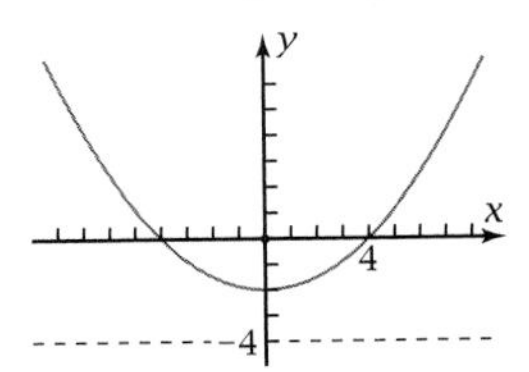

 b. $y = \frac{1}{8}x^2 - 2$
 c. The graphs agree.

25. a. Ellipse

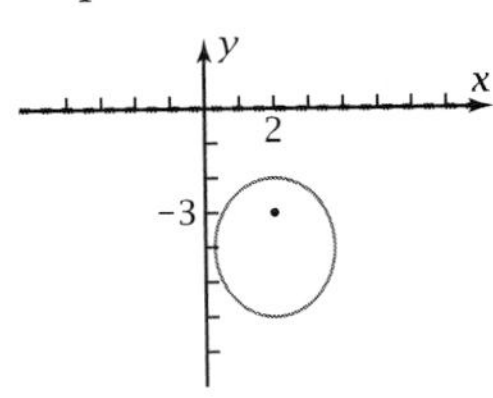

 b. $\left(\dfrac{x-2}{\sqrt{3}}\right)^2 + \left(\dfrac{y+4}{2}\right)^2 = 1$
 c. The graphs agree.

27. a.

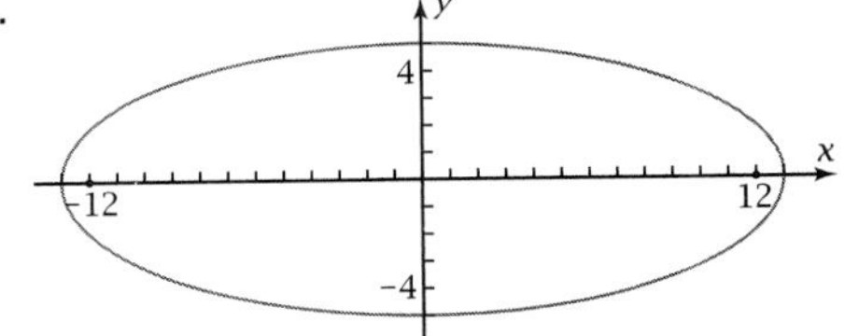

b. $\left(\frac{x}{13}\right)^2 + \left(\frac{y}{5}\right)^2 = 1$

c. The graphs agree.

29. a.

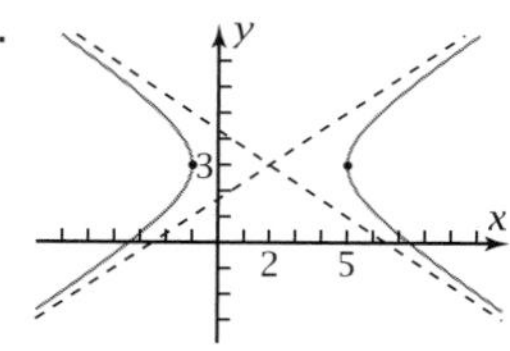

b. $\frac{(x-2)^2}{3} - \frac{(y-3)^2}{2} = 1$

c. The graphs agree.

31. a.

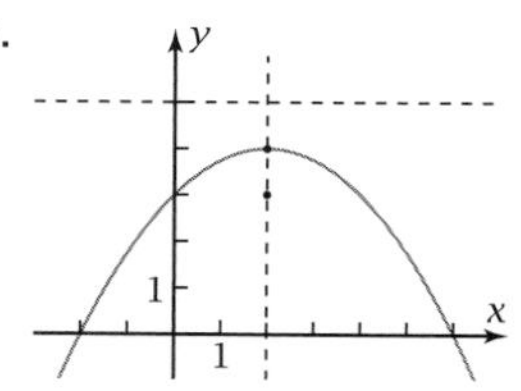

b. $y = \frac{1}{4}x^2 + x + 3$

c. The graphs agree.

33. For Problem 9, $LR = 280.4240\ldots$ million miles. For the ellipse of Problem 11, $LR = \frac{18}{5} = 3.6$. For the hyperbola of Problem 17,

$LR = \frac{6\sqrt{10}}{5} = 3.7947\ldots$

Problem Set 12-5

1. $x = -4 + 3\cos 20° \cos t + 5\cos 110° \sin t$
   $y = 2 + 3\sin 20° \cos t + 5\sin 110° \sin t$

3. $x = -2 + 7\cos(-10°) \tan t + 3\cos 80° \sec t$
   $y = -1 + 7\sin(-10°) \tan t + 3\sin 80° \sec t$

5. $x = 2 + \frac{1}{32}t^2 \cos 25° + t\cos 115°$
   $y = -1 + \frac{1}{32}t^2 \sin 25° + t\sin 115°$

7. $x = 8 + 25\cos(-30°) \cos t + 7\cos 60° \sin t$
   $y = 5 + 25\sin(-30°) \cos t + 7\sin 60° \sin t$

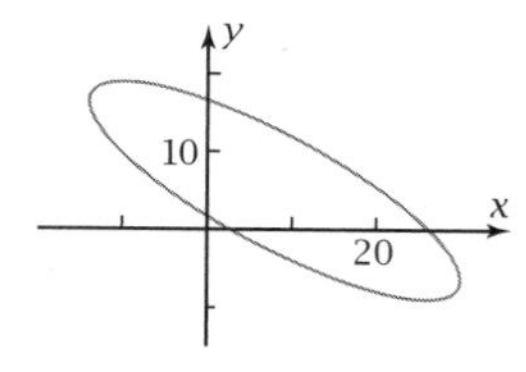

9. $x = 4\cos 15° \sec t + 3\cos 105° \tan t$
   $y = 4\sin 15° \sec t + 3\sin 105° \tan t$

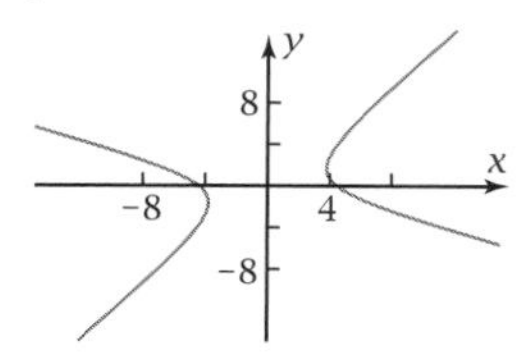

11. $x = -8 + t^2 \cos(-30°) + t\cos 60°$
    $y = 5 + t^2 \sin(-30°) + t\sin 60°$

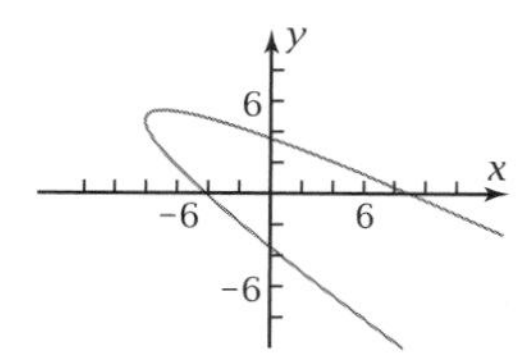

13. $B^2 - 4AC = (-5)^2 - 4(3)(9) = -83$; ellipse

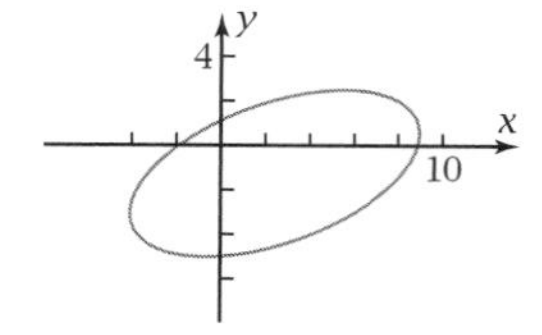

15. $B^2 - 4AC = (-10)^2 - 4(3)(6) = 28$; hyperbola

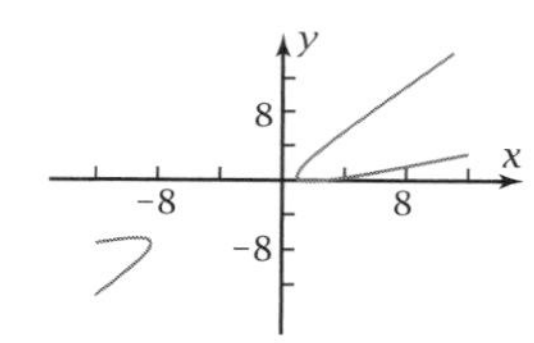

17. $B^2 - 4AC = 6^2 - 4(1)(9) = 0$; parabola

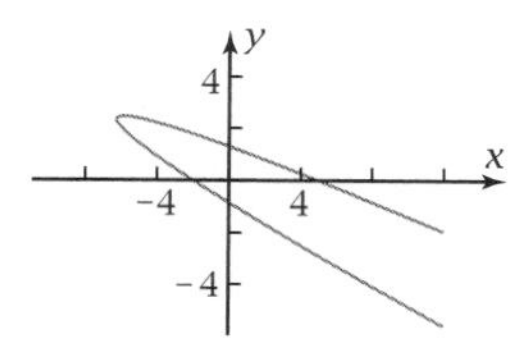

19. If $B = 0$, then $B^2 - 4AC = -4AC$. A and C have the same sign (ellipse) $\Leftrightarrow -4AC < 0$. A and C have opposite signs (hyperbola) $\Leftrightarrow -4AC > 0$. Either $A = 0$ or $C = 0$ (parabola) $\Leftrightarrow -4AC = 0$.

21. a. $\begin{bmatrix} \cos\alpha & \cos(\alpha + 90°) \\ \sin\alpha & \sin(\alpha + 90°) \end{bmatrix} \begin{bmatrix} a\cos t \\ b\sin t \end{bmatrix}$

$= \begin{bmatrix} (a\cos\alpha)\cos t + (b\cos(\alpha + 90°))\sin t \\ (a\sin\alpha)\cos t + (b\sin(\alpha + 90°))\sin t \end{bmatrix}$

b. $a = \sqrt{34} = 5.8309\ldots$ units

$b = \dfrac{2\sqrt{34}}{5} = 2.3323\ldots$ units

$\alpha = \tan^{-1}\frac{5}{3} = 59.0362\ldots°$

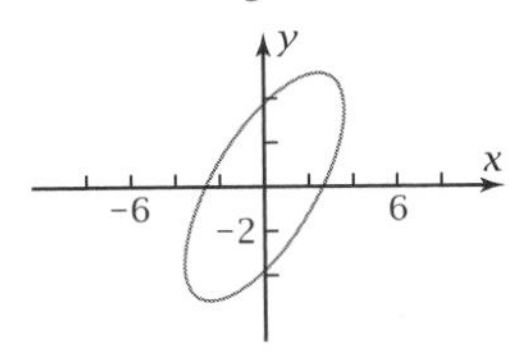

23. a. $B^2 - 4AC = 0^2 - 4(1)(1) = -4$; ellipse

$y = 4 \pm \sqrt{10x - x^2}$

b. $B^2 - 4AC = 1^2 - 4(1)(1) = -3$; ellipse

$y = \dfrac{-x + 8 \pm \sqrt{24x - 3x^2}}{2}$

c. $B^2 - 4AC = 2^2 - 4(1)(1) = 0$; parabola

$y = -x + 4 \pm \sqrt{2x}$

d. $B^2 - 4AC = 4^2 - 4(1)(1) = 12$; hyperbola

$y = -2x + 4 \pm \sqrt{3x^2 - 6x}$

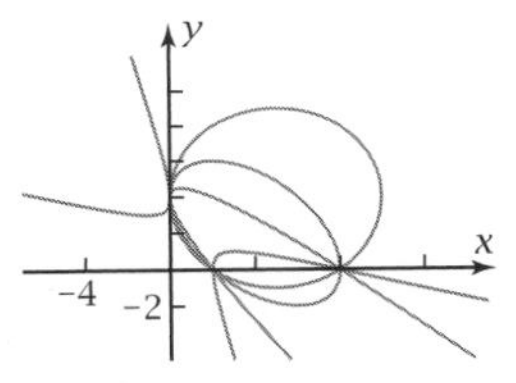

The x- and y-intercepts are the same for all the graphs. To see this algebraically, set either x or $y = 0$. Then the xy-term is 0 and the remaining equation in one variable is the same for all curves.

Problem Set 12-6

1. Assume the major (52-in.) and minor (26-in.) axes are already drawn, perpendicularly bisecting each other. Drive a nail at one end of the minor axis. Tie one end of a string to the nail and the other end of the string to a pencil so that the pencil is 26 (half of 52) in. from the nail. Use this as a compass to draw an arc of a circle with radius 26 in., intersecting the major axis in two points, which will be the foci. Now drive nails at the two foci and tie a 52-in. string between them. Use a pencil to pull the string taut, and slide the pencil back and forth, always keeping the string taut, to draw the ellipse. (Actually, this will draw *half* the ellipse; the string and pencil will have to be flipped to the other side of the nails to draw the other half.)

$e = \dfrac{\sqrt{3}}{2} = 0.8660\ldots$

3. a. $y = -\frac{1}{1000}x^2$

b. The columns divide the bridge into 20 equal sections, so they are $\frac{1000 \text{ ft}}{20} = 50$ ft apart.
Y1 = 2/1000 * X² + 20 (the *positive* value) was used in a grapher to get the following table:

x	y
−500	250
−450	202.5
−400	160
−350	122.5
−300	90
−250	62.5
−200	40
−150	22.5
−100	10
−50	2.5
0	0
50	2.5
100	10
150	22.5
200	40
250	62.5
300	90
350	122.5
400	160
450	202.5
500	250

c. By adding up all of the y-values in the table in part b and multiplying the total by 2, you get 3850 ft.

5. a. Parabola. The meteor's path doesn't intersect Earth's surface.

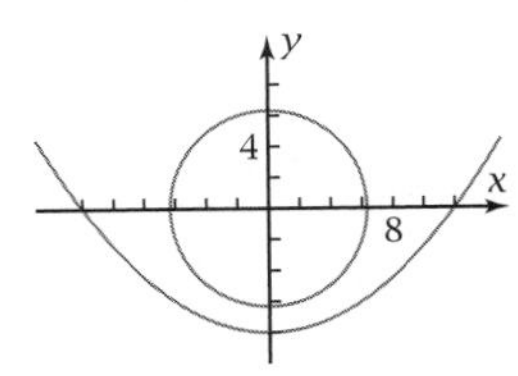

b. $x^2 + y^2 = 40 \Rightarrow x^2 = 40 - y^2$

$x^2 - 18y = 144 \Rightarrow (40 - y^2) - 18y = 144$

$\Rightarrow y^2 + 18y + 104 = 0$; the discriminant is $(18)^2 - 4(1)(104) = -92$; no real solution.

c. Hyperbola; the branch with the positive square root does not intersect, but the branch with the negative square root does:

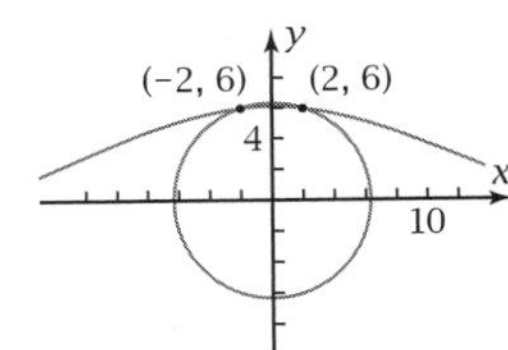

d. $x^2 + y^2 = 40 \Rightarrow x^2 = 40 - y^2$
$x^2 - 4y^2 + 80y = 340$
$\Rightarrow (40 - y^2) - 4y^2 + 80y = 340$
$\Rightarrow 5y^2 - 80y + 300 = 0 \Rightarrow y^2 - 16y + 60 = 0$
$\Rightarrow (y - 6)(y - 10) = 0 \Rightarrow y = 6$ or $y = 10$
$y = 10$ is an extraneous solution because it does not satisfy $x^2 + y^2 = 40$.
$y = 6 \Rightarrow x = \pm\sqrt{40 - (6)^2} = \pm 2$
The asteroid strikes at (−2000 km, 6000 km) or (2000 km, 6000 km), depending on which way it is traveling.

7. a. $(x - 8)^2 + (y - 0)^2 = 4^2$, a circle with center (8, 0) and radius 4.
b.

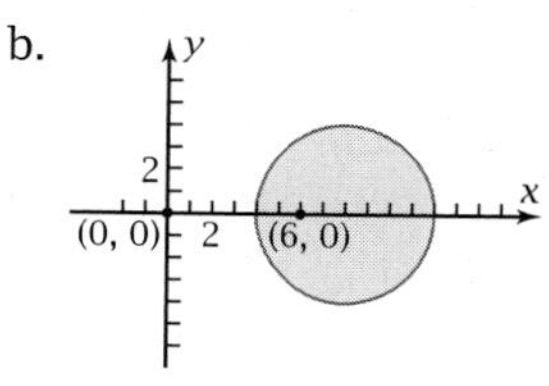

c. (15, 0) is closer to Supplier 2 but outside the shaded region, so Supplier 1 is cheaper.

9. Hyperbola: $x = 10 + 4 \sec t$, $y = \frac{23}{2} + \frac{26}{3} \tan t$
Top ellipse: $x = 10 + 5 \cos t$, $y = 18 + \sin t$
Bottom ellipse: $x = 10 + 6 \cos t$, $y = 2 + \frac{6}{5} \sin t$
To make the graph come out correctly, use a t-range of $-47 \le t \le 37$. Split the equation of the hyperbola in two. Use the original and
$x = 10 - 4 \sec t$
$y = \frac{23}{2} + \frac{26}{3} \tan t$
For the ellipse, you want the arguments to span 360°, so you need to change them. Multiplying t by $\frac{360}{84}$ accomplishes this. So you have:
Top ellipse: $x = 10 + 5 \cos\left(\frac{360}{84} \cdot t\right)$
$y = 18 + \sin\left(\frac{360}{84} \cdot t\right)$
Bottom ellipse: $x = 10 + 6 \cos\left(\frac{360}{84} \cdot t\right)$
$y = 2 + \frac{6}{5} \sin\left(\frac{360}{84} \cdot t\right)$

Problem Set 12-7

R1. a. Circle
b. Ellipse
c. Hyperbola
d. Hyperbola
e. Parabola

R2. a. i. $x^2 + y^2 = 1$
ii. $x = \cos t$, $y = \sin t$
iii. $x = \sec t$, $y = \tan t$
iv. $-x^2 + y^2 = 1$
v. $x = y^2$
vi. $y = x^2$
b. i.

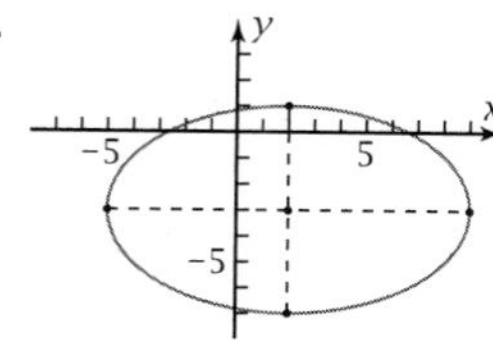

ii. $x^2 + y^2 - 4x + 6y - 771 = 0$
iii. The graph agrees with the sketch.
iv. $x = 2 + 7 \cos t$, $y = -3 + 4 \sin t$
c. i. $-\left(\frac{x + 4}{2}\right)^2 + \left(\frac{y - 1}{3}\right)^2 = 1$
ii. $x = -4 + 2 \sec t$, $y = 1 + 3 \tan t$
iii. The graph agrees.

R3. a. i. $4x^2 + y^2 = 16 \Rightarrow \left(\frac{x}{2}\right)^2 + \left(\frac{y}{4}\right)^2 = 1$

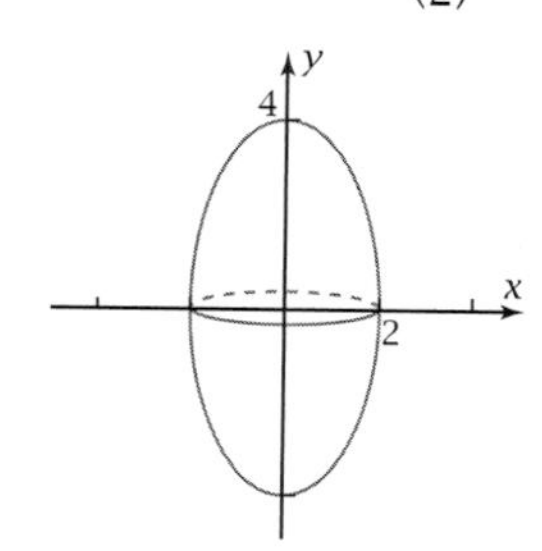

ii. $x^2 - 9y^2 = 9 \Rightarrow \left(\frac{x}{3}\right)^2 - \left(\frac{y}{1}\right)^2 = 1$

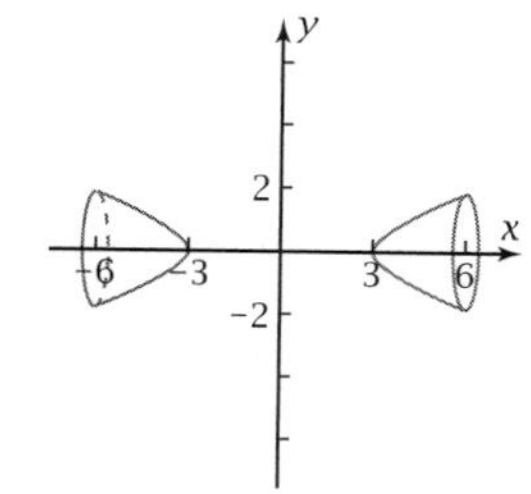

iii. $4x^2 - y^2 = 4 \Rightarrow \left(\frac{x}{1}\right)^2 - \left(\frac{y}{2}\right)^2 = 1$

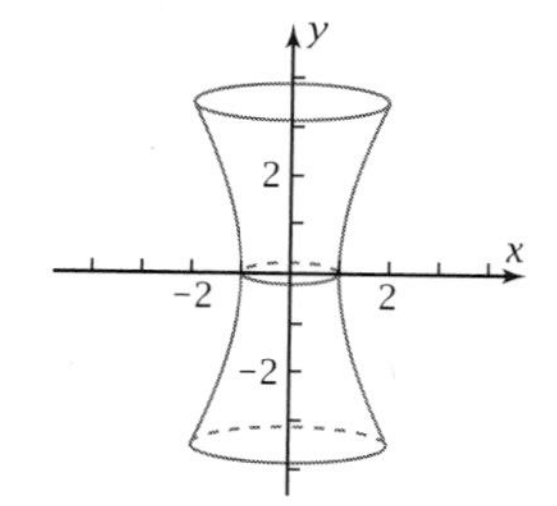

b. $V = \pi r^2 h = \pi y^2 x = \pi y^2(4 - y^2)$
$= \pi(4y^2 - y^4),\ 0 < y < 2$

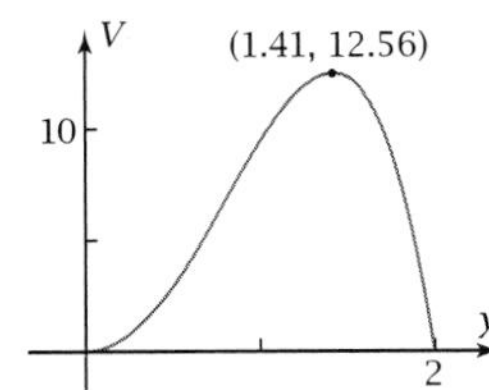

$h = y = 1.4142\ldots;\ r = x = 4 - y^2 = 2.0000\ldots;$
$V = 12.5663\ldots$

R4. a. $d_1 = 7,\ d_2 = 5.6,\ d_3 = 10.4$

b. $c = 6.4$

$$a = \frac{c}{e} = \frac{6.4}{0.8} = 8$$

$$b = \sqrt{a^2 - c^2} = \sqrt{23.04} = 4.8$$

$$\left(\frac{x}{8}\right)^2 + \left(\frac{y}{4.8}\right)^2 = 1$$

When $x = 3$, $y = \sqrt{19.8}$. The measurements agree.

c. $c = 81$ cm

$$a = \frac{c}{e} = \frac{81}{0.9} = 90 \text{ cm}$$

$$b = \sqrt{a^2 - c^2} = \sqrt{1539} = 39.2300\ldots \text{ cm}$$

$x = 39.2300\ldots \cos t,\ y = 90 \sin t$

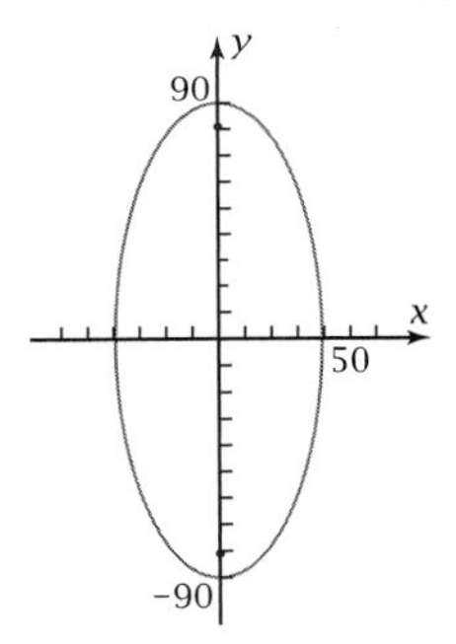

The ball will pass through the other focus.

d. Hyperbola opening horizontally

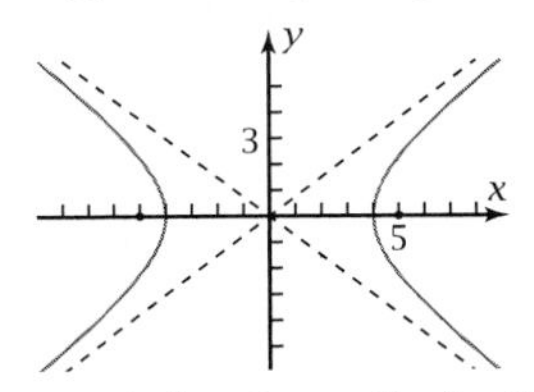

$a = 4,\ b = 3,\ c = 5,\ d = 3.2,\ e = 1.25$

e.

$x = 2 - \frac{1}{8}y^2$

f. $x = 2 + 1.8 \tan t,\ y = 4 + 2.4 \sec t$

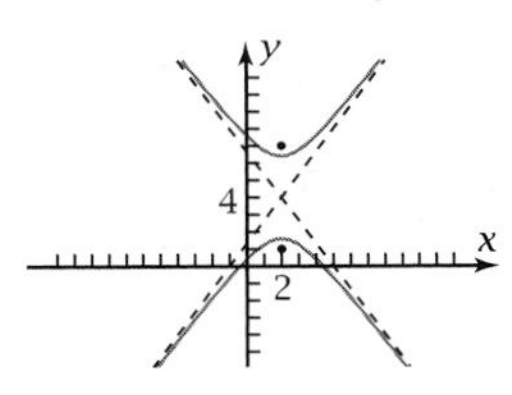

R5. a. $x = 10 + 9\cos 20° \cos t + 3\cos 110° \sin t,$
$y = 7 + 9\sin 20° \cos t + 3\sin 110° \sin t,$
$0° \le t \le 360°$. The graph is correct.

b. $x = 3 + 6\cos 35° \sec t + 6\sqrt{3}\cos 125° \tan t,$
$y = -4 + 6\sin 35° \sec t + 6\sqrt{3}\sin 125° \tan t,$
$0° \le t \le 360°$

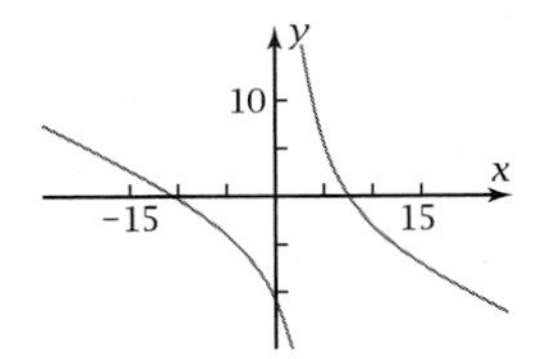

c. $x = 1 + \dfrac{\sqrt{2}}{24}t^2 \cos 45° + t\cos 135°,$

$y = 2 + \dfrac{\sqrt{2}}{24}t^2 \sin 45° + t\sin 135°,\ -24 \le t \le 24$

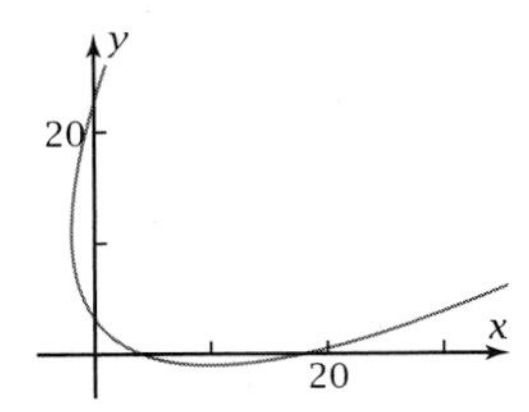

d. i. $B^2 - 4AC = (2)^2 - 4(4)(9) = -140 < 0$; ellipse

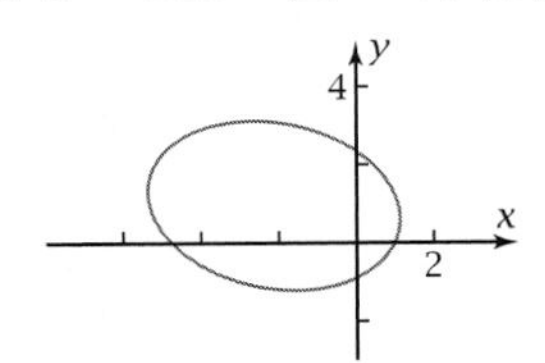

ii. $B^2 - 4AC = (12)^2 - 4(4)(9) = 0$; parabola

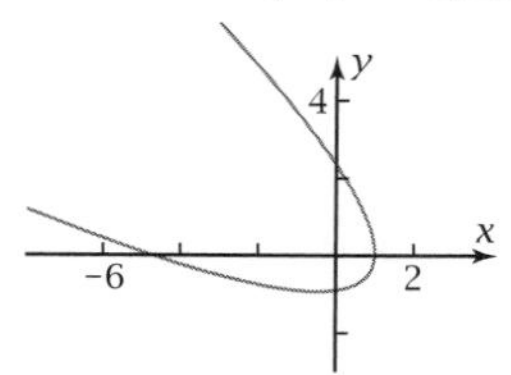

iii. $B^2 - 4AC = (22)^2 - 4(4)(9) = 340 > 0$; hyperbola

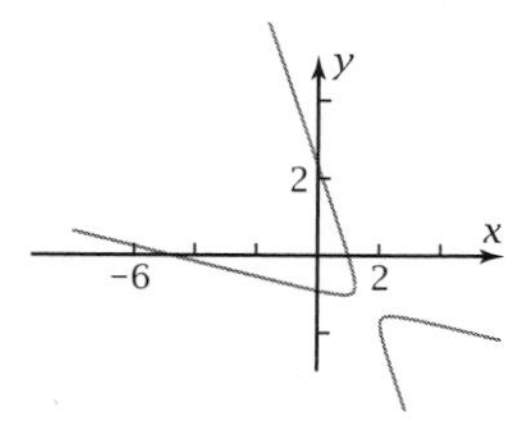

e. They all have the same x- and y-intercepts.

R6. a. $x = \frac{1}{100}y^2$; domain: 0 in. $\le x \le$ 16 in.

b.

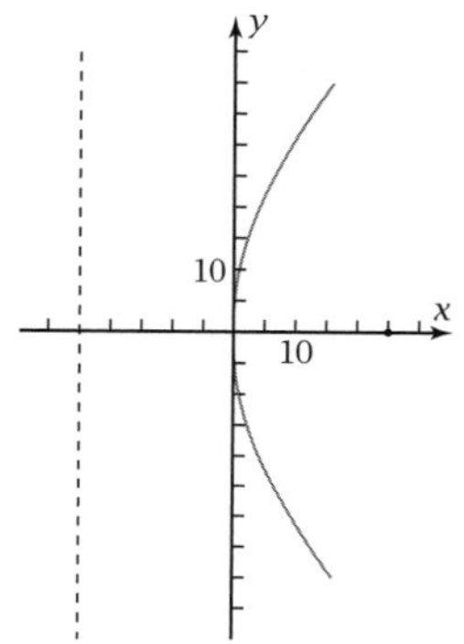

c. Yes, as discovered in part a and shown in the graph of part b, the focus is at $x = 25$, while the dish only extends to $x = 16$.

CHAPTER 13

Problem Set 13-1

1.

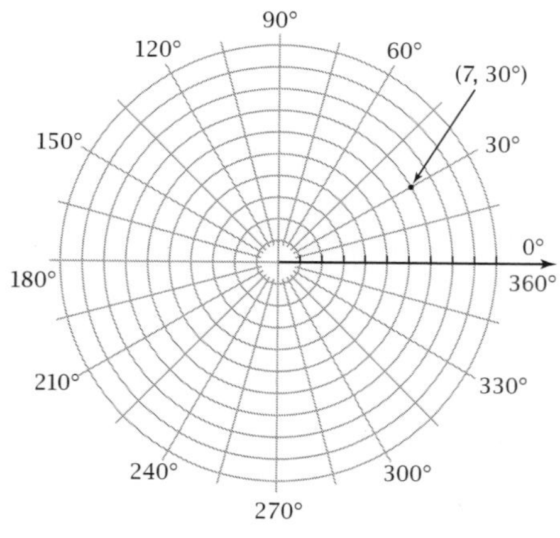

3.

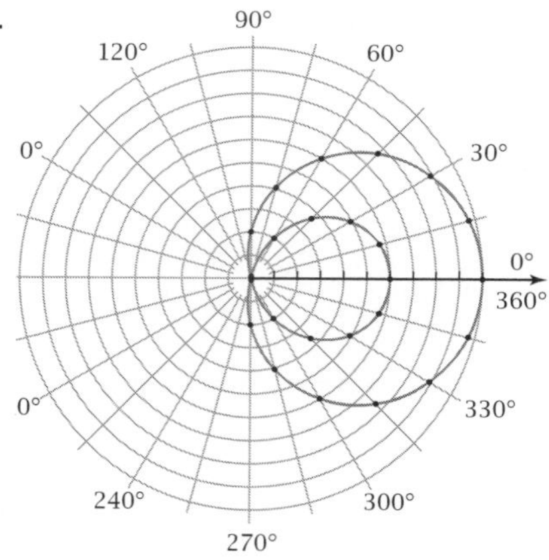

5.

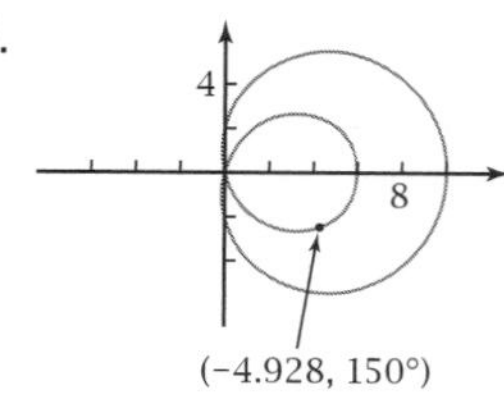

The coordinates agree.

Problem Set 13-2

1.

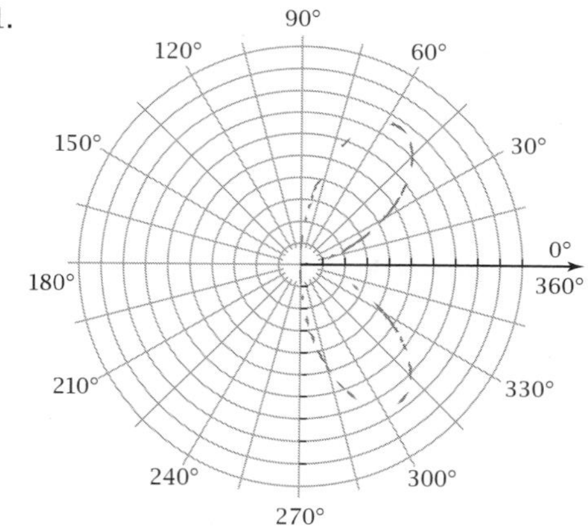

3.

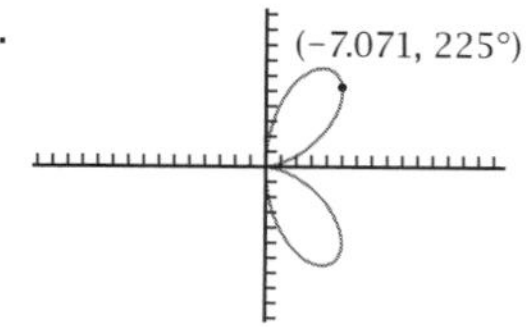

The graph is being retraced between 180° and 360°. The figure has two ("bi-") "leaves" ("folium").

5.

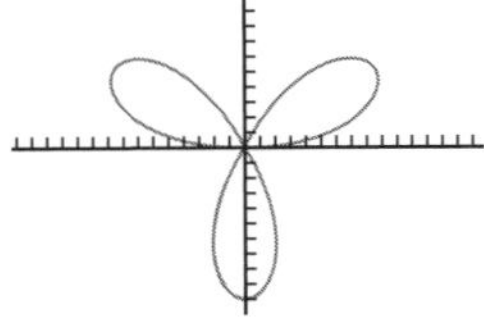

$10 \sin 3\theta = 0 \Leftrightarrow 3\theta = 0° + 180n° \Leftrightarrow \theta = 0° + 60n°$
$r < 0$ for $60° < \theta < 120°$

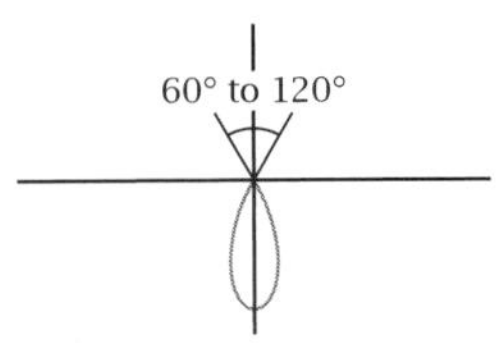

7.

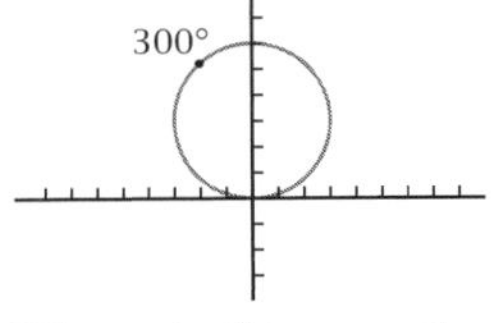

When $\sin \theta$ is negative, i.e., $180° < \theta < 360°$, the points are plotted in the opposite direction.

9. $r = 6 \sin \theta$
$r^2 = 6r \sin \theta$
$x^2 + y^2 = 6y$
$y^2 - 6y + 9 + x^2 = 9$
$(y - 3)^2 + x^2 = 3^2$
A circle with center (0, 3) and radius 3

11. $r < 0$ for $180° < \theta < 540°$

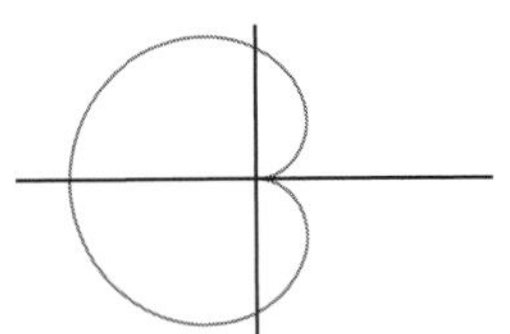

13.

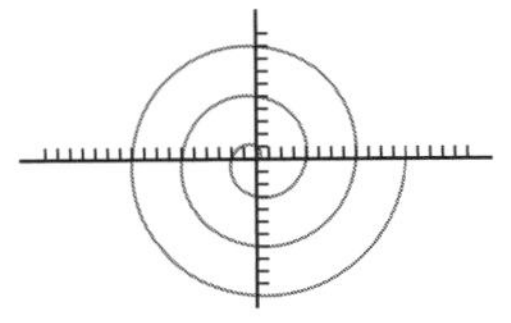

$r = 4, 8, 12$

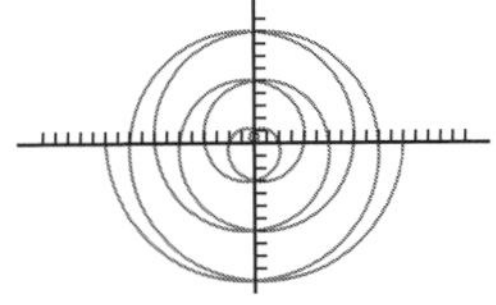

The graph is the original spiral with its mirror image across the y-axis.

15. a. $r = \cos \theta$
b. $r = \sin \theta$
c. $r = -\cos \theta$
d. $r = -\sin \theta$

17. a.

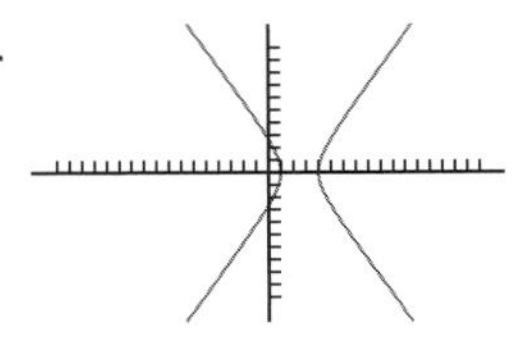

b. $\left(\dfrac{x - \frac{5}{2}}{\frac{3}{2}}\right)^2 - \left(\dfrac{y - 0}{2}\right)^2 = 1$

Hyperbola with center $\left(\frac{5}{2}, 0\right)$, horizontal semitransverse axis $\frac{3}{2}$, and vertical semiconjugate axis 2

c. One focus is at the pole. In the polar equation $r = \dfrac{8}{3 + 5 \cos \theta}$, $a = 3$ and $b = 5$, so $e = \left|\dfrac{b}{a}\right| = \dfrac{5}{3} > 1$, confirming that the graph is a hyperbola.

19. a.

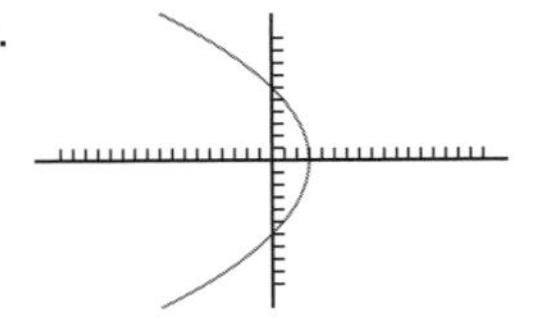

b. $x = -\frac{1}{12}y^2 + 3$
Parabola opening left, with vertex (0, 3)

c. The focus is at the pole. In the polar equation $r = \dfrac{6}{1 + \cos \theta}$, $a = b = 1$, so $e = \left|\dfrac{b}{a}\right| = 1$, confirming that the graph is a parabola.

21. Let the polar axis be where the two branches of the loop cross.

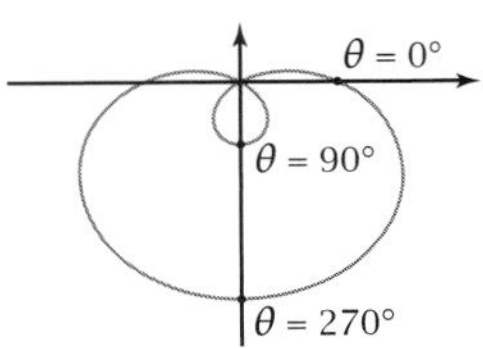

Starting at $\theta = 0°$ and moving to $\theta = 90°$ brings you to the bottom point of the inner loop, as shown. At this point, $r = -LL = -60$, its most negative.

Continuing on to $\theta = 270°$, $r = \text{D} - \text{SI} = 210$. From these points you can sketch a sinusoid in Cartesian coordinates, where the lower bound is $y = -60$, the upper bound is $y = 210$, and the sinusoidal axis is

$$y = \frac{210 + (-60)}{2} = 75$$

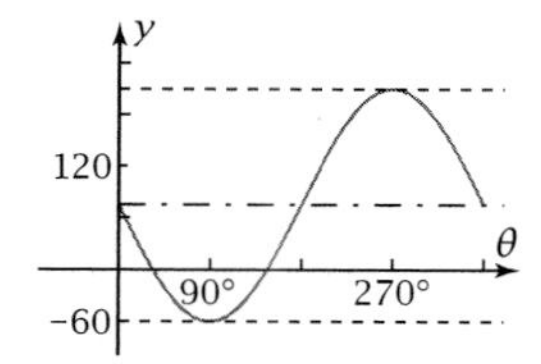

From this you find the amplitude, which is $210 - 75 = 135$. Because the graph is on the sinusoidal axis going down at $\theta = 0°$, the function is $r = 75 - 135 \sin \theta$.

Problem Set 13-3

1. (1.9230..., 78.4630...°)
(1.9230..., −78.4630...°)

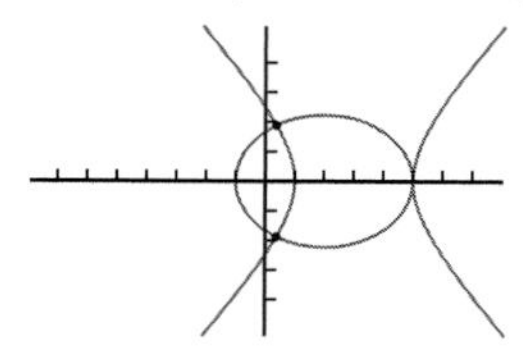

3. $r_1 = r_2 \Rightarrow \cos \theta = \frac{4}{3}$; no true intersections

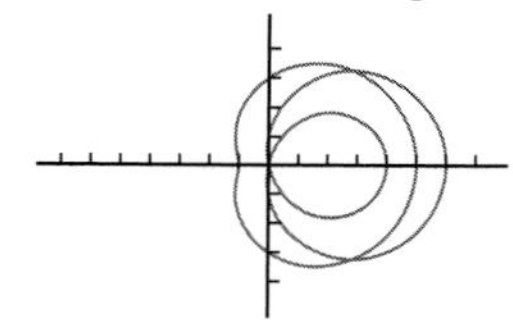

5. $\left(\frac{8 + \sqrt{19}}{3}, \sin^{-1} \frac{1 + \sqrt{19}}{6}\right) = (4.1196..., 63.2717...°)$

$\left(\frac{8 + \sqrt{19}}{3}, 180° - \sin^{-1} \frac{1 + \sqrt{19}}{6}\right)$
$= (4.1196..., 116.7282...°)$

$\left(\frac{8 - \sqrt{19}}{3}, \sin^{-1} \frac{1 - \sqrt{19}}{6}\right) = (1.2137..., -34.0431...°)$

$\left(\frac{8 - \sqrt{19}}{3}, 180° - \sin^{-1} \frac{1 - \sqrt{19}}{6}\right)$
$= (1.2137..., 214.0431...°)$

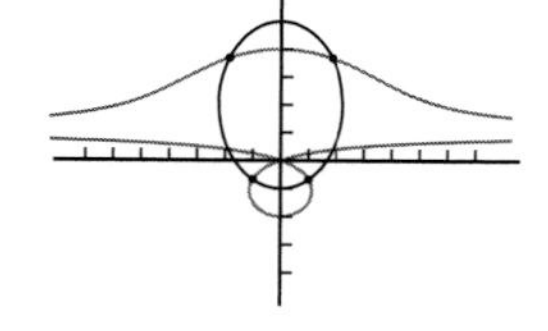

7. (0.3739..., 0.7479...)
(1.2407..., 2.4815...)
(1.8677..., 3.7355...)
(2.9038..., 5.8076...)
(3.3506..., 6.7013...)
(4.6134..., 9.2268...)
(4.7858..., 9.5716...)

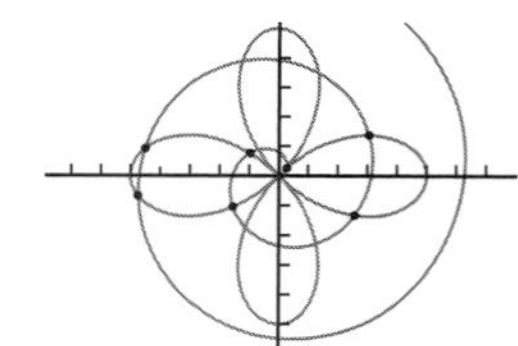

9.

For a given θ, r_2 is the opposite of the value of r_1 at $\theta + 180°$.

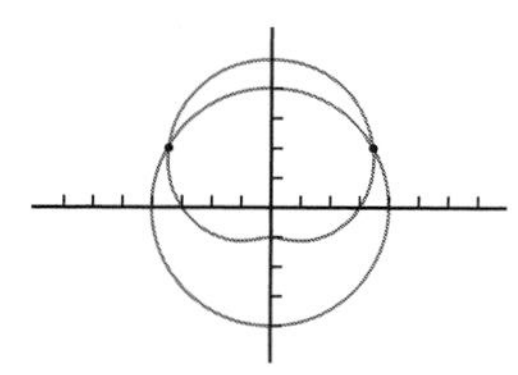

$r_1 = r_3 \Rightarrow (r, \theta) = (4, 30°)$ or $(4, 150°)$, while $r_2 = r_3 \Rightarrow \sin \theta = \frac{7}{2}$, which is impossible. However, the graphs of r_1 and r_2 coincide, so for r_2 those same two points are false intersections with r_3. To find the true intersections of $r_1 = f(\theta)$ and $r_2 = g(\theta)$, set $f(\theta) = g(\theta)$ and solve. To find the false intersections, set $f(\theta) = -g(180° - \theta)$ and solve.

Problem Set 13-4

1. $-1 + i = \sqrt{2} \text{ cis } 135°$
3. $\sqrt{3} - i = 2 \text{ cis } 330°$
5. $-4 - 3i = 5 \text{ cis } 216.8698...°$
7. $5 + 7i = \sqrt{74} \text{ cis } 54.4623...°$
9. $1 = 1 \text{ cis } 0°$
11. $-i = 1 \text{ cis } 270°$
13. $8(\cos 34° + i \sin 34°) = 6.6323... + 4.4735...i$
15. $6(\cos 120° + i \sin 120°) = -3 + 3i\sqrt{3}$
17. $\sqrt{2}(\cos 225° + i \sin 225°) = -1 - i$
19. $5(\cos 180° + i \sin 180°) = -5$

21. $3(\cos 270° + i \sin 270°) = -3i$

23. a. $3 \cdot 5 \text{ cis}(47° + 36°) = 15 \text{ cis } 83°$
    b. $\frac{3}{5} \text{ cis}(47° - 36°) = 0.6 \text{ cis } 11°$
    c. $3^2 \text{ cis}(2 \cdot 47°) = 9 \text{ cis } 94°$
    d. $5^3 \text{ cis}(3 \cdot 36°) = 125 \text{ cis } 108°$

25. a. $4 \cdot 2 \text{ cis}(238° + 51°) = 8 \text{ cis } 289°$
    b. $\frac{4}{2} \text{ cis}(238° - 51°) = 2 \text{ cis } 187°$
    c. $4^2 \text{ cis}(2 \cdot 238°) = 16 \text{ cis } 476° = 16 \text{ cis } 116°$
    d. $2^3 \text{ cis}(3 \cdot 51°) = 8 \text{ cis } 153°$

27. 3 cis 40°, 3 cis 160°, 3 cis 280°

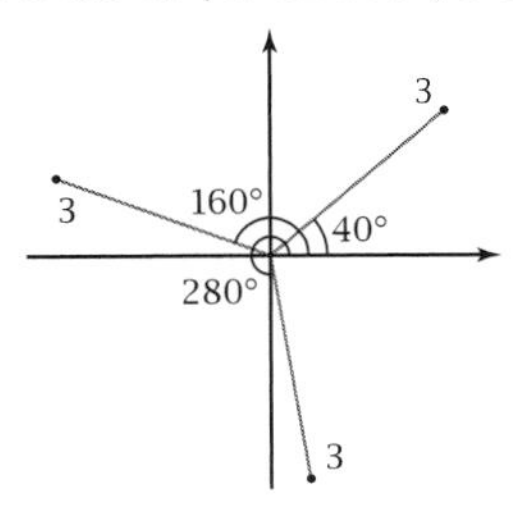

29. 2 cis 20°, 2 cis 110°, 2 cis 200°, 2 cis 290°

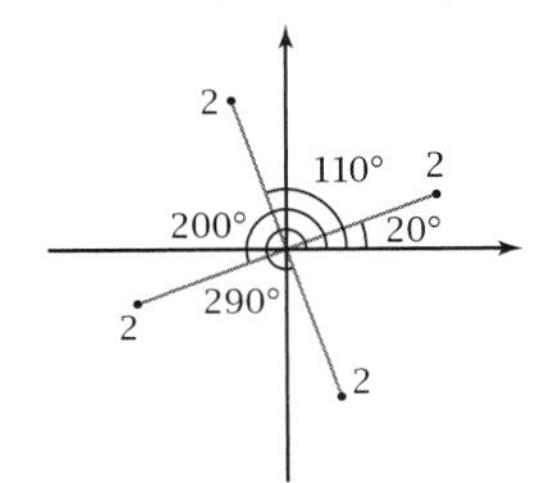

31. cis 45°, cis 225°

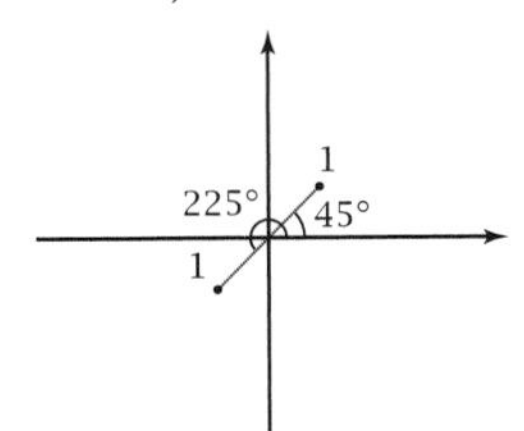

33. 2 cis 0° = 2, 2 cis 120°, 2 cis 240°

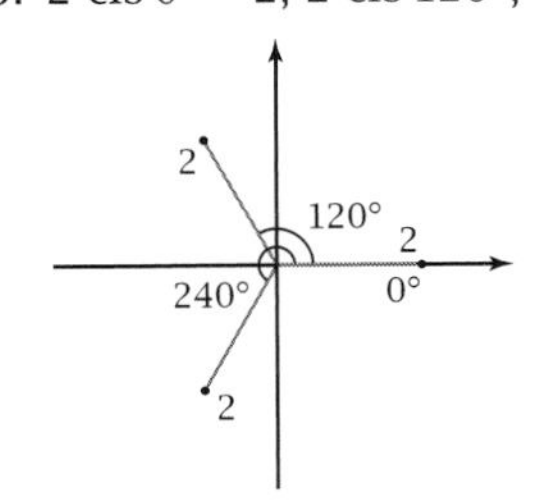

35. cis 30°, cis 90° = i, cis 150°, cis 210°, cis 270° = $-i$, cis 330°

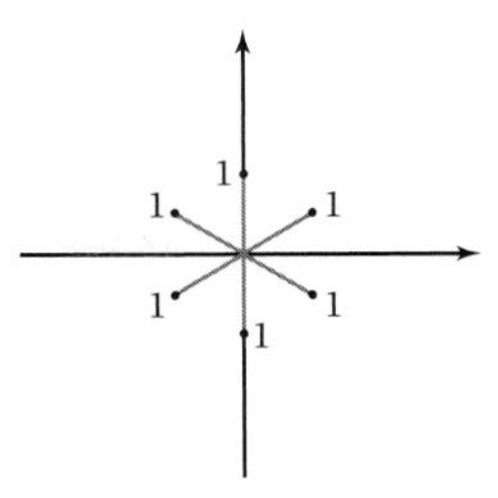

37. $(\cos\theta + i \sin\theta)^3 = \cos^3\theta - 3\cos\theta\sin^2\theta + i(3\cos^2\theta\sin\theta - \sin^3\theta)$

But by De Moivre's theorem,
$(\text{cis } \theta)^3 = \text{cis } 3\theta = \cos 3\theta + i \sin 3\theta$.
Equating real parts gives
$\cos 3\theta = \cos\theta(\cos^2\theta - 3\sin^2\theta)$.
Equating imaginary parts gives
$\sin 3\theta = \sin\theta(3\cos^2\theta - \sin^2\theta)$.

Problem Set 13-5

1. a. $x = 473 - 300t$
   $y = 155 + 100t$
   b. $x = 1.5766...$ hr
   At this time, $y = 312.6666...$ km
   c. Speed = 316.2277... km/hr

3. a. $x = 200t \cos 20°$
   $y = 10 + 200t \sin 20° - 16t^2$
   b.

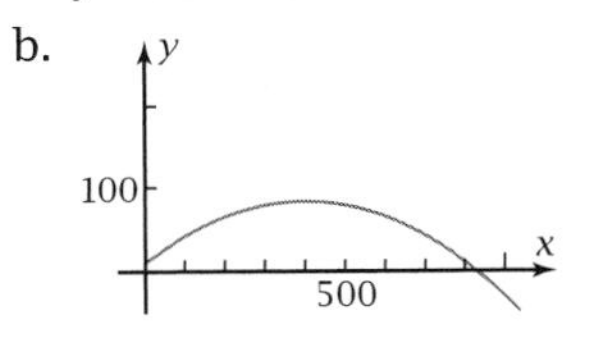

   c. $x = 900 \text{ ft} \Leftrightarrow t = \dfrac{900}{200 \cos 20°} = 4.7888... \text{ sec}$

   At this time,

   $$y = 10 + 200 \frac{900}{200 \cos 20°} \sin 20° - 16\left(\frac{900}{200 \cos 20°}\right)^2 = -29.3484... \text{ ft}$$

   The cannonball will fall short.

   d. $\theta = \tan^{-1} \dfrac{25 + \sqrt{311}}{18} = 67.1111...°$ or

   $\theta = \tan^{-1} \dfrac{25 - \sqrt{311}}{18} = 22.2522...°$

   e. Answers may vary.

5. a. The graph appears to be an ellipse. (See part c.)
   b. $x = 5 \cos t$, $y = 3 \sin t$. This is the parametric description of an ellipse.

c. 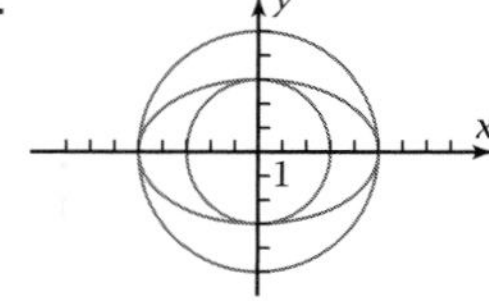

Yes, the circles have the same relationship.

d. $\left(\frac{x}{5}\right)^2 + \left(\frac{y}{3}\right)^2 = 1$

The variables have unequal coefficients but the same sign.

7. a. $\vec{v}_1$, $\vec{v}_2$, and $\vec{v}_3$ placed head-to-tail connect the origin with $P(x, y)$. Therefore, $\vec{r} = \vec{v}_1 + \vec{v}_2 + \vec{v}_3$.
   b. $\vec{v}_1 = (50t)\vec{i} + 0\vec{j}$ cm; $\vec{v}_2 = 0\vec{i} + 50\vec{j}$ cm
   c. $\vec{v}_3 = -(70 \sin t)\vec{i} - (70 \cos t)\vec{j}$;
   $\therefore \vec{r} = (50t - 70 \sin t)\vec{i} + (50 - 70 \cos t)\vec{j}$
   d. The graph is correct.
   e. $x(0.1) - x(0) = x(0.1) = 50(0.1) - 70 \sin 0.1$
   $= -1.9883\ldots$ cm
   Because P is below the track as the wheel rotates clockwise, P moves backward.

9. a. $\vec{r} = (5 \cos t + 5t \sin t) \cdot \vec{i} + (5 \sin t - 5t \cos t) \cdot \vec{j}$
   As parametric equations,
   $x = 5 \cos t + 5t \sin t$
   $y = 5 \sin t - 5t \cos t$

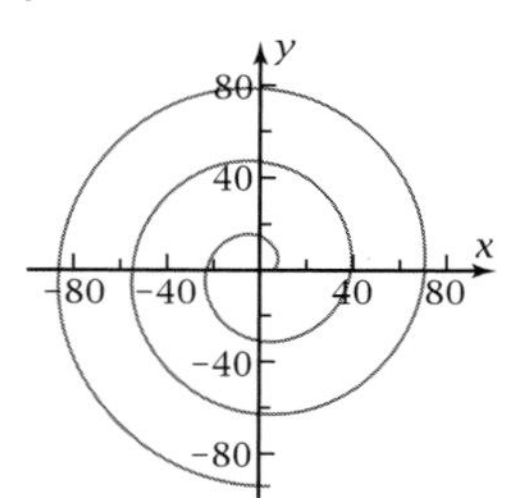

   b. Student project. The drawing should resemble the graph.

11. a. The graphs match.
   b. $x = \dfrac{9 \cos \theta}{5 - 4 \cos \theta}$
   $y = \dfrac{9 \sin \theta}{5 - 4 \cos \theta}$
   The equations give the same ellipse.
   c. $x = 4 + 5 \cos t$
   $y = 3 \sin t$
   d. The point for $t = 0.5$ is different from the point for $\theta = 0.5$. Note that θ is measured from the origin, but t is measured from the center, (4, 0).

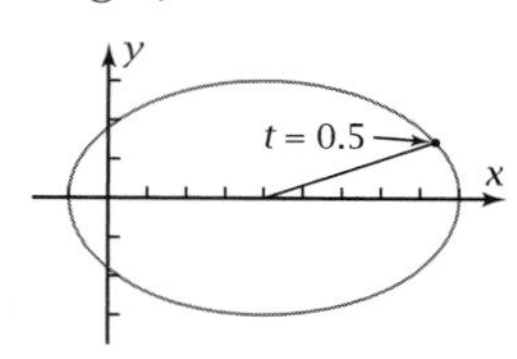

Problem Set 13-6

R1. 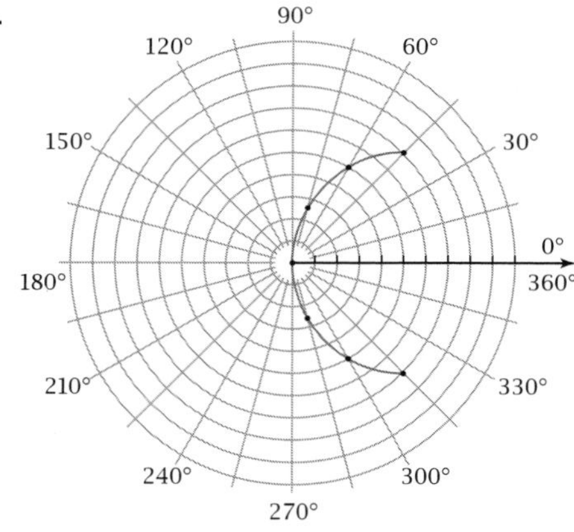

R2. a. 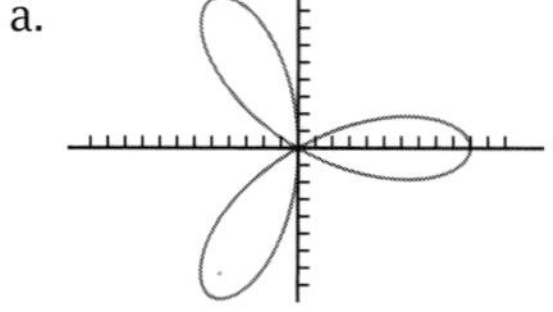

b. 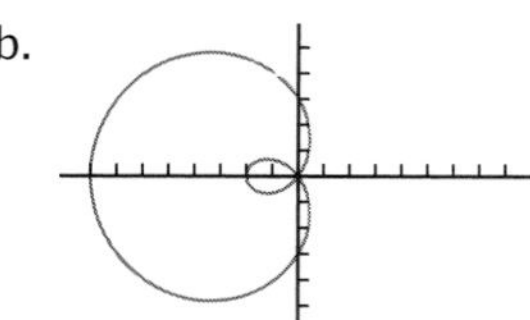

c. 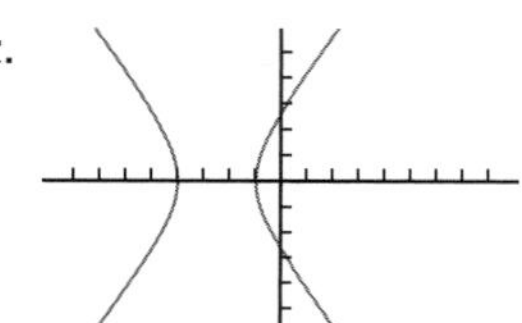

d. 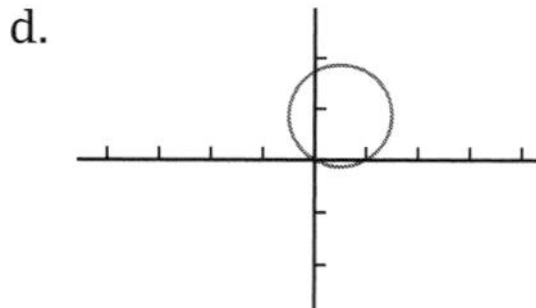

e. 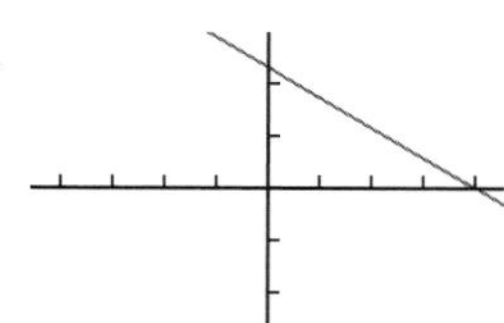

f. $r = \dfrac{8}{3 - 5 \cos \theta}$
$3r = 8 + 5r \cos \theta = 8 + 5x$
$9r^2 = 9x^2 + 9y^2 = 64 + 80x + 25x^2$
$16x^2 + 80x + 100 - 9y^2 = 36$
$\dfrac{(4x + 10)^2}{36} - \dfrac{9y^2}{36} = \left(\dfrac{x + 2.5}{1.5}\right)^2 - \left(\dfrac{y}{2}\right)^2 = 1$
Hyperbola with center (−2.5, 0), horizontal semitransverse axis 1.5, and vertical semiconjugate axis 2

One focus is at the origin.

$e = \left|\frac{b}{a}\right| = \frac{5}{3}$, confirming that this is a hyperbola.

R3. True intersections:

$\left(4\frac{2}{3}, \cos^{-1}\frac{1}{9}\right) = (4.6666..., 83.6206...°)$

$\left(4\frac{2}{3}, -\cos^{-1}\frac{1}{9}\right) = (4.6666..., -83.6206...°)$

False intersection:

$(-2, 180°), (2, 0°)$

R4. a. $-5 + 12i = 13 \text{ cis } 112.6198...°$

b. $7 \text{ cis } 234° = 7 \cos 234° + 7i \sin 234°$

c. $(2 \text{ cis } 52°)(5 \text{ cis } 38°) = 10 \text{ cis } 90° = 10i$

d. $(51 \text{ cis } 198°) \div (17 \text{ cis } 228°) = 3 \text{ cis } 330°$

e. $(2 \text{ cis } 27°)^5 = 32 \text{ cis } 135°$

f. $(8 \text{ cis } 120°)^{1/3} = 2 \text{ cis } 40°, 2 \text{ cis } 160°, 2 \text{ cis } 280°$

g.

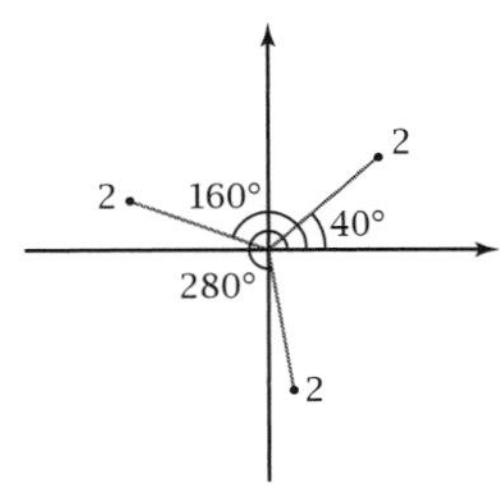

R5. a. $6.9097... + 15.0501...i$

b. $16.5605... \text{ cis } 65.3393...°$

R6. a. $x = 60t \cos 24°$

$y = 60t \sin 24°$

$y = 22.2614...$ mi

b. i. The drawing seems to match the description.

ii. $\vec{r} = \left(21 \cos t + 9 \cos \frac{7}{3}t\right) \cdot \vec{i} + \left(21 \sin t + 9 \sin \frac{7}{3}t\right) \cdot \vec{j}$

iii. Graph for $0 \le t \le 6\pi$. The dime makes three revolutions about the quarter.

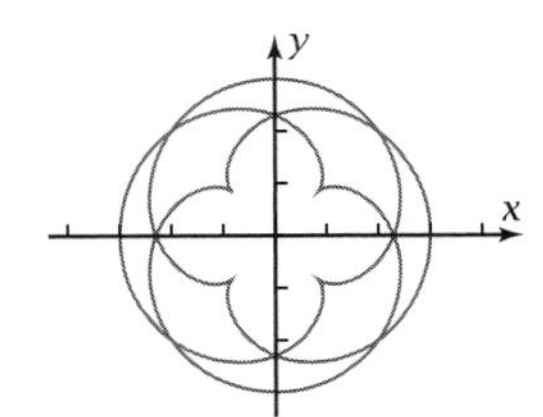

CHAPTER 14

Problem Set 14-1

1. You would have to add nine 2s; $5 + 9 \cdot 2 = 23$; $5 + 99 \cdot 2 = 203$

3. Linear

5. $\frac{5 + 23}{2} \cdot 10 = 140$ = the partial sum

$\frac{5 + 203}{2} \cdot 100 = 10{,}400$

7. Quadratic: $y = x^2 + 4x$

$100^2 + 4(100) = 10{,}400$

9. $6 + 12 + 24 + \cdots + 1536 + 3072 = 6138$

Problem Set 14-2

1. a. Geometric, with common ratio $\frac{4}{3}$

b. $64, 85.\overline{3}$

c. $t_{100} = 6.3139... \times 10^{13}$

d. 37th term

3. a. Arithmetic, with common difference -13

b. 19, 6

c. $t_{100} = -1229$

d. 50th term

5. a. Geometric, with common ratio 2.5

b. 856.25, 2140.625

c. $t_{100} = 1.3640... \times 10^{41}$

d. 13th term

7. a. Geometric, with common ratio $-\frac{9}{10}$

b. $-36.45, 32.805$

c. $t_{100} = -0.001475...$

d. 12th term

9. a. Neither; $t_n = n^2 - 1$

b. 120, 143

c. $t_{100} = 9999$

d. 57th term

11. Geometric, with common ratio 2;
$n = 10.9657... \Rightarrow$ the 11th square;
$1 \cdot 2^{64-1} = 9.2233... \times 10^{18}$ grains. The king was upset because the number of grains of rice was so large.

13. a. \$1,267,500, \$1,235,000, \$1,202,500, \$1,170,000, . . . , $t_n = 1{,}300{,}000 - n \cdot 32{,}500$

$1{,}300{,}000 - 30 \cdot 32{,}500 = 325{,}000$

$1{,}300{,}000 - n \cdot 32{,}500 = 0$

$\Rightarrow n = \frac{1{,}300{,}000}{32{,}500} = 40$ yr.

The depreciation function is linear, and the scatter plot points lie on a straight line.

b. \$1,170,000, \$1,053,000, \$947,700, \$852,930, . . . , $t_n = 1{,}300{,}000 \cdot (0.9)^n$

The business can deduct \$130,000 the first year, \$117,000 the second year, and \$105,300 the third year. After 15 years, the business deducts less than \$32,500.

15. 91%, ≈ 82.8%, ≈ 75.4%; geometric, with common ratio 0.91
$t_n = 0.91^n$; $t_{20} = 0.91^{20} = 0.1516\ldots \approx 15.2\%$
$0.91^n = 0.10$; only 10% will be left after 25 washings.

17. a. $t_n = t_{n-2} + t_{n-1}$, that is, each term is the sum of the preceding two.
$t_{11} = 34 + 55 = 89$
$t_{12} = 55 + 89 = 144$
$t_{20} = 6765$

b. 1, 2, 1.5, $1.\overline{6}$, 1.6, 1.625, 1.6153..., 1.6190..., 1.6176..., $1.6\overline{18}$

c. Answers will vary. The spirals in each direction will usually be consecutive Fibonacci numbers.

d. Answers will vary. Leonardo Fibonacci was an Italian mathematician of the late 12th and early 13th century. The Fibonacci term t_n is the number of pairs of rabbits there will be in the nth month if you start with one pair and if every pair that has reached the age of two months or more produces another pair every month.

19. a. To get to step 3, she can take 1 step from step 2 or 2 steps from step 1. So the number of ways to get to step 3 is the number of ways to get to step 1 plus the number of ways to get to step 2. Similarly, the number of ways to get to step 4 is the number of ways to get to step 2 plus the number of ways to get to step 3.

b. $t_n = t_{n-1} = t_{n-2}$, where $t_2 = 2$ and $t_1 = 1$;
$t_{20} = 10{,}946$

c. If you let $t_0 = 1$, then this is the same sequence.

d. $t_{91} = 7.5 \times 10^{18}$

21. a. 56.25, 65.50, 74.75, where common difference is 9.25

b. 6, 12, 24, where common ratio is 2

c. −6, 12, −24, where common ratio is −2

Problem Set 14-3

1. a. $3 + 8 + 13 + \cdots + 48$. There is a common difference (5).

b. $S_{10} = 255 = \dfrac{3 + 48}{2} \cdot 10$

Yes, the answers are the same.

c. $S_{100} = 25{,}050 = \dfrac{3 + 498}{2} \cdot 100$

3. a. $t_{10} = 13.4217\ldots$ cm; $S_{10} = 446.3129\ldots$ cm

b. The answers are getting closer and closer to 500. We can make the answer as close to 500 as we want by taking the sum of enough terms.

c. $\lim\limits_{n\to\infty} S_n = 100 \cdot \dfrac{1}{1 - 0.8} = 500$ cm

5. a. The series is geometric because there is a common ratio, 1.01. The amount at the fifth month is the fifth partial sum because it is the sum of the first five terms of the series.

$$100 \cdot \frac{1 - 1.01^5}{1 - 1.01} = \$510.10$$

b. $100 \cdot \dfrac{1 - 1.01^{120}}{1 - 1.01} = \$23{,}003.87$

$\$23{,}003.87 - 120 \cdot \$100 = \$11{,}003.87$

c. $n = 241$ months

7. a.

n	P	I_n	P_n	B_n
4	1050.00	998.48	51.52	99,796.97

b.
$$\begin{aligned}
B_1 &= B_0(1 + i) - P \\
B_2 &= B_1(1 + i) - P = [B_0(1 + i) - P](1 + i) - P \\
&= B_0(1 + i)^2 - P(1 + i) - P \\
B_3 &= B_2(1 + i) - P \\
&= [B_0(1 + i)^2 - P(1 + i) - P](1 + i) - P \\
&= B_0(1 + i)^3 - P(1 + i)^2 - P(1 + i) - P \\
B_4 &= B_3(1 + i) - P \\
&= B_0(1 + i)^4 - P(1 + i)^3 - P(1 + i)^2 - P(1 + i) - P \\
B_4 &= B_0(1 + i)^4 - P(1 + i)^3 - P(1 + i)^2 - P(1 + i) - P \\
&= B_0(1 + i)^4 - P[1 + (1 + i) + (1 + i)^2 + (1 + i)^3] \\
&= B_0(1 + i)^4 - P\frac{1 - (1 + i)^4}{1 - (1 + i)} \\
&= B_0(1 + i)^4 - P\frac{1 - (1 + i)^4}{-i} \\
&= B_0(1 + i)^4 + \frac{P}{i}(1 - (1 + i)^4)
\end{aligned}$$

c. $n = -\dfrac{\log\left(1 - \frac{iB_0}{P}\right)}{\log(1 + i)} = 305.9719\ldots$

so it takes 306 months.

d.

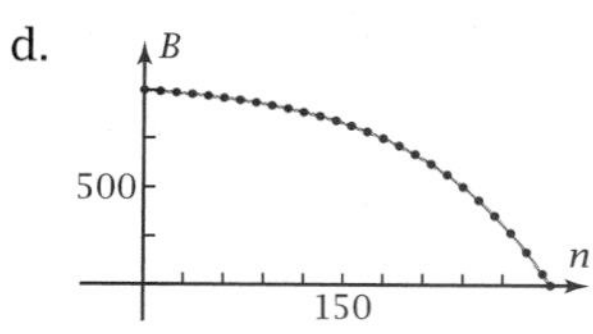

False. Halfway through the mortgage (month 153), the balance is still \$82,083.93, about 82% of the original amount.

9. a. $(0.4 + 0.6)^5 = 0.01024 + 0.07680 + 0.23040 + 0.34560 + 0.25920 + 0.07776$

b. ${}_5C_2(0.4)^2(0.6)^3 = 0.34560$, about 34.6%

c. Letting X = the number of "point-ups" out of five flips,
$$\begin{aligned}
P(X \le 3) &= P(X = 0) + P(X = 1) + P(X = 2) + P(X = 3) \\
&= 0.01024 + 0.07680 + 0.23040 + 0.34560 \\
&= 0.66304, \text{ about } 66.3\%
\end{aligned}$$

d. Letting X = the number of "point-ups" out of ten flips,

$P(X \le 6) = 1 - [(P(X=7) + P(X=8) + P(X=9) + P(X=10)]$
$= 1 - {}_{10}C_3(0.4)^3(0.6)^7 - {}_{10}C_2(0.4)^2(0.6)^8 - {}_{10}C_1(0.4)^1(0.6)^9 - {}_{10}C_0(0.4)^0(0.6)^{10}$
$= 0.6177\ldots$, about 61.8%, not the same probability

11. $9 + 11 + 13 + 15 + 17 = 65$

13. $3 + 9 + 27 + 81 + 243 + 729 = 1092$

15. Geometric, $r = 5$; $S_{11} = 2 \cdot \dfrac{1 - 5^{11}}{1 - 5} = 24{,}414{,}062$

17. Arithmetic, $d = 7.6$; $t_{54} = 24 + (54 - 1)(7.6) = 426.8$; $S_{54} = \dfrac{54}{2}(24 + 426.8) = 12{,}171.6$

19. Arithmetic, $d = -40$; $t_{78} = 1000 + (78 - 1)(-40) = -2080$; $S_{78} = \dfrac{78}{2}(1000 + (-2080)) = -42{,}120$

21. Geometric, $r = -3$; $S_{10} = 50 \cdot \dfrac{1 - (-3)^{10}}{1 - (-3)} = 738{,}100$

23. Arithmetic, $d = 11$; $n = 26$

25. Geometric, $r = \frac{5}{3}$; $n = 19$

27. Arithmetic, $d = -6$; $n = 31$

29. $r = \frac{90}{100} = \frac{81}{90} = \frac{9}{10}$; converges, because $|r| < 1$; $\lim\limits_{n\to\infty} S_n = 100 \cdot \dfrac{1}{1 - \frac{9}{10}} = 1000$

31. $r = \frac{50}{40} = \frac{62.5}{50} = 1.25$; diverges, because $|r| \ge 1$

33. $r = \frac{90}{300} = \frac{27}{90} = \frac{3}{10}$; converges, because $|r| < 1$; $\lim\limits_{n\to\infty} S_n = 300 \cdot \dfrac{1}{1 - \frac{3}{10}} = 428\frac{4}{7}$

35. $r = \frac{-950}{1000} = \frac{902.5}{-950} = -\frac{19}{20}$; converges, because $|r| < 1$; $\lim\limits_{n\to\infty} S_n = 1000 \cdot \dfrac{1}{1 - (-\frac{19}{20})} = 512\frac{32}{39}$

37. $(x - y)^3 = x^3 - 3x^2y + 3xy^2 - y^3$

39. $(2x - 3)^5 = 32x^5 - 240x^4 + 720x^3 - 1080x^2 + 810x - 243$

41. $(x^2 + y^3)^6 = x^{12} + 6x^{10}y^3 + 15x^8y^6 + 20x^6y^9 + 15x^4y^{12} + 6x^2y^{15} + y^{18}$

43. ${}_8C_3x^3y^5 = 56x^3y^5$

45. ${}_{15}C_4p^4(-j)^{11} = -1365p^4j^{11}$

47. ${}_{13}C_6(x^3)^6(-y^2)^7 = -1716x^{18}y^{14}$

49. ${}_8C_3(3x)^3(2y)^5 = 48{,}384x^3y^5$

51. ${}_{15}C_4r^4(-q)^{11} = -1365r^4q^{11}$

Problem Set 14-4

R1. a. 17, 20

b. 75

c. $75 = 6 \cdot \dfrac{5 + 20}{2}$

d. 80, 160

e. 315

R2. a. Neither; there is no common difference or common ratio.

b. $t_{200} = 1843$

c. $n = 392$

d. $t_{100} = 3731.7732\ldots$

e. $n = 114$

f. 0, 3, 8, 15, 24 $= 1^2 - 1, 2^2 - 1, 3^2 - 1, 4^2 - 1, 5^2 - 1, \ldots$; $6^2 - 1 = 35$; $7^2 - 1 = 48$; $8^2 - 1 = 63$

g. $t_{100} = 100^2 - 1 = 9999$

h. i. Every time you add one month, you multiply the amount by 1.005. The sequence is more appropriate because the interest compounds monthly, not continuously.

ii. 104th month

iii. $59,807.87

R3. a. $2 + 5 + 8 + 11 + 14 + 17 = 57$

b. There is a common difference, 3.

c. $6 \cdot \dfrac{2 + 17}{2} = 57$

d. 528th term

e. Because $t_n = 4000(0.95)^{n-1}$, you use

```
Y1 = sum(seq(4000*0.95^(N-1),N,1,200,1))
```

$= 79{,}997.1957\ldots$

f. $4000 \cdot \dfrac{1 - 0.95^{200}}{1 - 0.95} = 79{,}997.1957\ldots$

g. $4000 \cdot \dfrac{1 - 0.95^n}{1 - 0.95} = 78{,}377.8762\ldots$

$\Rightarrow n = \dfrac{\log\left(1 - \dfrac{(78{,}377.8762\ldots)(1 - 0.95)}{4000}\right)}{\log 0.95} = 76$; the 76th term

h. $4000 \cdot \dfrac{1}{1 - 0.95} = 80{,}000$

i. ${}_{13}C_6a^6b^7 = \dfrac{13!}{6! \cdot 7!}a^6b^7 = 1716a^6b^7$

j. i. Arithmetic; geometric

ii. Maya: $t_{21} = 22$ in.; Vincent: $t_{21} = 4.3767\ldots$ in.

iii. Maya: $S_{21} = 357$ in.; Vincent: $S_{21} = 320.6091\ldots$; Maya is ahead by 36.3908... in.

iv. Yes. Vincent approaches $36 \cdot \frac{1}{0.1} = 360$ in. as the number of steps increases to infinity, so he must at some point pass 357 in. In fact, Vincent passes Maya's stopping point on the 46th step.

k. $500(0.6) + 500 = 800$ mg;
$800(0.6) + 500 = 980$ mg;
$980(0.6) + 500 = 1088$ mg;
$980(0.6) + 500$
$= (500(0.6)^2 + 500(0.6) + 500)(0.6) + 500$
$= 500(0.6)^3 + 500(0.6)^2 + 500(0.6) + 500 \Rightarrow$
$S_n = 500 \cdot \dfrac{1 - 0.6^n}{1 - 0.6}$;
$500 \cdot \dfrac{1 - 0.6^{16}}{1 - 0.6} = 1249.6473\ldots$ mg;
$500 \cdot \dfrac{1}{1 - 0.6} = 1250$ mg

CHAPTER 15

Problem Set 15-1

1. The graphs match.
3. $f(-1) = (-1)^3 - 4(-1)^2 - 3(-1) + 2 = 0$
5. The graph of g is the same as the graph of f translated upward by 16 units. However, note that f crosses the axis in three distinct points, while g crosses the axis only once and is tangent at one point.
7. Two of the roots are identical. The graph intersects the horizontal axis twice; at one point the graph only touches the axis without crossing.
9. $h(-3) = (-3)^3 - 4(-3)^2 - 3(-3) + 54 = 0$
$h(x) = (x + 3)(x^2 - 7x + 18)$
$x^2 - 7x + 18 = 0 \Rightarrow x = \dfrac{7 \pm \sqrt{-23}}{2} = 3.5 \pm 2.3979\ldots i$

Problem Set 15-2

1. a.

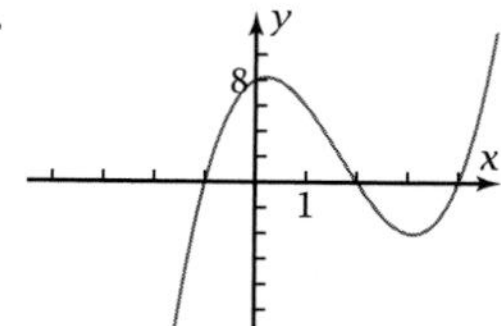

Three (two up and one down), the same as the degree of the polynomial

b. $x \approx -1, 2, 4$

c.

-1	1	-5	2	8
		-1	6	-8
	1	-6	8	0

$p(x) = (x + 1)(x^2 - 6x + 8) = (x + 1)(x - 2)(x - 4)$

d. The opposites of the zeros of the polynomial appear as the linear terms in the factors: If c is a zero, then $x - c$ is a factor.

3. Sixth-degree polynomial; four real zeros; two nonreal complex zeros
5. Eighth-degree polynomial; six real zeros (three double zeros); two nonreal complex zeros
7. Example: $f(x) = -(x + 3)(x + 1)(x - 1)$

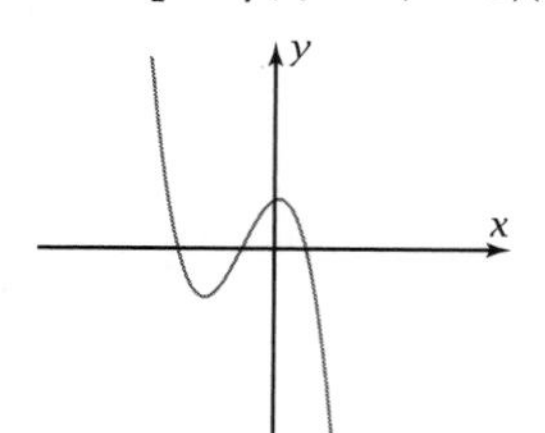

9. Example: $f(x) = (x + 1)(x^2 - 4x + 5)$

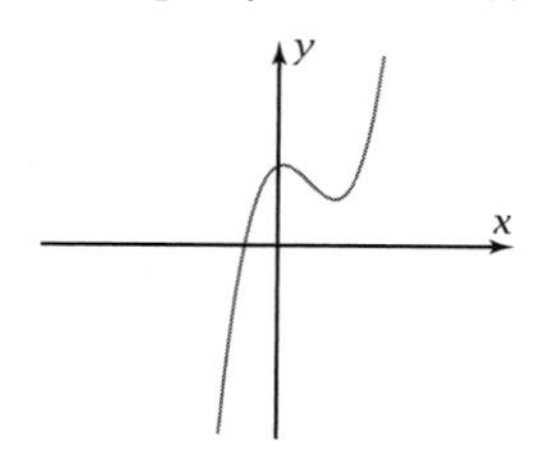

11. Example: $f(x) = x^3 + x$

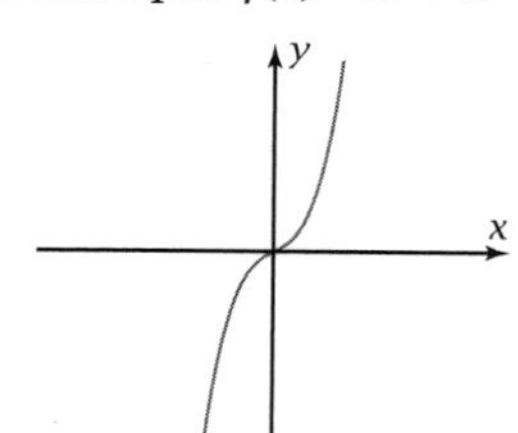

13. Example: $f(x) = x^4 + 1$

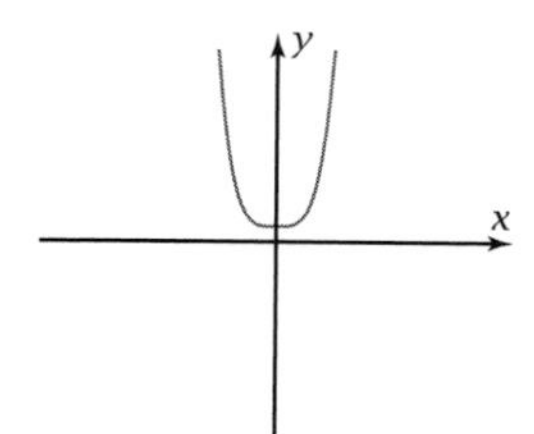

15. Example: $f(x) = (x + 1)^2(x - 2)^2$

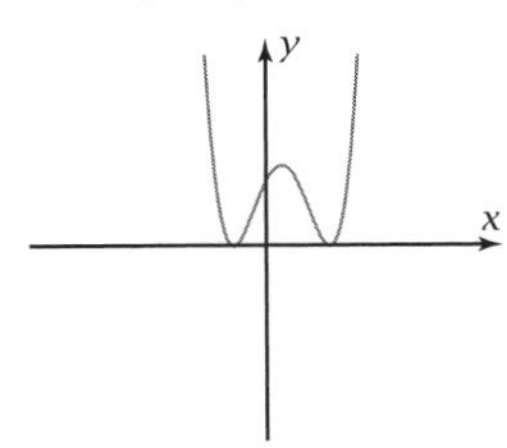

17. No such polynomial exists; a polynomial cannot have more zeros than its degree.

19. Sum = 1, product = −40, sum of pairwise products = −22, $x = -5, 2, 4$

21. Sum = $-\frac{18}{5}$, product = $\frac{156}{5}$, sum of pairwise products = $-\frac{7}{5}$, $x = \frac{12}{5}, -3 + 2i, -3 - 2i$

23. $f(x) = x^3 - 4x^2 - 11x + 30 = (x + 3)(x - 2)(x - 5)$

25. $f(x) = x^3 - 8x^2 + 29x - 52 = (x - 4)(x^2 - 4x + 13)$
$= (x - 4)(x - 2 - 3i)(x - 2 + 3i)$

27. a.

2	1	−7	5	4
		2	−10	−10
	1	−5	−5	−6

−3	1	−7	5	4
		−3	30	−105
	1	−10	35	−101

b. $p(2) = -6, \frac{P(x)}{x - 2} = x^2 - 5x - 5 - \frac{6}{x - 2}$

$p(-3) = -101, \frac{P(x)}{x + 3} = x^2 - 10x + 35 - \frac{101}{x + 3}$

29. If $p(x)$ is a polynomial, then $p(c)$ equals the remainder when $p(x)$ is divided by the quantity $(x - c)$.

31. A polynomial function has at least one zero in the set of complex numbers.

33. The following program works on a TI-83. You must enter the list of coefficients as a list.

```
Input "DEGREE: " , N
N+1 → dim(L1)
N+1 → dim(L2)
Input "COEFFS (LIST) : " , L1
While 1=1
Input "X: " , C
L1 (1) → L2 (1)
For (I, 2, N+1, 1)
L1 (I) +C*L2 (I–1) → L2 (I)
End
Disp L2
End
```

35. $f(x) = 2(x + 3)\left(x + \frac{3}{2}\right)(x - 1)(x - 2)$

Sum of zeros: $-3 - \frac{3}{2} + 1 + 2 = -\frac{3}{2} = -\frac{b}{a}$

Sum of products of two zeros:

$-3 \cdot \left(-\frac{3}{2}\right) - 3 \cdot 1 - 3 \cdot 2 - \frac{3}{2} \cdot 1 - \frac{3}{2} \cdot 2 + 1 \cdot 2$

$= -7 = \frac{-14}{2} = \frac{c}{a}$

Sum of products of three zeros:

$-3 \cdot \left(-\frac{3}{2}\right) \cdot 1 - 3 \cdot \left(-\frac{3}{2}\right) \cdot 2 - 3 \cdot 1 \cdot 2 - \frac{3}{2} \cdot 1 \cdot 2$

$= \frac{9}{2} = -\frac{-9}{2} = -\frac{d}{a}$

Product of zeros: $-3 \cdot \left(-\frac{3}{2}\right) \cdot 1 \cdot 2 = 9 = \frac{18}{2} = \frac{e}{a}$

37. $g(x) = x^3 - 8x^2 + 20x - 25$

$f(x) = (x - 4)\left(x - \frac{1}{2} - \frac{1}{2}i\sqrt{11}\right)\left(x - \frac{1}{2} + \frac{1}{2}i\sqrt{11}\right)$

$x = 4, \frac{1}{2} \pm \frac{1}{2}i\sqrt{11}$

$g(x) = (x - 5)\left(x - \frac{3}{2} - \frac{1}{2}i\sqrt{11}\right)\left(x - \frac{3}{2} + \frac{1}{2}i\sqrt{11}\right)$

$x = 5, \frac{3}{2} \pm \frac{1}{2}i\sqrt{11} = 4 + 1, \left(\frac{1}{2} + \frac{1}{2}i\sqrt{11}\right) + 1$

Problem Set 15-3

1. a.

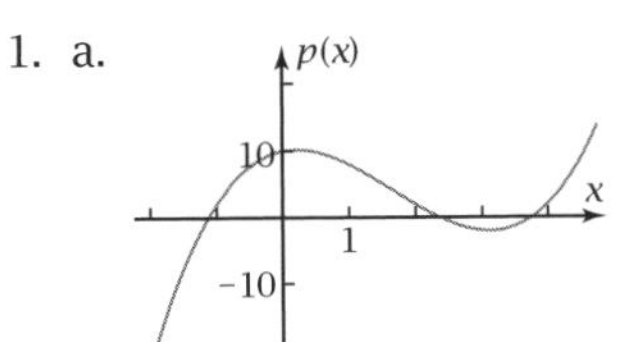

b. Function has three zeros (equals the degree of the equation) and two extreme points (one less than the degree of the equation).

c.

x	$p(x)$	1st Diff.	2nd Diff.	3rd Diff.
4	2			
		18		
5	20		20	
		38		6
6	58		26	
		64		6
7	122		32	
		96		6
8	218		38	
		134		
9	352			

3. a.

b.

x	$f(x)$	1st Diff.	2nd Diff.	3rd Diff.
2	25.4			
		−12.3		
3	13.1		−4.6	
		−16.9		1.8
4	−3.8		−2.8	
		−19.7		1.8
5	−23.5		−1.0	
		−20.7		1.8
6	−44.2		0.8	
		−19.9		
7	−64.1			

c. $f(x) = ax^3 + bx^2 + cx + d$

$8a + 4b + 2c + d = 25.4$

$27a + 9b + 3c + d = 13.1$

$64a + 16b + 4c + d = -3.8$

$125a + 25b + 5c + d = -23.5$

$$\begin{bmatrix} 8 & 4 & 2 & 1 \\ 27 & 9 & 3 & 1 \\ 64 & 16 & 4 & 1 \\ 125 & 25 & 5 & 1 \end{bmatrix}^{-1} \begin{bmatrix} 25.4 \\ 13.1 \\ -3.8 \\ -23.5 \end{bmatrix} = \begin{bmatrix} 0.3 \\ -5.0 \\ 7.0 \\ 29.0 \end{bmatrix}$$

so $f(x) = 0.3x^3 - 5x^2 + 7x + 29$

Cubic regression gives the same result, with $R^2 = 1$ (curve passes through all the points).

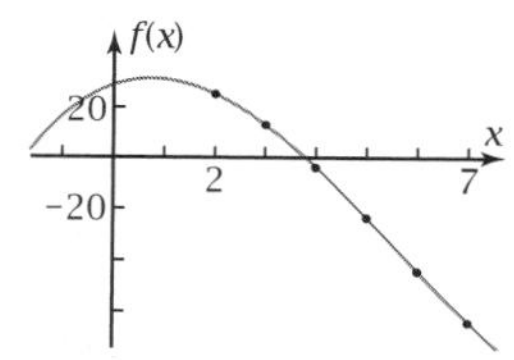

5. a. $f(x) = ax^3 + bx^2 + cx + d$

$0a + 0b + 0c + d = 0$

$a + b + c + d = 116$

$8a + 4b + 2c + d = 448$

$27a + 9b + 3c + d = 972$

$$\begin{bmatrix} 0 & 0 & 0 & 1 \\ 1 & 1 & 1 & 1 \\ 8 & 4 & 2 & 1 \\ 27 & 9 & 3 & 1 \end{bmatrix}^{-1} \begin{bmatrix} 0 \\ 116 \\ 448 \\ 972 \end{bmatrix} = \begin{bmatrix} -4 \\ 120 \\ 0 \\ 0 \end{bmatrix}$$

$f(x) = -4x^3 + 120x^2$

b. $f(10) = 8$ in.

c. $f(x) = -4x^3 + 120x^2 = -4x^2(x - 30)$

$f(x) = 0 \Rightarrow x = 0$ ft or 30 ft

d.

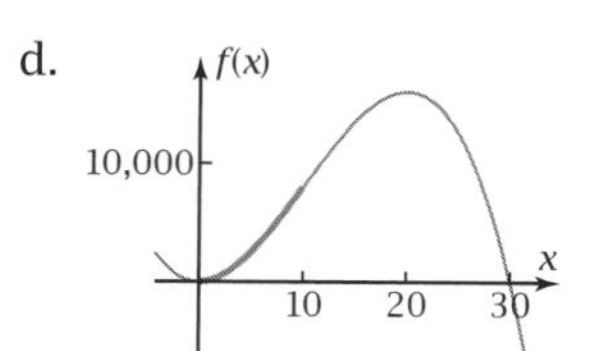

7. a.

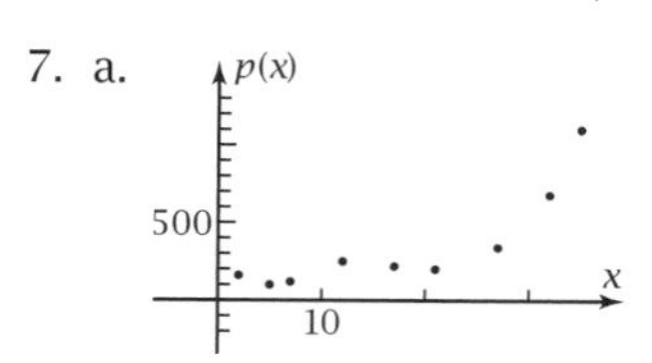

Function appears to have three vertices, so it must have degree at least four.

b. $p(x) = 0.0058\ldots x^4 - 0.3289\ldots x^3 + 6.1803\ldots x^2 - 36.6237\ldots x + 193.0963\ldots$

with $R^2 = 0.9931\ldots$

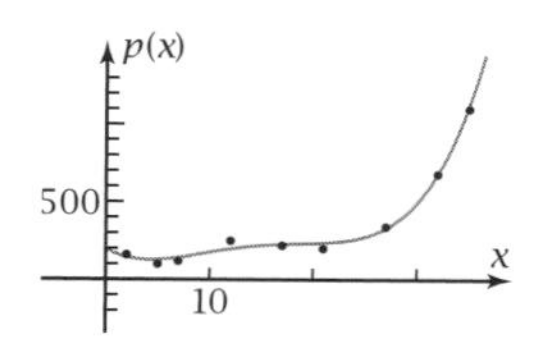

c. The predicted values $\hat{y} = p(x)$ and residuals $y - \hat{y}$ are

x	$\hat{y}$	$y - \hat{y}$
2	142.0312...	17.9687...
5	126.9978...	−26.9978...
7	140.6908...	−20.6908...
12	195.7265...	54.2734...
17	226.2139...	−6.2139...
21	234.2633...	−34.2633...
27	326.2134...	13.7865...
32	670.2425...	9.7574...
35	1107.6199...	−7.6199...

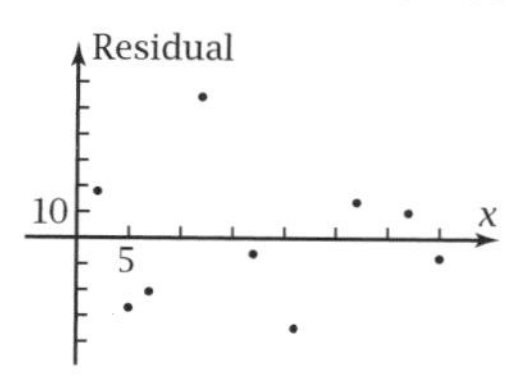

The 12-in. model is the most overpriced.

d. The demand for the smaller sets is greater, so the price is higher.

9. a. If $y = x^3 + bx^2 + cx + d$, then

$-b = 1.7 + 3.8 + 5.5 \Rightarrow b = -11.0$

$c = (1.7)(3.8) + (1.7)(5.5) + (3.8)(5.5) \Rightarrow c = 36.71$

$-d = (1.7)(3.8)(5.5) \Rightarrow d = -35.53$

so

$y = x^3 - 11x^2 + 36.71x - 35.53$

The y-intercept is −35.53 mi.

b. We must multiply by $\frac{4.1\text{ mi}}{-35.53\text{ mi}}$:

$$f(x) = \frac{-4.1}{35.53}x^3 + \frac{45.1}{35.53}x^2 - \frac{150.511}{35.53}x + 4.1$$

c. 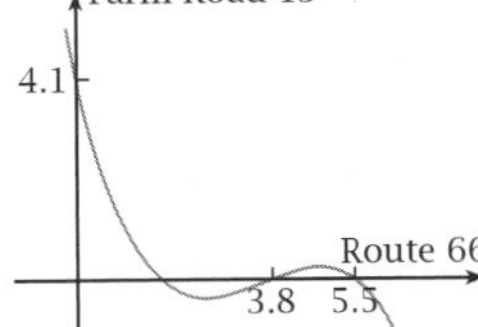

The river goes south before the first crossing and after the last.

d. Farthest south for 1.7 mi ≤ x ≤ 3.8 mi is $y = -0.3618...$ mi at $x = 2.5676...$ mi. Farthest north for 3.8 mi ≤ x ≤ 5.5 mi is $y = 0.2508...$ mi at $x = 4.7656...$ mi. The river is 10 miles south of Route 66 at $x = 2.8568...$ mi east of the zero at 5.5 mi.

11. a. The graphs match.

b. 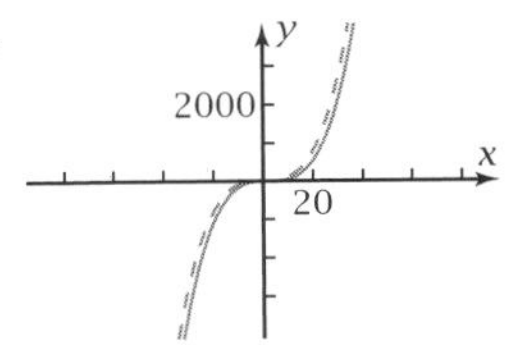

c. Both graphs look similar to $y = x^3$. The vertices and intercepts of the f graph are hard to see. The terms of lower degree do not affect the graph much for large x.

13. a. $R^2 = 0.9611...$ b. $SS_{\text{res}} = 67.0324...$

c. $\bar{y} = 10.875$
$SS_{\text{dev}} = 1724.875$

d. $R^2 = \frac{1724.875 - 67.0324...}{1724.875} = 0.9611...$

Problem Set 15-4

1. 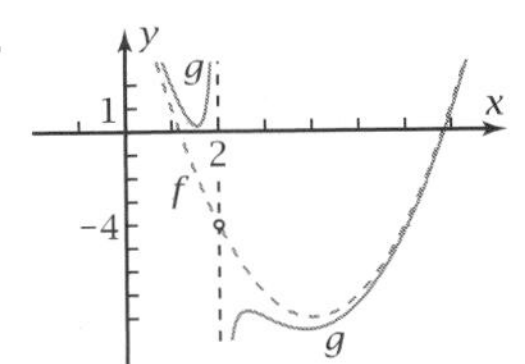

3.

x	$f(x)$	$g(x)$
1.9	-3.59	6.41
1.99	-3.9599	96.0401
1.999	-3.995999	996.004001
2	Undefined	Undefined
2.001	-4.003999	-1004.003999
2.01	-4.0399	-104.0399
2.1	-4.39	-14.39

They show the discontinuity and the asymptote.

5.

$$\begin{array}{r|rrrr} 2 & 1 & -10 & 24 & -16 \\ & & 2 & -16 & 16 \\ \hline & 1 & -8 & 8 & 0 \end{array}$$

$f(x) \cong x^2 - 8x + 8$

$$\begin{array}{r|rrrr} 2 & 1 & -10 & 24 & -17 \\ & & 2 & -16 & 16 \\ \hline & 1 & -8 & 8 & -1 \end{array}$$

$$g(x) = x^2 - 8x + 8 - \frac{1}{x-2}$$

7. $\lim_{x\to 2} f(x) = \lim_{x\to 2} x^2 - 8x + 8 = -4$

$\lim_{x\to 2} g(x) = \lim_{x\to 2} x^2 - 8x + 8 - \frac{1}{x-2} = \infty$

9. Horizontal translation by 2, vertical dilation by −1 (reflection over the x-axis)

11. 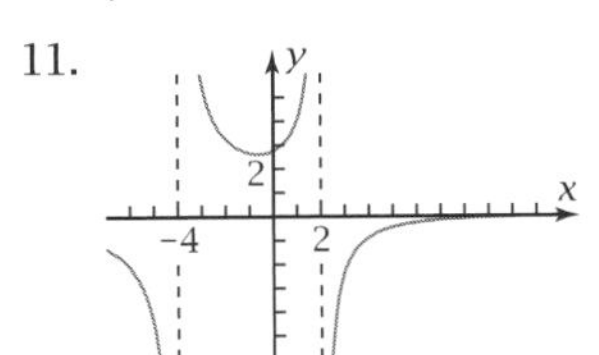

13. $f(x) = \frac{5}{x+4} + \frac{-3}{x-2}$

15. $\frac{5}{x+4}$: horizontal translation by −4, vertical dilation by 5

$\frac{-3}{x-2}$: horizontal translation by 2, vertical dilation by −3

17. An indeterminate form looks like $\frac{0}{0}$ and may or may not have a removable discontinuity; an infinite form looks like $\frac{a}{0}$, where $a \neq 0$, and has a vertical asymptote.

19. a. If $x < 2$, then $\lim_{x\to 2^-} f(x) = 6$. If $x > 2$, then $\lim_{x\to 2^+} f(x) = 2$. The graph should look like Figure 15-4c.

b. $g(2.99) = 34 + 23\text{ int}(2.99) = 34 + 23(2) = 70$
$g(3) = 34 + 23\text{ int}(3) = 34 + 23(3) = 70$
The graph should look like Figure 15-4d.

Problem Set 15-5

1. a. Near $x = 2$,

$$v \approx \frac{f(2.001) - f(2)}{2.001 - 2} = \frac{-0.003999999}{0.001}$$
$$= -3.999999 \approx -4$$

Near $x = 4$,

$$v \approx \frac{f(4.001) - f(4)}{4.001 - 4} = \frac{0.008006001}{0.001}$$
$$= 8.006001 \approx 8$$

b. $\lim_{x\to 2} \frac{f(x) - f(2)}{x - 2} = \lim_{x\to 2} \frac{(x^3 - 6x^2 + 8x + 5) - 5}{x - 2}$

$= \lim_{x\to 2} \frac{x^3 - 6x^2 + 8x}{x - 2} = \lim_{x\to 2} \frac{(x - 2)(x^2 - 4x)}{x - 2}$

$= \lim_{x\to 2} (x^2 - 4x) = -4$

$\lim_{x\to 4} \frac{f(x) - f(4)}{x - 4} = \lim_{x\to 4} \frac{(x^3 - 6x^2 + 8x + 5) - 5}{x - 4}$

$= \lim_{x\to 4} \frac{x^3 - 6x^2 + 8x}{x - 4} = \lim_{x\to 4} \frac{(x - 4)(x^2 - 2x)}{x - 4}$

$= \lim_{x\to 4} (x^2 - 2x) = 8$

The answers agree.

c. $y = 8x - 27$

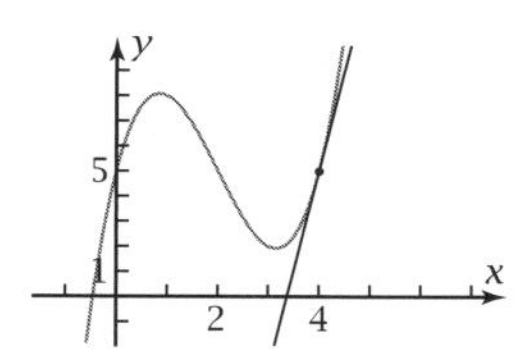

The line is tangent to the graph.

3. a.

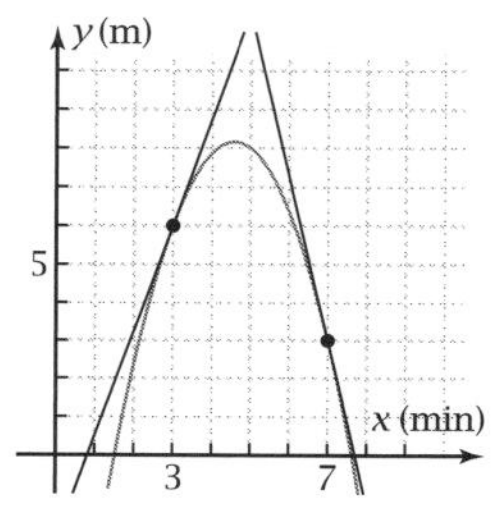

b. $y'(3) \approx 3$ m/min

$y'(7) \approx -4$ m/min

c. In the everyday sense of speed (without reference to direction), the object is slowing down at 3 min and speeding up at 7 min.

5. a.

1	1	−17	105	−89
		1	−16	89
	1	−16	89	0

$f(x) = (x - 1)(x^2 - 16x + 89)$

$= (x - 1)(x - 8 - 5i)(x - 8 + 5i)$

This means that a 1-ft board is free.

b. $f(20) = \$32.11$; $f(14.8647\ldots) = \$10.00$

A \$10 board would be about 14 ft $10\frac{3}{8}$ in.

c. $\frac{f(x) - f(8)}{x - 8} = \frac{(x^3 - 17x^2 + 105x - 89) - 175}{x - 8}$

$= \frac{x^3 - 17x^2 + 105x - 264}{x - 8} = \frac{(x - 8)(x^2 - 9x + 33)}{x - 8}$

$= x^2 - 9x + 33$ if $x \neq 8$

$\lim_{x\to 8} \frac{f(x) - f(8)}{x - 8} = \lim_{x\to 8} (x^2 - 9x + 33) = 25¢$

d. Line $y = 5x - 25$ is tangent to the graph.

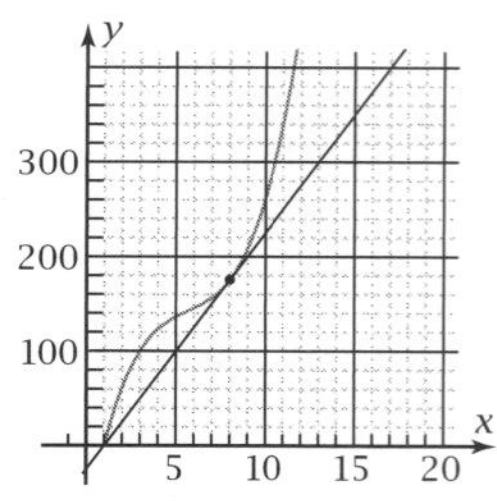

7. a. $\lim_{x\to 2} \frac{f(x) - f(2)}{x - 2} = \lim_{x\to 2} \frac{(x^3 - 11x^2 + 36x - 26) - 10}{x - 2}$

$= \lim_{x\to 2} (x^2 - 9x + 18) = 4$

$\lim_{x\to 3} \frac{f(x) - f(3)}{x - 3} = \lim_{x\to 3} \frac{(x^3 - 11x^2 + 36x - 26) - 10}{x - 3}$

$= \lim_{x\to 3} (x^2 - 8x + 12) = -3$

b. $g(2) = 3(2)^2 - 22(2) + 36 = 4$

$g(3) = 3(3)^2 - 22(3) + 36 = -3$

c. Multiply each term by the exponent of x, then reduce the exponent by 1. In the case of the constant term, the exponent is zero, so when the term is multiplied by the exponent, the term disappears.

d. $x \approx 2.5, 4.9$

e. $3x^2 - 22x + 36 = 0 \Rightarrow x$

$= \frac{-(-22) \pm \sqrt{(-22)^2 - 4(3)(36)}}{2(3)}$

$= \frac{11 \pm \sqrt{13}}{3} = 2.4648\ldots, 4.8685\ldots$

f. $g(5) = 3(5)^2 - 22(5) + 36 = 1$. This is just after the rate of change is zero, when the function is increasing relatively slowly.

g. $g(x) = 1776x^{1775}$

9. $f'(x) = 7x^6$

11. $f'(x) = 48x^5$

13. $f'(x) = 27x^2 - 10x + 2$

15. $f'(x) = 6x^5$

17. $f'(x) = 3x^2 - 24x + 36 = 3(x - 2)(x - 6)$

$\Rightarrow$ vertices at $x = 2, 6$

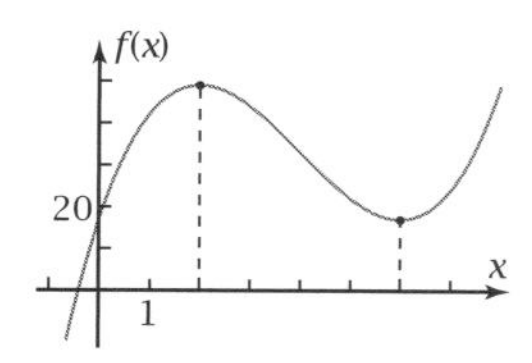

19. $f'(x) = 3x^2 - 8x + 1 = \left(x - \frac{4 - \sqrt{13}}{3}\right)\left(x - \frac{4 + \sqrt{13}}{3}\right)$

$\Rightarrow$ vertices at $x = \frac{4 \pm \sqrt{13}}{3} = 0.1314..., 2.5251...$

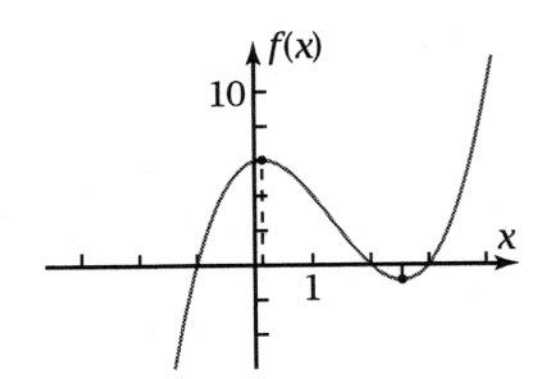

21. $f'(x) = 3x^2 + 1 > 0$ for all $x \Rightarrow$ no vertices

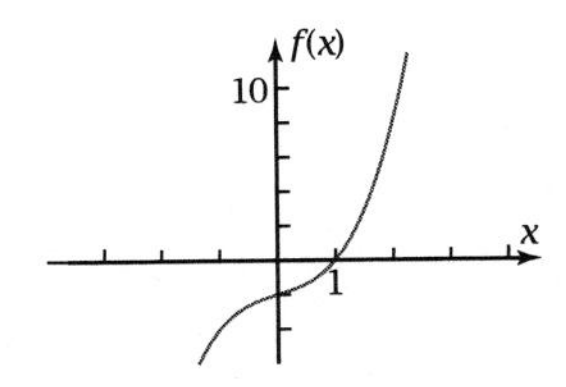

23. $f(3) = 44$, $f'(3) = -9$
$y = -9(x - 3) + 44 = -9x + 71$

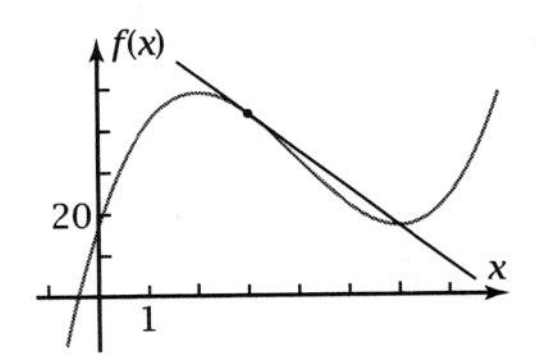

25. a. If the rule worked, then you would have $f'(x) = x \cdot 2^{x-1}$.

b. According to part a, $f'(0) = 0 \cdot 2^{-1} = 0$. However, the function is increasing at this point, so the derivative (rate of change) is in fact greater than zero.

c. The derivative of an exponential function cannot be found using the "power function shortcut" because part b shows that using this shortcut gives at least one answer that is wrong.

Problem Set 15-6

R1. a. $f(-2) = 8 + 8 - 10 - 6 = 0$
$g(-2) = 8 + 8 - 8 - 8 = 0$
$h(-2) = 8 + 8 - 6 - 10 = 0$

b. $f(x) = (x + 2)(-x^2 + 4x - 3)$
$g(x) = (x + 2)(-x^2 + 4x - 4)$
$h(x) = (x + 2)(-x^2 + 4x - 5)$

c. $f(x)$ crosses the horizontal axis at $x = 1, 3$.
$g(x)$ is tangent to the horizontal axis at $x = 2$.
$h(x)$ does not intercept the horizontal axis.

d. A double zero of a function is a number that makes two (identical) factors of the function equal zero. A complex zero of a function is a nonreal complex number that makes the function equal zero.

R2. a. $f(x) = -x^3 + 17x^2 - 32x + 30$

b. $f(5) = 170$; $f(6) = 234$

c. $y = -x^3 + 17x^2 - 32x + 30$
$R^2 = 1$ indicates perfect fit.

d.

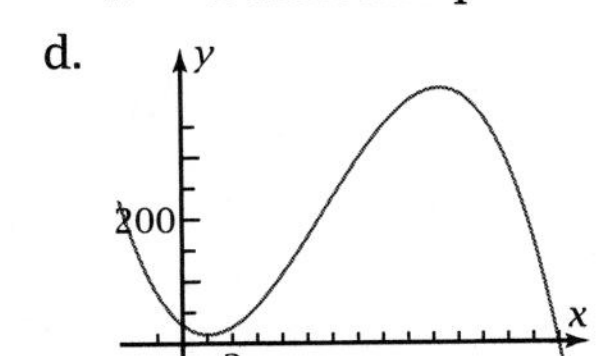

e. The share price goes up for a while, then plummets to zero. Reasons may vary.

R3. a.

5	120	−1200	3480	−2400
		600	−3000	2400
	120	−600	480	0

$d(x) = 120(x - 5)(x - 4)(x - 1)$
Zeros at $x = 1, 4, 5$. The number of zeros equals the degree of the function. The function is dominated by the cubic term, which means the value will eventually behave like $120x^3$. This eventually is greater than 700.

b. $\frac{x^3 - 4x^2 + 7x + 11}{x - 2} = x^2 - 2x + 3 + \frac{17}{x - 2}$
$d(2) = 2^3 - 4(2)^2 + 7(2) + 11 = 17$

c.

2	1	−10	57	−82
		2	−16	82
	1	−8	41	0

$f(x) = (x - 2)(x - 4 - 5i)(x - 4 + 5i)$
Zeros at $x = 2, 4 \pm 5i$.

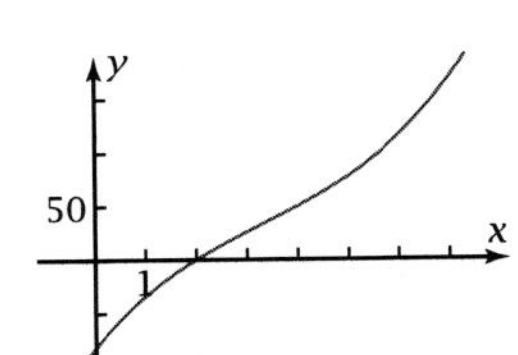

d. Fifth-degree polynomial; three real zeros; two nonreal complex zeros

R4. a.

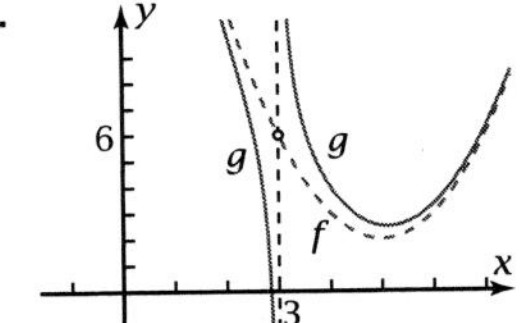

b. $f(x) \cong x^2 - 10x + 27$
$\lim_{x \to 3} f(x) = 6$
(3, 6) is the location of the "hole" in the graph.

c. $g(x) = x^2 - 10x + 27 + \frac{1}{x-3}$

The term $\frac{1}{x-3}$ has the infinite form $\frac{\neq 0}{0}$.

d. $\frac{0}{0}$: indeterminate form

$\frac{\neq 0}{0}$: infinite form

e. $h(x) = \frac{3}{x-1} + \frac{6}{x-4}$

$\frac{3}{x-1}$: horizontal translation by 1, vertical dilation by 3

$\frac{6}{x-4}$: horizontal translation by 4, vertical dilation by 6

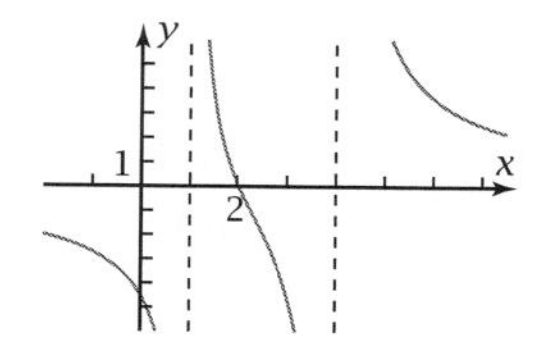

$0 = h(x) \Rightarrow x = 2$

R5. a. $v_{av} = 73.2$ ft/min

b. $v_{av} = \dfrac{d(x) - d(2)}{x - 2} = \dfrac{120x^3 - 1200x^2 + 3480x - 3120}{x - 2}$

$= 120\dfrac{(x-2)(x^2 - 8x + 13)}{x-2}$ ft/min

c. Removable discontinuity at $x = 2$

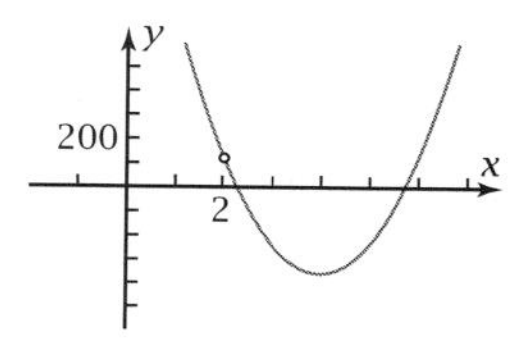

d. $v_{av} = 120(x^2 - 8x + 13)$

e. The derivative is $\lim_{x \to 2} \dfrac{d(x) - d(2)}{x - 2} = 120$ ft/min

f. $y = 120x + 480$

The line is tangent to the graph.

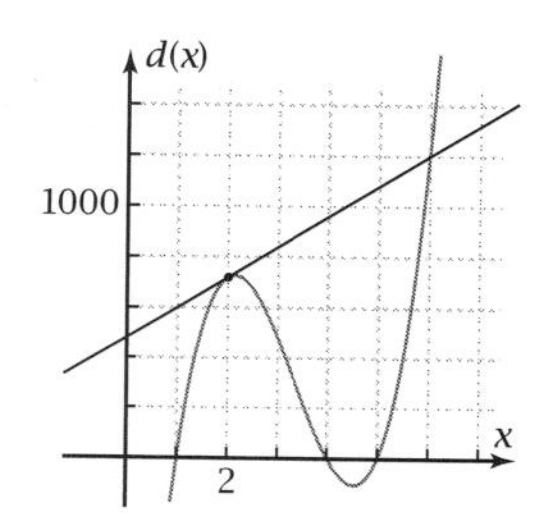

g. The train is going north; the derivative is positive.

h. $f'(2) = 120$ ft/min. For each term, multiply the term by the existing exponent, then subtract 1 from the exponent.

Problem Set 15-7

1.

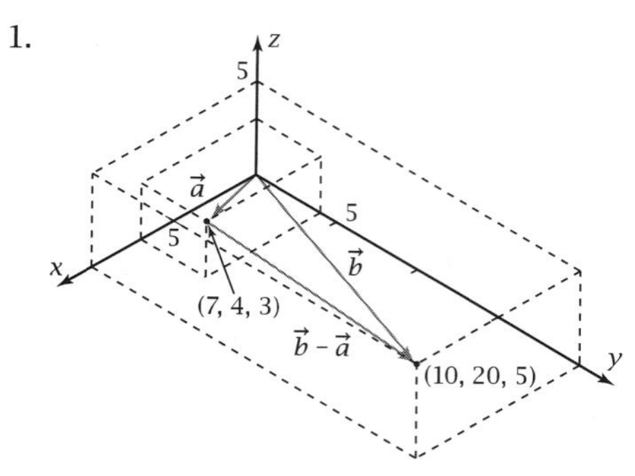

2. $\vec{b} - \vec{a} = 3\vec{i} + 16\vec{j} + 2\vec{k}$

3. $|\vec{b} - \vec{a}| = \sqrt{269} = 16.4012\ldots = 1640.1219\ldots$ m

4. The plane was higher by 200 m.

5. $\cos\theta = 0.8371\ldots$

$\theta = 33.1626\ldots°$

6. proj = 720.1190... m

7. $(\vec{a} \times \vec{b}) \cdot \vec{a} = 0$

$$\cos\theta = \frac{(\vec{a} \times \vec{b}) \cdot \vec{a}}{|\vec{a} \times \vec{b}||\vec{a}|} = 0 \Rightarrow (\vec{a} \times \vec{b}) \perp \vec{a}$$

8. $A = 539{,}096.4663\ldots$ m^2

9. $\theta = 60°$

10. $P \approx (3.5, 23.2)$

11. $\begin{bmatrix} 0.45 & -0.78 & 20 \\ 0.78 & 0.45 & 10 \\ 0 & 0 & 1 \end{bmatrix}\begin{bmatrix} x \\ y \\ 1 \end{bmatrix} = \begin{bmatrix} 0.45x - 0.78y + 20 \\ 0.78x + 0.45y + 10 \\ 1 \end{bmatrix}$

12. $P = (3.5130\ldots, 23.1639\ldots)$

13.

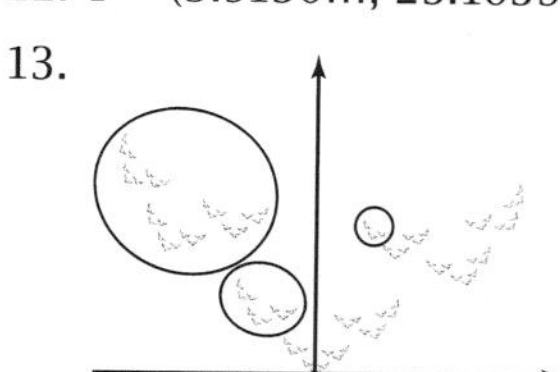

14. $[B] = \begin{bmatrix} 0.4\cos(-20°) & 0.4\cos 70° & -5 \\ 0.4\sin(-20°) & 0.4\sin 70° & 4 \\ 0 & 0 & 1 \end{bmatrix}$

15. $[C] = \begin{bmatrix} 0.4 & 0 & 0 \\ 0 & 0.4 & 0 \\ 0 & 0 & 1 \end{bmatrix}$

16. The graphs should match.

17.

Iteration	N	Perimeter of Each	Total P
0	1	36	36
1	4	12	48
2	16	4	64
3	64	$\frac{4}{3}$	$\frac{256}{3}$

18. Total $P = 63{,}568{,}114.6772\ldots$

19.

Iteration	A
0	81
1	36
2	16
3	$\frac{64}{9}$

20. $D = 1.2618\ldots$

21. $\lim_{n\to\infty} A = 0$

$\lim_{n\to\infty} P = \infty$

This makes sense because an object of dimension less than 2 has zero area, whereas an object of dimension greater than 1 has infinite length.

22. The signs of the x^2- and y^2-terms are opposite.

$25x^2 - 9y^2 - 200x + 18y = -391$

$$\Leftrightarrow \left(\frac{x-4}{3}\right)^2 - \left(\frac{y-1}{5}\right)^2 = 0$$

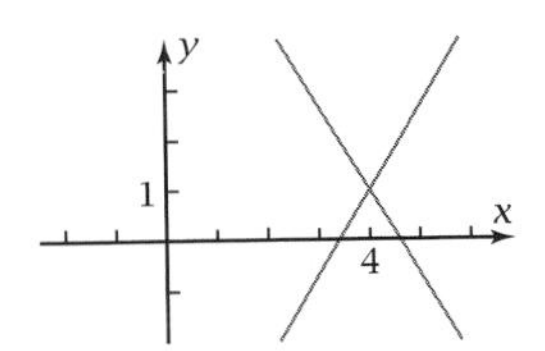

23. $c = 4$

$$e = \frac{c}{a} = \frac{4}{5} = 0.8$$

$$d = \frac{a}{e} = \frac{5}{0.8} = 6.25$$

24.

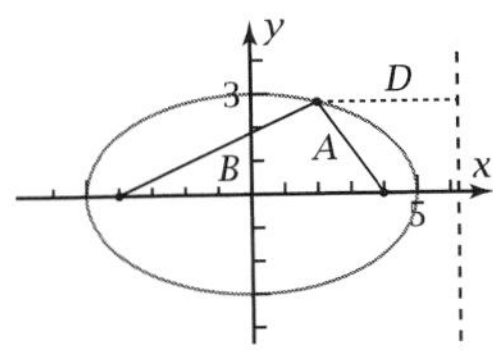

$A + B = 2a = 10$ = length of major axis

$D = eA = 6.25A$

25. $\begin{cases} x = 5\cos t \\ y = 3\sin t \end{cases}$

$\begin{cases} x' = 5\cos t + 2 \\ y' = 5\sin t + 1 \end{cases}$

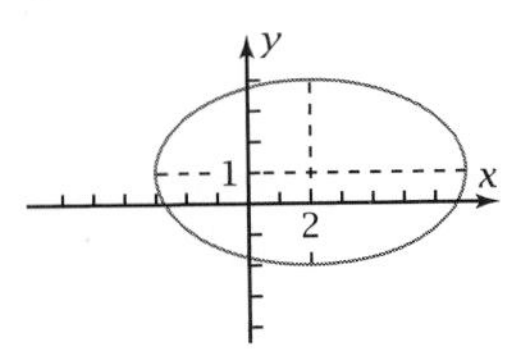

26. $V = 18\pi x - \dfrac{18}{25}\pi x^3$

$r = \sqrt{6} = \sqrt{2.4494\ldots}$; $h = \dfrac{10\sqrt{3}}{3} = 5.7735\ldots$

27. The crash occurs at $(r, \theta) = (3{,}960{,}325.9227\ldots°)$.

28. $(5 \text{ cis } 70°)(8 \text{ cis } 40°) = 40(\cos 110° + i\sin 110°)$
$= -13.6808\ldots + 37.5877\ldots i$

29. $\sqrt[3]{64} = 4 \text{ cis } 0°, 4 \text{ cis } 120°, 4 \text{ cis } 240°$

30. The figure is called a cardioid because it resembles a heart shape.

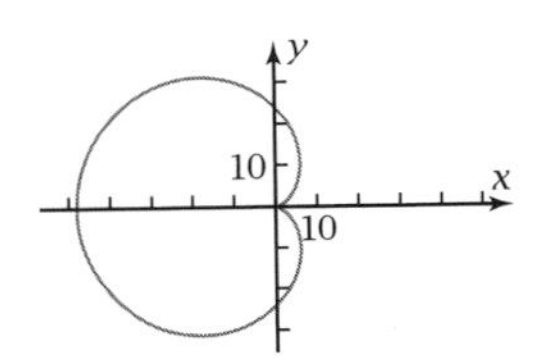

31.

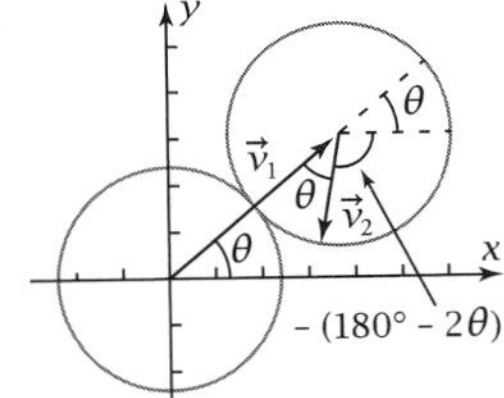

$\begin{cases} x = 24\cos\theta - 12\cos 2\theta \\ y = 24\sin\theta - 12\sin 2\theta \end{cases}$

The point on the rolling quarter that was touching the fixed quarter along the x-axis traces out the cardioid.

32. 7, 12, 17, 22

33. $8 + 24 + 72 + 216 = 320$

34. $12 + 15 + 18 + \cdots + 36 + 39 = \sum_{n=1}^{10} (9 + 3n) = \$255$

35. At the end of 36 months, the balance is $500 \cdot 1.0075^{36} = 654.3226... = \$654.32$. \$154.32 of that is interest. If the interest were compounded yearly, the balance would be

$500 \cdot 1.09^3 = 647.5145... = \$647.51$

Compound interest gives you an extra \$6.81. The value increases in a series of jumps (one jump every compounding period) rather than continuously.

36. In sequence mode on your grapher, enter
nMin = 0
u(n) = .9u(n − 1) + 30
u(nMin) = {100}

n	t_n	Rounded
0	100	100
1	120	120
2	138	138
3	154.2	154
4	168.78	169
5	181.902	182

and the series eventually levels off at 300. Solving $0.9x + 30 = x$ gives $x = \frac{30}{1 - 0.9} = 300$.

37. $t_n = n^2 + 1$; $t_{100} = 10{,}001$

38. $n = 71$

39. $\sum_{n=1}^{\infty} \frac{1}{n^{1.2}} = 1 + 0.4352... + 0.2675... + 0.1894... + \cdots$

$S_{100} = 3.6030...$

40.
$$\begin{array}{r|rrrr} 1 & 1 & -25 & 249 & -225 \\ & & 1 & -24 & 225 \\ \hline & 1 & -24 & 225 & 0 \end{array}$$

This means that 1-ft trees are free.

41. $f(x) = (x - 1)(x^2 - 24x + 225)$
$= (x - 1)(x - 12 - 9i)(x - 12 + 9i)$

42. $z_1 + z_2 + z_3 = 1 + (12 + 9i) + (12 - 9i) = 25$

43. $f(20) = \$2755.00$

44. $\frac{f(x) - f(10)}{x - 10} = \frac{(x - 10)(x^2 - 15x + 99)}{x - 10} = x^2 - 15x + 99$

if $x \neq 10$

$\lim_{x \to 10} \frac{f(x) - f(10)}{x - 10} = \lim_{x \to 10} (x^2 - 15x + 99) = \$49/\text{ft}$

45. Line $y = 49x + 275$ is tangent to the graph.

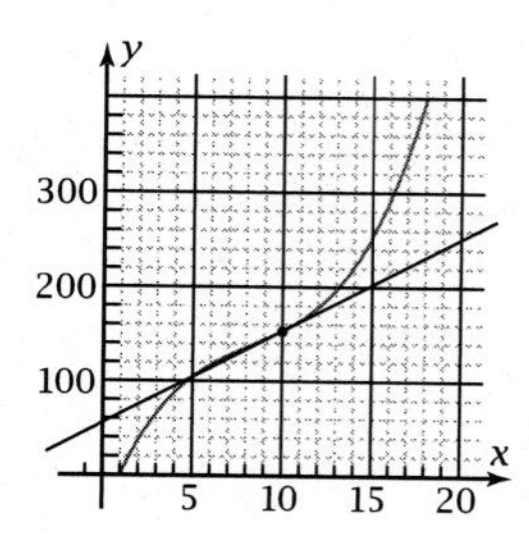

Glossary

Algebraic: Involving only the operations of algebra performed on the variable—namely, addition, subtraction, multiplication (including integer powers), division, and roots.

Amplitude (of a sinusoidal graph) (p. 85): The vertical distance from the sinusoidal axis of a graph to its maximum or minimum.

Argument (p. 8): In $f(x)$, the variable x or any expression substituted for x.

Arithmetic sequence (pp. 589, 592): A sequence in which each term is formed recursively by adding a constant to the previous term.

Arithmetic series (pp. 598, 604): The indicated sum of the terms of an arithmetic sequence.

Asymptote (p. 4): A line that the graph of a function approaches and to which it gets arbitrarily close as x or y approaches positive or negative infinity.

Barnsley's method (pp. 478–479): A means of generating fractal figures that involve iterations of multiple transformations; at each stage, one of a given set of transformations is selected at random and performed on the point that is the image of the preceding transformation.

Binomial distribution (p. 387): Probabilities that are associated with r successes in n repeated trials for an event with two outcomes.

Binomial experiment (p. 387): A random experiment in which the same action is repeated a number of times and in which only two results of the action are possible.

Binomial series (pp. 602–603): A series that comes from expanding a positive integer power of a binomial expression.

Circular functions (pp. 106–107): Periodic functions whose independent variable is in radians, a real number without units.

Coefficient of determination (pp. 327, 329): The fraction of SS_{dev} that has been removed by the linear regression $r^2 = \dfrac{SS_{\text{dev}} - SS_{\text{res}}}{SS_{\text{dev}}}$.

Combination (pp. 370–371): A subset of elements selected from a set without regard to the order in which the selected elements are arranged.

Common difference (pp. 589, 592): The constant added to a term to get the next term in an arithmetic sequence or series.

Common ratio (pp. 590, 592): The constant multiplier of a term to get the next term in a geometric sequence or series.

Completing the square: Reversing the process of squaring a binomial to go from $y = ax^2 + bx + c$ back to the form $y = a(x - h)^2 + k$ that shows translations and dilations.

Complex conjugates (p. 564): Two complex numbers of the forms $a + bi$ and $a - bi$, where a and b are real numbers and $i = \sqrt{-1}$.

Complex number (pp. 563–564): The sum of a real number and an imaginary number.

Complex plane (p. 563): The representation of complex numbers by points in a Cartesian coordinate system.

Composite function (pp. 22, 25): A function of the form $f(g(x))$, where function g is performed on x and then function f is performed on $g(x)$. The domain of the composite function is those x-values in the domain of g for which $g(x)$ is an element of the domain of f.

Composition of ordinates (p. 190): The method by which sinusoids or other functions are combined by adding or multiplying the corresponding ordinates for each value of x.

Concave upward (or downward) (p. 87): The graph of a function (or a portion of the graph between two asymptotes or two points of inflection) is called concave upward when its "hollowed out" side faces upward; it is called concave downward when its "hollowed out" side faces downward.

Conjugate axis (pp. 507–508): For a hyperbola, the name given to the axis of symmetry perpendicular to the transverse axis.

Constant: A number with a fixed value, as opposed to a *variable,* which can take on different values.

Correlation coefficient (pp. 327–329): The value r, which is the positive or negative square root of the coefficient of determination.

Cosine function (pp. 59–60): One of the trigonometric functions; abbreviated "cos." Let θ be the angle of rotation of the positive u-axis, terminating in a position containing the point (u, v), then $\cos \theta = \frac{u}{r}$, where r is the distance of the point (u, v) from the origin.

Coterminal angles (p. 55): Two angles in standard position are coterminal if and only if their degree measures differ by a multiple of 360°.

Critical point (pp. 87, 619): A point where the tangent line to the graph of a function is either horizontal or vertical.

Cross product (p. 434): The cross product of two vectors is a vector with the following properties: $\vec{a} \times \vec{b}$ is perpendicular to the plane containing $\vec{a}$ and $\vec{b}$; the magnitude of $\vec{a} \times \vec{b}$ is $|\vec{a} \times \vec{b}| = |\vec{a}||\vec{b}| \sin \theta$, where θ is the angle between the two vectors when they are placed tail-to-tail; the direction of $\vec{a} \times \vec{b}$ is determined by the right-hand rule.

Cycle (p. 85): The part of a graph from a point on it to another point where the graph first starts repeating itself.

De Moivre's theorem (p. 568): The "short way" to raise a complex number to a power. Raise the modulus to the power and multiply the argument by the exponent; for example, if $z = r(\cos \theta + i \sin \theta) = r \operatorname{cis} \theta$, then $z^n = r^n \operatorname{cis} n\theta$.

Degree (of a polynomial) (p. 618): For a polynomial expression in one variable, the greatest exponent of that variable.

Dependent variable (pp. 3–4): The output of a function, an equation, or a formula.

Derivative (p. 644): The instantaneous rate of change of a function.

Derivative function (p. 647): The function that gives the instantaneous rate of change of a given function at any x-value.

Deviation (pp. 327, 329): For a data point (x, y), the directed distance of its y-value from $\bar{y}$, $y - \bar{y}$, where $\bar{y}$ is the average of the y-values.

Dimension (the Hausdorff dimension) (p. 485): If an object is transformed into N self-similar pieces, the ratio of the length of each piece to the length of the original object is r, and the subdivisions can be carried on infinitely, then the dimension D of the object is $D = \dfrac{\log N}{\log \frac{1}{r}}$.

Direction angles (of a position vector) (p. 441): α, from the x-axis to the vector; β, from the y-axis to the vector; γ, from the z-axis to the vector.

Direction cosines (p. 441): The cosines of the direction angles of a position vector.

Direct variation function (p. 10): A function with the general equation $f(x) = ax$, where a stands for a constant.

Discrete function (p. 588): A function whose domain is a set of disconnected values.

Displacement (p. 37): A vector that indicates an object moving from one point to another.

Displacement vector (p. 411): The difference between an object's initial and final positions.

Distance formula (p. 176): A form of the Pythagorean theorem giving the distance D between two points (x_1, y_1) and (x_2, y_2) in a plane in terms of their coordinates: $D = \sqrt{(x_1 - x_2)^2 + (y_1 - y_2)^2}$.

Diverge (p. 601): A series diverges if the partial sums do not approach a finite limit as the number of terms becomes infinite.

Domain (p. 4): The set of values that the independent variable of a function can have.

Dot product (p. 422): $\vec{a} \cdot \vec{b} = |\vec{a}||\vec{b}| \cos \theta$, where θ is the angle between the two vectors when they are translated tail-to-tail. Also called the *scalar product* or *inner product.*

Eccentricity (pp. 514–515): Eccentricity, e, is the ratio of the distance between a point on a conic section and the focus, d_2, and the distance between the same point and the directrix, d_1: $e = \dfrac{d_2}{d_1}$.

Ellipsis format: A three-dot (...) notation that indicates something has been left out.

Even function (pp. 38, 182): The function f is an even function if and only if $f(-x) = f(x)$ for all x in the domain.

Event (p. 358): In probability, a set of outcomes.

Explicit formula (p. 591): For a sequence, specifies t_n as a function of n.

Exponential function (pp. 10, 272–273): A function in which the independent variable appears as an exponent.

Extrapolation (p. 6): Using a function to estimate a value *outside* the range of the given data.

Factorial (pp. 366–367): For any positive integer n, $n! = 1 \cdot 2 \cdot 3 \cdot \cdots \cdot n$ or, equivalently, $n! = n \cdot (n-1) \cdot (n-2) \cdot \cdots \cdot 2 \cdot 1$. 0! is defined to be equal to 1.

Fibonacci sequence (pp. 594–595): The sequence of integers 1, 1, 2, 3, 5, 8, 13, 21, 34, 55, 89, 144, 233, . . . , formed according to the rule that each integer is the sum of the preceding two.

Fractal (pp. 478, 484): A geometric figure that is composed of self-similar parts.

Frequency (pp. 86–87): The reciprocal of the period.

Function (pp. 3, 8): A relationship between two variable quantities for which there is exactly one value of the dependent variable for each value of the independent variable in the domain.

Geometric sequence (pp. 590, 592): A sequence in which each term is formed recursively by multiplying the previous term by a constant.

Geometric series (pp. 599, 604): The indicated sum of the terms of a geometric sequence.

Harmonic analysis (pp. 190, 192–193): The reverse of the composition of ordinates; a method by which parent sinusoid functions are found from the resultant function.

Harmonic series (p. 606): A series such as $1 + \frac{1}{2} + \frac{1}{3} + \frac{1}{4} + \cdots + \frac{1}{n} + \cdots$ in which successive terms are the reciprocals of the terms in an arithmetic sequence.

Hausdorff dimension: See **Dimension.**

Identity (p. 138): An equation that is true for all values of the variable in the domain.

Imaginary number (p. 563): A number that is the square root of a negative number.

Independent events (p. 361): Events are called independent if the way one event occurs does not affect the ways the other event(s) could occur.

Independent variable (pp. 3–4): The input of an equation or formula.

Indeterminate form (p. 636): An expression that has no direct meaning as a number, for example, $\frac{0}{0}$ or $0 \cdot \infty$.

Infinite form (p. 636): A form such as $\frac{1}{0}$ that indicates a value that is larger than any real number.

Inner product: See **Dot product.**

Interpolation (p. 6): Using a function to estimate a value *within* the range of given data.

Intersection (pp. 379–380): If A and B are two events, then the intersection of A and B, $A \cap B$, is the set of all outcomes that are found in both event A *and* event B.

Inverse of a function (pp. 30, 32, 71): The inverse of a function is the relation formed by interchanging the variables of the function. If the inverse relation is a function, then it is called the *inverse function* and is denoted $f^{-1}(x)$.

Inverse variation function (p. 10): A function with the general equation $f(x) = \frac{a}{x}$, where a stands for a constant and $x \neq 0$.

Iterate (p. 457): To perform an operation over and over again, each time operating on the image from the preceding step.

Law of cosines (p. 227): In triangle ABC, with sides a, b, and c opposite angles A, B, and C, respectively, $a^2 = b^2 + c^2 - 2bc \cos A$.

Law of sines (p. 235): In triangle ABC, with sides a, b, and c opposite angles A, B, and C, respectively,

$$\frac{a}{\sin A} = \frac{b}{\sin B} = \frac{c}{\sin C}.$$

Leading coefficient (p. 618): In a polynomial in one variable, the coefficient of the highest-degree term.

Limit (pp. 471, 636): A number that a function value $f(x)$ approaches, becoming arbitrarily close to it, as x approaches either a specific value or infinity.

Linear combination (p. 174): Of two expressions u and v, a sum of the form $au + bv$, where a and b stand for constants.

Linear function (pp. 9, 270): A function with the general equation $y = ax + b$, where a and b stand for constants.

Linear regression (pp. 325, 327, 329): The process of finding the best-fitting linear equation for a given set of data.

Logarithm (pp. 291, 296): $y = \log_a x$ is an exponent for which $a^y = x$.

Logarithmic function (pp. 290, 301): The function that is the inverse of an exponential function.

Logistic function (pp. 308, 310): A function of the form $f(x) = \dfrac{c}{1 + ae^{-bx}}$ or $f(x) = \dfrac{c}{1 + ab^{-x}}$, where a, b, and c stand for constants, e is the base of the natural logarithm, and the domain is all real numbers.

Major axis (p. 500): For an ellipse, the longer of its two perpendicular axes; for an ellipsoid, the largest of its three axes.

Mathematical expectation (p. 393): The weighted average of the values for a random experiment each time it is run. The "weights" are the probabilities of each outcome.

Mathematical models (p. 3): Functions that are used to make predictions and interpretations about a phenomenon in the real world.

Matrix (p. 458): A rectangular array of terms called *elements.*

Minor axis (p. 500): For an ellipse, the shorter of its two perpendicular axes; for an ellipsoid, the smallest of its three axes.

Mutually exclusive (pp. 361–362): When the occurrence of an event excludes the possibility that another event will also occur.

Numerical: Referring to constants, such as 2 or π, rather than to parameters or variables.

Oblique triangle (p. 225): A triangle none of whose angles are right angles.

Odd function (pp. 38, 182): The function f is an odd function if and only if $f(-x) = -f(x)$ for all x in the domain.

One-to-one function (p. 31): A function f is a one-to-one function if there are no y-values that correspond to more than one x-value.

Outcome (p. 358): A result of a random experiment.

Parameter (p. 153): A variable used as an arbitrary constant, or an independent variable on which two or more other variables depend.

Parametric equations (of a plane curve) (p. 153): The two equations that express the coordinates x and y of points on the curve as separate functions of a common variable (the parameter).

Parametric function (p. 152): A function specified by parametric equations.

Period (pp. 59, 87): For a periodic function, the difference between the horizontal coordinates of points at the ends of a single cycle.

Periodic function (p. 59): A function whose values repeat at regular intervals.

Permutation (p. 365): An arrangement in a definite order of some or all of the elements in a set.

Phase displacement (p. 85): The directed horizontal distance from the vertical axis to the point where the argument of a periodic function equals zero. For example, if $y = \cos(x - 2)$, the phase displacement is 2; or if $y = \sin\left(x + \frac{\pi}{3}\right)$, the phase displacement is $-\frac{\pi}{3}$.

Point of inflection (p. 87): The point where a graph switches its direction of concavity.

Point-slope form (p. 270): $y - y_1 = a(x - x_1)$, where (x_1, y_1) is a fixed point on the line.

Polar axis (p. 550): In the polar coordinate system, a fixed ray, usually in a horizontal position.

Polar coordinates (pp. 549, 551): A method of representing points in a plane by ordered pairs of (r, θ). The value r represents the distance of the point from a fixed point (pole), and θ represents the angle of rotation of the polar axis to a position that contains the point.

Pole (p. 550): In the polar coordinate system, the origin.

Polynomial (p. 618): An expression involving only the operations of addition, subtraction, and multiplication performed on the variable (including nonnegative integer powers).

Polynomial function (p. 9): A function of the form $y = p(x)$, where $p(x)$ is a polynomial expression. The general equation is $p(x) = a_nx^n + a_{n-1}x^{n-1} + \cdots + a_2x^2 + a_1x + a_0$.

Polynomial operations (p. 618): The operations of addition, subtraction, and multiplication.

Position vector (p. 409, 411): For a point (x, y), its position vector starts at the origin and ends at the point.

Power function (pp. 10, 271): A function of the form $y = ax^n$, where $a \neq 0$.

Power series (p. 612): A series in which each term is a constant coefficient times a power of the variable.

Probability (p. 358): If the outcomes of a random experiment are equally likely, then the probability that a particular event will occur is equal to the number of outcomes in the event divided by the number of outcomes in the sample space.

Probability distribution (p. 387): A probability function that tells how the 100% probability is distributed among the various possible events.

Proportionality constant (p. 271): The constant a in the equation of a power function $y = ax^n$. In general, a constant that equals the ratio of two quantities.

Quadratic function (p. 270): A function of the form $y = ax^2 + bx + c$, where $a \neq 0$.

Radian (pp. 99–100): A central angle of one radian intercepts an arc of the corresponding circle equal in length to the radius of the circle.

Radian measure (p. 101): The angle measure of a central angle in a circle defined as the arc length corresponding to the angle divided by the radius of the circle. Radian angle measures are real numbers with no units.

Random experiment (p. 358): In probability, an experiment in which there is no way of telling the outcome beforehand.

Range (p. 4): The set of all values of the dependent variable that correspond to values of the independent variable in the domain.

Rational function (p. 11): A function that can be expressed as the quotient of two polynomial expressions.

Recursion formula (p. 591): A formula for terms of a sequence that specifies t_n as a function of the preceding term, t_{n-1}.

Reference angle (p. 55): For an angle in standard position, the positive acute angle (measured counterclockwise) between the horizontal axis and the terminal side.

Relation (p. 8): Any set of ordered pairs.

Residual (pp. 325, 329): The residual deviation of a data point from the line $\hat{y} = mx + b$ is $y - \hat{y}$, the vertical directed distance of its y-value from the line.

Residual plot (p. 341): A scatter plot of residuals.

Sample space (p. 358): The set of all outcomes of an experiment.

Scalar (pp. 409, 411): A quantity, such as time, speed, or volume, that has magnitude but no direction.

Scalar product: See **Dot product.**

Scalar projection (p. 425): If θ is the angle between $\vec{a}$ and $\vec{b}$ when they are placed tail-to-tail, then the scalar projection of $\vec{a}$ on $\vec{b}$ is $p = |\vec{a}| \cos \theta$.

Scalar quantity (pp. 409, 411): A quantity, such as distance, time, or volume, that has magnitude but no direction.

Self-similar (p. 478): When an object or figure can be broken into smaller parts that are similar to one another and to the original object or figure.

Sequence (p. 591): A function whose domain is the set of positive integers. The independent variable is the term number, n, and the dependent variable is the term value, t_n.

Series (pp. 587, 596): The indicated sum of the terms of a sequence.

Simple event (p. 358): A single outcome of an experiment.

Sine function (pp. 59–60): One of the trigonometric functions; abbreviated "sin." Let θ be the angle of rotation of the positive u-axis, terminating in a position containing the point (u, v); then $\sin \theta = \frac{v}{r}$, where r is the distance of the point (u, v) from the origin.

Sinusoid (pp. 53, 86): Any translation or dilation of the parent function $y = \sin x$ or $y = \cos x$.

Sinusoidal axis (pp. 85–86): The axis that runs along the middle of the graph of a sinusoid.

Slope-intercept form (p. 270): The general equation $y = ax + b$, where a and b are constants and the domain is all real numbers.

Standard position (of an angle) (p. 54): The position of an angle with the vertex at the origin and the initial side along the positive horizontal axis.

Tangent function (p. 65): One of the trigonometric functions; abbreviated "tan." Let θ be the angle of rotation of the positive u-axis, terminating in a position containing the point (u, v); then $\tan \theta = \frac{v}{u}$.

Tangent line (p. 644): The tangent line of a curve at a point is the line whose slope equals the slope of the curve at that point.

Transformation (pp. 16, 19, 37): An operation that maps points in the plane or space uniquely. Some examples are dilation, translation, or rotation, which can change the shape or the position of the graph of a relation or of some other figure.

Transverse axis (pp. 507–508): For a hyperbola, the name given to the axis of symmetry that runs from vertex to vertex.

Trigonometric functions (pp. 59, 60, 65): The six functions sine, cosine, tangent, cotangent, secant, and cosecant.

Union (pp. 380–381): If A and B are two events, then the union of A and B, $A \cup B$, is the set of all outcomes that are in event A *or* event B (or both).

Unit circle (pp. 64, 154): A circle of radius 1.

Unit vector (pp. 409, 411): A vector that is one unit long.

Variable: A symbol used to stand for any one of a set of numbers, points, or other entities; the set is called the *domain* of the variable, and any member of the set is called a *value* of the variable.

Vector (pp. 242–243, 411): A directed line segment, $\vec{v}$.

Vector projection (p. 425): If $\vec{u}$ is a unit vector in the direction of $\vec{b}$, then the vector projection of $\vec{a}$ on $\vec{b}$ is $\vec{p} = |\vec{a}| \cos \theta \, \vec{u}$, where θ is the angle between $\vec{a}$ and $\vec{b}$.

Vector quantity (pp. 242, 411): A quantity, such as force, velocity, or displacement, that has both magnitude (size) and direction.

Vertex form (p. 271): For a quadratic function, $y - k = a(x - h)^2$, where a, h, and k stand for constants, the vertex of the parabola is at (h, k).

Zero of a function (pp. 617, 619): A value of $x = c$ for which $f(c) = 0$.

Index of Problem Titles

A

B

C

D

E

F

G

H

I

J

K

L

M

N

O

P

Q

R

S

T

U

V

W

X

Z

General Index

C

D

E

F

G

H

I

J

K

L

M

N

O

P

Q

R

S

T

U

V

W

X

Y

Z

Photograph Credits

Chapter 1

1: DG/Creatas. **3:** Ryan McVay/Getty Images. **6 (top):** Fototeca Storica Nazionale/Getty Images. **6 (bottom):** Ryan McVay/Getty Images. **8:** AP/Wide World Photos. **14:** Getty Images. **17:** Ken Karp Photography. **21:** Photolink/Getty Images. **26:** Hank Morgan/Rainbow/PictureQuest. **34 (left):** Jeremy Woodhouse/Getty Images. **34 (right):** Barros & Barros/The Image Bank. **37:** © Jacksonville Journal Courier/The Image Works. **45:** Yoshiaki Nagashima/Stockphoto.com. **46:** Brian Bahr/Allsport/Getty Images. **47:** © 1991 Ron Behrmann. **51:** John Running/Stockphoto.com.

Chapter 2

51: Brian Bahr/Allsport/Getty Images. **53:** Ben Wood/CORBIS. **55:** Gail Mooney/CORBIS. **57:** Michael Newman/PhotoEdit, Inc. **59:** Jason Reed/Getty Images. **63:** Allsport. **74:** David Portnoy/Stockphoto.com. **76 (left):** © CORBIS Images/PictureQuest. **76 (right):** Courtesy of NASA/JPL/CalTech. **77:** Bob Daemmrich/The Image Works. **78:** Roger Wood/CORBIS. **79 (left):** Michael Newman/PhotoEdit, Inc. **79 (right):** Ian Cartwright/Getty Images. **80:** Stephen Frisch/Stock Boston.

Chapter 3

83: Getty Images. **85:** Ken Karp Photography. **90:** Ken Karp Photography. **92:** Ken Karp Photography. **99:** Chad Ehlers/PictureQuest. **100:** © The British Museum. **106:** Index Stock Imagery, Inc./BSIP Agency. **112:** John Cetrino/PictureQuest. **116:** Michael Seigel/PictureQuest. **119 (top):** Courtesy of Nick Karanovich. **119 (bottom):** David W. Hamilton/Getty Images. **120:** Kwame Zikomo/SuperStock, Inc. **121 (left):** Courtesy of NASA. **121 (right):** Ryan McVay/Getty Images. **123:** Geostock/Getty Images. **124:** Mark Antman/The Image Works. **127:** Cousteau Society/Getty Images.

Chapter 4

131: Courtesy of Bing Quock. **140:** Ken Karp Photography. **142:** Ken Karp Photography. **154:** UNEP/Topham/The Image Works. **156:** William S. Helsel/Getty Images. **158:** Courtesy of the Kepler-Gesellschaft Society. **159:** Ken Karp Photography. **161:** IFA/eStock Photo. **165:** J. A. Hampton/Getty Images. **168:** Bettmann/CORBIS.

Chapter 5

171: www.comstock.com. **174:** Paul Arthur/Getty Images. **180:** Ken Karp Photography. **192:** Bob Cranston/Animals Animals/Earth Scenes. **195:** AP/Wide World Photos. **199:** Topham/The Image Works. **202:** Ken Karp Photography. **205:** AP/Wide World Photos. **219:** AIP/Emilio Segre Archives. **220:** Courtesy of BK Precision.

Chapter 6

223: Bill Bachmann/The Image Works. **232:** Provided by IIHR—Hydroscience and Engineering, The University of Iowa, Iowa City, IA, from 1688 translation of Hero of Alexander's *Pneumatics* in the History of Hydraulics Rare Book Collection, UI Libraries (Special Collections). **243:** Carl & Ann Purcell/CORBIS. **244 (top):** Richard Bickel/CORBIS. **244 (bottom):** AP/Wide World Photos. **246:** Chlaus Lotscher/Stock Boston. **248:** Bob Daemmrich/Stock Boston. **250:** Bettmann/CORBIS. **252 (top):** Erik Simonsen/Getty Images. **252 (bottom):** Daniel Wray/The Image Works. **253:** Steven Starr/Stock Boston. **254:** www.comstock.com. **255:** Bob Daemmrich/Stock Boston. **259:** Alex Calder/Helispot. **262:** Space Frontiers/Getty Images. **265:** AP/Wide World Photos.

Chapter 7

267: Erwin & Peggy Bauer/Wildstock. **271:** Michael Smith. **272:** AIP/Emilio Segre Archives. **275:** Richard Hutchings/PhotoEdit, Inc. **280:** Ken Karp Photography. **285:** © Science Photo Library/Photo Researchers, Inc. **291:** Courtesy of Eric Marcotte, Ph.D., www.sliderule.ca.

294: Michael Newman/PhotoEdit, Inc. **296:** Siede Preis/Getty Images. **299 (left):** NASA.
299 (right): Richard Laird/Getty Images. **304:** Sinclair Stammers/Photo Researchers, Inc.
305: Reuters NewMedia, Inc./CORBIS. **309:** AP/Wide World Photos. **312:** AP/Wide World Photos.
313: Neil McIntre/Getty Images. **316 (top):** Ron Dahlquist/SuperStock, Inc.
316 (bottom): Index Stock Imagery, Inc./Inga Spence **320:** Dominique Braud/Animals Animals.

Chapter 8

323: Kevin Fleming/CORBIS. **328:** Computer History Museum. **333:** Mike Robinson/SuperStock, Inc.
335: Michael Newman/PhotoEdit, Inc. **337:** JPL/NASA. **338:** Tom Bean/Getty Images.
341: Barry Runk/Grant Heilman Photography, Inc. **342:** Graham Neden/CORBIS.
345 (top): Spencer Grant/PhotoEdit, Inc. **345 (bottom):** AP/Wide World Photos.
347: AP/Wide World Photos. **350:** Malcolm Dunbar/Getty Images. **351:** AP/Wide World Photos.
353: Dan Suzio Photography.

Chapter 9

355: Vince Streano/Getty Images. **357:** Laura Murray. **359:** David Young-Wolff/PhotoEdit, Inc.
362: Laura Murray. **363:** Yves AEF Debay/Getty Images. **364:** Laura Murray.
368: Myrleen Cate/PhotoEdit, Inc. **369:** www.comstock.com. **373:** Don Tremain/Getty Images.
376: Collection, The Supreme Court Historical Society. **377:** Ken Karp Photography.
378: Lonny Kalfus/Getty Images. **382:** Bettmann/CORBIS. **384 (left):** Jeff Greenberg/PhotoEdit, Inc.
384 (right): Securtec. **385:** Michael J. Botos. **386:** SuperStock, Inc. **389:** Laura Murray.
390: Mark Richards/PhotoEdit, Inc. **394:** Mark Richards/PhotoEdit, Inc. **396:** The Sporting News.
399 (left): Ken Karp Photography. **399 (right):** Keoki Stender.

Chapter 10

407: Donald Corner and Jenny Young/www.greatbuildings.com.
411: AP/Wide World Photos. **415:** James L. Ames/CORBIS. **420:** NASA. **421 (left):** Stock Montage.
421 (right): AIP/Emilio Segre Archives. **424:** Ken Karp Photography. **427:** Look GMBH/eStock Photo.
432: Collection of Emilio Ellena/Kactus Foto/SuperStock, Inc. **439:** Laura Murray.
445: Reuters NewMedia, Inc./CORBIS.

Chapter 11

455: Kevin Schafer/Getty Images. **459:** Index Stock Imagery, Inc./Scott Smith.
465: AP/Wide World Photos. **467:** National Archives. **473:** Bettmann/CORBIS.
478 (left): Jack Fields/CORBIS. **478 (right):** Eric Crichton/CORBIS.
479: African Fractals by Ron Eglash. **483:** Jim Zuckerman/CORBIS. **487:** NASA.

Chapter 12

495 (left): Kenneth Hamm/Photo Japan. **495 (right):** NASA.
500: Ken Karp Photography. **505:** AP/Wide World Photos.
507: AP/Wide World Photos. **509 (top left):** Courtesy of Mr. Mom's All Weather Wicker.
509 (top right): Mark Burnett/Stock Boston. **509 (bottom):** Michael Rougier/TimePix.
514: Kelly/Mooney/CORBIS. **521:** Yann Arthus-Bertrand/CORBIS.
526 (left): Mark Antman/The Image Works. **526 (right):** Mark Keller/SuperStock, Inc.
528: AP/Wide World Photos. **532:** Courtesy of the Kepler-Gesellschaft Society.
536 (top): Beebe; Morton/CORBIS. **536 (bottom):** ©2002 Eames Office (www.eamesoffice.com).
537: Randy Wells/Getty Images. **542:** Don Baccus Photography.

Chapter 13

547: The Studio Dog/PictureQuest. **550:** Roger Ressmeyer/CORBIS. **557:** Mike Zens/CORBIS.
559: Courtesy USA Roller Sports. **570 (bottom):** Tony Freeman/PhotoEdit, Inc.
571: Bill Aron/PhotoEdit, Inc. **575:** Explorer, Paris/SuperStock, Inc.
576: Werner H. Miller/CORBIS. **578:** David Young-Wolff/PhotoEdit, Inc.
584: Michael Melford/Getty Images.

Chapter 14

585: NASA. **595:** Index Stock Imagery, Inc./ChromaZone Images.
596 (top): Richard A. Cook III/Getty Images. **596 (bottom):** S. Lowry/Univ. Ulstra/Getty Images.
600: Ken Karp Photography. **609:** Ken Karp Photography. **611:** Ken Karp Photography.
613: James Pickerell/The Image Works.

Chapter 15

615: AP/Wide World Photos. **628:** AP/Wide World Photos. **632:** Bob Torrez/Getty Images.
634: Steve Allen/Brand X Pictures/PictureQuest. **641:** Dean Abramson/Stock Boston.
646 (left): Kevin Fleming/CORBIS. **646 (right):** Jeff Lawrence/Stock Boston.
649: Reuters NewMedia, Inc./CORBIS. **650:** Dean Abramson/Stock Boston.
654: AP/Wide World Photos. **655:** Art Resource NY. **657:** Courtesy of the United States Mint.